Fungal Families of the World

Fungal Families of the World

Paul F. Cannon

And

Paul M. Kirk

CABI UK Centre (Egham)
Bakeham Lane
Egham, Surrey
TW20 9TY
United Kingdom

CABI is a trading name of CAB International

CABI Head Office
Nosworthy Way
Wallingford
Oxfordshire OX10 8DE
UK

CABI North American Office
875 Massachusetts Avenue
7th Floor
Cambridge, MA 02139
USA

Tel: +44 (0)1491 832111
Fax: +44 (0)1491 833508
E-mail: cabi@cabi.org
Website: www.cabi.org

Tel: +1 617 395 4056
Fax: +1 617 354 6875
E-mail: cabi-nao@cabi.org

© CAB International 2007. All rights reserved. No part of this publication may be reproduced in any form or by any means, electronically, mechanically, by photocopying, recording or otherwise, without the prior permission of the copyright owners.

A catalogue record for this book is available from the British Library, London, UK.

A catalogue record for this book is available from the Library of Congress, Washington, DC.

ISBN: 978 0 85199 827 5

Printed and bound in Singapore from copy supplied by the authors to Kyodo Press in association with MRM Graphics.

INTRODUCTION

The *Dictionary of the Fungi* has been published by CABI continuously from its outset in 1943 to the latest (ninth) edition in 2001. It has evolved over the years in many ways under the guidance of many editors, but has remained as the basic reference work for systematic mycology. Its primary feature is an authoritative consensus classification of the fungi, that has been widely accepted as an ennabling and informing framework for research into pure and applied mycology.

The purpose of this new book is three-fold. Firstly, it is intended as an illustrative and perhaps more approachable companion to the *Dictionary*. It places in context the recent seismic changes in classification driven by DNA sequence analyses, and forms a link between phenetic and phylogenetic approaches to fungal systematics. Secondly, it provides substantial further information on the 536 currently accepted families of *Fungi*, with more detailed descriptions and notes on ecology, economic uses etc. And thirdly (and perhaps most importantly), it depicts the extraordinary range of morphological structures found in fungi, celebrating myco-diversity and hopefully stimulating interest in mycology by those outside of the inner circle of fungal systematists.

The gestation of this book had its roots in an unpublished taxonomy manual on the *Ascomycota*, produced in the late 1980s by David Minter, David Hawksworth and Paul Cannon (CABI) for training courses in the UK, Africa and South America. The overall format has remained, but the text has been completely remodelled and the scope extended to cover the entire Kingdom *Fungi*. The original intention was to include fungal analogues now treated in different kingdoms (Protozoa and Chromista). However, except for a few well-known groups (mostly plant pathogens) recent knowledge of these organisms is scanty, and their inclusion would have required substantial extra research. Their exclusion also has the advantage that non-specialists need no longer be confused between the fungi (as traditionally circumscribed) and the *Fungi* in their strict phylogenetic sense.

The taxonomic framework used for this publication is based on that of the ninth edition of the *Dictionary of the Fungi* (2001), but substantially updated to conform with the findings of two major US-led research projects on fungal systematics, popularly referred to as *Deep Hypha* and AFTOL (*Assembling the Fungal Tree of Life*). These have operated primarily at ordinal level and above and have focused almost exclusively on molecular phylogenetic methods, leaving quite large numbers of 'orphan' families for which little or no sequence information has been acquired. Our practice has therefore been to assign provisional ordinal placements for these where possible, based on existing morphological and biological knowledge. We hope that new phylogenetic research will help to place the orphan families more securely. With a very few exceptions, the names used in this book conform to those accepted in the paper synthesizing fungal classification, a major output of the two programmes (Hibbett et al., 2007).

Fungal Families of the World could not have been produced without the very generous support of many people providing images of key genera (they are acknowledged more fully below), or without access to the CABI fungal collections. CABI's activities in mycology began in 1920 with inauguration of the Imperial Bureau of Mycology – soon to be renamed Imperial Mycological Institute (IMI). Despite its imperial name IMI was a highly innovative collaborative enterprise with its operational funds derived from a consortium of seven colonial governments. This pioneering spirit of cooperation has continued to this day with CABI operating as a successful intergovernmental organization, which is currently owned by no less than 45 member nations.

A major part of IMI's role was identification of fungi (especially those causing plant disease), and these duties led to establishment of world-leading reference collections of living and dried specimens. The dried collection (herbarium) was instigated in the early 1920s by Edmund Mason, and now contains around 400,000 specimens – each with its own unique 'IMI number'. The living culture collection now comprises around 27,000 accessions, stored as lyophilized samples or frozen in liquid nitrogen – some derived from living cultures dating back nearly sixty years to the collection's original inauguration. Many of the collections derive from the identification service, the samples being stored as reference material. However, the number and diversity of the samples were expanded greatly over the years through material collected by IMI staff. Especially valuable contributions were made by Colin Booth, Frederick Deighton, Martin Ellis, David Hawksworth, Stan Hughes, Paul Kirk, Edmund Mason, David Minter, Kris Pirozynski and Brian Sutton, and their labours within and outside formal work duties have ennabled images of many species to be included within this book.

Providing high-quality images for all 536 families included in this book has proved to be a mammoth task. This has been made made substantially easier as increasing numbers of images are placed on the Internet, but a significant number of families remain without illustration. Many of these have been hardly, if at all, researched in the era of digital colour photography. Nevertheless we have been able to provide images for over 400 families of *Fungi*, representing substantially wider fungal diversity than has been achieved before in a single publication. The images are necessarily of variable quality but, where practical, illustrations of both macroscopic and microscopic features have been included. Many of the photomicrographs are montages generated using Automontage software (Synoptics Ltd) which results in a greatly increased depth of field, at the expense of some 'flattening' of the image. Most are of material mounted in lactic acid (often with fuchsin or cotton blue dye) which accentuates morphological structures, though sometimes results in minor shrinkage. Where scale bars are present, these measure 20µm unless stated otherwise. Further details for individual images can be obtained from the authors.

Information provided in some of the categories within the accounts of individual families is sparse, to the extent that we debated whether to include them at all. This was especially true for economic uses, where much information (especially in industrial applications) is not in the public domain. Doubtless we have missed some data, but we have not been able to establish economic significance even in its widest sense for nearly 350 of the 536 families currently accepted. The potential for exploitation of these hardly-explored organisms must be substantial. The references provided for each family focus on phylogenetic studies where available, and should be taken in conjunction with the general bibliography below, as well as those provided in the *Dictionary of the Fungi*.

Future plans are to update this book at regular intervals, following production of new editions of the *Dictionary* (the tenth is due for publication in 2008). Whether or not both publications migrate onto the Internet, we will be very grateful to receive notes of omissions, corrections etc., and especially images of families for which illustrations are lacking or inadequate. All such contributions will be fully acknowledged, and copyright of images will remain with the original author.

Acknowledgements

We would like to record our gratitude to the many people listed below that have allowed us to use their images in this book. Their generous contributions have added immeasurably to the utility and visual appeal of *Fungal Families of the World*. We would like to record our especial thanks to an elite group of myco-photographers that not only have provided substantial freely accessible high-quality image resources on the Internet, but have allowed us to use many of their particularly striking images. The people concerned (each of whom has contributed more than ten images) include:

George Barron, Canada
Yves Deneyer, Belgium
David Ellis, Australia
Robert Lücking, USA
Bill Malcolm, Australia
Jens Petersen, Denmark
Einar Timdal, Norway
Chris Walker, UK

The complete list of people who have provided images is as follows:

Alex Weir, *Ceratomyces mirabilis, Cochliomyces, Zodiomyces vorticellarius*
Alfonso Spielmann, *Letrouitia domingensis*
Andrej Kunca, National Forest Centre – Slovakia, Slovak Republic, *Pucciniastrum areolatum*
Andrew Methven, *Boletinellus merulioides*
Andy Overall, http://www.fungitobewith.org, *Torrendia pulchella*
Ari Jumpponen, *Zelleromyces claridgei*
Barry Sullender, *Heterodermia echinata*

Bill Malcolm, *Bellemerea alpina, Biatorella, Calycidium cuneatum, Fellhanera bouteillei, Porina subapplanata, Porpidia macrocarpa, Roccella fuciformis, Vezdaea aestivalis, Wawea fruticulosa*

Bob Lichtwardt, http://www.nhm.ku.edu/~fungi, *Asellaria ligiae, Asellaria unguiformis, Harpella amazonica, Harpellomyces montanus, Smittium biforme, Smittium precipitiorum, Stachylina acutibasilaris*

Brian Ecott, http://www.hainaultforest.co.uk, *Podoscypha multizonata, Protomyces pachydermus*

Brian Spooner, http://www.rbgkew.org.uk/scihort/mycolexp.html, *Aseroë rubra*

Brian Sutton, *Ceuthospora lauri, Phacidium coniferarum*

Brigitte & Jan Kohlmeyer, *Koralionastes ellipticus, Koralionastes ovalis, Lulworthia calcicola, Spathulospora adelpha, Spathulospora adelpha, Spathulospora lanata*

Bryce Kendrick, http://www.mycolog.com, *Endogone, Entomophthora, Pilobolus*

Carol Shearer, *Anguillospora, Dipodascopsis uninucleata*

Chris Walker, http://www.lrz-muenchen.de/~schuessler/amphylo, *Acaulospora capsicula, Acaulospora mellea, Appendicispora fennica, Archaeospora trappei, Archaeospora trappei, Glomus epigaeum, Glomus epigaeum, Glomus macrocarpum, Glomus mosseae, Pacispora chimonobambusae, Pacispora scintillans, Paraglomus occultum, Paraglomus occultum, Scutellospora erythropa*

Claudio Carrai, *Phaeochora steinheilii*

David Ellis, http://www.mycology.adelaide.edu.au, *Apophysomyces elegans, Basidiobolus, Conidiobolus, Conidiobolus, Cunninghamella, Geotrichum candidum, Mortierella wolfii, Phaeoacremonium parasiticum, Saksenaea vasiformis, Syncephalastrum racemosum*

David Fischer, http://www.AmericanMushrooms.com, *Syzygospora mycetophila*

David Fry (dec.), *Coccomyces dentatus, Rhytisma acerinum*

David Larkin, *Hericium erinaceus*

David Minter, *Arthrobotrys superba, Arthrographis cuboidea, Ascodichaena rugosa, Ascosphaera apis, Aureobasidium pullulans, Aureobasidium pullulans, Bertia moriformis, Bertia moriformis, Botryobasidium chilense, Calicium trabinellum, Cladrrhinum foecundissimum, Cordyceps gracilis, Corynelia tropica, Corynelia uberata, Elaphomyces muricatus, Elaphomyces muricatus, Exophiala jeanselmei, Helvella lacunosa, Laboulbenia, Laboulbenia trichognati, Lichenopeltella pinophylla, Lophodermium baculiferum, Melanospora damnosa, Microascus cinereus, Microascus cinereus, Nematospora coryli, Nematospora coryli, Podostroma alutaceum, Rhachomyces zuphii, Sporothrix*

Dianna Smith, Photograph by Eleanor Yarrow, courtesy of the Connecticut-Westchester Mycological Association, *Peniophora rufa*

Donald Barr, *Allomyces, Blastocladia, Olpidium brassicae, Olpidium brassicae, Piromyces, Spizellomyces*

Einar Timdal, http://www.nhm.uio.no/botanisk/lav/Photo_Gallery, *Arctocetraria nigricascens, Arthrorhaphis citrinella, Baeomyces placophyllus, Buellia epigaea, Bunodophoron melanocarpum, Catillaria chalybeia, Chrysothrix chlorina, Cyphelium tigillare, Diploschistes muscorum, Gyalecta ulmi, Hypocenomyce scalaris, Lasallia rubiginosa, Lecidea silacea, Lopadium disciforme, Nephroma arcticum, Arctocetraria nigricascens, Arthrorhaphis citrinella, Baeomyces placophyllus, Buellia epigaea, Bunodophoron melanocarpum, Catillaria chalybeia, Chrysothrix chlorina, Cyphelium tigillare, Diploschistes muscorum, Gyalecta ulmi, Hypocenomyce scalaris, Lasallia rubiginosa, Lecidea silacea, Lopadium disciforme, Nephroma arcticum, Peltula omphaliza, Physcia stellaris, Placopsis gelida, Placynthium nigrum, Pseudocyphellaria crocata, Psoroma hypnorum, Solorina saccata, Teloschistes chrysophthalmus, Toninia bumamma, Trapeliopsis granulosa*

Ern Emmett, *Clavaria zollingeri, Taphrina johansonii*

Etienne Charles, http://champignons.moselle.free.fr, *Basidioradulum radula*

Eugene Bossenmaier (dec.) with permission from Mrs Alice Bossenmaier, *Cudonia circinans*

Franco Mondello, http://xoomer.alice.it/micologiamessinese/, *Battarrea phalloides*

Franz Lang, *Rhizophydium*

Fumio Ihara, *Podonectria cicadellidicola*

George Barron, http://www.uoguelph.ca/~gbarron, *Amoebophilus, Amoebophilus, Ballocephala, Cochlonema, Cotylidia, Cyathus olla, Cyathus striatus, Entomophthora, Favolaschia calocera, Helicocephalum oligosporum, Nematoctonus, Neolecta irregularis, Physalacria inflata, Thamnidium elegans, Tremellodendropsis semivestita, Urnula craterium, Zoophagus*

Gerald Benny, http://www.zygomycetes.org, *Radiomyces mexicanus*

Götz Palfner, http://www.chilefungi.cl, *Plectania chilensis*

Gunnar Hensel, http://www.pilzbestimmung.de, *Gastrosporium simplex, Hymenogaster arenarius, Hymenogaster rehsteineri*

Harrie Sipman, http://www.bgbm.org/BGBM/STAFF/Wiss/Sipman, *Mycoporum lacteum, Myeloconis*

Hege Gundersen, *Chaenotheca chrysocephala, Chaenotheca ferruginea, Chaenothecopsis fennica, Sphinctrina turbinata*

Hiroshi Takahashi, http://kinoko-ya.jp, *Albatrellus confluens, Spathulariopsis velutipes*

Iain Munro, *Anthracobia melaloma, Ascobolus crenulatus, Diatrype bullata, Geopyxis carbonaria, Helvella lacunosa, Peziza cerea*

Jan Vesterholt, http://www.mycokey.com, *Tubulicrinis subulatus*

Jeanne Lauwen, http://www.jeannelauwen.nl, *Coniophora puteana*

Jens Petersen, http://www.mycokey.com, *Calostoma cinnabarinum, Dasyscyphella nivea, Epithele typhae, Grammothele fuligo, Gyrophanopsis polonensis, Lachnum clandestinum, Lentinula edodes, Onygena corvina, Onygena equina*

Jerry Cooper, http://nzfungi.landcareresearch.co.nz, *Gomphidius glutinosus, Hohenbuehelia atrocoerulea*

Jim Trappe, http://www.natruffling.org/photo.htm, *Hydnangium carneum, Leucogaster citrinus, Rhizopogon truncatus*

Jim Worrall, *Serpula lacrymans*

John Weir, *Entoloma mougeotii*

Jørgen Koch, http://www.mycokey.com, *Nia vibrissa*

Joy Roberts, reproduced with permission from Susan K. Hagle (USDA Forest Service): http://www.fs.fed.us/r1-r4/spf/fhp/field_guide/index.htm to refs, *Rasutoria abietis*

Joyce Longcore, http://bama.ua.edu/~nsfpeet, *Chytridium lagenaria, Cladochytrium replicatum, Harpochytrium, Nephrochytrium aurantium, Nowakowskiella*

Karl Keck, http://www.karlkeck.de, *Inonotus hispidus, Tranzschelia anemones*

Lars Erik Lindell, Börge Pettersson, The Swedish Biodiversity Centre (CBM), *Tuburcinia leimbachii*

Laurens Sparrius, *Enterographa anguinella*

Lori Carris, *Hericium abietis*

Lu Guo-zhong, *Schizophyllum commune*

Malcolm Storey, http://www.bioimages.org.uk, *Physoderma asphodeli*

Marcel Lecomte, http://users.skynet.be/Champignons_passion/, *Prosthemium stellare*

Marilyn R.N. Mollicone, *Gonapodya, Monoblepharis, Monoblepharis, Oedogoniomyces lymnaeae*

Mark Holderness, *Marasmiellus scandens*

Markus Wilhelm, http://www.pilze-basel.ch/, *Gyroporus cyanescens*

Martin Ellis (dec.), *Acrogenospora sphaerocephala, Arthrinium caricicola, Botryobasidium conspersum, Capnobotrys neesii, Chaetosphaeria cupulifera, Corynespora smithii, Helminthosporium solani, Magnaporthe grisea, Pachnocybe ferruginea, Wallemia sebi*

Matt Trappe, http://www.natruffling.org/photo.htm, *Endogone lactiflua, Truncocolumella citrina*

Matthias Lutz, *Tuberculina persicina*

Meike Piepenbring, *Mycosyrinx cissi*

Meredith Blackwell, *Pyxidiophora*

Michael Beug, http://www.evergreen.edu/mushrooms, *Albatrellus ellisii*

Michael E. Hood, *Microbotryum violaceum*

Mike Dodd, http://www.amanita-photolibrary.co.uk/photo_library/page10.html, *Amanita phalloides*

Milan Zelenay, http://www.nahuby.sk, *Mycenastrum corium*

Mirek Junek, *Rutstroemia bulgarioides*

Patrick Lindsey, http://popgen.unimaas.nl/~jlindsey/commanster/Mushrooms/mushrooms.html, *Tulasnella violea, Veluticeps abietina*

Paul Busselen, *Peniophora quercina*

Paul Diederich, http://www.lichenology.info, *Arthrorhaphis grisea, Athelia arachnoidea, Pyrenocollema halodytes*

Peter Buchanan, http://nzfungi.landcareresearch.co.nz, *Entoloma hochstetteri*

Peter Roberts, http://www.rbgkew.org.uk/scihort/mycolexp.html, *Chionosphaera coppinsii, Oliveonia pauxilla, Phragmoxenidium mycophilum, Sirobasidium magnum, Spiculogloea occulta, Tulasnella tomaculum*

Randy Currah, *Ajellomyces capsulatus*

Richard Benjamin (dec.), reproduced with permission from Aliso, 1959: http://www.rsabg.org/content/view/49/59, *Dimargaris cristalligena, Spirodactylon aureum, Tieghemiomyces californicus*

Robert Lücking, http://www.fieldmuseum.org/research_collections/botany/botany_sites/ticolichen, *Aspidothelium cinerascens, Astrothelium galbineum, Badimia dimidiata, Brigantiaea leucoxantha, Byssoloma leucoblepharum, Chapsa platycarpella, Coccocarpia stellata, Coenogonium linkii, Cresponea melanocheloides,*

Crocynia gossypina, Cryptothecia rubrocincta, Dictyonema glabratum, Diploschistes cinereocaesius, Gomphillus, Gyalidea hyalinescens, Gyalideopsis lambinonii, Laurera purpurina, Leptogium burgessii, Megalospora tuberculosa, Microtheliopsis uleana, Mycomicrothelia hemisphaerica, Myeloconis guyanensis, Normandina pulchella, Pertusaria tetrathalamia, Phyllobaeis imbricata, Placomaronea candelarioides, Porina simulans, Pyrenula ochraceoflava, Sarcographa heteroclita, Sphinctrina tubiformis, Sticta humboldtii, Strigula macrocarpa, Trapeliopsis flexuosa, Trichothelium argenteum, Vezdaea dawsoniae

Roger Phillips, http://www.rogersmushrooms.com, *Echinodontium tinctorium*
Ross Beever, http://nzfungi.landcareresearch.co.nz, *Gallacea scleroderma*
Roy Halling, http://www.nybg.org/bsci/res/hall, *Alpova diplophloeus, Gloeomucro chlorinus*
Sharron Hicks Crane, *Dipodascopsis uninucleata*
Shirley Michelle Dulnuan, *Trypethelium aeneum*
Thomas Læssøe, http://www.mycokey.com, *Helicobasidium brebissonii, Stephanospora caroticolor*
Trude Vrålstad, *Cudonia circinans*
Ulrich Kirschbaum, http://kmubserv.tg.fh-giessen.de/pm/kirschbaum, *Protopannaria pezizoides, Thrombium epigaeum*
W. Wallace Martin, *Coelomomyces*
Yoshitaka Ono, *Phakopsora vitis*
Yousuke Degawa, *Helicoma*
Yves Deneyer, http://users.skynet.be/bs133881, *Astraeus hygrometricus, Ceriporia purpurea, Chroogomphus rutilus, Clathrus ruber, Crucibulum laeve, Exobasidium vaccinii, Geastrum pectinatum, Geastrum saccatum, Gymnopilus penetrans, Gyrodon lividus, Gyroporus castaneus, Hapalopilus rutilans, Helminthosphaeria clavariarum, Hydnum repandum, Hygrocybe calyptriformis, Hygrocybe pratensis, Hygrocybe virginea, Hygrophoropsis aurantiaca, Hymenochaete rubiginosa, Inocybe rimosa, Lachnellula suecica, Lactarius chrysorrheus, Melanogaster ambiguus, Orbilia luteorubella, Paxillus involutus, Phleogena faginea, Pisolithus tinctorius, Pterula multifida, Scytinostroma portentosum, Sebacina incrustans, Sistotrema confluens, Thelephora anthocephala, Thelephora palmata, Tulostoma brumale, Typhula erythropus, Typhula setipes, Verpa digitaliformis*

Classification of the *Fungi*

For many years from the days of Micheli and Linneaus, fungi were shoehorned into classifications as 'cryptogamic plants', and separated into lichenized and non-lichenized groups. The major assemblages were simplistically subdivided using a very restricted character suite, and basic distinctions were upheld between sexually and non-sexually reproducing species. This approach aided identification, but did little to advance understanding of relationships.

The biological continuity between meiotic and mitotic species was understood as early as the mid nineteenth century, but it was not until the work of Mason, Hughes and Ellis in the mid to late twentieth century that classifications began to reflect this reality. Successive editions of Ainsworth & Bisby's *Dictionary of the Fungi* (the first appeared in 1943) provided powerful impetus for this work, and these volumes also pioneered the integration of lichenized and non-lichenized taxa into a unified framework.

Fungi were not generally recognized as belonging to a separate kingdom from the plants until Whittaker's influential five-kingdom classification of living beings (Whittaker, 1959, 1969), and subsequent phylogenetic research has repeatedly confirmed that the *Fungi* are a sister group to the animals rather than the plants. For many years now, groups of organisms traditionally classified within the *Fungi* have been recognized as belonging to separate kingdoms. These include the *Hyphochytriomycota*, *Labyrinthulomycota* and *Oomycota* within chromistan clades of the Kingdom *Plantae*, and the *Acrasiomycota*, *Myxomycota* and *Plasmodiophoromycota* within the Kingdom *Protozoa*. These groups are therefore omitted from this publication.

Mycologists were quick to take advantage of the systematics revolution using molecular methods from the early 1990s, and there has been a series of major changes to the traditional morphology-based classifications. Publication of a major new classification of the Kingdom *Fungi* (Hibbett *et al.*, 2007) provides an opportune time to describe and illustrate at family level the existing classification, using an amalgamation of phenetic and phylogenetic data derived from multiple gene loci.

Fungi are eukaryotic heterotrophs that lack photosynthetic capacity (though many have symbiotic or commensal relationships with photosynthetic organisms) and gain nutrition by absorption rather than ingestion (Ainsworth, 1973; Webster & Weber, 2007). They typically form complex non-motile networks of hyphae within or on substrata, with cell walls composed primarily of glucans and chitin, with motile reproductive structures in some groups. Cells may be uninucleate or multinucleate (sometimes with large numbers of nuclei), and the mycelium may be homokaryotic or heterokaryotic, haploid, dikaryotic or diploid (usually only briefly in the last case). Reproduction may be sexual (following nuclear fusion and meiosis), parasexual (with nuclear fusion followed by de-diploidization) or asexual – following mitotic division. Reproductive structures and propagules are immensely varied – as illustrated throughout this book – but frequently of short duration. They are ubiquitous in terrestrial habitats and are important components of both marine and freshwater ecosystems. They are associated with an enormous range of plants, animals and their products in saprobic, mutualistic and parasitic relationships. Around 90,000 species have been described to date, and it has been estimated that in the order of 1.5 million species (including non-fungal analogues) exist (Hawksworth, 2001) – making the *Fungi* the second most diverse major group of organisms on Earth.

The current systematic arrangement of the *Fungi* (Hibbett *et al.*, 2007) includes one subkingdom *Dikarya*, into which the *Ascomycota* and *Basidiomycota* are placed due to their shared possession of dikaryotic hyphae. At phylum level, seven assemblages are recognized, the *Ascomycota*, *Basidiomycota*, *Blastocladiomycota*, *Chytridiomycota*, *Glomeromycota* and *Neocallimastigomycota*. The *Zygomycota* is not accepted as a formal phylum but the term is used here as a holding group pending more research to elucidate the relationships in these basla fungi. Other groups are included without formal placement, most notably the *Microsporidia* (excluded from this work due to the paucity of data on family level groupings and the lack of adequate illustrations), enigmatic organisms that lack mitochondria and are now considered as fungi in most phylogenetic studies.

The *Ascomycota* (characterized in morphological terms by production of sexual spores within asci) and *Basidiomycota* (with sexual spores produced on basidia) are by far the most diverse and well-studied. The *Chytridiomycota* have motile zoospores with a single posteriorly-directed flagellum, and are now restricted to the *Chytridiomycetes* and *Monoblepharidomycetes* due to separation of the *Blastocladiomycetes* into a separate phylum. The rumen-inhabiting *Neocallimastigomycota* were traditionally considered as with chytrid affinities, but molecular studies have confirmed their separation at phylum level. Major changes have taken place in our understanding of the evolution of species traditionally treated as zygomycetes, with the arbuscular mycorrhizal fungi being given their own phylum – *Glomeromycota*. The *Mucoromycotina*, *Kickxellomycotina*, *Zoopagomycotina* and *Entomophthoromycotina* are accepted as subphyla without placement within a specific phylum, as their evolutionary origins are still not fully understood.

General bibliography

A number of key works have general application in this book, and should be consulted in conjunction with the references provided in the individual family accounts. These include:

General works (including those on fungal analogues)

Adl, S.M., Simpson, A.G.B., Farmer, M.A., Andersen, R.A., Anderson, O.R., Barta, J.R., Bowser, S.S., Brugerolle, G., Fensome, R.A., Fredericq, S., James, T.Y., Karpov, S., Kugrens, P., Krug, J., Lane, C.E., Lewis, L.A., Lodge, J., Lynn, D.H., Mann, D.G., McCourt, R.M., Mendoza, L., Moestrup, Ø., Mozley-Standridge, S.E., Nerad, T.A., Shearer, C.A., Smirnov, A.V., Spiegel, F.W. & Taylor, M.F.J.R. (2005). The new higher level classification of eukaryotes with emphasis on the taxonomy of protists. *Journal of Eukaryotic Microbiology* **52**: 399–451.

Ainsworth, G.C., Sparrow, F.K. & Sussman, A.S. (1973). *The Fungi. An Advanced Treatise* vols 4A and 4B. New York etc.: Academic Press.

Alexopoulos, C.J., Mims, C.W. & Blackwell, M. (1996). *Introductory Mycology*. Edn 4. New York: John Wiley.

Baldauf, S.L., Roger, A.J., Wenk-Siefert, I. & Doolittle, W.F. (2000). A kingdom-level phylogeny of eukaryotes based on combined protein data. *Science* **290**: 972–977.

Carlile, M.J. & Watkinson, S.C. (1994). *The Fungi*. London etc.: Academic Press.

Cavalier-Smith, T. (1981). Eukaryote kingdoms: seven or nine? *BioSystems* **14**: 461–481.

Cavalier-Smith, T. (1998). A revised six-kingdom system of Life. *Biological Reviews* **73**: 203–266.

Cracraft, J. & Donoghue, M.J. (eds) (2004). *Assembling the Tree of Life*. Oxford: Oxford University Press.

David, J.C. (2002). A preliminary catalogue of the names of fungi above the rank of order. *Constancea* **83**: 1–30.

Ellis, M.B. & Ellis, J.P. (1997). *Microfungi on Land Plants.* Slough, UK: Richmond Publishing.

Hawksworth, D.L. (1991). The fungal dimension of biodiversity: magnitude, significance and conservation. *Mycological Research* **95**: 641–655.

Hawksworth, D.L. (2001). The magnitude of fungal diversity: the 1.5 million species estimate revisited. *Mycological Research* **105**: 1422–1432.

Kendrick, B. (2001). *The Fifth Kingdom.* Edn 3. Sidney, Canada: Mycologue.

Kirk, P.M., Cannon, P.F., David, J.C. & Stalpers, J.A. (eds) (2001). *Ainsworth & Bisby's Dictionary of the Fungi* edn 9. 655 pp. Wallingford, UK: CAB International.

Margulis, L., Corliss, J.O., Melkonian, M. & Chapman, D.J. (1990). *Handbook of Protoctista: The structure, cultivation, habitats and life histories of the eukaryotic micro-organisms and their descendants.* Boston, USA: Jones & Bartlett.

McLaughlin, D.J., McLaughlin, E.G. & Lemke, P.A. (eds) (2001). *The Mycota.* Vol. VII. Part A. *Systematics and Evolution.* Berlin: Springer-Verlag.

McLaughlin, D.J., McLaughlin, E.G. & Lemke, P.A. (eds) (2001). *The Mycota.* Vol. VII. Part B. *Systematics and Evolution.* Berlin: Springer-Verlag.

Steenkamp, E.T., Wright, J. & Baldauf, S.L. (2006). The protistan origins of animals and fungi. *Molecular Biology and Evolution* **23**: 93–106.

Van der Peer, Y., Baldauf, S.L., Doolittle, W.F. & Meyer, A. (2000). An updated and comprehensive rRNA phylogeny of (crown) Eukaryotes based on rate-calibrated evolutionary distances. *Journal of Molecular Evolution* **51**: 565–756.

Webster, J. & Weber, R. (2007). *Introduction to Fungi.* Edn 3. Cambridge, UK: Cambridge University Press.

Whittaker R.H. (1959). On the broad classification of organisms. *Quarterly Review of Biology* **34**: 210–226.

Whittaker, R.H. (1969). New concepts of kingdoms of organisms. *Science* **163**: 150–160.

Fungi

Anon. (ed.) (1996). *Fungi of Australia.* Vol. 1A. Introduction – Classification. 435 pp. Canberra: CSIRO Publishing/Australian Biological Resources Study.

Anon. (ed.) (1996). *Fungi of Australia.* Vol. 1B. Fungi in the Environment. 405 pp. Canberra: CSIRO Publishing/Australian Biological Resources Study.

Carroll, G.C. & Wicklow, D.T. (1992). *The Fungal Community. Its Organization and Role in the Ecosystem.* Edn 2. New York etc.: Marcel Dekker.

Domsch, K.H., Gams, W. & Anderson, T.-H. (1980). *Compendium of Soil Fungi.* 2 vols, London etc.: Academic Press.

Fréalle, E., Noël, C., Nolard, N., Symoens, F., Felipe, M.-S., Dei-Cas, E., Camus, D., Viscogliosi, E. & Delhaes, L. (2006). Manganese superoxide dismutase based phylogeny of pathogenic fungi. *Molecular Phylogenetics and Evolution* **41**: 28–39.

Galagan, J.E., Henn, M.R., Ma, L.-J., Cuomo, C.A. & Birren, B. (2005). Genomics of the fungal kingdom: insights into eukaryotic biology. *Genome Research* **15**: 1620–1631.

Germot, A., Philipe, H. & Le Guyader, H. (1997). Evidence for loss of mitochondria in microsporidia from a mitochondrial HSP70 in *Nosema locustae*. *Molecular and Biochemical Parasitology* **87**: 159–168.

Gill, E.E. & Fast, N.M. (2006). Assessing the microsporidia-fungi relationship: combined phylogenetic analysis of eight genes. *Gene* **375**: 103–109.

Hibbett, D.S., Binder, M., Bischoff, J.F., Blackwell, M., Cannon, P.F., Eriksson, O., Huhndorf, S., James, T., Kirk, P.M., Lücking, R., Lumbsch, T., Lutzoni, F., Matheny, P.B., Mclaughlin, D.J., Powell, M.J., Redhead, S., Schoch, C.L., Spatafora, J.W., Stalpers, J.A., Vilgalys, R., Aime, M.C., Aptroot, A., Bauer, R., Begerow, D., Benny, G.L., Castlebury, L.A., Crous, P.W., Dai, Y.-C., Gams, W., Geiser, D.M., Griffith, G.W., Gueidan, C., Hawksworth, D.L., Hestmark, G., Hosaka, K., Humber, R.A., Hyde, K., Kõljalg, U., Kurtzman, C.P., Larsson, K.-H., Lichtwardt, R., Longcore, J., Miądlikowska, J., Miller, A., Moncalvo, J.-M., Mozley-Standridge, S., Oberwinkler, F., Parmasto, E., Reeb, V., Rogers, J.D., Roux, C., Ryvarden, L., Sampaio, J.P., Schuessler, A., Sugiyama, J., Thorn, R.G., Tibell, L., Untereiner, W.A., Walker, C., Wang, Z.., Weir, A., Weiss, M., White, M., Winka, K., Yao, Y.-J. & Zhang, N. (2007). A higher-level phylogenetic classification of the *Fungi*. *Mycological Research* **111**: 509–548.

Hibbett, D.S. & Donoghue, M.J. (1998). Integrating phylogenetic analysis and classification in fungi. *Mycologia* **90**: 347–356

Hirt, R.P., Healy, B., Vossbrinck, C.R., Canning, E.U. & Embley, T.M. (1997). A mitochondrial Hsp70 orthologue in *Vairimorpha necatrix*: molecular evidence that *Microsporidia* once contained mitochondria. *Current Biology* **7**: 995–998.

Hoog, G.S. de, Guarro, J., Gené, J. & Figueras, M.J. (2000). *Atlas of Clinical Fungi.* Edn 2, 1126 pp. Utrecht: CBS and Reus: Universitat Rovire I Virgili.

James, T.Y., Kauff, F., Schoch, C.L., Matheny, P.B., Hofstetter, V., Cox, C., Celio, G., Gueidan, C., Fraker, E., Miądlikowska, J., Lumbsch, H.T., Rauhut, A., Reeb, V., Arnold, E.A., Amtoft, A., Stajich, J.E., Hosaka, K., Sung, G.-H., Johnson, D., O'Rourke, B., Crockett, M., Binder, M., Curtis, J.M., Slot, J.C., Wang, Z., Wilson, A.W., Schüßler, A., Longcore, J.E., O'Donnell, K., Mozley-Standridge, S., Porter, D., Letcher, P.M., Powell, M.J., Taylor, J.W., White, M.M., Griffith, G.W., Davies, D.R., Humber, R.A., Morton, J., Sugiyama, J., Rossman, A.Y., Rogers, J.D., Pfister, D.H., Hewitt, D., Hansen, K., Hambleton, S., Shoemaker, R.A., Kohlmeyer, J., Volkmann-Kohlmeyer, B., Spotts, R.A., Serdani, M., Crous, P.W., Hughes, K.W., Matsuura, K., Langer, E., Langer, G.,

Untereiner, W.A., Lücking, R., Büdel, B., Geiser, D.M., Aptroot, A., Diederich, P., Schmitt, I., Schultz, M., Yahr, R., Hibbett, D.S., Lutzoni, F., McLaughlin, D., Spatafora, J. & Vilgalys, R. (2006). Reconstructing the early evolution of the fungi using a six gene phylogeny. *Nature* **443**: 818–822.

Keeling, P.J. (2003). Congruent evidence for alpha-tubulin and beta-tubulin gene phylogenies for a zygomycete origin of *Microsporidia*. *Fungal Genetics and Biology* **38**: 298–309.

Keeling, P.J., Luker, M.A. & Palmer, J.D. (2000). Evidence from beta-tubulin phylogeny that *Microsporidia* evolved from within the fungi. *Molecular Biology and Evolution* **17**: 23–31.

Kendrick, B. (1979). *The Whole Fungus. The Sexual-Asexual Synthesis*. 793 pp. Ottawa: National Museum of Natural Sciences.

Kohlmeyer, J. & Kohlmeyer, E. (1979). *Marine Mycology. The Higher Fungi*. 690 pp. New York etc.: Academic Press.

Kurtzman, C.P. & Fell, J.W. (2000). *The Yeasts. A Taxonomic Study*. Edn 4, 1055 pp. Amsterdam: Elsevier.

Larsson, J.I.R. (2000). The hyperparasitic microsporidium *Amphiacantha longa* Caullery et Mesnil, 1914 (*Microspora*: Metchnikovellidae) – description of the cytology, redescription of the species, emended diagnosis of the genus *Amphiacantha* and establishment of the new family Amphiacanthidae. *Folia Parasitologica* **47**: 241–256.

Liu, Y.J., Hodson, M.C. & Hall, B.D. (2006). Loss of the flagellum happened only once in the fungal lineage: phylogenetic structure of kingdom *Fungi* inferred from RNA polymerase II subunit genes. *BMC Evolutionary Biology* **6**: 74.

Lutzoni, F., Kauff, F., Cox, C.J., McLaughlin, D., Celio, G., Dentinger, B., Padamsee, M., Hibbett, D.S., James, T.Y., Baloch, E., Grube, M., Reeb, V., Hofstetter, V., Schoch, C., Arnold, A.E., Miądlikowska, J., Spatafora, J., Johnson, D., Hambleton, S., Crockett, M., Shoemaker, R., Sung, G.-H., Lücking, R., Lumbsch, T., O'Donnell, K., Binder, M., Diederich, P., Ertz, D., Gueidan, C., Hansen, K., Harris, R.C., Hosaka, K., Lim, Y.-W., Matheny, B., Nishida, H., Pfister, D., Rogers, J., Rossman, A., Schmitt, I., Sipman, H., Stone, J., Sugiyama, J., Yahr, R. & Vilgalys, R. (2004). Assembling the fungal tree of life: progress, classification, and evolution of subcellular traits. *American Journal of Botany* **91**: 1446–1480.

Moore, R.T. (1980). Taxonomic proposals for the classification of marine yeasts and other yeast-like fungi including the smuts. *Botanica Marina* **23**: 361–373.

Reynolds, D.R. & Taylor, J.W. (eds) (1993). *The Fungal Holomorph. Mitotic, Meiotic and Pleomorphic Speciation in Fungal Systematics*. 375 pp. Wallingford, UK: CAB International.

Spooner, B.M. & Roberts, P. (2005). *Fungi*. 594 pp. London: Collins.

Subramanian, C.V. (1983). *Hyphomycetes. Taxonomy and Biology*. 502 pp. London etc.: Academic Press.

Taylor, J.W., Spatafora, J., O'Donnell, K., Lutzoni, F., James, T., Hibbett, D.S., Geiser, D., Bruns, T.D. & Blackwell, M. (2004). The fungi. In: Cracraft J, Donoghue MJ (eds), *Assembling the Tree of Life*, Oxford University Press, Oxford, pp. 171–196.

Tehler, A. (1988). A cladistic outline of the *Eumycota*. *Cladistics* **4**: 227–277.

Tehler, A., Little, D.P. & Farris, J.S. (2003). The full-length phylogenetic tree from 1551 ribosomal sequences of chitinous fungi, Fungi. *Mycological Research* **107**: 901–916.

Vossbrinck, C.R. & Debrunner-Vossbrinck, B.A. (2005). Molecular phylogeny of the *Microsporidia*: ecological, ultrastructural and taxonomic considerations. *Folia Parasitologica* **52**: 131–142.

Ascomycota

Arx, J.A. von & Müller, E. (1954). Die Gattungen von amerosporen Ascomyceten. *Beiträge zur Kryptogamenflora der Schweiz* **11**(1): 434 pp.

Barr, M.E. (1983). The ascomycete connection. *Mycologia* **75**: 1–13.

Barr, M.E. (1987). *Prodromus to class Loculoascomycetes*. Amherst, USA: University of Massachusetts.

Berbee, M.L. (1996). Loculoascomycete origins and evolution of filamentous ascomycete morphology based on 18S rDNA sequence data. *Molecular Biology and Evolution* **13**: 462–470.

Berbee, M.L. (2001). The phylogeny of plant and animal pathogens in the *Ascomycota*. *Physiological and Molecular Plant Pathology* **59**: 165–187.

Brodo, I.M., Sharfnoff, S.D. & Sharnoff, S. (2001). *Lichens of North America*. 795 pp. New Haven, USA: Yale University Press.

Ellis, M.B. (1971). *Dematiaceous Hyphomycetes*. 608 pp. Kew, UK: CABI.

Ellis, M.B. (1976). *More Dematiaceous Hyphomycetes*. 507 pp. Kew, UK: CABI.

Eriksson, O.E. (1981). The families of bitunicate ascomycetes. *Opera Botanica* **60**: 220 pp.

Eriksson, O. (1982). Outline of the ascomycetes. *Mycotaxon* **15**: 203–248 [continued as *Myconet*; now accessible at http://www.fieldmuseum.org/myconet/].

Eriksson, O.E. (1994). *Pneumocystis carinii*, a parasite in lungs of mammals, referred to a new family and order (*Pneumocystidaceae, Pneumocystidales, Ascomycota*). *Systema Ascomycetum* **13**: 165–180.

Eriksson, O.E., Svedskog, A. & Landvik, S. (1993). Molecular evidence for the evolutionary hiatus between *Saccharomyces cerevisiae* and *Schizosaccharomyces pombe*. *Systema Ascomycetum* **11**: 119–162.

Eriksson, O.E. & Winka, K. (1997). Supraordinal taxa of *Ascomycota*. *Myconet* **1**: 1–16.

Geiser, D.M., Gueidan, C., Miądlikowska, J., Lutzoni, F., Kauff, F., Hofstetter, V., Fraker, E., Schoch, C.L., Tibell, L., Untereiner, W.A. & Aptroot, A. (2007).

Eurotiomycetes: Eurotiomycetidae and *Chaetothyriomycetidae*. *Mycologia* **98**: 1053–1064.

Gilbert, O. (2000). *Lichens*. 288 pp. London: Collins.

Hansen, L. & Knudsen, H. (eds) (2000). *Nordic Macromycetes*. Vol. 1. *Ascomycota*. 309 pp. Copenhagen: Nordsvamp.

Henssen, A. & Jahns, H.M. (2003) *Lichenes: Eine Einführung in die Flechtenkunde*. Stuttgart: Georg Thieme.

Kohlmeyer, J. & Volkmann-Kohlmeyer, B. (1991). Illustrated key to the filamentous higher marine fungi. *Botanica Marina* **34**: 1–61.

Kurtzman, C.P. & Sugiyama, J. (2001). Ascomycetous yeasts and yeastlike taxa. In: McLaughlin DJ, McLaughlin EJ, Lemke P (eds), *The Mycota*. Vol. VII. Part A. *Systematics and Evolution,* Springer-Verlag, Berlin, pp. 179–200.

Liew, E.C.Y., Aptroot, A. & Hyde, K.D. (2000). Phylogenetic significance of the pseudoparaphyses in loculoascomycete taxonomy. *Molecular Phylogenetics and Evolution* **16**: 392–402.

Lindemuth, R., Wirtz, N. & Lumbsch, H.T. (2001). Phylogenetic analysis of nuclear and mitochondrial rDNA sequences supports the view that loculoascomycetes (*Ascomycota*) are not monophyletic. *Mycological Research* **105**: 1176–1181.

Liu, Y.J. & Hall, B.D. (2004). Body plan evolution of ascomycetes, as inferred from an RNA polymerase II phylogeny. *Proceedings of the National Academy of Sciences, USA* **101**: 4507–4512.

Lopandic, K., Molnár, O., Suzuki, M., Pinsker, W. & Prillinger, H. (2005). Estimation of phylogenetic relationships within the *Ascomycota* on the basis of 18S rDNA sequences and chemotaxonomy. *Mycological Progress* **4**: 205–214.

Lumbsch, H.T. & Lindemuth, R. (2001). Major lineages of *Dothideomycetes* (*Ascomycota*) inferred from SSU and LSU rDNA sequences. *Mycological Research* **105**: 901–908.

Lumbsch, T., Palice, Z., Wiklund, E., Ekman, S. & Wedin, M. (2004). Supraordinal phylogenetic relationships of *Lecanoromycetes* based on Bayesian analysis of combined nuclear and mitochondrial sequences. *Molecular Phylogenetics and Evolution* **31**: 822–832.

Lumbsch, H.T., Schmitt, I., Lindemuth, R., Miller, A., Mangold, A., Fernandez, F. & Huhndorf, S. (2005). Performance of four ribosomal DNA regions to infer higher-level phylogenetic relationships of inoperculate euascomycetes (*Leotiomyceta*). *Molecular Phylogenetics and Evolution* **34**: 512–524.

Lumbsch, H.T., Wirtz, N., Lindemuth, R. & Schmitt, I. (2002). Higher level phylogenetic relationships of euascomycetes (*Pezizomycotina*) inferred from a combined analysis of nuclear and mitochondrial sequence data. *Mycological Progress* **1**: 57–70.

Lutzoni, F., Pagel, M. & Reeb, V. (2001). Major fungal lineages are derived from lichen symbiotic ancestors. *Nature* **411**: 937–940.

McCarthy, P.M. (ed.) (1992). *Flora of Australia*. Vol. 54. Lichens: Introduction, *Lecanorales* I. 349 pp. Canberra: Australian Government Publishing Service.

McCarthy, P.M. (ed.) (1994). *Flora of Australia*. Vol. 55. *Lecanorales* 2, *Parmeliaceae*. 360 pp. Canberra: CSIRO Publishing/Australian Biological Resources Study.

McCarthy, P.M. (ed.) (2001). *Flora of Australia*. Vol. 58A. Lichens 3. 242 pp. Canberra: Australian Government Publishing Service.

McCarthy, P.M. (ed.) (2004). *Flora of Australia*. Vol. 56A. Lichens 4. 240 pp. Canberra: Australian Government Publishing Service.

Miądlikowska, J. Kauff, F., Hofstetter, V., Fraker, E., Reeb, V., Grube, M., Hafellner, J., Kukwa, M., Lücking, R., Hestmark, G., Otalora, M.G., Rauhut, A., Büdel, B., Scheidegger, C., Timdal, E., Stenroos, S., Brodo, I., Perlmutter, G.B., Ertz, D., Diederich, P., Lendemer, J.C., May, P., Schoch, C.L., Arnold, A.E., Hodkinson, B.P., Gueidan, C., Tripp, E., Yahr, R., Robertson, C. & Lutzoni, F. (2007). New insights into classification and evolution of the *Lecanoromycetes* (*Pezizomycotina, Ascomycota*) from phylogenetic analyses of three ribosomal RNA- and two protein-coding genes. *Mycologia* **98**: 1088–1103.

Müller, E. & Arx, J.A. von (1962). Die Gattungen von didymosporen Ascomyceten. *Beiträge zur Kryptogamenflora der Schweiz* **11**(2): 922 pp.

Nag Raj, T.R. (1993). *Coelomycetous Anamorphs with Appendage-Bearing Conidia*. 1101 pp. Sidney, Canada: Mycologue.

Nannfeldt, J.A. (1932). Studien über die Morphologie und Systematik der nicht-lichenisierten inoperculaten Discomyceten., *Nova Acta Regiae Societatis Scientarum Upsaliensis, ser.* 4 **8**(2): 1–368.

Nishida, H. & Sugiyama, J. (1994). *Archiascomycetes*: detection of a major new lineage within the *Ascomycota*. *Mycoscience* **35**: 361–366.

Peršoh, D., Beck, A. & Rambold, G. (2004). The distribution of ascus types and photobiontal selection in *Lecanoromycetes* (*Ascomycota*) against the background of a revised SSU nrDNA phylogeny. *Mycological Progress* **3**: 103–121.

Purvis, O.W., Coppins, B.J., Hawksworth, D.L., James, P.W. & Moore, D.M. (1992). *Lichen Flora of Great Britain and Ireland*. London: Natural History Museum.

Robbertse, B., Reeves, J., Schoch, C. & Spatafora, J.W. (2006). A phylogenomic analysis of the Ascomycota. *Fungal Genetics and Biology* **43**: 715–725.

Santesson. R. (1952). Foliicolous lichens I. A revision of the taxonomy of the obligately foliicolous, lichenized fungi. *Symbolae Botanicae Upsalienses* **12**(1): 590 pp.

Schoch, C.L., Shoemaker, R.A., Seifert, K.A., Hambleton, S., Spatafora, J.W. & Crous, P.W. (2007). A multigene phylogeny of the Dothideomycetes using four nuclear loci. *Mycologia* **98**: 1041–1052.

Seifert, K.A., Gams, W., Crous, P.W. & Samuels, G.J. (2000). Molecules, morphology and classification: towards monophyletic genera in the *Ascomycetes*. *Studies in Mycology* no. 45: 230 pp. Baarn: CBS.

Sivanesan, A. (1984). *The Bitunicate Ascomycetes and their Anamorphs*. 701 pp. Vaduz: J. Cramer.

Spatafora, J.W., Johnson, D., Sung, G.-H., Hosaka, K., O'Rourke, B., Serdani, M., Spotts, R., Lutzoni, F., Hofstetter, V., Fraker, E., Gueidan, C., Miądlikowska, J., Reeb, V., Lumbsch, T., Lücking, R., Schmitt, I., Aptroot, A., Roux, C., Miller, A., Geiser, D.,

Hafellner, J., Hestmark, G., Arnold, A.E., Büdel, B., Rauhut, A., Hewitt, D., Untereiner, W., Cole, M.S., Scheidegger, C., Schultz, M., Sipman, H. & Schoch, C.L. (2007). A five-gene phylogenetic analysis of the *Pezizomycotina*. *Mycologia* **98**: 1018–1028.

Sugiyama, J., Hosaka, K. & Suh, S.-O. (2007). Early diverging *Ascomycota*: phylogenetic divergence and related evolutionary enigmas. *Mycologia* **98**: 996–1005.

Suh, S.-O., Blackwell, M., Kurtzman, C.P. & Lachance, M.-A. (2007). Phylogenetics of *Saccharomycetales*, the ascomycete yeasts. *Mycologia* **98**: 1006–1017.

Sutton, B.C. (1980). *The Coelomycetes. Fungi Imperfecti with Pycnidia, Acervuli and Stromata*. 696 pp. Kew, UK: CABI.

Wang, Z., Johnston, P.R., Takamatsu, S., Spatafora, J.W. & Hibbett, D.S. (2007). Phylogenetic classification of the *Leotiomycetes* based on rDNA data. *Mycologia* **98**: 1065–1075.

Wedin, M., Wiklund, E., Crewe, A., Döring, H., Ekman, S., Nyberg, Å., Schmitt, I. & Lumbsch, H.T. (2005). Phylogenetic relationships of *Lecanoromycetes* (*Ascomycota*) as revealed by analyses of mtSSU and nLSU rDNA sequence data. *Mycological Research* **109**: 159–172.

Zhang, N., Castlebury, L.A., Miller, A.N., Huhndorf, S., Schoch, C.L., Seifert, K., Rossman, A.Y., Rogers, J.D., Kohlmeyer, J., Volkmann-Kohlmeyer, B. & Sung, G.-H. (2007). An overview of the systematics of the Sordariomycetes based on a four-gene phylogeny. *Mycologia* **98**: 1076–1087.

Basidiomycota

Aime, M.C. (2006). Towards resolving family-level relationships in rust fungi (*Uredinales*). *Mycoscience* **47**: 112–122.

Aime, M.C., Matheny, P.B., Henk, D.A., Frieders, E.M., Nilsson, R.H., Piepenbring, M., McLaughlin, D.J., Szabo, L.J., Begerow, D., Sampaio, J.P., Bauer, R., Weiß, M., Oberwinkler, F. & Hibbett, D.S. (2007). An overview of the higher-level classification of *Pucciniomycotina* based on combined analyses of nuclear large and small subunit rDNA sequences. *Mycologia* **98**: 896–905.

Bauer, R., Begerow, D., Oberwinkler, F., Piepenbring, M. & Berbee, M.L. (2001). Ustilaginomycetes. In: McLaughlin DJ, McLaughlin EJ, Lemke P, eds. *The Mycota. Vol. VIII: Part B. Systematics and Evolution.* Springer-Verlag, Berlin, pp. 57–84.

Bauer, R., Begerow, D., Sampaio, J.P., Weiß, M. & Oberwinkler, F. (2006). The simple-septate basidiomycetes: a synopsis. *Mycological Progress* **5**: 41–66.

Bauer, R., Oberwinkler, F. & Vánky, K. (1997). Ultrastructural markers and systematics in smut fungi and allied taxa. *Canadian Journal of Botany* **75**: 1273–1314.

Begerow, D., Bauer, R. & Boekhout, T. (2000). Phylogenetic placements of ustilaginomycetous anamorphs as deduced from nuclear LSU rDNA sequences. *Mycological Research* **104**: 53–60.

Begerow, D., Bauer, R. &Oberwinkler, F. (1997). Phylogenetic studies on nuclear large subunit ribosomal DNA sequences of smut fungi and related taxa. *Canadian Journal of Botany* **75**: 2045–2056.

Begerow, D. Stoll, M. & Bauer, R. (2007). A phylogenetic hypothesis of *Ustilaginomycotina* based on multiple gene analyses and morphological data. *Mycologia* **98**: 906–916.

Binder, M. & Hibbett, D.S. (2002). Higher-level phylogenetic relationships of *Homobasidiomycetes* (mushroom-forming fungi) inferred from four rDNA regions. *Molecular Phylogenetics and Evolution* **22**: 76–90.

Binder, M., Hibbett, D.S., Larsson, K.-H., Larsson, E., Langer, E. & Langer, G. (2005). The phylogenetic distribution of resupinate forms across the major clades of mushroom-forming fungi (*Homobasidiomycetes*). *Systematics and Biodiversity* **3**: 1–45.

Binder, M. & Hibbett, D.S. (2007). Molecular systematics and biological diversification of *Boletales*. *Mycologia* **98**: 971–981.

Bodensteiner, P., Binder, M., Moncalvo, J.-M., Agerer, R. & Hibbett, D.S. (2004). Phylogenetic relationships of cyphelloid homobasidiomycetes. *Molecular Phylogenetics and Evolution* **33**: 501–515.

Boidin, J., Mugnier, J. & Canales, R. (1998). Taxonomie moleculaire des *Aphyllophorales*. *Mycotaxon* **66**: 445–492.

Bruns, T.D., Szaro, T.M., Gardes, M., Cullings, K.W., Pan, J.J., Taylor, D.L., Horton, T.R., Kretzer, A., Garbelotto, M. & Li, Y. (1998). A sequence database for the identification of ectomycorrhizal basidiomycetes by phylogenetic analysis. *Molecular Ecology* **7**: 257–272.

Buchanan, P.K. (2001). *Aphyllophorales* in Australia. *Australian Systematic Botany* **14**: 417–437.

Corner, E.J.H. (1950). *A Monograph of Clavaria and Allied Genera*. 740 pp. Annals of Botany Memoirs no. 1. Oxford: Oxford University Press.

Donk, M.A. (1964). A conspectus of the families of *Aphyllophorales*. *Persoonia* **3**: 199–324.

Fell, J.W., Boekhout, T., Fonseca, A. & Sampaio, J.P. (2001). Basidiomycetous yeasts. In: McLaughlin DJ, McLaughlin EJ, Lemke P (eds). *The Mycota*. Vol. VII. Part B., *Systematics and Evolution,*. Springer-Verlag, Berlin, pp. 3–35.

Gilbertson, R. & Ryvarden, L. (1986–7). *North American Polypores*. Parts 1 and 2. 885 pp. Oslo: Fungiflora.

Grgurinovic, C.A. (1997). *Larger Fungi of South Australia*. 725 pp. Adelaide: Botanic Gardens of Adelaide and State Herbarium.

Hansen, L. & Knudsen, H. (eds) (1992). *Nordic Macromycetes*. Vol. 2. *Polyporales, Boletales, Agaricales, Russulales*. 474 pp. Copenhagen: Nordsvamp.

Hansen, L. & Knudsen, H. (eds) (1997). *Nordic Macromycetes*. Vol. 3. Heterobasidioid, Aphyllophoroid and gastromycetoid *Basidiomycetes*. 444 pp. Copenhagen: Nordsvamp.

Hibbett, D.S. (2007). A phylogenetic overview of the *Agaricomycotina*. *Mycologia* **98**: 917–925.

Hibbett, D.S. & Binder, M. (2002). Evolution of complex fruiting-body morphologies in homobasidiomycetes. *Proceedings of the Royal Society of London* **B**, **269**: 1963–1969.

Hibbett, D.S. & Thorn, R.G. (2001). *Homobasidiomycetes*. In: McLaughlin DJ, McLaughlin EJ, Lemke P (eds), *The Mycota*. Vol. VII. Part B., *Systematics and Evolution*, Springer-Verlag, Berlin, pp. 121–168.

Hosaka, K., Bates, S.T., Beever, R.T., Castellano, M.A., Colgan, W. III, Domínguez, L.S., Nouhra, E.R., Geml, J., Giachini, A.J., Kenney, S.R., Simpson, N.B., Spatafora, J.W. & Trappe, J.M. 2007. Molecular phylogenetics of the gomphoid-phalloid fungi with an establishment of the new subclass *Phallomycetidae* and two new orders. *Mycologia* **98**: 949–959.

Jülich, W. (1981). Higher taxa of basidiomycetes. *Bibliotheca Mycologica* **85**: 1–485.

Larsson, E. & Larsson, K.-H. (2003). Phylogenetic relationships of russuloid basidiomycetes with emphasis on aphyllophoralean taxa. *Mycologia* **95**: 1037–1065.

Larsson, K.-H., Larsson, E. & Kõljalg, U. (2004). High phylogenetic diversity among corticioid homobasidiomycetes. *Mycological Research* **108**: 983–1002.

Larsson, K.-H., Parmasto, E., Fischer, M., Langer, E., Nakasone, K.K. & Redhead, S.A. (2007). *Hymenochaetales*: a molecular phylogeny for the hymenochaetoid clade. *Mycologia* **98**: 926–936.

Matheny, P.B. (2005). Improving phylogenetic inference of mushrooms with RPB1 and RPB2 nucleotide sequences. *Molecular Phylogenetics and Evolution* **35**: 1–20.

Matheny, P.B., Curtis, J.M., Hofstetter, V., Aime, M.C., Moncalvo, J.-M., Ge, Z.-W., Yang, Z.-L., Slot, J.C., Ammirati, J.F., Baroni, T.J., Bougher, N.L., Hughes, K.W., Lodge, D.J., Kerrigan, R.W., Seidl, M.T., Aanen, D.K., DeNitis, M., Daniele, G.M., Desjardin, D.E., Kropp, B.R., Norvell, L.L., Parker, A., Vellinga, E.C., Vilgalys, R. & Hibbett, D.S. (2007). Major clades of *Agaricales*: a multi-locus phylogenetic overview. *Mycologia* **98**: 982–995.

Matheny, P.B., Gossman, J.A., Zalar, P., Arun Kumar, T.K. & Hibbett, D.S. (2007). Resolving the phylogenetic position of the Wallemiomycetes: an enigmatic major lineage of Basidiomycota. *Canadian Journal of Botany*: in press.

May, T.W., Milne, J., Shingles, S. & Jones, R.H. (2003). *Fungi of Australia*. Vol. 2B. Catalogue and Bibliography of Australian Macrofungi 2. *Basidiomycota* p.p. and *Myxomycota*. 484 pp. Canberra: CSIRO Publishing/Australian Biological Resources Study.

May, T.W. & Wood, A.E. (1997). *Fungi of Australia*. Vol. 2A. Catalogue and Bibliography of Australian Macrofungi 1. *Basidiomycota* p.p. 358 pp. Canberra: CSIRO Publishing/Australian Biological Resources Study.

Moncalvo, J.-M., Lutzoni, F.M., Rehner, S.A., Johnson, J. & Vilgalys, R. (2000). Phylogenetic relationships of agaric fungi based on nuclear large subunit ribosomal DNA sequences. *Systematic Biology* **49**: 278–305.

Moncalvo, J.-M., Nilsson, R.H., Koster, B., Dunham, S.M., Bernauer, T., Matheny, P.B., McLenon, T., Margaritescu, S., Weiß, M., Garnica, S., Danell, E., Langer, G., Langer, E., Larsson, E., Larsson, K.-H. & Vilgalys, R. (2007). The cantharelloid clade: dealing with incongruent gene trees and phylogenetic reconstruction methods. *Mycologia* **98**: 937–948.

Moncalvo, J.-M., Vilgalys, R., Redhead, S.A., Johnson, J.E., James, T.Y., Aime, M.C., Hofstetter, V., Verduin, S.J.W., Larsson, E., Baroni, T.J., Thorn, R.G., Jacobsson, S., Clémençon, H. & Miller, O.K. (2002). One hundred seventeen clades of euagarics. *Molecular Phylogenetics and Evolution* **23**: 357–400.

Nuñez, M. & Ryvarden, L. (2000). *East Asian Polypores*. Vol. 1. *Ganodermataceae* and *Hymenochaetaceae*. 168 pp. Oslo: Fungiflora.

Nuñez, M. & Ryvarden, L. (2001). *East Asian Polypores*. Vol. 2. *Polyporaceae* s. lato. 352 pp. Oslo: Fungiflora.

Oberwinkler, F. (1977). Das neue System der Basidiomyceten. In: Frey W, Hurka H, Oberwinkler F (eds), *Beiträge zur Biologie der niederen Pflanzen*, Gustav Fischer Verlag, Stuttgart, pp. 59–105.

Oberwinkler, F. & Bandoni, R.J. (1982). A taxonomic survey of the gasteroid, auricularioid *Heterobasidiomycetes*. *Canadian Journal of Botany* **60**: 1726–1750.

Oberwinkler, F. & Bauer, R. (1989). The systematics of gasteroid, auricularioid *Heterobasidiomycetes*. *Sydowia* **41**: 224–256.

Pegler, D.N. (1977). *A Preliminary Agaric Flora of East Africa*. Kew Bulletin Additional Series no. 6, 615 pp. London: HMSO.

Pegler, D.N. (1983). *Agaric Flora of the Lesser Antilles*. Kew Bulletin Additional Series no. 9, 668 pp. + 27 pl. London: HMSO.

Petersen, R.H. (ed.) (1971). *Evolution in the Higher Basidiomycetes*. 562 pp. Knoxville, USA: University of Knoxville Press.

Ryvarden, L. (1992). *Genera of Polypores, Nomenclature and Taxonomy*. 363 pp. Oslo: Fungiflora.

Ryvarden, L. & Johansen, I. (1980). *Preliminary Polypore Flora of East Africa*. 630 pp. Oslo: Fungiflora.

Ryvarden, L. & Gilbertson, R.L. (1993). *European Polypores*. Part 1. *Abortiporus – Lindtneria*. 387 pp. Oslo: Fungiflora.

Ryvarden, L. & Gilbertson, R.L. (1994). *European Polypores*. Part 2. *Meripilus – Tyromyces*. pp. 388–743. Oslo: Fungiflora.

Sampaio, J.P. (2004). Diversity, phylogeny and classification of basidiomycetous yeasts. In: Agerer R, Piepenbring M, Blanz P, eds. *Frontiers in basidiomycote mycology*. IHW Verlag, Eching, pp. 49–80.

Singer, R. (1986). *The Agaricales in Modern Taxonomy*. 981 pp. Königstein: Koeltz.

Swann, E.C., Frieders, E.M., McLaughlin, D.J. (2001). *Urediniomycetes*. In: McLaughlin DJ, McLaughlin EJ, Lemke P (eds), *The Mycota*. Vol. VII. Part B., *Systematics and Evolution*, Springer-Verlag, Berlin, pp. 37–56.

Swann, E.C. & Taylor, J.W. (1995). Phylogenetic perspectives on basidiomycete systematics: evidence from the 18S rRNA gene. *Canadian Journal of Botany* **73**(Suppl): S862–S868.

Vánky, K. (1999). The new classificatory system for smut fungi, and two new genera. *Mycotaxon* **70**: 35–49.

Weiß, M., Bauer, R. & Begerow, D. (2004). Spotlights on heterobasidiomycetes. In: Agerer R, Piepenbring M, Blanz P (eds), *Frontiers in Basidiomycote Mycolog,*. IHW Verlag, Eching, pp. 7–48.

Wells, K. & Bandoni, R.J. (2001). *Heterobasidiomycetes*. In: McLaughlin DJ, McLaughlin EJ, Lemke P (eds), *The Mycota*. Vol. VII. Part B., *Systematics and Evolution,* Springer-Verlag, Berlin, pp. 85–120.

Wingfield, B.D., Ericson, L., Szaro, T. & Burdon, J.J. (2004). Phylogenetic patters in the *Uredinales*. *Australasian Plant Pathology* **33**: 327–335.

Zalar, P., de Hoog, G.S., Schroers, H.-J., Frank, J.M. & Gunde-Cimerman, N. (2005). Taxonomy and phylogeny of the xerophilic genus *Wallemia* (*Wallemiomycetes* and *Wallemiales*, cl. et ord. nov.). *Antonie van Leeuwenhoek* **87**: 311–328.

Blastocladiomycota, *Chytridiomycota* and *Neocallimastigomycota*

Barr, D.J.S. (1980). An outline for the reclassification of the *Chytridiales*, and for a new order, the *Spizellomycetales*. *Canadian Journal of Botany* **58**: 2380–2394.

Bullerwell, C.E., Forget, L. & Lang, B.F. (2003). Evolution of monoblepharidalean fungi based on complete mitochondrial genome sequences. *Nucleic Acids Research* **31**: 1614–1623.

James, T.Y., Letcher, P.M., Longcore, J.E., Mozley-Standridge, S.E., Porter, D., Powell, M.J., Griffith, G.W. & Vilgalys, R. (2007). A molecular phylogeny of the flagellated *Fungi* (*Chytridiomycota*) and a proposal for a new phylum (*Blastocladiomycota*). *Mycologia* **98**: 860–871.

James, T.Y., Porter, D., Leander, C.A., Vilgalys, R. & Longcore, J.E. (2000). Molecular phylogenetics of the *Chytridiomycota* supports the utility of ultrastructural data in chytrid systematics. *Canadian Journal of Botany* **78**: 336–350.

Kanouse, B.B. (1927). A monographic study of special groups of the water molds I. *Blastocladiaceae*. *American Journal of Botany* **14**: 287–306.

Letcher, P.M., Powell, M.J., Churchill, P.F., Chambers, J.G. (2006). Ultrastructural and molecular phylogenetic delineation of a new order, the *Rhizophydiales* (*Chytridiomycota*). *Mycological Research* **110**: 898–915.

Li, J.L., Heath, I.B. & Packer, L. (1993). The phylogenetic relationships of the anaerobic chytridiomycetous gut fungi (*Neocallimasticaceae*) and the *Chytridiomycota*. II. Cladistic analysis of structural data and description of *Neocallimasticales* ord. nov. *Canadian Journal of Botany* **71**: 393–407.

Sparrow, F.K. (1943). *Aquatic Phycomycetes Exclusive of the Saprolegniaceae and Pythium*. Ann Arbor, USA: University of Michigan Press.

Sparrow F.K. (1958). Interrelationships and phylogeny of the aquatic *Phycomycetes*. *Mycologia* **50:** 797–813.

Tanabe, Y., Watanabe, M.M. & Sugiyama, J. (2005). Evolutionary relationships among basal fungi (*Chytridiomycota* and *Zygomycota*): Insights from molecular phylogenetics. *Journal of General and Applied Microbiology* **51**: 267–276.

Glomeromycota

Helgason, T., Watson, I.J. & Young, P.W. (2003). Phylogeny of the *Glomerales* and *Diversisporales* (*Fungi: Glomeromycota*) from actin and elongation factor 1-alpha sequences. *FEMS Microbiology Letters* **229**: 127–132.

Kuhn, G., Hijri, M. & Sanders, I.R. (2001). Evidence for the evolution of multiple genomes in arbuscular mycorrhizal fungi. *Nature* **414**: 745–748.

Morton, J.B. & Benny, G.L. (1990). Revised classification of arbuscular mycorrhizal fungi (*Zygomycetes*): a new order, *Glomales*, two new suborders, *Glomineae* and *Gigasporineae*, and two new families, *Acaulosporaceae* and *Gigasporaceae*, with an emendation of *Glomaceae*. *Mycotaxon* **37**: 471–491.

Morton, J.B. & Redecker, D. (2001). Two new families of *Glomales*, *Archaeosporaceae* and *Paraglomaceae*, with two new genera *Archaeospora* and *Paraglomus*, based on concordant molecular and morphological characters. *Mycologia* **93**: 181–195.

Oehl, F. & Sieverding, E. (2004). *Pacispora*, a new vesicular arbuscular mycorrhizal fungal genus in the *Glomeromycetes*. *Journal of Applied Botany* **78**: 72–82.

Redecker, D., Morton, J.B. & Bruns, T.D. (2000). Ancestral lineages of arbuscular mycorrhizal fungi (*Glomales*). *Molecular Phylogenetics and Evolution* **14**: 276–284.

Sanders, I.R. (2002). Ecology and evolution of multigenomic arbuscular mycorrhizal fungi. *American Naturalist* **160**: S128–141.

Schüßler, A. Schwarzott, D. & Walker, C. (2001). A new fungal phylum, the *Glomeromycota*: phylogeny and evolution. *Mycological Research* **105**: 1413–1421.

Walker, C. & Schüßler, A. (2004). Nomenclatural clarifications and new taxa in the *Glomeromycota*. *Mycological Research* **108**: 981–982.

'Zygomycota'

Benjamin, R.K. (1979). Zygomycetes and their spores. In: Kendrick B (ed.) *The Whole Fungus The Sexual-Asexual Synthesis.* National Museum of Natural Sciences, Ottawa, pp. 573–616.

Benny, G.L. & White, M.M. (2001). The classification and phylogeny of *Trichomycetes* and *Zygomycetes.* In: Misra, JK, Horn BW (eds), *Trichomycetes and Other Fungal Groups..* Science Publishers, Enfield, NeH, pp. 39–53.

Lichtwardt, R.W., Manier, J.F. (1978). Validation of the *Harpellales* and *Asellariales. Mycotaxon* **7**: 441–442.

Moss, S.T. (1975). Septal structure in the *Trichomycetes* with special reference to *Astreptonema gammari* (*Eccrinales*). *Transactions of the British Mycological Society* **65**: 115–127

Nagahama, T., Sato, H., Shimazu, M. & Sugiyama, J. (1995). Phylogenetic divergence of the entomophtoralean fungi: evidence from nuclear 18S ribosomal RNA gene sequences. *Mycologia* **87**: 203–209.

O'Donnell, K., Cigelnik, E. & Benny, G.L. (1998). Phylogenetic relationships among the *Harpellales* and *Kickxellales. Mycologia* **90**: 286–297.

O'Donnell, K., Lutzoni, F.M., Ward, T.J. & Benny, G.L. (2001). Evolutionary relationships among mucoralean fungi (*Zygomycota*): evidence for family polyphyly on a large scale. *Mycologia* **93**: 286–297.

Seif, E., Leigh, L., Roewer, I., Forget, L. & Lang, B.F. (2005). Comparative mitochondrial genomics in zygomycetes: bacteria-like RNase P RNAs, mobile elements and a close source of the group I intron invasion in angiosperms. *Nucleic Acids Research* **33**: 734–744.

Tanabe, Y., O'Donnell, K., Saikawa, M. & Sugiyama, J. (2000). Molecular phylogeny of parasitic *Zygomycota* (*Dimargaritales, Zoopagales*) based on nuclear small subunit ribosomal DNA sequences. *Molecular Phylogenetics and Evolution* **16**: 253–262.

Tanabe, Y., Saikawa, M., Watanabe, M.M. & Sugiyama, J. (2004). Molecular phylogeny of *Zygomycota* based on EF-1 and RPB1 sequences: limitations and utility of alternative markers to rDNA. *Molecular Phylogenetics and Evolution* **30**: 438–449.

Tanabe, Y., Watanabe, M.M. & Sugiyama, J. (2005). Evolutionary relationships among basal fungi (*Chytridiomycota* and *Zygomycota*): Insights from molecular phylogenetics. *Journal of General and Applied Microbiology* **51**: 267–276.

Voigt, K. & Wöstemeyer, J. (2001). Phylogeny and origin of 83 *Zygomycetes* from all 54 genera of the *Mucorales* and *Mortierellales* based on combined analysis of actin and translation elongation factor EF-1α genes. *Gene* **270**: 113–120.

Walker, C. (1983). Taxonomic concepts in the *Endogonaceae*: spore wall characteristics in species descriptions. *Mycotaxon* **18**: 443–455.

White, M.M., James, T.Y., O'Donnell, K., Cafaro, M.J., Tanabe, Y. & Sugiyama, J. (2007). Phylogeny of the *Zygomycota* based on nuclear ribosomal sequence data. *Mycologia* **98**: 872–884.

Acarosporaceae Zahlbr. 1906
Acarosporales: Ascomycota

Thallus usually crustose or squamulose, rarely foliose or umbilicate. Ascomata usually deeply immersed, rarely almost perithecial, with a variably developed margin. Interascal tissue of simple or branched and anastomosed paraphyses, immersed in gel. Asci usually with a well-developed apical dome, usually J– though sometimes with an outer layer of J+ gel, with a well-developed ocular chamber, polysporous. Ascospores small, hyaline, usually aseptate, without a gelatinous sheath. Conidiomata pycnidial, immersed in the thallus, producing small, usually ellipsoidal, hyaline conidia.

Acarospora contigua, Arizona, USA; thallus

Significant Genera: *Acarospora*, *Sarcogyne*.

Distribution: Widespread.

Acarospora fuscata, Massachusetts, USA; thallus

Economic Significance: None is known.

Ecology: Lichenized with green algae, usually on rocks and often in nutrient-enriched habitats.

Notes: The *Acarosporaceae* form a separate well-defined clade to the *Lecanoromycetes* with a sister group that includes the *Eurotiomycetes* and the *Chaetothyriomycetes*. *Thelocarpon* appears not to belong within the *Acarosporaceae*.

References: **Bellemère, A.** (1994). Documents et commentaires sur l'ultrastructure des asques polysporés des *Acarospora*, de quelques genres de la famille des *Acarosporaceae* et de genres similaires. *Bull. Soc. linn. Provence* 45: 355–388; **Crewe, A.T.; Purvis, O.W.; Wedin, M.** (2006). Molecular phylogeny of *Acarosporaceae* (*Ascomycota*) with focus on the proposed genus *Polysporinopsis*. *Mycol. Res.* 110: 521–526; **David, J.C.; Coppins, B.J.** (1997). *Thelocarpon opertum* (*Acarosporaceae*), a new species from the British Isles. *Lichenologist* 29: 291–299; **Hafellner, J.** (1995). Towards a better circumscription of the *Acarosporaceae* (Lichenized *Ascomycotina*, *Lecanorales*). *Cryptog. bot.* 5: 99–104; **Kocourková-Horáková, J.** (1998). Distribution and ecology of the genus *Thelocarpon* (*Lecanorales*, *Thelocarpaceae*) in the Czech Republic. *Czech Mycol.* 50: 271–302; **Navarro-Rosinés, P.; Roux, C.; Bellemère, A.** (1999). *Thelocarpella gordensis* gen. et sp. nov. (*Ascomycetes* lichenisati *Acarosporaceae*). *Can. J. Bot.* 77. 035–042; **Reeb, V.; Lutzoni, F.; Roux, C.** (2004). Contribution of *RPB2* to multilocus phylogenetic studies of the euascomycetes (*Pezizomycotina*, *Fungi*) with special emphasis on the lichen-forming *Acarosporaceae* and evolution of polyspory. *Mol. Phylogen. Evol.* 32: 1036–1060; **Seppelt, R.D.; Nimis, P.L.; Castello, M.** (1998). The genus *Sarcogyne* (*Acarosporaceae*) in Antarctica. *Lichenologist* 30: 249–258; **Stenroos, S.K.; DePriest, P.T.** (1998). SSU rDNA phylogeny of cladoniiform lichens. *Am. J. Bot.* 85: 1548–1559; **Wedin, M.; Wiklund, E.; Crewe, A.; Döring, H.; Ekman, S.; Nyberg, A.; Schmitt, I.; Lumbsch, H.T.** (2005). Phylogenetic relationships of *Lecanoromycetes* (*Ascomycota*) as revealed by sequences of mtSSU and nLSU rDNA sequence data. *Mycol. Res.* 109: 159–172.

Acaulosporaceae J.B. Morton & Benny 1990
Diversisporales: Glomeromycota

Mycorrhizal species forming arbuscules and vesicles in roots (intraradical), anamorphic; the penetrating hyphae producing finely branched haustorial branches (arbuscules) or coils (pelotons) and vesicles. Spores hypogeous or partly epigeous, formed from or in the neck of a sporiferous saccule, auxillary cells not produced; spore wall multilayed, with a thin bilayered and flexible inner layer and germination orb formed on the innermost layer.

Significant Genera: *Acaulospora*.

Distribution: Widespread, although undoubtedly more so than current records suggest.

Economic Significance: Enhancement of phosphorus (P) supply to the 'host' is a characteristic of AM (arbuscular mycorrhizal) infections. As such they have a significant influence, either directly or indirectly, on land plants including ecosystem variability and productivity.

Ecology: As mycorrhizal partners of land plants forming arbuscular mycorrhizas; the hosts of AM fungi are primarily found in P-limited tropical and temperate grasslands.

Notes: As with some other families of the *Glomeromycota*, sequence diversity even within single spores makes phylogenetic analysis complex.

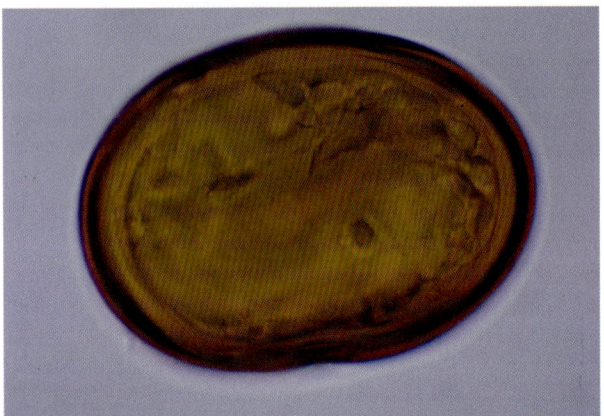
Acaulospora mellea; spore

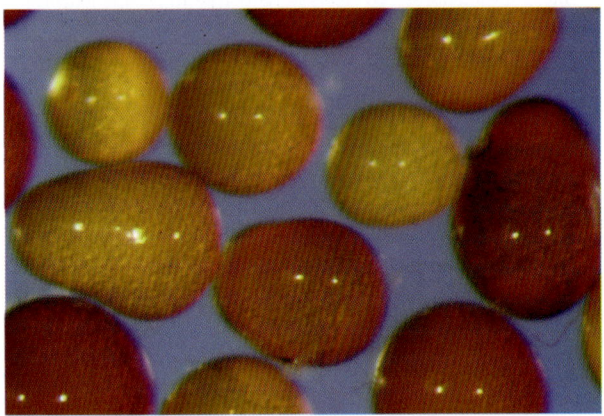
Acaulospora capsicula; spores

References: **Azcon-Aguilar, C.; Barea, J.M.** (1996). Arbuscular mycorrhizas and biological control of soil-borne plant pathogens – an overview of the mechanisms involved. *Mycorrhiza* 6: 457–464; **Fracchia, S.; Scervino, J.M.; Menéndez, A.; Godeas, A.** (2003). Isolation, culture and host colonization of *Entrophospora schenckii* (*Glomales*), an arbuscular mycorrhizal fungus. *Nova Hedwigia* 77: 383–388; **Maia, L.C.; Kimbrough, J.W.** (1993). Ultrastructural studies of spore walls of *Acaulospora morrowiae* and *A. scrobiculata*. *Mycol. Res.* 97: 1183–1189; **Morton, J.B.; Benny, G.L.** (1990). Revised classification of arbuscular mycorrhizal fungi (*Zygomycetes*): a new order, *Glomales*, two new suborders, *Glomineae* and *Gigasporineae* and two new families, *Acaulosporaceae* and *Gigasporaceae*, with an emendation of *Glomaceae*. *Mycotaxon* 37: 471–491; **Morton, J.B.; Bever, J.D.; Pfleger, F.L.** (1997). Taxonomy of *Acaulospora gerdemannii* and *Glomus leptotichum*, synanamorphs of an arbuscular mycorrhizal fungus in *Glomales*. *Mycol. Res.* 101: 625–631; **Pringle, A.; Moncalvo, J.-M.; Vilgalys, R.** (2003). Revisiting the rDNA sequence diversity of a natural population of the arbuscular mycorrhizal fungus *Acaulospora colossica*. *Mycorrhiza* 13: 227–231; **Redecker, D.** (2005). *Glomeromycota* Arbuscular mycorrhizal fungi and their relative(s). Version 01 July 2005. http://tolweb.org/Glomeromycota/28715/2005.07.01 in The Tree of Life Web Project, http://tolweb.org; **Rodriguez, A.; Dougall, T.; Dodd, J.C.; Clapp, J.P.** (2001). The large subunit ribosomal RNA genes of *Entrophospora infrequens* comprise sequences related to two different glomalean families. *New Phytol.* 152: 159–167; **Sawaki, H.; Sugawara, K.; Saito, M.** (1998). Phylogenetic position of an arbuscular mycorrhizal fungus, *Acaulospora gerdemannii*, and its synanamorph *Glomus leptotichum*, based upon 18S rRNA gene sequence. *Mycoscience* 39: 477–480; **Schüßler, A.; Schwarzott, D.; Walker, C.** (2001). A new fungal phylum, the *Glomeromycota*: phylogeny and evolution. *Mycol. Res.* 105: 1413–1421; **Stürmer, S.L.; Morton, J.B.** (1999). Taxonomic reinterpretation of morphological characters in *Acaulosporaceae* based on developmental patterns. *Mycologia* 91: 849–857; **van der Heijden, M.A.G.; Klironomos, J.N.; Ursic, M.; Moutoglis, P.; Streitwolf-Engel, R.; Boller, T.; Wiemken, A.; Sanders, I.R.** (1998). Mycorrhizal fungal diversity determines plant biodiversity, ecosystem variability and productivity. *Nature* Lond. 396: 69–72.

Acrospermaceae Fuckel 1870
Uncertain position within Dothideomycetes, Ascomycota

Stromata ± superficial, hyaline to brown, pulvinate or stipitate, composed of gelatinous pseudoparenchymatous tissue, often containing only a single ascoma. Ascomata perithecial, thin-walled, the ostiole large, not periphysate. Interascal tissue composed of narrow paraphyses. Asci cylindrical, elongated, ? with separable wall layers, not fissitunicate, usually with a capitate apical thickening with a narrow pore. Ascospores hyaline, elongate, multiseptate, not fragmenting, without a sheath. Anamorphs ? hyphomycetous, connections need confirmation.

Significant Genera: *Acrospermum*.

Acrospermum compressum, Austria; ascomata

Distribution: Widespread in temperate regions.

Economic Significance: None is known.

Ecology: Presumably saprobic, on wood and herbaceous stems.

Notes: Probably part of the *Dothideomycetes*, but more phylogenetic studies are needed.

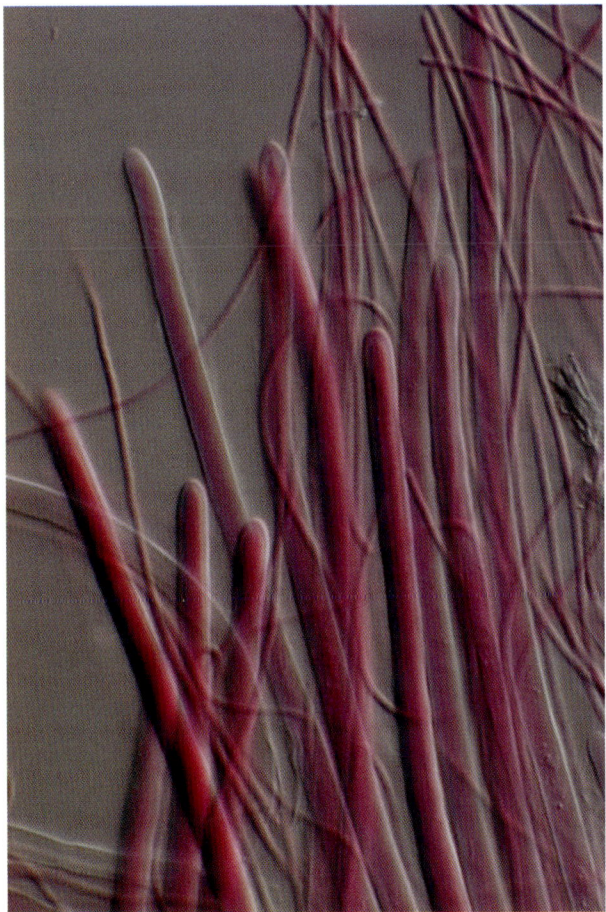

Acrospermum compressum, Suffolk, UK; asci and ascospores

References: **Eriksson, O.E.** (1967). On graminicolous pyrenomycetes from Fennoscandia. 2. Phragmosporous and scolecosporous species. *Arkiv för Botanik* ser 2, **6**(9): 381-440; **Nograsek, A.** (1990). Ascomyceten auf Gefässpflanzen der Polsterseggenrasen in der Ostalpen. *Bibliotheca Mycologica* **133**: 271 pp.; **Webster, J.** (1956). Conidia of *Acrospermum compressum* and *A. graminum*. *Trans. Br. mycol. Soc.* **39**: 361–366; **Winka, K.; Eriksson, O.E.** (2000). Adding to the bitunicate puzzle – studies on the systematic position of five aberrant ascomycete taxa. *Phylogenetic Relationships Within the Ascomycota Based on 18S rDNA Sequences, Akademisk Avhandling* [Thesis (PhD), Department of Ecology and Environmental Science, Umeå University] (Umeå): [17] pp.

Adelococcaceae Triebel 1993
Verrucariales: Ascomycota

Thallus absent. Ascomata perithecial, not clypeate, ± globose; peridium dark brown, especially at the apex, composed of pseudoparenchymatous cells. Hymenium gelatinous, blueing in iodine. Interascal tissue of narrow anastomosing paraphysis-like hyphae and apical paraphyses, the ostiole periphysate. Asci ± clavate, thick-walled.

Significant Genera: *Sagediopsis*.

Distribution: Primarily north temperate, but poorly studied.

Economic Significance: None is known.

Ecology: Biotrophic or necrotrophic on lichens.

Notes: No molecular data are available.

References: **Hoffmann, N.; Hafellner, J.** (2000). Eine Revision der lichenicoler Arten der Sammelgattungen *Guignardia* und *Physalospora* (Ascomycotina). *Biblthca Lichenol.* **77**: 181 pp.; **Matzer, M.; Pelzmann, B.** (1991). REM-Studien an Ascosporen der lichenicolen Gattungen *Adelococcus, Reconditella, Roselliniella* und *Roselliniopsis*. *Nova Hedwigia* **52**: 1–9; **Triebel, D.** (1989). Lecideicole Ascomyceten. Eine Revision der obligat lichenicolen Ascomyceten auf lecideoiden Flechten. *Biblthca Lichenol.* **35**: 278 pp.; **Triebel, D.** (1993). Notes on the genus *Sagediopsis* (*Verrucariales, Adelococcaceae*). *Sendtnera* **1**: 273–280.

Agaricaceae Chevall. 1826
Agaricales: Basidiomycota

Basidioma pileate, secotioid or gasteroid. Clamp connections present or absent. Pileate basidiomata with thin, usually free gills, often large, the cap often scurfy or scaly, ± flat or umbonate, the stipe central with a membranous annular veil, volva present or absent. Basidia usually small and 4-spored, sometimes accompanied by cystidia, rarely deliquescing. Basidiospores variable in colour from hyaline to greenish, ochraceous, pink or sepia, but never rusty-brown or cinnamon-brown; smooth or ornamented, with or without a germ pore, sometimes staining in iodine, usually binucleate. Sequestrate basidiomata stalked or not, with a chambered, fertile region (gleba) that either remains completely enclosed or becomes exposed following rupture of the surrounding peridial tissues. Anamorph hyphomycetous, with sometimes swollen conidia seceding rhexolytically from undifferentiated fertile hyphae.

Significant Genera: *Agaricus, Coprinus, Lepiota, Leucoagaricus, Leucocoprinus*.

Distribution: Widespread.

Economic Significance: *Agaricus bisporus* (*A. brunnescens*) is the most widely cultivated edible mushroom in the world, with global sales reaching US$millions annually. Other species are toxic, e.g. some *Chlorophyllum* and *Lepiota* species.

Ecology: Primarily saprobes in grassland and woodland situations, with some species commonly forming typical 'fairy rings'. Some species of *Leucoagaricus* and *Attamyces* have coevolved relationships with ants. Species of *Leucocoprinus* are well known as colonizers of compost and woodchips in horticultural situations, and appear to be resistant to steam sterilization.

Agaricus urinascens, Devon, UK; basidiomata

Notes: Molecular studies are revolutionizing understanding of the evolution of the *Agaricaceae*, and the classification is still in flux. Indications are that the family circumscription should include *Coprinus* in its newly restricted sense, and an expanding series of gasteroid genera that should now probably include the *Lycoperdaceae* as well as *Tulostoma*, *Battarrea* etc. In the light of this uncertainty, the classification of the *Dictionary of the Fungi* edn 9 is largely followed.

Macrolepiota procera, Suffolk, UK; basidioma

Coprinus comatus, Surrey, UK; basidioma

References: Breitenbach, J.; Kränzlin, F. (1995). *Fungi of Switzerland*. 4, Agarics, 2nd part: *Entolomataceae, Pluteaceae, Amanitaceae, Agaricaceae, Coprinaceae, Bolbitiaceae, Strophariaceae* (Lucerne): 368 pp.; **Didukh, M.; Vilgalys, R.; Wasser, S.P.; Islkhuemhen, O.S.; Nevo, E.** (2005). Notes on *Agaricus* section *Duploannulati* using molecular and morphological data. *Mycol. Res.* 109: 729–740; **Geml, J.; Geiser, D.M.; Royse, D.J.** (2004). Molecular evolution of *Agaricus* species based on ITS and LSU rDNA sequences. *Mycol. Prog.* 3: 157–176; **Grgurinovic, C.A.** (1997). *Larger Fungi of South Australia* (Adelaide): 725 pp.; **Hibbett, D.S.; Pine, E.M.; Langer, E.; Langer, G.; Donoghue, M.J.** (1996). Evolution of gilled mushrooms and puffballs inferred from ribosomal DNA sequences. *Proc. natn Acad. Sci. U.S.A.* 94: 12002–12006; **Hopple, J.S.; Vilgalys, R.** (1999). Phylogenetic relationships in the mushroom genus *Coprinus* and dark-spored allies based on sequence data from the nuclear gene coding for the large ribosomal subunit RNA: divergent domains, outgroups, and monophyly. *Mol. Phylogen. Evol.* 13: 1–19; **Johnson, J.** (1999). Phylogenetic relationships within *Lepiota sensu lato* based on morphological and molecular data. *Mycologia* 91: 443–458; **Kerrigan, R.W.** (2005). *Agaricus subrufescens*, a cultivated edible and medicinal mushroom, and its synonyms. *Mycologia* 97: 12–24; **Lebel, T.; Thompson, D.K.; Udovicic, F.** (2006). Description and affinities of a new sequestrate fungus, *Barcheria willisiana* gen. et sp. nov. (*Agaricales*) from Australia. *Mycol. Res.* 109: 206–213; **Malloch, D.; Castle, A.; Hintz, W.** (1988). Further evidence for *Agaricus brunnescens* Peck as the preferred name for the cultivated *Agaricus*. *Mycologia* 79: 839–846; **Mitchell, A.D.; Bresinsky, A.** (1999). Phylogenetic relationships of *Agaricus* species based on ITS-2 and 28S ribosomal DNA sequences. *Mycologia* 91: 811–819; **Moncalvo, J.-M.; Vilgalys, R.; Redhead, S.A.; Johnson, J.E.; James, T.Y.; Aime, M.C.; Hofstetter, V.; Verduin, S.J.W.; Larsson, E.; Baroni, T.J.; Thorn, R.G.; Jacobsson, S.; Clemencon, H.; Miller, O.K.** (2002). One hundred and seventeen clades of euagarics. *Mol. Phylogen. Evol.* 23: 357–400; **Pegler, D.N.** (1986). Agaric Flora of Sri Lanka. *Kew Bull. Addit. Ser.* 12: 519 pp.; **Redhead, S.A.; Vilgalys, R.; Moncalvo, J.-M.; Johnson, J.; Hopple, J.S.** (2001). *Coprinus* Pers. and the disposition of *Coprinus* species *sensu lato*. *Taxon* 50: 203–241; **Singer, R.** (1986). *Agaricales in Modern Taxonomy* Edn 4 (Koenigstein): 981 pp.; **Vellinga, E.C.** (2004). Genera in the family *Agaricaceae*: evidence from nrITS and nrLSU sequences. *Mycol. Res.* 108: 354–377; **Vellinga, E.C.; de Kok, R.P.J.; Bruns, T.D.** (2003). Phylogeny and taxonomy of *Macrolepiota* (*Agaricaceae*). *Mycologia* 95: 442–456; **Walther, G.; Garnica, S.; Weiss, M.** (2005). The systematic relevance of conidiogenesis modes in the gilled *Agaricales*. *Mycol. Res.* 109: 525–544; **Xu, J.P.; Kerrigan, R.W.; Sonnenberg, A.S.; Callac, P.; Horgen, P.A.; Anderson, J.B.** (1998). Mitochondrial DNA variation in natural populations of the mushroom *Agaricus bisporus*. *Mol. Ecol.* 7: 19–33.

Agaricostilbaceae Oberw. & R. Bauer 1989

Agaricostilbales: Basidiomycota

Basidiomata small, stalked, capitate, not gelatinous, the outer surface covered in basidia, cystidia absent. Hyphal system monomitic, composed of ± hyaline hyphae with simple septa. Basidia elongate, thick-walled, becoming transversely septate, repeatedly producing basidiospores from short sterigma-like outgrowths. Basidiospores not violently discharged, hyaline, not staining in iodine. Anamorph yeast-like, with cells formed both by budding and as endospores.

Significant Genera: *Agaricostilbum*.

Distribution: Scattered throughout subtropical and tropical zones, but certainly under-recorded.

Economic Significance: None is known.

Ecology: Probably saprobic, usually associated with dead palm tissues.

Notes: The phylogeny of the family has not been intensively studied, but it appears to occupy a basal position within the *Urediniomycetes*, perhaps close to the *Platygloeaceae*. For many years fruit bodies were assumed to be anamorphic, with spores produced mitotically rather than meiotically.

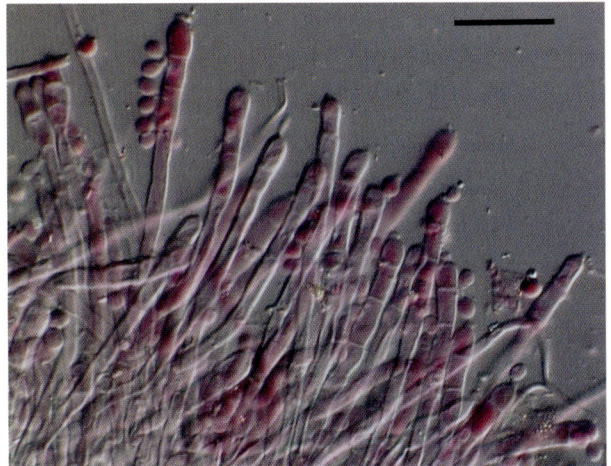

Agaricostilbum pulcherrimum, India; basidia and basidiospores

References: Bandoni, R.J.; Boekhout, T. (1998). *Agaricostilbum* Wright. *The Yeasts, A Taxonomic Study* (Amsterdam): 639–640; **Frieders, E.M.; McLaughlin, D.J.** (1996). Mitosis in the yeast phase of *Agaricostilbum pulcherrimum* and its evolutionary significance. *Can. J. Bot.* 74: 1392–1406; **Kendrick, B.; Gong, X.-d.** (1995). *Agaricostilbum nova-zelandica*, a new auricularoid fungus on *Rhopalostylis sapida* from New Zealand. *Mycotaxon* 54: 19–25; **Oberwinkler, F.; Bauer, R.** (1989). The systematics of gasteroid, auricularioid *Heterobasidiomycetes*. *Sydowia* 41: 224–256; **Scorzetti, G.; Fell, J.W.; Fonseca, A.; Statzell-Tallman, A.** (2002). Systematics of basidiomycetous yeasts: a comparison of large subunit D1/D2 and internal transcribed spacer rDNA regions. *FEMS Yeast Res.* 2: 495–517; **Swann, E.C.; Taylor, J.W.** (1995). Phylogenetic diversity of yeast-producing basidiomycetes. *Mycol. Res.* 99: 1205–1210.

Agyriaceae Corda 1838

Agyriales: Ascomycota

Thallus crustose, lobed or squamulose, sometimes absent or with weakly developed immersed stromatic tissue. Ascomata apothecial, brightly coloured, brown or black, not stalked, sometimes elongated or contorted, often domed, with marginal tissue absent or weakly to strongly developed, if present composed of globose cells, hymenium usually gelatinous, often blueing in iodine. Interascal tissue of branched and anastomosing paraphyses, sometimes with a well-developed pigmented epithecial layer. Asci usually with a strongly thickened J+ or J– apical cap but sometimes without an ocular chamber, rarely with a non-thickened apex, rarely with an internal more strongly staining apical region, sometimes with a J+ outer gelatinous layer, opening by eversion through a vertical split, sometimes polysporous. Ascospores small, hyaline, aseptate, sometimes with a gelatinous sheath. Cephalodia sometimes present. Anamorphs pycnidial.

Significant Genera: *Agyrium, Placopsis, Trapeliopsis*.

Distribution: Widespread, especially in temperate boreal and austral regions.

Economic Significance: None is known.

Ecology: Lichenized with green algae (with cyanobacteria in the cephalodia) on rotten wood or soil, or saprobic on bark and wood, especially on conifers.

Notes: The *Agyriaceae* was removed from the *Lecanorales* and given its own order in 2001, with molecular evidence suggesting a closer relationship with the *Ostropales* and *Pertusariales*.

Placopsis gelida, Iceland; thalli and ascomata

Trapeliopsis pseudogranulosa, La Palma; thallus

Trapeliopsis flexuosa; thallus and ascoma

Ajellomycetaceae Unter., J.A. Scott & Sigler 2004
Onygenales: Ascomycota

Stromata absent. Ascomata cleistothecial, sometimes aggregated, ± globose, small, yellowish or brown; peridium formed of loosely intertwined branched and anastomosed hyphae, sometimes sinuous and/or constricted at the septa and inflated between; also with conspicuous brown helical appendages arising from the central ascogonium. Asci globose to pyriform, thin-walled and evanescent, 8-spored. Ascospores globose to oblate, hyaline, aseptate, lacking germ pores. Anamorphs hyphomycetous, forming simple thick-walled lateral conidia or irregular intercalary arthroconidia.

Significant Genera: *Ajellomyces*. Anamorphs: *Blastomyces, Histoplasma, Paracoccidioides*.

Distribution: Primarily found in warm temperate and tropical regions of the New World.

Economic Significance: This family contains most of the serious (non-opportunistic) human pathogenic fungi, sometimes affecting healthy as well as immunocompromised people especially in the neotropics. A wide variety of symptoms may present, and some species can kill if not treated.

Ecology: Facultative pathogens of humans and other mammals, with saprobic phases colonizing soil, dung, dead wood etc., especially in arid conditions.

Notes: Until recently this group was assumed to belong within the *Arthrodermataceae*, but molecular studies show that the families occupy quite separate clades.

References: **Bellemère, A.** (1994). Documents et commentaires sur l'ultrastructure des asques polysporés des *Acarospora*, de quelques genres de la famille des *Acarosporaceae* et de genres similaires. *Bull. Soc. linn. Provence* 45: 355–388; **Brodo, I.M.** (1995). Notes on the lichen genus *Placopsis* (*Ascomycotina, Trapeliaceae*) in North America. *Biblthca Lichenol.* 57: 59–70; **Lumbsch, H.T.** (1997). Systematic studies in the suborder *Agyriineae* (*Lecanorales*). *J. Hattori bot. Lab.* 83: 1–73; **Lumbsch, H.T.; Schmitt, I.; Döring, H.; Wedin, M.** (2001). ITS sequence data suggest variability of ascus types support ontogenetic characters as phylogenetic discriminators in the *Agyriales* (*Ascomycota*). *Mycol. Res.* 105: 265–274; **Lumbsch, H.T.; Schmitt, I.; Döring, H.; Wedin, M.** (2001). Molecular systematics supports the recognition of an additional order of *Ascomycota*: the *Agyriales*. *Mycol. Res.* 105: 16–23; **Moberg, R.; Carlin, G.** (1996). The genus *Placopsis* (*Trapeliaceae*) in Norden. *Symb. bot. upsal.* 31: 319–325; **Reeb, V.; Lutzoni, F.; Roux, C.** (2004). Contribution of *RPB2* to multilocus phylogenetic studies of the euascomycetes (*Pezizomycotina, Fungi*) with special emphasis on the lichen-forming *Acarosporaceae* and evolution of polyspory. *Mol. Phylogen. Evol.* 32: 1036–1060; **Schmitt, I.; Lumbsch, H.T.; Søchting, U.** (2003). Phylogeny of the lichen genus *Placopsis* and its allies based on Bayesian analyses of nuclear and mitochondrial sequences. *Mycologia* 95: 827–835; **Wedin, M.; Wiklund, E.; Crewe, A.; Döring, H.; Ekman, S.; Nyberg, A.; Schmitt, I.; Lumbsch, H.T.** (2005). Phylogenetic relationships of *Lecanoromycetes* (*Ascomycota*) as revealed by sequences of mtSSU and nLSU rDNA sequence data. *Mycol. Res.* 109: 159–172.

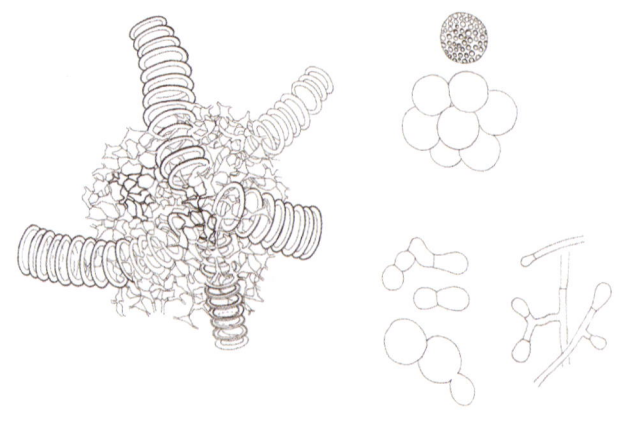
Ajellomyces dermatitidis; ascomata, asci, ascospores and anamorph

References: **Bialek, R.; Ibricevic, A.; Fothergill, A.; Begerow, D.** (2000). Small subunit ribosomal DNA sequence shows *Paracoccidioides brasiliensis* closely related to *Blastomyces dermatitidis*. *J. Clin. Microbiol.* 38: 3190–3193; **Currah, R.S.** (1985). Taxonomy of the *Onygenales*: *Arthrodermataceae, Gymnoaceae, Myxotrichaceae* and *Onygenaceae*. *Mycotaxon* 24: 1–216; **Feitosa, L. dos S.; Cisalpino, P.S.; Santos, M.R. Machado dos; Mortara, R.A.; Barros, T.F.; Morais, F.V.; Puccia, R.; Silveira, J.F. da; Camargo, Z.P. de** (2003). Chromosomal polymorphism, syntenic relationships, and ploidy in the pathogenic fungus *Paracoccidioides brasiliensis*. *Fungal Genetics Biol.* 39: 60–69; **Fukushima, K.; Takeo, K.; Takizawa, K.; Nishimura, K.; Miyaji, M.** (1991). Reevaluation of the teleomorph of the genus *Histoplasma* by ubiquinone systems. *Mycopathologia* 116: 151–154; **Guého, E.; Leclerc, M.C.; Hoog, G.S. de; Dupont, B.** (1997). Molecular taxonomy and epidemiology of *Blastomyces* and *Histoplasma* species. *Mycoses* 40: 69–81; **Larone, D.H.; Mitchell, T.G.; Walsh, T.J.** (1999). *Histoplasma, Blastomyces, Coccidioides*, and other dimorphic fungi causing systemic mycosis. *Manual of Clinical Microbiology* (Washington): 1259–1274; **San-Blas, G.; Niño-Vega, G.; Iturriaga, T.** (2002). *Paracoccidioides brasiliensis* and paracoccidioidomycosis: molecular approaches to morphogenesis, diagnosis, epidemiology, taxonomy and genetics. *Medical Mycol.* 40: 225–242; **Sano, A.; Tanaka, R.; Yokoyama, K.; Franco, M.; Bagagli, E.; Montenegro, M.R.; Mikami, Y.; Miyaji, M.; Nishimura, K.** (1999). Comparison between human and armadillo *Paracoccidioides brasiliensis* isolates by random amplified polymorphic DNA analysis. *Mycopathologia* 143: 165–169; **Semighini, C.P.; Camargo, Z.P. de; Puccia, R.; Goldman, M.H.S.; Goldman, G.H.** (2002). Molecular identification of *Paracoccidioides brasiliensis* by 5' nuclease assay. *Diagn. Microbiol. Infect. Dis.* 44: 383–386; **Sigler, L.** (1996). *Ajellomyces crescens* sp. nov., taxonomy of *Emmonsia* spp., and relatedness with *Blastomyces dermatitidis* (teleomorph *Ajellomyces dermatitidis*). *J. Med. Vet. Mycol.* 34: 303–314; **Sigler, L.** (2003). Ascomycetes. The *Onygenaceae* and other fungi from the order *Onygenales*. *Mycology Series* (New York) 16: 195–236; **Sugiyama, M.; Summerbell, R.C.; Mikawa, T.** (2002). Molecular phylogeny of onygenalean fungi based on small subunit (SSU) and large subunit (LSU) ribosomal DNA sequences. *Stud. Mycol.* 47: 5–23; **Ueda, Y.; Sano, A.; Tamura, M.; Inomata, T.; Kamei, K.; Yokoyama, K.; Kishi, F.; Ito, J.; Mikami, Y.; Miyaji, M.; Nishimura, K.** (2003). Diagnosis of histoplasmosis by detection of the internal transcribed spacer region of fungal rRNA gene from a paraffin-embedded skin sample from a dog in Japan. *Veter. Pathol.* 94: 219–224; **Untereiner, W.A.; Scott, J.A.; Naveau, F.A.; Currah, R.S.; Bachewich, J.** (2002). Phylogeny of *Ajellomyces, Polytolypa* and *Spiromastix* (*Onygenaceae*) inferred from rDNA sequence and non-molecular data. *Stud. Mycol.* 47: 25–35; **Untereiner, W.A.; Scott, J.A.; Naveau, F.A.; Sigler, L.; Bachewich, J.; Angus, A.** (2004). The *Ajellomycetaceae*, a new family of vertebrate-associated *Onygenales*. *Mycologia* 96: 812–821.

Albatrellaceae Nuss 1980
Polyporales: Basidiomycota

Basidioma annual, fleshy, stipitate, pileate; the stem central or eccentric, the cap medium-sized, becoming ± flat or convoluted, smooth, scaly or scurfy, usually white, yellowish or orange, pigments developing with age or after handling. Hyphae monomitic or mostly so, with or without clamp connections, thin-walled or thick-walled, often inflated, sometimes blueing in iodine. Hymenophore poroid, usually minutely and irregularly so, sometimes with isolated spines, whiteish or concolorous with the cap. Basidia ± cylindrical to clavate, with 4 sterigmata. Basidiospores ellipsoidal to subglobose, rarely lacrimiform or fusiform, smooth or rarely pitted, blueing in iodine in some species.

Albatrellus confluens; basidiomata

Significant Genera: *Albatrellus*.

Distribution: Widespread, primarily from temperate regions.

Economic Significance: Ningyoutake (*Albatrellus confluens*) is edible, and is claimed to contain cholesterol-lowering compounds, but is not exploited commercially.

Ecology: Most species are associated with conifers and considered to be ectomycorrhizal, but some may be saprobic and gain nutrition from rotten buried wood. The ectomycorrhizal association might not be stable.

Albatrellus ellisii; basidiomata

Notes: The *Albatrellaceae* has been considered to be related to the *Hericiaceae* and *Thelephoraceae* using morphological and mycorrhizal criteria respectively. Molecular studies indicate the closest relative to be the *Bondarzewiaceae*, within the russuloid clade of homobasidiomycetes. The principal genus *Albatrellus* is grossly polyphyletic. There are documented conservation concerns for some European and North American species, and some species have been investigated as old-growth forest indicators.

References: **Agerer, R.; Klostermeyer, D.; Steglich, W.; Franz, F.; Acker, G.** (1996). Ectomycorrhizae of *Albatrellus ovinus* (*Scutigeraceae*) on Norway spruce with some remarks on the systematic position of the family. *Mycotaxon* 59: 289–307; **Binder, M.; Hibbett, D.S.** (2002). Higher-level phylogenetic relationships of homobasidiomycetes (mushroom-forming fungi) inferred from four rDNA regions. *Mol. Phylogen. Evol.* 22: 76–90; **Binder, M.; Hibbett, D.S.; Larsson, K.-H.; Larsson, E.; Langer, E.; Langer, G.** (2005). The phylogenetic distribution of resupinate forms across the major clades of mushroom-forming fungi (*Homobasidiomycetes*). *Syst. Biodiv.* 3: 113–157; **Bruns, T.D.; Szaro, T.M.; Gardes, M.; Cullings, K.W.; Pan, J.J.; Taylor, D.L.; Horton, T.R.; Kretzer, A.; Garbelotto, M.; Li, Y.** (1998). A sequence database for the identification of ectomycorrhizal basidiomycetes by phylogenetic analysis. *Mol. Ecol.* 7: 257–272; **Corner, E.J.H.** (1989). Ad Polyporaceas V. The genera *Albatrellus, Boletopsis, Coriolopsis* (dimitic), *Cristelloporia, Diacanthodes, Elmerina, Fomitopsis* (dimitic), *Gloeoporus, Grifola, Hapalopilus, Heterobasidion, Hydnopolyporus, Ischnoderma, Loweporus, Parmastomyces, Perenniporia, Pyrofomes, Steccherinium, Trechispora, Truncospora* and *Tyromyces*. *Beih. Nova Hedwigia* 96: 218 pp.; **Dai, Y.C.; Zeng, X.L.** (1999). [*Jahnoporus*, a new record genus for the Chinese polypore flora]. *Mycosystema* 18: 226–227; **Ginns, J.** (1997). The taxonomy and distribution of rare or uncommon species of *Albatrellus* in western North America. *Can. J. Bot.* 75: 261–273; **Keller, J.** (1986). Ultrastructure des parois sporiques de quelques Aphyllophorales. *Mycol. helv.* 2: 1–33; **Larsson, E.; Larsson, K.-H.** (2003). Phylogenetic relationships of russuloid basidiomycetes with emphasis on aphyllophoralean taxa. *Mycologia* 95: 1037–1065; **Ryman, S.; Fransson, P.; Johanneson, H.; Danell, E.** (2003). *Albatrellus citrinus* sp. nov., connected to *Picea abies* on lime rich soils. *Mycol. Res.* 107: 1243–1246; **Stalpers, J.A.** (1992). *Albatrellus* and the *Hericiaceae*. *Persoonia* Suppl. 14: 537–541; **Thorn, R.G.** (2000). Some polypores misclassified in *Piptoporus*. *Karstenia* 40: 181–197; **Valenzuela, R.; Nava, R.; Cifuentes, J.** (1994). El género *Albatrellus* en México I. *Revta Mex. Micol.* 10: 113–152; **Zheng, G.Y.; Zhang, W.M.; Li, T.H.; Lai, J.P.** (1992). [The genus *Albatrellus* of Guangdong and Hainan provinces]. *Acta Mycol. Sin.* 11: 107–110.

Aliquandostipitaceae Inderb. 2001
Jahnulales: Ascomycota

Stroma absent. Ascomata (presumably ascostromata) perithecial, papillate, pale to dark brown, sessile or with a long stalk composed of a single broad thick-walled hypha. Ascomatal wall single-layered, membranous, smooth. Interascal tissue of septate sparsely branched pseudoparaphyses. Asci fissitunicate, clavate, the apex thickened with a small ocular chamber, not staining in iodine, 8-spored. Ascospores ellipsoidal to clavate, pale brown, 1- to 3-septate, thin-walled and smooth-walled, with a wide gelatinous sheath and also apical appendages.

Significant Genera: *Aliquandostipite, Jahnula*.

Distribution: *Aliquandostipite* is apparently pantropical, recorded from South-East Asia and Central America, with one record from the USA. *Jahnula* is more widely distributed.

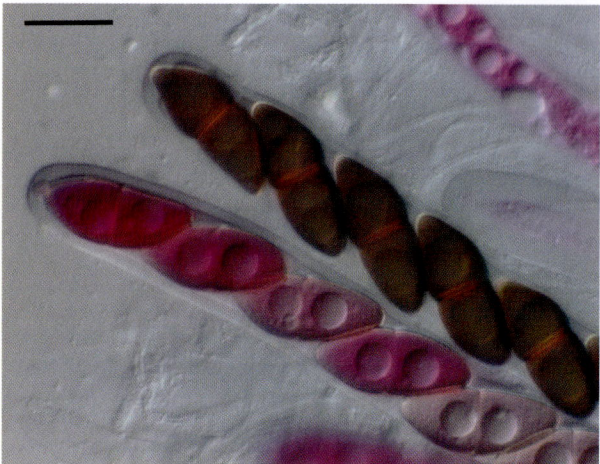

Jahnula aquatica, Devon, UK; asci and ascospores

Economic Significance: None is known.

Ecology: Saprobic, on decaying wood in aquatic habitats.

Notes: This family occupies a very isolated phylogenetic position, and may comprise a sister group to the entire *Dothideomycetes*. Curiosities include the extremely wide hyphal stalk subtending the ascoma in *Aliquandostipite*, which may be well developed or entirely absent within individual species.

References: **Hawksworth, D.L.** (1984). Observations on *Jahnula* Kirschst., a remarkable aquatic pyrenomycete. *Sydowia* 37: 43–46; **Inderbitzin, P.; Landvik, S.; Abdel-Wahab, M.A.; Berbee, M.L.** (2001). Aliquandostipitaceae, a new family for two new tropical ascomycetes with unusually wide hyphae and dimorphic ascomata. *Am. J. Bot.* 88: 52–61; **Pang, K.-L.; Abdel-Wahab, M.A.; Sivichai, S.; El-Sharouney, H.M.; Jones, E.B.G.** (2002). Jahnulales (Dothideomycetes, Ascomycota): a new order of lignicolous freshwater ascomycetes. *Mycol. Res.* 106: 1031–1042; **Raja, H.A.; Ferrer, A.; Shearer, C.A.** (2005). *Aliquandostipite crystallinus*, a new ascomycete species from wood submerged in freshwater habitats. *Mycotaxon* 91: 207–215; **Raja, H.A.; Shearer, C.A.** (2006). *Jahnula* species from North and Central America, including three new species. *Mycol.* 98: 319–332.

Amorphothecaceae Parbery 1969
Uncertain position within Eurotiomycetes, Ascomycota

Stromata absent. Ascomata cleistothecial, dark brown to black, ± globose; peridium amorphous, sometimes incorporating a few hyphae, often variable in thickness, evanescent. Interascal tissue absent. Asci irregularly arranged, ± globose, very thin-walled, evanescent. Ascospores pale brown, aseptate, thin-walled, without germ pores, without a sheath. Anamorph hyphomycetous, with pigmented conidiophores and sympodially proliferating denticulate conidiogenous cells, producing often-branched chains of brown septate or aseptate conidia.

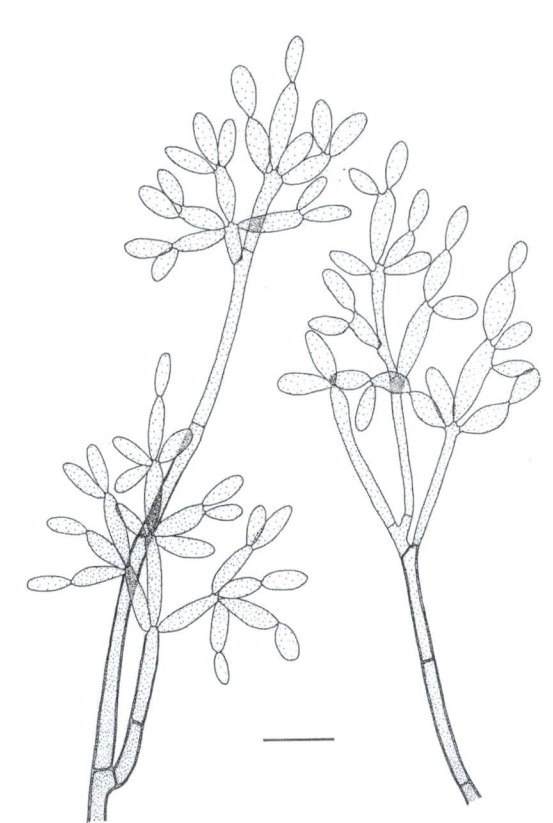

Amorphotheca resinae; conidiophores and conidia

Significant Genera: *Amorphotheca*. Anamorph: *Hormoconis*.

Distribution: Widespread.

Economic Significance: The single included species (better known as its anamorph *Hormoconis resinae*) is a major problem in the aviation industry. It grows at the interface between kerosene and contaminating water in fuel, and produces acids that erode fuel tanks and lines.

Ecology: In soil, or associated with hydrocarbons; isolated from substrata such as creosote-coated utility poles as well as fuel.

Notes: Affinities are uncertain with few molecular data available, but the family appears to be most closely allied to the *Herpotrichiellaceae* (*Chaeothyriomycetes*). The teleomorph is rarely formed and has only been described on one occasion.

Amorphotheca resinae; conidiophores and conidia

References: **Abliz, P.; Fukushima, K.; Takizawa, K.; Nishimura, K.** (2004). Identification of pathogenic dematiaceous fungi and related taxa based on large subunit ribosomal DNA D1/D2 domain sequence analysis. *FEMS Immunol. Med. Microbiol.* 40: 41–49; **Braun, U.; Crous, P.W.; Dugan, F.; Hoog, G.S. de** (2003). Phylogeny and taxonomy of *Cladosporium*-like hyphomycetes, including *Davidiella* gen. nov., the teleomorph of *Cladosporium* s. str. *Mycol. Prog.* 2: 3–18; **Parbery, D.G.** (1969). *Amorphotheca resinae* gen. nov., sp. nov.: the perfect state of *Cladosporium resinae*. *Aust. J. Bot.* 17: 331–357.

Amphisphaeriaceae G. Winter 1885
Xylariales: Ascomycota

Stromata crustose, often clypeate, rarely absent. Ascomata perithecial, immersed or erumpent, ± globose, the ostiole papillate, periphysate. Interascal tissue of numerous thin-walled usually gelatinized true paraphyses. Asci cylindrical, persistent, with a small, usually J+ apical ring. Ascospores hyaline to brown, ellipsoidal to fusiform, sometimes elongate, transversely septate, rarely ornamented, often with germ pores. Anamorphs coelomycetous, usually acervular, often with versicoloured appendaged conidia.

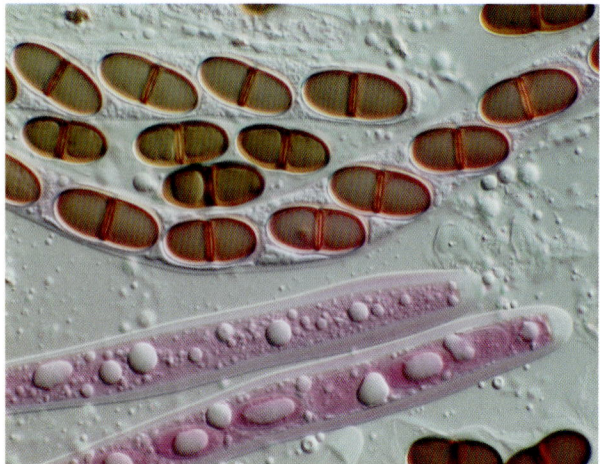

Amphisphaeria fallax, Italy; asci and ascospores

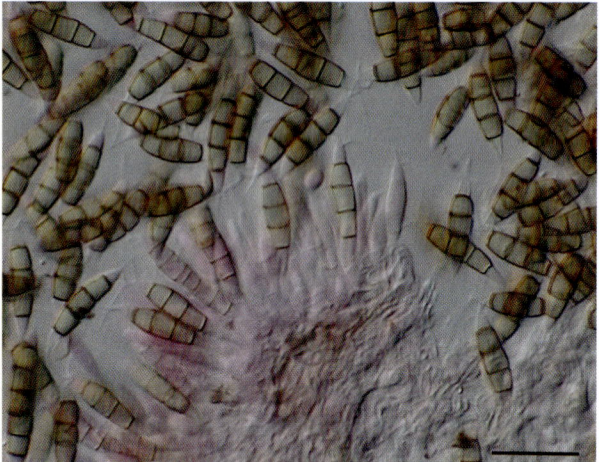

Pestalotiopsis aff. *uvicola* from mango skin; conidia and conidiogenous cells

Significant Genera: *Amphisphaeria*, *Pestalosphaeria*. Anamorphs: *Pestalotiopsis*, *Seimatosporium*, *Seiridium*.

Distribution: Widespread in both temperate and tropical zones.

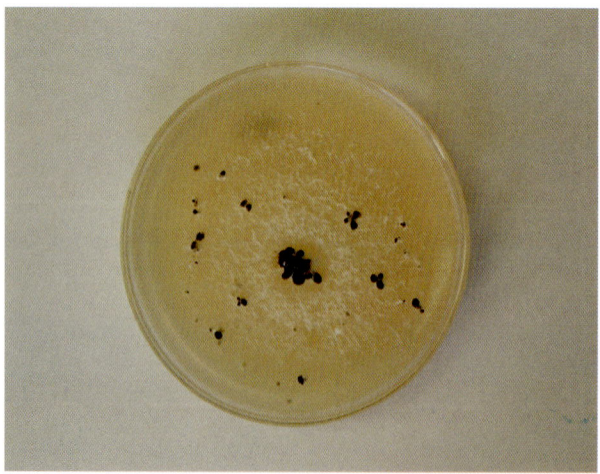

Pestalotiopsis aff. *uvicola* from mango skin, unlocalized; colony

Economic Significance: Some species are parasites of crop plants and trees, though few are significant primary pathogens. Some endophytic taxa produce taxol and other pharmaceutically active compounds.

Ecology: Saprobes or weak parasites of leaves, stems etc., often appearing as secondary colonizers following damage to plants by other agents.

Notes: A rather heterogenous family, despite recent monographic treatment of several of its constituent groups and a number of phylogenetic analyses. Placement in the *Xylariales* appears appropriate for the time being.

References: Goh, T.K.; Hyde, K.D. (1997). *Lepteutypa hexagonalis* sp. nov. from *Pinanga* sp. in Ecuador. *Mycol. Res.* 101: 85–88; **Graniti, A.** (1998). Cypress canker: a pandemic in progress. *A. Rev. Phytopath.* 36: 91–114; **Hyde, K.D.** (1997). The genus *Roussoëlla*, including two new species from palms in Cuyabeno, Ecuador. *Mycol. Res.* 101: 609–616; **Jeewon, R.; Liew, E.C.Y.; Hyde, K.D.** (2002). Phylogenetic relationships of *Pestalotiopsis* and allied genera inferred from ribosomal DNA sequences and morphological characters. *Mol. Phylogen. Evol.* 25: 378–392; **Jeewon, R.; Liew, E.C.Y.; Hyde, K.D.** (2004). Phylogenetic evaluation of species nomenclature of *Pestalotiopsis* in relation to host association. *Fungal Diversity* 17: 39–55; **Kang, J.C.; Hyde, K.D.; Kong, R.Y.C.** (1999). Studies on *Amphisphaeriales*: the *Amphisphaeriaceae* (*sensu stricto*). *Mycol. Res.* 103: 53–64; **Kang, J.C.; Kong, R.Y.C.; Hyde, K.D.** (1998). Studies on the *Amphisphaeriales* 1. *Amphisphaeriaceae* (*sensu stricto*) and its phylogenetic relationships inferred from 5.8S rDNA and ITS2 sequences. *Fungal Diversity* 1: 147–157; **Nag Raj, T.R.; Mel'nik, V.** (1994). Redisposals and redescriptions in the *Monochaetia* – *Seiridium*, *Pestalotia* – *Pestalotiopsis* complexes. X. *Pestalotia granati* and *Pestalozzina punicae*. *Mycotaxon* 50: 435–440; **Okane, I.; Nakagiri, A.; Ito, T.** (1996). *Discostroma tricellulare*, a new endophytic ascomycete with a *Seimatosporium* anamorph isolated from *Rhododendron*. *Can. J. Bot.* 74: 1338–1344; **Strobel, G.A.; Ford, E.; Li, J.Y.; Sears, J.; Sidhu, R.S.; Hess, W.M.** (1999). *Seimatoantlerium tepuiense* gen. nov., a unique epiphytic fungus producing taxol from the Venezuelan Guyana. *Syst. Appl. Microbiol.* 22: 426–433.

Anamylopsoraceae Lumbsch & Lunke 1995
Agyriales: Ascomycota

Thallus squamulose, orange to brown. Ascomata apothecial, without a distinct margin, dark brown, distinctly stalked, the excipular tissue weakly annular, hymenium not blueing in iodine. Interascal tissue composed of sparingly branched paraphyses, slightly thickened at the apex and with a brownish gelatinized epithecium. Asci with a weakly J+ outer wall and a thickened domed J– tholus. Ascospores hyaline, aseptate. Anamorph pycnidial.

Significant Genera: *Anamylopsora*.

Distribution: Widespread.

Economic Significance: None is known.

Ecology: Lichenized with green algae.

Notes: Only a single species has been described.

References: **Döring, H.; Lumbsch, H.T.** (1998). Ascoma ontogeny: is this character set of any use in the systematics of lichenized ascomycetes?. *Lichenologist* 30: 489–500; **Huneck, S.; Elix, J.A.** (1993). The chemistry of the lichens *Anamylopsora pulcherrima* and *Tephromela armeniaca*. *Herzogia* 9: 647–651; **Lumbsch, H.T.** (1997). Systematic studies in the suborder *Agyriineae* (*Lecanorales*). *J. Hattori bot. Lab.* 83: 1–73; **Lumbsch, H.T.; Lunke, T.; Feige, T.B.; Huneck, S.** (1995). *Anamylopsoraceae* – a new family of lichenized ascomycetes with stipitate apothecia (*Lecanorales*: *Agyriineae*). *Pl. Syst. Evol.* 198: 275–286; **Lumbsch, H.T.; Schmitt, I.; Döring, H.; Wedin, M.** (2001). ITS sequence data suggest variability of ascus types support ontogenetic characters as phylogenetic discriminators in the *Agyriales* (*Ascomycota*). *Mycol. Res.* 105: 265–274; **Lumbsch, H.T.; Schmitt, I.; Döring, H.; Wedin, M.** (2001). Molecular systematics supports the recognition of an additional order of *Ascomycota*: the *Agyriales*. *Mycol. Res.* 105: 16–23; **Timdal, E.** (1991). *Anamylopsora*, a new genus in the *Lecideaceae*. *Mycotaxon* 42: 249–254.

Ancylistaceae J. Schröt. 1893
Entomophthorales: Zygomycota

Colonies growing rapidly, initially flat but becoming radially folded, cream but darkening to tan or brown with age, appearing powdery due to surface mycelium and sporophores (conidiophores). Sporophores simple, solitary, without subsporangial vesicle, arising directly from the vegetative mycelium. Spores (conidia) terminal, spherical, single-celled, with a prominent papilla and sometimes with hair-like appendages, forcibly discharged; germinating to form either hyphae or one to several short sporophores, each bearing a small secondary spore; nucleus difficult to observe during mitosis, nucleolus prominent. Zygospores formed by hyphal conjugation.

Significant Genera: *Conidiobolus*.

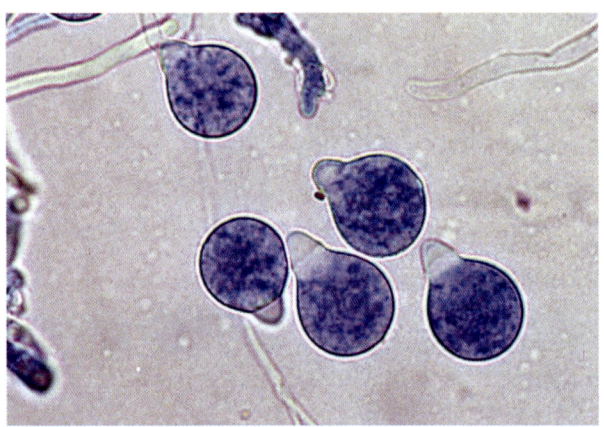

Conidiobolus; conidia

Distribution: Widespread but more commonly encountered in subtropical and tropical areas.

Economic Significance: Species of *Conidiobolus* cause entomophthoromycosis in man. The primary species is *C. coronatus* with a world-wide distribution, but mainly found in tropical rain forests, especially in Africa.

Ecology: Mainly saprobes in the soil or on decaying leaves and other dead plant material; some pathogens of insects, soil invertebrates, desmid algae or facultative mycotic agents in man.

Notes: Very poorly known; no recent information is available.

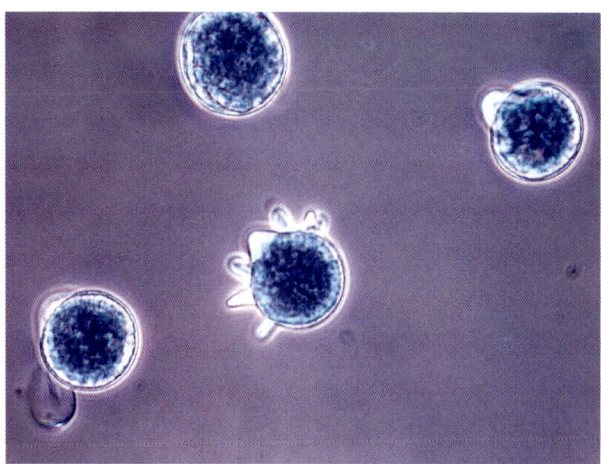

Conidiobolus; conidia

References: **Humber, R.A.** (1989). Synopsis of a revised classification for the *Entomophthorales* (*Zygomycotina*). *Mycotaxon* 34: 441–460; **Kędra, E.; Boguś, M.I.** (2006). The influence of *Conidiobolus coronatus* on phagocytic activity of insect hemocytes. *J. Invert. Path.* 91: 50–52; **Keller, S.; Petrini, O.** (2005). Keys to the identification of the arthropod pathogenic genera of the families *Entomophthoraceae* and *Neozygitaceae* (*Zygomycetes*), with descriptions of three new subfamilies and a new genus. *Sydowia* 57: 23–53; **Tadano, T.; Paim, N.P.; Hueb, M.; Fontes, C.J.** (2005). Entomophthoramycosis (zygomycosis) caused by *Conidiobolus coronatus* in Mato Grosso (Brazil): case report. *Revta Soc. Bras. Med. Trop.* 38: 188–190; **Tanabe, Y.; Saikawa, M.; Watanabe, M.M.; Sugiyama, J.** (2004). Molecular phylogeny of *Zygomycota* based on EF-1α and RPB1 sequences: limitations and utility of alternative markers to rDNA. *Mol. Phylogen. Evol.* 30: 438–449; **Voigt, K.; Cigelnik, E.; O'Donnell, K.** (1999). Phylogeny and PCR identification of clinically important *Zygomycetes* based on nuclear ribosomal-DNA sequence data. *J. Clin. Microbiol.* 37: 3957–3964; **Wolf, F.T.** (1988). *Entomophthorales* and their parasitism of insects. *Nova Hedwigia* 46: 121–142.

Annulatascaceae S.W. Wong, K.D. Hyde & E.B.G. Jones 1998
Uncertain position within Sordariomycetes, Ascomycota

Stromata absent. Ascomata solitary or aggregated, immersed to erumpent, dark brown to black, sometimes long-necked, periphysate; peridium coriaceous, composed of dark brown angular cells, paler and thinner-walled towards the cavity. Interascal tissue of wide persistent thin-walled gelatinized true paraphyses. Asci cylindrical, fairly thin-walled, not fissitunicate, with a massive refractive usually J– apical ring. Ascospores hyaline or brown, usually septate, rarely with a germ pore, sometimes versicoloured or with hyaline apical chambers, smooth or verruculose, often with polar appendages or a gelatinous sheath.

Significant Genera: *Annulatascus*, *Ceratosphaeria*.

Distribution: Widespread; many species have tropical distributions.

Economic Significance: None is known.

Ecology: Saprobes in rotten wood, often in freshwater habitats.

Notes: This may well be a polyphyletic family, and many of its constituent genera need revision. The family appears to occupy an isolated position within the *Sordariomycetes*, and there is now doubt as to whether *Ceratosphaeria* should be placed here rather than with the *Magnaporthaceae*.

References: **Campbell, J.; Shearer, C.A.** (2004). *Annulusmagnus* and *Ascitendus*, two new genera in the *Annulatascaceae*. *Mycologia* 96: 822–833; **Ho, W.H.; Hyde, K.D.** (2000). A new family of freshwater ascomycetes. *Fungal Diversity* 4: 21–36; **Ho, W.H.; Ranghoo, V.M.; Hyde, K.D.; Hodgkiss, I.J.** (1999). Ascal ultrastructural study in *Annulatascus hongkongensis* sp. nov., a freshwater ascomycete. *Mycologia* 91: 885–892; **Ho, W.W.H.; Tsui, C.K.M.; Hodgkiss, I.J.; Hyde, K.D.** (1999). *Aquaticola*, a new genus of *Annulatascaceae* from freshwater habitats. *Fungal Diversity* 3: 87–97; **Inderbitzin, P.** (2000). *Ceratostomella hyalocoronata*, a new pyrenomycete from a stream in southern China. *Mycoscience* 41: 167–169; **Ranghoo, V.M.; Hyde, K.D.; Liew, E.C.Y.** (1999). Family placement of *Ascotaiwania* and *Ascolacicola* based on DNA sequences from the large subunit rRNA gene. *Fungal Diversity* 2: 159–168; **Réblová, M.** (2006). Molecular systematics of *Ceratostomella* sensu lato and morphologically similar fungi. *Mycologia* 98: 68–93; **Wong, S.W.; Hyde, K.D.; Jones, E.B.G.** (1998). *Annulatascaceae*, a new ascomycete family from the tropics. *Syst. Ascom.* 16: 17–25.

Antennulariellaceae Woron. 1925
Capnodiales: Ascomycota

Superficial mycelium irregular, dark, smooth or rough-walled, adpressed or erect. Ascomata small, perithecial, ± globose, stalked or sessile, opening by a small poorly defined lysigenous pore, sometimes with hyphal appendages. Interascal tissue absent or of inconspicuous apical periphysoids. Asci small, ovoid, fissitunicate, not blueing in iodine, 8-spored. Ascospores hyaline to brown, usually 1-septate, sheath lacking. Anamorphs coelomycetous and hyphomycetous, if the latter then conidia elongate, multiseptate.

Significant Genera: *Achaetobotrys*, *Antennulariella*. Anamorphs: *Antennariella*, *Capnodendron*.

Distribution: Widespread; warm temperate to tropical.

Economic Significance: Little is known, though colonies may affect photosynthetic potential of leaves.

Ecology: Saprobic on exudates, usually epiphytic on leaves.

Notes: One of the families of sooty moulds, which form conspicuous black communities of fungi on living plant parts in areas of high humidity. No molecular research is available.

References: **Barr, M.E.; Rogerson, C.T.** (1999). Some loculoascomycete species from the Great Basin, USA. *Mycotaxon* 71: 473–480; **Hughes, S.J.** (1976). Sooty moulds. *Mycologia* 68: 693–820; **Hughes, S.J.** (2000). *Antennulariella batistae* n. sp. and its *Capnodendron* and *Antennariella* synanamorphs, with notes on *Capnodium capsuliferum*. *Can. J. Bot.* 78: 1215–1226; **Reynolds, D.R.** (1986). Foliicolous ascomycetes 7. Phylogenetic systematics of the *Capnodiaceae*. *Mycotaxon* 27: 377–403.

Anthracoideaceae Denchev 1997
Ustilaginales: Basidiomycota

Anthracoidea heterospora, Sweden; infected inflorescence

Sori globose to ovoid, composed of host ovarian tissue surrounded by a dry black spore mass, initially covered by a thin greyish layer of fungal cells. Teliospores formed singly, globose, ovoid or polygonal, often flattened, dark brown (rarely with pale, thinner-walled polar regions), usually ornamented with spines, warts or granules, surrounded by a gelatinous sheath that breaks down at maturity. Spores germinate to form 2-spored aerial basidia, in most cases each cell producing a single basidium from a well-developed sterigma, terminal in the upper cell and lateral in the lower. Basidiospores ovoid to cylindrical, hyaline, not discharged explosively, apparently homothallic.

Significant Genera: *Anthracoidea*.

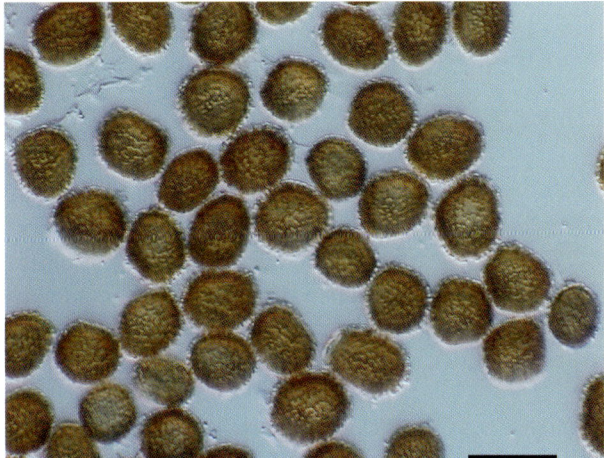
Anthracoidea echinospora, Sweden; teliospores

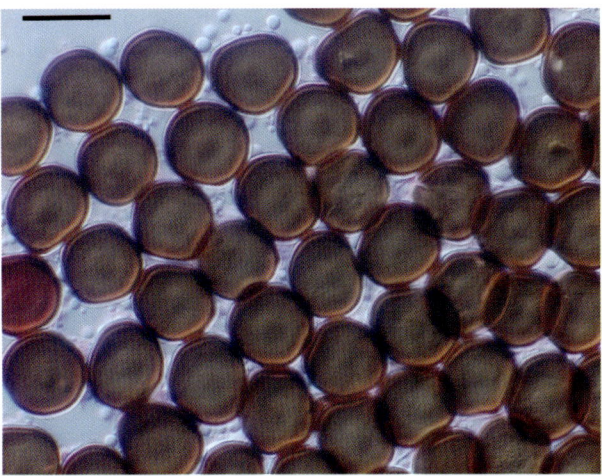
Anthracoidea elynae, Ellesmere Island, Canada; teliospores

Distribution: Widespread, primarily in north and south temperate ecosystems.

Economic Significance: None is known.

Ecology: Biotrophic in ovaries of the *Cyperaceae*.

Notes: Diagnostic features of the family are primarily the 2-celled aerial basidia, a feature not known elsewhere in the smut fungi. Molecular phylogenetic studies have not been carried out to establish the position of the *Anthracoideaceae* within the *Ustilaginales*.

References: **Bauer, R.; Oberwinkler, F.; Vánky, K.** (1997). Ultrastructural markers and systematics in smut fungi and allied taxa. *Can. J. Bot.* 75: 1273–1314; **Denchev, C.M.** (1997). *Anthracoideaceae*, a new family in the *Ustilaginales*. *Mycotaxon* 65: 411–417; **Hendrichs, M.; Begerow, D.; Bauer, R.; Oberwinkler, F.** (2005). The genus *Anthracoidea* (*Basidiomycota, Ustilaginales*): a molecular phylogenetic approach using LSU rDNA sequences. *Mycol. Res.* 109: 31–40; **Ingold, C.T.** (1989). The basidium of *Anthracoidea inclusa* in relation to smut taxonomy. *Mycol. Res.* 92: 245–246; **Vánky, K.** (1987). Illustrated genera of smut fungi. *Cryptog. Stud.* 1: 159 pp.; **Vánky, K.** (1994). *European Smut Fungi* (Stuttgart): 570 pp.; **Vánky, K.** (1995). Taxonomical studies on *Ustilaginales*. XII. *Mycotaxon* 54: 215–238; **Vánky, K.; Websdane, K.** (1995). *Ustilaginales* of *Schoenus* (*Cyperaceae*). *Mycotaxon* 56: 217–229.

Aphanopsidaceae Printzen & Rambold 1995
Lecanorales: Ascomycota

Thallus crustose. Ascomata apothecial, ± flat, without a well-developed wall. Interascal tissue of rarely branched thin-walled paraphyses, the apices not swollen. Asci thick-walled, with a well-developed apical dome, ocular chamber poorly developed, with a well-developed J+ plug or tube and an outer J+ gelatinized layer. Ascospores hyaline, aseptate, without a sheath. Anamorphs pycnidial.

Significant Genera: *Aphanopsis*.

Distribution: Only reported from Europe.

Economic Significance: None is known.

Ecology: Lichenized with green algae; on soil, rotten wood and similar substrata.

Notes: Affinities are uncertain and no sequences are available, so placement within the *Lecanorales* relies only on morphological comparison.

References: **Kantvilas, G.; McCarthy, P.M.** (1999). *Steinia australis*, a new species in the lichen family *Aphanopsidaceae*. *Lichenologist* 31: 555–558; **Printzen, C.; Rambold, G.** (1995). *Aphanopsidaceae* – a new family of lichenized *Ascomycetes*. *Lichenologist* 27: 99–103.

Aphelariaceae Corner 1970
Cantharellales: Basidiomycota

Basidioma ramarioid, caespitose, repeatedly branched and occasionally anastomosed, small, slender, cartilaginous, usually whiteish, pink or buff, the basal part sometimes fused into a short stipe and the upper branches usually flattened and cristate. Hyphae monomitic, ± thin-walled, not inflated between septa, clamp connections present or absent, cystidia absent. Basidia clavate, mostly with 3 or 4 elongate sterigmata, lacking clamp connections, not forming secondary septa. Basidiospores ellipsoidal, ovoid or ± cylindrical, hyaline, smooth, not staining in iodine, with a papillate hilum.

Significant Genera: *Aphelaria*.

Distribution: Widespread in tropical regions.

Economic Significance: None is known.

Ecology: Little is known, but species are assumed to be saprobic, rotting dead wood and litter.

Notes: One of two families of the *Cantharellales* with elongate fruit-bodies; the *Clavulinaceae* differs by having clavarioid basidiomata and curved, 2-spored basidia. No molecular data are available.

References: **Corner, E.J.H.** (1950). A monograph of *Clavaria* and allied genera. *Ann. Bot. Mem.* 1: 1–740; **Corner, E.J.H.** (1966). The clavarioid complex of *Aphelaria* and *Tremellodendropsis*. *Trans. Br. mycol. Soc.* 49: 205–211; **Petersen, R.H.; Zang, M.** (1986). New or interesting clavarioid fungi from Yunnan, China. *Acta bot. Yunn.* 8: 281–294; **Roberts, P.** (1999). Clavarioid fungi from Korup National Park, Cameroon. *Kew Bull.* 54: 517–539.

Apiosporaceae K.D. Hyde, J. Fröhl., Joanne E. Taylor & M.E. Barr 1998
Uncertain position within Sordariomycetes, Ascomycota

Stromata poorly developed, though ascomata are frequently strongly aggregated in a linear arrangement. Ascomata perithecial, immersed in the substratum, black, thin-walled, papillate, the ostiole periphysate. Interascal tissue of thin-walled unbranched hypha-like true paraphyses. Asci clavate, thin-walled, not fissitunicate, without distinct apical structures, 8-spored. Ascospores hyaline, 1-septate with the septum near the base, thin-walled, smooth, often with a mucous sheath. Anamorphs hyphomycetous with basauxic conidiophores (often with thickened, pigmented septa) formed from generative cells, and dark, variably shaped and often ornamented conidia with germ pores or slits formed laterally from cells of the conidiophore.

Significant Genera: *Apiospora*. Anamorphs: *Arthrinium*, *Cordella*.

Distribution: Widespread.

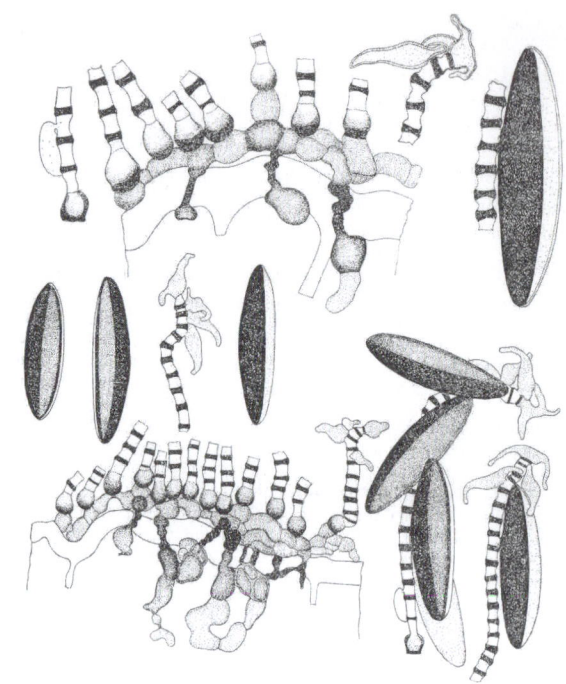

Arthrinium caricicola; conidiophores and conidia

Economic Significance: Little is known, although some species have been implicated as allergens and mycotoxin producers.

Ecology: Saprobic on dead plant parts, especially of palms, grasses etc., also isolated from soil.

Notes: At one time considered to be part of the *Lasiosphaeriaceae*, but anamorphic features are completely different to those of that family, and molecular studies have confirmed the separate nature of the two families. Further research is needed, but a position in the *Xylariales* appears to be justified.

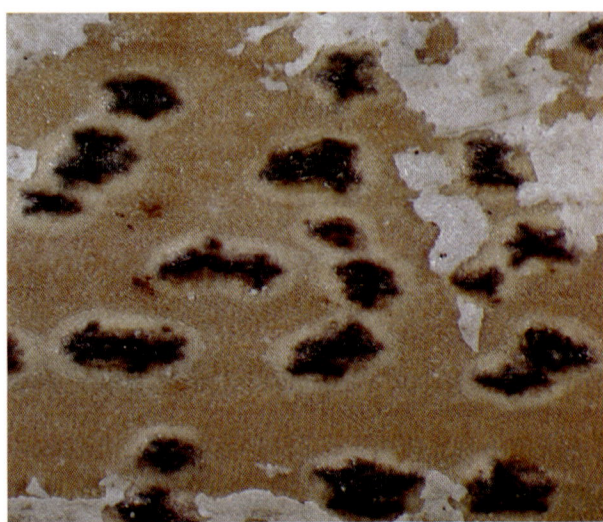

Apiospora bambusae, Gloucestershire, UK; ascomata on bamboo stem

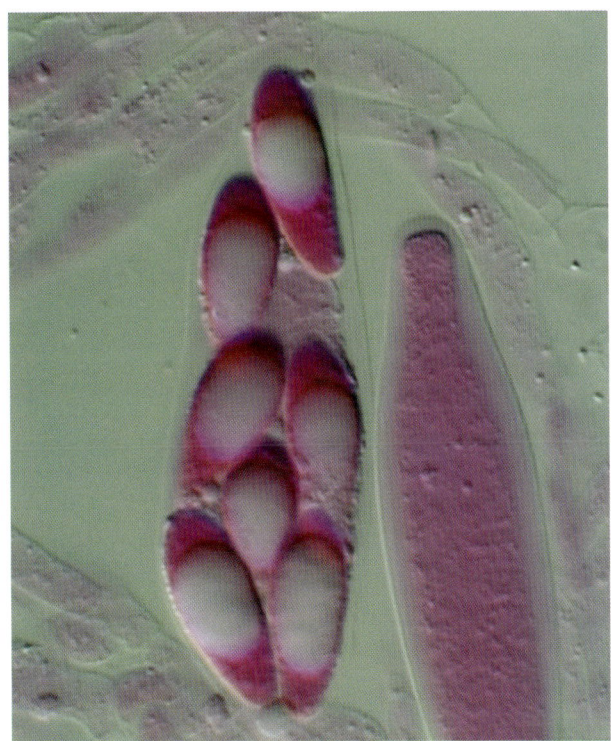

Apiospora bambusae, Gloucestershire, UK; ascus

References: **Huhndorf, S.M.; Miller, A.N.; Fernández, F.A.** (2004). Molecular systematics of the *Sordariales*: the order and the family *Lasiosphaeriaceae* revisited. *Mycologia* 96: 368–387; **Hyde, K.D.; Fröhlich, J.; Taylor, J.E.** (1998). Fungi from palms. XXXVI. Reflections on unitunicate ascomycetes with apiospores. *Sydowia* 50: 21–80; **Müller, E.** (1992). A new parasitic species of *Apiospora*. *Boln Soc. argent. Bot.* 28: 201–203; **Samuels, G.S.; McKenzie, E.H.C.; Buchanan, D.E.** (1981). Ascomycetes of New Zealand. 3. Two new species of *Apiospora* and their *Arthrinium* anamorphs on bamboo. *N.Z. Jl Bot.* 19: 137–149; **Wang, Y.Z.; Hyde, K.D.** (1999). *Hyponectria buxi* with notes on the *Hyponectriaceae*. *Fungal Diversity* 3: 159–172.

Aporpiaceae Bondartsev & Bondartseva 1960
Tremellales: Basidiomycota

Basidiomata pileate or resupinate, coriaceous, poroid; when pileate upper surface smooth or tomentose. Hyphal structure usually dimitic, with thin-walled to thick-walled, sometimes encrusted skeletal hyphae dominating, and thin-walled generative hyphae usually with clamp connections. Hymenium cream, pale brown or pinkish, with angular pores varying considerably in size, often with conspicuous hyphal pegs fringing the pore mouths. Cystidia present or absent. Basidia 4-spored, varied in shape, becoming cruciately and longitudinally septate, usually dividing off a small basal cell with a clamp connection. Basidiospores variably shaped, thin-walled, hyaline, not staining in iodine.

Significant Genera: *Elmerina*.

Distribution: Widespread, with most species occurring in tropical climates.

Economic Significance: None is known.

Ecology: Little information is available, but species are probably saprobic on rotten wood.

Notes: Molecular data are sparse, but it appears that the *Aporpiaceae* is an isolated group within the *Tremellomycetes*. It is unusual in this group in having well-developed coriaceous polypore-like basidiomata.

References: **Corner, E.J.H.** (1989). Ad Polyporaceas V. The genera *Albatrellus*, *Boletopsis*, *Coriolopsis* (dimitic), *Cristelloporia*, *Diacanthodes*, *Elmerina*, *Fomitopsis* (dimitic), *Gloeoporus*, *Grifola*, *Hapalopilus*, *Heterobasidion*, *Hydnopolyporus*, *Ischnoderma*, *Loweporus*, *Parmastomyces*, *Perenniporia*, *Pyrofomes*, *Steccherinium*, *Trechispora*, *Truncospora* and *Tyromyces*. *Beih. Nova Hedwigia* 96: 218 pp.; **Larsson, K.-H.; Larsson, E.; Koljalg, U.** (2004). High phylogenetic diversity among corticioid homobasidiomycetes. *Mycol. Res.* 108: 983–1002; **Núñcz, M.** (1997). *Protodaedalea*, a synonym of *Elmerina* (*Heterobasidiomycetes*). *Mycotaxon* 61: 177–183; **Núñez, M.** (1998). The genus *Elmerina* (*Heterobasidiomycetes*) in Japan. *Folia cryptog. Estonica* 33: 99–101; **Reid, D.A.** (1992). The genus *Elmerina* (*Tremellales*), with accounts of two species from Queensland, Australia. *Persoonia* Suppl. 14: 465–474; **Ryvarden, L.** (1991). Genera of Polypores. Nomenclature and Taxonomy. *Syn. Fung.* (Oslo) 5: 363 pp.; **Weiss, M.; Oberwinkler, F.** (2001). Phylogenetic relationships in *Auriculariales* and related groups – hypotheses derived from nuclear ribosomal DNA sequences. *Mycol. Res.* 105: 403–415.

Appendicisporaceae C. Walker, Vestberg & A. Schüssler 2007
Archaeosporales: Glomeromycota

Mycorrhizal species forming arbuscules and vesicles in roots (intraradical), anamorphic; the penetrating hyphae producing finely branched haustorial branches (arbuscules) or coils (pelotons) and vesicles. Spores hypogeous or partly epigeous, small, hyaline, monomorphic and acaulosporoid (developing from the neck of a saccule) and/or dimorphic and glomeroid (developing blastically from the tip of a sporogenous hypha); spore wall multilayed, with a thick and flexible inner layer (thin bi-layered flexible inner walls and germination orb absent).

Significant Genera: *Appendicispora*.

Distribution: Widespread, although undoubtedly more so than current records suggest.

Economic Significance: Enhancement of phosphorus (P) supply to the 'host' is a characteristic of AM (arbuscular mycorrhizal) infections. As such they have a significant influence, either directly or indirectly, on land plants including ecosystem variability and productivity.

Ecology: As mycorrhizal partners of land plants forming arbuscular mycorrhizas; the hosts of AM fungi are primarily found in P-limited tropical and temperate grasslands.

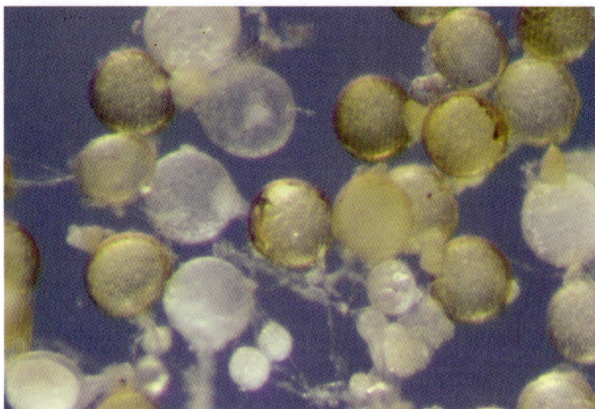

Appendicispora fennica; spores

References: Walker, C.; Vestberg, M.; Demircik, F.; Stockingler, H.; Saito, M.; Sawaki, H.; Nishmura, I; Schüssler, A. (2007). Molecular phylogeny and new taxa in the *Archaeosporales* (*Glomeromycota*: *Ambispora fennica* gen. sp. nov., *Ambisporaceae* fam. nov., and emendation of *Archaeospora* and *Archaeosporaceae*. *Mycol. Res.* 111: 137-153; **Walker, C.; Vestberg, M.; Schüssler, A.** (2007). Nomenclatural clarifications in *Glomeromycota*. *Mycol. Res.* 111: [in press].

Archaeosporaceae Morton & Redecker 2001
Archaeosporales: Glomeromycota

Mycorrhizal species forming arbuscules and vesicles in roots (intraradical), anamorphic; the penetrating hyphae producing finely branched haustorial branches (arbuscules) or coils (pelotons) and vesicles. Spores hypogeous or partly epigeous, monomorphic and acaulosporoid (developing from the neck of a saccule or from a hypha or pedicel branching from the saccule neck) and/or dimorphic and glomeroid (developing blastically from the tip of a sporogenous hypha); spore wall multilayed, with a thick and flexible inner layer (thin bi-layered flexible inner walls and germination orb absent).

Significant Genera: *Archaeospora*.

Distribution: Widespread, although undoubtedly more so than current records suggest.

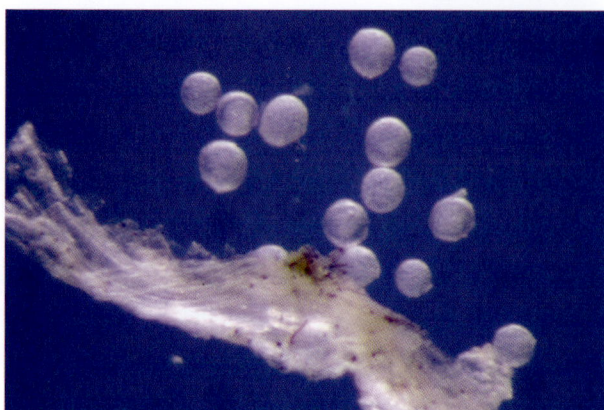

Archaeospora trappei; spores

Economic Significance: Enhancement of phosphorus (P) supply to the 'host' is a characteristic of AM (arbuscular mycorrhizal) infections. As such they have a significant influence, either directly or indirectly, on land plants including ecosystem variability and productivity.

Ecology: As mycorrhizal partners of land plants forming arbuscular mycorrhizas; the hosts of AM fungi are primarily found in P-limited tropical and temperate grasslands.

Notes: An unusual feature of the family is the formation of dimorphic spores, with atypical *Acaulospora*-like propagules formed as well as *Glomus*-like structures. Considered (with *Paraglomeraceae*) to form the 'ancestral' lineage within the *Glomeromycota*.

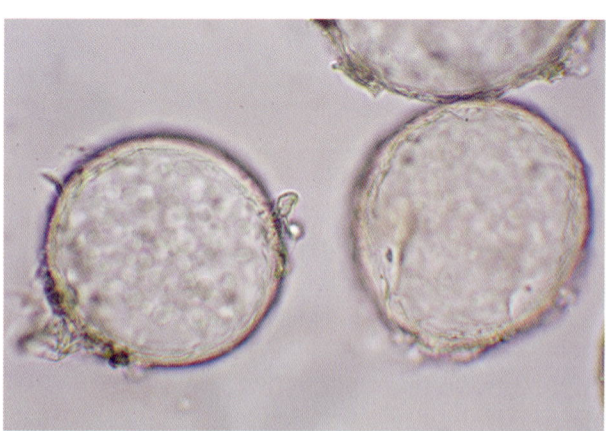

Archaeospora trappei; spores

References: Azcon-Aguilar, C.; Barea, J.M. (1996). Arbuscular mycorrhizas and biological control of soil-borne plant pathogens – an overview of the mechanisms involved. *Mycorrhiza* 6: 457–464; **Hafeel, K.M.** (2004). Spore ontogeny of the arbuscular mycorrhizal fungus *Archaeospora trappei* (Ames & Linderman) Morton & Redecker (*Archaeosporaceae*). *Mycorrhiza* 14: 213–219; **Morton, J.B.; Redecker, D.** (2001). Two new families of *Glomales*, *Archaeosporaceae* and *Paraglomaceae*, with two new genera *Archaeospora* and *Paraglomus*, based on concordant molecular and morphological characters. *Mycologia* 93: 181–195; **Redecker, D.** (2005). *Glomeromycota* Arbuscular mycorrhizal fungi and their relative(s).

Version 01 July 2005. http://tolweb.org/Glomeromycota/28715/ 2005.07.01 in The Tree of Life Web Project, http://tolweb.org; **Schüßler, A.; Schwarzott, D.; Walker, C.** (2001). A new fungal phylum, the *Glomeromycota*: phylogeny and evolution. *Mycol. Res.* 105: 1413–1421; **Spain, J.L.** (2003). Emendation of *Archaeospora* and of its type species, *Archaeospora trappei*. *Mycotaxon* 87: 109–112; **van der Heijden, M.A.G.; Klironomos, J.N.; Ursic, M.; Moutoglis, P.; Streitwolf-Engel, R.; Boller, T.; Wiemken, A.; Sanders, I.R.** (1998). Mycorrhizal fungal diversity determines plant biodiversity, ecosystem variability and productivity. *Nature* Lond. 396: 69–72; **Walker, C.; Vestberg, M.; Demircik, F.; Stockinger, H.; Saito, M.; Sawaki, H.; Nishmura, I; Schüssler, A.** (2007). Molecular phylogeny and new taxa in the *Archaeosporales* (*Glomeromycota*: *Ambispora fennica* gen. sp. nov., *Ambisporaceae* fam. nov., and emendation of *Archaeospora* and *Archaeosporaceae*. *Mycol. Res.* 111: 137-153;

Arctomiaceae Th. Fr. 1860
Lecanorales: Ascomycota

Thallus crustose or fruticose, gelatinized, with rhizoids. Ascomata apothecial, the outer wall often not well-developed, sometimes becoming compound. Interascal tissue of paraphyses or anastomosing pseudoparaphysis-like hyphae. Asci cylindrical, with a well-developed apical cap but no ocular chamber, 8-spored, with a J+ mucous outer layer. Ascospores hyaline, elongate, transversely septate, often with attenuated apices. Anamorph pycnidial.

Significant Genera: *Arctomia, Wawea*.

Distribution: Arctic and subarctic.

Economic Significance: None is known.
Ecology: Associated with bryophytes, lichenized with cyanobacteria.

Notes: Preliminary molecular data suggest placement in the *Ostropomycetidae*.

Wawea fruticulosa, Tasmania; thallus and ascomata

References: **Jørgensen, P.M.** (2003). A new species of *Arctomia* from Sichuan province, China. *Lichenologist* 35: 287–289; **Lumbsch, H.T.; Prado, R. del; Kantvilas, G.** (2005). *Gregorella*, a new genus to accommodate *Moelleropsis humida* and a molecular phylogeny of *Arctomiaceae*. *Lichenologist* 37: 291–302.

Argynnaceae Shearer & J.L. Crane 1980
Uncertain position within Dothideomycetes, Ascomycota

Ascomata initially formed from clusters of pseudoparenchymatous cells, sessile, cleistothecial, the peridium parenchymatous or thick and fragmenting into well-defined plates. Interascal tissue composed of branched paraphyses or pseudoparaphyses. Asci thin-walled or thick-walled, if the latter then apparently fissitunicate but not discharging actively. Ascospores brown, 1-septate, the septa sometimes thickened, inequilateral, smooth. Anamorphs unknown.

Significant Genera: *Argynna, Lepidopterella*.

Distribution: Known from North America.

Economic Significance: None is known.

Ecology: Saprobic on wood, sometimes aquatic.

Notes: The two genera currently included may not be closely related.

References: **Shearer, C.A.; Crane, J.L.** (1980). Taxonomy of two cleistothecial ascomycetes with papilionaceous ascospores. *Trans. Br. mycol. Soc.* 75: 193–200.

Arthoniaceae Rchb. 1841
Arthoniales: Ascomycota

Thallus usually absent or poorly developed, rarely crustose. Ascomata with rudimentary walls, basically apothecial in form but often elongated and branched; hymenium reddish or brownish. Interascal tissue composed of branched cellular pseudoparaphyses in a gelatinous matrix. Asci thick-walled, ± fissitunicate, usually with a large apical dome, blueing in iodine. Ascospores simple or septate, sometimes brown and/or ornamented, without a gelatinous sheath. Anamorph coelomycetous.

Significant Genera: *Arthonia*.

Distribution: Widespread, especially diverse in the tropics.

Economic Significance: None is known.

Ecology: Very varied, species are lichenized with green algae or lichenicolous, on a wide range of substrata including leaves, bark and rock.

Notes: An under-researched family with very little molecular phylogenetic information available. There is some evidence that the Arthoniales represent a completely separate clade from other apothecial lichenized fungi.

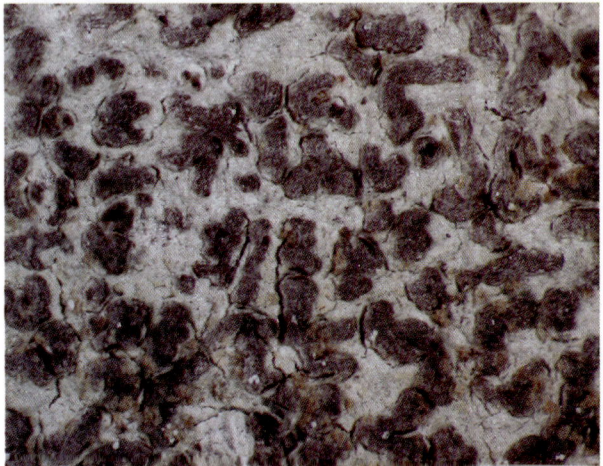

Arthonia radiata, Devon, UK; ascomata on bark

Cryptothecia rubrocincta; crustose thallus on bark

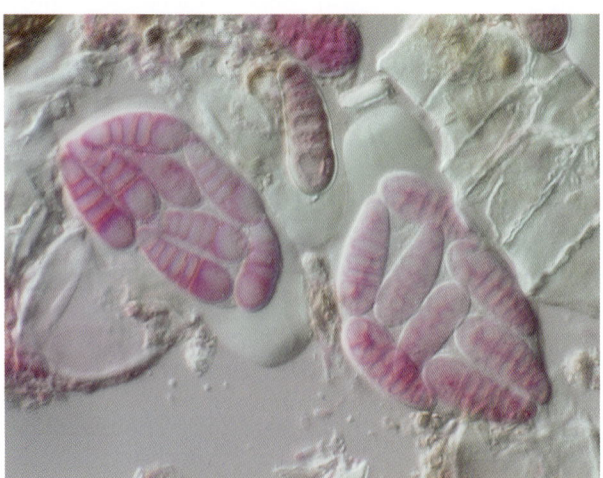

Arthonia ilicina, Argyll, Scotland; asci and ascospores

References: **Coppins, B.J.** (1989). Notes on the Arthoniaceae in the British Isles. *Lichenologist* 21: 195–216; **Diederich, P.** (1995). New or interesting lichenicolous fungi. VI. *Arthonia follmanniana*, a new species from the Galápagos Islands. *Flechten Follmann, Contributions to Lichenology in Honour of Gerhard Follmann* (Cologne): 179–182; **Grube, M.** (1998). Classification and phylogeny in the *Arthoniales* (Lichenized Ascomycetes). *Bryologist* 101: 377–391; **Grube, M.; Matzer, M.** (1997). Taxonomic concepts of lichenicolous *Arthonia* species. *Biblthca Lichenol.* 68: 1–17; **Lücking, R.** (1995). Additions and corrections to the foliicolous lichen flora of Costa Rica. The family Arthoniaceae, with notes on the genus *Stirtonia*. *Lichenologist* 27: 127–153; **Lutzoni, F.M. et al.** (2004). Assembling the fungal tree of life: progress, classification, and evolution of subcellular traits. *Am. J. Bot.* 91: 1446–1480; **Makhija, U.; Patwardhan, P.G.** (1998). The lichen genus *Stirtonia* (family Arthoniaceae). *Mycotaxon* 67: 287–311; **Myllys, L.; Källersjö, M.; Tehler, A.** (1998). A comparison of SSU rDNA data and morphological data in Arthoniales (Euascomycetes) phylogeny. *Bryologist* 101: 70–85; **Sundin, R.; Tehler, A.** (1998). Phylogenetic studies of the genus *Arthonia*. *Lichenologist* 30: 381–413; **Thor, G.** (1997). The genus *Cryptothecia* in Australia and New Zealand and the circumscription of the genus. *In* Tibell, L. & Hedberg, I. (eds), Lichen Studies Dedicated to Rolf Santesson. *Symb. bot. upsal.* 32: 267–289; **Wedin, M.; Hafellner, J.** (1998). Lichenicolous species of *Arthonia* on Lobariaceae with notes on excluded species. *Lichenologist* 30: 59–91.

Arthopyreniaceae Walt. Watson 1929
Pleosporales: Ascomycota

Thallus or stroma poorly developed, where present often immersed in bark. Ascomata clypeate, with a single perithecial locule, black, spherical or flattened, composed of pseudoparenchymatous cells, with a broad ostiole, the wall at least sometimes staining green in KOH. Interascal tissue of narrow ? cellular pseudoparaphyses, sometimes immersed in gel, sometimes evanescent. Asci clavate or elongate, fissitunicate, with an ocular chamber but otherwise without clear apical structures. Ascospores hyaline or brown, sometimes transversely septate and/or weakly ornamented, thin-walled. Anamorph coelomycetous, probably spermatial.

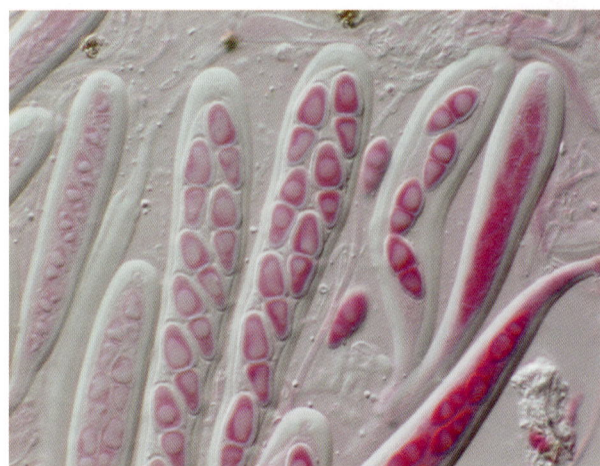

Arthopyrenia cinchonae, Louisiana, USA; asci and ascospores

Significant Genera: *Arthopyrenia*.

Distribution: Widespread.

Economic Significance: None is known.

Ecology: Saprobic, especially in smooth bark, or lichenized with green algae.

Notes: No molecular data are available, and the position of this family remains rather enigmatic.

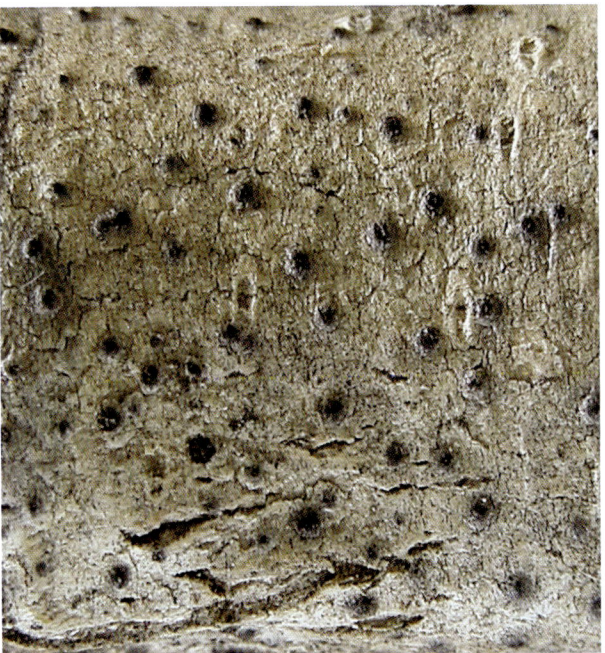

Arthopyrenia cinchonae, Louisiana, USA; ascomata on bark

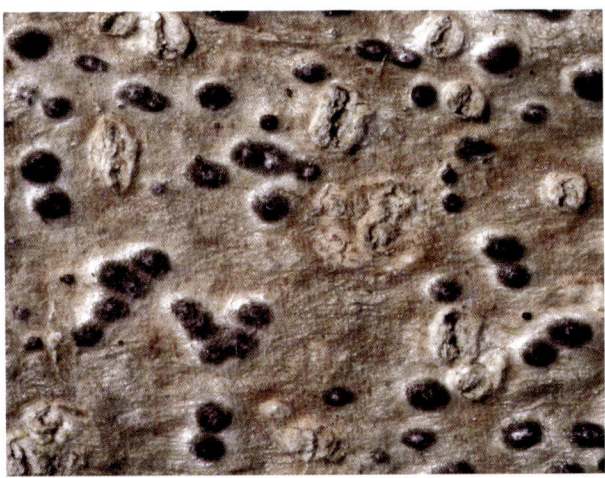

Mycomicrothelia hemisphaerica; ascomata on bark

References: **Aptroot, A.** (1997). Corticolous pyrenocarpous ascomycetes (lichenized and non-lichenized) from the Sonoran Desert (Arizona and Mexico). *Nova Hedwigia* 64: 169–176; **Coppins, B.J.** (1988). Notes on the genus *Arthopyrenia* in the British Isles. *Lichenologist* 20: 305–325; **Sérusiaux, E.; Aptroot, A.** (1998). *Mycomicrothelia striguloides* sp. nov. from New Zealand. *Bryologist* 101: 144–146; **Upreti, D.K.; Pant, G.** (1993). Notes on *Arthopyrenia* species from India. *Bryologist* 96: 226–232.

Arthrodermataceae Locq. ex Currah 1985
Onygenales: Ascomycota

Stromata absent. Ascomata cleistothecial, small, not stipitate, thin-walled, with a ± loose hyphal peridium, usually with complex appendages. Interascal tissue absent. Asci small, globose or saccate, thin-walled, evanescent. Ascospores aseptate, ± oblate, smooth, without a gelatinous sheath. Anamorphs hyphomycetous, thallic, with thick-walled conidia that are sometimes septate and ornamented.

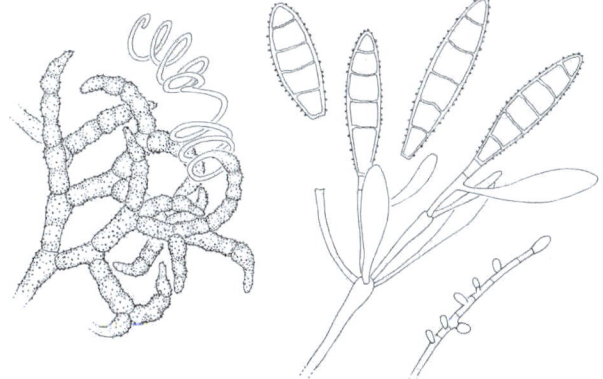

Arthroderma gypseum; ascomatal appendages and anamorph

Significant Genera: *Arthroderma*. Anamorphs: *Chrysosporium*, *Microsporum*, *Trichophyton*.

Distribution: Widespread.

Economic Significance: The *Arthrodermataceae* contains the principal dermatophyte fungi, causing ringworm and tinea of keratinous tissues (hair, skin and nails).

Ecology: Species principally affect mammals (including humans). Colonies are present in substrata containing keratin, such as soil and birds' nests, and can be isolated using baiting techniques.

Notes: A small but economically important family, well-defined in phylogenetic and nutritional terms.

References: **Amer, M.A.; Taha, M.; Diab, N.A.; El-Moughith, A.; El-Harras, M.** (1993). Ultrastructure of *Trichophyton violaceum*. *Int. J. Dermat.* 32: 97–99; **Bastert, J.; Korting, H.C.; Traenkle, P.; Schmalreck, A.F.** (1999). Identification of dermatophytes by Fourier transform infrared spectroscopy (FT-IR). *Mycoses* 42: 525–528; **Chandler, F.W.** (1998). Histopathological diagnosis of mycotic diseases. *Topley & Wilson's Microbiology and Microbial Infections* Edn 9. Vol. 4 Medical Mycology (London): 111–131; **Currah, R.S.** (1985). Taxonomy of the *Onygenales*: Arthrodermataceae, Gymnoaceae, Myxotrichaceae and Onygenaceae. *Mycotaxon* 24: 1–216; **Currah, R.S.** (1988). An annotated key to the genera of the *Onygenales*. *Syst. Ascom.* 7: 1–12; **Gräser, Y.; El Fari, M.; Vilgalys, R.; Kuijpers, A.F.A.; Hoog, G.S. de; Presber, W.; Tietz, H.-J.** (1999). Phylogeny and taxonomy of the family *Arthrodermataceae* (dermatophytes) using sequence analysis of the ribosomal ITS region. *Medical Mycol.* 37: 105–114; **Gräser, Y.; Kuijpers, A.F.A.; El Fari,**

M.; Presber, W.; Hoog, G.S. de (2000). Molecular and conventional taxonomy of the *Microsporum canis* complex. *Medical Mycol.* 38: 143–153; **Guillamón, J.M.; Cano, J.; Ramón, D.; Guarro, J.** (1996). Molecular differentiation of *Keratinomyces* (*Trichophyton*) species. *Antonie van Leeuwenhoek* 69: 223–227; **Harmsen, D.; Schwinn, A.; Bröcker, E.-B.; Frosch, M.** (1999). Molecular differentiation of dermatophyte fungi. *Mycoses* 42: 67–70; **Hoog, G.S. de; Bowman, B.; Graser, Y.; Haase, G.; El Fari, M.; Gerrits van den Ende, A.H.G.; Melzer-Krick, B.; Untereiner, W.A.** (1998). Molecular phylogeny and taxonomy of medically important fungi. *Medical Mycol.* 36: 52–56; **Kano, R.; Nakamura, Y.; Watari, T.; Watanabe, S.; Takahashi, H.; Tsujimoto, H.; Hasegawa, A.** (1999). Species-specific primers of chitin synthase 1 gene for the differentiation of the *Trichophyton mentagrophytes* complex. *Mycoses* 42: 71–74; **Makimura, K.; Tamura, Y.; Mochizuki, T.; Hasegawa, A.; Tajiri, Y.; Hanazawa, R.; Uchida, K.; Saito, H.; Yamaguchi, H.** (1999). Phylogenetic classification and species identification of dermatophyte strains based on DNA sequences of nuclear ribosomal internal transcribed spacer 1 regions. *J. Clin. Microbiol.* 37: 807–811; **Simpanya, M.F.** (2000). Dermatophytes: their taxonomy, ecology and pathogenicity. *Revta Iberoamer. Micol.* 17: 1–12; **Sugiyama, M.; Ohara, A.; Mikawa, T.** (1999). Molecular phylogeny of onygenalean fungi based on small subunit ribosomal DNA (SSU rDNA) sequences. *Mycoscience* 40: 251–258; **Summerbell, R.C.** (2000). Form and function in the evolution of dermatophytes. *Revta Iberoamer. Micol.* 17: 30–43; **Weitzman, I.; McGinnis, M.R.; Padhye, A.A.; Ajello, L.** (1986). The genus *Arthroderma* and its later synonym *Nannizzia*. *Mycotaxon* 25: 505–518.

Arthrorhaphidaceae Poelt & Hafellner 1976
Patellariales: Ascomycota

Thallus crustose or immersed within the host thallus, sometimes eventually free and squamulose. Ascomata apothecial, discoid or strongly cupulate, ± sessile; peridium poorly developed, composed of hyphae with swollen walls. Interascal tissue of narrow branched and anastomosing paraphyses. Asci clavate, hardly thickened at the apex but with an ocular chamber, not blueing in iodine. Ascospores hyaline, elongated, multiseptate. Anamorphs not known.

Arthrorhaphis grisea; thallus and ascomata

Significant Genera: *Arthrorhaphis*.

Distribution: Widespread in temperate and montane regions.

Economic Significance: None is known.

Ecology: Lichenized with green algae or parasitic on other lichens, sometimes eventually free-living.

Notes: Molecular data suggest that *Anzina* may belong here, but there are many morphological characters separating the two groups.

Arthrorhaphis citrinella; squamulose thallus and ascomata

References: **Galloway, D.J.; Bartlett, J.K.** (1986). *Arthrorhaphis* Th. Fr. (lichenized *Ascomycotina*) in New Zealand. *N.Z. Jl Bot.* 24: 393–402; **Hansen, E.S.; Obermayer, W.** (1999). Notes on *Arthrorhaphis* and its lichenicolous fungi in Greenland. *Bryologist* 102: 104–107; **Obermayer, W.** (1994). Die Flechtengattung *Arthrorhaphis* (*Arthrorhaphidaceae*, *Ascomycotina*) in Europa und Grönland. *Nova Hedwigia* 58: 275–333; **Santesson, R.; Tønsberg, T.** (1994). *Arthrorhaphis aeruginosa* and *A. olivaceae*, two new lichenicolous fungi. *Lichenologist* 26: 295–299; **Wedin, M.; Wiklund, E.; Crewe, A.; Döring, H.; Ekman, S.; Nyberg, A.; Schmitt, I.; Lumbsch, H.T.** (2005). Phylogenetic relationships of *Lecanoromycetes* (*Ascomycota*) as revealed by sequences of mtSSU and nLSU rDNA sequence data. *Mycol. Res.* 109: 159–172.

Ascobolaceae Boud. ex Sacc. 1884
Pezizales: Ascomycota

Stromata absent. Ascomata apothecial, rarely cleistothecial, usually pulvinate, fleshy, brightly coloured, not setose. Interascal tissue of simple paraphyses. Asci broad, operculate, protruding above the hymenium when mature, usually with a J+ wall. Ascospores biseriately arranged, thick-walled, usually ornamented, often with purple/brown epispore, sometimes with a sheath, sometimes dispersed as a compound unit. Anamorphs not known.

Significant Genera: *Ascobolus*.

Ascobolus crenulatus, Scotland; fruiting on the ground

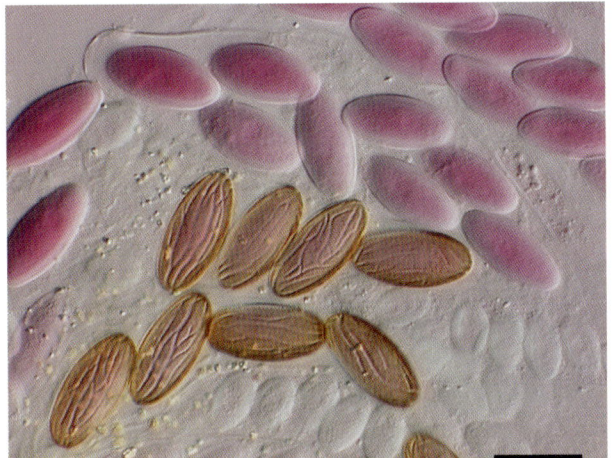
Ascobolus stercorareus, Devon, UK; asci and ascospores

Distribution: Widespread, especially in temperate zones.

Economic Significance: None is known.

Ecology: Most species are saprobes and occur on dung, some occur on soil or rotten wood.

Notes: A well-defined family with distinctively ornamented ascospores.

References: **Brummelen, J. van** (1990). Notes on cup-fungi – 4. On two rare species of *Ascobolus*. *Persoonia* 14: 203–207; **Brummelin, J. van** (1998). Reconsideration of relationships within the *Thelebolaceae* based on ascus ultrastructure. *Persoonia* 16: 425–469; **Dissing, H.** (1989). Four new coprophilous species of *Ascobolus* and *Saccobolus* from Greenland (*Pezizales*). *Op. bot.* 100: 43–50; **Gargas, A.** (1991). Molecular systematics of fungal orders within the ascomycotina using PCR amplified 18s rDNA [abstract]. *Fungal Genetics Newsl.* Suppl. 38: 26; **Jahn, E.; Benkert, D.; Schmidt, A.; Unger, H.-G.** (1997). [Coprophilous *Pezizales* on dung cultures from northern Germany and different parts of the world]. *Z. Mykol.* 63: 133–148; **Landvik, S.; Kristiansen, R.; Schumacher, T.** (1998). Phylogenetic and structural studies in the *Thelebolaceae* (*Ascomycota*). *Mycoscience* 39: 49–56; **Prokhorov, V.P.** (1997). Genera *Iodophanus*, *Thecotheus* and *Ascodesmis*: species identification keys. *Mikol. Fitopatol.* 31: 27–30; **Ranalli, M.E.; Mercuri, O.A.** (1998). The genus *Thecotheus* in Argentina. *Mycotaxon* 67: 505–518; **Wu, C.G.; Kimbrough, J.W.** (1996). Comparative studies on the septal development in *Ascobolaceae* and *Ciliarieae* (*Humariaceae*). *Taiwania* 41: 7–16.

Ascocorticiaceae J. Schröt. 1893
Helotiales: Ascomycota

Stromata absent. Ascomata effuse, forming an indefinite palisade, white or pinkish; peridium ± absent. Interascal tissue not well defined, composed of irregular undifferentiated hyphae interspersed with the asci. Asci with a J– apical ring. Ascospores small, hyaline, aseptate. Anamorphs unknown.

Significant Genera: *Ascocorticium*.

Distribution: Widespread; temperate.

Economic Significance: None is known.

Ecology: Saprobic, especially on conifer bark.

Notes: A poorly studied family containing a single genus. Its position within the *Leotiomycetes* is uncertain and no molecular data are available.

References: **Vellinga, E.C.; Vries, B. de** (1987). Weinig opvallend en veel voorkomend: twee makkelijk herkenbare ascomyceten. *Coolia* 30: 50–52

Ascodesmidaceae J. Schröt. 1893
Pezizales: Ascomycota

Ascomata ± globose, formed from coiled antheridial and ascogonial branches; peridium absent. Interascal tissue poorly developed. Asci saccate, with a very large operculum. Ascospores pigmented, with ornamentation derived directly from the secondary wall layer. Anamorphs unknown.

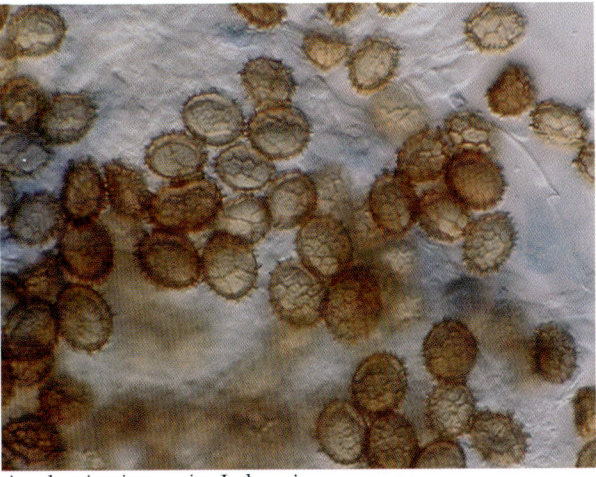
Ascodesmis microscopica, Indonesia; ascospores

Significant Genera: *Ascodesmis*.

Distribution: Widespread.

Economic Significance: None is known.

Ecology: Species are saprobic, coprophilous or soil-dwelling.

Notes: Molecular studies indicate that the family clusters within the *Pyronemataceae*, but the latter family is widely circumscribed and may need division.

References: **Brummelen, J. van** (1989). Ultrastructure of the ascus and the ascospore wall in *Eleutherascus* and *Ascodesmis* (*Ascomycotina*). *Persoonia* 14: 1–17; **Currah, R.** (1986). A new species of *Ascodesmis* from Alberta. *Mycologia* 78: 198–201; **Hansen, K.; Perry, B.A.; Pfister, D.H.** (2006). Phylogenetic origins of two cleistothecial fungi, *Orbicula parietina* and *Lasiobolidium orbiculoides*, within the operculate discomycetes. *Mycologia* 97: 1023–1033; **Landvik, S.; Kristiansen, R.; Schumacher, T.** (1998). Phylogenetic and structural studies in the *Thelebolaceae* (*Ascomycota*). *Mycoscience* 39: 49–56.

Ascodichaenaceae D. Hawksw. & Sherwood 1982
Rhytismatales: Ascomycota

Ascomata apothecial, erumpent in clusters, elongate, opening by a longitudinal split, the wall tissue stromatal in origin, composed of vertically oriented pseudoparenchymatous tissue, black. Interascal tissue of simple paraphyses. Asci clavate or saccate, thin-walled, with a wide J– apical ring. Ascospores ellipsoidal, hyaline, aseptate, without a sheath. Anamorphs coelomycetous, disseminative.

Significant Genera: *Ascodichaena*.

Distribution: Widespread in temperate regions.

Economic Significance: None is known.

Ecology: Saprobic on bark, possibly endophytic.

Notes: The placement of this family is not clear; no sequences are available. *Ascodichaena rugosa* is extremely common in Europe on beech and oak bark in less polluted areas.

References: **Butin, H.; Marmolejo, J.G.** (1990). *Ascodichaena mexicana* sp. nov. (*Rhytismatales*), Erreger der 'Warzenkrankheit' mexikanischer Eichen. *Sydowia* 42: 8–16; **Minter, D.W.** (1995). The *Rhytismatales* on conifers from Europe. *Shoot and Foliage Diseases in Forest Trees* Proceedings of a Joint Meeting of the Working Parties: Canker & Shoot Blight of Conifers, Foliage Diseases (Firenze): 65–84; **Yuan, Z.Q.; Rudman, T.; Mohammed, C.** (2000). *Pseudophacidium diselmae* sp. nov. isolated from stem chankers on *Diselma archeri* in Tasmania, Australia. *Australas. Pl. Path.* 29: 215–221.

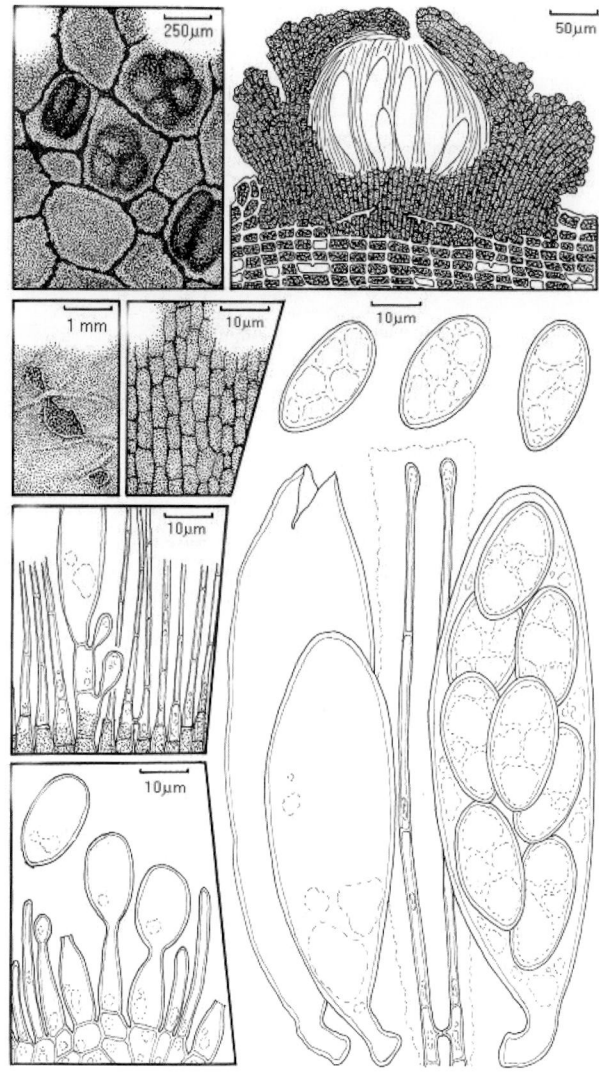

Ascodichaena rugosa; asci, ascospores and conidia

Ascoideaceae J. Schröt. 1894
Saccharomycetales: Ascomycota

Mycelium well-developed. Asci multispored, proliferating internally to form phialide-like structures. Ascospores ± ellipsoidal to hat-shaped with an eccentric sheath, not blueing in iodine, sometimes discharged in a mucous cirrus, at least sometimes conjugating in pairs within the ascus.

Significant Genera: *Ascoidea*.

Distribution: Widespread.

Economic Significance: None is known.

Ecology: Found typically in beetle galleries in dead wood, apparently transmitted by insects.

Notes: A poorly known family in an isolated phylogenetic position. The percurrent proliferation of asci is unusual.

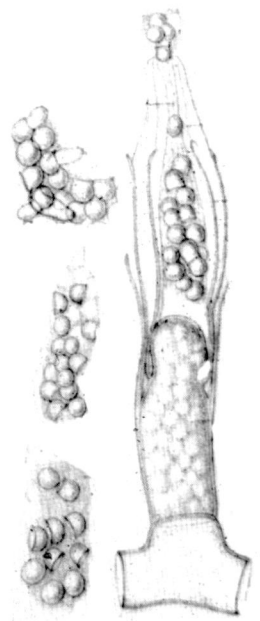

Ascoidea rubescens; derived from Brefeld, *Unters. Gesammtgeb. Mykol.* **9**: 94–108 (1891)

References: **Arx, J.A. von; Müller, E.** (1984). Notes on some ascomycetes. *Sydowia* 37: 6–10; **Batra, L.R.** (1987). Insect-associated, filamentous *Endomycetes* – their growth and strategies for survival. *Stud. Mycol.* 30: 415–428; **Hoog, G.S. de** (1998). *Ascoidea* Brefeld & Lindau. *The Yeasts, A Taxonomic Study* (Amsterdam): 136–140; **Kurtzman, C.P.; Blanz, P.A.** (1998). Ribosomal RNA/DNA sequence comparisons for assessing phylogenetic relationships. *The Yeasts, A Taxonomic Study* (Amsterdam): 69–74; **Kurtzman, C.P.; Robnett, C.J.** (1998). Identification and phylogeny of ascomycetous yeasts from analysis of nuclear large subunit (26S) ribosomal DNA partial sequences. *Antonie van Leeuwenhoek* 73: 331–371.

Ascoporiaceae Kutorga & D. Hawksw. 1997
Uncertain position within Dothideomycetes, Ascomycota

Stromata multiloculate, superficial, discoid to pulvinate, on a basal subiculum, leathery, with a pseudoparenchymatous surface layer covering numerous locules separated by sterile regions, the locules initially ovoid with a narrow ostiole but becoming apothecial. Interascal tissue of branched trabecular pseudoparaphyses, forming an epithecial layer above. Asci cylindric-clavate, fissitunicate with an ocular chamber, not blueing in iodine. Ascospores 1-septate, smooth, brown. Anamorph not clearly known.

Significant Genera: *Ascoporia*.

Distribution: Only known from Brazil.

Economic Significance: None is known.

Ecology: Found on decorticated wood, presumably saprobic.

Notes: Very poorly known; no recent information is available.

References: **Kutorga, E.; Hawksworth, D.L.** (1997). A reassessment of the genera referred to the family *Patellariaceae* (*Ascomycota*). *Syst. Ascom.* 15: 1–110; **Samuels, G.J.; Romero, A.I.** (1993). *Ascoporia* (*Fungi, Loculoascomycetes*), a new genus from Pará. *Bolm Mus. paraense 'Emílio Goeldi'* sér. bot. 7: 263–268.

Ascosphaeraceae L.S. Olive & Spiltoir 1955
Ascosphaerales: Ascomycota

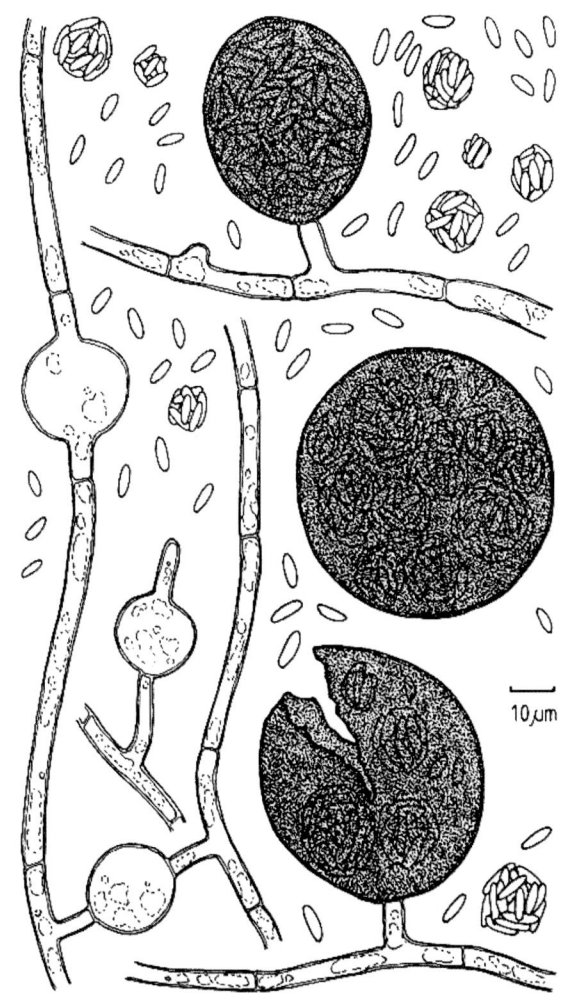

Ascosphaera apis; hyphae, ascomata, spore balls and ascospores

Teleomorph. Ascomata small, acellular, consisting of a brown, hollow, non-ostiolate cyst-like organ formed from multinucleate ascogenous cells, lacking appendages. Interascal tissue absent. Asci evanescent at a very early stage (the wall possibly never formed), enclosed in clusters within a new communal structure, so that a number of 'spore balls' probably composed of ascospores from several different asci can be observed. Ascospores hyaline, smooth, aseptate, lacking pores. Anamorph *Chrysosporium*-like. Conidiomata absent. Conidiophores hardly differentiated. Conidiogenous cells little different in morphology from the vegetative hyphae. Conidia delimited after all or most cell wall growth has finished, colourless, aseptate, smooth or minutely roughened.

Significant Genera: *Ascosphaera*, *Bettsia*.

Distribution: Widespread, especially in north temperate regions.

Economic Significance: *Ascosphaera apis* is of some importance as the cause of chalkbrood disease of honeybees. Chalkbrood was historically thought to occur solely in Europe, but it has now spread to almost all regions where honeybees are prevalent.

Ecology: Associated with bees, many on pollen in hives, some parasitic on larvae and killing them (chalkbrood disease). Ascospores are ingested by young larvae, germinating in the gut and the subsequent growth breaking out of the hindquarters when the larva is sealed in its cell prior to pupation. This fungus is heterothallic and the presence or absence of suitable strains may affect the manner in which the larvae are colonized. Ascospores may persist for many years, providing periodic outbreaks when conditions are adverse for the bees. Transmission occurs within the cells of the comb to young larvae, and from hive to hive probably by robber bees. Chalkbrood occurs almost exclusively in honeybees, but other species of the family are associated with bumble bees and leaf cutting bees.

Notes: The anamorph is often not observed, and may not play a major role in the life cycle. In literature on this order the ascomata are often referred to as cysts. Recent molecular evidence has suggested links with the *Trichocomaceae*, but the family has very distinctive morphological features.

References: **Anderson, D.L.; Gibbs, A.J.; Gibson, N.L.** (1998). Identification and phylogeny of spore-cyst fungi (*Ascosphaera* spp.) using ribosomal DNA sequences. *Mycol. Res.* 102: 541–547; **Anderson, D.L.; Gibson, N.L.** (1998). New species and isolates of spore-cyst fungi (*Plectomycetes*, *Ascosphaerales*) from Australia. *Aust. Syst. Bot.* 11: 53–72; **Bissett, J.** (1988). Contribution toward a monograph of the genus *Ascosphaera*. *Can. J. Bot.* 66: 2541–2560; **Brady, B.L.** (1979). *Ascosphaera apis*. *IMI Descr. Fungi Bact.* 62: [1–2]; **Geiser, D.M.; LoBuglio, K.F.** (2001). The monophyletic plectomycetes: *Ascosphaeriales*, *Onygenales*, *Eurotiales*. *The Mycota* (Berlin) 7: 201–219; **Kish, L.P.; Bowers, N.A.; Benny, G.L.; Kimbrough, J.W.** (1988). Cytological development of *Ascosphaera atra*. *Mycologia* 80: 312–319; **Landvik, S.; Shailer, N.F.J.; Eriksson, O.E.** (1997). SSU rDNA sequence support for a close relationship between the *Elaphomycetales* and the *Eurotiales* and *Onygenales*. *Mycoscience* 37: 237–241; **Skou, J.P.** (1982). *Ascosphaerales* and their unique ascomata. *Mycotaxon* 15: 487–499.

Asellariaceae Manier ex Manier & Lichtw. 1968
Asellariales: Zygomycota

Thallus branched, septate, with a polymorphic basal cell secreting, or functioning as, a holdfast. Asexual reproduction by uninucleate arthrospores (? sporangia) or long-fusiform sporangia produced in chains, each bearing one cylindrical spore. Sexual reproduction unknown.

Significant Genera: *Asellaria*.

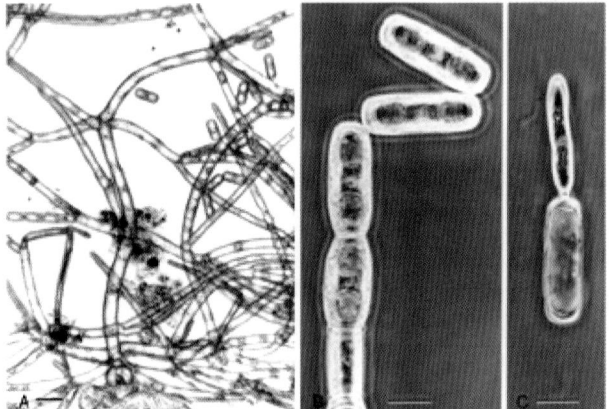
Asellaria ligiae; thallus and sporangia

Distribution: Widespread but relatively rare.

Economic Significance: None is known.

Ecology: Endocommensals (obligate) of aquatic, marine and terrestrial *Isopoda* (Crustacea) and *Collembola* (Insecta).

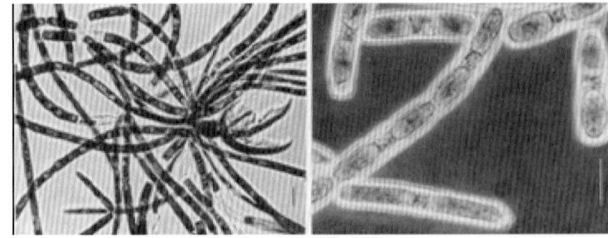
Asellaria unguiformis; thalli and sporangia

Notes: The only family of the *Asellariales* with just 11 species. The anamorph is arthrosporic and the teleomorph is not known although what appears to be conjugation has been reported. A close affinity with the *Kickxellales* has been suggested based first on ultrastructural and serological evidence and augmented by analysis of molecular data. The exclusion of the *Amoebidiales* and *Eccrinales* (*Protozoa*) was suggested as early as 1975 (Moss, 1975), a conclusion subsequently confirmed by molecular data.

References: **Benny, G.L.** (2001). *Zygomycota*: *Trichomycetes*. *The Mycota* (Berlin) 7: 147–160; **Cafaro, M.J.** (1999). *Baltomyces*, a new genus of gut-inhabiting fungus in an isopod. *Mycologia* 91: 517–519; **Lichtwardt, R.W.** (1986). *The Trichomycetes. Fungal associates of arthropods* (New York): 343 pp.; **Lichtwardt, R.W.** (2002). *Trichomycetes*: fungi in relationship with insects and other arthropods. *Cellular Origin and Life in Extreme Habitats* (Dordrecht) 4: 577–588; **Moss, S.T.** (1975). Septal ultrastructure in the *Trichomycetes* with special reference to *Astreptonema* (*Eccrinales*). *Trans. Br. mycol. Soc.* 65: 115–127; **Moss, S.T.; Young, T.W.K.** (1978). Phyletic considerations of the *Harpellales* and *Asellariales* (*Trichomycetes, Zygomycotina*) and *Kickxellales* (*Zygomycetes, Zygomycotina*). *Mycologia* 70: 944–963.

Aspidotheliaceae Räsänen ex J.C. David & D. Hawksw. 1991
Uncertain position within Ascomycetes, Ascomycota

Thallus crustose. Ascomata solitary, perithecial, ± globose, sometimes clypeate, the wall sometimes ornamented, the ostiole periphysate. Interascal tissue of sometimes branched and anastomosing paraphyses. Asci clavate, persistent, with a multilayered wall, ? fissitunicate, the apex thickened, not blueing in iodine, with a small ocular chamber. Ascospores hyaline, transversely septate or muriform, with a mucous sheath. Anamorphs pycnidial.

Significant Genera: *Aspidothelium*.

Distribution: Widespread, mainly pantropical.

Economic Significance: None is known.

Ecology: Lichenized with green algae, on leaves and bark.

Notes: Affinities are not clear and little research is available.

Aspidothelium cinerascens; thallus and ascomata

References: **Aptroot, A.; Sipman, H.** (1993). *Musaespora*, a genus of pyrenocarpous lichens with campylidia, and other additions to the foliicolous lichen flora of New Guinea. *Lichenologist* 25: 121–135; **Lücking, R.** (1999). Foliicolous lichens and their lichenicolous fungi from Ecuador, with a comparison of lowland and montane rain forest. *Willdenowia* 29: 299–335; **Lücking, R.; Sérusiaux, E.** (1996). *Musaespora kalbii* (lichenized *Ascomycetes*: *Melanommatales*), a new foliicolous lichen with a pantropical distribution. *Nordic Jl Bot.* 16: 661–668; **Malcolm, W.M.; Vězda, A.** (1995). Additional lichen records from New Zealand. 13: *Aspidothelium cinerascens* Vain. *Australas. Lichenol. Newsl.* 37: 13–15.

Asterinaceae Hansf. 1946
Uncertain position within Dothideomycetes, Ascomycota

Mycelium superficial, brown, thick-walled, frequently branching, often hyphopodiate. Ascomata round or elongate, strongly flattened, dark brown, usually opening by radiating or longitudinal splits, but in some species by a lysigenous pore; peridium (upper wall) composed of a single layer or a few layers of radiating thick-walled isodiametric cells. Lower wall thin, pale, at least in some species also composed of radiating cells. Hymenium usually gelatinous, blueing in iodine. Interascal tissue absent, at least at maturity. Asci ovoid to cylindrical, thick-walled, with rostrate dehiscence. Ascospores brown, thick-walled, transversely septate. Anamorph coelomycetous. Conidiomata similar in form to (though rather smaller than) the ascomata, the conidia formed from the under-surface of the upper wall, which sometimes breaks away completely leaving a ring of tissue surrounding the exposed fertile layer. Conidiogenous cells small and inconspicuous, possibly not proliferating. Conidia dark, thick-walled, smooth or finely roughened, aseptate, usually ovoid with a wide base, sometimes with a transverse pale region.

Significant Genera: *Asterina, Lembosia*. Anamorph: *Asterostomella*.

Distribution: Widespread, especially in the tropics.

Economic Significance: None is known.

Ecology: Assumed to be biotrophic on leaves, rarely causing any clear damage to the host, but nutritional studies have yet to be carried out.

Notes: The family and its limits are poorly known, it is probably artificial in its delimitation and species concepts generally assume a high degree of host specificity. There may be close evolutionary links with the *Parmulariaceae* and/or the *Meliolaceae*, but no molecular data are available.

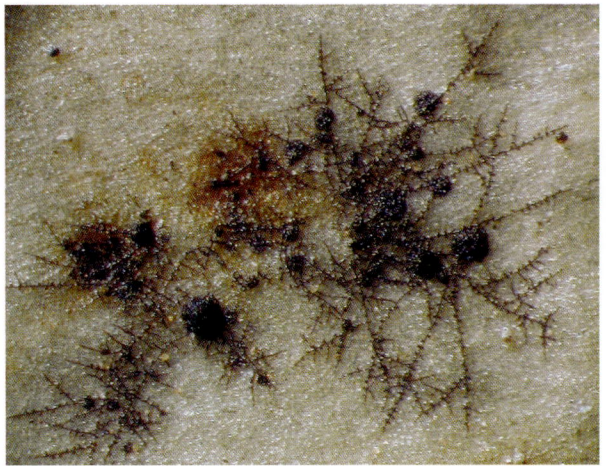

Asterina mahoniae, Nepal; ascomata and superficial mycelium

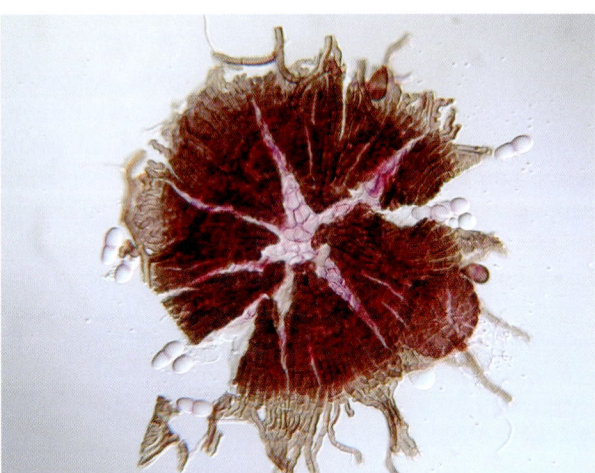

Asterina alchorneicola, Kenya; ascoma

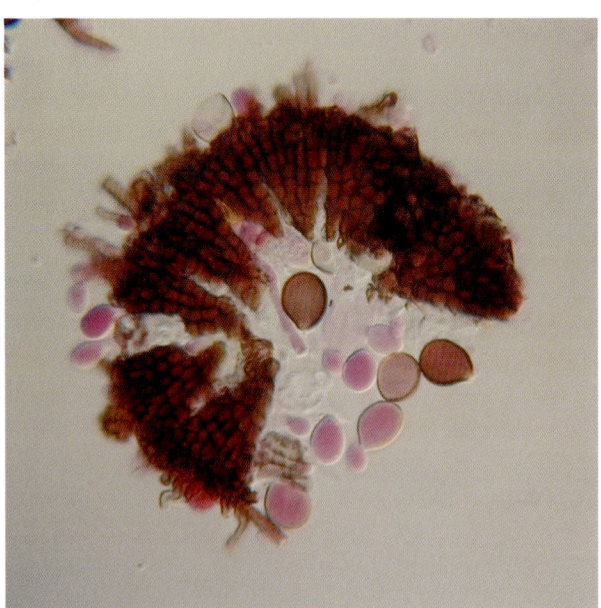

Asterina nyanzae, Uganda; conidioma

References: **Doidge, E.M.** (1942). A revision of the South African *Microthyriaceae*. *Bothalia* 4: 273–420; **Farr, M.L.** (1986). Amazonian foliicolous fungi. II. *Deuteromycotina*. *Mycologia* 78: 269–286; **Farr, M.L.** (1987). Amazonian foliicolous fungi. IV. Some new and critical taxa in *Ascomycotina* and associated anamorphs. *Mycologia* 79: 97–116; **Goos, R.D.** (1999). Notes on the genus *Echidnodella* (*Asterinaceae*). *Mycotaxon* 73: 455–464; **Hosagoudar, V.B.; Balakrishnan, M.P.; Goos, R.D.** (1996). Some *Asterinella*, *Asterostomella* and *Echidnodella* species fom southern India. *Mycotaxon* 58: 489–498; **Hosagoudar, V.B.; Goos, R.D.** (1994). Some *Asterina*, *Asterostomella* and *Lembosia* species from southern India. *Mycotaxon* 52: 467–473; **Mibey, R.K.; Hawksworth, D.L.** (1997). *Meliolaceae* and *Asterinaceae* of the Shimba Hills, Kenya. *Mycol. Pap.* 174: 108 pp.; **Rahayu, G.; Parbery, I.H.** (1991). Revision of Australian *Asterinaceae*: *Asterina* species on *Winteraceae* and *Eupomatiaceae*. *Mycol. Res.* 95: 731–740; **Reynolds, D.R.** (1987). A nonbitunicate ascus in the ascostromatic genus *Asterina*. *Cryptog. Mycol.* 8: 251–268; **Swart, H.J.** (1986). Australian leaf-inhabiting fungi XXII. *Microthyrium*-like fungi on *Eucalyptus*. *Trans. Br. mycol. Soc.* 87: 81–91.

Asterothyriaceae Walt. Watson ex R. Sant. 1952
Ostropales: Ascomycota

Thallus crustose, usually effuse. Ascomata often deeply immersed, apothecial, with rudimentary to well-developed walls of thick gelatinized hyphae. Hymenium not blueing in iodine. Paraphyses gelatinized, simple or anastomosing. Asci ± cylindrical, with a slightly thickened apical cap, 1- to 8-spored. Ascospores transversely septate to muriform, hyaline, sometimes with a sheath. Anamorphs pycnidial.

Significant Genera: *Asterothyrium*, *Gyalidea*.

Distribution: Widespread, especially in tropical forests.

Economic Significance: None is known.

Ecology: Lichenized with green algae, mostly foliicolous.

Notes: Molecular studies confirm phenotypic evidence that the *Gomphillaceae* and *Asterothyriaceae* are closely related, within the major Ostropalean clade, but more sequences are required.

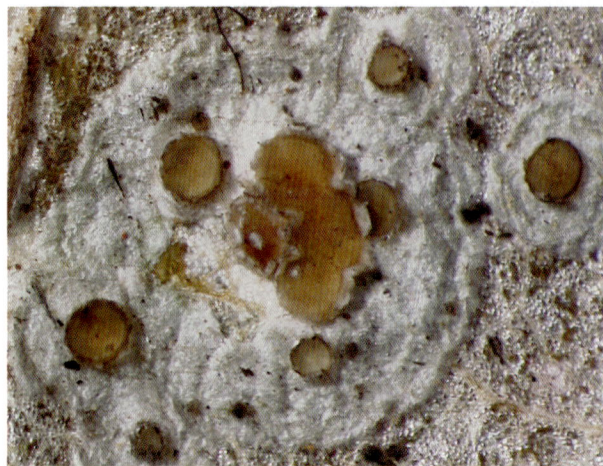

Asterothyrium octomerum, Togo; thalli and ascomata

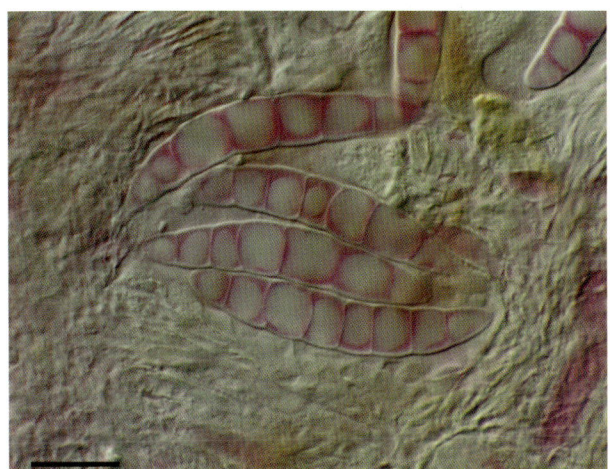
Asterothyrium octomerum, Togo; ascospores

Gyalidea hyalinescens; thallus and ascomata

References: **Boom, P.P.G. van den; Vězda, A.** (1995). A new species and a new variety of the lichen genus *Gyalidea* from western Europe. *Mycotaxon* 54: 421–426; **Etayo, J.; Vězda, A.** (1994). Two new species of *Gyalidea* from Europe. *Lichenologist* 26: 333–335; **Hansen, E.S.; Poelt, J.; Vězda, A.** (1987). The lichen genera *Gyalecta*, *Gyalidea* and *Sagiolechia* in Greenland. *Herzogia* 7: 367–374; **Henssen, A.; Lücking, R.** (2002). Morphology, ontogeny and anatomy in the *Asterothyriaceae* (*Ascomycetes*: *Ostropales*), a greatly misunderstood group of lichenized fungi. *Ann. bot. fenn.* 39: 273–299; **Lücking, R.** (1999). Foliicolous lichens and their lichenicolous fungi from Ecuador, with a comparison of lowland and montane rain forest. *Willdenowia* 29: 299–335; **Lücking, R.; Stuart, B.L.; Lumbsch, H.T.** (2004). Phylogenetic relationships of *Gomphillaceae* and *Asterothyriaceae*: evidence from a combined Bayesian analysis of nuclear and mitochondrial sequences. *Mycologia* 96: 283–294; **Sérusiaux, E.; Sloover, J.R. de** (1986). Taxonomical and ecological observations on foliicolous lichens in northern Argentina, with notes on the hyphophores of *Asterothyriaceae*. *Veröff. geobot. Inst.* Zürich 91: 260–292; **Vězda, A.; Poelt, J.** (1991). Die Flechtengattung *Gyalidea* Lett. ex Vězda (*Solorinellaceae*). Eine Übersicht mit Bestimmungsschlüssel. *Nova Hedwigia* Beih. 53: 99–113.

Astraeaceae Zeller ex Jülich 1982
Boletales: Basidiomycota

Basidiomata gasteroid, epigeous, ± globose, with an outer peridial layer that splits into hygroscopic stellate rays to expose a thin-walled felty inner fruit body that opens with an irregular apical tear. Columella absent, gleba initially divided into chambers by narrow mycelial layers, becoming dark brown and powdery, with a capillitium composed of septate, hyaline, thick-walled, branched hyphae with clamp connections. Basidia clavate, with 2–8 sterigmata. Basidiospores globose, dark brown, thick-walled, verrucose or echinulate, not staining in iodine.

Significant Genera: *Astraeus*.

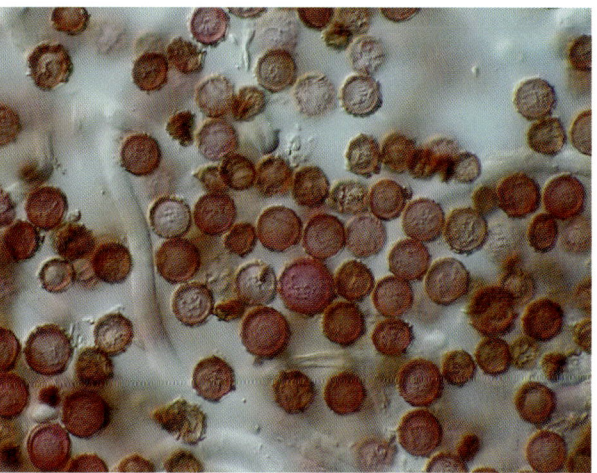
Astraeus hygrometricus, Czech Republic; basidiospores

Astraeus hygrometricus; basidiomata

Distribution: Cosmopolitan.

Economic Significance: None is known.

Ecology: On the ground, ectomycorrhizal with woody plants.

Notes: Combined with the *Sclerodermataceae* by some authors, but clearly distinguishable on both morphology and molecular characters. The similarity with *Geastrum* is a product of parallel evolution.

References: **Binder, M.; Bresinsky, A.** (2002). Derivation of a polymorphic lineage of gasteromycetes from boletoid ancestors. *Mycologia* 94: 85–98; **Hughey, B.D.; Adams, G.C.; Bruns, T.D.; Hibbett, D.S.** (2000). Phylogeny of *Calostoma*, the gelatinous-stalked puffball, based on nuclear and mitochondrial ribosomal DNA sequences. *Mycologia* 92: 94–104; **Pegler, D.N.; Læssøe, T.; Spooner, B.M.** (1995). *British Puffballs, Earthstars and Stinkhorns* An Account of the British Gasteroid Fungi (Kew): 255 pp.

Atheliaceae Jülich 1982
Atheliales: Basidiomycota

Basidiomata poorly developed, usually white or yellowish, covered with a thin, web-like, occasionally crystalline, subiculum and a pale, smooth to rugose or merulioid, rarely weakly poroid hymenium. Hyphal system monomitic, very rarely dimitic, clamp connections present or absent. Cystidia absent or smooth and thin-walled or finely encrusted. Basidia small, clavate, sometimes stalked, with 2 or 4 sterigmata. Basidiospores small, almost always hyaline, ellipsoidal to cylindrical, navicular, rarely warted or lobed, usually thin-walled, rarely staining in iodine. Anamorph either sporodochial and hyphomycetous or *Rhizoctonia*-like, forming long-lived infective sclerotia in the soil.

Significant Genera: *Amylocorticium, Athelia*.

Athelia arachnoidea, Nova Scotia, Canada; basidiomata

Distribution: Widespread in all climatic zones.

Economic Significance: Some sclerotioid species of *Athelia* are important plant pathogens, especially *A. rolfsii* and *A. arachnoidea*. Species have an extremely wide host range, and the sclerotia may remain viable in the soil for long periods. Lichenized species may not belong here.

Ecology: Many species form ectomycorrhizal associations, often occupying a dominant position in mycorrhizal communities. Other species are pathogens of many different plants and even lichens, and one species has even been suggested to form a lichen association with cyanobacteria. Yet more species are saprobes and active decomposer and a few are aquatic.

Dictyonema glabratum; thalli

Notes: The family occupies a well-defined clade and seems to belong to a major grouping of the homobasidiomycetes that also includes the euagaric and boletoid clades, and relationships with the true polypores now seem less likely. More research on phylogenetics is needed, and it is likely that the family will require remodelling.

Athelia rolfsii; sclerotia in culture

Athelia arachnoidea; byssoid basidiomata

References: **Adams, G.C.; Kropp, B.R.** (1996). *Athelia arachnoidea*, the sexual state of *Rhizoctonia carotae*, a pathogen of carrot in cold storage. *Mycologia* 88: 459–472; **Binder, M.; Hibbett, D.S.; Larsson, K.-H.; Larsson, E.; Langer, E.; Langer, G.** (2005). The phylogenetic distribution of resupinate forms across the major clades of mushroom-forming fungi (*Homobasidiomycetes*). *Syst. Biodiv.* 3: 113–157; **Boidin, J.; Mugnier, J.; Canales, R.** (1998). Taxonomie moleculaire des *Aphyllophorales*. *Mycotaxon* 66: 445–491; **Gilbertson, R.L.; Lindsey, J.P.** (1989). North American species of *Amylocorticium* (*Aphyllophorales, Corticiaceae*), a genus of brown rot fungi. *Mem. N. Y. bot. Gdn* 49: 138–146; **Ginns, J.** (1998). Genera of the North American *Corticiaceae sensu lato*. *Mycologia* 90: 1–35; **Harlton, C.E.; Lévesque, C.A.; Punja, Z.K.** (1995). Genetic diversity in *Sclerotium* (*Athelia*) *rolfsii* and related species. *Phytopathology* 85: 1269–1281; **Hibbett, D.S.; Gilbert, L.B.; Donoghue, M.J.** (2000). Evolutionary instability of ectomycorrhizal symbioses in basidiomycetes. *Nature Lond.* 407: 506–508; **Kirschner, R.; Oberwinkler, F.** (1999). A new basidiomycetous anamorph genus with cruciform conidia. *Mycoscience* 40: 345–348; **Larsson, K.-H.; Larsson, E.; Koljalg, U.** (2004). High phylogenetic diversity among corticioid homobasidiomycetes. *Mycol. Res.* 108: 983–1002; **Nakasone, K.K.** (1990). Cultural studies and identification of wood-inhabiting *Corticiaceae* and selected *Hymenomycetes* from North America. *Mycol. Mem.* 15: 412 pp.; **Ryvarden, L.** (1991). Genera of Polypores. Nomenclature and Taxonomy. *Syn. Fung.* (Oslo) 5: 363 pp.; **Stalpers, J.A.; Andersen, T.F.** (1996). A synopsis of the taxonomy of teleomorphs connected with *Rhizoctonia s.l. Rhizoctonia Species, Taxonomy, Molecular Biology, Ecology, Pathology and Disease Control* (Dordrecht): 49–63.

Atractogloeaceae Oberw. & R. Bauer 1989
Atractiellales: Basidiomycota

Basidiomata pulvinate, gelatinous. Hyphal system monomitic, hyphae with simple, swollen, septal pores and diglobular spindle pole bodies, clamp connections present. Cystidia absent. Basidia elongate, cylindrical and transversely septate, sterigmata absent, producing spores sequentially from terminal or lateral buds. Basidiospores not discharged actively, hyaline, thin-walled, smooth, not staining in iodine. Anamorph yeast-like, proliferating by budding.

Significant Genera: *Atractogloea*.

Distribution: Apparently the only known collection is from California.

Economic Significance: None is known.

Ecology: On dead palm fronds, presumably saprobic.

Notes: An isolated family within the auricularioid fungi. No molecular studies have been completed.

References: **Oberwinkler, F.** (1987). *Heterobasidiomycetes* with ontogenetic yeast-stages – systematic and phylogenetic aspects. *Stud. Mycol.* 30: 61–74; **Oberwinkler, F.; Bandoni, R.J.** (1982). *Atractogloea*: a new genus in the *Hoehnelomycetaceae* (*Heterobasidiomycetes*). *Mycologia* 74: 634–639; **Oberwinkler, F.; Bauer, R.** (1989). The systematics of gasteroid, auricularioid *Heterobasidiomycetes*. *Sydowia* 41: 224–256.

Aulographaceae Luttr. ex P.M. Kirk, P.F. Cannon & J.C. David 2001
Microthyriales: Ascomycota

Mycelium superficial, brown, without hyphopodia, often inconspicuous. Stromata absent. Ascomata strongly flattened, elongate, opening by an irregular split, composed of epidermoid cells, merging at the edge with the vegetative mycelium by irregularly branched hyphae. Interascal tissue absent. Asci numerous, fissitunicate, clavate to broadly cylindrical, not blueing in iodine. Ascospores transversely septate, hyaline. Anamorph unknown.

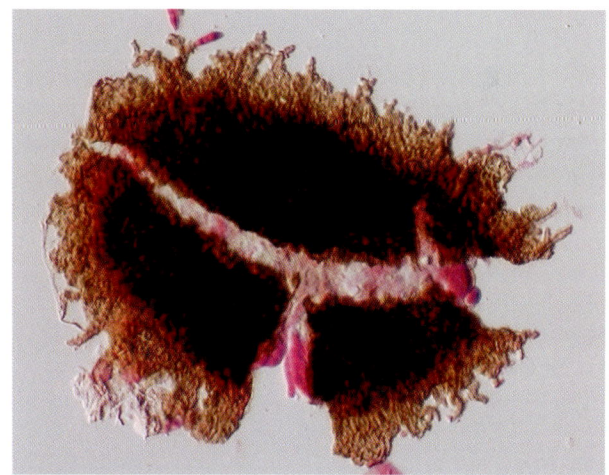

Aulographum hederae, Devon, UK; ascoma showing epidermoid peridium

Significant Genera: *Aulographum*.

Distribution: Widespread.

Economic Significance: None is known.

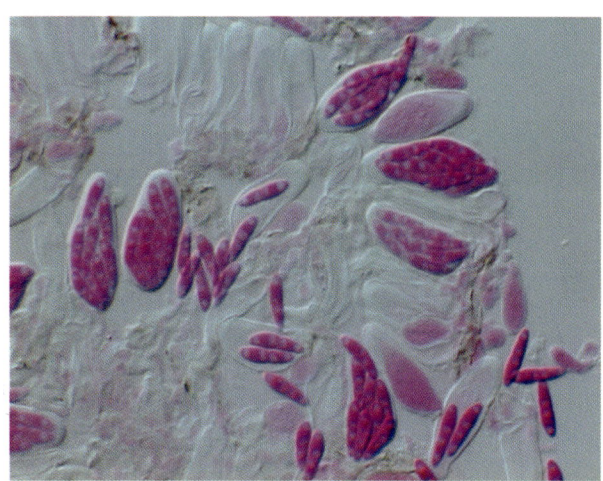

Aulographum hederae, Devon, UK; asci and ascospores

Ecology: Saprobic, usually on dead leaves and stems.

Notes: One of a number of families of the *Ascomycota* with thyriothecia. Generic limits and relationships with other familes are poorly understood. Most are very inconspicuous.

References: **Batista, A.C.** (1959). Monografia dos fungos *Micropeltaceae*. *Publções Inst. Micol. Recife* 56: 1–88; **McKenzie, E.H.C.; Foggo, M.N.** (1989). Fungi of New Zealand subantarctic islands. *N.Z. Jl Bot.* 27: 91–100; **Petrini, O.; Candoussau, F.; Petrini, L.E.** (1989). Bambusicolous fungi collected in southwestern France 1982–1989. *Mycol. helv.* 3: 263–279.

Auricularia fuscosuccinea, Galapagos Islands, Ecuador; basidioma

Auriculariaceae Fr. 1838
Auriculariales: Basidiomycota

Basidioma resupinate or applanate, rarely stipitate, gelatinous, usually pallid or brownish. Sterile surface smooth or hairy, hymenial surface smooth to coarsely reticulate. Hyphae with clamp connections, varied in form. Basidia interspersed with hyphidia, cylindrical, transversely septate, producing spores from well-developed terminal or lateral sterigmata. Basidiospores discharged actively, hyaline, smooth, reniform to allantoid, not staining in iodine, commonly becoming septate, the daughter cells forming conidia repeatedly from short tubular extensions.

Significant Genera: *Auricularia*.

Distribution: Widespread in both temperate and tropical regions.

Economic Significance: *Auricularia* species are prized as food, especially in the Orient, and are widely cultivated both on natural logs and on woodchip-based substrata. Cholesterol-lowering properties have been claimed.

Auricularia auricula-judae, Surrey, UK; basidiomata in frosty conditions

Auricularia auricula-judae, Surrey, UK; basidiomata

Ecology: Saprobic, occasionally weakly parasitic on wood of both gymnosperms and angiosperms.

Notes: Preliminary phylogenetic studies suggest the *Exidiaceae* is closely related to this family; the two share many morphological traits. The *Exidiaceae* is traditionally separated using basidial characters – having subglobose basidia with both transverse and longitudinal septa.

References: **Begerow, D.; Bauer, R.; Oberwinkler, F.** (1998). Phylogenetic studies on nuclear large subunit ribosomal DNA sequences of smut fungi and related taxa. *Can. J. Bot.* 75: 2045–2056; **Lü, H.; McLaughlin, D.J.** (1991). Ultrastructure of the septal pore apparatus and early septum initiation in *Auricularia auricula-judae*. *Mycologia* 83: 322–334; **Lü, H.S.; McLaughlin, D.J.** (1995). A light and electron microscopic study of mitosis in the clamp connection of *Auricularia auricula-judae*. *Can. J. Bot.* 73: 315–332; **Weiss, M.; Oberwinkler, F.** (2001). Phylogenetic relationships in *Auriculariales* and related groups – hypotheses derived from nuclear ribosomal DNA sequences. *Mycol. Res.* 105: 403–415; **Wong, G.J.; Wells, K.** (1988). Comparative morphology, compatibility, and infertility of *Auricularia cornea*, *A. polytricha*, and *A. tenuis*. *Mycologia* 79: 847–856; **Yan, P.S.; Luo, X.C.; Zhou, Q.** (1999). [RFLP analysis of amplified nuclear ribosomal DNA in the genus *Auricularia*]. *Mycosystema* 18: 206–213.

Auriscalpiaceae Maas Geest. 1963
Russulales: Basidiomycota

Basidioma pileate, dimidiate or effuse, tough, coriaceous, brownish, sessile or stipitate, the stalk sometimes eccentric, usually tomentose or hispid at least initially. Hyphal system monomitic or dimitic with pigmented skeletal hyphae, clamp connections usually present. Hymenium spinose or lamellar with serrate edges, with well-developed gloeocystidia abundant in the spines. Basidia clavate, hyaline, thin-walled, with 2–4 sterigmata. Basidiospores discharged actively, ovoid to ellipsoidal, hyaline, thin-walled, warted or echinulate, sometimes staining in iodine.

Significant Genera: *Auriscalpium*, *Lentinellus*.

Lentinellus cochleatus, Surrey, UK; basidiomata on rotten stump

Distribution: Widespread, primarily in temperate zones.

Auriscalpium vulgare, Surrey, UK; basidioma

Economic Significance: None is known.

Ecology: Saprobic or parasitic on wood of gymnosperms and angiosperms, producing white rots.

Notes: *Clavicorona* and *Artomyces* have been included in this family, but differ significantly in morphological and molecular characters and should be excluded. The remaining genera can be divided into two groups supported by molecular data with spiny or lamellar hymenia; *Gloiodon* should also be included.

Auriscalpium vulgare, Surrey, UK; hymenium

References: **Berbee, M.L.; Wells, K.** (1989). Light and electron microscopic studies of meiosis and basidium ontogeny in *Clavicorona pyxidata*. *Mycologia* 81: 20–41; **Binder, M.; Hibbett, D.S.; Larsson, K.-H.; Larsson, E.; Langer, E.; Langer, G.** (2005). The phylogenetic distribution of resupinate forms across the major clades of mushroom-forming fungi (Homobasidiomycetes). *Syst. Biodiv.* 3: 113–157; **Desjardin, D.E.; Ryvarden, L.** (2003). The genus *Gloiodon*. *Sydowia* 55: 153–161; **Ginns, J.** (1998). Genera of the North American *Corticiaceae* sensu lato. *Mycologia* 90: 1–35; **Hibbett, D.S.; Donoghue, M.J.** (1995). Progress toward a phylogenetic classification of the *Polyporaceae* through parsimony analysis of mitochondrial ribosomal DNA sequences. *Can. J. Bot.* 73: S853–S861; **Larsson, E.; Larsson, K.-H.** (2003). Phylogenetic relationships of russuloid basidiomycetes with emphasis on aphyllophoralean taxa. *Mycologia* 95: 1037–1065; **Lickey, E.B.; Hughes, K.W.; Petersen, R.H.** (2003). Phylogenetic and taxonomic studies in *Artomyces* and *Clavicorona* (Homobasidiomycetes: Auriscalpiaceae). *Sydowia* 55: 181–254; **Miller, A.N.; Methven, A.S.** (2000). Biological species concepts in eastern North American populations of *Lentinellus*. *Mycologia* 92: 792–800; **Petersen, R.H.; Cifuentes, J.** (1994). Notes on mating systems of *Auriscalpium vulgare* and *A. villipes*. *Mycol. Res.* 98: 1427–1430; **Pine, E.M.; Hibbett, D.S.; Donoghue, M.J.** (1999). Phylogenetic relationships of cantharelloid and clavarioid Homobasidiomycetes based on mitochondrial and nuclear rDNA sequences. *Mycologia* 91: 944–963; **Stalpers, J.A.** (1996). The Aphyllophoraceous Fungi – II. Keys to the Species of the *Hericiales*. *Stud. Mycol.* 40: 185 pp.; **Wu, Q.X.; Petersen, R.H.** (1991). Morphological and mating studies on Asian *Clavicorona*. *Mycosystema* Suppl. 4: 33–44.

Bacidiaceae Walt. Watson 1929
Lecanorales: Ascomycota

Thallus crustose, squamulose or granular. Ascomata apothecial, pale to black, concave to flat, usually without a well-developed thalline margin, ± superficial on the thallus. Interascal tissue of true paraphyses, usually branched and/or swollen at the apex, sometimes anastomosing, often with a well-developed epithecium. Asci cylindrical to cylindric-clavate, with a well-developed apical cap, strongly blueing in iodine apart from the apical cushion, ocular chamber small, often with an outer J+ mucous layer, rarely polyspored. Ascospores ellipsoidal to elongated, usually septate, usually without a gelatinous sheath. Anamorph pycnidial, usually immersed in the thallus, producing hyaline aseptate or septate conidia from percurrently proliferating conidiogenous cells.

Significant Genera: *Bacidia, Lecania, Phyllopsora, Squamarina, Tephromela*.

Tephromela atra, Outer Hebrides, UK; thallus and ascomata

Distribution: Widespread in all climatic zones.

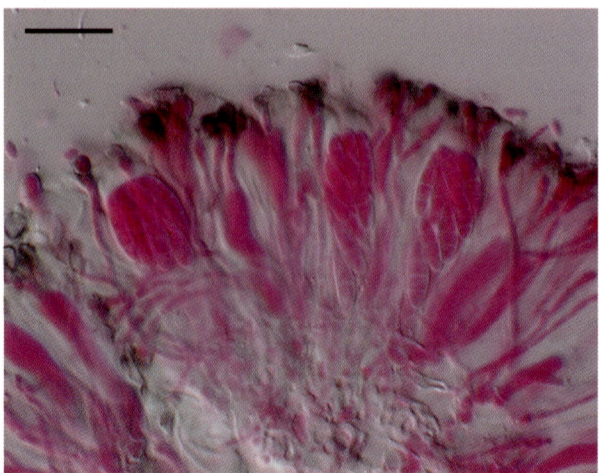

Bacidia naegelii, Jersey, UK; asci and paraphyses

Economic Significance: None is known.

Ecology: Lichenized with green algae, with occasional lichenicolous species, on rocks and bark.

Notes: Molecular studies suggest that the *Bacidiaceae* may need to be subsumed into the *Ramalinaceae*, but more sequences from the latter family are needed to establish a clear phylogeny. Many of the genera need revision.

References: **Aptroot, A.; Herk, C.M. van** (1999). *Bacidia neosquamulosa*, a new and rapidly spreading corticolous lichen species from western Europe. *Lichenologist* 31: 121–127; **Bellemère, A.; Letrouit-Galinou, M.-A.** (1987). Differentiation of lichen asci including dehiscence and sporogenesis: an ultrastructural survey. *In* Peveling, E. (ed.), Progress and Problems in Lichenology in the Eighties. *Biblthca Lichenol.* 25: 137–161; **Boom, P.P.G. van den** (1992). The saxicolous species of the lichen genus *Lecania* in the Netherlands, Belgium and Luxemburg. *Nova Hedwigia* 54: 229–254; **Brako, L.** (1989). Reevaluation of the genus *Phyllopsora* with taxonomic notes and introduction of *Squamacidia*, gen. nov. *Mycotaxon* 35: 1–19; **Brako, L.** (1991). *Phyllopsora* (*Bacidiaceae*). *Fl. Neotrop. Monogr.*: 67 pp.; **Ekman, S.** (1996). The Corticolous and Lignicolous Species of *Bacidia* and *Bacidina* in North America. *Op. bot.* 127: 148 pp.; **Ekman, S.** (1997). The genus *Cliostomum* revisited. *In* Tibell, L. & Hedberg, I. (eds), Lichen Studies Dedicated to Rolf Santesson. *Symb. bot. upsal.* 32: 17–28; **Ekman, S.** (2001). Molecular phylogeny of the *Bacidiaceae* (*Lecanorales*, lichenized *Ascomycota*). *Mycol. Res.* 105: 783–797; **Ekman, S.; Wedin, M.** (2000). The phylogeny of the families *Lecanoraceae* and *Bacidiaceae* (lichenized *Ascomycota*) inferred from nuclear SSU rDNA sequences. *Pl. Biol.* 2: 350–360; **Feige, G.B.; Röser, G.; Lumbsch, H.T.** (1997). Chemotaxonomic studies on European *Squamarina* species (*Ascomycotina, Lecanorales*). *Biblthca Lichenol.* 67: 25–31; **Kalb, K.; Elix, J.A.** (1995). The lichen genus *Physcidia*. *Biblthca Lichenol.* 57: 265–296; **Kalb, K.; Lücking, R.; Sérusiaux, E.** (2000). Studies in *Bacidia sensu lato* (lichenized *Ascomycetes*; *Lecanorales*) I. The genus *Bapalmuia*. *Mycotaxon* 75: 281–309; **Kantvilas, G.; Elix, J.A.** (1995). The lichen genus *Cliostomum* in Australia. *Biblthca Lichenol.* 58: 199–212; **Lumbsch, H.T.; Schmitt, I.; Palice, Z.; Wiklund, E.; Ekman, S.; Wedin, M.** (2004). Supraordinal phylogenetic relationships of *Lecanoromycetes* based on a Bayesian analysis of combined nuclear and mitochondrial sequences. *Mol. Phylogen. Evol.* 31: 822–832; **Rambold, G.** (1993). Further species of the genus *Tephromela* (*Lecanorales*). *Sendtnera* 1: 281–288.

Baeomycetaceae Dumort. 1829
Baeomycetales: Ascomycota

Thallus varied, usually crustose or squamulose. Ascomata sessile or shortly stipitate, sometimes clustered, formed on specialized generally non-lichenized thalline branches, flat or convex, pink or brown, the wall of interwoven hyphae. Interascal tissue of simple or sparingly branched paraphyses, often swollen at the apices. Hymenial gel J+ or J–. Asci thin-walled, not thickened at the apex, with a J+ or J– apical pore. Ascospores hyaline, simple or transversely septate. Anamorph pycnidial.

Significant Genera: *Baeomyces*.

Distribution: Widespread.

Economic Significance: None is known.

Ecology: Lichenized with green algae, on soil, rock and similar substrata.

Notes: The family has recently been placed in a separate order, the *Baeomycetales*, but more research is needed and the name is so far invalidly published. The *Anamylopsoraceae* may be closely related. Some molecular data suggest a relationship with the *Icmadophilaceae* and the *Pertusariales*.

Baeomyces placophyllus; thallus and ascomata

Phyllobaeis imbricatus; thallus and ascomata

References: **Christensen, S.N.; Alstrup, V.** (1990). Chemical and morphological variation in *Baeomyces rufus*, including *B. speciosus* (*Baeomycetaceae, Lecanorales, Ascomycotina*) with special reference to Denmark. *Nova Hedwigia* 51: 469–474; **Galloway, D.J.** (2000). *Knightiella* belongs in *Icmadophila* (*Helotiales, Icmadophilaceae*). *Lichenologist* 32: 294–297; **Gierl, C.; Kalb, K.** (1993). Die Flechtengattung *Dibaeis*. Eine Übersicht über die rosafrüchtigen Arten von *Baeomyces* sens. lat. nebst Anmerkungen zu *Phyllobaeis* gen. nov. *Herzogia* 9: 593–645; **Hibbett et al.** (2007). A higher-level phylogenetic classification of the *Fungi*. *Mycol. Res.* 111(5): 509–548. **Ihlen, P.G.** (1997). The lichen genus *Baeomyces* (*Leotiales, Ascomycotina*) in Norway. *Nova Hedwigia* 64: 137–146; **Kauff, F.; Lutzoni, F.** (2002). Phylogeny of the *Gyalectales* and *Ostropales* (*Ascomycota, Fungi*): among and within order relationships based on nuclear ribosomal RNA small and large subunits. *Mol. Phylogen. Evol.* 25: 138–156; **Lumbsch, H.T.; Schmitt, I.; Palice, Z.; Wiklund, E.; Ekman, S.; Wedin, M.** (2004). Supraordinal phylogenetic relationships of *Lecanoromycetes* based on a Bayesian analysis of combined nuclear and mitochondrial sequences. *Mol. Phylogen. Evol.* 31: 822–832; **Miądlikowska et al.** (2007). New insights into classification and evolution of the *Lecanoromycetes* (*Pezizomycotina, Ascomycota*) from phylogenetic analyses of three ribosomal RNA- and two protein-coding genes. *Mycologia* 98(6): 1088–1103; **Peršoh, D.; Beck, A.; Rambold, G.** (2004). The distribution of ascus types and photobiontal selection in *Lecanoromycetes* (*Ascomycota*) against the background of a revised SSU nrDNA phylogeny. *Mycol. Prog.* 3: 103–121; **Platt, J.L.; Spatafora, J.W.** (1999). A re-examination of generic concepts of baeomycetoid lichens based on phylogenetic analyses of nuclear SSU and LSU ribosomal DNA. *Lichenologist* 31: 409–418; **Wedin, M.; Wiklund, E.; Crewe, A.; Döring, H.; Ekman, S.; Nyberg, A.; Schmitt, I.; Lumbsch, H.T.** (2005). Phylogenetic relationships of *Lecanoromycetes* (*Ascomycota*) as revealed by sequences of mtSSU and nLSU rDNA sequence data. *Mycol. Res.* 109: 159–172.

Bankeraceae Donk 1961
Thelephorales: Basidiomycota

Basidiomata annual, pileate and centrally or eccentrically stipitate, fleshy or tough and woody, white, cream or brownish, more rarely yellow or red, at first tomentose above, sometimes becoming scaly. Hyphal system monomitic, rarely with some skeletal hyphae, generative hyphae often inflated, with or without clamp connections. Cystidia occasionally present. Hymenium poroid, spiny or ridged, usually white, cream or grey, rarely brown or purplish. Basidia clavate, with or without a basal clamp connection, with 4 sterigmata. Basidiospores small, subglobose to ellipsoidal, thin-walled, hyaline and echinulate or brown and warted, not staining in iodine.

Significant Genera: *Hydnellum, Phellodon, Sarcodon*.

Bankera fuligineoalba, Glen Affric, Scotland; basidioma

Distribution: Widespread, mainly known from north temperate zones.

Economic Significance: None is known.

Ecology: Terrestrial, ectomycorrhizal with gymnosperm and angiosperm trees, typically *Pinaceae* or *Fagaceae*.

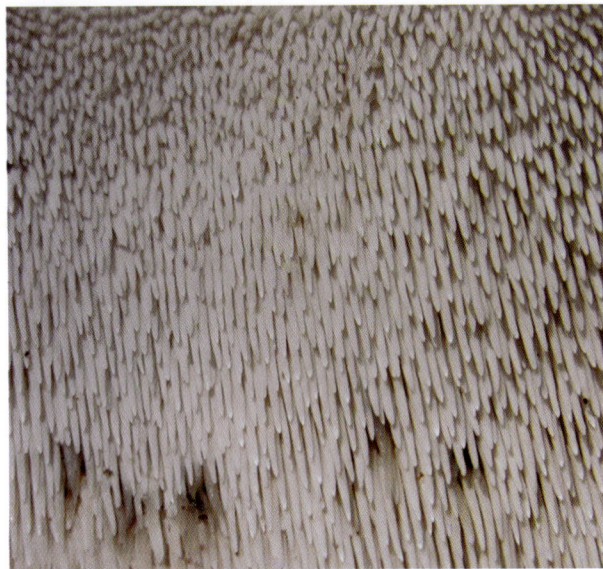

Bankera fuligineoalba, Glen Affric, Scotland; toothed hymenium

Notes: A number of these stipitate hydnoid species are of conservation concern in northern Europe and North America. The family appears to be well defined in molecular terms as part of the thelephoroid clade of the homobasidiomycetes.

Sarcodon scabrosus, Andorra; basidioma

References: **Agerer, R.** (1992). Studies on ectomycorrhizae XLIV. Ectomycorrhizae of *Boletopsis leucomelaena* (*Thelephoraceae*, *Basidiomycetes*) and their relationship to an unidentified ectomycorrhiza. *Nova Hedwigia* 55: 501–518; **Arnolds, E.** (1989). Changes in frequency and distribution of macromycetes in the Netherlands in relation to a changing environment. *Atti del IV Convegno Internazionale di Micologia, Borgo val di Taro – I, Funghi Atque Loci Natura (Funghi ed Ambiente)* (Italy): 163–232; **Baird, R.E.** (1986). Study of the stipitate hydnums from the Southern Appalachian Mountains – Genera: *Bankera*, *Hydnellum*, *Phellodon*, *Sarcodon*. *Biblthca Mycol.* 104: 156 pp.; **Baird, R.E.; Khan, S.R.** (1986). The stipitate Hydnums (*Thelephoraceae*) of Florida. *Brittonia* 30: 171–184; **Harrison, K.A.; Grund, D.W.** (1987). Preliminary keys to the terrestrial stipitate Hydnums of North America. *Mycotaxon* 28: 419–426; **Hibbett, D.S.; Binder, M.** (2002). Evolution of complex fruiting-body morphologies in homobasidiomycetes. *Proc. R. Soc. Lond.* B. Biol. Sci. 269: 1963–1969; **Hibbett, D.S.; Gilbert, L.B.; Donoghue, M.J.** (2000). Evolutionary instability of ectomycorrhizal symbioses in basidiomycetes. *Nature* Lond. 407: 506–508; **Kõljalg, U.; Renvall, P.** (2000). *Hydnellum gracilipes* – a link between stipitate and resupinate hymenomycetes. *Karstenia* 40: 71–77; **Larsson, K.-H.; Larsson, E.; Koljalg, U.** (2004). High phylogenetic diversity among corticioid homobasidiomycetes. *Mycol. Res.* 108: 983–1002; **Stalpers, J.A.** (1993). The aphyllophoraceous fungi I. Keys to the species of the *Thelephorales*. *Stud. Mycol.* 35: 168 pp.

Basidiobolaceae Engl. & E. Gilg 1924
Basidiobolales: Zygomycota

Colonies growing rapidly, formed from typically broad hyphae, initially flat but becoming radially folded, grey with cream or yellow tints, appearing powdery due to surface mycelium and sporophores (conidiophores); satellite colonies formed from discharged spores. Sporophores dimorphic; primary sporophores with a subsporangial vesicle, primary spores globose, solitary, forcibly discharged from the sporophore; secondary sporophores not basally swollen, secondary spores clavate, with a knob-like adhesive tip, passively released. Spherical, smooth, thick-walled spores (zygospores) with two closely appressed beak-like appendages formed in the mycelium.

Significant Genera: *Basidiobolus*.

Distribution: Widespread.

Economic Significance: Cause of subcutaneous zygomycosis in humans.

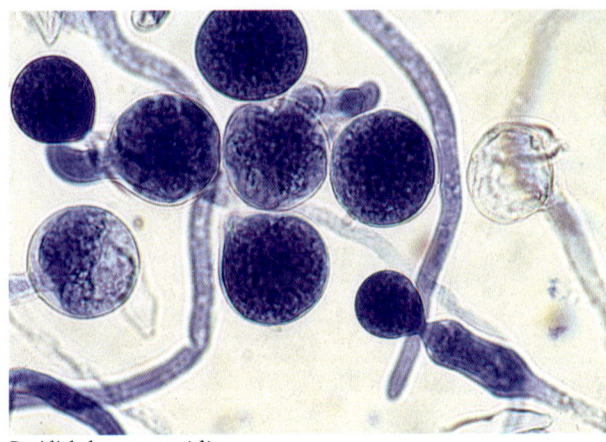

Basidiobolus sp.; conidia.

Ecology: *Basidiobolus ranarum* is commonly present in decaying fruit and vegetable matter, and as a commensal in the intestinal tract of frogs, toads and lizards. It has been reported from tropical Africa, Australia, India, Indonesia and South-East Asia.

Notes: Previously classified in the *Entomophthorales* but based on an analysis of molecular data, *Basidiobolaceae* clusters with the *Chytridiales* and *Neocallimasticales* (*Chytridiomycota*).

References: **Dykstra, M.J.** (1994). Ballistosporic conidia in *Basidiobolus ranarum*; the influence of light and nutrition on the production of conidia and endospores (sporangiospores). *Mycologia* 86: 494–501; **Dykstra, M.J.; Bradley-Kerr, B.** (1994). The adhesive droplet of capilliconidia of *Basidiobolus ranarum* exhibits unique ultrastructural features. *Mycologia* 86: 336–342; **Jensen, A.B.; Gargas, A.; Eilenberg, J.; Rosendahl, S.** (1998). Relationships of the insect-pathogenic order *Entomophthorales* (*Zygomycota, Fungi*) based on phylogenetic analyses of nuclear small subunit ribosomal DNA sequences (SSU rDNA). *Fungal Genetics Biol.* 24: 325–334; **Nagayama, Y.; Sato, H.; Shimazu, M.; Sugiyama, J.** (1995). Phylogenetic divergence of the entomophthoralean fungi: evidence from nuclear 18S ribosomal RNA gene sequences. *Mycologia* 87: 203–209; **Tanabe, Y.; Saikawa, M.; Watanabe, M.M.; Sugiyama, J.** (2004). Molecular phylogeny of *Zygomycota* based on EF-1α and RPB1 sequences: limitations and utility of alternative markers to rDNA. *Mol. Phylogen. Evol.* 30: 438–449.

Batistiaceae Samuels & K.F. Rodrigues 1989
Uncertain position within Ascomycetes, Ascomycota

Stromata absent (or not distinguishable from ascoma). Ascomata non-ostiolate, black, thick-walled, cephalothecoid, with a long transversely ridged stalk. Interascal tissue absent. Asci minute, deliquescent. Ascospores translucent, brown, simple, without germ pores. Anamorph hyphomycetous, synnematal.

Batistia annulipes, Brazil; ascomata on rotten wood

Significant Genera: *Batistia*. Anamorph: *Acrostroma*.

Distribution: Only known from the neotropics.

Economic Significance: None is known.

Ecology: Saprobic on rotten wood.

Notes: An isolated family, at one time thought to belong to the *Sordariales*. Molecular data suggest that it may be a sister group to the *Sordariomycetes* as a whole.

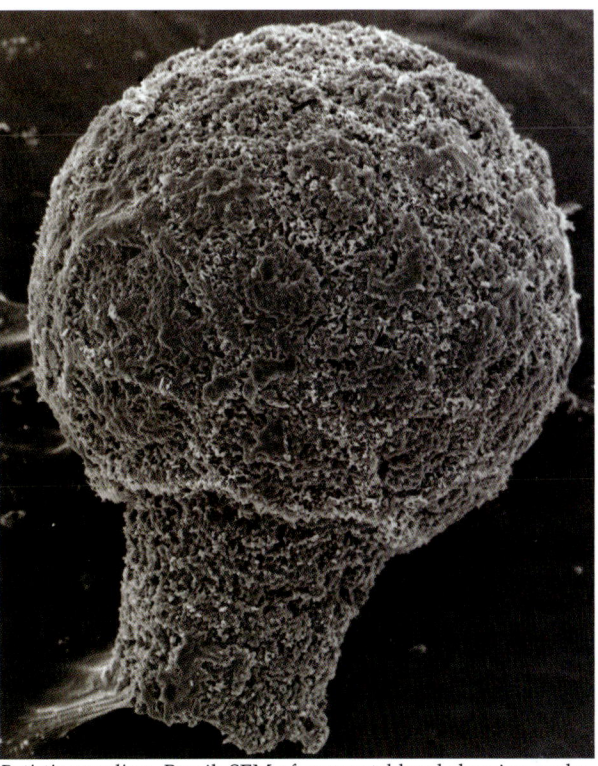

Batistia annulipes, Brazil; SEM of ascomatal head showing cephalothecoid plates

References: **Huhndorf, S.M.; Miller, A.N.; Fernández, F.A.** (2004). Molecular systematics of the *Sordariales*: the order and the family *Lasiosphaeriaceae* revisited. *Mycologia* 96: 368–387; **Samuels, G.J.; Rodrigues, K.F.** (1989). *Batistia annulipes* and its anamorph, *Acrostroma annellosynnema*. *Mycologia* 81: 52–56.

Biatorellaceae M. Choisy ex Hafellner & Casares 1992
Lecanorales: Ascomycota

Thallus crustose, often poorly developed. Ascomata apothecia, developing ± superficially, pale yellowish to brown, without a well-developed wall, usually slightly domed; hymenium blueing in iodine. Interascal tissue of paraphyses, branched near the apex and forming an epithecium. Asci cylindrical to clavate, with a well-developed usually weakly J+ apical dome which has a darker staining basal layer, no ocular chamber, and a outer gelatinous layer which blues strongly in iodine, multispored. Ascospores small, hyaline, aseptate. Anamorph not known.

Significant Genera: *Biatorella*.

Biatorella sp., Australia; thallus and ascomata

Distribution: Spread throughout the north temperate region, especially in Europe.

Economic Significance: None is known.

Ecology: Lichenized with green algae, on soil or bark.

Notes: The family is of uncertain position within the *Lecanorales*, with no molecular data available and little recent research.

References: **Hafellner, J.; Casares-Porcel, M.** (1992). Untersuchungen an den Typusarten der lichenisierten Ascomycetengattungen *Acarospora* und *Biatorella* und die daraus entstehenden Konsequenzen. *Nova Hedwigia* 55: 309–323; **Weber, W.A.; Nash, T.H.** (1992). *Biatorella clauzadeana* in North America. *Lichenologist* 24: 101–103.

Bionectriaceae Samuels & Rossman 1999
Hypocreales: Ascomycota

Stromata absent or if present pulvinate. Ascomata perithecial, superficial or rarely immersed, often hardly papillate, the ostiole periphysate; peridium membranous, white, pale yellow, orange or brown, not changing colour in KOH or lactic acid, often ornamented and sometimes setose. Interascal tissue of apical paraphyses, sometimes absent or deliquescent. Asci cylindrical, thin-walled, not fissitunicate, often with a minute J– apical ring. Ascospores varied in shape, usually transversely septate, not disarticulating, hyaline to yellow or pale brown, sometimes ornamented, without gelatinous sheaths or appendages. Anamorphs hyphomycetous, hyaline or brightly coloured, often morphologically simple, proliferating sympodially.

Significant Genera: *Bionectria, Nectriella, Nectriopsis*. Anamorph: *Clonostachys*.

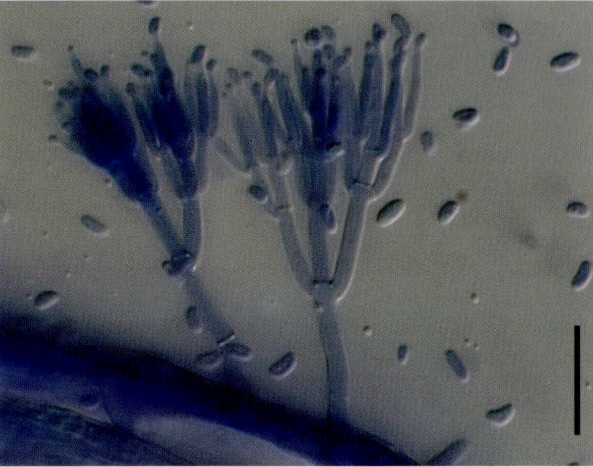

Clonostachys solani, Singapore; conidiophores and conidia

Distribution: Cosmopolitan.

Economic Significance: A few species are pathogens of crop plants (especially those with *Clonostachys* anamorphs), causing stem and root rots, damping off etc. Novel antibiotics from these species are being investigated.

Ecology: Associated with plant material, especially on wood and herbaceous debris, or with algae, bryophytes or other fungi, rarely pathogenic.

Notes: A polymorphic family, which may need further division when more genera have been sequenced.

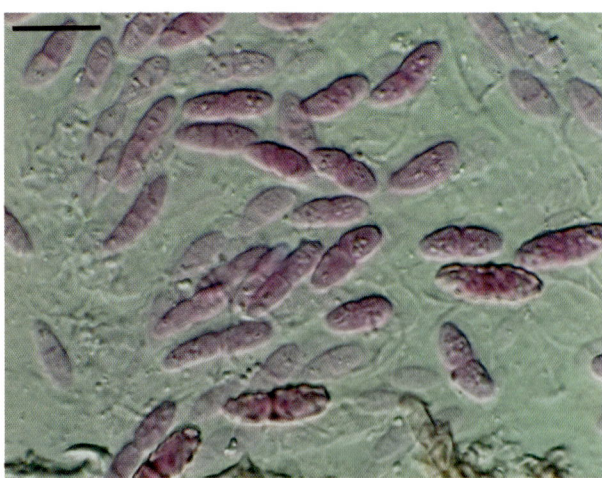

Xenonectriella ornamentata; ascospores

References: **Castlebury, L.A.; Rossman, A.Y.; Sung, G.-H.; Hyten, A.S.; Spatafora, J.W.** (2004). Multigene phylogeny reveals new lineage for *Stachybotrys chartarum*, the indoor air fungus. *Mycol. Res.* 108: 864–872; **Kohlmeyer, J.; Volkmann-Kohlmeyer, B.** (1993). Observations on *Hydronectria* and *Kallichroma* gen. nov. *Mycol. Res.* 97: 753–761; **Lowen, R.** (1991). A Monograph of the Genera *Nectriella*, Nitschke and *Pronectria* Clements. With Reference to *Charonectria*, *Cryptonectriella*, *Hydronectria* and *Pseudonectria* [Thesis (PhD) Graduate Faculty of Biology, City University,

New York]: 331 pp.; **Okuda, T.; Kohno, J.; Kishi, N.; Asai, Y.; Nishio, M.; Komatsubara, S.** (2000). Production of TMC-151, TMC-154 and TMC-171, a new class of antibiotics, is specific to 'Gliocladium roseum' group. *Mycoscience* 41: 239–253; **Rehner, S.A.; Samuels, G.J.** (1995). Molecular systematics of the *Hypocreales*: a teleomorph gene phylogeny and the status of their anamorphs. *Can. J. Bot.* 73: S816–S823; **Rossman, A.Y.** (2000). Towards monophyletic genera in the holomorphic *Hypocreales*. *Stud. Mycol.* 45: 27–34; **Rossman, A.Y.; Samuels, G.J.; Rogerson, C.T.; Lowen, R.** (1999). Genera of *Bionectriaceae*, *Hypocreaceae* and *Nectriaceae*. *Stud. Mycol.* 42: 248 pp.; **Schroers, H.-J.** (2000). Generic delimitation of *Bionectria* (*Bionectriaceae*, *Hypocreales*) based on holomorph characters and rDNA sequences. *Stud. Mycol.* 45: 63–82; **Schroers, H.-J.** (2001). A monograph of *Bionectria* (*Ascomycota*, *Hypocreales*, *Bionectriaceae*) and its *Clonostachys* anamorphs. *Stud. Mycol.* 46: 1–214; **Schroers, H.-J.; Samuels, G.J.; Seifert, K.A.; Gams, W.** (1999). Classification of the mycoparasite *Gliocladium roseum* in *Clonostachys* as *C. rosea*, its relationship to *Bionectria ochroleuca*, and notes on other *Gliocladium*-like fungi. *Mycologia* 91: 365–385.

Blastocladiaceae H.E. Petersen 1909
Blastocladiales: Blastocladiomycota

Thallus monocentric (comprising a single cell) or polycentric (typically branched with two or more centres of development), non-colonial, usually with a ± prominent basal part bearing rhizoids and one or more reproductive structures; zoosporangia sessile, with a single, apical, discharge papilla; zoospores posteriorly uniflagellate, with an often conspicuous nuclear cap; resting spores with a think, minutely punctate, wall and a truncate base.

Significant Genera: *Allomyces*, *Blastocladia*.

Distribution: Widespread but especially common in waterlogged tropical soils.

Economic Significance: None is known.

Ecology: Saprobic in soil or water.

Notes: *Allomyces* appears to be closer to *Zygomycetes* than *Chytridiomycetes* although a link to the *Entomophthorales* has also been demonstrated, based on an analysis of molecular data. Some of the oldest fossil forms recognized as fungi appear to belong here.

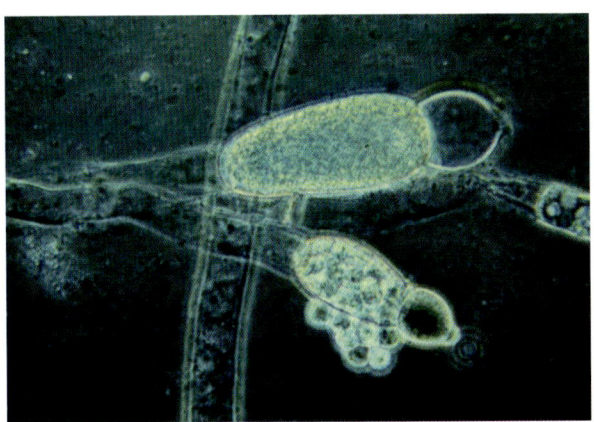
Allomyces sp.; thallus

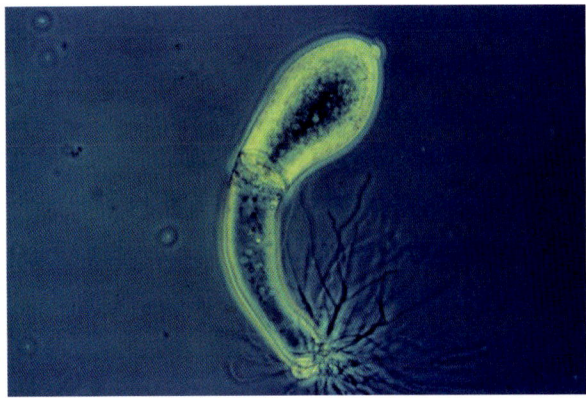
Blastocladia sp.; thallus

References: **Bullerwell, C.E.; Forget, L.; Langa, B.F.** (2003). Evolution of monoblepharidalean fungi based on complete mitochondrial genome sequences. *Nucl. Acids Res.* 31: 1614–1623; **Olson, L.W.** (1984). *Allomyces* – a different fungus. *Op. bot.* 73: 1–96; **Remy, W.; Taylor, T.N.; Hass, H.** (1994). Early Devonian fungi: a Blastocladalean fungus with sexual reproduction. *Am. J. Bot.* 81: 690–702; **Tanabe, Y.; Watanabe, M.M.; Sugiyama, J.** (2005). Evolutionary relationships among basal fungi (*Chytridiomycota* and *Zygomycota*): Insights from molecular phylogenetics. *J. gen. appl. Microbiol.* Tokyo 51: 267–276.

Bolbitiaceae Singer 1948
Agaricales: Basidiomycota

Basidiomata usually small, annual, sometimes ephemeral, agaricoid with a central stipe, rarely gasteroid or secotioid and gelatinous, mostly brownish or yellowish, sometimes staining blue on contact, not staining in iodine, the upper surface usually glistening or appearing water-soaked, sometimes with cystidia. Veil rarely present. Clamp connections present in most species. Gills brown to reddish, with cystidia on the edges and sometimes also on the face, rarely deliquescing. Basidia often short and broad, 1- to 4-spored. Basidiospores normally discharged actively, ochraceous to rusty brown, rarely ornamented, with a usually broad, truncate germ pore. Anamorph hyphomycetous, chains of conidia seceding rhexolytically from poorly differentiated fertile hyphae.

Significant Genera: *Agrocybe*, *Bolbitius*, *Conocybe*.

Distribution: Widespread, in all climatic zones.

Economic Significance: Some species are toxic, containing psilocybins, phallotoxins and cyclic peptide amatoxins.

Ecology: On the soil, rotting vegetation etc., sometimes associated with dung, animal nests etc. Most species are found in grassland, some are drought-adapted.

Panaeolus sphinctrinus, Inverness-shire, Scotland; basidioma

Notes: The glistening basidiomata and dark basidiospores with germ pores are the most important diagnostic features. The gasteroid/secotioid forms were at one time segregated into the *Galeropsidaceae*. There has been speculation that this characteristic is a result of bacterial infection rather than evolutionary radiation. Ectomycorrhizal taxa formerly assigned to this family (e.g. *Hebeloma*, *Naucoria*) are now placed in the related family *Cortinariaceae*; more work is needed on the boundary between these two groups.

References: **Aanen, D.K.; Kuyper, T.W.; Boekhout, T.; Hoekstra, R.F.** (2000). Phylogenetic relationships in the genus *Hebeloma* based on ITS1 and 2 sequences, with special emphasis on the *Hebeloma crustuliniforme* complex. *Mycologia* 92: 269–281; **Fukihara, T.; Hongo, T.** (1995). Ammonia fungi of Iriomote Island in the southern Ryukyus, Japan and a new ammonia fungus, *Hebeloma luchuense*. *Mycoscience* 36: 425–430; **Gerhardt, E.** (1996). Taxonomische Revision der Gattungen *Panaeolus* und *Panaeolina* (*Fungi*, *Agaricales*, *Coprinaceae*). *Biblthca Bot.* 147: 149 pp.; **Guidot, A.; Lumini, E.; Debaud, J.-C.; Marmeisse, R.** (1999). The nuclear ribosomal DNA intergenic spacer as a target sequence to study intraspecific diversity of the ectomycorrhizal basidiomycete *Hebeloma cylindrosporum* directly on *Pinus* root systems. *Appl. Environm. Microbiol.* 65: 903–909; **Hallen, H.E.; Watling, R.; Adams, G.C.** (2003). Taxonomy and toxicity of *Conocybe lactea* and related species. *Mycol. Res.* 107: 969–979; **Moncalvo, J.-M.; Lutzoni, F.M.; Rehner, S.A.; Johnson, J.; Vilgalys, R.** (2000). Phylogenetic relationships of agaric fungi based on nuclear large subunit ribosomal DNA sequences. *Syst. Biol.* 49: 278–305; **Pegler, D.N.** (1986). Agaric Flora of Sri Lanka. *Kew Bull.* Addit. Ser. 12: 519 pp.; **Peintner, U.; Bougher, N.L.; Castellano, M.A.; Moncalvo, M.-C.; Moser, M.M.; Trappe, J.M.; Vilgalys, R.** (2001). Multiple origins of sequestrate fungi related to *Cortinarius* (*Cortinariaceae*). *Am. J. Bot.* 88: 2168–2179; **Singer, R.** (1986). *Agaricales in Modern Taxonomy* Edn 4 (Koenigstein): 981 pp.; **Singer, R.; Hausknecht, A.** (1988). Notes on *Conocybe* (*Bolbitiaceae*). *Pl. Syst. Evol.* 159: 107–121; **Singer, R.; Hausknecht, A.** (1992). The group of *Conocybe mesospora* in Europe (*Bolbitiaceae*). *Pl. Syst. Evol.* 180: 77–104; **Walther, G.; Garnica, S.; Weiss, M.** (2005). The systematic relevance of conidiogenesis modes in the gilled *Agaricales*. *Mycol. Res.* 109: 525–544; **Walther, G.; Weiss, M.** (2006). Anamorphs of the *Bolbitiaceae* (*Basidiomycota*, *Agaricales*). *Mycologia* 98: 792–800; **Watling, R.** (1989). The natural history of the *Bolbitiaceae* (*Agaricales*): ecological strategies. *Op. bot.* 100: 259–265; **Watling, R.; Taylor, G.M.** (1987). Observations on the *Bolbitiaceae*: 27. Preliminary account of the *Bolbitiaceae* of New Zealand. *Biblthca Mycol.* 117: 61 pp.; **Young, A.M.** (1989). The *Panaeoloideae* (*Fungi*, *Basidiomycetes*) of Australia. *Aust. Syst. Bot.* 2: 75–97.

Boletaceae Chevall. 1826
Boletales: Basidiomycota

Basidioma pileate or gasteroid. Pileate forms typically robust, stipe stout, solid or hollow and often bulbous, with a scabrous or reticulate surface, the cap viscid or dry, veil sometimes present, brownish or brightly coloured. Clamp connections normally absent. Hymenium tubulate or with wide irregular anastomosing gills, often brightly coloured and sometimes changing colour (often rapidly and dramatically) when damaged. Cystidia common in most tissues but not well differentiated. Basidia usually small, clavate, 4-spored. Basidiospores either ovoid to fusoid-cylindric and smooth or subglobose with a reticulately ridged ornamentation, spore-print olivaceous or brown, rarely purplish.

Significant Genera: *Boletus*, *Leccinum*, *Tylopilus*.

Distribution: Widespread; the gasteroid forms are mainly tropical. Increasing numbers of species are found outside their natural range as a result of introductions with their mycorrhizal (tree) associates.

Economic Significance: Some are edible and highly prized, while others are dangerously toxic. The cep is the edible *Boletus edulis* ('fungi suilli' of Pliny and other classical writers), a major commercial crop, all of which is collected from the wild.

Ecology: Most species form ectomycorrhizal associations with gymnosperm or angiosperm trees. Some are saprobes on decaying wood and leaf litter, a few are parasitic on other fungi.

Notes: The main family of the well-known boletes. The gasteroid forms, previously classified separately, are increasingly shown to belong within the main pileate taxon, and the *Sclerodermataceae* seems to be a sister group. A number of species are of conservation concern in Europe and North America.

Boletus edulis, Suffolk, UK; basidioma

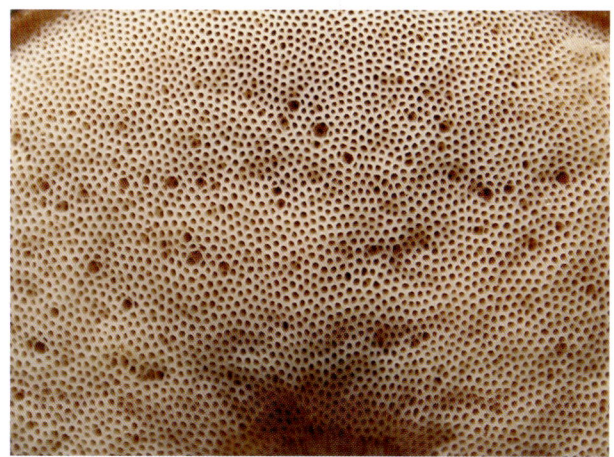
Leccinum scabrum, Glen Affric, Scotland; poroid hymenium

References: **Alessio, C.L.** (1985). *Boletus Dill. ex L.* (Saronno): 712 pp.; **Binder, M.; Besl, H.** (2000). 28S rDNA sequence data and chemotaxonomical analyses on the generic concept of *Leccinum* (*Boletales*). *Micologia 2000* (Trento): 75–86; **Binder, M.; Bresinsky, A.** (2002). Derivation of a polymorphic lineage of gasteromycetes from boletoid ancestors. *Mycologia* 94: 85–98; **Binder, M.; Bresinsky, A.** (2002). *Retiboletus*, a new genus for a species complex in the *Boletaceae* producing retipolides. *Feddes Repert.* Beih. 113: 30–40; **Den Bakker, H.C.; Gravendeel, B.; Kuyper, T.W.** (2004). An ITS phylogeny of *Leccinum* and an analysis of the evolution of minisatellite-like sequences within ITS1. *Mycologia* 96: 102–118; **Den Bakker, H.C.; Zuccarello, G.C.; Kuyper, T.W.; Noordeloos, M.E.** (2004). Evolution of host specificity in the ectomycorrhizal genus *Leccinum*. *New Phytol.* 163: 201–215; **Kretzer, A.M.; Bruns, T.D.** (1999). Use of *atp6* in fungal phylogenetics: an example from the *Boletales*. *Mol. Phylogen. Evol.* 13: 483–492; **Lakhanpal, T.N.** (1996). Mushrooms of India. Boletaceae. *Stud. Cryptog. Bot.* 1: 170 pp.; **Lannoy, G.; Estadès, A.** (1995). *Monographie des Leccinum d'Europe* (France): 229 pp.; **Mello, A.; Ghignone, S.; Vizzini, A.; Sechi, C.; Ruiu, P.; Bonfante, P.** (2006). ITS primers for the identification of marketable boletes. *J. Biotechn.* 121: 318–329; **Peintner, U.; Ladurner, H.; Simonini, G.** (2003). *Xerocomus cisalpinus* sp. nov., and the delimitation of species in the *X. chrysenteron* complex based on morphology and rDNA-LSU sequences. *Mycol. Res.* 107: 659–679; **Singer, R.** (1986). *Agaricales in Modern Taxonomy* Edn 4 (Koenigstein): 981 pp.; **Singer, R.; García, J.; Gomez, L.D.** (1991). The *Boletineae* of Mexico and Central America III. *Beih. Nova Hedwigia* 102: 99 pp.; **Thiers, H.D.** (1989). *Gastroboletus* revisited. *Mem. N. Y. bot. Gdn* 49: 355–359; **Watling, R.** (2001). Australian boletes: their diversity and possible origins. *Aust. Syst. Bot.* 14: 407–416; **Watling, R.; Hollands, R.** (1990). Boletes from Sarawak. *Notes R. bot. Gdn Edinb.* 46: 405–422.

Boletinellaceae P.M. Kirk, P.F. Cannon & J.C. David 2001
Boletales: Basidiomycota

Basidioma boletoid, stipitate and pileate. Cap glabrous to tomentose, yellowish or brown, stipe similarly pigmented, ± smooth, insertion variable, sometimes eccentric. Hyphae with clamp connections. Hymenophore tubulate and decurrent, sometimes depressed around the stipe, yellowish to brown. Basidiospores short-ellipsoid, smooth, not staining in iodine, spore-print olivaceous-brown. Sclerotia produced abundantly in culture.

Significant Genera: *Boletinellus*, *Phlebopus*.

Distribution: Widespread, primarily in tropical zones.

Economic Significance: None is known.

Ecology: Saprobic on buried wood and roots, normally not forming ectomycorrhizas. Facultative associations with legume trees have been reported.

Notes: The family appears most closely allied to the primarily gasteroid *Scleroderma* lineage within the *Boletales*.

Boletinellus merulioides; basidiomata

References: **Corner, E.J.H.** (1970). Supplement to a monograph of *Clavaria* and allied genera. *Beih. Nova Hedwigia* 33: 1–299; **Cotter, H. van T.; Miller, O.K.** (1985). Sclerotia of *Boletinellus merulioides* in nature. *Mycologia* 77: 927–931; **Deschamps, J.; Moreno, G.** (1999). *Phlebopus bruchii* (*Boletales*): an edible fungus from Argentina with possible commercial value. *Mycotaxon* 72: 205–213; **Kretzer, A.M.; Bruns, T.D.** (1999). Use of *atp6* in fungal phylogenetics: an example from the *Boletales*. *Mol. Phylogen. Evol.* 13: 483–492; **Watling, R.; Meijer, A.R. de** (1997). Macromycetes from the state of Paraná, Brazil. 5. Poroid and lamellate boletes. *Edinb. J. Bot.* 54: 231–251.

Boliniaceae Rick 1931
Boliniales: Ascomycota

Stromata immersed to erumpent, crustose or pulvinate, sometimes absent, usually soft-textured, composed of thin-walled hyphal tissue. Ascomata perithecial, long-necked, sometimes vertically elongate, the ostiole periphysate. Interascal tissue of narrow thin-walled paraphyses, sometimes evanescent. Asci cylindrical, persistent, thin-walled, with a small, J apical ring. Ascospores brown, sometimes transversely septate, sometimes with germ pores. Anamorphs not known.

Significant Genera: *Camarops*, *Endoxyla*.

Distribution: Widespread in temperate regions.

Economic Significance: None is known.

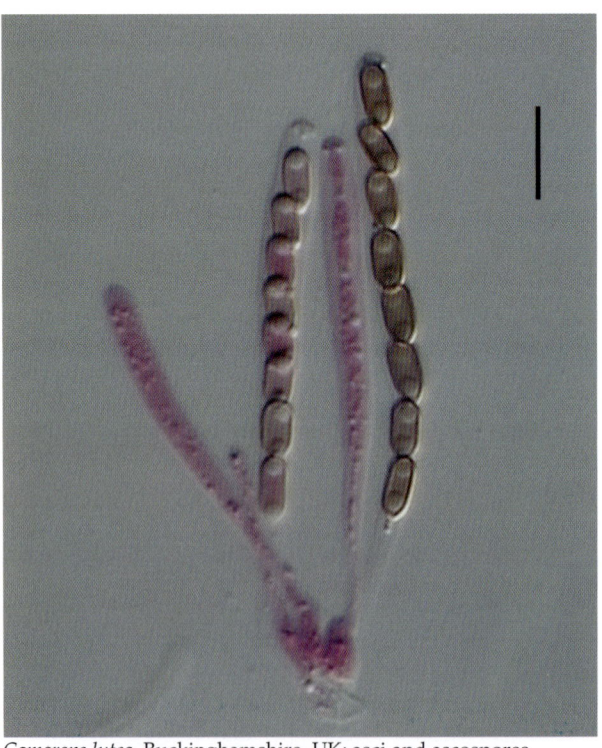

Camarops lutea, Buckinghamshire, UK; asci and ascospores

Ecology: Saprobic on and in wood and bark.

Notes: Placement in the separate order *Boliniales* within the *Sordariomycetes* has been confirmed using molecular studies, and the family is clearly more closely related to the *Sordariales* than the *Xylariales*.

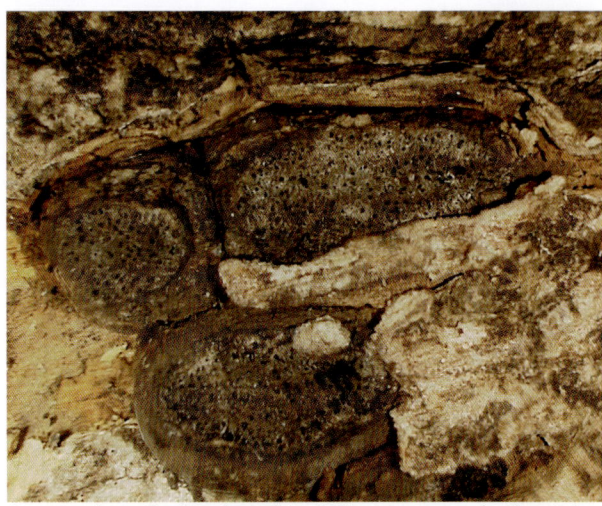

Camarops lutea, Buckinghamshire, UK; stromata on dead wood

References: **Andersson, K.; Eriksson, O.E.; Landvik, S.** (1995). *Boliniaceae* transferred to *Sordariales* (*Ascomycota*). *Syst. Ascom.* 14: 1–16; **Huhndorf, S.M.; Miller, A.N.; Fernández, F.A.** (2004). Molecular systematics of the *Sordariales*: the order and the family *Lasiosphaeriaceae* revisited. *Mycologia* 96: 368–387; **Park, H.G.; Jong, S.C.** (2003). Molecular characterization of *Monascus* strains based on the D1/D2 regions of LSU rRNA genes. *Mycoscience* 44: 25–32; **Rogers, J.D.; Samuels, G.J.** (1988). *Apiocamarops cryptocellula*, a new species from Guyana. *Mycologia* 80: 738–741; **Samuels, G.J.; Rogers, J.D.** (1987). *Camarops flava* sp. nov., *Apiocamarops alba* gen. et sp. nov., and notes on *Camarops scleroderma* and *C. ustulinoides*. *Mycotaxon* 28: 45–59; **Untereiner, W.A.** (1988). A taxonomic re-evaluation of the pyrenomycete genus *Endoxyla* Fuckel [abstract]. *Mycol. Soc. Amer. Newsl.* 39: 52; **Vasilyeva, L.N.** (1997). *Camarops pugillus* (Schw.: Fr.) Shear from the Russian Far East. *Mikol. Fitopatol.* 31: 5–7.

Bondarzewiaceae Kotl. & Pouzar 1957
Russulales: Basidiomycota

Basidiomata usually annual, effused or fan-shaped, rarely stipitate and club-shaped or pulvinate, generally tomentose, soft-textured to tough and woody, often zoned, pale. Hyphal system usually dimitic, skeletal hyphae not or sparingly branched, clamp connections present or absent (but then often present in culture). Hymenium smooth, spiny, toothed or poroid. Gloeocystidia typically present. Basidia hyaline, thin-walled, clavate to urniform, 2- to 4-spored. Basidiospores globose to ellipsoidal, rarely ± cylindrical, usually ornamented, thin-walled to thick-walled, staining in iodine. Anamorph hyphomycetous, with small thin-walled to thick-walled conidia formed simultaneously from multiple denticles on clavate conidiogenous cells.

Significant Genera: *Heterobasidion*, *Wrightoporia*. Anamorph: *Spiniger*.

Distribution: Widespread, especially in temperate regions.

Economic Significance: *Heterobasidion annosum* is a major pathogen of conifer plantations, causing rots of roots and stem bases. Control is problematic, and biological methods have been extensively researched.

Heterobasidion annosum, Surrey, UK; basidioma

Ecology: Parasitic or saprobic on wood of gymnosperms and angiosperms, causing white rots.

Notes: Initial studies indicate that the family is paraphyletic, with the *Albatrellaceae* nesting within. The *Echinodontaceae* is confirmed as separate based on molecular as well as morphological characters. The fertile structures of the anamorph appear very similar morphologically to the basidia and basidiospores, and are doubtless generated using similar or identical developmental control processes.

Heterobasidion annosum, Surrey, UK; old basidioma showing poroid hymenium

References: **Binder, M.; Hibbett, D.S.** (2002). Higher-level phylogenetic relationships of homobasidiomycetes (mushroom-forming fungi) inferred from four rDNA regions. *Mol. Phylogen. Evol.* 22: 76–90; **Binder, M.; Hibbett, D.S.; Larsson, K.-H.; Larsson, E.; Langer, E.; Langer, G.** (2005). The phylogenetic distribution of resupinate forms across the major clades of mushroom-forming fungi (Homobasidiomycetes). *Syst. Biodiv.* 3: 113–157; **Garbelotto, M.; Otrosina, W.J.; Cobb, F.W.; Bruns, T.D.** (1998). The European S and F intersterility groups of *Heterobasidion annosum* may represent sympatric protospecies. *Can. J. Bot.* 76: 397–409; **Ginns, J.** (1998). Genera of the North American *Corticiaceae sensu lato*. *Mycologia* 90: 1–35; **Korhonen, K.** (1987). Breeding units in the forest pathogens *Armillaria* and *Heterobasidion*. *Evolutionary Biology of the Fungi* Symposium of the British Mycological Society held at the University of Bristol, April 1986 (Cambridge): 301–310; **Niemelä, T.; Korhonen, K.** (1998). Taxonomy of the genus *Heterobasidion*. *Heterobasidion annosum, Biology, Ecology, Impact and Control* (Wallingford): 27–33; **Redhead, S.A.; Norvell, L.L.** (1993). Notes on *Bondarzewia*, *Heterobasidion* and *Pleurogala*. *Mycotaxon* 48: 371–380; **Ryvarden, L.** (2000). Studies in neotropical polypores 7. *Wrightoporia* (*Hericiaceae*, *Basidiomycetes*) in tropical America. *Karstenia* 40: 153–158; **Schulze, S.** (1999). Rapid detection of European *Heterobasidion annosum* intersterility groups and inter-group gene flow using taxon-specific competitive-priming PCR (TSCP-PCR). *J. Phytopath.* 147: 125–127; **Stalpers, J.A.** (1996). The Aphyllophoraceous Fungi – II. Keys to the Species of the *Hericiales*. *Stud. Mycol.* 40: 185 pp.; **Stenlid, J.; Karlsson, J.-O.; Högberg, N.** (1994). Intraspecific genetic variation in *Heterobasidion annosum* revealed by amplification of minisatellite DNA. *Mycol. Res.* 98: 57–63.

Boreostereaceae Jülich 1982
Gloeophyllales: Basidiomycota

Basidioma annual or perennial, resupinate or dimidiate, rarely with a short undifferentiated stipe, leathery to woody, dark brown to black. Context complex, brown, usually staining darker or yellowish in KOH. Hyphal system interpreted as either monomitic or dimitic with skeletal hyphae, with or without clamp connections. Hymenophore brown or grey, smooth to warted, often velvety due to projecting cystidia and rarely hydnoid with teeth composed of fasciculate cystidia. Basidia cylindrical to clavate, hyaline to yellowish, sometimes thick-walled, with 4 sterigmata, with or without clamp connections. Basidiospores ellipsoidal to cylindrical, smooth, hyaline or pale yellow. Mating system tetrapolar or homothallic.

Significant Genera: *Boreostereum*, *Veluticeps*.

Distribution: Widespread, known especially from north temperate zones.

Economic Significance: *Veluticeps berkeleyi* causes heart rot of living conifers, especially *Pinus ponderosa* in North America.

Ecology: Species are active wood decayers, inducing brown and white rots of conifers and broadleaved trees.

Notes: The family is poorly defined and may need to be combined with the *Gloeophyllaceae*, together forming an isolated clade within the homobasidiomycetes. *Columnocystis*, previously considered a synonym of *Veluticeps*, may not belong in this family, but more studies are needed.

Veluticeps abietina; basidioma

References: **Binder, M.; Hibbett, D.S.; Larsson, K.-H.; Larsson, E.; Langer, E.; Langer, G.** (2005). The phylogenetic distribution of resupinate forms across the major clades of mushroom-forming fungi (*Homobasidiomycetes*). *Syst. Biodiv.* 3: 113–157; **Chamuris, G.P.** (1988). The non-stipitate stereoid fungi in the northeastern United States and adjacent Canada. *Mycol. Mem.* 14: 247 pp.; **Ginns, J.** (1998). Genera of the North American *Corticiaceae sensu lato*. *Mycologia* 90: 1–35; **Hibbett et al.** (2007). A higher-level phylogenetic classification of the *Fungi*. *Mycol. Res.* 111(5): 509–548; **Hjortstam, K.; Tellería, M.T.** (1990). *Columnocystis*, a synonym of *Veluticeps*. *Mycotaxon* 37: 53–56; **Jung, H.S.** (1987). Wood-rotting *Aphyllophorales* of the southern Appalachian spruce-fir forest. *Biblthca Mycol.* 119: 260 pp.; **Nakasone, K.K.** (1990). Cultural studies and identification of wood-inhabiting *Corticiaceae* and selected *Hymenomycetes* from North America. *Mycol. Mem.* 15: 412 pp.; **Nakasone, K.K.** (1990). Taxonomic study of *Veluticeps* (*Aphyllophorales*). *Mycologia* 82: 622–641.

Botryobasidiaceae Jülich 1982
Cantharellales: Basidiomycota

Basidioma resupinate, poorly developed, smooth or floccose, composed of broad, monomitic, effuse, reticulate, net-like or thread-like hyphae, frequently branched at right angles, with clamp connections rare or lacking, rarely with septate cystidia. Basidia ± cylindrical to obconical, sometimes constricted in the mid portion, with 4–8 sterigmata, often arranged in botryose clusters. Basidiospores hyaline, ellipsoidal, elongate or navicular, smooth, tuberculate or echinate, not staining in iodine. Anamorph hyphomycetous, with poorly differentiated conidiogenous cells and hyaline aseptate conidia formed from multiple denticles, and/or dark thick-walled aseptate chlamydospores, sometimes in chains.

Significant Genera: *Botryobasidium*. Anamorphs: *Alysidium, Allescheriella, Haplotrichum*.

Distribution: Widespread in both temperate and tropical environments.

Economic Significance: None is known.

Ecology: Saprobic, typically on rotten wood.

Notes: The *Botryobasidiaceae* appears to occupy an isolated position at the base of the cantharelloid clade within the homobasidiomycetes, allied to the *Sistotremataceae* which has somewhat better developed basidiomata and more prominent clamp connections. The anamorphic forms are prominent.

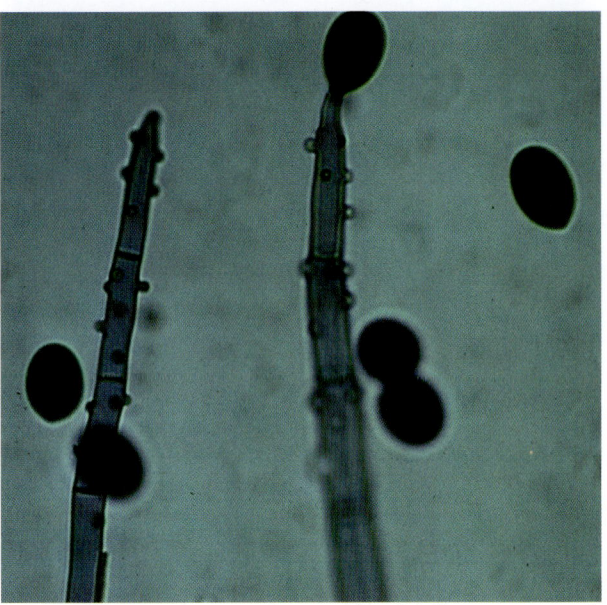

Botryobasidium conspersum (anamorph: *Acladium conspersum*); conidiophores and conidia

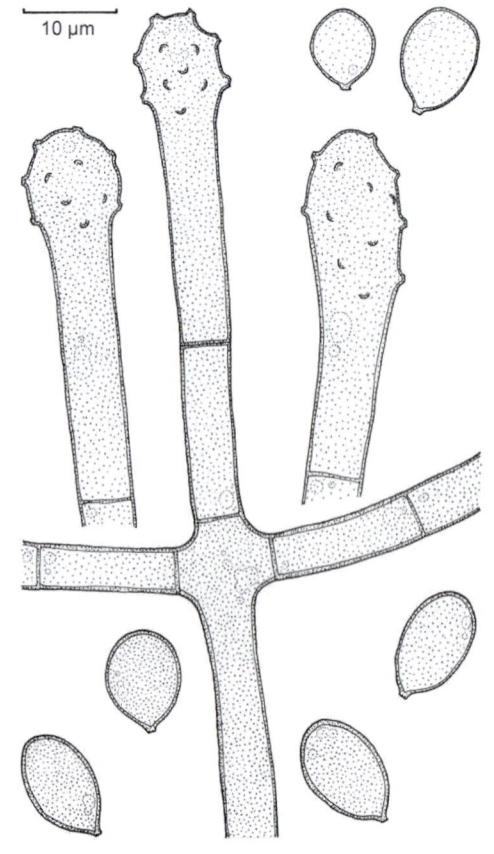

Botryobasidium chilense; anamorph (conidiophores and conidia)

References: **Binder, M.; Bresinsky, A.** (2002). Derivation of a polymorphic lineage of gasteromycetes from boletoid ancestors. *Mycologia* 94: 85–98; **Ginns, J.** (1998). Genera of the North American *Corticiaceae sensu lato. Mycologia* 90: 1–35; **Langer, G.** (1994). Die Gattung *Botryobasidium* Donk (*Corticiaceae, Basidiomycetes*). *Biblthca Mycol.* 158: 459 pp.; **Langer, G.; Langer, E.; Chen, C.J.** (2000). *Botryobasidium musaisporum* sp. nov. collected in Taiwan. *Mycol. Res.* 104: 510–512; **Langer, G.; Langer, E.; Oberwinkler, F.; Chen, J.** (2000). Speciation of *Botryobasidium subcoronatum* (*Basidiomycota*) collected in Taiwan: morphology, mating tests, and molecular data. *Mycoscience* 41: 201–210; **Maekawa, N.** (1992). The genus *Botryohypochnus* (*Corticiaceae, Aphyllophorales*) from Japan. *Trans. Mycol. Soc. Japan* 33: 317–324.

Botryosphaeriaceae Theiss. & P. Syd. 1918

Botryosphaeriales: Ascomycota

Stromata varied in development, ± immersed in or erumpent from plant material. Ascostromata perithecial, usually uniloculate, sometimes strongly aggregated, with a well-developed ostiole, the peridium black, thick-walled, composed of large pseudoparenchymatous cells. Interascal tissue of wide thin-walled cellular pseudoparaphyses, often deliquescent at maturity. Asci widely clavate, fissitunicate, with a well-developed ocular chamber but lacking well-developed apical structures. Ascospores hyaline to pale yellow, usually aseptate, smooth or slightly verrucose, sometimes with mucilaginous appendages. Anamorphs coelomycetous. Conidiomata pycnidial, often aggregated in or on stromatic tissue, thick-walled, ostiolate. Conidiogenous cells doliiform or hyphal in form, either non-proliferating or forming annellides. Conidia hyaline to dark brown, aseptate or eventually 1-septate, smooth or ornamented, often with a mucilaginous coat, sometimes with an apical mucilaginous appendage. A spermatial conidial morph is also present in some species.

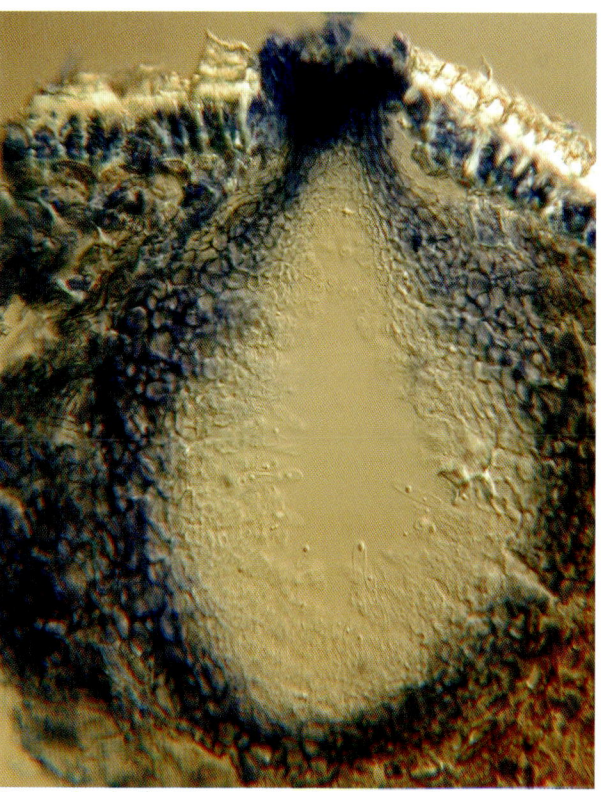

Botryosphaeria ribis, Niger; section through conidioma

Significant Genera: *Botryosphaeria, Guignardia.* Anamorphs: *Diplodia, Fusicoccum, Phyllosticta, Lasiodiplodia, Macrophomina.*

Distribution: Very widespread, both in the tropics and temperate zones.

Economic Significance: Frequently implicated as economically significant crop and forest pathogens, including both host-specific and generalist species. However, others are weak wound or secondary parasites, and are frequently found as saprobes or endophytes.

Ecology: Biotrophic, necrotrophic or saprobic, especially associated with woody tissues.

Notes: The *Botryosphaeriaceae* is now well-defined using molecular characters, and includes a number of well-defined lineages that correspond well to anamorph type. *Botryosphaeria* and *Guignardia* have been confused in the past, and at one stage the former genus was equated with the quite unrelated *Physalospora* (*Hyponectriaceae*). The *Lasiodiplodia theobromae* aggregate (often referred to incorrectly as *Botryodiplodia*) is one of the most ubiquitous and non-specific tropical pathogens, though it rarely causes serious diseases. Its striate conidia are diagnostic, but recent research suggests that several closely related species exist rather than one polymorphic assemblage, and that the anamorph genus cannot be unequivocally separated from *Diplodia*.

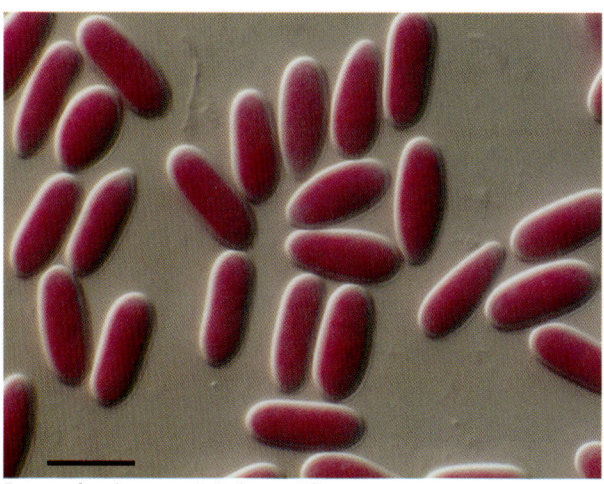

Botryosphaeria stevensii, Italy; conidia

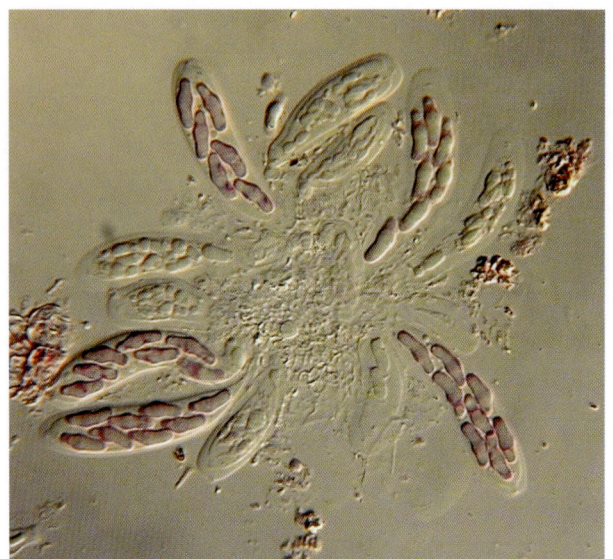

Guignardia sp. on leaves of *Ancistrocladus*, Kenya; asci and ascospores

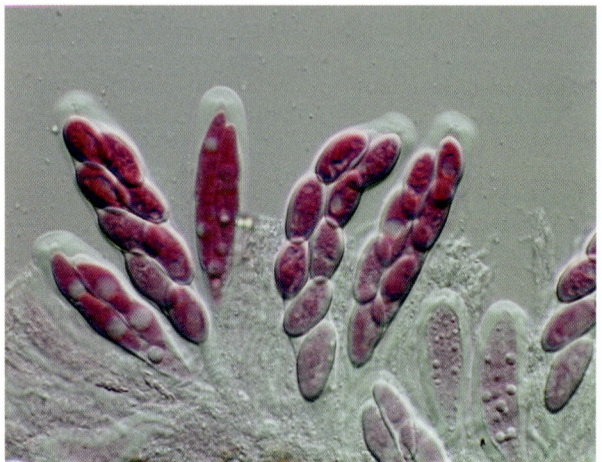

Botryosphaeria rhodina, India; asci and ascospores

References: Barber, P.A.; Burgess, T.J.; Hardy, G.E.StJ.; Slippers, B.; Keane, P.J.; Wingfield, M.J. (2005). *Botryosphaeria* species from *Eucalyptus* in Australia are pleoanamorphic, producing *Dichomera* synanamorphs in culture. *Mycol. Res.* 109: 1347–1363; **Bissett, J.** (1986). A note on the typification of *Guignardia*. *Mycotaxon* 25: 519–522; **Bissett, J.; Palm, M.E.** (1989). Species of *Phyllosticta* on conifers. *Can. J. Bot.* 67: 3378–3385; **Crous, P.W.; Slippers, B.; Wingfield, M.J.; Rheeder, J.; Marasas, W.F.O.; Philips, A.J.L.; Alves, A.; Brugess, T.; Barber, P.; Groenewald, J.Z.** (2006). Phylogenetic lineages in the *Botryosphaeriaceae*. *Stud. Mycol.* 55: 235–253; **Denman, S.; Crous, P.W.; Taylor, J.E.; Kang, J.C.; Pascoe, I.; Wingfield, M.J.** (2000). An overview of the taxonomic history of *Botryosphaeria*, and a re-evaluation of its anamorphs based on morphology and ITS rDNA phylogeny. *Stud. Mycol.* 45: 129–140; **Hibbett et al.** (2007). A higher-level phylogenetic classification of the *Fungi*. *Mycol. Res.* 111(5): 509–548; **Hyde, K.D.** (1995). Fungi from palms. XX. The genus *Guignardia*. *Sydowia* 47: 180–198; **Jacobs, K.A.; Rehner, S.A.** (1998). Comparison of cultural and morphological characters and ITS sequences in anamorphs of *Botryosphaeria* and related taxa. *Mycologia* 90: 601–610; **Jacobs, K.A.; Rehner, S.A.; Johnson, G.R.** (1995). Differentiating *Botryosphaeria* species utilizing morphological, cultural and molecular characters [abstract]. *Phytopathology* 85: 1206; **Morgan-Jones, G.; White, J.F.** (1987). Notes on coelomycetes. II. Concerning the *Fusicoccum* anamorph of *Botryosphaeria ribis*. *Mycotaxon* 30: 117–125; **Pennycook, S.R.; Samuels, G.J.** (1985). *Botryosphaeria* and *Fusicoccum* species associated with ripe fruit rot of *Actinidia deliciosa* (kiwi fruit) in New Zealand. *Mycotaxon* 24: 445–458; **Rodrigues, K.F.; Sieber, T.N.; Grünig, C.R.; Holdenrieder, O.** (2004). Characterization of *Guignardia mangiferae* isolated from tropical plants based on morphology, ISSR-PCR amplifications and ITS1-5.8S-ITS2 sequences. *Mycol. Res.* 108: 45–52; **Schoch, C.L.; Shoemaker, R.A.; Seifert, K.A.; Hambleton, S.; Spatafora, J.W.; Crous, P.W.** (2007). A multigene phylogeny of the Dothideomycetes using four nuclear loci. *Mycologia* 98(6): 1041–1052; **Slippers, B.; Crous, P.W.; Denman, S.; Coutinho, T.A.; Wingfield, B.D.; Wingfield, M.J.** (2004). Combined multiple gene genealogies and phenotypic characters differentiate several species previously identified as *Botryosphaeria dothidea*. *Mycologia* 96: 83–101; **Van Niekerk, J.M.; Crous, P.W.; Groenewald, J.Z.; Fourie, P.H.; Halleen, F.** (2004). DNA phylogeny, morphology and pathogenicity of *Botryosphaeria* species on grapevines. *Mycologia* 96: 781–798.

Brachybasidiaceae Gäum. 1926
Exobasidiales: Basidiomycota

Basidiomata minutely pustulate to discoid, gelatinous, composed of a reduced internal stroma from which basidia emerge in clusters through stomata or epidermal cells of the host. Haustoria apparently absent. Basidia initially clavate with the terminal section becoming elongate, aseptate, thin-walled to thick-walled, with 2 well-developed apical sterigmata. Gloeocystidia sometimes present. Basidiospores formed with their hilar appendages oriented adaxially, discharged actively, thin-walled, smooth, cylindrical to allantoid, sometimes becoming septate.

Kordyana celebensis; parasitic on leaves of *Commelina*

Significant Genera: *Dicellomyces*, *Kordyana*.

Distribution: Widespread, including tropical and temperate zones.

Economic Significance: None is known.

Ecology: Biotrophic or necrotrophic on leaves of monocots, including palms, grasses, sedges and *Commelinaceae*.

Notes: The family is poorly known, may be paraphyletic, and appears to be related to the *Graphiolaceae*, also parasitic on monocotyledonous plants.

References: **Barreto, R.W.; Evans, H.C.** (1988). Taxonomy of a fungus introduced into Hawaii for biological control of *Ageratina riparia* (*Eupatorieae*; *Compositae*), with observations on related weed pathogens. *Trans. Br. mycol. Soc.* 91: 81–97; **Bauer, R.; Oberwinkler, F.; Vánky, K.** (1997). Ultrastructural markers and systematics in smut fungi and allied taxa. *Can. J. Bot.* 75: 1273–1314; **Begerow, D.; Bauer, R.; Oberwinkler, F.** (2002). The *Exobasidiales*: an evolutionary hypothesis. *Mycol. Prog.* 1: 187–199; **Berndt, R.; Sharma, N.D.** (1998). *Dicellomyces calami* sp. nov., from India. *Mycol. Res.* 102: 1484–1486; **Gruèzo, W.S.** (1990). The genus *Kordyana* Rac. (*Exobasidiaceae*) in the Philippines. *Nat. Hist. Bull. Siam Soc.* 38: 89–92; **Ingold, C.T.** (1985). *Dicellomyces scirpia*: its conidial stage and taxonomic position. *Trans. Br. mycol. Soc.* 84: 542–545; **Oberwinkler, F.** (1987). *Heterobasidiomycetes* with ontogenetic yeast-stages – systematic and phylogenetic aspects. *Stud. Mycol.* 30: 61–74.

Brefeldiellaceae E. Müll. & Arx 1962
Uncertain position within Dothideomycetes, Ascomycota

Stromata strongly flattened, skin-like, often irregularly shaped, composed of irregular cells with pores in their walls. Ascomatal locules flattened, thin-walled, circular, opening by a lysigenous pore, not periphysate. Interascal tissue of ? pseudoparaphyses. Asci ± clavate, short-stalked, with a thick apical cap and a narrow ocular chamber, ? fissitunicate. Ascospores hyaline, 1-septate, thin-walled, without a sheath. Anamorphs unknown.

Significant Genera: *Brefeldiella*.

Distribution: Only known from South America.

Economic Significance: None is known.

Ecology: Biotrophic on leaves.

Notes: No recent research is available.

References: **Eriksson, O.** (1981). The families of bitunicate ascomycetes. *Op. bot.* no. 60: 220 pp.

Brigantiaeaceae Hafellner & Bellem. 1982
Lecanorales: Ascomycota

Thallus crustose or lobed. Ascomata apothecial, with well-developed margins. Interascal tissue of long narrow anastomosing paraphyses, often with a well-developed epithecium. Asci thick-walled, with a well-developed J+ apical cap, and J+ inner and outer ascus walls, ocular chamber hardly developed, 1- to 2-spored. Ascospores thin-walled, hyaline, muriform, sometimes germinating to form conidia.

Brigantiaea leucoxantha; ascomata

Significant Genera: *Brigantiaea*.

Distribution: Very widespread.

Brigantiaea fuscolutea, Louisiana, USA; ascomata on bark

Economic Significance: None is known.

Ecology: Lichenized with green algae, usually on bark.

Notes: The systematic position of this family is uncertain, and no molecular phylogenetic studies are available.

References: **Awasthi, D.D.; Srivastava, P.** (1989). Lichen genera *Brigantiaea* and *Letrouitia* from India. *Proc. Indian Acad. Sci. Pl. Sci.* 99: 165–177; **Hafellner, J.** (1997). A world monograph of *Brigantiaea* (lichenized *Ascomycotina*, *Lecanorales*). *In* Tibell, L. & Hedberg, I. (eds), Lichen Studies Dedicated to Rolf Santesson. *Symb. bot. upsal.* 32: 35–74.

which has smooth basidiomata and ostioles that lack a well-defined peristome.

References: **Bottomley, A.M.** (1948). *Gasteromycetes* of South Africa. *Bothalia* 4: 473–810; **Jacobson, K.M.** (1997). Macrofungal ecology in the Namib Desert: a fruitful or futile study?. *McIlvainea* 12: 21–32; **Ryvarden, L.; Piearce, G.D.; Masuka, A.J.** (1994). *An Introduction to the Larger Fungi of South Central Africa* (Harare): 200 pp.; **Sharp, C.; Piearce, G.** (1999). Some interesting gasteroid fungi from Zimbabwe. *Kew Bull.* 54: 739–746.

Broomeiaceae Zeller 1948
Agaricales: Basidiomycota

Fruit-body consisting of a large whitish effuse or columnar stroma with a thick, corky base from which multiple basidiomata are emergent simultaneously, divided into round or angular cavities from which the basidiomata emerge. Basidiomata gasteroid, ± globose to turbinate, brown, covered with minute balls of hyphae that presumably function as sclerotia, the ostiole surrounded by a peristome formed from dark brown conical fimbriate processes. Columella absent. Capillitium pale to dark brown, consisting of sparsely septate sparingly branched, irregularly thickened hyphae. Basidia not described. Basidiospores globose to ellipsoidal, brown, initially smooth but becoming punctate or echinulate.

Significant Genera: *Broomeia*.

Distribution: Known from dry areas of southern Africa.

Economic Significance: None is known.

Broomeia congregata; illustration on a postage stamp

Ecology: Little is known, but fruit bodies occur as apparent vagrants on sandy soil and seem to be associated with living *Acacia* roots. A parasitic relationship has been postulated but not demonstrated.

Notes: No molecular data have been gathered. Similarities have been noted with the *Diplocystaceae*,

Bulgariaceae Fr. 1849
Helotiales: Ascomycota

Stromata poorly developed, immersed. Ascomata large, with a brownish carbonaceous verrucose ectal excipulum and a black highly gelatinized hymenium; excipulum composed of an outer layer of parallel or interwoven hyphae, and a central strongly gelatinized area, ± glabrous. Interascal tissue of simple paraphyses. Asci cylindrical, thin-walled and not fissitunicate, with a complex J+ apical ring. Ascospores dark brown, ellipsoidal, aseptate, with a longitudinal germ slit, sometimes dimorphic. Anamorph hyphomycetous, *Aureobasidium*-like.

Significant Genera: *Bulgaria*. Anamorph: *Endomelanconium*.

Distribution: Known from north and south temperate zones.

Economic Significance: None is known.

Ecology: Saprobic on wood and bark, a primary colonizer and probably endophytic.

Bulgaria inquinans, Wiltshire, UK; ascomata on fallen beech tree

Notes: Phylogenetic research on *Bulgaria* is preliminary, but some research indicates that the family is distinct from the *Leotiaceae*. That has similar large gelatinized ascomata, but many other morphological features are distinct.

References: **Döring, H.; Triebel, D.** (1998). Phylogenetic relationships of *Bulgaria* inferred by 18S rDNA sequence analysis. *Cryptog. Bryol.-Lichénol.* 19: 123–136; **Fenwick, G.A.** (1992). A conidial form of *Bulgaria inquinans*. *Mycologist* 6: 177–179; **Gamundí, I.J.; Arambarri, A.M.** (1983). *Bulgaria nana* Cash y su anamorpho *Endomelanconium* (Helotiaceae, Ombrophiloidea) (X). *Revta Fac. Agron. Univ. nat. La Plata* 59: 17–23; **Triebel, D.; Rambold, G.** (1992). On the ascus apex and ascospore germ slit in *Bulgaria inquinans*. *Arnoldia* 4: 15–16; **Wang, Z.; Binder, M.; Schoch, C.L.; Johnston, P.R.; Spatafora, J.W.; Hibbett, D.S.** (2006). Evolution of helotialean fungi (*Leotiomycetes*, *Pezizomycotina*): a nuclear rDNA phylogeny. *Mol. Phylogen. Evol.* 41: 295–312.

Cainiaceae J.C. Krug 1978
Xylariales: Ascomycota

Stromata absent or reduced to clypeate structures. Ascomata perithecial, ± globose to lenticular, black, immersed, the ostiole papillate, periphysate, scattered or aggregated in small groups; peridium usually thin-walled, brown, composed of 1–2 layers of angular cells. Interascal tissue composed of simple persistent filiform paraphyses, but of varied ontogeny. Asci cylindrical, usually short-stalked, fairly thin-walled, not fissitunicate, the apex rounded, with a complex J+ cylindrical or funnel-shaped apical structure formed from a series of rings, 8-spored. Ascospores ellipsoidal, brown, 1-septate, striate or with a series of longitudinal germ slits, with a mucous sheath. Anamorphs poorly known, with a possible link to the coelomycetous genus of uncertain affinity *Rhabdospora*.

Significant Genera: *Cainia*.

Cainia graminis, Ireland; habit view of ascomata

Distribution: Widespread.

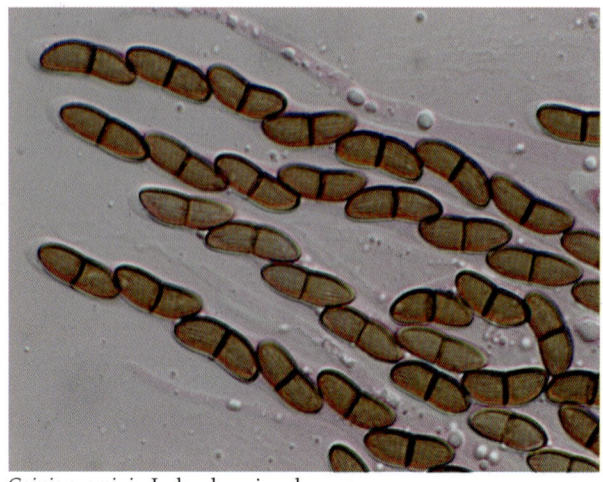

Cainia graminis, Ireland; asci and ascospores

Economic Significance: None is known.

Ecology: Saprobic in grasses and other monocotyledons.

Notes: *Atrotorquata* and *Cainia* may well be congeneric. Their 2-celled ascospores with longitudinal germ pores are almost unparalleled within the Ascomycota.

References: Kang, J.C.; Hyde, K.D.; Kong, R.Y.C. (1999). Studies on *Amphisphaeriales*: the Cainiaceae. *Mycol. Res.* 103: 1621–1627; Kohlmeyer, J.; Volkmann-Kohlmeyer, B. (1993). *Atrotorquata* and *Loratospora*: new ascomycete genera on *Juncus roemerianus*. *Syst. Ascom.* 12: 7–22; Lumbsch, H.T.; Wirtz, N.; Lindemuth, R.; Schmitt, I. (2002). Higher-level phylogenetic relationships of euascomycetes (*Pezizomycotina*) inferred from a combined analysis of nuclear and mitochondrial sequence data. *Mycol. Prog.* 1: 57–70.

Caliciaceae Chevall. 1826
Lecanorales: Ascomycota

Thallus crustose where present; verrucose, areolate or granular, generally grey, yellow or greenish, sometimes immersed within the outer layers of dead wood or bark. Ascomata apothecial, with a hemispherical to ± spherical head and a usually well-developed ascoma wall, long-stalked or sessile, the stalk where present usually strongly melanized. Interascal tissue of simple paraphyses where present. Asci cylindrical to broadly clavate, formed from croziers, thin-walled, without apical structures, evanescent, releasing spores passively in a mazaedial mass. Ascospores 0- to 1-septate, dark brown, smooth or with ornamentation formed from rupturing of the outer wall layers, sometimes with a gelatinous coat. Anamorph sometimes present. Conidiomata pycnidial, sessile or partially immersed, dark-walled, globose, the ostiole apiculate or inconspicuous. Conidiogenous cells formed on short branched conidiophores, ± cylindrical, proliferating percurrently. Conidia ellipsoidal to cylindrical, hyaline, aseptate.

Significant Genera: *Calicium*, *Cyphelium*, *Thelomma*.

Distribution: Cosmopolitan, especially in temperate and tropical montane regions.

Economic Significance: None is known.

Ecology: Almost all lichen-forming with *Trebouxia* photobionts on bark and dead wood, rarely saprobic or lichenicolous.

Notes: Some species are good indicator species for ancient woodlands. Many are drought-tolerant.

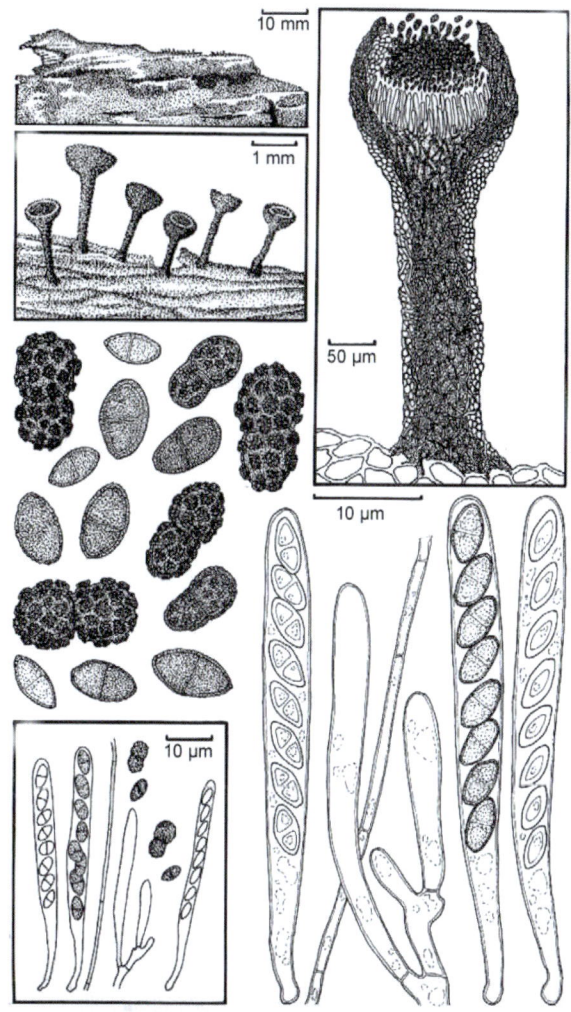

Calicium trabinellum; ascomata

Cyphelium tigillare, Norway; ascomata

References: **McCune, B.; Rosentreter, R.** (1992). *Texosporium sancti-jacobi*, a rare western North American lichen. *Bryologist* 95: 329–333; **Sarrión, F.; Aragón, G.; Burgaz, A.R.** (1999). Studies on mazaediate lichens and calicioid fungi of the Iberian Peninsula. *Mycotaxon* 71: 169–198; **Tibell, L.** (1984). A reappraisal of the taxonomy of *Caliciales*. *Nova Hedwigia* Beih. 79: 597–713; **Tibell, L.** (1987). Australasian *Caliciales*. *Symb. bot. upsal.* 27: 279 pp.; **Tibell, L.** (1997). Anamorphs in mazaediate lichenized fungi and the *Mycocaliciaceae* ('*Caliciales s. lat.*'). *In* Tibell, L. & Hedberg, I. (eds), Lichen Studies Dedicated to Rolf Santesson. *Symb. bot. upsal.* 32: 291–322; **Tibell, L.** (1998). Crustose Mazaediate Lichens and the *Mycocaliciaceae* in Temperate South America. *Biblthca Lichenol.* 71: 107 pp.; **Tibell, L.** (1999). Calicioid lichens and fungi. *Nordic Lichen Flora* 1. Introductory Parts; Calicioid Lichens and Fungi (Uddevalla): 20–94; **Tibell, L.; Kalb, K.** (1992). *Calicium* in the tropical and subtropical Americas. *Nova Hedwigia* 55: 11–36; **Wedin, M.; Döring, H.; Nordin, A.; Tibell, L.** (2000). Small subunit rDNA phylogeny shows the lichen families *Caliciaceae* and *Physciaceae* (*Lecanorales*, *Ascomycotina*) to form a monophyletic group. *Can. J. Bot.* 78: 246–254; **Wedin, M.; Tibell, L.** (1997). Phylogeny and evolution of *Caliciaceae*, *Mycocaliciaceae*, and *Sphinctrinaceae* (*Ascomycota*), with notes on the evolution of the prototunicate ascus. *Can. J. Bot.* 75: 1236–1242.

Calosphaeriaceae Munk 1957
Calosphaeriales: Ascomycota

Stromata immersed in bark, varied in development, composed of fungal cells mixed with host tissue. Ascomata perithecial, often aggregated in valsoid or eutypoid clusters, the ostioles usually converging, papillate or long-necked and lined with periphyses; peridium reddish brown to black, composed of several rows of compressed cells. Interascal tissue of broad thin-walled tapering paraphyses. Asci formed in a small fascicle from a short ascogenous hypha which sometimes proliferates to form a spike-like cluster, thin-walled with one functional wall-layer, the apex often thickened and refractive with a J– apical ring. Ascospores hyaline or pale yellow, allantoid, cylindrical or ellipsoidal, thin-walled, mostly aseptate, budding within the ascus in a few species. Anamorphs hyphomycetous, *Acremonium*-like or *Phialophora*-like.

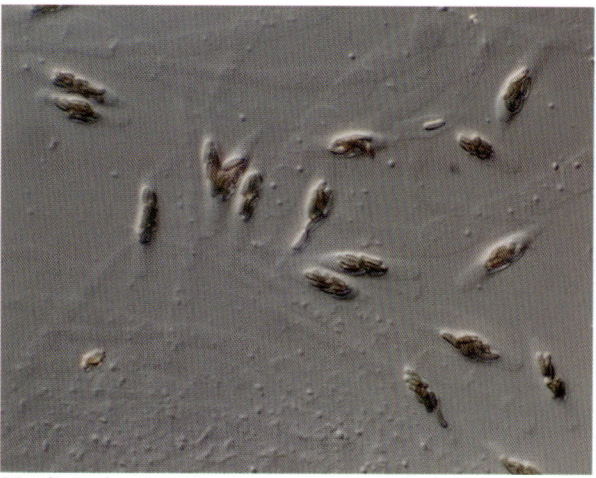

Wegelina polyporina, Germany; asci and paraphyses

Significant Genera: *Calosphaeria, Togniniella*.

Distribution: Widespread in temperate regions.

Economic Significance: None is known.

Ecology: Little information is available, but most species appear to be saprobes in dead bark.

Notes: A poorly known family, both in terms of its constituent species and its interrelationships. Genera with coelomycetous anamorphs (e.g. *Pachytrype*) probably do not belong here, but their ascus arrangement in spicate clusters is considered to be generally diagnostic of the group.

Calosphaeria pulchella, South Africa; long-necked ascomata

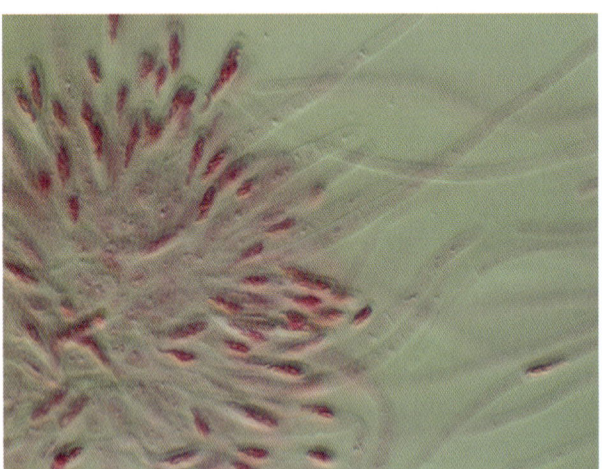
Calosphaeria pulchella, South Africa; asci, ascospores and paraphyses

References: **Barr, M.E.** (1985). Notes on the *Calosphaeriales*. *Mycologia* 77: 549–565; **Barr, M.E.** (1998). *Wegelina*, a reinstated genus in the *Calosphaeriales*. *Cryptog. Bryol.-Lichénol.* 19: 169–173; **Mostert, L.; Crous, P.W.; Groenewald, J.Z.; Gams, W.; Summerbell, R.C.** (2003). *Togninia* (*Calosphaeriales*) is confirmed as teleomorph of *Phaeoacremonium* by means of morphology, sexual compatibility and DNA phylogeny. *Mycologia* 95: 646–659; **Réblová, M.; Mostert, L.; Gams, W.; Crous, P.W.** (2004). New genera in the *Calosphaeriales*: *Togniniella* and its anamorph *Phaeocrella*, and *Calosphaeriophora* as anamorph of *Calosphaeria*. *Stud. Mycol.* 50: 533–550; **Romero, A.I.; Minter, D.W.** (1988). Fluorescence microscopy: an aid to the elucidation of ascomycete structures. *Trans. Br. mycol. Soc.* 90: 457–470; **Samuels, G.J.; Candoussau, F.** (1996). Heterogeneity in the *Calosphaeriales*: a new *Calosphaeria* with *Ramichloridium*- and *Sporothrix*-like synanamorphs. *Nova Hedwigia* 62: 47–60.

Calostomataceae E. Fisch. 1900
Boletales: Basidiomycota

Basidiomata gasteroid, epigeal, consisting of a ± globose fruit body upon a wide, strongly gelatinized, hygroscopic stipe that becomes hard and horny and persists long after spore dispersal. Fruit body ± globose, with a well-developed apical ostiole surrounded by an ornamented and often brightly coloured peristome; peridium 3-layered; the outer layer gelatinous, the median layer membranous and the inner layer thin and parchment-like. Capillitium absent, the gleba powdery. Basidia clavate or inflated, with 4–12 apical or lateral sterigmata. Basidiospores globose to ellipsoidal, hyaline or yellowish, with a reticulate epispore.

Significant Genera: *Calostoma*.

Distribution: Primarily in tropical and warm temperate regions.

Economic Significance: None is known.

Ecology: On the ground, presumably saprobic.

Notes: Molecular data confirm that the family is part of the *Scleroderma* clade within the *Boletales*, but its closest relative is not clear. Its placement over the years has not been stable, with some authors postulating links with the *Tulostomataceae*.

Calostoma cinnabarinum; basidiomata

References: **Binder, M.; Bresinsky, A.** (2002). Derivation of a polymorphic lineage of gasteromycetes from boletoid ancestors. *Mycologia* 94: 85–98; **Cunningham, G.H.** (1942). *Gast. Austr. N.Z.* (Dunedin): 236 pp.; **Hughey, B.D.; Adams, G.C.; Bruns, T.D.; Hibbett, D.S.** (2000). Phylogeny of *Calostoma*, the gelatinous-stalked puffball, based on nuclear and mitochondrial ribosomal DNA sequences. *Mycologia* 92: 94–104.

Calycidiaceae Elenkin 1929
Uncertain position within Lecanoromycetidae, Ascomycota

Thallus foliose, usually grey or white. Ascomata sessile, marginal, the hymenium dark brown to black. Asci thin-walled, evanescent. Ascospores spherical, aseptate, brown, forming a dry mazaedial mass.

Significant Genera: *Calycidium*.

Distribution: Only known from New Zealand.

Economic Significance: None is known.

Ecology: Lichenized with green algae.

Notes: Very poorly known; no recent information is available.

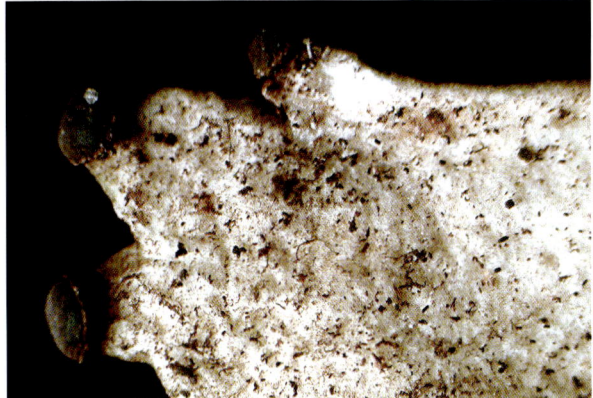
Calycidium cuneatum, New Zealand; thallus and ascomata

References: **Tibell, L.** (1987). Australasian *Caliciales*. *Symb. bot. upsal.* 27: 279 pp.; **Tibell, L.** (1997). Anamorphs in mazaediate lichenized fungi and the *Mycocaliciaceae* ('Caliciales s. lat.'). In Tibell, L. & Hedberg, I. (eds), Lichen Studies Dedicated to Rolf Santesson. *Symb. bot. upsal.* 32: 291–322; **Wedin, M.** (1994). Taxonomic Studies in *Sphaerophoraceae* (*Caliciales, Ascomycotina*). *Taxonomic Studies in Sphaerophoraceae* (Caliciales, Ascomycotina), Acta Universitatis Upsaliensis (Comprehensive Summaries of Uppsala Dissertations from the Faculty of Science and Technology no. 77) (Uppsala): 168 pp..

Candelariaceae Hakul. 1954
Candelariales: Ascomycota

Thallus crustose, fruticose or foliose, yellow-green to orange. Ascomata apothecial, with or without a clear margin, usually orange or yellow, the ascomatal wall formed from closely septate twisted hyphae. Interascal tissue of simple paraphyses, often swollen at the apex. Asci cylindrical to cylindric-clavate, with a thick non-amyloid cap above a blue-staining region with a central less strongly stained cushion, ocular chamber ± absent, often multispored. Ascospores small, hyaline, usually aseptate. Conidiomata pycnidial, immersed within the thallus, forming small ellipsoidal conidia from percurrently proliferating conidiogenous cells.

Placomaronea candelarioides; thallus and ascomata

Candelina submexicana, Arizona, USA; thallus and ascomata

Candelariella vitellina, Cornwall, UK; thallus and ascomata

Significant Genera: *Candelaria*, *Candelariella*.

Distribution: Widespread, especially in polar and cold temperate regions.

Economic Significance: None is known.

Ecology: Lichenized with green algae, usually associated with nitrogen-rich substrata.

Notes: A poorly known family. Recent data suggest that the *Candelariaceae* are separate from the *Lecanorales*, but positioning within the *Lecanoromycetes* is uncertain.

References: **Castello, M.; Nimis, P.L.** (1994). Critical notes on the genus *Candelariella* (lichenes) in Antarctica. *Acta bot. fenn*. 150: 5–10; **Hibbett et al.** (2007). A higher-level phylogenetic classification of the *Fungi*. *Mycol. Res.* 111(5): 509–548; **Hofstetter, V.; Miądlikowska, J.; Kauff, F.; Lutzoni, F.** 2007. Phylogenetic comparison of protein-coding versus ribosomal RNA-coding sequence data: a case study of the *Lecanoromycetes* (*Ascomycota*). *Mol. Phylogen. Evol.* 16(3): [in press]; **Jørgensen, P.M.; Galloway, D.J.** (1992). Notes on *Candelaria crawfordii*. *Lichenologist* 24: 407–410; **LaGreca, S.; Lumbsch, H.T.** (2001). The phylogenetic position of the *Candelariaceae* (*Lecanorales*) inferred from anatomical and molecular data. *Biblthca Lichenol.* 78: 211–222; **Miądlikowska et al.** (2007). New insights into classification and evolution of the *Lecanoromycetes* (*Pezizomycotina, Ascomycota*) from phylogenetic analyses of three ribosomal RNA- and two protein-coding genes. *Mycologia* 98(6): 1088–1103.

Cantharellaceae J. Schröt. 1888
Cantharellales: Basidiomycota

Basidiomata pileate, often funnel-shaped and sinuate or lobate, appearing agaricoid, fleshy or thin-fleshed, centrally or laterally stalked, dry or slightly waxy, variously coloured. Hyphal system monomitic, hyphae often strongly inflated, usually thin-walled, clamp connections present or absent. Hymenium merging with the stem, with radial folds and interveining, wrinkles or ± smooth, fertile layer extending over much of the surface of the radial folds, usually concolorous with the cap. Cystidia absent. Basidia (2–) 4 (–8)-spored, initially cylindrical, becoming elongate-clavate, with longitudinally oriented nuclear spindles, thin-walled, smooth, basal clamp sometimes present. Basidiospores discharged actively, subglobose to ellipsoidal, usually thin-walled, hyaline, not staining in iodine.

Significant Genera: *Cantharellus, Craterellus.*

Distribution: Widespread, known best from north temperate regions but also occurring in the tropics.

Economic Significance: *Cantharellus cibarius* is the edible chanterelle and *Craterellus cornucopioides* is the edible 'Horn of Plenty'. Species are collected commercially from the wild, with estimated annual global production $1.67 billion in 1997. Cultivation has not so far proved a reliable alternative and there are understandably concerns over sustainability.

Cantharellus cibarius, Hampshire, UK; basidiomata

Ecology: Species form ectomycorrhizas with coniferous and broadleaved trees.

Notes: One of the best-known families of *Basidiomycota*, probably close to the *Hydnaceae* which has spiny rather than ridged hymenia. Delimitation of the two principal genera has proved troublesome and molecular data are not conclusive.

Cantharellus tubaeformis, Argyll, Scotland; basidiomata

References: **Corner, E.J.H.** (1957). *Craterellus* Pers., *Cantharellus* Fr. and *Pseudocraterellus* gen. nov. *Beih. Sydowia* 1: 266–276; **Dahlman, M.; Danell, E.; Spatafora, J.W.** (2000). Molecular systematics of *Craterellus*: cladistic analysis of nuclear LSU rDNA sequence analysis. *Mycol. Res.* 104: 388–394; **Danell, E.** (1994). Formation and growth of the ectomycorrhiza of *Cantharellus cibarius*. *Mycorrhiza* 5: 89–97; **Dunham, S.M.; Kretzer, A.; Pfrender, M.E.** (2003). Characterization of Pacific golden chanterelle (*Cantharellus formosus*) genet size using co-dominant microsatellite markers. *Mol. Ecol.* 12: 1607–1618; **Dunham, S.M.; O'Dell, T.E.; Molina, R.** (2003). Analysis of nrDNA sequences and microsatellite allele frequencies reveals a cryptic chanterelle species *Cantharellus cascadensis* sp. nov. from the American Pacific Northwest. *Mycol. Res.* 107: 1163–1177; **Feibelman, T.P.;**

Doudrick, R.L.; Cibula, W.G.; Bennett, J.W. (1997). Phylogenetic relationships within the *Cantharellaceae* inferred from sequence analysis of the nuclear large subunit rDNA. *Mycol. Res.* 101: 1423–1430; **Li, T.H.; Zhuang, L.; Chen, Y.Q.; Qu, L.H.** (1999). Partial 25S rDNA sequences and phylogenetic relationships of *Cantharellus*, *Craterellus* and *Gomphus*. *Acta Sci. nat. Univ. Sunyats.* 38: 29–33; **Petersen, R.H.; Mueller, G.M.** (1992). New South American taxa of *Cantharellus*, *C. nothofagorum*, *C. xanthoscyphus* and *C. lateritius* var. *colombianus*. *Boln Soc. argent. Bot.* 28: 195–200; **Pine, E.M.; Hibbett, D.S.; Donoghue, M.J.** (1999). Phylogenetic relationships of cantharelloid and clavarioid *Homobasidiomycetes* based on mitochondrial and nuclear rDNA sequences. *Mycologia* 91: 944–963; **Redhead, S.A.; Norvell, L.L.; Danell, E.** (1997). *Cantharellus formosus* and the Pacific golden chanterelle harvest in western North America. *Mycotaxon* 65: 285–322; **Watling, R.** (1997). The business of fructification. *Nature* Lond. 365: 299–300.

Capnodiaceae Höhn. ex Theiss. 1916
Capnodiales: Ascomycota

Mycelium superficial, well-developed, dark, composed of ± cylindrical hyphae with mucous coating. Ascomata small, sometimes vertically elongated, thin-walled, covered in a mucous layer, sometimes setose, usually with a clearly defined ostiole. Interascal tissue absent. Asci saccate, fissitunicate. Ascospores brown, septate, sometimes muriform. Anamorphs pycnidial, elongate, sometimes stipitate.

Significant Genera: *Capnodium*.

Distribution: Widespread; especially so in the tropics and subtropics but also in temperate rainforest.

Economic Significance: Heavy infestation must reduce growth in crop plants, due to reduction in photosynthetic activity.

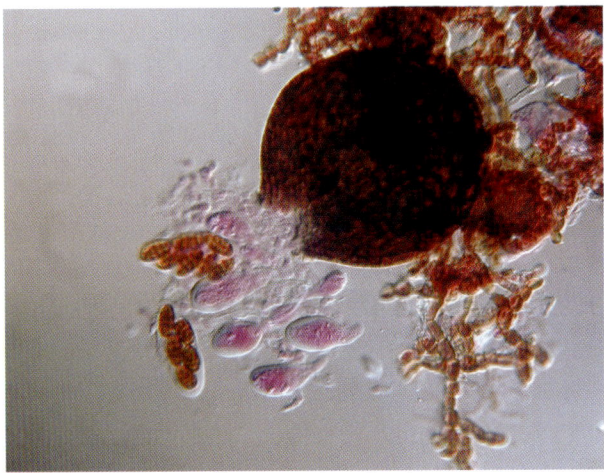

Capnodium citri, Australia; ascomata and asci

Ecology: Most species are apparently saprobic, usually on insect exudates on leaves and branches. The heavily pigmented mycelium is presumably an adaptation to high light levels.

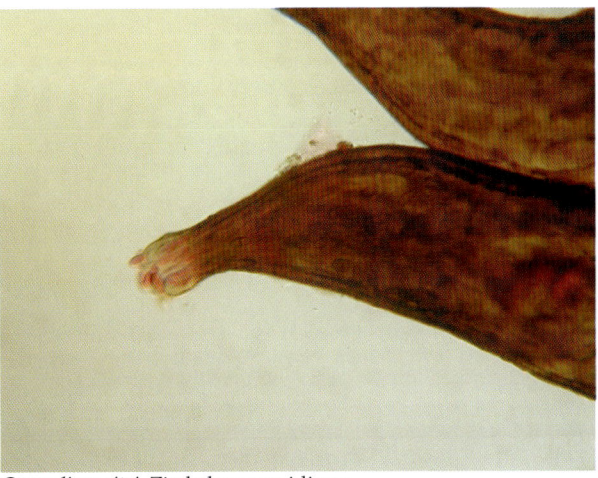

Capnodium citri, Zimbabwe; conidioma

Notes: One of several families of 'sooty moulds', which comprise complex communities that are often parasitized by other fungi. Species concepts and host specificity have hardly been addressed.

References: **Inácio, C.A.; Dianese, J.C.** (1998). Some foliicolous fungi on *Tabebuia* species. *Mycol. Res.* 102: 695–708; **Lindemuth, R.; Wirtz, N.; Lumbsch, H.T.** (2001). Phylogenetic analysis of nuclear and mitochondrial rDNA sequences supports the view that loculoascomycetes (*Ascomycota*) are not monophyletic. *Mycol. Res.* 105: 1176–1181; **Lumbsch, H.T.; Lindemuth, R.** (2001). Major lineages of *Dothideomycetes* (*Ascomycota*) inferred from SSU and LSU rDNA sequences. *Mycol. Res.* 105: 901–908; **Mibey, R.K.** (1997). Sooty moulds. *Soft Scale Insects. Their Biology, Natural Enemies and Control World Crop Pests* 7A (Amsterdam): 275–290; **Olejnik, I.M.; Ingrouille, M.** (1999). Numerical taxonomy of the sooty moulds *Leptoxyphium*, *Caldariomyces* and *Aithaloderma* based on micromorphology and physiology. *Mycol. Res.* 103: 333–346; **Parbery, I.H.; Brown, J.F.** (1986). Sooty moulds and black mildews in extra-tropical rainforests. *Microbiology of the Phyllosphere* (Cambridge): 101–120; **Reynolds, D.R.** (1999). Capnodiaceous sooty mold phylogeny. *Can. J. Bot.* 76: 2125–2130; **Reynolds, D.R.** (2000). *Capnodium citri*: the sooty mold fungi comprising the taxon concept. *Mycopathologia* 148: 141–147; **Rodríguez Hernández, M.** (1985). Clave para los hongos de la fumagina en Cuba. *Revta Jardín bot. Nac., Univ. Habana* 6: 53–62; **Schoch, C.L.; Shoemaker, R.A.; Seifert, K.A.; Hambleton, S.; Spatafora, J.W.; Crous, P.W.** (2007). A multigene phylogeny of the Dothideomycetes using four nuclear loci. *Mycologia* 98(6): 1041–1052; **Sterflinger, K.; Hoog, G.S. de; Haase, G.** (1999). Phylogeny and ecology of meristematic ascomycetes. *Stud. Mycol.* 43: 5–22.

Carbomycetaceae Trappe 1971
Pezizales: Ascomycota

Ascomata large, cleistothecial, ± globose, smooth; peridium of intertwining hyphae with strongly inflated cells; interior solid, with fertile pockets surrounded by sterile veins. Asci ± globose, randomly arranged, the wall not blueing in iodine, ? usually evanescent at maturity. Ascospores ellipsoidal, hyaline or pale brown, ornamented. Anamorphs unknown.

Significant Genera: *Carbomyces*.

Distribution: Known from the USA.

Economic Significance: None is known.

Ecology: Hypogeous but frequently emergent, in deserts.

Notes: Very poorly known; no recent information is available.

References: **Zak, J.C.; Whitford, W.G.** (1986). The occurrence of a hypogeous ascomycete in the Northern Chihuahuan Desert. *Mycologia* 78: 840–841.

Catabotrydaceae Petr. ex M.E. Barr 1990
Boliniales: Ascomycota

Stromata large, pulvinate, erumpent to almost superficial, composed of thin-walled angular cells. Ascomata perithecial, deeply immersed, globose, long-necked, with periphysate ostioles. Interascal tissue of thin-walled true paraphyses, sometimes deliquescent. Asci cylindrical, thin-walled, with a small J– ring. Ascospores ellipsoidal, aseptate, hyaline, without a mucous sheath or appendages. Anamorph unknown.

Significant Genera: *Catabotrys*.

Distribution: Apparently pantropical.

Economic Significance: None is known.

Ecology: Saprobic on palm fronds, possibly with an initial endophytic phase.

Notes: The systematic position of this family is in doubt; there are morphological similarities with the *Boliniaceae* but these are not strongly supported by molecular data.

References: **Huhndorf, S.M.; Miller, A.N.; Fernández, F.A.** (2004). Molecular systematics of the *Sordariales*: the order and the family *Lasiosphaeriaceae* revisited. *Mycologia* 96: 368–387; **Hyde, K.D.; Cannon, P.F.** (1999). Fungi Causing Tar Spots on Palms. *Mycol. Pap.* 175: 114 pp.; **Rossman, A.Y.; Samuels, G.J.; Rogerson, C.T.; Lowen, R.** (1999). Genera of *Bionectriaceae, Hypocreaceae* and *Nectriaceae*. *Stud. Mycol.* 42: 248 pp..

Catenariaceae Couch 1945
Blastocladiales: Blastocladiomycota

Thallus polycentric (typically branched with two or more centres of development), rhizomycelial with catenulate swellings separated by sterile isthmi; zoosporangia intercalary, formed from the aforementioned swellings; zoospores posteriorly uniflagellate, with a conspicuous nuclear cap, germinating directly to form a new thallus; resting spores formed within a zoosporangium-like structure, thick-walled, smooth or minutely pitted and typically brownish, germinating via a discharge tube and functioning as a zoosporangium, the uniflagellate zoospores either forming a new thallus or encysting, the cyst then serving as a gametangium to release four, uniflagellate, isogamous gametes which fuse in pairs to form a biflagellate zygote which upon germination forms a new thallus.

Significant Genera: *Catenaria*.

Distribution: Widespread.

Economic Significance: None is known.

Ecology: Saprobic in water-logged soil or in water, or parasitic on various soil-inhabiting invertebrates.

Notes: Ultrastructure has suggested a relationship with *Chytridiomycetes* but recent analysis of molecular data show that *Blastocladiales* is closer to *Entomophthorales*.

References: **Manier, J.F.** (1977). Life cycle, ultrastructure of a *Catenaria* (*Phycomycetes Blastocladiales*) parasite of *Crustacea*: *Cyclopoida*. *Annls Parasit. hum. comp.* 52: 363–376; **Tanabe, Y.; Watanabe, M.M.; Sugiyama, J.** (2005). Evolutionary relationships among basal fungi (*Chytridiomycota* and *Zygomycota*): Insights from molecular phylogenetics. *J. gen. appl. Microbiol.* Tokyo 51: 267–276.

Catillariaceae Hafellner 1984
Lecanorales: Ascomycota

Thallus crustose, sometimes poorly developed. Ascomata sessile, dark, small, with a poorly developed proper margin and no thalline margin. Interascal tissue of sparsely branched paraphyses, often pigmented at the apices. Asci with a J+ apical cap, ocular chamber absent or poorly developed, with an outer J+ gelatinous layer, 8- or 16-spored. Ascospores hyaline, aseptate or transversely septate, without a sheath. Anamorphs pycnidial, usually inconspicuous.

Significant Genera: *Catillaria, Toninia*.

Distribution: Widespread, especially in temperate regions.

Economic Significance: None is known.

Ecology: Lichenized with green algae on bark or rock, a few lichenicolous.

Notes: Several of the constituent genera (including *Catillaria*) are polyphyletic, and the identity and circumscription of the family remains uncertain. There is confusion especially over links with the *Bacidiaceae* and *Ramalinaceae*.

Catillaria chalybeia, Iceland; thallus and ascomata

Catillaria lenticularis, Fife, Scotland; asci, ascospores and paraphyses with pigmented apices

Toninia bumamma, South Africa; thallus and ascomata

References: **Coppins, B.J.** (1989). On some species of *Catillaria* s. lat. and *Halecania* in the British Isles. *Lichenologist* 21: 217–227; **Ekman, S.** (2001). Molecular phylogeny of the *Bacidiaceae* (*Lecanorales*, lichenized *Ascomycota*). *Mycol. Res.* 105: 783–797; **Friday, A.M.; Coppins, B.J.** (1996). Three new species in the *Catillariaceae* from the Central Highlands of Scotland. *Lichenologist* 28: 507–512; **Lumbsch, H.T.** (1997). Systematic studies in the suborder *Agyriineae* (*Lecanorales*). *J. Hattori bot. Lab.* 83: 1–73; **Pant, G.; Awasthi, D.D.** (1989). Lichen genus *Catillaria* s. lat. in India. *Proc. Indian Acad. Sci. Pl. Sci.* 99: 369–384; **Reeb, V.; Lutzoni, F.; Roux, C.** (2004). Contribution of *RPB2* to multilocus phylogenetic studies of the euascomycetes (*Pezizomycotina*, *Fungi*) with special emphasis on the lichen-forming *Acarosporaceae* and evolution of polyspory. *Mol. Phylogen. Evol.* 32: 1036–1060.

Caulochytriaceae Subram. 1974
Spizellomycetales: Chytridiomycota

Thallus monocentric, comprising a single cell; zoosporangia sessile, forming posteriorly uniflagellate zoospores, without a nuclear cap, which are liberated via 1–7 exit papillae and which subsequently develop vegetatively to form similar sporangia or function as isogametes, fusing in pairs to produce zygotes; zygotes giving rise to slender-stalked aerial sporangia (resembling a protostelid); this aerial sporangium (where meiosis occurs) germinates to liberate eight zoospores. Homothallic.

Significant Genera: *Caulochytrium*.

Distribution: Currently the only two known species are recorded from the USA.

Economic Significance: None is known.

Ecology: Reported parasitizing the conidia of *Cladosporium* and *Gloeosporium*.

Notes: The species are noted for their alternation of generations.

References: **Barr, D.J.S.** (1990). Phylum *Chytridiomycota*. *Handbook of Protoctista* (Boston): 454–466; **Voos, J.R.** (1969). Morphology and life cycle of a new chytrid with aerial sporangia. *Am. J. Bot.* 56: 898–909.

Cephaloascaceae L.R. Batra 1973
Saccharomycetales: Ascomycota

Vegetative cells ± ellipsoidal, with multilateral budding. Ascomata absent. Asci evanescent, formed in branched chains at the apex of an erect pigmented diploid hyphal seta-like stalk, ellipsoidal. Ascospores hyaline, oblate, with an eccentric thickening.

Significant Genera: *Cephaloascus*.

Distribution: Known from Japan, Canada and the UK.

Economic Significance: None is known.

Ecology: On coniferous wood or on other fungi, at least sometimes insect-associated.

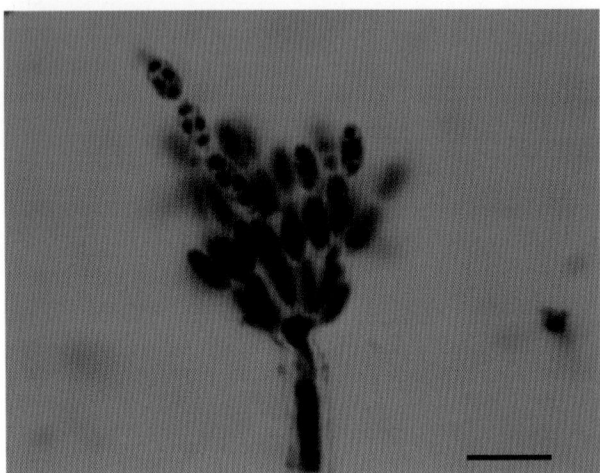

Cephaloascus fragrans, Ontario, Canada; fertile stalk and chains of asci

Notes: *Cephaloascus* is an enigmatic fungus and probably not closely allied to the *Saccharomycetales*, but morphological similarities between this genus and some anamorphs of the *Ophiostomatales* are now considered to be a result of convergent evolution.

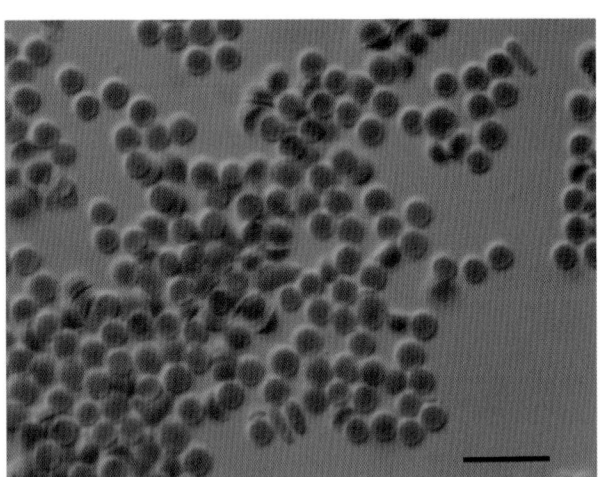

Cephaloascus fragrans, Ontario, Canada; hat-shaped ascospores

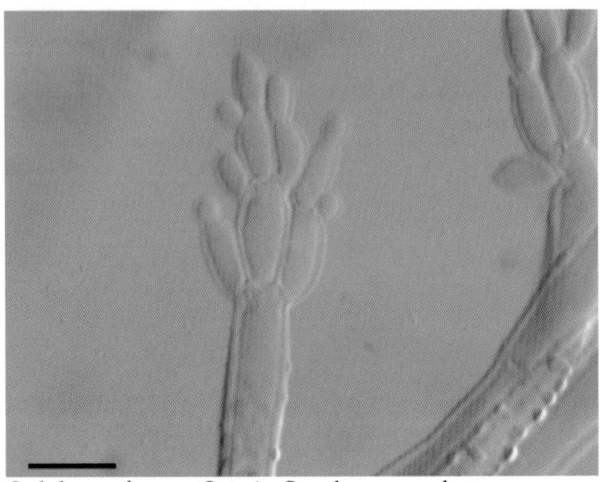

Cephaloascus fragrans, Ontario, Canada; anamorph

References: **Batra, L.R.** (1987). Insect-associated, filamentous *Endomycetes* – their growth and strategies for survival. *Stud. Mycol.* 30: 415–428; **Hoog, G.S. de; Kurtzman, C.P.** (1998). *Cephaloascus* Hanawa. *The Yeasts, A Taxonomic Study* (Amsterdam): 143–145; **Kurtzman, C.P.; Robnett, C.J.** (1995). Molecular relationships among hyphal ascomycetous yeasts and yeastlike taxa. *Can. J. Bot.* 73: S824–S830; **Read, N.D.; Beckett, A.** (1996). Ascus and ascospore morphogenesis. *Mycol. Res.* 100: 1281–1314; **Spatafora, J.W.; Blackwell, M.** (1993). Molecular systematics of unitunicate perithecial ascomycetes: the *Clavicipitales-Hypocreales* connection. *Mycologia* 85: 912–922.

Cephalothecaceae Höhn. 1917
Sordariales: Ascomycota

Stromata absent, though ascomata sometimes clustered. Ascomata cleistothecial, small, thick-walled, sometimes covered with dense yellow hyphae when young, the wall fragmenting into preformed polygonal plates. Interascal tissue absent. Asci irregularly disposed, ± globose, thin-walled, evanescent. Ascospores brown, ellipsoidal, aseptate, smooth, without a gelatinous sheath or appendages. Anamorph hyphomycetous, *Chalara*-like, *Paecilomyces*-like or *Acremonium*-like.

Significant Genera: *Cephalotheca*.

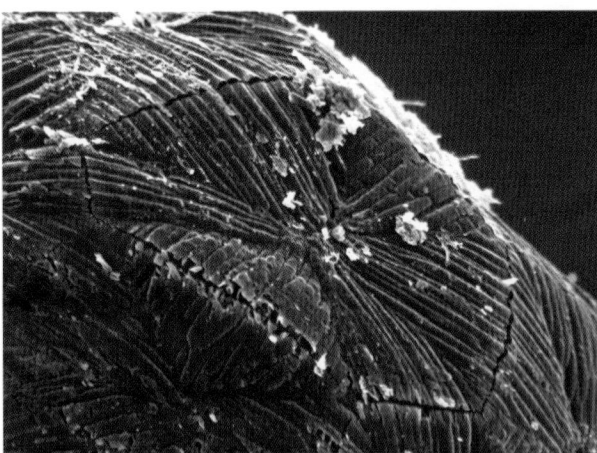

Cephalotheca sulfurea; ascoma

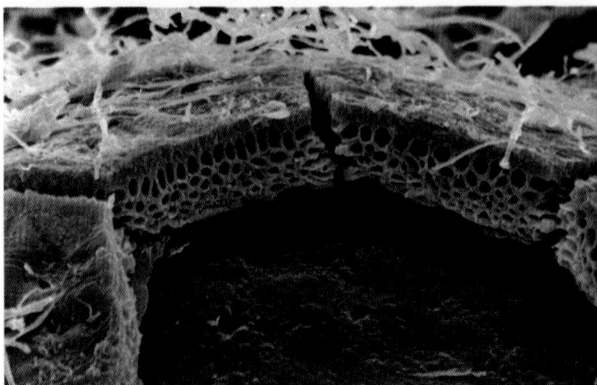

Fragosphaeria reniformis; detail of peridial structure

Distribution: Known from north temperate zones.

Economic Significance: Little is known, but there is a report of one species as a human pathogen.

Ecology: Typically saprobic, fruiting on very rotten wood or other fungi.

Notes: The cephalothecoid peridium (where a cleistothecium fragments into small plates along predetermined fracture lines) has evolved on several occasions. *Fragosphaeria* may not belong in this family; a sequence identified as *F. purpurea* clusters with the Ophiostomataceae and confirmation of this position is required.

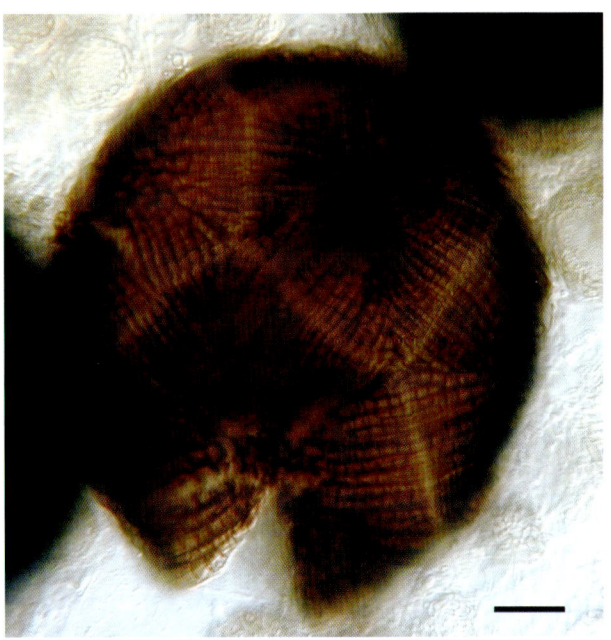

Cryptendoxyla hypophloia; ascoma

References: **Fortey, R.A.; Yao, Y.J.; Spooner, B.M.** (1997). Profiles of Fungi. 86. *Cephalotheca sulfurea* Fuckel; 87. *Lamprospora carbonicola* Boud. *Mycologist* 11: 132–133; **Huhndorf, S.M.; Miller, A.N.; Fernández, F.A.** (2004). Molecular systematics of the *Sordariales*: the order and the family Lasiosphaeriaceae revisited. *Mycologia* 96: 368–387; **Lumley, T.C.; Abbott, S.P.; Currah, R.S.** (2000). Microscopic ascomycetes isolated from rotting wood in the boreal forest. *Mycotaxon* 74: 395–414; **Lundqvist, N.** (1992). *Albertiniella polyporicola*, en askomycet på plattika, funnen i Sverige. *Svensk bot. Tidskr.* 86: 261–270; **Paulin, A.E.; Harrington, T.C.** (2000). Phylogenetic placement of anamorphic species of *Chalara* among *Ceratocystis* species and other ascomycetes. *Stud. Mycol.* 45: 209–222; **Suh, S.O.; Blackwell, M.** (1999). Molecular phylogeny of the cleistothecial fungi placed in Cephalothecaceae and Pseudeurotiaceae. *Mycologia* 91: 836–848; **Yaguchi, T.; Sano, A.; Yarita, K.; Suh, M.-K.; Nishimura, K.; Udagawa, S.-i.** (2006). A new species of *Cephalotheca* isolated from a Korean patient. *Mycotaxon* 96: 309–322.

Ceratobasidiaceae G.W. Martin 1948
Ceratobasidiales: Basidiomycota

Basidioma annual, poorly developed, composed of effuse or net-like hyphae, sometimes pellicular or waxy in form. Hyphal system monomitic, with thick-walled hyaline to brownish hyphae usually branching at right angles and constricted around the nodes. Basidia hyaline, thin-walled, ± globose, ovoid, clavate or broadly cylindrical, with usually 2–4 large broad-based sterigmata. Basidiospores ballistosporic, hyaline to yellowish, smooth, thin-walled, ± globose to ellipsoidal or cylindrical, not staining in iodine, often producing secondary spores and sometimes becoming septate. Anamorphs forming a sterile mycelium or producing sclerotia.

Significant Genera: *Ceratobasidium*, *Thanatephorus*. Anamorph: *Rhizoctonia*.

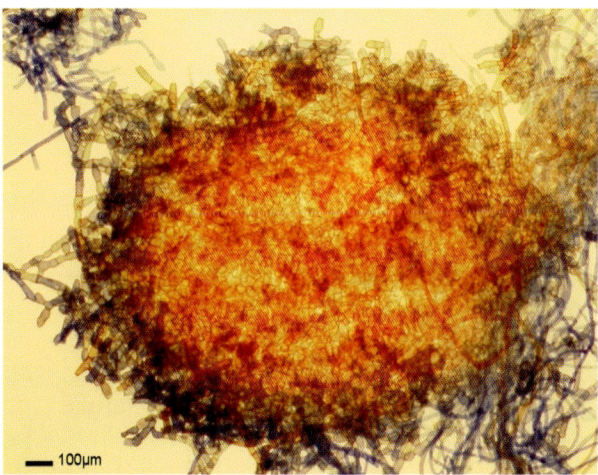

Rhizoctonia aff. *solani*, Singapore; sclerotium

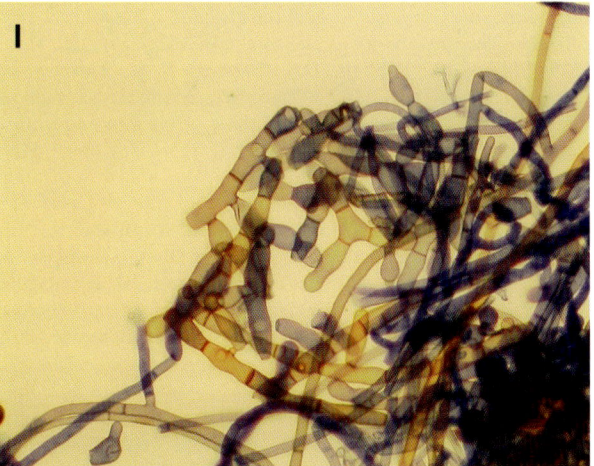

Rhizoctonia aff. *solani*, Singapore; detail of sclerotial hyphae

Distribution: Extremely widespread, prominent in both temperate and tropical ecosystems.

Economic Significance: Many species (especially the *Rhizoctonia solani* aggregate) cause economically significant plant diseases, primarily rots of root and stem bases. Sclerotia may remain viable in the soil for long periods.

Ecology: Best known as plant pathogens, but also saprobic. Species are endomycorrhizal with a wide range of plant roots (especially of Orchidaceae but also even bryophytes).

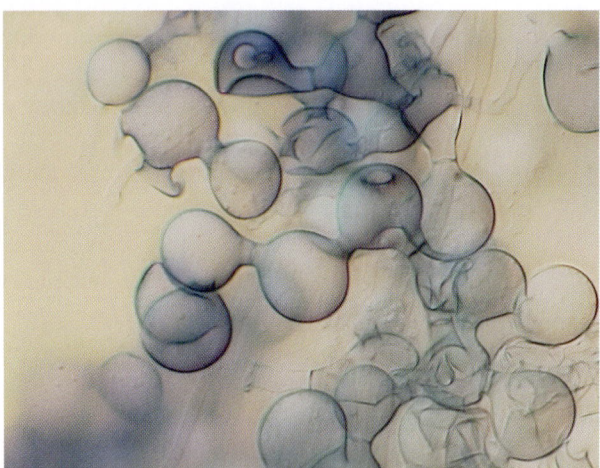

Rhizoctonia oryzae-sativae, Côte d'Ivoire; swollen cells surrounding sclerotium

Notes: The pathogenic species aggregates are divided largely through analysis of hyphal anastomosis groups. Molecular studies have largely confirmed these as monophyletic entities. The family occupies an isolated position within the cantharelloid clade of the basal homobasidiomycetes, with the Sistotremataceae probably its closest relative.

References: **Andersen, T.B.; Rasmussen, H.N.** (1996). The mycorrhizal species of *Rhizoctonia*. *Rhizoctonia Species, Taxonomy, Molecular Biology, Ecology, Pathology and Disease Control* (Dordrecht): 379–390; **Anderson, T.F.** (1996). A comparative taxonomic study of *Rhizoctonia sensu lato* employing morphological, ultrastructural and molecular methods. *Mycol. Res.* 100: 1117–1128; **Binder, M.; Hibbett, D.S.; Larsson, K.-H.; Larsson, E.; Langer, E.; Langer, G.** (2005). The phylogenetic distribution of resupinate forms across the major clades of mushroom-forming fungi (Homobasidiomycetes). *Syst. Biodiv.* 3: 113–157; **Currah, R.S.; Sigler, L.; Hambleton, S.** (1987). New records and taxa of fungi from the mycorrhizae of terrestrial orchids of Alberta. *Can. J. Bot.* 65: 2473–2482; **Gonzalez, D.; Carling, D.E.; Kuninaga, S.; Vilgalys, R.; Cubeta, M.A.** (2001). Ribosomal DNA systematics of *Ceratobasidium* and *Thanatephorus* with *Rhizoctonia* anamorphs. *Mycologia* 93: 1138–1150; **González, D.; Cubeta, M.A.; Vilgalys, R.** (2006). Phylogenetic utility of indels within ribosomal DNA and β-tubulin sequences from fungi in the *Rhizoctonia solani* species complex. *Mol. Phylogen. Evol.* 40: 459–470; **Gunnell, P.S.; Webster, R.K.** (1987). *Ceratobasidium oryzae-sativae* sp. nov., the teleomorph of *Rhizoctonia oryzae-sativae* and *Ceratobasidium setariae* comb. nov., the probable teleomorph of *Rhizoctonia fumigata* comb. nov. *Mycologia* 79: 731–736; **Jin, M.S.; Korpradiskul, V.** (1998). Isozyme analysis of genetic diversity among isolates of *Rhizoctonia solani* anastomosis group 1-IA (AG1-IA). *Mycosystema* 17: 331–338; **Kataria, H.R.; Hoffmann, G.M.** (1988). A critical review of plant pathogenic species of *Ceratobasidium* Rogers. *Z. PflKrankh. PflPath. PflSchutz* 95: 81–107; **Kuninaga, S.; Natsuaki, T.; Takeuchi, T.; Yokosawa, R.** (1997). Sequence variation of the rDNA ITS regions within and between anastomosis groups in *Rhizoctonia solani*. *Curr. Genet.* 32: 237–243; **Larsson, K.-H.; Larsson, E.; Koljalg, U.** (2004). High phylogenetic diversity among corticioid homobasidiomycetes. *Mycol. Res.* 108: 983–1002; **Leiner, R.H.; Carling, D.E.** (1991). Characterization of isolates of *Waitea circinata* collected from Alaskan agricultural soils. *Phytopathology* 81: 1242 (no. 820); **Maekawa, N.** (1997). *Ypsilonidium bananisporum* sp. nov. (Ceratobasidiales) from Iriomote Island, Japan. *Mycoscience* 38: 71–73; **Matsumoto, M.; Furuya, N.; Takanami, Y.; Matsuyama, N.** (1997). Rapid detection of *Rhizoctonia* species, causal agents of rice sheath diseases, by PCR-RFLP analysis using an alkaline DNA extraction method. *Mycoscience* 38: 451–454; **Mordue, J.E.M.; Currah, R.S.; Bridge, P.D.** (1989). An integrated approach to *Rhizoctonia* taxonomy: cultural, biochemical and numerical techniques. *Mycol. Res.* 92: 78–90; **Otero, J.T.; Ackerman, J.D.; Bayman, P.** (2004). Differences in mycorrhizal preferences between two tropical orchids. *Mol. Ecol.* 13: 2393–2404; **Pope, E.J.; Carter, D.A.** (2001). Phylogenetic placement and host specificity of mycorrhizal isolates belonging to AG-6 and AG-12 in the *Rhizoctonia solani* species complex. *Mycologia* 93: 712–719; **Roberts, P.** (1998). *Ceratobasidium obscurum*: an atypical *Thanatephorus* species, misinterpreted as an orchid associate. *Mycol. Res.* 102: 1074–1076; **Roberts, P.** (1998). Synonymy of *Tofispora* & *Thanatephorus*, with notes on a new collection from Puerto Rico. *Mycotaxon* 69: 35–38; **Roberts, P.** (1999). *Rhizoctonia-Forming Fungi* (Richmond): 239 pp.; **Roberts, P.J.** (1998). *Thanatephorus ochraceus*: a saprotrophic and orchid endomycorrhizal species. *Sydowia* 50: 252–256; **Salazar, O.; Schneider, J.H.M.; Julián, M.C.; Keijer, J.; Rubio, V.** (1999). Phylogenetic subgrouping of *Rhizoctonia solani* AG 2 isolates based on ribosomal ITS sequences. *Mycologia* 91: 459–467; **Saunders, G.C.; Owens, S.J.** (1998). RAPD and ITS analysis of orchid mycorrhizal fungi. *Mycorrhiza Manual* Springer Lab Manual (Berlin): 413–424; **Sharon, M.; Kuninaga, S.; Hyakumachi, M.; Sneh, B.** (2006). The advancing identification and classification of *Rhizoctonia* spp. using molecular and biotechnological methods compared with the classical anastomosis grouping. *Mycoscience* 47: 299–316; **Stalpers, J.A.; Andersen, T.F.** (1996). A synopsis of the taxonomy of teleomorphs connected with *Rhizoctonia s.l.*. *Rhizoctonia Species, Taxonomy, Molecular Biology, Ecology, Pathology and Disease Control* (Dordrecht): 49–63.

Ceratocystidaceae Locq. 1972
Microascales: Ascomycota

Stromata absent. Ascomata perithecial, often strongly aggregated, dark brown to black, thin-walled, usually long-necked, usually with divergent ostiolar setae. Interascal tissue absent. Asci at least usually formed in chains from a fertile layer lining the ascomatal cavity, ± saccate, very thin-walled, evanescent, 8-spored. Ascospores small, hyaline, varied in shape, aseptate, often with eccentric wall thickening or sheaths. Anamorphs hyphomycetous, often dimorphic and very variable in morphology. Conidiophores single or in small clusters, usually poorly developed and subtending a single conidiogenous cell. Conidiogenous cells phialidic, with a wall-building ring, the neck elongated, periclinal thickening absent, with a cylindrical collarette extending beyond the wall-building ring, usually forming small hyaline aseptate cylindrical conidia. Chains of dark-walled, often ornamented conidia ('chlamydospores') are also formed from undifferentiated fertile hyphae.

Significant Genera: *Ceratocystis*. Anamorph: *Chalara*.

Distribution: Widespread, especially in the tropics.

Ceratocystis fimbriata, Oman; ascomatal neck

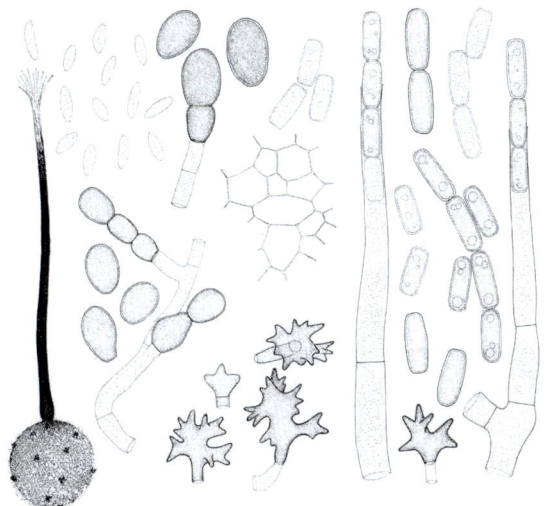

Ceratocystis paradoxa; ascoma, ascospores, conidiophores, conidia

Economic Significance: Several species, especially *Ceratocystis fimbriata* and *C. paradoxa*, are economically significant wide-spectrum pathogens, causing stem, root and fruit rots and cankers of tropical crops including pineapple, mango, sugar cane and banana. Some species cause staining of wood following blockage of the vascular system.

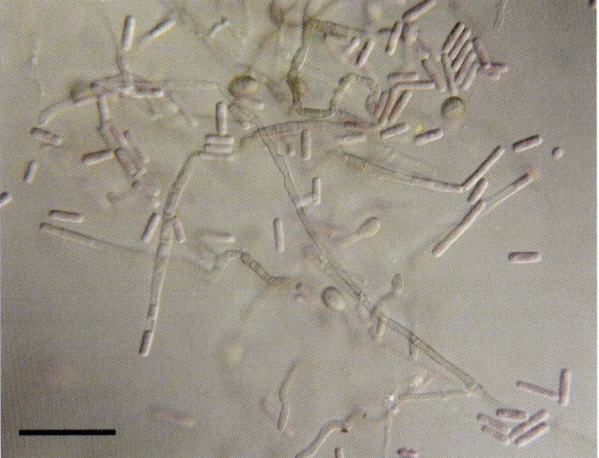

Ceratocystis fimbriata, Oman; conidiogenous cells and conidia

Ecology: Most species are necrotrophs, infecting plant tissues through wounds caused by human intervention, sap-feeding insects etc. Some species produce volatiles which play a part in insect dispersal, but spread through the soil and by water-splash is also important. Most temperate species are associated with bark beetles in a similar manner to the *Ophiostomataceae*.

Ceratocystis fimbriata, Oman; chlamydospores formed in chains

Notes: The teleomorphs of the *Ceratocystidaceae* are almost indistinguishable morphologically from those of the *Ophiostomataceae*, although the anamorphic forms are quite distinct. Molecular evidence has confirmed that the two families are not closely related, and the similarities are likely to be the result of convergent evolution.

References: Baker Engelbrecht, C.J.; Harrington, T.C. (2005). Intersterility, morphology and taxonomy of *Ceratocystis fimbriata* on sweet potato, cacao and sycamore. *Mycologia* 97: 57–69; **Beer, C. de; Wyk, P.W.J. van; Wingfield, M.J.; Kemp, G.H.J.** (1995).

The fine structure of ascospore shape and development in *Ceratocystis fimbriata*. *Antonie van Leeuwenhoek* 67: 325–332; **Brasier, C.M.** (1993). *Ceratocystis* or *Ophiostoma*? The status of *Ophiostoma* and *Ceratocystis* species on *Quercus*. *Recent Advances in Studies on Oak Decline* Proceedings of an International Congress, Selva di Fasano (Brindisi), Italy, September 13-18, 1992 (Bari): 241–245; **Grylls, B.T.; Seifert, K.A.** (1993). A synoptic key to species of *Ophiostoma*, *Ceratocystis* and *Ceratocystiopsis*. *Ceratocystis* and *Ophiostoma* Taxonomy, Ecology and Pathogenicity (St. Paul): 261–268; **Harrington, T.C.; Wingfield, M.J.** (1999). The *Ceratocystis* species on conifers. *Can. J. Bot.* 76: 1446–1457; **Hausner, G.; Reid, J.; Klassen, G.R.** (2000). On the phylogeny of members of *Ceratocystis* s.s. and *Ophiostoma* that possess different anamorphic states, with emphasis on the anamorph genus *Leptographium*, based on partial ribosomal DNA sequen. *Can. J. Bot.* 78: 903–916; **Johnson, J.A.; Harrington, T.C.; Engelbrecht, C.J.B.** (2006). Phylogeny and taxonomy of the North American clade of the *Ceratocystis fimbriata* complex. *Mycologia* 97: 1067–1092; **Marin, M.; Preisig, O.; Wingfield, B.D.; Kirisits, T.; Yamaoka, Y.; Wingfield, M.J.** (2005). Phenotypic and DNA sequence data comparisons reveal three discrete species in the *Ceratocystis polonica* species complex. *Mycol. Res.* 109: 1137–1148; **Paulin, A.E.; Harrington, T.C.** (2000). Phylogenetic placement of anamorphic species of *Chalara* among *Ceratocystis* species and other ascomycetes. *Stud. Mycol.* 45: 209–222, **Samuels, G.J.** (1993). The case for distinguishing *Ceratocystis* and *Ophiostoma*. *Ceratocystis* and *Ophiostoma* Taxonomy, Ecology and Pathogenicity (St. Paul): 15–20; **Seifert, K.A.; Wingfield, M.J.; Kendrick, W.B.** (1993). A nomenclator for described species of *Ceratocystis*, *Ophiostoma*, *Ceratocystiopsis*, *Ceratostomella* and *Sphaeronaemella*. *Ceratocystis* and *Ophiostoma* Taxonomy, Ecology and Pathogenicity (St. Paul): 269–287; **Witthuhn, R.C.; Wingfield, B.D.; Wingfield, M.J.; Harrington, T.C.** (1999). PCR-based identification and phylogeny of species of *Ceratocystis sensu stricto*. *Mycol. Res.* 103: 743–749; **Wyk, P.W.J. van; Wingfield, M.J.; Wyk, P.S. van** (1993). Ultrastructure of centrum and ascospore development in selected *Ceratocystis* and *Ophiostoma* species. *Ceratocystis* and *Ophiostoma* Taxonomy, Ecology and Pathogenicity (St. Paul): 133–138.

Ceratomyces mirabilis; ascomata

References: **Bameul, F.** (1993). *Drepanomyces malayanus* Thaxter (*Laboulbeniales*, *Ceratomycetaceae*) and its hosts of the genus *Psalitrus* d'Orchymont (*Coleoptera*, *Hydrophilidae*, *Sphaeridiinae*). *Nouv. Rev. Entomol.* Nouv. sér. 10: 19–30; **Majewski, T.** (1999). New and rare *Laboulbeniales* (*Ascomycetes*) from the Białowiża Forest (NE Poland). *Acta Mycologica* Warszawa 34: 7–39; **Santamaría, S.** (1999). New and interesting Iberian *Laboulbeniales* (*Fungi*, *Ascomycota*). *Nova Hedwigia* 68: 351–364; **Tavares, I.I.** (1985). *Laboulbeniales* (*Fungi*, *Ascomycetes*). *Mycol. Mem.* 9: 627 pp.; **Weir, A.; Hughes, M.** (2002). The taxonomic status of *Corethromyces bicolor* from New Zealand, as inferred from morphological, developmental, and molecular studies. *Mycologia* 94: 483–493.

Ceratomycetaceae S. Colla 1934
Laboulbeniales: Ascomycota

Stroma (thallus) present. Ascomata formed directly from successive intercalary cells of the primary thallus; outer wall layer of ascoma composed of many short usually ± equal cells. Asci 4-spored. Ascospores with a submedian septum. Usually monoecious.

Significant Genera: *Autoicomyces*, *Ceratomyces*.

Distribution: Widespread.

Economic Significance: None is known.

Ecology: Epibiotic or weakly parasitic on insect cuticles.

Notes: Also referred to as the subfamily *Ceratomycetoideae* of the *Laboulbeniaceae*, but more work is required to establish the correct rank for this group.

Ceratostomataceae G. Winter 1885
Melanosporales: Ascomycota

Stromata absent. Ascomata perithecial or cleistothecial, usually translucent, yellow to pale brown, ostiolate or not, often long-necked and with smooth ostiolar setae. Interascal tissue absent. Asci clavate, thin-walled, without apical apparatus, deliquescing. Ascospores usually large, brown, usually 2-pored, non-septate, smooth to strongly ornamented, sheath absent. Anamorphs hyphomycetous.

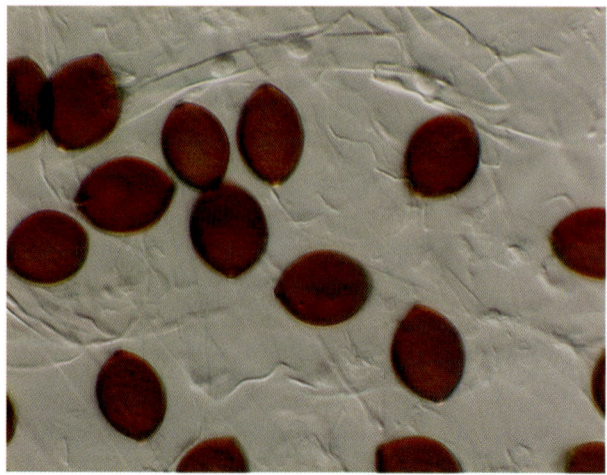

Melanospora zamiae, Martinique; ascospores

Significant Genera: *Melanospora*.

Distribution: Widespread.

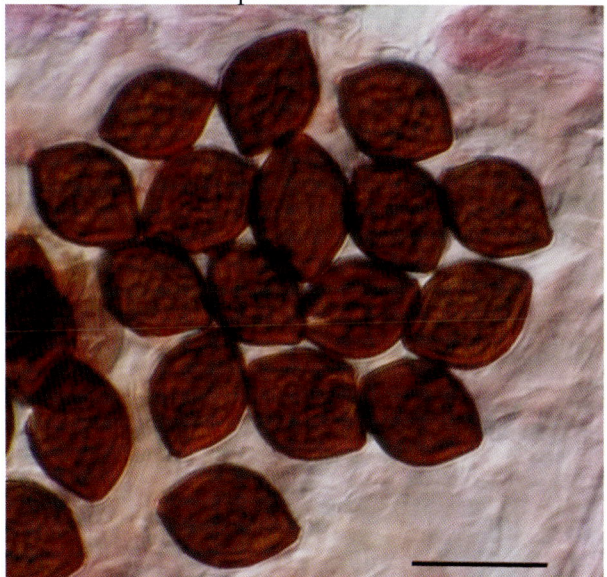

Sphaerodes fimicola, Scotland; ascospores

Economic Significance: None is known.

Ecology: Often fungicolous, but frequently isolated also from soil, rotting vegetation etc. Saprobic or weakly parasitic.

Notes: The correct name for the family is in doubt; it is probably not monophyletic, though molecular data are equivocal.

References: **Goh, T.K.; Hanlin, R.T.** (1998). Ultrastructural observations of ascomal development in *Melanospora zamiae*. *Mycologia* 90: 655–666; **Goh, T.K.; Hyde, K.D.; Hanlin, R.T.** (1998). Spore germination, hyphal morphology, homothallism, and conidial state in *Melanospora zamiae*. *Fungal Science* Taipei 13: 1–9; **Hibbett et al.** (2007). A higher-level phylogenetic classification of the *Fungi*. *Mycol. Res.* 111(5): 509–548; **Horie, Y.; Udagawa, S.-i.; Cannon, P.F.** (1986). Four new Japanese species of *Ceratostomataceae* (*Ascomycetes*). *Mycotaxon* 25: 229–245; **Huhndorf, S.M.; Miller, A.N.; Fernández, F.A.** (2004). Molecular systematics of the *Sordariales*: the order and the family *Lasiosphaeriaceae* revisited. *Mycologia* 96: 368–387; **Rehner, S.A.; Samuels, G.J.** (1995). Molecular systematics of the *Hypocreales*: a teleomorph gene phylogeny and the status of their anamorphs. *Can. J. Bot.* 73: S816–S823; **Stchigel, A.M.; Cano, J.; Guarro, J.** (1999). A new species of *Melanospora* from Easter Island. *Mycol. Res.* 103: 1305–1308; **Vakili, N.G.** (1989). *Gonatobotrys simplex* and its teleomorph, *Melanospora damnosa*. *Mycol. Res.* 93: 67–74; **Zhang, N.; Blackwell, M.** (2002). Molecular phylogeny of *Melanospora* and similar pyrenomycetous fungi. *Mycol. Res.* 106: 148–155; **Zhang et al.** (2007). An overview of the systematics of the *Sordariomycetes* based on a four-gene phylogeny. *Mycologia* 98(6): 1076–1087.

Cerinomycetaceae Jülich 1982
Dacrymycetales: Basidiomycota

Basidiomata resupinate, membranous or waxy, originating as fertile patches on a loose subiculum, effuse. Hyphal system monomitic, hyphae thin-walled or thick-walled, with or without clamp connections. Hymenial surface smooth or with sparse spines due to projecting hyphal pegs. Basidia initially cylindrical to clavate, at first thin-walled but becoming stouter with age, with a basal clamp connection, with 2 very large dacrymycetoid sterigmata. Basidiospores large, weakly to strongly curved, apiculate, aseptate or septate, smooth, thin-walled, not staining in iodine.

Significant Genera: *Cerinomyces*.

Distribution: Widespread but probably under-recorded.

Economic Significance: None is known.

Ecology: Little is known, species are likely to be saprobic on rotten wood.

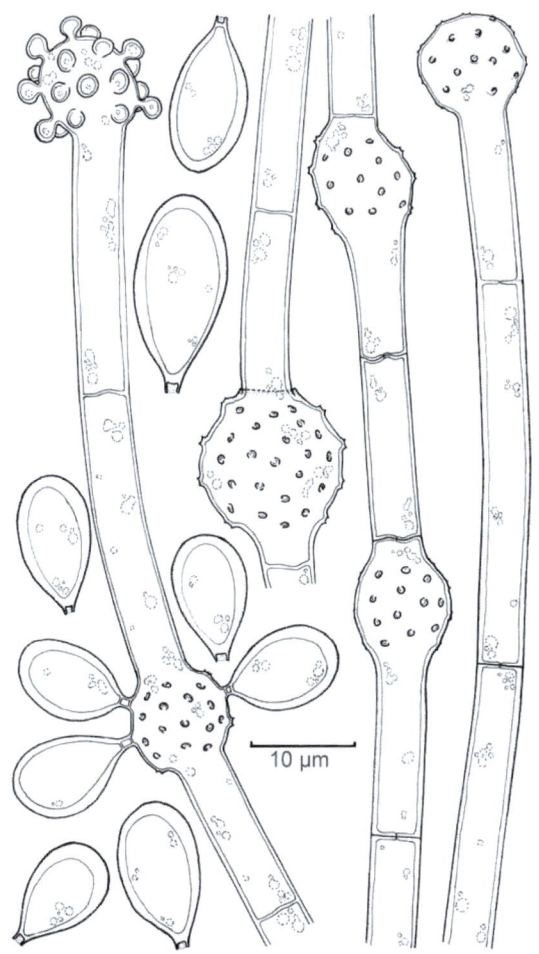

Melanospora damnosa (*Gonatobotrys* anamorph); conidiophores and conidia

Notes: Molecular data confirm placement of this family in the *Dacrymycetales*, but more work is needed.

References: **Duhem, B.** (1998). *Cerinomyces megalosporus*, sp. nov., et *C.* aff. *pallidus* Martin, 1949 (Dacrymycétales corticioïdes). *Bull. trimest. Soc. mycol. Fr.* 114: 1–9; **Larsson, K.-H.; Larsson, E.; Koljalg, U.** (2004). High phylogenetic diversity among corticioid homobasidiomycetes. *Mycol. Res.* 108: 983–1002; **Maekawa, N.** (1987). A new species of the genus *Cerinomyces*. *Can. J. Bot.* 65: 583–588; **Maekawa, N.; Zang, M.** (1997). *Cerinomyces curvisporus* sp. nov. (*Dacrymycetales*) from Yunnan, China. *Mycotaxon* 61: 343–346.

Chaconiaceae Cummins & Y. Hirats. 1983
Uredinales: Basidiomycota

Spermogonia discoid to conical with a discrete wall, subepidermal or subcuticular. Aecia *Aecidium*-like or *Uredo*-like, with or without a well-developed wall, irregular or cupulate. Aeciospores stalked, sometimes in chains, sometimes interspersed with paraphyses, usually echinulate, germ pores variously oriented. Uredinia *Uredo*-type, with or without paraphyses. Urediniospores not catenate, usually echinulate, germ pores variously oriented. Telia erumpent. Teliospores 1-celled, sessile or stalked, thin-walled, germ pore when present poorly defined, the spore forming basidia externally or internally by septation, germinating without dormancy.

Significant Genera: *Hemileia, Maravalia*.

Distribution: Almost all species are tropical in distribution.

Economic Significance: Most species are not economically significant pathogens, but *Hemileia vastatrix* is the causal agent of coffee rust. It devastated plantations in South Asia in the nineteenth century and led to domination of the industry in the western hemisphere. Recently, *Maravalia cryptostegiae* has been used for control of rubber vine in Australia, resulting in one of the most spectacular success stories in biocontrol globally.

Ecology: Autoecious or heteroecious, biotrophic on leaves and green stems of a variety of plant families.

Notes: A poorly studied family that appears to be one of the most basal in rust phylogenies. *Hemileia* was recently added to the family based on morphological criteria. Some preliminary molecular studies confirm this move, but placement within the *Mikronegeriaceae* has also been proposed.

Hemileia vastatrix, Galápagos Islands, Ecuador; aecia on living leaves, parasitized by *Simplicillium lanosoniveum* (*Clavicipitaceae*)

Chaconia ingae, Galápagos Islands, Ecuador; teliospores

References: **Aime, M.C.** (2006). Towards resolving family-level relationships in rust fungi (*Uredinales*). *Mycoscience* 47: 112–122; **Berndt, R.** (1999). Neotropical rust fungi: new species and observations. *Mycologia* 91: 1045–1059; **Cummins, G.B.; Hiratsuka, Y.** (2003). *Illustrated Genera of Rust Fungi* (St Paul): 225 pp.; **Dai, Y.C.; Shen, R.X.** (1993). A numerical taxonomic study on the position of *Maravalia* and other genera of *Uredinales*. *Mycotaxon* 48: 193–200; **Evans, H.C.** (1993). Studies on the rust *Maravalia cryptostegiae*, a potential biological control agent of rubber-vine weed (*Cryptostegia grandiflora*, Asclepiadaceae: Periplocoideae) in Australia, I: Life-cycle. *Mycopathologia* 124: 163–174; **Maier, W.; Begerow, D.; Weiss, M.; Oberwinkler, F.** (2003). Phylogeny of the rust fungi: an approach using nuclear large subunit ribosomal DNA sequences. *Can. J. Bot.* 81: 12–23; **Ono, Y.; Adhikari, M.K.; Kaneko, R.** (1995). An annotated list of the rust fungi (*Uredinales*) of Nepal. *Cryptogams of the Himalayas, 3. Nepal and Pakistan* (Tsukuba): 69–125; **Ono, Y.; Harada, Yu.** (1994). *Aplopsora corni* sp. nov. on *Cornus controversa* from Hokkaido, Japan. *Mycoscience* 35: 179–181; **Payak, M.M.** (1997). Cytology, life cycles and spore development in some Indian rust fungi (*Urediniomycetes*). *Indian Phytopath.* 49: 307–318; **Ritschel, A.** (2005). Monograph of the genus *Hemileia* (*Uredinales*). *Biblthca Mycol.* 200: 132 pp.; **Wingfield, B.D.; Ericson, L.; Szaro, T.; Burdon, J.J.** (2004). Phylogenetic patterns in the *Uredinales*. *Australas. Pl. Path.* 33: 327–335.

Chadefaudiellaceae Faurel & Schotter ex Benny & Kimbr. 1980

Microascales: Ascomycota

Stromata absent. Ascomata perithecial, the upper part formed from anastomosing setae. Interascal tissue of undifferentiated hyphae. Asci ± spherical, formed in vertical chains. Ascospores striate, pale brown, ? formed in a mazaedial mass. Anamorphs hyphomycetous, forming arthrospores.

Significant Genera: *Chadefaudiella, Faurelina*.

Distribution: Widespread, especially Africa and Arabia.

Economic Significance: None is known.

Ecology: Coprophilous, in desert environments.

Notes: Morphologically rather distinctive but with unclear relationships.

References: **Arx, J.A. von; Figueras, M.J.; Guarro, J.** (1988). Sordariaceous ascomycetes without ascospore ejaculation. *Beih. Nova Hedwigia* 94: 104 pp.; **Valldosera, M.; Guarro, J.; Figueras, M.J.** (1987). Coprophilous fungi from Spain VII. *Faurelina hispanica* sp. nov. *Mycotaxon* 30: 5–7.

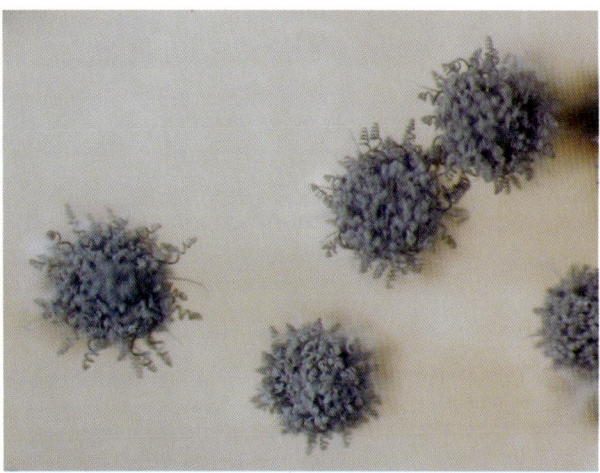

Chaetomium robustum, UK; ascomata with coiled hairs

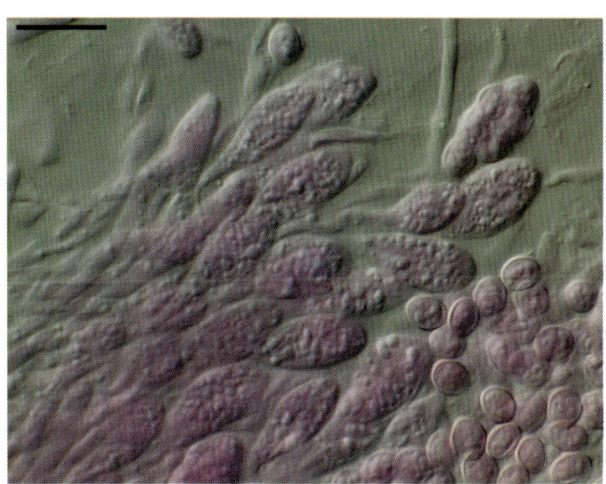

Chaetomium globosum; immature asci

Chaetomiaceae G. Winter 1885

Sordariales: Ascomycota

Stromata absent. Ascomata perithecial, pale or dark, thin-walled, membranous, occasionally with cells arranged in shields, often with complex ornamented hairs, ostiolate or not. Interascal tissue apparently absent. Asci clavate to saccate, very thin-walled, lacking apical structures, deliquescing when mature. Ascospores small, pale to mid-brown, thin-walled, often discharged as a cirrus, usually 1-pored, non-septate, smooth, gelatinous sheath and appendages absent. Anamorph usually absent but when present hyphomycetous. Conidia thick-walled and sometimes ornamented, produced singly from undifferentiated hyphae, usually with a survival function; a few species producing small hyaline spermatia from percurrently proliferating simple phialidic conidiogenous cells.

Significant Genera: *Chaetomium, Corynascus, Thielavia*. Anamorph: *Myceliophthora*.

Distribution: Extremely widespread, wherever suitable substrata are available.

Economic Significance: Many species are highly competent enzymatically and there has been substantial research into utilization of cellulases and other key enzymes. They are commonly found as contaminants in food production and laboratory systems, as deteriogenic organisms of cellulose-rich substrata such as paper and canvas, and occasionally as opportunistic pathogens. A few are thermophilic.

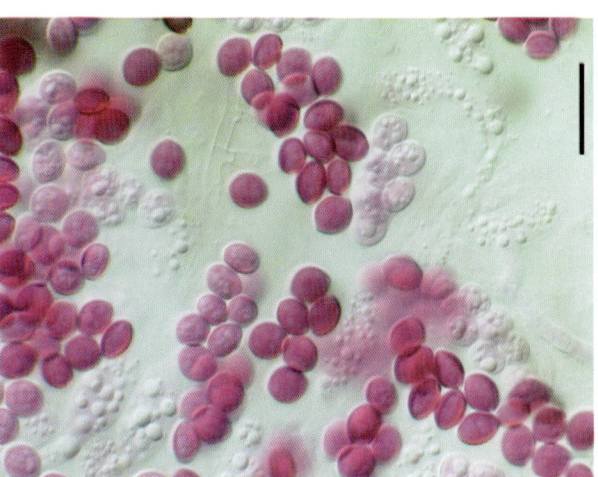

Chaetomium globosum, UK; immature ascospores

Ecology: Saprobic, acting as key decay organisms of plant materials and plant-derived commodities.

Notes: The *Chaetomiaceae* appears to be a well-defined family within the *Sordariales*. The thin-walled ascomata and evanescent asci are its most diagnostic characters in morphological terms, but a few species in other families within this order appear to have also developed these characteristics.

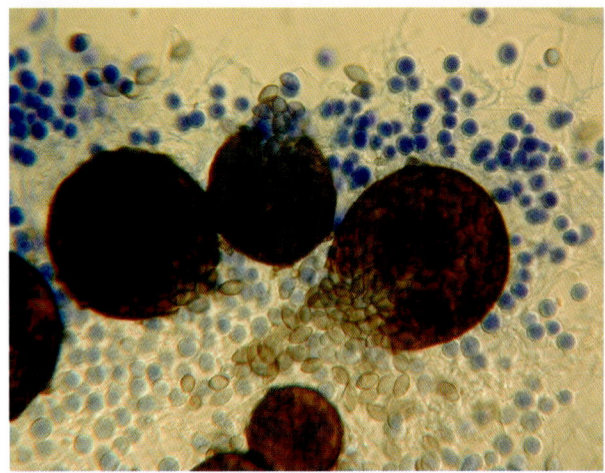

Corynascus sepedonium, India; ascomata, ascospores and conidia

References: **Arx, J.A. von; Figueras, M.J.; Guarro, J.** (1988). Sordariaceous ascomycetes without ascospore ejaculation. *Beih. Nova Hedwigia* 94: 104 pp.; **Arx, J.A. von; Guarro, J.; Figueras, M.J.** (1986). The ascomycete genus *Chaetomium*. *Beih. Nova Hedwigia* 84: 162 pp.; **Cannon, P.F.** (1986). A revision of *Achaetomium*, *Achaetomiella* and *Subramaniula*, and similar species of *Chaetomium*. *Trans. Br. mycol. Soc.* 87: 45–76; **Domsch, K.H.; Gams, W.; Anderson, T.-H.** (1993). *Compendium of Soil Fungi and Supplement* (Germany): 406 pp.; **Huhndorf, S.M.; Miller, A.N.; Fernández, F.A.** (2004). Molecular systematics of the *Sordariales*: the order and the family *Lasiosphaeriaceae* revisited. *Mycologia* 96: 368–387; **Lee, S.J.; Hanlin, R.T.** (1999). Phylogenetic relationships of *Chaetomium* and similar genera based on ribosomal DNA sequences. *Mycologia* 91: 434–442; **Mouchacca, J.** (1997). Thermophilic fungi: biodiversity and taxonomic status. *Cryptog. Mycol.* 18: 19–69; **Silva, D.M.W.; Hanlin, R.T.** (1997). *Chaetomidium heterotrichum* from Venezuela, with a key to species and cladistic analysis of the genus *Chaetomidium*. *Mycoscience* 37: 261–267; **Stchigel, A.M.; Figuera, L.; Cano, J.; Guarro, J.** (2002). New species of *Thielavia*, with a molecular study of representative species of the genus. *Mycol. Res.* 106: 975–983; **Stchigel, A.M.; Sagués, M.; Cano, J.; Guarro, J.** (2000). Three new thermotolerant species of *Corynascus* from soil, with a key to the known species. *Mycol. Res.* 104: 879–887; **Udagawa, S.-i.** (1986). Taxonomy of mycotoxin-producing *Chaetomium*. *Toxigenic Fungi: their toxins and health hazard* (Tokyo): 139–147; **Untereiner, W.A.; Débois, V.; Naveau, F.A.** (2001). Molecular systematics of the ascomycete genus *Farrowia* (*Chaetomiaceae*). *Can. J. Bot.* 79: 321–333.

Chaetosphaerellaceae Huhndorf, A.N. Mill. & F.A. Fernández 2004
Coronophorales: Ascomycota

Stromata absent or composed of an often setose hyphal subiculum. Ascomata perithecial, superficial, ± globose to ovoid, ostiolate, glabrous or hairy, peridium dark brown and leathery, cells with Munk pores. Interascal tissue composed of wide thin-walled paraphyses, inflated between the septa. Asci cylindrical to clavate, thin-walled, not fissitunicate, sometimes with a small refractive apical ring, not staining in iodine, 8-spored. Ascospores ellipsoidal to fusiform, usually multiseptate, brown or with ± hyaline end cells, smooth, without a gelatinous sheath or appendages. Anamorphs hyphomycetous, usually with conidia formed in short chains from polyblastic percurrently proliferating conidiogenous cells.

Significant Genera: *Chaetosphaerella*, *Crassochaeta*.

Distribution: Widespread in both temperate and tropical regions.

Economic Significance: None is known.

Ecology: Saprobic on fallen wood and similar substrata.

Notes: *Chaetosphaerella* has previously been placed in various families. Its relationships with the *Nitschkiaceae* as revealed by rDNA sequences are unexpected, but both included genera have Munk pores in their peridial cells as are typical of that family.

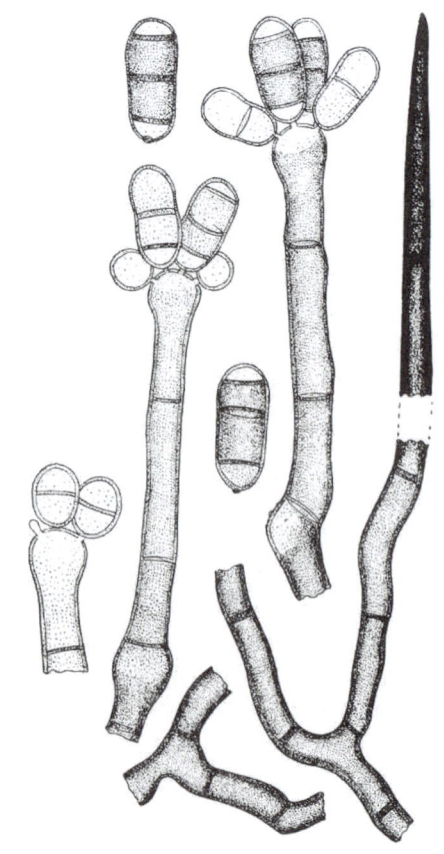

Chaetosphaerella phaeostroma; anamorph

Chaetosphaerella phaeostroma, Devon, UK; ascomata and conidiophores

Chaetosphaeria myriocarpa, Surrey, UK; ascomata

References: Huhndorf, S.M.; Miller, A.N.; Fernández, F.A. (2004). Molecular systematics of the *Coronophorales* and new species of *Bertia*, *Lasiobertia* and *Nitschkia*. Mycol. Res. 108: 1384–1398; Huhndorf, S.M.; Miller, A.N.; Fernández, F.A. (2004). Molecular systematics of the *Sordariales*: the order and the family *Lasiosphaeriaceae* revisited. Mycologia 96: 368–387; **Réblová, M.** (1999). Studies in *Chaetosphaeria* sensu lato I. The genera *Chaetosphaerella* and *Tengiomyces* gen. nov. of the *Helminthosphaeriaceae*. Mycotaxon 70: 387–420; **Réblová, M.** (1999). Studies in *Chaetosphaeria* sensu lato IV. *Crassochaeta* gen. nov., a new lignicolous genus of the *Trichosphaeriaceae*. Mycotaxon 71: 45–67.

Economic Significance: Species of *Stachybotrys* produce potent mycotoxins and are one of the key fungi associated with indoor air quality concerns in North America.

Ecology: Saprobes associated with decaying woody and herbaceous plant material, including timber and timber products in damp buildings.

Notes: The separation of the *Chaetosphaeriaceae* from the *Lasosphaeriaceae* has now been confirmed using molecular methods and it has recently been segregated into its own order.

Chaetosphaeriaceae Réblová, M.E. Barr & Samuels 1999
Chaetosphaeriales: Ascomycota

Stroma absent or restricted to a thin sometimes subiculate crust. Ascomata perithecial, superficial, often aggregated, black, ± globose to conical, usually rather thin-walled but often carbonaceous, glabrous or setose, the ostiole papillate, periphysate. Interascal tissue of copious persistent true paraphyses. Asci cylindrical, persistent, thin-walled, not fissitunicate, with a well-developed refractive J– apical ring. Ascospores ellipsoidal to fusiform, transversely septate, sometimes fragmenting at the septa, hyaline or brown. Anamorphs hyphomycetous, varied, with pigmented conidiophores and percurrently proliferating conidiogenous cells often with widely flared collarettes, conidia very varied in form but mostly small and hyaline.

Significant Genera: *Chaetosphaeria*, *Melanopsamma*. Anamorphs: *Cylindrotrichum*, *Custingophora*, *Dictyochaeta*, *Chloridium*, *Gonytrichum*, *Menispora*, *Stachybotrys*.

Distribution: Cosmopolitan, very frequent in temperate as well as tropical climates.

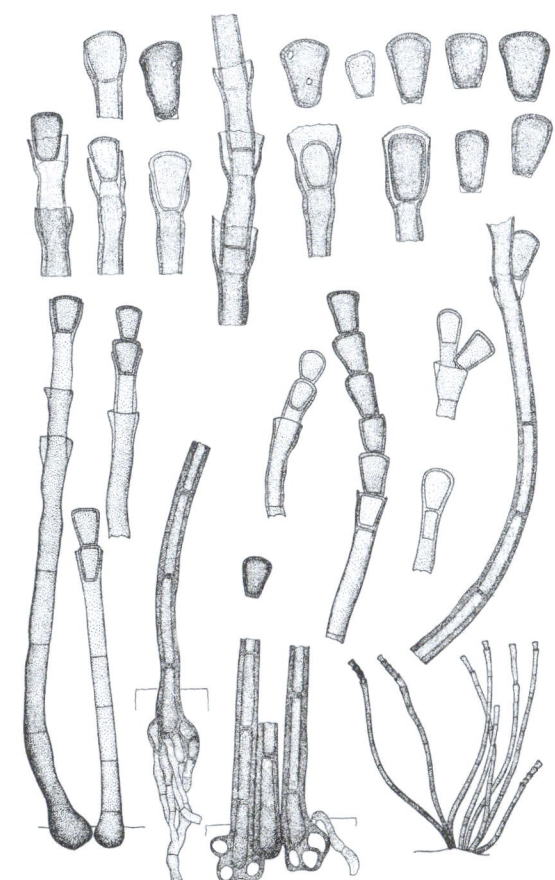

Chaetosphaeria cupulifera; conidiophores and conidia

Chaetosphaeria chloroconia, Czech Republic; conidiophores and conidia

References: **Fernández, F.A.; Miller, A.N.; Huhndorf, S.M.; Lutzoni, F.M.; Zoller, S.** (2006). Systematics of the genus *Chaetosphaeria* and its allied genera: morphological and phylogenetic diversity in north temperate and neotropical taxa. *Mycologia* 98: 121–130; **Gams, W.** (2000). *Phialophora* and some similar morphologically little-differentiated anamorphs of divergent ascomycetes. *Stud. Mycol.* 45: 187–199; **Huhndorf, S.M.; Miller, A.N.; Fernández, F.A.** (2004). Molecular systematics of the *Sordariales*: the order and the family *Lasiosphaeriaceae* revisited. *Mycologia* 96: 368–387; **Hyde, K.D.; Goh, T.K.; Taylor, J.E.; Fröhlich, J.** (1999). *Byssosphaeria*, *Chaetosphaeria*, *Niesslia* and *Ornatispora* gen. nov., from palms. *Mycol. Res.* 103: 1423–1439; **Koster, B.; Scott, J.; Wong, B.; Malloch, D.; Straus, N.** (2003). A geographically diverse set of isolates indicates two phylogenetic lineages within *Stachybotrys chartarum*. *Can. J. Bot.* 81: 633–643; **Kuthubutheen, A.J.; Nawawi, A.** (1991). Key to *Dictyochaeta* and *Codinaea* species. *Mycol. Res.* 95: 1224–1229; **Réblová, M.** (1997). Revision and reclassification of some *Chaetosphaeria* species. *Czech Mycol.* 50: 73–83; **Réblová, M.** (1999). Studies in *Chaetosphaeria* sensu lato IV. *Crassochaeta* gen. nov., a new lignicolous genus of the *Trichosphaeriaceae*. *Mycotaxon* 71: 45–67; **Réblová, M.** (1999). Teleomorph-anamorph connections in *Ascomycetes* 2. *Ascochalara gabretae* gen. et sp. nov. and its *Chalara*-like anamorph. *Sydowia* 51: 210–222; **Réblová, M.** (2000). The genus *Chaetosphaeria* and its anamorphs. *Stud. Mycol.* 45: 149–168; **Réblová, M.; Barr, M.E.; Samuels, G.J.** (1999). *Chaetosphaeriaceae*, a new family for *Chaetosphaeria* and its relatives. *Sydowia* 51: 49–70; **Réblová, M.; Winka, K.** (2000). Phylogeny of *Chaetosphaeria* and its anamorphs based on morphological and molecular data. *Mycologia* 92: 939–954; **Samuels, G.J.; Candoussau, F.; Magni, J.-F.** (1997). Fungicolous pyrenomycetes 2. *Ascocodinaea*, gen. nov., and reconsideration of *Litschaueria*. *Mycologia* 89: 156–162.

Chaetothyriaceae Hansf. ex M.E. Barr 1979
Chaetothyriales: Ascomycota

Mycelium largely superficial, forming a thin layer of tissue composed of hyaline to mid-brown broad thin-walled hyphae, sometimes clypeus-like, sometimes with long dark brown thick-walled setose appendages. Ascomata often formed beneath the external hyphal mat, perithecial, spherical or flattened, often collapsing when dry, the apex ± papillate, with a periphysate ostiole; peridium dark brown, composed of a few layers of thin-walled angular cells; hymenium usually J+. Interascal tissue of short apical periphysoids. Asci formed in a basal fascicle, clavate to saccate, short-stalked, fairly thick-walled, fissitunicate, ocular chamber absent, with a small refractive apical ring, 8-spored. Ascospores hyaline or pale, transversely septate or muriform, constricted at the septa, the primary septum not clearly defined, thin-walled and smooth-walled. Anamorphs hyphomycetous. Conidiophores absent. Conidiogenous cells undifferentiated, intercalary on pigmented vegetative hyphae, holoblastic, proliferation where occurring sympodial. Conidia brown or olivaceous, multiseptate, usually tapering, strongly constricted at the primary septa, often branched at the basal cell.

Significant Genera: *Ceramothyrium*, *Chaetothyrium*. Anamorph: *Stanhughesia*.

Distribution: Widespread, mainly tropical but also with cold temperate representatives.

Economic Significance: None is known.

Ecology: Epiphytic or biotrophic on living leaves; very little is known.

Chaetothyrium javanicum, Ghana; asci and ascospores

Notes: Until quite recently *Chaetothyrium* and its relatives were assumed to belong to the *Dothideomycetidae* due to their possession of fissitunicate asci. However, several recent molecular studies have concluded that the group is only distantly related. The other family of the *Chaetothyriomycetidae*, the

Herpotrichiellaceae, lacks superficial mycelium and is mostly saprobic, and has varied black yeast-like anamorphs.

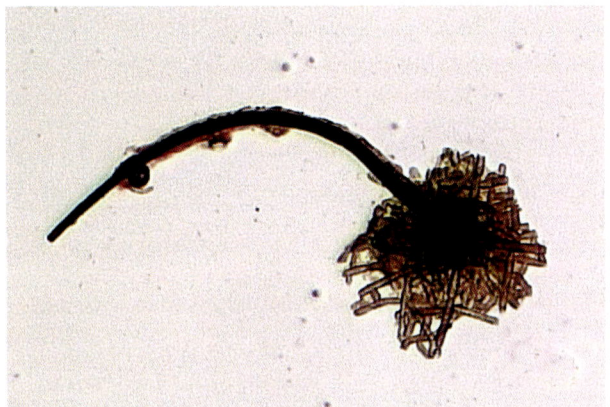

Chaetothyrium sp., Galápagos Islands, Ecuador; mycelial seta

References: **Cannon, P.F.** (2001). *Asterina nyanzae*; *Capnodium salicinum*; *Chaetothyrium guaraniticum*; *Cucurbitaria laburni*; *Dothidea sambuci*; *Hysterographium fraxini*; *Micropeltis ugandae*; *Parodiella hedysari*; *Tubeufia cerea*; *Vizella gomphispora*. IMI Descr. Fungi Bact. 141: [22] pp.; **Constantinescu, O.; Holm, K.; Holm, L.** (1989). Teleomorph-anamorph connections in ascomycetes. 1–3. *Stanhughesia* (*Hyphomycetes*) new genus, the anamorph of *Ceramothyrium*. Stud. Mycol. 31: 69–84; **Eriksson, O.; Yue, J.** (1985). Studies on Chinese ascomycetes. I. *Phaeosaccardinula dictyospora*. Mycotaxon 22: 269–280; **Haase, G.; Sonntag, L.; Melzer-Krick, B.; Hoog, G.S. de** (1999). Phylogenetic inference by SSU-gene analysis of members of the *Herpotrichiellaceae* with special reference to human pathogenic species. Stud. Mycol. 43: 80–97; **Panwar, A.B.; Jagtap, S.S.** (1990). Studies in foliicolous fungi-X. Genus – *Chaetothyrium* Speg. Geobios New Rep. 9: 121–127; **Pohlad, B.R.** (1988). *Rhombostilbella* parasitizing *Chaetothyriaceae* and *Capnodiaceae*. Mycologia 80: 757–759; **Winka, K.; Eriksson, O.E.; Bång, Å.** (1998). Molecular evidence for recognizing the *Chaetothyriales*. Mycologia 90: 822–830.

Chionosphaeraceae Oberw. & Bandoni 1982
Atractiellales: Basidiomycota

Basidiomata small, stipitate, capitate, synnema-like, pale and gelatinous or waxy, the stipe and fertile region not strongly differentiated. Hyphal system monomitic, the hyphae of the stipe ± parallel, thick-walled, sparingly septate, septal pores simple, without clamp connections. Basidia elongate-clavate, aseptate or with a single transverse septum, with 1–2 apiculate sterigmata or a cluster of 6–8 in a crown-like arrangement. Basidiospores discharged actively, hyaline, aseptate, smooth, thin-walled or thick-walled, germinating by budding. Anamorph yeast-like.

Significant Genera: *Chionosphaera*.

Distribution: Most species are known from north temperate regions.

Economic Significance: None is known.

Ecology: Species appear to be mycoparasitic, associated with fungi of bark beetle galleries, wood-inhabiting *Xylariaceae* or lichens. Others are found erumpent from bark or decaying wood and may be saprobic.

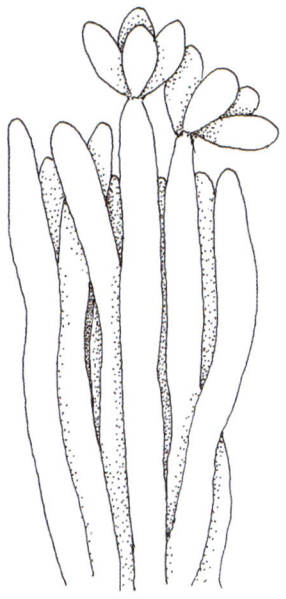

Chionosphaera coppinsii, Wester Ross, Scotland; basidia and basidiospores

Notes: A small family of uncertain taxonomic position; it appears to be amongst the basal groups of *Urediniomycetes*. The *Agaricostilbaceae* seems to be its closest relative.

References: **Boekhout, T.** (1998). Diagnostic descriptions and key to presently accepted heterobasidiomycetous genera. *The Yeasts, A Taxonomic Study* (Amsterdam): 627–634; **Diederich, P.** (1996). The lichenicolous heterobasidiomycetes. Biblthca Lichenol. 61: 198 pp.; **Kwon-Chung, K.J.** (1998). *Chionosphaera* Cox. *The Yeasts, A Taxonomic Study* (Amsterdam): 643–645; **Oberwinkler, F.; Bandoni, R.J.** (1982). A taxonomic survey of the gasteroid, auricularioid *Heterobasidiomycetes*. Can. J. Bot. 60: 1726–1750; **Oberwinkler, F.; Bauer, R.** (1989). The systematics of gasteroid, auricularioid *Heterobasidiomycetes*. Sydowia 41: 224–256; **Scorzetti, G.; Fell, J.W.; Fonseca, A.; Statzell-Tallman, A.** (2002). Systematics of basidiomycetous yeasts: a comparison of large subunit D1/D2 and internal transcribed spacer rDNA regions. FEMS Yeast Res. 2: 495–517; **Seifert, K.A.; Oberwinkler, F.; Bandoni, R.** (1992). Notes on *Stilbum vulgare* and *Fibulostilbum phylacicola* gen. et sp. nov. (*Atractiellales*). Boln Soc. argent. Bot. 28: 213–217.

Choanephoraceae J. Schröt. 1894
Mucorales: Zygomycota

Vegetative thallus a mycelium, complex fruit bodies absent. Sporophores arising from substratum mycelium or aerial hyphae, simple or branched, each sporophore producing either a sporangium or sporangiola, never both. Sporangia usually large, multispored and columellate, or infrequently smaller, few-spored and acolumellate; wall persistent, echinulate, rupturing into two hemispherical portions via a line of weakness, or into three or more irregular portions via two or more lines of weakness; sporangiospores from sporangia with a brown, longitudinally striate wall (sometimes indistinct) and long, hyaline, polar appendages. Sporangiola unispored or multispored, wall persistent, of two types, indehiscent or dehiscent; indehiscent sporangiola with wall inseparable from sporangiospore wall, sporangiospores brown, wall with longitudinal striations but without polar appendages; dehiscent sporangiola with a separable wall, rupturing via one or more zones of weakness, sporangiospores brown, wall with longitudinal striations and with long, hyaline, polar appendages. Zygospores globose to subglobose, of two types: (1) with a thin, hyaline, relatively smooth zygosporangial wall enclosing a thick-walled, brown, striate zygospore; suspensors apposed or tongs-like, basally entwined, without appendages; and (2) with a brown to dark brown wall, opaque, ornamented with more or less coarse projections; suspensors opposed, smooth, more or less swollen, equal or unequal. Heterothallic.

Significant Genera: *Choanephora, Blakeslea, Gilbertella*.

Distribution: Tropics or subtropics (Mediterranean-type), rarely occurring in temperate zones and then only under unusual circumstances (warm summers and suitable hosts).

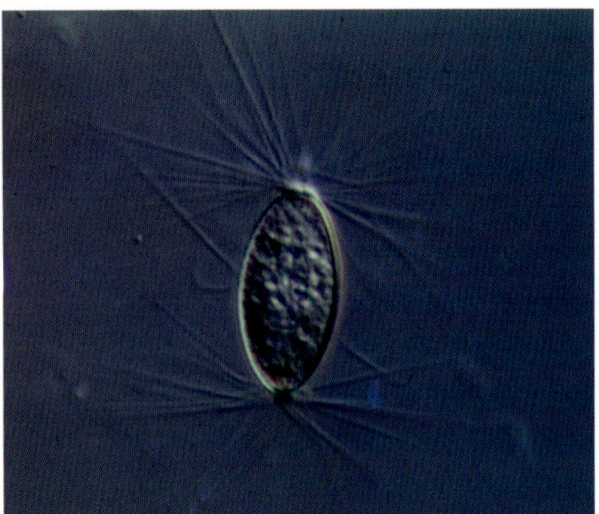
Choanephora cucurbitarum; mature sporangiolum bearing filamentous appendages

Economic Significance: As facultative pathogens of a range of plants, traditionally associated with members of the *Cucurbitaceae* (hence *Choanephora cucurbitarum*) where it causes a blossom blight and fruit rot but also found on other commercially important crops although usually as a recurring minor disease. The species *G. persicaria* can cause a dry rot of peach (*Prunus persica*) which is now less significant due to improved methods of harvesting and transport resulting in less damage to the fruit.

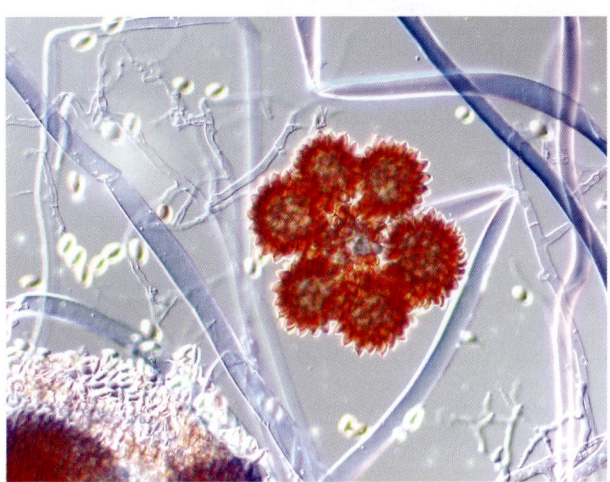
Choanephora cucurbitarum; sporangiophore bearing sporangiola

Ecology: Essentially soil fungi from where they are infrequently isolated; can often be found with fallen and decaying flowers of *Hibiscus rosa-sinensis* where this has been planted as an ornamental.

Notes: Recent molecular work has show that *Gilbertella*, previously segregated into a separate family, belongs in *Choanephoraceae*, from where it was removed. Apparently the characters of the anamorph – persistent sporangial wall with dehiscence suture, appendaged and differentially thickened sporangiospores – are a better indication of relationships than the non *Choanephora*-like zygospores.

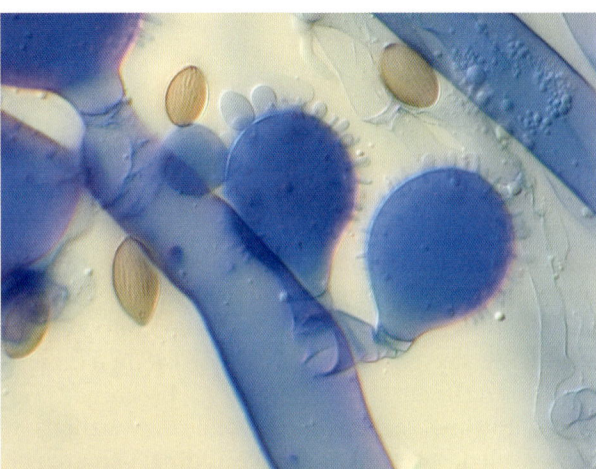
Choanephora cucurbitarum; immature fertile vesicles and sporangiola

References: **Benny, G.L.** (1991). *Gilbertellaceae*, a new family of the *Mucorales* (Zygomycetes). *Mycologia* 83: 150–157; **Higham, M.T.; Cole, K.M.** (1982). Fine structure of sporangiole development in *Choanephora cucurbitarum* (Mucorales). *Can. J. Bot.* 60:

2313–2324; **Kirk, P.M.** (1984). A Monograph of the *Choanephoraceae*. *Mycol. Pap.* 152: 61 pp.; **Sakai, S; Kato, M.; Nagamasu, H.** (2000). *Artocarpus* (*Moraceae*) – gall midge pollination mutualism mediated by a male-flower parasitic fungus. *Am. J. Bot.* 87: 440–445; **Tanabe, Y.; Saikawa, M.; Watanabe, M.M.; Sugiyama, J.** (2004). Molecular phylogeny of *Zygomycota* based on EF-1α and RPB1 sequences: limitations and utility of alternative markers to rDNA. *Mol. Phylogen. Evol.* 30: 438–449; **Yu, M.Q.; Ko, W.H.** (1999). Azygospore formation following protoplast fusion in *Choanephora cucurbitarum*. *Mycol. Res.* 103: 684–688.

Christianseniaceae F. Rath 1991
Tremellales: Basidiomycota

Basidiomata gelatinous, smooth, thin, folded or cerebriform, superficial. Hyphal system monomitic, hyphae hyaline, nodulose, ± thin-walled, with or without clamp connections, haustoria present. Cystidia absent. Basidia clavate to ± urniform, with 2–6 sterigmata, aseptate or incompletely longitudinally septate at the apex. Basidiospores discharged actively and lacrimiform or passively and cylindrical to ellipsoidal and radially symmetrical, hyaline, smooth, thin-walled, germinating by budding. Conidia formed in chains from poorly differentiated fertile hyphae (blastoconidia), or dikaryotically from transversely septate basidium-like structures with spores derived from each cell fusing (zygoconidia).

Significant Genera: *Christiansenia*. Anamorph: *Syzygospora*.

Distribution: Widespread, known primarily from north temperate zones but also recorded from the neotropics.

Economic Significance: None is known.

Ecology: Mycoparasitic on agarics, corticioid fungi and *Leotiales*, often forming galls.

Syzygospora mycetophila; colony parasitizing an agaric fungus

Notes: There has been much disagreement over structural interpretation and nomenclature in this family. There is doubt as to whether it should be referred to as the *Christianseniaceae* or *Syzygosporaceae*, depending on whether the so-called zygoconidia are considered to be teleomorphic or anamorphic structures.

References: **Ginns, J.** (1986). The genus *Syzygospora* (*Heterobasidiomycetes*: *Syzygosporaceae*). *Mycologia* 78: 619–636; **Gottschalk, M.; Blanz, P.A.** (1985). Untersuchungen an 5S ribosomalen Ribonukleinsäuren als Beitrag zur Klärung von Systematik und Phylogenie der Basidiomyceten. *Z. Mykol.* 51: 205–243; **Hauerslev, K.** (1969). *Christiansenia pallida*, gen. nov., sp. nov. A new parasitic homobasidiomycete from Denmark. *Friesia* 9: 43–45; **Kotiranta, H.; Larsson, K.-H.** (1990). New or little collected corticolous fungi from Finland (*Aphyllophorales*, *Basidiomycetes*). *Windahlia* 18: 1–14; **Oberwinkler, F.** (1987). Heterobasidiomycetes with ontogenetic yeast-stages – systematic and phylogenetic aspects. *Stud. Mycol.* 30: 61–74; **Rath, F.** (1991). *Christianseniales*, *Christianseniaceae*, ordine e famiglia nuovi. *Atti Soc. ital. Sci. nat. Mus. Civico Storia nat. Milano* 132: 13–24.

Chrysothricaceae Zahlbr. 1905
Arthoniales: Ascomycota

Thallus granular, without a cortex, usually poorly developed; ascomata apothecial, ± circular, with rudimentary walls, the hymenium usually yellowish.

Chrysothrix candelaris, British Columbia, Canada; bright yellow farinose thallus on conifer bark

Significant Genera: *Byssoloma*, *Chrysothrix*.

Distribution: Widespread, especially in the tropics.

Economic Significance: None is known.

Ecology: Lichenized with green algae, especially on wood.

Notes: Very poorly known; no recent information is available.

Chrysothrix chlorina, Norway; thallus

Ecology: Saprobic or more frequently parasitic on 'algae' (*Chromista* and *Platae*: *Chlorophyta*) and other chytrids.

Notes: The largest of the families of *Chytridiales* with about 35 accepted genera (12 monotypic) and over 300 species.

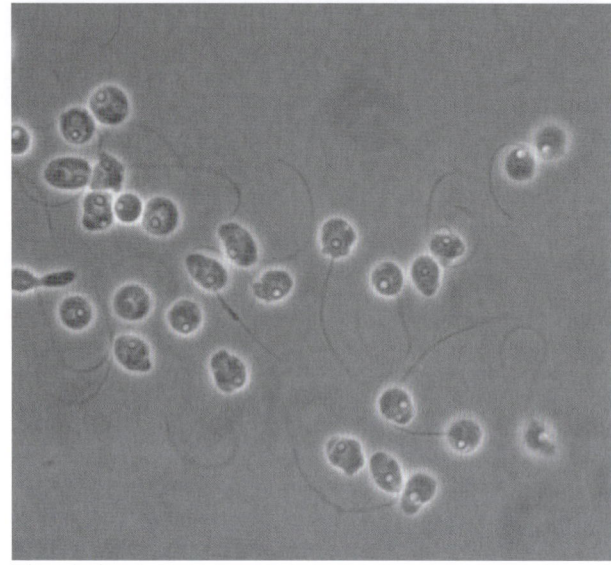

Rhizophydium sp.; zoospores

References: Grube, M. (1998). Classification and phylogeny in the *Arthoniales* (Lichenized *Ascomycetes*). *Bryologist* 101: 377–391; Thor, G. (1988). Two new species of *Chrysothrix* from South America. *Bryologist* 91: 360–363; Tønsberg, T. (1994). *Chrysothrix flavovirens* sp. nov. – the sorediate counterpart of *C. chrysophthalma*. *Graphis Scripta* 6: 31–33.

Chytridiaceae Nowak. 1878
Chytridiales: Chytridiomycota

Thallus eucarpic (with rhizoids) and monocentric (comprising a single cell), epibiotic and either expanding to form an operculate zoosporangium or an evanescent cyst, or endobiotic and forming the reproductive organs, rhizoids endobiotic or interbiotic; zoosporangium endogenous to the zoospore cyst, operculate or inoperculate; zoospores posteriorly uniflagellate, usually with a single globule; resting spore endogenous or exogenous, germinating to form an epibiotic zoosporangium; zygote, where known, formed by fusion of aplanogametes.

Significant Genera: *Chytridium*.

Distribution: Widespread and relatively common.

Economic Significance: None is known.

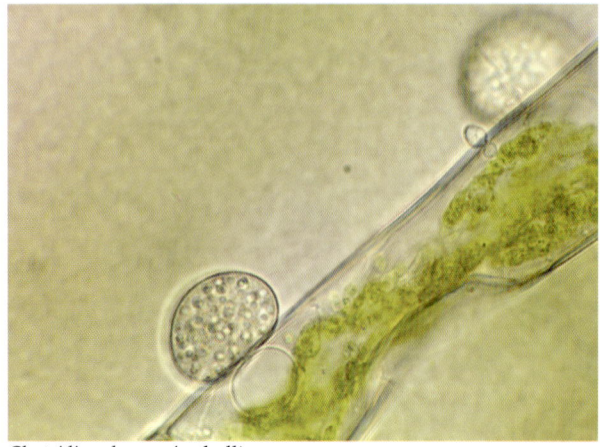

Chytridium lagenaria; thalli

References: Barr, D.J.S. (1990). Phylum *Chytridiomycota*. *Handbook of Protoctista* (Boston): 454–466; James, T.Y.; Porter, D.; Leander, C.A.; Vilgalys, R.; Longcore, J.E. (2000). Molecular phylogenetics of the *Chytridiomycota* supports the utility of ultrastructural data in chytrid systematics. *Can. J. Bot.* 78: 336–350; Sparrow, F.K. (1960). *Aquatic Phycomycetes* Edn 2 (Ann Arbor): 1187 pp..

Cintractiaceae Vánky 2000
Ustilaginales: Basidiomycota

Sori formed on the surface of plant tissues, usually initially enclosed by a network of gelatinous hyphae. Hyphae intracellular, with septa lacking pores when mature. Spore mass black, sometimes mixed with sterile cells, composed of single reddish or brownish, usually ornamented teliospores germinating to produce transversely septate basidia usually with 4 cells, rarely with 1 or more than 4. Basidial cells fusing in pairs via clamp connections or copulation bridges giving rise to a large, probably dikaryotic, blastoconidium on a sterigma, monokaryotic basidiospores also produced.

Significant Genera: *Cintractia*, *Tolyposporium*.

Distribution: Very widespread, with a similar range to its host family.

Economic Significance: None is known; no important crop plants are affected.

Cintractia axicola, Peru; habit

Ecology: Parasitic on inflorescences of species of Cyperaceae and Juncaceae, sometimes also spreading systemically through the plant.

Notes: Molecular evidence suggests that the *Cintractiaceae* is paraphyletic; *Leucocintractia* and *Ustanciosporium* may belong in the *Ustilaginaceae*, but more research is required.

Cintractia axicola, Sarawak; teliospores

References: **Begerow, D.; Bauer, R.; Oberwinkler, F.** (1998). Phylogenetic studies on nuclear large subunit ribosomal DNA sequences of smut fungi and related taxa. *Can. J. Bot.* 75: 2045–2056; **Ingold, C.T.** (1995). Blastoconidium production from conjugated basidial cells in *Cintractia*. *Mycol. Res.* 99: 140–142; **Ingold, C.T.** (1999). The *Cintractia* pattern of teliospore germination. *Mycol. Res.* 103: 1071–1072; **Piepenbring, M.** (1995). *Trichocintractia*, a new genus for *Cintractia utriculosa* (*Ustilaginales*). *Can. J. Bot.* 73: 1089–1096; **Piepenbring, M.; Begerow, D.; Oberwinkler, F.** (1999). Molecular sequence data assess the value of morphological characteristics for a phylogenetic classification of species of *Cintractia*. *Mycologia* 91: 485–498; **Piepenbring, M.; Oberwinkler, F.** (2003). Integrating morphological and molecular characteristics for a phylogenetic system of smut fungi. *Bot. Jb.* 24: 241–253; **Vánky, K.** (1987). Illustrated genera of smut fungi. *Cryptog. Stud.* 1: 159 pp.; **Vánky, K.** (1997). *Heterotolyposporium*, a new genus of Ustilaginales. *Mycotaxon* 63: 143–154; **Vánky, K.** (1999). New smut fungi from South Africa. *Mycotaxon* 70: 17–34; **Vánky, K.** (2000). New taxa of *Ustilaginomycetes*. *Mycotaxon* 74: 343–356.

Cladochytriaceae J. Schröt. 1892
Chytridiales: Chytridiomycota

Thallus eucarpic and polycentric (typically branched with two or more centres of development), differentiated into a well-developed rhizomycelium system bearing numerous reproductive structures, typically extensive; zoosporangia inoperculate, terminal or intercalary; zoospores posteriorly uniflagellate; resting spores thick-walled, smooth or spiny, borne on the rhizoidal system, germinating to function as a prosporangium or zoosporangium.

Significant Genera: *Cladochytrium*, *Nowakowskiella*.

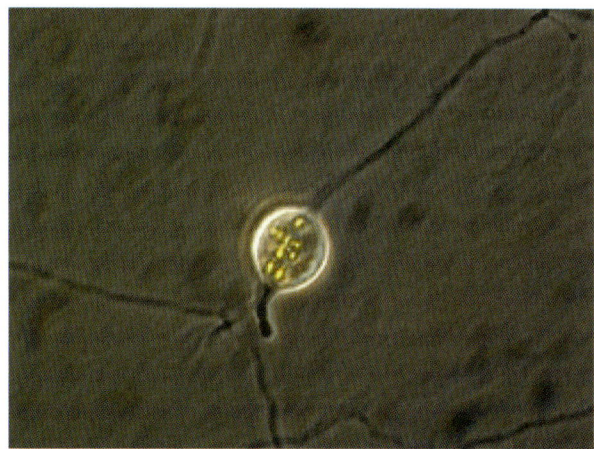

Cladochytrium replicatum; thallus

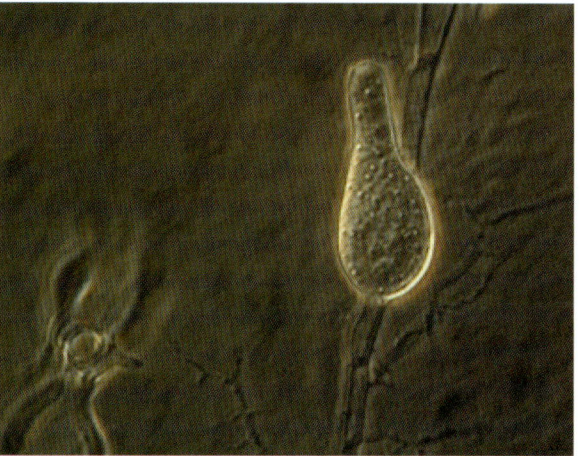

Nowakowskiella sp.; thallus

Economic Significance: None is known.

Ecology: Aquatic; one species found in *Elodea* ('Canadian pondweed'), others in ascidians (sea squirts), some are keratinophilic.

Notes: Very poorly known; no recent information is available.

Distribution: Widespread but clearly under-recorded.

References: **Blackwell, W.H.; Letcher, P.M.; Powell, M.J.** (2004). Synopsis and systematic reconsideration of *Karlingiomyces* (*Chytridiomycota*). *Mycotaxon* 89: 259–276; **Longcore, J.E.** (1993). Morphology and zoospore ultrastructure of *Lacustromyces hiemalis* gen. et sp. nov. (*Chytridiales*). *Can. J. Bot.* 71: 414–425; **Sparrow, F.K.** (1960). *Aquatic Phycomycetes* Edn 2 (Ann Arbor): 1187 pp..

Cladoniaceae Zenker 1827
Lecanorales: Ascomycota

Thallus usually dimorphic, composed of an evanescent to persistent, crustose, foliose or squamulose primary thallus lacking rhizoids and a ± vertical, usually branched, solid or hollow, secondary thallus which bears the ascomata. Soredia sometimes present. Ascomata apothecial, entire to lobate, without a fully differentiated margin, often brightly coloured, sometimes proliferating. Interascal tissue of sparsely branched and anastomosed paraphyses. Asci cylindric-clavate, thickened at the apex, with a J+ apical cap and usually a J+ gelatinous outer layer. Ascospores usually aseptate and hyaline. Cephalodia sometimes present. Anamorphs coelomycetous, pycnidial, with simple ± filiform spermatial conidia. Lichenized with green algae.

Significant Genera: *Cladonia*, *Pilophorus*.

Distribution: Cosmopolitan, prominent in all climatic zones but especially in boreal and austral regions.

Economic Significance: There is a lucrative export trade in some species for decorative purposes, with markets measured in thousands of kg. Many species produce a range of distinctive bioactive compounds and feature as traditional medicines in many parts of the world.

Cladonia pyxidata, Surrey, UK; habit

Cladonia rangiferina, Bhutan; habit

Cladonia macilenta, Surrey, UK; habit

Cladonia chlorophaea, Surrey, UK; habit

Ecology: On soil, bark or wood, mainly on acid substrata. *Cladonia* species form a major component of the biomass in taiga and tundra ecosystems and a large part of the diet of large mammals.

Notes: The family is dominated by *Cladonia*, which has around 400 species. The circumscription of that genus and its constituent sections has been the subject of recent debate.

References: **Ahti, T.** (1998). The lichen family *Cladoniaceae* in the neotropics. *Lichenology in Latin America. History, Current Knowledge and Applications, [Proceedings of GLAL-3, Terceiro Encontro do Grupo Latino-Americano de Liquenólogos, São Paulo, Brazil, 24-28 September, 1997]* (São Paulo): 109–115; **Burgaz, A.R.; Ahti, T.** (1994). Contribution to the study of the genera *Cladina* and *Cladonia* in Spain. II. *Nova Hedwigia* 59: 399–440; **DePriest, P.T.** (1995). Phylogenetic analyses of the variable ribosomal DNA of the *Cladonia chlorophaea* complex. *Cryptog. bot.* 5: 60–70; **Galloway, D.J.; James, P.W.** (1987). *Metus*, a new austral lichen genus and notes on an Australasian species of *Pycnothelia*. *Notes R. bot. Gdn Edinb.* 44: 561–579; **Goward, T.; Ahti, T.** (1997). Notes on the distributional ecology of the *Cladoniaceae* (Lichenized *Ascomycetes*) in temperate and boreal western North America. *J. Hattori bot. Lab.* 82: 143–155; **Hammer, S.** (2000). Meristem growth dynamics and branching patterns in the *Cladoniaceae*. *Am. J. Bot.* 87: 33–47; **Stenroos, S.** (1998). Configuration and location of pycnidia in the lichen genus *Cladonia* section *Perviae*. *Nova Hedwigia* 66: 457–462; **Stenroos, S.; Ahti, T.; Hyvönen, J.** (1997). Phylogenetic analysis of the genera *Cladonia* and *Cladina* (*Cladoniaceae*, lichenised *Ascomycota*). *Pl. Syst. Evol.* 207: 43–58; **Stenroos, S.K.; DePriest, P.T.** (1998). SSU rDNA phylogeny of cladoniiform lichens. *Am. J. Bot.* 85: 1548–1559; **Wedin, M.; Döring, H.; Ekman, S.** (2000). Molecular phylogeny of the lichen families *Cladoniaceae*, *Sphaerophoraceae*, and *Stereocaulaceae* (*Lecanorales*, *Ascomycotina*). *Lichenologist* 32: 171–187.

Clavariaceae Chevall. 1826
Agaricales: Basidiomycota

Basidioma cylindrical to clavate, simple, fasciculate or branched, ± erect, fleshy to waxy, rarely ± fibrous, often brightly coloured with carotenoid pigments, the hymenium covering the entire upper surface. Hyphal system monomitic, hyphae usually hyaline and inflated, clamp connections present or absent. Cystidia usually absent. Basidia 2- to 4- (6-) spored, usually short, with horizontally or vertically arranged spindle pole bodies, sometimes with clamp connections, then sometimes appearing bifurcate at the base due to their fracture. Basidiospores smooth or echinulate, thin-walled, sometimes with a prominent hilar apiculus, sometimes guttulate, not staining in iodine.

Significant Genera: *Clavaria*, *Clavulinopsis*, *Ramariopsis*.

Distribution: Widespread, best known from the north temperate regions.

Economic Significance: None is known.

Clavulinopsis corniculata, Cornwall, UK; basidiomata

Ecology: Most species are presumed saprobic on the ground or rotten wood, but many *Multiclavula* species are lichenized with green algae.

Clavulinopsis helvola, Surrey, UK; basidiomata

Notes: The *Clavariaceae* at one time included all basidiomycete fungi with cylindrical or club-shaped fruit bodies, but these have been demonstrated to have multiple evolutionary origins and the family is now recognized with a much restricted circumscription. Even so it remains polyphyletic; *Macrotyphula* seems to belong to a separate subgroup of the euagaric clade, and *Clavaria* itself needs revision. *Mucronella* may belong in the family.

Clavaria zollingeri; basidiomata

References: **Corner, E.J.H.** (1950). A monograph of *Clavaria* and allied genera. *Ann. Bot. Mem.* 1: 1–740; **Corner, E.J.H.** (1970). Supplement to a monograph of *Clavaria* and allied genera. *Beih. Nova Hedwigia* 33: 1–299; **Dentinger, B.T.M.; McLaughlin, D.J.** (2006). Reconstructing the *Clavariaceae* using nuclear large subunit rDNA sequences and a new genus segregated from *Clavaria*. *Mycologia* 98: 746–762; **Garcia-Sandoval, R.; Cifuentes, J.; Luna, E. de; Estrada-Torres, A.; Villegas, M.** (2006). A phylogeny of *Ramariopsis* and allied taxa. *Mycotaxon* 94: 265–292; **Pegler, D.N; Young, T.W.K.** (1985). Basidiospore structure in *Rumariopsis* (*Clavariaceae*). *Trans. Br. mycol. Soc.* 84: 207–214; **Petersen, R.H.** (1988). The clavarioid fungi of New Zealand. *DSIR Bulletin* 236: 170 pp.; **Petersen, R.H.** (1989). Some clavarioid fungi from northern China. *Mycosystema* 2: 159–174; **Pine, E.M.; Hibbett, D.S.; Donoghue, M.J.** (1999). Phylogenetic relationships of cantharelloid and clavarioid *Homobasidiomycetes* based on mitochondrial and nuclear rDNA sequences. *Mycologia* 91: 944–963; **Rodríguez-Armas, J.L.; Beltrán Tejera, E.; Bañares Baudet, A.** (1992). Contribucion al estudio de *Clavariaceae* y familias affines (*Aphyllophorales*) de las Islas Canarias. *Docums Mycol.* mém. hors sér. 22: 21–38; **Thind, K.S.; Sharda, R.M.** (1987). The genera *Clavulinopsis* and *Ramariopsis* in the Eastern Himalayas. *Kavaka* 14: 9–16; **Villarreal, L.; Pérez-Moreno, J.** (1991). The clavarioid fungi from Mexico, I.; Addition of the genera *Macrotyphula* and *Typhula*. *Micol. Neotrop. Aplic.* 4: 119–126.

Clavicipitaceae O.E. Erikss. 1982
Hypocreales: Ascomycota

Stromata conspicuous, usually brightly coloured, fleshy and elongate, with a distinct stalk and fertile head, sometimes developing from a sclerotium [ergot]. Ascomata sometimes superficial but more usually immersed in stromatic tissue, perithecial, usually brightly coloured. Interascal tissue of apical paraphyses. Asci cylindrical, elongate, with a prominent apical cap with a narrow pore. Ascospores hyaline, filiform, multiseptate, usually fragmenting at the septa, without gelatinous appendages or sheath. Anamorphs hyphomycetous or coelomycetous, varied in form, with conidial development often in or on stromata.

Significant Genera: *Balansia, Claviceps, Cordyceps, Epichloë, Hypocrella*. Anamorphs: *Aschersonia, Beauveria, Ephelis, Hirsutella, Metarhizium, Neotyphodium, Sphacelia, Tolypocladium.*

Cordyceps sinensis, Bhutan; stroma

Distribution: Widespread but especially prominent in wet tropical regions.

Gibellula pulchra, Thailand; stroma with perithecial ascomata and conidiophores

Economic Significance: Many metabolically active compounds are produced, including ergot alkaloids of notable medical significance and the psychotropic drug LSD. Cyclosporin, used in transplant surgery, is derived from *Tolypocladium inflatum*, the anamorph of *Cordyceps subsessilis*. The lolitrem and pyrrolizidine alkaloids from grass endophytes are toxic to cattle and sheep. *Cordyceps sinensis* is highly prized in Eastern traditional medicine and the sustainability of its harvest is in doubt. Species of *Beauveria* and *Metarhizium* are widely used in biological control of insect pests.

Metarhizium cylindrosporae Kenya; on cicada

Ecology: Parasitic, endophytic or epiphytic on Insecta, other fungi or *Gramineae*. There has been much research on the potential symbiotic relationship between the grass endophytes and their hosts, with suggested ecological benefits including resistance to herbivory and disease in addition to increased drought tolerance.

Notes: This is a well-circumscribed family within the *Hypocreales*, though it was at one time allied with the *Hypomycetaceae* in a separate order.

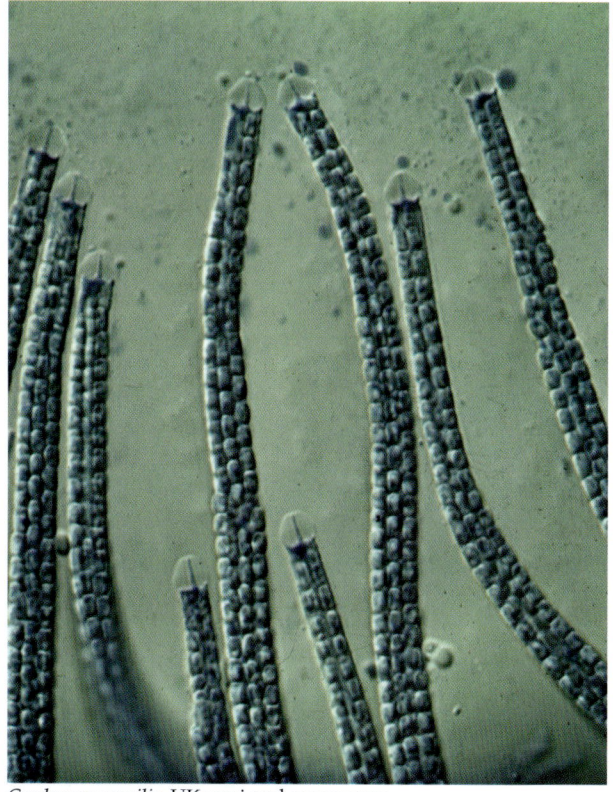

Cordyceps gracilis, UK; asci and ascospores

References: **Bacon, C.W.; White, J.F.** (2000). *Microbial Endophytes* (New York): 600 pp.; **Clay, K.** (1988). Fungal endophytes of grasses: a defensive mutualism between plants and fungi. *Ecology* 69: 10–16; **Glenn, A.E.; Bacon, C.W.** (1997). Distribution of ergot alkaloids within the family *Clavicipitaceae*. *Neotyphodium/Grass Interactions* Proceedings of the Third International Symposium on Acremonium/Grass Interactions, held May 28–31, 1997, in Athens, Georgia (New York): 53–56; **Glenn, A.E.; Bacon, C.W.; Price, R.; Hanlin, R.T.** (1996). Molecular phylogeny of *Acremonium* and its taxonomic implications. *Mycologia* 88: 369–383; **Hodge, K.T.; Krasnoff, S.B.; Humber, R.A.** (1996). *Tolypocladium inflatum* is the anamorph of *Cordyceps subsessilis*. *Mycologia* 88: 715–719; **Hywel-Jones, N.L.; Evans, H.C.** (1993). Taxonomy and ecology of *Hypocrella discoidea* and its anamorph, *Aschersonia samoensis*. *Mycol. Res.* 97: 871–876; **Kobayasi, Y.; Shimizu, D.** (1983). *[Iconography of Vegetable Wasps and Plant Worms]* (Osaka): 280 pp.; **Pažoutová, S.; Kolařík, M.; Kolínská, R.** (2004). Pleomorphic conidiation in *Claviceps*. *Mycol. Res.* 108: 126–135; **Rehner, S.A.; Buckley, E.** (2005). A *Beauveria* phylogeny inferred from nuclear ITS and EFI-α sequences: evidence for cryptic diversification and links to *Cordyceps* teleomorphs. *Mycologia* 97: 84–98; **Samson, R.A.; Evans, H.C.; Latgé, J.-P.** (1988). *Atlas of Entomopathogenic Fungi* (Berlin): 187 pp.; **Spatafora, J.W.; Blackwell, M.** (1993). Molecular systematics of unitunicate perithecial ascomycetes: the *Clavicipitales-Hypocreales* connection. *Mycologia* 85: 912–922; **Spatafora, J.W.; Sung, G.-H.; Sung, J.-M.; Hywel-Jones, N.L.; White, J.F.** (2007). Phylogenetic evidence for an animal pathogen origin of ergot and the grass endophytes. *Molecular Ecology* 16: [in press]; **Stensrud, Ø.; Hywel-Jones, N.L.; Schumacher, T.** (2005). Towards a phylogenetic classification of *Cordyceps*: ITS nrDNA sequence data confirm divergent lineages and paraphyly. *Mycol. Res.* 109: 41–56; **White, J.F.; Reddy, P.V.; Bacon, C.W.** (2000). Biotrophic endophytes of grasses: a systematic appraisal. *Microbial Endophytes* (New York): 49–62; **White, J.F.; Sullivan, R.; Moy, M.; Patel, R.; Duncan, R.** (2000). An overview of problems in the classification of plant-parasitic *Clavicipitaceae*. *Stud. Mycol.* 45: 95–105.

Clavulinaceae Donk 1970
Cantharellales: Basidiomycota

Basidiomata resupinate or club-shaped to cylindrical and often strongly branched, often furrowed, pale to brownish or pinkish, membranous to ± coriaceous, the hymenium covering much of the outer surface of the fruit body. Hyphal system monomitic, hyphae hyaline, thin-walled, inflated, clamp connections present or absent. Cystidia present or absent. Basidia elongate, sometimes sinuous, with 2 to 4 strongly curved sterigmata. Basidiospores ellipsoidal to subglobose, smooth, thin-walled, sometimes guttulate, not staining in iodine.

Significant Genera: *Clavulina*.

Distribution: Widespread, especially in temperate zones.

Economic Significance: None is known.

Ecology: Some species form ectomycorrhizas, with fruit bodies in leaf litter and on soil. Others are probably saprobes, associated with rotten wood.

Notes: Superficially similar to the *Clavariaceae*, but basidia are elongate with well-developed sterigmata and molecular studies link the family to the cantharelloid clade of the homobasidiomycetes.

Clavulina coralloides, Inverness-shire, Scotland; basidiomata

References: **Corner, E.J.H.** (1950). A monograph of *Clavaria* and allied genera. *Ann. Bot. Mem.* 1: 1–740; **Corner, E.J.H.** (1986). The genus *Clavulina* (*Basidiomycetes*) in south-eastern Australia. *Aust. J. Bot.* 34: 103–105; **Henkel, T.W.; Meszaros, R.; Aime, M.C.; Kennedy, A.** (2005). New *Clavulina* species from the Pakaraima Mountains of Guyana. *Mycol. Prog.* 4: 343–350; **Koide, R.T.; Xu, B.; Sharda, J.; Lekberg, Y.; Ostiguy, N.** (2005). Evidence of species interactions within an ectomycorrhizal fungal community. *New Phytol.* 165: 305–316; **Petersen, R.H.** (1985). Notes on clavarioid fungi. XX. New taxa and distributional records in *Clavulina* and *Ramaria*. *Mycologia* 77: 903–919; **Petersen, R.H.** (1988). The clavarioid fungi of New Zealand. *DSIR Bulletin* 236: 170 pp.; **Pine, E.M.; Hibbett, D.S.; Donoghue, M.J.** (1999). Phylogenetic relationships of cantharelloid and clavarioid *Homobasidiomycetes* based on mitochondrial and nuclear rDNA sequences. *Mycologia* 91: 944–963; **Villegas, M.; De Luna, E.; Cifuentes, J.; Estrada Torres, A.** (1999). Phylogenetic studies in *Gomphaceae sensu lato* (*Basidiomycetes*). *Mycotaxon* 70: 127–147.

Clintamraceae Vánky 2001
Ustilaginales: Basidiomycota

Sori developing externally on plant surfaces, forming a blackish-brown granular spore mass. Hyphae intracellular, with an electron-opaque coating, mature septa without pores. Teliospores formed singly or in small groups, irregular in shape, thick-walled, faintly verrucose, brown, germinating to produce a short bifurcate aseptate 2-spored basidium. Basidiospores apical, multiseptate.

Significant Genera: *Clintamra*.

Distribution: The only species occurs in the southern USA and Mexico.

Economic Significance: None is known.

Ecology: Biotrophic on infloresences and young leaves of *Nolina* (*Nolinaceae*).

Notes: No molecular data are available. The *Clintamraceae* is a recent segregate from the *Ustilaginaceae*, distinguished especially by the bifurcate basidia.

References: **Cordas, D.I.; Durán, R.** (1977). *Clintamra*, a new smut genus. *Mycologia* 68: 1239–1245; **Durán, R.** (1987). *Ustilaginales of Mexico. Taxonomy, Symptomatology, Spore Germination and Basidial Cytology* (Pullman): 331 pp.; **Vánky, K.** (1987). Illustrated genera of smut fungi. *Cryptog. Stud.* 1: 159 pp.; **Vánky, K.** (1991). Spore morphology in the taxonomy of *Ustilaginales*. *Trans. Mycol. Soc. Japan* 32: 381–400; **Vánky, K.** (1998). A survey of the sporeball-forming smut fungi. *Mycol. Res.* 102: 513–526; **Vánky, K.** (2001). The emended *Ustilaginaceae* of the modern classificatory system for smut fungi. *Fungal Diversity* 6: 131–147.

Clypeosphaeriaceae G. Winter 1886
Xylariales: Ascomycota

Stromatal tissues poorly developed, often clypeate, composed of a mixture of host and fungal cells. Ascomata perithecial, immersed or erumpent, sometimes aggregated, usually thick-walled, the ostiolar region thick-walled, widely papillate, periphysate; peridium dark, composed of flattened cells. Interascal tissue of numerous thin-walled gelatinized paraphyses. Asci cylindrical, persistent, short-stalked, fairly thin-walled, not fissitunicate, with a simple usually J– apical ring. Ascospores hyaline or brown, transversely septate, sometimes with a germ pore, sometimes with a mucous sheath. Anamorphs not known.

Clypeosphaeria mamillana, Argyll, Scotland; habit

Significant Genera: *Clypeosphaeria*.

Distribution: Widespread.

Economic Significance: None is known.

Ecology: Saprobic in woody stems and similar substrata.

Notes: Probably polyphyletic; more molecular data are required.

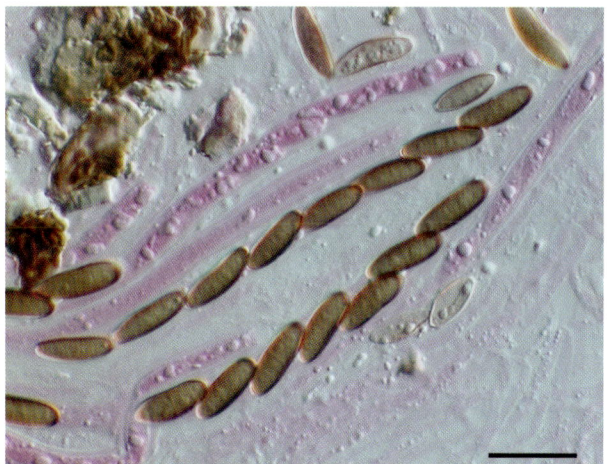

Clypeosphaeria mamillana, Kent, UK; asci and ascospores

References: **Barr, M.E.** (1989). *Clypeosphaeria* and the *Clypeosphaeriaceae*. *Syst. Ascom.* 8: 1–8; **Hyde, K.D.; Fröhlich, J.; Taylor, J.E.** (1998). Fungi from palms. XXXVI. Reflections on unitunicate ascomycetes with apiospores. *Sydowia* 50: 21–80; **Jeewon, R.; Liew, E.C.; Hyde, K.D.** (2003). Molecular systematics of the *Amphisphaeriaceae* based on cladistic analyses of partial LSU rDNA gene sequences. *Mycol. Res.* 107: 1392–1402; **Kang, J.C.; Hyde, K.D.; Kong, R.Y.C.** (1999). Studies on the *Amphisphaeriales* I. The *Clypeosphaeriaceae*. *Mycoscience* 40: 151–164.

Coccocarpiaceae Henssen 1986
Lecanorales: Ascomycota

Thallus varied but most often foliose, well-developed, loosely attached. Ascomata apothecial, formed from the aggregation of short cells derived from adjacent thalline hyphae, flat to strongly convex, typically dark or concolorous with the thallus; hymenium not gelatinous. Interascal tissue of sparsely branched paraphyses, sometimes pigmented at the apices or with a well-developed epithecium. Asci with a well-developed apical cap, either wholly J+ or with a ring structure, usually with a J+ gelatinized outer layer. Ascospores simple or transversely septate. Anamorphs coelomycetous, pycnidial.

Significant Genera: *Coccocarpia*.

Distribution: Very widespread, including boreal and austral regions.

Economic Significance: None is known.

Ecology: Lichenized with cyanobacteria.

Notes: A poorly known family, probably most closely related to the *Collemataceae* and *Pannariaceae*. The association with cyanobacteria seems to have some phylogenetic significance.

Coccocarpia albida, Martinique; thallus

Coccocarpia stellata; thallus and ascomata

References: **Awasthi, D.D.** (1987). Lichen genus *Coccocarpia* from India. *Kavaka* 13: 83–86; **Ekman, S.; Joergensen, P.M.** (2002). Towards a molecular phylogeny for the lichen family *Pannariaceae* (*Lecanorales, Ascomycota*). *Can. J. Bot.* 80: 625–634; **Henssen, A.; Tønsberg, T.** (2000). *Spilonemella*, a new genus of cyanophilic lichens with species from North America and Japan (*Coccocarpiaceae*). *Bryologist* 103: 108–116; **Lumbsch, H.T.; Kothe, H.W.** (1992). Thallus surfaces in *Coccocarpiaceae* and *Pannariaceae* (lichenized *Ascomycetes*) viewed with scanning electron microscopy. *Mycotaxon* 43: 277–282; **Makhija, U.; Adawadkar, B.; Patwardhan, P.G.** (1999). The lichen genus *Coccocarpia* from the Andaman and Nicobar Islands, India. *Trop. Bryol.* 17: 47–55; **Marcano, V.; Morales Méndez, A.; Mohali, S.; Galiz, L.; Palacios-Prü, E.** (1995). El genero *Coccocarpia* Pers. (Ascomicetes liquenizados) en Venezuela. *Trop. Bryol.* 10: 215–227; **Wagner, B.** (1995). *Spilonema paradoxum* a *Thermutis velutina* – dvě překvapení v Herbáři PRC. *Bryonora* 15: 8–9; **Wedin, M.; Wiklund, E.** (2004). The phylogenetic relationships of *Lecanorales* suborder *Peltigerineae* revisited. *Symb. bot. upsal.* 34: 469–475.

Coccodiniaceae Höhn. ex O.E. Erikss. 1981
Capnodiales: Ascomycota

Mycelium at least partially superficial, with brown hyphae. Ascomata ± globose but frequently collapsing when dry, thin-walled, sometimes setose, with a periphysate ostiole. Interascal tissue of short periphysoids. Asci saccate, fissitunicate, with a J+ outer gelatinous layer. Ascospores hyaline or brown, usually muriform. Anamorphs hyphomycetous, often poorly developed.

Significant Genera: *Dennisiella*.

Distribution: Widespread.

Economic Significance: None is known.

Ecology: Epiphytic or biotrophic on leaves and stems.

Notes: Molecular data are sparse, but placement within the *Chaetothyriomycetidae* along with the *Capnodiaceae* seems pragmatic.

References: **Liu, Y.J.; Hall, B.D.** (2004). Body plan evolution of ascomycetes, as inferred from an RNA polymerase II phylogeny. *Proc. natn Acad. Sci. U.S.A.* 101: 4507–4512; **Reynolds, D.R.** (1986). Foliicolous ascomycetes 7. Phylogenetic systematics of the *Capnodiaceae*. *Mycotaxon* 27: 377–403; **Winka, K.** (2000). *Phylogenetic Relationships Within the Ascomycota Based on 18S rDNA Sequences*, Akademisk Avhandling [Thesis (PhD), Department of Ecology and Environmental Science, Umeå University] (Umeå): [91] pp.; **Winka, K.; Eriksson, O.E.; Bång, Å.** (1998). Molecular evidence for recognizing the *Chaetothyriales*. *Mycologia* 90: 822–830.

Coccoidella perseae, Costa Rica; ascostromata

Notes: Very poorly known; no recent information is available.

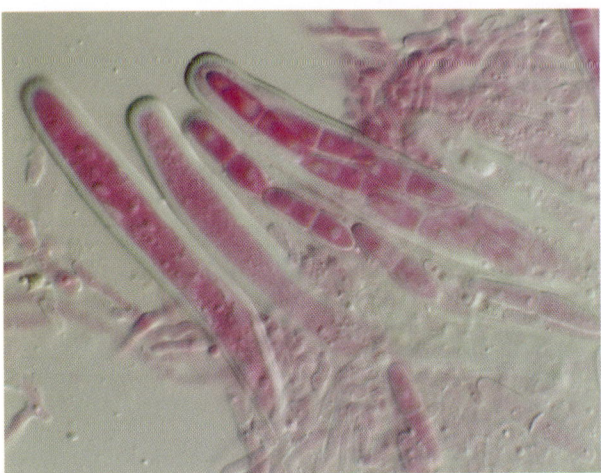
Coccoidella perseae, Costa Rica; asci and ascospores

References: **Inácio, C.A.; Cannon, P.F.** (2002). Re-interpretation of *Cocconia palmae*, with description of the genus *Dianesea* (*Ascomycota*: *Dothideomycetidae*). *Fungal Diversity* 9: 71–79; **Sivanesan, A.** (1987). *Coccoidella perseae* sp. nov. and its anamorph *Colletogloeum perseae* sp. nov. *Trans. Br. mycol. Soc.* 89: 265–270; **Yuan, Z.Q.; Mohammed, C.; Wardlaw, T.J.** (1996). *Coccoidella exocarpi* sp. nov. on *Exocarpos* spp. from Australia. *Mycotaxon* 60: 175–180.

Coccoideaceae Henn. ex Sacc. & D. Sacc. 1905
Dothideales: Ascomycota

Ascomata formed as locules in an erumpent pulvinate or peltate stroma with a short stalk, ± globose, the peridium spongy, two-layered, ostiole lysigenous, not periphysate. Interascal tissue composed of branched cellular pseudoparaphyses, sometimes evanescent; asci ± cylindrical, fissitunicate; ascospores relatively small, 1-septate, without sheath. Anamorph absent or acervular.

Significant Genera: *Coccoidella*.

Distribution: Apparently pantropical.

Economic Significance: None is known.

Ecology: Biotrophic on leaves.

Coccotremataceae Henssen ex J.C. David & D. Hawksw. 1991
Uncertain position within Ascomycetes, Ascomycota

Thallus crustose or lobate, sometimes becoming ± foliose; sometimes with cephalodia. Ascomata sessile or immersed in thalline warts, originating as a cavity within the primordium, opening by a pore, often perithecium-like. Interascal tissue of apical and basal paraphyses, the basal paraphyses sometimes evanescent; hymenium non-amyloid. Asci short, cylindrical, the wall multilayered, the apex thickened and staining blue in iodine, apparently dehiscing by an apical split. Ascospores large, aseptate, hyaline. Anamorphs pycnidial, with bacilliform conidia.

Significant Genera: *Coccotrema*.

Distribution: Widespread.

Economic Significance: None is known.

Ecology: Lichenized with green algae, on bark or rocks especially in maritime zones.

Notes: Very poorly known; no recent information is available.

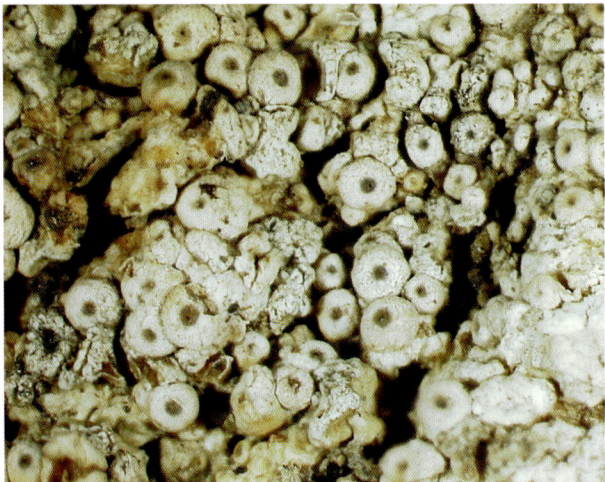

Coccotrema cucurbitula, Chile; habit

References: **Galloway, D.J.; Watson-Gandy, L.A.** (1992). *Lepolichen coccophorus* (Lichenized *Ascomycotina*, *Coccotremataceae*) in South America. *Bryologist* 95: 227–232; **Messuti, M.I.** (1996). Notes on the lichen genus *Coccotrema* in southern South America. *N.Z. Jl Bot.* 34: 57–64; **Schmitt, I.; Messuti, M.I.; Feige, G.B.; Lumbsch, H.T.** (2001). Molecular data support rejection of the generic concept in the *Coccotremataceae* (*Ascomycota*). *Lichenologist* 33: 315–321.

Cochlonemataceae Dudd. 1974
Zoopagales: Zygomycota

Thallus comprising an internal spiral hyphal filament, fertile hyphae forming outside the host, usually unbranched. Spores (conidia) simple or appendaged, either single and formed successively or in short or long chains on unbranched or branched hyphae. Zygospores minute, polyhedral to warty, borne on somewhat parallel, randomly arranged or spirally twisted suspensors.

Significant Genera: *Amoebophilus, Cochlonema*.

Distribution: Widespread but monumentally under-recorded.

Economic Significance: None is known.

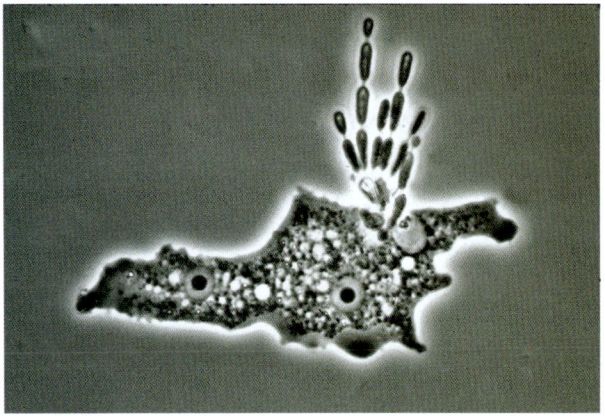

Amoebophilus sp.; fertile hyphae emerging from an amoeba

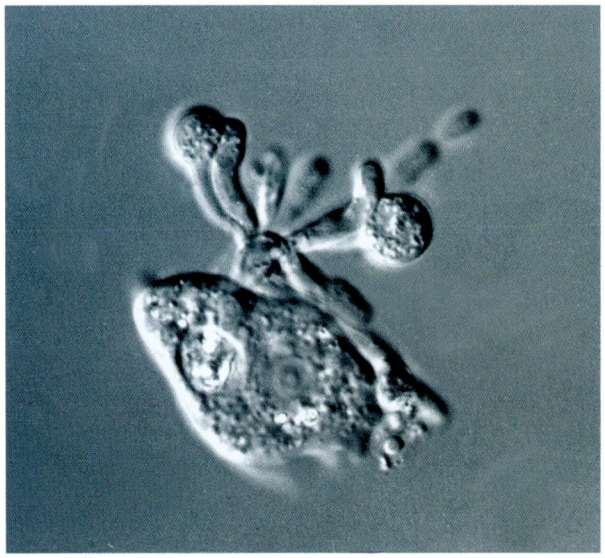

Amoebophilus sp.; zygospores (immature)

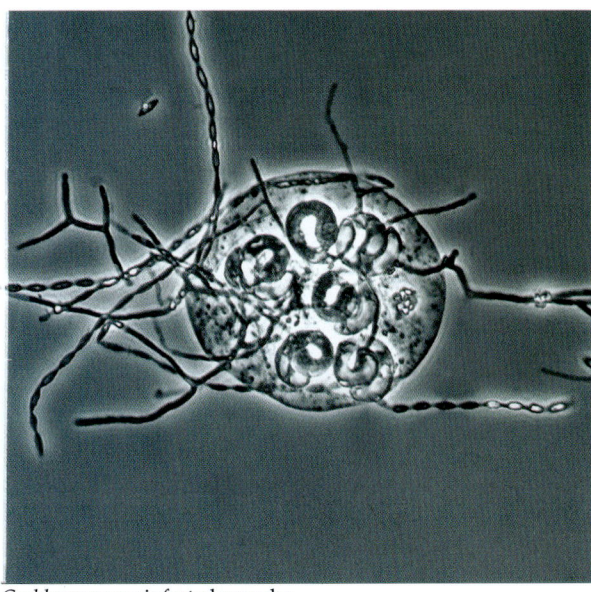

Cochlonema sp.; infected amoeba

Ecology: Ectoparasites or endoparasites of amoebae and other soil-inhabiting invertebrates.

Notes: Molecular data are lacking for this family, the order usually represented by the better-known species of *Piptocephalidaceae*.

References: **Barron, G.L.** (2004). Fungal parasites and predators of rotifers, nematodes, and other invertebrates. *Biodiversity of Fungi, Inventory and Monitoring Methods* (Amsterdam): 435–450; **Saikawa, M.; Aoki, Y.** (1995). Taxonomic studies on *Euryancale marsipospora* and *E. saccispora*. *Nova Hedwigia* 60: 571–581; **Saikawa, M.; Katsurashima, E.** (1993). Light and electron microscopy of a new species of *Euryancale* producing phallus-shaped conidia. *Mycologia* 85: 24–29; **Saikawa, M.; Sato, H.** (1986). Zygosporic structures in two species of *Euryancale* (*Zoopagales*). *Trans. Br. mycol. Soc.* 87: 337–340; **Saikawa, M.; Sato, H.** (1991). Ultrastructure of *Cochlonema odontosperma*, an endoparasite in amoebae. *Mycologia* 83: 403–408; **Tanabe, Y.; O'Donnell, K.; Saikawa, M.; Sugiyama, J.** (2000). Molecular phylogeny of parasitic *Zygomycota* (*Dimargaritales*, *Zoopagales*) based on nuclear small subunit ribosomal DNA sequences. *Mol. Phylogen. Evol.* 16: 253–262.

Coelomomycetaceae Couch ex Couch 1962
Blastocladiales: Blastocladiomycota

Sporophyte thallus in dipterans, unwalled, branched or lobed, without rhizoids, converting into a mass of thick-walled, variously ornamented, typically brownish, resting sporangia which function as zoosporangia, these dehiscing by fracture of the outer wall along a preformed groove and subsequent gelatinization and disintegration of the exposed surface of the extruded inner wall; zoospores posteriorly uniflagellate, with a nuclear cap; gametophyte thallus in microcrustaceans.

Significant Genera: *Coelomomyces*.

Distribution: Widespread, all continents except, apparently, Antarctica.

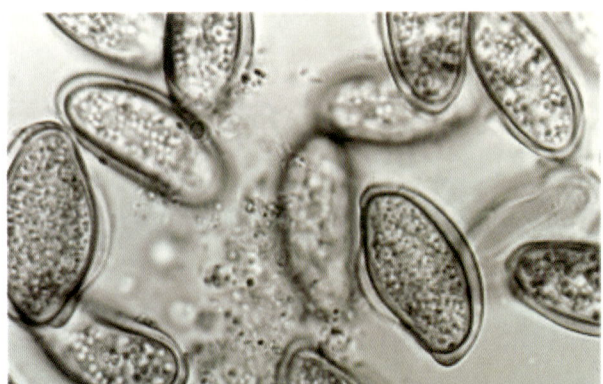

Coelomomyces sp.; resting spores

Economic Significance: There has been much research on species of *Coelomomyces* as potential biological control agents for aquatic insect larvae of, for example, mosquitoes. The complexity of the life cycle and the inability to produce large volumes of inoculum has not yet resulted in the development of sucessful applications.

Ecology: Aquatic.

Notes: Very poorly known; little recent information is available.

References: **Couch, J.N.; Bland, C.E. (eds)** (1985). *The Genus Coelomomyces* (Orlando): 399 pp.; **Scholte, E.J.; Knols, B.G.; Samson, R.A.; Takken W.** (2004). Entomopathogenic fungi for mosquito control: a review. *J. Insect Sci.* 4: 1943; **Whisler, H.C.; Zebold, S.L.; Shemanchuk, J.A.** (1975). Life history of *Coelomomyces psorophorae*. *Proc. natn Acad. Sci. U.S.A.* 72: 693–696.

Coenogoniaceae (Fr.) Stizenb. 1862
Ostropales: Ascomycota

Thallus variably developed, filamentous, felt-like or crustose, rarely dimidiate or beard-like, grey, green or yellowish. Ascomata apothecial, discoid to urceolate, yellow to brownish, sometimes with a minutely lobed or pilose margin that sometimes incorporates algal elements. Interascal tissue composed of simple paraphyses, often with swollen apices. Asci cylindrical, thin-walled, the wall staining blue in iodine, without a thickened apex or apical structures, 8-spored. Ascospores ± ellipsoidal, hyaline, septate or aseptate, without a sheath or appendages. Conidiomata pycnidial, tubular, clavate, globose or flattened, sometimes with complex chambers, conidia cylindrical, usually aseptate.

Significant Genera: *Coenogonium*.

Dimerella lutea, Ireland; ascomata

Distribution: Very widespread in both Old and New Worlds, especially in tropical zones.

Dimerella pineti, Louisiana, USA; ascomata

Economic Significance: None is known.

Ecology: Lichenized with green algae, on bark, leaves, overgrowing mosses etc..

Notes: *Coenogonium* is now accepted to include the perhaps better-known genus *Dimerella*, that was distinguished primarily by its crustose rather than filamentous thallus. The family is separated from the *Gyalectaceae* primarily based on molecular data, but ascoma ontogeny may apparently also be used to separate the two taxa.

References: **Davis, J.S.** (1994). *Coenogonium missouriense*, a new lichen species from Missouri. *Bryologist* 97: 186–189; **Kauff, F.; Lutzoni, F.** (2002). Phylogeny of the *Gyalectales* and *Ostropales* (*Ascomycota, Fungi*): among and within order relationships based on nuclear ribosomal RNA small and large subunits. *Mol. Phylogen. Evol.* 25: 138–156; **Lücking, R.; Kalb, K.** (2000). Foliikole Flechten aus Brasilien (vornehmlich Amazonien), inklusive einer Checkliste und Bemerkungen zu *Coeonomgonium* und *Dimerella* (*Gyalectaceae*). *Bot. Jb.* 122: 1–61; **Rivas Plata, E.; Lücking, R.; Aptroot, A.; Sipman, H.J.M.; Chaves, J.L.; Umaña, L.; Lizano, D.** (2006). A first assessment of the Ticolichen biodiversity inventory in Costa Rica: the genus *Coenogonium* (*Ostropales: Coenogoniaceae*), with a world-wide key and checklist and a phenotype-based cladistic analysis. *Fungal Diversity* 23: 255–321; **Stocker-Worgötter, E.** (1997). Investigations on the photobiont and resynthesis of the tropical lichen *Coenogonium leprieurii* (Mont.) Nyl. From the NE coast of Brazil in culture. *Symbiosis* 23: 117–124.

Coleosporiaceae Dietel 1900
Uredinales: Basidiomycota

Spermogonia discoid to conical, without a discrete wall, usually subepidermal. Aecia *Peridermium*-like, with a strongly developed peridium. Aeciospores in chains, verrucose. Uredinia *Caeoma*-like. Urediniospores in chains, verrucose or echinulate, germ pores poorly defined and usually scattered. Telia erumpent, waxy or gelatinous, pulvinate or columnar. Teliospores unicellular, sessile, in a single layer or in chains, thin-walled, germ pores absent, germinating to form basidia within teliospores followed by development of sterigmata, or directly externally to produce ellipsoidal or globose basidiospores.

Significant Genera: *Chrysomyxa, Coleosporium*.

Distribution: Widespread.

Economic Significance: Some economic damage in conifer plantations results, especially in seed orchards, though species are generally not as noxious as in the related *Cronartiaceae*.

Ecology: Mostly heteroecious, biotrophic on various plant parts, the aecial stages typically developing on needles, stems or cones of conifers (primarily *Picea* species), sometimes stimulating witches' brooms. The *Ericaceae* and *Asteraceae* (*Compositae*) are typical alternate host families.

Notes: Preliminary molecular data indicate that the family may be polyphyletic, and the divisions between it and the *Cronartiaceae* and *Pucciniastraceae* need further research.

Chrysomyxa ledi, Alberta, Canada; uredinium

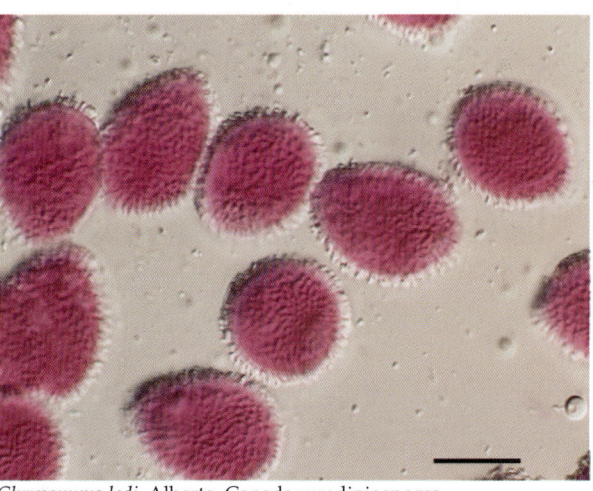

Chrysomyxa ledi, Alberta, Canada; urediniospores

References: Aime, M.C. (2006). Towards resolving family-level relationships in rust fungi (*Uredinales*). *Mycoscience* 47: 112–122; **Berndt, R.** (1999). *Chrysomyxa* rust: morphology and ultrastructure of D-haustoria, uredinia, and telia. *Can. J. Bot.* 77: 1469–1484; **Crane, P.E.** (2001). Morphology, taxonomy, and nomenclature of the *Chrysomyxa ledi* complex and related rust fungi on spruce and *Ericaceae* in North America and Europe. *Can. J. Bot.* 79: 957–982; **Cummins, G.B.; Hiratsuka, Y.** (2003). *Illustrated Genera of Rust Fungi* (St Paul): 225 pp.; **Hernández, J.R.; Hennen, J.F.** (2003). Rust fungi causing galls, witches' brooms, and other abnormal plant growths in northwestern Argentina. *Mycologia* 95: 728–755; **Maier, W.; Begerow, D.; Weiss, M.; Oberwinkler, F.** (2003). Phylogeny of the rust fungi: an approach using nuclear large subunit ribosomal DNA sequences. *Can. J. Bot.* 81: 12–23; **Ono, Y.** (2002). The diversity of nuclear cycle in microcyclic rust fungi (*Uredinales*) and its ecological and evolutionary implications. *Mycoscience* 43: 421–439; **Wingfield, B.D.; Ericson, L.; Szaro, T.; Burdon, J.J.** (2004). Phylogenetic patterns in the *Uredinales*. *Australas. Pl. Path.* 33: 327–335.

Collemataceae Zenker 1827
Peltigerales: Ascomycota

Thallus usually foliose to fruticose, dark grey, brown or green to black, gelatinized, corticate or not, often with isidia. Ascomata sessile, apothecial, usually strongly concave, the margin varied in form. Interascal tissue of simple or branched paraphyses, sometimes pigmented, immersed in a J+ gelatinous matrix. Asci clavate, with a well-developed J+ apical cap, often with a distinct more heavily staining apical ring. Ascospores varied, often elongate, hyaline, thin-walled, without a sheath. Anamorph pycnidial, with small ± bacilliform or cylindrical conidia.

Significant Genera: *Collema*, *Leptogium*.

Collema coccophorum, France; habit

Leptogium saturninum, Alberta, Canada; habit

Collema cristatum, Var, France; habit

Leptogium burgessii; ascomata

Distribution: Widespread.

Economic Significance: None is known.

Ecology: Lichenized with cyanobacteria, on rocks, soil and bark, in habitats ranging from xeric to ± inundated.

Notes: The *Collemataceae* represents a very well-defined major clade within the *Lecanoromycetidae*, and appears to be most closely allied to the *Placynthiaceae* and *Pannariaceae*, and should be classified within the *Peltigerales* rather than the *Lecanorales* as in the *Dictionary* edn 9.

References: **Degelius, G.** (1994). Studies in the lichen family *Collemataceae* VII. Two new tropical *Collema* species. *Nordic Jl Bot.* 14: 229–233; **Galloway, D.J.** (1999). Notes on the lichen genus *Leptogium* (*Collemataceae*, *Ascomycota*) in New Zealand. *Nova Hedwigia* 69: 317–355; **Galloway, D.J.; Jørgensen, P.M.** (1995). The lichen genus *Leptogium* (*Collemataceae*) in southern Chile, South America. *Flechten Follmann, Contributions to Lichenology in Honour of Gerhard Follmann* (Cologne): 227–247; **Galloway, D.J.; Knight, A.** (1999). *Leptogium australe* (*Collemataceae*), new to New Zealand. *Lichenologist* 31: 642–646; **Guttová, A.** (2000). Three *Leptogium* species new to central Europe. *Lichenologist* 32: 291–293; **Jørgensen, P.M.; Henssen, A.** (1993). *Physma omphalarioides* – its taxonomic position and phytogeography. *Graphis Scripta* 5: 12–17; **Jørgensen, P.M.; Tønsberg, T.** (1999). Notes on small species of *Leptogium* from Pacific North America. *Bryologist* 102: 412–417; **Miadlikowska, J.; Lutzoni, F.** (2004). Phylogenetic classification of peltigeralean fungi (*Peltigerales*, *Ascomycota*) based on ribosomal RNA small and large subunits. *Am. J. Bot.* 91: 449–464; **Upreti, D.K.; Singh, A.** (1988). Lichen family *Collemataceae* from Andaman Islands, India. *J. Bombay nat. Hist. Soc.* 85: 234–237; **Verdon, D.; Elix, J.A.** (1994). A new species and new records of *Physma* from Australasia. *Acta bot. fenn.* 150: 209–215; **Wiklund, E.; Wedin, M.** (2003). The phylogenetic relationships of the cyanobacterial lichens in the *Lecanorales* suborder *Peltigerineae*. *Cladistics* 19: 419–431..

Completoriaceae Humber 1989
Entomophthorales: Zygomycota

Vegetative structures small, irregular hyphal bodies probably lacking cell walls. Conidiophores short, unbranched, arising directly from vegetative cells, without a conidiogenous cell. Conidia unitunicate, more or less globose, with a rounded papilla; release by papillar eversion. Resting spores globose (ontogeny and germination unknown), formed in the axis of parental cell.

Significant Genera: *Completoria*.

Distribution: Widespread in north temperate regions (USA and central Europe) but very rarely reported; apparently no record since 1895.

Economic Significance: None is known.

Ecology: Obligate intercellular parasites of fern gametophytes (prothalli).

Notes: A monospecific family, the single species occurring in the prothalli of *Pteridophyta*.

References: **Atkinson, G.F.** (1894). *Completoria complens*. *Bot. Gaz.* 19: 467–468; **Humber, R.A.** (1989). Synopsis of a revised classification for the *Entomophthorales* (*Zygomycotina*). *Mycotaxon* 34: 441–460.

Coniochaetaceae Malloch & Cain 1971
Coniochaetales: Ascomycota

Stromata absent. Ascomata usually perithecial, solitary or aggregated, sometimes on a poorly-developed subiculum; peridium occasionally organized into plate-like structures, but not strictly cephalothecoid. Interascal tissue inconspicuous, of simple paraphyses. Asci usually cylindrical, rather thin-walled, with a small undifferentiated apical ring, sometimes J+ blue, sometimes polysporous. Ascospores aseptate, usually dark brown and with a germ slit, rarely ornamented, sheath lacking. Anamorphs hyphomycetous, in most species *Acremonium*-like or *Phialophora*-like.

Significant Genera: *Coniochaeta*.

Distribution: Widespread, in all climatic zones.

Economic Significance: None is known.

Ecology: Saprobic on wood and bark (especially in dry habitats), coprophilous or isolated from soil and similar substrata.

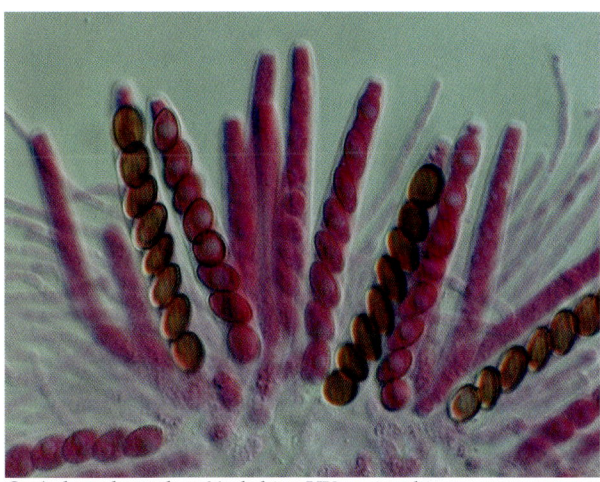

Coniochaeta leucoplaca, Yorkshire, UK; asci and ascospores

Notes: Some species previously referred to *Coniochaeta* have *Nodulisporium*-like anamorphs and have recently been shown to belong in the *Xylariales*. The family is now placed in its own order *Coniochaetales*.

References: **Checa, J.; Barrasa, J.M.; Moreno, G.; Fort, F.; Guarro, J.** (1988). The genus *Coniochaeta* (Sacc.) Cooke (*Coniochaetaceae*, *Ascomycotina*) in Spain. *Cryptog. Mycol.* 9: 1–34; **Crane, J.L.; Shearer, C.A.** (1995). A new *Coniochaeta* from fresh water. *Mycotaxon* 54: 107–110; **García, D.; Stchigel, A.M.; Cano, J.; Calduch, M.; Hawksworth, D.L.; Guarro, J.** (2006). Molecular phylogeny of *Coniochaetales*. *Mycol. Res.* 110: 1271–1289; **Guarro, J.; Gené, J.; Al-Bader, S.M.; Abdullah, S.K.** (1997). A new species of *Coniochaetidium* from soil. *Mycoscience* 38: 123–125; **Huhndorf, S.M.; Fernández, F.; Candoussau, F.** (1999). Two new species of *Synatospora*. *Sydowia* 51: 176–182; **Huhndorf, S.M.; Miller, A.N.; Fernández, F.A.** (2004). Molecular systematics of the *Sordariales*: the order and the family *Lasiosphaeriaceae* revisited. *Mycologia* 96: 368–387; **Kamiya, S.; Uchiyama, S.; Udagawa, S.** (1995). Two new species of *Coniochaeta* with a cephalothecoid peridium wall. *Mycoscience* 36: 377–383; **Lee, S.J.; Hanlin, R.T.** (1999). Phylogenetic relationships of *Chaetomium* and similar genera based on ribosomal DNA sequences. *Mycologia* 91: 434–442; **Ramaley, A.W.** (1997). *Barrina*, a new genus with polysporous asci. *Mycologia* 89: 962–966; **Romero, A.I.; Carmarán, C.C.; Lorenzo, L.E.** (1999). A new species of *Coniochaeta* with a key to the species known in Argentina. *Mycol. Res.* 103: 689–695; **Weber, E.; Görke, C.; Bege-**

row, D. (2002). The *Lecythophora-Coniochaeta* complex. II. Molecular studies based on sequences of the large subunit of ribosomal DNA. *Nova Hedwigia* 74: 187–200.

Coniocybaceae Rchb. 1837
Uncertain position within Lecanoromycetes, Ascomycota

Thallus crustose, often inconspicuous but brightly coloured in some species. Ascomata stalked, the stalk composed of parallel hyphae. Interascal tissue absent. Asci thin-walled, evanescent. Ascospores usually spherical and aseptate, hyaline or pale brown, smooth or with inconspicuous ornamentation, forming a dry mazaedial mass.

Chaenotheca chrysocephala; thallus and ascomata

Chaenotheca ferruginea; thallus and ascomata

Significant Genera: *Chaenotheca, Coniocybe*.

Distribution: Widespread.

Economic Significance: None is known.

Ecology: Mostly lichenized with green algae, on dry wood and similar substrata.

Notes: Very poorly known; no recent information is available.

References: **Honegger, R.** (1985). The hyphomycetous anamorph of *Coniocybe furfuracea*. *Lichenologist* 17: 273–279; **Honegger, R.** (1986). Ascus structure and ascospore formation in the lichen-forming *Chaenotheca chrysocephala* (*Caliciales*). *Sydowia* 38: 146–157; **Rikkinen, J.** (1998). *Chaenotheca olivaceorufa* (*Caliciales*) new to North America. *Bryologist* 101: 558–559; **Selva, S.B.; Tibell, L.** (1999). Lichenized and non-lichenized calicioid fungi from North America. *Bryologist* 102: 377–397; **Tibell, L.** (1987). Australasian Caliciales. *Symb. bot. upsal.* 27: 279 pp.; **Tibell, L.** (1996). *Caliciales*. *Fl. Neotrop. Monogr.* 69: 78 pp.; **Tibell, L.** (1997). Anamorphs in mazaediate lichenized fungi and the *Mycocaliciaceae* ('Caliciales s. lat.'). In Tibell, L. & Hedberg, I. (eds), Lichen Studies Dedicated to Rolf Santesson. *Symb. bot. upsal.* 32: 291–322; **Tibell, L.** (1998). Crustose Mazaediate Lichens and the *Mycocaliciaceae* in Temperate South America. *Biblthca Lichenol.* 71: 107 pp.; **Tibell, L.** (1999). Calicioid lichens and fungi. *Nordic Lichen Flora* 1. Introductory Parts; Calicioid Lichens and Fungi (Uddevalla): 20–94; **Tibell, L.** (2001). Photobiont association and molecular phylogeny of the lichen genus *Chaenotheca*. *Bryologist* 104: 191–198; **Tibell, L.** (2001). *Cybebe gracilenta* in an ITS/5.8S rDNA based phylogeny belongs to *Chaenotheca* (*Coniocybaceae*, lichenized ascomycetes). *Lichenologist* 33: 519–525; **Tibell, L.; Koffman, A.** (2002). *Chaenotheca nitidula*, a new species of calicioid lichen from northeastern North America. *Bryologist* 105: 353–357; **Wedin, M.; Tibell, L.** (1997). Phylogeny and evolution of *Caliciaceae, Mycocaliciaceae*, and *Sphinctrinaceae* (*Ascomycota*), with notes on the evolution of the prototunicate ascus. *Can. J. Bot.* 75: 1236–1242.

Coniophoraceae Ulbr. 1928
Boletales: Basidiomycota

Basidiomata effuse or resupinate, fleshy, membranous, pruinose or granular, often with cottony margins, producing spherical or elongate sclerotia in some species. Hyphal system usually monomitic, rarely dimitic with skeletal hyphae, hyphae hyaline to yellowish, sometimes encrusted, with normally simple septa and sometimes verticillate clamp connections. Hymenium smooth to folded, rarely ridged or spine-like, generally brown or yellowish. Cystidia present or absent. Basidia clavate-cylindrical to ± urniform, with a basal clamp connection and 4 short sterigmata. Basidiospores yellowish-brown, smooth, thick-walled, with a germ pore, staining brown in iodine.

Significant Genera: *Coniophora, Serpula*.

Distribution: Very widespread.

Economic Significance: *Serpula lacrymans* ('dry rot') is an aggressive brown-rot fungus that causes very significant economic losses through degradation of wooden parts of buildings. It is particularly difficult to eradicate and may persist, for example, in brickwork and plaster, if not carefully treated. Other members of the family also cause rots of construction timber, though with lesser economic impact.

Serpula lacrymans; basidioma

Ecology: Saprobic on wood, soil etc., some species cause damage to living trees through heart-wood rot. One species is lichenicolous.

Notes: A fairly well-defined family of resupinate fungi within the bolete clade of the homobasidiomycetes, though some genera (e.g. *Hydnomerulius*, *Jaapia* and *Pseudomerulius*) appear not to belong judging from molecular studies.

Coniophora puteana; basidioma

References: **Bigelow, D.M.; Gilbertson, R.L.; Matheron, M.E.** (1998). Cultural studies of fungi causing brown rot in heartwood of living lemon trees in Arizona. *Mycol. Res.* 102: 257–262; **Binder, M.; Hibbett, D.S.; Larsson, K.-H.; Larsson, E.; Langer, E.; Langer, G.** (2005). The phylogenetic distribution of resupinate forms across the major clades of mushroom-forming fungi (Homobasidiomycetes). *Syst. Biodiv.* 3: 113–157; **Gilbertson, R.L.; Hemmes, D.E.** (1997). Notes on Hawaiian Coniophoraceae. *Mycotaxon* 65: 427–442; **Ginns, J.** (1998). Genera of the North American Corticiaceae sensu lato. *Mycologia* 90: 1–35; **Hjortstam, K.** (1987). A check-list to genera and species of corticioid fungi (Hymenomycetes). *Windahlia* 17: 55–85; **Kauserud, H.** (2004). Widespread vegetative compatibility groups in the dry rot fungus *Serpula lacrymans*. *Mycologia* 96: 232–239; **Kauserud, H.; Hogberg, N.; Knudsen, H.; Elborne, S.A.; Schumacher, T.** (2004). Molecular phylogenetics suggest a North American link between the anthropogenic dry rot fungus *Serpula lacrymans* and its wild relative *S. himantioides*. *Mol. Ecol.* 13: 3137–3146; **Kauserud, H.; Schmidt, O.; Elfstrand, M.; Högberg, N.** (2004). Extremely low AFLP variation in the European dry rot fungus (*Serpula lacrymans*): implications for self/nonself-recognition. *Mycol. Res.* 108: 1264–1270; **Kauserud, H.; Stensrud, O.; Decock, C.; Shalchian-Tabrizi, K.; Schumacher, T.** (2006). Multiple gene genealogies and AFLPs suggest cryptic speciation and long-distance dispersal in the basidiomycete *Serpula himantioides* (Boletales). *Mol. Ecol.* 15: 421–431; **Moreth, U.; Schmidt, O.** (2005). Investigations on ribosomal DNA of indoor wood decay fungi for their characterization and identification. *Holzforschung* 59: 90–93; **Palfreyman, J.W.; Gartland, J.S.; Sturrock, C.J.; Lester, D.; White, N.A.; Low, G.A.; Bech-Andersen, J.; Cooke, D.E.** (2003). The relationship between 'wild' and 'building' isolates of the dry rot fungus *Serpula lacrymans*. *FEMS Microbiol. Lett.* 228: 281–286; **Pegler, D.N.** (1991). Taxonomy, identification and recognition of *Serpula lacrymans*. *Serpula lacrymans. Fundamental Biology and Control Strategies*: 1–7; **Schmidt, O.; Moreth, U.** (2000). Species-specific PCR primers in the rDNA-ITS region as a diagnostic tool for *Serpula lacrymans*. *Mycol. Res.* 104: 69–72; **Singh, J.; Bech-Andersen, J.; Elborne, S.A.; Singh, S.; Walker, B.; Goldie, F.** (1993). The search for wild dry rot fungus (*Serpula lacrymans*) in the Himalayas. *Mycologist* 7: 124–130; **Thorn, R.G.; Malloch, D.W.; Ginns, J.** (1998). *Leucogyrophana lichenicola* sp. nov., and a comparison with basidiomes and cultures of the similar *Leucogyrophana romellii*. *Can. J. Bot.* 76: 686–693.

Cookellaceae Höhn. ex Sacc. & Trotter 1913

Myriangiales: Ascomycota

Ascomata small, crustose or pulvinate, opening by irregular tears in the upper surface, consisting of thin-walled cells in a gelatinous matrix. Interascal tissue absent. Asci ± globose, scattered irregularly throughout the ascoma, fissitunicate, without an ocular chamber or other apical structures. Ascospores hyaline to brown, muriform, with a gelatinous sheath. Anamorphs unknown.

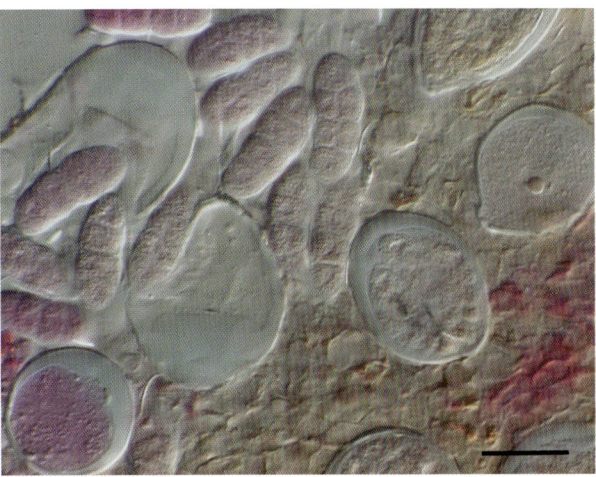

Uleomyces decipiens, Japan; asci and ascospores

Significant Genera: *Cookella*, *Uleomyces*.

Distribution: Pantropical.

Economic Significance: None is known.

Ecology: Fungicolous, on leaf-inhabiting species.

Notes: A small and poorly known family. No molecular data are available.

Uleomyces decipiens, Japan; ascomata

References: Eriksson, O.E.; Yue, J.Z. (1990). Notes on bambusicolous pyrenomycetes. Nos. 1-10. *Mycotaxon* 38: 201–220; **Tehler, A.** (1990). A new approach to the phylogeny of *Euascomycetes* with a cladistic outline of *Arthoniales* focussing on *Roccellaceae*. *Can. J. Bot.* 68: 2458–2492.

Corticiaceae Herter 1910
Corticiales: Basidiomycota

Basidiomata resupinate, effuse or discoid and then centrally attached, crustose, cottony, waxy or membranous, generally whitish, cream or pink, in one species dark blue. Hyphal system usually monomitic, septa simple, clamp connections mostly absent, cystidia usually present. Hymenium smooth to tuberculate. Basidia urniform, clavate or more usually elongate, 2- to 4-spored, rarely staining in iodine, interspersed with sterile, usually branched hyphae. Basidiospores subglobose to ellipsoidal or allantoid, smooth or occasionally ornamented, thin-walled, not staining in iodine. Sclerotia sometimes produced.

Significant Genera: *Corticium*, *Cytidia*, *Dendrothele*.

Distribution: Very widespread.

Terana caerulea, Sussex, UK; basidioma

Economic Significance: *Erythricium salmonicolor* (causal agent of pink disease of rubber, tea and other tropical plants) should probably be placed in the *Corticiaceae* based on molecular studies. Some species (e.g. *Laetisaria fuciformis*) are turfgrass pathogens.

Ecology: Saprobes or pathogens on a wide range of mainly woody substrata; a few species are lichenicolous and some are grass pathogens.

Notes: Formerly used for many resupinate basidiomycetes; now restricted to a small group of genera centred on *Corticium* in its original sense and probably still polyphyletic. *Corticium solani* (anamorph *Rhizoctonia solani*) is now treated as *Thanatephorus cucumeris* (*Ceratobasidiaceae*).

Cytidia salicina, Bhutan; basidiomata

Lindtneria trachyspora; basidiomata

References: **Binder, M.; Hibbett, D.S.; Larsson, K.-H.; Larsson, E.; Langer, E.; Langer, G.** (2005). The phylogenetic distribution of resupinate forms across the major clades of mushroom-forming fungi (*Homobasidiomycetes*). *Syst. Biodiv.* 3: 113–157; **Ginns, J.** (1998). Genera of the North American *Corticiaceae* sensu lato. *Mycologia* 90: 1–35; **Greslebin, A.; Rajchenberg, M.** (1999). Corticioid *Aphyllophorales* (*Basidiomycota*) from the Patagonian Andes forests of Argentina 4. *Nothocorticium patagonicum* gen. et sp. nov. *Mycotaxon* 70: 371–375; **Hallenberg, N.; Parmasto, E.** (1998). Phylogenetic studies in species of *Corticiaceae* growing on branches. *Mycologia* 90: 640–654; **Hibbett et al.** (2007). A higher-level phylogenetic classification of the *Fungi*. *Mycol. Res.* 111(5): 509–548; **Hjortstam, K.** (1998). A Checklist to Genera and Species of Corticioid Fungi (*Basidiomycotina*, *Aphyllophorales*). *Windahlia* 23: 1–54; **Hjortstam, K.; Larsson, K.-H.; Ryvarden, L.** (1988). *The Corticiaceae of North Europe, Introduction and keys* (Oslo) 1: 59 pp.; **Larsson, K.-H.; Larsson, E.; Koljalg, U.** (2004). High phylogenetic diversity among corticioid homobasidiomycetes. *Mycol. Res.* 108: 983–1002; **Maekawa, N.** (1993). Taxonomic study of Japanese *Corticiaceae* (*Aphyllophorales*) I. *Rep. Tottori Mycol. Inst.* 31: 1–149; **Maekawa, N.** (1994). Taxonomic study of Japanese *Corticiaceae* (*Aphyllophoraceae*) II. *Rep. Tottori Mycol. Inst.* 32: 1–123; **Parmasto, E.** ([undated]). *CORTBASE: A Nomenclatural Database of Corticioid Fungi* Version 1 (Tartu): [unpaginated].

Cortinariaceae R. Heim ex Pouzar 1983
Agaricales: Basidiomycota

Basidiomata pileate or gasteroid, membranous to fleshy, often brightly coloured, typically yellowish, reddish or brownish, the cap dry, silky, fibrose, scaly, or slimy and glutinous, sometimes striate. Pileate taxa lamellate, with the stipe central or eccentric and sometimes reduced, dry or viscid, sometimes bulbous at the base, often annular, veil often present protecting the young gills but breaking up in many species at an early stage of development, the gills concolorous with or differently pigmented to the cap, sometimes fimbriate due to presence of cheilocystidia. Hyphae often staining in iodine, clamp connections present or absent, cystidia usually present, varied in form. Basidia 2- or 4-spored. Basidiospores usually some shade of brown, variously shaped, thin-walled or thick-walled, usually without a germ pore, smooth or more usually ornamented, not staining in iodine.

Significant Genera: *Cortinarius*, *Crepidotus*, *Galerina*, *Gymnopilus*.

Galerina mniophila, Devon, UK; basidiomata

Distribution: Very widespread, especially in temperate climates. The family constitutes the dominant ectomycorrhizal group in many temperate forest ecosystems.

Gymnopilus penetrans; basidiomata

Economic Significance: The family must make a major contribution to forest productivity, but no financial estimates of their value have been made. Many species are poisonous, containing amatoxins, orellanine, muscarine, hallucinogens etc., and the North American species *Galerina sulciceps* has been claimed as the most toxic mushroom known to man.

Ecology: Ectomycorrhizal with woody plants, especially *Pinaceae*, *Fagaceae* and *Salicaceae*.

Notes: A highly speciose family; *Cortinarius* alone has more than 4000 species-level epithets described. Attempts to subdivide this genus based on morphological characteristics have met with mixed success. Some preliminary work has been carried out to identify clades based on molecular data, and the *Crepidotaceae* and *Inocybaceae* have been identified as independent units.

References: **Ammirati, J.F.; Traquair, J.A.; Horgen, P.A.** (1986). *Poisonous Mushrooms of the Northern United States and Canada* (Minneapolis): 396 pp.; **Bidaud, A.; Moënne-Loccoz, P.; Reumaux, P.; Henry, R.** (1996). Sous-Genre *Phlegmacium* (Fr.) Trog, Section *Patibiles* Moënne-L. & Reum., Sous-Section *Patibiles* Bid. & al.; Sous-Section *Cyanipedes* Bid. & al.; Sous-Section *Balteati* R. Hry. ex Moënne-L. & Reum. *Atlas des Cortinaires* (Meyzieu) 8: 239–304; **Bougher, N.L.; Lebel, T.** (2001). Sequestrate (truffle-like) fungi of Australia and New Zealand. *Aust. Syst. Bot.* 14: 439–484; **Boyle, H.; Zimdars, B.; Renker, C.; Buscot, F.** (2006). A molecular phylogeny of *Hebeloma* species from Europe. *Mycol. Res.* 110: 369–380; **Brandrud, T.E.** (1998). *Cortinarius* subgenus *Phlegmacium* section *Phlegmacioides* (= *Variecolores*) in Europe. *Edinb. J. Bot.* 55: 65–156; **Frøslev, T.G.; Matheny, P.B.; Hibbett, D.S.** (2005). Lower level relationships in the mushroom genus *Cortinarius* (*Basidiomycota, Agaricales*): a comparison of RPB1, RPB2 and ITS phylogenies. *Mol. Phylogen. Evol.* 37: 602–618; **Garnica, S.; Weiss, M.; Oberwinkler, F.** (2003). Morphological and molecular phylogenetic studies in South American *Cortinarius* species. *Mycol. Res.* 107: 1143–1156; **Garnica, S.; Weiss, M.; Oertel, B.; Oberwinkler, F.** (2003). Phylogenetic relationships of European *Phlegmacium* species (*Cortinarius, Agaricales*). *Mycologia* 95: 1155–1170; **Garnica, S.; Weiss, M.; Oertel, B.; Oberwinkler, F.** (2005). A framework for a phylogenetic classification in the genus *Cortinarius* (*Basidiomycota, Agaricales*) derived from morphological and molecular data. *Can. J. Bot.* 83: 1457–1477; **Høiland, K; Holst-Jensen, A.** (2000). *Cortinarius* phylogeny and possible taxonomic implications of ITS rDNA sequences. *Mycologia* 92: 694–710; **Liu, Y.J.; Rogers, S.O.; Ammirati, J.F.** (1997). Phylogenetic relationships in *Dermocybe* and related *Cortinarius* taxa based on nuclear ribosomal DNA internal transcribed spacers. *Can. J. Bot.* 75: 519–532; **Matheny, P.B.** (2005). Improving phylogenetic inference of mushrooms with RPB1 and RPB2 nucleotide sequences (*Inocybe; Agaricales*). *Mol. Phylogen. Evol.* 35: 1–20; **Moreau, P.-A.; Peintner, U.; Gardes, M.** (2006). Phylogeny of the ectomycorrhizal mushroom genus *Alnicola* (*Basidiomycota, Cortinariaceae*) based on rDNA sequences with special emphasis on host specificity and morphological characters. *Mol. Phylogen. Evol.* 38: 794–807; **Peintner, U.; Bougher, N.L.; Castellano, M.A.; Moncalvo, M.-C.; Moser, M.M.; Trappe, J.M.; Vilgalys, R.** (2001). Multiple origins of sequestrate fungi related to *Cortinarius* (*Cortinariaceae*). *Am. J. Bot.* 88: 2168–2179; **Peintner, U.; Horak, E.; Moser, M.M.; Vilgalys, R.** (2002). Phylogeny of *Rozites*, *Cuphocybe* and *Rapacea* inferred from ITS and LSU rDNA sequences. *Mycologia* 94: 620–629; **Peintner, U.; Moncalvo, J.-M.; Vilgalys, R.** (2004). Towards a better understanding of the infrageneric relationships in *Cortinarius* (*Agaricales, Basidiomycota*). *Mycologia* 96: 1042–1058; **Peintner, U.; Moser, M.M.; Thomas, K.A.; Manimohan, P.** (2003). First records of ectomycorrhizal *Cortinarius* species from tropical India and their phylogenetic position based on rDNA ITS sequences. *Mycol. Res.* 107: 485–494.

Coryneliaceae Sacc. ex Berl. & Voglino 1886
Coryneliales: Ascomycota

Stromata absent or small and cushion-like. Ascomata clustered, basically perithecial but sometimes with a mazaedial chamber above the ascus-bearing part, black, thick-walled and rough-walled, opening by an ostiole or irregular split. Interascal tissue absent. Asci initially double-walled, the outer layer deliquescing early, becoming long-stalked and very thin-walled, without apical structures, evanescent. Ascospores usually aseptate, dark, varied and usually strongly ornamented, collecting in a mass above the ascus layer. Anamorph coelomycetous, probably spermatial in function.

Significant Genera: *Caliciopsis*, *Corynelia*.

Distribution: Mainly tropical and south temperate.

Corynelia tropica, Chile; ascomata

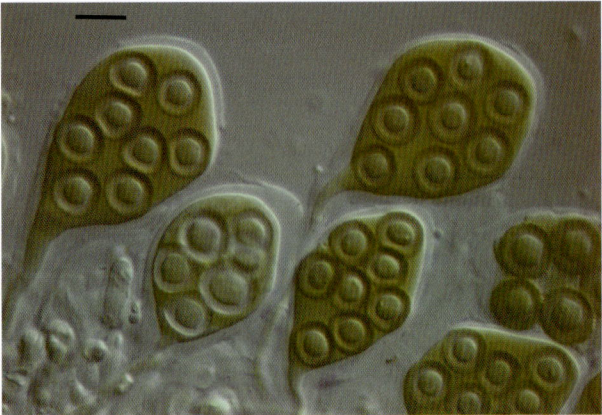

Corynelia uberata, South Africa; asci and immature ascospores

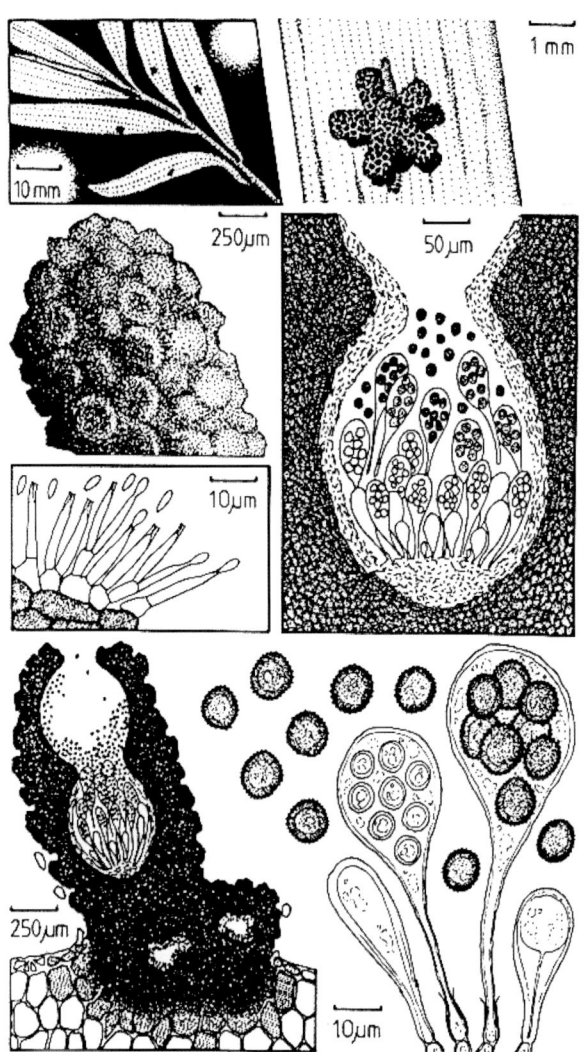

Corynelia tropica; ascomata, asci, conidiogenous cells, conidia

Economic Significance: Little is known. Presence of the parasite does not appear to cause significant damage to the host.

Ecology: Biotrophic on leaves and stems of *Podocarpaceae* and other conifers.

Notes: The family has a highly distinctive ascus structure, and its taxonomic position is uncertain; preliminary molecular evidence suggests placement within the *Chaetothyriomycetidae/Eurotiomycetidae* clade, but morphological analysis does not support this.

References: **Benny, G.L.; Samuelson, D.A.; Kimbrough, J.W.** (1985). Studies on the *Coryneliales*. III. Taxa parasitic on *Podocarpaceae*: *Lagenulopsis* and *Tripospora*. *Bot. Gaz.* 146: 431–436; **Benny, G.L.; Samuelson, D.A.; Kimbrough, J.W.** (1985). Studies on the *Coryneliales*. II. Taxa parasitic on *Podocarpaceae*: *Corynelia*. *Bot. Gaz.* 146: 238–251; **Benny, G.L.; Samuelson, D.A.; Kimbrough, J.W.** (1985). Studies on the *Coryneliales*. IV. *Caliciopsis*, *Coryneliopsis* and *Coryneliospora*. *Bot. Gaz.* 146: 437–448; **Benny, G.L.; Samuelson, D.A.; Kimbrough, J.W.** (1985). Studies on the *Coryneliales*. I. *Fitzpatrickella*, a monotypic genus on the fruits of *Drimys*. *Bot. Gaz.* 146: 232–237; **Inderbitzin, P.; Lim, S.R.; Volkmann-Kohlmeyer, B.; Kohlmeyer, J.; Berbee, M.L.** (2004). The phylogenetic position of *Spathulospora* based on DNA sequences from dried herbarium material. *Mycol. Res.* 108: 737–748; **Johnston, P.R.; Minter, D.W.** (1989). Structure and taxonomic significance of the ascus in the *Coryneliaceae*. *Mycol. Res.* 92: 422–430; **Marmolejo, J.G.** (1999). A new species of *Caliciopsis* on oaks from Mexico. *Mycotaxon* 72: 195–197; **Rikkinen, J.** (2000). Two new species of *Caliciopsis* (*Coryneliaceae*) from Hunan Province, China. *Karstenia* 40: 147–151; **Winka, K.; Eriksson, O.E.** (2000). Adding to the bitunicate puzzle – studies on the systematic position of five aberrant ascomycete taxa. *Phylogenetic Relationships Within the Ascomycota Based on 18S rDNA Sequences, Akademisk Avhandling* [Thesis (PhD), Department of Ecology and Environmental Science, Umeå University] (Umeå): [17] pp..

Corynesporascaceae Sivan. 1996
Pleosporales: Ascomycota

Ascomata cleistothecial, brown, thin-walled. Interascal tissue of narrow thin-walled deliquescing pseudoparaphyses. Asci ovoid, bitunicate, deliquescing, without clear apical structures. Ascospores large, brown, transversely euseptate and distoseptate, thick-walled, without a gelatinous sheath.

Significant Genera: *Corynesporasca*. Anamorph: *Corynespora*.

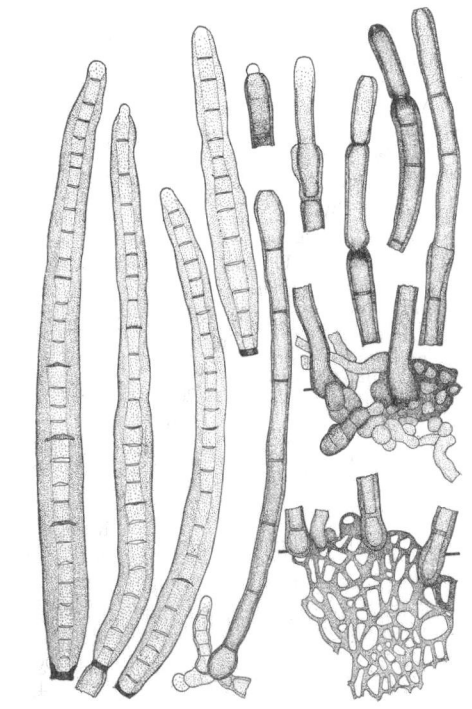

Corynespora smithii; conidiophores and conidia

Distribution: Tropical.

Economic Significance: A number of economically important plant diseases are caused by this family, especially leaf spots and fruit disorders; many are ascribed to the *Corynespora cassiicola* complex.

Ecology: Saprobes or weak parasites on a large number of plant species.

Notes: Species concepts and host specificity are uncertain. A link with the *Massarinaceae* (which includes *Helminthosporium*) is possible.

References: **Carris, L.M.** (1987). A new species of *Corynespora* forming secondary conidia [abstract]. *Mycol. Soc. Amer. Newsl.* 38: 19; **Goh, T.K.; Hyde, K.D.; Lee, D.K.L.** (1998). Generic distinction in the *Helminthosporium*-complex based on restriction analysis of the nuclear ribosomal RNA gene. *Fungal Diversity* 1: 85–107; **Morgan-Jones, G.** (1988). Notes on hyphomycetes. LVII. *Corynespora biseptata*, reclassified in *Corynesporopsis*. *Mycotaxon* 31: 511–515; **Sivanesan, A.** (1996). *Corynesporasca caryotae* gen. et sp. nov. with a *Corynespora* anamorph, and the family *Corynesporascaceae*. *Mycol. Res.* 100: 783–788.

Crepidotaceae Singer 1951
Agaricales: Basidiomycota

Basidiomata with lateral attachment, lamellate. Hyphae with or without clamp connections. Upper surface of cap white, yellow, red or brown, with an indefinite cuticle or covered with erect thin-walled hypha-like hairs which may be pigment-encrusted, below which a gelatinized layer is often present. Lamellae usually ± hyaline, narrow or broad, sometimes decurrent, not anastomosing, cystidia usually present on the edges. Basidia small, clavate, 2- or 4-spored. Basidiospores globose to cylindrical, pale to dark brown, smooth or with short spines, warts or ridges, usually without a germ pore.

Significant Genera: *Crepidotus*, *Simocybe*.

Distribution: Cosmopolitan.

Crepidotus variabilis, Surrey, UK; undersurface of young basidiomata

Economic Significance: None is known.

Ecology: On wood, palm fronds, herbaceous stems etc., rarely parasitic on living woody tissues.

Notes: Recently considered part of the *Cortinariaceae* in a wide sense, molecular studies have confirmed that *Crepidotus* occupies a clearly demarcated clade but that the family is less widely demarcated than originally proposed by Singer.

References: **Aime, M.C.; Vilgalys, R.; Miller, O.K.** (2005). The *Crepidotaceae* (*Basidiomycota*, *Agaricales*): phylogeny and taxonomy of the genera and revision of the family based on molecular evidence. *Am. J. Bot.* 92: 74-82; **Binder, M.; Hibbett, D.S.; Larsson, K.-H.; Larsson, E.; Langer, E.; Langer, G.** (2005). The phylogenetic distribution of resupinate forms across the major clades of mushroom-forming fungi (*Homobasidiomycetes*). *Syst. Biodiv.* 3: 113–157; **Matheny, P.B.** (2005). Improving phylogenetic inference of mushrooms with RPB1 and RPB2 nucleotide sequences (*Inocybe*; *Agaricales*). *Mol. Phylogen. Evol.* 35: 1–20; **Matheny, P.B.; Bougher, N.L.** (2006). The new genus *Auritella* from Africa and Australia (*Inocybaceae*, *Agaricales*>: molecular systematics, taxonomy and historical biogeography. *Mycol. Prog.* 5: 2–17; **Nordstein, S.** (1990). The genus *Crepidotus* (*Basidiomycotina*, *Agaricales*) in Norway. *Syn. Fung.* (Oslo) 2: 1-115; **Senn-Irlet, B.** (1995). The genus *Crepidotus* (Fr.) Staude in Europe. *Persoonia* 16: 1-80; **Singer, R.** (1986). *Agaricales in Modern Taxonomy* Edn 4 (Koenigstein): 981 pp..

Crocyniaceae M. Choisy ex Hafellner 1984
Lecanorales: Ascomycota

Thallus crustose, often spongy. Ascomata apothecial, sessile, pale, convex, with a poorly-developed margin. Interascal tissue of simple paraphyses, the apices hardly swollen, in a gelatinous matrix. Asci with a well-developed J+ apical cap, a darker-staining ring, a small ocular chamber and an outer J+ gelatinous layer. Ascospores simple, hyaline, without a sheath.

Significant Genera: *Crocynia*.

Crocynia gossypina; thallus and ascomata

Distribution: Primarily tropical.

Economic Significance: None is known.

Ecology: Lichenized with green algae.

Notes: Very poorly known; no recent information is available.

References: **Aptroot, A.; Sipman, H.J.M.** (1991). New lichens and lichen records from New Guinea. *Willdenowia* 20: 221–256; **Hafellner, J.** (1984). Studien in Richtung einer natürlicheren Gliederung der Sammel-familien *Lecanoraceae* und *Lecideaceae*. *Nova Hedwigia* Beih. 79: 241–371.

Cronartiaceae Dietel 1900
Uredinales: Basidiomycota

Spermogonia intracortical, without discrete walls, growth indeterminate. Aecia *Peridermium*-like, large and blister-like, with well-developed walls that rupture irregularly to release spores. Aeciospores formed in chains, usually strongly verrucose. Uredinia discoid or domed, with discrete walls and a well-defined ostiole. Urediniospores formed singly, pedicellate, echinulate, with germ pores in two rings, sometimes accompanied by paraphyses. Telia columnar, not gelatinous. Teliospores 1-celled with 1–3 germ pores, produced in basipetal succession, adhering to each other and forming thread-like columns; or telia peridermioid with telial aeciospores, germinating with external sterigmata to produce ellipsoidal or globose basidiospores.

Significant Genera: *Cronartium*.

Distribution: Widespread, especially in temperate regions.

Cronartium coleosporioides, Nova Scotia, Canada; blister rust on pine

Economic Significance: Several species are important pathogens of pines, especially *Cronartium flaccidum*, *C. ribicola* and *C. quercuum*, which cause blister rusts in north temperate zones and elevated regions in the tropics.

Ecology: Most species are heteroecious, with the aecial stages forming on stems and cones of *Pinus* species and uredinia and telia on a range of dicotyledonous plants.

Notes: Separated from the *Coleosporiaceae* primarily by the urediniospores being formed singly rather than in chains and differences in spermogonial morphology. Preliminary molecular studies indicate that the two families may have a shared ancestry.

References: **Aime, M.C.** (2006). Towards resolving family-level relationships in rust fungi (*Uredinales*). *Mycoscience* 47: 112–122; **Crane, P.E.; Jakubec, K.M.; Hiratsuka, Y.** (1995). Production of spermogonia and aecia of *Endocronartium harknessii* on artificially inoculated lodgepole pine and their significance in the life cycle. *Proceedings of the Fourth IUFRO Rusts of Pines Working Party Conference* Tsukuba (Tsukuba): 101–107; **Cummins, G.B.; Hiratsuka, Y.** (2003). *Illustrated Genera of Rust Fungi* (St Paul): 225 pp.; **Ettouil, K.; Bernier, L.; Beaulieu, J.; Bérubé, J.A.; Hopkin, A.; Hamelin, R.C.** (1999). Genetic structure of *Cronartium ribicola* populations in eastern Canada. *Phytopathology* 89: 915–919; **Hantula, J.; Kasanen, R.; Kaitera, J.; Moricca, S.** (2002). Analyses of genetic variation suggest that pine rusts *Cronartium flaccidum* and *Peridermium pini* belong to the same species. *Mycol. Res.* 106: 203–209; **Hiratsuka, Y.** (1995). Pine stem rusts of the world – framework for a monograph. *Proceedings of the Fourth IUFRO Rusts of Pines Working Party Conference* Tsukuba (Tsukuba): 1–8; **Kubisiak, T.L.; Roberds, J.H.; Spaine, P.C.; Doudrick, R.L.** (2004). Microsatellite DNA suggests regional structure in the fusiform rust fungus *Cronartium quercuum* f. sp. *fusiforme*. *Heredity* 92: 41–50; **Maier, W.; Begerow, D.; Weiss, M.; Oberwinkler, F.** (2003). Phylogeny of the rust fungi: an approach using nuclear large subunit ribosomal DNA sequences. *Can. J. Bot.* 81: 12–23; **Mims, C.W.; Liljebjelke, K.A.; Covert, S.F.** (1996). Ultrastructure of telia and teliospores of the rust fungus *Cronartium quercuum* f. sp. *fusiforme*. *Mycologia* 88: 47–56; **Vogler, D.R.** (2000). Coevolution of *Cronartium* with its hosts. *HortTechnol.* 10: 518; **Vogler, D.R.; Bruns, T.D.** (1998). Phylogenetic relationships among the pine stem rust fungi (*Cronartium* and *Peridermium* spp.). *Mycologia* 90: 244–257; **Wingfield, B.D.; Ericson, L.; Szaro, T.; Burdon, J.J.** (2004). Phylogenetic patterns in the *Uredinales*. *Australas. Pl. Path.* 33: 327–335.

Cryphonectriaceae Gryzenh. & M.J. Wingf. 2006
Diaporthales: Ascomycota

Stromata well-developed, composed of a combination of degraded plant cells and fungal tissue, small to large, erumpent, the ectostromatic elements usually with conspicuous orange pigmentation. Ascomata perithecial, brown to black-walled, with long slender necks and usually immersed in orange to dark brown stromatic tissue. Interascal tissue absent or deliquescent at an early stage of development. Asci obovoid to fusiform, with short deliquescent stalks, the apex truncate with a conspicuous refractive apical ring that does not blue in iodine, usually

8-spored. Ascospores ellipsoidal, fusiform or allantoid, hyaline, aseptate to multiseptate, lacking gelatinous sheath or appendages. Conidiomata eustromatic, sometimes formed in common stromata with ascomata, sometimes long-necked. Conidiogenous cells ± cylindrical, proliferating percurrently. Conidia very small, usually ovoid or cylindrical, hyaline, aseptate.

Significant Genera: *Cryphonectria, Chrysoporthe, Endothia*.

Distribution: Worldwide, in tropical and temperate zones, primarily in the Old World.

Economic Significance: Several species are important plant pathogens. *Cryphonectria parasitica*, apparently native to southern Europe, caused devastation of *Castanea* populations in North America in the early 1900s. *Chrysoporthe cubensis* is an important pathogen of *Myrtaceae* including eucalyptus and clove, and has been widely introduced in the tropics and semitropics along with its host plant; it has been confused in the past with *Microthia havanensis*.

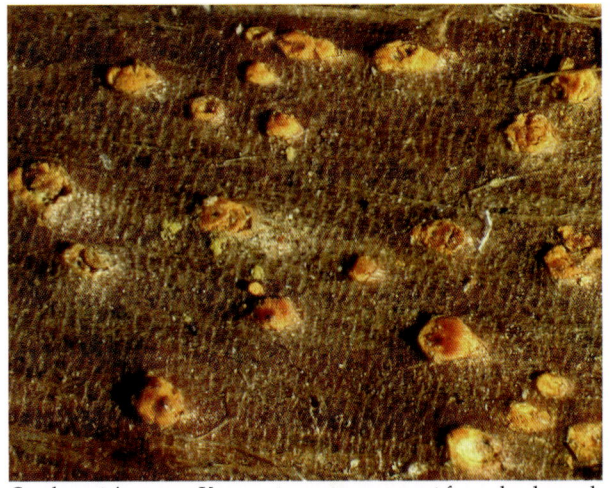

Cryphonectria gyrosa, Kenya; stromata erumpent from dead wood

Ecology: Many species are parasitic on woody dicotyledonous plants, causing cankers and wilt due to disruption of cambial tissues. Species on plants in their native ecosystems are not necessarily aggressive pathogens.

Notes: One of about seven major monophyletic lineages within the *Diaporthales*, characterized especially by its usually orange stromatic tissues. Its circumscription still requires some study, in particular the possible inclusion of *Cryptodiaporthe corni* which is an important pathogen of *Cornus* in the USA.

References: **Anagnostakis, S.L.** (1987). Chestnut blight: the classical problem of an introduced pathogen. *Mycologia* 79: 23–37; **Barr, M.E.** (1978). The *Diaporthales* in North America with emphasis on *Gnomonia* and its segregates. *Mycol. Mem.* 7: 1–232; **Castlebury, L.A.; Rossman, A.Y.; Jaklitsch, W.J.; Vasilyeva, L.N.** (2002). A preliminary overview of the *Diaporthales* based on large subunit nuclear ribosomal DNA sequences. *Mycologia* 94: 1017–1031; **Gryzenhout, M.; Myburg, H.; Merwe, N.A. van der; Wingfield, B.D.; Wingfield, M.J.** (2004). *Chrysoporthe*, a new genus to accommodate *Cryophonectria cubensis*. *Stud. Mycol.* 50: 119–142; **Gryzenhout, M.; Myburg, H.; Wingfield, B.D.; Wingfield, M.J.** (2006). Cryophonectriaceae (*Diaporthales*), a new family including *Cryphonectria, Chrysoporthe, Endothia* and allied genera. *Mycologia* 98: 239–249; **Myburg, H.; Gryzenhout, M.; Wingfield, B.D.; Stipes, R.J.; Wingfield, M.J.** (2004). Phylogenetic relationships of *Cryphonectria* and *Endothia* species, based on DNA sequence data and morphology. *Mycologia* 96: 990–1001; **Myburg, H.; Gryzenhout, M.; Wingfield, B.D.; Wingfield, M.J.** (2003). Conspecificity of *Endothia eugeniae* and *Cryphonectria cubensis*: a re-evaluation based on morphology and DNA sequence data. *Mycoscience* 104: 187–196; **Redlin, S.C.; Rossman, A.Y.** (1991). *Cryptodiaporthe corni* (*Diaporthales*), cause of Cryptodiaporthe canker of pagoda dogwood. *Mycologia* 83: 200–209.

Cryptobasidiaceae Malençon ex Donk 1956
Exobasidiales: Basidiomycota

Basidiomata poorly developed, fruit bodies formed as gall-like malformations of host tissue, spores dispersed following rupture of the host tissue. Septal pores simple, haustoria present in host cells. Basidia usually formed in a hymenial layer lining cavities of the host gall, rarely sporulating on the external surface, in most cases mixed with cylindrical or clavate thick-walled paraphysis-like hyphal elements. Basidia ± cylindrical, 2- to 8-spored, sterigmata poorly developed or ± absent. Basidiospores not discharged actively, formed singly or in clusters of 6–8, globose to ellipsoidal or falcate, sometimes septate, often very thick-walled, hyaline, often ornamented.

Significant Genera: *Botryoconis*.

Distribution: Widespread, studied especially in the neotropics.

Economic Significance: None is known.

Ecology: Parasitic on leaves, stems and fruits of plants, primarily *Lauraceae*. Many species cause hypertrophy and other deformations.

Notes: A small family, separated from the *Exobasidiaceae* primarily due to passively rather than actively discharged basidiospores and parasitism of the *Lauraceae* rather than the *Ericaceae*. Initial molecular studies confirm this split.

References: **Bauer, R.; Oberwinkler, F.; Vánky, K.** (1997). Ultrastructural markers and systematics in smut fungi and allied taxa. *Can. J. Bot.* 75: 1273–1314; **Begerow, D.; Bauer, R.; Oberwinkler, F.** (2002). The *Exobasidiales*: an evolutionary hypothesis. *Mycol.*

Prog. 1: 187–199; **Ciccarone, C.** (1989). *Acaulopage tigrina* sp. nov. e *Drepanoconis divertigastra* sp. nov.: due componenti nuovi di generi microfungi poco noti. *Micol. Ital.* 18: 29–32; **Gómez, L.D.; Kisimova-Horovitz, L.** (1998). [Basidiomycetes of Costa Rica. *Exobasidiales, Cryptobasidiales.* Historic, taxonomic and phytogeographic notes]. *Revta Biol. trop.* 45: 1293–1310; **Hendrichs, M.; Bauer, R.; Oberwinkler, F.** (2003). The *Cryptobasidiaceae* of tropical Central and South America. *Sydowia* 55: 33–64.

Cryptomycocolacaceae Oberw. & R. Bauer 1990
Cryptomycocolacales: Basidiomycota

Basidiomata absent, the basidia formed in an exposed hymenial layer on the host surface. Hyphae hyaline, thin-walled, smooth, dikaryotic, septa simple in structure, Woronin bodies common, clamp connections usually present. Cystidia absent. Basidia initially ± globose, the apical part extending to form an elongate structure delimited by a septum, which is released; the basal cell then itself elongates to form a ± lanceolate structure from the apex of which basidiospores are formed sequentially in a cluster. Basidiospores released passively, ± cylindrical, hyaline, smooth, thin-walled, not staining in iodine, sometimes budding in a yeast-like fashion. Chlamydospores common in old cultures.

Significant Genera: *Cryptomycocolax*.

Distribution: Only reported from Costa Rica.

Economic Significance: None is known.

Ecology: Parasitic on sclerotia of *Ascomycota* in dead stems.

Notes: Relationships of this family are obscure, but are assumed to fall within the auriculariod homobasidiomycetes. Cytological features of basidium development appear to be unique within the *Basidiomycota*.

References: **Oberwinkler, F.; Bauer, R.** (1990). *Cryptomycocolax*: a new mycoparasitic heterobasidiomycete. *Mycologia* 82: 671–692.

Cucurbitariaceae G. Winter 1885
Pleosporales: Ascomycota

Ascomata often strongly aggregated, formed from a basal stromatic structure, perithecial, erumpent, variable in shape, with a well-developed periphysate ostiole, thick-walled especially at the base, often roughened or hairy. Interascal tissue composed of cellular pseudoparaphyses. Asci ± cylindrical, fissitunicate, with a well-developed ocular chamber, sometimes with a distinct J– apical ring. Ascospores pigmented, transversely septate or muriform, usually without a gelatinous sheath or appendages. Anamorph coelomycetous, pycnidial, strongly aggregated, formed from the same basal stroma, thick-walled, weakly papillate. Conidiogenous cells irregular, where proliferating percurrent or annellidic. Conidia pigmented, transversely septate or muriform; spermatial microconidia also formed.

Significant Genera: *Cucurbitaria, Syncarpella*. Anamorphs: *Camarosporium, Pyrenochaeta, Syntholus*.

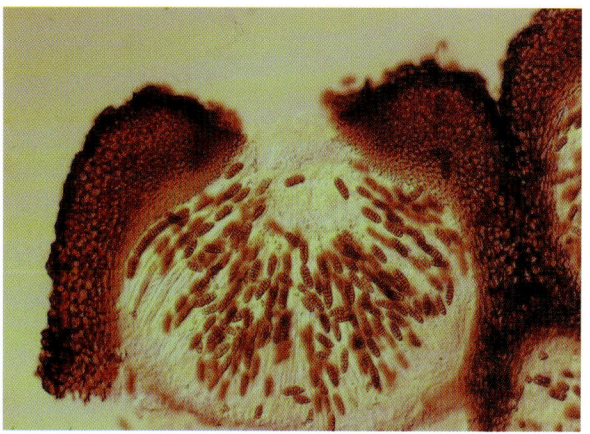
Cucurbitaria laburni, Gloucestershire, UK; vertical section through ascoma

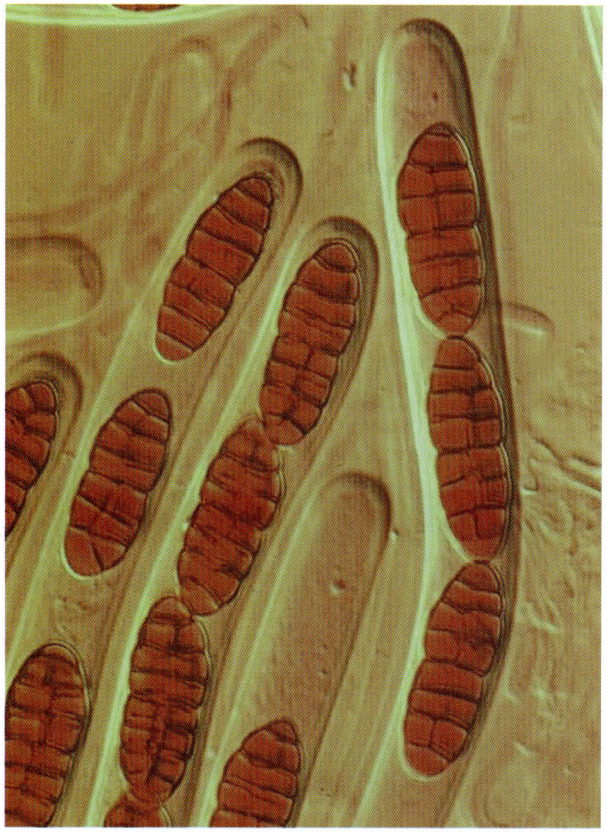
Cucurbitaria laburni, Gloucestershire, UK; asci and ascospores

Distribution: Widespread in temperate regions.

Economic Significance: No economically significant pathogens are known.

Ecology: Necrotrophic or saprobic on woody plants.
Notes: A small but well-defined family of the *Pleosporales*. It is perhaps most closely related to the *Leptosphaeriaceae* and *Phaeosphaeriaceae*, but molecular data are sparse.

References: **Barr, M.E.** (1990). Some dictyosporous genera and species of *Pleosporales* in North America. *Mem. N. Y. bot. Gdn* 62: 92 pp.; **Barr, M.E.; Boise, J.R.** (1989). *Syncarpella* (*Pleosporales, Cucurbitariaceae*). *Mem. N. Y. bot. Gdn* 49: 298–304; **Berbee, M.L.** (1996). Loculoascomycete origins and evolution of filamentous ascomycete morphology based on 18S rRNA gene sequence analysis. *Mol. Biol. Evol.* 13: 462–470; **Mirza, F.** (1968). Taxonomic investigations on the ascomycetous genus *Cucurbitaria* S.F. Gray. *Nova Hedwigia* 16: 161–213; **Ramaley, A.W.; Barr, M.E.** (1997). *Syncarpella ribis* sp. nov. and its anamorph, *Syntholus ribis* gen. et sp. nov. *Mycotaxon* 65: 499–506; **Silva-Hanlin, D.M.; Hanlin, R.T.** (1999). Small subunit ribosomal RNA gene phylogeny of several loculoascomycetes and its taxonomic implications. *Mycol. Res.* 103: 153–160.

Cudoniaceae P.F. Cannon 2001
Helotiales: Ascomycota

Stromata absent. Ascomata stipitate, brightly coloured, usually with a distinct stipe and a laterally flattened or irregular hymenial region. Hymenium brownish to yellow. Interascal tissue composed of narrow branched paraphyses which are curved towards the tip. Asci cylindric-clavate, thin-walled, with a small J+ or J– pore. Ascospores elongate, hyaline, multiseptate, arranged fasciculately within the ascus. Anamorphs unknown.

Significant Genera: *Cudonia*, *Spathularia*.

Cudonia circinans, Saskatchewan, Canada; ascomata

Cudonia circinans, Norway; ascomata

Distribution: Widespread in north temperate regions.

Spathulariopsis velutipes, Japan; ascomata

Economic Significance: None is known.

Ecology: Soil-dwelling, perhaps associated with buried wood.

Notes: Possibly related to the *Rhytismataceae*, though more research is needed.

References: **Döring, H.; Triebel, D.** (1998). Phylogenetic relationships of *Bulgaria* inferred by 18S rDNA sequence analysis. *Cryptog. Bryol.-Lichénol.* 19: 123–136; **Gernandt, D.S.; Platt, J.L.; Stone, J.K.; Spatafora, J.W.; Holst-Jensen, A.; Hamelin, R.C.; Kohn, L.M.** (2001). Phylogenetics of *Helotiales* and *Rhytismatales* based on partial small subunit nuclear ribosomal DNA sequences. *Mycologia* 93: 915–933; **Landvik, S.; Shailer, N.F.J.; Eriksson, O.E.** (1997). SSU rDNA sequence support for a close relationship between the *Elaphomycetales* and the *Eurotiales* and *Onygenales*. *Mycoscience* 37: 237–241; **Wang, Z.; Binder, M.; Hibbett, D.S.** (2002). A new species of *Cudonia* based on morphological and molecular data. *Mycologia* 94: 641–650.

Cuniculitremaceae J.P. Samp., R. Kirschner & M. Weiss 2001
Tremellales: Basidiomycota

Basidiomata absent. Hyphae hyaline, septa with dolipores, clamp connections present, haustoria composed of narrow, thin-walled, sometimes branched, hypha-like structures growing from a swollen base. Basidia ± globose, becoming longitudinally septate, not formed in chains, sterigmata strongly elongate. Basidiospores allantoid, hyaline, thin-walled, aseptate, germinating to produce secondary spores. Conidiogenous cells formed in clusters on slender hyphae, sometimes on hyphal swellings, with an ellipsoidal sterile base and a septate, fertile apical part with numerous clamp connections that form persistent peg-like structures after conidium dehiscence. Conidia ± globose, hyaline, ± thin-walled, smooth, aseptate. A second anamorphic stage has conidiogenous cells formed singly on septate hyphae, with or without a basal clamp connection, giving rise to one or more apical sterigma-like processes from which conidia secede.

Significant Genera: *Cuniculitrema*. Anamorphs: *Fellomyces*, *Sterigmatosporidium*.

Distribution: The teleomorphic taxon is known from southern Germany and Switzerland, anamorphic taxa are more widely distributed.

Economic Significance: None is known.

Ecology: Parasitic on other fungi, probably *Ophiostoma* or similar species, in bark beetle galleries in coniferous bark.

Notes: This is the only family of tremelloid heterobasidiomycetes that have anamorphs producing conidia on synnema-like structures and that lack a recognizable fruit-body. Molecular data confirm the family as distinct.

References: **Kirschner, R.; Sampaio, J.P.; Gadanho, M.; Weiss, M.; Oberwinkler, F.** (2001). *Cuniculitrema polymorpha* (Tremellales, gen. nov. and sp. nov.), a heterobasidiomycete vectored by bark beetles, which is the teleomorph of *Sterigmatosporidium polymorphum*. *Antonie van Leeuwenhoek* 80: 149–161.

Cunninghamellaceae Naumov ex R.K. Benj. 1959
Mucorales: Zygomycota

Vegetative thallus a mycelium, complex fruit bodies absent. Sporophores arising from the substratum mycelium or from stolons, erect, ascending or repent, branched, usually verticillately terminating in fertile vesicles. Fertile vesicles bearing pedicellate sporangiola over their entire surface; sporangia absent. Fertile vesicles globose to subglobose or somewhat clavate, smooth. Pedicels tapered, very short. Sporangiola globose to subglobose, unispored, lacking an obvious columella; wall thin, spinulose, persistent. Sporangiospores like the sporangiola in size and shape; wall thin, smooth. Zygospores globose to subglobose; wall brown to dark brown, opaque, ornamented with more or less coarse projections; suspensors opposed, smooth, more or less swollen, equal or unequal; heterothallic or rarely homothallic.

Significant Genera: *Cunninghamella*.

Distribution: Widespread, more common in the tropics.

Cunninghamella; fertile vesicle and developing sporangiola

Economic Significance: None is known.

Ecology: Soil fungi. One species (*C. bertholletii*) is a recognized cause of a zygomycosis but the correct name for this may be *C. elegans*, a common and widespread species.

Notes: Very poorly known; no recent information is available.

References: **Dermoumi, H.** (1993). A rare zygomycosis due to *Cunninghamella bertholletiae*. *Mycoses* 36: 293–294; **Su, Y.C.; Huang, H.; Liu, X.Y.; Zheng, R.Y.** (1999). Systematic relationship of several controversial *Cunninghamella* taxa inferred from sequence comparisons of ITS2 of rDNA. *Mycol. Res.* 103: 805–810; **Zheng, R.Y.; Chen, G.Q.** (1994). *Cunninghamella phaeospora* var. *multiverticillata* var. nov. and its mating with var. *phaeospora*. *Mycosystema* 7: 1–11; **Zheng, R.Y.; Chen, G.Q.** (1996). *Cunninghamella echinulata* (Thaxt.) Thaxt. ex Blakeslee var. *echinulata* and var. *verticillata* (Paine) comb. nov. *Mycosystema* 8-9: 1–13; **Zheng, R.Y.; Chen, G.Q.** (1998). *Cunninghamella clavata* sp. nov., a fungus with an unusual type of branching of sporophore. *Mycotaxon* 69: 187–198.

Cystofilobasidiaceae Well & Bandoni 2001
Cystofilobasidiales: Basidiomycota

Basidiomata absent, usually restricted to a free-living yeast-like phase, heterothallic or secondarily homothallic. Cell walls containing xylose, hyphal septa usually with dolipores lacking a parenthesome. Teliospores present, producing aseptate basidia. Coenzyme 8 or 10 present, D-glucuronate, nitrate, nitrite and usually *myo*-inositol assimilated.

Significant Genera: *Cystofilobasidium*, *Mrakia*. Anamorph: *Cryptococcus*-like.

Distribution: Widespread, especially in cold climates.

Economic Significance: Few studies are available, but there has been some interest in cold-tolerant enzymes in the family.

Ecology: Presumably saprobic, little is known.

Notes: The *Cystofilobasidiaceae* are clearly separate from the other yeast-forming families of the *Tremellomycetidae* based on phylogenetic analysis of rDNA.

References: **Diaz, M.R.; Fell, J.W.** (2000). Molecular analyses of the IGS & ITS regions of rDNA of the psychrophilic yeasts in the genus *Mrakia*. *Antonie van Leeuwenhoek* 77: 7–12; **Fell, J.W.; Blatt, G.M.** (1999). Separation of strains of the yeasts *Xanthophyllomyces dendrorhus* and *Phaffia rhodozyma* based on rDNA IGS and ITS sequence analysis. *J. Industr. Microbiol. Biotechnol.* 23: 677–681; **Fell, J.W.; Roeijmans, H.; Boekhout, T.** (1999). Cystofilobasidiales, a new order of basidiomycetous yeasts. *Int. J. Syst. Bacteriol.* 49: 907–913; **Fell, J.W.; Statzell-Tallman, A.** (1998). *Mrakia* Y. Yamada & Komagata. *The Yeasts, A Taxonomic Study* (Amsterdam): 676–677; **Kwon-Chung, K.J.** (1987). Filobasidiaceae – a taxonomic survey. *Stud. Mycol.* 30: 75–85; **Kwon-Chung, K.J.** (1998). *Cystofilobasidium* Oberwinkler & Bandoni. *The Yeasts, A Taxonomic Study* (Amsterdam): 646–653; **Miller, M.W.; Phaff, H.J.** (1998). *Phaffia* M.W. Miller, Yoneyama & Soneda. *The Yeasts, A Taxonomic Study* (Amsterdam): 789; **Sampaio, J.P.; Gadanho, M.; Bauer, R.** (2001). Taxonomic studies on the genus *Cystofilobasidium*: description of *Cystofilobasidium ferigula* sp. nov. and clarification of the status of *Cystofilobasidium lari-marini*. *Int. J. Syst. Evol. Microbiol.* 51: 221–229; **Wery, J.; Dalderup, M.J.M.; Linde, J.T.; Boekhout, T.; Ooyen, A.J.J. van** (1996). Structural and phylogenetic analysis of the actin gene from the yeast *Phaffia rhodozyma*. *Yeast* (Chichester) 12: 641–651.

Cystostereaceae Jülich 1982
Polyporales: Basidiomycota

Basidiomata annual or perennial, effuse or pileate, membranous, woody or crustose, the hymenium grey, ochraceous or pinkish, smooth to tuberculate or spiny (odontioid). Hyphal system usually dimitic, clamp connections present, hyphae hyaline or coloured, skeletal hyphae not encrusted, thick-walled hymenial cystidia sometimes present. Basidia ± urniform or narrowly clavate, not elongate, 4-spored. Basidiospores ellipsoidal to cylindrical, smooth, not staining in iodine.

Significant Genera: *Cystostereum*.

Distribution: Widespread, especially in temperate regions.

Economic Significance: None is known.

Ecology: On rotting coniferous or angiosperm wood, probably saprobic.

Notes: Initial studies suggest this family belongs in the phlebioid clade of the homobasidiomycetes, but more phylogenetic research is needed before the limits of the family and its relationships are elucidated.

References: **Chamuris, G.P.** (1986). The *Cystostereum pini-canadense* complex in North America. *Mycologia* 78: 380–390; **Chamuris, G.P.** (1988). The non-stipitate stereoid fungi in the northeastern United States and adjacent Canada. *Mycol. Mem.* 14: 247 pp.; **Ginns, J.** (1998). Genera of the North American Corticiaceae sensu lato. *Mycologia* 90: 1–35; **Hjortstam, K.** (1991). *Athelopsis* instead of *Pteridomyces* (Corticiaceae, Basidiomycetes). *Mycotaxon* 42: 149–154; **Larsson, K.-H.; Larsson, E.; Koljalg, U.** (2004). High phylogenetic diversity among corticioid homobasidiomycetes. *Mycol. Res.* 108: 983–1002.

Cyttariaceae Speg. 1887
Cyttariales: Ascomycota

Stromata large and conspicuous, usually ± globose, fleshy or leathery, orange or brown, often containing numerous ascomata which give the stroma the appearance of a golf ball. Ascomata large, apothecial, strongly concave to nearly globose with a wide opening, the wall often brightly coloured. Interascal tissue of numerous filiform paraphyses. Asci developing synchronously or sequentially, cylindrical, with active discharge, the apex often with a J+ apical ring, opening irregularly. Ascospores ± ellipsoidal, aseptate, pale grey, thin-walled, without a sheath. Anamorph pycnidial, with fruit bodies usually predating the teleomorph and formed on the surface of developing stromata, the wall blackish and fleshy. Conidiogenous cells lining the entire inner surface of the conidioma, proliferating percurrently. Conidia small, hyaline, aseptate, probably spermatial in function.

Significant Genera: *Cyttaria*.

Distribution: Known from southern South America and Australasia, more or less coexistent with the host trees.

Economic Significance: Wood quality of infected trees is greatly diminished due to disruption of growth processes leading to extensive gall formation. Some species are edible and sold in local markets.

Cyttaria berteroi, Chile; pycnidia on stroma surface

Cyttaria darwinii, Tierra del Fuego, Argentina; stromata

Cyttaria hariotii, Tierra del Fuego, Argentina; habit on *Nothofagus betuloides*

Ecology: Restricted to and biotrophic on wood of *Nothofagus* species.

Notes: At one time treated in its own order. Molecular evidence indicates a position within the *Helotiales*, though that order may be polyphyletic. Coevolution of *Cyttaria* with both its plant host and with fungus-feeding insects has been studied in some detail.

References: **Crisci, J.V.; Gamundí, I.J.; Cabello, M.N.** (1988). A cladistic analysis of the genus *Cyttaria* (*Fungi – Ascomycotina*). *Cladistics* 4: 279–290; **Döring, H.; Triebel, D.** (1998). Phylogenetic relationships of *Bulgaria* inferred by 18S rDNA sequence analysis. *Cryptog. Bryol.-Lichénol.* 19: 123–136; **Gamundí, I.J.** (1991). Review of recent advances in the knowledge of the *Cyttariales*. *Syst. Ascom.* 10: 69–77; **Humphries, C.J.; Cox, J.M.; Nielsen, E.S.** (1986). *Nothofagus* and its parasites: a cladistic approach to coevolution. *Coevolution and Systematics* (Oxford) 32: 55–76; **Landvik, S.; Kristiansen, R.; Schumacher, T.** (1998). Phylogenetic and structural studies in the *Thelebolaceae* (*Ascomycota*). *Mycoscience* 39: 49–56; **Minter, D.W.; Cannon, P.F.; Peredo, H.L.** (1987). South American species of *Cyttaria* (a remarkable and beautiful group of edible ascomycetes). *Mycologist* 21: 7–11.

Dacampiaceae Körb. 1855
Pleosporales: Ascomycota

Ascomata perithecial, globose to obpyriform, black, thick-walled, sometimes with hyphal appendages, with a well-developed periphysate ostiole; peridium thick, composed of three distinct layers. Interascal tissue poorly to well-developed, composed of narrow cellular pseudoparaphyses, sometimes immersed in gel. Asci ± cylindrical, fissitunicate, without clear apical structures. Ascospores hyaline or brown, usually septate, sometimes muriform, sometimes with a gelatinous sheath. Anamorphs coelomycetous.

Significant Genera: *Polycoccum, Weddellomyces*.

Distribution: Widespread.

Economic Significance: None is known.

Ecology: Saprobic, necrotrophic or biotrophic, many lichenicolous, rarely lichenized.

Notes: Probably polyphyletic, but almost no molecular data are available.

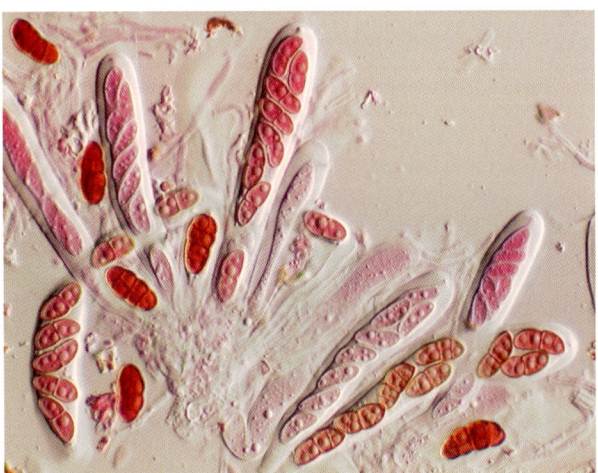

Dacampia rufescentis, Yorkshire, UK; asci and ascospores

References: Alstrup, V.; Hawksworth, D.L. (1990). The lichenicolous fungi of Greenland. *Meddr Grønland* Biosc. 31: 90 pp.; Hawksworth, D.L.; Diederich, P. (1988). A synopsis of the genus *Polycoccum* (*Dothideales*), with a key to accepted species. *Trans. Br. mycol. Soc.* 90: 293–312.

Dacrymycetaceae J. Schröt. 1888
Dacrymycetales: Basidiomycota

Basidioma pustulate or hemispherical to clavate or elongate, rarely cupulate, sometimes branched, gelatinous, usually yellow or orange, hymenium covering much of the external surface, glabrous or rarely with cortical hairs. Hyphal system generally monomitic, hyphal walls sometimes thickened and glassy, with a gelatinous outer layer, often staining brown in iodine, clamp connections present or absent, sessile or loop-like. Basidia varied in shape, elongate to urniform, with 2 well-developed sterigmata making the basidium appear bifurcate. Basidiospores ellipsoidal, ± cylindrical or allantoid, sometimes becoming multiseptate, hyaline, thin-walled, not staining in iodine. Anamorph hyphomycetous, conidia generally formed by budding from vegetative hyphae or basidiospores, thick-walled resting spores also sometimes present.

Significant Genera: *Calocera, Dacrymyces, Guepiniopsis*.

Dacrymyces stillatus, Vaucluse, France; basidiomata

Calocera viscosa, Surrey, UK; basidioma

Distribution: Widespread.

Economic Significance: None is known. Species are known to produce antibiotics, but these are not exploited commercially.

Ecology: Saprobic on rotten coniferous or broad-leaved wood and bark.

Notes: A distinctive family considered to represent a basal lineage of the *Basidiomycota*, recognizable morphologically especially by its bifurcate basidia. The *Cerinomycetaceae* should perhaps be merged with this group.

References: **Binder, M.; Hibbett, D.S.** (2002). Higher-level phylogenetic relationships of homobasidiomycetes (mushroom-forming fungi) inferred from four rDNA regions. *Mol. Phylogen. Evol.* 22: 76–90; **Ing, B.** (1990). Profiles of Fungi 25. *Calocera pallidospathulata* Reid. *Mycologist* 4: 34; **McNabb, R.F.R.** (1965). Taxonomic studies on *Dacrymycetaceae* III. *Dacryopinax* Martin. *N.Z. Jl Bot.* 3: 59–72; **McNabb, R.F.R.** (1965). Taxonomic studies on *Dacrymycetaceae* II. *Calocera* (Fries) Fries. *N.Z. Jl Bot.* 3: 31–58; **Mossebo, D.C.; Akoa, A.; Atangana Eteme, R.** (2001). Ultrastructural studies of the patterns of conidiogenesis in *Dacrymyces stillatus* Nees: Fries (basidiomycete). *Cryptog. Mycol.* 22: 119–138; **Oberwinkler, F.** (1987). *Heterobasidiomycetes* with ontogenetic yeast-stages – systematic and phylogenetic aspects. *Stud. Mycol.* 30: 61–74; **Reid, D.A.** (1974). A monograph of the British *Dacrymycetales*. *Trans. Br. mycol. Soc.* 62: 433–494; **Weiss, M.; Oberwinkler, F.** (2001). Phylogenetic relationships in *Auriculariales* and related groups – hypotheses derived from nuclear ribosomal DNA sequences. *Mycol. Res.* 105: 403–415.

Significant Genera: *Dactylospora*.

Distribution: Widespread.

Economic Significance: None is known.

Ecology: Saprobic on wood, on liverworts, or commensalistic on lichens.

Notes: Very poorly known; no recent information is available.

References: **Au, D.W.T.; Vrijmoed, L.L.P.; Jones, E.B.G.** (1996). Ultrastructure of asci and ascospores of the mangrove ascomycete *Dactylospora haliotrepha*. *Mycoscience* 37: 129–135; **Döbbeler, P.; Triebel, D.** (1985). Hepaticole Vertreter der Gattungen *Muellerella* und *Dactylospora* (*Ascomycetes*). *Bot. Jb.* 107: 503–519; **Jones, E.B.G.; Abdel-Wahab, M.A.; Alias, S.A.; Hsieh, S.Y.** (1999). *Dactylospora mangrovei* sp. nov. (*Discomycetes, Ascomycota*) from mangrove wood. *Mycoscience* 40: 317–320; **Kohlmeyer, J.; Volkmann-Kohlmeyer, B.** (1998). *Dactylospora canariensis* sp. nov. and notes on *D. haliotrepha*. *Mycotaxon* 67: 247–250; **Wedin, M.; Kondratyuk, S.Y.** (1997). *Dactylospora plectocarpoides*, a gall-forming species of *Arthonia* on *Pseudocyphellaria*. *Lichenologist* 29: 97–102.

Dactylosporaceae Bellem. & Hafellner 1982
Lecanorales: Ascomycota

Stromata absent. Ascomata apothecial, ± superficial, sometimes shortly stipitate, black, the wall composed of dark brown pseudoparenchymatous tissue, sometimes poorly developed. Interascal tissue of sparsely branched narrow paraphyses, the apices swollen and pigmented. Asci cylindric-clavate, persistent, thick-walled especially at the apex, not fissitunicate, with an outer J+ gelatinized layer, without a well-developed apical cap or ocular chamber. Ascospores brown, septate, occasionally ornamented. Anamorphs not known.

Delitschiaceae M.E. Barr 2000
Pleosporales: Ascomycota

Stromata absent. Ascomata perithecial, pyriform, pale brown to black, membranous or coriaceous, with a well-developed lysigenous periphysate ostiole. Interascal tissue copious, of cellular pseudoparaphyses. Asci cylindrical, fissitunicate, with refractive apical structures, not staining in iodine. Ascospores dark brown, smooth, with a ± median transverse or oblique strongly constricted septum, sometimes fragmenting, with germ slits in each cell, usually with a conspicuous gelatinous sheath. Anamorphs unknown.

Significant Genera: *Delitschia*.

Dactylospora parasitica, Devon, UK; ascomata on thallus of *Pertusaria* sp.

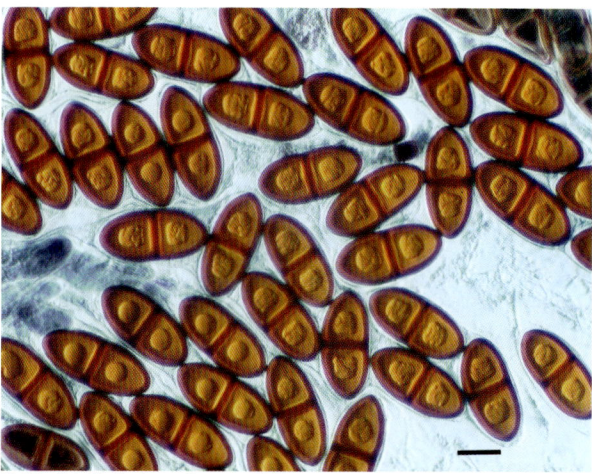

Delitschia furfuracea, Surrey, UK; ascospores

Distribution: Widespread, particularly in temperate zones.

Economic Significance: None is known.

Ecology: Saprobic, in herbivore dung.

Notes: Until recently placed within the *Sporormiaceae* but this family differs primarily in ascus and ostiolar structure and provides an excellent example of convergent evolution. It appears to be one of the most basal families of the *Dothideomycetes*.

References: **Kruys, Å.; Eriksson, O.E.; Wedin, M.** (2006). Phylogenetic relationships of coprophilous *Pleosporales* (*Dothideomycetes, Ascomycota*), and the classification of some bitunicate taxa of unknown position. *Mycol. Res.* 110: 527–536; **Liew, E.C.Y.; Aptroot, A.; Hyde, K.D.** (2000). Phylogenetic significance of the pseudoparaphyses in loculoascomycete taxonomy. *Mol. Phylogen. Evol.* 16: 392–402; **Winka, K.; Eriksson, O.E.** (2000). Adding to the bitunicate puzzle – studies on the systematic position of five aberrant ascomycete taxa. *Phylogenetic Relationships Within the Ascomycota Based on 18S rDNA Sequences, Akademisk Avhandling* [Thesis (PhD), Department of Ecology and Environmental Science, Umeå University] (Umeå): [17] pp.

Notes: A polymorphic and poorly defined family that is in urgent need of revision. Molecular data suggest that *Mollisia* and its relatives may be placed more appropriately in the *Vibrisseaceae*, but more studies are needed.

Mollisia cinerea, Hampshire, UK; ascomata

Dermateaceae Fr. 1849
Helotiales: Ascomycota

Stroma (sclerotia) absent. Ascomata apothecial, small, flat or concave, usually sessile, usually grey-brown or black, occasionally immersed in plant tissue, then sometimes with a specialized opening mechanism; usually with a well-defined margin which is often downy but rarely with distinct hairs; excipulum composed of ± brown, thin-walled or thick-walled, isodiametric cells. Interascal tissue of simple paraphyses. Asci cylindrical to cylindric-clavate, small, usually with a well-developed J+ or J– ring. Ascospores small, hyaline, septate or aseptate, often elongated. Anamorphs very varied, including both hyphomycetous and coelomycetous forms.

Significant Genera: *Diplocarpon, Mollisia, Oculimacula, Pezicula, Pyrenopeziza*. Anamorphs: *Helgardia, Marssonina*.

Distribution: Cosmopolitan.

Economic Significance: *Oculimacula* species are economically significant cereal pathogens and *Pezicula* and its relatives cause damage to roots and induce stem cankers of woody plants. *Diplocarpon rosae* causes black spot of roses.

Ecology: Saprobic or parasitic on herbaceous and woody material.

Diplocarpon rosae, Shenyang, China; symptoms on rose leaves

References: **Abeln, E.C.A.; Pagter, M.A. de; Verkley, G.J.M.** (2000). Phylogeny of *Pezicula, Dermea* and *Neofabraea* inferred from partial sequences of the nuclear ribosomal RNA gene cluster. *Mycologia* 92: 685–693; **Baral, H.O.** (1987). Der Apikalapparat der *Helotiales*. Eine lichtmikroskopische Studie über Arten mit Amyloidring. *Z. Mykol.* 53: 119–136; **Cunnington, J.H.** (2004). Three *Neofabraea* species on pome fruit in Australia. *Australas. Pl. Path.* 33: 453–454; **Dyer, P.S.; Furneaux, P.A.; Douhan, G.; Murray, T.D.** (2001). A multiplex PCR test for determination of mating type applied to the plant pathogens *Tapesia yallundae* and *Tapesia acuformis. Fungal Genetics Biol.* 33: 173–180; **Dyer, P.S.; Nicholson, P.; Lucas, J.A.; Peberdy, J.F.** (1996). *Tapesia acuformis* as a causal agent of eyespot disease of cereals and evidence for a heterothallic mating system using molecular markers. *Mycol. Res.* 100: 1219–1226; **Gamundí, I.J.** (1987). Fungi, Ascomycetes. Cyttariales, Helotiales: Geoglossaceae, Dermateaceae. *Fl. criptog. Tierra del Fuego* 10: 126 pp.; **Jong, S.N. de; Lévesque, C.A.; Verkley, G.J.M.; Abeln, E.C.A.; Rahe, J.E.; Braun, P.G.** (2001). Phylogenetic relationships among *Neofabraea* species causing tree cankers and bull's-eye rot of apple based on DNA sequencing of ITS nuclear rDNA, mitochondrial rDNA, and the β-tubulin gene. *Mycol. Res.* 105: 658–669; **Nauta, M.M.; Spooner, B.** (2000). British *Dermateaceae*: 4B. *Dermateoideae* genera B-E. *Mycologist* 14: 21–28; **Pärtel, K.; Raitviir, A.** (2005). The ultrastructure of the ascus apical apparatus of some *Dermateaceae* (*Helotiales*). *Mycol. Prog.* 4:

149–159; **Spiers, A.G.; Hopcroft, D.H.** (1998). Morphology of *Drepanopeziza* species pathogenic to poplars. *Mycol. Res.* 102: 1025–1037; **Stewart, E.L.; Liu, Z.W.; Crous, P.W.; Szabo, L.J.** (1999). Phylogenetic relationships among some cercosporoid anamorphs of *Mycosphaerella* based on rDNA sequence analysis. *Mycol. Res.* 103: 1491–1499; **Verkley, G.J.M.** (1999). A Monograph of the Genus *Pezicula* and its Anamorphs. *Stud. Mycol.* 44: 180 pp.; **Verkley, G.J.M.; Zijlstra, J.D.; Summerbell, R.C.; Berendse, F.** (2003). Phylogeny and taxonomy of root-inhabiting *Cryptosporiopsis* species, and *C. rhizophila* sp. nov., a fungus inhabiting roots of several *Ericaceae*. *Mycol. Res.* 107: 689–698; **Wang, Z.; Binder, M.; Schoch, C.L.; Johnston, P.R.; Spatafora, J.W.; Hibbett, D.S.** (2006). Evolution of helotialean fungi (*Leotiomycetes, Pezizomycotina*): a nuclear rDNA phylogeny. *Mol. Phylogen. Evol.* 41: 295–312.

Dermatosoraceae Ványk 2001
Ustilaginales: Basidiomycota

Sori formed in swollen plant ovaries with a central columella, at first covered by a thick black peridial membrane that ruptures irregularly to release the spores. Hyphae intracellular, the septa lacking pores. Teliospores formed in many-spored, irregular, multifaceted, persistent, dark brown spore balls that are surrounded by a cortex of sterile cells and eventually break down irregularly; globose to ovoid, smooth or reticulate, germinating directly to form basidia. Basidia cylindrical, with three transverse septa, producing basidiospores from the distal ends of each cell, laterally in the three lower cells and terminal in the apical cell.

Significant Genera: *Dermatosorus*.

Distribution: Known from south and east Asia and Australia.

Dermatosorus eleocharidis, Queensland, Australia; sori

Economic Significance: None is known.

Ecology: Parasitic in ovaries of members of the *Cyperaceae*.

Notes: One of a series of families with *Ustilago*-like teliospore germination, perhaps most closely related to the *Cintractiaceae*. Only one genus is known.

References: **Begerow, D.; John, B.; Oberwinkler, F.** (2004). Evolutionary relationships among β-tubulin gene sequences of basidiomycetous fungi. *Mycol. Res.* 108: 1257–1263; **Piepenbring, M.; Begerow, D.; Oberwinkler, F.** (1999). Molecular sequence data assess the value of morphological characteristics for a phylogenetic classification of species of *Cintractia*. *Mycologia* 91: 485–498; **Ványk, K.** (1987). The genus *Dermatosorus* (*Ustilaginales*). *Trans. Br. mycol. Soc.* 89: 61–65; **Ványk, K.** (1987). Illustrated genera of smut fungi. *Cryptog. Stud.* 1: 159 pp.; **Ványk, K.** (2001). The emended *Ustilaginaceae* of the modern classificatory system for smut fungi. *Fungal Diversity* 6: 131–147.

Diademaceae Shoemaker & C.E. Babc. 1992
Pleosporales: Ascomycota

Ascomata perithecial, globose to ellipsoid-elongate, typically opening by a disc-like operculum but also by a lysigenous pore; peridium composed of large thick-walled pseudoparenchyma. Interascal tissue composed of cellular pseudoparaphyses. Asci cylindrical to saccate, fissitunicate. Ascospores large, brown, muriform, usually radially flattened.

Significant Genera: *Clathrospora*, *Nimbya*.

Distribution: Widespread, especially in temperate regions.

Economic Significance: A number of species of *Nimbya* (with their *Embellisia* anamorph) are economically significant plant pathogens.

Ecology: Saprobic or parasitic in leaves and stems.

Notes: More research on relationships is needed. *Lewia* may be placed more appropriately in the *Pleosporaceae*.

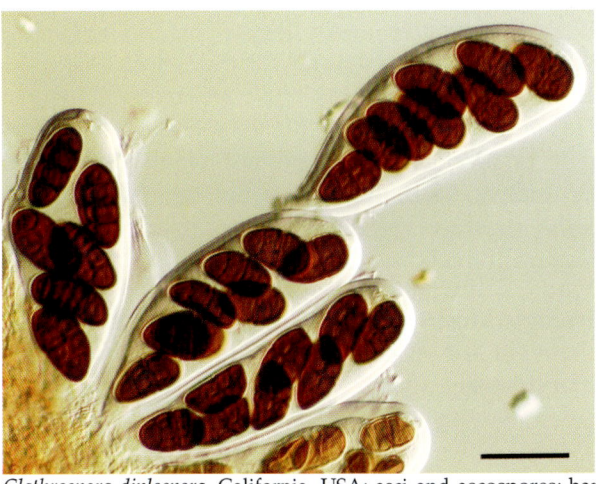

Clathrospora diplospora, California, USA; asci and ascospores; bar = 40 μm

References: **Ahn, Y.M.; Shearer, C.A.** (1998). Reexamination of taxa in *Leptosphaeria* originally described on host species in *Ranunculaceae, Papaveraceae,* and *Magnoliaceae. Can. J. Bot.* 76: 258–280; **Barreto, R.W.; Torres, A.N.L.** (1999). *Nimbya alternantherae* and *Cercospora alternantherae*: two new records of fungal pathogens on *Alternanthera philoxeroides* (alligatorweed) in Brazil. *Australas. Pl. Path.* 28: 103–107; **Dong, J.W.; Chen, W.D.; Crane, J.L.** (1998). Phylogenetic studies of the *Leptosphaeriaceae, Pleosporaceae* and some other *Loculoascomycetes* based on nuclear ribosomal DNA sequences. *Mycol. Res.* 102: 151–156; **Pryor, B.M.; Bigelow, D.M.** (2003). Molecular characterization of *Embellisia* and *Nimbya* species and their relationship to *Alternaria, Ulocladium* and *Stemphylium. Mycologia* 95: 1141–1154; **Shoemaker, R.A.; Babcock, C.E.** (1989). *Diadema. Can. J. Bot.* 67: 1349–1355; **Simmons, E.G.** (2000). *Alternaria* themes and variations (244-286). Species on *Solanaceae. Mycotaxon* 75: 1–115.

Diaporthaceae Höhn. ex Wehm. 1926
Diaporthales: Ascomycota

Stromata usually well-developed, formed from a mixture of fungal and plant tissues. Ascomata perithecial, often long-necked, usually in a ring formation surrounding the conidioma, dark brown to black, membranous to leathery, the necks often protruding from the substratum surface. Interascal tissue absent or with rapidly evanescent, wide, thin-walled paraphysis-like structures, periphyses present. Asci clavate, short-stalked, evanescent at the base, thin-walled, with a large and conspicuous refractive apical ring, usually 8-spored. Ascospores ± hyaline, usually septate, thin-walled, usually without a gelatinous sheath or appendages. Anamorph coelomycetous, stromatic, the fertile chamber often convoluted. Conidiogenous cells formed in clusters on short conidiophores, ± cylindrical, proliferating percurrently. Conidia usually dimorphic, alpha conidia fusiform to fusiform-cylindrical, hyaline, aseptate and conspicuously guttulate, beta conidia filiform and usually hamate.

Significant Genera: *Diaporthe.* Anamorph: *Phomopsis.*

Distribution: Cosmopolitan.

Economic Significance: Many species are pathogens of economically important plants, especially legumes and sunflowers and are often seedborne. Some produce metabolites with toxic effects on grazing animals.

Ecology: Most species are weak pathogens or endophytes in leaf or stem tissues, growing saprobically after the plant dies.

Notes: For many years the family was treated as a synonym of the *Valsaceae*, but recent molecular studies indicate that the two groups should be separated. *Mazzantia* has been shown to belong within the *Diaporthaceae*, and other genera previously placed within the *Valsaceae* in its wide sense may also belong here.

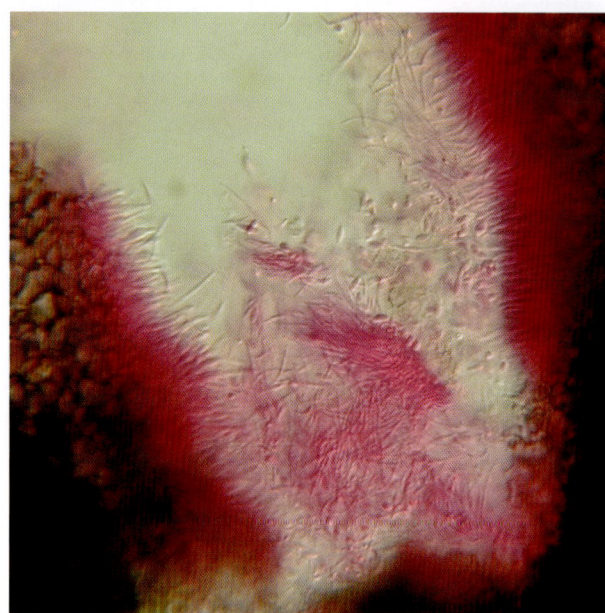

Phomopsis sp. from *Cinnamomum* bark; conidiogenous cells and conidia

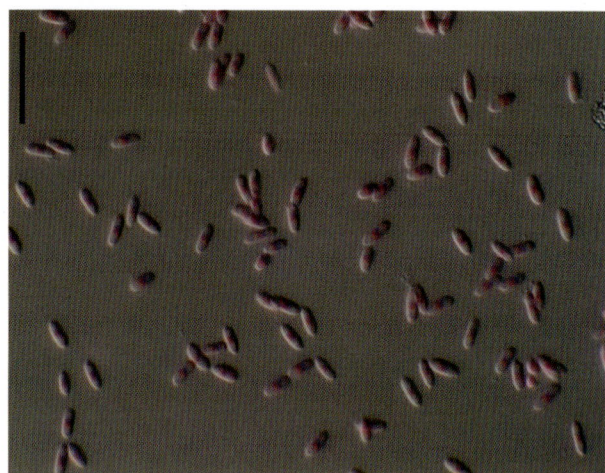

Diaporthe phaseolorum, Mauritius; alpha conidia

References: **Castlebury, L.A.; Farr, D.F.; Rossman, A.Y.; Jaklitsch, W.** (2003). *Diaporthe angelicae* comb. nov., a modern description and placement of *Diaporthopsis* in *Diaporthe. Mycoscience* 44: 203–208; **Castlebury, L.A.; Rossman, A.Y.; Jaklitsch, W.J.; Vasilyeva, L.N.** (2002). A preliminary overview of the *Diaporthales* based on large subunit nuclear ribosomal DNA sequences. *Mycologia* 94: 1017–1031; **Mostert, L.; Crous, P.W.; Kang, J.C.; Phillips, A.J.L.** (2001). Species of *Phomopsis* and a *Libertella* sp. occurring on grapevines with specific reference to South Africa: morphological, cultural, molecular and pathological characterization. *Mycologia* 93: 146–167; **Myburg, H.; Gryzenhout, M.; Wingfield, B.D.; Stipes, R.J.; Wingfield, M.J.** (2004). Phylogenetic relationships of *Cryphonectria* and *Endothia* species, based on DNA sequence data and morphology. *Mycologia* 96: 990–1001; **Rekab, D.; Del Sorbo, G.; Reggio, C.; Zoina, A.; Firrao, G.** (2004). Polymorphisms in nuclear rDNA and mtDNA reveal the polyphyletic nature of isolates of *Phomopsis* pathogenic to sunflower and a tight monophyletic clade of defined geographic origin. *Mycol. Res.* 108: 393–402; **Says-Lesage, V.; Roeckel-Drevet, P.; Viguié, A.; Tourvieille, J.; Nicolas, P.; Tourvieille de

Labrouhe, D. (2002). Molecular variability within *Diaporthe/Phomopsis helianthi* from France. *Phytopathology* 92: 308–313; **Uecker, F.A.** (1989). A timed sequence of development of *Diaporthe phaseolorum* (*Diaporthaceae*) from *Stokesia laevis*. *Mem. N. Y. bot. Gdn* 49: 38–50; **Zhang, A.W.; Hartman, G.L.; Curio-Penny, B.; Pedersen, W.L.; Becker, K.B.** (1999). Molecular detection of *Diaporthe phaseolorum* and *Phomopsis longicolla* from soybean seeds. *Phytopathology* 89: 796–804.

Diatrypaceae Nitschke 1869
Xylariales: Ascomycota

Ascomata perithecial, black, immersed in a well-developed eu- or pseudostroma, often long-necked, usually with sulcate ostioles. Interascal tissue composed of thin-walled, unbranched, cylindrical paraphyses. Asci long-stalked, the apex ± truncate, with a small often J+ apical ring, sometimes polyspored. Ascospores usually small, pale brown, allantoid, without a gelatinous sheath or appendages. Anamorphs coelomycetous, formed as locules within the stroma. Conidiogenous cells ± cylindrical, usually proliferating sympodially. Conidia hyaline, ± filiform.

Significant Genera: *Diatrype*. Anamorph: *Libertella*.

Diatrype bullata, Scotland; stromata

Diatrypella quercina, Surrey, UK; asci and ascospores

Distribution: Widespread, especially in temperate regions.

Economic Significance: Little is known, though components of the *Eutypa lata* complex and similar species are pathogens of woody crops including grapes and apricots.

Ecology: Saprobes and (usually) weak parasites in wood and bark.

Notes: Genus and species concepts are very uncertain. Some genera of the *Xylariaceae* are intermediate between the main body of that family and the *Diatrypaceae*. Some morphological features, e.g. the polysporous asci, appear to have evolved several times within this family.

Eutypa spinosa, Surrey, UK; stromata

References: **Acero, F.J.; González, V.; Sánchez-Ballesteros, J.; Rubio, V.; Checa, J.; Bills, G.F.; Salazar, O.; Platas, G.; Peláez, F.** (2004). Molecular phylogenetic studies on the *Diatrypaceae* based on rDNA-ITS sequences. *Mycologia* 96: 249–259; **Carmarán, C.C.; Romero, A.I.; Giussani, L.M.** (2006). An approach towards a new phylogenetic classification in *Diatrypaceae*. *Fungal Diversity* 23: 67–87; **Chacón, S.; Medel, R.** (1988). *Ascomycetes* lignicolas de México, I: *Diatrypales*. *Revta Mex. Micol.* 4: 323–331; **Chlebicki, A.** (1986). Variability in *Diatrypella favacea* in Poland. *Trans. Br. mycol. Soc.* 86: 441–449; **Dargan, J.S.; Bhatia, M.** (1989). The genus *Diatrype* in western Himalayas. *Nova Hedwigia* 48: 405–418; **DeScenzo, R.A.; Engel, S.R.; Gomez, G.; Jackson, E.L.; Munkvold, G.P.; Weller, J.; Irelan, N.A.** (1999). Genetic analysis of *Eutypa* strains from California supports the presence of two pathogenic species. *Phytopathology* 89: 884–893; **Glawe, D.A.** (1989). Variable proliferation of conidiogenous cells in *Diatrypaceae* and other fungi. *Sydowia* 41: 122–135; **Glawe, D.A.; Rogers, J.D.** (1986). Conidial states of some species of *Diatrypaceae* and *Xylariaceae*. *Can. J. Bot.* 64: 1493–1498; **Ju, Y.M.; Glawe, D.A.; Rogers, J.D.** (1991). Conidial germination in *Eutypa armeniacae* and selected other species of *Diatrypaceae*: implications for the systematics and biology of Diatrypaceous fungi. *Mycotaxon* 41: 311–320; **Lardner, R.; Stummer, B.E.; Sosnowski, M.R.; Scott, E.S.** (2005). Molecular identification and detection of *Eutypa lata* in grapevine. *Mycol. Res.* 109: 799–808; **Peros, J.-P.; Jamaux-Despréaux, I.; Berger, G.; Gerba, D.** (1999). The potential importance of diversity in *Eutypa lata* and co-colonising fungi in explaining variation in development of grapevine dieback. *Mycol. Res.* 103: 1385–1390; **Rappaz, F.**

(1987). Taxonomie et nomenclature des Diatrypacées à asques octosporés. *Mycol. helv.* 2: 285–648; **Trouillas, F.P.; Gubler, W.D.** (2004). Identification and characterization of *Eutypa leptoplaca*, a new pathogen of grapevine in Northern California. *Mycol. Res.* 108: 1195–1204.

Didymosphaeriaceae Munk 1953
Pleosporales: Ascomycota

Ascomata perithecial, globose to flattened, immersed or erumpent, sometimes clypeate, with a clearly defined periphysate ostiole; peridium hyaline to black, thick-walled especially in the apical region, composed of closely interspersed hyphal cells. Interascal tissue composed of thick-walled trabeculate pseudoparaphyses in a gelatinous matrix. Asci cylindrical, fissitunicate, with a small ocular chamber. Ascospores brown, usually 1-septate, the septa often thickened, often thick-walled, sometimes ornamented. Anamorphs *Fusicladium*-like and *Phoma*-like.

Significant Genera: *Didymosphaeria*.

Distribution: Cosmopolitan.

Economic Significance: None is known.

Ecology: Saprobes in both woody and herbaceous material, also parasitic on other fungi.

Notes: The limits and relationships of this family are uncertain; only minimal molecular data are available.

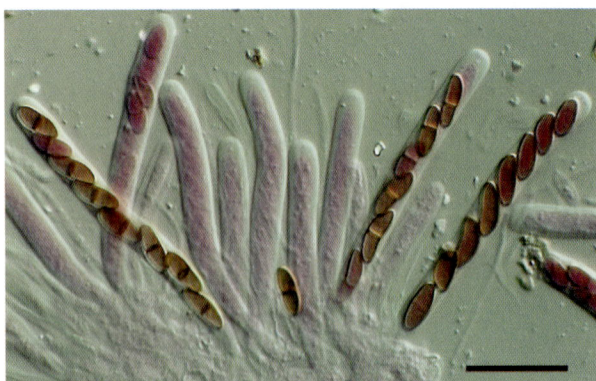

Didymosphaeria futilis, India; asci and ascospores

References: **Aptroot, A.** (1995). A monograph of *Didymosphaeria*. *Stud. Mycol.* 37: 160 pp.; **Aptroot, A.** (1995). Redisposition of some species excluded from *Didymosphaeria* (Ascomycotina). *Nova Hedwigia* 60: 325–379; **Hyde, K.D.; Aptroot, A.; Fröhlich, J.; Taylor, J.E.** (1999). Fungi from palms. XLII. *Didymosphaeria* and similar ascomycetes from palms. *Nova Hedwigia* 69: 449–471; **Kohlmeyer, J.; Volkmann-Kohlmeyer, B.** (1990). Revision of marine species of *Didymosphaeria* (Ascomycotina). *Mycol. Res.* 94: 685–690.

Dimargaritaceae R.K. Benj. 1959
Dimargaritales: Zygomycota

Vegetative thallus a mycelium, complex fruit bodies absent. Sporophores arising from the substratum mycelium, erect, ascending or repent, simple or branched, sometimes setiform, septate, the septa with a unique plug bearing two refractive spheres, one on each side of the septum; sporogenous region characterized by either complex branching, a helical portion or fertile vesicles; bearing 2-spored merosporangia. Merosporangia broadly ellipsoid, bisporous, lacking an obvious columella; wall thin, persistent; forming a liquid droplet or remaining dry at maturity. Zygospores globose; wall pale brown, translucent, smooth or ornamented and appearing punctate; suspensors opposed, smooth, more or less indistinguishable for the vegetative hyphae; heterothallic or rarely homothallic.

Significant Genera: *Dimargaris*, *Tieghemiomyces*.

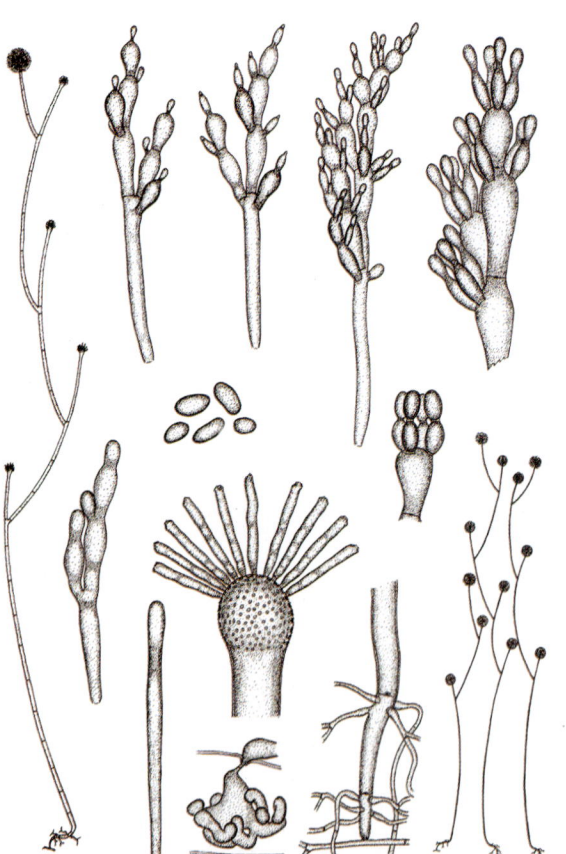

Dimargaris cristalligena; sporophores, merosporangia and zygospores. Reproduced with permission from *Aliso*, 1959.

Distribution: Widespread but very rarely reported; some presumed to be restricted to warm dry areas, others with a temperate range.

Economic Significance: None is known.

Ecology: Frequently isolated as a component of the early coprophilous succession (possibly obligately so); all are obligate haustorial mycoparasites with a host range restricted to members of the *Mucorales* or, in one case, *Chaetomium* and related ascomycetes.

Notes: Some morphological characters appear to have evolved to facilitate spore liberation in the specialized ecosystem associated with rodent dung. The regularly formed septa with the unique plug bearing two sphaerical refractive structures (from which the family name is derived) have suggested a close affinity with the *Kickxellales* and *Harpellales* and this is supported by some molecular studies.

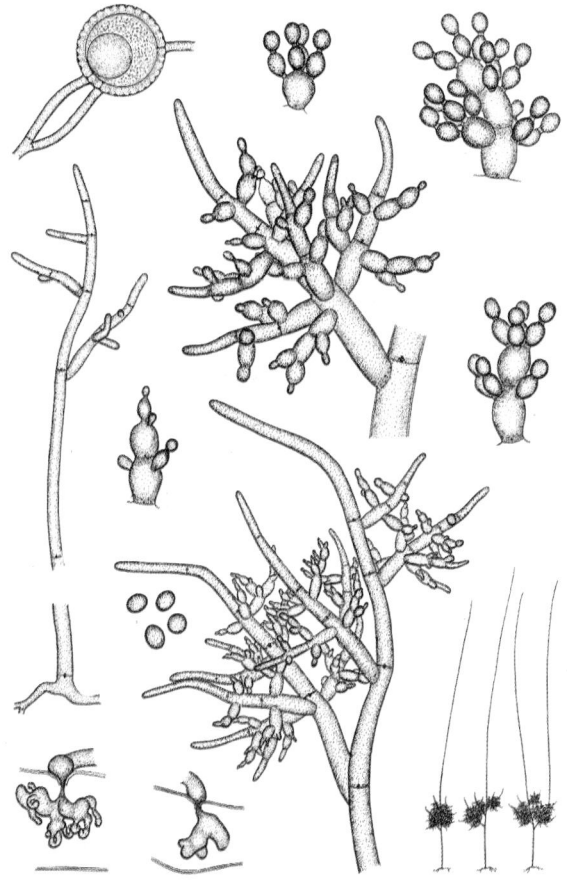

Tieghemiomyces californicus; sporophores, merosporangia and zygospores. Reproduced with permission from *Aliso*, 1959

References: **Bebłowska, M.** (1992). Four noteworthy coprophilic *Mucorales* from eastern Poland. *Acta Mycologica* Warszawa 27: 271–276; **Benjamin, R.K.** (1959). The Merosporangiferous *Mucorales*. *Aliso* 4: 321–433; **Benjamin, R.K.** (1961). Addenda to 'The Merosporangiferous *Mucorales*'. *Aliso* 5: 11–19; **Benjamin, R.K.** (1963). Addenda to 'The Merosporangiferous *Mucorales*' II. *Aliso* 5: 273–288; **Benjamin, R.K.** (1965). Addenda to 'The Merosporangiferous *Mucorales*' III Dimargaris. *Aliso* 6: 1–10; **Tanabe, Y.; Saikawa, M.; Watanabe, M.M.; Sugiyama, J.** (2004). Molecular phylogeny of *Zygomycota* based on EF-1α and RPB1 sequences: limitations and utility of alternative markers to rDNA. *Mol. Phylogen. Evol.* 30: 438–449; **Wrzosek, M.; Gajowniczek, Z.** (1998). Some zygomycetous fungi new to Poland. *Acta Mycologica* Warszawa 33: 265–271.

Diplocystidiaceae Kreisel 1974
Boletales: Basidiomycota

Fruit-body consisting of a thin effuse stroma from which multiple basidiomata emerge simultaneously, each forming from a separate cupulate exoperidium. Basidiomata gasteroid, ± globose to turbinate, cartilaginous, yellowish, smooth, opening by a small irregularly defined aperture. Basidia not described. Basidiospores globose to ellipsoidal, brown, initially smooth but becoming punctate or echinulate.

Significant Genera: *Diplocystis*.

Distribution: Known from the Caribbean.

Economic Significance: None is known.

Ecology: On dead wood, presumably saprobic.

Notes: A poorly known family, with only one species now accepted. It might be related to the *Broomeiaceae*, but molecular data are not available.

References: **Miller, O.K.; Miller, H.H.** (1988). *Gasteromycetes. Morphological and Development Features with Keys to the Orders, Families, and Genera* (Eureka): 157 pp.

Dipodascaceae Engl. & E. Gilg 1924
Saccharomycetales: Ascomycota

Mycelium well-developed, lacking a polysaccharide sheath, the septa with clusters of minute pores, fragmenting to produce thallic conidia. Asci formed by fusion of gametangia from adjacent cells or separate mycelia, usually elongate, ± persistent, 1-spored or multispored. Ascospores released from the attenuated apex, usually ± ellipsoidal, rarely ornamented, usually with a mucous sheath, not blueing in iodine.

Significant Genera: *Dipodascus*. Anamorph: *Geotrichum*.

Distribution: Widespread.

Economic Significance: Species of *Geotrichum* are important spoilage organisms in the food industry.

Ecology: In decaying tissues and exudates of plants, and plant-derived and animal-derived commodities, often associated with high-nutrient environments.

Notes: Very poorly known; no recent information is available.

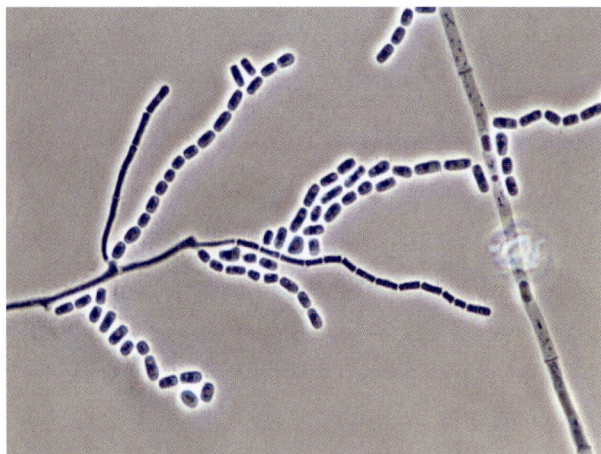

Geotrichum candidum; hyphae fragmenting to form conidia

References: **Hoog, G.S. de; Smith, M.T.; Guého, E.** (1986). A revision of the genus *Geotrichum* and its teleomorphs. *Stud. Mycol.* 29: 131 pp.; **Hoog, G.S. de; Smith, M.T.; Guého, E.** (1998). *Geotrichum* Link: Fries. *The Yeasts, A Taxonomic Study* (Amsterdam): 574–579; **Hoog, G.S. de; Smith, M.T.; Guého, E.** (1998). *Dipodascus* de Lagerheim *The Yeasts, A Taxonomic Study* (Amsterdam): 181–193; **Hoog, G.S. de; Smith, M.T.; Guého, E.** (1998). *Galactomyces*. *The Yeasts, A Taxonomic Study* (Amsterdam): 209–213; **Smith, M.T.; Poot, G.A.; de Cock, A.W.A.M.** (2000). Re-examination of some species of the genus *Geotrichum* Link: Fr. *Antonie van Leeuwenhoek* 77: 71–81; **Ueda-Nishimura, K.; Mikata, K.** (2000). Two distinct 18S rRNA secondary structures in *Dipodoscus* (*Hemiascomycetes*). *Microbiology* (Reading) 146: 1045–1051.

Diporothecaceae Mibey & D. Hawksw. 1995

Uncertain position within Ascomycetes, Ascomycota

Mycelium superficial, dark, not setose, with hyphopodia. Stromata absent. Ascomata perithecial, thin-walled, dark. Interascal tissue absent, the ostiole periphysate. Asci ± clavate, very thin-walled, without apical structures, evanescent, 8-spored. Ascospores dark brown, 2-septate, the septa near the apices, with a separable perispore with poroid ends. Anamorph unknown.

Significant Genera: *Diporotheca*.

Distribution: Only known from the north-western USA.

Economic Significance: None is known.

Ecology: Biotrophic or necrotrophic on roots.

Notes: Very poorly known; no recent information is available.

References: **Mibey, R.K.; Hawksworth, D.L.** (1995). *Diporothecaceae*, a new family of ascomycetes, and the term 'hyphopodium'. *Syst. Ascom.* 14: 25–31.

Discinaceae Benedix 1961

Pezizales: Ascomycota

Ascomata variable, epigeous or hypogeous, if epigeous irregularly cupulate to stipitate, the hymenial surface brown and often somewhat contorted, the stipe pale and also contorted. Interascal tissues of simple branched paraphyses, often swollen at the tip, without a mucous coating, irregular when hypogeous. Asci cylindrical to saccate, the apex rounded, not blueing in iodine. Ascospores hyaline to brown, aseptate, verrucose, tetranucleate, usually with apical gelatinous appendages. Anamorphs unknown.

Significant Genera: *Discina*, *Gyromitra*.

Distribution: Widespread, especially in temperate zones.

Discina ancilis, Scotland; ascomata

Gyromitra tasmanica, Tierra del Fuego, Argentina; ascoma

Economic Significance: Species of *Gyromitra* are toxic and cause poisoning incidents, although some cultures eat the fruit bodies after appropriate cooking procedures.

Ecology: In or on soil and forest litter.

Notes: Very poorly known; no recent information is available.

References: **Abbott, S.P.; Currah, R.S.** (1997). The *Helvellaceae*: systematic revision and occurrence in northern and northwestern North America. *Mycotaxon* 62: 1–125; **Gibson, J.L.; Kimbrough, J.W.** (1988). Ultrastructural observations on *Helvellaceae* (*Pezizales*). II. Ascosporogenesis of *Gyromitra esculenta*. *Can. J. Bot.* 66: 1743–1749; **Kimbrough, J.W.** (1991). Ultrastructural observations on *Helvellaceae* (*Pezizales, Ascomycetes*). V. Septal structures in *Gyromitra*. *Mycol. Res.* 95: 421–426; **Kimbrough, J.W.; Wu, C.-g.; Gibson, J.L.** (1990). Ultrastructural observations on *Helvellaceae* (*Pezizales, Ascomycetes*). IV. Ascospore ontogeny in selected species of *Gyromitra* subgenus *Discina*. *Can. J. Bot.* 68: 317–328; **Landvik, S.; Egger, K.N.; Schumacher, T.** (1997). Towards a subordinal classification of the *Pezizales* (*Ascomycota*): phylogenetic analyses of SSU rDNA sequences. *Nordic Jl Bot.* 17: 403–418; **O'Donnell, K.; Cigelnik, E.; Weber, N.S.; Trappe, J.M.** (1997). Phylogenetic relationships among ascomycetous truffles and the true and false morels inferred from 18S and 28S ribosomal DNA sequence analysis. *Mycologia* 89: 48–65.

Diversisporaceae C. Walker & A. Schüssler 2004
Diversisporales: Glomeromycota

Mycorrhizal species forming arbuscules and vesicles in roots (intraradical), anamorphic; the penetrating hyphae producing finely branched haustorial branches (arbuscules) or coils (pelotons) and vesicles. Spores hypogeous or partly epigeous, glomoid; spore wall comprising a thin outer wall, a laminated structural wall and a flexible inner wall (I–); germination not accompanied by the formation of a germination shield.

Significant Genera: *Diversispora*.

Distribution: Widespread, although undoubtedly more so than current records suggest; apparently restricted to warmer regions.

Diversispora (*Glomus epigaeum*); spores on root

Economic Significance: Enhancement of phosphorus (P) supply to the 'host' is a characteristic of AM (arbuscular mycorrhizal) infections. As such they have a significant influence, either directly or indirectly, on land plants including ecosystem variability and productivity.

Ecology: As mycorrhizal partners of land plants forming arbuscular mycorrhizas; the hosts of AM fungi are primarily found in P-limited tropical and temperate grasslands.

Notes: Only a single species is presently known.

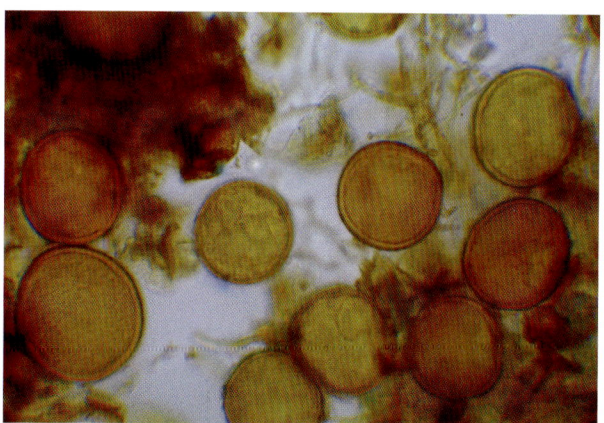

Diversispora (*Glomus epigaeum*); spore

References: **Azcon-Aguilar, C.; Barea, J.M.** (1996). Arbuscular mycorrhizas and biological control of soil-borne plant pathogens – an overview of the mechanisms involved. *Mycorrhiza* 6: 457–464; **Redecker, D.** (2005). *Glomeromycota* Arbuscular mycorrhizal fungi and their relative(s). Version 01 July 2005. http://tolweb.org/Glomeromycota/28715/2005.07.01 in The Tree of Life Web Project, http://tolweb.org; **Schüßler, A.; Schwarzott, D.; Walker, C.** (2001). A new fungal phylum, the *Glomeromycota*: phylogeny and evolution. *Mycol. Res.* 105: 1413–1421; **van der Heijden, M.A.G.; Klironomos, J.N.; Ursic, M.; Moutoglis, P.; Streitwolf-Engel, R.; Boller, T.; Wiemken, A.; Sanders, I.R.** (1998). Mycorrhizal fungal diversity determines plant biodiversity, ecosystem variability and productivity. *Nature Lond.* 396: 69–72; **Walker, C.; Schüssler, A.** (2004). Nomenclatural clarifications and new taxa in the *Glomeromycota*. *Mycol. Res.* 108: 981–982.

Doassansiaceae R.T. Moore ex P.M. Kirk, P.F. Cannon & J.C. David 2001
Doassansiales: Basidiomycota

Sori developing in leaves, petioles and stems, visible as pale green, yellowish or brownish lesions with spore balls as minute brown dots. Hyphae intercellular, haustoria absent. Spore balls ± persistent, globose or irregular, composed of a central mass of spores, central sterile region absent, sometimes interspersed with sterile cells and surrounded or not by a sterile cortex. Teliospores globose or irregularly faceted, hyaline or yellowish, smooth, thin-walled, germinating to form a short hypha-like promycelium from which a cluster of basidiospores are produced apically.

Significant Genera: *Burrillia*, *Doassansia*.

Distribution: Widespread.

Economic Significance: None is known; no economically important plants are parasitized.

Ecology: Parasitic on leaves and stems of aquatic monocotyledons.

Notes: An isolated family, perhaps most closely related to the *Entylomataceae* or *Microstromataceae*. A diagnostic feature is the pale persistent spore balls immersed in host tissue.

Burrillia pustulata, Wisconsin, USA; sori in leave of *Sagittaria latifolia*

References: **Begerow, D.; Bauer, R.; Boekhout, T.** (2000). Phylogenetic placements of ustilaginomycetous anamorphs as deduced from nuclear LSA rDNA sequences. *Mycol. Res.* 104: 53–60; **Begerow, D.; Bauer, R.; Oberwinkler, F.** (1998). Phylogenetic studies on nuclear large subunit ribosomal DNA sequences of smut fungi and related taxa. *Can. J. Bot.* 75: 2045–2056; **Piepenbring, M.; Bauer, R.; Oberwinkler, F.** (1998). Teliospores of smut fungi. General aspects of teliospore walls and sporogenesis. *Protoplasma* 204: 155–169; **Piepenbring, M.; Oberwinkler, F.** (2003). Integrating morphological and molecular characteristics for a phylogenetic system of smut fungi. *Bot. Jb.* 24: 241–253; **Vánky, K.** (1987). Illustrated genera of smut fungi. *Cryptog. Stud.* 1: 159 pp.; **Vánky, K.** (1998). A survey of the spore-ball-forming smut fungi. *Mycol. Res.* 102: 513–526; **Vánky, K.** (2001). The new classification of the smut fungi, exemplified by Australasian taxa. *Aust. Syst. Bot.* 14: 385–394.

Doassansiopsidaceae Begerow, R. Bauer & Oberw. 1998
Urocystidales: Basidiomycota

Sori developing in leaves, visible as pale yellowish or brownish spots or swellings with spore balls as minute brown dots. Hyphae intercellular, septal pores simple with two membrane caps and two non-membranous inner plates closing the pore, haustoria absent. Spore balls ± persistent, globose or irregular, composed of a central sterile region of pseudoparenchymatous cells surrounded by a layer of spores and a thin sterile cortex. Teliospores globose or irregularly faceted, hyaline or yellowish, smooth, thin-walled, germinating to form a short hypha-like promycelium from which a cluster of basidiospores are produced apically.

Significant Genera: *Doassansiopsis*.

Distribution: Widespread, in both temperate and tropical zones.

Economic Significance: None is known.

Ecology: Parasitic on plants in aquatic habitats.

Notes: At one time assumed to be allied to the *Doassansiaceae*, a family with very similar ecological characteristics and parallels in morphology. However, molecular and ultrstructural data indicate a relationship to the *Ustilaginomycetes* rather than the *Exobasidiomycetes*.

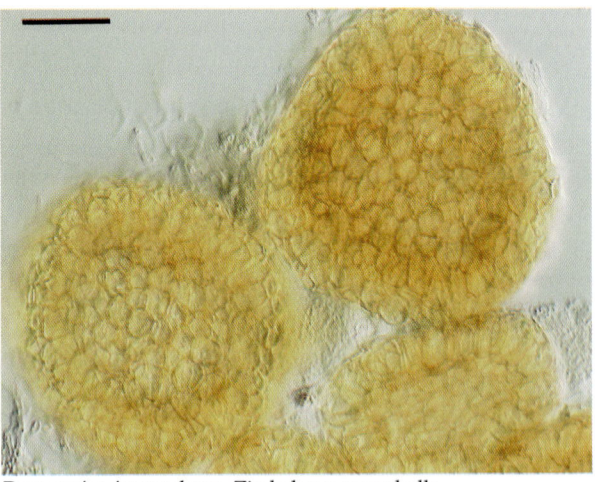

Doassansiopsis nymphaeae, Zimbabwe; spore balls

References: **Bauer, R.; Oberwinkler, F.; Vánky, K.** (1997). Ultrastructural markers and systematics in smut fungi and allied taxa. *Can. J. Bot.* 75: 1273–1314; **Begerow, D.; Bauer, R.; Oberwinkler, F.** (1998). Phylogenetic studies on nuclear large subunit ribosomal DNA sequences of smut fungi and related taxa. *Can. J. Bot.* 75: 2045–2056; **Begerow, D.; John, B.; Oberwinkler, F.** (2004). Evolutionary relationships among β-tubulin gene sequences of basidiomycetous fungi. *Mycol. Res.* 108: 1257–1263; **Piepenbring, M.; Oberwinkler, F.** (2003). Integrating morphological and molecular characteristics for a phylogenetic system of smut fungi. *Bot. Jb.* 24: 241–253; **Vánky, K.** (1987). Illustrated genera of smut fungi. *Cryptog. Stud.* 1: 159 pp.; **Vánky, K.** (2001). The new classification of the smut fungi, exemplified by Australasian taxa. *Aust. Syst. Bot.* 14: 385–394.

Dothideaceae Chevall. 1826

Dothideales: Ascomycota

Stromata multiloculate, immersed or erumpent, usually pulvinate or crustose; black, thick-walled, opening by an apical, usually lysigenous pore. Interascal tissue lacking. Asci saccate or clavate, fissitunicate. Ascospores small, hyaline or brown, transversely septate.

Significant Genera: *Auerswaldia*, *Dothidea*.

Distribution: Widespread, especially in the tropics.

Economic Significance: A few species (primarily belonging to the genus *Scirrhia*) cause economically significant damage in conifer plantations.

Ecology: Biotrophic, necrotrophic or saprobic on plant tissue.

Notes: Almost certainly polyphyletic. The *Dothioraceae* and *Mycosphaerellaceae* may be closely related, but more molecular data are needed.

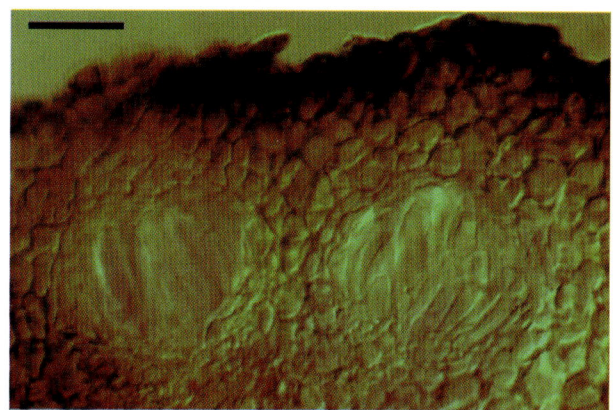
Dothidea sambuci, Bedfordshire, UK; vertical section through stroma

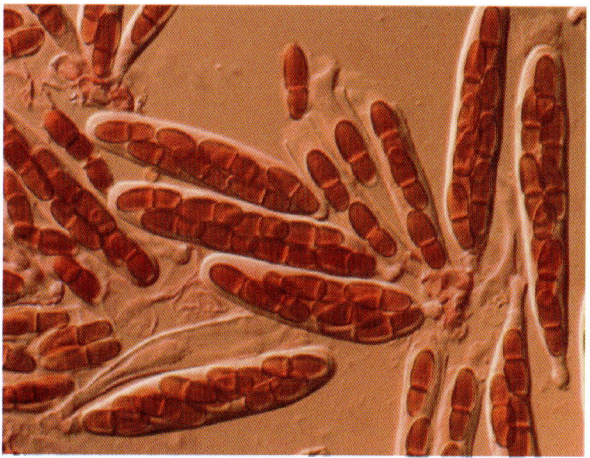
Dothidea sambuci, Czech Republic; asci and ascospores

References: **Butin, H.** (1986). Teleomorph- und Anamorph-Entwicklung von *Scirrhia pini* Funk & Parker auf Nadeln von *Pinus nigra* Arnold. *Sydowia* 38: 20–27; **Hyde, K.D.; Cannon, P.F.** (1999). Fungi Causing Tar Spots on Palms. *Mycol. Pap.* 175: 114 pp.; **Lindemuth, R.; Wirtz, N.; Lumbsch, H.T.** (2001). Phylogenetic analysis of nuclear and mitochondrial rDNA sequences supports the view that loculoascomycetes (*Ascomycota*) are not monophyletic. *Mycol. Res.* 105: 1176–1181; **Silva-Hanlin, D.M.; Hanlin, R.T.** (1999). Small subunit ribosomal RNA gene phylogeny of several loculoascomycetes and its taxonomic implications. *Mycol. Res.* 103: 153–160.

Dothioraceae Theiss. & P. Syd. 1918

Dothideales: Ascomycota

Ascomata immersed to erumpent, globose to pulvinate, uniloculate, sometimes with a central sterile column, opening by a large irregular pore; peridium thick, of large-celled pseudoparenchymatous cells. Interascal tissue absent but asci sometimes separated by stromatic tissue. Asci clavate, fissitunicate, sometimes polysporous. Ascospores aseptate or septate, sometimes muriform. Anamorphs very varied, including coelomycetous and 'black yeast' forms.

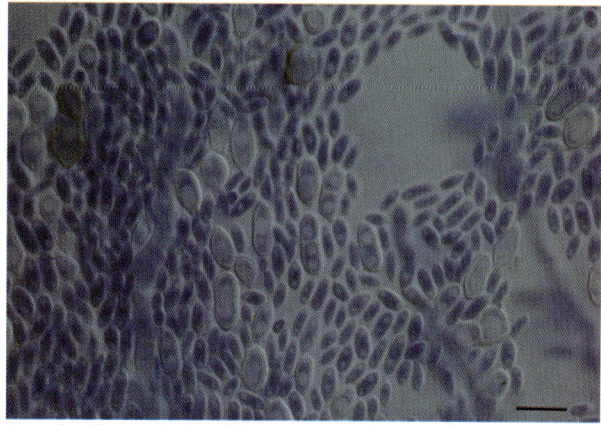
Aureobasidium pullulans, Western Australia; conidia

Significant Genera: *Dothiora*, *Discosphaerina*. Anamorphs: *Aureobasidium*, *Hormonema*, *Kabatiella*.

Distribution: Widespread.

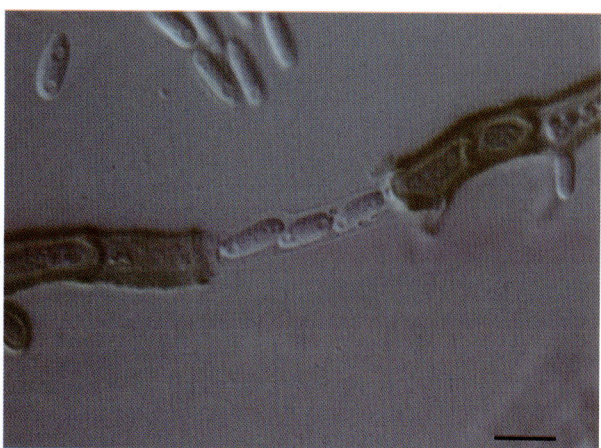
Aureobasidium pullulans, Western Australia; hypha showing internal conidiation

Economic Significance: Some species are significant plant pathogens. *Aureobasidium* and *Hormonema* species are important spoilage organisms and are also implicated as human pathogens.

Ecology: Biotrophic or necrotrophic, usually associated with woody plants.

Notes: Teleomorphs very poorly known; no recent information is available.

References: **Dupont, J.; Laloui, W.; Roquebert, M.F.** (1998). Partial ribosomal DNA sequences show an important divergence between *Phaeoacremonium* species isolated from *Vitis vinifera*. *Mycol. Res.* 102: 631–637; **Hoog, G.S. de; Zalar, P.; Urzì, C.; Leo, F. de; Yurlova, N.A.; Sterflinger, K.** (1999). Relationships of dothideaceous black yeasts and meristematic fungi based on 5.8S and ITS2 rDNA sequence comparison. *Stud. Mycol.* 43: 31–37; **Simon, L.; Bouchet, B.; Caye-Vaugien, C.; Gallant, D.J.** (1995). Pullulan elaboration and differentiation of the resting forms in *Aureobasidium pullulans*. *Can. J. Microbiol.* 41: 35–45; **Sivanesan, A.; Hsieh, W.H.** (1995). A re-appraisal of the systematic status of the ascomycete genus *Yoshinagaia*. *Mycol. Res.* 99: 1295–1298; **Untereiner, W.A.; Naveau, F.A.** (1999). Molecular systematics of the *Herpotrichiellaceae* with an assessment of the phylogenetic positions of *Exophiala dermatitidis* and *Phialophora americana*. *Mycologia* 91: 67–83; **Urzí, C.; De Leo, F.; Lo Passo, C.; Criseo, G.** (1999). Intra-specific diversity of *Aureobasidium pullulans* strains isolated from rocks and other habitats assessed by physiological methods and by random amplified polymorphic DNA (RAPD). *J. Microbiol. Meth.* 36: 95–105; **Verkely, G.J.M.; Starink-Willemse, M.; Iperen, A. van; Abelh, E.C.A.** (2004). Phylogenetic analyses of *Septoria* species based on the ITS and LSU-D2 region of nuclear ribosomal DNA. *Mycologia* 96: 558–571; **Yurlova, N.A.; Hoog, G.S. de; Gerrits van den Ende, A.H.G.** (1999). Taxonomy of *Aureobasidium* and allied genera. *Stud. Mycol.* 43: 63–69.

Eballistraceae R. Bauer, Begerow, A. Nagler & Oberw. 2001
Georgefischeriales: Basidiomycota

Sori forming in leaves of host plants, usually ± linear, initially immersed beneath the epidermis, visible from the surface as agglutinated grey or brown spore masses. Teliospores formed singly (not in spore balls), ± globose or angular due to compression, the wall fairly thin and homogenous, smooth. Basidia formed directly from germinating teliospores, ± cylindrical, aseptate, producing a cluster of 4–5 basidiospores at the apex. Basidiospores passively discharged, triradiate or cylindrical, sometimes becoming septate, mid-brown or olivaceous, thick-walled, smooth. Yeast phase also produced.

Significant Genera: *Eballistra*.

Distribution: Widespread in the tropics.

Economic Significance: One species is a rice pathogen, but does not cause economically significant losses.

Ecology: Parasitic on grasses.

Notes: A sister group to the *Gjaerumiaceae* in the *Georgefischeriales*, this family is characterized by the formation of holobasidia and basidiospores that are passively rather than actively discharged.

References: **Bauer, R.; Begerow, D.; Nagler, A.; Oberwinkler, F.** (2001). The *Georgefischeriales*: a phylogenetic hypothesis. *Mycol. Res.* 105: 416–424; **Bauer, R.; Lutz, M.; Oberwinkler, F.** (2005). *Gjaerumia*, a new genus in the *Georgefischeriales* (*Ustilaginomycetes*). *Mycol. Res.* 109: 1250–1258; **Vánky, K.** (2005). Taxonomic studies on *Ustilaginomycetes* – 25. *Mycotaxon* 91: 217–272.

Significant Genera: *Echinodontium*. Anamorph: *Spiniger*.

Distribution: Widespread, especially in north temperate zones.

Economic Significance: *Echinodontium* species can attack valuable living trees, especially *Chamaecyparis* and *Taxodium* species. Pigments derived from ground-up basidiomata have been used by native traditional artists in North America.

Ecology: Parasitic or saprobic on wood, causing sometimes aggressive white rots of both angiosperms and gymnosperms.

Notes: Relationships of the *Echinodontiaceae* are poorly understood and the type genus appears to be polyphyletic. A link with the *Bondarzewiaceae* is indicated by some molecular data, confirming anatomical studies of the two families. However, other studies suggest a linkage with the *Amylostereaceae*.

Echinodontium tinctorium; basidioma

Echinodontiaceae Donk 1961
Russulales: Basidiomycota

Basidiomata perennial, resupinate, effuse or sessile and pileate, ± glabrous or tomentose, often zonate, woody or corky in texture, the inner tissues often reddish or brown. Hyphal system usually interpreted as dimitic or trimitic, though skeletal hyphae are sometimes not well differentiated, clamp connections present and thick-walled encrusted cystidia often also present. Hymenium smooth to warted or spinose. Basidia clavate, with 4 sterigmata. Basidiospores hyaline to pale yellow, ± ellipsoidal, smooth or echinulate, usually thin-walled, staining blue in iodine.

References: **Binder, M.; Hibbett, D.S.** (2002). Higher-level phylogenetic relationships of homobasidiomycetes (mushroom-forming fungi) inferred from four rDNA regions. *Mol. Phylogen. Evol.* 22: 76–90; **Ginns, J.** (1998). Genera of the North American *Corticiaceae* sensu lato. *Mycologia* 90: 1–35; **Gross, H.L.** (1964). The *Echinodontiaceae*. *Mycopath. Mycol. appl.* 24: 1–26; **Larsson, E.; Larsson, K.-H.** (2003). Phylogenetic relationships of russuloid basidiomycetes with emphasis on aphyllophoralean taxa. *Mycologia* 95: 1037–1065; **Maijala, P.; Harrington, T.C.; Raudaskoski, M.** (2003). A peroxidase gene family and gene trees in *Heterobasidion* and related genera. *Mycologia* 95: 209–221; **Stalpers, J.A.** (1979). *Heterobasidion* (*Fomes*) *annosum* and the *Bondarzewiaceae*. *Taxon* 28: 414–417; **Stalpers, J.A.** (1996). The Aphyllophoraceous Fungi – II. Keys to the Species of the *Hericiales*. *Stud. Mycol.* 40: 185 pp.; **Tabata, M.; Harrington, T.C.; Chen, W.; Abe, Y.** (2000). Molecular phylogeny of species in the genera *Amylostereum* and *Echinodontium*. *Mycoscience* 41: 585–593; **Wilson, A.D.** (1990). The genetics of sexual incompatibility in the Indian paint fungus, *Echinodontium tinctorium*. *Mycologia* 82: 332–341.

Ectolechiaceae Zahlbr. 1905
Lecanorales: Ascomycota

Thallus usually foliose or crustose. Ascomata apothecial, pale, convex, without a well-developed margin. Interascal tissue of very narrow anastomosing pseudoparaphysis-like hyphae, variously pigmented, the apices thin-walled. Asci with a J+ apical cap and a very well-developed ocular chamber, with an outer J+ gelatinous layer, mostly 1- or 2-spored. Ascospores muriform, usually thin-walled. Sometimes with cephalodia. Anamorphs pycnidial, often with campylidia.

Significant Genera: *Calopadia*, *Lopadium*.

Distribution: Primarily tropical.

Economic Significance: None is known.

Ecology: Lichenized with green algae, usually foliicolous.

Notes: Very poorly known; no recent information is available.

Lopadium disciforme, Norway; thallus and ascomata on bryophytes

References: **Kalb, K.; Vězda, A.** (1987). Einige nicht-foliicole Arten der Familie Ectolechiaceae (*Lichenes*) aus Brasilien. *Folia geobot. phytotax.* 22: 287–312; **Lücking, R.** (1998). Foliicolous lichens and their lichenicolous fungi collected during the Smithsonian International Cryptogamic Expedition to Guyana 1996. *Trop. Bryol.* 15: 45–76; **Lücking, R.** (1999). [Additions and corrections to the knowledge of the foliicolous lichen flora of Costa Rica. The family Ectolechiaceae]. *Phyton Horn* 39: 131–165; **Lücking, R.; Lumbsch, H.T.; Elix, J.A.** (1994). Chemistry, anatomy and morphology of foliicolous species of *Fellhanera* and *Badimia* (lichenized *Ascomycotina*: Lecanorales). *Bot. Acta* 107: 393–401; **McCarthy, P.M.; Elix, J.A.; Sérusiaux, E.** (2000). *Kantvilasia* (Lecanorales, Ectolechiaceae), a new foliicolous lichen genus from Tasmania. *Lichenologist* 32: 317–324; **Sérusiaux, E.** (1997). *Sporopodiopsis*, a new genus of lichens (Ectolechiaceae) from S-E Asia. *Abstracta Botanica* 21: 145–152.

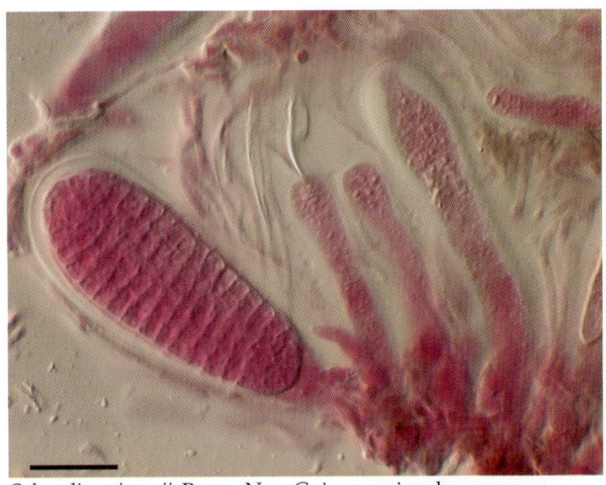
Calopadia puiggarii, Papua New Guinea; asci and ascospore

Badimia dimidiata; ascomata on leaf

Elaphomycetaceae Tul. ex Paol. 1889
Eurotiales: Ascomycota

Stromata absent. Ascomata large, cleistothecial, pale to black, subterranean; peridium very thick, smooth or warted, sometimes covered by a mycelial layer, homogenous or marbled in appearance. Interascal tissue (gleba) of copious unordered hyphae, sometimes developing in locule-like structures, breaking down at maturity. Asci thin-walled, ± globose, evanescent, 2- to 8-spored. Ascospores large, globose, aseptate, at first with a mucous sheath, becoming strongly ornamented and pigmented. Anamorphs unknown.

Significant Genera: *Elaphomyces*.

Distribution: Most species have temperate distributions.

Economic Significance: None is known.

Ecology: Ectomycorrhizal hypogeous fungi, associated with conifers or broadleaved trees.

Notes: The *Elaphomycetales* are remotely related to other ascomycetous truffles, and appear to be aligned with the *Eurotiales*. The genus *Pseudotulostoma* was recently described as a member of the *Elaphomycetaceae*, but it may well deserve its own family. It is immediately distinguishable from *Elaphomyces* by its aerial stalked ascomata growing from a partly buried volva-like structure.

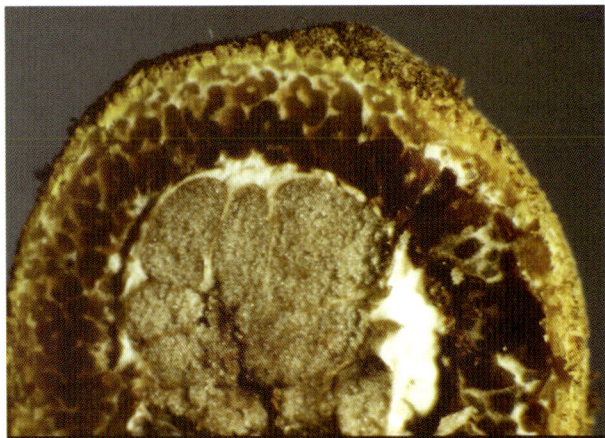

Elaphomyces muricatus, England; section through ascoma

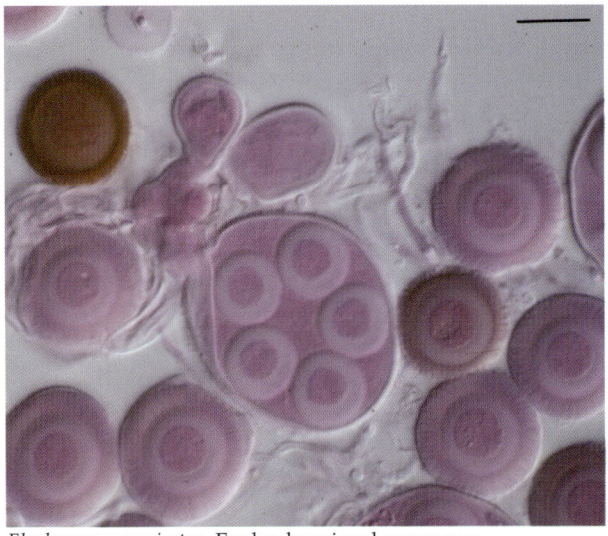

Elaphomyces muricatus, England; asci and ascospores

References: **Landvik, S.; Eriksson, O.E.** (1994). Relationships of *Tuber, Elaphomyces,* and *Cyttaria* (*Ascomycotina*), inferred from 18S rDNA studies. *Ascomycete Systematics. Problems and Perspectives in the Nineties* NATO ASI Series vol. 269 (New York): 225–231; **Landvik, S.; Shailer, N.F.J.; Eriksson, O.E.** (1997). SSU rDNA sequence support for a close relationship between the *Elaphomycetales* and the *Eurotiales* and *Onygenales*. *Mycoscience* 37: 237–241; **Miller, O.K.; Henkel, T.W.; James, T.Y.; Miller, S.L.** (2001). *Pseudotulostoma*, a remarkable new volvate genus in the *Elaphomycetaceae* from Guyana. *Mycol. Res.* 105: 1268–1272; **Samuelson, D.A.; Benny, G.L.; Kimbrough, J.W.** (1987). Ultrastructure of ascospore ornamentation in *Elaphomyces* (*Ascomycetes*). *Mycologia* 79: 571–577; **Xu, A S.** (1999). [The genus *Elaphomyces* and its ecological distribution in Xizang]. *Mycosystema* 18: 238–242; **Zhang, B.C.** (1991). Revision of Chinese species of *Elaphomyces* (*Ascomycotina, Elaphomycetales*). *Mycol. Res.* 95: 973–985; **Zhang, B.C.; Minter, D.W.** (1989). *Elaphomyces spinoreticulatus* sp. nov., with notes on Canadian species of *Elaphomyces*. *Can. J. Bot.* 67: 909–914.

Elixiaceae Lumbsch 1997
Agyriales: Ascomycota

Thallus crustose, indistinct, consisting of dark brown granules on or just under the substratum surface. Ascomata apothecial, sessile, constricted at the base, round, angular or elongate, ± black, with a slit-like hymenial disk and a cupulate outer wall. Interascal tissue of simple branched and anastomosing paraphyses, slightly thickened and pigmented at the apex, immersed in a pigmented epithecial gel. Asci clavate, with an outer J+ gel layer and a narrow J+ tholus, 8-spored. Ascospores hyaline, aseptate, ellipsoidal, without a gelatinous sheath. Anamorph unknown.

Significant Genera: *Elixia*.

Distribution: Known from Europe.

Economic Significance: None is known.

Ecology: Lichenized with green algae.

Notes: Probably more closely related to the *Umbilicariaceae* than the *Agyriaceae* as originally supposed.

References: **Lumbsch, H.T.** (1997). Systematic studies in the suborder *Agyriineae* (*Lecanorales*). *J. Hattori bot. Lab.* 83: 1–73; **Lumbsch, H.T.; Schmitt, I.; Palice, Z.; Wiklund, E.; Ekman, S.; Wedin, M.** (2004). Supraordinal phylogenetic relationships of *Lecanoromycetes* based on a Bayesian analysis of combined nuclear and mitochondrial sequences. *Mol. Phylogen. Evol.* 31: 822–832.

Elsinoaceae Höhn. ex Sacc. & Trotter 1913
Myriangiales: Ascomycota

Ascomata immersed or erumpent, round or elongate, usually crustose, composed of pale gelatinous thin-walled hyphal or pseudoparenchymatous cells, opening by unordered breakdown of the surface layers. Specialized interascal tissue absent. Asci arranged in individual locules, in a single layer or irregularly disposed, saccate to globose, fissitunicate, not blueing in iodine. Ascospores hyaline to brown, septate, sometimes muriform, without a gelatinous sheath or appendages. Anamorphs coelomycetous, acervular, often coalescing, with poorly developed conidiophores and short polyphialidic conidiogenous cells producing hyaline aseptate conidia.

Significant Genera: *Elsinoë*. Anamorph: *Sphaceloma*.

Distribution: Widespread, especially in the tropics.

Economic Significance: Some species cause economically significant scab diseases, especially of citrus.

Elsinoë fawcettii; symptoms on living leaves

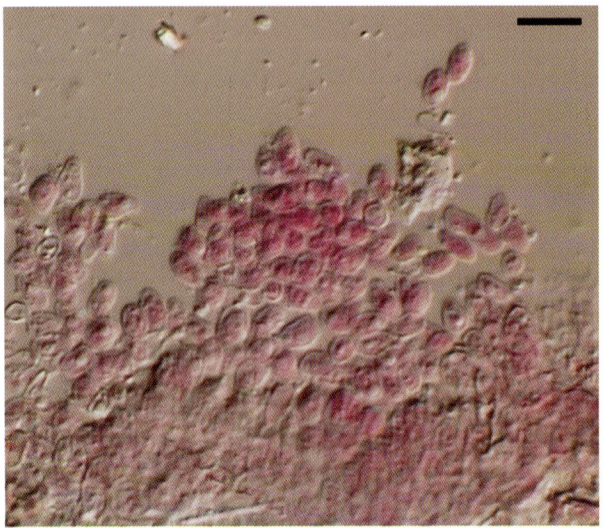
Elsinoë ampelina, Vietnam; conidiogenous cells and conidia

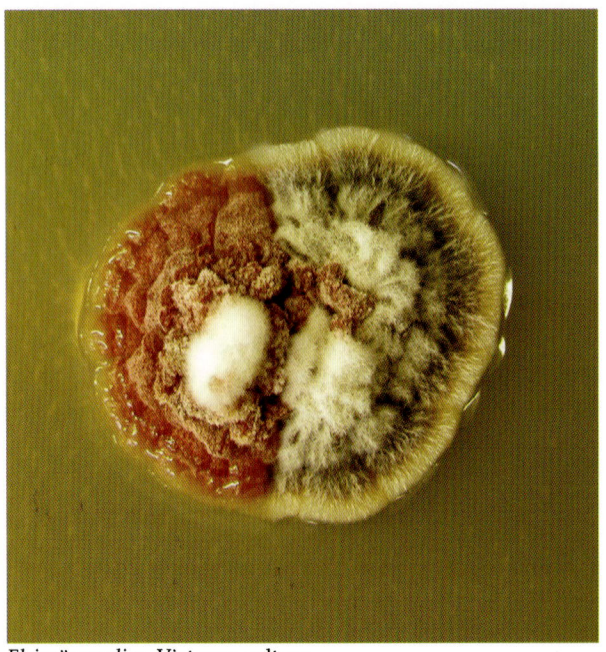
Elsinoë ampelina, Vietnam; culture

Ecology: Biotrophic or necrotropic on leaves, stems and fruits of a wide range of plants.

Notes: A poorly known family, probably with no close relatives.

References: **Alvarez, E.; Mejia, J.F.; Valle, T.L.** (2003). Molecular and pathogenicity characterization of *Sphaceloma manihoticola* isolates from South-central Brazil. *Pl. Dis.* 87: 1322–1328; **Gabel, A.W.; Tiffany, L.H.** (1987). Host-parasite relations and development of *Elsinoë panici*. *Mycologia* 79: 737–744; **Gardner, D.E.; Hodges, C.S.** (1986). Hawaiian forest fungi. VII. A new species of *Elsinoë* on native *Vaccinium*. *Mycologia* 78: 506–508; **Johnston, P.R.; Beever, R.E.** (1994). *Elsinoë dracophylli* sp. nov. *N.Z. Jl Bot.* 32: 519–520; **Palm, M.E.** (1999). Mycology and world trade: a view from the front line. *Mycologia* 91: 1–12; **Ridley, G.S.; Ramsfield, T.D.** (2006). *Elsinoë takoropuku* sp. nov. infects twigs of *Pittosporum tenuifolium* in New Zealand. *Mycologia* 97: 1362–1364; **Swart, L.; Crous, P.W.; Kang, J.C.; Mchau, G.R.A.; Pascoe, I.; Palm, M.E.** (2001). Differentiation of species of *Elsinoë* associated with scab disease of Proteaceae based on morphology, symptomatology, and ITS sequence phylogeny. *Mycologia* 93: 366–379.

Endochytriaceae Sparrow ex D.J.S. Barr 1980
Chytridiales: Chytridiomycota

Thallus eucarpic or holocarpic, monocentric (comprising a single cell), endogenous; zoosporangium endogenous with the exception of the discharge tube, operculate or inoperculate; zoospores posteriorly uniflagellate, with a single globule; resting spore thick-walled, endogenous, germinating to function as an exogenous prosporangium.

Significant Genera: *Diplophlyctis*, *Entophlyctis*.

Distribution: Widespread and apparently temperate.

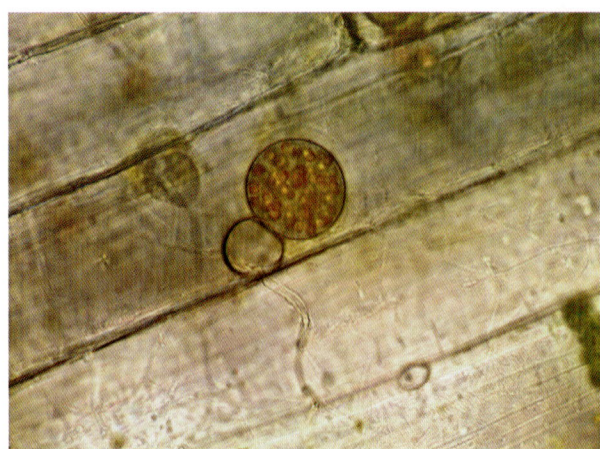
Nephrochytrium aurantium; zoosporangium

Economic Significance: None is known.

Ecology: Saprobic on characeans, sometimes chitinophitic.

Notes: Very poorly known; no recent information is available.

References: **Barr, D.J.S.; Désaulniers, N.L.; Knox, J.S.** (1987). *Catenochytridium hemicysti* n. sp.: morphology, physiology and zoospore ultrastructure. *Mycologia* 79: 587–594; **Chen, S.F.; Chien, C.Y.** (1995). Some chytrids of Taiwan (I). *Bot. Bull. Acad. Sinica* 36: 235–241; **Longcore, J.E.** (1995). Morphology and zoospore ultrastructure of *Entophlyctis luteolus* sp. nov. (*Chytridiales*): implications for chytrid taxonomy. *Mycologia* 87: 25–33; **Shin, W.G.; Boo, S.M.; Longcore, J.E.** (2001). *Entophlyctis apiculata*, a chytrid parasite of *Chlamydomonas* sp. (*Chlorophyceae*). *Can. J. Bot.* 79: 1083–1089.

Endogonaceae Paol. 1889
Endogonales: Zygomycota

Sporocarps mainly hypogeous (where present), up to *ca* 25mm diam, ± globose to irregular in shape, rather thin-walled, forming only zygospores. Zygospores often very large, subglobose to ovoid, thick-walled, reddish to brown and aseptate, formed as bud-like outgrowths from the larger of two unequal gametangia.

Significant Genera: *Endogone*.

Distribution: Widespread.

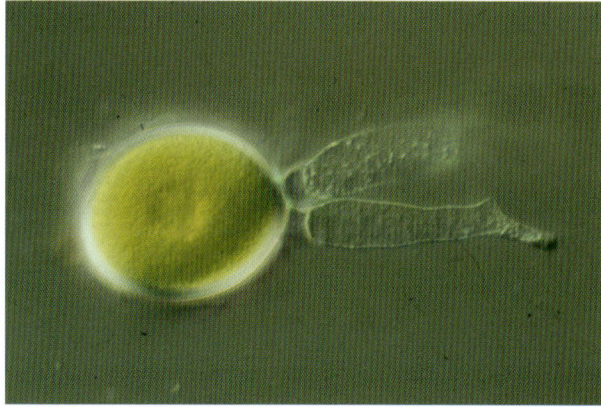

Endogone sp.; zygospore

Endogone lactiflua; sporocarp

Economic Significance: None is known, but the significance of mycorrhizal association is not yet fully understood.

Ecology: Saprobic or forming ectomycorrhiza with roots of annual and perennial plants.

Notes: Very poorly known; no recent information is available. The very simple morphological structures make interpretation difficult, and few molecular data are so far available.

References: **Benny, G.L.; Humber, R.A.; Morton, J.B.** (2001). Zygomycota: Zygomycetes. *The Mycota* (Berlin) 7: 113–146; **Gleason, F.H.; McGee, P.A.** (2004). The ultrastructure of walls of some sporocarpic species of *Densospora*, *Glomus* and *Endogone*. *Australas. Mycol.* 22: 73–78; **Jeffries, P.; Dodd, J.C.** (2000). Molecular ecology of mycorrhizal fungi. *Appl. Microb. System.*: 73–105; **Kyde, M.M.; Gould, A.B.** (2000). Mycorrhizal endosymbiosis. *Microbial Endophytes* (New York): 161–198; **Yao, Y.J.; Pegler, D.N.; Young, T.W.K.** (1996). *Genera of Endogonales* (Kew): 229 pp.

Endomycetaceae J. Schröt. 1893
Saccharomycetales: Ascomycota

Mycelium ranging from well-developed to ± absent, then vegetative cells proliferating by multilateral budding. Asci formed singly or in short irregular chains, directly from vegetative cells, usually ± globose, usually evanescent, sometimes multispored. Ascospores usually with asymmetrical flanges or sheaths. Fermentation sometimes present, coenzyme system usually Q-8.

Significant Genera: *Endomyces*, *Phialoascus*.

Distribution: Cosmopolitan.

Economic Significance: None is known.

Ecology: Associated with other fungi, at least some species are parasitic.

Notes: The family is poorly understood and the application of many names is uncertain.

References: **Hoog, G.S. de** (1998). *Endomyces* Reess. *The Yeasts, A Taxonomic Study* (Amsterdam): 194–196; **Suh, S.O.; Kurtzman, C.P.; Blackwell, M.** (2001). The status of *Endomyces scopularum* – a filamentous fungus and two yeasts. *Mycologia* 93: 317–322.

Englerulaceae Henn. 1904
Uncertain position within Dothideomycetes, Ascomycota

Mycelium superficial, dark, usually hyphopodiate. Ascomata superficial, sometimes stalked, globose to pulvinate; peridium thin-walled, deliquescing in the upper part to expose the asci; hymenium gelatinous. Interascal tissue absent. Asci ovoid to saccate, fissitunicate, with a J+ outer layer. Ascospores brown, 1-septate, usually with a mucous sheath. Anamorphs hyphomycetous or coelomycetous, usually prominent.

Significant Genera: *Englerula, Schiffnerula*. Anamorphs: *Questieriella, Sarcinella*.

Distribution: Primarily tropical.

Economic Significance: None is known.

Ecology: Biotrophic on leaves.

Notes: Possibly polyphyletic, but no molecular data are available.

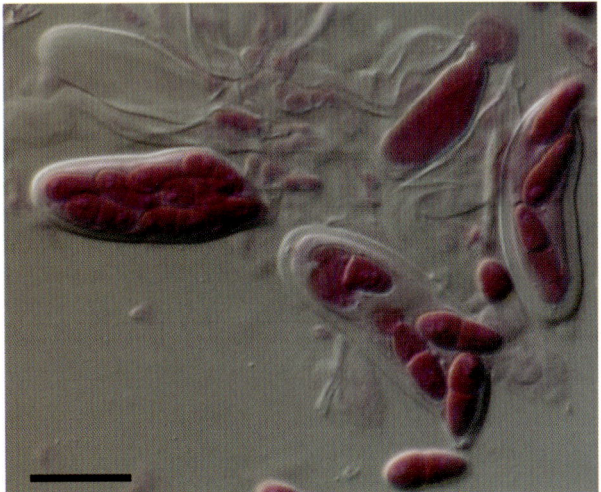

Schiffnerula spectabilis, Uganda; asci and ascospores

References: **Castlebury, L.A.; Crane, J.L.; Huhndorf, S.M.** (1995). A new species of *Rhytidenglerula* from Everglades National Park, Florida. *Mycotaxon* 54: 461–463; **Hughes, S.J.** (1986). *Questieriella quercina* n. sp. and notes on an additional collection of *Schiffnerula oyedaeae. Can. J. Bot.* 64: 1591–1593; **Hughes, S.J.** (1990). *Schiffnerula corni* n. sp., and its *Sarcinella* and *Questieriella* synanamorphs from Quebec. *Mycologia* 82: 657–658.

Entolomataceae Kotl. & Pouzar 1972
Agaricales: Basidiomycota

Basidiomata pileate and agaricoid, gasteroid in a few species, varied in form, ± glabrous, not viscid, usually pale, stipe central, veil lacking. Hyphal system monomitic, hyphae with or without clamp connections, not staining in iodine. Hymenium lamellate, the gills often decurrent, cystidia often present. Basidia usually 4-spored. Basidiospores pink, ± thin-walled, apiculate or nodose, lacking a germ pore, faceted, angular or longitudinally ridged, not staining in iodine.

Significant Genera: *Clitopilus, Entoloma, Rhodocybe*.

Entoloma mougeotii, UK; basidiomata

Distribution: Cosmopolitan, frequent in both temperate and tropical climates.

Economic Significance: Some species (especially of *Entoloma*) are toxic and cause regular poisoning incidents (including fatalities) in North America and Europe. Other species (e.g. *Clitopilus prunulus*) are prized as edible mushrooms.

Entoloma hochstetteri, New Zealand; basidiomata

Ecology: Saprobic or ectomycorrhizal, in both broadleaved and coniferous forests; a few species associated with other substrata such as dung and mosses.

Notes: A highly speciose family, recognized particularly by their pink, angular or striate, basidiospores. Molecular data indicate a probable common ancestor with the *Tricholomataceae*, but more research is required.

References: Baroni, T.J. (1999). *Rhodocybe pegleri* sp. nov. with notes on *Rhodocybe pseudonitellina* from East Africa. *Kew Bull.* 54: 777–782; **Baroni, T.J.; Horak, E.** (1994). *Entolomataceae* in North America III: new taxa, new combinations and notes on species of *Rhodocybe*. *Mycologia* 86: 138–145; **Eyssartier, G.; Buyck, B.; Courtecuisse, R.** (2001). New species and combinations in cuboid-spored *Entoloma* species from Madagascar. *Mycol. Res.* 105: 1144–1148; **Hofstetter, V.; Clémençon, H.; Vilgalys, R.; Moncalvo, J.-M.** (2002). Phylogenetic analyses of the *Lyophylleae* (*Agaricales*, *Basidiomycota*) based on nuclear and mitochondrial rDNA sequences. *Mycol. Res.* 106: 1043–1059; **Manimohan, P.; Joseph, A.V.; Leelavathy, K.M.** (1995). The genus *Entoloma* in Kerala State, India. *Mycol. Res.* 99: 1083–1097; **Moncalvo, J.-M.; Vilgalys, R.; Redhead, S.A.; Johnson, J.E.; James, T.Y.; Aime, M.C.; Hofstetter, V.; Verduin, S.J.W.; Larsson, E.; Baroni, T.J.; Thorn, R.G.; Jacobsson, S.; Clemencon, H.; Miller, O.K.** (2002). One hundred and seventeen clades of euagarics. *Mol. Phylogen. Evol.* 23: 357–400; **Noordeloos, M.E.** (1987). *Entoloma* (*Agaricales*) in Europe. Synopsis and keys to all species and a monograph of the subgenera *Trichopilus*, *Inocephalus*, *Alboleptonia*, *Leptonia*, *Paraleptonia*, and *Omphaliopsis*. *Beih. Nova Hedwigia* 91: 419 pp.; **Noordeloos, M.E.; Gulden, G.** (1989). *Entoloma* (*Basidiomycetes*, *Agaricales*) of alpine habitats on the Hardangervidda near Finse, Norway, with a key including species from Northern Europe and Greenland. *Can. J. Bot.* 67: 1727–1738; **Noordeloos, M.E.; Gulden, G.** (2004). The genus *Entoloma* (*Basidiomycetes*, *Agaricales*) on Svalbard. *Mem. N. Y. bot. Gdn* 89: 97–106.

Entomophthoraceae Nowak. 1877
Entomophthorales: Zygomycota

Vegetative growth mycelial or as hyphal bodies, with or without a cell wall, usually spherical to rounded or fusoid to irregular. Sporophores (conidiophores) either simple or branched (dichotomously or digitately), solitary, cylindrical to slightly clavate, bearing terminal conidiogenous cells. Spores (conidia) terminal, spherical, single-celled, with a truncate or small papilla, forcibly discharged; germinating to form either hyphae or one to several short sporophores, each bearing a small secondary spore; nucleus small, visible during mitosis and interphase, nucleolus prominent. Zygospores formed by hyphal conjugation, zygosporangium (epispore) hyaline or slightly pigmented, smooth or ornamented, zygospore (endospore) ovoid and smooth or globose to subglobose and roughened; germinating directly to produce secondary spores or a sporophore.

Significant Genera: *Entomophthora*, *Entomophaga*, *Pandora*.

Distribution: Widespread, perhaps more so in the tropics.

Economic Significance: None noted, except perhaps as potential agents of biological control or as areas of research for developing biological control strategies.

Ecology: Obligate pathogens of arthropods – mainly insects (esp. Diptera in *Entomopthora*; others, for example, Orthoptera and Coleoptera, in *Entomophaga*, and Cicadidae in *Massospora*) and spiders (Phalangiidae in *Entomophaga*).

Notes: Three subfamilies have been distinguished; *Entomophthoroideae* for species where the forcibly discharged and binucleate to multinucleate conidia are produced on unbranched conidiophores, *Erynoideae* differing by uninucleate conidia on branched conidiophores and *Massosporoideae* where the conidia are produced within a mycelial mass and are passively discharged.

Entomophthora sp.; colonies on fly

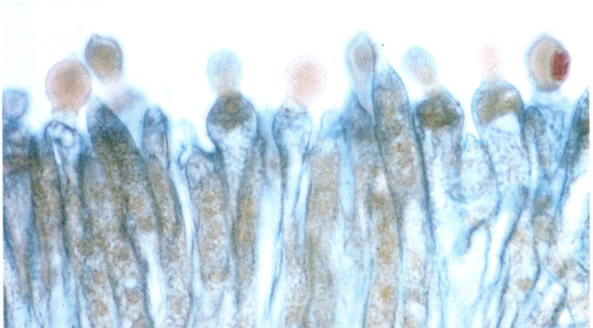

Entomophthora; sporophores bearing conidia

References: Bidochka, M.J.; Walsh, S.R.A.; Ramos, M.E.; St Leger, R.J.; Carruthers, R.I.; Silver, J.C.; Roberts, D.W. (1997). Cloned DNA probes distinguish endemic and exotic *Entomophaga grylli* fungal pathotype infections in grasshopper life stages. *Mol. Ecol.* 6: 303–308; **Bidochka, M.J.; Walsh, S.R.A.; Ramos, M.E.; St Leger, R.J.; Silver, J.C.; Roberts, D.W.** (1995). Pathotypes in the *Entomophaga grylli* species complex of grasshopper pathogens differentiated with random amplification of polymorphic DNA and cloned-DNA probes. *Appl. Environm. Microbiol.* 61: 556–560; **Freimoser, F.M.; Jensen, A.B.; Tuor, U.; Aebi, M.; Eilenberg, J.** (2001). Isolation and *in vitro* cultivation of the aphid pathogenic fungus *Entomophthora planchoniana*. *Can. J. Microbiol.* 47: 1082–1087; **Humber, R.A.** (1989). Synopsis of a revised classification for the *Entomophthorales* (*Zygomycotina*). *Mycotaxon* 34: 441–460; **Jensen, A.B.; Eilenberg, J.** (2001). Genetic variation within the insect-pathogenic genus *Entomophthora*, focusing on the *E. muscae* complex, using PCR-RFLP of the ITS II and the LSU rDNA. *Mycol. Res.* 105: 307–312; **Jensen, A.B.; Gargas, A.; Eilenberg, J.; Rosendahl, S.** (1998). Relationships of the insect-pathogenic order *Entomophthorales* (*Zygomycota, Fungi*) based on phylogenetic analyses of nuclear small subunit ribosomal DNA sequences (SSU rDNA). *Fungal Genetics Biol.* 24: 325–334; **Keller, S.; Kalsbeek, V.; Eilenberg, J.** (1999). Redescription of *Entomophthora muscae* (Cohn) Fresenius. *Sydowia* 51: 197–209; **Keller, S.; Petrini, O.** (2005). Keys to the identification of the arthropod pathogenic genera of the families *Entomophthoraceae* and *Neozygitaceae* (*Zygomycetes*), with descriptions of three new subfamilies and a new genus. *Sydowia* 57: 23–53; **Murrin, F.; Nolan, R.A.** (1989). Ultrastructure of conidial germ-tube development *in vitro* by the insect pathogen *Entomophaga aulicae*. *Can. J. Bot.* 67: 754–762; **Perry, D.F.; Fleming, R.A.** (1989). *Erynia crustosa* zygospore germination. *Mycologia* 81: 154–158; **Tymon, A.M.; Pell, J.K.** (2005). ISSR, ERIC and RAPD techniques to detect genetic diversity in the aphid pathogen *Pandora neoaphidis*. *Mycol. Res.* 109: 285–293; **Tymon, A.M.; Shah, P.A.; Pell, J.K.** (2004). PCR-based molecular discrimination of *Pandora neoaphidis* isolates from related entomopathogenic fungi and development of species-specific diagnostic primers. *Mycol. Res.* 108: 419–433.

Entorrhizaceae R. Bauer & Oberw. 1997
Entorrhizales: Basidiomycota

Sori forming galls on roots of host plants, irregularly shaped but often elongate and furcate, dark brown. Septa of dolipore type, but lacking membranous pore caps or bands. Haustoria present. Teliospores formed intracellularly, usually solitary, ± ovoid or ellipsoidal, brown or yellowish, thick-walled, the wall usually ornamented, sometimes with a hyaline outer layer, germinating to form a 4-celled cruciate [presumed] basidium within the host cell, from which develop frequently branched fertile hyphae with filiform conidia formed in basipetal succession from apiculate fertile loci.

Significant Genera: *Entorrhiza*.

Distribution: Widespread, primarily in temperate regions.

Economic Significance: None is known.

Ecology: Biotrophic and gall-forming on roots of *Cyperaceae* and *Juncaceae*.

Notes: The family occupies a basal position within the *Ustilaginomycetes*, with unique septal structures for the class.

References: **Bauer, R.; Oberwinkler, F.; Vánky, K.** (1997). Ultrastructural markers and systematics in smut fungi and allied taxa. *Can. J. Bot.* 75: 1273–1314; **Begerow, D.; Bauer, R.; Oberwinkler, F.** (1998). Phylogenetic studies on nuclear large subunit ribosomal DNA sequences of smut fungi and related taxa. *Can. J. Bot.* 75: 2045–2056; **Piepenbring, M.; Bauer, R.; Oberwinkler, F.** (1998). Teliospores of smut fungi. Teliospore walls and the development of ornamentation studied by electron microscopy. *Protoplasma* 204: 170–201; **Piepenbring, M.; Hagedorn, G.; Oberwinkler, F.** (1998). Spore liberation and dispersal in smut fungi. *Bot. Acta* 111: 444–460; **Vánky, K.** (1987). Illustrated genera of smut fungi. *Cryptog. Stud.* 1: 159 pp.; **Vánky, K.** (2004). Taxonomic studies on *Ustilaginomycetes* – 24. *Mycotaxon* 89: 55–118; **Vánky, K.; McKenzie, E.H.C.** (2002). Smut Fungi of New Zealand. *Fungal Diversity Res. Ser.* (Hong Kong) 8: 259 pp.

Entylomataceae R. Bauer & Oberw. 1997
Entylomatales: Basidiomycota

Sori formed in leaves and stems, causing galls, swellings or spots. Septa with simple pores. Haustoria absent, host interaction structures poorly differentiated with homogenous contents. Teliospores immersed in host tissues, solitary or adhering in irregular groups, hyaline or pale yellowish, usually smooth, often with a hyaline gelatinous outer layer, germinating to form a transversely septate hypha-like basidium, from which 4–8 basidiospores are produced in an apical cluster. Basidiospores elongate, copulating in pairs to produce mycelia or secondary sporidia. Anamorphs yeast-like.

Significant Genera: *Entyloma*. Anamorph: *Tilletiopsis*.

Distribution: Cosmopolitan, frequently encountered in both tropical and temperate climates.

Economic Significance: Some species cause diseases of ornamental *Asteraceae*, while others have been researched for their potential in biological control of weeds. Some yeast-like members of this family cause postharvest deterioration of apples.

Ecology: Biotrophic, forming galls and spots in living leaves, occurring on a wide range of dicot plant families, primarily from the *Asteridae, Rosidae* and *Ranunculales*.

Notes: Species on asterid plants appear to have undergone a recent major evolutionary radiation, and most have highly restricted host ranges. The family appears to be a sister group of the *Microstromataceae* and *Doassansiaceae*.

References: **Adejumo, T.O.; Ikotun, T.; Florini, D.A.** (2001). Identification and survival of the causal organism of leaf smut disease of cowpea in Nigeria. *Mycopathologia* 150: 85–90; **Bauer, R.; Oberwinkler, F.; Vánky, K.** (1997). Ultrastructural markers and systematics in smut fungi and allied taxa. *Can. J. Bot.* 75: 1273–1314; **Begerow, D.; Bauer, R.; Boekhout, T.** (2000). Phylogenetic placements of ustilaginomycetous anamorphs as deduced from nuclear LSA rDNA sequences. *Mycol. Res.* 104: 53–60; **Begerow, D.; Bauer, R.; Oberwinkler, F.** (1998). Phylogenetic studies on nuclear large subunit ribosomal DNA sequences of smut fungi and related taxa. *Can. J. Bot.* 75: 2045–2056; **Begerow, D.; Lutz, M.; Oberwinkler, F.** (2002). Implications of molecular characters for the phylogeny of the genus *Entyloma. Mycol. Res.* 106: 1392–1399; **Boekhout, T.; Fell, J.W.; O'Donnell, K.** (1995). Molecular systematics of some yeast-like anamorphs belonging to the *Ustilaginales* and *Tilletiales. Stud. Mycol.* 38: 175–183; **Boekhout, T.; Gildemacher, P.; Theelen, B.; Müller, W.H.; Heijne, B.; Lutz, M.** (2005). Extensive colonization of apples by smut anamorphs causes a new postharvest disorder. *FEMS Yeast Res.* 6: 63–76; **Jackson, A.P.** (2004). A reconciliation analysis of host switching in plant-fungal symbiosis. *Evolution, Lancaster, Pa.* 58: 1909–1923; **Piepenbring, M.; Oberwinkler, F.** (2003). Integrating morphological and molecular characters for a phylogenetic system of smut fungi. *Bot. Jb.* 24: 241–253; **Zwetko, P.; Blanz, P.** (2004). Die Brandpilze Österreichs Doassansiales, Entorrhizales, Entylomatales, Georgefischeriales, Microbotryales, Tilletiales, Urocystales, Ustilaginales. *Biosystem. Ecol. Ser.* 21: 241 pp.

Eoterfeziaceae G.F. Atk. 1902
Uncertain position within Ascomycetes, Ascomycota

Ascomata small, cleistothecial, ± globose; the peridium thin, membranous, either composed of coalescing hyphae or with a narrow layer of pseudoparenchymatous cells covered by a hyaline granular layer, interior sometimes divided into locules by radiating mycelial strands. Interascal tissue absent. Asci ? formed from croziers, saccate, very thin-walled, evanescent. Ascospores small, hyaline, aseptate. Anamorphs unknown.

Significant Genera: *Eoterfezia*.

Distribution: Only known from North America.

Economic Significance: None is known.

Ecology: Fungicolous, but the nutritional status is uncertain.

Notes: Very poorly known; no recent information is available.

References: Jeng, R.S.; Cain, R.F. (1976). A new species of *Eoterfezia* from Mexico and Venezuela. *Mycotaxon* 3: 387–390.

Epigloeaceae Zahlbr. 1903
Uncertain position within Ascomycetes, Ascomycota

Thallus inconspicuous or absent. Ascomata immersed in a thin gelatinous algal film, perithecial, dark green to black, superficial, composed of thin-walled periclinally arranged hyphae immersed in a gelatinous matrix. Interascal tissue of narrow, rarely branched, thin-walled paraphyses, the apices not swollen. Asci cylindrical, persistent, at first thick-walled but becoming thin-walled, the wall blueing in iodine, without apical structures, releasing ascospores through a vertical split. Ascospores hyaline, transversely septate, thin-walled, without a sheath. Anamorphs pycnidial.

Significant Genera: *Epigloea*.

Distribution: Known from Europe and Antarctica.

Economic Significance: None is known.

Ecology: Either lichenized or parasitic on algae.

Notes: Very poorly known; no recent information is available.

References: David, J.C. (1987). Studies on the genus *Epigloea. Syst. Ascom.* 6: 217–221; **Döbbeler, P.** (1994). *Epigloea urosperma* (*Ascomycetes*) – ein neuer Flechtenparasit. *Sendtnera* 2: 277–282.

Epitheliaceae Jülich 1982
Polyporales: Basidiomycota

Basidiomata resupinate and effuse, membranous to waxy, white or yellowish, the margin indeterminate. Hyphal system dimitic, sometimes with clamp connections. Hymenium appearing powdery or dentiform, with hymenial teeth composed of hyphae surrounding cystidia or aggregations of dendritic hyphae. Basidia sessile or borne on short hyphae, initially ± globose but becoming elongate, with or without a basal clamp connection, with 2 or 4 well-developed sterigmata. Basidiospores variously shaped, hyaline, thin-walled, with a well-developed apicular hilum, sometimes staining in iodine.

Significant Genera: *Epithele*.

Distribution: Widespread, represented in both tropical and temperate zones.

Epithele typhae; basidioma

Economic Significance: None is known.

Ecology: Saprobic or weakly parasitic on stems, petioles etc. of plants, especially palms and ferns.

Notes: No molecular data are available and the systematic position and delimitation of this family is uncertain.

References: **Boquiren, D.T.** (1971). The genus *Epithele*. *Mycologia* 63: 937–957; **De, A.B.** (1989). *Epitheleopsis bosei* De, sp. nov. *Int. J. Mycol. Lichenol.* 4: 59–63; **Ginns, J.** (1998). Genera of the North American *Corticiaceae* sensu lato. *Mycologia* 90: 1–35; **Greslebin, A.G.; Rajchenberg, M.** (2003). Diversity of *Corticiaceae* sens. lat. in Patagonia, southern Argentina. *N.Z. Jl Bot.* 41: 437–446.

Eremascaceae Engl. & E. Gilg 1924
Uncertain position within Eurotiomycetes, Ascomycota

Mycelium copious. Ascomata absent. Asci formed by anastomosis of equal cells on short coiled ascogenous hyphae, ± globose, thin-walled, evanescent. Ascospores hyaline, smooth, aseptate. Anamorph hyphomycetous, thallic, *Chrysosporium*-like.

Significant Genera: *Eremascus*.

Distribution: Widespread.

Economic Significance: Occasionally implicated in deterioration of stored food.

Ecology: Saprobic, found especially on substrata with low water content.

Notes: Probably close to the *Ascosphaeraceae*.

References: **Berbee, M.L.; Taylor, J.W.** (1992). Two ascomycete classes based on fruiting-body characters and ribosomal DNA sequence. *Mol. Biol. Evol.* 9: 278–284; **Hocking, A.D.** (1991). Xerophilic fungi in intermediate and low moisture foods. *Handbook of Applied Mycology* Vol. 3. Foods and Feeds: 69–97; **Kurtzman, C.P.; Robnett, C.J.** (1998). Identification and phylogeny of ascomycetous yeasts from analysis of nuclear large subunit (26S) ribosomal DNA partial sequences. *Antonie van Leeuwenhoek* 73: 331–371; **Landvik, S.; Shailer, N.F.J.; Eriksson, O.E.** (1997). SSU rDNA sequence support for a close relationship between the *Elaphomycetales* and the *Eurotiales* and *Onygenales*. *Mycoscience* 37: 237–241; **Lumbsch, H.T.; Lindemuth, R.; Schmitt, I.** (2000). Evolution of filamentous *Ascomycetes* inferred from LSU rDNA sequence data. *Pl. Biol.* 2: 525–529.

Eremomycetaceae Malloch & Cain 1971
Uncertain position within Dothideomycetes, Ascomycota

Ascomata cleistothecial, formed from solid pseudoparenchymatous initials, spherical, small, sometimes setose; peridium brown to black, pseudoparenchymatous or cephalothecoid. Interascal tissue absent. Asci irregularly disposed within the ascoma, clavate, thin-walled, not fissitunicate, evanescent. Ascospores small, hyaline to yellow-brown, often reniform, aseptate. Anamorph hyphomycetous, conidiogenesis thallic.

Significant Genera: *Eremomyces*. Anamorph: *Arthrographis*.

Distribution: Widespread.

Economic Significance: Some species are implicated as opportunistic human pathogens.

Ecology: Saprobic, isolated primarily from soil and dung.

Notes: Very poorly known; no recent information is available.

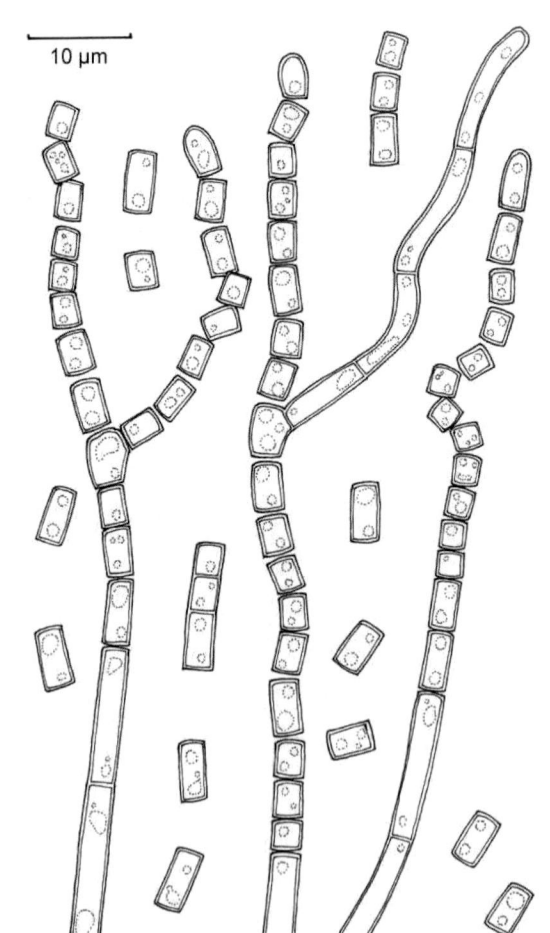

Arthrographis cuboidea; conidia of anamorph

References: **Arx, J.A. von; Figueras, M.J.; Guarro, J.** (1988). Sordariaceous ascomycetes without ascospore ejaculation. *Beih. Nova Hedwigia* 94: 104 pp.; **Lumley, T.C.; Abbott, S.P.; Currah, R.S.** (2000). Microscopic ascomycetes isolated from rotting wood in the boreal forest. *Mycotaxon* 74: 395–414; **Malloch, D.W.; Sigler, L.** (1988). The *Eremomycetaceae* (*Ascomycotina*). *Can. J. Bot.* 66: 1929–1932.

Eremotheciaceae Kurtzman 1995
Saccharomycetales: Ascomycota

Hyphae septate, composed of globose to cylindrical cells which proliferate multilaterally. Asci ellipsoidal to elongate, evanescent, 8- to 32-spored. Ascospores acicular, septate or aseptate, smooth or ornamented.

Significant Genera: *Nematospora*.

Distribution: Widespread, especially in the tropics.
Economic Significance: Some species have been implicated as crop pathogens, but no major diseases are caused.

Ecology: Biotrophic or necrotrophic on plant tissues.

Notes: An outline of the entire genome of *Eremothecium ashbyi* is available on the Internet.

Nematospora coryli; asci and ascospores

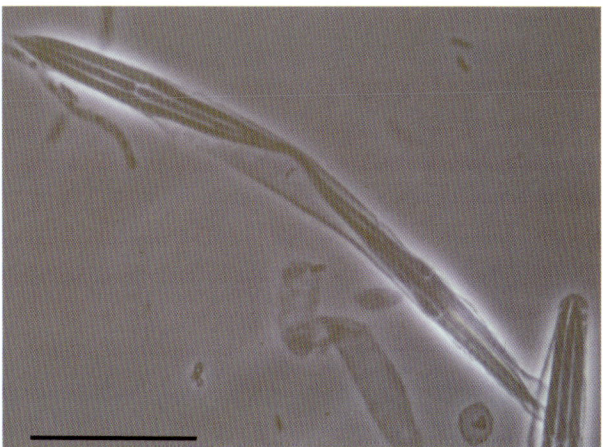
Nematospora coryli; ascus

References: **Kurtzman, C.P.; Robnett, C.J.** (2003). Phylogenetic relationships among yeasts of the 'Saccharomyces complex' determined from multigene sequence analyses. *FEMS Yeast Res.* 3: 417–432; **Prillinger, H.; Schweigkofler, W.; Breitenbach, M.; Briza, P.; Staudacher, E.; Lopandic, K.; Molnar, O.; Weigang, F.; Ibl, M.; Ellinger, A.** (1997). Phytopathogenic filamentous (*Ashbya, Eremothecium*) and dimorphic fungi (*Holleya, Nematospora*) with needle-shaped ascospores as new members within the Saccharomycetaceae. *Yeast* (Chichester) 13: 945–960; **Wendland, J.; Pohlmann, R.; Dietrich, F.; Steiner, S.; Mohr, C.; Philippsen, P.** (1999). Compact organization of rRNA genes in the filamentous fungus *Ashbya gossypii*. *Curr. Genet.* 35: 618–625; **Yamada, Y.; Nagahama, T.; Banno, I.** (1994). The phylogenic relationships among species of the genus *Metschnikowia* Kamienski and its related genera based on the partial sequences of 18S and 26S ribosomal RNAs (*Metschnikowiaceae*). *Bull. Fac. Agric. Shizuoka Univ.* 44: 9–20.

Erysiphaceae Tul. & C. Tul. 1861
Erysiphales: Ascomycota

Mycelium superficial, ± hyaline, conspicuous, infiltrating surface host cells with appressoria and internal haustoria. Ascomata cleistothecial, globose, solitary or aggregated, peridium one-layered, pale, sometimes multilayered, translucent to dark, often with complex appendages. Interascal tissue absent. Asci solitary to few, broadly clavate, thin-walled, with two wall layers at the base but the inner layer absent towards the apex, dehiscence forcible via an apical slit. Ascospores hyaline to yellowish, subglobose to ellipsoidal-ovoid, aseptate, without a sheath. Anamorphs hyphomycetous, prominent. Conidiophores formed from external mycelium, conidiogenous cells not differentiated, sometimes with a swollen foot cell. Conidia often formed in chains by ring wall-building with the oldest at the apex, secession schizolytic, large, hyaline, aseptate, thin-walled, ellipsoidal to doliiform, sometimes distinctly swollen in the mid portion, sometimes inconspicuously ornamented, sometimes containing fibrosin (crystalline) bodies.

Microsphaera diffusa, Galápagos Islands; symptoms on leaves of *Senna obtusifolia*

Significant Genera: *Blumeria, Erysiphe, Phyllactinia, Podosphaera*. Anamorphs: *Oidium, Oidiopsis*.

Distribution: Cosmopolitan.

Economic Significance: The powdery mildews. Many species are important plant pathogens, affecting leaves, fruits and herbaceous stems. Yield reductions can be very significant in some temperate crops, including cereals, beet and sunflower.

Ecology: Biotrophic, with affected plant tissues damaged but usually not killed outright.

Notes: An isolated group that has in the past been linked with quite diverse ascomycete clades, with recent molecular studies indicating a link with the inoperculate discomycetes.

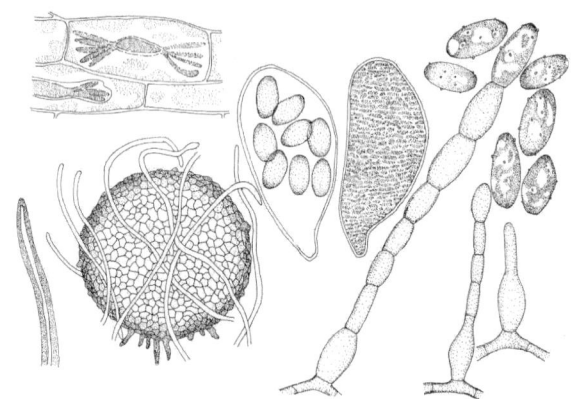

Blumeria graminis; showing haustoria in host cells, ascoma, asci, ascospores and anamorph

References: **Braun, U.** (1987). A monograph of the *Erysiphales* (powdery mildews). *Beih. Nova Hedwigia* 89: 700 pp.; **Braun, U.** (1995). *The Powdery Mildews (Erysiphales) of Europe* (Jena): 337 pp.; **Braun, U.** (1999). Some critical notes on the classification and the generic concept of the *Erysiphaceae*. *Schlechtendalia* 3: 48–54; **Braun, U.; Takamatsu, S.** (2000). Phylogeny of *Erysiphe, Microsphaera, Uncinula* (*Erysipheae*) and *Cystotheca, Podosphaera, Sphaerotheca* (*Cystotheceae*) inferred from rDNA. *Schlechtendalia* 4: 1–33; **Cook, R.T.A.; Inman, A.J.; Billings, C.** (1997). Identification and classification of powdery mildew anamorphs using light and scanning electron microscopy and host range data. *Mycol. Res.* 101: 975–1002; **Hirose, S.; Tanda, S.; Kiss, L.; Grigaliunaite, B.; Havrylenko, M.; Takamatsu, S.** (2005). Molecular phylogeny and evolution of the maple powdery mildew (*Sawadaea, Erysiphaceae*) inferred from nuclear rDNA sequences. *Mycol. Res.* 109: 912–922; **Mori, Y.; Sato, Y.; Takamatsu, S.** (2000). Evolutionary analysis of the powdery mildew fungi using nucleotide sequences of the nuclear ribosomal DNA. *Mycologia* 92: 74–93; **Saenz, G.S.** (1998). Evolutionary relationships of the powdery mildews (*Erysiphales*) inferred from ribosomal DNA sequences. *McIlvainea* 13: 33–44; **Saenz, G.S.; Taylor, J.W.** (1999). Phylogeny of the *Erysiphales* (powdery mildews) inferred from internal transcribed spacer ribosomal DNA sequences. *Can. J. Bot.* 77: 150–168; **Takamatsu, S.** (2004). Phylogeny and evolution of the powdery mildew fungi (*Erysiphales, Ascomycota*) inferred from nuclear ribosomal DNA sequences. *Mycoscience* 45: 147–157; **Takamatsu, S.; Hirata, T.; Sato, Y.** (1998). Phylogenetic analysis and predicted secondary structures of the rDNA internal transcribed spacers of the powdery mildew fungi (*Erysiphaceae*). *Mycoscience* 39: 441–453; **Takamatsu, S.; Matsuda, S.** (2004). Estimation of molecular clocks for ITS and 28S rDNA in *Erysiphales*. *Mycoscience* 45: 340–344; **Takamatsu, S.; Matsuda, S.; Niinomi, S.; Havrylenko, M.** (2006). Molecular phylogeny supports a Northern Hemisphere origin of *Golovinomyces* (*Ascomycota: Erysiphales*). *Mycol. Res.* 110: 1093–1101.

Euantennariaceae S. Hughes & Corlett ex S. Hughes 1972
Capnodiales: Ascomycota

Mycelium superficial, dark, forming a flattened mat but frequently with erect branches. Ascomata cleistothecial, ± spherical, small, superficial, with a small lysigenous pore, the peridium dark, with hyphal appendages. Interascal tissue absent. Asci saccate, fissitunicate. Ascospores brown, transversely septate or muriform, sometimes attenuated at the apices. Anamorphs hyphomycetous.

Significant Genera: *Euantennaria*. Anamorph: *Antennatula*.

Distribution: Widespread, especially in wet forest zones.

Economic Significance: None is known.

Ecology: Epiphytic, presumably gaining nutrition predominantly from plant exudates.

Notes: Very poorly known; no recent information is available.

References: **Hughes, S.J.; Arnold, G.** (1989). *Antennatula cubensis* sp. nov. and its *Hormisciomyces* synanamorph. *Mem. N. Y. bot. Gdn* 49: 198–201; **Hughes, S.J.; Seifert, K.A.** (1998). The hyphomycete genus *Heterosporiopsis* Petrak. *Sydowia* 50: 192–199; **Parbery, I.H.; Brown, J.F.** (1986). Sooty moulds and black mildews in extra-tropical rainforests. *Microbiology of the Phyllosphere* (Cambridge): 101–120; **Sugiyama, J.; Amano, N.; Yokoyama, K.** (1984). *Euantennaria mucronata*, an euantennariaceous sooty mould with *Antennatula* and *Hormisciomyces* synanamorphs from southern Chile. *Studies on Cryptogams in Southern Chile* (Tokyo): 169–173.

Euceratomycetaceae I.I. Tav. 1980
Laboulbeniales: Ascomycota

Stroma (thallus) present. Ascomata formed from successive cells of a lateral appendage of the primary thallus, the appendage extending beyond the base of the ascoma; outer wall cells of ascoma usually small, ± equal. Asci 4-spored. Ascospores with a submedian septum. Usually monoecious.

Significant Genera: *Euceratomyces, Euzodiomyces*.

Distribution: Primarily temperate.

Economic Significance: None is known.

Ecology: Parasitic or epibiotic on insect exoskeletons.

Notes: Very poorly known; no recent information is available.

Cochliomyces sp.; ascoma

References: **Santamaria i del Campo, S.** (1990). *L'ordre Laboulbenials (Fungi, Ascomycotina) a la Península Ibèrica i Illes Ballears* [Thesis] (Barcelona): 669 pp.; **Santamaría, S.** (1995). New and interesting *Laboulbeniales* (*Fungi, Ascomycotina*) from Spain, III. *Nova Hedwigia* 61: 65–83; **Santamaría, S.; Rossi, W.** (1999). New or interesting *Laboulbeniales* (*Ascomycota*) from the Mediterranean region. *Pl. Biosystems* 133: 163–171; **Tavares, I.I.** (1985). *Laboulbeniales* (*Fungi, Ascomycetes*). *Mycol. Mem.* 9: 627 pp.

Exidiaceae R.T. Moore 1978
Tremellales: Basidiomycota

Basidiomata resupinate, pustulate or discoid, membranous, waxy or ± gelatinous, varied in coloration, sometimes inconspicuous. Hyphal system monomitic, hyphae with clamp connections, often gelatinized. Hymenium smooth or with fertile pegs or spines, composed of basidia and thin branched, often with nodular sterile hyphal structures (hyphidia). Basidia large, ovoid to ellipsoidal, vertically (cruciately) septate, with a basal clamp connection and elongate, often sinuous, tubular sterigmata. Basidiospores usually allantoid, thin-walled, smooth, hyaline, often germinating to form curved microconidia.

Significant Genera: *Exidia, Exidiopsis, Heterochaete*.

Distribution: Very widespread.

Economic Significance: None is known.

Exidia glandulosa, Surrey, UK; basidiomata on dead branch

Eichleriella deglubens, Surrey, UK; basidioma on dead wood

Ecology: Saprobic or possibly weakly parasitic on wood and bark.

Guepinia helvelloides, Vancouver Island, Canada; basidiomata

Notes: The *Exidiaceae* is likely to be a close relative of the *Auriculariaceae*, the phylogenetic significance of the cruciate versus transversely septate basidium appearing to be less than is traditionally surmised. Further revision is needed. *Sebacina* and its allies have recently been separated from the *Exidiaceae* into a separate order *Sebacinales* on morphological, molecular and ecological evidence.

References: **Larsson, K.-H.; Larsson, E.; Koljalg, U.** (2004). High phylogenetic diversity among corticioid homobasidiomycetes. *Mycol. Res.* 108: 983–1002; **Roberts, P.** (1993). *Exidiopsis* species from Devon, including the new segregate genera *Ceratosebacina*, *Endoperplexa*, *Microsebacina*, and *Serendipita*. *Mycol. Res.* 97: 467–478; **Weiss, M.; Oberwinkler, F.** (2001). Phylogenetic relationships in *Auriculariales* and related groups – hypotheses derived from nuclear ribosomal DNA sequences. *Mycol. Res.* 105: 403–415; **Weiss, M.; Selosse, M.-A.; Rexer, K.-H.; Urban, A.; Oberwinkler, F.** (2004). *Sebacinales*: a hitherto overlooked cosm of heterobasidiomycetes with a broad mycorrhizal potential. *Mycol. Res.* 108: 1003–1010.

Exobasidiaceae J. Schröt. 1888
Exobasidiales: Basidiomycota

Basidiomata absent or effuse and poorly defined, immersed in plant tissues and frequently causing hypertrophy. Hyphal system monomitic, hyphae intercellular and intracellular, haustoria sometimes formed, clamp connections present or absent. Hymenium whitish, forming on the under-surface of leaves, hyphae emerging through stomata or by breaking through the cuticle. Basidia cylindrical to clavate or urniform, with variable numbers of elongate sterigmata. Basidiospores discharged actively or rarely passively, small, hyaline, thin-walled, clavate to cylindrical, often curved, becoming septate, with a conspicuous abaxial apiculate hilum, germinating to form bacillar, fusiform or filiform conidia.

Significant Genera: *Exobasidium*.

Distribution: Widespread, especially in temperate regions.

Economic Significance: Species on economic plants, such as *Exobasidium japonicum*, *E. oxycocci* and *E. vaccinii* on azalea and cranberry and *E. camelliae* on tea, can cause significant losses.

Ecology: Biotrophic and gall-forming in leaves, primarily of the *Ericaceae*.

Notes: The *Exobasidiaceae* appears to be a sister group of the *Graphiolaceae*, parasites of palms with cupulate basidiomata. Gall formation can be remarkably rapid, with globose 'gall apples' formed within 24 hours in some circumstances.

Exobasidium vaccinii; galls on leaves of *Vaccinium*

References: **Bauer, R.; Oberwinkler, F.; Vánky, K.** (1997). Ultrastructural markers and systematics in smut fungi and allied taxa. *Can. J. Bot.* 75: 1273–1314; **Begerow, D.; Bauer, R.; Oberwinkler, F.** (1998). Phylogenetic studies on nuclear large subunit ribosomal DNA sequences of smut fungi and related taxa. *Can. J. Bot.* 75: 2045–2056; **Begerow, D.; Bauer, R.; Oberwinkler, F.** (2001). *Muribasidiospora*: Microstromatales or Exobasidiales?. *Mycol. Res.* 105: 798–810; **Begerow, D.; Bauer, R.; Oberwinkler, F.** (2002). The *Exobasidiales*: an evolutionary hypothesis. *Mycol. Prog.* 1: 187–199; **Crous, P.W.; Groenewald, J.Z.; Carroll, G.** (2003). *Muribasidiospora indica* causing a prominent leaf spot disease on *Rhus lancea* in South Africa. *Australas. Pl. Path.* 32: 313–316; **Nagao, H.; Sato, T.; Kakishima, M.** (2004). Three species of *Exobasidium* causing exobasidium leaf blight on subgenus *Hymenanthes*, *Rhododendron* spp., in Japan. *Mycoscience* 45: 85–95.

Farysiaceae Vánky 2001
Ustilaginales: Basidiomycota

Sori formed in ovaries, initially covered by a pale hyphal membrane, bursting at maturity to form a brown or blackish powdery mass of spores, interspersed with capillitium-like clusters of sterile hyphae. Hyphae lacking septal pores. Teliospores formed in chains following division of the sporogenous hyphae, fairly small, variable in shape and size, often angular, thick-walled, dark brown, usually verrucose, germinating to form a hypha-like transversely septate basidium from which fusiform, ovoid or cylindrical basidiospores are formed.

Significant Genera: *Farysia*.

Distribution: Cosmopolitan, perhaps with a distribution almost as extensive as that of its host genus.

Economic Significance: None is known.

Ecology: Biotrophic in ovaries of *Carex* species.

Farysia merrillii, Papua New Guinea; spore masses in infected flowers

Notes: The family is isolated within the *Ustilaginales*, though molecular evidence suggests that *Schizonella* might also belong here. That genus has teliospores formed in pairs and aggregated in clusters.

References: **Bauer, R.; Oberwinkler, F.; Vánky, K.** (1997). Ultrastructural markers and systematics in smut fungi and allied taxa. *Can. J. Bot.* 75: 1273–1314; **Begerow, D.; Bauer, R.; Oberwinkler, F.** (1998). Phylogenetic studies on nuclear large subunit ribosomal DNA sequences of smut fungi and related taxa. *Can. J. Bot.* 75: 2045–2056; **Piepenbring, M.; Bauer, R.; Oberwinkler, F.** (1998). Teliospores of smut fungi. General aspects of teliospore walls and sporogenesis. *Protoplasma* 204: 155–169; **Stoll, M.; Piepenbring, M.; Begerow, D.; Oberwinkler, F.** (2003). Molecular phylogeny of *Ustilago* and *Sporisorium* species (*Basidiomycota, Ustilaginales*) based on internal transcribed spacer (ITS) sequences. *Can. J. Bot.* 81: 976–984; **Vánky, K.** (2001). The emended *Ustilaginaceae* of the modern classificatory system for smut fungi. *Fungal Diversity* 6: 131–147; **Vánky, K.** (2004). Taxonomic studies on *Ustilaginomycetes* – 24. *Mycotaxon* 89: 55–118; **Vánky, K.; McKenzie, E.H.C.** (2002). Smut Fungi of New Zealand. *Fungal Diversity Res. Ser.* (Hong Kong) 8: 259 pp.

Fenestellaceae M.E. Barr 1979
Pleosporales: Ascomycota

Ascomata perithecial, immersed or erumpent, often aggregated or in stromata with convergent ostioles, large, thick-walled, ovoid, with a well-developed periphysate ostiole; peridium dark, two-layered, consisting of small pseudoparenchymatous cells. Interascal tissue of trabeculate pseudoparaphyses in a gelatinous matrix. Asci cylindrical, fissitunicate, with a wide ocular chamber. Ascospores large, complex, dark brown, transversely septate or muriform, constricted at the septa, sometimes ornamented, often with a mucous sheath. Anamorphs not known.

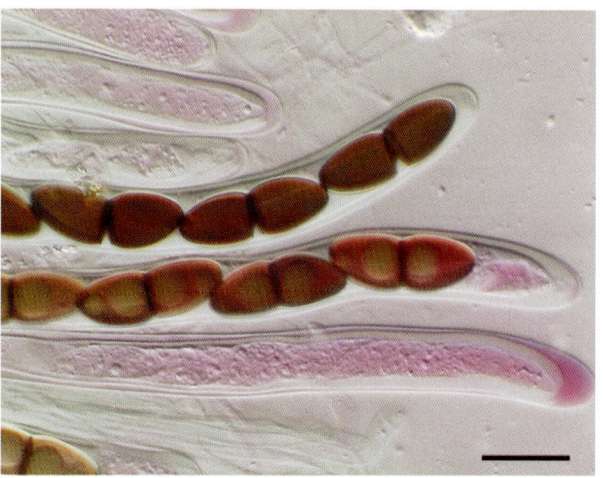

Lojkania separans, India; asci and ascospores

Significant Genera: *Fenestella, Lojkania*.

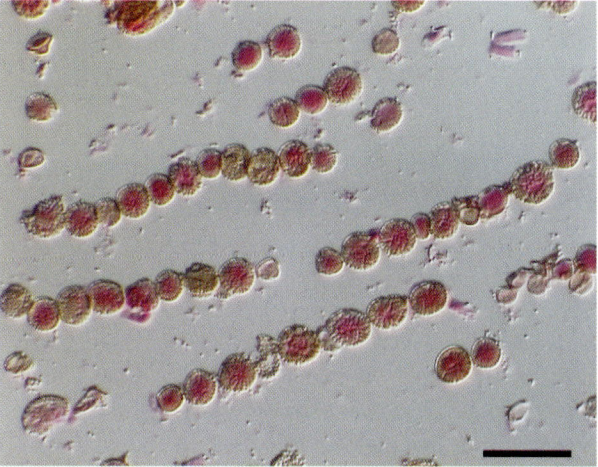

Farysia catenata, New South Wales, Australia; chains of teliospores

Distribution: Widespread.

Economic Significance: None is known.

Ecology: Saprobic or necrotrophic in woody tissue.

Notes: Molecular data confirm placement of the family in the *Pleosporales* as it is currently circumscribed. Links with the *Delitschiaceae* have been proposed. Some taxa have ascospores with germ slits; it is not clear whether they are true members of the family.

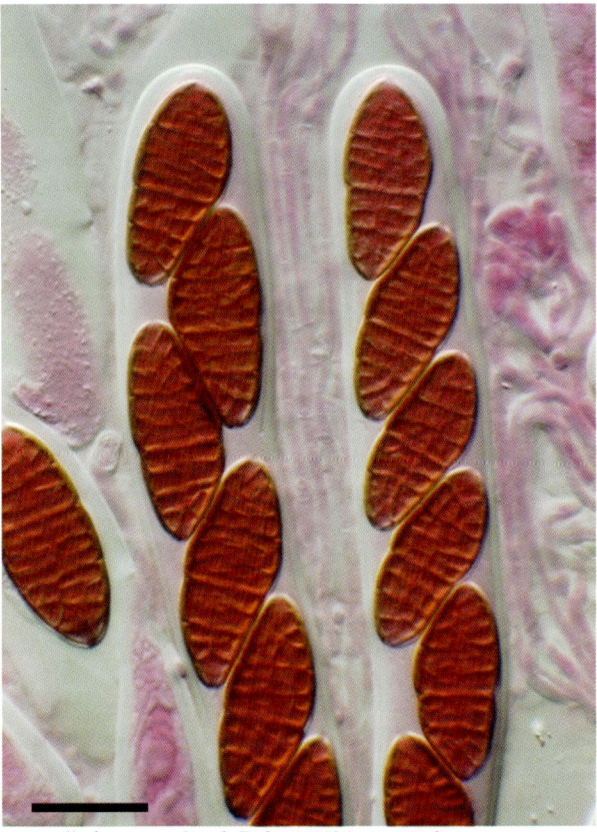

Fenestella fenestrata, South Dakota, USA; asci and ascospores

References: **Barr, M.E.** (1990). Melanommatales (Loculoascomycetes). *N. Amer. Fl.* Ser. 2 13: 129 pp.; **Huhndorf, S.M.; Glawe, D.A.** (1990). Pycnidial development from ascospores of *Fenestella princeps*. *Mycologia* 82: 541–548; **Lumbsch, H.T.; Lindemuth, R.** (2001). Major lineages of *Dothideomycetes* (*Ascomycota*) inferred from SSU and LSU rDNA sequences. *Mycol. Res.* 105: 901–908; **Vasil'eva, L.N.** (1987). Semeistvo *Fenestellaceae* Barr [The family Fenestellaceae Barr]. *Mikol. Fitopatol.* 21: 403–408; **Yuan, Z.Q.; Barr, M.E.** (1994). Three new species of *Lojkania* from Xinjiang, China. *Sydowia* 46: 338–343.

Filobasidiaceae L.S. Olive 1968
Filobasidiales: Basidiomycota

Basidiomata absent, colonies cream or white. Hyphae often with haustoria, septa with dolipores, clamp connections often present. Teliospores not formed. Basidia formed laterally or terminally on undifferentiated hyphae with basal clamp connections, ± cylindrical, bearing a cluster of basidiospores at the apex, not discharging actively. Basidiospores hyaline, ellipsoidal, fusiform or reniform, ± thick-walled, smooth, budding to produce hyaline, thin-walled, ± ellipsoidal yeast-like cells.

Significant Genera: *Filobasidium*. Anamorph: *Cryptococcus*.

Distribution: Widespread; some species are psychrophilic and have been isolated from Antarctic ecosystems.

Economic Significance: *Filobasidium uniguttulatum* is frequently associated with human environments, but does not appear to be a major pathogen; it is unable to grow at physiological temperatures.

Ecology: Most strains appear to be saprobic and are isolated from a very wide range of habitats.

Notes: *Filobasidiella* (including the important human pathogen *F. neoformans*: anamorph *Cryptococcus neoformans*) does not belong in this family, but in an isolated position at the base of the *Tremellales*.

References: **Fell, J.W.; Statzell-Tallman, A.** (1998). *Cryptococcus* Vuillemin. *The Yeasts, A Taxonomic Study* (Amsterdam): 742–767; **Guffogg, S.P.; Thomas-Hall, S.; Holloway, P.; Watson, K.** (2004). A novel psychrotolerant member of the hymenomycetous yeasts from Antarctica: *Cryptococcus watticus* sp. nov. *Int. J. Syst. Evol. Microbiol.* 54: 275–277; **Kwon-Chung, K.J.** (1998). *Filobasidium* Olive. *The Yeasts, A Taxonomic Study* (Amsterdam): 663–669; **Mitchell, T.G.; White, T.J.; Taylor, J.W.** (1992). Comparison of 5.8S ribosomal DNA sequences among the basidiomycetous yeast genera *Cystofilobasidium*, *Filobasidium* and *Filobasidiella*. *J. Med. Vet. Mycol.* 30: 207–218; **Scorzetti, G.; Fell, J.W.; Fonseca, A.; Statzell-Tallman, A.** (2002). Systematics of basidiomycetous yeasts: a comparison of large subunit D1/D2 and internal transcribed spacer rDNA regions. *FEMS Yeast Res.* 2: 495–517; **Sivakumaran, S.; Bridge, P.; Roberts, P.** (2003). Genetic relatedness among *Filobasidiella* species. *Mycopathologia* 156: 157–162; **Suh, S.O.; McHugh, J.V.; Pollock, D.D.; Blackwell, M.** (2005). The beetle gut: a hyperdiverse source of novel yeasts. *Mycol. Res.* 109: 261–265.

Fistulinaceae Lotsy 1907
Agaricales: Basidiomycota

Basidioma pileate, sessile or stipitate, the stalk when present lateral, fleshy, brown or reddish, smooth or papillate, sometimes with red exudates; flesh soft or fibrous. Hyphal system monomitic, hyphae hyaline, gelatinized, sometimes inflated, clamp connections present or absent. Hymenium initially pale yellowish or pink, becoming darker with age, with long crowded but not conjoined tubes. Basidia 4-spored, interspersed in some species with clavate, sometimes branched, sterile hyphae with numerous pin-like outgrowths. Basidiospores small, ± ellipsoidal, hyaline, thin-walled, smooth.

Significant Genera: *Fistulina*.

Distribution: Widespread in temperate regions and the tropics.

Economic Significance: *Fistulina hepatica*, the beefsteak fungus is eaten in north temperate regions, though not exploited economically. *Pseudofistulina radicata* is also edible and wild-collected fruit bodies are sold in Central American countries.

Ecology: Weakly parasitic on broadleaved trees, causing discoloration of the wood.

Notes: The *Fistulinaceae* is part of the core euagaric clade according to recent molecular studies, and is possibly related to the *Schizophyllaceae*. More research is needed, bearing in mind the major morphological differences between these groups. *Porodisculus* may be a further member of the family.

Fistulina hepatica, Suffolk, UK; basidiomata growing from dying tree

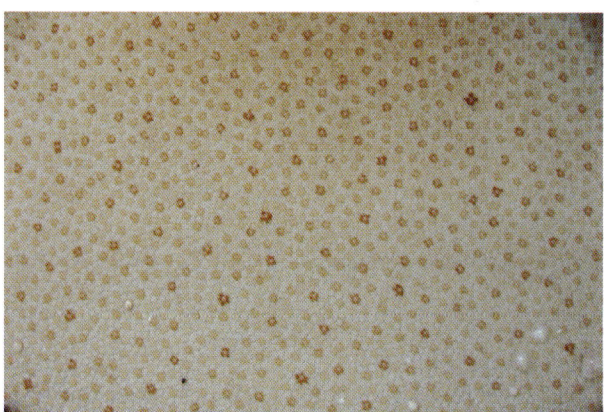

Fistulina hepatica, Devon, UK; hymenial surface

References: **Binder, M.; Hibbett, D.S.; Larsson, K.-H.; Larsson, E.; Langer, E.; Langer, G.** (2005). The phylogenetic distribution of resupinate forms across the major clades of mushroom-forming fungi (*Homobasidiomycetes*). *Syst. Biodiv.* 3: 113–157; **Guzmán, G.** (1987). Distribucion y etnomicologia de *Pseudofistulina radicata* en Mesoamerica, con nuevas localidades en Mexico y su primer registro en Guatemala. *Revta Mex. Micol.* 3: 29–38; **Hibbett, D.S.; Pine, E.M.; Langer, E.; Langer, G.; Donoghue, M.J.** (1996). Evolution of gilled mushrooms and puffballs inferred from ribosomal DNA sequences. *Proc. natn Acad. Sci. U.S.A.* 94: 12002–12006; **Niemelä, T.; Kotiranta, H.** (1986). Polypore survey of Finland 4. *Phaeolus, Fistulina, Ganoderma* and *Ischnoderma*. *Karstenia* 26: 57–64; **Quanten, E.** (1997). The Polypores (*Polyporaceae s.l.*) of Papua New Guinea: A Preliminary Prospectus. *Op. bot. Belg.* 11: 352 pp.; **Rajchenberg, M.; Greslebin, A.** (1995). Cultural characters, compatibility tests and taxonomic remarks of selected polypores of the Patagonian Andes forests of Argentina. *Mycotaxon* 56: 325–346.

Fomitopsidaceae Jülich 1982
Polyporales: Basidiomycota

Basidiomata annual or perennial, resupinate, applanate, effuse or pileate, woody, leathery or corky in texture, smooth to rugose or crustose. Hyphal system dimitic or trimitic, skeletal hyphae often well-developed, clamp connections usually present, cystidia absent. Hymenium poroid, the surface usually pale or brownish, the pores sometimes convolute or lamellate. Basidiospores ellipsoidal to cylindric or allantoid, hyaline, smooth, thin-walled, not staining in iodine.

Significant Genera: *Daedalea, Fomitopsis, Piptoporus, Postia*.

Fomitopsis pinicola, Vancouver Island, Canada; basidioma

Distribution: Widespread, especially north temperate.

Economic Significance: Some species affect economically important trees, e.g. *Daedalea quercina* (wood rot of various broadleaved trees, especially oak) and *Fomitopsis pinicola* (brown rot of conifers). *Piptoporus betulinus* was at one time used for tinder, and carved basidiomata of *Fomitopsis officinalis* were used as grave guardians by indigenous peoples of the Pacific coast of Canada. A number of species have had medicinal applications in the past, including styptics and antiperspirants.

Ecology: Most species are parasitic on woody plants, causing brown rots.

Notes: The family requires redefinition; generic delimitation appears to be confused and not congruous with molecular phylogenetic data. Merging with the *Polyporaceae* has been suggested, but that family also requires redefinition.

Piptoporus betulinus, Surrey, UK; basidioma

References: **Blanchette, R.A.; Compton, B.D.; Turner, N.J.; Gilbertson, R.L.** (1992). Nineteenth century shaman grave guardians are carved *Fomitopsis officinalis* sporophores. *Mycologia* 84: 119–124; **Carranza-Morse, J.; Gilbertson, R.L.** (1986). Taxonomy of the *Fomitopsis rosea* complex (*Aphyllophorales*; *Polyporaceae*). *Mycotaxon* 25: 469–486; **Chang, T.T.; Chiu, W.H.; Hua, J.** (1996). The Cultural Atlas of Wood-Inhabiting *Aphyllophorales* in Taiwan, vol. 1. *Mycol. Monogr.* 10: 126 pp.; **Hibbett, D.S.; Donoghue, M.J.** (1995). Progress toward a phylogenetic classification of the *Polyporaceae* through parsimony analysis of mitochondrial ribosomal DNA sequences. *Can. J. Bot.* 73: S853–S861; **Hjortstam, K.; Ryvarden, L.** (1987). A note on *Anomoporia* Pouz. (*Polyporaceae*, *Aphyllophorales*). *Mycotaxon* 28: 553–555; **Högberg, N.; Stenlid, J.** (1999). Population genetics of *Fomitopsis rosea* – a wood-decay fungus of the old-growth European taiga. *Mol. Ecol.* 8: 703–710; **Kauserud, H.; Schumacher, T.** (2003). Genetic structure of Fennoscandian populations of the threatened wood-decay fungus *Fomitopsis rosea* (*Basidiomycota*). *Mycol. Res.* 107: 155–163; **Kim, K.M.; Yoon, Y.-G.; Jung, H.S.** (2005). Evaluation of the monophyly of *Fomitopsis* using parsimony and MCMC methods. *Mycologia* 97: 812–822; **Kim, S.Y.; Park, S.Y.; Ko, K.S.; Jung, H.S.** (2003). Phylogenetic analysis of *Antrodia* and related taxa based on partial mitochondrial SSU rDNA sequences. *Antonie van Leeuwenhoek* 83: 81–88; **Larsen, M.J.; Lombard, F.F.** (1986). New combinations in the genus *Postia* Fr. (*Polyporaceae*). *Mycotaxon* 26: 271–273; **Larsson, K.-H.; Larsson, E.; Koljalg, U.** (2004). High phylogenetic diversity among corticioid homobasidiomycetes. *Mycol. Res.* 108: 983–1002; **Parmasto, E.** (2001). *Gilbertsonia*, a new genus of polypores (*Hymenomycetes*, *Basidiomycota*). *Harvard Pap. Bot.* 6: 179–182; **Rajchenberg, M.** (1995). A taxonomic study of the subantarctic *Piptoporus* (*Polyporaceae*, *Basidiomycetes*) II. *Nordic Jl Bot.* 15: 105–119; **Thorn, R.G.** (2000). Some polypores misclassified in *Piptoporus*. *Karstenia* 40: 181–197.

Fuscideaceae Hafellner 1984
Teloschistales: Ascomycota

Thallus crustose, areolate. Ascomata apothecial, blackish, lacking anthraquinone pigments, without a well-developed thalline margin. Interascal tissue of sparsely branched paraphyses, the apices often pigmented. Asci with a well-developed outer J+ gelatinous layer and usually a conspicuous inner J+ layer surrounding the often poorly defined ocular chamber. Ascospores usually aseptate, sometimes with attenuated apices.

Fuscidea kochiana, Northumberland, UK; thallus

Significant Genera: *Fuscidea*, *Maronea*.

Distribution: Widespread.

Economic Significance: None is known.

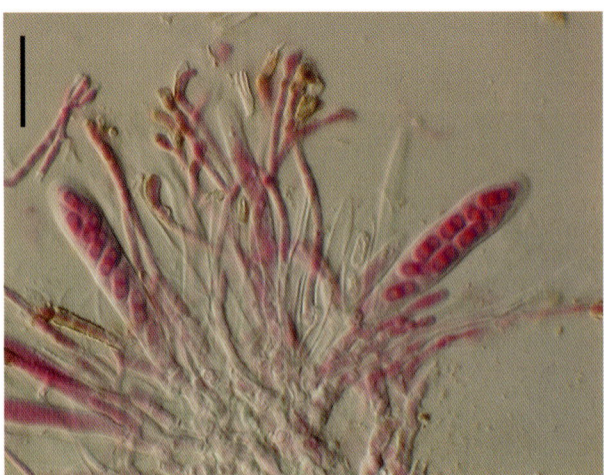
Fuscidea lightfootii, Devon, UK; asci, ascospores and paraphyses

Ecology: Lichenized with green algae, mostly on bark; a few species lichenicolous.

Notes: Previous links with the *Teloschistaceae* have not been supported by molecular data; the family appears to be more closely related to the *Ostropales* clade.

References: **Farkas, E.É.** (1995). Notes on the genus *Sarrameana* Vězka & P. James and some black fruited species of *Bacidia* s.l. (Lichenized *Ascomycetes*). *Biblthca Lichenol.* 58: 97–106; **Ihlen, P.G.; Ekman, S.** (2002). Outline of phylogeny and character evolution in *Rhizocarpon* (*Rhizocarpaceae*, lichenized *Ascomycota*) based on nuclear ITS and mitochondrial SSU ribosomal DNA sequences. *Biol. J. Linn. Soc.* 77: 535–546; **Ihlen, P.G.; Tønsberg, T.** (1996). The lichenicolous genus *Lettauia* in North America. *Bryologist* 99: 32–33; **Kantvilas, G.; Vězda, A.** (1996). The lichen genus *Sarrameana*. *Nordic Jl Bot.* 16: 325–333; **Oberhollenzer, H.; Wirth, V.** (1990). Contributions to a revision of the lichen genus *Fuscidea*. III. *Fuscidea recensa* (Stirton) Hertel, V. Wirth & Vězda. *Biblthca Lichenol.* 38: 363–375; **Reeb, V.; Lutzoni, F.; Roux, C.** (2004). Contribution of *RPB2* to multilocus phylogenetic studies of the euascomycetes (*Pezizomycotina*, *Fungi*) with special emphasis on the lichen-forming *Acarosporaceae* and evolution of polyspory. *Mol. Phylogen. Evol.* 32: 1036–1060.

Ganodermataceae Donk 1948
Polyporales: Basidiomycota

Basidioma pileate, annual or perennial, sessile to stipitate, large, brown or reddish, woody or coriaceous with a zoned, waxy and often glossy crust. Hyphal system trimitic; generative hyphae hyaline, skeletal and binding hyphae hyaline or brown, clamp connections present. Hymenium usually pale, poroid, thick, the pores round in section. Basidia large, clavate to pyriform, with 4 sterigmata. Basidiospores brown, ovoid to ellipsoidal, the apex truncate and usually thicker-walled than the basal region, with a 2-layered wall, the outer (exospore) layer relatively thin and hyaline, and the inner layer strongly pigmented, thick and often ornamented, the surface typically punctate, rarely reticulate or longitudinally crested.

Significant Genera: *Amauroderma*, *Ganoderma*.

Distribution: Widespread, especially in the tropics.

Economic Significance: *Ganoderma* species cause decay of standing timber; *G. philippii* causes a root rot of, for example, cacao, coffee, rubber and tea whilst *G. orbiformum* (syn. *G. boninense*, *G. lucidum* sensu auct.) causes a root rot on palms (oil palm, coconut palm). *Ganoderma* is also highly prized in Oriental traditional medicine, with numerous preparations commercially available.

Ganoderma adspersum, Berkshire, UK; basidiomata

Ecology: Lignicolous, causing sometimes aggressive white rots of timber.

Notes: The classification of the principal genus, *Ganoderma*, is undergoing change, with numerous taxa now recognized as insignificant morphological variants. Particular types of fruit bodies are highly prized and treated as ornaments, presumably a reflection of their perceived medicinal value. The family could be subsumed into the *Polyporaceae*.

Ganoderma sp., Khao Yai National Park, Thailand; basidiomata

References: **Gottlieb, A.M.; Ferrer, E.; Wright, J.E.** (2000). rDNA analyses as an aid to the taxonomy of species of *Ganoderma*. *Mycol. Res.* 104: 1033–1045; **Hong, S.G.; Jung, H.S.** (2004). Phylogenetic analysis of *Ganoderma* based on nearly complete mitochondrial small-subunit ribosomal DNA sequences. *Mycologia* 96: 742–755; **Kim, S.Y.; Jung, H.S.** (2001). Phylogenetic relationships of the *Polyporaceae* based on gene sequences of nuclear small subunit ribosomal RNAs. *Mycobiology* 29: 73–79; **Miller, R.N.G.; Holderness, M.; Bridge, P.D.** (2000). Molecular and morphological characterization of *Ganoderma* in oil-palm plantings. *Ganoderma Diseases of Perennial Crops* (Wallingford): 159–182; **Moncalvo, J.-M.** (2000). Systematics of *Ganoderma*. *Ganoderma Diseases of Perennial Crops* (Wallingford): 23–45; **Moncalvo, J.-M.; Ryvarden, L.** (1997). A Nomenclatural Study of the *Ganodermataceae* Donk. *Syn. Fung.* (Oslo) 11: 114 pp.; **Moncalvo, J.-M.; Wang, H.F.; Hseu, R.S.** (1995). Gene phylogeny of the *Ganoderma lucidum* complex based on ribosomal DNA sequences. Comparison with traditional taxonomic characters. *Mycol. Res.* 99: 1489–1499; **Moncalvo, J.-M.; Wang, H.F.; Wang, H.H.; Hseu, R.S.** (1995). The use of ribosomal DNA sequence data for species identification and phylogeny in the *Ganodermataceae*. *Ganoderma: Systematics, Phytopathology and Pharmacology* Proceedings of Contributed Symposium 59A, B, 5th International Mycological Congress, Vancouver, August 14-21, 1994 (Taipei): 31–44; **Moncalvo, J.-M.; Wang, H.-h.; Hseu, R.-s.** (1995). Phylogenetic relationships in *Ganoderma* inferred from the internal transcribed spacers and 25S ribosomal DNA sequences. *Mycologia* 87: 223–238; **Pilotti, C.A.; Sanderson, F.R.; Aitken, E.A.B.; Bridge, P.D.** (2000). Genetic variation in *Ganoderma* spp. from Papua New Guinea as revealed by molecular (PCR) methods. *Ganoderma Diseases of Perennial Crops* (Wallingford): 195–204; **Ryvarden, L.** (2004). Studies in Neotropical polypores 19. Two wood-inhabiting *Amauroderma* species. *Syn. Fung.* (Oslo) 18: 57–61; **Smith, B.J.; Sivasithamparam, K.** (2003). Morphological studies of *Ganoderma* (*Ganodermataceae*) from the Australasian and Pacific regions. *Aust. Syst. Bot.* 16: 487–503; **Wang, D.M.; Yao, Y.J.** (2005). Intrastrain internal transcribed spacer heterogeneity in *Ganoderma* species. *Can. J. Microbiol.* 51: 113–121; **Zhao, J.D.** (1989). The *Ganodermataceae* in China. *Biblthca Mycol.* 132: 176 pp.

Gasterellaceae Zeller 1948
Boletales: Basidiomycota

Basidiomata gasteroid, minute, pulvinate, lacking a stipe, initially white but becoming darker with age, hollow, the surface cottony or fibrose, dry. Hyphal system monomitic, the fruit body wall composed of loosely woven hyphae. Hymenium lining the inner surface of the basidioma, of basidia interspersed with sterile hyphae and capitate cystidia with dark encrusted heads. Basidia evanescent, clavate, with well-developed sterigmata, 2- or 4-spored, the spore forming symmetrically on the sterigma. Basidiospores ovoid to subglobose, with a distinctly apiculate distal region, dark brown, thick-walled, verrucose.

Significant Genera: *Gasterella*.

Distribution: Only known from North America.

Economic Significance: None is known.

Ecology: Apparently saprobic, occurring on soil.

Notes: Relationships of this family are uncertain and no molecular data are available. Only one species has been described. Its placement is doubtless influenced by comparison with *Hymenogaster* (*Hymenogasteraceae*, *Boletales*) in the original paper.

References: **Zeller, S.M.** (1948). Notes on certain gasteromycetes, including two new orders. *Mycologia* 40: 639–668; **Zeller, S.M.; Walker, L.B.** (1935). *Gasterella*, a new uniloculate gasteromycete. *Mycologia* 27: 573–579.

Gastrosporiaceae Pilát 1934
Boletales: Basidiomycota

Basidiomata gasteroid, hypogeous, ± globose, pale, soft-textured, dry, composed of two easily separable layers. Outer wall layer crystalline, hyphal in composition, the inner layer tough, cartilaginous, gelatinous. Gleba filling the entire basidiomatal cavity, initially white but becoming yellowish or olivaceous, without a true capillitium. Basidia evanescent, irregular in form, clavate to widely cylindrical, with a basal clamp connection, with a cluster of up to 12 sterigmata at the apex. Basidiospores globose, pale brown or ochraceous, minutely warted. Rhizomorphs well-developed, composed of hyphae with ampullate inflations at the septa.

Significant Genera: *Gastrosporium*.

Distribution: Typically found in xerothermic areas, widespread but known especially from southern Europe.

Economic Significance: None is known.

Ecology: In soil, associated with *Poaceae* (*Gramineae*), nutrition and mycorrhizal status unknown.

Notes: Molecular study and anatomical features suggest this family is related to the *Phallaceae*, but more research is needed.

Gastrosporium simplex; basidiomata

References: **Domínguez de Toledo, L.S.; Castellano, M.A.** (1997). First report of *Gastrosporium simplex* (*Gasteromycetes*) from South America. *Mycotaxon* 64: 443–448; **Iosifidou, P.; Agerer, R.** (2002). [The rhizomorphs of *Gastrosporium simplex* and some ideas on the systematic position of the *Gastrosporiaceae* (*Hymenomycetes, Basidiomycota*)]. *Feddes Repert.* 113: 11–23; **Kreisel, H.; Hausknecht, A.** (2002). The gasteral *Basidiomycetes* of Mascarenes and Seychelles. *Öst. Z. Pilzk.* 11: 191–211; **Zeller, S.M.** (1948). Notes on certain gasteromycetes, including two new orders. *Mycologia* 40: 639–668.

Geastraceae Corda 1842
Geastrales: Basidiomycota

Basidiomata gasteroid and epigeous, rarely hypogeous, thick-walled, solitary or in small clusters on a basal stroma, usually multilayered. Peridium soft-textured, leathery or parchment-like, sometimes gelatinous, primarily hyphal in construction with up to 4 distinct layers, usually opening with a wide stellate split to expose the inner wall; clamp connections usually present. Inner wall when present quite separate in construction, thin and pliable, persistent, usually with a single large circular ostiole. Sporogenous tissue (gleba) composed of a single convoluted or multiple fertile chambers radiating from a central columella, mostly breaking down to a powdery mass but in one genus discharged forcibly from the basidioma as a discrete unit. Basidia evanescent at an early stage, clavate or hypha-like, 4- or 8-spored, interspersed with unbranched thick-walled capillitial hyphae. Basidiospores globose or ± ellipsoidal, brown, thin-walled, finely warted or spiny, not staining in iodine.

Significant Genera: *Geastrum*, *Sphaerobolus*.

Distribution: Cosmopolitan.

Economic Significance: Little is known. *Sphaerobolus* species can cause cosmetic damage through adhesion of discharged glebae to house walls etc.

Ecology: Terrestrial, rarely on wood or coprophilous. Most species are presumed saprobes, but at least some may be ectomycorrhizal.

Notes: A distinctive family within the phalloid clade of euagarics. Indiacations are that the family diverged from the euagarics at an early stage, and that in some cases there is substantial sequence variation between morphologically similar taxa. Species of *Geastrum* are known to non-specialists as the earth-stars.

Geastrum pectinatum; basidiomata

Geastrum saccatum; basidiomata

References: **Baseia, I.G.; Cavalcanti, M.A.; Milanez, A.I.** (2003). Additions to our knowledge of the genus *Geastrum* (*Phallales*: *Geastraceae*) in Brazil. *Mycotaxon* 85: 409–415; **Binder, M.; Bresinsky, A.** (2002). Derivation of a polymorphic lineage of gasteromycetes from boletoid ancestors. *Mycologia* 94: 85–98; **Calonge, F.D.** (1998). *Gasteromycetes*, I. *Lycoperdales*, *Nidulariales*, *Phallales*, *Sclerodermatales*, *Tulostomatales*. *Fl. Mycol. Iberica* 3: 271 pp.; **Coetzee, J.C.; Eicker, A.; Van Wyck, A.E.** (1997). Taxonomic notes on the *Geastraceae*, *Tulostomataceae*, *Nidulariaceae* and *Sphaerobolaceae* (*Gasteromycetes*) *sensu* Bottomly, in southern Africa. *Bothalia* 27: 117–123; **Domínguez de Toledo, L.S.; Castellano, M.A.** (1996). A revision of the genera *Radiigera* and *Pyrenogaster*. *Mycologia* 88: 863–884; **Esqueda, M.; Herrera, T.; Pérez-Silva, E.; Sánchez, A.** (2003). Distribution of *Geastrum* species from some priority regions for conservation of biodiversity of Sonora, Mexico. *Mycotaxon* 87: 445–456; **Estrada-Torres, A.; Gaither, T.W.; Miller, D.L.; Lado, C.; Keller, H.W.** (2005). The myxomycete genus *Schenella*: morphological and DNA sequence evidence for synonymy with the gasteromycete genus *Pyrenogaster*. *Mycologia* 97: 139–149; **Geml, J.; Davis, D.D.; Geiser, D.M.** (2005). Phylogenetic analyses reveal deeply divergent species lineages in the genus *Sphaerobolus* (*Phallales*, *Basidiomycota*). *Mol. Phylogen. Evol.* 35: 313–322; **Geml, J.; Davis, D.D.; Geiser, D.M.** (2005). Systematics of the genus *Sphaerobolus* based on molecular and morphological data, with the description of *Sphaerobolus ingoldii* sp. nov. *Mycologia* 97: 680–694; **Hibbett, D.S.; Binder, M.** (2002). Evolution of complex fruiting-body morphologies in homobasidiomycetes. *Proc. R. Soc. Lond. B. Biol. Sci.* 269: 1963–1969; **Hibbett, D.S.; Pine, E.M.; Langer, E.; Langer, G.; Donoghue, M.J.** (1996). Evolution of gilled mushrooms and puffballs inferred from ribosomal DNA sequences. *Proc. natn Acad. Sci. U.S.A.* 94: 12002–12006; **Hibbett et al.** (2007). A higher-level phylogenetic classification of the *Fungi*. *Mycol. Res.* 111(5): 509–548; **Hosaka et al.** (2007). Molecular phylogenetics of the gomphoid-phalloid fungi with an establishment of the new subclass *Phallomycetidae* and two new orders. *Mycologia* 98(6): 949–959; **Ingold, C.T.** (1972). *Sphaerobolus*: the story of a fungus. *Trans. Br. mycol. Soc.* 58: 179–195; **Kreisel, H.; Hausknecht, A.** (2002). The gasteral *Basidiomycetes* of Mascarenes and Seychelles. *Öst. Z. Pilzk.* 11: 191–211; **Krüger, D.; Binder, M.; Fischer, M.; Kreisel, H.** (2001). The *Lycoperdales*. A molecular approach to the systematics of some gasteroid mushrooms. *Mycologia* 93: 947–957; **Miller, O.K.; Miller, H.H.** (1988). *Gasteromycetes. Morphological and Development Features with Keys to the Orders, Families, and Genera* (Eureka): 157 pp.

Geminaginaceae Vánky 2001
Ustilaginales: Basidiomycota

Sori immersed in floral tissues, causing hypertrophy and deformation, containing ovoid to irregular cavities in an irregular honeycomb arrangement, in which dark brown, powdery spore masses form. Hyphae intracellular, septa lacking pores. Teliospores formed in pairs, ± hemispherical, dark brown, the wall uneven in thickness, ± smooth, adhering with the flattened surfaces adjacent, often eventually separating, germinating to form septate branched basidia. Basidiospores produced laterally or terminally, often in chains, ovoid to fusiform, brown, thin-walled.

Significant Genera: *Geminago*.

Distribution: Only known from Central Africa.

Economic Significance: The host is exploited as a timber tree, but the impact on reproduction is poorly known.

Ecology: Parasitic on floral tissues of *Triplochiton* (*Sterculiaceae*), causing hypertrophy.

Notes: The only species, *Geminago noveilleri*, is one of the few smuts to affect woody plants. Its systematic position remains uncertain, and no molecular studies have been carried out.

References: **Piepenbring, M.; Bauer, R.; Oberwinkler, F.** (1998). Teliospores of smut fungi. Teliospore connections, appendages, and germ pores studied by electron microscopy; phylogenetic discussion of characteristics of teliospores. *Protoplasma* 204: 202–218; **Vánky, K.** (1996). *Mycosyrinx* and other pair-spored *Ustilaginales*. *Mycoscience* 37: 173–185; **Vánky, K.** (1998). A survey of the spore-ball-forming smut fungi. *Mycol. Res.* 102: 513–526; **Vánky, K.** (2001). The emended *Ustilaginaceae* of the modern classificatory system for smut fungi. *Fungal Diversity* 6: 131–147.

Geoglossaceae Corda 1838
Helotiales: Ascomycota

Stromata absent. Ascomata large, clavate, spathulate or strongly stipitate, the hymenium usually not clearly separable from the stalk, usually dark. Interascal tissue of paraphyses, often strongly pigmented, often with complex branching and swollen apices and covered with agglutinated melanized material; the hymenium sometimes setose. Asci large, cylindric-clavate, thin-walled, 4- or 8-spored, with a J+ or J– pore. Ascospores clavate or elongate, usually brown at maturity, usually with many transverse septa. Anamorphs unknown.

Significant Genera: *Geoglossum*, *Trichoglossum*.

Geoglossum cookeanum, St Kilda, Scotland; asci, ascospores and paraphyses

Distribution: Widespread, especially in temperate zones.

Economic Significance: None is known.

Ecology: Probably saprobic on soil or rotten vegetation, especially in grassland ecosystems; little is known.

Notes: An isolated family which is now recognized as completely separate from the *Helotiales* where it has traditionally been classified. Older systematic treatments used a rather wider concept including many stalked genera that are now placed in the *Cudoniaceae*, *Helotiaceae* and *Sclerotiniaceae*.

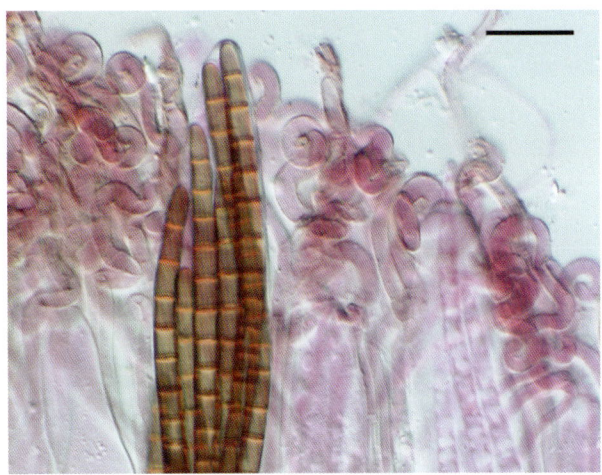

Geoglossum difforme, North Carolina, USA; ascus, ascospores and paraphyses

References: **Maas Geesteranus, R.A.** (1965). *Geoglossaceae* of India and adjacent countries. *Persoonia* 4: 19–46; **Mains, E.B.** (1954). North American species of *Geoglossum* and *Trichoglossum*. *Mycologia* 46: 586–631; **Nannfeldt, J.A.** (1942). The *Geoglossaceae* of Sweden. *Ark. Bot.* 30A: 1–67; **Olsen, S.** (1986). Jordtunger i Norge. *Agarica* 7: 120–168; **Spooner, B.M.** (1987). Helotiales of Australasia: Geoglossaceae, Orbiliaceae, Sclerotiniaceae, Hyaloscyphaceae. *Biblthca Mycol.* 116: 711 pp.; **Verkley, G.J.M.** (1994). Ultrastructure of the ascus apical apparatus in *Leotia lubrica* and some Geoglossaceae (Leotiales, Ascomycotina). *Persoonia* 15: 405–430; **Wang, Z.; Binder, M.; Schoch, C.L.; Johnston, P.R.; Spatafora, J.W.; Hibbett, D.S.** (2006). Evolution of helotialean fungi (Leotiomycetes, Pezizomycotina): a nuclear rDNA phylogeny. *Mol. Phylogen. Evol.* 41: 295–312.

Georgefischeriaceae R. Bauer, Begerow & Oberw. 1997
Georgefischeriales: Basidiomycota

Sori formed in leaves and stems, often causing hypertrophy or witches' brooms, mycelium intracellular and intercellular. Spore masses formed in lysigenous cavities within the plant tissues, the epidermis splitting away or fragmenting to expose the teliospores. Teliospores globose to irregular, brownish, smooth, the wall sometimes lamellate, with a well-developed mucous coating, aggregating in groups, germinating to form a simple hypha-like aseptate basidium. Basidiospores discharged actively, 2–4 per basidium, cylindrical to ellipsoidal, formed in an apical cluster.

Significant Genera: *Georgefischeria*, *Jamesdicksonia*.

Distribution: Widespread in the Old World tropics and warm temperate zones.

Economic Significance: None is known.

Ecology: Parasitic in leaves and stems of grasses, sedges and *Convolvulaceae*.

Notes: Few sequence data are available, but the family appears to be close to the *Entylomatales*. Other families of the *Georgefischeriales* either have septate basidia or lack actively discharged basidiospores.

References: **Bauer, R.; Begerow, D.; Nagler, A.; Oberwinkler, F.** (2001). The *Georgefischeriales*: a phylogenetic hypothesis. *Mycol. Res.* 105: 416–424; **Bauer, R.; Oberwinkler, F.; Vánky, K.** (1997). Ultrastructural markers and systematics in smut fungi and allied taxa. *Can. J. Bot.* 75: 1273–1314; **Begerow, D.; Bauer, R.; Oberwinkler, F.** (1998). Phylogenetic studies on nuclear large subunit ribosomal DNA sequences of smut fungi and related taxa. *Can. J. Bot.* 75: 2045–2056; **Begerow, D.; Bauer, R.; Oberwinkler, F.** (2002). The *Exobasidiales*: an evolutionary hypothesis. *Mycol. Prog.* 1: 187–199; **Begerow, D.; John, B.; Oberwinkler, F.** (2004). Evolutionary relationships among β-tubulin gene sequences of basidiomycetous fungi. *Mycol. Res.* 108: 1257–1263.

Geosiphonaceae Engl. & E. Gilg 1924
Archaeosporales: Glomeromycota

Thallus hyphal, coenocytic, forming an endosymbiotic association with cyanobacteria (*Nostoc punctiforme*), the hyphal tip swelling to form a unicellular 'bladder'. Spores similar to those of *Glomeraceae*.

Significant Genera: *Geosiphon*.

Distribution: Widespread, although undoubtedly more so than current records suggest.

Economic Significance: None is known.

Ecology: Occurring in the soil but rarely recorded.

Notes: *Geosiphon pyriforme* is the only known fungus forming an endosymbiosis with cyanobacteria, specifically *Nostoc punctiforme*. At the hyphal tips unicellular, multinucleated 'bladders', about 1–2 mm in size, harbouring the cyanobacteria, are formed. There have been only five reports of it being found in nature and presently a region in the Spessart mountains (Germany) is the only known natural habitat worldwide.

References: **Azcon-Aguilar, C.; Barea, J.M.** (1996). Arbuscular mycorrhizas and biological control of soil-borne plant pathogens – an overview of the mechanisms involved. *Mycorrhiza* 6: 457–464; **Gehrig, H.; Schüßler, A.; Kluge, M.** (1996). *Geosiphon pyriforme*, a fungus forming endocytobiosis with *Nostoc* (*Cyanobacteria*), is an ancestral member of the *Glomales*: evidence by SSU rRNA analysis. *J. Mol. Evol.* 43: 71–81; **Redecker, D.** (2005). *Glomeromycota* Arbuscular mycorrhizal fungi and their relative(s). Version 01 July 2005. http://tolweb.org/Glomeromycota/28715/2005.07.01 in The Tree of Life Web Project, http://tolweb.org; **Schüßler, A.; Schwarzott, D.; Walker, C.** (2001). A new fungal phylum, the *Glomeromycota*: phylogeny and evolution. *Mycol. Res.* 105: 1413–1421; **Schüßler, A.; Wolf, E.** (2005). *Geosiphon pyriformis*: a glomeromycotan soil fungus forming endosymbiosis with cyanobacteria. in Declerck, Strullu & Fortin (eds). *Root-organ culture of mycorrhizal fungi* (Heidelberg): [in press]; **Schüßler. A.; Kluge, M.** (2001). *Geosiphon pyriformis*, an endocytosymbiosis between fungus and cyanobacteria, and its meaning as a model system for arbuscular mycorrhizal research. in Hock (ed.). *The Mycota* (Berlin) 9: 151–161; **van der Heijden, M.A.G.; Klironomos, J.N.; Ursic, M.; Moutoglis, P.; Streitwolf-Engel, R.; Boller, T.; Wiemken, A.; Sanders, I.R.** (1998). Mycorrhizal fungal diversity determines plant biodiversity, ecosystem variability and productivity. *Nature* Lond. 396: 69–72.

Gigaspermaceae Jülich 1982
Agaricales: Basidiomycota

Basidiomata hypogeous, ± globose, fragile, white or yellowish. Outer wall (peridium) hyphal in construction, the hyphae narrow, not encrusted or gelatinized, with clamp connections sometimes present. Fertile zone (gleba) labyrinthiform with irregular locules, grey or brownish, not gelatinous, columella lacking. Basidia clavate, hyaline, thin-walled, 4-spored or with a single well-developed thick-walled sterigma, interspersed with inflated hyaline sterile hypha-like cells, cystidia absent. Basidiospores large, ± globose, hyaline or brown, extremely thick-walled, smooth, sometimes staining brown in iodine.

Significant Genera: *Gigasperma*.

Distribution: Known from Australasia and North America.

Economic Significance: None is known.

Ecology: Species are presumed to be ectomycorrhizal with woody plants (*Nothofagus* and *Cercocarpus*).

Notes: Inclusion within the *Tricholomataceae* has been suggested, but sequence studies have not been attempted. The extremely thick-walled basidiospores are characteristic of the family.

References: Castellano, M.A.; Trappe, J.M. (1992). Australasian truffle-like fungi. VI. *Gigasperma* (Basidiomycotina, Tricholomataceae) and *Horakiella* gen. nov. (Basidiomycotina, Sclerodermataceae). *Aust. Syst. Bot.* 5: 639–643; **Kropp, B.R.; Hutchison, L.J.** (1996). *Gigasperma americanum*, a new hypogeous member of the *Basidiomycota* associated with *Cercocarpus ledifolius* in Utah. *Mycologia* 88: 662–665.

Gigasporaceae J.B. Morton & Benny 1990
Diversisporales: Glomeromycota

Mycorrhizal species forming arbuscules in roots (intraradical), anamorphic; the penetrating hyphae producing finely branched haustorial branches (arbuscules) or coils (pelotons), vesicles absent. Spores hypogeous or partly epigeous, formed at the apex of a sporogenous cell on a fertile hypha, auxillary cells produced.

Significant Genera: *Gigaspora, Scutellospora*.

Distribution: Widespread, although undoubtedly more so than current records suggest.

Economic Significance: Enhancement of phosphorus (P) supply to the 'host' is a characteristic of AM (arbuscular mycorrhizal) infections. As such they have a significant influence, either directly or indirectly, on land plants including ecosystem variability and productivity.

Ecology: As mycorrhizal partners of land plants forming arbuscular mycorrhizas; the hosts of AM fungi are primarily found in P-limited tropical and temperate grasslands.

Notes: The two genera are sufficiently similar that they might be merged but are currently maintained based on minor differences in germination type; a germinal wall in *Gigaspora* and a germination shield in *Scutellospora*.

Scutellospora erythropa; spores

References: **Azcon-Aguilar, C.; Barea, J.M.** (1996). Arbuscular mycorrhizas and biological control of soil-borne plant pathogens – an overview of the mechanisms involved. *Mycorrhiza* 6: 457–464; **Jeffries, P.; Dodd, J.C.** (2000). Molecular ecology of mycorrhizal fungi. *Appl. Microb. System.*: 73–105; **Redecker, D.** (2005). *Glomeromycota* Arbuscular mycorrhizal fungi and their relative(s). Version 01 July 2005. http://tolweb.org/Glomeromycota/28715/2005.07.01 in The Tree of Life Web Project, http://tolweb.org; **Schüßler, A.; Schwarzott, D.; Walker, C.** (2001). A new fungal phylum, the *Glomeromycota*: phylogeny and evolution. *Mycol. Res.* 105: 1413–1421; **Souza, F.A. de; Declerck, S.; Smit, E.; Kowalchuk, G.A.** (2005). Morphological, ontogenetic and molecular characterization of *Scutellospora reticulata* (*Glomeromycota*). *Mycol. Res.* 109: 697–706; **van der Heijden, M.A.G.; Klironomos, J.N.; Ursic, M.; Moutoglis, P.; Streitwolf-Engel, R.; Boller, T.; Wiemken, A.; Sanders, I.R.** (1998). Mycorrhizal fungal diversity determines plant biodiversity, ecosystem variability and productivity. *Nature Lond.* 396: 69–72; **Walker, C.; Schüssler, A.** (2004). Nomenclatural clarifications and new taxa in the *Glomeromycota*. *Mycol. Res.* 108: 981–982.

Gjaerumiaceae R. Bauer, M. Lutz & Oberw. 2005
Georgefischeriales: Basidiomycota

Colonies visible as greyish patches on living leaves. Mycelium intercellular, hyphal septa with simple dolipores, haustoria absent. Teliospores ± globose, brown, smooth-walled, germinating to produce a single cylindrical hypha-like basidium that is often septate. Basidiospores formed in a cluster at the apex of the basidium, released passively, fusiform, ± radially symmetrical, hyaline and thin-walled. Anamorph *Tilletiopsis*-like.

Significant Genera: *Gjaerumia*.

Distribution: Known from northern Europe.

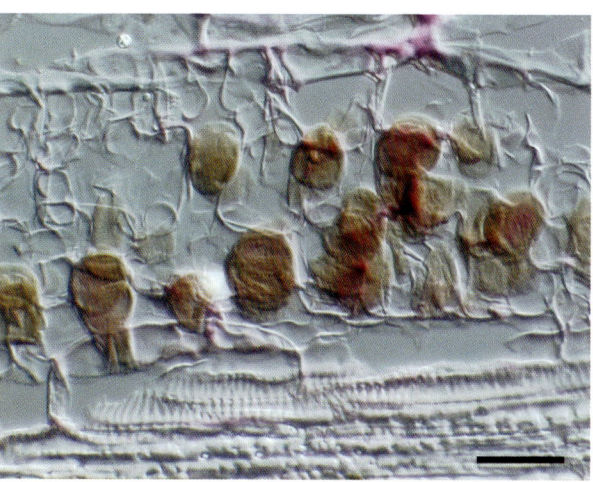

Gjaerumia ossifragi, Sutherland, Scotland; spore balls in leaf tissue

Economic Significance: None is known.

Ecology: Biotrophic in leaves of *Narthecium ossifragum*.

Notes: This is the only family of the *Georgefischeriales* with dolipore septa. Molecular phylogenetic studies indicate a possible relationship with the *Eballistraceae*.

References: **Bauer, R.; Lutz, M.; Oberwinkler, F.** (2005). *Gjaerumia*, a new genus in the *Georgefischeriales* (*Ustilaginomycetes*). *Mycol. Res.* 109: 1250–1258.

Glaziellaceae J.L. Gibson 1986
Pezizales: Ascomycota

Ascomata hollow, cleistothecial, lobed, with basal opening, orange to red; peridium thin and gelatinous. Interascal tissue absent. Asci embedded in ascomatal wall, clavate to subglobose, deliquescent, 1-spored. Ascospores globose, large, smooth, orange. Anamorph unknown.

Significant Genera: *Glaziella*.

Distribution: Widespread.

Economic Significance: None is known.

Ecology: Epigeous, forming mycorrhizas.

Notes: At one time assumed to be allied to the *Endogonales*, the monospecific *Glaziella* (the only representative of this family) is now known to be a cleistothecial member of the *Pezizales* with 1-spored asci. Molecular studies indicate that the family clusters within the *Pyronemataceae*, but the latter family is widely circumscribed and may need division.

References: **Gibson, J.L.; Kimbrough, J.W.; Benny, G.L.** (1986). Ultrastructural observations on *Endogonaceae* (*Zygomycetes*). II. *Glaziellales* ord. nov. and *Glaziellaceae* fam. nov.: new taxa based upon light and electron microscopic observations of *Glaziella aurantiaca*. *Mycologia* 78: 941–954; **Hansen, K.; Perry, B.A.; Pfister, D.H.** (2006). Phylogenetic origins of two cleistothecial fungi, *Orbicula parietina* and *Lasiobolidium orbiculoides*, within the operculate discomycetes. *Mycologia* 97: 1023–1033; **Harrington, F.A.; Pfister, D.H.; Potter, D.; Donoghue, M.J.** (1999). Phylogenetic studies within the *Pezizales*. I. 18S rRNA sequence data and classification. *Mycologia* 91: 41–50; **Landvik, S.; Egger, K.N.; Schumacher, T.** (1997). Towards a subordinal classification of the *Pezizales* (*Ascomycota*): phylogenetic analyses of SSU rDNA sequences. *Nordic Jl Bot.* 17: 403–418; **Landvik, S.; Eriksson, O.E.** (1994). Relationship of the genus *Glaziella* (*Ascomycota*) inferred from 18S rDNA sequences. *Syst. Ascom.* 13: 13–23.

Gloeoheppiaceae Henssen 1995
Lichinales: Ascomycota

Thallus squamulose to somewhat peltate, dark, attached by rhizoidal strands, homoiomerous, hyphae forming a reticulum surrounding cyanobacteria cells. Apothecia laminal, immersed to adnate. Interascal tissue composed of sparsely branched or anastomosed paraphyses, the tips with enlarged cells. Asci 8- to 16-spored. Ascospores ellipsoidal, aseptate to 1-septate.

Significant Genera: *Gloeoheppia, Pseudopeltula*.

Distribution: Primarily in desert areas, in both the Old and New World.

Economic Significance: None is known.

Ecology: Lichenized with cyanobacteria.

Notes: Very poorly known; no recent information is available.

Stenhammarella turgida, Saudi Arabia; thallus on sandy soil

References: **Henssen, A.** (1995). The new lichen family *Gloeoheppiaceae* and its genera *Gloeoheppia, Pseudopeltula* and *Gudelia* (*Lichinales*). *Lichenologist* 27: 261–290.

Gloeophyllaceae Jülich 1982
Gloeophyllales: Basidiomycota

Basidiomata annual or perennial, resupinate or pileate, coriaceous or woody, the upper surface dark brown or greyish, smooth, velutinous or hirsute, often zoned, the flesh dark brown. Hyphal system ditrimitic or trimitic, generative hyphae with clamp connections, skeletal hyphae often dominant. Cystidia present or absent, when present smooth or with a crystalline apex. Hymenium poroid or with straight or sinuous lamellae, reddish or dark brown. Basidia cylindric-clavate, thin-walled, with 4 sterigmata and a basal clamp connection. Basidiospores cylindrical or allantoid, hyaline, thin-walled, smooth, not staining in iodine.

Significant Genera: *Gloeophyllum*.

Distribution: Widespread.

Economic Significance: Some species are implicated in deterioration of buildings.

Ecology: Saprobic on wood of broadleaved and coniferous trees, causing aggressive brown rots.

Notes: Considered to be a monotypic family in the *Dictionary of the Fungi* edn 9, but molecular studies indicate that *Gloeophyllum* may belong with the *Boreostereaceae* within the *Polyporales*. Further revisionary work is required.

Gloeophyllum odoratum, Vancouver Island, Canada; young basidioma

References: **Adair, S.; Kim, S.H.; Breuil, C.** (2002). A molecular approach for early monitoring of decay basidiomycetes in wood chips. *FEMS Microbiol. Lett.* 211: 117–122; **Binder, M.; Hibbett, D.S.** (2002). Higher-level phylogenetic relationships of homobasidiomycetes (mushroom-forming fungi) inferred from four rDNA regions. *Mol. Phylogen. Evol.* 22: 76–90; **Corner, E.J.H.** (1987). Ad Polyporaceas IV. *Beih. Nova Hedwigia* 86: 265 pp.; **Hibbett, D.S.; Donoghue, M.J.** (2001). Analysis of character correlations among wood decay mechanisms, mating systems and substrate ranges in *Homobasidiomycetes*. *Syst. Biol.* 50: 215–242; **Hibbett et al.** (2007). A higher-level phylogenetic classification of the Fungi. *Mycol. Res.* 111(5): 509–548; **Hof, T.** (1981). *Gloeophyllum abietinum* (Bull. ex Fr.) Karst. *Some Wood-Destroying Basidiomycetes* (Papua New Guinea) 1: 55–66; **Moreth, U.; Schmidt, O.** (2005). Investigations on ribosomal DNA of indoor wood decay fungi for their characterization and identification. *Holzforschung* 59: 90–93; **Ryvarden, L.; Gilbertson, R.L.** (1993). European polypores. Part 1. *Abortiporus – Lindtneria*. *Syn. Fung.* (Oslo) 6: 387 pp.; **Schmidt, O.; Grimm, K.; Moreth, U.** (2002). [Molecular and biological characterization of *Gloeophyllum* species in buildings]. *Z. Mykol.* 68: 141–152; **Thorn, R.G.; Moncalvo, J.-M.; Reddy, C.A.; Vilgalys, R.** (2000). Phylogenetic analyses and the distribution of nematophagy support a monophyletic *Pleurotaceae* within the polyphyletic pleurotoid-lentinoid fungi. *Mycologia* 92: 241–252; **Yuan, H.S.; Dai, Y.C.; Wei, Y.L.** (2004). *Gloeophyllum* (*Basidiomycota*, *Aphyllophorales*) in China. *Mycosystema* 23: 173–176.

Glomeraceae Piroz. & Dalpé 1989
Glomerales: Glomeromycota

Mycorrhizal species forming arbuscules and vesicles in roots (intraradical), anamorphic; the penetrating hyphae producing finely branched haustorial branches (arbuscules) or coils (pelotons) and vesicles. Spores hypogeous or partly epigeous, arising terminally from fertile hyphae (auxillary cells not formed), solitary or formed in clusters or in sporocarps.

Significant Genera: *Glomus*.

Distribution: Widespread, although undoubtedly more so than current records suggest.

Economic Significance: Enhancement of phosphorus (P) supply to the 'host' is a characteristic of AM (arbuscular mycorrhizal) infections. As such they have a significant influence, either directly or indirectly, on land plants including ecosystem variability and productivity.

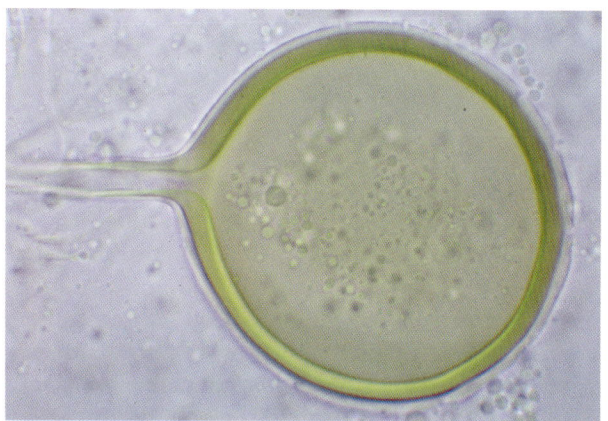

Glomus macrocarpum; spore

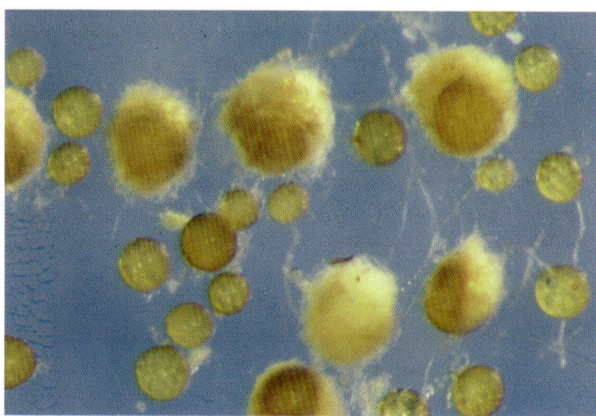

Glomus mosseae; spores

Ecology: As mycorrhizal partners of land plants forming arbuscular mycorrhizas; the hosts of AM fungi are primarily found in P-limited tropical and temperate grasslands.

Notes: The phylogeny of this family (in its restricted sense) has only recently been investigated in a more than cursory context.

References: **Azcon-Aguilar, C.; Barea, J.M.** (1996). Arbuscular mycorrhizas and biological control of soil-borne plant pathogens – an overview of the mechanisms involved. *Mycorrhiza* 6: 457–464; **Redecker, D.** (2005). *Glomeromycota* Arbuscular mycorrhizal fungi and their relative(s). Version 01 July 2005. http://tolweb.org/Glomeromycota/28715/2005.07.01 in The Tree of Life Web Project, http://tolweb.org; **Schüßler, A.; Schwarzott, D.; Walker, C.** (2001). A new fungal phylum, the *Glomeromycota*: phylogeny and evolution. *Mycol. Res.* 105: 1413–1421; **van der Heijden, M.A.G.; Klironomos, J.N.; Ursic, M.; Moutoglis, P.; Streitwolf-Engel, R.; Boller, T.; Wiemken, A.; Sanders, I.R.** (1998). Mycorrhizal fungal diversity determines plant biodiversity, ecosystem variability and productivity. *Nature Lond.* 396: 69–72; **Walker, C.; Schüssler, A.** (2004). Nomenclatural clarifications and new taxa in the *Glomeromycota*. *Mycol. Res.* 108: 981–982.

Glomerellaceae Locq. 1984
Uncertain position within Sordariomycetes, Ascomycota

Stromata absent. Ascomata perithecial, black, thick-walled and often irregular and ± sclerotial in form, the ostiole periphysate. Interascal tissue of copious thin-walled tapering true paraphyses. Asci clavate, short-stalked, thin-walled and not fissitunicate, with a small J– apical ring. Ascospores hyaline, almost all aseptate, smooth, often curved, without a mucous sheath or appendages. Anamorphs prominent, coelomycetous. Conidiomata acervular, often poorly developed or absent in culture, usually with well-developed dark setae. Conidiophores poorly developed, bearing small clusters of conidiogenous cells. Conidiogenous cells flask-shaped, proliferating percurrently, collarette usually inconspicuous. Conidia cylindrical, fusiform or lunate, hyaline, thin-walled, aseptate, rarely with one cellular attenuated apex, lacking a mucous sheath or appendages, germinating to produce dark brown ± circular to irregular appressoria.

Significant Genera: *Glomerella*. Anamorph: *Colletotrichum*.

Distribution: Widespread, but most frequent in tropical and warm temperate areas.

Economic Significance: Many species are highly significant plant pathogens, causing disease to growing crops and post-harvest problems, including cereals, fruits and vegetables. There has been widespread interest in use of highly host-specialized genotypes as mycoherbicides, although commercialization has proved difficult.

Ecology: Species are hemibiotrophic, with an initial endophytic (symptomless) phase that leads to a necrotic phase where significant damage may be caused to the host. Appressoria appear to function as survival propagules.

Colletotrichum dematium, Kenya; setae, conidiogenous cells and conidia

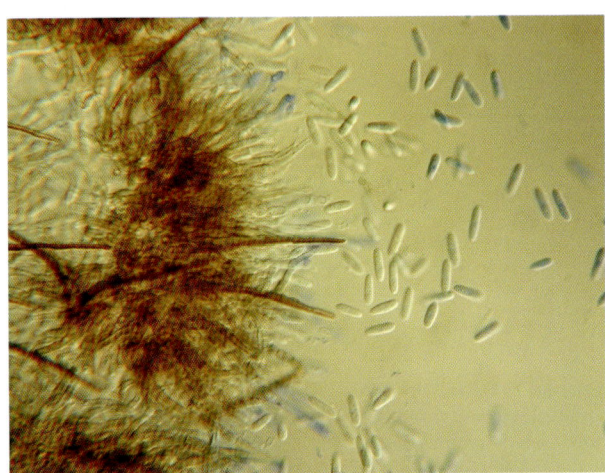
Glomerella cingulata, Guyana; setae, conidiogenous cells and conidia

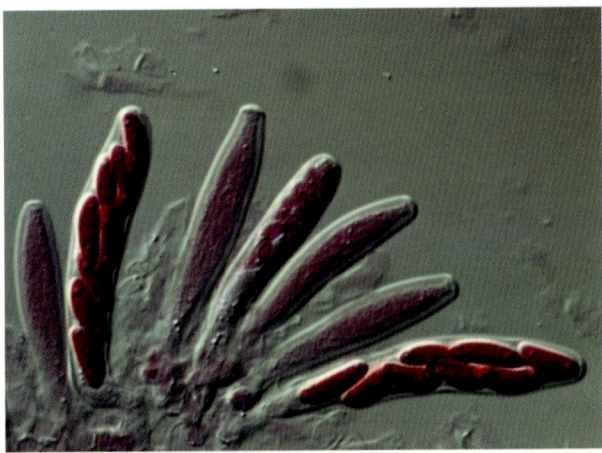
Glomerella cingulata; asci and ascospores

Notes: There is conflicting evidence of the extent of host specialization. Many genotypes and infraspecific groups appear to be generalists and at best weak pathogens, while others are highly host-specific and may represent recently evolved pathogens. Until recently, *Glomerella* and *Colletotrichum* were assumed to be members of the *Phyllachoraceae*, biotrophic fungi with appressoria and broadly similar (though stromatic) teleomorphs, but molecular data indicate that the two groups are distantly related.

References: **Abang, M.M.; Winter, S.; Green, K.R.; Hoffmann, P.; Mignouna, H.D.; Wolf, G.A.** (2002). Molecular identification of *Colletotrichum gloeosporioides* causing yam anthracnose in Nigeria. *Pl. Path.* 51: 63–71; **Du, M.; Schardl, C.L.; Nuckles, E.M.; Vaillancourt, L.J.** (2005). Using mating-type sequences for improved phylogenetic resolution of *Colletotrichum* species complexes. *Mycologia* 97: 641–658; **Farr, D.F.; Aime, M.C.; Rossman, A.Y.; Palm, M.E.** (2006) Species of *Colletotrichum* on *Agavaceae*. *Mycol. Res.* 110: 1395–1408; **García Muñoz, J.-A.; Belén Suárez, M.; Grondona, I.; Monte, E.; Buddie, A.G.; Bridge, P.D.; Cannon, P.F.** (2000). A physiological and biochemical approach to the systematics of *Colletotrichum* species pathogenic to strawberry. *Mycologia* 92: 488–498; **Guerber, J.C.; Correll, J.C.** (2001). Characterization of *Glomerella acutata*, the teleomorph of *Colletotrichum acutatum*. *Mycologia* 93: 216–229; **Guerber, J.C.; Liu, B.; Correll, J.C.; Johnston, P.R.** (2003). Characterization of diversity in *Colletotrichum acutatum sensu lato* by sequence analysis of two gene introns, mtDNA and intron RFLPs, and mating compatibility. *Mycologia* 95: 872–895; **Johnston, P.R.; Jones, D.** (1997). Relationships among *Colletotrichum* isolates from fruit-rots assessed using rDNA sequences. *Mycologia* 89: 420–430; **Lu, G.Z.; Cannon, P.F.; Reid, A.; Simmons, C.M.** (2004). Diversity and molecular relationships of endophytic *Colletotrichum* isolates from the Iwokrama Forest Reserve, Guyana. *Mycol. Res.* 108: 53–63; **Lubbe, C.M.; Denman, S.; Cannon, P.F.; Groenewald, J.Z.; Lamprecht, S.C.; Crous, P.W.** (2004). Characterization of *Colletotrichum* species associated with diseases of Proteaceae. *Mycologia* 96: 1268–1279; **Prusky, D.; Freeman, S.; Dickman, M.B. (eds)** (2000). *Host specificity, Pathology, and Host–Pathogen Interactions of Colletotrichum* (St Paul): 400 pp.; **Silva-Hanlin, D.M.; Hanlin, R.T.** (1999). Small subunit ribosomal RNA gene phylogeny of several loculoascomycetes and its taxonomic implications. *Mycol. Res.* 103: 153–160; **Sivanesan, A.; Hsieh, W.H.** (1993). A new ascomycete, *Glomerella septospora* sp. nov. and its coelomycete anamorph, *Colletotrichum taiwanense* sp. nov. from Taiwan. *Mycol. Res.* 97: 1523–1529; **Sutton, B.C.** (1992). The genus *Glomerella* and its anamorph *Colletotrichum*. *Colletotrichum: Biology, Pathology and Control* (Wallingford): 1–26; **Uecker, F.A.** (1994). Ontogeny of the ascoma of *Glomerella cingulata*. *Mycologia* 86: 82–88.

Glomosporiaceae Cif. 1963
Ustilaginales: Basidiomycota

Sori developing in ovaries, sometimes hardly causing deformation, hyphae intracellular. Interior of the sori granular, filled with globose to ovoid persistent or eventually fragmenting spore balls, each often composed of numerous spores. Teliospores rounded or angular due to compression, brown, thick-walled, verrucose on exposed surfaces, germinating to produce basidia, or septate or aseptate hyphae. Basidia aseptate or septate, cylindrical or hypha-like, with an apical cluster of basidiospores. Basidiospores elongate, ellipsoidal to ovoid, sometimes with long spine-like appendages that have been interpreted as prematurely developed conjugation branches.

Significant Genera: *Thecaphora*.

Distribution: Widespread, especially in temperate regions.

Economic Significance: *Thecaphora solani* causes an economically important disease of potatoes.

Ecology: Parasitic in ovaries of a range of dicotyledonous herbaceous plants, especially legumes.

Notes: A well-defined family, probably a sister group of the *Urocystales* rather than a member of the *Ustilaginales* as indicated in the *Dictionary of the Fungi* edn 9. Some older literature can be accessed under *Sorosporium*, now recognized as a synonym of *Thecaphora*.

Thecaphora deformans, Hungary; parasitized ovaries

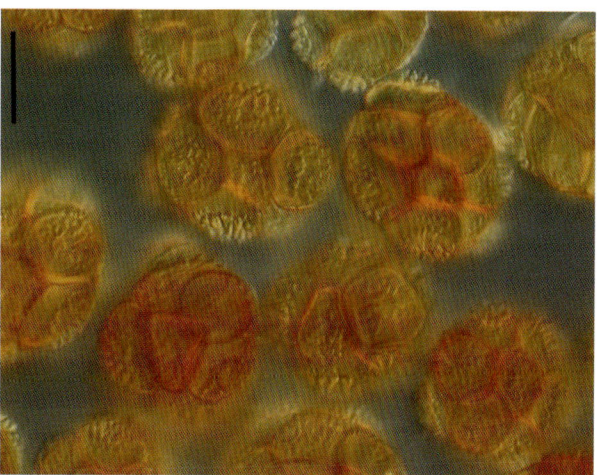

Thecaphora deformans, Hungary; spore balls; bar = 40 μm

References: **Andrade, O.; Muñoz, G.; Galdames, R.; Durán, P.; Honorato, R.** (2004). Characterization, *in vitro* culture, and molecular analysis of *Thecaphora solani*, the causal agent of potato smut. *Phytopathology* 94: 875–882; **Bauer, R.; Oberwinkler, F.; Vánky, K.** (1997). Ultrastructural markers and systematics in smut fungi and allied taxa. *Can. J. Bot.* 75: 1273–1314; **Begerow, D.; Bauer, R.; Boekhout, T.** (2000). Phylogenetic placements of ustilaginomycetous anamorphs as deduced from nuclear LSA rDNA sequences. *Mycol. Res.* 104: 53–60; **Begerow, D.; Bauer, R.; Oberwinkler, F.** (1998). Phylogenetic studies on nuclear large subunit ribosomal DNA sequences of smut fungi and related taxa. *Can. J. Bot.* 75: 2045–2056; **Piepenbring, M.; Bauer, R.; Oberwinkler, F.** (1998). Teliospores of smut fungi. Teliospore connections, appendages, and germ pores studied by electron microscopy; phylogenetic discussion of characteristics of teliospores. *Protoplasma* 204: 202–218; **Snetselaar, K.M.; Tiffany, L.H.** (1990). Light and electron microscopy of sorus development in *Sorosporium provinciale*, a smut of big bluestem. *Mycologia* 82: 480–492; **Vánky, K.** (1991). *Thecaphora* (*Ustilaginales*) on *Leguminosae*. *Trans. Mycol. Soc. Japan* 32: 145–159; **Vánky, K.** (1999). The new classificatory system for smut fungi, and two new genera. *Mycotaxon* 70: 35–49; **Vánky, K.; Berbee, M.L.** (1988). Are there *Thecaphora* species (*Ustilaginales*) on *Gramineae*?. *Mycotaxon* 33: 281–282.

Gnomoniaceae G. Winter 1886
Diaporthales: Ascomycota

Stromata absent or poorly developed, when present composed of a mixture of fungal and plant tissues. Ascomata perithecial, solitary or in clusters, black, erect or oblique, often long-necked, erumpent singly. Asci numerous, clavate, with a single wall layer, the apex thickened and truncate with a large and conspicuous refractive J– ring, evanescent at the base, becoming free within the ascomatal cavity, 8- or poly-spored. Ascospores usually ± ellipsoidal or fusiform, ± hyaline, often with mucous appendages. Anamorphs coelomycetous, the conidia spermatial in function.

Significant Genera: *Gnomonia*, *Cryptodiaporthe*, *Ophiovalsa*. Anamorph: *Discula*.

Distribution: Very widespread, especially in temperate zones.

Gnomonia setacea, Sutherland, Scotland; ascomata in dead leaf

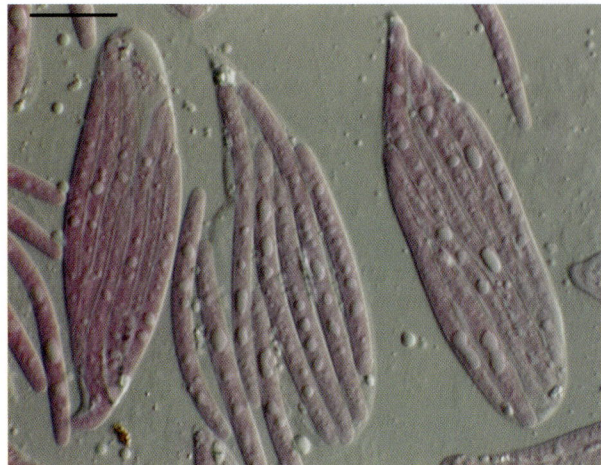
Ophiovalsa suffusa, Yorkshire, UK; asci and ascospores

Economic Significance: *Discula destructiva* (teleomorph *Cryphonectria corni*) causes an important disease of *Cornus* in North America and *Gnomonia comari* and *G. radiciciola* are minor pathogens of strawberry and rose respectively. *Cryptodiaporthe* species cause dieback of willow stems and other woody plant tissues and may be better placed in the *Cryohonectriaceae*.

Ecology: Most species are saprobes or endophytes in leaves and bark, but a few are necrotrophic pathogens.

Notes: This is currently a poorly circumscribed family, containing a rather disparate assemblage of genera in morphological terms. Further studies are needed to establish its boundaries.

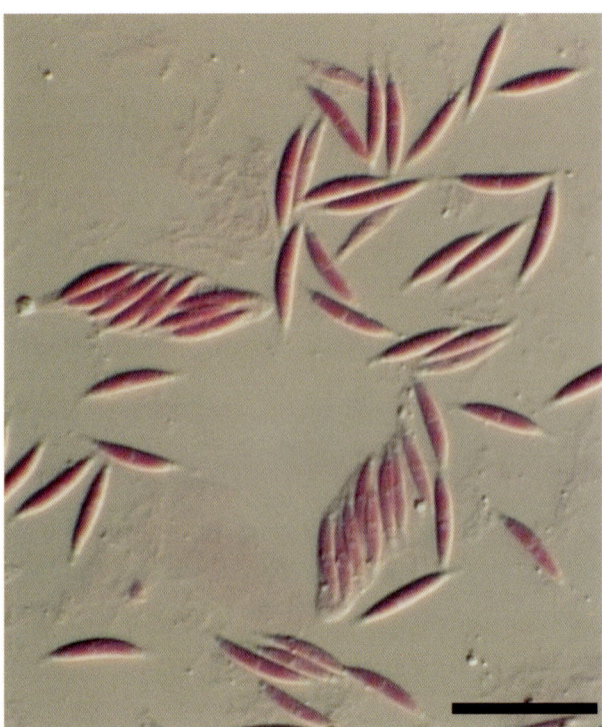
Gnomonia setacea, Ontario, Canada; asci and ascospores

GOMPHACEAE

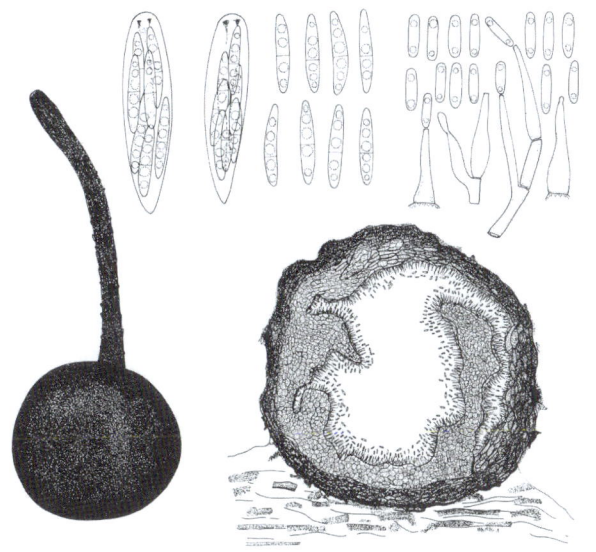

Gnomonia comari; ascoma, asci, ascospores and anamorph

References: **Barr, M.E.** (1978). The *Diaporthales* in North America with emphasis on *Gnomonia* and its segregates. *Mycol. Mem.* 7: 1–232; **Castlebury, L.A.; Rossman, A.Y.; Jaklitsch, W.J.; Vasilyeva, L.N.** (2002). A preliminary overview of the *Diaporthales* based on large subunit nuclear ribosomal DNA sequences. *Mycologia* 94: 1017–1031; **Lappalainen, J.H.; Yli-Mattila, T.** (1999). Genetic diversity in Finland of the birch endophyte *Gnomonia setacea* as determined by RAPD-PCR markers. *Mycol. Res.* 103: 328–332; **Monod, M.** (1983). Monographie taxonomique des *Gnomoniaceae* (Ascomycètes de l'ordre des *Diaporthales*). I. *Beih. Sydowia* 9: 1–314; **Noordeloos, M.E.; Kesteren, H.A. van; Veenbaas-Rijks, J.W.** (1989). Studies in plant pathogenic fungi – I. *Gnomonia radicicola*, spec. nov., a new pathogen of roses. *Persoonia* 14: 47–49; **Redlin, S.C.; Rossman, A.Y.** (1991). *Cryptodiaporthe corni* (*Diaporthales*), cause of Cryptodiaporthe canker of pagoda dogwood. *Mycologia* 83: 200–209; **Sogonov, M.V.; Castlebury, L.A.; Rossman, A.Y.; Farr, D.F.; White, J.F.** (2005). The type species of the genus *Gnomonia*, *G. gnomon*, and the closely related *G. setacea*. *Sydowia* 57: 102–119; **Zhang, N.; Blackwell, M.** (2001). Molecular phylogeny of dogwood anthracnose fungus (*Discula destructiva*) and the *Diaporthales*. *Mycologia* 93: 355–365.

Gomphaceae Donk 1961
Gomphales: Basidiomycota

Basidiomata funnel-shaped or club-shaped or coralloid and repeatedly branched, rarely resupinate or sequestrate, often brightly coloured, usually robust, fleshy, cartilaginous or gelatinous. Hyphal system mostly monomitic, the hyphae thin-walled or thick-walled, most species with clamp connections. Hymenium smooth, rarely gilled (cantharelloid), hydnoid or reticulate, sometimes setose. Basidia aseptate, ± cylindrical, generally with a basal clamp and with 4 sterigmata. Basidiospores discharged actively and asymmetric in most taxa, ellipsoidal or elongate, hyaline to yellowish-brown, smooth or ornamented with warts, spines or longitudinal ribs, the ornamentation usually staining in cotton blue and similar mountants.

Significant Genera: *Clavariadelphus, Gautieria, Gomphus, Ramaria, Ramariopsis*.

Distribution: Cosmopolitan, widespread in both temperate and tropical climatic zones.

Economic Significance: Some species are edible, e.g. oniusutake (*Gomphus kauffmanii*). A number of taxa have been identified as of conservation concern, particularly in the Pacific North West region of North America.

Ecology: Terrestrial or lignicolous, either saprobic or forming ectomycorrhizal relationships with woody plants.

Notes: The family as currently circumscribed contains a bewildering range of fruit-body types. Phylogenetic analyses suggest that the *Gomphaceae* cannot be readily separated from the *Ramariaceae*, but the combined group is paraphyletic with the *Phallaceae* clustering within. Further studies are needed.

Gomphus floccosus, Vancouver Island, Canada; basidiomata

Ramaria aff. *subviolacea*, Glacier National Park, Canada; basidioma

References: **Binder, M.; Hibbett, D.S.; Larsson, K.-H.; Larsson, E.; Langer, E.; Langer, G.** (2005). The phylogenetic distribution of resupinate forms across the major clades of mushroom-forming fungi (*Homobasidiomycetes*). *Syst. Biodiv.* 3: 113–157; **Corner, E.J.H.** (1950). A monograph of *Clavaria* and allied genera. *Ann. Bot.*

Mem. 1: 1–740; **Corner, E.J.H.** (1970). Supplement to a monograph of *Clavaria* and allied genera. *Beih. Nova Hedwigia* 33: 1–299; **Hibbett, D.S.; Binder, M.** (2002). Evolution of complex fruiting-body morphologies in homobasidiomycetes. *Proc. R. Soc. Lond. B. Biol. Sci.* 269: 1963–1969; **Hibbett et al.** (2007). A higher-level phylogenetic classification of the *Fungi*. *Mycol. Res.* 111(5): 509–548; **Hosaka et al.** (2007). Molecular phylogenetics of the gomphoid-phalloid fungi with an establishment of the new subclass *Phallomycetidae* and two new orders. *Mycologia* 98(6): 949–959; **Humpert, A.J.; Muench, E.L.; Giachini, A.J.; Castellano, M.A.; Spatafora, J.W.** (2001). Molecular phylogenetics of *Ramaria* and related genera: evidence from nuclear large subunit and mitochondrial small subunit rDNA sequences. *Mycologia* 93: 465–477; **Marr, C.D.; Stuntz, D.E.** (1973). *Ramaria* of western Washington. *Biblthca Mycol.* 38: 232 pp.; **Nouhra, E.R.; Horton, T.R.; Cazares, E.; Castellano, M.** (2005). Morphological and molecular characterization of selected *Ramaria* mycorrhizae. *Mycorrhiza* 15: 55–59; **Núñez, M.; Ryvarden, L.** (1994). A note on the genus *Beenakia*. *Sydowia* 46: 321–328; **Parmasto, E.; Ryvarden, L.** (1990). The genus *Beenakia* (*Gomphaceae, Aphyllophorales*). *Windahlia* 18: 35–42; **Petersen, R.H.** (1975). *Ramaria* subg. *Lentoramaria*, with emphasis on North American taxa. *Biblthca Mycol.* 43: 161 pp.; **Petersen, R.H.** (1981). *Ramaria* subg. *Echinoramaria*. *Biblthca Mycol.* 79: 261 pp.; **Pine, E.M.; Hibbett, D.S.; Donoghue, M.J.** (1999). Phylogenetic relationships of cantharelloid and clavarioid *Homobasidiomycetes* based on mitochondrial and nuclear rDNA sequences. *Mycologia* 91: 944–963; **Villegas, M.; De Luna, E.; Cifuentes, J.; Estrada Torres, A.** (1999). Phylogenetic studies in *Gomphaceae sensu lato* (*Basidiomycetes*). *Mycotaxon* 70: 127–147.

Gomphidiaceae Maire ex Jülich 1982
Boletales: Basidiomycota

Basidiomata pileate, rarely secotioid, small to medium, dry, viscid or glutinous, usually reddish or brown, veil dry or viscid, sometimes partial or lacking, stipe slender, not annulate or reticulate. Hymenium lamellate, the gills adnate to decurrent, broad, becoming dark brown or black. Hyphal system monomitic, clamp connections only on hyphae at the base of the stipe, flesh sometimes staining in iodine; thin or thick-walled elongate cylindric cystidia present, protruding from the hymenium. Basidia elongate, clavate to cylindrical, hyaline or pigmented, usually with 4 sterigmata. Basidiospores mostly discharged actively and radially asymmetrical, fuscous, elongate and fusoid-cylindric, smooth or longitudinally striate to ridged.

Significant Genera: *Chroogomphus, Gomphidius*.

Distribution: Widespread, primarily in north temperate zones.

Economic Significance: None is known.

Ecology: Ectomycorrhizal in soil and leaf mould, always in association with *Pinaceae*.

Notes: The family is close to the *Suillaceae*, which has a poroid rather than lamellate hymenium, and several workers have suggested their synonymy. More molecular work is required.

Gomphidius glutinosus; basidiomata

Chroogomphus rutilus; basidiomata

References: **Agerer, R.** (1991). Studies on ectomycorrhizae XXXIV. Mycorrhizae of *Gomphidius glutinosus* and of *G. roseus* with some remarks on *Gomphidiaceae* (*Basidiomycetes*). *Nova Hedwigia* Beih. 53: 127–170; **Besl, H.; Bresinsky, A.** (1997). Chemosystematics of *Suillaceae* and *Gomphidiaceae* (suborder *Suillineae*). *Pl. Syst. Evol.* 206: 223–242; **Greselin, A.** (1991). Il genere *Gomphidius*. *Boll. Gruppo Micol. 'G. Bresadola'* (Trento) 34: 56–74; **Miller, O.K.** (1971). The genus *Gomphidius* with a revised description of the *Gomphidiaceae* and a key to the genera. *Mycologia* 63: 1129–1163; **Miller, O.K.** (2003). The *Gomphidiaceae* revisited: a worldwide perspective. *Mycologia* 95: 176–183; **Miller, O.K.; Aime, M.C.** (2001). Systematics, ecology and world distribution in the genus *Chroogomphus* (*Gomphidiaceae*). *Trichomycetes and Other Fungal Groups* Robert W. Lichwardt Commemoration Volume (Enfield): 315–333; **Miller, O.K.; Aime, M.C.; Camacho, F.J.; Peintner, U.** (2002). Two new species of *Gomphidius* from the western United States and eastern Siberia. *Mycologia* 94: 1044–1050; **Watling, R.** (2004). New combinations in *Boletaceae* and *Gomphidiaceae*. *Edinb. J. Bot.* 61: 41–47.

Gomphillaceae Walt. Watson ex Hafellner 1984
Ostropales, Ascomycota

Thallus crustose, very varied in form. Ascomata apothecial, usually sessile, discoid, without a well-defined margin. Interascal tissue of branched and anastomosing filiform paraphyses. Asci cylindrical, with only a slight apical thickening, 1- to 8-spored. Ascospores ellipsoidal to filiform, with transverse septa, sometimes muriform. Anamorphs coelomycetous, spine-like to brush-like or peltate; forming hyphophores.

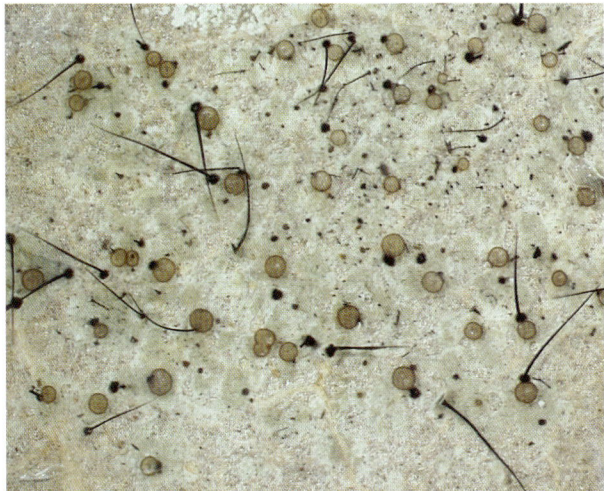

Tricharia vainioi, Togo; ascomata and mycelial setae

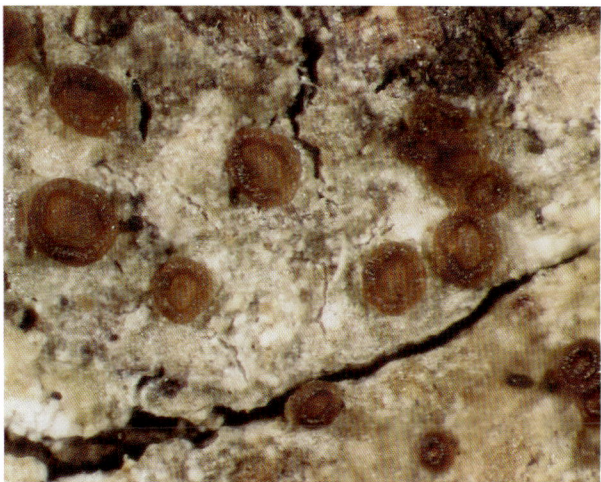

Gyalideopsis anastomosans, Surrey, UK; ascomata

Gyalideopsis lambinonii; ascomata

Gomphillus sp.; anamorph

Significant Genera: *Gomphillus, Gyalideopsis, Tricharia*.

Distribution: Most species occur in the tropics.

Economic Significance: None is known.

Ecology: Lichenized with green algae, mostly foliicolous but also on bark, bryophytes, soil etc., occasionally lichenicolous.

Notes: Closely related to the *Asterothyriaceae*, but more sequences are needed.

References: **Brusse, F.** (1992). A new species of *Bullatina* from the Transkei Wild Coast. *Bothalia* 22: 44–46; **Buck, W.R.** (1998). Lichen flora of eastern North America: the genus *Gomphillus* (*Gomphillaceae*). *Lichenographia Thomsoniana, North American Lichenology in Honor of John W. Thomson* (Ithaca): 71–76; **Buck, W.R.; Sérusiaux, E.** (2000). *Gyalectidium yahriae*, sp. nov. (lichenized *Ascomycetes, Gomphillaceae*) from Florida and Papua New Guinea. *Bryologist* 103: 134–138; **Dennetière, B.; Péroni, J.** (1998). [Phylogenetic approach to the *Gomphillaceae*]. *Cryptog. Bryol.-Lichénol.* 19: 105–121; **Ferraro, L.I.** (2004). Morphological diversity in the hyphophores of *Gomphillaceae* (Ostropales, lichenized *Ascomycetes*). *Fungal Diversity* 15: 153–169; **Hartmann, C.** (1996). Two new foliicolous species of *Calenia* (lichens, *Gomphillaceae*) from Costa Rica. *Mycotaxon* 59: 483–488; **Kalb, K.; Vězda, A.** (1988). Neue oder bemerkenswerte Arten der Flechtenfamilie *Gomphillaceae* in der Neotropis. *Biblthca Lichenol.* 29: 80 pp.; **Lücking, R.** (1998). Foliicolous lichens and their lichenicolous fungi collected during the Smithsonian International Cryptogamic Expedition to Guyana 1996. *Trop. Bryol.* 15: 45–76; **Lücking, R.; Stuart, B.L.; Lumbsch, H.T.** (2004). Phylogenetic relationships of *Gomphillaceae* and *Asterothyriaceae*: evidence from a combined Bayesian analysis of nuclear and mitochondrial sequences. *Mycologia* 96: 283–294; **Sérusiaux, E.** (1998). Notes on the *Gomphillaceae* (Lichens) from Guadeloupe (West Indies), with four new species of *Gyalideopsis*. *Nova Hedwigia* 67: 381–402.

Gonapodyaceae H.E. Petersen ex P.M. Kirk, P.F. Cannon & J.C. David 2001
Monoblepharidales: Chytridiomycota

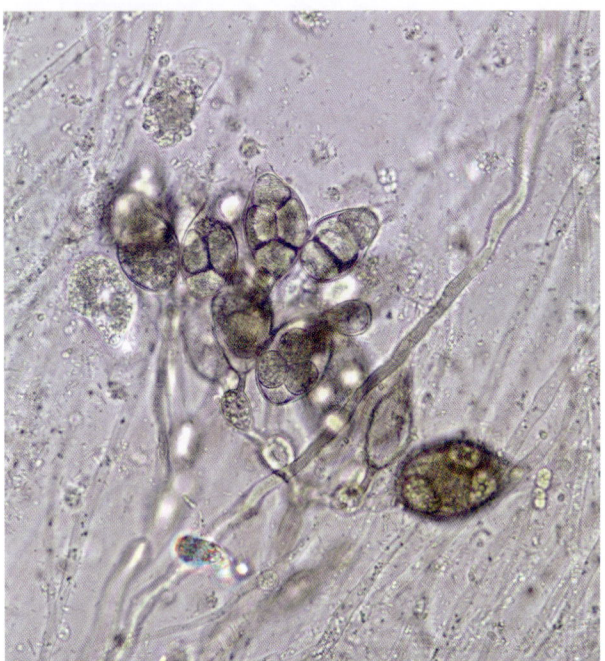

Gonapodya sp.; zoosporangia and zoospores

Thallus a well-developed, differentiated, hyphal system bearing numerous reproductive structures, hyphae irregularly or dichotomously branched, often moniliform; zoosporangia terminal, single or occuring in fascicles, variable; zoospores formed completely inside the sporangium, escaping by sporangium deliquescence at the apex; zoospores ovoid to cylindrical, posteriorly uniflagellate; zygote undergoing a period of motility before encystment, propelled by flagellum of the male gamete, oogamous, the gametangia borne terminally in fascicles, fertilization internal or external.

Significant Genera: *Gonapodya*.

Distribution: Widespread, mainly temperate.

Economic Significance: None is known.

Ecology: Occurring in a variety of aquatic environments.

Notes: Very poorly known; no recent information is available.

References: Bullerwell, C.E.; Forget, L.; Langa, B.F. (2003). Evolution of monoblepharidalean fungi based on complete mitochondrial genome sequences. *Nucl. Acids Res.* 31: 1614–1623; **Noyes Mollicone, M.R.; Longcore, J.E.** (1999). Zoospore ultrastructure of *Gonapodya polymorpha*. *Mycologia* 91: 727–734; **Steciow, M.M.; Elíades, L.A.; Arambarri, A.M.** (2001). [The genus *Gonapodya* (*Monoblepharidales, Chytridiomycota*) in polluted environments of Ensenada (Buenos Aires, Argentina)]. *Boln Soc. argent. Bot.* 36: 203–208.

Grammotheleaceae Jülich 1982
Polyporales: Basidiomycota

Basidioma resupinate, adnate or effuse, annual or perennial, woody, whiteish or cream. Hyphal system monomitic or dimitic, generative hyphae with clamp connections, hyaline, thin-walled or thick-walled. Hymenium composed of irregular tubes or irregular anastomosing ridges, the basidia present only at the base of the cavities, sometimes spinose due to the presence of dark hyphal pegs. Basidia clavate, with basal clamp connections and usually 4 sterigmata, usually interspersed with simple thin-walled cystidia, dendritic hyphidia sometimes present. Basidiospores ellipsoidal, hyaline, thin-walled, smooth, not staining in iodine.

Significant Genera: *Grammothele, Porogramme*.

Distribution: Widespread in both the Old World and New World tropics.

Grammothele fuligo, West Africa; detail of hymenial surface

Economic Significance: None is known.

Ecology: Little is known; species are probably saprobic on woody substrata.

Notes: The family appears to be isolated within the *Polyporales*, and few molecular data are available to assess phylogenetic relationships. A diagnostic character appears to be the hymenial arrangement, where basidia are confined to the basal part of the hymenium.

References: Kotlaba, F.; Pouzar, Z. (2003). Polypores (*Polyporales s.l.*) collected in Cuba. *Czech Mycol.* 55: 7–50; **Quanten, E.** (1997). The Polypores (*Polyporaceae s.l.*) of Papua New Guinea: A Preliminary Prospectus. *Op. bot. Belg.* 11: 352 pp.; **Ryvarden, L.** (1979). *Porogramme* and related genera. *Trans. Br. mycol. Soc.* 73: 9–19.

Graphidaceae Dumort. 1822
Ostropales: Ascomycota

Thallus crustose, often immersed or inconspicuous. Ascomata apothecial, elongated, curved or branched; marginal tissues well-developed, black, carbonized. Interascal tissue of simple paraphyses. Asci cylindrical, with a single functional wall layer and a thickened J– apex. Ascospores mostly large, transversely and sometimes longitudinally septate, hyaline or brown. Anamorphs pycnidial.

Significant Genera: *Graphina*, *Graphis*, *Phaeographis*.

Dyplolabia afzelii, Sierra Leone; ascomata

Graphis elegans, Hampshire, UK; ascomata

Distribution: Most species have tropical distributions, but species of *Graphis* in particular are prominent members of the temperate mycota.

Economic Significance: None is known.

Ecology: Lichenized with green algae (usually *Trentopohliaceae*), especially on bark.

Notes: Superficially similar to *Opegrapha* and allies, but those species have fissitunicate asci. Recent studies suggest that the family may cluster within the *Thelotremataceae* as currently circumscribed and the two taxa may be combined once further research is completed.

Sarcographa heteroclita; ascomata

Graphis scripta, Hampshire, UK; ascomata

References: **Archer, A.W.** (1999). The lichen genera *Graphis* and *Graphina* (*Graphidaceae*) in Australia 1. Species based on Australian type specimens. *Telopea* 8: 273–295; **Archer, A.W.** (2000). The lichen genera *Phaeographis* and *Phaeographina* (*Graphidaceae*) in Australia 1: species based on Australian type specimans. *Telopea* 8: 461–475; **Kalb, K.; Staiger, B.; Elix, J.A.** (2004). A monograph of the lichen genus *Diorygma* – a first attempt. *Symb. bot. upsal.* 34: 133–181; **Kauff, F.; Lutzoni, F.** (2002). Phylogeny of the *Gyalectales* and *Ostropales* (*Ascomycota, Fungi*): among and within order relationships based on nuclear ribosomal RNA small and large subunits. *Mol. Phylogen. Evol.* 25: 138–156; **Kauff, F.; Lutzoni, F.** (2002). Phylogeny of the *Gyalectales* and *Ostropales* (*Ascomycota, Fungi*): among and within order relationships based on nuclear ribosomal RNA small and large subunits. *Mol. Phylogen. Evol.* 25: 138–156; **Nakanishi, M.; Harada, H.** (1999). Five new species of the lichen genus *Graphis* (lichenized *Ascomycota*; *Graphidaceae*) from the Mariana Islands, Micronesia. *Nat. Hist. Res.* 5: 63–71; **Staiger, B.; Kalb, K.** (1999). *Acanthothecis* and other graphidioid lichens with warty periphysoids or paraphysis-tips. *Mycotaxon* 73: 69–134; **Staiger, B.; Kalb, K.; Grube, M.** (2006). Phylogeny and phenotypic variation in the lichen family *Graphidaceae* (*Ostropomycetidae, Ascomycota*). *Mycol. Res.* 110: 765–772; **Winka, K.; Ahlberg, C.; Eriksson, O.E.** (1998). Are there lichenized *Ostropales?*. *Lichenologist* 30: 455–462.

Graphiolaceae Clem. & Shear 1931
Exobasidiales: Basidiomycota

Sori stromatic, black, simple or compound, cup-shaped, composed of an outer peridium surrounding a central fertile region and usually also clusters of thick-walled hyphae. Hyphae intracellular, haustoria present, septal pores simple. Fertile hyphae repeatedly septate, producing small, pale, basipetally formed cells that develop into basidia and in most species fragment into primary propagules. Basidia thin-walled, ± ellipsoidal; after meiosis the nuclei migrate into lateral whorls of readily detached propagules (basidiospores). Basidiospores small, irregular in shape, thick-walled and verrucose, becoming 1-septate, germinating to form yeast-like cells (or hyphae) from which reinfection is established.

Significant Genera: *Graphiola*.

Distribution: Widespread in the tropics and warm temperate regions.

Economic Significance: Species can cause damage to ornamental palm trees, but do not cause significant mortality.

Ecology: Biotrophic on leaves of *Palmae*.

Notes: An isolated family that has been classified in almost every fungal group at one time or another. Molecular and ultrastructural research has confirmed its position within the *Exobasidiales*, with its closest relative provisionally the *Brachybasidiaceae*.

Graphiola phoenicis, India; sori on living palm leaf

References: **Bauer, R.; Oberwinkler, F.; Vánky, K.** (1997). Ultrastructural markers and systematics in smut fungi and allied taxa. *Can. J. Bot.* 75: 1273–1314; **Begerow, D.; Bauer, R.; Oberwinkler, F.** (1998). Phylogenetic studies on nuclear large subunit ribosomal DNA sequences of smut fungi and related taxa. *Can. J. Bot.* 75: 2045–2056; **Begerow, D.; Bauer, R.; Oberwinkler, F.** (2002). The *Exobasidiales*: an evolutionary hypothesis. *Mycol. Prog.* 1: 187–199; **Blanz, P.A.; Gottschalk, M.** (1986). Systematic position of *Septobasidium*, *Graphiola* and other basidiomycetes as deduced on the basis of their 5S ribosomal RNA nucleotide sequences. *Syst. Appl. Microbiol.* 8: 121–127; **Oberwinkler, F.; Bandoni, R.J.; Blanz, P.; Deml, G.; Kisimova-Horowitz, L.** (1982). *Graphiolales*: basidiomycetes parasitic on palms. *Pl. Syst. Evol.* 140: 251–277; **Sjamsuridzal, W.; Sugiyama, J.** (1998). Detection of multiple insertions of group I introns in the nuclear small subunit ribosomal RNA gene from the ustilaginomycete *Graphiola phoenicis* parasitic on palm. *J. gen. appl. Microbiol.* Tokyo 44: 355–360.

Graphostromataceae M.E. Barr, J.D. Rogers & Y.M. Ju 1993
Xylariales: Ascomycota

Stroma flat, effuse, dark, two-layered. Ascomata in a single layer, perithecial, short-necked with separate ostioles, the ostioles not furrowed. Interascal tissue of simple paraphyses, swollen towards the base. Asci ± cylindrical, small, one-layered, the apex thickened with a faint J+ apical ring. Ascospores hyaline, aseptate, allantoid, smooth, without a mucous sheath or appendages. Anamorph *Nodulisporium*-like.

Significant Genera: *Graphostroma*.

Distribution: Known from temperate North America.

Economic Significance: None is known.

Ecology: Saprobic on bark of deciduous trees.

Notes: Probably more closely related to the *Xylariales* rather than the *Calosphaeriales*, as was previously the case, but more research is needed.

References: **Barr, M.E.** (1985). Notes on the *Calosphaeriales*. *Mycologia* 77: 549–565; **Barr, M.E.; Rogers, J.D.; Ju, Y.M.** (1993). Revisionary studies in the *Calosphaeriales*. *Mycotaxon* 48: 529–535; **Glawe, D.A.; Rogers, J.D.** (1986). Conidial states of some species of *Diatrypaceae* and *Xylariaceae*. *Can. J. Bot.* 64: 1493–1498; **Pirozynski, K.A.** (1974). *Xenotypa* Petrak and *Graphostroma* gen. nov., segregates from *Diatrypaceae*. *Can. J. Bot.* 52: 2129–2135.

Gyalectaceae Stizenb. 1862
Ostropales: Ascomycota

Thallus crustose, often effuse or inconspicuous. Ascomata apothecial, flat or cup-shaped to almost perithecial in shape, partly immersed in the thallus, pallid or brown. Interascal tissue of paraphyses, usually unbranched, with knob-like apices. Asci cylindrical, thin-walled at the apex, the walls blueing in iodine, without a tholus or apical structure. Ascospores usually hyaline, transversely septate or muriform, without a sheath or appendages. Anamorph pycnidial, not prominent.

Significant Genera: *Gyalecta*, *Petractis*.

Distribution: Mainly in the tropics; also in humid environments within the temperate zone.

Gyalecta ulmi, Norway; thallus and ascomata

Gyalecta jenensis, Devon, UK; ascomata

Coenogonium linkii; thalli

Economic Significance: None is known.

Ecology: Mainly lichen-forming with green, especially trentepohlioid, photobionts. Some are foliicolous or associated with bryophytes.

Notes: Recently linked with the *Ostropales* within a separate subclass *Ostropomycetidae*, but molecular data are sparse in both groups and the *Gyalectaceae* as currently circumscribed is probably polyphyletic. *Dimerella* and *Coenogonium* are now combined, and placed in the *Coenogoniaceae*.

References: **Coppins, B.; Thor, G.; Nordin, A.** (1994). The genus *Ramonia* in Sweden. *Graphis Scripta* 6: 89–92; **Hansen, E.S.; Poelt, J.; Vězda, A.** (1987). The lichen genera *Gyalecta*, *Gyalidea* and *Sagiolechia* in Greenland. *Herzogia* 7: 367–374; **Kauff, F.; Lutzoni, F.** (2002). Phylogeny of the *Gyalectales* and *Ostropales* (*Ascomycota*, *Fungi*): among and within order relationships based on nuclear ribosomal RNA small and large subunits. *Mol. Phylogen. Evol.* 25: 138–156; **Lücking, R.** (1998). Foliicolous lichens and their lichenicolous fungi collected during the Smithsonian International Cryptogamic Expedition to Guyana 1996. *Trop. Bryol.* 15: 45–76; **Lücking, R.** (1999). Additions and corrections to the foliicolous lichen flora of Costa Rica. The family *Gyalectaceae*. *Lichenologist* 31: 359–374; **Lumbsch, H.T.; Schmitt, I.; Palice, Z.; Wiklund, E.; Ekman, S.; Wedin, M.** (2004). Supraordinal phylogenetic relationships of *Lecanoromycetes* based on a Bayesian analysis of combined nuclear and mitochondrial sequences. *Mol. Phylogen. Evol.* 31: 822–832; **Messuti, M.I.; Vězda, A.; Lumbsch, H.T.** (1999). *Belonia uncinata* (*Gyalectales*, *Ascomycotina*) new to South America. *Bryologist* 102: 314–316; **Reeb, V.; Lutzoni, F.; Roux, C.** (2004). Contribution of *RPB2* to multilocus phylogenetic studies of the euascomycetes (*Pezizomycotina*, *Fungi*) with special emphasis on the lichen-forming *Acarosporaceae* and evolution of polyspory. *Mol. Phylogen. Evol.* 32: 1036–1060; **Wiklund, E.; Wedin, M.** (2003). The phylogenetic relationships of the cyanobacterial lichens in the *Lecanorales* suborder *Peltigerineae*. *Cladistics* 19: 419–431.

Gymnoascaceae Baran. 1872
Onygenales: Ascomycota

Stromata absent. Ascomata cleistothecial, ± globose, not stipitate; peridium formed from intertwined hyphae, sometimes with complex thick-walled appendages. Interascal tissue absent. Asci small, ± globose, thin-walled, evanescent, probably produced from croziers. Ascospores ± oblate, often with polar and/or equatorial thickenings, often minutely ornamented. Anamorphs unknown or arthric, usually inconspicuous.

Significant Genera: *Gymnascella*, *Gymnoascus*.

Distribution: Cosmopolitan.

Economic Significance: A few species are implicated in biodeterioration of cellulose-rich substrata, but are likely to have little impact compared with more ubiquitous and enzymatically competent groups. There are also occasional reports of mycoses caused by members of the family.

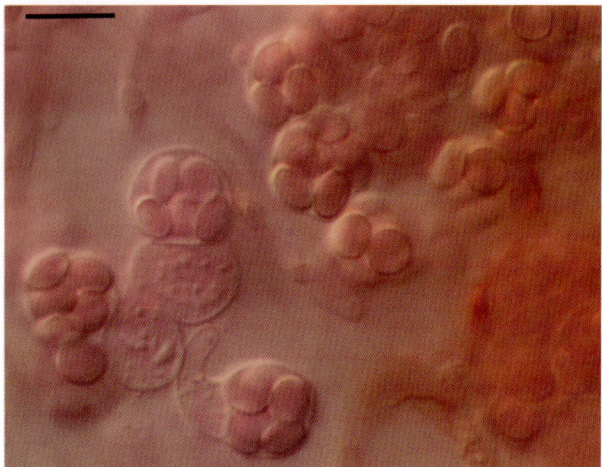

Gymnascella aff. *devroeyi*, England; asci and ascospores

Ecology: Keratinophilic or cellulolytic, usually encountered as isolations from soil, dung etc. Baiting techniques are useful to improve recovery rates.

Notes: The *Gymnoascaceae* forms a well-supported clade within the *Onygenales*, though some of the genera are polyphyletic and ascospore morphology appears to be unreliable as a phylogenetic marker.

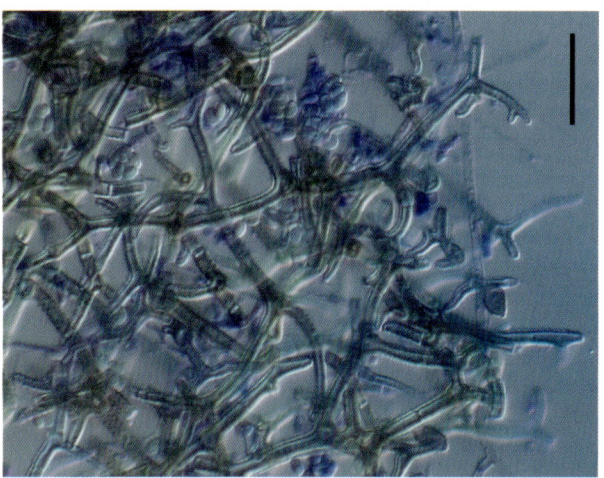

Gymnoascus reessii, Cleveland, UK; peridial elements and ascospores

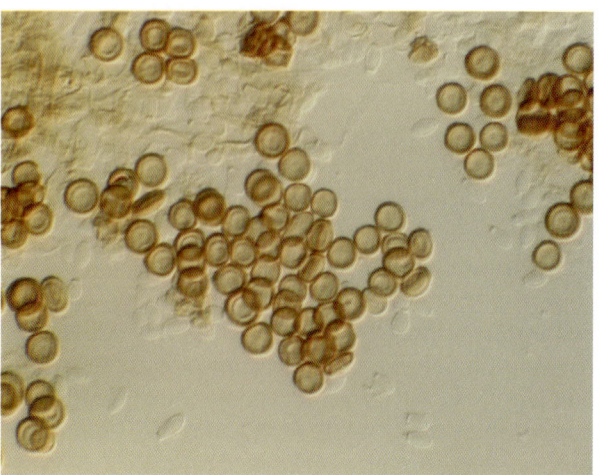

Gymnoascus arxii, Spain; ascospores

References: **Bowman, B.H.; White, T.J.; Taylor, J.W.** (1996). Human pathogenic fungi and their close nonpathogenic relatives. *Mol. Phylogen. Evol.* 6: 89–96; **Currah, R.S.** (1985). Taxonomy of the Onygenales: Arthrodermataceae, Gymnoacaceae, Myxotrichaceae and Onygenaceae. *Mycotaxon* 24: 1–216; **Currah, R.S.** (1988). An annotated key to the genera of the Onygenales. *Syst. Ascom.* 7: 1–12; **Dalpé, Y.** (1989). Ericoid mycorrhizal fungi in the *Myxotrichaceae* and *Gymnoascaceae*. *New Phytol.* 113: 523–527; **Iwen, P.C.; Sigler, L.; Tarantolo, S.; Sutton, D.A.; Rinaldi, M.G.; Lackner, R.P.; McCarthy, D.I.; Hinrichs, S.H.** (2000). Pulmonary infection caused by *Gymnascella hyalinospora* in a patient with acute myelogenous leukemia. *J. Clin. Microbiol.* 38: 375–381; **Kuraishi, H.; Itoh, M.; Katayama, Y.; Ito, T.; Hasegawa, A.; Sugiyama, J.** (2000). Ubiquinone systems in fungi. V. Distribution and taxonomic implications of ubiquinones in *Eurotiales*, *Onygenales* and the related plectomycete genera, except for *Aspergillus*, *Paecilomyces*, *Penicillium*, and their related teleomorphs. *Antonie van Leeuwenhoek* 77: 179–186; **Sugiyama, M.; Mikawa, T.** (2001). Phylogenetic analysis of the non-pathogenic genus *Spiromastix* (*Onygenaceae*) and related onygenalean taxa based on large subunit ribosomal DNA sequences. *Mycoscience* 42: 413–421; **Sugiyama, M.; Ohara, A.; Mikawa, T.** (1999). Molecular phylogeny of onygenalean fungi based on small subunit ribosomal DNA (SSU rDNA) sequences. *Mycoscience* 40: 251–258; **Sugiyama, M.; Summerbell, R.C.; Mikawa, T.** (2002). Molecular phylogeny of onygenalean fungi based on small subunit (SSU) and large subunit (LSU) ribosomal DNA sequences. *Stud. Mycol.* 47: 5–23; **Untereiner, W.A.; Scott, J.A.; Naveau, F.A.; Currah, R.S.; Bachewich, J.** (2002). Phylogeny of *Ajellomyces*, *Polytolypa* and *Spiromastix* (*Onygenaceae*) inferred from rDNA sequence and non-molecular data. *Stud. Mycol.* 47: 25–35.

Gypsoplacaceae Timdal 1990
Lecanorales: Ascomycota

Thallus foliose, squamiform, with well-developed rhizoids. Ascomata apothecial, effuse, irregular, without a margin or distinct wall, the asci and interascal tissue developing directly from cortical hyphae. Interascal tissue of branched and anastomosing paraphyses. Asci with a well-developed apical cap and a stronger-staining J+ ring around the apical cushion, usually with an ocular chamber. Ascospores simple, hyaline, without a sheath. Anamorph pycnidial.

Significant Genera: *Gypsoplaca*.

Distribution: Widespread.

Economic Significance: None is known.

Ecology: Lichenized with green algae.

Notes: Very poorly known; no recent information is available.

References: **Goward, T.; Breuss, O.; Ryan, B.; McCune, B.; Sipman, H.; Scheidegger, C.** (1996). Notes on the lichens and allied fungi of British Columbia. III. *Bryologist* 99: 439–449; **Timdal, E.**

(1990). *Gypsoplacaceae* and *Gypsoplaca*, a new family and genus of squamiform lichens. *Biblthca Lichenol.* 38: 419–427.

Gyroporaceae Locq. 1984
Boletales: Basidiomycota

Basidiomata pileate, stipitate, yellowish, brownish or reddish, the upper surface glabrous, scaly or weakly tomentose, not viscid. Stipe glabrous to fibrous, veil absent, not reticulate, solid or partly hollow, with an outer layer of transversely oriented hyphae at least in some species. Internal hyphae with clamp connections, remaining hyaline or blueing on exposure to the air. Hymenium tubular, white or yellowish, depressed around the stipe, cystidia present. Basidiospores ellipsoidal to subglobose, thin-walled, ± smooth, not staining in iodine.

Significant Genera: *Gyroporus*.

Gyroporus cyanescens; basidiomata

Distribution: Widespread in the tropics and both north and south temperate zones.

Economic Significance: Some species are edible, but are not commercially developed. At least one species is reported to be toxic.

Ecology: Ectomycorrhizal at least in most circumstances, associated with both coniferous and broad-leaved trees.

Notes: The family is clearly separate from the *Boletaceae* and its relatives, according to phylogenetic analysis, and nests within the suborder *Sclerodermatineae*. *Rubinoboletus* was placed in the family in the *Dictionary of the Fungi* edn 9, but preliminary data suggest that this genus belongs within the *Boletineae*.

Gyroporus castaneus; basidiomata

References: **Binder, M.; Bresinsky, A.** (2002). Derivation of a polymorphic lineage of gasteromycetes from boletoid ancestors. *Mycologia* 94: 85–98; **Bruns, T.D.; Gardes, M.** (1993). Molecular tools for the identification of ectomycorrhizal fungi – taxon-specific oligonucleotide probes for suilloid fungi. *Mol. Ecol.* 2: 233–242; **Bruns, T.D.; Shefferson, R.P.** (2004). Evolutionary studies of ectomycorrhizal fungi: recent advances and future directions. *Can. J. Bot.* 82: 1122–1132; **Buchanan, P.K.; May, T.W.** (2003). Conservation of New Zealand and Australian fungi. *N.Z. Jl Bot.* 41: 407–421; **Castro, M.L.; Freire, L.** (1995). *Gyroporus ammophilus*, a new poisonous bolete from the Iberian Peninsula. *Persoonia* 16: 123–126; **Nagasawa, E.** (2001). Taxonomic studies of Japanese boletes. I. The genera *Boletinellus*, *Gyrodon* and *Gyroporus*. *Rep. Tottori Mycol. Inst.* 39: 1–27; **Watling, R.** (2001). Australian boletes: their diversity and possible origins. *Aust. Syst. Bot.* 14: 407–416.

Haematommataceae Hafellner 1984
Lecanorales: Ascomycota

Thallus crustose, often sorediate. Ascomata apothecial, sessile or immersed, with a thalline margin, the disc often red; hymenium blueing in iodine. Interascal tissue of sparingly branched and anastomosing paraphyses. Asci with a large J+ apical cap, sometimes with a very large conical ocular chamber, and an outer J+ gelatinized layer. Ascospores elongate, multiseptate, hyaline, without a sheath. Anamorphs coelomycetous, pycnidial, immersed in the thallus, producing filiform or bacilliform conidia from ampulliform, percurrently proliferating, conidiogenous cells.

Significant Genera: *Haematomma*.

Distribution: Widespread in both tropical and temperate regions.

Economic Significance: None is known.

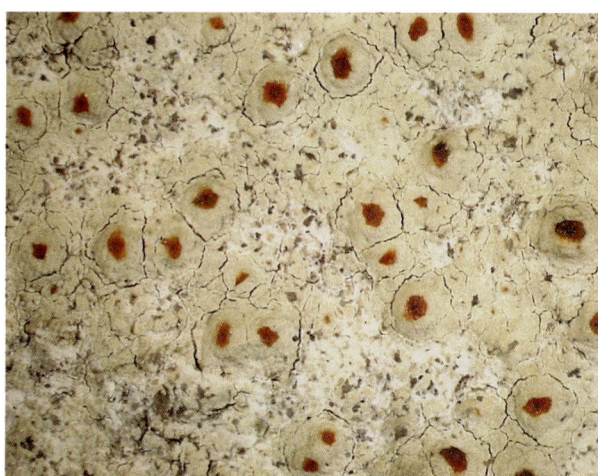
Haematomma ochroleucum, Wales; thallus and ascomata

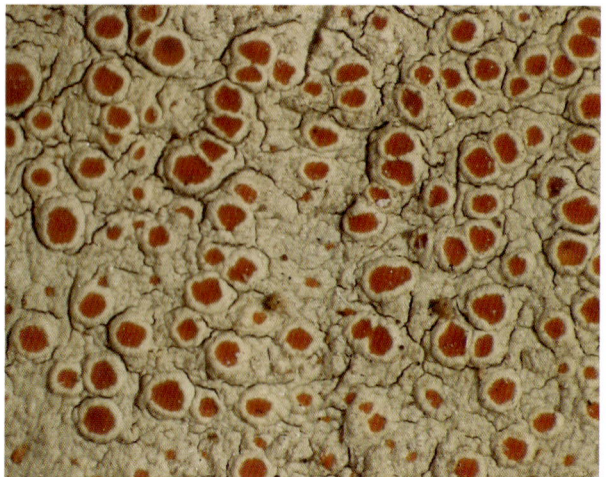
Haematomma puniceum, Louisiana, USA; thallus and ascomata

Ecology: Lichenized with green algae, on rocks, tree bark and similar substrata.

Notes: Perhaps allied to the *Ophioparmaceae* and/or the *Loxosporaceae*, but molecular data are lacking.

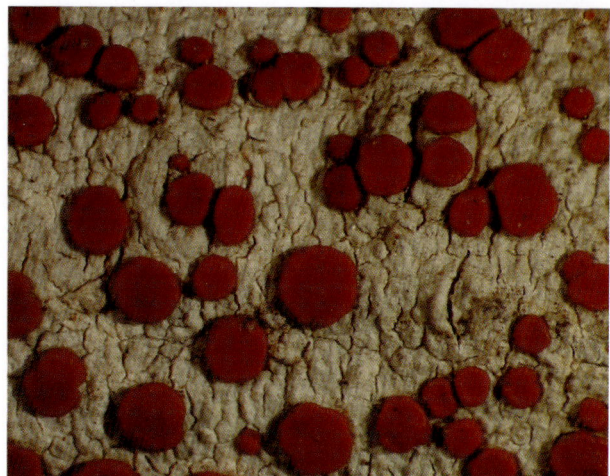
Haematomma similis, Zimbabwe; thallus and ascomata

References: **Brodo, I.M.; Culberson, W.L.** (1987). *Haematomma pustulatum*, sp. nov. (*Ascomycotina, Haematommataceae*): a common, widespread, sterile lichen of Eastern North America. *Bryologist* 89: 203–205; **Honegger, R.** (1986). Ultrastructural studies in lichens I. Haustorial types and their frequences in a range of lichens with trebouxioid photobionts. *New Phytol.* 103: 785–795; **Kalb, K.; Hafellner, J.; Staiger, B.** (1995). *Haematomma*-Studien, II. Lichenicole Pilze auf Arten der Flechtengattung *Haematomma*. *Biblthca Lichenol.* 59: 199–222; **Rogers, R.W.** (1985). Additional notes on *Haematomma* in Australia. *Lichenologist* 17: 307–309; **Rogers, R.W.; Bartlett, J.K.** (1986). The lichen genus *Haematomma* in New Zealand. *Lichenologist* 18: 247–255; **Rogers, R.W.; Hafellner, J.** (1988). *Haematomma* and *Ophioparma*: two superficially similar genera of lichenized fungi. *Lichenologist* 20: 167–174; **Staiger, B.; Kalb, K.** (1995). *Haematomma*-Studien, I. Die Flechtengattung *Haematomma*. *Biblthca Lichenol.* 59: 3–198.

Halosphaeriaceae E. Müll. & Arx ex Kohlm. 1972
Microascales: Ascomycota

Stromata absent. Ascomata perithecial, usually solitary, immersed; peridium usually thin, membranous, hyaline or black. Interascal tissue absent but central pseudoparenchyma characteristic, often breaking up to form catenophyses. Asci clavate, thin-walled, evanescent, without apical structures. Ascospores varied, hyaline, usually septate, occasionally ornamented, usually with complex mucous or cellular appendages. Anamorphs hyphomycetous, often absent.

Significant Genera: *Corollospora, Halosarpheia*.

Distribution: Cosmopolitan. Many species have extensive distributions, perhaps reflecting the relative ease of transmission in aquatic environments.

Economic Significance: Very little is known, though there have been studies on the potential value of their metabolites.

Ecology: Found in aquatic habitats, mostly in marine environments; saprobic on submerged or intertidal wood, occasionally associated with sand grains and similar substrata.

Notes: Classification has relied heavily on morphological and ultrastructural aspects of ascospore development, but some recent molecular studies suggest that these features may not be reliable indicators of phylogeny.

Halosarpheia spartinae, Wales; ascospores

References: **Abdel-Wahab, M.A.; Jones, E.B.G.; Vrijmoed, L.L.P.** (1999). *Halosarpheia kandeliae* sp. nov. on intertidal bark of the mangrove tree *Kandelia candel* in Hong Kong. *Mycol. Res.* 103: 1500–1504; **Anderson, J.L.; Chen, W.; Shearer, C.A.** (2001). Phylogeny of *Halosarpheia* based on 18S rDNA. *Mycologia* 93: 897–906; **Campbell, J.; Anderson, J.L.; Shearer, C.A.** (2003). Systematics of *Halosarpheia* based on morphological and molecular data. *Mycologia* 95: 530–552; **Chen, W.; Shearer, C.A.; Crane, J.L.** (1999). Phylogeny of *Ophioceras* spp. based on morphological and molecular data. *Mycologia* 91: 84–94; **Hyde, K.D.; Ho, W.H.; Tsui, C.K.M.** (1999). The genera *Aniptodera*, *Halosarpheia*, *Nais* and *Phaeonectriella* from freshwater habitats. *Mycoscience* 40: 165–183; **Hyde, K.D.; Jones, E.B.G.** (1989). Observations on ascospore morphology in marine fungi and their attachment to surfaces. *Bot. Mar.* 32: 205–218; **Jones, E.B.G.** (1995). Ultrastructure and taxonomy of the aquatic ascomycetous order *Halosphaeriales*. *Can. J. Bot.* 73: S790–S801; **Jones, E.B.G.; Moss, S.T.** (1987). Key and notes on genera of the *Halosphaeriaceae* examined at the ultrastructural level. *Syst. Ascom.* 6: 179–200; **Kohlmeyer, J.; Volkmann-Kohlmeyer, B.** (1991). Illustrated key to the filamentous higher marine fungi. *Bot. Mar.* 34: 1–61; **Kong, R.Y.C.; Chan, J.Y.C.; Mitchell, J.I.; Vrijmoed, L.L.P.; Jones, E.B.G.** (2000). Relationships of *Halosarpheia*, *Lignincola* and *Nais* inferred from partial 18S rDNA. *Mycol. Res.* 104: 35–43; **Nakagiri, A.; Tubaki, K.** (1987). Pleomorphy in marine fungi: Teleomorph-anamorph connections in the *Halosphaeriaceae*. *Pleomorphic Fungi: The Diversity and its Taxonomic Implications* (Tokyo): 79–101; **Pang, K.L.; Vrijmoed, L.L.P.; Kong, R.Y.C.; Jones, E.B.G.** (2003). Polyphyly of *Halosarpheia* (Halosphaeriales, Ascomycota): implications on the use of unfurling ascospore appendage as a systematic character. *Nova Hedwigia* 77: 1–18; **Sakayaroj, J.; Pang, K.-l.; Phongpaichit, S.; Jones, E.B.G.** (2005). A phylogenetic study of the genus *Haligena* (Halosphaeriales, Ascomycota). *Mycologia* 97: 804–811; **Spatafora, J.W.; Volkmann-Kohlmeyer, B.; Kohlmeyer, J.** (1998). Independent terrestrial origins of the *Halosphaeriales* (marine Ascomy-

cota). *Am. J. Bot.* 85: 1569–1580; **Yusoff, M.; Jones, E.B.G.; Moss, S.T.** (1994). A taxonomic reappraisal of the genus *Ceriosporopsis* based on ultrastructure. *Can. J. Bot.* 72: 1550–1559.

Hapalopilaceae Jülich 1982
Polyporales: Basidiomycota

Basidiomata resupinate or pileate and laterally attached, annual, initially soft-textured, hard and resinous when dry, highly varied in colour, glabrous, tomentose or scurfy. Hyphal system mostly monomitic, generative hyphae usually with clamp connections and sometimes crystalline, cystidia absent. Hymenium poroid, the pores round to angular, sometimes brightly coloured. Basidia clavate, usually with a basal clamp connection and 4 sterigmata. Basidiospores ellipsoidal to cylindrical or allantoid, smooth, hyaline, thin-walled, not staining in iodine.

Significant Genera: *Ceriporia*, *Ceriporiopsis*, *Hapalopilus*, *Ischnoderma*.

Hapalopilus rutilans; basidiomata

Ceriporia purpurea; basidioma

Distribution: Widespread, mainly north temperate.

Economic Significance: *Hapalopilus* species have been invesigated for their potential as tumour inhibitors. Species are occasionally pathogenic to living woody plants.

Ecology: Generally saprobic, causing white rots of hardwoods and conifers.

Notes: Initial molecular studies places the family within the phlebioid clade of the homobasidiomycetes, but the family is clearly polyphyletic and needs reshaping. *Bjerkandera* in particular appears to be misplaced and *Porostereum* should be included.

References: **Binder, M.; Hibbett, D.S.** (2002). Higher-level phylogenetic relationships of homobasidiomycetes (mushroom-forming fungi) inferred from four rDNA regions. *Mol. Phylogen. Evol.* 22: 76–90; **Chang, T.T.; Chou, W.N.** (1999). *Ceriporiopsis microporus* and *Tyromyces formosanus*, two new polypore species from Taiwan. *Mycol. Res.* 103: 674–676; **Hibbett, D.S.; Donoghue, M.J.** (1995). Progress toward a phylogenetic classification of the *Polyporaceae* through parsimony analysis of mitochondrial ribosomal DNA sequences. *Can. J. Bot.* 73: S853–S861; **Ko, K.S.; Jung, H.S.; Ryvarden, L.** (2001). Phylogenetic relationships of *Hapalopilus* and related genera inferred from mitochondrial small subunit ribosomal DNA sequences. *Mycologia* 93: 270–276; **Loguercio-Leite, C.; Gonçalves, G.V. de C.** (2001). Studies in neotropical polypores 13. *Ceriporiopsis cystidiata* sp. nov. *Mycotaxon* 79: 285–288; **Piątek, M.; Seta, D.; Szczepkowski, A.** (2004). Notes on Polish polypores 5. Synopsis of the genus *Spongipellis*. *Acta Mycologica* Warszawa 39: 25–32; **Pouzar, Z.** (1990). Additional notes on the taxonomy and nomenclature of *Ischnoderma* (*Polyporaceae*). *Česká Mykol.* 44: 92–100; **Rajchenberg, M.** (2000). The genus *Ceriporia* Donk (*Polyporaceae, Basidiomycota*) in the Patagonian Andes forests of Argentina. *Karstenia* 40: 143–146; **Suhara, H.; Maekawa, N.; Kaneko, S.; Hattori, T.; Sakai, K.; Kondo, R.** (2003). A new species, *Ceriporia lacerata*, isolated from white-rotted wood. *Mycotaxon* 86: 335–347; **Vampola, P.; Pouzar, Z.** (1996). Notes on some species of genera *Ceriporia* and *Ceriporiopsis* (*Polyporaceae*). *Czech Mycol.* 48: 315–324.

Harpellaceae L. Léger & Duboscq ex P.M. Kirk & P.F. Cannon 2007[1]
Harpellales: Zygomycota

Thallus unbranched, attached to the periotrophic membrane (usually) or hindgut lining of the host by a non-cellular holdfast. Sporagangia (trichospores) unispored, each usually bearing one or more long basal appendages. Zygospores conical or biconical, formed by hyphal conjugation; heterothallic.

Significant Genera: *Harpella, Stachylina*.

Distribution: Widespread though infrequently recorded.

Economic Significance: None is known.

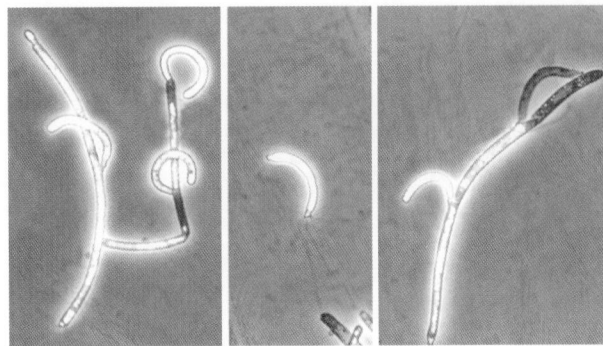
Harpella amazonica; thalli and sporangia

Ecology: Endocommensals (obligate) in aquatic larvae of Diptera; usually found in fast-flowing streams.

Notes: Six genera containing 38 species (one genus, *Stachylina*, with 25). Attached to the peritrophic membrane, or to the hindgut lining of the host (two of the four genera are known to attach to the hindgut lining). A close affinity with the *Kickxellales* has been suggested based first on ultrastructural and serological evidence and augmented by analysis of molecular data. The exclusion of the Amoebidiales and Eccrinales (Protozoa) was suggested as early as 1975, a conclusion confirmed by molecular data.

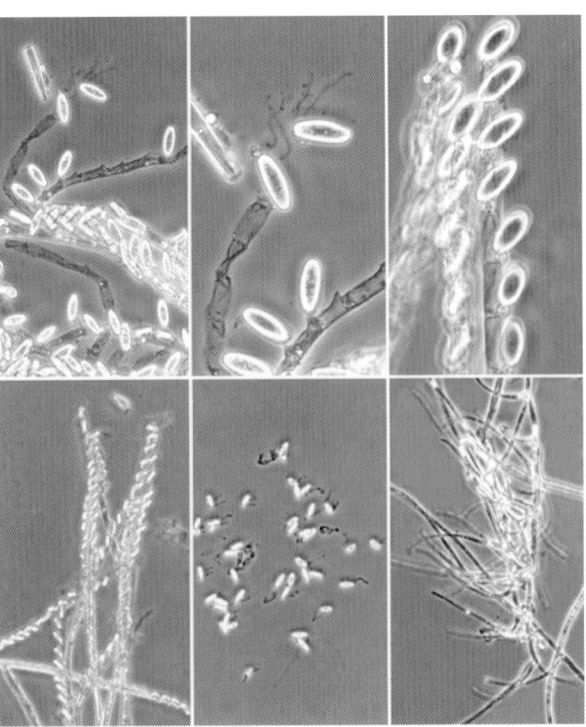
Harpellomyces montanus; thalli and sporangia

---

[1] Thallus non ramosus; trichosporae basipetaliter enatae, basi appendiculis; zygosporae biconicae; intra insectas amborum bono mutuo habitans. Typus *Harpella* L. Léger & Duboscq 1929

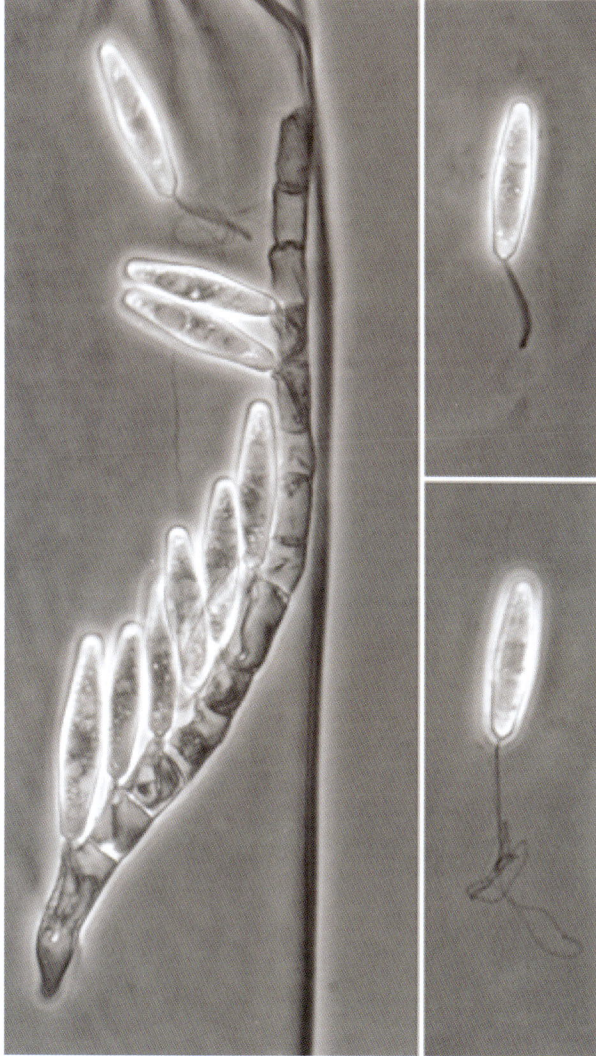

Stachylina acutibasilaris; thallus and sporangia

References: **Benny, G.L.** (2001). Zygomycota: Trichomycetes. *The Mycota* (Berlin) 7: 147–160; **Lichtwardt, R.W.** (1986). *The Trichomycetes. Fungal associates of arthropods* (New York): 343 pp.; **Lichtwardt, R.W.** (2002). Trichomycetes: fungi in relationship with insects and other arthropods. *Cellular Origin and Life in Extreme Habitats* (Dordrecht) 4: 577–588; **Moss, S.T.** (1975). Septal ultrastructure in the *Trichomycetes* with special reference to *Astreptonema* (Eccrinales). *Trans. Br. mycol. Soc.* 65: 115–127; **Moss, S.T.; Young, T.W.K.** (1978). Phyletic considerations of the *Harpellales* and *Asellariales* (*Trichomycetes, Zygomycotina*) and *Kickxellales* (*Zygomycetes, Zygomycotina*). *Mycologia* 70: 944–963.

Harpochytriaceae Wille 1900
Monoblepharidiales: Chytridiomycota

Thallus eucarpic with a sub-basal holdfast and either sessile or with a long slender stalk, the upper region forming the zoosporangium, this capable of repeated sporulation; zoospores usually formed in a single row.

Significant Genera: *Harpochytrium*.

Distribution: Widespread, mainly north temperate.

Economic Significance: None is known.

Ecology: Occurring in a variety of aquatic environments.

Notes: Very poorly known; no recent information is available.

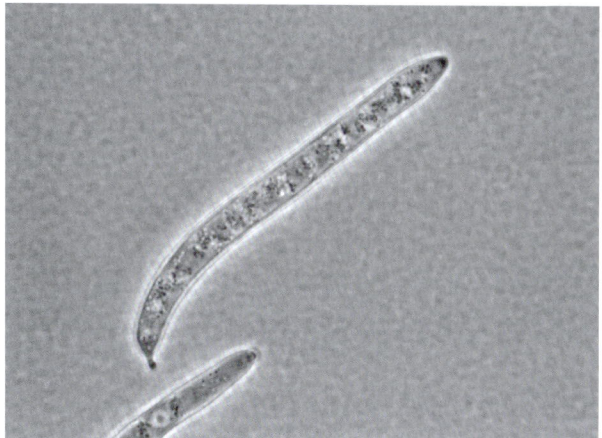

Harpochytrium sp.; thallus

References: **Einax, E.; Voigt, K.** (2003). Oligonucleotide primers for the universal amplification of β-tubulin genes facilitate phylogenetic analyses in the regnum *Fungi*. *Organ. Divers. Evol.* 3: 185–194; **Gaurilof, L.P.; Delay, R.J.; Fuller, M.S.** (1980). Comparative ultrastructure and biochemistry of chytridiomycetous fungi and the future of the *Harpochytriaceae*. *Can. J. Bot.* 58: 2098–2109; **Gaurilof, L.P.; Delay, R.J.; Fuller, M.S.** (1980). The fine structure of zoospores of *Harpochytrium hedinii*. *Can. J. Bot.* 58: 2090–2097.

Helicocephalidaceae Boedijn 1959
Zoopagales: Zygomycota

Sporophores unbranched, rhizoidal. Spores small or large, pigmented, formed in often helical chains. Zygospores unknown.

Helicocephalum oligosporum; spores

Significant Genera: *Helicocephalum*.

Distribution: Widespread although not well recorded.

Economic Significance: None is known.

Ecology: Obligate parasites of nematodes, rotifers and their eggs.

Notes: Very poorly known; little recent information is available. *Helicocephalum* appears to be polyphyletic.

References: **Benny, G.L.; Humber, R.A.; Morton, J.B.** (2001). *Zygomycota: Zygomycetes. The Mycota* (Berlin) 7: 113–146; **Borowska, A.A.** (1997). A new species of *Helicocephalum* from Poland. *Acta Mycologica* Warszawa 32: 129–131.

Distribution: Most known species have temperate distributions.

Economic Significance: None is known.

Ecology: Saprobic on wood or old basidiomycete fruit bodies.

Notes: The family occupies a well-defined clade within the *Sordariomycetes*, but more work is required before an ordinal position is established. *Chaetosphaerella* has recently been given its own family within the *Coronophorales*.

Helminthosphaeriaceae Samuels, Cand. & Magni 1997
Trichosphaeriales: Ascomycota

Stromata absent. Ascomata perithecial, superficial or with the base immersed, black, carbonaceous, usually distinctly setose, ostiolate, the ostiole ± papillate, periphysate; peridium composed of a thick outer layer of dark angular cells and an inner layer of small hyaline angular tissue. Interascal tissue of copious, persistent, rarely branched, true paraphyses. Asci cylindrical, persistent, thin-walled, not fissitunicate, without visible apical structures. Ascospores grey-brown to olivaceous, aseptate or transversely septate, sometimes versicoloured, sometimes with an inconspicuous pore at one or both ends, without a gelatinous sheath or appendages. Anamorphs hyphomycetous, with pigmented conidiophores and dark, poorly differentiated, conidiogenous cells with conidia formed through pores.

Significant Genera: *Helminthosphaeria*.

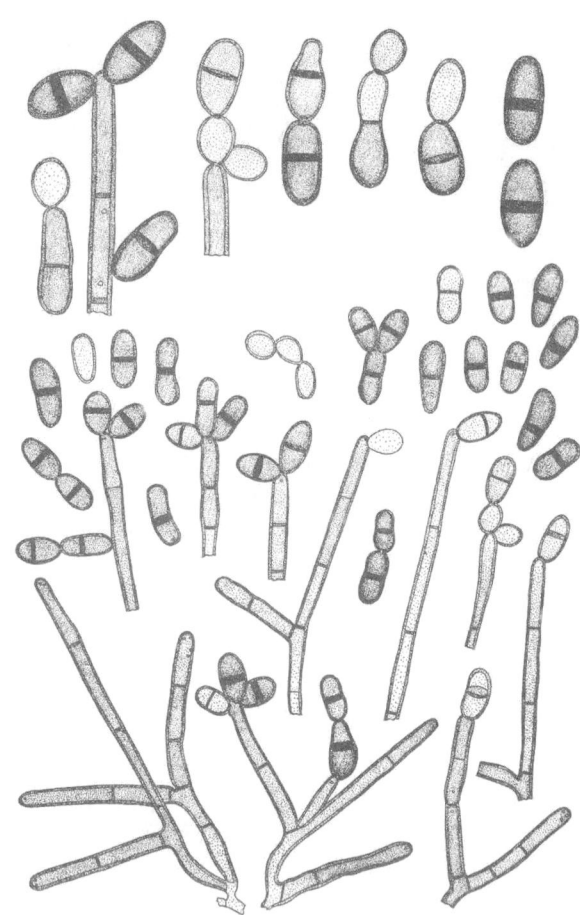

Helminthosphaeria clavariarum; conidiophores and conidia

Helminthosphaeria clavariarum; ascomata on old clavarioid fungus

References: **Goh, T.K.; Hyde, K.D.** (1998). A synopsis of and a key to *Diplococcium* species, based on the literature, with a description of a new species. *Fungal Diversity* 1: 65–83; **Goh, T.K.; Hyde, K.D.** (1999). *Spadicoides palmicola* sp. nov. on *Licuala* sp. from Brunei, and a note on *Spadicoides heterocolorata* comb. nov. *Can. J. Bot.* 76: 1698–1702; **Huhndorf, S.M.; Miller, A.N.; Fernández, F.A.** (2004). Molecular systematics of the *Sordariales*: the order and the family *Lasiosphaeriaceae* revisited. *Mycologia* 96: 368–387; **Miller, A.N.; Huhndorf, S.M.** (2004). A natural classification of *Lasiosphaeria* based on nuclear LSU rDNA sequences. *Mycol. Res.* 108: 26–34; **Park, H.G.; Jong, S.C.** (2003). Molecular characterization of *Monascus* strains based on the D1/D2 regions of LSU rRNA genes. *Mycoscience* 44: 25–32; **Réblová, M.** (1999). Studies in *Chaetosphaeria sensu lato* I. The genera *Chaetosphaerella* and *Tengiomyces* gen. nov. of the *Helminthosphaeriaceae*. *Mycotaxon* 70: 387–420; **Réblová, M.** (1999). Teleomorph-anamorph connections in Ascomycetes 3. Three new lignicolous species of *Helminthosphaeria*. *Sydowia* 51: 223–244; **Samuels, G.J.; Candoussau, F.;**

Magni, J.-F. (1997). Fungicolous pyrenomycetes 1. *Helminthosphaeria* and the new family *Helminthosphaeriaceae*. *Mycologia* 89: 141–155; **Wang, C.J.K.; Sutton, B.C.** (1999). *Diplococcium hughesii* sp. nov. with a *Selenosporella* synanamorph. *Can. J. Bot.* 76: 1608–1613.

Helotiaceae Rehm 1886
Helotiales: Ascomycota

Stromata usually absent. Ascomata apothecial, small to medium-sized, discoid or cupulate, often brightly coloured; excipulum usually composed of parallel or interwoven hyphae with ± remote septa, sometimes gelatinized but not in the outer layer, almost always glabrous or downy (with poorly differentiated hairs). Interascal tissue of simple paraphyses. Asci cylindrical, usually thin-walled, the apical part not significantly thickened, with a small but usually distinct J+ or J– apical ring. Ascospores usually small, hyaline, ellipsoidal or elongate, septate or not. Anamorphs very varied, only known for a few taxa.

Significant Genera: *Calycina, Cenangium, Crocicreas, Godronia, Hymenoscyphus, Tympanis*.

Hymenoscyphus herbarum, Devon, UK; ascomata

Distribution: Widespread in temperate and tropical areas.

Economic Significance: A few species are pathogenic on woody plants, but little is known for almost all taxa.

Ecology: Usually saprobic on herbaceous or woody tissues, some species fungicolous. A few species are known to form mycorrhizas.

Notes: A poorly known family that is varied in form, ecology and almost certainly polyphyletic; little molecular systematic research has been carried out. The family contains many species that are prominent microfungal members of the decomposer community. *Neobulgaria* and *Ascocoryne* can be separated from *Hymenoscyphus* and its relatives using molecular data and may merit the introduction of a separate family.

Bisporella citrina, Surrey, UK; ascomata

Ascocoryne sarcoides, Devon, UK; ascomata

Neobulgaria pura, Devon, UK; ascomata

References: **Baral, H.O.** (1987). Der Apikalapparat der *Helotiales*. Eine lichtmikroskopische Studie über Arten mit Amyloidring. *Z. Mykol.* 53: 119–136; **Baral, H.-O.; Marson, G.** (2000). Monographic revision of *Gelatinopsis* and *Calloriopsis* (*Calloriopsideae, Leotiales*). *Micologia 2000* (Trento): 23–46; **Döbbeler, P.** (1997). Biodiversity of bryophilous ascomycetes. *Biodiv. Cons.* 6: 721–738; **Gamundí, I.; Romero, A.I.** (1998). *Fungi, Ascomycetes. Helotiales: Helotiaceae. Fl. criptog. Tierra del Fuego* 10: 130 pp.; **Monreal, M.; Berch, S.M.; Berbee, M.** (1999). Molecular diversity of ericoid mycorrhizal fungi. *Can. J. Bot.* 77: 1580–1594; **Petrini, O.; Petrini, L.E.; LaFlamme, G.; Ouellette, G.B.** (1989). Taxonomic position of *Gremmeniella abietina* and related species: a reappraisal. *Can. J. Bot.* 67: 2805–2814; **Triebel, D.; Baral, H.-O.** (1996). Notes on the ascus types in *Crocicreas* (*Leotiales, Ascomycetes*) with a characterization of selected taxa. *Sendtnera* 3: 199–218; **Verkley, G.J.M.** (1995). The Ascus Apical Apparatus in Leotiales, An Evaluation of Ultrastructural Characters as Phylogenetic Markers in the Families *Sclerotiniaceae, Leotiaceae* and *Geoglossaceae* (Leiden): 209 pp.; **Verkley, G.J.M.** (1999). A Monograph of the Genus *Pezicula* and its Anamorphs. *Stud. Mycol.* 44: 180 pp.; **Vrålstad, T.; Myhre, E.; Schumacher, T.** (2002). Molecular diversity and phylogenetic affinities of symbiotic root-associated ascomycetes of the *Helotiales* in burnt and metal polluted habitats. *New Phytol.* 155: 131–148; **Wang, Z.; Binder, M.; Schoch, C.L.; Johnston, P.R.; Spatafora, J.W.; Hibbett, D.S.** (2006). Evolution of helotialean fungi (*Leotiomycetes, Pezizomycotina*): a nuclear rDNA phylogeny. *Mol. Phylogen. Evol.* 41: 295–312.

Helvellaceae Fr. 1823
Pezizales: Ascomycota

Stromata absent. Ascomata large, apothecial or cleistothecial; if apothecial varied in form, stalked, the hymenium cupulate or everted, either saddle-shaped or corrugated, pale to dark brown or grey; if cleistothecial solid or chambered, often convoluted, with pockets of fertile tissue separated by sterile undifferentiated cells. Interascal tissue, where present, composed of simple unbranched paraphyses. Asci cylindrical, persistent, operculate, not blueing in iodine (globose to saccate and indehiscent in cleistothecial taxa). Ascospores hyaline or brown, smooth, lacking appendages, tetranucleate. Anamorphs unknown.

Significant Genera: *Helvella, Hydnotrya*.

Distribution: Widespread, especially in temperate, boreal and austral regions.

Economic Significance: None is known.

Ecology: Epigeous or hypogeous, primarily in woodlands and grassy habitats. Probably at least most species form ectomycorrhizas.

Notes: A rather polymorphic family, largely due to the variation in growth habit, for which rearrangement may be needed following further molecular studies. *Gyromitra* was until recently placed in the *Helvellaceae* due to a superficial morphological similarity between that genus and *Helvella*.

Helvella lacunosa, Scotland; ascomata

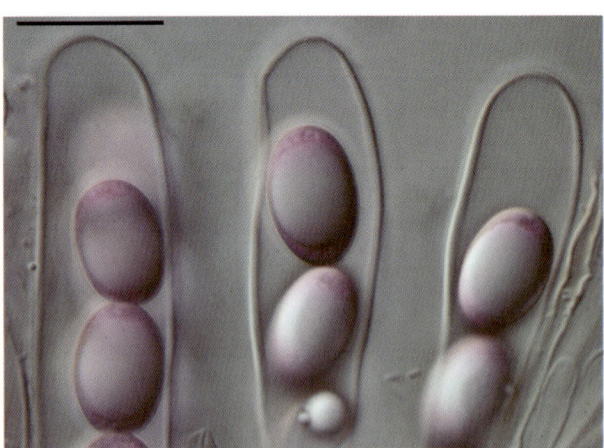

Helvella lacunosa; asci and ascospores

References: **Abbott, S.P.; Currah, R.S.** (1997). The *Helvellaceae*: systematic revision and occurrence in northern and northwestern North America. *Mycotaxon* 62: 1–125; **Calonge, F.D.; Arroyo, I.** (1990). Notes on the genus *Helvella* in Spain. *Mycotaxon* 39: 203–217; **Donadini, J.C.** (1987). Les Balsamiacées sont des Helvellacées: cytologie et scanning de *Balsamia vulgaris* Vitt. et de *Balsamia platyspora* Berk. et Br. *Bull. trimest. Soc. mycol. Fr.* 102: 373–387; **Gibson, J.L.; Kimbrough, J.W.** (1988). Ultrastructural observations on *Helvellaceae* (*Pezizales*). Ascosporogenesis of selected species of *Helvella*. *Can. J. Bot.* 66: 771–783; **Harrington, F.A.; Pfister, D.H.; Potter, D.; Donoghue, M.J.** (1999). Phylogenetic studies within the *Pezizales*. I. 18S rRNA sequence data and classification. *Mycologia* 91: 41–50; **Kimbrough, J.W.; Gibson, J.L.** (1990). Ultrastructural observations on *Helvellaceae* (*Pezizales*; *Ascomycetes*). III. Septal structures in *Helvella*. *Mycologia* 81: 914–920; **Kimbrough, J.W.; Li, L.T.; Wu, C.G.** (1996). Ultrastructural evidence for the placement of the truffle *Barssia* in the *Helvellaceae* (*Pezizales*). *Mycologia* 88: 38–46; **Landvik, S.; Egger, K.N.; Schumacher, T.** (1997). Towards a subordinal classification of the *Pezizales* (*Ascomycota*): phylogenetic analyses of SSU rDNA sequences. *Nordic Jl Bot.* 17: 403–418; **Landvik, S.; Kristiansen, R.; Schumacher, T.** (1999). *Pindara*: a miniature *Helvella*. *Mycologia* 91: 278–285; **O'Donnell, K.; Cigelnik, E.; Weber, N.S.; Trappe, J.M.** (1997). Phylogenetic relationships among ascomycetous truffles and the true and false morels inferred from 18S and 28S ribosomal DNA sequence analysis. *Mycologia* 89: 48–65; **Percudani, R.; Trevisi, A.; Zambonelli, A.; Ottonello, S.** (1999). Molecular phylogeny of truffles (*Pezizales*: *Terfeziaceae*, *Tuberaceae*) derived from nuclear rDNA sequence analysis. *Mol. Phylogen. Evol.* 13: 169–180; **Zhang, B.C.** (1991). Morphology, cytology and taxonomy of *Hydnotrya cerebriformis* (*Pezizales*). *Mycotaxon* 42: 155–162.

Hemigasteraceae Gäum. & C.W. Dodge 1928
Agaricales: Basidiomycota

Basidiomata initially gasteroid but later becoming angiocarpous, the stipe developing into a columella. Basidia discharging spores passively, produced in a closed hymenial cavity in which hyphae producing chlamydospores also occur.

Significant Genera: *Hemigaster*.

Distribution: Only known from a single collection in Sweden.

Economic Significance: None is known.

Ecology: Saprobic on rabbit dung.

Notes: The affinities of *Hemigaster* are uncertain; it could represent an early or abortive developmental stage of a *Coprinus*-like fungus.

References: **Reijnders, A.F.M.** (2000). A morphogenetic analysis of the basic characters of the gasteromycetes and their relation to other basidiomycetes. *Mycol. Res.* 104: 900–910; **Singer, R.** (1986). *Agaricales in Modern Taxonomy* Edn 4 (Koenigstein): 981 pp.

Hemiphacidiaceae Korf 1962
Helotiales: Ascomycota

Stromata absent. Ascomata apothecial, immersed, becoming exposed by circumscissile or laciniate rupture of the overlying host tissues; peridium hardly developed. Interascal tissue of simple paraphyses, sometimes swollen towards the apices. Asci with a J+ or J– ring. Ascospores usually septate, hyaline or brown, sometimes with a sheath. Anamorphs coelomycetous.

Significant Genera: *Didymascella*, *Rhabdocline*.

Distribution: Confined to the north temperate zone.

Economic Significance: Some species are pathogens, though of fairly minor economic importance.

Ecology: Saprobic (endophytic) or parasitic, especially on conifers.

Notes: The family has many morphological parallels with the *Rhytismataceae*.

Didymascella thujina, Vancouver Island, Canada; ascomata

References: **Gernandt, D.S.; Camacho, F.J.; Stone, J.K.** (1997). *Meria laricis*, an anamorph of *Rhabdocline*. *Mycologia* 89: 735–744; **Gernandt, D.S.; Platt, J.L.; Stone, J.K.; Spatafora, J.W.; Holst-Jensen, A.; Hamelin, R.C.; Kohn, L.M.** (2001). Phylogenetics of *Helotiales* and *Rhytismatales* based on partial small subunit nuclear ribosomal DNA sequences. *Mycologia* 93: 915–933; **Minter, D.W.** (1995). The *Rhytismatales* on conifers from Europe. *Shoot and Foliage Diseases in Forest Trees* Proceedings of a Joint Meeting of the Working Parties: Canker & Shoot Blight of Conifers, Foliage Diseases (Firenze): 65–84; **Sherwood-Pike, M.A.; Stone, J.K.; Carroll, G.C.** (1986). *Rhabdocline parkeri*, a ubiquitous foliar endophyte of Douglas fir. *Can. J. Bot.* 64: 1849–1855; **Stone, J.K.; Gernandt, D.S.** (2005). A reassessment of *Hemiphacidium*, *Rhabdocline*, and *Sarcotrochila* (*Hemiphacidiaceae*). *Mycotaxon* 91: 115–126.

Heppiaceae Zahlbr. 1906
Lichinales: Ascomycota

Thallus foliose to fruticose, sometimes peltate, not gelatinized, with rather wide hyphae. Ascomata apothecial, deeply immersed, sometimes initially cleistothecial. Interascal tissue of thick-walled anastomosing paraphyses, swollen at the apices. Asci ± thin-walled, with a J+ apical cap. Ascospores aseptate, hyaline, without a sheath. Anamorphs pycnidial.

Significant Genera: *Heppia*, *Solorinaria*.

Distribution: Primarily tropical, often in arid ecosystems.

Economic Significance: None is known.

Ecology: Lichenized with cyanobacteria.

Notes: Very poorly known; no recent information is available.

References: **Büdel, B.** (1987). Zur Biologie und Systematik der Flechtengattungen *Heppia* und *Peltula* im südlichen Afrika. *Biblthca Lichenol.* 23: 105 pp.; **Henssen, A.** (1994). Contribution to the morphology and species delimitation in *Heppia* sensu stricto (lichenized *Ascomycotina*). *Acta bot. fenn.* 150: 57–73; **Schultz, M.; Arendholz, W.-R.; Büdel, B.** (2001). Origin and evolution of the lichenized ascomycete order *Lichinales*: monophyly and systematic relationships inferred from ascus, fruiting body and SSU rDNA evolution. *Pl. Biol.* 3: 116–123; **Schultz, M.; Büdel, B.** (2003). On the systematic position of the genus *Heppia* in the *Lichinales*. *Lichenologist* 35: 151–156; **Upreti, D.K.; Büdel, B.** (1990). The lichen genera *Heppia* and *Peltula* in India. *J. Hattori bot. Lab.* 68: 279–284.

evolved on numerous occasions. The non-hydnoid genus *Laxitextum* appears to belong here.

Hericium erinaceus, Sussex, UK; basidioma

Hericiaceae Donk 1964
Russulales: Basidiomycota

Basidiomata effuse, clavate or shortly stipitate and strongly branched, frequently with fleshy downward-pointing spines, the upper surface smooth, tomentose or hairy, pale or brightly coloured. Hyphal system monomitic, flesh generally soft-textured, hyphae often inflated and clamp connections usually present, some hyphae irregular and with refractive contents, not staining in iodine. Gloeocystidia present, not darkening in sulphoaldehydes. Basidia clavate, with 4 sterigmata. Basidiospores globose to ellipsoidal, hyaline, thin-walled or moderately thick-walled, hyaline, usually ornamented, staining blue in iodine.

Significant Genera: *Hericium*.

Distribution: Mainly known from north temperate regions.

Economic Significance: Little is known. Some species are edible, though a number of species are of conservation concern and species of *Hericium* are prized for their medicinal properties in Oriental medicine.

Ecology: Saprobic on rotten wood, sometimes associated with heart rots and found growing on dead parts of living trees.

Notes: The *Hericiaceae* appear to belong to the russuloid clade of homobasidiomycetes, with morphological and molecular evidence linking it with the *Auriscalpiaceae*, *Bondarzewiaceae* and *Echinodontiaceae*. The spiny (hydnoid) hymenium appears to have

Hericium abietis; basidioma

References: **Ginns, J.** (1986). The genus *Dentipellis* (*Hericiaceae*). *Windahlia* 16: 35–45; **Hibbett, D.S.; Donoghue, M.J.** (2001). Analysis of character correlations among wood decay mechanisms, mating systems and substrate ranges in *Homobasidiomycetes*. *Syst. Biol.* 50: 215–242; **Koski-Kotiranta, S.; Niemelä, T.** (1988). Hydnaceous fungi of the *Hericiaceae*, *Auriscalpiaceae* and *Climacodontaceae* in northwestern Europe. *Karstenia* 27: 43–70; **Larsson, E.; Larsson, K.-H.** (2003). Phylogenetic relationships of russuloid basidiomycetes with emphasis on aphyllophoralean taxa. *Mycologia* 95: 1037–1065; **Stalpers, J.A.** (1996). The Aphyllophoraceous Fungi – II. Keys to the Species of the *Hericiales*. *Stud. Mycol.* 40: 185 pp.

Herpomycetaceae I.I. Tav. 1981
Laboulbeniales: Ascomycota

Stroma (thallus) present; with a usually 4-celled primary thallus and ascomata formed from secondary thalli. Ascomata perithecial, pyriform, thin-walled. Asci thin-walled, evanescent, 8-spored. Ascospores hyaline, thin-walled, with median septa. Dioecious, with separate thalli forming antheridia and gynoecia.

Significant Genera: *Herpomyces*.

Distribution: Widespread.

Economic Significance: None is known. There has been some interest in the use of species as biological control agents, but commercial exploitation has not yet happened.

Ecology: Ectoparasitic or epibiotic on cockroaches (Blattidae), parasitic status unknown but not causing significant damage.

Notes: Very poorly known; no recent information is available.

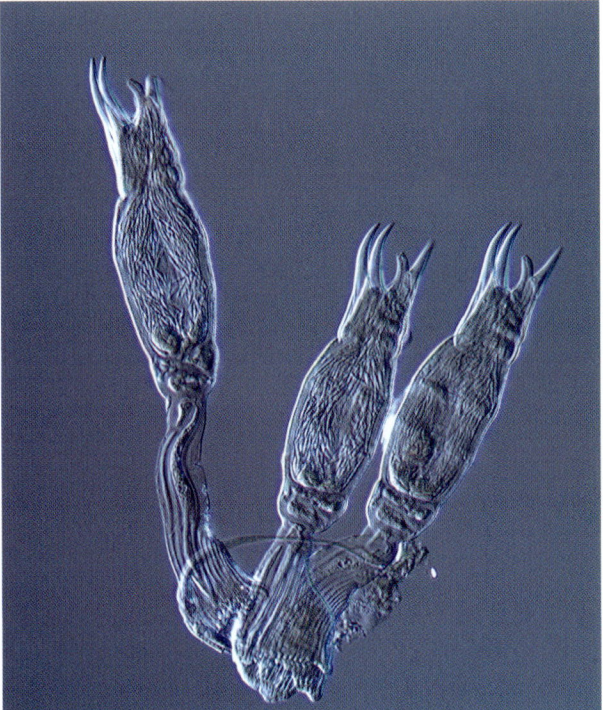

Herpomyces sp.; ascomata

References: **Majewski, T.** (1988). Some *Laboulbeniales* (*Ascomycotina*) collected in Japan I. Species from Shizuoka Prefecture. *Trans. Mycol. Soc. Japan* 29: 33–54; **Majewski, T.; Sugiyama, K.** (1985). The *Laboulbeniomycetes* of eastern Asia. IV. On ten species from Japan including four new species. *Trans. Mycol. Soc. Japan* 26: 295–313; **Santamaria, S.; Balazuc, J.; Tavares, I.I.** (1991). Distribution of the European *Laboulbeniales* (*Fungi, Ascomycotina*). An annotated list of species. *Treb. Inst. Bot. Barcelona* 14: 1–123; **Tavares, I.I.** (1985). *Laboulbeniales* (*Fungi, Ascomycetes*). *Mycol. Mem.* 9: 627 pp.

Herpotrichiellaceae Munk 1953
Chaetothyriales: Ascomycota

Mycelium immersed. Ascomata perithecial, erumpent or superficial, small, ± globose, occasionally aggregated in small stromata, with a well-developed often periphysate ostiole, often setose; peridium thin, composed of compressed pseudoparenchymatous cells, varied in pigmentation. Interascal tissue of short periphysoids. Asci saccate to clavate, the inner wall layer often conspicuously thickened in the apical region, fissitunicate, sometimes polysporous. Ascospores hyaline to greyish, septate, sometimes muriform. Anamorphs hyphomycetous, sometimes yeast-like (commonly referred to as 'black yeasts').

Significant Genera: *Capronia*. Anamorphs: *Cladophialophora*, *Exophiala*.

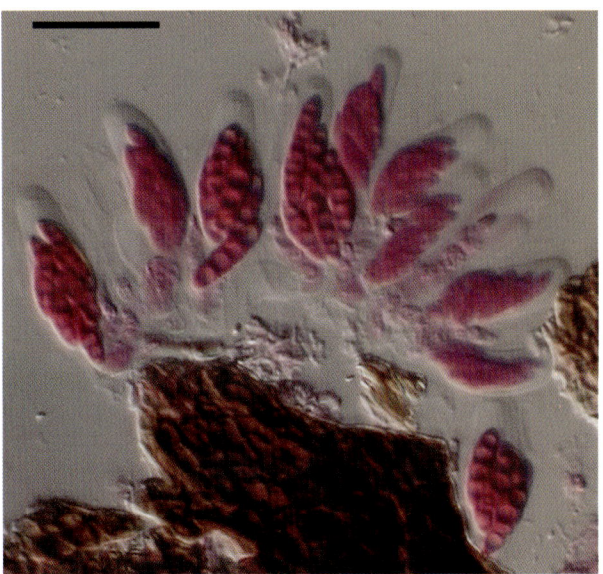

Capronia spinifera, Malawi; asci and ascospores

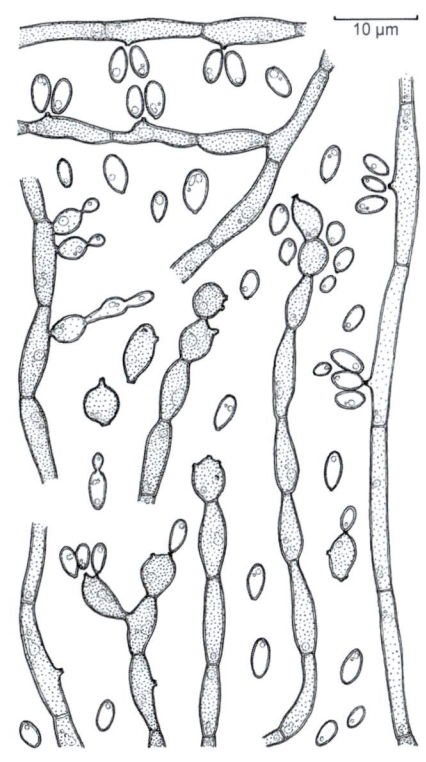

Exophiala jeanselmei; conidiophores and conidia

Distribution: Cosmopolitan, often in extreme environments.

Economic Significance: Species are implicated in human disease, causing chromatoblastomycoses and phaeohyphomycoses.

Ecology: Mostly saprobic on plants or other fungi, isolated from soil etc., and from medical sources.

Notes: The ecological requirements of the family are diverse, and merit further investigation. *Coniosporium* and *Phaeococcomyces*, found in the surface layers of rock amongst other habitats, may also belong here, but the genera need revision.

References: **Abliz, P.; Fukushima, K.; Takizawa, K.; Nishimura, K.** (2004). Identification of pathogenic dematiaceous fungi and related taxa based on large subunit ribosomal DNA D1/D2 domain sequence analysis. *FEMS Immunol. Med. Microbiol.* 40: 41–49; **Au, D.W.T.; Jones, E.B.G.; Vrijmoed, L.L.P.** (1999). The ultrastructure of *Capronia ciliomaris*, an intertidal marine fungus from San Juan Island. *Mycologia* 91: 326–333; **Braun, U.; Feiler, U.** (1995). *Cladophialophora* and its teliomorph. *Microbiol. Res.* 150: 81–91; **Gams, W.** (2000). *Phialophora* and some similar morphologically little-differentiated anamorphs of divergent ascomycetes. *Stud. Mycol.* 45: 187–199; **Gerrits van den Ende, A.H.G.; Hoog, G.S. de** (1999). Variability and molecular diagnostics of the neurotropic species *Cladophialophora bantiana*. *Stud. Mycol.* 43: 151–162; **Haase, G.; Sonntag, L.; Melzer-Krick, B.; Hoog, G.S. de** (1999). Phylogenetic inference by SSU-gene analysis of members of the *Herpotrichiellaceae* with special reference to human pathogenic species. *Stud. Mycol.* 43: 80–97; **Haase, G.; Spatafora, J.W.; Mitchell, T.G.; Vilgalys, R.** (1996). Analysis of genes coding for small-subunit rRNA sequences in studying phylogenetics of dematiaceous fungal pathogens. *J. Clin. Microbiol.* 34: 2049–2050; **Hoog, G.S. de; Bowman, B.; Graser, Y.; Haase, G.; El Fari, M.; Gerrits van den Ende, A.H.G.; Melzer-Krick, B.; Untereiner, W.A.** (1998). Molecular phylogeny and taxonomy of medically important fungi. *Medical Mycol.* 36: 52–56; **Hoog, G.S. de; Poonwan, N.; Gerrits van den Ende, A.H.G.** (1999). Taxonomy of *Exophiala spinifera* and its relationship to *E. jeanselmei*. *Stud. Mycol.* 43: 133–142; **McKemy, J.M.; Rogers, S.O.; Wang, C.J.K.** (1999). Emendation of the genus *Wangiella* and a new combination, *W. heteromorpha*. *Mycologia* 91: 200–205; **Müller, E.; Petrini, O.; Fisher, P.J.; Samuels, G.J.; Rossman, A.Y.** (1987). Taxonomy and anamorphs of the *Herpotrichiellaceae* with notes on generic synonymy. *Trans. Br. mycol. Soc.* 88: 63–74; **Rogers, S.O.; McKemy, J.M.; Wang, C.J.K.** (1999). Molecular assessment of *Exophiala* and related hyphomycetes. *Stud. Mycol.* 43: 122–133; **Uijthof, J.M.J.; Hoog, G.S. de** (1996). Molecular biodiversity in the ascomycete family *Herpotrichiellaceae*. *Culture Collections to Improve the Quality of Life, Proceedings of the Eighth International Congress for Culture Collections, Veldhoven, The Netherlands, 25-29 August 1996* (Baarn): 389–392; **Untereiner, W.A.** (1997). Taxonomy of selected members of the ascomycete genus *Capronia* with notes on anamorph-teleomorph connections. *Mycologia* 89: 120–131; **Untereiner, W.A.** (2000). *Capronia* and its anamorphs: exploring the value of morphological and molecular characters in the systematics of the *Herpotrichiellaceae*. *Stud. Mycol.* 45: 141–149; **Untereiner, W.A.; Naveau, F.A.** (1999). Molecular systematics of the *Herpotrichiellaceae* with an assessment of the phylogenetic positions of *Exophiala dermatitidis* and *Phialophora americana*. *Mycologia* 91: 67–83; **Untereiner, W.A.; Straus, N.A.; Malloch, D.** (1995). A molecular-morphotaxonomic approach to the systematics of the *Herpotrichiellaceae* and allied black yeasts. *Mycol. Res.* 99: 897–913.

Heterodeaceae Filson 1978

Lecanorales: Ascomycota

Thallus foliose, irregular, sometimes becoming erect, the lower surface ecorticate, rhizoidal. Ascomata apothecial, irregular, domed, without a well-developed margin. Interascal tissue of unbranched paraphyses. Asci with a well-developed J+ apical cap, lacking an ocular chamber. Ascospores ellipsoidal, hyaline, aseptate, thin-walled, without a sheath. Anamorph pycnidial, superficial.

Significant Genera: *Heterodea*.

Distribution: Australia.

Economic Significance: None is known.

Ecology: Lichenized with green algae.

Notes: The family should probably be placed within the *Cladoniaceae*, though more data are needed before this change can be justified.

Heterodea muelleri, Tasmania; thallus on dry soil.

References: **Wedin, M.; Döring, H.; Ekman, S.** (2000). Molecular phylogeny of the lichen families *Cladoniaceae*, *Sphaerophoraceae*, and *Stereocaulaceae* (*Lecanorales*, *Ascomycotina*). *Lichenologist* 32: 171–187; **Wiklund, E.; Wedin, M.** (2003). The phylogenetic relationships of the cyanobacterial lichens in the *Lecanorales* suborder *Peltigerineae*. *Cladistics* 19: 419–431.

Heterogastridiaceae Oberw. & R. Bauer 1990

Heterogastridiales: Basidiomycota

Basidiomata flask-shaped, long-necked. Basidioma wall composed of a single layer of vertically oriented, loosely adhering, parallel thin-walled hyphae, flared and ± setose at the apex. Hyphae with simple septa, clamp connections and cystidia absent. Hymenium enclosed, basal, composed of a loose cluster of fertile hyphae bearing basidia and interspersed with branched conidium-bearing hyphae. Basidia cylindrical, becoming transversely septate, each daughter cell bearing a single basidiospore. Basidiospores discharged passively, initially cylindrical but becoming tetraradiate, hyaline, thin-walled, smooth, not staining in iodine. Conidia formed from undifferentiated non-proliferating hyphae, ellipsoidal, hyaline, thin-walled, smooth, aseptate.

Significant Genera: *Heterogastridium*. Anamorph: *Hyalopycnis*.

Distribution: Widespread in north temperate zones.

Economic Significance: None is known.

Ecology: Saprobic or fungicolous, isolated from fungal fruit bodies, rotten plant material and leaf litter.

Notes: Preliminary molecular studies indicate that this family may represent a basal lineage of the *Microbotryum* clade of *Urediniomycetes*, but more research is needed. The morphology is certainly unique amongst the *Basidiomycota*.

References: **Bandoni, R.J.; Oberwinkler, F.** (1981). *Hyalopycnis blepharistoma*: a pycnidial basidiomycete. *Can. J. Bot.* 59: 1613–1620; **Oberwinkler, F.; Bauer, R.; Bandoni, R.J.** (1990). *Heterogastridiales*: a new order of basidiomycetes. *Mycologia* 82: 45–58; **Scorzetti, G.; Fell, J.W.; Fonseca, A.; Statzell-Tallman, A.** (2002). Systematics of basidiomycetous yeasts: a comparison of large subunit D1/D2 and internal transcribed spacer rDNA regions. *FEMS Yeast Res.* 2: 495–517; **Seeler, E.V.** (1943). Several fungicolous fungi. *Farlowia* 1: 119–133; **Swann, E.C.; Frieders, E.M.; McLaughlin, D.J.** (1999). *Microbotryum*, *Kriegeria* and the changing paradigm in basidiomycete classification. *Mycologia* 91: 51–66.

Hispidicarpomycetaceae Nakagiri 1993
Spathulosporales: Ascomycota

Thallus composed of hyphal cells. Ascomata perithecial, without sterile setose hairs; peridium very thick, 3-layered. Interascal tissue of simple paraphysis-like hyphae, the ostiole not periphysate. Hymenium extending over the whole of the inner surface of the ascoma. Asci thin-walled, not fissitunicate, deliquescing, without apical structures. Ascospores without mucous appendages. Anamorph spermatial, with verticillately branched conidiophores.

Significant Genera: *Hispidocarpomyces*.

Distribution: Known from the Japanese sector of the Pacific Ocean.

Economic Significance: None is known.

Ecology: Apparently parasitic on marine red algae.

Notes: Very poorly known; no recent information is available.

References: **Nakagiri, A.** (1993). A new marine ascomycete in *Spathulosporales*, *Hispidicarpomyces galaxauricola* gen. et sp. nov. (Hispidicarpomycetaceae fam. nov.), inhabiting a red alga, *Galaxaura falcata*. *Mycologia* 85: 638–652; **Nakagiri, A.; Ito, T.** (1997). *Retrostium umphiroae* gen. et sp. nov. inhabiting a marine red alga, *Amphiroa zonata*. *Mycologia* 89: 484–493.

Hoehnelomycetaceae Jülich 1982
Atractiellales: Basidiomycota

Basidiomata stipitate, capitate, fleshy to gelatinous, ± hyaline, composed of a synnema-like cluster of parallel adherent thin-walled or thick-walled hyphae with simple septal pores. Hyphae of upper section repeatedly branched, the apices developing into basidia and hyphidia. Basidia narrowly clavate to cylindrical, transversely septate, usually forming 4 daughter cells that each produce a lateral basidiospore at the distal end. Basidiospores ovoid to ellipsoidal, hyaline, smooth, thin-walled, germinating to form a short hypha from which minute conidia develop.

Significant Genera: *Atractiella*.

Distribution: Scattered throughout north temperate and tropical regions.

Economic Significance: None is known.

Ecology: Presumably saprobic, known from decaying plant material and plant exudates.

Notes: *Agaricostilbum* was once included within the *Hoehnelomycetaceae*, but that has now been separated into its own family. Little molecular work has been done, but *Helicogloea* and *Platygloea* have recently been described as belonging to the *Atractiella* group.

References: **Bandoni, R.J.; Inderbitzin, P.** (2002). On a new *Atractiella*. *Czech Mycol.* 53: 265–273; **Bauer, R.; Begerow, D.; Oberwinkler, F.; Marvanová, L.** (2003). *Classicula*: the teleomorph of *Naiadella fluitans*. *Mycologia* 95: 756–764; **Kirschner, R.; Bauer, R.; Oberwinkler, F.** (1999). *Atractocolax*, a new heterobasidiomycetous genus based on a species vectored by coniferous bark beetles. *Mycologia* 91: 538–543; **Moore, R.T.** (1990). Order *Platygloeales* ord. nov. *Mycotaxon* 39: 245–248; **Oberwinkler, F.; Bandoni, R.J.** (1982). A taxonomic survey of the gasteroid, auricularioid Heterobasidiomycetes. *Can. J. Bot.* 60: 1726–1750; **Oberwinkler, F.; Bandoni, R.J.** (1982). *Atractogloea*: a new genus in the Hoehnelomycetaceae (Heterobasidiomycetes). *Mycologia* 74: 634–639;

Hyaloriaceae Lindau 1897
Tremellales: Basidiomycota

Basidiomata small, clavate, pulvinate or synnema-like, gelatinous or waxy, whiteish, solitary or formed in clusters on a common subiculum, the hymenium covering its entire upper surface. Hyphal system monomitic, composed of loosely woven gelatinized hyphae with clamp connections, cystidia present or absent. Hyphidia commonly present, with complex branching patterns interspersed with basidia. Basidia initially hypha-like, the distal part becoming swollen and divided from the stalk cell by a septum, the stalk cell with a basal clamp connection and becoming enucleate. Distal cell ovoid to fusiform, separated into 2 or 4 cells with longitudinal or oblique septa, elongated into a long sterigma-like process. Basidiospores discharged passively, ellipsoidal to cylindrical or allantoid, hyaline, smooth, with thick gelatinous walls, not staining in iodine. Conidiomata sporodochial, gelatinous, fertile hyphae with clamp connections, forming reniform to C-shaped conidia apically.

Significant Genera: *Hyaloria, Myxarium*. Anamorph: *Helicomyxa*.

Distribution: Known from Europe and North America, and both New World and Old World tropics.

Economic Significance: None is known.

Ecology: Presumably saprobic, little research is available. Species are found on decaying wood, palm fronds and similar substrata.

Notes: Content of the *Hyaloriaceae* has changed several times in the past thirty years. Molecular studies indicate that the family should also contain *Myxarium*, previously considered to be a synonym of *Exidia* (*Exidiaceae*). It is largely typified by the so-called 'myxarioid' basidium, where the cell subtending the basidium becomes enucleate at maturity.

References: Bandoni, R.J. (1984). The *Tremellales* and *Auriculariales*: an alternative classification. *Trans. Mycol. Soc. Japan* 25: 489–530; **Bononi, V.L.; Capelari, M.** (1984). Basidiomicetos do Parque Estadual da Ilha do Cardoso: *Tremellales*. *Rickia* 11: 109–114; **Kirschner, R.; Chen, C.J.** (2004). *Helicomyxa everhartioides*, a new helicosporous sporodochial hyphomycete from Taiwan with relationships to the *Hyaloriaceae* (*Auriculariales, Basidiomycota*). *Stud. Mycol.* 50: 337–342; **Roberts, P.** (1998). A revision of the genera *Heterochaetella, Myxarium, Protodontia*, and *Stypella* (*Heterobasidiomycetes*). *Mycotaxon* 69: 209–248; **Weiss, M.; Oberwinkler, F.** (2001). Phylogenetic relationships in *Auriculariales* and related groups – hypotheses derived from nuclear ribosomal DNA sequences. *Mycol. Res.* 105: 403–415; **Wells, K.** (1969). New or noteworthy *Tremellales* from southern Brazil. *Mycologia* 61: 77–86.

Hyaloscyphaceae Nannf. 1932
Helotiales: Ascomycota

Stromata absent. Ascomata apothecial, usually small, flat or concave, the excipulum soft, fleshy, usually composed of prismatic or isodiametric cells, almost always with conspicuous, sometimes ornamented hairs surrounding the disc, often brightly coloured. Interascal tissue of simple paraphyses, sometimes lanceolate. Asci cylindric-clavate, small, thin-walled, with a J+ or J– ring. Ascospores small, usually elongated, sometimes septate. Anamorphs rarely noted, when so hyphomycetous, varied in morphology, with some aero-aquatic.

Significant Genera: *Cistella, Hyaloscypha, Lachnum*.

Lachnum clandestinum; ascomata

Distribution: Cosmopolitan.

Economic Significance: None is known.

Ecology: Saprobic on woody and herbaceous material.

Notes: Many species are brightly coloured and may have contrasting ascomatal hairs, making them amongst the most aesthetically attractive of the smaller *Ascomycota*. The hairs are highly varied in form and are used as taxonomic markers for many of the constituent genera, although not all are monophyletic judging from the limited molecular data available.

Lachnellula suecica; ascomata

Dasyscyphella nivea; ascomata

References: Baral, H.O. (1987). Der Apikalapparat der *Helotiales*. Eine lichtmikroskopische Studie über Arten mit Amyloidring. *Z. Mykol.* 53: 119–136; **Baral, H.O.** (1989). Beiträge zur Taxonomie der Discomyceten II. Die *Calycellina*-Arten mit 4sporigen Asci. *Beitr. Kenntn. Pilze Mitteleur.* 5: 209–236; **Cantrell, S.A.; Hanlin, R.T.** (1997). Phylogenetic relationships in the family *Hyaloscyphaceae* inferred from sequences of ITS regions, 5.8S ribosomal DNA and morphological characters. *Mycologia* 89: 745–755; **Galán, R.; Raitviir, A.; Ayala, N.; Ochoa, C.** (1994). First contribution to the knowledge of the *Leotiales* of Baja California and adjacent areas. *Mycol. Res.* 98: 1137–1152; **Haines, J.H.** (1992). Studies in the *Hyaloscyphaceae* VI: the genus *Lachnum* (ascomycetes) of the Guayana Highlands. *Nova Hedwigia* 54: 97–112; **Huhtinen, S.** (1990). A monograph of *Hyaloscypha* and allied genera. *Karstenia* 29: 45–252; **Leenurm, K.; Raitviir, A.; Raid, R.** (2000). Studies on the ultrastructure of *Lachnum* and related genera (*Hyaloscyphaceae, Helotiales, Ascomycetes*). *Sydowia* 52: 30–45; **Spooner, B.M.** (1987). Helotiales of Australasia: *Geoglossaceae, Orbiliaceae, Sclerotiniaceae, Hyaloscyphaceae*. *Biblthca Mycol.* 116: 711 pp.; **Verkley, G.J.M.** (1996). Ultrastructure of the ascus in the genera *Lachnum* and *Trichopeziza* (*Hyaloscyphaceae, Ascomycotina*). *Nova Hedwigia* 63: 215–228; **Wang, Z.; Binder, M.; Schoch, C.L.; Johnston, P.R.; Spatafora, J.W.; Hibbett, D.S.** (2006). Evolution of helotialean fungi (*Leotiomycetes, Pezizomycotina*): a nuclear rDNA phylogeny. *Mol. Phylogen. Evol.* 41: 295–312; **Zhang, Y.H.; Zhuang, W.Y.** (2004). Phylogenetic relationships of some members in the genus *Hymenoscyphus* (*Ascomycetes, Helotiales*). *Nova Hedwigia* 78: 475–484; **Zhuang, W.Y.** (1998). Discomycetes of tropical China. III. Hyaloscyphaceous fungi from tropical Guangxi. *Mycotaxon* 69: 359–376.

Hybogasteraceae Jülich 1982
Russulales: Basidiomycota

Basidiomata gasteroid, epigeous, tuber-like and often lobed, with a small stipe-like base; peridium thin, ± glabrous, brownish or orange, glebal chambers sometimes visible from the surface. Internal structure exuding latex from all parts, composed of a central columella arising from the stipe, which is branched repeatedly to anastomose with the peridial tissue, between which multiple glebal chambers are formed. Hyphal system composed of thick-walled generative and latex-bearing hyphae, staining brown in sulfobenzaldehyde, cystidia absent. Hymenium lining the chambers, composed of basidia interspersed with branched hyphae. Basidia clavate, 4-spored. Basidiospores globose to ellipsoidal, hyaline to pale yellow, symmetrical or with an eccentric hilum, thick-walled, with a prominent central oil droplet and ornamented with scattered spines that stain blue in iodine.

Significant Genera: *Hybogaster*.

Distribution: Only known from southern Chile.

Economic Significance: None is known.

Ecology: Associated with living *Nothofagus* trees, probably ectomycorrhizal.

Notes: The genus was contrasted with *Zelleromyces* and *Arcangeliella* (*Russulaceae*) and it is likely that the families are closely related. No molecular data are available.

References: Singer, R. (1964). New genera of fungi – XII: *Hybogaster*. *Sydowia* 17: 13–16.

Hydnaceae Chevall. 1826
Cantharellales: Basidiomycota

Basidiomata pileate, stipitate, the stalk central or eccentric, usually fleshy, whiteish to brown or orange, or thin and somewhat resupinate, velutinous or glabrescent. Hyphal system monomitic, hyphae with clamp connections, inflated between the septa, cystidia absent. Hymenium spinose, white to yellowish, fleshy and fragile. Basidia narrowly clavate, with 2–6 sterigmata, frequently aborting. Basidiospores subglobose to ovoid or ellipsoidal, hyaline, fairly thin-walled, smooth, not staining in iodine.

Significant Genera: *Hydnum*.

Distribution: Widespread.

Economic Significance: Some species (e.g. *Hydnum repandum*) are well-known edible fungi, others are reported as a cause of heartwood-rot in living trees.

Ecology: Saprobic or facultative parasites; most species form ectomycorrhizas.

Hydnum repandum; basidiomata

Gloeomucro chlorinus; basidiomata

Notes: The *Hydnaceae* occupies a consistent but unstable position within the cantharelloid clade of the homobasidiomycetes, characterized in part by the stichic arrangement of meiotic nuclei, with the spindle oriented vertically rather than transversely. Some molecular studies indicate that *Sistotrema* may belong here.

References: **Agerer, R.; Kraigher, H.; Javornik, B.** (1996). Identification of ectomycorrhizae of *Hydnum rufescens* on Norway spruce and the variability of the ITS region of *H. rufescens* and *H. repandum* (*Basidiomycetes*). *Nova Hedwigia* 63: 183–194; **Binder, M.; Hibbett, D.S.** (2002). Higher-level phylogenetic relationships of homobasidiomycetes (mushroom-forming fungi) inferred from four rDNA regions. *Mol. Phylogen. Evol.* 22: 76–90; **Gulden, G.; Hanssen, E.W.** (1992). Distribution and ecology of stipitate hydnaceous fungi in Norway, with special reference to the question of decline. *Sommerfeltia* 13: 58 pp.; **Harrison, K.A.; Grund, D.W.** (1987). Differences in European and North American stipitate Hydnums. *Mycotaxon* 28: 427–435; **Harrison, K.A.; Grund, D.W.** (1987). Preliminary keys to the terrestrial stipitate Hydnums of North America. *Mycotaxon* 28: 419–426; **Larsson, K.-H.; Larsson, E.; Koljalg, U.** (2004). High phylogenetic diversity among corticioid homobasidiomycetes. *Mycol. Res.* 108: 983–1002; **Pine, E.M.; Hibbett, D.S.; Donoghue, M.J.** (1999). Phylogenetic relationships of cantharelloid and clavarioid *Homobasidiomycetes* based on mitochondrial and nuclear rDNA sequences. *Mycologia* 91: 944–963.

Hydnangiaceae Gäum. & C.W. Dodge 1928
Agaricales: Basidiomycota

Basidiomata stipitate and pileate or gasteroid, epigeous or hypogeous; when pileate the cap glabrous to scaly, sometimes striate, generally orange-brown or violet in coloration, the gills widely spaced, thick and waxy; when gasteroid irregular, thin-walled, the peridium sometimes evanescent, columella absent or present, the hymenium not gelatinized, formed in locules. Hyphal system monomitic, hyphae often inflated, clamp connections present, cystidia generally absent. Basidia clavate, with 2 or 4 sterigmata, sometimes accompanied by cheilocystidia. Basidiospores discharged actively or passively, globose to cylindrical, rarely elongate, mostly echinulate but occasionally slightly roughened, the spines composed of radially oriented microtubules, hyaline, not staining in iodine.

Significant Genera: *Hydnangium, Laccaria*.

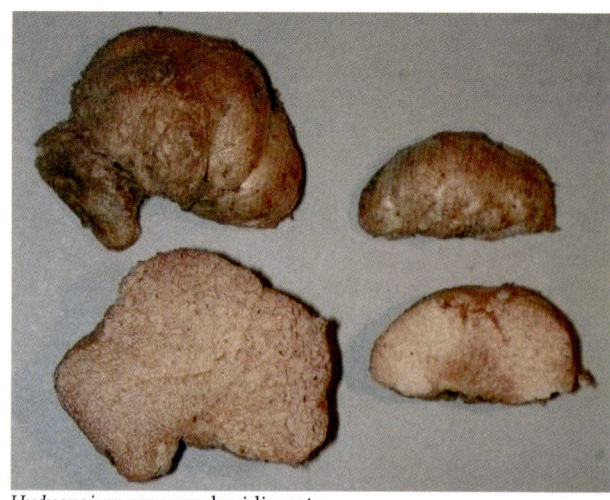

Hydnangium carneum; basidiomata

Distribution: Widespread in both temperate and tropical zones.

Economic Significance: None is known.

Ecology: Studied species are ectomycorrhizal, in soil of both coniferous and deciduous forests.

Laccaria amethystina, Surrey, UK; basidioma

(*Agaricales*) in the Continental United States and Canada, with discussions on extralimital taxa and descriptions of extant types. *Fieldiana* Bot. 30: 158 pp.; **Mueller, G.M.** (1997). Distribution and species composition of *Laccaria* (*Agaricales*) in tropical and subtropical America. *Revta Biol. trop.* 44: 131–135; **Mueller, G.M.** (1997). Designation of epitypes for *Laccaria proxima* and *Laccaria tortilis* (*Agaricales*). *Mycotaxon* 61: 205–207; **Mueller, G.M.; Ammirati, J.F.** (1993). Cytological studies in *Laccaria* (*Agaricales*). II. Assessing phylogenetic relationships among *Laccaria*, *Hydnangium*, and other *Agaricales*. *Am. J. Bot.* 80: 322–329; **Mueller, G.M.; Gardes, M.** (1991). Intra- and interspecific relations within *Laccaria bicolor sensu lato*. *Mycol. Res.* 95: 592–601; **Mueller, G.M.; Hofsten, A. von; Axén, A.; Strack, B.A.** (1994). Basidiospore wall ultrastructure of the false-truffle *Hydnangium* and its phylogenetic significance. *Mycologia* 85: 890–893; **Sweeney, M.; Harmey, M.A.; Mitchell, D.T.** (1996). Detection and identification of *Laccaria* species using a repeated DNA sequence from *Laccaria proxima*. *Mycol. Res.* 100: 1515–1521; **Wang, L.; Yang, Z.L.; Liu, J.H.** (2004). Two new species of *Laccaria* (*Basidiomycetes*) from China. *Nova Hedwigia* 79: 511–517.

Hygrophoraceae Lotsy 1907
Agaricales: Basidiomycota

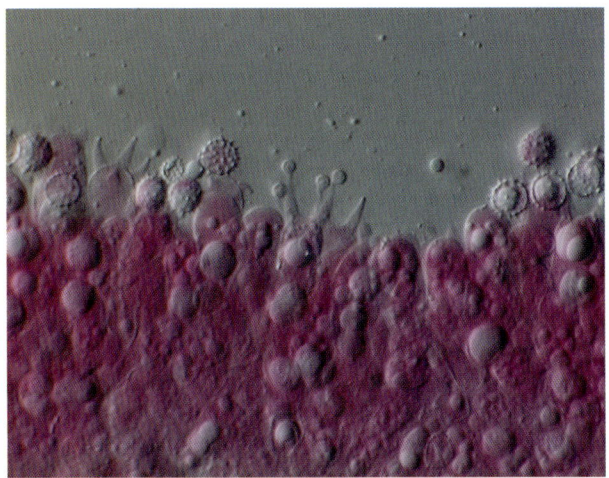
Laccaria amethystina, Surrey, UK; basidia and basidiospores

Hygrocybe conica, Surrey, UK; basidioma

Notes: Ultrastructural and molecular examination confirm the close relationship between the pileate genus *Laccaria* and the gasteroid *Hydnangium*, and some authors place them in synonymy.

References: **Bougher, N.L.; Lebel, T.** (2001). Sequestrate (trufflelike) fungi of Australia and New Zealand. *Aust. Syst. Bot.* 14: 439–484; **Fiore-Donno, A.-M.; Martin, F.** (2001). Populations of ectomycorrhizal *Laccaria amethystina* and *Xerocomus* spp. show contrasting colonization patterns in a mixed forest. *New Phytol.* 152: 533–542; **Gherbi, H.; Delaruelle, C.; Selosse, M.-A.; Martin, F.** (1999). High genetic diversity in a population of the ectomycorrhizal basidiomycete *Laccaria amethystina* in a 150-year-old beech forest. *Mol. Ecol.* 8: 2003–2013; **Henrion, B.; Battista C. di; Bouchard, D.; Vairelles, D.; Thompson, B.D.; Le Tacon, F.; Martin, F.** (1994). Monitoring the persistence of *Laccaria bicolor* as an ectomycorrhizal symbiont of nursery-grown Douglas fir by PCR of the rDNA intergenic spacer. *Mol. Ecol.* 3: 571–580; **Martin, F.; Selosse, M.-A.; Le Tacon, F.** (1999). The nuclear rDNA intergenic spacer of the ectomycorrhizal basidiomycete *Laccaria bicolor*: structural analysis and allelic polymorphism. *Microbiology* (Reading) 145: 1605–1611; **Mueller, G.M.** (1992). Systematics of *Laccaria*

Basidiomata pileate, stipitate, often brightly coloured with carotenoid pigments. Stipe fleshy to cartilaginous, sometimes with a veil, smooth or fibrose. Cap usually viscid to glutinous, glabrous or hairy, with an outer layer of radially arranged hyphae. Hyphal system monomitic, usually with clamp connections, hyphae not staining in iodine, cystidia normally absent. Hymenium lamellar, the gills waxy and thick. Basidia elongate, filamentous to narrowly clavate, with 2 or 4 sterigmata. Basidiospores varied in form, small to large, globose to cylindrical or ellipsoidal, thin-walled, almost always smooth, rarely staining in iodine.

Significant Genera: *Cuphophyllus*, *Hygrocybe*, *Hygrophorus*.

Distribution: Widespread in all ecological zones apart from in highly xeric conditions.

number of species have been investigated due to conservation concern in various European countries.

References: **Arnolds, E.** (1989). Notes on *Hygrophoraceae* – XI. Observations on some species of *Hygrocybe* subgenus *Cuphophyllus*. *Persoonia* 14: 43–46; **Beisenherz, M.** (2002). [On the ecology and taxonomy of *Hygrocybe*, *Agaricales*]. *Regensb. Mykol. Schr.* 10: 3–65; **Boertmann, D.** (1990). The identity of *Hygrocybe vitellina* and related species. *Nordic Jl Bot.* 10: 311–317; **Boertmann, D.** (1995). The genus *Hygrocybe*. *Fungi of Northern Europe* (Greve) 1: 184 pp.; **Boertmann, D.** (2002). *Index Hygrocybearum* A Catalogue to Names and Potential Names in Tribus *Hygrocybeae* Kühner (*Tricholomatales, Fungi*). *Biblthca Mycol.* 192: 168 pp.; **Bougher, N.L.; Young, A.M.** (1997). *Hygrophoraceae* of Western Australia. *Mycotaxon* 63: 25–35; **Cantrell, S.A.; Lodge, D.J.** (2004). *Hygrophoraceae* (*Agaricales*) of the Greater Antilles: *Hygrocybe* subgenus *Pseudohygrocybe* sections *Coccineae* and *Neohygrocybe*. *Mycol. Res.* 108: 1301–1314; **Chandrasrikul, A.; Saifa, Y.; Panichpol, D.** (1985). The Amanitas of Thailand. *Thai J. Agric. Sci.* 18: 287–295; **Horak, E.** (1990). Monograph of the New Zealand *Hygrophoraceae* (*Agaricales*). *N.Z. Jl Bot.* 28: 255–306; **Larsson, E.; Jacobsson, S.** (2004). Controversy over *Hygrophorus cossus* settled using ITS sequence data from 200 year-old type material. *Mycol. Res.* 108: 781–786; **Moncalvo, J.-M.; Vilgalys, R.; Redhead, S.A.; Johnson, J.E.; James, T.Y.; Aime, M.C.; Hofstetter, V.; Verduin, S.J.W.; Larsson, E.; Baroni, T.J.; Thorn, R.G.; Jacobsson, S.; Clemencon, H.; Miller, O.K.** (2002). One hundred and seventeen clades of euagarics. *Mol. Phylogen. Evol.* 23: 357–400; **Young, A.M.** (1999). A field key to the *Hygrophoraceae* of South-eastern Australia. *Australas. Mycol.* 18: 63–69; **Young, A.M.** (2003). Brief notes on the status of family *Hygrophoraceae* Lotsy. *Australas. Mycol.* 21: 114–116; **Young, A.M.** (2005). *Fungi of Australia:* Hygrophoraceae (Canberra): 188 pp.; **Young, A.M.; Mills, A.K.** (2002). The *Hygrophoraceae* of Tasmania. *Muelleria* 16: 3–28.

Hygrocybe calyptriformis; basidiomata

Hygrocybe pratensis; basidioma

Hygrocybe virginea; basidioma

Economic Significance: None is known.

Ecology: Ectomycorrhizal or saprobic, on the ground in woods, fields and moorland.

Notes: A mostly well-defined family, subsumed into the *Tricholomataceae* in the *Dictionary of the Fungi* edn 9 but clearly separable using molecular data. A

Hygrophoropsidaceae Kühner 1980
Boletales: Basidiomycota

Basidiomata pileate, usually stipitate, the cap flat to funnel-shaped, with an incurved margin at least when young, ± tomentose, reddish to yellowish, fleshy or soft-textured, the stipe central or eccentric, rarely absent, veil absent. Hymenium lamellar, the gills decurrent, repeatedly forked, thin. Hyphae with clamp connections, cystidia absent. Basidia small, clavate to cylindric-clavate, usually 4-spored. Basidiospores ellipsoidal to cylindrical, hyaline, smooth, thin-walled, not staining in iodine.

Significant Genera: *Hygrophoropsis*.

Distribution: Widespread.

Economic Significance: None is known.

Ecology: Apparently saprobic, not obligately mycorrhizal.

Notes: Separated from the *Paxillaceae*, primarily on chemical and anatomical features, the *Hygrophoropsidaceae* has been found to be polyphyletic based on preliminary molecular studies. *Austropaxillus* and *Tapinella* may not belong here as indicated in the *Dictionary of the Fungi* edn 9.

Hygrophoropsis aurantiaca; basidiomata

References: **Binder, M.; Besl, H.; Bresinsky, A.** (1997). [*Agaricales* or *Boletales*? Molecular evidence towards the classification of some controversial taxa]. *Z. Mykol.* 63: 189–196; **Bresinsky, A.; Jarosch, M.; Fischer, M.; Schönberger, I.; Wittmann-Bresinsky, B.** (1999). Phylogenetic relationships within *Paxillus* s.l. (*Basidiomycetes, Boletales*): separation of a southern hemisphere genus. *Pl. Biol.* 1: 327–333; **Šutara, J.** (1992). The genera *Paxillus* and *Tapinella* in Central Europe. *Česká Mykol.* 46: 50–56; **Tuthill, D.E.; Frisvad, J.C.** (2002). *Eupenicillium bovifimosum*, a new species from dry cow manure in Wyoming. *Mycologia* 94: 240–246; **Watling, R.** (2001). The relationships and possible distributional patterns of boletes in South-East Asia. *Mycol. Res.* 105: 1440–1448.

Hymeneliaceae Körb. 1855
Lecanorales: Ascomycota

Thallus varied but usually crustose, without rhizoids, sometimes evanescent. Ascomata apothecial, deeply immersed, with a poorly to well-developed margin, the disc pale or dark. Interascal tissue of paraphyses, often branched and moniliform at the apex. Asci with a well-developed usually J+ or J– apical cap, usually without an ocular chamber, with an outer layer of J+ gelatinous material. Ascospores large, aseptate, hyaline, thin-walled. Anamorphs pycnidial.

Significant Genera: *Hymenelia*.

Distribution: Widespread.

Economic Significance: None is known.

Ecology: Lichenized with green algae, usually on rocks.

Notes: Very poorly known; no recent information is available. *Aspicilia* is now accepted as a member of the *Megasporaceae* rather than this family.

References: **Buschbom, J.; Mueller, G.** (2004). Resolving evolutionary relationships in the lichen-forming genus *Porpidia* and related allies (*Porpidiaceae, Ascomycota*). *Mol. Phylogen. Evol.* 32: 66–82; **Lumbsch, H.T.** (1997). A comparison of ascoma ontogeny supports the inclusion of the *Eigleraceae* in the *Hymeneliaceae* (*Lecanorales*). *Bryologist* 100: 180–192; **Lutzoni, F.M.; Brodo, I.M.** (1995). A generic redelimitation of the *Ionaspis-Hymenelia* complex (lichenized *Ascomycotina*). *Syst. Bot.* 20: 224–258; **Reeb, V.; Lutzoni, F.; Roux, C.** (2004). Contribution of *RPB2* to multilocus phylogenetic studies of the euascomycetes (*Pezizomycotina, Fungi*) with special emphasis on the lichen-forming *Acarosporaceae* and evolution of polyspory. *Mol. Phylogen. Evol.* 32: 1036–1060; **Schmitt, I.; Yamamoto, Y.; Lumbsch, H.T.** (2006). Phylogeny of *Pertusariales* (*Ascomycotina*): resurrection of *Ochrolechiaceae* and new circumscription of *Megasporaceae*. *J. Hattori bot. Lab.* 100: 753–764.

Hymenochaetaceae Imazeki & Toki 1954
Hymenochaetales: Basidiomycota

Basidioma resupinate to pileate, rarely clavarioid, annual or perennial, stipe when present usually lateral or eccentric, often tomentose or hispid, usually brown, staining black with KOH, usually woody or fibrous in texture. Hyphal system monomitic or dimitic, clamp connections absent, cystidia absent. Hymenium smooth to poroid. Basidia clavate to cylindrical or widely ellipsoidal, 4-spored, interspersed with dichotomously branched sterile filaments or simple or rarely stellate setae. Basidiospores hyaline to yellowish or brown, usually smooth, rarely staining blue or brown in iodine.

Phellinus ferruginosus, Surrey, UK; basidioma

Significant Genera: *Hymenochaete, Inonotus, Phellinus*.

Distribution: Cosmopolitan.

Phellinus ferreus, Devon, UK; basidioma

Inonotus hispidus; basidioma on living tree

Hymenochaete rubiginosa; basidiomata

Notes: One of the most prominent families of polypores. It may be a sister family of the *Schizoporaceae*, though the latter group is polyphyletic according to recent studies.

References: **Binder, M.; Hibbett, D.S.; Larsson, K.-H.; Larsson, E.; Langer, E.; Langer, G.** (2005). The phylogenetic distribution of resupinate forms across the major clades of mushroom-forming fungi (Homobasidiomycetes). *Syst. Biodiv.* 3: 113–157; **Dai, Y.C.; Niemelä, T.; Zang, M.** (1997). Synopsis of the genus *Inonotus* (Basidiomycetes) sensu lato in China. *Mycotaxon* 65: 273–283; **Fischer, M.; Binder, M.** (2004). Species recognition, geographic distribution and host-pathogen relationships: a case study in a group of lignicolous basidiomycetes, *Phellinus* s.l. *Mycologia* 96: 799–811; **Gatica, M.; Césari, C.; Escoriaza, G.** (2004). *Phellinus* species inducing *hoja de malvón* symptoms on leaves and wood decay in mature field-grown grapevines. *Phytopath. Mediterr.* 43: 59–65; **Germain, H.; Laflamme, G.; Bernier, L.; Boulet, B.; Hamelin, R.C.** (2002). DNA polymorphism and molecular diagnosis in *Inonotus* spp. *Can. J. Pl. Path.* 24: 194–199; **Góes-Neto, A.; Loguercio-Leite, C.; Guerrero, R.T.** (2002). Molecular phylogeny of tropical *Hymenochaetales* (Basidiomycota). *Mycotaxon* 84: 337–354; **Gottlieb, A.M.; Wright, J.E.; Moncalvo, J.-M.** (2002). *Inonotus* s.l. in Argentina – morphology, cultural characters and molecular analysis. *Mycol. Prog.* 1: 299–313; **Kauserud, H.; Schumacher, T.** (2002). Population structure of the endangered wood decay fungus *Phellinus nigrolimitatus* (Basidiomycota). *Can. J. Bot.* 80: 597–606; **Küffer, N.; Senn-Irlet, B.** (2005). Diversity and ecology of wood-inhabiting aphyllophoroid basidiomycetes on fallen woody debris in various forest types in Switzerland. *Mycol. Prog.* 4: 77–86; **Müller, W.H.; Stalpers, J.A.; Aelst, A.C. van; Jong, M.D.M. de; Krift, T.P. van der; Boekhout, T.** (2000). The taxonomic position of *Asterodon*, *Asterostroma* and *Coltricia* inferred from the septal pore cap ultrastructure. *Mycol. Res.* 104: 1485–1492; **Nam, B.H.; Lee, J.Y.; Kim, G.Y.; Jung, H.H.; Park, H.S.; Kim, C.Y.; Jo, W.S.; Jeong, S.J.; Lee, T.H.; Lee, J.D.** (2003). Phylogenetic analysis and rapid detection of genus *Phellinus* using the nucleotide sequences of 18S ribosomal RNA. *Mycobiology* 31: 133–138; **Parmasto, E.** (2001). Hymenochaetoid fungi (Basidiomycota) of North America. *Mycotaxon* 79: 107–176; **Parmasto, E.** (2005). New data on rare species of *Hydnochaete* and *Hymenochaete* (Hymenochaetales). *Mycotaxon* 91: 137–163; **Wagner, T.** (2001). Phylogenetic relationships of *Asterodon* and *Asterostroma* (Basidiomycetes), two genera with asterosetae. *Mycotaxon* 79: 235–246; **Wagner, T.; Fischer, M.** (2001). Natural groups and a revised system for the European poroid *Hymenochaetales* (Basidiomycota) supported by nLSU rDNA sequence data. *Mycol. Res.* 105: 773–782; **Wagner, T.; Fischer, M.** (2002). Classification and phylogenetic relationships of *Hymenochaete* and allied genera of the *Hymenochaetales*, inferred from rDNA sequence data and nuclear behaviour of vegetative mycelium. *Mycol. Prog.* 1: 93–104; **Wagner, T.; Ryvarden, L.** (2002). Phylogeny and taxonomy of the genus *Phylloporia* (Hymenochaetales). *Mycol. Prog.* 1: 105–116.

Economic Significance: Species of *Inonotus* and *Phellinus* are implicated in many diseases of broadleaved and coniferous forest trees, causing heart rot, canker and root disease, also esca disease of grapevine.

Ecology: Lignicolous, causing white rots, rarely ectomycorrhizal.

Hymenogasteraceae Vittad. 1831
Boletales: Basidiomycota

Basidiomata gasteroid, globose to irregular, multiloculate, hypogeous, whiteish to brown or olivaceous; peridium usually smooth, dry or viscid, rarely fibrose, thin and not easily separable from internal tissues. Columella variably developed, sometimes almost absent, in other genera present as a basal pad or rarely dendritic veins. Hyphal system typically monomitic with inflated hyphae, clamp connections usually present, cystidia normally absent. Hymenium lining the locules. Basidia clavate to cylindric-clavate, 2- to 4-spored, discharging spores passively. Basidiospores symmetrical or asymmetrical, ellipsoidal, cylindrical, fusiform or citriform, brown, initially smooth but in most species developing verrucose, reticulate or irregularly ridged ornamentation, sometimes with an external sheath, not staining in iodine.

Significant Genera: *Hymenogaster*.

Hymenogaster rehsteineri; basidiomata

Hymenogaster arenarius; basidiomata

Distribution: The group has a centre of diversity in Australasia, but overall a widespread distribution especially in temperate zones. *Descomyces albellus* has been widely introduced with its *Eucalyptus* mycorrhizal partners.

Economic Significance: None is known.

Ecology: Ectomycorrhizal with a range of woody plant species, especially *Myrtaceae*.

Notes: The family should probably be united with the *Cortinariaceae*, but links with the *Bolbitiaceae* also need to be explored.

References: **Binder, M.; Bresinsky, A.** (2002). Derivation of a polymorphic lineage of gasteromycetes from boletoid ancestors. *Mycologia* 94: 85–98; **Bougher, N.L.; Castellano, M.A.** (1993). Delimitation of *Hymenogaster sensu stricto* and four new segregate genera. *Mycologia* 85: 273–293; **Bougher, N.L.; Lebel, T.** (2001). Sequestrate (truffle-like) fungi of Australia and New Zealand. *Aust. Syst. Bot.* 14: 439–484; **Fogel, R.** (1985). Studies on *Hymenogaster* (*Basidiomycotina*): A re-evaluation of the subgenus *Dendrogaster*. *Mycologia* 77: 72–82; **Francis, A.A.; Bougher, N.L.** (2004). Cortinarioid sequestrate (truffle-like) fungi of Western Australia. *Australas. Mycol.* 23: 1–26; **Moreau, P.-A.; Peintner, U.; Gardes, M.** (2006). Phylogeny of the ectomycorrhizal mushroom genus *Alnicola* (*Basidiomycota*, *Cortinariaceae*) based on rDNA sequences with special emphasis on host specificity and morphological characters. *Mol. Phylogen. Evol.* 38: 794–807; **Pegler, D.N.; Young, T.W.K.** (1987). A reassessment of the British species of *Hymenogaster* (*Basidiomycota*: *Cortinariales*). *Notes R. bot. Gdn Edinb.* 44: 437–458; **Peintner, U.; Bougher, N.L.; Castellano, M.A.; Moncalvo, M.-C.; Moser, M.M.; Trappe, J.M.; Vilgalys, R.** (2001). Multiple origins of sequestrate fungi related to *Cortinarius* (*Cortinariaceae*). *Am. J. Bot.* 88: 2168–2179; **Peintner, U.; Moncalvo, J.-M.; Vilgalys, R.** (2004). Towards a better understanding of the infrageneric relationships in *Cortinarius* (*Agaricales*, *Basidiomycota*). *Mycologia* 96: 1042–1058; **Reijnders, A.F.M.** (2000). A morphogenetic analysis of the basic characters of the gasteromycetes and their relation to other basidiomycetes. *Mycol. Res.* 104: 900–910.

Hyphodermataceae Jülich 1982
Polyporales: Basidiomycota

Basidioma resupinate or effuse, thin or thick, membranous, gelatinous or waxy, smooth, pilose or warted, usually pale, yellowish or brownish. Hyphal system monomitic, clamp connections present or lacking, cystidia usually present, often thick-walled and rarely encrusted. Hymenium smooth to tuberculate or spinose. Basidia usually large, cylindrical, clavate or urniform, sometimes becoming septate, usually 4-spored. Basidiospores globose, allantoid, cylindrical or lacrimiform, smooth or rarely verrucose, not staining in iodine.

Significant Genera: *Hyphoderma*, *Hypochnicium*.

Distribution: Widespread, especially in temperate regions.

Economic Significance: None is known apart from occasional reports of damage to amenity timber.

Ecology: Most are saprobes causing white rots on decaying wood, herbaceous stems etc., though some affect living trees. Some species appear to be nematophagous, possessing special capture structures termed stephanocysts.

Notes: The family is polyphyletic and *Hyphoderma* itself, as currently accepted, includes species in both the phlebioid and hymenochaetoid clades of the homobasidiomycetes. Ultrastructural differences in septal structure provide further evidence of the heterogeneity of the family. Anamorphic forms referred to *Aegerita* sometimes accompany the teleomorph.

Gyrophanopsis polonensis; basidioma

References: **Ginns, J.** (1998). Genera of the North American *Corticiaceae sensu lato*. *Mycologia* 90: 1–35; **Hallenberg, N.** (1990). Ultrastructure of stephanocysts and basidiospores in *Hyphoderma praetermissum*. *Mycol. Res.* 94: 1090–1095; **Hjortstam, K.** (2001). Two new species of *Brevicellicium* and a survey of tropical and subtropical species in the genus (*Basidiomycotina, Aphyllophorales*). *Mycotaxon* 79: 181–187; **Küffer, N.; Senn-Irlet, B.** (2005). Diversity and ecology of wood-inhabiting aphyllophoroid basidiomycetes on fallen woody debris in various forest types in Switzerland. *Mycol. Prog.* 4: 77–86; **Langer, E.** (1998). Evolution of *Hyphodontia* (*Corticiaceae, Basidiomycetes*) and related *Aphyllophorales* inferred from ribosomal DNA sequences. *Folia cryptog. Estonica* 33: 57–62; **Larsson, K.-H.** (2007). Molecular phylogeny of *Hyphoderma* and the reinstatement of *Peniophorella*. *Mycological Research* 111: 186–195; **Larsson, K.-H.; Larsson, E.; Koljalg, U.** (2004). High phylogenetic diversity among corticioid homobasidiomycetes. *Mycol. Res.* 108: 983–1002; **Nilsson, R.H.; Hallenberg, N.; Nordén, B.; Maekawa, N.; Wu, S.H.** (2003). Phylogeography of *Hyphoderma setigerum* (*Basidiomycota*) in the Northern Hemisphere. *Mycol. Res.* 107: 645–652; **Tzean, S.S.; Liou, J.Y.** (1993). Nematophagous resupinate basidiomycetous fungi. *Phytopathology* 83: 1015–1020.

Hypocreaceae De Not. 1844
Hypocreales: Ascomycota

Stromata discrete or effuse, rarely club-shaped, usually brightly coloured, fleshy or formed from loose subicular hyphae. Ascomata brightly coloured, perithecial, rarely non-ostiolate. Asci clavate to cylindrical, thin-walled, not fissitunicate, often with a J– apical ring. Ascospores hyaline or green, 1-septate, often disarticulating, sometimes ornamented. Anamorphs prominent, hyphomycetous, varied in morphology, with different taxa forming resting spores and disseminative conidia. Resting spores hyaline or pigmented, thick-walled, often ornamented. Disseminative conidia small, often rough-walled, produced from elongate percurrently proliferating conidiogenous cells on branched conidiophores.

Significant Genera: *Hypocrea*, *Hypomyces*. Anamorphs: *Sepedonium*, *Trichoderma*, *Verticillium*.

Distribution: Cosmopolitan, common in all environmental zones.

Economic Significance: A number of mycoparasitic species of *Trichoderma* and related genera are used in biocontrol programmes and there is currently great interest in the use of proteins derived from these fungi as novel fungicides. A number of species of *Trichoderma* produce antibiotics and some are serious pests in the commercial mushroom industry.

Hypocrea aff. *crassa*, Thailand; stromata

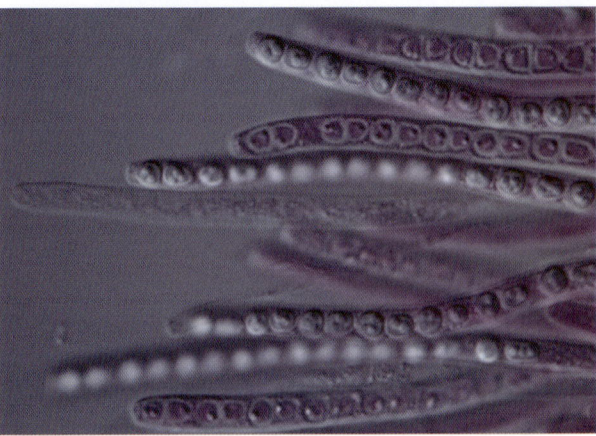

Podostroma alutaceum; asci and ascospores

Ecology: Saprobic on rotting wood and other vegetation or parasitic on other fungi, rarely coprophilous. Many species are prominent components of the soil biota and are frequently encountered in isolation programmes.

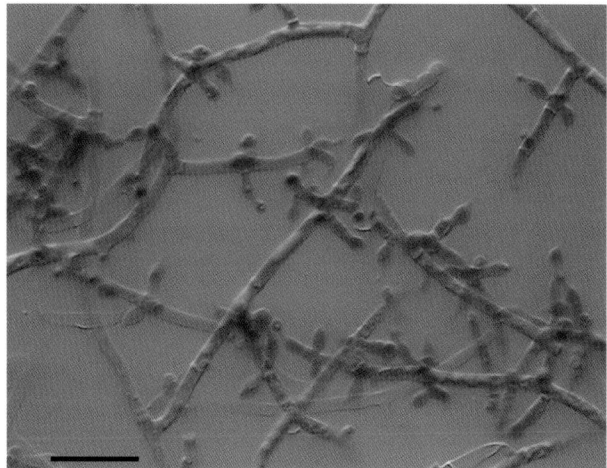

Trichoderma atroviride, UK; conidiophores and conidia

Notes: Phylogenetic studies have concentrated on relationships within genera, rather than the systematics of the family as a whole. The web-based keys to *Hypomyces* and *Trichoderma* (USDA, ARS, Beltsville) are models of modern phenetic classification systems.

References: **Bissett, J.** (1992). A revision of the genus *Trichoderma*. II. Infrageneric classification. *Can. J. Bot.* 69: 2357–2372; **Bissett, J.** (1992). A revision of the genus *Trichoderma*. III. Section *Pachybasium*. *Can. J. Bot.* 69: 2373–2417; **Bulat, S.A.; Lübeck, M.; Mironenko, N.; Jensen, D.F.; Lübeck, P.S.** (1998). UP-PCR analysis and ITS1 ribotyping of strains of *Trichoderma* and *Gliocladium*. *Mycol. Res.* 102: 933–943; **Castle, A.; Speranzini, D.; Rghei, N.; Alm, G.; Rinker, D.; Bissett, J.** (1998). Morphological and molecular identification of *Trichoderma* isolates on North American mushroom farms. *Appl. Environm. Microbiol.* 64: 133–137; **Chen, X.; Romaine, C.P.; Ospina-Giraldo, M.D.; Royse, D.J.** (1999). A polymerase chain reaction-based test for the identification of *Trichoderma harzianum* biotypes 2 and 4, responsible for the worldwide green mold epidemic in cultivated *Agaricus bisporus*. *Appl. Microbiol. Biotechn.* 52: 246–250; **Douhan, G.W.; Rizzo, D.M.** (2004). Host-parasite relationships among bolete infecting *Hypomyces* species. *Mycol. Res.* 107: 1342–1349; **Druzhinina, I.S.; Kopchinskiy, A.G.; Kubicek, C.P.** (2006). The first 100 *Trichoderma* species characterized by molecular data. *Mycoscience* 47: 55–64; **Gams, W. (ed.)** (2006). *Hypocrea* and *Trichoderma* studies marking the 90th birthday of Joan M. Dingley. *Stud. Mycol.* 56: 177 pp.; **Gams, W.; Bissett, J.** (1998). Morphology and identification of *Trichoderma*. *Trichoderma and Gliocladium* Vol. 1. Basic Biology, Taxonomy and Genetics (London): 3–34; **Grondona, I.; Hermosa, R.; Tejada, M.; Gomis, M.D.; Mateos, P.F.; Bridge, P.D.; Monte, E.; García-Acha, I.** (1997). Physiological and biochemical characterization of *Trichoderma harzianum*, a biological control agent against soilborne fungal plant pathogens. *Appl. Environm. Microbiol.* 63: 3189–3198; **Harman, G.E.; Hayes, C.K.; Ondik, K.L.** (1998). Asexual genetics in *Trichoderma* and *Gliocladium*: mechanisms and implications. *Trichoderma and Gliocladium* Vol. 1. Basic Biology, Taxonomy and Genetics (London): 243–270; **Hermosa, M.R.; Grondona, I.; Iturriaga, E.A.; Diaz-Minguez, J.M.; Castro, C.; Monte, E.; Garcia-Acha, I.** (2000). Molecular characterization and identification of biocontrol isolates of *Trichoderma* spp. *Appl. Environm. Microbiol.* 66: 1890–1898; **Kindermann, J.; El-Ayouti, Y.; Samuels, G.J.; Kubicek, C.P.** (1998). Phylogeny of the genus *Trichoderma* based on sequence analysis of the internal transcribed spacer region 1 of the rDNA cluster. *Fungal Genetics Biol.* 24: 298–309; **Kuhls, K.; Lieckfeldt, E.; Samuels, G.J.; Meyer, W.; Kubicek, C.P.; Börner, T.** (1997). Revision of *Trichoderma* sect. *Longibrachiatum* including related teleomorphs based on analysis of ribosomal DNA internal transcribed spacer sequences. *Mycologia* 89: 442–460; **Lieckfeldt, E.; Cavignac, Y.; Fekete, C.; Börner, T.** (2000). Endochitinase gene-based phylogenetic analysis of *Trichoderma*. *Microbiol. Res.* 155: 7–15; **Lieckfeldt, E.; Kuhls, K.; Muthumeenakshi, S.** (1998). Molecular taxonomy of *Trichoderma* and *Gliocladium* and their teleomorphs. *Trichoderma and Gliocladium* Vol. 1. Basic Biology, Taxonomy and Genetics (London): 35–56; **Lieckfeldt, E.; Samuels, G.J.; Börner, T.; Gams, W.** (1999). *Trichoderma koningii*: neotypification and *Hypocrea* teleomorph. *Can. J. Bot.* 76: 1507–1522; **Lu, B.; Druzhinina, I.S.; Fallah, P.; Chaverri, P.; Gradinger, C.; Kubicek, C.P.; Samuels, G.J.** (2004). *Hypocrea/Trichoderma* species with *Pachybasium*-like conidiophores: teleomorphs for *T. minutisporum* and *T. polysporum* and their newly discovered relatives. *Mycologia* 96: 310–342; **McKay, G.; Egan, D.; Morris, E.; Scott, C.; Brown, A.E.** (1999). Genetic and morphological characterization of *Cladobotryum* species causing cobweb disease of mushrooms. *Appl. Environm. Microbiol.* 65: 606–610; **Põldmaa, K.** (2000). Generic delimitation of the fungicolous *Hypocreaceae*. *Stud. Mycol.* 45: 83–94; **Põldmaa, K.; Larsson, E.; Kõljalg, U.** (1999). Phylogenetic relationships in *Hypomyces* and allied genera, with emphasis on species growing on wood-decaying homobasidiomycetes. *Can. J. Bot.* 77: 1756–1768; **Rossman, A.Y.** (2000). Towards monophyletic genera in the holomorphic Hypocreales. *Stud. Mycol.* 45: 27–34; **Sahr, T.; Ammer, H.; Besl, H.; Fischer, M.** (1999). Infrageneric classification of the boleticolous genus *Sepedonium*: species delimitation and phylogenetic relationships. *Mycologia* 91: 935–943; **Samuels, G.J.** (1996). *Trichoderma*: a review of biology and systematics of the genus. *Mycol. Res.* 100: 923–935.

Hyponectriaceae Petr. 1923
Xylariales: Ascomycota

Stromatic tissues reduced, usually either clypeate or absent. Ascomata immersed or erumpent, perithecial, usually thin-walled, the ostiole papillate, sometimes periphysate. Interascal tissue of narrow or wide thin-walled paraphyses. Asci cylindrical, persistent, thin-walled, with a small J+ or J– apical ring. Ascospores variously shaped, hyaline to pale brown, simple or transversely septate, sometimes thick-walled, sometimes with a mucous sheath. Anamorphs generally unknown.

Significant Genera: *Hyponectria, Physalospora*.

Distribution: Cosmopolitan.

Economic Significance: None is known.

Ecology: Saprobic or necrotrophic in herbaceous and woody plant material.

Notes: Many of the included genera are poorly known, especially *Physalospora*. That genus has been used in several senses and confused with *Botryosphaeria* (Botryosphaeriaceae). *Monographella* (anamorph *Microdochium*) species are well-known, *Fusarium*-like, cereal pathogens; they may belong to this family but molecular data are needed to confirm the relationship.

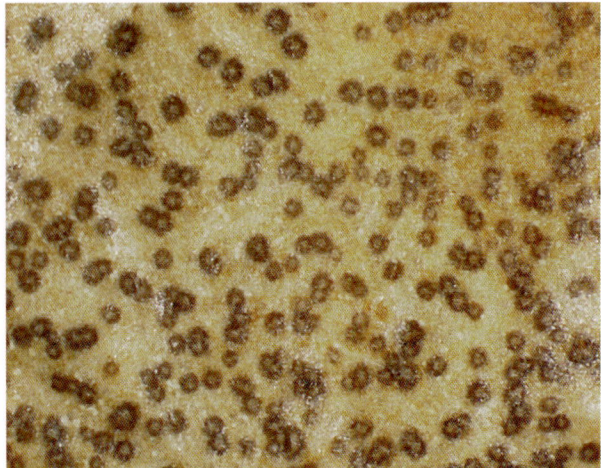

Hyponectria buxi, Derbyshire, UK; ascomata in dead leaf

Hyponectria buxi, Surrey, UK; asci and ascospores

References: **Hyde, K.D.; Fröhlich, J.; Taylor, J.E.** (1998). Fungi from palms. XXXVI. Reflections on unitunicate ascomycetes with apiospores. *Sydowia* 50: 21–80; **Jeewon, R.; Liew, E.C.; Hyde, K.D.** (2003). Molecular systematics of the *Amphisphaeriaceae* based on cladistic analyses of partial LSU rDNA gene sequences. *Mycol. Res.* 107: 1392–1402; **Mahuku, G.S.; Hsiang, T.; Yang, L.** (1998). Genetic diversity of *Microdochium nivale* isolates from turfgrass. *Mycol. Res.* 102: 559–567; **Samuels, G.J.; Müller, E.; Petrini, O.** (1987). Studies in the *Amphisphaeriaceae* (*sensu lato*) 3. New species of *Monographella* and *Pestalosphaeria*, and two new genera. *Mycotaxon* 28: 473–499; **Wang, Y.Z.; Hyde, K.D.** (1999). *Hyponectria buxi* with notes on the *Hyponectriaceae*. *Fungal Diversity* 3: 159–172; **Winka, K.; Eriksson, O.E.** (2000). *Papulosa amerospora* accommodated in a new family (*Papulosaceae, Sordariomycetes, Ascomycota*) inferred from morphological and molecular data. *Mycoscience* 41: 97–103.

Hypsostromataceae Huhndorf 1994
Uncertain position within Dothideomycetes, Ascomycota

Ascomata large, perithecial, vertically elongate, superficial, sometimes on a basal subiculum; wall soft-textured, of pseudoparenchymatous cells. Interascal tissue of trabeculate pseudoparaphyses. Asci long-stalked, with an apical chamber and fluorescing ring, dehiscence unknown. Ascospores fusiform to cylindrical, septate, sometimes fragmenting, brown, sometimes with a germ slit. Anamorph coelomycetous.

Significant Genera: *Manglicola, Hypsostroma*.

Distribution: Restricted to the neotropics.

Economic Significance: None is known.

Ecology: Saprobic on wood.

Notes: Very poorly known; no molecular information is available.

References: **Huhndorf, S.M.** (1992). Neotropical ascomycetes 2. *Hypsostroma*, a new genus from the Dominican Republic and Venezuela. *Mycologia* 84: 750–758; **Huhndorf, S.M.** (1994). Neotropical *Ascomycetes* 5. *Hypsostromataceae*, a new family of *Loculoascomycetes* and *Manglicola samuelsii*, a new species from Guyana. *Mycologia* 86: 266–269.

Hysterangiaceae E. Fisch. 1899
Hysterangiales: Basidiomycota

Basidiocarp gasteroid, usually remaining hypogeous at maturity, globose to tuberiform, often cartilaginous. Hyphal system monomitic, the hyphae often gelatinized and inflated, clamp connections usually absent; peridium usually pale, smooth, clearly differentiated from the fertile part (gleba), composed of hyphal or pseudoparenchymatous tissue, rarely two-layered. Gleba olivaceous grey with radiating tramal plates, gelatinous, with cavities or sterile locules sometimes formed, columella cartilaginous, finally deliquescent and foetid. Basidia cylindrical to clavate, sometimes elongate, 2- to 8-spored. Basidiospores discharged passively, ellipsoidal to cylindrical or fusiform, hyaline to greenish or olivaceous, smooth or with an often wrinkled epispore, staining brown or not in iodine.

Significant Genera: *Gallacea, Hysterangium*.

Distribution: Widespread in both temperate zones and the tropics.

Economic Significance: None is known.

Ecology: Ectomycorrhizal or saprobic, found in the surface layers of forest soil.

Notes: Few molecular data are available, but the family is confirmed as a member of the *Phallales*.

Gallacea scleroderma, New Zealand; basidiomata

References: **Beaton, G.; Pegler, D.N.; Young, T.W.K.** (1985). Gasteroid *Basidiomycota* of Victoria State, Australia: 4. *Hysterangium*. *Kew Bull.* 40: 435–444; **Binder, M.; Bresinsky, A.** (2002). Derivation of a polymorphic lineage of gasteromycetes from boletoid ancestors. *Mycologia* 94: 85–98; **Calonge, F.D.** (1998). *Gasteromycetes*, I. *Lycoperdales, Nidulariales, Phallales, Sclerodermatales, Tulostomatales. Fl. Mycol. Iberica* 3: 271 pp.; **Castellano, M.A.** (1990). The new genus *Trappea* (*Basidiomycotina, Hysterangiaceae*), a segregate from *Hysterangium*. *Mycotaxon* 38: 1–9; **Castellano, M.A.** (1999). *Hysterangium*. *Ectomycorrhizal Fungi* (Berlin): 311–323; **Castellano, M.A.; Beever, R.E.** (1994). Truffle-like *Basidiomycotina* of New Zealand: *Gallacea, Hysterangium, Phallobata*, and *Protubera*. *N.Z. Jl Bot.* 32: 305–328; **Castellano, M.A.; Muchovej, J.J.** (1996). Truffle-like fungi from South America: *Hysterangium sensu lato. Mycotaxon* 57: 329–345; **Hibbett et al.** (2007). A higher-level phylogenetic classification of the *Fungi*. *Mycol. Res.* 111(6): 949–959; **Hosaka et al.** (2007). Molecular phylogenetics of the gomphoid-phalloid fungi with an establishment of the new subclass *Phallomycetidae* and two new orders. *Mycologia* 98(6): 949–959; **Humpert, A.J.; Muench, E.L.; Giachini, A.J.; Castellano, M.A.; Spatafora, J.W.** (2001). Molecular phylogenetics of *Ramaria* and related genera: evidence from nuclear large subunit and mitochondrial small subunit rDNA sequences. *Mycologia* 93: 465–477; **Lebel, T.; Castellano, M.A.** (1999). Australasian truffle-like fungi. IX. History and current trends in the study of the taxonomy of sequestrate macrofungi from Australia and New Zealand. *Aust. Syst. Bot.* 12: 803–817.

Hysteriaceae Chevall. 1826
Hysteriales: Ascomycota

Ascostromata perithecial, erumpent or superficial, often aggregated, elongated, sometimes branched, wider than tall, opening by a longitudinal split; peridium black, very thick-walled, composed of small pseudoparenchymatous cells. Interascal tissue of narrow cellular pseudoparaphyses. Asci cylindrical, fissitunicate, with a distinct ocular chamber. Ascospores hyaline to brown, variously septate, sometimes with a mucous sheath. Anamorphs varied.

Significant Genera: *Hysterium, Hysterographium*. Anamorph: *Acrogenospora*.

Distribution: Widespread, especially in temperate regions.

Economic Significance: None is known.

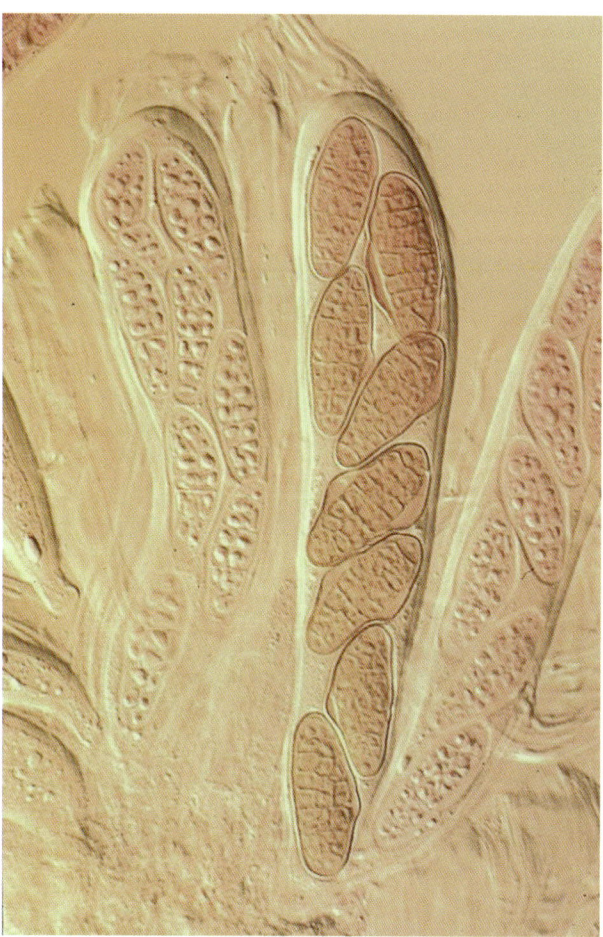

Hysterographium fraxini, Dyfed, UK; asci and ascospores

Hysterium angustatum, Cornwall, UK; ascostromata on bare wood

Ecology: Most species are saprobic on wood and bark; a few are lichenicolous.

Notes: Ascostromata are long-lived, making the family easy to survey. The limited molecular data available suggest a relationship with the *Dothideales*, but more research is needed.

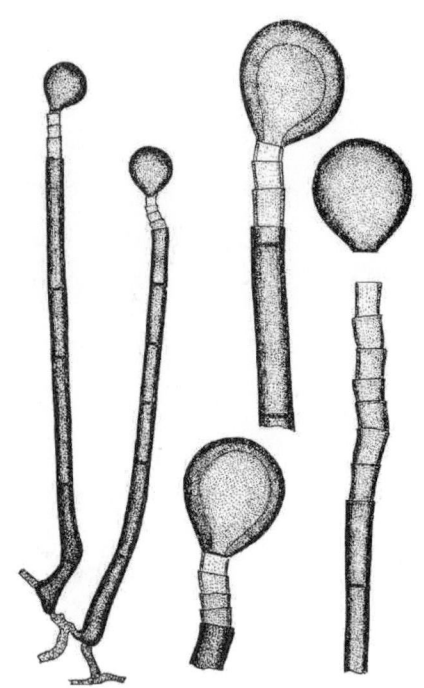

Acrogenospora sphaerocephala; conidiophores and conidia

References: **Checa, J.** (1997). Annotated list of the *Hysteriaceae* (*Dothideales, Ascomycotina*) reported from the Iberian Peninsula and Balearic Islands. *Mycotaxon* 62: 349–374; **Diederich, P.; Wedin, M.** (2000). The species of *Hemigrapha* (lichenicolous Ascomycetes, Dothideales) on *Peltigerales*. *Nordic Jl Bot.* 20: 203–214; **Goh, T.K.; Hyde, K.D.; Tsui, K.M.** (1998). The hyphomycete genus *Acrogenospora*, with two new species and two new combinations. *Mycol. Res.* 102: 1309–1315; **Larios, J.M.; Honrubia, M.** (1988). *Hysteriaceae* Chev. y *Lophiaceae* Zogg ex Arx & Müller (*Ascomycotina*) en el sudeste de la Península Ibérica. *Revta Ibér. Micol.* 5: 111–117; **Liew, E.C.Y.; Aptroot, A.; Hyde, K.D.** (2000). Phylogenetic significance of the pseudoparaphyses in loculoascomycete taxonomy. *Mol. Phylogen. Evol.* 16: 392–402; **Lorenzo, L.E.; Messuti, M.I.** (1998). Noteworthy *Hysteriaceae* from southern South America. *Mycol. Res.* 102: 1101–1107; **Renobales, G.; Aguirre, B.** (1990). The nomenclature and systematic position of the genus *Encephalographa*. *Syst. Ascom.* 8: 87–92; **Sivanesan, A.; Rajak, R.C.; Gupta, R.C.** (1988). New species of *Gloniella* from India. *Trans. Br. mycol. Soc.* 90: 665–670; **Tretiach, M.; Modenesi, P.** (1999). Critical notes on the lichen genus *Encephalographa* A. Massal. (?*Hysteriaceae*). *Nova Hedwigia* 68: 527–544; **Vasil'eva, L.N.** (1999). Hysteriaceous fungi in the Russian Far East. II. The genus *Hysterographium*. *Mikol. Fitopatol.* 33: 297–301.

Icmadophilaceae Triebel 1993
Uncertain position within Ascomycetes, Ascomycota

Thallus usually crustose or squamulose. Ascomata apothecial, sessile or shortly stipitate, sometimes clustered, formed on specialized often non-lichenized thalline branches, flat or convex. Interascal tissue of simple or sparingly branched paraphyses, often swollen at the apices. Asci persistent, cylindrical, with a J+ apical cap. Ascospores hyaline, simple or transversely septate. Anamorph pycnidial.

Significant Genera: *Dibaeis, Icmadophila, Siphula*.

Distribution: Widespread.

Economic Significance: None is known.

Ecology: Lichenized with green algae, generally on soil or wood.

Notes: Molecular data suggest that the family occupies an isolated position within the *Lecanoromycetes*. The *Baeomycetaceae* may be related to the *Icmadophilaceae* as originally supposed.

Icmadophila ericetorum, Inverness-shire, Scotland; thallus and ascomata

References: **Galloway, D.J.** (2000). *Knightiella* belongs in *Icmadophila* (Helotiales, Icmadophilaceae). *Lichenologist* 32: 294–297; **Kantvilas, G.** (1996). Studies on the lichen genus *Siphula* in Tasmania I. *S. complanata* and its allies. *Herzogia* 12: 7–22; **Lumbsch, H.T.; Schmitt, I.; Palice, Z.; Wiklund, E.; Ekman, S.; Wedin, M.** (2004). Supraordinal phylogenetic relationships of *Lecanoromycetes* based on a Bayesian analysis of combined nuclear and mitochondrial sequences. *Mol. Phylogen. Evol.* 31: 822–832; **Platt, J.L.; Spatafora, J.W.** (1999). A re-examination of generic concepts of baeomycetoid lichens based on phylogenetic analyses of nuclear SSU and LSU ribosomal DNA. *Lichenologist* 31: 409–418; **Platt, J.L.; Spatafora, J.W.** (2000). Evolutionary relationships of nonsexual lichenized fungi: molecular phylogenetic hypotheses for the genera *Siphula* and *Thamnolia* from SSU and LSU rDNA. *Mycologia* 92: 475–487; **Reeb, V.; Lutzoni, F.; Roux, C.** (2004). Contribution of *RPB2* to multilocus phylogenetic studies of the euascomycetes (Pezizomycotina, Fungi) with special emphasis on the lichen-forming Acarosporaceae and evolution of polyspory. *Mol. Phylogen. Evol.* 32: 1036–1060; **Stenroos, S.K.; DePriest, P.T.** (1998). SSU rDNA phylogeny of cladoniiform lichens. *Am. J. Bot.* 85: 1548–1559; **Wedin, M.; Wiklund, E.; Crewe, A.; Döring, H.; Ekman, S.; Nyberg, A.; Schmitt, I.; Lumbsch, H.T.** (2005). Phylogenetic relationships of *Lecanoromycetes* (Ascomycota) as revealed by sequences of mtSSU and nLSU rDNA sequence data. *Mycol. Res.* 109: 159–172.

Inocybaceae Jülich 1982
Agaricales: Basidiomycota

Basidiomata pileate or rarely sequestrate, membranous to fleshy, often brightly coloured, typically yellowish or brownish, the cap where present usually fibrose or scaly, rarely glutinous, sometimes striate. Pileate taxa lamellate, the cap conical to convex, the gills sometimes fimbriate due to presence of cheilocystidia, the stipe cylindrical or bulbose, sometimes with a partial veil. Hyphae often staining in iodine, clamp connections present, cystidia usually present, varied in form. Basidia 2- or 4-spored. Basidiospores brown, variously shaped, thin-walled or thick-walled, usually without a germ pore, smooth or ornamented, sometimes staining in iodine. Hyphal system monomitic, clamp connections usually present.

Significant Genera: *Auritella, Inocybe*.

Inocybe rimosa; basidiomata

Distribution: Widespread in both tropical and temperate regions.

Economic Significance: None is known.

Ecology: Forming ectomycorrhizas with both angiosperm and gymnosperm trees, terrestrial or on very rotten wood.

Notes: Separated from the *Cortinariaceae* primarily using molecular evidence; many morphological features are shared with that family or at least do not allow effective separation. Some species of *Inocybe* are poisonous, containing muscarine compounds; at least one is considered edible in countries in south east Africa.

References: **Kropp, B.R.; Matheny, P.B.** (2004). Basidiospore homoplasy and variation in the *Inocybe chelanensis* group in North America. *Mycologia* 96: 295–309; **Kupyer, T.W.** (1986). A revision of the genus *Inocybe* in Europe: I. Subgenus *Inosperma* and the smooth-spored species of subgenus *Inocybe*. *Persoonia* Suppl. 3: 1–247; **Matheny, P.B.** (2005). Improving phylogenetic inference of mushrooms with RPB1 and RPB2 nucleotide sequences (*Inocybe*; *Agaricales*). *Mol. Phylogen. Evol.* 35: 1–20; **Matheny, P.B.; Bougher, N.L.** (2006). The new genus *Auritella* from Africa and Australia (*Inocybaceae, Agaricales*>: molecular systematics, taxonomy and historical biogeography. *Mycol. Prog.* 5: 2–17.

Karstenellaceae Harmaja 1974
Pezizales: Ascomycota

Ascomata apothecial, flat, very thin, formed on a thick-walled, papillate, hyaline, cyanophilic, hyphal mat; peridium composed of intertwined hyphae. Interascal tissue of simple paraphyses, reddish-brown at the apices. Asci cylindric, operculate, not blueing in iodine. Ascospores hyaline, smooth, without oil droplets or appendages, binucleate. Anamorph not known.

Significant Genera: *Karstenella*.

Distribution: Only known from Finland.

Economic Significance: None is known.

Ecology: Saprobic on leaf litter.

Notes: Very poorly known; no recent information is available.

References: **Harmaja, H.** (1969). *Karstenella vernalis* Harmaja, a new genus and species of discomycetes from Finland. *Karstenia* 9: 20–22; **Harmaja, H.** (1974). Two new families of the *Pezizales*: *Karstenellaceae* and *Pseudorhizinaceae*. *Karstenia* 14: 109–112; **Kimbrough, J.W.** (1989). Arguments towards restricting the limits of the *Pyronemataceae* (*Ascomycetes*, *Pezizales*). *Mem. N. Y. bot. Gdn* 49: 326–335.

Kathistaceae Malloch & M. Blackw. 1990
Ophiostomatales: Ascomycota

Stromata absent. Ascomata small, perithecial, long-necked, hyaline, with ostiolar setae. Interascal tissue absent. Asci formed in a basal fascicle, ± ellipsoidal, evanescent. Ascospores elongated, septate, hyaline to pale brown, without a sheath. Anamorphs yeast-like, formed directly from ascospores; putative sporidiomata also formed.

Significant Genera: *Kathistes*.

Distribution: Known from north temperate zones.

Economic Significance: None is known.

Ecology: Coprophilous, dispersed by arthropods.

Notes: Very poorly known; no recent information is available.

References: **Blackwell, M.; Jones, K.** (1997). Taxonomic diversity and interactions of insect-associated ascomycetes. *Biodiv. Cons.* 6: 689–699; **Malloch, D.W.; Blackwell, M.** (1990). *Kathistes*, a new genus of pleomorphic ascomycetes. *Can. J. Bot.* 68: 1712–1721.

Kickxellaceae Linder 1943
Kickxellales: Zygomycota

Vegetative thallus a mycelium, complex fruit bodies absent. Sporophores arising from the substratum mycelium, erect, ascending or repent, simple or branched, sometimes setiform; sporogenous region characterized by either complex branching, a spiral or helical portion or fertile vesicles; bearing unisporous sporangiola on sporogenous cells. Sporangiola dry or wet spored. Sporangiospores variously shaped but typically elongate, narrowly ellipsoid, fusiform to acicular. Zygospores formed from conjugation of undifferentiated hyphae, appearing chlamydospore-like, zygosporangium thin and unpigmented, zygospore wall smooth.

Significant Genera: *Coemansia, Kickxella, Spirodactylon*.

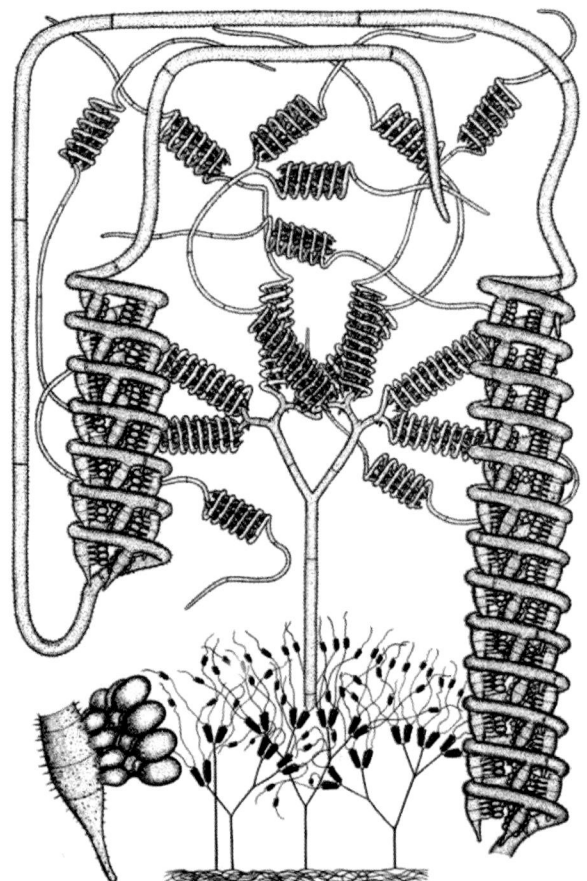

Spirodactylon aureum; sporophore with sporangiola; reproduced with permission from *Aliso* (1967)

Distribution: Widespread, apparently with distinct distribution patterns, although rarely recorded; often known from single collections.

Economic Significance: None is known.

Ecology: Saprobes from soil and coprophilous (rarely mycoparasites on *Isaria* and *Vesiculomyces*).

Notes: Recent molecular studies have shown strong support for a monophyletic clade including *Kickxellaceae*, *Dimargaritaceae* and *Harpellaceae* (*Harpellales*: *Trichomycetes*), which augments the results from previous ultrastructural studies on septal type. It has also been suggested that *Meristracaceae* (*Entomophthorales*) might be related, based on similar septal ultrastructure, but this has yet to be demonstrated with molecular data.

Coemansia reversa; sporophore

References: **Benjamin, R.K.** (1959). The Merosporangiferous *Mucorales*. *Aliso* 4: 321–433; **Benjamin, R.K.** (1961). Addenda to 'The Merosporangiferous *Mucorales*'. *Aliso* 5: 11–19; **Benjamin, R.K.** (1963). Addenda to 'The Merosporangiferous *Mucorales*' II. *Aliso* 5: 273–288; **Moss, S.T.; Young, T.W.K.** (1978). Phyletic considerations of the *Harpellales* and *Asellariales* (*Trichomycetes, Zygomycotina*) and *Kickxellales* (*Zygomycetes, Zygomycotina*). *Mycologia* 70: 944–963; **Saikawa, M.** (1989). Ultrastructure of the septum in *Ballocephala verrucospora* (*Entomophthorales, Zygomycetes*). *Can. J. Bot.* 67: 2484–2488; **Tanabe, Y.; Saikawa, M.; Watanabe, M.M.; Sugiyama, J.** (2004). Molecular phylogeny of *Zygomycota* based on EF-1α and RPB1 sequences: limitations and utility of alternative markers to rDNA. *Mol. Phylogen. Evol.* 30: 438–449.

Koralionastetaceae Kohlm. & Volkm.-Kohlm. 1987

Uncertain position within Ascomycetes, Ascomycota

Stromata absent. Ascomata formed from a dark hyphal layer, perithecial, superficial, black, thick-walled, the ostiole papillate, periphysate; peridium composed of pseudoparenchymatous tissue. Interascal tissue of simple thin-walled paraphyses. Asci cylindrical, thin-walled, without apical structures, evanescent. Ascospores hyaline, transversely septate, thick-walled, without a mucous sheath. Anamorphs hyphomycetous, probably spermatial.

Significant Genera: *Koralionastes*.

Distribution: Caribbean & Australasia.

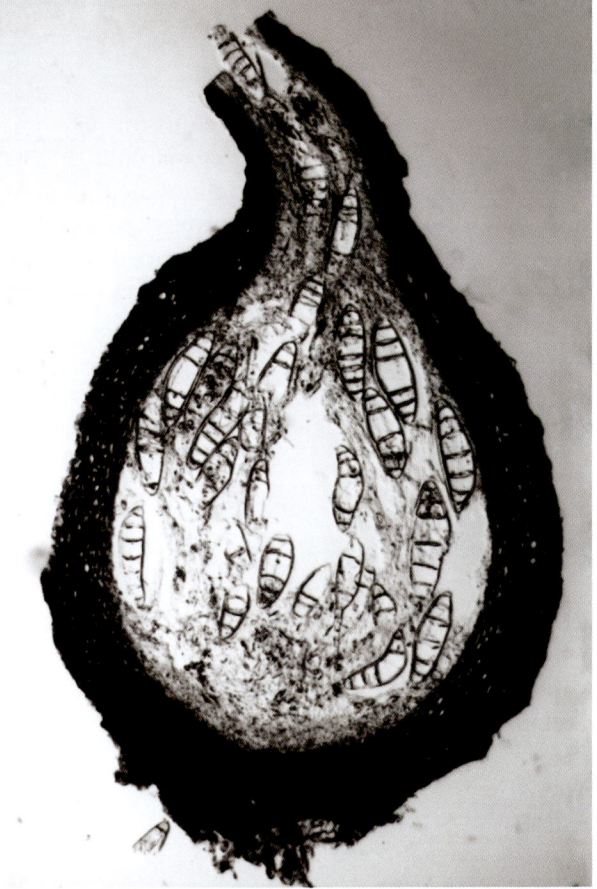

Koralionastes ellipticus; ascoma

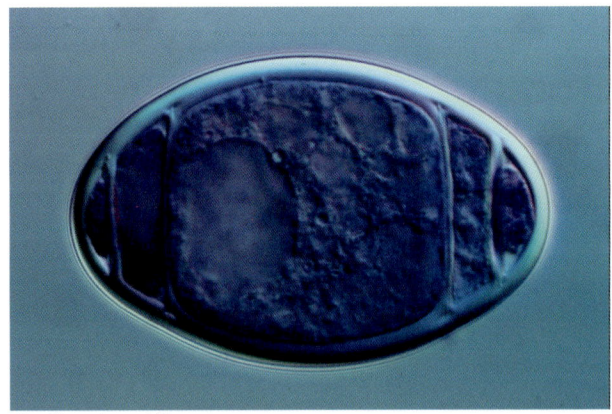

Koralionastes ovalis; ascospore

Economic Significance: None is known.

Ecology: On dead corals, associated with marine sponges.

Notes: Very poorly known; no recent information is available.

References: **Kohlmeyer, J.; Volkmann-Kohlmeyer, B.** (1987). *Koralionastetaceae* fam. nov. (*Ascomycetes*) from coral rock. *Mycologia* 79: 764–778; **Kohlmeyer, J.; Volkmann-Kohlmeyer, B.** (1990). New species of *Koralionastes* (*Ascomycotina*) from the Caribbean and Australia. *Can. J. Bot.* 68: 1554–1559.

Laboulbeniaceae G. Winter 1886
Laboulbeniales: Ascomycota

Stroma (thallus) present, composed of a basal black haustorial structure that anchors the fungus to its host, and a dark cellular secondary thallus. Ascomata formed directly from successive cells of a lateral appendage of the primary thallus, the lateral appendage very varied in form but not extending beyond the base of the ascoma, perithecial, pyriform to ovoid; outer wall cells of ascoma usually large and unequal, thin and translucent. Interascal tissue absent. Asci 4-spored, thin-walled, evanescent, without apical structures. Ascospores hyaline, usually elongate, thin-walled, with a submedian septum, surrounded by a gelatinous sheath and with an adhesive patch that allows attachment to the host. Usually monoecious. Anamorphs hyphomycetous, spermatial in function. Conidiogenous cells (antheridia) formed directly from branches of the secondary thallus, singly or in defined series, pyriform to flask-shaped, proliferating percurrently to form small hyaline aseptate conidia, which fertilize elongate hypha-like trichogynes.

Laboulbenia sp.; on dead fly abdomen

Distribution: Very widely distributed, probably more or less congruent with the arthropod host groups.

Economic Significance: None is known.

Rhachomyces zuphii; ascoma and thallus hairs

Significant Genera: *Dimeromyces, Laboulbenia, Rickia, Stigmatomyces*.

Laboulbenia trichognati; asci inside ascoma

Ecology: Associated with various orders of insects, millipedes and mites, especially on beetles; presumably parasitic but apparently causing little or no damage. At least many species appear not to penetrate the exoskeleton, but there is a lack of detailed studies on host/parasite relations.

Notes: Along with most other families of the *Laboulbeniomycetes*, the *Laboulbeniaceae* is unusual in having a determinate developmental process in which cell number and shape are tightly defined. The group is highly diverse and many species appear to be highly host-specific. Very little molecular analysis has been carried out, but the data available largely support the family and subfamily divisions based on morphological information.

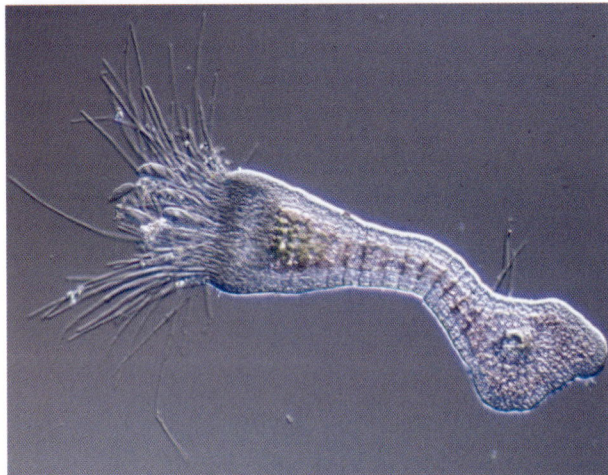

Zodiomyces vorticellarius; ascoma

References: **Bánhegyi, J.; Tóth, S.; Ubrizsy, G.; Vörös, J.** (1985). *Magyarország Mikroszkopikus Gombáinak Határozókönyve* 2. Kötet Eumycotina (Ascomycetes: A Discomycetestöl, Basidiomycetes, Deuteromycetes) (Budapest): 635 pp.; **Hughes, M.; Weir, A.; Gillen, B.; Rossi, W.** (2004). *Stigmatomyces* from New Zealand and New Caledonia: new records, new species and two new host families. *Mycologia* 96: 834–844; **Huldén, L.** (1985). Floristic notes on palaearctic *Laboulbeniales* (*Ascomycetes*). *Karstenia* 25: 1–16; **Lee, Y.B.** (1986). Taxonomy and geographical distribution of the *Laboulbeniales* in Asia. *Korean J. Pl. Taxon.* 16: 1–97; **Rossi, W.; Santamaria, S.** (2000). New *Laboulbeniales* parasitic on *Staphylinidae*. *Mycologia* 92: 786–791; **Santamaría, S.** (1996). Dioecism in two species of *Laboulbenia* (Fungi, Ascomycotina, Laboulbeniales). *Nova Hedwigia* 63: 63–70; **Santamaría, S.; Rossi, W.** (1999). New or interesting *Laboulbeniales* (*Ascomycota*) from the Mediterranean region. *Pl. Biosystems* 133: 163–171; **Tavares, I.I.** (1985). *Laboulbeniales* (Fungi, Ascomycetes). *Mycol. Mem.* 9: 627 pp.; **Weir, A.; Beakes, G.W.** (1996). Correlative light- and scanning electron microscope studies on the developmental morphology of *Hesperomyces virescens*. *Mycologia* 88: 677–693; **Weir, A.; Hammond, P.M.** (1997). *Laboulbeniales* on beetles: host utilization patterns and species richness of the parasites. *Biodiv. Cons.* 6: 701–719; **Weir, A.; Hughes, M.** (2002). The taxonomic status of *Corethromyces bicolor* from New Zealand, as inferred from morphological, developmental, and molecular studies. *Mycologia* 94: 483–493.

Lachnocladiaceae D.A. Reid 1965
Russulales: Basidiomycota

Basidiomata effuse and resupinate, or erect, pileate, spathulate or coralloid, annual or perennial, soft-textured or ± leathery, smooth or tomentose, flesh typically yellowish or brown. Hyphal system dimitic, with generative hyphae possessing or lacking clamp connections and branched skeletal hyphae that stain brown in iodine, cystidia usually present, rarely encrusted. Hymenium smooth, warted or ridged. Basidia cylindrical to clavate and elongate, 2- or 4-spored, in most genera interspersed with dichotomously branched sterile hyphae (dichohyphidia) that stain brown in iodine, and sometimes also stellate setae. Basidiospores variously shaped, hyaline, usually thin-walled, smooth or ornamented, sometimes blueing in iodine.

Significant Genera: *Asterostroma*, *Lachnocladium*, *Scytinostroma*, *Vararia*.

Distribution: Widespread, in both the tropics and temperate zones.

Economic Significance: None is known.

Ecology: On decaying coniferous or broadleaved wood, sometimes on herbaceous stems, presumably saprobic.

Notes: The family as currently circumscribed is extremely variable in gross morphology and is united primarily by the possession of branched dextrinoid hyphidia. Molecular studies indicate that the *Peniophoraceae* (which lack these structures) should probably be combined with the *Lachnocladiaceae*, but the affinities of *Lachnocladium* itself and similar tropical coralloid allies need further investigation.

Scytinostroma portentosum; basidioma

References: **Boidin, J.; Lanquetin, P.** (1987). Le Genre *Scytinostroma* Donk (Basidiomycètes, Lachnocladiaceae). *Biblthca Mycol.* 114: 130 pp.; **Ginns, J.** (1998). Genera of the North American Corticiaceae sensu lato. *Mycologia* 90: 1–35; **Hallenberg, N.** (1985). *The Lachnocladiaceae and Coniophoraceae of North Europe* (Oslo): 96 pp.; **Hibbett, D.S.; Donoghue, M.J.** (2001). Analysis of character correlations among wood decay mechanisms, mating systems and substrate ranges in *Homobasidiomycetes*. *Syst. Biol.* 50: 215–242; **Larsson, E.; Larsson, K.-H.** (2003). Phylogenetic relationships of russuloid basidiomycetes with emphasis on aphyllophoralean taxa. *Mycologia* 95: 1037–1065; **Müller, W.H.; Stalpers, J.A.; Aelst, A.C. van; Jong, M.D.M. de; Krift, T.P. van der; Boekhout, T.** (2000). The taxonomic position of *Asterodon*, *Asterostroma* and

Coltricia inferred from the septal pore cap ultrastructure. *Mycol. Res.* 104: 1485–1492; **Nakasone, K.K.; Micales, J.A.** (1988). *Scytinostroma galactinum* species complex in the United States. *Mycologia* 80: 546–559; **Stalpers, J.A.** (1996). The Aphyllophoraceous Fungi – II. Keys to the Species of the *Hericiales*. *Stud. Mycol.* 40: 185 pp.; **Wagner, T.** (2001). Phylogenetic relationships of *Asterodon* and *Asterostroma* (*Basidiomycetes*), two genera with asterosetae. *Mycotaxon* 79: 235–246; **Welden, A.L.** (1997). Colombian and Costa Rican species of stipitate stereoid fungi. *Revta Biol. trop.* 44: 91–102.

Lahmiaceae O.E. Erikss. 1986
Lahmiales: Ascomycota

Stromata absent. Ascomata cleistothecial, scattered, turbinate, shortly stipitate, black, uniloculate, opening by irregular splits in the upper part of the wall; peridium hyphal, gelatinized. Interascal tissue of trabecular pseudoparaphyses, secondarily producing paraphysis-like elements in the upper part of the locule. Asci clavate, long-stalked, probably with separate wall layers but not fissitunicate, without distinct apical structures, not blueing in iodine. Ascospores hyaline, falcate, transversely septate, without a sheath. Anamorph unknown.

Significant Genera: *Lahmia*.

Distribution: Known from Europe and North America, with most records from higher latitudes.

Economic Significance: None is known.

Ecology: Probably saprobic on rough bark of *Populus* species, often encountered with species of *Rhytidiella*.

Notes: The only genus, *Lahmia*, occupies a very isolated position within the *Dothideomycetes*; no molecular data are available.

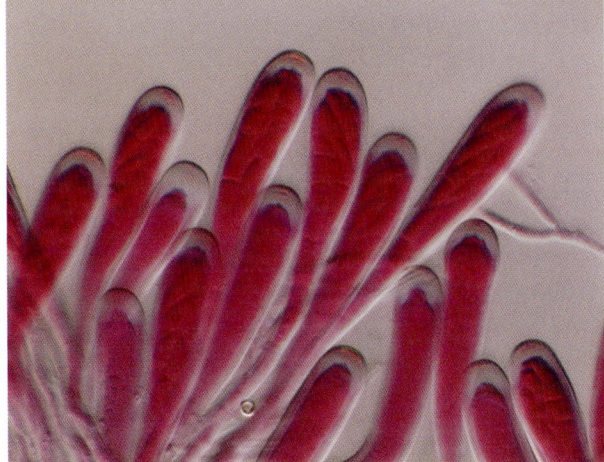

Lahmia kunzei, Sweden; asci and ascospores

References: **Eriksson, O.** (1986). *Lahmia* Körber (= *Parkerella* A.Funk) a misinterpreted genus with isolated position. *Mycotaxon* 27: 347–360; **Ostry, M.E.** (1986). Association of *Parkerella populi* with declining hybrid aspen in Wisconsin. *Can. J. Bot.* 64: 1834–1835.

Lasiosphaeriaceae Nannf. 1932
Sordariales: Ascomycota

Stromata absent, rarely with a basal subiculum. Ascomata perithecial, dark, often thick-walled, ostiolate or not, often hairy or ornamented, the ostiole if present periphysate. Interascal tissue often present, of wide thin-walled paraphyses, often inconspicuous and usually evanescent. Asci cylindric-clavate to clavate, fairly thin-walled and not fissitunicate, rarely evanescent, usually with a small J– apical ring. Ascospores variable, usually with at least one dark brown and one hyaline cell, occasionally ornamented, with gelatinous appendages or caudae (often long), normally lacking a sheath. Anamorphs varied but rarely prominent, hyphomycetous, probably spermatial in function.

Significant Genera: *Cercophora*, *Lasiosphaeria*, *Podospora*, *Zopfiella*. Anamorphs: *Cladorrhinum*.

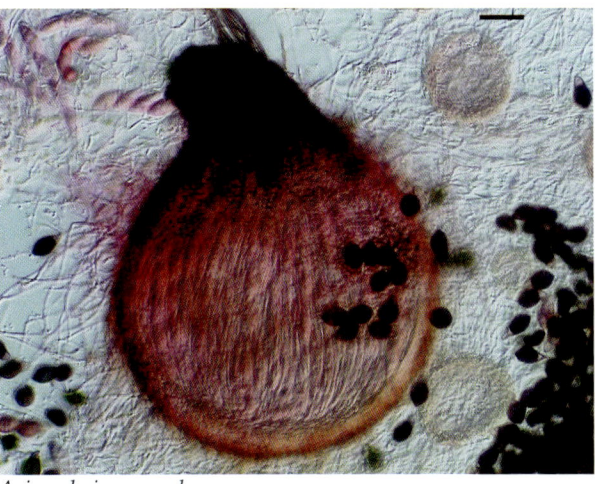

Apiosordaria verruculosa; ascoma

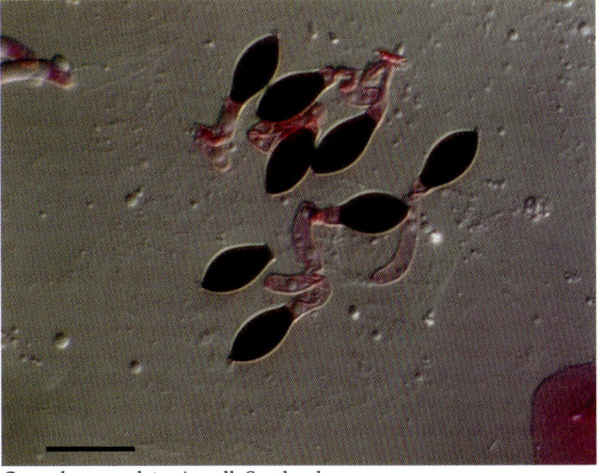

Cercophora caudata, Argyll, Scotland; ascospores

Distribution: Widespread, especially in temperate regions.

Economic Significance: Little is known; a few species are model organisms for genetic research.

Ecology: Saprobes found in dung, soil and rotting vegetation.

Notes: A basically well-defined family within the *Sordariales*, though some recent molecular evidence suggests that it is polyphyletic with the *Sordariaceae* and *Chaetomiaceae* clustered within and that the type genus as traditionally circumscribed contains disparate elements. The wide concept of the family previously advocated has now been discredited. The *Nitschkiaceae* was at one time considered to be close to the *Lasiosphaeriaceae* due to the shared attribute of conspicuous pores in the peridial cells, but these are now known to be widely distributed within the *Ascomycota* and molecular data clearly separate the two families.

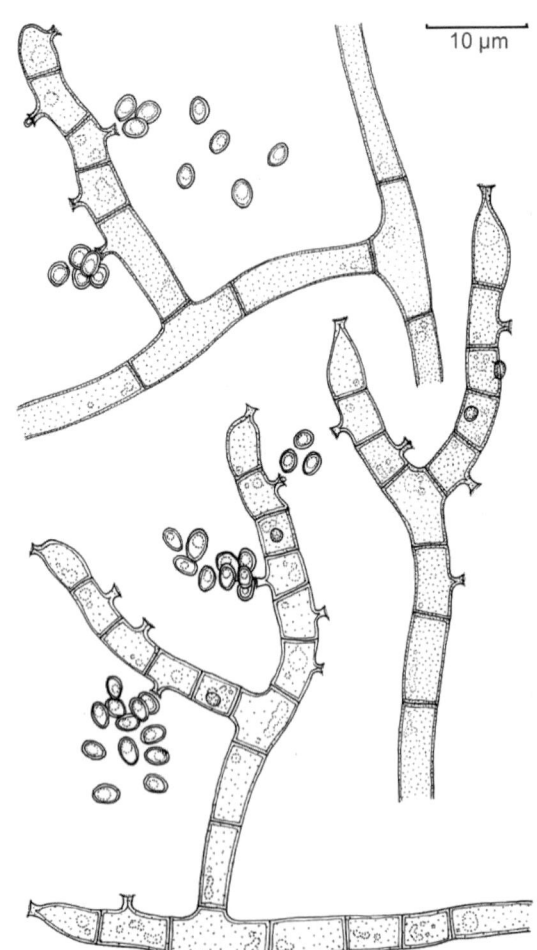

Cladorrhinum foecundissimum; conidiophores and conidia

References: **Barr, M.E.** (1990). Prodromus to nonlichenized, pyrenomycetous members of class *Hymenoascomycetes*. *Mycotaxon* 39: 43–184; **Bell, A.; Mahoney, D.P.** (1995). Coprophilous fungi in New Zealand. I. *Podospora* species with swollen agglutinated perithecial hairs. *Mycologia* 87: 375–396; **Bell, A.; Mahoney, D.P.** (1996). Perithecium development in *Podospora tetraspora* and *Podospora vesticola*. *Mycologia* 88: 163–170; **Bell, A.; Mahoney, D.P.** (1997). Coprophilous fungi in New Zealand. II. *Podospora* species with coriaceous perithecia. *Mycologia* 89: 908–915; **Cai, L.; Jeewon, R.; Hyde, K.D.** (2005). Phylogenetic evaluation and taxonomic revision of *Schizothecium* based on ribosomal DNA and protein coding genes. *Fungal Diversity* 19: 1–21; **Cai, L.; Jeewon, R.; Hyde, K.D.** (2006). Phylogenetic investigations of *Sordariaceae* based on multiple gene sequences and morphology. *Mycol. Res.* 110: 137–150; **Cai, L.; Jeewon, R.; Hyde, K.D.** (2006). Molecular systematics of *Zopfiella* and allied genera: evidence from multigene sequence analyses. *Mycol. Res.* 110: 359–368; **Chen, W.; Shearer, C.A.; Crane, J.L.** (1999). Phylogeny of *Ophioceras* spp. based on morphological and molecular data. *Mycologia* 91: 84–94; **Coppin, E.; Debuchy, R.; Arnaise, S.; Picard, M.** (1997). Mating types and sexual development in filamentous ascomycetes. *Microbiol. Mol. Biol. Rev.* 61: 411–428; **Gams, W.** (2000). *Phialophora* and some similar morphologically little-differentiated anamorphs of divergent ascomycetes. *Stud. Mycol.* 45: 187–199; **Guarro, J.; Cannon, P.F.; Aa, H.A. van der** (1991). A synopsis of the genus *Zopfiella* (Ascomycetes, Lasiosphaeriaceae). *Syst. Ascom.* 10: 79–112; **Huhndorf, S.M.; Miller, A.N.; Fernández, F.A.** (2004). Molecular systematics of the *Sordariales*: the order and the family *Lasiosphaeriaceae* revisited. *Mycologia* 96: 368–387; **Khan, R.S.; Krug, J.C.** (1990). New records of the *Sordariaceae* from East Africa. *Mycologia* 81: 862–869; **Krug, J.C.; Scott, J.A.** (1994). The genus *Bombardioidea*. *Can. J. Bot.* 72: 1302–1310; **Lee, S.J.; Hanlin, R.T.** (1999). Phylogenetic relationships of *Chaetomium* and similar genera based on ribosomal DNA sequences. *Mycologia* 91: 434–442; **Lundqvist, N.** (1972). Nordic *Sordariaceae* s. lat. *Symb. bot. upsal.* 20: 374 pp.; **Miller, A.N.; Huhndorf, S.M.** (2004). Using phylogenetic species recognition to delimit species boundaries within *Lasiosphaeria*. *Mycologia* 96: 1106–1127; **Miller, A.N.; Huhndorf, S.M.** (2004). A natural classification of *Lasiosphaeria* based on nuclear LSU rDNA sequences. *Mycol. Res.* 108: 26–34; **Taylor, J.W.; White, T.J.** (1991). Molecular evolution of nuclear and mitochondrial ribosomal DNA regions in *Neurospora* and other *Sordariaceae* and comparison with the mushroom, *Laccaria* [abstract]. *Fungal Genetics Newsl.* Suppl. 38: 26.

Lautosporaceae Kohlm., Volkm.-Kohlm. & O.E. Erikss. 1995

Uncertain position within Ascomycetes, Ascomycota

Ascomata perithecioid, immersed, with an elongated neck; peridium brown, coriaceous, composed of thick-walled pseudoparenchymatous cells. Interascal tissue of simple septate tapering pseudoparaphyses. Asci ± cylindrical, thick-walled with a large ocular chamber, apparently not fissitunicate, 4-spored. Ascospores muriform, with numerous Δ-trans-septa but no primary euseptum, all septa developing simultaneously to form a mesh-like pattern, with an outer thick-walled layer which is not divided by septa.

Significant Genera: *Lautospora*.

Distribution: Known from North America and Brunei.

Economic Significance: None is known.

Ecology: On plant material in marine environments.

Notes: Very poorly known; no recent information is available.

References: **Hyde, K.D.; Jones, E.B.G.** (1989). Intertidal mangrove fungi from Brunei. *Lautospora gigantea* gen. et sp. nov., a new loculoascomycete from prop roots of *Rhizophora* spp. *Bot. Mar.* 32: 479–482; **Kohlmeyer, J.; Volkmann-Kohlmeyer, B.; Eriksson, O.E.** (1995). Fungi on *Juncus roemerianus* 2. New dictyosporous *Ascomycetes*. *Bot. Mar.* 38: 165–174.

Lecanoraceae Körb. 1855
Lecanorales: Ascomycota

Thallus crustose, granular or scurfy, sometimes areolate and/or with radiating margins, sometimes internal, rarely lacking (in parasitic or commensal species), rhizoids absent or poorly developed, sometimes sorediate, isidia and cephalodia absent. Ascomata apothecial, sessile or short-stalked, rarely ± immersed, usually with a well-developed lecanorine margin concolorous with the thallus, the disc often strongly pigmented. Interascal tissue of simple or anastomosing paraphyses, often with clavate tips and a pigmented epithecium. Asci cylindrical to clavate, thick-walled, with a well-developed J+ apical cap, a usually well-developed ocular chamber and apical cushion, with an outer layer of J+ gelatinized material, 8- to 32-spored. Ascospores clavate or elongate, rarely acicular, almost always hyaline, smooth, aseptate or septate, without a gelatinous sheath or appendages. Anamorphs pycnidial, varied in form, ± immersed in thallus tissue, colourless or brown. Conidiogenous cells often borne on short branched conidiophores, elongate, proliferating percurrently with well-developed collarettes. Conidia usually bacilliform or filamentous, hyaline, aseptate.

Significant Genera: *Lecanora*, *Lecidella*.

Lecanora garovaglii, Arizona, USA; thallus and ascomata

Distribution: Very widespread.

Economic Significance: None is known, although many species produce unusual metabolites that may have economic value.

Ecology: Lichenized with green algae, occasionally lichenicolous, on rocks, mosses and wood.

Notes: The central family of the *Lecanorales* is dominated by the large genus *Lecanora* and is defined largely using ascus morphology. There have been few phylogenetic studies, but Ekman & Wedin (2000) suggest that the family is polyphyletic and that the *Lecanora* ascus type has evolved on more than one occasion.

Lecanora muralis, Salamanca, Spain; thallus and ascomata

Lecanora chlarotera, Ross-shire, Scotland; thallus and ascomata

References: **Arup, U.; Grube, M.** (1998). Molecular systematics of *Lecanora* subgenus *Placodium*. *Lichenologist* 30: 415–425; **Arup, U.; Grube, M.** (2000). Is *Rhizoplaca* (*Lecanorales*, lichenized *Ascomycota*) a monophyletic genus?. *Can. J. Bot.* 78: 318–327; **Ekman, S.; Wedin, M.** (2000). The phylogeny of the families *Lecanoraceae* and *Bacidiaceae* (lichenized *Ascomycota*) inferred from nuclear SSU rDNA sequences. *Pl. Biol.* 2: 350–360; **Gargas, A.; DePriest, P.T.; Taylor, J.W.** (1995). Positions of multiple insertions in SSU rDNA of lichen-forming fungi. *Mol. Biol. Evol.* 12: 208–218; **Grube, M.;**

Baloch, E.; Arup, U. (2004). A phylogenetic study of the *Lecanora rupicola* group (*Lecanoraceae*, *Ascomycota*). *Mycol. Res.* 108: 506–514; **Hafellner, J.** (1984). Studien in Richtung einer natürlicheren Gliederung der Sammel-familien *Lecanoraceae* und *Lecideaceae*. *Nova Hedwigia* Beih. 79: 241–371; **Kärnefelt, E.I.** (1997). On the nature of species in lichenized *Ascomycotina*. *Abstracta Botanica* 21: 21–29; **Lumbsch, H.T.; Plümper, M.; Guderley, R.; Feige, G.B.** (1997). The corticolous species of *Lecanora sensu stricto* with pruinose apothecial discs. *In* Tibell, L. & Hedberg, I. (eds), Lichen Studies Dedicated to Rolf Santesson. *Symb. bot. upsal.* 32: 131–162; **Rambold, G.; Hagedorn, G.** (1998). The distribution of selected diagnostic characters in the *Lecanorales*. *Lichenologist* 30: 473–487; **Ryan, B.D.; Nash, T.H.** (1997). Placidioid taxa of *Lecanoraceae sensu* Zahlbr. (lichenized *Ascomycotina*) in North America: taxa excluded from *Lecanora* subg. *Placodium*. *Nova Hedwigia* 64: 393–420.

Lecideaceae Chevall. 1826
Lecanorales: Ascomycota

Thallus crustose to squamulose, sometimes areolate, sometimes immersed, rarely absent, with poorly developed rhizoids, soredia sometimes present. Ascomata apothecial, immersed to sessile, the margin absent or weakly developed, usually domed, light or dark; hymenium usually blueing in iodine. Interascal tissue of simple or sparsely branched and anastomosed paraphyses, usually swollen at the apices, often pigmented or with an epithecium. Asci cylindrical to clavate, thick-walled, with a J+ apical cap, often with a more strongly staining shallow or tube-like apical ring and an outer J+ gelatinized layer. Ascospores cylindrical to ellipsoidal, hyaline, aseptate, thin-walled, without a sheath. Anamorphs coelomycetous. Conidiomata pycnidial, immersed, dark-walled. Conidiogenous cells flask-shaped, proliferating percurrently. Conidia cylindrical to bacilliform or filiform, hyaline, aseptate.

Significant Genera: *Lecidea*, *Hypocenomyce*.

Distribution: Cosmopolitan.

Economic Significance: None is known.

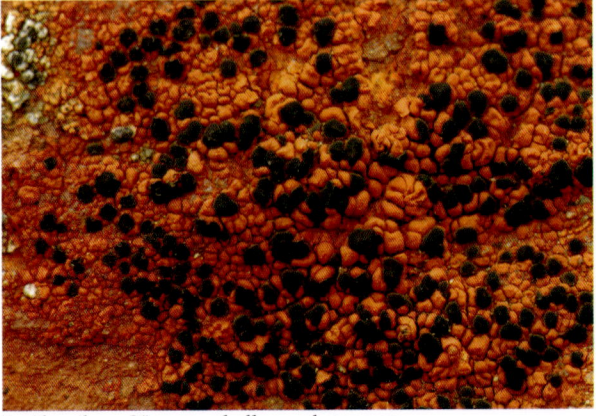

Lecidea silacea, Norway; thallus and ascomata

Ecology: Lichenized with green algae, rarely lichenicolous, on rock, wood and soil.

Notes: The speciose genus *Lecidea* was traditionally loosely circumscribed, including almost any crustose lichens with hyaline ascospores that lacked a thalline outer wall to the ascomata. Many groups have been split off, but the taxonomy still requires attention. Very few molecular data are available.

Hypocenomyce scalaris, Norway; thallus and ascomata

References: **Hertel, H.** (1987). Progress and problems in taxonomy of Antarctic saxicolous lichens. *In* Peveling (ed.), Progress and Problems in Lichenology in the Eighties. *Biblthca Lichenol.* 25: 219–242; **Hertel, H.** (1995). Schlüssel für die Arten der Flechtenfamilie *Lecideaceae* in Europa. *Biblthca Lichenol.* 58: 137–180; **Hertel, H.** (1997). On the genus *Lecidea* (*Lecanorales*) in southern Chile and Argentina. *In* Tibell, L. & Hedberg, I. (eds), Lichen Studies Dedicated to Rolf Santesson. *Symb. bot. upsal.* 32: 95–111; **Hertel, H.; Rambold, G.** (1985). *Lecidea* sect. *Armeniaceae*: lecideoide Arten der Flechtengattungen *Lecanora* und *Tephromela* (*Lecanorales*). *Bot. Jb.* 107: 469–501; **Kantvilas, G.; McCarthy, P.M.** (1999). *Steinia australis*, a new species in the lichen family *Aphanopsidaceae*. *Lichenologist* 31: 555–558; **Pietschmann, M.** (1990). Morphometrics of tubiform apical apparatus in *Lecideaceae*, *Micareaceae*, *Porpidiaceae* and allied families (lichenized *Ascomycetes*, *Lecanorales*): limitations and perspectives of statistical inference. *Nova Hedwigia* 51: 521–549; **Rambold, G.** (1989). A monograph of the saxicolous lecideoid lichens of Australia (excl. Tasmania). *Biblthca Lichenol.* 34: 345 pp.; **Rambold, G.; Hagedorn, G.** (1998). The distribution of selected diagnostic characters in the *Lecanorales*. *Lichenologist* 30: 473–487; **Triebel, D.** (1989). Lecideicole Ascomyceten. Eine Revision der obligat lichenicolen Ascomyceten auf lecideoiden Flechten. *Biblthca Lichenol.* 35: 278 pp.

Legeriomycetaceae Pouzar 1972
Harpellales: Zygomycota

Thallus branched, attached to the hindgut cuticle of aquatic insect larvae. Sporangia (trichospores) unispored, each usually bearing one or more long basal appendages. Zygospores conical or biconical, formed by hyphal conjugation; heterothallic.

Significant Genera: *Smittium*.

Distribution: Widespread and sometimes rather common.

Economic Significance: None is known.

Ecology: Attached to the hindgut cuticle of aquatic insect larvae, especially Diptera but also Ephemeroptera and Plecoptera, where they are obligate endocommensals.

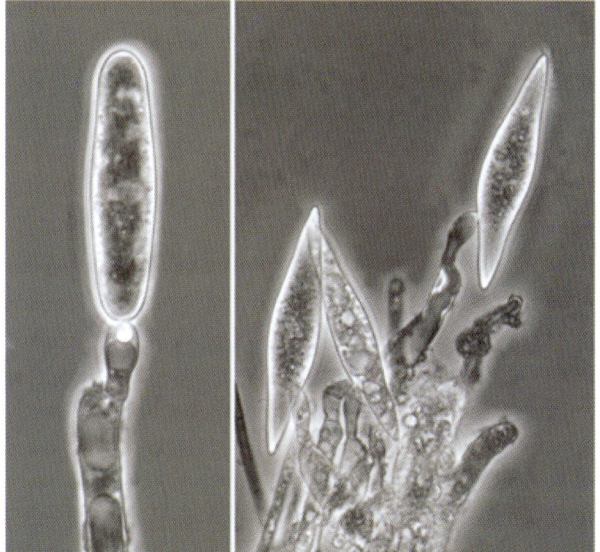

Smittium biforme; thallus bearing sporangia and zygospores

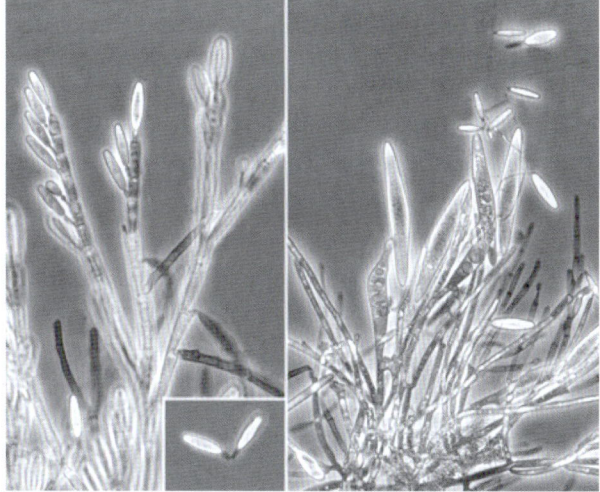

Smittium precipitiorum; thallus bearing sporangia and zygospores

Notes: By far the largest family of *Trichomycetes* with 30 recognized genera (but most either monotypic or with few species) and 135 species (almost half of these, 66, in *Smittium*). A close affinity with the *Kickxellales* has been suggested based first on ultrastructural and serological evidence and augmented by analysis of molecular data. The exclusion of the Amoebidiales and Eccrinales (Protozoa) was suggested as early as 1975, a conclusion confirmed by molecular data.

References: **Beard, C.E.; Adler, P.H.** (2003). Zygospores of selected *Trichomycetes* in larval *Diptera* of the families *Chironomidae* and *Simuliidae*. *Mycologia* 95: 317–320; **Benny, G.L.** (2001). *Zygomycota*: *Trichomycetes*. The *Mycota* (Berlin) 7: 147–160; **Lichtwardt, R.W.** (1986). *The Trichomycetes. Fungal associates of arthropods* (New York): 343 pp.; **Lichtwardt, R.W.** (2002). *Trichomycetes*: fungi in relationship with insects and other arthropods. *Cellular Origin and Life in Extreme Habitats* (Dordrecht) 4: 577–588; **Moss, S.T.** (1975). Septal ultrastructure in the *Trichomycetes* with special reference to *Astreptonema* (*Eccrinales*). *Trans. Br. mycol. Soc.* 65: 115–127; **Moss, S.T.; Young, T.W.K.** (1978). Phyletic considerations of the *Harpellales* and *Asellariales* (*Trichomycetes, Zygomycotina*) and *Kickxellales* (*Zygomycetes, Zygomycotina*). *Mycologia* 70: 944–963; **Valle, L.G.; Santamaría, S.** (2004). The genus *Smittium* (*Trichomycetes, Harpellales*) in the Iberian peninsula. *Mycologia* 96: 682–701.

Leotiaceae Corda 1842
Helotiales: Ascomycota

Stromata absent. Ascomata apothecial, medium-sized to large, usually stalked, often brightly coloured, glabrous; excipulum composed of an outer layer of parallel or interwoven gelatinized hyphae with ± remote septa, an inner layer of non-gelatinized hyphal tissue, and an intermediate hyphal layer which may or may not be gelatinized. Interascal tissue of simple paraphyses. Asci cylindrical, with a thickened apex and a usually diffuse J+ apical ring. Ascospores hyaline, ellipsoidal or elongate, septate or not. Anamorphs unknown.

Leotia lubrica, Argyll, Scotland; ascomata

Significant Genera: *Leotia*.

Distribution: Widespread, especially in temperate zones.

Economic Significance: None is known.

Ecology: Saprobic on soil, especially in damp habitats, and on rotting vegetation.

Notes: The family is accepted in a restricted sense and occupies a clearly separate clade from the *Helotiaceae*, with which it has traditionally been allied. The *Bulgariaceae* may belong here.

References: **Döring, H.; Triebel, D.** (1998). Phylogenetic relationships of *Bulgaria* inferred by 18S rDNA sequence analysis. *Cryptog. Bryol.-Lichénol.* 19: 123–136; **Gamundí, I.; Romero, A.I.** (1998). *Fungi, Ascomycetes. Helotiales: Helotiaceae. Fl. criptog. Tierra del Fuego* 10: 130 pp.; **Gargas, A.; Taylor, J.W.** (1995). Phylogeny of *Discomycetes* and early radiations of the apothecial *Ascomycotina* inferred from SSU rDNA sequence data. *Exp. Mycol.* 19: 7–15; **Korf, R.P.** (1999). *Pezoloma kathiae* sp. nov. (Ascomycetes: *Leotiales, Leotiaceae*) and its placement in a new subgenus, *Phaeopezoloma*. *Mycotaxon* 73: 493–497; **Landvik, S.; Kristiansen, R.; Schumacher, T.** (1998). Phylogenetic and structural studies in the *Thelebolaceae* (*Ascomycota*). *Mycoscience* 39: 49–56; **Liu, Y.J.J.; Whelen, S.; Hall, B.D.** (1999). Phylogenetic relationships among ascomycetes: evidence from an RNA polymer[a]se II subunit. *Mol. Biol. Evol.* 16: 1799–1808; **Verkley, G.J.M.** (1994). Ultrastructure of the ascus apical apparatus in *Leotia lubrica* and some *Geoglossaceae* (*Leotiales, Ascomycotina*). *Persoonia* 15: 405–430; **Wang, Z.; Binder, M.; Schoch, C.L.; Johnston, P.R.; Spatafora, J.W.; Hibbett, D.S.** (2006). Evolution of helotialean fungi (*Leotiomycetes, Pezizomycotina*): a nuclear rDNA phylogeny. *Mol. Phylogen. Evol.* 41: 295–312; **Zhong, Z.H.; Pfister, D.H.** (2004). Phylogenetic relationships among species of *Leotia* (*Leotiales*) based on ITS and RPB2 sequences. *Mycol. Prog.* 3: 237–246.

Leptopeltidaceae Höhn. ex Trotter 1928
Microthyriales: Ascomycota

Ascomata formed within a small, strongly flattened, subcuticular stroma, elongate, opening by a longitudinal split; covering wall brown, composed of a single layer of radially arranged, thick-walled, isodiametric cells. Interascal hyphae present but inconspicuous. Asci small, clavate, apparently not functionally fissitunicate, probably with rostrate dehiscence, not blueing in iodine. Ascospores hyaline, septate, without a sheath. Anamorphs coelomycetous, conidiomata morphologically similar to the ascomata.

Significant Genera: *Leptopeltis*.

Distribution: Most species have temperate distributions.

Economic Significance: None is known.

Ecology: Saprobic, especially on ferns.

Notes: Very poorly known; no recent information is available.

Leptopeltis holmorum, Devon, UK; ascomata

References: **Cannon, P.F.** (1997). Two new genera of *Ascomycota*, and other new or interesting fungi from Slapton Ley National Nature Reserve and its environs. *Syst. Ascom.* 15: 121–138; **Holm, L.; Holm, K.** (1977). A study of the *Leptopeltidaceae*. *Bot. Notiser* 130: 215–229; **Spooner, B.M.; Kirk, P.M.** (1990). Observations on some genera of *Trichothyriaceae*. *Mycol. Res.* 94: 223–230.

Leptosphaeriaceae M.E. Barr 1987
Pleosporales: Ascomycota

Ascomata perithecial, often ± conical, papillate, immersed or erumpent, sometimes aggregated into small stromata, with a well-developed, usually periphysate, ostiole; peridium black, well-developed, sometimes thicker at the base, composed of thick-walled pseudoparenchymatous cells. Interascal tissue of cellular pseudoparaphyses. Asci cylindrical, relatively narrow and thin-walled but fissitunicate, with a distinct ocular chamber. Ascospores hyaline to brown, transversely septate, sometimes elongated, sometimes with a sheath. Anamorphs coelomycetous. Conidiomata pycnidial. Conidiogenous cells small, doliiform, proliferating percurrently. Conidia small, aseptate, thin-walled.

Significant Genera: *Leptosphaeria, Ophiobolus*. Anamorphs: *Coniothyrium, Phoma*.

Distribution: Cosmopolitan, but especially prominent in temperate regions.

Economic Significance: Little is known. A number of species are known as crop pathogens, especially *Leptosphaeria maculans* on crucifers.

Ecology: Saprobic or weakly necrotrophic on stems or leaves.

Notes: Some of the constituent genera are polyphyletic, including the type genus *Leptosphaeria*. Several groups have been segregated from the family; the most significant of these is the *Phaeosphaeriaceae* that includes important cereal pathogens.

Leptosphaeria acuta, Isle of Man, UK; ascomata

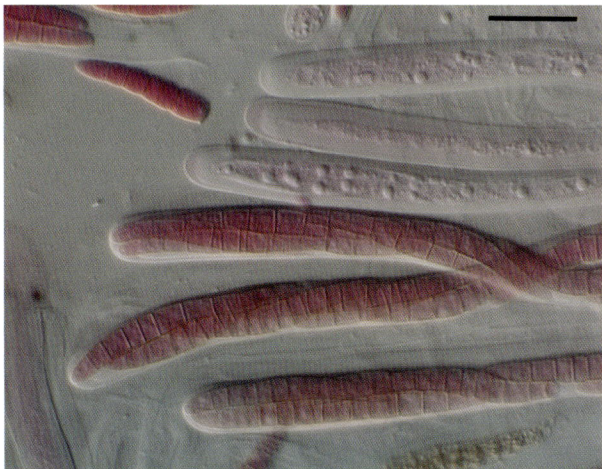

Leptosphaeria acuta, Isle of Man, UK; asci

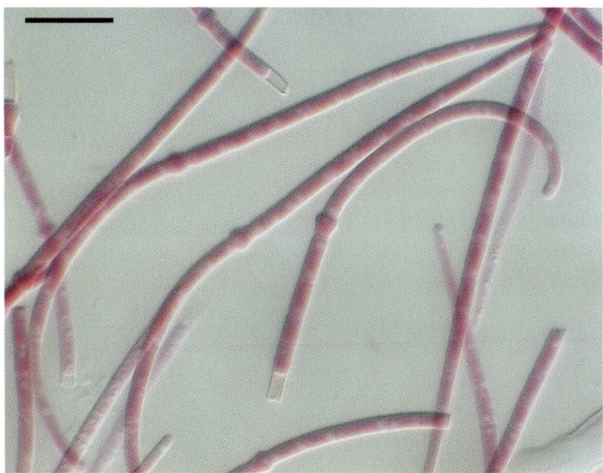

Ophiobolus acuminatus, Shetland Isles, Scotland; ascospores

References: **Ahn, Y.M.; Shearer, C.A.** (1998). Reexamination of taxa in *Leptosphaeria* originally described on host species in Ranunculaceae, Papaveraceae, and Magnoliaceae. *Can. J. Bot.* 76: 258– 280; **Ahn, Y.M.; Shearer, C.A.** (1999). Taxonomic revision of *Leptosphaeria vagabunda* and four infraspecific taxa. *Mycologia* 91: 684–693; **Barr, M.E.** (1987). *Prodromus to Class Loculoascomycetes* (Amherst): 168 pp.; **Boerema, G.H.; Gruyer, J. de.; Noordeloos, M.E.; Hamers, M.E.C.** (2004). *Phoma Identification Manual* (Wallingford): 470 pp.; **Câmara, M.P.S.; Palm, M.E.; Berkum, P. van; O'Neill, N.R.** (2002). Molecular phylogeny of *Leptosphaeria* and *Phaeosphaeria*. *Mycologia* 94: 630–640; **Crane, J.L.; Shearer, C.A.** (1991). A nomenclator of *Leptosphaeria* V. Cesati & G. de Notaris (*Mycota – Ascomycotina – Loculoascomycetes*). *Bull. Ill. St. nat. Hist. Surv.* 34: 1–355; **Dong, J.W.; Chen, W.D.; Crane, J.L.** (1998). Phylogenetic studies of the *Leptosphaeriaceae, Pleosporaceae* and some other *Loculoascomycetes* based on nuclear ribosomal DNA sequences. *Mycol. Res.* 102: 151–156; **Hill, C.B.; Williams, P.H.** (1988). *Leptosphaeria maculans*, cause of blackleg of crucifers. *Phytophthora infestans, the Cause of Late Blight of Potato* Advances in Plant Pathology vol. **7** (London): 169–174; **Huhndorf, S.M.** (1992). Systematics of *Leptosphaeria* species found on the Rosaceae. *Bull. Ill. nat. Hist. Surv.* 34: 475–534; **Khashnobish, A.; Shearer, C.A.; Crane, J.L.** (1995). Reexamination of species of *Leptosphaeria* on asteraceous hosts. *Mycotaxon* 54: 91–106; **Kodsueb, R.; Dhanasekraran, V.; Aptroot, A.; Lumyong, S.; McKenzie, E.H.C.; Hyde, K.D.; Jeewon, R.** (2006). The family *Pleosporaceae*: intergeneric relationships and phylogenetic perspectives based on sequence analyses of partial 28S rDNA. *Mycologia* 98: 571–583; **Mahuku, G.S.; Goodwin, P.H.; Hall, R.; Hsiang, T.** (1997). Variability in the highly virulent type of *Leptosphaeria maculans* within and between oilseed rape fields. *Can. J. Bot.* 75: 1485–1492; **Morales, V.M.; Jasalavich, C.A.; Pelcher, L.E.; Petrie, G.A.; Taylor, J.L.** (1995). Phylogenetic relationships among several *Leptosphaeria* species based on their ribosomal DNA sequences. *Mycol. Res.* 99: 593–603; **Olivier, C.; Berbee, M.L.; Shoemaker, R.A.; Loria, R.** (2000). Molecular phylogenetic support from ribosomal DNA sequences for origin of *Helminthosporium* from *Leptosphaeria*-like loculoascomycete ancestors. *Mycologia* 92: 736–746; **Purwantara, A.; Barrins, J.M.; Cozijnsen, A.J.; Ades, P.K.; Howle, B.J.** (2000). Genetic diversity of isolates of the *Leptosphaeria maculans* species complex from Australia, Europe and North America using amplified fragment length polymorphism analysis. *Mycol. Res.* 104: 772–781; **Shearer, C.A.; Crane, J.L.; Swofford, D.L.** (1990). A preliminary cladistic analysis of *Leptosphaeria* and allied genera [abstract]. *Fourth International Mycological Congress Abstracts* (Regensburg): 156; **Shoemaker, R.A.** (1985). Canadian and some extralimital *Leptosphaeria* species. *Can. J. Bot.* 62: 2688–2729; **Yuan, Z.Q.; Barr, M.E.** (1994). Species of *Chaetoplea* on desert plants in China. *Mycotaxon* 52: 495–499.

Letrouitiaceae Hafellner & Bellem. 1982
Teloschistales: Ascomycota

Thallus crustose, green or brownish, lacking anthraquinone pigments. Ascomata apothecial, without a well-developed thalline margin. Asci with a diffuse J+ outer apical layer and a well-developed internal apical structure which blues diffusely in iodine. Ascospores hyaline, multiseptate, sometimes muriform, with strongly thickened internal walls, often very large.

Significant Genera: *Letrouitia*.

Distribution: Widespread.

Economic Significance: None is known.

Ecology: Lichenized with green algae.

Notes: Very poorly known; no recent information is available.

Letrouitia domingensis; thallus and ascomata

References: **Awasthi, D.D.; Srivastava, P.** (1989). Lichen genera *Brigantiaea* and *Letrouitia* from India. *Proc. Indian Acad. Sci.* Pl. Sci. 99: 165–177; **Kärnefelt, I.** (1989). Morphology and phylogeny in the *Teloschistales*. *Cryptog. bot.* 1: 147–203; **Kärnefelt, I.** (1990). Isidiate taxa in the *Teloschistales* and their ecological and evolutionary significance. *Lichenologist* 22: 307–320; **Kasalicky, T.; Döring, H.; Rambold, G.; Wedin, M.** (2000). A comparison of ITS and LSU nrDNA phylogenies of *Fulgensia* (*Teloschistaceae, Lecanorales*), a genus of lichenised ascomycetes. *Can. J. Bot.* 78: 1580–1589.

Leucogastraceae Moreau ex Fogel 1979
Boletales: Basidiomycota

Basidiomata gasteroid, loculate, hypogeous or becoming emergent, globose to irregular; peridium well-developed, white to yellow, thin or thick. Hyphal system monomitic, clamp connections absent. Gleba lacunose, tending to be compact and gelatinous at maturity, columella absent. Hymenium lining the globose or lobed locules, composed of basidia only. Basidia ± globose to clavate, 2- to 4-spored, evanescent. Basidiospores globose to ellipsoidal, hyaline, thick-walled, ornamented with spines or pits, not staining in iodine, with a gelatinous sheath or a separating perispore layer.

Significant Genera: *Leucogaster, Leucophleps*.

Distribution: Widespread throughout north temperate regions.

Economic Significance: None is known.

Ecology: Found in broadleaved and coniferous forests, presumably ecomycorrhizal.

Notes: Molecular studies have not yet been published, but there are indications that the family occupies a basal position within the homobasidiomycetes.

Leucogaster citrinus; basidiomata

References: **Beaton, G.; Pegler, D.N.; Young, T.W.K.** (1985). Gasteroid basidiomycota of Victoria State, Australia. 8-9. *Kew Bull.* 40: 827–842; **Fogel, R.** (1979). The genus *Leucophleps* (*Basidiomycotina, Leucogastrales. Can. J. Bot.* 57: 1718–1728; **Fogel, R.** (1990). *Leucogaster columellatus* is a *Sclerogaster*. *Mycologia* 82: 655–657; **Montecchi, A.; Sarasini, M.** (2000). *Funghi Ipogei d'Europa* (Trento): 714 pp.

Leucosporidiaceae Jülich 1982
Leucosporidiales: Basidiomycota

Colonies colourless to cream, yeast-like. Hyphae lacking haustoria, with colacosomes and septal pores. Teliospores ± globose, thin-walled or thick-walled, germinating to form elongate hypha-like transversely septate basidia with basidiospores formed at the distal end of each cell. Basidiospores small, hyaline, thin-walled, discharged passively. Anamorph hyphomycetous, proliferating by budding and rarely also ballistoconidia.

Significant Genera: *Leucosporidium*. Anamorph: *Leucosporidiella*.

Distribution: Widespread, including aquatic and psychrophilic habitats.

Economic Significance: None is known.

Ecology: Isolated from a wide range of substrata.

Notes: The family has been placed in its own order *Leucosporidiales*, but it is paraphyletic as currently described in some analyses and separation from the *Microbotryales* may not be sufficient to justify this move.

References: **DePriest, P.T.; Ivanova, N.V.; Fahselt, D.; Alstrup, V.; Gargas, A.** (2000). Sequences of psychrophilic fungi amplified from glacier-preserved ascolichens. *Can. J. Bot.* 78: 1450–1459; **Joo, W.H.** (1991). Ubiquinone compounds of the strains in *Leucosporidium scottii* and its related taxa. *Korean J. Mycol.* 19: 258–265; **Sampaio, J.P.; Gadanho, M.; Bauer, R.; Weiss, M.** (2003). Taxonomic studies in the *Microbotryomycetidae*: *Leucosporidium golubevii* sp. nov., *Leucosporidiella* gen. nov. and the new orders *Leucosporidiales* and *Sporidiobolales*. *Mycol. Prog.* 2: 53–68; **Statzell-Tallman, A.; Fell, J.W.** (1998). *Leucosporidium* Fell, Statzell, I.L. Hunter & Phaff. *The Yeasts, A Taxonomic Study* (Amsterdam): 670–675; **Suh, S.O.; Sugiyama, J.** (1993). Septal spore ultrastructure of *Leucosporidium lari-marini*, a basidiomycetous yeast, and its taxonomic implications. *J. gen. appl. Microbiol.* Tokyo 39: 257–260; **Swann, E.C.; Frieders, E.M.; McLaughlin, D.J.** (1999). *Microbotryum*, *Kriegeria* and the changing paradigm in basidiomycete classification. *Mycologia* 91: 51–66.

Lichenotheliaceae Henssen 1986
Uncertain position within Dothideomycetes, Ascomycota

Ascomata formed in a dark crustose pseudoparenchymatous stroma, uniloculate, perithecial (without a clearly defined ostiole) or apothecial, globose to pulvinate; peridium of dark pseudoparenchymatous cells. Interascal tissue present (? pseudoparaphyses). Asci clavate, ? fissitunicate. Ascospores dark brown, septate, thick-walled, sometimes ornamented, with a mucous sheath. Anamorphs hyphomycetous or coelomycetous.

Significant Genera: *Lichenothelia*.

Distribution: Widespread.

Economic Significance: None is known.

Ecology: Saprobic or biotrophic, on rocks, lichens, or cyanobacterial mats.

Notes: Very poorly known; no recent information is available.

Lichenothelia sp., Castellón, Spain; ascostromata

References: **Hafellner, J.; Calatayud, V.** (1999). *Lichenostigma cosmopolites*, a common lichenicolous fungus on *Xanthoparmelia* species. *Mycotaxon* 72: 107–114; **Henssen, A.** (1987). *Lichenothelia*, a genus of microfungi on rocks. In Peveling, E. (ed.), Progress and Problems in Lichenology in the Eighties. *Biblthca Lichenol.* 25: 257–293; **Navarro-Rosinés, P.; Hafellner, J.** (1996). *Lichenostigma elongata* spec. nov. (*Dothideales*), a lichenicolous ascomycete on *Lobothallia* and *Aspicilia* species. *Mycotaxon* 57: 211–225.

Lichinaceae Nyl. 1854
Lichinales: Ascomycota

Thallus crustose, foliose or fruticose with rounded or flattened lobes, generally dark brown, greenish or black, gelatinous, without a well-defined cortex, sometimes producing vegetative propagules (hormocysts) from specialized parts of the thallus. Ascomata apothecial, ± immersed in thallus branches, usually strongly cupulate, the opening narrow or pore-like, with a well-developed wall. Interascal tissue of simple or branched and anastomosed paraphyses, often swollen at the apices, the hymenium often J+. Asci clavate to cylindrical, thin-walled at all stages, evanescent or persistent, the outer layer often blueing in iodine. Ascospores hyaline, aseptate, thin-walled, sometimes with a gelatinous coat. Anamorphs coelomycetous, pycnidial or stromatic. Conidiogenous cells flask-shaped, proliferating percurrently. Conidia small, hyaline, aseptate, usually ± globose to bacilliform.

Significant Genera: *Lempholemma*, *Lichina*, *Psorotichia*, *Pyrenopsis*.

Lichina confinis, Barra, Scotland; thallus

Distribution: Widespread in north and south temperate regions.

Economic Significance: None is known.

Ecology: Lichenized with cyanobacteria, on rocks, soil or mosses. Some species are marine, occurring around the high tide mark or in the intertidal zone.

Notes: The central family of the *Lichinales*, which molecular data suggest is quite separate from the *Lecanorales*. Species with functionally bitunicate asci (e.g. in *Pyrenopsis*) are likely to be ancestral, while the prototunicate (thin-walled and evanescent) taxa are derived. Lichenization with cyanobacteria is relatively uncommon amongst the *Ascomycota*.

References: **Henssen, A.; Dobelmann, A.** (1987). *Euopsis* and *Harpidium*, genera of *Lichinaceae* (*Lichenes*). In Peveling, E. (ed.), Progress and Problems in Lichenology in the Eighties. *Biblthca Lichenol.* 25: 103–108; **Henssen, A.; Jørgensen, P.M.** (1990). New combinations and synonyms in the *Lichinaceae*. *Lichenologist* 22: 137–147; **Jørgensen, P.M.; Henssen, A.** (1999). Further species of the lichen genus *Staurolemma* (*Collemataceae*, lichenized ascomycetes). *Bryologist* 102: 22–25; **Kantvilas, G.; Jørgensen, P.M.** (1998). Observations on the lichen genus *Lempholemma* Körb. in Australia. *Muelleria* 11: 45–50; **Lutzoni, F.; Pagel, M.; Reeb, V.** (2001). Major fungal lineages are derived from lichen symbiotic ancestors. *Nature* Lond. 411: 937–940; **Moreno, P.; Egea, J.M.** (1990). Revision de las especies de la familia *Lichinaceae* incluidas en el herbario Werner (BC). *Acta bot. Malac.* 15: 19–26; **Moreno, P.P.; Egea, J.M.** (1991). Biología y Taxonomía de la Familia *Lichinaceae*, con Especial Referencia a las Especies del S.E. Español y Norte de África (Murcia): 87 pp.; **Moreno, P.P.; Egea, J.M.** (1994). El género *Psorotichia* y especies próximas en el sureste de España y norte de Africa. *Bull. Soc. linn. Provence* 45: 291–308; **Schultz, M.; Arendholz, W.-R.; Büdel, B.** (2001). Origin and evolution of the lichenized ascomycete order *Lichinales*: monophyly and systematic relationships inferred from ascus, fruiting body and SSU rDNA evolution. *Pl. Biol.* 3: 116–123; **Schultz, M.; Porembski, S.; Büdel, B.** (2000). Diversity of rock-inhabiting cyanobacterial lichens: studies on granite inselbergs along the Orinoco and in Guyana. *Pl. Biol.* 2: 482–494; **Wedin, M.; Wiklund, E.; Crewe, A.; Döring, H.; Ekman, S.; Nyberg, A.; Schmitt, I.; Lumbsch, H.T.** (2005). Phylogenetic relationships of *Lecanoromycetes* (*Ascomycota*) as revealed by sequences of mtSSU and nLSU rDNA sequence data. *Mycol. Res.* 109: 159–172.

Lipomycetaceae E.K. Novák & Zsolt 1961
Saccharomycetales: Ascomycota

Mycelium usually poorly developed; vegetative cells usually proliferating by multilateral budding, rarely by thallic fragmentation, covered with a J+ gelatinous matrix. Asci formed either directly by budding from daughter cells or from conjugation, often multispored, sometimes elongate, usually evanescent. Ascospores usually cylindrical or allantoid, smooth or ornamented. Fermentation negative; coenzyme system Q8–Q10.

Significant Genera: *Dipodascopsis, Lipomyces*.

Distribution: Widespread.

Economic Significance: None is known.

Ecology: Mostly isolated from soil, though some species are associated with insects.

Notes: Very poorly known; no recent information is available.

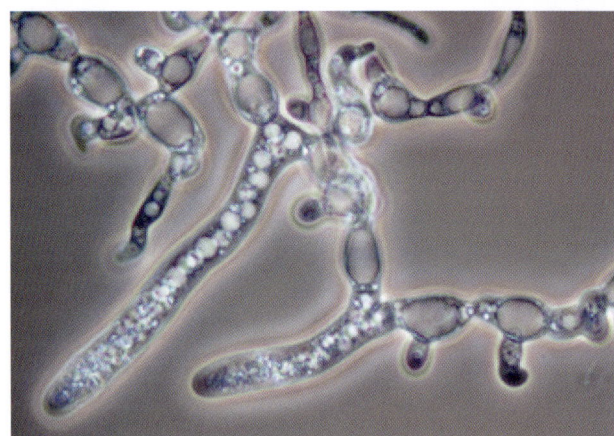
Dipodascopsis uninucleata; ascus forming from vegetative hyphae

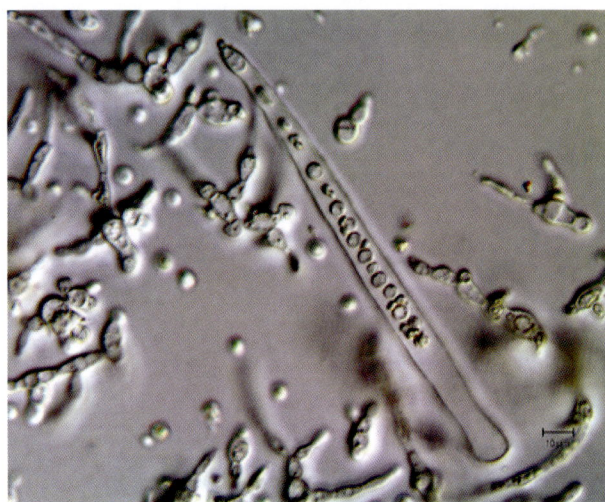
Dipodascopsis uninucleata; asci

References: **Cottrell, M.; Kock, J.L.F.** (1989). The yeast family *Lipomycetaceae* Novák et Zsolt emend. van der Walt *et al.* and the genus *Myxozyma* van der Walt *et al.* 1. A historical account of its delimitation and 2. The taxonomic relevance of cellular long-chain fatty acid composition and other phenotypic characters. *Syst. Appl. Microbiol.* 12: 291–305; **Gouliamova, D.E.; Hennebert, G.L.; Smith, M.T.; Walt, J.P. van der** (1998). Diversity and affinities among species and strains of *Lipomyces*. *Antonie van Leeuwenhoek* 74: 283–291; **Hoog, G.S. de; Smith, M.T.; Guého, E.** (1986). A revision of the genus *Geotrichum* and its teleomorphs. *Stud. Mycol.* 29: 131 pp.; **Jansen van Rensburg, E.L.; Kock, J.L.F.; Coetzee, D.J.; Botha, A.; Botes, P.J.** (1995). Lipid composition and DNA band patterns in the yeast family *Lipomycetaceae*. *Syst. Appl. Microbiol.* 18: 410–424; **Kurtzman, C.P.** (1998). Discussion of teleomorphic and anamorphic ascomycetous yeasts and a key to the genera. *The Yeasts, A Taxonomic Study* (Amsterdam): 111–121; **Kurtzman, C.P.; Liu, Z.-w.** (1990). Evolutionary affinities of species assigned to *Lipomyces* and *Myxozyma* estimated from ribosomal RNA sequence divergence. *Curr. Microbiol.* 21: 387–393; **Lomascolo, A.; Dubreucq, E.; Perrier, V.; Galzy, P.** (1994). Study of lipids in *Lipomyces* and *Waltomyces*. *Can. J. Microbiol.* 40: 724–729; **Smith, D.P.; Kock, J.L.F.; Wyk, P.W.J. van; Venter, P.; Coetzee, D.J.; Van Heerden, E.; Linke, D.; Nigam, S.** (2000). The occurrence of 3-hydroxy oxylipins in the ascomycetous yeast family *Lipomycetaceae*. *S. Afr. J. Sci.* 96: 247–249; **Smith, M.T.** (1998). Li-

pomyces Lodder & Kreger-van Rij. *The Yeasts, A Taxonomic Study* (Amsterdam): 248–253; **Walt, J.P. van der; Smith, M.T.; Roeijmans, H.J.** (1999). Four new species in *Lipomyces*. *Syst. Appl. Microbiol.* 22: 229–236; **Yamada, Y.; Nogawa, C.** (1995). *Kawasakia* gen. nov. for *Zygozyma arxii*, the Q9-equipped species in the genus *Zygozyma* (*Lipomycetaceae*). *Bull. Fac. Agric. Shizuoka Univ.* 45: 31–34.

Lobariaceae Chevall. 1826
Peltigerales: Ascomycota

Thallus foliose, irregularly spreading, lobed, the lobes often branched; upper surface smooth or wrinkled, sometimes ridged or reticulate; lower surface corticate or not, sometimes with cyphellae or pseudocyphellae (punctate or areolate openings exposing the medulla), sometimes tomentose, rhizoids present. Soredia and isidia sometimes present. Ascomata apothecial, rarely developed in many species, sessile or ± stalked, initially deeply immersed, the covering layer becoming stretched and eventually fragmenting; hymenium strongly concave when immature, often with a conspicuous margin. Interascal tissue of simple paraphyses immersed in a gelatinous matrix, often with a very well-developed pigmented epithecial layer. Asci with a thickened apex, with a large J+ ring, ocular chamber often poorly developed, sometimes with an outer J+ gelatinized layer. Ascospores often elongated, hyaline or brown, transversely septate. Anamorphs coelomycetous. Conidiomata pycnidial, at least the ostiolar region pigmented. Conidiogenous cells elongate or bacilliform, formed from percurrently proliferating conidiogenous cells. Conidia small, hyaline.

Significant Genera: *Lobaria, Pseudocyphellaria, Sticta*.

Lobaria pulmonaria, Vaucluse, France; thallus

Distribution: Widespread, with good representation especially in the tropics and southern temperate regions.

Lobaria virens, Agyll, Scotland; thallus and ascomata

Pseudocyphellaria crocata, Norway; thallus with yellow soredia

Sticta humboldtii; thallus

Economic Significance: None is known.

Ecology: Lichenized with green algae or cyanobacteria; if the former then usually with cyanobacteria in cephalodia.

Notes: Many species are large and conspicuous and have been used as indicator taxa for pollution and woodland management. There has been some research on their conservation status, focusing especially on genetic variation.

References: **Galloway, D.J.** (1988). Studies in *Pseudocyphellaria* (lichens) I. The New Zealand species. *Bull. Br. Mus. nat. Hist.* Bot. 17: 1–267; **Galloway, D.J.** (1991). Chemical evolution in the order *Peltigerales*: triterpenoids. *Symbiosis* 11: 327–344; **Galloway, D.J.** (1998). Studies on the lichen genus *Sticta* (Schreber) Ach.: V. Australian species. *Trop. Bryol.* 15: 117–160; **Galloway, D.J.; Stenroos, S.; Ferraro, L.I.** (1995). *Lichenes Peltigerales: Lobariaceae y Stictacea. Fl. criptog. Tierra del Fuego* 13: 78 pp.; **McDonald, T.; Miadlikowska, J.; Lutzoni, F.** (2003). The lichen genus *Sticta* in the Great Smoky Mountains: a phylogenetic study of morphological, chemical, and molecular data. *Bryologist* 106: 61–79; **Miadlikowska, J.; Lutzoni, F.** (2004). Phylogenetic classification of peltigeralean fungi (*Peltigerales, Ascomycota*) based on ribosomal RNA small and large subunits. *Am. J. Bot.* 91: 449–464; **Sillet, S.C.; Goward, T.** (1998). Ecology and conservation of *Pseudocyphellaria rainierensis*, a Pacific Northwest endemic lichen. *Lichenographia Thomsoniana, North American Lichenology in Honor of John W. Thomson* (Ithaca): 377–388; **Stenroos, S.; Stocker-Wörgötter, E.; Yoshimura, I.; Myllys, L.; Thell, A.; Hyvönen, J.** (2003). Culture experiments and DNA sequence data confirm the identity of *Lobaria* photomorphs. *Can. J. Bot.* 81: 232–247; **Walser J.-C.; Sperisen, C.; Soliva, M.; Scheidegger, C.** (2003). Fungus-specific microsatellite primers of lichens: application for the assessment of genetic variation on different spatial scales in *Lobaria pulmonaria*. *Fungal Genetics Biol.* 40: 72–82; **Wiklund, E.; Wedin, M.** (2003). The phylogenetic relationships of the cyanobacterial lichens in the *Lecanorales* suborder *Peltigerineae*. *Cladistics* 19: 419–431; **Zoller, S.; Lutzoni, F.; Scheidegger, C.** (1999). Genetic variation within and among populations of the threatened lichen *Lobaria pulmonaria* in Switzerland and implications for its conservation. *Mol. Ecol.* 8: 2049–2059.

Lophiostomataceae Sacc. 1883
Pleosporales: Ascomycota

Ascostromata perithecial, immersed or erumpent, often strongly flattened, often aggregated, black, usually papillate, with a round or slit-like, sometimes periphysate, lysigenous pore; peridium of small pseudoparenchymatous cells, usually thickened above, sometimes clypeate. Interascal tissue copious, of narrow cellular pseudoparaphyses. Asci ± widely cylindrical, fissitunicate, with a distinct ocular chamber, not blueing in iodine. Ascospores transversely septate, occasionally muriform, constricted at the primary and secondary septa, often with a sheath or appendages. Anamorphs mostly coelomycetous, poorly known.

Significant Genera: *Lophiostoma, Massariosphaeria*.

Distribution: Widespread, especially in temperate zones.

Economic Significance: None is known.

Ecology: Saprobic or necrotrophic on herbaceous and woody stems.

Lophiostoma angustilabrum, Derbyshire, UK; ascomata with elongate ostioles

Notes: *Lophiostoma* species have distinctive slit-like ostioles. The delimitation of the family has been unstable in recent years and its limits are still not properly understood. The *Massarinaceae* has been synonymized with the *Lophiostomataceae*, but *Massarina* itself is highly polyphyletic and its type species does not cluster with the main *Lophiostomataceae* clade.

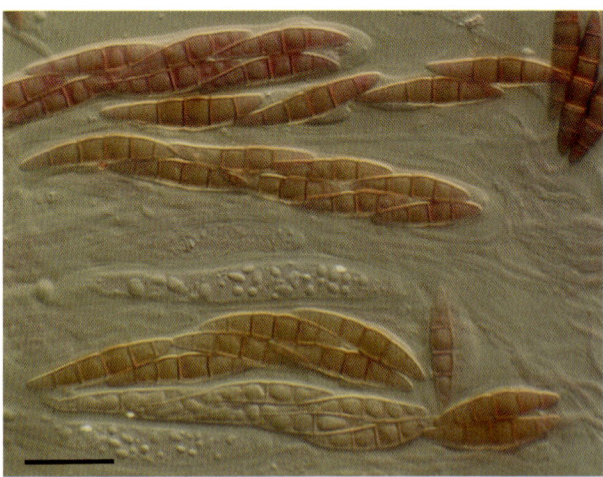

Lophiostoma arundinis, Devon, UK; asci, ascospores and interascal tissue

References: **Aptroot, A.** (1998). A world revision of *Massarina* (*Ascomycota*). *Nova Hedwigia* 66: 89–162; **Barr, M.E.** (1990). *Melanommatales* (*Loculoascomycetes*). *N. Amer. Fl.* Ser. 2 13: 129 pp.; **Berbee, M.L.** (1996). Loculoascomycete origins and evolution of filamentous ascomycete morphology based on 18S rRNA gene sequence analysis. *Mol. Biol. Evol.* 13: 462–470; **Checa, J.** (1997). Annotated list of the *Lophiostomataceae* and *Mytilinidiaceae* (*Dothideales, Ascomycotina*) reported from the Iberian Peninsula and Balearic Islands. *Mycotaxon* 63: 467–491; **Eriksson, O.E.; Hawksworth, D.L.** (2003). *Saccharicola*, a new genus for two *Leptosphaeria* species on sugar cane. *Mycologia* 95: 426–433; **Holm, L.** (1986). A note on *Byssolophis ampla*. *Windahlia* 16: 49–52; **Holm, L.; Holm, K.** (1988). Studies in the *Lophiostomataceae* with emphasis on the

Swedish species. *Symb. bot. upsal.* 28: 50 pp.; **Hyde, K.D.; Aptroot, A.; Fröhlich, J.; Taylor, J.E.** (2000). Fungi from palms. XLIII. *Lophiostoma* and *Astrosphaeriella* species with slit-like ostioles. *Nova Hedwigia* 70: 143–160; **Kohlmeyer, J.** (1986). *Ascocratera manglicola* gen. et sp. nov. and key to the marine loculoascomycetes on mangroves. *Can. J. Bot.* 64: 3036–3042; **Liew, E.C.Y.; Aptroot, A.; Hyde, K.D.** (2000). Phylogenetic significance of the pseudoparaphyses in loculoascomycete taxonomy. *Mol. Phylogen. Evol.* 16: 392–402; **Liew, E.C.Y.; Aptroot, A.; Hyde, K.D.** (2002). An evaluation of the monophyly of *Massarina* based on ribosomal DNA sequences. *Mycologia* 94: 803–813.

Loramycetaceae Dennis ex Digby & Goos 1988
Helotiales: Ascomycota

Stromata absent. Ascomata deeply cupulate, almost perithecial, formed within an exogenous gelatinous matrix; peridium hyaline, thin-walled. Interascal tissue of thin-walled simple paraphyses. Asci with a wide pore surrounded by a poorly developed J– ring. Ascospores 2-septate, hyaline, with a long basal cellular appendage, with a gelatinous sheath. Anamorphs hyphomycetous, *Anguillospora*-like.

Significant Genera: *Loramyces*.

Distribution: Known from North America and Europe.

Economic Significance: None is known.

Ecology: Saprobic on submerged decaying plant tissue.

Notes: Possibly related to *Phialocephala*, a group of root endophytic fungi placed by some authors in the *Vibrisseaceae*, but more data are required.

Loramyces juncicola, Cumbria, UK; ascomata on dead leaves

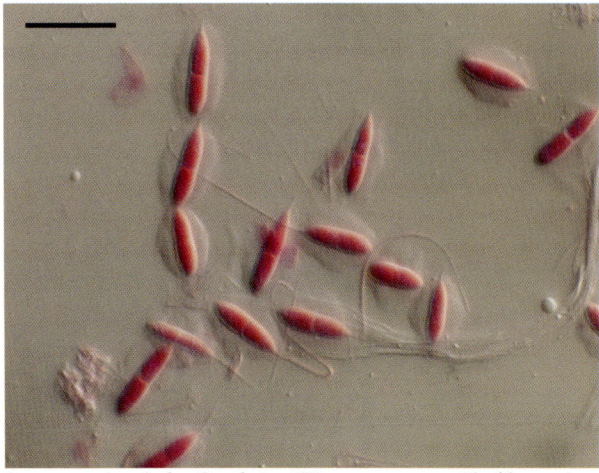
Loramyces juncicola, Cumbria, UK; ascospores with filamentous appendage and gelatinous sheath

References: **Digby, S.; Goos, R.D.** (1988). Morphology, development and taxonomy of *Loramyces*. *Mycologia* 79: 821–831; **Gernandt, D.S.; Platt, J.L.; Stone, J.K.; Spatafora, J.W.; Holst-Jensen, A.; Hamelin, R.C.; Kohn, L.M.** (2001). Phylogenetics of *Helotiales* and *Rhytismatales* based on partial small subunit nuclear ribosomal DNA sequences. *Mycologia* 93: 915–933; **Wang, Z.; Binder, M.; Schoch, C.L.; Johnston, P.R.; Spatafora, J.W.; Hibbett, D.S.** (2006). Evolution of helotialean fungi (*Leotiomycetes, Pezizomycotina*): a nuclear rDNA phylogeny. *Mol. Phylogen. Evol.* 41: 295–312.

Loxosporaceae Kalb & Staiger 1995
Lecanorales: Ascomycota

Thallus crustose, often with soredia. Ascomata apothecial, sessile, the disc reddish, often pruinose, with a well-defined thickened margin, sometimes also with a thalline margin. Interascal tissue of relatively wide, sparingly branched and anastomosed paraphyses, not gelatinized. Asci with a well-developed J+ tholus with a distinct ocular chamber and an outer J+ mucous layer. Ascospores hyaline, fusiform, aseptate or septate, thin-walled, without a mucus sheath. Anamorph pycnidial.

Significant Genera: *Loxospora*.

Distribution: Widespread, primarily north temperate.

Economic Significance: None is known.

Ecology: Lichenized with green algae, generally on bark.

Notes: At one time treated as part of the *Haematommataceae*, doubtless due to the red coloration of the ascomata. Little recent research is available.

Loxospora elatina, Norway; thallus and soredia

References: Staiger, B.; Kalb, K. (1995). *Haematomma*-Studien, I. Die Flechtengattung *Haematomma*. *Biblthca Lichenol.* 59: 3–198.

Lulworthiaceae Kohlm., Spatafora & Volkm.-Kohlm. 2000
Lulworthiales: Ascomycota

Stromata absent. Ascomata perithecial, subglobose to cylindrical, immersed or superficial, brown to black, glabrous or with a subiculum; peridium coriaceous to carbonaceous. Interascal tissue absent, the ascomatal cavity initially filled with deliquescing, thin-walled, parenchymatous cells. Asci cylindrical to fusiform, thin-walled and deliquescing, without apical structures, 8-spored. Ascospores mostly filiform, often many-septate, hyaline, sometimes with small apical mucus-secreting chambers. Anamorphs hyphomycetous. Conidiophores undifferentiated, non-proliferating. Conidia hyaline, filamentous or dark coiled; sometimes producing resting spores.

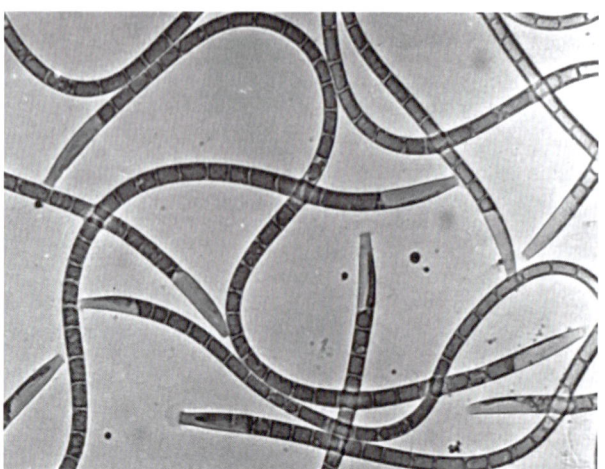

Lulworthia lignoarenaria, Denmark; ascospores

Significant Genera: *Lulworthia*.

Distribution: Widespread in both temperate and tropical oceans.

Economic Significance: None is known.

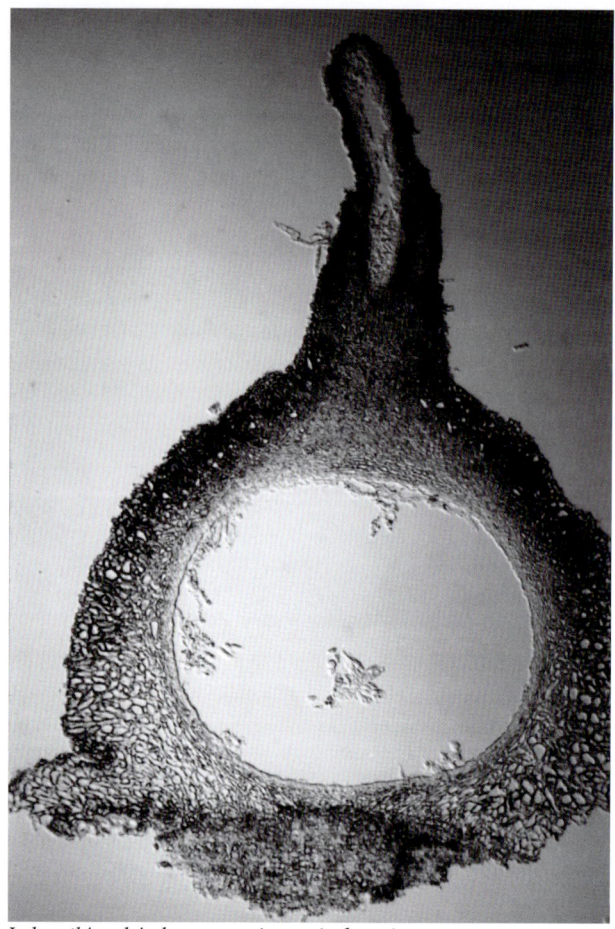

Lulworthia calcicola; ascoma in vertical section

Ecology: Marine, usually on immersed wood or marine algae.

Notes: A phylogenetically isolated family that appears to be basal to the *Sordariomycetes*; certainly not closely related to the *Halosphaeriales* where it formerly resided. *Kohlmeyeriella* is a recent addition but *Lulworthia* itself is polyphyletic.

References: Campbell, J.; Shearer, C.A.; Mitchell, J.I.; Eaton, R.A. (2002). *Corollospora* revisited: a molecular approach. *Fungi in Marine Environments*: 15–33; **Campbell, J.; Volkmann-Kohlmeyer, B.; Gräfenhan, T.; Spatafora, J.W.; Kohlmeyer, J.** (2005). A re-evaluation of *Lulworthiales*: relationships based on 18S and 28S rDNA. *Mycol. Res.* 109: 556–568; **Inderbitzin, P.; Lim, S.R.; Volkmann-Kohlmeyer, B.; Kohlmeyer, J.; Berbee, M.L.** (2004). The phylogenetic position of *Spathulospora* based on DNA sequences from dried herbarium material. *Mycol. Res.* 108: 737–748; **Kohlmeyer, J.; Volkmann-Kohlmeyer, B.** (1989). A new *Lulworthia* (*Ascomycotina*) from corals. *Mycologia* 81: 289–292; **Kohlmeyer, J.W.; Spatafora, J.W.; Volkmann-Kohlmeyer, B.** (2000). *Lulworthiales*, a new order of marine *Ascomycota*. *Mycologia* 92: 453–458; **Nakagiri, A.** (1984). Two new species of *Lulworthia* and evaluation of genera-delimiting characters between *Lulwor-*

thia and *Lindra* (*Halosphaeriaceae*). *Trans. Mycol. Soc. Japan* 25: 377–388; **Spatafora, J.W.; Volkmann-Kohlmeyer, B.; Kohlmeyer, J.** (1998). Independent terrestrial origins of the *Halosphaeriales* (marine *Ascomycota*). *Am. J. Bot.* 85: 1569–1580; **Yusoff, M.; Jones, E.B.G.; Moss, S.T.** (1995). Ascospore ultrastructure in the marine genera *Lulworthia* Sutherland and *Lindra* Wilson. *Cryptog. bot.* 5: 307–315; **Zuccaro, A.; Schulz, B.; Mitchell, J.I.** (2003). Molecular detection of ascomycetes associated with *Fucus serratus*. *Mycol. Res.* 107: 1451–1466.

Ecology: Primarily saprobes in grassland and woodland, with some species commonly forming typical 'fairy rings'.

Notes: The family clusters within the *Agaricaceae* in molecular analyses and may well be placed into synonymy in future classifications. Family limits remain to be agreed.

Lycoperdaceae Chevall. 1826
Agaricales: Basidiomycota

Basidiomata gasteroid, sometimes very large, epigeous at maturity, ± globose or pyriform to irregular, sessile or shortly stalked, mostly white or pale brown, smooth, granular or weakly spinose, releasing spores through an apical pore or irregular opening. Outer peridial layer thin, often inconspicuous, persistent or evanescent. Inner peridium usually with multiple layers. Hyphae lacking clamp connections. Gleba composed of a single chamber with a well-developed capillitial structure, columella lacking, degenerating into a powdery mass. Capillitial hyphae long, aseptate or remotely septate, sclerified, branched or unbranched, sometimes ornamented with circular or linear pits. Basidia clavate with distinct sterigmata, evanescent. Basidiospores globose to ovoid, sometimes remaining attached to the broken-off sterigma, pale to medium brown, smooth to verrucose, often staining brown in iodine.

Significant Genera: *Bovista, Calvatia, Lycoperdon, Vascellum*.

Distribution: Cosmopolitan.

Handkea excipuliformis, Surrey, UK; basidioma

Economic Significance: Some species (e.g. *Calvatia gigantea*) are prized as edible fungi. *Lycoperdon* species have been investigated as sources of antibiotics.

Lycoperdon perlatum, Hampshire, UK; basidiomata

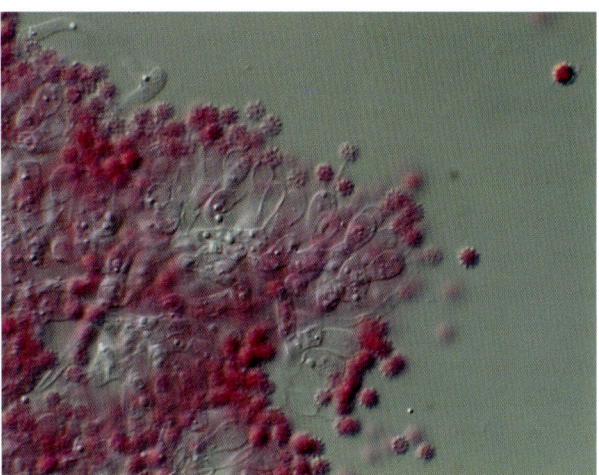

Lycoperdon molle, Surrey, UK; basidia and basidiospores

References: **Agerer, R.** (2002). Rhizomorph structures confirm the relationship between *Lycoperdales* and *Agaricaceae* (*Hymenomycetes, Basidiomycota*). *Nova Hedwigia* 75: 367–385; **Baseia, I.G.** (2003). Contribution to the study of the genus *Calvatia* (*Lycoperdaceae*) in Brazil. *Mycotaxon* 88: 107–112; **Hibbett, D.S.; Pine, E.M.; Langer, E.; Langer, G.; Donoghue, M.J.** (1996). Evolution of gilled mushrooms and puffballs inferred from ribosomal DNA sequences. *Proc. natn Acad. Sci. U.S.A.* 94: 12002–12006; **Kreisel, H.** (1998). Die Gattungen *Calvatia* und *Handkea* in Europa und der Arktis. *Öst. Z. Pilzk.* 7: 215–225; **Kreisel, H.; Moreno, G.** (1996). The genus *Handkea* Kreisel (*Basidiomycetes, Lycoperdaceae*) in the southern hemisphere. *Feddes Repert.* 107: 83–87; **Krüger, D.; Binder, M.; Fischer, M.; Kreisel, H.** (2001). The *Lycoperdales*. A molecular approach to the systematics of some gasteroid mushrooms. *Mycologia* 93: 947–957; **Krüger, D.; Kreisel, H.** (2003). Proposing *Morganella* subgen. *Apioperdon* subgen. nov. for the puffball *Lycoperdon pyriforme*. *Mycotaxon* 86: 169–177; **Lebel, T.; Thompson, D.K.; Udovicic, F.** (2004). Description and affinities of a new sequestrate fungus, *Barcheria willisiana* gen. et sp. nov. (*Agaricales*) from Australia. *Mycol. Res.* 108: 206–213; **Pegler, D.N.;**

Young, T.W.K. (1994). *Lycoperdopsis*, a genus belonging to Lycoperdaceae. *Mycol. Res.* 98: 904–906; **Portman, R.; Moseman, R.; Levetin, E.** (1997). Ultrastructure of basidiospores in North American members of the genus *Calvatia*. *Mycotaxon* 62: 435–443; **Suárez, V.L.; Wright, J.E.** (1996). South American gasteromycetes V: The genus *Morganella*. *Mycologia* 88: 655–661; **Terashima, Y.; Fukiharu, T.; Fujiie, A.** (2004). Morphology and comparative ecology of the fairy ring fungi, *Vascellum curtisii* and *Bovista dermoxantha*, on turf of bentgrass, bluegrass, and Zoysiagrass. *Mycoscience* 45: 251–260; **Vellinga, E.C.** (2004). Genera in the family *Agaricaceae*: evidence from nrITS and nrLSU sequences. *Mycol. Res.* 108: 354–377.

Magnaporthaceae P.F. Cannon 1994
Uncertain position within Sordariomycetes, Ascomycota

Stromata absent. Sclerotia sometimes formed, lobed appressoria usually present. Ascomata perithecial, black, immersed in plant tissues, usually with long hairy necks, sometimes accompanied by setae. Interascal tissue of simple, thin-walled, tapering paraphyses. Asci ± cylindrical, persistent, fairly thick-walled, without separable layers, with a large apical pore surrounded by an often J+ apical ring. Ascospores septate, often filiform, often with the middle cells pigmented, without germ pores, without a sheath. Anamorphs varied, hyphomycetous, sometimes formed directly from sclerotium-like structures. Conidiophores simple, often undifferentiated. Conidiogenous cells hyphal or much reduced, usually pigmented, proliferating percurrently with flared collarettes. Conidia hyaline, often elongate and transversely septate, usually curved, lacking gelatinous sheaths or appendages.

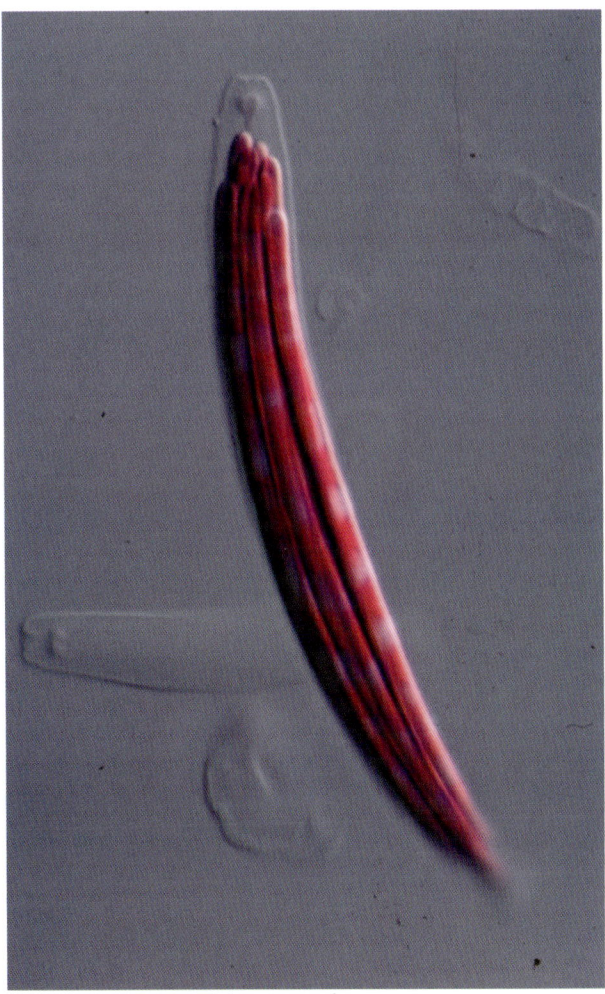

Gaeumannomyces graminis; ascus and ascospores

Significant Genera: *Gaeumannomyces, Magnaporthe*. Anamorph: *Pyricularia*.

Distribution: Widespread especially in temperate regions.

Economic Significance: Some species cause economically important diseases of grasses, especially the rice blast pathogen *Pyricularia grisea* (anamorph of *Magnaporthe grisea*) and the take-all fungi, species of *Gaeumannomyces*, on wheat and other cereals.

Ecology: Usually necrotrophic on roots.

Notes: *Magnaporthe* has been used widely as a model organism for genetics research, especially into aspects of pathogenicity, and the whole genome sequence of *M. grisea* is available on the Internet. *Ceratosphaeria* appears to belong to this family based on molecular data, but its life strategy is very different.

Magnaporthe grisea (*Pyricularia* anamorph); conidiophores and conidia

References: Augustin, C.; Ulrich, K.; Ward, E.; Werner, A. (1999). RAPD-based inter- and intravarietal classification of fungi of the *Gaeumannomyces-Phialophora* complex. *J. Phytopath.* 147: 109–117; **Bryan, G.T.; Labourdette, E.; Melton, R.E.; Nicholson, P.; Daniels, M.J.; Osbourn, A.E.** (1999). DNA polymorphism and host range in the take-all fungus, *Gaeumannomyces graminis*. *Mycol. Res.* 103: 319–327; **Bunting, T.E.; Plumley, K.A.; Clarke, B.B.; Hillman, B.I.** (1994). Phylogenic comparison of *Magnaporthe* spp. using ITS-1 sequence analysis. *Phytopathology* 84: 1097; **Bussaban, B.; Lumyong, S.; Lumyong, P.; Seelanan, T.; Park, D.C.; McKenzie, E.H.C.; Hyde, K.D.** (2006). Molecular and morphological characterization of *Pyricularia* and allied genera. *Mycologia* 97: 1002–1011; **Cannon, P.F.** (1994). The newly recognized family Magnaporthaceae and its interrelationships. *Syst. Ascom.* 13: 25–42; **Chen, W.; Shearer, C.A.; Crane, J.L.** (1999). Phylogeny of *Ophioceras* spp. based on morphological and molecular data. *Mycologia*

91: 84–94; **Kumar, J.; Nelson, R.J.; Zeigler, R.S.** (1999). Population structure and dynamics of *Magnaporthe grisea* in the Indian Himalayas. *Genetics* Bethesda 152: 971–984; **Kusaba, M.; Eto, Y.; Don, L.D.; Nishimoto, N.; Tosa, Y.; Nakayashiki, H.; Mayama, S.** (2000). Genetic diversity in *Pyricularia* isolates from various hosts revealed by polymorphisms of nuclear ribosomal DNA and the distribution of the MAGGY retrotransposon. *Ann. phytopath. Soc. Japan* 65: 588–596; **Réblová, M.** (2006). Molecular systematics of *Ceratostomella* sensu lato and morphologically similar fungi. *Mycologia* 98: 68–93; **Ulrich, K.; Augustin, C.; Werner, A.** (2000). Identification and characterization of a new group of root-colonizing fungi within the *Gaeumannomyces-Phialophora* complex. *New Phytol.* 145: 127–135; **Ward, E.; Bateman, G.L.** (1999). Comparison of *Gaeumannomyces-* and *Phialophora*-like fungal pathogens from maize and other plants using DNA methods. *New Phytol.* 141: 323–331.

Marasmiaceae Roze ex Kühner 1980
Agaricales: Basidiomycota

Basidiomata pileate and usually stipitate, often drought-tolerant, the cap convex to campanulate, often appearing pleated, sometimes inrolled, usually verrucose, hairy or setose, the hairs often browning in iodine, the stipe central or eccentric, fleshy, cartilaginous or wiry, sometimes hairy, veil absent. Hyphal system monomitic, often staining brown in iodine, clamp connections usually present, cystidia often present. Hymenium lamellate, the gills sometimes narrower at the point of attachment. Basidia cylindrical to clavate, usually with 4 sterigmata, often interspersed with thick-walled irregularly branched, sometimes ± spinose hyphae. Basidiospores ellipsoidal to cylindric or clavate, hyaline, thin-walled, smooth, not staining in iodine. Rhizomorphs often prominent.

Significant Genera: *Crinipellis*, *Marasmius*, *Moniliophthora*.

Distribution: Cosmopolitan, but with the centre of variation in the tropics.

Economic Significance: Species of *Moniliophthora* are devastating pathogens of tree crops, causing witches' broom disease and frosty pod rot of cacao.

Ecology: Saprobic on litter and dead wood, or parasitic on woody plants. The family is often one of the most prominent fungal components of tropical forests, playing a major role in nutrient cycling.

Notes: The classification of this family in the *Dictionary of the Fungi* edn 9 requires restructuring; molecular data indicates the genera accepted there belong to a number of distinct clades. Most prominent amongst these is *Armillaria*, a highly important tree pathogen group which has been provisionally transferred to the *Physalacriaceae* based on molecular evidence. *Favolaschia* should probably belong in the *Mycenaceae*. The genus *Marasmius* itself is polyphyletic with some species apparently belonging to the *Omphalotaceae*.

Marasmius rotula, Argyll, Scotland; basidiomata

Marasmiellus candidus, Aude, France; basidiomata

Marasmiellus scandens; disease symptoms on *Theobroma cacao*

References: **Abesha, E.; Caetano-Anollés, G.; Høiland, K.** (2003). Population genetics and spatial structure of the fairy ring fungus *Marasmius oreades* in a Norwegian sand dune ecosystem. *Mycologia* 95: 1021–1031; **Antonín, V.** (1999). [An annotated key to European species of the genera *Chaetocalathus* and *Crinipellis*]. *Mykol. Listy* 68: 13–15; **Antonín, V.** (2003). New species of marasmioid genera (*Basidiomycetes*, *Tricholomataceae*) from tropical Africa – II. *Gloiocephala*, *Marasmius*, *Setulipes* and two new combinations. *Mycotaxon* 88: 53–78; **Antonín, V.; Halling, R.E.; Noordeloos, M.E.** (1997). Generic concepts within the groups of *Marasmius* and *Collybia sensu lato*. *Mycotaxon* 63: 359–368; **Arruda, M.C.C. de; Ferreira, M.A.S.V.; Miller, R.N.G.; Resende, M.L.V.; Felipe, M.S.S.** (2003). Nuclear and mitochondrial rDNA variability in *Crinipellis perniciosa* from different geographic origins and hosts. *Mycol. Res.* 107: 25–37; **Bodensteiner, P.; Binder, M.; Moncalvo, J.-M.; Agerer, R.; Hibbett, D.S.** (2004). Phylogenetic relationships of cyphelloid homobasidiomycetes. *Mol. Phylogen. Evol.* 33: 501–515; **Corner, E.J.H.** (1996). The agaric genera *Marasmius, Chaetocalathus, Crinipellis, Heimiomyces, Resupinatus, Xerula* and *Xerulina* in Malesia. *Nova Hedwigia* Beih. 111: 1–164; **Desjardin, D.E.; Horak, E.** (1997). *Marasmius* and *Gloiocephala* in the South Pacific Region: Papua New Guinea, New Caledonia, and New Zealand taxa. *In* Petrini, O.; Petrini, L.E. & Horak, E. (eds), Taxonomic Monographs of *Agaricales* II. *Biblthca Mycol.* 168: 152 pp.; **Desjardin, D.E.; Retnowati, A.; Horak, E.** (2000). *Agaricales* of Indonesia. 2. A preliminary monograph of *Marasmius* from Java and Bali. *Sydowia* 52: 92–193; **Gordon, S.A.; Desjardin, D.E.; Petersen, R.H.** (1994). Mating systems in *Marasmius*: additional evidence to support sectional consistency. *Mycol. Res.* 98: 200–204; **Griffith, G.W.; Nicholson, J.; Nenninger, A.; Birch, R.N.; Hedger, J.N.** (2003). Witches' brooms and frosty pods: two major pathogens of cacao. *N.Z. Jl Bot.* 41: 423–435; **Moncalvo, J.-M.; Vilgalys, R.; Redhead, S.A.; Johnson, J.E.; James, T.Y.; Aime, M.C.; Hofstetter, V.; Verduin, S.J.W.; Larsson, E.; Baroni, T.J.; Thorn, R.G.; Jacobsson, S.; Clemencon, H.; Miller, O.K.** (2002). One hundred and seventeen clades of euagarics. *Mol. Phylogen. Evol.* 23: 357–400; **Mossebo, D.C.; Antonín, V.** (2004). *Marasmius* species (*Tricholomataceae*) found in man-influenced habitats in the vicinity of Yaoundé, Cameroon. *Czech Mycol.* 56: 85–111; **Rincones, J.; Mazotti, G.D.; Griffith, G.W.; Pomela, A.; Figueira, A.; Leal, G.A.; Queiroz, M.V.; Pereira, J.F.; Azevedo, R.A.; Pereira, G.A.; Meinhardt, L.W.** (2006). Genetic variability and chromosome-length polymorphisms of the witches' broom pathogen *Crinipellis perniciosa* from various plant hosts inSouth America. *Mycol. Res.* 110: 821–832; **Segedin, B.P.** (1993). Studies in the *Agaricales* of New Zealand: some new and revised species of *Campanella* (*Tricholomataceae: Collybieae*). *N.Z. Jl Bot.* 31: 375–384; **Takahashi, H.** (2002). Four new species of *Crinipellis* and *Marasmius* in eastern Honshu, Japan. *Mycoscience* 43: 343–350; **Wilson, A.W.; Desjardin, D.E.** (2005). Phylogenetic relationships in the gymnopoid and marasmioid fungi (*Basidiomycetes*, euagarics clade). *Mycologia* 97: 667–679.

Massalongiaceae Wedin, P.M. Jørg. & Wiklund 2007
Peltigerales: Ascomycota

Thallus large, squamulose or fruticulose, lobed or rosette-shaped, brown, the upper surface corticate, the lower surface tomentose. Ascomata apothecial, formed on the upper surface, laminar or marginal, with a thin covering layer which breaks apart at an early stage to expose the hymenium, brown or black. Interascal tissue of simple or branched paraphyses. Asci fissitunicate with a thickened apex and a J+ gelatinized outer layer, lacking an amyloid apical tube. Ascospores hyaline, ellipsoidal to fusiform, transversely septate. Anamorphs pycnidial.

Significant Genera: *Massalongia*.

Distribution: Widespread, especially in north temperate zones.

Economic Significance: None is known.

Ecology: Lichenized with cyanobacteria, on rock, overgrowing mosses etc.

Notes: A recently recognized family that is clearly distinct from other members of the *Peltigerales* in molecular characters and with a distinctive ascus structure.

References: **Henssen, A.** (1963). The North American species of *Massalongia* and generic relationships. *Can. J. Bot.* 41: 1331-1346; **Miadlikowska, J.; Lutzoni, F.** (2004). Phylogenetic classification of peltigeralean fungi (*Peltigerales, Ascomycota*) based on ribosomal RNA small and large subunits. *Am. J. Bot.* 91: 449–464; **Wedin, M.; Jorgensen, P.M.; Wiklund, E.** (2007). *Massalongiaceae* fam. nov., an overlooked monophyletic group among the cyanobacterial lichens (*Peltigerales, Lecanoromycetes, Ascomycota*). *Lichenologist* 39: 61-67; **Wedin, M.; Wiklund, E.** (2004). The phylogenetic relationships of *Lecanorales* suborder *Peltigerineae* revisited. *Symb. bot. upsal.* 34: 469–475.

Massariaceae Nitschke 1869
Pyrenulales: Ascomycota

Ascomata perithecial, immersed to erumpent, often aggregated within pseudostromatic tissues or beneath a clypeus, ± globose, black; peridium composed of small-celled pseudoparenchymatous tissue. Interascal tissue of trabeculate pseudoparaphyses, true paraphyses not formed. Asci cylindrical, persistent, with a conspicuous refractive apical ring. Ascospores hyaline to dark brown, transversely septate or muriform, often with thickened septa, often with a sheath. Anamorphs unknown.

Significant Genera: *Decaisnella, Massaria*.

Massaria inquinans, Gwynedd, Wales; ascospore

Distribution: Cosmopolitan, though more well-known in temperate areas.

Economic Significance: None is known.

Ecology: Probably saprobic in wood and bark. A few species are weak pathogens.

Notes: A poorly known family that has hardly been examined using molecular techniques and placement within the *Pyrenulales* needs re-examining. The large and complex ascospores are diagnostic.

References: **Barr, M.E.** (1986). On *Julella, Delacourea,* and *Decaisnella,* three dictyosporous genera described by J.H.Fabre. *Sydowia* 38: 11–19; **Barr, M.E.** (1990). Melanommatales (Loculoascomycetes). *N. Amer. Fl.* Ser. 2 13: 129 pp.; **Hyde, K.D.** (1992). *Aigialus striatispora* sp. nov. from intertidal mangrove wood. *Mycol. Res.* 96: 1044–1046; **Hyde, K.D.; Wong, S.W.; Jones, E.B.G.** (1996). Tropical Australian freshwater fungi. XI. *Mamillisphaeria dimorphospora* gen. et sp. nov. and notes on freshwater ascomycetes with dimorphic ascospores. *Nova Hedwigia* 62: 513–520; **Liew, E.C.Y.; Aptroot, A.; Hyde, K.D.** (2000). Phylogenetic significance of the pseudoparaphyses in loculoascomycete taxonomy. *Mol. Phylogen. Evol.* 16: 392–402.

Mastodiaceae Zahlbr. 1907
Uncertain position within Ascomycetes, Ascomycota

Stromata absent or reduced to a small clypeus-like structure. Ascomata perithecial, ± globose, black, thick-walled, weakly papillate, ostiole periphysate. Interascal tissue of simple paraphyses, sometimes evanescent, or of apical paraphyses. Asci clavate, thin-walled, without apical structures, evanescent or ? persistent. Ascospores hyaline, simple or transversely septate, thin-walled, without a mucous sheath. Anamorphs unknown.

Significant Genera: *Mastodia*.

Distribution: Widespread, especially in austral regions.

Economic Significance: None is known.

Ecology: Parasitic or lichenized with large marine algae.

Notes: Very poorly known; no recent information is available.

References: **Kohlmeyer, J.; Volkmann-Kohlmeyer, B.** (1991). Illustrated key to the filamentous higher marine fungi. *Bot. Mar.* 34: 1–61; **Sancho, L.G.; Valladares, F.** (1993). Lichen colonization of recent moraines on Livingston Island (South Shetland I., Antarctica). *Polar Biol.* 13: 227–233.

Medeolariaceae Korf 1982
Medeolariales: Ascomycota

Stroma absent. Ascomata ± absent, the hymenium exposed in an indefinite palisade on surface of the host, originating from epidermal tissues. Interascal tissues well-developed, of simple or branched pigmented rather irregular paraphyses. Asci clavate, thin-walled, without apical structures, persistent, 8-spored. Ascospores aseptate, brown, fusiform, flattened on one side, striate. Anamorph unknown.

Significant Genera: *Medeolaria*.

Distribution: Only known from eastern North America.

Economic Significance: None is known.

Ecology: Biotrophic and gall-forming on *Medeola* (*Liliaceae*).

Notes: Apparently a very isolated family; the single sequence currently available does not match any other closely.

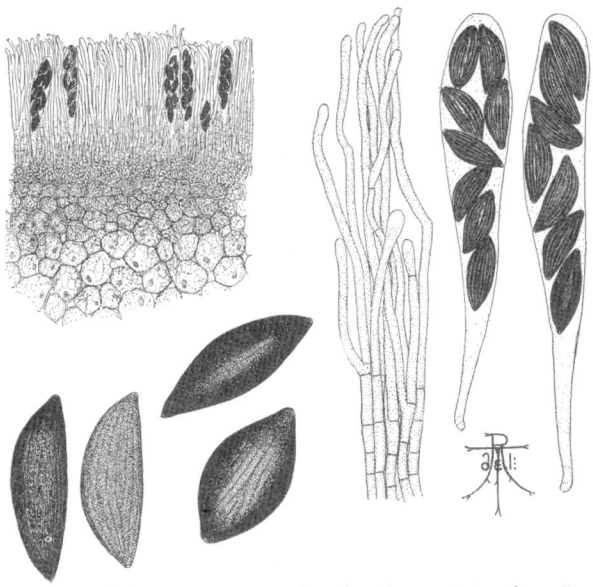

Medeolaria farlowii; image reproduced with permission from Proc. Amer. Acad. Arts Sci 57: 431, 1922

References: **Korf, R.P.** (1990). Discomycete systematics today: a look at some unanswered questions in a group of unitunicate ascomycetes. *Mycosystema* 3: 19–27; **Pfister, D.H.; Kimbrough, J.W.** (2001). Discomycetes. *The Mycota* (Berlin) 7: 257–281.

Megalariaceae Hafellner 1984
Lecanorales: Ascomycota

Thallus crustose, soredia sometimes present. Ascomata apothecial, with a well-developed margin of hyphal tissue, often bluish; hymenial gel mostly J–. Interascal tissue of sparingly branched and anastomosing paraphyses. Asci with a well-developed J+ tholus with a non-reactive axial body and a distinct ocular chamber, covered with a J+ mucus layer. Ascospores septate or aseptate, the wall thick and two-layered, without a mucous sheath. Anamorph pycnidial.

Significant Genera: *Megalaria*.

Distribution: Widespread.

Economic Significance: None is known.

Ecology: Lichenized with chlorococcoid algae.

Notes: *Megalaria* clusters within the *Bacidiaceae* in its traditional circumscription and also close to *Ramalina*. A reassessment of the classification is required.

Megalaria pulverea, Devon, UK; thallus and ascomata

References: **Ekman, S.** (2001). Molecular phylogeny of the *Bacidiaceae* (*Lecanorales*, lichenized *Ascomycota*). *Mycol. Res.* 105: 783–797; **Ekman, S.; Tønsberg, T.** (1996). A new species of *Megalaria* from the North American West Coast, and notes on the generic circumscription. *Bryologist* 99: 34–40; **Kantvilas, G.; Hafellner, J.; Elix, J.A.** (1999). *Tasmidella*, a new lichen genus from Tasmania, with a revised circumscription of the family *Megalariaceae*. *Lichenologist* 31: 213–225.

Megalosporaceae Vězda ex Hafellner & Bellem. 1982
Lecanorales: Ascomycota

Thallus crustose. Ascomata apothecial, sessile, with a wide margin, the disc brown or black. Interascal tissue of very narrow anastomosing pseudoparaphysis-like hyphae, often with a pigmented epithecium. Asci with a well-developed J+ apical cap and an outer J+ gelatinized layer, 1- to 8-spored. Ascospores large, hyaline, thick-walled, without a sheath. Anamorphs pycnidial.

Significant Genera: *Megalospora*.

Distribution: Most species have tropical distributions.

Economic Significance: None is known.

Ecology: Lichenized with green algae.

Notes: Very poorly known; no recent information is available.

Megalospora sulphurata, Rio de Janeiro, Brazil; thallus and ascomata

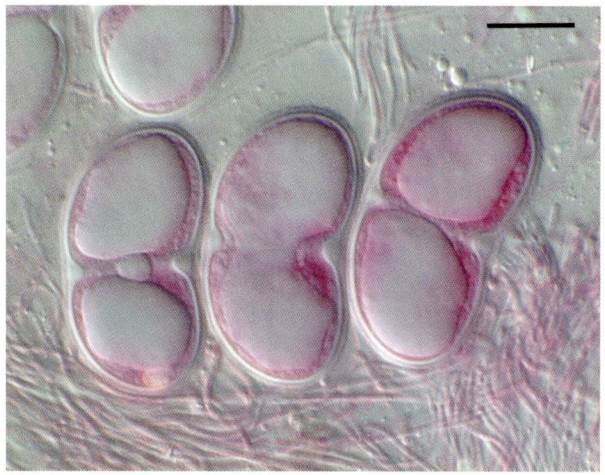

Megalospora pulverata, New Zealand; ascospores

Megalospora tuberculosa; thallus and ascomata

References: **Kantvilas, G.** (1994). Additions to the family *Megalosporaceae* in Tasmania and mainland Australia. *Lichenologist* 26: 349–366; **Sipman, H.J.M.** (1986). Additional notes on the lichen family *Megalosporaceae*. *Willdenowia* 15: 557–564; **Stenroos, S.K.; DePriest, P.T.** (1998). SSU rDNA phylogeny of cladoniiform lichens. *Am. J. Bot.* 85: 1548–1559.

Megasporaceae Lumbsch 1994
Pertusariales: Ascomycota

Thallus crustose. Ascomata deeply immersed within verrucose swellings of the thallus, apothecial, cupulate, with a differentiated ascomatal wall and a separate thalline margin. Interascal tissue of narrow anastomosing pseudoparaphyses. Asci thick-walled, with a strongly thickened apical cap without an ocular chamber, the outer layer weakly J+, 8-spored. Ascospores large, hyaline, aseptate, with a two-layered wall, lacking a gelatinous sheath. Anamorphs pycnidial.

Significant Genera: *Aspicilia*, *Megaspora*.

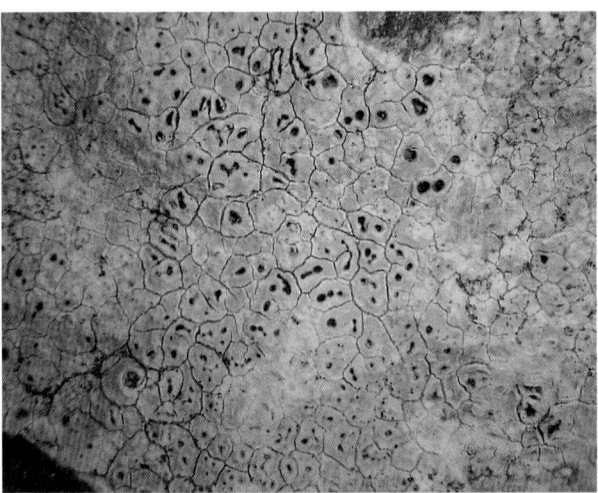
Aspicilia calcarea, Derbyshire, UK; thallus

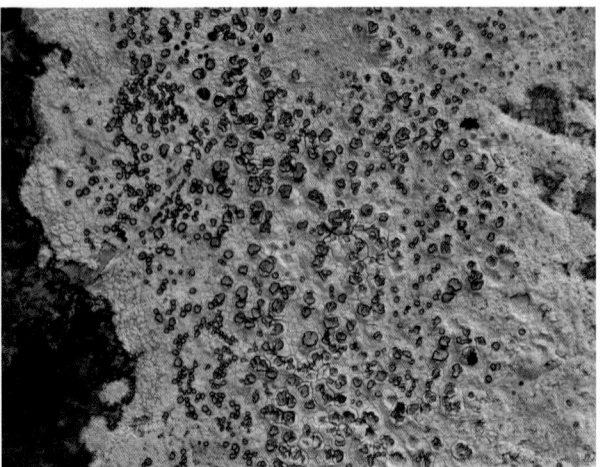
Aspicilia candida, Vancouver Island, Canada; thallus and ascomata

Distribution: Widespread in north temperate regions, also reported from Venezuela.

Economic Significance: None is known.

Ecology: Lichenized with green algae, on rocks, often maritime or in/close to fresh water.

Notes: Very poorly known; little recent information is available on *Megaspora*. Molecular studies indicate that the famiy is related to the *Pertusariaceae*, and that *Aspicilia* belongs here rather than in the *Hymeneliaceae*.

References: **Hafellner, J.** (1991). Die Gattung *Aspicilia*, ihre Ableitungen nebst Bemerkungen über cryptolecanorine Ascocarporganisation bei anderen Genera der *Lecanorales* (*Ascomycetes* lichenisati). *Acta bot. Malac.* 16: 133–140; **Ivanova, N.V.; Hafellner, J.** (2002). Searching for the correct placement of *Megaspora* by use of ITS1, 5.8S and ITS2 rDNA data. *Biblthca Lichenol.* 82: 113–122; **Lumbsch, H.T.; Feige, G.B.; Schmitz, K.E.** (1994). Systematic studies in the *Pertusariales* I. *Megasporaceae*, a new family of lichenized ascomycetes. *J. Hattori bot. Lab.* 75: 295–304; **Schmitt, I.; Yamamoto, Y.; Lumbsch, H.T.** (2006). Phylogeny of *Pertusariales* (*Ascomycotina*): resurrection of *Ochrolechiaceae* and new circumscription of *Megasporaceae*. *J. Hattori bot. Lab.* 100: 753–764; **Stenroos, S.K.; DePriest, P.T.** (1998). SSU rDNA phylogeny of cladoniiform lichens. *Am. J. Bot.* 85: 1548–1559.

Melampsoraceae Dietel 1897
Uredinales: Basidiomycota

Spermogonia discoid to conical, with discrete walls, subepidermal or subcuticular. Aecia *Caeoma*-like, wall structures absent or poorly developed. Aeciospores formed in chains, verrucose. Uredinia with a poorly developed or evanescent wall. Urediniospores formed singly on pedicels, echinulate, with germ pores scattered or in two zones, accompanied by capitate paraphyses. Telia usually subepidermal. Teliospores unicellular, sessile, pigmented, with a single germ pore, laterally adherent and forming single-layered crusts, germinating externally to form basidiospores.

Significant Genera: *Melampsora*.

Distribution: Cosmopolitan.

Economic Significance: Willow rusts on *Populus* and *Salix* cause significant damage and *Melampsora lini* causes losses in flax production.

Ecology: Biotrophic, heteroecious or autoecious. A prominent group of species produces aecia on conifer needles and telia on leaves of *Salicaceae*. Many species appear to have partially or completely lost the requirement for alternation of hosts.

Notes: The diverse genus *Melampsora* (with around 90 species) is the only representative of this family. Molecular studies indicate that the family is well-defined and perhaps forms a basal lineage within the *Uredinales*.

Melampsora medusae, Montana, USA; aecia on needles of *Pseudotsuga mucronata*

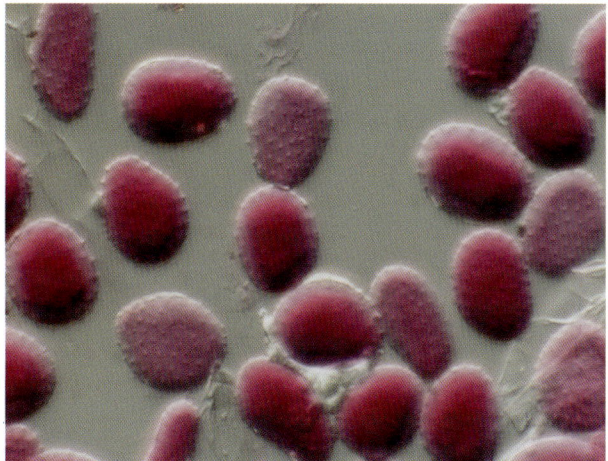

Melampsora populnea, Malta; urediniospores from leaves of *Populus alba*

References: **Aime, M.C.** (2006). Towards resolving family-level relationships in rust fungi (*Uredinales*). *Mycoscience* 47: 112–122; **Brasier, C.M.** (2001). Rapid evolution of introduced plant pathogens via interspecific hybridization. *BioScience* 51: 123–133; **Burdon, J.J.; Roberts, J.K.** (1995). The population genetic structure of the rust fungus *Melampsora lini* as revealed by pathogenicity, isozyme and RFLP markers. *Pl. Path.* 44: 270–278; **Cummins, G.B.; Hiratsuka, Y.** (2003). *Illustrated Genera of Rust Fungi* (St Paul): 225 pp.; **Maier, W.; Begerow, D.; Weiss, M.; Oberwinkler, F.** (2003). Phylogeny of the rust fungi: an approach using nuclear large subunit ribosomal DNA sequences. *Can. J. Bot.* 81: 12–23; **Nakamura, H.; Kaneko, S.; Yamaoka, Y.; Kakishima, M.** (2003). Correlation between pathogenicity and molecular characteristics in the willow leaf rusts *Melampsora epitea* and *M. humilis* in Japan. *Mycoscience* 44: 253–256; **Pei, M.H.** (2005). A brief review of *Melampsora* rusts on *Salix*. *Rust Diseases of Willow and Poplar* (Wallingford): 11–28; **Pei, M.H.; Bayon, C.; Ruiz, C.** (2005). Phylogenetic relationships in some *Melampsora* rusts on *Salicaceae* assessed using rDNA sequence information. *Mycol. Res.* 109: 401–409; **Pei, M.H.; Bayon, C.; Ruiz, C.** (2005). Phylogenetic position of *Melampsora* in rust fungi inferred from ribosomal DNA sequences. *Rust Diseases of Willow and Poplar* (Wallingford): 1–9; **Samils, B.; Lagercrantz, U.; Lascoux, M.; Gullberg, U.** (2001). Genetic structure of *Melampsora epitea* populations in Swedish *Salix viminalis* plantations. *Eur. J. Pl. Path.* 107: 399–409; **Smith, J.A.; Blanchette, R.A.; Newcombe, G.** (2004). Molecular and morphological characterization of the willow rust fungus, *Melampsora epitea*, from arctic and temperate hosts in North America. *Mycologia* 96: 1330–1338; **Spiers, A.G.; Hopcroft, D.H.** (1996). Morphological and host range studies of *Melampsora* rusts attacking *Salix* species in New Zealand. *Mycol. Res.* 100: 1163–1175; **Tian, C.M.; Shang, Y.Z.; Zhuang, J.Y.; Wang, Q.; Kakishima, M.** (2004). Morphological and molecular phylogenetic analysis of *Melampsora* species on poplars in China. *Mycoscience* 45: 56–66; **Wingfield, B.D.; Ericson, L.; Szaro, T.; Burdon, J.J.** (2004). Phylogenetic patterns in the *Uredinales*. *Australas. Pl. Path.* 33: 327–335.

Melanconidaceae G. Winter 1886
Diaporthales: Ascomycota

Ascomata often large, perithecial, erect, oblique or horizontal, necks central, oblique or lateral, erumpent singly or converging and erumpent through a stromatic disc or immersed in a stroma. Asci persistent, thick-walled, often remaining attached. Ascospores mainly broad and brown, simple or septate, sometimes distoseptate, end cells sometimes hyaline. Anamorphs coelomycetous, varied.

Significant Genera: *Melanconis*. Anamorph: *Melanconium*.

Distribution: Widespread, especially in temperate regions.

Economic Significance: A few species are pathogens of economically important trees, including *Melanconis alni* on *Alnus* spp.

Ecology: Saprobic or weakly parasitic in wood and bark, possibly endophytic.

Notes: The family is certainly polyphyletic as it was circumscribed in the *Dictionary of the Fungi* edn 9, but more work is needed to establish its boundaries. The *Pseudovalsaceae* may belong here, but sequences are not available.

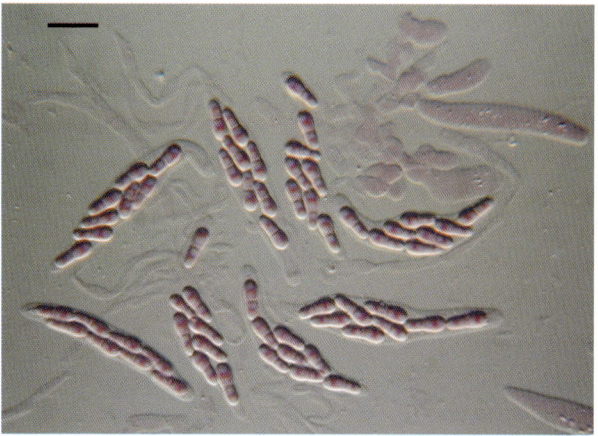

Uleoporthe orbiculata, Guyana; asci and ascospores

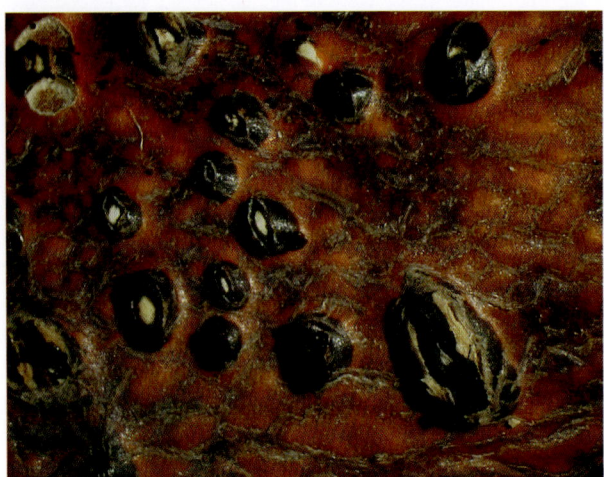

Melanconis alni, Somerset, UK; conidiomata

References: **Belisario, A.** (1999). Cultural characteristics and pathogenicity of *Melanconium juglandinum*. *Eur. J. For. Path.* 29: 317–322; **Belisario, A.; Onofri, S.** (1995). Conidiogenesis and morphology of *Melanconium juglandinum*. *Mycol. Res.* 99: 1059–1062; **Castlebury, L.A.; Rossman, A.Y.; Jaklitsch, W.J.; Vasilyeva, L.N.** (2002). A preliminary overview of the *Diaporthales* based on large subunit nuclear ribosomal DNA sequences. *Mycologia* 94: 1017–1031; **Orsenigo, M.; Rodondi, G.; Sutton, B.C.** (1998). A short note on three critical *Coryneum* species: *Coryneum populinum*, *C. elongatum* and *C. corni-albae*. *Mycotaxon* 67: 257–263; **Sieber, T.N.; Sieber-Canavesi, F.; Petrini, O.; Ekramoddoullah, A.K.M.; Dorworth, C.E.** (1992). Characterization of Canadian and European *Melanconium* from some *Alnus* species by morphological, cultural, and biochemical studies. *Can. J. Bot.* 69: 2170–2176; **Yanna; Hyde, K.D.; Goh, T.K.** (1999). *Endomelanconium phoenicicola* sp. nov., a new coelomycete from *Phoenix hanceana* in Hong Kong. *Fungal Diversity* 2: 199–204.

Melaniellaceae R. Bauer, Vánky, Begerow & Oberw. 1999
Doassansiales: Basidiomycota

Sori forming irregular black spots in leaves and stems, not breaking down. Hyphae with clamp connections, exclusively intercellular, septal pore simple with membrane caps, haustoria absent but specialized interaction apparatus present with non-homogenous contents. Teliospores embedded in host tissue, variable in shape and size, often polyhedral due to compression, dark brown, thick-walled, smooth or tuberculate. Basidia formed directly from germinating teliospores, aseptate (*Exobasidium*-type), hypha-like, thin-walled, usually with a cluster of 4 apical sterigmata. Basidiospores discharged actively, fusiform to cylindrical, 2-celled, hyaline, germinating in turn to produce narrow cylindrical yeast-like cells.

Significant Genera: *Melaniella*.

Distribution: Known from South and South-East Asia.

Economic Significance: None is known.

Ecology: Biotrophic in leaves and stems of *Selaginella* species.

Notes: The family appears to be a basal assemblage within the *Doassansiales*, though more molecular studies are needed. The ancient origins of the lycopods invites speculation that plant and fungus share a very long-established coevolutionary relationship, though some authors favour host-jumping as an evolutionary mechanism.

References: **Bauer, R.; Oberwinkler, F.; Vánky, K.** (1997). Ultrastructural markers and systematics in smut fungi and allied taxa. *Can. J. Bot.* 75: 1273–1314; **Bauer, R.; Vánky, K.; Begerow, D.; Oberwinkler, F.** (1999). *Ustilaginomycetes* on *Selaginella*. *Mycologia* 91: 475–484; **Vánky, K.** (2004). Biodiversity and conservation of smut fungi (*Ustilaginomycetes* p.p., *Microbotryales*) reflected in Vánky, *Ustilaginales* exsiccata no. 1-1250. *Mycol. Balcanica* 1: 175–187.

Melanogastraceae E. Fisch. 1933
Boletales: Basidiomycota

Basidiomata gasteroid, hypogeous or ± emergent, globose or irregular, generally brownish in coloration, often blackening when damaged, the peridium ± smooth or felty, persistent. Hyphae generally brown, inflated or not, with or without clamp connections. Gleba dark, marbled, gelatinous and deliquescent at maturity. Hymenium varied in development, columella absent, with basidia irregularly arranged and developing simultaneously. Basidia clavate, elongate, evanescent, with 2–6 sterigmata. Basidiospores radially ± symmetrical, ovoid, ellipsoidal, ± cylindrical or citriform, pale yellow or becoming dark brown or black, thick-walled, smooth or faintly roughened, not staining in iodine.

Significant Genera: *Alpova*, *Melanogaster*.

Distribution: Widespread in temperate regions.

Economic Significance: None is known.

Ecology: Ectomycorrhizal with coniferous and broadleaved trees, in the upper soil layers and under litter and similar substrata.

Notes: At one time assumed to be closely related to the *Rhizopogonaceae*, molecular data confirm that the two families have evolved quite separately. The *Melanogastraceae* belongs to the boletoid lineage of the *Boletales*, perhaps as part of a widely circumscribed *Paxillaceae*, while the *Rhizopogonaceae* is part of the suilloid lineage. However, several of the genera are polyphyletic.

Alpova diplophloeus; basidiomata

Melanogaster ambiguus; basidioma

References: **Beaton, G.; Pegler, D.N.; Young, T.W.K.** (1985). Gasteroid *Basidiomycota* of Victoria State, Australia: 5–7. *Kew Bull.* 40: 573–598; **Besl, H.; Dorsch, R.; Fischer, M.** (1996). [On the systematic position of genus *Melanogaster* (*Melanogastraceae*, *Basidiomycetes*)]. *Z. Mykol.* 62: 195–199; **Calonge, F.D.; Siquier, J.L.** (1998). *Alpova pseudostipitatus,* sp. nov. (gasteromycetes), from Majorca (Spain). *Boln Soc. Micol. Madrid* 23: 91–96; **Grubisha, L.C.; Trappe, J.M.; Molina, R.; Spatafora, J.W.** (2001). Biology of the ectomycorrhizal genus *Rhizopogon.* V. Phylogenetic relationships in the *Boletales* inferred from LSU rDNA sequences. *Mycologia* 93: 82–89; **Liu, B.; Tao, K.; Chang, M.C.** (1989). Two new species of *Melanogaster* from China. *Acta Mycol. Sin.* 8: 210–213; **Nouhra, E.R.; Domínguez, L.S.; Becerra, A.G.; Trappe, J.M.** (2005). Morphological, molecular and ecological aspects of the South American hypogeous fungus *Alpova austroalnicola* sp. nov. *Mycologia* 97: 598–604.

Melanommataceae G. Winter 1885
Pleosporales: Ascomycota

Ascomata formed from an immersed stroma, perithecial, erumpent, often becoming superficial, often strongly aggregated, black, variable in form, with a well-developed lysigenous pore; peridium composed of small, thick-walled, pseudoparenchymatous cells, roughly equal in thickness. Interascal tissue of trabeculate pseudoparaphyses (paraphysoids), usually immersed in gel. Asci cylindrical, fissitunicate, with a distinct ocular chamber, not blueing in iodine. Ascospores brown, septate, sometimes muriform, fairly thick-walled, often with a sheath. Anamorphs coelomycetous, pycnidial, with small aseptate conidia formed from percurrently proliferating conidiogenous cells, often on branched conidiophores.

Significant Genera: *Astrosphaeriella, Byssosphaeria, Melanomma.* Anamorphs: *Aposphaeria, Pyrenochaeta.*

Distribution: Widespread in temperate and subtropical regions.

Economic Significance: None is known.

Ecology: Saprobic on wood and bark.

Notes: The *Melanommataceae,* along with several other families, was considered to belong to a separate clade within the *Dothideomycetes* due to the presence of very narrow and stretched (so-called trabeculate) pseudoparaphyses, but molecular data do not strongly support this arrangement. More sequences are needed to confirm the phylogenetic position of this family.

Melanomma pulvis-pyrius, Surrey, UK; ascomata

References: **Barr, M.E.** (1990). *Melanommatales* (*Loculoascomycetes*). *N. Amer. Fl.* Ser. 2 13: 129 pp.; **Boise, J.** (1985). An amended description of *Trematosphaeria. Mycologia* 77: 230–237; **Hawksworth, D.L.; Boise, J.R.** (1986). Some additional species of *Astrosphaeriella,* with a key to the members of the genus. *Sydowia* 38: 114–124; **Huhndorf, S.M.** (1998). Neotropical ascomycetes 6. A new species of *Ostropella* from French Guiana. *Mycologia* 90: 527–530; **Hyde, K.D.; Fröhlich, J.** (1998). Fungi from palms XXXVII. The genus *Astrosphaeriella,* including ten new species. *Sydowia* 50: 81–132; **Hyde, K.D.; Goh, T.K.** (1999). Tropical Australian freshwater fungi. XVI. Some new melanommataceous fungi from woody substrata and a key to genera of lignicolous loculoascomycetes in freshwater. *Nova Hedwigia* 68: 251–272; **Kodsueb, R.; Dhanasekraran, V.; Aptroot, A.; Lumyong, S.; McKenzie,**

E.H.C.; Hyde, K.D.; Jeewon, R. (2006). The family *Pleosporaceae*: intergeneric relationships and phylogenetic perspectives based on sequence analyses of partial 28S rDNA. *Mycologia* 98: 571–583; **Liew, E.C.Y.; Aptroot, A.; Hyde, K.D.** (2000). Phylogenetic significance of the pseudoparaphyses in loculoascomycete taxonomy. *Mol. Phylogen. Evol.* 16: 392–402; **Lindemuth, R.; Wirtz, N.; Lumbsch, H.T.** (2001). Phylogenetic analysis of nuclear and mitochondrial rDNA sequences supports the view that loculoascomycetes (*Ascomycota*) are not monophyletic. *Mycol. Res.* 105: 1176–1181; **Lumbsch, H.T.; Lindemuth, R.** (2001). Major lineages of *Dothideomycetes* (*Ascomycota*) inferred from SSU and LSU rDNA sequences. *Mycol. Res.* 105: 901–908; **Réblová, M.** (1998). Revision of three *Melanomma* species described by L. Fuckel. *Czech Mycol.* 50: 161–179.

Melanopsichiaceae Vánky 2001
Ustilaginales: Basidiomycota

Sori forming galls in inflorescences, leaves and stems, ± black, irregular, the surface gelatinous, with multiple chambers filled with teliospores in a gelatinous matrix. Hyphae intracellular, lacking septal pores. Teliospores formed singly, globose to irregular, yellow or brown, thick-walled, smooth or minutely verrucose, with a gelatinous sheath, germinating to form 1 or 2 basidia. Basidia cylindrical, elongate, variably septate, with basidiospores formed laterally and terminally. Basidiospores ovoid to cylindrical, hyaline, thin-walled, smooth.

Significant Genera: *Melanopsichium*.

Distribution: Widespread in north and south temperate zones.

Melanopsichium pennsylvanicum; sori in inflorescence of *Polygonum setulosum*, Kenya

Economic Significance: None is known.

Ecology: Biotrophic on species of *Polygonaceae*; taxa associated with other plant families should probably belong elsewhere.

Notes: *Melanopsichium* is the only genus of *Ustilaginales* that does not parasitize grasses, suggesting that a host jump may have occurred.

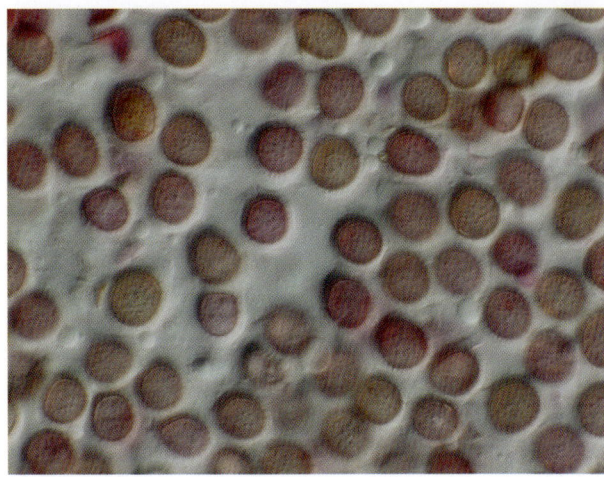

Melanopsichium pennsylvanicum, Kenya; teliospores

References: **Bauer, R.; Oberwinkler, F.; Vánky, K.** (1997). Ultrastructural markers and systematics in smut fungi and allied taxa. *Can. J. Bot.* 75: 1273–1314; **Begerow, D.; Bauer, R.; Oberwinkler, F.** (1998). Phylogenetic studies on nuclear large subunit ribosomal DNA sequences of smut fungi and related taxa. *Can. J. Bot.* 75: 2045–2056; **Begerow, D.; John, B.; Oberwinkler, F.** (2004). Evolutionary relationships among β-tubulin gene sequences of basidiomycetous fungi. *Mycol. Res.* 108: 1257–1263; **Spooner, B.M.** (1985). *Melanopsichium* (*Ustilaginales*), a genus new to the British Isles. *Trans. Br. mycol. Soc.* 85: 540–544; **Stoll, M.; Begerow, D.; Oberwinkler, F.** (2005). Molecular phylogeny of *Ustilago*, *Sporisorium*, and related taxa based on combined analyses of rDNA sequences. *Mycol. Res.* 109: 342–356; **Vánky, K.** (2001). The emended *Ustilaginaceae* of the modern classificatory system for smut fungi. *Fungal Diversity* 6: 131–147; **Vánky, K.** (2004). Biodiversity and conservation of smut fungi (*Ustilaginomycetes* p.p., *Microbotryales*) reflected in Vánky, *Ustilaginales* exsiccata no. 1-1250. *Mycol. Balcanica* 1: 175–187; **Vánky, K.; Oberwinkler, F.** (1994). *Ustilaginales* on *Polygonaceae* – a taxonomic revision. *Nova Hedwigia* Beih. 107: 96 pp.

Melanotaeniaceae Begerow, R. Bauer & Oberw. 1998
Urocystidales: Basidiomycota

Sori formed in leaves, stems or roots, ± black, irregularly swollen, the host tissue hypertrophied, stunted or with witches' broom (hyperplasia) symptoms. Hyphae intracellular, haustoria present, septal pores simple, with membrane caps and without non-membranous bands. Spore mass agglutinated and not powdery. Teliospores formed singly, ± globose to irregular, the shape often affected by compression, dark brown, thick-walled, two-layered, smooth, germinating to form a single basidium. Basidia cylindrical, hypha-like, transversely septate, with a cluster of 4–7 apical persistent basidiospores that fuse in pairs to form a septate mycelium.

Significant Genera: *Melanotaenium*.

Distribution: Widespread.

Economic Significance: None is known.

Ecology: Biotrophic in various tissues of plants, including dicotyledonous groups, *Liliaceae* and ferns. Species on grasses appear to be misplaced.

Notes: Family delimitation requires further research; *Melanotaenium* has been shown to be polyphyletic. It is currently unclear whether placement in the *Ustilaginales* or *Urocystales* is preferable.

Melanotaenium endogenum, Germany; sori in stems of *Galium mollugo*

References: **Bauer, R.; Oberwinkler, F.; Vánky, K.** (1997). Ultrastructural markers and systematics in smut fungi and allied taxa. *Can. J. Bot.* 75: 1273–1314; **Begerow, D.; Bauer, R.; Oberwinkler, F.** (1998). Phylogenetic studies on nuclear large subunit ribosomal DNA sequences of smut fungi and related taxa. *Can. J. Bot.* 75: 2045–2056; **Boekhout, T.; Fell, J.W.; O'Donnell, K.** (1995). Molecular systematics of some yeast-like anamorphs belonging to the *Ustilaginales* and *Tilletiales*. *Stud. Mycol.* 38: 175–183; **Ingold, C.T.** (1988). Ballistospores in *Melanotaenium endogenum*. *Trans. Br. mycol. Soc.* 91: 712–714; **Piepenbring, M.; Bauer, R.; Oberwinkler, F.** (1998). Teliospores of smut fungi. Teliospore connections, appendages, and germ pores studied by electron microscopy; phylogenetic discussion of characteristics of teliospores. *Protoplasma* 204: 202–218; **Vánky, K.** (2004). Biodiversity and conservation of smut fungi (*Ustilaginomycetes p.p.*, *Microbotryales*) reflected in Vánky, *Ustilaginales* exsiccata no. 1-1250. *Mycol. Balcanica* 1: 175–187; **Walker, J.** (2001). *Yelsemia arthropodii* gen. et sp. nov. (*Tilletiales*) on *Arthropodium* in Australia. *Mycol. Res.* 105: 225–232.

Melaspileaceae Walt. Watson 1929
Uncertain position within Arthoniomycetes, Ascomycota

Thallus crustose or immersed, often evanescent. Ascomata apothecial, immersed to superficial, elongate, sometimes branched, the hymenium exposed permanently or by a longitudinal slit; peridium black, well-developed. Interascal tissue of narrow, sparsely branched and anastomosing, paraphyses sometimes pigmented at the apex. Asci cylindrical, persistent, with a poorly developed apical cap and an ocular chamber, usually not blueing in iodine. Ascospores usually 1-septate, becoming brown, sometimes ornamented. Anamorphs unknown.

Significant Genera: *Melaspilea*.

Distribution: Widespread, especially in the tropics.

Economic Significance: None is known.

Ecology: Mostly lichenized with green algae; occasional species are lichenicolous.

Notes: The only genus is poorly understood and relationships are obscure.

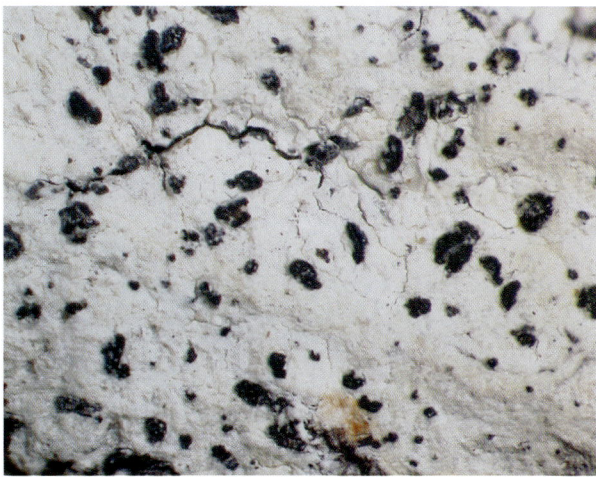

Melaspilea lentiginosa, Ireland; ascomata

References: **Vrijmoed, L.L.P; Hyde, K.D.; Jones, E.B.G.** (1996). *Melaspilea mangrovei* sp. nov., from Australian and Hong Kong mangroves. *Mycol. Res.* 100: 291–294.

Meliolaceae G.W. Martin ex Hansf. 1946
Meliolales: Ascomycota

Mycelium usually superficial, copious, dark, thick-walled, frequently branched, often rough-walled or with mucous coating, usually with appressoria ('capitate hyphopodia') and producing conidiogenous cells ('mucronate hyphopodia'). Stromata absent. Ascomata superficial on lateral branches of hyphae, ± spherical or flattened, black, thick-walled, cleistothecial, breaking down irregularly or opening with a large irregular apical hole. Interascal tissue poorly developed, of wide thin-walled often evanescent paraphyses. Asci very thin-walled, clavate or ± globose, without apical structures, evanescent, usually 2-spored. Ascospores brown, usually 4-septate, constricted at the septa, often with a narrow sheath. Anamorphs inconspicuous, probably spermatial, with flask-shaped conidiogenous cells formed directly from vegetative hyphae. Conidia small, hyaline.

Significant Genera: *Meliola*.

Meliola sp. on *Psychotria rufipes*, Galápagos Islands, Ecuador; ascomata and setae

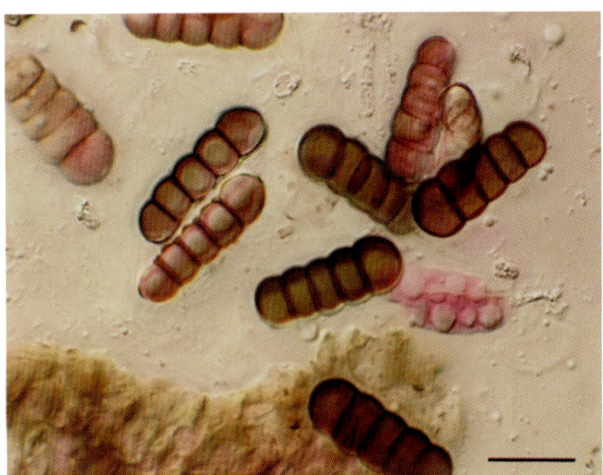

Meliola sp. on *Psychotria rufipes*, Galápagos Islands, Ecuador; ascospores

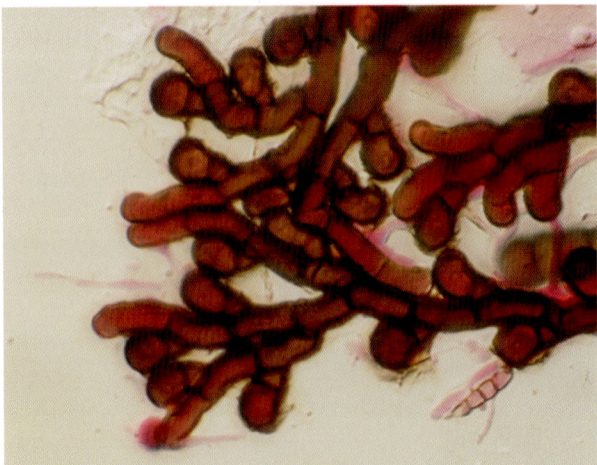

Meliola sp. on *Psychotria rufipes*, Galápagos Islands, Ecuador; superficial mycelium with appressoria

Distribution: Most species are tropical in distribution, but a few species occur in humid temperate zones.

Economic Significance: The family is not generally considered to be of economic significance, but extensive colonization of perennial crop plants must result in loss of photosynthetic capacity and probably also nutrients.

Ecology: Biotrophic on leaves and stems. Many species appear to cause only minimal damage to the host and nutrition may also be obtained from exudates and cuticular waxes.

Notes: The *Meliolales* is an isolated order of fungi and its closest relatives are still uncertain. The single phylogenetic study available to date suggests a placement within the *Sordariomycetes* rather than allied to the *Erysiphales* as commonly supposed, but there is insufficient data to fix the position of the order.

References: **Hosagoudar, V.B.** (1996). *Meliolales of India* (Calcutta): 363 pp.; **Hosagoudar, V.B.; Abraham, T.K.; Pushpangadan, P.** (1997). *The Meliolineae* A Supplement (Thiurvananthapuram): 201 pp.; **Hu, Y.X.; Ouyang, Y.S.; Song, B.; Jiang, G.Z.** (1996). *Flora Fungorum Sinicorum* 4. *Meliolales* (Beijing): 270 pp.; **Mibey, R.K.; Hawksworth, D.L.** (1997). Meliolaceae and Asterinaceae of the Shimba Hills, Kenya. *Mycol. Pap.* 174: 108 pp.; **Mueller, W.C.; Goos, R.D.; Quainoo, J.; Morgham, A.T.** (1991). The structure of the phialides (mucronate hyphopodia) of the *Meliolaceae*. *Can. J. Bot.* 69: 803–807; **Mueller, W.C.; Morgham, A.T.; Goos, R.D.** (1991). Ultrastructure of the haustoria of *Meliola sandwicensis*. *Mycol. Res.* 95: 1208–1210; **Saenz, G.S.; Taylor, J.W.** (1999). Phylogenetic relationships of *Meliola* and *Meliolina* inferred from nuclear small subunit rRNA sequences. *Mycol. Res.* 103: 1049–1056.

Meliolinaceae S. Hughes 1993
Uncertain position within Ascomycetes, Ascomycota

Colonies superficial, black, velutinous, composed of dark-walled external hyphae with stomatopodia (appressoria forming within stomata) and pale internal hyphae. Ascomata black, superficial, cleistothecial, ± globose, ostiolate, sometimes setose. Interascal tissue persistent, of broad paraphysis-like hyphae, always free at the apex. Asci thick-walled but not fissitunicate, J–, with a broad opening when discharged. Ascospores 3-septate, constricted at the septa, usually with transverse hyaline bands. Anamorphs hyphomycetous.

Significant Genera: *Meliolina*.

Distribution: Widespread, especially in tropical regions; also occurring in wet temperate forests.

Economic Significance: None is known.

Ecology: Apparently biotrophic on leaves; the nutritional status needs further analysis.

Notes: Very poorly known; no recent information is available.

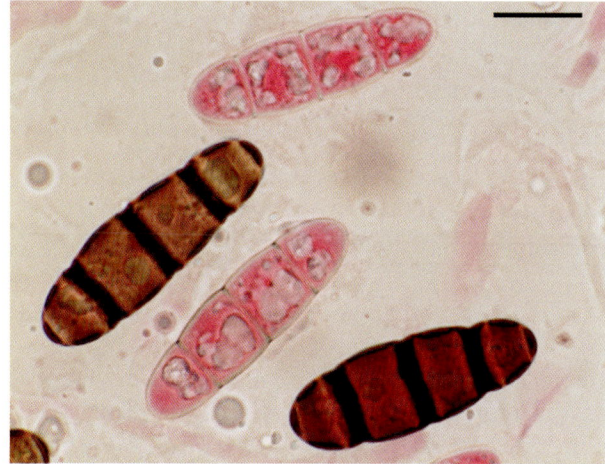

Meliolina stevensii, Hawaii; mature and immature ascospores

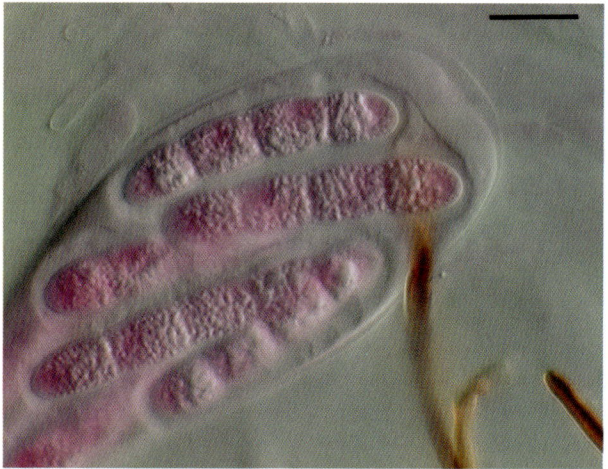

Meliolina pulcherrima, Assam, India; ascus

References: **Hughes, S.J.** (1993). *Meliolina* and its excluded species. *Mycol. Pap.* 166: 255 pp.; **Johnston, P.R.** (1999). Some hyperparasites of *Meliolina* in New Zealand. *N.Z. Jl Bot.* 37: 289–295; **Reynolds, D.R.** (1989). *Briania* gen. nov. and *Briania fruticetum* sp. nov. *Pacific Sci.* 43: 161–165; **Reynolds, D.R.** (1989). An extending ascus in the ascostromatic genus *Meliolina. Cryptog. Mycol.* 10: 305–320; **Saenz, G.S.; Taylor, J.W.** (1999). Phylogenetic relationships of *Meliola* and *Meliolina* inferred from nuclear small subunit rRNA sequences. *Mycol. Res.* 103: 1049–1056.

Meripilaceae Jülich 1982
Polyporales: Basidiomycota

Basidiomata pileate, with clusters of fruit bodies produced from a common base, sometimes very large, fan-shaped or spathulate, fleshy to coriaceous, the upper surface brown, ± smooth, velvety or squamulose. Hyphal system monomitic, hyphae thin-walled in the internal tissues and thick-walled and almost skeletal in the outer layers, clamp connections sparse, cystidia absent. Hymenium poroid, white or grey, becoming almost black when damaged, the pores angular and tubes often elongate. Basidia clavate, lacking clamp connections, with 4 sterigmata. Basidiospores ellipsoidal to subglobose or ovoid, smooth, thin-walled, not staining in iodine.

Significant Genera: *Meripilus*.

Distribution: Widespread in north temperate ecosystems.

Economic Significance: A number of important timber trees are affected, including *Fagus* and *Quercus* species, but extensive economic damage does not appear to occur.

Ecology: Saprobic or parasitic on hardwood trees, often growing from buried roots and causing white rots.

Notes: The *Meripilaceae* as delimited in the *Dictionary of the Fungi* edn 9 is poorly defined in morphology and clearly polyphyletic when molecular data are examined. Pending further study, the family is here reduced to the type genus *Meripilus*. Of the well-known genera previously included, *Antrodia* has been given clade status along with *Grifola* and several further genera not previously placed in the *Meripilaceae*.

Meripilus giganteus, Surrey, UK; basidiomata

References: **Binder, M.; Hibbett, D.S.; Larsson, K.-H.; Larsson, E.; Langer, E.; Langer, G.** (2005). The phylogenetic distribution of resupinate forms across the major clades of mushroom-forming fungi (*Homobasidiomycetes*). *Syst. Biodiv.* 3: 113–157; **Corner, E.J.H.** (1989). Ad Polyporaceas V. The genera *Albatrellus, Boletopsis, Coriolopsis* (dimitic), *Cristelloporia, Diacanthodes, Elmerina, Fomitopsis* (dimitic), *Gloeoporus, Grifola, Hapalopilus, Heterobasidion, Hydnopolyporus, Ischnoderma, Loweporus, Parmastomyces, Perenniporia, Pyrofomes, Stecchericium, Trechispora, Truncospora* and *Tyromyces*. *Beih. Nova Hedwigia* 96: 218 pp.; **Kim, S.Y.; Park, S.Y.; Ko, K.S.; Jung, H.S.** (2003). Phylogenetic analysis of *Antrodia* and related taxa based on partial mitochondrial SSU rDNA sequences. *Antonie van Leeuwenhoek* 83: 81–88; **Larsen, M.J.; Lombard, F.F.**

(1988). The status of *Meripilus giganteus* (*Aphyllophorales, Polyporaceae*) in North America. *Mycologia* 80: 612–621; **Shen, Q.; Geiser, D.M.; Royse, D.J.** (2002). Molecular phylogenetic analysis of *Grifola frondosa* (aaitake) reveals a species partition separating eastern North American and Asian isolates. *Mycologia* 94: 472–482; **Wang, Z.; Binder, M.; Dai, Y.C.; Hibbett, D.S.** (2004). Phylogenetic relationships of *Sparassis* inferred from nuclear and mitochondrial ribosomal DNA and RNA polymerase sequences. *Mycologia* 96: 1015–1029.

References: **Humber, R.A.** (1989). Synopsis of a revised classification for the *Entomophthorales* (*Zygomycotina*). *Mycotaxon* 34: 441–460; **Saikawa, M.** (1989). Ultrastructure of the septum in *Ballocephala verrucospora* (*Entomophthorales, Zygomycetes*). *Can. J. Bot.* 67: 2484–2488; **Saikawa, M.; Oguchi, M.; Castañeda Ruíz, R.F.** (1997). Electron microscopy of two nematode-destroying fungi, *Meristacrum asterospermum* and *Zygnemomyces echinulatus* (*Meristacraceae, Entomophthorales*). *Can. J. Bot.* 75: 762–768; **Saikawa, M.; Sakuramata, H.** (1992). Observation on zygospores in *Ballocephala verrucospora*. *Trans. Mycol. Soc. Japan* 33: 237–241.

Meristacraceae Humber 1989
Entomophthorales: Zygomycota

Vegetative growth mycelial at first then forming hyphal bodies, these usually spherical to rounded. Sporophores (conidiophores) simple, solitary, cylindrical to slightly clavate, bearing terminal conidiogenous cells. Spores (conidia) terminal, spherical, single-celled, forcibly or passively discharged; secondary spores present or absent, either similar to primary spores and forcibly discharged or passively discharged from strongly narrowed and elongate sporophores; nucleus small, not clearly visible during mitosis and interphase, nucleolus prominent. Zygospores formed by hyphal conjugation, zygosporangium (epispore) hyaline or slightly pigmented, smooth or ornamented, zygospore (endospore) ovoid and smooth or globose to subglobose and roughened; germinating directly to produce secondary spores or a sporophore.

Significant Genera: *Ballocephala, Meristacrum*.

Distribution: Widespread although undoubtedly under-recorded.

Economic Significance: None is known.

Ecology: Obligate pathogens of soil invertebrates, particularly nematodes and tardigrades.

Notes: Very poorly known; no recent information is available.

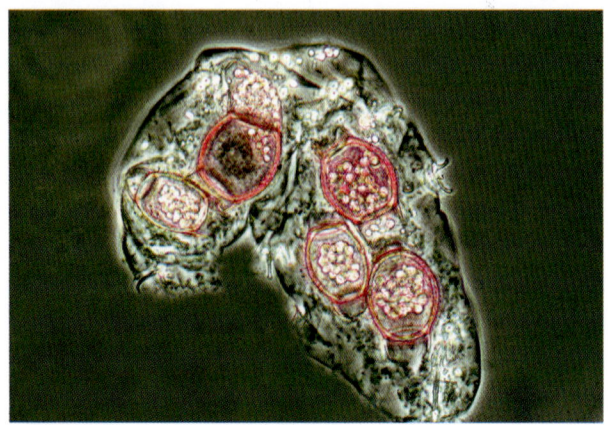

Ballocephala sp.; zygospores in body of tardigrade

Meruliaceae P. Karst. 1881
Polyporales: Basidiomycota

Basidioma resupinate to effuse, pileate or not, waxy or soft-textured, whiteish to brown, smooth or tomentose, rarely reduced to a subiculum. Hyphal system monomitic or rarely dimitic, hyphae gelatinized in some species, clamp connections often present, thick-walled cystidia usually present. Hymenium smooth, radially folded, spinose or ± poroid. Basidia narrow, cylindrical or clavate, 2- to 4-spored. Basidiospores ellipsoidal, cylindrical or allantoid, hyaline, smooth, not staining in iodine. Anamorphs unknown.

Significant Genera: *Byssomerulius, Gloeoporus, Mycoacia, Phlebia*.

Merulius tremellosus, Surrey, UK; basidiomata

Distribution: Widespread, best studied in north temperate zones.

Economic Significance: Some species cause decay of timber trees, though damage frequently is of minor impact.

Ecology: Saprobic or parasitic on woody plants, causing a white rot.

Notes: A poorly defined assemblage that is not reliably delimited in molecular studies. Here it is restricted in extent compared with the *Dictionary of the*

Fungi edn 9. The important pathogen genus *Chondrostereum* is removed, along with *Cylindrobasidium* and *Gloeostereum*, to the euagaric clade; *Dacryobolus* and *Resinicium* are also excluded. The principal genus *Phlebia* is clearly polyphyletic.

Phlebia radiata, Surrey, UK; basidioma

References: **Binder, M.; Hibbett, D.S.; Larsson, K.-H.; Larsson, E.; Langer, E.; Langer, G.** (2005). The phylogenetic distribution of resupinate forms across the major clades of mushroom-forming fungi (Homobasidiomycetes). *Syst. Biodiv.* 3: 113–157; **Corner, E.J.H.** (1989). Ad Polyporaceas V. The genera *Albatrellus, Boletopsis, Coriolopsis* (dimitic), *Cristelloporia, Diacanthodes, Elmerina, Fomitopsis* (dimitic), *Gloeoporus, Grifola, Hapalopilus, Heterobasidion, Hydnopolyporus, Ischnoderma, Loweporus, Parmastomyces, Perenniporia, Pyrofomes, Stecchericium, Trechispora, Truncospora* and *Tyromyces*. *Beih. Nova Hedwigia* 96: 218 pp.; **Ginns, J.** (1998). Genera of the North American *Corticiaceae* sensu lato. *Mycologia* 90: 1–35; **Greslebin, A.; Nakasone, K.K.; Rajchenberg, M.** (2004). *Rhizochaete*, a new genus of phanerochaetoid fungi. *Mycologia* 96: 260–271; **Hjortstam, K.; Ryvarden, L.** (2004). Tropical species of *Mycoaciella* (*Basidiomycotina, Aphyllophorales*). *Syn. Fung.* (Oslo) 18: 14–16; **Koker, T.H. de; Nakasone, K.K.; Haarhof, J.; Burdsall, H.H.; Janse, B.J.H.** (2003). Phylogenetic relationships of the genus *Phanerochaete* inferred from the internal transcribed spacer region. *Mycol. Res.* 107: 1032–1040; **Larsson, K.-H.; Larsson, E.; Koljalg, U.** (2004). High phylogenetic diversity among corticioid homobasidiomycetes. *Mycol. Res.* 108: 983–1002; **Nakasone, K.K.** (2002). *Mycoaciella*, a synonym of *Phlebia*. *Mycotaxon* 81: 477–490; **Nakasone, K.K.; Burdsall Jr., H.H.** (1995). *Phlebia* species from eastern and southeastern United States. *Mycotaxon* 54: 335–359; **Nakasone, K.K.; Sytsma, K.J.** (1994). Biosystematic studies on *Phlebia acerina, P. rufa*, and *P. radiata* in North America. *Mycologia* 85: 996–1016; **Parmasto, E.; Hallenberg, N.** (2000). A taxonomic study of phlebioid fungi (*Basidiomycota*). *Nordic Jl Bot.* 20: 105–118.

Mesnieraceae Arx & E. Müll. 1975
Uncertain position within Dothideomycetes, Ascomycota

Ascomata perithecial, thin-walled, brightly coloured, ostiolate but the upper wall layers sometimes also degenerating; peridium consisting of thin-walled ± flattened cells. Interascal tissue of cellular pseudoparaphyses. Asci cylindrical, ? fissitunicate, often polysporous. Ascospores aseptate or septate, often ornamented, without a sheath. Anamorphs unknown.

Significant Genera: *Mesniera*.

Distribution: Primarily tropical.

Economic Significance: None is known.

Ecology: Necrotrophic or saprobic on leaves.

Notes: Very poorly known; no recent information is available.

References: **Hyde, K.D.** (1996). Fungi from palms. XXIV. The genus *Bondiella*. *Mycotaxon* 57: 347–352.

Mesophelliaceae Jülich 1982
Agaricales: Basidiomycota

Basidiomata gasteroid, sequestrate, hypogeous or epigeous, irregular in shape but usually ± globose, the peridium with several layers, incorporating mycorrhizal rootlets of associated trees, externally dark brown to black, sometimes surrounded by gelatinized hyphae. Gleba lacking a columella or with a tough, rubbery to corky, central region connected to the peridium by narrow capillitial threads or narrow to wide columns of gelatinized hyphae, spores formed in a powdery mass or in discrete locules, white to yellowish, pink or brown. Basidiospores ellipsoidal to fusiform, sometimes irregular, smooth, verrucose or ridged, ± hyaline, sometimes with an evanescent outer layer.

Significant Genera: *Malaczjukia, Mesophellia*.

Distribution: Apparently confined to Australasia.

Economic Significance: None is known.

Ecology: Ectomycorrhizal with various woody plants. Some species are important components of native mammal diets.

Notes: No molecular data are available and the affinities of the family are unclear. There is apparently no northern hemisphere counterpart.

References: **Bougher, N.L.; Lebel, T.** (2001). Sequestrate (truffle-like) fungi of Australia and New Zealand. *Aust. Syst. Bot.* 14: 439–484; **Cribb, J.W.** (1990). An occurrence of *Castoreum* (*Lycoperdaceae*) on Cape York. *Qd Nat.* 30: 25–26; **Dell, B.; Malajczuk, N.; Grove, T.S.; Thomson, G.** (1990). Ectomycorrhiza formation in *Eucalyptus*. IV. Ectomycorrhizas in the sporocarps of the hypogeous fungi *Mesophellia* and *Castoreum* in eucalypt forests of western Australia. *New Phytol.* 114: 449–456; **Trappe, J.M.; Bougher, N.L.** (2002). Australasian sequestrate (truffle-like) fungi. XI. *Gummivena potorooi* gen. & sp. nov. (*Basidiomycota*,

Mesophelliaceae), with a key to the 'gummy' genera and species of the *Mesophelliaceae*. *Australas. Mycol.* 21: 9–11; **Trappe, J.M.; Castellano, M.A.; Amaranthus, M.P.** (1996). Australasian truffle-like fungi. VIII. *Gummiglobus* and *Andebbia* gen. nov. (*Basidiomycotina, Mesophelliaceae*) and a supplement to the 'Nomenclatural Bibliography of Basidiomycotina'. *Aust. Syst. Bot.* 9: 803–811; **Trappe, J.M.; Castellano, M.A.; Malajczuk, N.** (1996). Australasian truffle-like fungi. VII. *Mesophellia* (*Basidiomycotina, Mesophelliaceae*). *Aust. Syst. Bot.* 9: 773–802; **Trappe, J.M.; Castellano, M.A.; Trappe, M.J.** (1992). Australasian truffle-like fungi. IV. *Malajczukia* gen. nov. (*Basidiomycotina, Mesophelliaceae*). *Aust. Syst. Bot.* 5: 617–630.

Metacapnodiaceae S. Hughes & Corlett 1972
Capnodiales: Ascomycota

Mycelium superficial, dark, copious, hyphae strongly constricted at the septa with cells ± spherical, tapering towards the apices. Ascomata superficial, small, cleistothecial, ± globose, black, thin-walled, with a periphysate ostiole; peridium of pseudoparenchyma. Interascal tissue of periphysoids. Asci ± saccate, fissitunicate. Ascospores brown, transversely septate, sometimes ornamented. Anamorphs hyphomycetous, conidiogenous cells tretic.

Significant Genera: *Metacapnodium*. Anamorph: *Capnobotrys*.

Distribution: Widespread in the tropics; also known from USA and New Zealand.

Economic Significance: None is known.

Ecology: Epiphytic, probably gaining nutrition from insect or plant exudates.

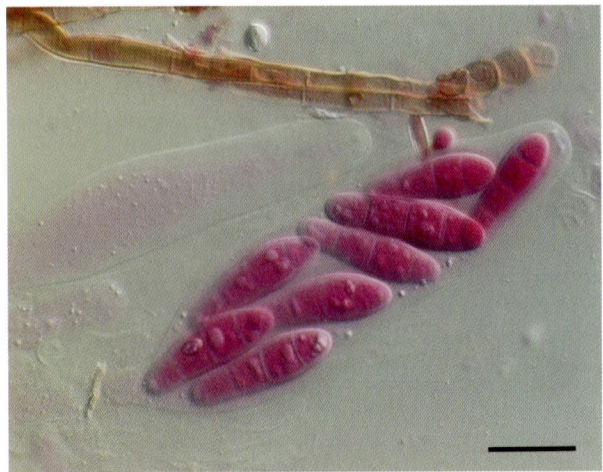

Metacapnodium dennisii, Sarawak, Malaysia; ascus and ascospores

Notes: Very poorly known; no recent information is available.

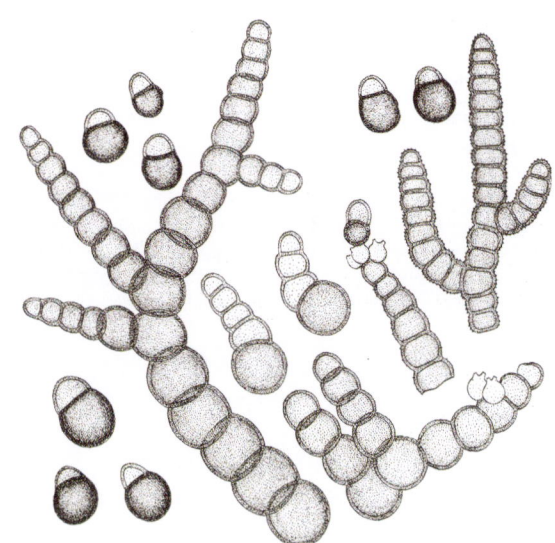

Capnobotrys neesii; superficial mycelium and conidia

References: **Parbery, I.H.; Brown, J.F.** (1986). Sooty moulds and black mildews in extra-tropical rainforests. *Microbiology of the Phyllosphere* (Cambridge): 101–120; **Sugiyama, J.; Amano, N.** (1987). Two Metacapnodiaceous sooty moulds from Japan: Their identity and behavior in pure culture. *Pleomorphic Fungi: The Diversity and its Taxonomic Implications* (Tokyo): 141–156.

Metschnikowiaceae T. Kamieński 1899
Saccharomycetales: Ascomycota

Mycelium often well-developed, usually remotely septate; vegetative cells usually proliferating by multilateral budding. Asci often formed on short lateral branches, elongate, sometimes curved or clavate, evanescent or persistent. Ascospores narrowly fusiform to filiform, sometimes curved, sometimes with filiform appendages; fermentation usually negative.

Significant Genera: *Clavispora, Metschnikowia*.

Distribution: Widespread, especially in the tropics.

Economic Significance: A few species cause food spoilage problems.

Ecology: Many are either marine or necrotrophic on plant tissue, others are found in nectar and are transmitted by insects.

Notes: Little recent phylogenetic information is available.

References: **Diezmann, S.; Cox, C.J.; Schonian, G.; Vilgalys, R.J.; Mitchell, T.G.** (2004). Phylogeny and evolution of medical species of *Candida* and related taxa: a multigenic analysis. *J. Clin. Microbiol.* 42: 5624–5635; **Gimenez-Jurado, G.; Kurtzman, C.P.; Starmer, W.T.; Spencer-Martins, I.** (2003). *Metschnikowia vanudenii* sp. nov. and *Metschnikowia lachancei* sp. nov., from flowers and associated insects in North America. *Int. J. Syst. Evol. Microbiol.* 53: 1665–1670; **Lachance, M.-A.; Bowles, J.M.; Starmer, W.T.**

(2003). *Metschnikowia santaceciliae, Candida hawaiiana*, and *Candida kipukae*, three new yeast species associated with insects of tropical morning glory. *FEMS Yeast Res.* 3: 97–103; **Lachance, M.A.; Daniel, H.M.; Meyer, W.; Prasad, G.S.; Gautam, S.P.; Boundy-Mills, K.** (2003). The D1/D2 domain of the large-subunit rDNA of the yeast species *Clavispora lusitaniae* is unusually polymorphic. *FEMS Yeast Res.* 4: 253–258; **Lachance, M.A.; Phaff, H.J.** (1998). *Clavispora* Rodrigues de Miranda. *The Yeasts, A Taxonomic Study* (Amsterdam): 148–152; **Marinoni, G.; Lachance, M.A.** (2004). Speciation in the large-spored *Metschnikowia* clade and establishment of a new species, *Metschnikowia borealis* comb. nov. *FEMS Yeast Res.* 4: 587–596; **Miller, M.W.; Phaff, H.J.** (1998). *Metschnikowia* Kamienski. *The Yeasts, A Taxonomic Study* (Amsterdam): 256–267.

Micareaceae Vězda ex Hafellner 1984
Lecanorales: Ascomycota

Thallus crustose. Ascomata apothecial, flat or domed, without a well-developed margin, pale or dark. Interascal tissue of relatively wide branched and anastomosing paraphyses, sometimes with a pigmented epithecium. Asci with a well-developed J+ apical cap, with a large tube-like more strongly staining ring surrounding a usually well-developed apical cushion, and an outer J+ gelatinized layer. Ascospores small, hyaline, simple or septate, without a sheath. Anamorphs pycnidial, with rather varied conidia.

Significant Genera: *Micarea*.

Distribution: Very widespread.

Economic Significance: None is known.

Micarea prasina, Devon, UK; thallus and ascomata

Ecology: Lichenized with green algae, primarily on wood and bark but also known on soil, rock and bryophytes. A few species are lichenicolous.

Notes: The family as currently circumscribed is polyphyletic (as is the genus *Micarea* itself) and should perhaps be merged with the *Pilocarpaceae*, but more research is needed.

References: **Andersen, H.L.; Ekman, S.** (2004). Phylogeny of the *Micareaceae* inferred from nrSSU DNA sequences. *Lichenologist* 36: 27–35; **Andersen, H.L.; Ekman, S.** (2005). Disintegration of the *Micareaceae* (lichenized *Ascomycota*): a molecular phylogeny based on mitochondrial rDNA sequences. *Mycol. Res.* 109: 21–30; **Brodo, I.M.; Tønsberg, T.** (1994). A new species of *Micarea* with stalked pycnidia from the west coast of North America. *Acta bot. fenn.* 150: 1–4; **Coppins, B.J.** (1999). Two new species of *Micarea* from South Africa. *Lichenologist* 31: 559–565; **Coppins, B.J.; Kantvilas, G.** (1990). Studies on *Micarea* in Australasia I. Four new species from Tasmania. *Lichenologist* 22: 277–288; **Coppins, B.J.; Purvis, O.W.** (1987). A review of *Psilolechia*. *Lichenologist* 19: 29–42; **Hawksworth, D.L.** (1988). Conidiomata, conidiogenesis, and conidia. *CRC Handbook of Lichenology* (Boca Raton): 181–193; **Pietschmann, M.** (1990). Morphometrics of tubiform apical apparatus in *Lecideaceae, Micareaceae, Porpidiaceae* and allied families (lichenized *Ascomycetes, Lecanorales*): limitations and perspectives of statistical inference. *Nova Hedwigia* 51: 521–549; **Triebel, D.; Wedin, M.; Rambold, G.** (1997). The genus *Scutula* (lichenicolous ascomycetes, *Lecanorales*): species on the *Peltigera canina* and *P. horizontalis* groups. *In* Tibell, L. & Hedberg, I. (eds), Lichen Studies Dedicated. *Symb. bot. upsal.* 32: 323–337.

Microascaceae Luttr. ex Malloch 1970
Microascales: Ascomycota

Stromata absent. Ascomata perithecial or cleistothecial, usually dark, fairly thin-walled, sometimes with coiled appendages, neck if present papillate or elongated. Ascomatal wall composed entirely of black, small-celled, pseudoparenchymatous cells. Interascal tissue absent. Asci small, ± globose, very thin-walled and evanescent, formed in chains, usually 8-spored. Ascospores small, bright yellow or copper-coloured, often curved (sometimes almost horseshoe-shaped), aseptate, smooth-walled, sometimes with minute germ pores. Anamorphs prominent, hyphomycetous. Conidiophores often elongated, sometimes synnematous, dark and usually smooth-walled. Conidiogenous cells proliferating percurrently, often annellidic, elongate or flask-shaped. Conidia small, ± hyaline, smooth or roughened. Synanamorphs with dark, thick-walled and sometimes ornamented; resting spores sometimes produced.

Significant Genera: *Microscus, Petriella, Pseudallescheria*. Anamorphs: *Scopulariopsis, Graphium, Scedosporium*.

Distribution: Cosmopolitan.

Economic Significance: *Pseudallecheria boydii* is sometimes associated with human and animal mycetomas, especially in infections of immunocompromised patients; other species may act as opportunists.

Ecology: Saprobic in soil or rotting vegetation, some species heat-tolerant and associated with silage and compost, a few associated with humans and animals.

Notes: There are strong morphological parallels between the *Microascaceae* and the *Ophiostomataceae* in both teleomorphic and anamorphic features, due primarily to similar spore dispersal strategies.

Microascus cinereus; ascomata with extruded ascospores

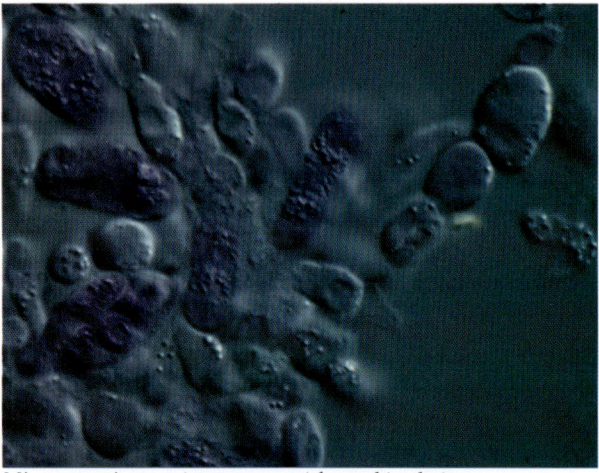
Microascus cinereus; immature asci formed in chains

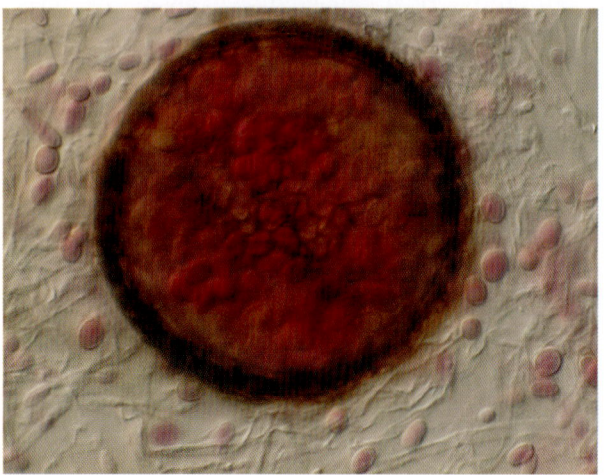
Pseudallescheria ellipsoidea; *Graphium* and *Scedosporium* synanamorphs

References: **Abbott, S.P.; Sigler, L.; Currah, R.S.** (1998). *Microascus brevicaulis* sp. nov., the teleomorph of *Scopulariopsis brevicaulis*, supports placement of *Scopulariopsis* with the *Microascaceae*. *Mycologia* 90: 297–302; **Arx, J.A. von; Figueras, M.J.; Guarro, J.** (1988). Sordariaceous ascomycetes without ascospore ejaculation. *Beih. Nova Hedwigia* 94: 104 pp.; **Dykstra, M.J.; Salkin, I.F.; McGinnis, M.R.** (1990). An ultrastructural comparison of conidiogenesis in *Scedosporium apiospermum*, *Scedosporium inflatum* and *Scopulariopsis brumptii*. *Mycologia* 81: 896–904; **Hausner, G.; Reid, J.; Klassen, G.R.** (2000). On the phylogeny of members of *Ceratocystis s.s.* and *Ophiostoma* that possess different anamorphic states, with emphasis on the anamorph genus *Leptographium*, based on partial ribosomal DNA sequen. *Can. J. Bot.* 78: 903–916; **Issakainen, J.; Jalava, J.; Campbell, C.K.** (1999). Relationship of *Scedosporium prolificans* with *Petriella* confirmed by partial LSU rDNA sequences. *Mycol. Res.* 103: 1179–1184; **Issakainen, J.; Jalava, J.; Eerola, E.; Campbell, C.K.** (1997). Relatedness of *Pseudallescheria*, *Scedosporium* and *Graphium pro parte* based on SSU rDNA sequences. *J. Med. Vet. Mycol.* 35: 389–398; **Issakainen, J.; Jalava, J.; Hyvönen, J.; Sahlberg, N.; Pirnes, T.; Campbell, C.K.** (2003). Relationships of *Scopulariopsis* based on LSU rDNA sequences. *Medical Mycol.* 41: 31–42; **Lee, S.J.; Hanlin, R.T.** (1999). Phylogenetic relationships of *Chaetomium* and similar genera based on ribosomal DNA sequences. *Mycologia* 91: 434–442; **Okada, G.; Seifert, K.A.; Takematsu, A.; Yamaoka, Y.; Miyazaki, S.; Tubaki, K.** (1999). A molecular phylogenetic reappraisal of the *Graphium* complex based on 18S rDNA sequences. *Can. J. Bot.* 76: 1495–1506; **Rainer, J.; de Hoog, G.S.** (2006). Molecular taxonomy and ecology of *Pseudallescheria*, *Petriella* and *Scedosporium prolificans* (*Microascaceae*) containing opportunistic agents on humans. *Mycol. Res.* 110: 151–160; **Rainer, J.; Hoog, G.S. de; Wedde, M.; Gräser, Y.; Gilges, S.** (2000). Molecular variability of *Pseudallescheria boydii*, a neurotropic opportunist. *J. Clin. Microbiol.* 38: 3267–3273; **Wedde, M.; Müller, D.; Tintelnot, K.; Hoog, G.S. de; Stahl, U.** (1998). PCR-based identification of clinically relevant *Pseudallescheria/Scedosporium* strains. *Medical Mycol.* 36: 61–67.

Microbotryaceae R.T. Moore 1996
Microbotryales: Basidiomycota

Mycelial septa poreless, hyphae intercellular, haustoria lacking. Sori developing in swollen anthers, filaments or ovaries, rarely in whole flowers or leaf tissue, typically resulting in replacement of generative tissue by a dark brown or violet spore mass, peridium distinct or ± lacking, columella usually absent. Teliospores globose to ovoid or angular due to compression, ± hyaline to brown or pale violet, smooth, reticulate or verrucose, rarely separated by disjunctors, germinating to form a 3- to 4-celled promycelium (basidium) which proliferates to produce a succession of sessile basidiospores apically and laterally, sterigmata not developing. Basidiospores small, hyaline, aseptate, thin-walled.

Significant Genera: *Microbotryum*, *Sphacelotheca*.

Distribution: Widespread in temperate zones.

Economic Significance: A few species parasitize ornamental plants (especially *Gypsophila*) but little economic damage is caused.

Ecology: Biotrophic, almost all species occurring in dicot flowers and effectively replacing anthers or ovaries.

Microbotryum violaceum; sori

Liroa emodensis, Tamil Nadu, India; sori in ovaries of *Polygonum chinense*

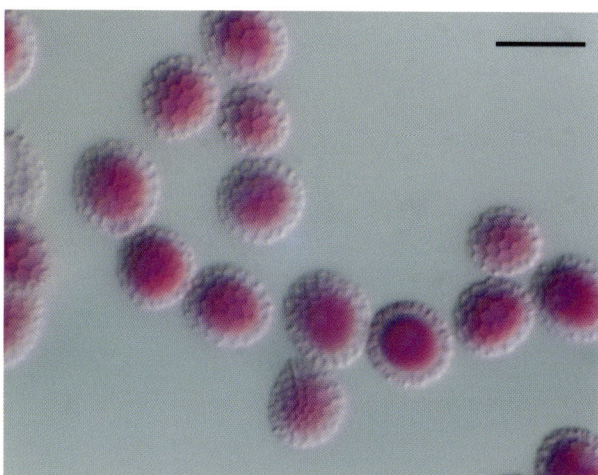
Microbotryum violaceum, Finland; teliospores

Notes: Much recent research has focused on speciation in the *Microbotryum* species parasitizing *Caryophyllaceae* and around 80 species are currently recognized. However, generic delimitation is uncertain and most *Microbotryum* species associated with other families (primarily *Polygonaceae*) may merit separation. *Bauerago* (on monocots) may belong in this family.

Microbotryum violaceum; sori in infected anthers

References: **Almaraz, T.; Roux, C.; Maumont, S.; Durrieu, G.** (2002). Phylogenetic relationships among smut fungi parasitizing dicotyledons based on ITS sequence analysis. *Mycol. Res.* 106: 541–548; **Begerow, D.; Bauer, R.; Oberwinkler, F.** (1998). Phylogenetic studies on nuclear large subunit ribosomal DNA sequences of smut fungi and related taxa. *Can. J. Bot.* 75: 2045–2056; **Begerow, D.; John, B.; Oberwinkler, F.** (2004). Evolutionary relationships among β-tubulin gene sequences of basidiomycetous fungi. *Mycol. Res.* 108: 1257–1263; **Berbee, M.L.; Bauer, R.; Oberwinkler, F.** (1991). The spindle pole body cycle, meiosis, and basidial cytology of the smut fungus *Microbotryum violaceum*. *Can. J. Bot.* 69: 1795–1803; **Bucheli, E.; Gautschi, B.; Shykoff, J.A.** (2001). Differences in population structure of the anther smut fungus *Microbotryum violaceum* on two closely related host species, *Silene latifolia* and *S. dioica*. *Mol. Ecol.* 10: 285–294; **Giraud, T.** (2004). Patterns of within population dispersal and mating of the fungus *Microbotryum violaceum* parasitising the plant *Silene latifolia*. *Heredity* 93: 559–565; **Hood, M.E.; Antonovics, J.; Heishman, H.** (2003). Karyotypic similarity identifies multiple host-shifts of a pathogenic fungus in natural populations. *Infect. Genet. Evol.* 2: 167–172; **Kemler, M.; Goker, M.; Oberwinkler, F.; Begerow, D.** (2006). Implications of molecular characters for the phylogeny of the *Microbotryaceae* (*Basidiomycota*: *Urediniomycetes*). *BMC Evol. Biol.* 6: 35; **Lutz, M.; Göker, M.; Piatek, M.; Kemler, M.; Begerow, D.; Oberwinkler, F.** (2005). Anther smuts of *Caryophyllaceae*: molecular characters indicate host-dependent species delimitation. *Mycol. Prog.* 4: 225–238; **Sampaio, J.P.; Gadanho, M.; Bauer, R.; Weiss, M.** (2003). Taxonomic studies in the *Microbotryomycetidae*: *Leucosporidium golubevii* sp. nov., *Leucosporidiella* gen. nov. and the new orders *Leucosporidiales* and *Sporidiobolales*. *Mycol. Prog.* 2: 53–68; **Swann, E.C.; Frieders, E.M.; McLaughlin, D.J.** (1999). *Microbotryum*, *Kriegeria* and the changing paradigm in basidiomycete classification. *Mycologia* 91: 51–66; **Van Putten,**

W.F.; Biere, A.; Van Damme, J.M.M. (2003). Intraspecific competition and mating between fungal strains of the anther smut *Microbotryum violaceum* from the host plants *Silene latifolia* and *S. dioica*. *Evolution* Lancaster, Pa. 57: 766–776; **Van Putten, W.F.; Biere, A.; Van Damme, J.M.M.** (2004). Host-related genetic differentiation in the anther smut fungus *Microbotryum violaceum* in sympatric, parapatric and allopatric populations of two host species *Silene latifolia* and *S. dioica*. *J. Evol. Biol.* 18: 203–212; **Vánky, K.** (1998). The genus *Microbotryum* (smut fungi). *Mycotaxon* 67: 33–60; **Vánky, K.** (2001). The new classification of the smut fungi, exemplified by Australasian taxa. *Aust. Syst. Bot.* 14: 385–394; **Vánky, K.** (2004). Biodiversity and conservation of smut fungi (*Ustilaginomycetes p.p., Microbotryales*) reflected in Vánky, *Ustilaginales* exsiccata no. 1-1250. *Mycol. Balcanica* 1: 175–187.

Microcaliciaceae Tibell 1984
Uncertain position within Ascomycetes, Ascomycota

Thallus absent. Ascomata small, usually long-stalked. Interascal tissue absent. Asci ellipsoidal, formed in chains, thin-walled and evanescent. Ascospores 1- to 7-septate, dark brown, with an ornamentation of helical ridges, forming a dry mazaedial mass.

Significant Genera: *Microcalicium*.

Distribution: Most species are European and/or boreal in distribution.

Economic Significance: None is known.

Ecology: Saprobic on wood or fungicolous.

Notes: No molecular data are available and the systematic position is unclear.

Microcalicium arenarium, Devon, UK; stalked ascomata

References: **Tibell, L.** (1998). Crustose Mazaediate Lichens and the *Mycocaliciaceae* in Temperate South America. *Biblthca Lichenol.* 71: 107 pp.; **Tibell, L.** (1999). Calicioid lichens and fungi. *Nordic Lichen Flora* 1. Introductory Parts; Calicioid Lichens and Fungi (Uddevalla): 20–94.

Micropeltidaceae Clem. & Shear 1931
Uncertain position within Dothideomycetes, Ascomycota

Mycelium at least partially superficial, often not strongly pigmented. Ascomata strongly flattened, occasionally coalescing, often greenish, opening by a wide irregular ostiole; peridium of one to several layers of pseudoparenchymatous cells, sometimes epidermoid, the lower wall indistinct. Interascal tissue of cellular pseudoparaphyses, often sparse or evanescent. Asci clavate to saccate, fissitunicate, sometimes faintly J+ blue at the apex, with a poorly defined apical ring. Ascospores hyaline or pale brown, transversely septate or muriform, sometimes attenuated at the base. Anamorphs poorly known, pycnothyrial.

Significant Genera: *Micropeltis, Stomiopeltis*.

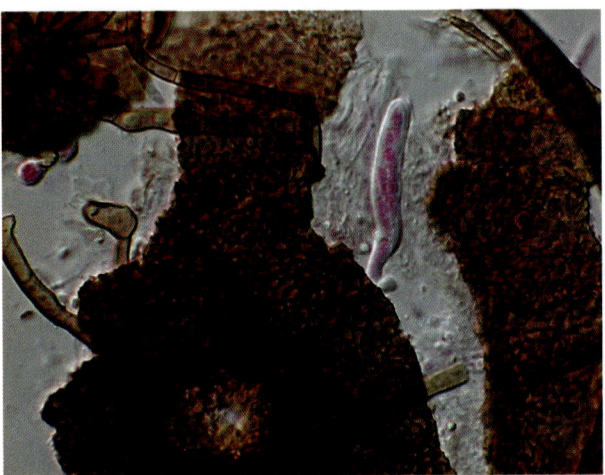

Stomiopeltis sp. on *Miconia robinsoniana*, Galápagos Islands, Ecuador; ascoma and ascus

Micropeltis ugandae, Kenya; ascus and ascospores

Distribution: Most species have tropical distributions.

Economic Significance: A few species are implicated in plant disease, causing 'sooty blotch' of fruits, but epidemiology is poorly understood.

Ecology: Superficial on leaves etc. The mode of nutrition is unknown for most species, but many are likely to be saprobic on plant exudates.

Notes: Little is known of this family. Species concepts are very uncertain and it seems likely that they are not strongly host-specific. Phylogenetic studies are lacking.

References: **Cannon, P.F.** (2001). *Asterina nyanzae*; *Capnodium salicinum*; *Chaetothyrium guaraniticum*; *Cucurbitaria laburni*; *Dothidea sambuci*; *Hysterographium fraxini*; *Micropeltis ugandae*; *Parodiella hedysari*; *Tubeufia cerea*; *Vizella gomphispora*. IMI Descr. Fungi Bact. 141: [22] pp.; **Eboh, D.O.** (1987). Further contribution to the East African foliicolous *Ascomycotina*. Sydowia 39: 37–49; **Farr, M.L.** (1987). Amazonian foliicolous fungi. IV. Some new and critical taxa in *Ascomycotina* and associated anamorphs. Mycologia 79: 97–116; **Hsieh, W.H.; Chen, C.Y.; Sivanesan, A.** (1995). Taiwan fungi: new species and new records of ascomycetes. Mycol. Res. 99: 917–931; **Hyde, K.D.** (1997). Ascomycetes described on *Freycinetia*. Sydowia 49: 1–20; **Panwar, A.B.; Bhoite, A.S.; Patil, M.S.** (1991). Studies in foliicolous fungi – XI. Genus *Chaetothyrina* Theiss. Geobios New Rep. 10: 41–46.

Microstromataceae Jülich 1982
Microstromatales: Basidiomycota

Sori formed in substomatal cavities, wall structures not defined. Hyphae intercellular, with simple pores. Telia absent. Basidia derived from fertile hyphae or sac-like structures that are sometimes surrounded by sterile hyphal tissue, erumpent through host stomata, clavate or elongate, with a cluster of 6–8 sterigmata at the apex. Basidiospores discharged passively, small, hyaline and aseptate. Anamorph yeast-like, cells globose to ellipsoidal, not ballistosporic.

Significant Genera: *Microstroma*.

Distribution: Widespread.

Microstroma juglandis, Pakistan; colony on living leaf of *Juglans regia*

Economic Significance: *Microstroma juglandis* causes downy leaf-spot of walnut (*Juglans*), but it is not a major pathogen.

Ecology: Biotrophic in living leaves of woody plants.

Notes: *Volvocisporium* has been separated into its own family, characterized by its septate rather than aseptate basidiospores, but it does not clearly fall into a separate clade based on existing molecular data.

References: **Bauer, R.; Oberwinkler, F.; Vánky, K.** (1997). Ultrastructural markers and systematics in smut fungi and allied taxa. Can. J. Bot. 75: 1273–1314; **Beer, Z.W. de; Begerow, D.; Bauer, R.; Pegg, G.S.; Crous, P.W.; Wingfield, M.J.** (2006). Phylogeny of the *Quambalariaceae* fam. nov., including important *Eucalyptus* pathogens in South Africa and Australia. Stud. Mycol. 55: 289–298; **Begerow, D.; Bauer, R.; Oberwinkler, F.** (1998). Phylogenetic studies on nuclear large subunit ribosomal DNA sequences of smut fungi and related taxa. Can. J. Bot. 75: 2045–2056; **Begerow, D.; Bauer, R.; Oberwinkler, F.** (2001). *Muribasidiospora*: *Microstromatales* or *Exobasidiales*?. Mycol. Res. 105: 798–810; **Begerow, D.; Bauer, R.; Oberwinkler, F.** (2002). The *Exobasidiales*: an evolutionary hypothesis. Mycol. Prog. 1: 187–199.

Microtheliopsidaceae O.E. Erikss. 1981
Uncertain position within Dothideomycetes, Ascomycota

Thallus crustose, effuse, non-corticate. Ascomata scattered, perithecial, immersed, conical, black, ostiolate. Interascal tissue absent. Asci ovoid to saccate, apparently fissitunicate, blue in KOH/IKI, the inner layer thickened towards the apex. Ascospores brown, thin-walled, septate, without a sheath. Anamorphs unknown.

Microtheliopsis uleana; thallus and ascomata

Significant Genera: *Microtheliopsis*.

Distribution: Pantropical.

Economic Significance: None is known.

Ecology: Lichenized with green algae (*Phycopeltis*), mostly foliicolous.

Notes: Very poorly known; no recent information is available.

References: **Aptroot, A.** (1998). Aspects of the integration of the taxonomy of lichenized and non-lichenized pyrenocarpous ascomycetes. *Lichenologist* 30: 501–514; **Lücking, R.** (1994). A new foliicolous species of *Microtheliopsis* (Lichens, *Microtheliopsidaceae*) from Costa Rica. *Mycotaxon* 51: 69–73; **Lücking, R.; Sérusiaux, E.; Maia, L.C.; Pereira, E.C.G.** (1998). A revision of the names of foliicolous lichenized fungi published by Batista and co-workers between 1960 and 1975. *Lichenologist* 30: 121–191.

Microthyriaceae Sacc. 1883
Microthyriales: Ascomycota

Superficial mycelium indistinct. Ascomata small, thyrothecial, strongly flattened, superficial, with a small central ostiole sometimes surrounded by thickened tissue; peridium brown, composed of radiating rows of irregular, epidermoid or ± isodiametric cells, the basal layer hyaline, poorly developed or absent. Interascal tissue of trabeculate pseudoparaphyses, often deliquescing. Asci saccate, fissitunicate, without a clear ocular chamber, not blueing in iodine. Ascospores hyaline or brown, transversely septate, sometimes ciliate, without a sheath. Anamorphs pycnothyrial, similar in structure to the ascomata.

Significant Genera: *Lichenopeltella, Microthyrium*.

Distribution: Cosmopolitan, especially diverse in the tropics.

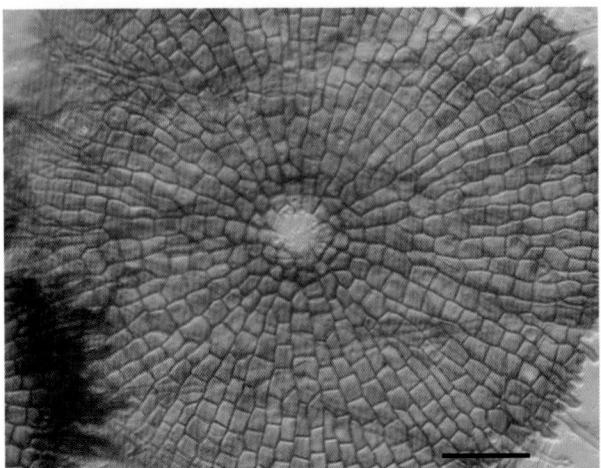

Microthyrium ciliatum, Suffolk, UK; cellular structure of ascoma

Economic Significance: None is known.

Ecology: Saprobic or epibiotic on leaves and stems, occasionally on lichens or other fungi, probably gaining nutrition primarily from exudates.

Notes: A polymorphic and poorly studied family, in common with most other thyriothecial ascomycetes.

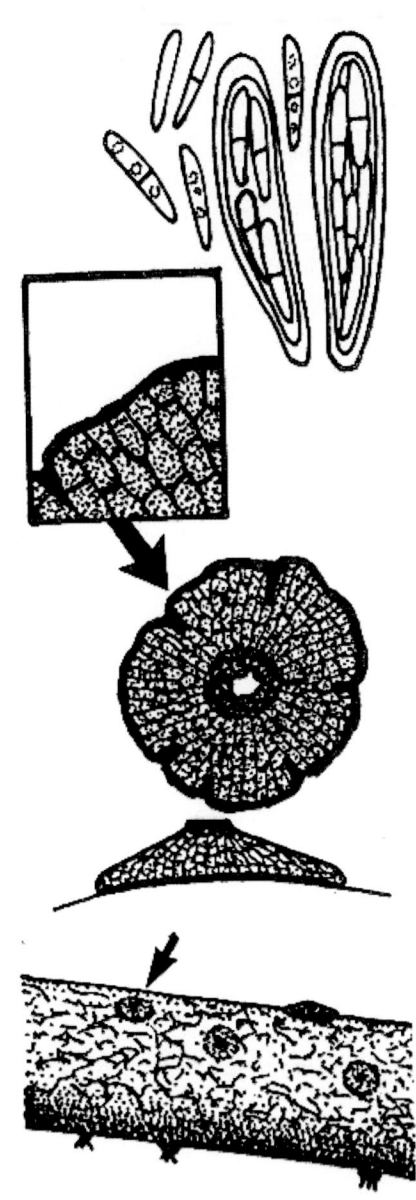

Lichenopeltella pinophylla; ascoma, asci and ascospores

References: **Crous, P.W.; Kendrick, W.B.** (1994). *Arnaudiella eucalyptorum* sp. nov. (*Dothideales, Ascomycetes*), and its hyphomycetous anamorph *Xenogliocladiopsis* gen. nov., from *Eucalyptus* leaf litter in South Africa. *Can. J. Bot.* 72: 59–64; **Hariharan, G.N.; Mibey, R.K.; Hawksworth, D.L.** (1996). A new species of *Lichenopeltella* on *Porina* in India. *Lichenologist* 28: 294–296; **Holm, K.; Holm, L.** (1991). Ascomycetes on *Myrica gale* in Sweden. *Nordic Jl Bot.* 11: 675–687; **Hosagoudar, V.B.; Balakrishnan, M.P.; Goos, R.D.** (1996). Some *Asterinella, Asterostomella* and *Echidnodella* species fom southern India. *Mycotaxon* 58: 489–498; **Kirk, P.M.; Spooner, B.M.** (1989). Ascomycetes on leaf litter of *Laurus nobilis* and *Hedera helix*. *Mycol. Res.* 92: 335–346; **Matzer, M.** (1996). Lichenicolous Ascomycetes with Fissitunicate Asci on Foliicolous Lichens. *Mycol. Pap.* 171: 202 pp.; **Ramaley, A.W.** (1999).

Three species of *Microthyrium* from *Nolina*. *Mycotaxon* 70: 7–16; **Spooner, B.M.; Kirk, P.M.** (1990). Observations on some genera of *Trichothyriaceae*. *Mycol. Res.* 94: 223–230; **Swart, H.J.** (1986). Australian leaf-inhabiting fungi XXII. *Microthyrium*-like fungi on *Eucalyptus*. *Trans. Br. mycol. Soc.* 87: 81–91.

Mikronegeriaceae Cummins & Y. Hirats. 1983
Uredinales: Basidiomycota

Spermogonia deeply immersed and irregularly ampulliform, often lobed or multiloculate. Aecia *Petersonia*-like, subepidermal with upper wall absent or reduced to a hyphal layer. Aeciospores formed in chains without intercalary cells, verrucose. Uredinia *Uredo*-like, lacking a discrete wall. Urediniospores borne singly, echinulate, apparently lacking germ pores, paraphyses absent. Telia waxy, lacking an outer wall, paraphyses absent. Teliospores unicellular, sessile, laterally free, germinating by prolongation of teliospore apex to form a septate basidium, from which ellipsoidal basidiospores develop on distinct sterigmata.

Significant Genera: *Mikronegeria*.

Distribution: Known from temperate South America.

Economic Significance: None is known.

Ecology: Biotrophic, heteroecious with aecia on conifers (*Araucaria* and *Austrocedrus*) and telia on *Nothofagus*.

Notes: *Blastospora* and *Chrysocelis* may also belong in this family and it may be synonymous with the *Chaconiaceae*, but further revision is required and molecular data are sparse. The group may represent the most basal family of the *Uredinales*.

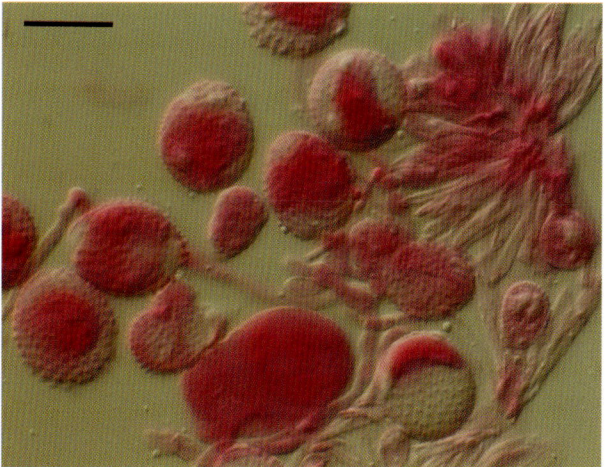

Mikronegeria fagi, Chile; urediniospores from living leaves of *Nothofagus obliqua*

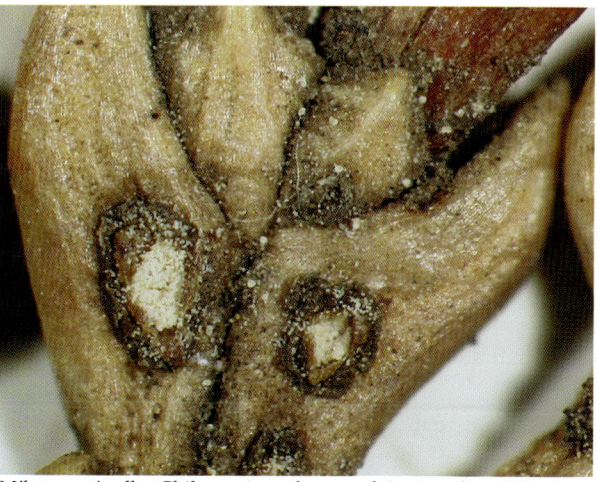

Mikronegeria alba, Chile; aecia on leaves of *Austrocedrus chilensis*

References: **Aime, M.C.** (2006). Towards resolving family-level relationships in rust fungi (*Uredinales*). *Mycoscience* 47: 112–122; **Cummins, G.B.; Hiratsuka, Y.** (2003). *Illustrated Genera of Rust Fungi* (St Paul): 225 pp.; **Sato, T.; Sato, S.** (1985). Morphology of aecia of the rust fungi. *Trans. Br. mycol. Soc.* 85: 223–238.

Miltideaceae Hafellner 1984
Lecanorales: Ascomycota

Thallus crustose. Ascomata apothecial, ± sessile, pale, without a well-developed margin. Interascal tissue of branched and anastomosing paraphyses, swollen at the apices, with a reddish crystalline epithecium. Asci with a weakly J+ apical cap and a well-developed ocular chamber, with a J+ gelatinized outer layer. Ascospores hyaline, aseptate, with a gelatinous sheath.

Significant Genera: *Miltidea*.

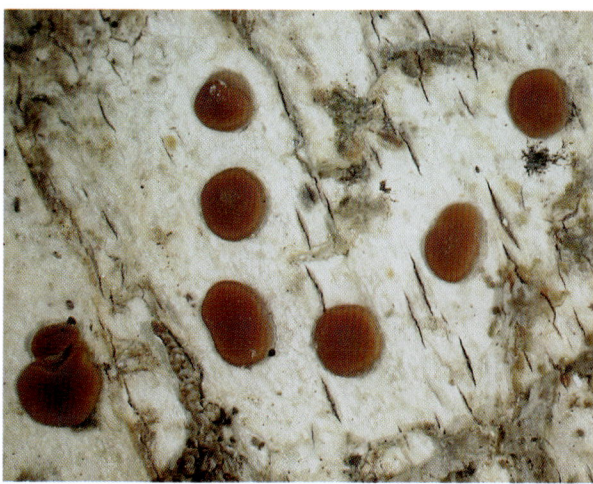

Miltidea ceroplasta, Tasmania; ascomata

Distribution: Known from Sri Lanka and Australasia.

Economic Significance: None is known.

Ecology: Lichenized with green algae, tropical and cool temperate rainforests.

Notes: Very poorly known; no phylogenetic information is available.

References: **Lumbsch, H.T.** (1997). Systematic studies in the suborder *Agyriineae* (*Lecanorales*). *J. Hattori bot. Lab.* 83: 1–73.

Mixiaceae C.L. Kramer 1987
Mixiales: Basidiomycota

Sporogenous cells ('spore sacs') formed directly from the germination of specialized cells of the internal mycelium, with a wall separating a peripheral layer of sporogenous multinucleate protoplasm; irregular in form but usually fusiform or pyriform, thick-walled, aseptate, producing spores from multiple small denticles covering the surface. Spores small, ellipsoidal, hyaline, aseptate. Anamorph yeast-like.

Significant Genera: *Mixia*.

Distribution: Known from scattered locations throughout the north temperate zone, including North America, Europe and Japan.

Economic Significance: None is known.

Ecology: Biotrophic in living leaves of the fern genus *Osmunda*.

Notes: *Mixia osmundae*, the only species in the *Mixiaceae*, was originally described as a species of *Taphrina* and subsequently assumed to be a member of the *Protomycetales* (*Ascomycota*). However, it is now recognized as having affinities with the *Basidiomycota* and although its closest relatives are uncertain it appears to occupy a basal position within that lineage.

References: **Bauer, R.; Begerow, D.; Sampaio, J.P.; Weiß, M.; Oberwinkler, F.** (2006). The simple-septate basidiomycetes: a synopsis. *Mycol. Progr.* 5: 41-66; **Kramer, C.L.** (1987). The *Taphrinales*. *Stud. Mycol.* 30: 151–166; **Nishida, H.; Ando, K.; Ando, Y.; Hirata, A.; Sugiyama, J.** (1995). *Mixia osmundae*: transfer from the *Ascomycota* to the *Basidiomycota* based on evidence from molecules and morphology. *Can. J. Bot.* 73: S660–S666; **Sjamsuridzal, W.; Nishida, H.; Ogawa, H.; Kakishima, M.; Sugiyama, J.** (1999). Phylogenetic positions of rust fungi parasitic on ferns: evidence from 18S rDNA sequence analysis. *Mycoscience* 40: 21–27; **Sjamsuridzal, W.; Nishida, H; Yokota, A.** (2002). Phylogenetic position of *Mixia osmundae* inferred from 28S rDNA comparison. *J. gen. appl. Microbiol.* Tokyo 48: 121–123.

Monascaceae J. Schröt. 1894
Uncertain position within Eurotiomycetes, Ascomycota

Stromata absent. Ascomata small, cleistothecial, globose, hyaline or brown, thin-walled; peridium composed of flattened thick-walled hyphae. Interascal tissue absent. Asci presumably evanescent at a very early stage, morphology unknown. Ascospores hyaline, aseptate, ellipsoidal, thick-walled. Anamorphs prominent, hyphomycetous. Conidia formed in chains from apparently undifferentiated hyphae, the oldest at the apex and with conidiogenesis regressive so the conidiophore becomes shorter, ellipsoidal to pyriform, usually brown, thick-walled and often roughened.

Significant Genera: *Monascus*. Anamorph: *Basipetospora*.

Distribution: Cosmopolitan, but more commonly encountered in the tropics.

Economic Significance: Some species are used to produce fermented foods in the Orient. The bright red pigments produced by some species in culture have also been investigated as food dyes.

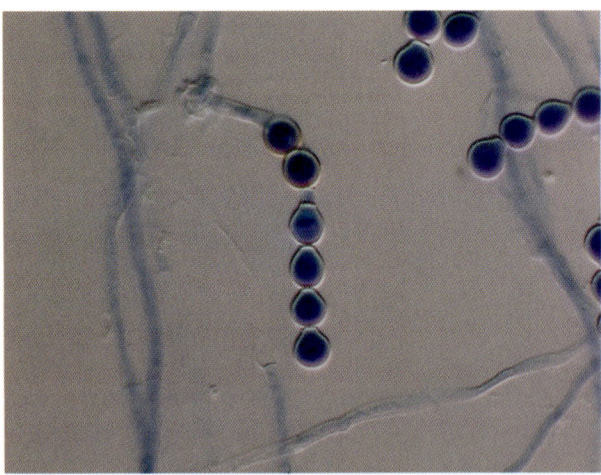

Monascus ruber; conidiophore and conidia

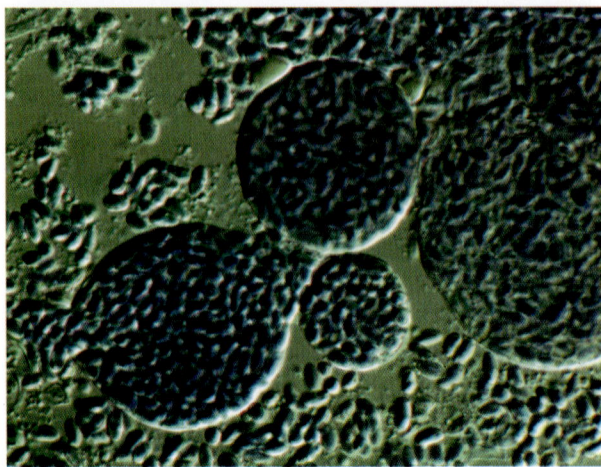

Xeromyces bisporus; ascomata and ascospores

Ecology: Saprobes, especially on substrata with high water tension, e.g. dried foods; also commonly encountered in compost, silage and similar substrata.

Notes: The unusual ascoma morphology has led to placement in a variety of major groups, but molecular data indicate a position within the *Eurotiomycetes*.

References: **Bridge, P.D.; Hawksworth, D.L.** (1985). Biochemical tests as an aid to the identification of *Monascus* species. *Lett. Appl. Microbiol.* 1: 25–29; **Cannon, P.F.; Abdullah, S.K.; Abbas, B.A.** (1995). Two new species of *Monascus* from Iraq, with a key to known species of the genus. *Mycol. Res.* 99: 659–662; **Chaisrisook, C.** (2002). Mycelial reactions and mycelial compatibility groups of red rice mould (*Monascus purpureus*). *Mycol. Res.* 106: 298–304; **Hocking, A.D.** (1991). Xerophilic fungi in intermediate and low moisture foods. *Handbook of Applied Mycology* Vol. 3. Foods and Feeds: 69–97; **Park, H.G.; Jong, S.C.** (2003). Molecular characterization of *Monascus* strains based on the D1/D2 regions of LSU rRNA genes. *Mycoscience* 44: 25–32; **Pitt, J.I.; Hocking, A.D.** (1997). *Fungi and Food Spoilage* Edn 2 (London): 593 pp.; **Udagawa, S.; Baba, H.** (1998). *Monascus lunisporas*, a new species isolated from mouldy feeds. *Cryptog. Mycol.* 19: 269–276.

Monoblastiaceae Walt. Watson 1929
Pyrenulales: Ascomycota

Thallus crustose, often immersed. Ascomata sometimes aggregated, perithecial, usually clypeate, ± globose, papillate, the ostiole sometimes lateral. Interascal tissue of narrow trabeculate pseudoparaphyses, true paraphyses not developing. Asci cylindrical, fissitunicate, with a large ocular chamber but without an apical ring. Ascospores usually 1-septate, hyaline, smooth or ornamented, sometimes with a sheath. Anamorphs coelomycetous, pycnidial.

Significant Genera: *Anisomeridium*.

Acrocordia gemmata, Argyll, Scotland; thallus and ascomata

Distribution: Widespread, especially in the tropics.

Economic Significance: None is known.

Ecology: Lichenized with green algae.

Notes: Poorly known; little recent information and no phylogenetic data are available.

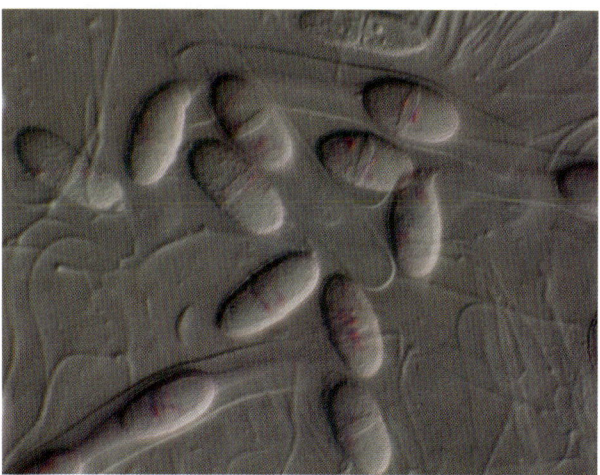

Acrocordia gemmata, Argyll, Scotland; asci and ascospores

References: **Aptroot, A.** (1991). *Monoblastia pellucida*: a remarkable new lichen species from tropical coastal areas, with comments on the *Monoblastiaceae*. *Bryologist* 94: 404–406; **Aptroot, A.** (1991). Tropical pyrenocarpous lichens. A phylogenetic approach. *Tropical Lichens: Their Systematics, Conservation, and Ecology* (Oxford) 43: 253–273; **Aptroot, A.** (1999). Notes on taxonomy, distribution and ecology of *Anisomeridium polypori*. *Lichenologist* 31: 641–642; **Etayo, J.; Lücking, R.** (1999). *Anisomeridium musaesporoides*, a new foliicolous lichen from tropical America. *Lichenologist* 31: 145–148; **Sérusiaux, E.; Aptroot, A.** (1998). A further new species of *Monoblastia* (lichenized ascomycetes: *Monoblastiaceae*) from Papua New Guinea. *Nova Hedwigia* 67: 259–265.

Monoblepharidaceae A. Fisch. 1892
Monoblepharidales: Chytridiomycota

Thallus differentiated into a well-developed, branched or unbranched, hyphal, vegetative system attached by rhizoids and bearing numerous reproductive structures; zoosporangia terminal, narrowly cylindrical or irregular, delimited by a septum, a succession formed by hyphal branching or internal proliferation; zoospores fully formed withing the zoosporangium, escaping by apical dissolution thereof; oogonia intercalary or terminal, usually narrowly pyriform or spherical, delimited by a septum, papillate at maturity, with a single oosphere; antheridia somewhat cylindrical, variously inserted; zygote remaining in oogonium or encysting on the opening of the oogonium; male gamete completely engulfed at fertilization, oogamous.

Significant Genera: *Monoblepharis*.

Distribution: Known from Europe and North America and possible restricted to temperate zones.

Economic Significance: None is known.

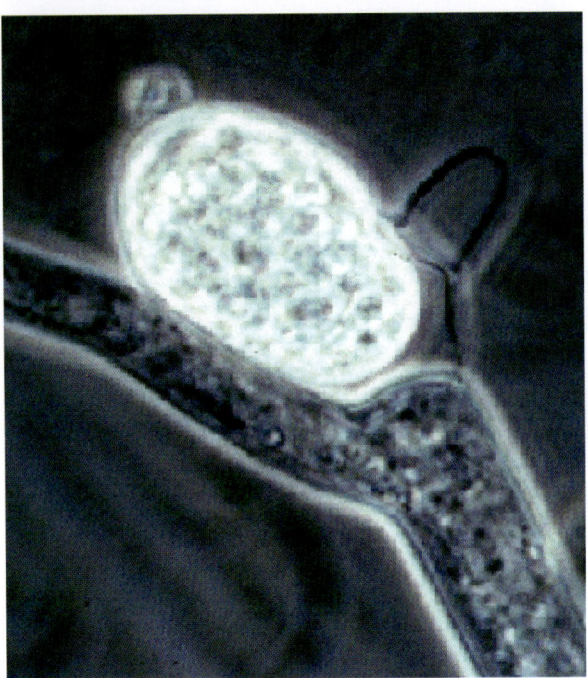

Monoblepharis sp.; young oogonium with attached antheridium

Ecology: Saprobic on submerged plant material, particularly waterlogged but not decorticated twigs of broadleaf trees (esp. *Quercus robur* in the British Isles). Apparently psychrophilic, with optimum temperature for growth at a few degrees above freezing, although this may be due to reduced competition.

Notes: Some ultrastructural characters indicate a possible relationship with choanoflagellates although this has not yet been corroborated with other evidence and the similarity may be due to independent origins.

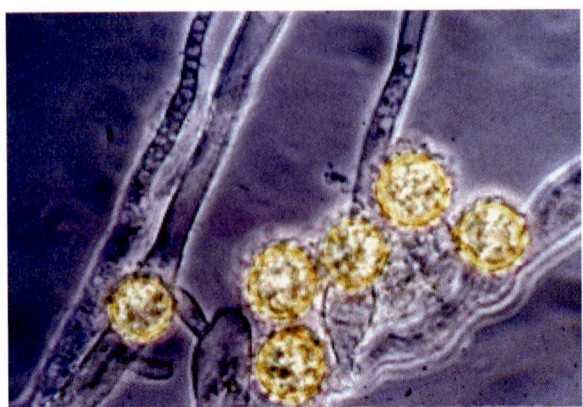

Monoblepharis sp.; mature oospores

References: **Noyes Mollicone, M.R.; Longcore, J.E.** (1994). Zoospore ultrastructure of *Monoblepharis polymorpha*. Mycologia 86: 615–625; **Perrott, P.E.T.** (1955). The genus *Monoblepharis*. Trans. Br. mycol. Soc. 38: 247–282.

Morchellaceae Rchb. 1834
Pezizales: Ascomycota

Ascomata formed from large subterranean sclerotium-like structures, large, apothecial, usually distinctly stalked, the hymenium usually everted, often lobed or corrugated, often differently coloured or darker than the stalk, which is often hollow. Interascal tissue well-developed, composed of simple, ± unbranched, true paraphyses. Asci cylindrical, persistent, not blueing in iodine, with an operculate apex. Ascospores ellipsoidal, hyaline, smooth, eguttulate but with a minutely guttulate capitate appendage at each pold, multinucleate. Anamorphs hyphomycetous, with branched or verticillate conidiophores arising from dark ornamented setose hyphae. Conidiogenous cells ± cylindrical, proliferating sympodially, minutely denticulate. Conidia small, hyaline, aseptate, ± spherical.

Significant Genera: *Morchella*, *Verpa*. Anamorph: *Costantinella*.

Morchella esculenta, Chile; ascoma

Distribution: Widespread, especially in temperate regions.

Economic Significance: *Morchella* species are highly prized gastronomically and there has been considerable interest in their cultivation in North America.

Verpa digitaliformis; ascomata

Ecology: Saprobic on soil, rotten wood and leaf mould primarily in broadleaved forests, also forming ectomycorrhizal associations.

Notes: Phylogenetic studies suggest that the hypogeous genera *Fischerula* and *Leucangium* might belong within the *Morchellaceae*, but more research is needed.

References: **Amir, R.; Levanon, D.; Hadar, Y.; Chet, I.** (1992). Formation of sclerotia by *Morchella esculenta*: relationship between media composition and turgor potential in the mycelium. *Mycol. Res.* 96: 943–948; **Bunyard, B.A.; Nicholson, M.S.; Royse, D.J.** (1995). Phylogenetic resolution of *Morchella*, *Verpa* and *Disciotis* [Pezizales: Morchellaceae] based on restriction enzyme analysis of the 28S ribosomal RNA gene. *Exp. Mycol.* 19: 223–233; **Bunyard, B.A.; Nicholson, M.S.; Royse, D.J.** (1995). A systematic assessment of *Morchella* using RFLP analysis of the 28S ribosomal RNA gene. *Mycologia* 86: 762–772; **Buscot, F.** (1994). Ectomycorrhizal types and endobacteria associated with ectomycorrhizas of *Morchella elata* (Fr.) Boudier with *Picea abies* (L.) Karst. *Mycorrhiza* 4: 223–232; **Guzmán, G.; Tapia, F.** (1998). The known morels in Mexico, a description of a new blushing species, *Morchella rufobrunnea*, and new data on *M. guatemalensis*. *Mycologia* 90: 705–714; **Harrington, F.A.; Pfister, D.H.; Potter, D.; Donoghue, M.J.** (1999). Phylogenetic studies within the *Pezizales*. I. 18S rRNA sequence data and classification. *Mycologia* 91: 41–50; **Jacquetant, E.** (1984). *Les Morilles* (Paris): 144 pp.; **Landvik, S.; Egger, K.N.; Schumacher, T.** (1997). Towards a subordinal classification of the *Pezizales* (Ascomycota): phylogenetic analyses of SSU rDNA sequences. *Nordic Jl Bot.* 17: 403–418; **O'Donnell, K.; Cigelnik, E.; Weber, N.S.; Trappe, J.M.** (1997). Phylogenetic relationships among ascomycetous truffles and the true and false morels inferred from 18S and 28S ribosomal DNA sequence analysis. *Mycologia* 89: 48–65; **Parguey-Leduc, A.; Janex-Favre, M.-C.; Bruxelles, G.** (1998). [Comparative study of the ascus and ascospores of some morels (genus *Morchella*, Ascomycetes)]. *Cryptog. Bryol.-Lichénol.* 19: 277–292; **Volk, T.J.; Leonard, T.J.** (1990). Cytology of the life-cycle of *Morchella*. *Mycol. Res.* 94: 399–406; **Wipf, D.; Fribourg, A.; Munch, J.C.; Botton, B.; Buscot, F.** (1999). Diversity of the internal transcribed spacer of rDNA in morels. *Can. J. Microbiol.* 45: 769–778; **Wipf, D.; Koschinsky, S.; Clowez, P.; Munch, J.C.; Botton, B.; Buscot, F.** (1997). Recent advances in ecology and systematics of morels. *Cryptog. Mycol.* 18: 95–109.

Moriolaceae Zahlbr. 1903
Uncertain position within Dothideomycetes, Ascomycota

Mycelium dark, superficial, ? sometimes associated with algal cells. Stromata absent. Ascomata perithecial, ± flattened to globose, superficial, black, ± thin-walled. Interascal tissue of ? narrow anastomosing pseudoparaphyses. Asci cylindrical, persistent, ± thick-walled, with separable wall layers, fissitunicate, with an indistinct ocular chamber surrounded by an inconspicuous J– ring. Ascospores brown, transversely septate, without a sheath. Anamorphs hyphomycetous.

Significant Genera: *Moriola*.

Distribution: Known from Europe.

Economic Significance: None is known.

Ecology: Saprobic on bark, or lichenized with green algae.

Notes: *Moriola*, the only genus, is in need of revision and its circumscription is uncertain.

References: **Clauzade, G.; Roux, C.** (1985). Likenoj de Okcidenta Europa. Ilustrita determinlibro. *Bull. Soc. bot. Centre-Ouest* Nouv. sér., num. spec. 7: 893 pp.

Mortierellaceae A. Fisch. 1892
Mortierellales: Zygomycota

Vegetative thallus a mycelium, complex fruit bodies absent. Sporophores (sporangiophore) arising from substratum mycelium or aerial hyphae, simple or more frequently branched (often distictly acrotonous or basitonous), typically erect and ascending, often with characteristic basal rhizoids. Sporangia and/or sporangiola formed; sporangia multispored, wall diffluent, the sporangium forming a liquid droplet at maturity, or persistent, echinulate, acolumellate or columella rudimentary; sporangiola few to 1-spored, persistent walled and acolumellate. Sporangiospores globose to narrowly ellipsoid or variable, smooth or ornamented, hyaline. Zygospores globose to subglobose, sometimes surrounded by hyphae; wall hyaline, smooth or angular; suspensors opposed, smooth, equal or unequal; heterothallic or homothallic.

Significant Genera: *Mortierella*.

Distribution: Widespread and very common as a component of the soil biota, including soils from Antarctica.

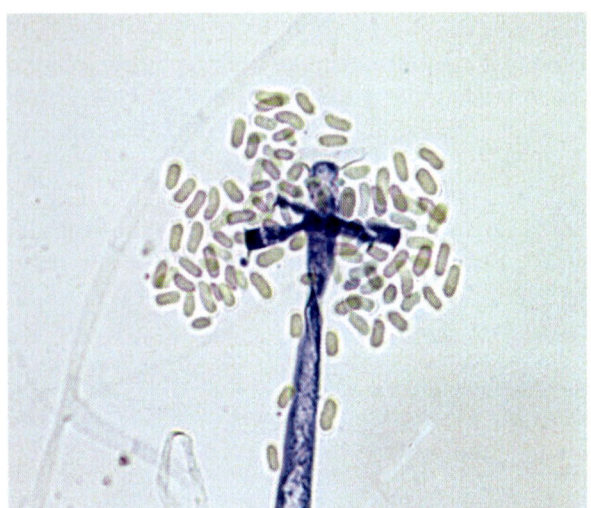

Mortierella wolfii; dehisced sporangium with acrotonous (terminal) branches

Economic Significance: At least one species (*Mortierella wolfii*) is an important causal agent of bovine mycotic abortion, pneumonia and systemic mycosis in New Zealand, Australia, Europe and USA. Confirmed infections in man have not, however, been documented.

Ecology: Saprobic in soil and thus frequently isolated from soil and material contaminated with soil such as hay or rotten silage and coal spoil tips.

Notes: Originally considered a family of the *Mucorales* but recognized as distinct from the core families due to fatty acid profiles and sterol analysis. The significance of these differences was augmented by results from the analysis of molecular data which confirmed their separation into a new order. They appear to be a sister group to the *Endogonales*.

References: **Corradi, N.; Kuhn, G.; Sanders, I.R.** (2004). Monophyly of β-tubulin and H+-ATPase gene variants in *Glomus intraradices*: consequences for molecular evolutionary studies of AM fungal genes. *Fungal Genetics Biol.* 41: 262–273; **Kwanśa, H.; Ward, E.; Bateman, G.L.** (2006). Phylogenetic relationships among *Zygomycetes* from soil based on ITS1/2 rDNA sequences. *Mycol. Res.* 110: 501–510; **O'Donnell, K.; Lutzoni, F.M.; Ward, T.J.; Benny, G.L.** (2001). Evolutionary relationships among mucoralean fungi (*Zygomycota*): evidence for family polyphyly on a large scale. *Mycologia* 93: 286–296; **Tanabe, Y.; Saikawa, M.; Watanabe, M.M.; Sugiyama, J.** (2004). Molecular phylogeny of *Zygomycota* based on EF-1α and RPB1 sequences: limitations and utility of alternative markers to rDNA. *Mol. Phylogen. Evol.* 30: 438–449; **Voigt, K.; Wöstemeyer, J.W.** (2001). Phylogeny and origin of 82 zygomycetes from all 54 genera of the *Mucorales* and *Mortierellales* based on combined analysis of actin and translation elongation factor EF-1α genes. *Gene* 270: 113–120.

Mucoraceae Dumort. 1822
Mucorales: Zygomycota

Vegetative thallus a mycelium, complex fruit bodies absent. Sporophores arising from substratum mycelium aerial hyphae or stolons, simple or more frequently branched, typically erect and ascending or rarely repent. Sporangia and/or sporangiola formed; sporangia multispored, wall persistent or diffluent and the sporangium forming a liquid droplet at maturity, echinulate, columellate; sporangiola few- to 1-spored, persistent walled and columellate, sometimes indehiscent, borne on the same or separate, morphologically similar sporangiophores, often with extensive dichotomously or verticillately branching systems. Sporangiospores globose to narrowly ellipsoid or variable, smooth or ornamented, hyaline or pale brown to brown. Zygospores globose to subglobose; wall light brown, brown or reddish-brown to dark brown, opaque, ornamented with more or less coarse projections; suspensors opposed, tongs-like or apposed, smooth, more or less swollen, equal or unequal, naked or appendaged; heterothallic or homothallic.

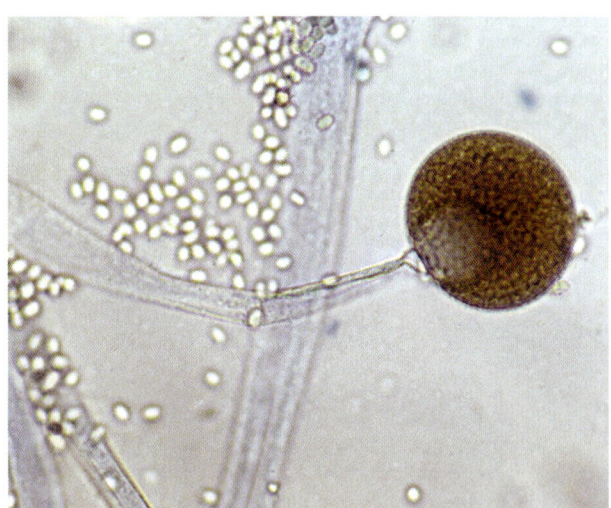

Mucor sp.; sporophore

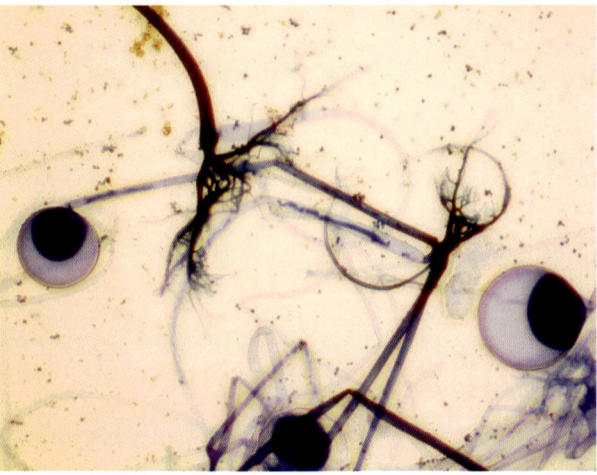

Rhizopus stolonifer; sporophores

Significant Genera: *Absidia, Amylomyces, Mucor, Rhizomucor, Rhizopus, Cokeromyces, Dicranophora, Thamnidium*.

Distribution: Widespread; species in some genera, for example *Rhizopus*, are particularly common in the tropics and are considered ubiquitous. Some with a distinctly warm to tropical distribution and others psychrophilic and thus restricted to temperate regions.

Economic Significance: Attenuated strains of *Amylomyces* and *Rhizopus* are used in the manufacture of some fermented products in South-East Asia.

Ecology: Mainly soil fungi, some readily isolated as a component of the early coprophilous succession. A few are facultative pathogens of animals, including man. *Rhizopus stolonifer* (syn. *R. nigricans*) is a common saprobe and facultative parasites of mature fruits and vegetables. One species causing mucormycosis; some are facultative parasites of fungi, for example, fruit bodies of *Boletaceae*.

Notes: The family as circumscribed is polyphyletic but is presently retained to provide a pragmatic framework for classification and identification.

Thamnidium elegans; sporophore and sporangiola

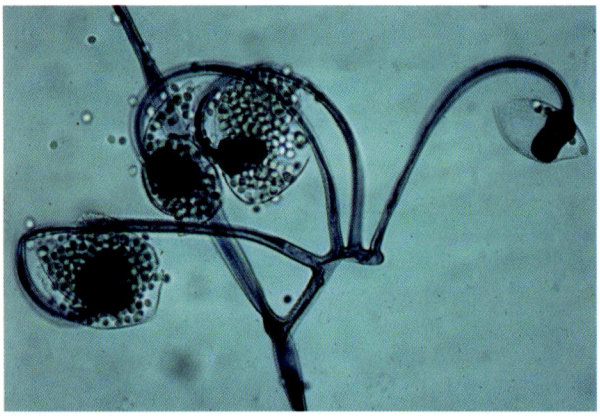
Circinella minor; sporophore

References: **Benny, G.L.; Benjamin, R.K.** (1975). Observations on *Thamnidiaceae* (*Mucorales*). New taxa, new combinations, and notes on selected species. *Aliso* 8: 301–351; **Benny, G.L.; Benja-min, R.K.** (1976). Observations on *Thamnidiaceae* (*Mucorales*). II. *Chaetocladium, Cokeromyces, Mycotypha,* and *Phascolomyces*. *Aliso* 8: 391–424; **Ellis, J.J.; Hesseltine, C.W.** (1966). Studies on *Absidia* with ovoid sporangiospores. II. *Sabouraudia* 5: 59–77; **Iwen, P.C.; Freifeld, A.G.; Sigler, L.; Tarantolo, S.R.** (2005). Molecular identification of *Rhizomucor pusillus* as a cause of sinus-orbital zygomycosis in a patient with acute myelogenous leukemia. *J. Clin. Microbiol.* 43: 5819–5821; **Kimura, M.; Udagawa, S.; Toyazaki, N.; Iimori, M.; Hashimoto, S.** (1995). Isolation of *Rhizopus microsporus* var. *rhizopodiformis* in the ulcer of human gastric carcinoma. *J. Med. Vet. Mycol.* 33: 137–139; **Kwanśa, H.; Ward, E.; Bateman, G.L.** (2006). Phylogenetic relationships among *Zygomycetes* from soil based on ITS1/2 rDNA sequences. *Mycol. Res.* 110: 501–510; **Liou, G.-Y.; Chen, S.-R.; Wei, Y.-H.; Lee, F.-L.; Fu, H.-M.; Yuan, G.-F.; Stalpers, J.A.** (2007). Polyphasic approach to the taxonomy of the *Rhizopus stolonifer* group. *Mycol. Res.* 111: 196-203; **Munipalli, B.; Rinaldi, M.G.; Greenberg, S.B.** (1996). *Cokeromyces recurvatus* isolated from pleural and peritoneal fluid: case report. *J. Clin. Microbiol.* 34: 2601–2603; **Schipper, M.A.A.** (1978). 1. On certain species of *Mucor* with a key to all accepted species. 2. On the genera *Rhizopus* and *Parasitella*. *Stud. Mycol.* 17: 1–71; **Schipper, M.A.A.** (1990). Notes on *Mucorales* – I. Observations on *Absidia*. *Persoonia* 14: 133–149; **Tanabe, Y.; Saikawa, M.; Watanabe, M.M.; Sugiyama, J.** (2004). Molecular phylogeny of *Zygomycota* based on EF-1α and RPB1 sequences: limitations and utility of alternative markers to rDNA. *Mol. Phylogen. Evol.* 30: 438–449; **Zycha, H.; Siepmann, R.; Linnemann, G.** (1969). *Mucorales* (Lehre): 355 pp.

Mycenaceae Roze 1876
Agaricales: Basidiomycota

Basidiomata pileate and stipitate, often small and delicate, usually with a central stipe, rarely polyporoid, sometimes brightly coloured; cap convex to campanulate, often appearing pleated, usually thin and pellucid, fleshy or gelatinous, smooth, veil absent. Hyphal system monomitic, often staining brown in iodine, clamp connections usually present, cystidia often present. Hymenium usually distinctly lamellate, the gills free or decurrent, or in an angular cupulate pore-like arrangement. Basidia cylindrical to clavate, usually with 4 sterigmata. Basidiospores ellipsoidal to cylindric or clavate, hyaline, thin-walled, smooth, usually not staining in iodine.

Significant Genera: *Favolaschia, Mycena, Panellus*.

Ecology: Saprobes, most species as early colonizers of dead wood and leaves. A few species of *Favolaschia* and *Mycena* have been described as mycorrizal with terrestrial orchids, but the associations need confirmation.

Notes: One of several families separated from the *Tricholomataceae* and further molecular studies may identify other constituent genera. *Mycena* itself is highly diverse with about 500 accepted species. It is polyphyletic in its traditional sense and not all species belong in this family. *Favolaschia* has been shown to belong here despite its unusual hymenial arrangement.

Mycena pura, Surrey, UK; basidiomata

Distribution: Cosmopolitan, diverse in almost all ecological zones.

Mycena sanguinolenta, Surrey, UK; basidiomata

Economic Significance: Little is known. The luminescent species *Mycena citricolor* is associated with leaf spot of coffee; its affinities have not been studied recently.

Mycena galericulata, Inverness-shire, Scotland; basidioma

Favolaschia calocera; basidiomata

References: **Corner, E.J.H.** (1986). The agaric genus *Panellus* Karst. (including *Dictyopanus* Pat.) in Malaysia. *Gdns' Bull.* Singapore 39: 103–147; **Desjardin, D.E.; Boonpratuang, T.; Hywel-Jones, N.L.** (2002). An accounting of the worldwide members of *Mycena* sect. *Longisetae*. *Fungal Diversity* 11: 69–85; **Grgurinovic, C.A.** (2003). The Genus *Mycena* in South-Eastern Australia. *Fungal Diversity Res. Ser.* (Hong Kong) 9: 329 pp.; **Jin, J.K.; Hughes, K.W.; Petersen, R.H.** (2001). Phylogenetic relationships of *Panellus* (*Agaricales*) and related species based on morphology and ribosomal large subunit DNA sequences. *Mycotaxon* 79: 7–21; **Jin, J.K.; Hughes, K.W.; Petersen, R.H.** (2001). Biogeographical patterns in *Panellus stypticus*. *Mycologia* 93: 309–316; **Lodge, D.J.; Perry, B.A.; Desjardin, D.E.** (2004). *Mycena* sect. *Hygrocyboideae* in the mountains of the Dominican Republic. *Mem. N. Y. bot. Gdn* 89: 131–139; **Maas Geesteranus, R.A.** (1992). Mycenas of the northern hemisphere. I. Studies in Mycenas and other papers. *Verh. K. Akad. Wet.* tweede sect. 1: 571 pp.; **Maas Geesteranus, R.A.; Horak, E.** (1995). *Mycena* and related genera from Papua New Guinea and New Caledonia. *Biblthca Mycol.* 159: 143–229; **Treu, R.; Agerer, R.** (1990). Culture characteristics of some *Mycena* species. *Mycotaxon* 38: 279–309.

Mycenastraceae Zeller 1948
Agaricales: Basidiomycota

Basidiomata gasteroid, epigeal, irregularly shaped but usually ± spherical or pyriform, attached to the underground mycelium with a fibrous root-like structure, breaking open irregularly or stellately to expose the spore mass; peridium with a fragmenting felty outer layer and a thick, tough, glabrous, leathery or corky inner layer. Clamp connections present. Gleba initially white, eventually becoming olive-brown or orange, with irregular chambers separated by poorly defined glebal columns. Capillitium of thick-walled, densely coiled, cells with spine-like projections. Basidia in fascicles, evanescent at an early stage. Basidiospores ± globose, thick-walled, aseptate and spinose or verrucose, yellow or yellowish-brown.

Significant Genera: *Mycenastrum*.

Distribution: Widespread in both north and south temperate zones.

Economic Significance: None is known.

Ecology: Occurring in fields or on bare soil, probably saprobic.

Mycenastrum corium; basidioma

Notes: The family originally included *Calbovista*, but that genus is now referred to the *Lycoperdaceae* leaving only the single species *Mycenastrum corium*. Molecular data confirm placement within the *Agaricales* but more studies are needed.

References: **Homrich, M.H.; Wright, J.E.** (1973). South American gasteromycetes. The genera *Gastropila*, *Lanopila* and *Mycenastrum*. *Mycologia* 65: 779–794; **Krüger, D.; Binder, M.; Fischer, M.; Kreisel, H.** (2001). The *Lycoperdales*. A molecular approach to the systematics of some gasteroid mushrooms. *Mycologia* 93: 947–957; **Miller, O.K., Jr; Brace, R.-L.; Evenson, V.** (2005). A new subspecies of *Mycenastrum corium* from Colorado. *Mycologia* 97: 530–533.

Mycoblastaceae Hafellner 1984
Lecanorales: Ascomycota

Thallus crustose. Ascomata apothecial, black, flat or domed, without a well-developed margin. Interascal tissue of branched and anastomosing paraphyses, with a pigmented epithecium. Asci clavate, with an enormous J+ apical cap and a very large ocular chamber, with a very thick (especially at the apex) outer J+ gelatinized layer, 1- or few-spored. Ascospores usually very large, thick-walled, hyaline, aseptate, without a sheath. Anamorphs pycnidial.

Significant Genera: *Mycoblastus*.

Distribution: Widespread in temperate and boreal zones.

Economic Significance: None is known.

Ecology: Lichenized with green algae, especially on soil or dead wood.

Notes: Little is known of the affinities of this monogeneric family.

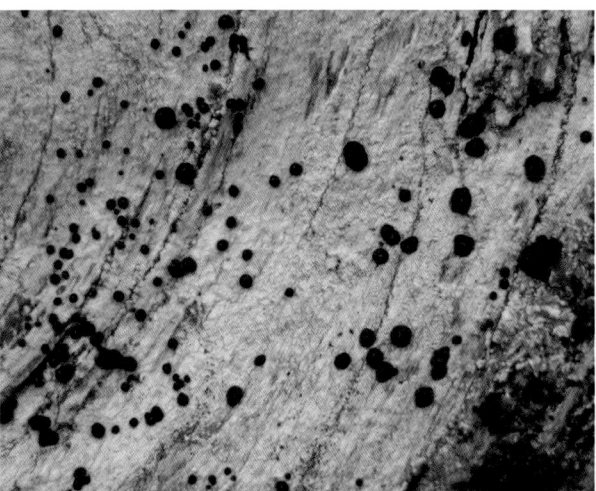

Mycoblastus sanguinarius, Inverness-shire, Scotland; thallus and ascomata

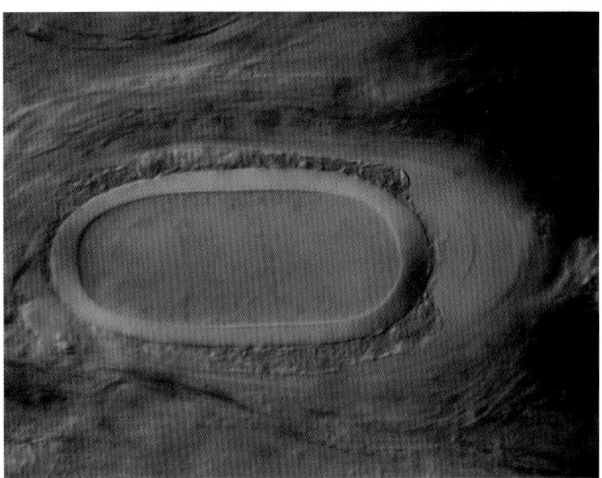

Mycoblastus sanguinarius, Inverness-shire, Scotland; ascus and ascospore

References: **Śliwa, L.** (1996). Note on *Mycoblastus fucatus* [= *M. sterilis*] (Lichenes, *Mycoblastaceae*) from Poland. *Fragm. flor. geobot.* (Kraków) 41: 491–492.

Mycocaliciaceae Alf. Schmidt 1970
Mycocaliciales: Ascomycota

Thallus immersed, often absent. Ascomata stalked, brown or black, the head discoid to subglobose. Interascal tissue absent. Asci small, cylindrical, thick-walled at least at the apex, not evanescent at an early stage. Ascospores ellipsoidal to cylindrical, pale to mid brown, smooth- and thin-walled, not liberated in a mazaedial mass.

Significant Genera: *Chaenothecopsis*, *Mycocalicium*, *Phaeocalicium*. Anamorph: *Asterophoma*.

Chaenothecopsis australis, Tierra del Fuego, Argentina; stalked ascomata

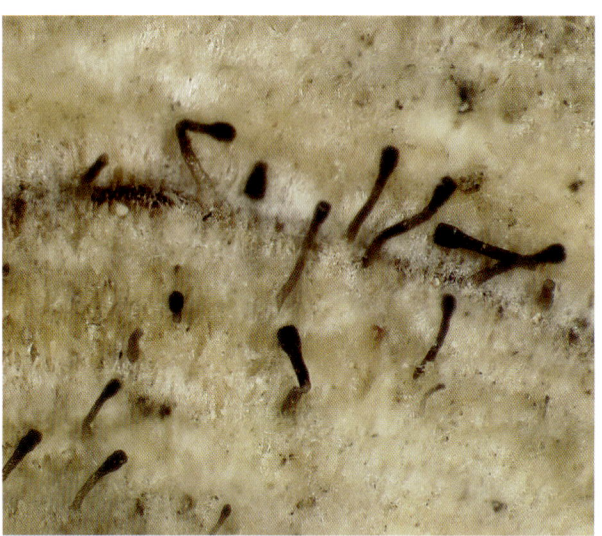
Phaeocalicium polyporaeum, Michigan, USA; stalked ascomata

Chaenothecopsis fennica; stalked ascomata

Distribution: Widespread, especially in boreal and austral regions.

Economic Significance: None is known.

Ecology: Mainly saprobic on wood or bark, some lichenicolous, fungicolous or possibly lichen-forming.

Notes: At one time assumed to be closely related to the *Caliciaceae*, the *Mycocaliciaceae* and *Sphinctrinaceae* are now considered to be isolated relatives of the *Eurotiomycetes* based on molecular analysis.

References: **Angeles Vinuesa, M. de los; Sanchez-Puelles, J.M.; Tibell, L.** (2001). Intraspecific variation in *Mycocalicium subtile* (*Mycocaliciaceae*) elucidated by morphology and the sequences of the ITS1-5.8S-ITS2 region of rDNA. *Mycol. Res.* 105: 323–330; **Hutchison, L.J.** (1987). Studies on *Phaeocalicium polyporaeum* in North America. *Mycologia* 79: 786–789; **Peterson, E.B.; Rikkinen, J.** (1998). *Stenocybe fragmenta*, a new species of *Mycocaliciaceae* with fragmenting spores. *Mycologia* 90: 1087–1090; **Rikkinen, J.; Poinar, G.O.** (2000). A new species of resinicolous *Chaenothecopsis* (*Mycocaliciaceae*, *Ascomycota*) from 20 million year old Bitterfeld amber, with remarks on the biology of resinicolous fungi. *Mycol. Res.* 104: 7–15; **Selva, S.B.; Tibell, L.** (1999). Lichenized and non-lichenized calicioid fungi from North America. *Bryologist* 102: 377–397; **Tibell, L.** (1990). Anamorphs in *Mycocalicium albonigrum* and *M. subtile* (*Caliciales*). *Nordic Jl Bot.* 10: 221–242; **Tibell, L.** (1993). The anamorphs of *Chaenothecopsis viridireagens*. *Nordic Jl Bot.* 13: 331–335; **Tibell, L.** (1996). *Phaeocalicium* (*Mycocaliciaceae*, *Ascomycetes*) in northern Europe. *Ann. bot. fenn.* 33: 205–221; **Tibell, L.** (1997). Anamorphs in mazaediate lichenized fungi and the *Mycocaliciaceae* ('*Caliciales s. lat.*'). *In* Tibell, L. & Hedberg, I. (eds), Lichen Studies Dedicated to Rolf Santesson. *Symb. bot. upsal.* 32: 291–322; **Tibell, L.** (1998). Crustose Mazaediate Lichens and the *Mycocaliciaceae* in Temperate South America. *Biblthca Lichenol.* 71: 107 pp.; **Tibell, L.; Wedin, M.** (2000). *Mycocaliciales*, a new order for nonlichenized calicioid fungi. *Mycologia* 92: 577–581; **Wedin, M.; Tibell, L.** (1997). Phylogeny and evolution of *Caliciaceae*, *Mycocaliciaceae*, and *Sphinctrinaceae* (*Ascomycota*), with notes on the evolution of the prototunicate ascus. *Can. J. Bot.* 75: 1236–1242.

Mycoporaceae Zahlbr. 1903
Uncertain position within Dothideomycetes, Ascomycota

Ascomata erumpent or superficial, perithecial, globose or aggregated into small pulvinate stromata, seated in a loose, copiously branched, pale mycelium; peridium of pseudoparenchymatous cells. Interascal tissue absent or scarcely differentiated from stromatal tissue; hymenium often gelatinous. Asci saccate, fissitunicate. Ascospores becoming brown, variably septate (usually muriform), often with a sheath. Anamorphs not prominent.

Significant Genera: *Mycoporum*.

Distribution: Widespread.

Economic Significance: None is known.

Ecology: Saprobic on woody tissue, many species apparently non-lichenized.

Notes: The only genus is poorly known and may well be polyphyletic. *Cyrtidula* may also belong here, but apparently has not been correctly typified.

Mycoporum lacteum; thallus

References: **Coppins, B.J.; James, P.W.** (1979). New or interesting British lichens III. *Lichenologist* 11: 27–45; **Lumbsch, H.T.** (1999). The ascoma development in *Mycoporum elabens* (Mycoporaceae, Dothideales). *Pl. Biol.* 1: 321–326.

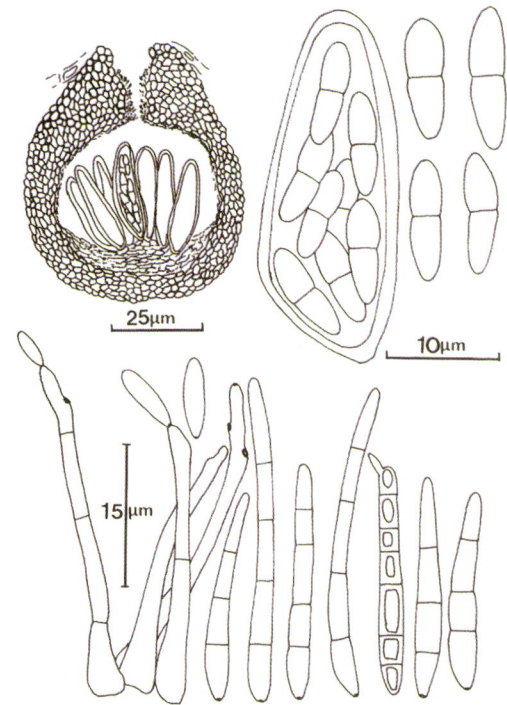

Mycosphaerella rosicola; ascoma, asci, ascospores and anamorph

Mycosphaerellaceae Lindau 1897
Capnodiales: Ascomycota

Ascomata perithecial, small, immersed in plant tissue, often becoming erumpent, often strongly aggregated or on a weakly developed basal stroma, black, papillate, with a well-developed lysigenous ostiole; peridium usually thin, composed of pseudoparenchymatous cells. Interascal tissue lacking. Asci ovoid to saccate, fissitunicate, with an ocular chamber but lacking other apical structures, not blueing in iodine. Ascospores usually hyaline and transversely septate, without a sheath. Anamorphs usually hyphomycetous, with cylindrical, often pigmented conidiophores forming in clusters from small basal stromata, proliferating sympodially and often with conspicuous scars, with conidia that are often elongate, pigmented and multiseptate; in some formed in fragmenting branched chains. Coelomycetous anamorphs pycnidial in form, usually with short sympodially proliferating conidiogenous cells producing elongate conidia.

Mycosphaerella henningsii, Galápagos Islands, Ecuador; conidiophores

Significant Genera: *Davidiella, Mycosphaerella, Microcyclus, Sphaerulina*. Anamorphs: *Cladosporium, Cercospora, Pseudocercospora, Ramularia, Septoria*.

Distribution: Cosmopolitan.

Economic Significance: Many species are important plant pathogens, in both tropical and temperate zones. *Cladosporium* species are also significant as saprobes, often forming major components of mould growth in buildings.

Ecology: Biotrophic, necrotrophic or saprobic on various plant tissues.

Notes: The family is dominated by the large genus *Mycosphaerella* with around 500 teleomorph species and more than double this number as only the anamorph. The teleomorphs are remarkably homogenous and the anamorphs extremely varied in form.

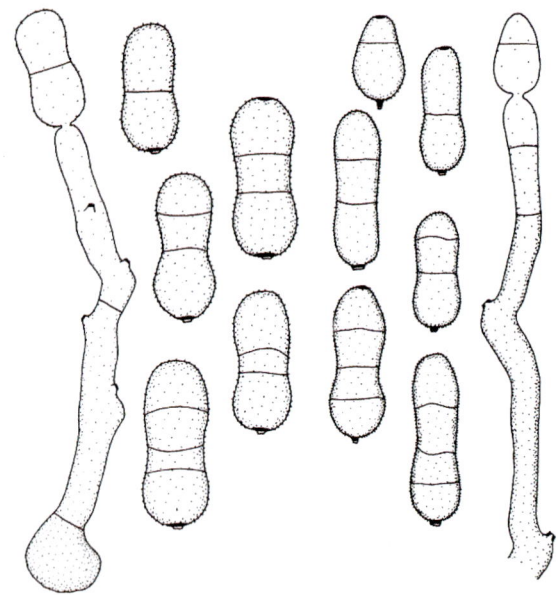

Davidiella allii-cepae; conidiophores and conidia

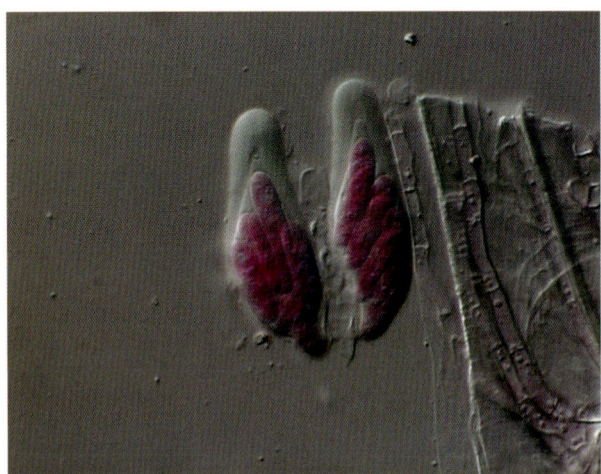

Davidiella tassiana, Galápagos Islands, Ecuador; asci and ascospores

References: **Braun, U.; Crous, P.W.; Dugan, F.; Hoog, G.S. de** (2003). Phylogeny and taxonomy of *Cladosporium*-like hyphomycetes, including *Davidiella* gen. nov., the teleomorph of *Cladosporium s. str. Mycol. Prog.* 2: 3–18; **Cannon, P.F.; Carmarán, C.C.; Romero, A.I.** (1995). Studies on biotrophic fungi from Argentina: *Microcyclus porlieriae*, with a key to South American species of *Microcyclus*. *Mycol. Res.* 99: 353–356; **Caten, C.E.** (1999). Molecular genetics of *Stagonospora* and *Septoria*. *Septoria on Cereals, A Study of Pathosystems* (Wallingford): 26–43; **Corlett, M.** (1991). An annotated list of the published names in *Mycosphaerella* and *Sphaerella*. *Mycol. Mem.* 18: 328 pp.; **Crous, P.W.** (1998). *Mycosphaerella* spp. and Their Anamorphs Associated with Leaf Spot Diseases of *Eucalyptus*. *Mycol. Mem.* 21: 170 pp.; **Crous, P.W.** (1999). Species of *Mycosphaerella* and related anamorphs occurring on *Myrtaceae* (excluding *Eucalyptus*). *Mycol. Res.* 103: 607–621; **Crous, P.W.; Aptroot, A.; Kang, J.C.; Braun, U.; Wingfield, M.J.** (2000). The genus *Mycosphaerella* and its anamorphs. *Stud. Mycol.* 45: 107–121; **Crous, P.W.; Braun, U.** (2003). *CBS Diversity Ser.* (Utrecht) 1: 571 pp.; **Crous, P.W.; Corlett, M.** (1999). Reassessment of *Mycosphaerella* spp. and their anamorphs occurring on *Platanus*. *Can. J. Bot.* 76: 1523–1532; **Crous, P.W.; Kang, J.C.; Braun, U.** (2001). A phylogenetic redefinition of anamorph genera in *Mycosphaerella* based on ITS rDNA sequence and morphology. *Mycologia* 93: 1081–1101; **Cunfer, B.M.; Ueng, P.P.** (1999). Taxonomy and identification of *Septoria* and *Stagonospora* species on small-grain cereals. *A. Rev. Phytopath.* 37: 267–284; **David, J.C.** (1997). A Contribution to the Systematics of *Cladosporium*: Revision of the Fungi Previously Referred to *Heterosporium*. *Mycol. Pap.* 172: 157 pp.; **Hunter, G.C.; Roux, J.; Wingfield, B.D.; Crous, P.W.; Wingfield, M.J.** (2004). *Mycosphaerella* species causing leaf disease in South African *Eucalyptus* plantations. *Mycol. Res.* 108: 672–681; **Johanson, A.; Crowhurst, R.N.; Rikkerink, E.H.A.; Fullerton, R.A.; Templeton, M.D.** (1994). The use of species-specific DNA probes for the identification of *Mycosphaerella fijiensis* and *M. musicola*, the causal agents of Sigatoka disease of banana. *Pl. Path.* 43: 701–707; **Scharen, A.L.; Sanderson, F.R.** (1985). Identification, distribution and nomenclature of the *Septoria* species that attack cereals. *Septoria of Cereals* Proceedings of the Workshop held August 2-4, 1983, at Montana State University, Bozeman, Montana (USA): 37–41; **Schoch, C.L.; Shoemaker, R.A.; Seifert, K.A.; Hambleton, S.; Spatafora, J.W.; Crous, P.W.** (2007). A multigene phylogeny of the Dothideomycetes using four nuclear loci. *Mycologia* 98(6): 1041–1052; **Starink-Willemse, M.; Iperen, A. van; Abelh, E.C.A.** (2004). Phylogenetic analyses of *Septoria* species based on the ITS and LSU-D2 region of nuclear ribosomal DNA. *Mycologia* 96: 558–571; **Stewart, E.L.; Liu, Z.W.; Crous, P.W.; Szabo, L.J.** (1999). Phylogenetic relationships among some cercosporoid anamorphs of *Mycosphaerella* based on rDNA sequence analysis. *Mycol. Res.* 103: 1491–1499; **Taylor, J.W.; Jacobson, D.J.; Fisher, M.C.** (1999). The evolution of asexual fungi: reproduction, speciation and classification. *A. Rev. Phytopath.* 37: 197–246; **Verkely, G.J.M.; Priest, M.J.** (2000). *Septoria* and similiar coelomycetous anamorphs of *Mycosphaerella*. *Stud. Mycol.* 45: 123–128 **Verkely, G.J.M.; Starinke-Willemse, M.; Iperen, A. van; Abelh, E.C.A.** (2004) Phylogenetic analysis of Septoria species based on the ITS and LSU-D2 region of nuclear ribosomal DNA. *Mycologia* 96: 558–571.

Mycosyringaceae R. Bauer & Oberw. 1997
Ustilaginales: Basidiomycota

Mycosyrinx cissi; witches' broom on *Cissus*

Sori formed in inflorescences and flowers, causing hypertrophy and distortion, producing a mass of dark brown spores. Hyphae intercellular, with septal pores of the *Ustilago*-type. Teliospores formed in pairs in the branches of sporogenous hyphae, ± hemispherical, dark brown, minutely verrucose on the outer surfaces. Basidia not developed, the teliospores germinating directly to produce basidiospores. Basidiospores elongate-fusiform, the upper part bent back sharply and the two half-spores becoming swollen.

Significant Genera: *Mycosyrinx*.

Distribution: Pantropical.

Mycosyrinx cissi; sori in infected inflorescence

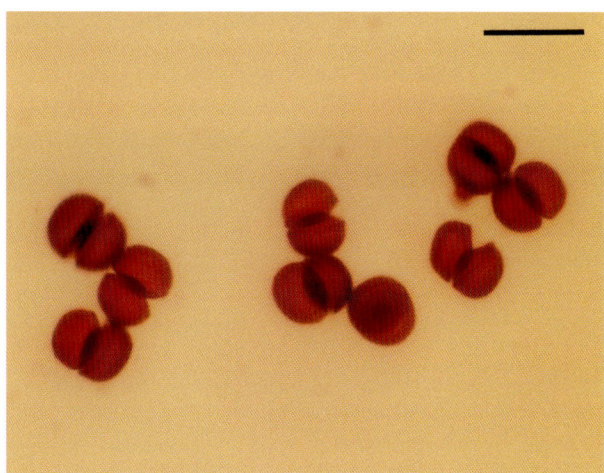

Mycosyrinx cissi, Trinidad; teliospores

Economic Significance: None is known.

Ecology: Biotrophic, associated with *Vitaceae*.

Notes: An isolated family of smuts with very distinctive morphological features. No sequences are available.

References: **Bauer, R.; Oberwinkler, F.** (1997). The *Ustomycota*: an inventory. *Mycotaxon* 64: 303–319; **Bauer, R.; Oberwinkler, F.; Vánky, K.** (1997). Ultrastructural markers and systematics in smut fungi and allied taxa. *Can. J. Bot.* 75: 1273–1314; **Piepenbring, M.; Bauer, R.; Oberwinkler, F.** (1998). Teliospores of smut fungi. Teliospore walls and the development of ornamentation studied by electron microscopy. *Protoplasma* 204: 170–201; **Vánky, K.** (1987). Illustrated genera of smut fungi. *Cryptog. Stud.* 1: 159 pp.; **Vánky, K.** (1996). *Mycosyrinx* and other pair-spored *Ustilaginales*. *Mycoscience* 37: 173–185; **Vánky, K.** (1998). A survey of the spore-ball-forming smut fungi. *Mycol. Res.* 102: 513–526; **Vánky, K.** (2004). Biodiversity and conservation of smut fungi (*Ustilaginomycetes p.p.*, *Microbotryales*) reflected in Vánky, *Ustilaginales* exsiccata no. 1-1250. *Mycol. Balcanica* 1: 175–187.

Mycotyphaceae Benny & R.K. Benj. 1985
Mucorales: Zygomycota

Vegetative thallus a mycelium, complex fruit bodies absent. Sporophores arising from submerged mycelium, simple or branched, stolons or rhizoids not produced; terminating in a fertile vesicle bearing pedicellate sporangiola. Sporangiola unispored or multispored, monomorphic or dimorphic, dehiscing by circumscissile fracture at a preformed zone of weakness leaving monomorphic or dimorphic, truncate denticles on the fertile vesicle. Sporangiospores variously shaped; those from the unispored sporangiola of similar shape and size to the intact sporangiolum. Zygospores globose to subglobose; zygosporangial wall ornamented with more or less conical projections; suspensors opposed, non-appendaged; homothallic, or presumed heterothallic.

Significant Genera: *Mycotypha*, *Benjaminiella*.

Distribution: Widespread but rarely reported, more common in warmer areas.

Economic Significance: None is known.

Ecology: Essentially soil fungi but most readily isolated as a component of the early coprophilous succession; characterized by a hyphal-yeast dimorphism.

Notes: Very poorly known; no recent information is available.

References: **Benny, G.L.; Kirk, P.M.; Samson, R.A.** (1985). Observations on *Thamnidaiceae* (*Mucorales*). III. *Mycotyphaceae* fam. nov. and a re-evaluation of *Mycotypha* sensu Benny & Benjamin illustrated by two new species. *Mycotaxon* 22: 119–148; **Forst, T.G.; Prillinger, H.** (1988). Vergleichende karyologische Untersuchungen an dimorphen Zygomyceten. *Z. Mykol.* 54: 139–153.

Myeloconidiaceae P.M. McCarthy 2001
Trichotheliales: Ascomycota

Thallus crustose, thinly corticate or ecorticate, yellowish or orange internally. Ascomata perithecial, solitary, lacking an involucrellum, immersed in the thallus but sometimes within prominent pustules. Interascal tissue composed of narrow paraphyses, often anastomosed near the base and rarely branched above, periphyses absent. Asci cylindrical, thin-walled and unitunicate, lacking apical structures, 8-spored, not staining in iodine. Ascospores fusiform to elongate, muriform, sometimes tapering to a narrow point at one or both ends, hyaline. Anamorphs unknown.

Significant Genera: *Myeloconis*.

Distribution: Widespread in tropical climates.

Myeloconis guyanensis; thallus

Economic Significance: None is known.

Ecology: Lichenized with green algae, on bark in humid environments.

Notes: Many microscopic features suggest links with the *Porinaceae*, but molecular studies have not been carried out.

References: **Anon.** (2001). *Flora of Australia* (Melbourne) 58: 242 pp.; **McCarthy, P.M.; Elix, J.A.** (1996). *Myeloconis*, a new genus of pyrenocarpous lichens from the tropics. *Lichenologist* 28: 401–414; **McCarthy, P.M.; Kantvilas, G.; Elix, J.A.** (2001). *Amphorothecium*, a new pyrenocarpous lichen genus from New South Wales, Australia. *Lichenologist* 33: 291–296.

Myelospermataceae K.D. Hyde & S.W. Wong 1999
Xylariales: Ascomycota

Stromata immersed in plant tissue, restricted, composed of a mixture of fungal and plant tissues, blistering and blackening the plant surface tissues. Ascomata perithecial, short-necked, arranged in a valsoid ring with ostioles convergent and exiting through a single pore; peridium narrow, composed of brown-walled compressed cells. Interascal tissue composed of thin-walled true paraphyses with globose basal cells, apparently evanescent in some circumstances. Asci cylindrical, short-stalked, thin-walled, not fissitunicate, the apex rounded with a J–refractive ring, 8-spored. Ascospores ellipsoidal, often slightly curved, hyaline, smooth-walled, aseptate, surrounded by a mucous sheath. Anamorph not known.

Significant Genera: *Myelosperma*.

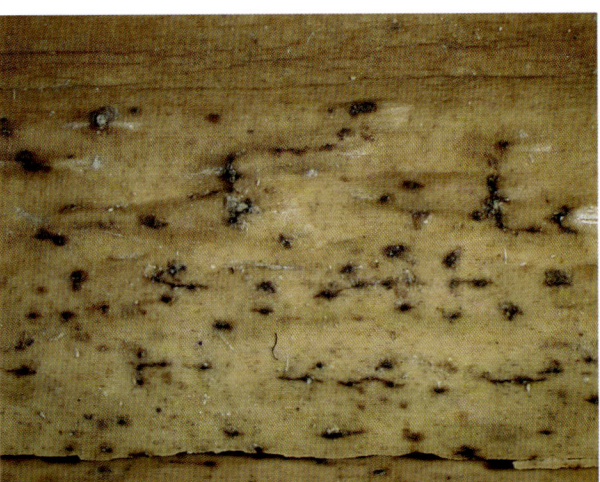

Myelosperma tumidum, Papua New Guinea; ascomata in dead palm frond

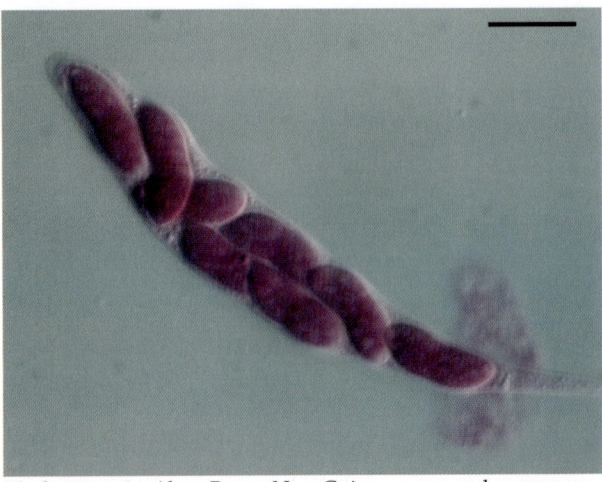

Myelosperma tumidum, Papua New Guinea; ascus and ascospores

Distribution: Pantropical.

Economic Significance: None is known.

Ecology: Saprobic and possibly endophytic on palms.

Notes: Very poorly known; no recent information is available.

References: **Hyde, K.D.** (1993). Fungi from palms. VIII. The genus *Myelosperma* (*Ascomycotina*). *Sydowia* 45: 241–245; **Hyde, K.D.; Wong, S.W.** (1999). Ultrastructural studies on the *Myelospermaceae* [sic] fam. nov., with a new species of *Myelosperma*. *Mycol. Res.* 103: 347–352; **Kang, J.C.; Kong, R.Y.C.; Hyde, K.D.** (1998). Studies on the *Amphisphaeriales* 1. *Amphisphaeriaceae* (sensu stricto) and its phylogenetic relationships inferred from 5.8S rDNA and ITS2 sequences. *Fungal Diversity* 1: 147–157.

Myriangiaceae Nyl. 1854
Myriangiales: Ascomycota

Myriangium duriaei, Thailand; ascomata on scale insects on *Prunus persica* twigs

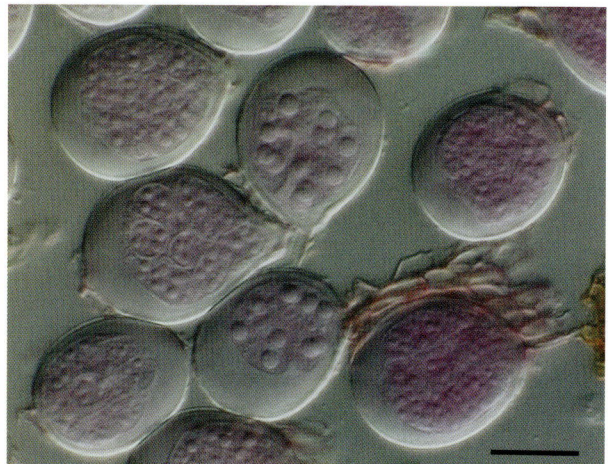

Myriangium duriaei, Thailand; immature asci

Stromata crustose or pulvinate, composed of subhyaline or brown thin-walled pseudoparenchymatous tissue, with ± globose sessile or short-stalked fertile outgrowths of similar tissue containing scattered asci in individual perithecial locules, becoming gelatinous at maturity. Interascal tissue absent. Asci ± globose, sessile, fissitunicate, with a poorly defined ocular chamber. Ascospores pale brown, transversely septate or muriform. Anamorphs unknown.

Significant Genera: *Anhellia*, *Myriangium*.

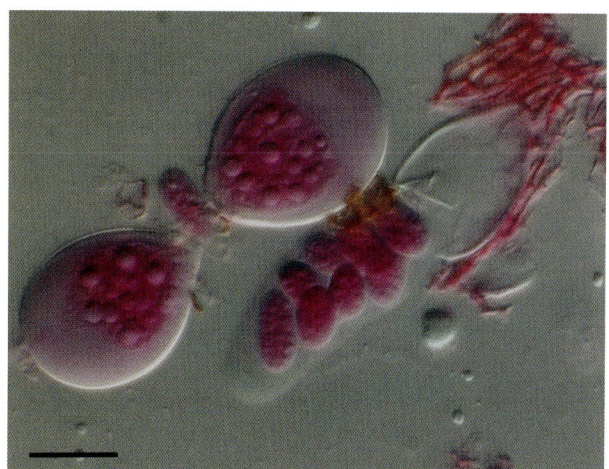

Myriangium duriaei, Thailand; mature and immature asci

Distribution: Widespread, especially in the tropics.

Economic Significance: None is known; the potential for use as biocontrol organisms appears not to be high.

Ecology: Associated with scale insects or resinous exudates, presumably parasitic.

Notes: An isolated group within the *Dothideomycetes*. The *Piedraiaceae* may be related as has traditionally been assumed, but more molecular data are needed.

References: **Inácio, C.A.; Dianese, J.C.** (1998). Some foliicolous fungi on *Tabebuia* species. *Mycol. Res.* 102: 695–708; **Lumbsch, H.T.; Lindemuth, R.** (2001). Major lineages of *Dothideomycetes* (*Ascomycota*) inferred from SSU and LSU rDNA sequences. *Mycol. Res.* 105: 901–908; **Rao, V.G.; Pande, A.** (1993). A new species of *Myriangium* (Ascomycete) on rose. *Cryptog. bot.* 3: 255–256; **Winka, K.; Eriksson, O.E.** (2000). Adding to the bitunicate puzzle – studies on the systematic position of five aberrant ascomycete taxa. *Phylogenetic Relationships Within the Ascomycota Based on 18S rDNA Sequences, Akademisk Avhandling* [Thesis (PhD), Department of Ecology and Environmental Science, Umeå University] (Umeå): [17] pp.

Mytilinidiaceae Kirschst. 1924
Hysteriales: Ascomycota

Ascomata superficial or with the base immersed, hysterothecial, elongate, sometimes branched, usually taller than wide, opening by a long slit or radial slits in the upper surface, black, thick-walled; peridium carbonaceous, composed of intertwined hyphal tissue. Interascal tissue of trabeculate pseudoparaphyses and/or periphysoids, sometimes evanescent. Asci ± cylindrical, long and narrow, fissitunicate in most cases. Ascospores becoming brown, septate, sometimes filiform or muriform. Anamorphs varied.

Lophium mytilinum, Inverness-shire, Scotland; ascomata

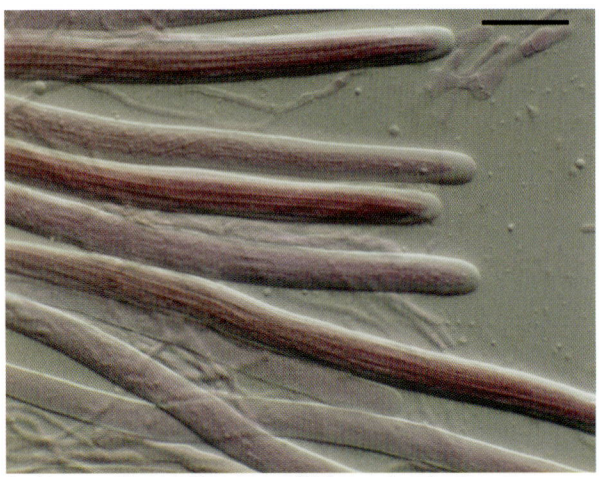
Lophium mytilinum, Glamorgan, Wales; asci and ascospores

Significant Genera: *Lophium*, *Mytilinidion*. Anamorphs: *Chalara*-like, *Taeniolella*.

Distribution: Widespread, especially in temperate zones.

Economic Significance: None is known.

Ecology: Usually saprobic on woody tissue, especially on gymnosperms; a few species lichenicolous.

Notes: Poorly known; little recent information is available. Molecular phylogenetic analysis of *Lophium mytilinum* indicates a shared evolutionary pathway with the *Chaetothyriomycetes*, but more data are needed.

References: **Blackwell, M.; Gilbertson, R.L.** (1985). *Quasiconcha reticulata* and its anamorph from conifer roots. *Mycologia* 77: 50–54; **Checa, J.** (1997). Annotated list of the *Lophiostomataceae* and *Mytilinidiaceae* (*Dothideales*, *Ascomycotina*) reported from the Iberian Peninsula and Balearic Islands. *Mycotaxon* 63: 467–491; **Vasilyeva, L.** (1997). *Ostropella luxurians* sp. nov. from the Russian Far East. *Mycoscience* 38: 341–342.

Myxotrichaceae Locq. ex Currah 1985
Uncertain position within Ascomycetes, Ascomycota

Ascomata cleistothecial, not stipitate; peridium composed of a loose network of hyphae, often with complex thick-walled appendages. Asci small, thin-walled, evanescent, probably originating from croziers. Ascospores pigmented, ± fusiform, smooth or striate, without a gelatinous sheath or appendages. Anamorphs hyphomycetous, arthric, with simple, thick-walled, conidia derived directly from short side branches or intercalary cells of vegetative hyphae.

Significant Genera: *Myxotrichum*. Anamorphs: *Geomyces*, *Malbranchea*, *Oidiodendron*.

Myxotrichum chartarum, Yorkshire, UK; ascomata and appendages

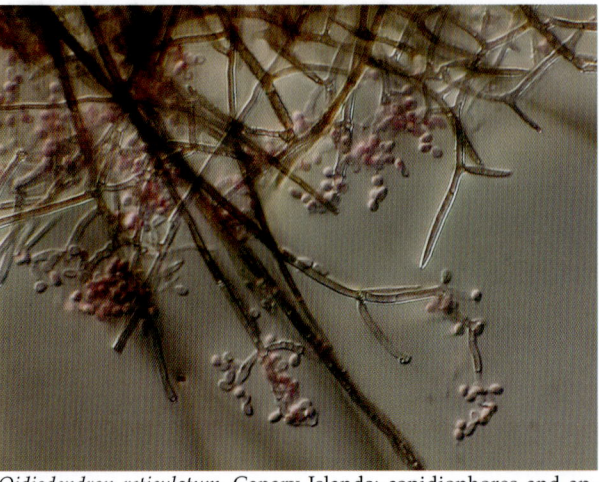
Oidiodendron reticulatum, Canary Islands; conidiophores and appendages

Distribution: Cosmopolitan.

Economic Significance: Species of *Myxotrichum* are occasionally implicated in biodeterioration of paper where storage conditions are poor.

Ecology: Many species are cellulolytic, occurring on paper, straw etc. Others are mycorrhizal, associated with roots of *Ericaceae* and *Epacridaceae*.

Notes: Until recently considered to be related to the *Onygenales*, but that group is mostly keratinolytic in contrast to the *Myxotrichaceae*. Molecular phylogenetic studies now place the family in an isolated position, probably within the *Leotiomycetes*.

References: **Caretta, G.; Piontelli, E.** (1998). Preserved ascomatal and other fungal structures on the remains of a ninth century Longobard abbess exhumed from a monastery in Pavia, Italy. *Mycopathologia* 140: 77–83; **Chambers, S.M.; Liu, G.; Cairney, J.W.G.** (2000). ITS rDNA sequence comparison of ericoid mycorrhizal endophytes from *Woollsia pungens*. *Mycol. Res.* 104: 168–174; **Currah, R.S.** (1985). Taxonomy of the *Onygenales*: *Arthrodermataceae*, *Gymnoacaceae*, *Myxotrichaceae* and *Onygenaceae*. *Mycotaxon* 24: 1–216; **Currah, R.S.** (1988). An annotated key to the genera of the *Onygenales*. *Syst. Ascom.* 7: 1–12; **Currah, R.S.; Niemi, M.; Huhtinen, S.** (1999). *Oidiodendron maius* and *Scytalidium vaccinii* from the mycorrhizas of *Ericaceae* in northern Finland. *Karstenia* 39: 65–68; **Gibas, C.F.C., Sigler, L., Summerbell, R.C.; Currah, R.S.** (2002). Phylogeny of the genus *Arachnomyces* and its anamorphs and the establishment of *Arachnomycetales*, a new eurotiomycete order in the *Ascomycota*. *Stud. Mycol.* 47: 131–139; **Hambleton, S.; Egger, K.N.; Currah, R.S.** (1998). The genus *Oidiodendron*: species delimitation and phylogenetic relationships based on nuclear ribosomal DNA analysis. *Mycologia* 90: 854–869; **McLean, C.B.; Cunnington, J.H.; Lawrie, A.C.** (1999). Molecular diversity within and between ericoid endophytes from the *Ericaceae* and *Epacridaceae*. *New Phytol.* 144: 351–358; **Monreal, M.; Berch, S.M.; Berbee, M.** (1999). Molecular diversity of ericoid mycorrhizal fungi. *Can. J. Bot.* 77: 1580–1594; **Rosing, W.C.** (1985). Fine structure of cleistothecia, asci, and ascospores of *Myxotrichum deflexum*. *Mycologia* 77: 920–926; **Sugiyama, M.; Ohara, A.; Mikawa, T.** (1999). Molecular phylogeny of onygenalean fungi based on small subunit ribosomal DNA (SSU rDNA) sequences. *Mycoscience* 40: 251–258; **Tsuneda, A.; Currah, R.S.** (2004). Ascomatal morphogenesis in *Myxotrichum arcticum* supports the derivation of the *Myxotrichaceae* from a discomycetous ancestor. *Mycologia* 96: 627–635.

Naetrocymbaceae Höhn. ex R.C. Harris 1995
Pleosporales: Ascomycota

Thallus or stroma absent or poorly developed, usually with a single perithecial locule, black, spherical or flattened, composed of pseudoparenchymatous cells, with a broad ostiole. Interascal tissue of narrow, short-celled, cellular pseudoparaphyses with refractive tips. Asci clavate or obpyriform, fissitunicate, without an ocular chamber or clear apical structure. Ascospores hyaline or brown, sometimes transversely septate, ornamented, thin-walled. Anamorph coelomycetous, probably spermatial.

Significant Genera: *Leptorhaphis*, *Tomasellia*.

Leptorhaphis atomaria, Finland; ascomata in bark

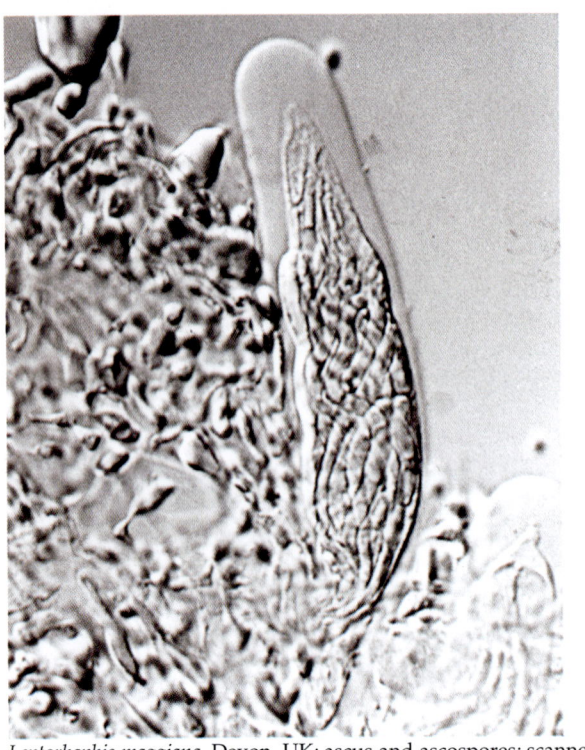

Leptorhaphis maggiana, Devon, UK; ascus and ascospores; scanned from image in IMI

Distribution: Widespread.

Economic Significance: None is known.

Ecology: Saprobic, especially in smooth bark, or lichenized with green algae.

Notes: The family was separated from the *Arthopyreniaceae*, another lichenized family of the *Pleosporales* with immersed thalli. It remains poorly known and no molecular phylogenetic data are available.

References: **Aguirre, B.; Hawksworth, D.L.** (1987). The circumscription, biology and relationships of the genus *Leptorhaphis* Körber. In Peveling, E. (ed.), Progress and Problems in Lichenology in the Eighties. *Biblthca Lichenol.* 25: 249–255; **Aguirre-Hudson, B.; Fiol, L.** (1993). A new species of *Leptorhaphis* (*Arthopyreniaceae*) on *Opuntia* from the Balearic Islands. *Lichenologist* 25: 207–210; **Aptroot, A.** (1991). Tropical pyrenocarpous lichens. A phylogenetic approach. *Tropical Lichens: Their Systematics, Conservation, and Ecology* (Oxford) 43: 253–273; **Harris, R.C.** (1995). *More Florida Lichens, Including the 10 Cent Tour of the Pyrenolichens* (Bronx): 192 pp.

Nectriaceae Tul. & C. Tul. 1865
Hypocreales: Ascomycota

Stromata absent or if present basal and pulvinate. Ascomata perithecial, superficial, often hardly papillate, the ostiole periphysate; peridium membranous, orange-red to purple, rarely pallid, changing colour in KOH and lactic acid, sometimes ornamented or setose. Interascal tissue of apical paraphyses, sometimes absent or deliquescent. Asci cylindrical, thin-walled, not fissitunicate, often with a minute J– apical ring. Ascospores varied in shape, usually transversely septate and rarely muriform, not disarticulating, hyaline to yellow or pale brown, sometimes ornamented, without gelatinous sheaths or appendages. Anamorphs hyphomycetous, hyaline or brightly coloured, sometimes sporodochial or synnematous, sometimes with sterile extensions to the conidiophore, the conidiogenous cells single or in clusters, proliferating percurrently. Conidia variable, ± hyaline, aseptate to elongate and multi-septate, often curved, sometimes dimorphic (with macroconidia and microconidia).

Significant Genera: *Calonectria*, *Cosmospora*, *Gibberella*, *Haematonectria*, *Nectria*. Anamorphs: *Fusarium*, *Cylindrocarpon*, *Cylindrocladium*.

Distribution: Cosmopolitan.

Economic Significance: *Fusarium* species are amongst the most economically significant plant pathogens, causing millions of dollar losses worldwide to a very wide range of crops. In addition to causing wilts and stem and root rots, many (especially those associated with cereals) produce highly toxic metabolites that render the grain poisonous and unsaleable. *Gibberella* species (which have *Fusarium* anamorphs) also produce gibberellins, hormone-like chemicals that have effects on plant growth. A non-toxin producing strain of *Fusarium venenatum* is used for industrial production of the mycoprotein Quorn, marketed as a vegetarian meat substitute.

Nectria cinnabarina, Suffolk, UK; ascomata and conidiomata

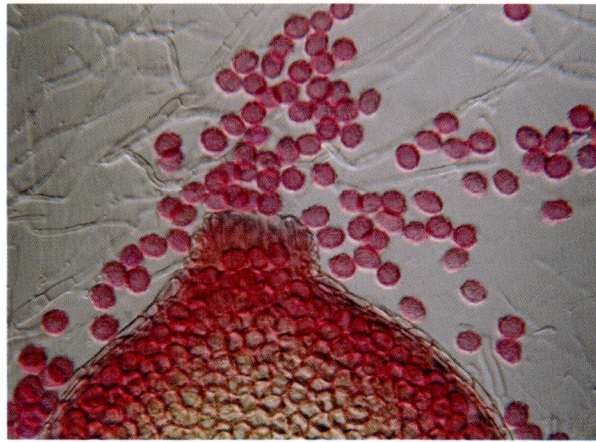

Neocosmospora vasinfecta; ascoma and ascospores

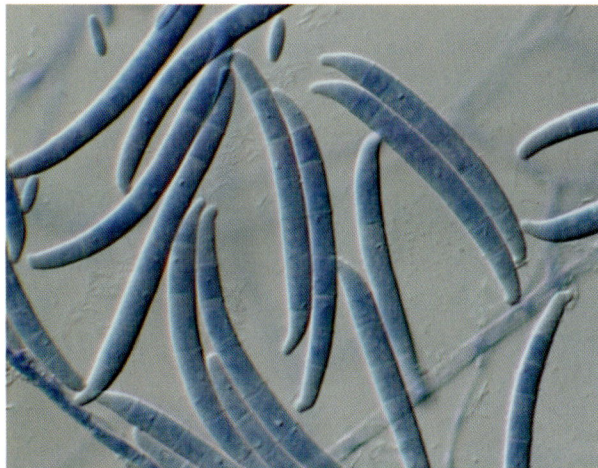

Albonectria rigidiuscula(*Fusarium* anamorph); macroconidia

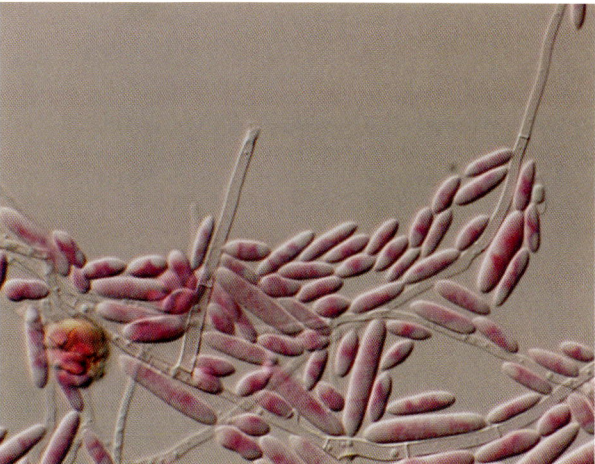

Cylindrocarpon didymum; conidiogenous cell, micro- and macroconidia

Ecology: Associated with dead plant material or other fungi, often pathogenic.

Notes: The *Nectriaceae* is separated from the other families of the *Hypocreales* principally using molecular data, although the brightly coloured ascomata that are not immersed in a stroma are largely diagnostic.

References: **Aoki, T.; O'Donnell, K.** (1999). Morphological and molecular characterization of *Fusarium pseudograminearum* sp. nov., formerly recognized as the Group 1 population of *F. graminearum*. *Mycologia* 91: 597–609; **Arie, T.; Yoshida, T.; Shimizu, T.; Kawabe, M.; Yoneyama, K.; Yamaguchi, I.** (1999). Assessment of *Gibberella fujikuroi* mating type by PCR. *Mycoscience* 40: 311–314; **Baayen, R.P.; O'Donnell, K.; Bonants, P.J.M.; Cigelnik, E.; Kroon, L.P.N.M.; Roebroeck, E.J.A.; Waalwijk, C.** (2000). Gene genealogies and AFLP analyses in the *Fusarium oxysporum* complex identify monophyletic and nonmonophyletic *formae speciales* causing wilt and rot disease. *Phytopathology* 90: 891–900; **Brayford, D.; Honda, B.M.; Mantiri, F.R.; Samuels, G.J.** (2004). *Neonectria* and *Cylindrocarpon*: the *Nectria mammoidea* group and species lacking microconidia. *Mycologia* 96: 572–597; **Brayford, D.; Samuels, G.J.** (1993). Some didymosporous species of *Nectria* with nonmicroconidial *Cylindrocarpon* anamorphs. *Mycologia* 85: 612–637; **Crous, P.W.** (2002). *Taxonomy and Pathology of Cylindrocladium (Calonectria) and Allied Genera* (St Paul): 278 pp.; **Crous, P.W.; Groenewald, J.Z.; Risède, J.-M.; Simoneau, P.; Hyde, K.D.** (2006). *Calonectria* species and their *Cylindrocladium* anamorphs: species with clavate vesicles. *Stud. Mycol.* 55: 213–226; **Leslie, J.F.** (1999). Genetic status of the *Gibberella fujikuroi* species complex. *Pl. Path. J.* 15: 259–269; **Lieckfeldt, E.; Seifert, K.A.** (2000). An evaluation of the use of ITS sequences in taxonomy of the *Hypocreales*. *Stud. Mycol.* 45: 35–44; **Möller, E.M.; Chełkowski, J.; Geiger, H.H.** (1999). Species-specific PCR assays for the fungal pathogens *Fusarium moniliforme* and *Fusarium subglutinans* and their application to diagnose maize ear rot disease. *J. Phytopath.* 147: 497–508; **Nirenberg, H.I.; O'Donnell, K.** (1998). New *Fusarium* species and combinations within the *Gibberella fujikuroi* species complex. *Mycologia* 90: 434–458; **O'Donnell, K.; Cigelnik, E.; Nirenberg, H.I.** (1998). Molecular systematics and phylogeography of the *Gibberella fujikuroi* species complex. *Mycologia* 90: 465–493; **O'Donnell, K.; Kistler, H.C.; Tacke, B.K.; Casper, H.H.** (2000). Gene genealogies reveal global phylogeographic structure and reproductive isolation among lineages of *Fusarium graminearum*, the fungus causing wheat scab. *Proc. natn Acad. Sci. U.S.A.* 97: 7905–7910; **Rehner, S.A.; Samuels, G.J.** (1995). Molecular systematics of the *Hypocreales*: a teleomorph gene phylogeny and the status of their anamorphs. *Can. J.*

Bot. 73: S816–S823; **Rossman, A.Y.** (2000). Towards monophyletic genera in the holomorphic *Hypocreales*. *Stud. Mycol.* 45: 27–34; **Rossman, A.Y.; Samuels, G.J.; Rogerson, C.T.; Lowen, R.** (1999). Genera of *Bionectriaceae, Hypocreaceae* and *Nectriaceae. Stud. Mycol.* 42: 248 pp.; **Samuels, G.J.; Brayford, D.** (1993). Phragmosporous *Nectria* species with *Cylindrocarpon* anamorphs. *Sydowia* 45: 55–80; **Schoch, C.L.; Crous, P.W.; Wingfield, M.J.; Wingfield, B.D.** (2000). Phylogeny of *Calonectria* and selected hyprocrealean genera with cylindrical macroconidia. *Stud. Mycol.* 45: 45–62; **Victor, D.; Crous, P.W.; Janse, B.J.H.; Van Zyl, W.H.; Wingfield, M.J.; Alfenas, A.C.** (1998). Systematic appraisal of species complexes within *Cylindrocladiella*. *Mycol. Res.* 102: 273–279.

Neocallimastigaceae I.B. Heath 1983
Neocallimastigales: Neocallimastigomycota

Thallus monocentric or polycentric. Zoospores monoflagellate or polyflagellate, without mitochondria, ribosome aggregates are mainly in the anterior, the nucleus in the centre and the hydrogenosome in the posterior part; kinetosomes with a skirt, spur, cylinder and microtubule root, lacking props. In place of props there is a complex, electron-opaque structure that partially surrounds the kinetosome; the flagellar root comprises an array of microtubules that extend without regular arrangement into the cytoplasm from a spur on the kinetosome(s), when polyflagellate the flagella beat as a unit. Zoospores encyst, develop rhizoids and can produce new zoospores at any stage in their subsequent development from a uninucleate to a multinucleate eucarpic thallus; zoospore production involves intracystoplasmic flagellar vesicles, organelle clustering and loss of the sporangium plasmalemma; zoospore release follows non-localized dissolution of the entire sporangium wall.

Significant Genera: *Neocallimastix, Piromyces*.

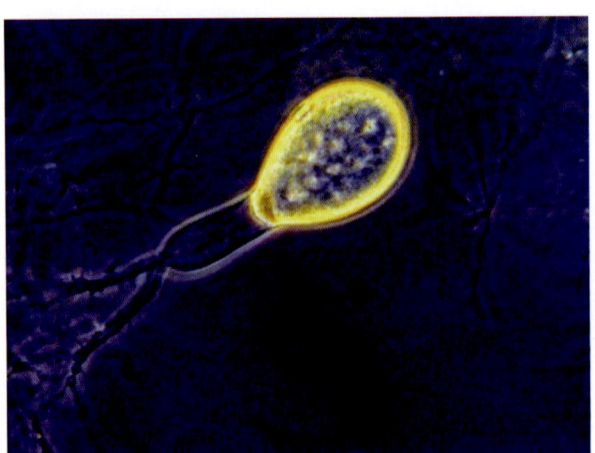

Piromyces sp.; thallus and sporangium

Distribution: Widespread, probably coextensive with their mammal hosts.

Economic Significance: None is known, apart from their important role in ruminant nutrition (see below).

Ecology: Obligate anaerobes that live in the rumen and hindgut of vertebrate herbivores and metabolize cellulose.

Notes: The family is the sole member of the order *Neocallimastigales* and contains five genera and approximately 16 spp. Molecular data from several representatives of this family are available and support the distinctness of the group which superficially resemble members of the *Spizellomycetales*.

References: **Heath, I.B.; Bauchop, T.; Skipp, R.A.** (1983). Assignment of the rumen anaerobe *Neocallimastix frontalis* to the *Spizellomycetales* (*Chytridiomycetes*) on the basis of its polyflagellate zoospore ultrastructure. *Can. J. Bot.* 61: 295–307; **Li, J.L.; Heath, I.B.; Packer, L.** (1993). The phylogenetic relationships of the anaerobic chytridiomycetous gut fungi (*Neocallimastigaceae*) and the *Chytridiomycota*. II. Cladistic analysis of structural data and description of *Neocallimasticales* ord.nov. *Can. J. Bot.* 71: 393–407.

Neolectaceae Redhead 1977
Neolectales: Ascomycota

Stromata absent. Ascomata pale yellowish, stalked, clavate or spathulate, the entire surface covered with asci; peridium and stalk composed of interwoven hyphae, often with large J+ septal plugs. Interascal tissue absent. Asci not formed from croziers, cylindrical, thin-walled, the walls blueing in iodine, slightly thickened at the apex, with a wide apical pore, persistent, 8-spored. Ascospores ellipsoidal, hyaline, aseptate, without gelatinous appendages or sheaths. Anamorphs unknown.

Significant Genera: *Neolecta*.

Distribution: Boreal forests and southern Brazil.

Economic Significance: None is known.

Neolecta irregularis; ascomata

Ecology: On soil and leaf mould, one species possibly parasitic on tree roots.

Notes: This is an extremely isolated family that is now considered to be amongst the most basal of the *Ascomycota*. Most phylogenetic trees place the *Neolectaceae* near the *Saccharomycetes* or *Pneumocystidomycetes*. However, morphological differences are extreme and it is entirely possible that the juxtaposition of these groups in phylogenetic trees is artifactual. Little is known of the biology of its species.

References: **Berbee, M.L.; Taylor, J.W.** (2001). Fungal molecular evolution: gene trees and geologic time. *The Mycota* (Berlin) 7: 229–245; **Landvik, S.** (1996). *Neolecta*, a fruit-body-producing genus of the basal ascomycetes, as shown by SSU and LSU rDNA sequences. *Mycol. Res.* 100: 199–202; **Landvik, S.; Eriksson, O.E.; Berbee, M.L.** (2001). *Neolecta* – a fungal dinosaur? Evidence from β-tubulin amino acid sequences. *Mycologia* 93: 1151–1163; **Landvik, S.; Eriksson, O.E.; Gargas, A.; Gustafsson, P.** (1993). Relationships of the genus *Neolecta* (*Neolectales* ordo nov., *Ascomycotina*) inferred from 18S rDNA sequences. *Syst. Ascom.* 11: 107–118; **Liu, Y.J.J.; Whelen, S.; Hall, B.D.** (1999). Phylogenetic relationships among ascomycetes: evidence from an RNA polymer[a]se II subunit. *Mol. Biol. Evol.* 16: 1799–1808; **Redhead, S.A.** (1977). The genus *Neolecta* (*Neolectaceae* fam. nov., *Lecanorales*, *Ascomycetes*) in Canada. *Can. J. Bot.* 55: 301–306; **Sugiyama, J.** (1998). Relatedness, phylogeny, and evolution of the fungi. *Mycoscience* 39: 487–511.

Neozygitaceae Ben Ze'ev, R.G. Kenneth & Uziel 1987
Entomophthorales: Zygomycota

Vegetative growth as globose to rod-shaped hyphal bodies, with or without a cell wall. Sporophores (conidiophores) simple, solitary, cylindrical to slightly clavate, bearing terminal conidiogenous cells, or conidiogenous cells arising directly from the hyphal bodies. Spores (conidia) terminal, pigmented pale grey, spherical, single-celled, with a truncate or small papilla, weakly forcibly discharged; germinating to form either hyphae or one to several short sporophores, each bearing a small secondary spore; nucleus small, nucleolus prominent. Zygospores formed by hyphal conjugation, zygosporangium (epispore) dark brown to black, readily detaching from the zygospore (endospore), ovoid and smooth or globose to subglobose and roughened; germinating directly to produce secondary spores or a sporophore.

Significant Genera: *Neozygites*.

Distribution: Widespread.

Economic Significance: None is known; may be useful adjuncts to biocontrol research projects.

Ecology: Obligately pathogenic in insects (especially Homoptera) and mites (Acari).

Notes: Very poorly known; no recent information is available.

References: **Delalibera, I.; Hajek, A.E.; Humber, R.A.** (2004). *Neozygites tanajoae* sp. nov., a pathogen of the cassava green mite. *Mycologia* 96: 1002–1009; **Delalibera, I.; Humber, R.A.; Hajek, A.E.** (2004). Preservation of *in vitro* cultures of the mite pathogenic fungus *Neozygites tanajoae*. *Can. J. Microbiol.* 50: 579–586; **Humber, R.A.** (1989). Synopsis of a revised classification for the *Entomophthorales* (*Zygomycotina*). *Mycotaxon* 34: 441–460; **Keller, S.** (1991). Arthropod-pathogenic *Entomophthorales* of Switzerland. II. *Erynia*, *Eryniopsis*, *Neozygites*, *Zoophthora* and *Tarichium*. *Sydowia* 43: 39–122; **Keller, S.; Petrini, O.** (2005). Keys to the identification of the arthropod pathogenic genera of the families *Entomophthoraceae* and *Neozygitaceae* (*Zygomycetes*), with descriptions of three new subfamilies and a new genus. *Sydowia* 57: 23–53.

Nephromataceae Wetmore ex J.C. David & D. Hawksw. 1991
Peltigerales: Ascomycota

Thallus foliose, corticate on both surfaces. Ascomata apothecial, initially immersed with a vegetative covering layer splitting open at a late stage of development, produced on the lower surface of the thallus which subsequently turns up to expose the hymenium. Interascal tissue of simple paraphyses. Asci with a J– apical cap, lacking an apical ring, with a well-developed ocular chamber but no J+ gelatinized outer layer. Ascospores brown, elongate, transversely septate. Anamorphs pycnidial.

Significant Genera: *Nephroma*.

Distribution: Primarily temperate.

Economic Significance: None is known.

Nephroma aff. *laevigatum*, Bhutan; thallus and ascomata

Ecology: Lichenized with green algae or cyanobacteria, the latter sometimes in cephalodia; on soil, bark and similar substrata.

Notes: This small family forms a well-supported clade within the *Peltigerales* in most recent phylogenetic analyses. Species are large and conspicuous, and are used as indicators of ecological continuity.

Nephroma arcticum, Norway; thallus and ascomata

References: **Burgaz, A.R.; Martínez, I.** (1999). The genus *Nephroma* Ach. in the Iberian Peninsula. *Cryptog. Mycol.* 20: 225–235; **Buschbom, J.; Mueller, G.** (2004). Resolving evolutionary relationships in the lichen-forming genus *Porpidia* and related allies (*Porpidiaceae, Ascomycota*). *Mol. Phylogen. Evol.* 32: 66–82; **Eriksson, O.E.; Strand, Å.** (1995). Relationships of the genera *Nephroma*, *Peltigera* and *Solorina* (*Peltigerales, Ascomycota*) inferred from 18S rDNA sequences. *Syst. Ascom.* 14: 33–39; **Galloway, D.J.** (1991). Chemical evolution in the order *Peltigerales*: triterpenoids. *Symbiosis* 11: 327–344; **Goffinet, B.; Goward, T.** (1998). Is *Nephroma silvae-veteris* the cyanomorph of *Lobaria oregana*? Insights from molecular, chemical and morphological characters. *Lichenographia Thomsoniana, North American Lichenology in Honor of John W. Thomson* (Ithaca): 41–52; **James, P.W.; White, F.J.** (1987). Studies on the genus *Nephroma* I. The European and Macronesian species. *Lichenologist* 19: 215–268; **Miadlikowska, J.; Lutzoni, F.** (2004). Phylogenetic classification of peltigeralean fungi (*Peltigerales, Ascomycota*) based on ribosomal RNA small and large subunits. *Am. J. Bot.* 91: 449–464; **White, F.J.; James, P.W.** (1988). Studies on the genus *Nephroma* II. The southern temperate species. *Lichenologist* 20: 103–166; **Wiklund, E.; Wedin, M.** (2003). The phylogenetic relationships of the cyanobacterial lichens in the *Lecanorales* suborder *Peltigerineae*. *Cladistics* 19: 419–431.

Niaceae Jülich 1982
Agaricales: Basidiomycota

Basidiomata small, gasteroid or cyphelloid, superficial, short-stalked, white, orange or brown; when gasteroid ± spherical, opening by irregular rupture; when cyphelloid initially turbinate but eventually becoming funnel-shaped with the hymenium lining the inner wall; thin-walled, glabrous or with straight, apically curved and/or bifurcate hairs. Clamp connections present. Basidia clavate or cylindrical, long-stalked, either evanescent or with 4–8 evanescent sterigmata. Basidiospores ± globose, thin-walled, smooth or with elongate thin-walled appendages, not staining in iodine.

Significant Genera: *Halocyphina*, *Nia*.

Distribution: Widespread in tropical and temperate zones.

Economic Significance: None is known.

Ecology: Marine, growing on submerged wood, roots and grass culms, often in mangroves.

Notes: Molecular studies have revealed *Nia* to be related to the terrestrial genera *Cyphellopsis*, *Lachnella* and *Merismodes*, all traditionally placed in the *Marasmiaceae*. The gasteroid habit in *Nia* appears to have been derived from common cyphelloid ancestors. It is not currently clear whether the *Niaceae* should be delimited to include all of these genera.

Nia vibrissa; basidiomata

References: **Barata, M.; Basilio, M.C.; Baptista-Ferreira, J.L.** (1997). *Nia globospora*, a new marine gasteromycete on baits of *Spartina maritima* in Portugal. *Mycol. Res.* 101: 687–690; **Binder, M.; Hibbett, D.S.; Molitoris, H.P.** (2001). Phylogenetic relationships of the marine gasteromycete *Nia vibrissa*. *Mycologia* 93: 679–688; **Bodensteiner, P.; Binder, M.; Moncalvo, J.-M.; Agerer, R.; Hibbett, D.S.** (2004). Phylogenetic relationships of cyphelloid homobasidiomycetes. *Mol. Phylogen. Evol.* 33: 501–515; **Jones, A.M.; Jones, E.B.G.** (1993). Observations on the marine gasteromycete *Nia vibrissa*. *Mycol. Res.* 97: 1–6; **Kohlmeyer, J.; Kohlmeyer, E.** (1979). *Marine Mycology, the Higher Fungi* (London): 690 pp.; **Sivichai, S.; Jones, E.B.G.** (2004). *Stauriella* gen. nov., proposed for a new lignicolous basidiomycetous anamorph from freshwater in Thailand. *Sydowia* 56: 131–136.

Nidulariaceae Dumort. 1822
Agaricales: Basidiomycota

Basidiocarp gasteroid, epigeous, small, often gregarious, the peridium obconical and resembling a nest or funnel, the top of which is first covered by a membrane that dehisces irregularly or by a circumscissile split, peridium one- or multilayered and sometimes with a striate inner surface, containing one to several peridioles ('eggs') either fixed to the wall by thread-like funiculi or free within the peridium and immersed in mucilage. Hyphal system dimitic, clamp connections present. Gleba consisting of individualized glebal chambers with hardened wall (peridioles), capillitium absent, hymenium not defined. Basidia borne singly or in groups, evanescent, with 4 to 8 sterigmata. Basidiospores usually large, ellipsoidal to cylindrical, smooth, hyaline and thin-walled.

Cyathus olla; basidiomata

Cyathus striatus; basidiomata

Crucibulum laeve; basidiomata

Significant Genera: *Crucibulum*, *Cyathus*, *Nidularia*.

Distribution: Widespread in most ecological zones.

Economic Significance: None is known.

Ecology: Saprobic on dead plant material or dung, especially common on dead grass stems.

Notes: The Bird's Nest Fungi. Distribution of the peridioles is effected by a splash-cup mechanism. Molecular studies place this family within the *Agaricales*, but its closest relatives are unclear.

Nidula candida, Vancouver Island, Canada; basidiomata

References: **Binder, M.; Bresinsky, A.** (2002). Derivation of a polymorphic lineage of gasteromycetes from boletoid ancestors. *Mycologia* 94: 85–98; **Brodie, H.J.** (1975). *The Bird's Nest Fungi* (Toronto): 199 pp.; **Brodie, H.J.** (1984). More Bird's Nest Fungi (*Nidulariaceae*) – A supplement to "The bird's nest fungi." (1975). *Lejeunia* n.s. 112: 1–72; **Calonge, F.D.** (1998). *Gasteromycetes*, I. *Lycoperdales, Nidulariales, Phallales, Sclerodermatales, Tulostomatales*. *Fl. Mycol. Iberica* 3: 271 pp.; **Coetzee, J.C.; Eicker, A.; Van Wyck, A.E.** (1997). Taxonomic notes on the *Geastraceae, Tulostomataceae, Nidulariaceae* and *Sphaerobolaceae* (*Gasteromycetes*) *sensu* Bottomly, in southern Africa. *Bothalia* 27: 117–123; **Diehl, P.** (2000). Anatomy of the peridium in the genus *Nidula* (*Nidulariales, Basidiomycetes*). *Sydowia* 52: 16–29; **Hibbett, D.S.; Pine, E.M.; Langer, E.; Langer, G.; Donoghue, M.J.** (1996). Evolution of gilled mushrooms and puffballs inferred from ribosomal DNA sequences. *Proc. natn Acad. Sci. U.S.A.* 94: 12002–12006; **Sarasini, M.; Pina, G.** (1995). Nidulariace[a]e. Prima parte. Ciclo vitale e caratteri generali: il genere *Crucibulum*. *Riv. Micol.* 38: 237–252; **Sarasini, M.; Pina, G.** (1996). *Nidulariaceae*. Seconda parte (seguito del genere *Crucibulum*). *Riv. Micol.* 39: 115–126; **Sarasini, M.; Pina, G.C.** (1997). *Nidulariaceae*. Terza parte: il genere *Cyathus*. *Riv. Micol.* 40: 19–35; **Shinners, T.C.; Tewari, J.P.** (1998). Morphological and RAPD analyses of *Cyathus olla* from crop residue. *Mycologia* 90: 980–989.

Niessliaceae Kirschst. 1939
Hypocreales: Ascomycota

Ascomata perithecial, superficial on or rarely immersed in a subiculum or crustose stroma, small, ± globose, sometimes collapsing when dry, pale to dark brown, often setose or ornamented, the ostiole periphysate; peridium thin, soft. Interascal tissue of apical paraphyses. Asci ± clavate, thin-walled, not fissitunicate, with a small refractive J– apical ring. Ascospores ± hyaline, 0- or 1-septate, sometimes fragmenting. Anamorphs hyphomycetous, *Acremonium*-like.

Significant Genera: *Niesslia*. Anamorph: *Monocillium*.

Distribution: Widespread.

Economic Significance: None is known.

Ecology: Saprobic on herbaceous and woody tissue.

Notes: *Myrothecium* appears to form a sister group to the *Niessliaceae* and the two clades are apparently ancestral within the *Hypocreales*.

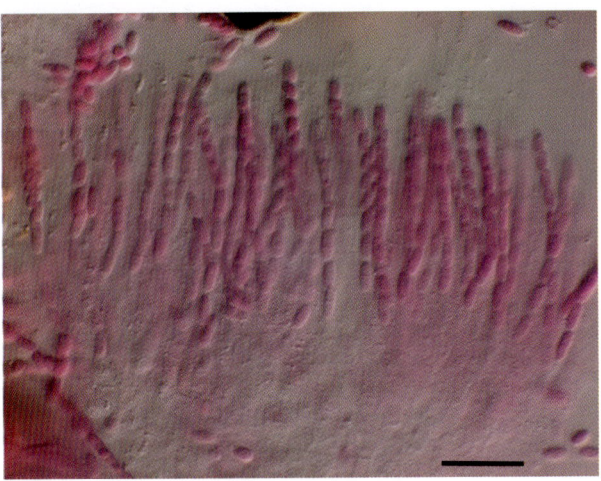
Niesslia ilicifolia, Devon, UK; asci and ascospores

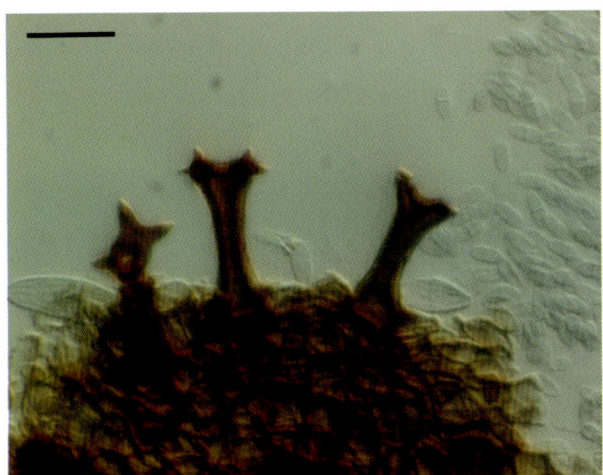
Valetoniella crucipila, Sierra Leone; ascomatal appendages

References: **Castlebury, L.A.; Rossman, A.Y.; Sung, G.-H.; Hyten, A.S.; Spatafora, J.W.** (2004). Multigene phylogeny reveals new lineage for *Stachybotrys chartarum*, the indoor air fungus. *Mycol. Res.* 108: 864–872; **Crous, P.W.; Gams, W.; Wingfield, M.J.; Wyk, P.S. van** (1996). *Phaeoacremonium* gen. nov. associated with wilt and decline diseases of woody hosts and human infections. *Mycologia* 88: 786–796; **Gams, W.; Turhan, G.** (1996). An extreme modification of *Monocillium*: *Monocillium curvisetosum* n. sp. *Mycotaxon* 59: 343–348; **Rossman, A.Y.; Samuels, G.J.; Rogerson, C.T.; Lowen, R.** (1999). Genera of *Bionectriaceae*, *Hypocreaceae* and *Nectriaceae*. *Stud. Mycol.* 42: 248 pp.; **Samuels, G.J.; Barr, M.E.** (1998). Notes on and additions to the *Niessliaceae* (*Hypocreales*). *Can. J. Bot.* 75: 2165–2176; **Scheuer, Ch.** (1993). *Cryptoniesslia setulosa* gen. et sp. nov. *Mycol. Res.* 97: 543–546.

Nitschkiaceae Nannf. 1932
Coronophorales: Ascomycota

Stromata absent, though ascomata often strongly clustered. Ascomata perithecial, dark, thick-walled and strongly verrucose, often collapsing when dry, opening with an irregular lysigenous pore, cells with 1 μm diam. perforations ('Munk pores'). Interascal tissue absent, at least at maturity. Asci clavate, long-stalked, without apical apparatus, sometimes multispored. Ascospores hyaline or brown, usually allantoid, sometimes septate, smooth, sheath absent. Anamorphs not known.

Significant Genera: *Bertia*, *Nitschkia*.

Distribution: Widespread, in both temperate and tropical regions.

Economic Significance: None is known.

Ecology: Saprobic on wood. A few species are lichenicolous.

Notes: The *Nitschkiaceae* was considered for many years to belong to its own order the *Coronophorales*, but was subsequently linked with the *Lasiosphaeriaceae* due to the presence of Munk pores in some species of that family. Molecular evidence now suggests that the original hypothesis is correct, but further work is required on the limits of the family. The *Bertiaceae* and the *Scortechiniaceae* are found in separate clades, but are not necessarily separable at family level.

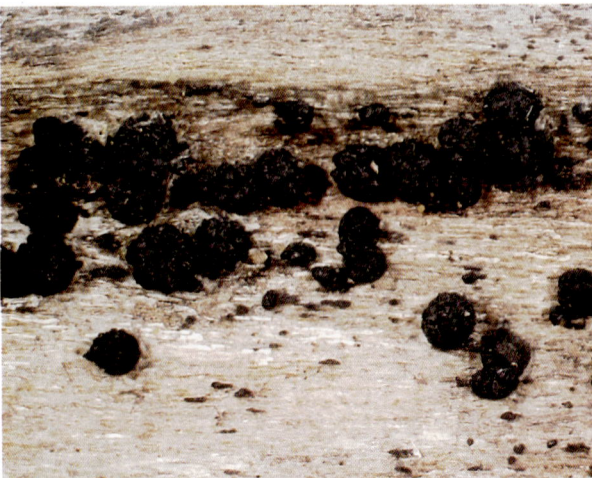
Bertia moriformis, Tierra del Fuego, Chile; ascomata on rotten wood

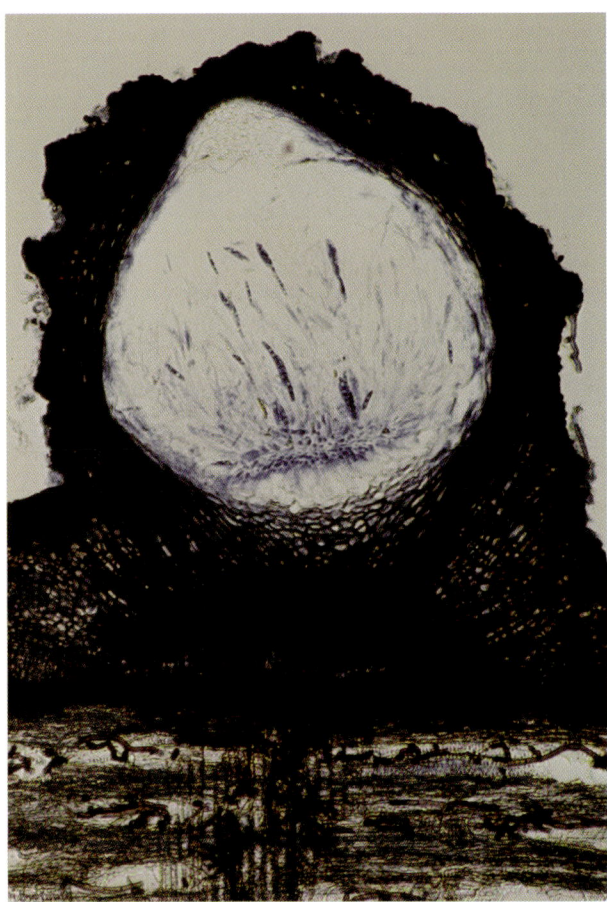

Bertia moriformis; vertical section through ascoma

Bertia moriformis; Munk pores in peridial cells

References: **Bianchinotti, M.V.** (2004). Two new lignicolous species of *Nitschkia* from Argentina. *Mycologia* 96: 911–916; **Corlett, M.; Krug, J.C.** (1985). *Bertia moriformis* and its varieties. *Can. J. Bot.* 62: 2561–2569; **Eriksson, O.; Santesson, R.** (1986). *Lasiosphaeriopsis stereocaulicola*. *Mycotaxon* 25: 569–580; **Huhndorf, S.M.; Miller, A.N.; Fernández, F.A.** (2004). Molecular systematics of the *Coronophorales* and new species of *Bertia*, *Lasiobertia* and *Nitschkia*. *Mycol. Res.* 108: 1384–1398; **Huhndorf, S.M.; Miller, A.N.; Fernández, F.A.** (2004). Molecular systematics of the *Sordariales*: the order and the family *Lasiosphaeriaceae* revisited. *Mycologia* 96: 368–387; **Hyde, K.D.** (1995). Tropical Australian freshwater fungi. VIII: *Bertia convolutispora* sp. nov. *Nova Hedwigia* 61: 141–146; **Jensen, J.D.** (1985). Peridial anatomy and pyrenomycete taxonomy. *Mycologia* 77: 688–701; **Linde, E.J. van der; Botha, A.J.** (1997). Profiles of South African ascomycetous fungi. I. *Nitschkia broomeiana*. *S. Afr. J. Bot.* 63: 66–67; **Navarro-Rosinés, P.; Etayo, J.; Calatayud, V.** (1999). *Rhagadostoma collematum* sp. nov. (ascomicetes liquenícolas, *Sordariales*) y nuevos datos para otras especies del género. *Bull. Soc. linn. Provence* 50: 233–241; **Petersen, K.R.L.** (1997). Ultrastructural studies of the marine ascomycete *Groenhiella bivestia*. *Bot. Mar.* 40: 71–76; **Subramanian, C.V.; Sekar, G.** (1993). *Coronophorales* from India – a monograph. *Kavaka* 18: 19–90.

Obryzaceae Körb. 1855
Uncertain position within Sordariomycetes, Ascomycota

Stromata absent. Ascomata perithecial, pyriform, immersed, black-walled near the ostiole, the ostiole periphysate. Interascal tissue absent. Asci clavate, short-stalked, thin-walled, not fissitunicate, apical structures lacking, evanescent at the base. Ascospores hyaline, aseptate, smooth, without a gelatinous sheath or appendages.

Significant Genera: *Obryzum*.

Distribution: Known from north temperate zones.

Economic Significance: None is known.

Ecology: Lichenicolous, reported as parasitic on cyanobacterial lichens from the genus *Leptogium*.

Notes: Very poorly known; little recent information is available and no molecular data are available. The type species has some similarities to *Guignardia* (*Botryosphaeriaceae*).

References: **Hoffmann, N.; Hafellner, J.** (2000). Eine Revision der lichenicoler Arten der Sammelgattungen *Guignardia* und *Physalospora* (*Ascomycotina*). *Biblthca Lichenol.* 77: 181 pp.

Ochrolechiaceae R.C. Harris ex Lumbsch & I. Schmitt 2006
Pertusariales: Ascomycota

Thallus well-developed, crustose and often thick and warted or rarely spinose, white to dark grey, the upper cortex thhin or absent. Ascomata apothecia, the disc often pruinose, the well-developed wall composed of fertile and thalline tissue and sometimes with an inner non-lichenized layer. Interascal tissue composed of narrow densely branched and anastomosed paraphyses. Asci fissitunicate, short, widely cylindrical, with a thick multilayered usually J+ wall, the apex often more strongly thickened, releasing spores through a ± vertical split, 2- to 8-spored. Ascospores large, very thick-walled, aseptate, smooth-walled, without gelatinous sheath or appendages. Conidiomata pycnidial, the conidia elongate-cylindrical.

Significant Genera: *Ochrolechia*.

Distribution: Widespread, especially in temperate zones.

Economic Significance: None is known.

Ecology: Lichenized with green algae; on rock or bark, rarely on moss mats and similar substrata.

Notes: Recently separated from the *Pertusariaceae*, which has very similar asci but ascomata which in many species are almost perithecial in form. *Varicellaria* and *Variolaria* may also belong here, but more study is needed.

Ochrolechia tartarea, Devon, UK; thallus and ascomata

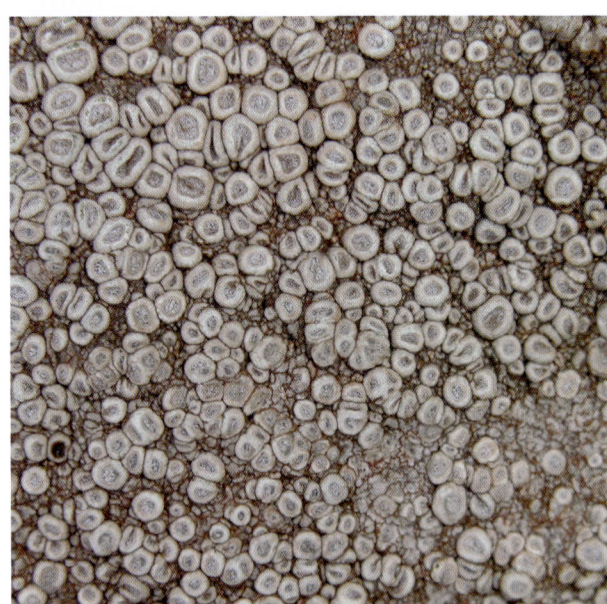
Ochrolechia parella, Devon; thallus and ascomata

References: **Brodo, I.M.** (1991). Studies in the lichen genus *Ochrolechia*. 2. Corticolous species of North America. *Can. J. Bot.* 69: 733–772; **Lumbsch, H.T.; Dickhäuser, A.; Feige, G.B.** (1995). Systematic studies in the *Pertusariales* III. Taxonomic position of *Thamnochrolechia* (lichenized *Ascomycetes*). *Biblthca Lichenol.* 57: 355–361; **Schmitt, I.; Martín, M.P.; Türk, R.; Lumbsch, H.T.** (2003). Phylogenetic position of the genera *Varicellaria, Melanaria* and *Thamnochrolechia* (*Pertusariales*). *Biblthca Lichenol.* 86: 147–154; **Schmitt, I.; Yamamoto, Y.; Lumbsch, H.T.** (2006). Phylogeny of *Pertusariales* (*Ascomycotina*): resurrection of *Ochrolechiaceae* and new circumscription of *Megasporaceae*. *J. Hattori bot. Lab.* 100: 753–764.

Octavianiaceae Locq. ex Pegler & T.W.K. Young 1979
Boletales: Basidiomycota

Basidiomata gasteroid, hypogeous or semi-epigeous, small, ± globose, without a stalk, thin-walled, sometimes embedded in a mycelial mat, the peridium not breaking down to allow spore release. Hyphal system monomitic, the generative hyphae thin-walled and sometimes inflated, clamp connections absent, lactifers and cystidia sometimes present. Hymenium labyrinthoid or loculate, columella absent or reduced to a small pulvinate base, capillitium lacking. Basidia thin-walled, clavate, with 2–8 sterigmata, deliquescing. Basidiospores globose, spinose, tuberculate or verrucose, sometimes with an epispore layer, sometimes staining brown but not blue in iodine.

Significant Genera: *Octaviania, Sclerogaster*.

Distribution: Widespread, in both temperate zones and the Old World tropics.

Economic Significance: None is known.

Ecology: Probably ectomycorrhizal, but little is known.

Notes: A poorly studied family with no molecular phylogenetics research, its position within the *Boletales* is questionable. *Stephanospora* was originally included but it is now given separate family status.

References: **Lebel, T.; Castellano, M.A.** (2002). Type studies of sequestrate *Russulales* II. Australian and New Zealand species related to *Russula*. *Mycologia* 94: 327–354; **Pegler, D.N.; Young, T.W.K.** (1979). The gasteroid *Russulales*. *Trans. Br. mycol. Soc.* 72: 353–388; **Trappe, J.M.; Lebel, T.; Castellano, M.A.** (2002). Nomenclatural revisions in the sequestrate russuloid genera. *Mycotaxon* 81: 195–214.

Odontotremataceae D. Hawksw. & Sherwood 1982
Ostropales: Ascomycota

Stroma sometimes present. Ascomata apothecial, immersed or erumpent, with a well-developed hyphal margin, often black and carbonized. Interascal tissue of simple paraphyses, sometimes branched at the base. Asci elongate, with a thickened apex, with or without a well-developed J– pore. Ascospores ellipsoidal to filiform, variously septate. Anamorphs not definitely known.

Significant Genera: *Odontotrema, Skyttea*.

Distribution: Widespread, primarily known from north temperate zones.

Economic Significance: None is known.

Ecology: Saprobic on wood, especially in xeric conditions, or lichenicolous. A few species are lichenized with green algae.

Notes: There is little research available on this family and its position is inferred largely through morphology rather than molecular comparison.

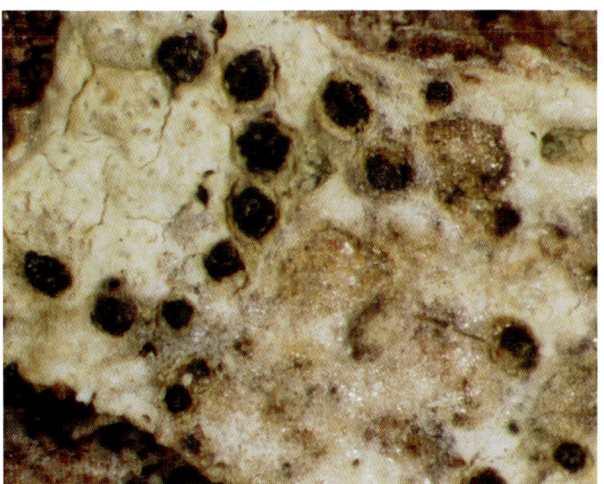

Skyttea nitschkei, Hampshire, UK; ascomata on thallus of *Thelotrema lepadinum*

References: **Döbbeler, P.** (1996). *Potriphila navicularis* gen. et sp. nov. (Ostropales, Ascomycetes), ein bipolar verbreiteter Parasit von *Polytrichum alpinum*. *Nova Hedwigia* 62: 61–77; **Holien, H.; Triebel, D.** (1996). *Spirographa vinosa*, a new odontotremoid fungus on *Ochrolechia* and *Pertusaria*. *Lichenologist* 28: 307–313; **Lumbsch, H.T.; Hawksworth, D.L.** (1990). The species of *Lethariicola* Grumm. (Odontotremataceae). *Biblthca Lichenol.* 38: 325–333; **Sherwood-Pike, M.A.** (1987). The Ostropalean fungi III: The Odontotremataceae. *Mycotaxon* 28: 137–177; **Triebel, D.** (1989). Lecideicole Ascomyceten. Eine Revision der obligat lichenicolen Ascomyceten auf lecideoiden Flechten. *Biblthca Lichenol.* 35: 278 pp.; **Wedin, M.; Döring, H.; Könberg, K.; Gilenstam, G.** (2005). Generic delimitations in the family Stictidaceae (Ostropales, Ascomycota): the *Stictis-Conotrema* problem. *Lichenologist* 37: 67–75.

Oedogoniomycetaceae D.J.S. Barr 1990
Monoblepharidales: Chytridiomycota

Thallus an unbranched filament attached to the substratum by a holdfast, single or fasciculate; zoosporangia maturing in basipetal succession, shot cylindrical with truncate ends, without a discharge tube or papilla but with an exit pore or cleavage line; zoospores posteriorly uniflagellate, without a nuclear cap.

Significant Genera: *Oedogoniomyces*.

Distribution: Widespread, mainly north tropical.

Economic Significance: None is known.

Ecology: Saprobic.

Notes: Very poorly known; no recent information is available.

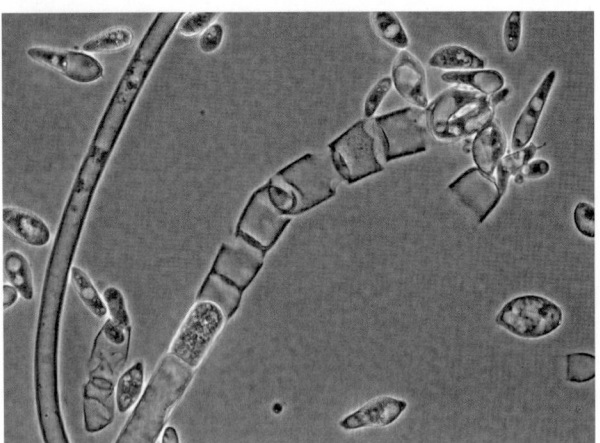

Oedogoniomyces lymnaeae; thallus and zoosporangia

References: **Barr, D.J.S.** (2001). *Chytridiomycota. The Mycota* (Berlin) 7: 93–112; **Emerson, R.; Whisler, H.C.** (1968). Cultural studies of *Oedogoniomyces* and *Harpochytrium*, and a proposal to place them in a new order of aquatic *Phycomycetes*. *Archs Microbiol.* 61: 195–211.

Oliveoniaceae P. Roberts 1998
Ceratobasidiales: Basidiomycota

Basidiomata thin and waxy, usually grey. Hyphae hyaline or becoming brown, clamp connections present or absent, thin-walled cystidia sometimes present. Basidia globose to widely clavate, sometimes becoming brown, not becoming septate, with 1–4 elongate sterigmata. Basidiospores discharged actively, subglobose to cylindrical or lemon-shaped, aseptate, smooth, germinating repeatedly to produce daughter spores.

Significant Genera: *Oliveonia*.

Distribution: Widespread, known from north temperate and tropical zones.

Economic Significance: None is known.

Ecology: Presumably saprobic on rotten wood etc. or associated with termite nests; little is known.

Notes: Only a single genus is known. Links with the *Ceratobasidiales* have been questioned and it has been postulated that the family belongs with the *Exidiales* but molecular data appear not to be publically available.

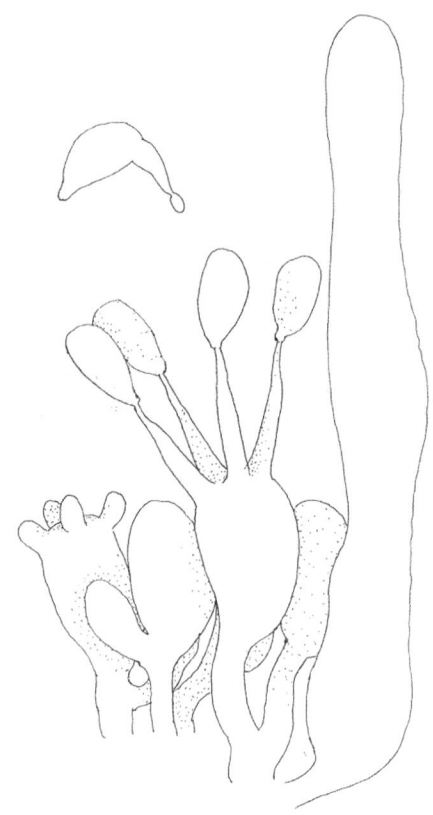

Oliveonia pauxilla, Bedfordshire, UK; basidia and basidiospores

References: **Ginns, J.** (1998). Genera of the North American *Corticiaceae sensu lato*. *Mycologia* 90: 1–35; **Grosse-Brauckmann, H.** (2002). [*Spiculogloea subminuta* and *Oliveonia fibrillosa* (*Heterobasidiomycetes*) – reported from Germany for the first time]. *Z. Mykol.* 68: 135–140; **Roberts, P.** (1998). *Oliveonia* and the origin of the holobasidiomycetes. *Folia cryptog. Estonica* 33: 127–132; **Roberts, P.** (1999). *Rhizoctonia-Forming Fungi* (Richmond): 239 pp.; **Roberts, P.** (2003). *Heterobasidiomycetes* from Rancho Grande, Venezuela. *Mycotaxon* 87: 25–41.

Olpidiaceae J. Schröt. 1889
Uncertain position within Chytridiomycetes, Chytridiomycota

Thallus bearing a single (monocentric) reproductive structure (holocarpic), a sporangium, or with a specialized rhizoidal element (eucarpic), converted as a whole into a single inoperculate sporangium or a resting spore (exogenous to the zoospore cyst), usually endobiotic; zoospores posteriorly uniflagellate; sexual reproduction by fusion of posteriorly uniflagellate planogametes, the zygote, after penetration of the host, forming an endobiotic resting spore which functions as a zoosporangium on germination.

Significant Genera: *Olpidium*.

Distribution: Widespread.

Economic Significance: *Olpidium brassicae* is a vector of lettuce big vein virus; *O. uredinis* is a common species found on rust spores.

Ecology: Typically found parasitizing algae, other aquatic fungi, rotifers and other soil or water-borne invertebrates.

Notes: Referred tentatively to *Spizellomycetales*; molecular data are scanty and it is unclear whether this position is supported.

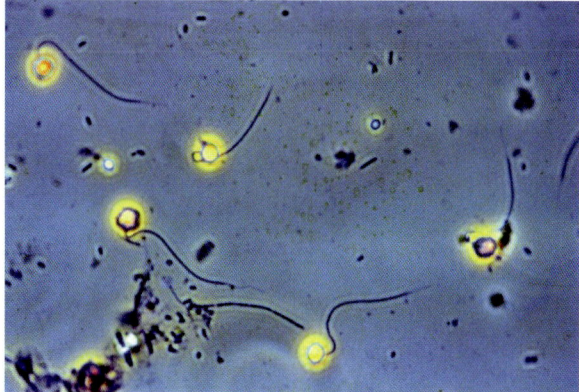
Olpidium brassicae; globose, uniflagellated zoospores

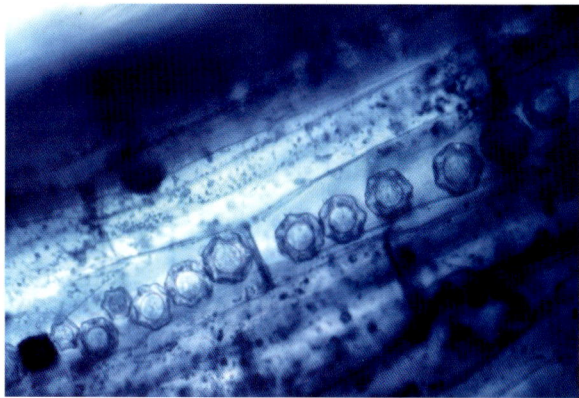
Olpidium brassicae; stellate resting spores inside root cells

References: **Barr, D.J.S.** (1980). An outline for the reclassification of the *Chytridiales*, and for a new order, the *Spizellomycetales*. *Can. J. Bot.* 58: 2380–2394; **Karling, J.S.** (1977). *Chytriomyc. Iconogr.* (Vaduz): 414 pp.; **Lange, L.; Insunza, V.** (1977). Root-inhabiting *Olpidium* species: The *O. radicale* complex. *Trans. Br. mycol. Soc.* 69: 377–384; **Sahtiyanci, S.** (1962). Studien über einige wurzelparasitäre Olpidiaceen. *Archs Microbiol.* 41: 187–228; **Sampson, K.** (1939). *Olpidium brassicae* (Wor.) Dang. and its connection with *Asterocystis radicis* De Wildeman. *Trans. Br. mycol. Soc.* 23: 199–205; **Sparrow, F.K.** (1960). *Aquatic Phycomycetes* Edn 2 (Ann Arbor): 1187 pp.

Omphalotaceae Bresinsky 1985
Agaricales: Basidiomycota

Basidiomata pileate and stipitate, fleshy, brightly coloured, often luminescent, the cap glabrous or fibrillose, not viscid, the stipe fibrous or fleshy, central or eccentric, usually without a veil. Hyphal system monomitic, usually with inflated generative hyphae, clamp connections present, not staining in iodine, cystidia usually lacking. Hymenium lamellar, the gills usually deeply decurrent. Basidia clavate, often with a proportion with smaller numbers of sterigmata or intergrading with cystidioles. Basidiospores small or large, globose or ellipsoidal, hyaline, smooth-walled, thin-walled or thick-walled, not staining in iodine.

Significant Genera: *Lentinula, Omphalotus*.

Distribution: Very widespread, cosmopolitan in forested regions.

Economic Significance: *Lentinula edodes*, better known as the shiitake mushroom, is a prized edible species and is widely cultivated by inoculation into hardwood logs. Several other species of the family are toxigenic, producing illudins and muscarin.

Ecology: Saprobic, causing white rots of dead wood.

Notes: A rather poorly defined assemblage that requires further molecular studies. *Lampteromyces*, previously placed in the *Tricholomataceae*, has been shown to cluster within *Omphalotus* based on molecular data, despite their rather different phenotypic appearance.

Lentinula edodes; basidiomata

References: **Hibbett, D.S.; Hansen, K.; Donoghue, M.J.** (1998). Phylogeny and biogeography of *Lentinula* inferred from an expanded rDNA dataset. *Mycol. Res.* 102: 1041–1049; **Hughes, K.W.; Petersen, R.H.** (1998). Relationships among *Omphalotus* species (*Paxillaceae*) based on restriction sites in the ribosomal ITS1-5.8S-ITS2 region. *Pl. Syst. Evol.* 211: 231–237; **Kämmerer, A.; Besl, H.; Bresinsky, A.** (1985). Omphalotaceae fam. nov. und Paxillaceae, ein chemotaxonomischer Vergleich zwier Pilzfamilien der *Boletales*. *Pl. Syst. Evol.* 150: 101–117; **Kirchmair, M.; Morandell, S.; Stolz, D.; Pöder, R.; Sturmbauer, C.** (2004). Phylogeny of the genus *Omphalotus* based on nuclear ribosomal DNA-sequences. *Mycologia* 96: 1253–1260; **Kirchmair, M.; Pöder, R.; Huber, C.G.; Miller, O.K.** (2002). Chemotaxonomical and morphological observations in the genus *Omphalotus* (*Omphalotaceae*). *Persoonia* 17: 583–600; **Mata, J.L.; Petersen, R.H.; Hughes, K.W.** (2001). The genus *Lentinula* in the Americas. *Mycologia* 93: 1102–1112; **Miller, O.K.**

(1994). Observations on the genus *Omphalotus* in Australia. *Mycol. helv.* 6: 91–100; **Moncalvo, J.-M.; Vilgalys, R.; Redhead, S.A.; Johnson, J.E.; James, T.Y.; Aime, M.C.; Hofstetter, V.; Verduin, S.J.W.; Larsson, E.; Baroni, T.J.; Thorn, R.G.; Jacobsson, S.; Clemencon, H.; Miller, O.K.** (2002). One hundred and seventeen clades of euagarics. *Mol. Phylogen. Evol.* 23: 357–400; **Nicholson, M.S.; Bunyard, B.A.; Royse, D.J.** (1997). Phylogeny of the genus *Lentinula* based on ribosomal DNA restriction fragment length polymorphism analysis. *Mycologia* 89: 400–407; **Petersen, R.H.; Hughes, K.W.** (1998). Mating systems in *Omphalotus* (*Paxillaceae, Agaricales*). *Pl. Syst. Evol.* 211: 217–229; **Saito, T.; Tanaka, N.; Shinozawa, T.** (2002). Characterization of subrepeat regions within rDNA intergenic spacers of the edible basidiomycete *Lentinula edodes*. *Biosc., Biotechn., Biochem.* 66: 2125–2133; **Wilson, A.W.; Desjardin, D.E.** (2005). Phylogenetic relationships in the gymnopoid and marasmioid fungi (*Basidiomycetes*, euagarics clade). *Mycologia* 97: 667–679.

Onygenaceae Berk. 1857
Onygenales: Ascomycota

Ascomata cleistothecial, mostly small and ± globose, in a few species large and stipitate; peridium thin, varied, hyaline or brightly coloured, usually hyphal and net-like but occasionally pseudoparenchymatous. Asci small, globose, thin-walled and evanescent, probably arising from croziers, 8-spored. Ascospores oblate or allantoid, pitted or reticulate, often yellowish, aseptate, without mucous sheath or appendages. Anamorphs hyphomycetous, mostly *Chrysosporium*-like or *Malbranchea*-like, with thick-walled, often ornamented thallic conidia formed as side branches or intercalary on hyaline fertile hyphae.

Significant Genera: *Onygena*.

Distribution: Widespread, especially in warmer regions of the world.

Onygena corvina; ascostromata on feathers

Economic Significance: A few species may cause skin infections, but most appear to be harmless.

Ecology: Generally saprobic and keratinophilic, in soil or coprophilous, occasionally found on hair or horn.

Notes: *Ajellomyces* and its close relatives are segregated from the *Onygenaceae* and placed in a separate family; *Spiromastix* appears to form a third major lineage.

Onygena equina; ascostromata on horn

Onygena corvina, Suffolk, UK; ascostromata on rotten felt

References: **Bialek, R.; Ibricevic, A.; Fothergill, A.; Begerow, D.** (2000). Small subunit ribosomal DNA sequence shows *Paracoccidioides brasiliensis* closely related to *Blastomyces dermatitidis*. *J. Clin. Microbiol.* 38: 3190–3193; **Cano, J.; Guarro, J.** (1990). The genus *Aphanoascus*. *Mycol. Res.* 94: 355–377; **Cano, J.; Vidal, P.; Guarro, J.; Castañeda Ruíz, R.F.; De Vroey, C.** (1996). Three new species of *Amauroascus* from tropical regions. *Mycol. Res.* 100: 343–348; **Currah, R.S.** (1985). Taxonomy of the *Onygenales*: *Arthrodermataceae, Gymnoacaceae, Myxotrichaceae* and *Onygenaceae*. *Mycotaxon* 24: 1–216; **Currah, R.S.** (1988). An annotated key to the genera of the *Onygenales*. *Syst. Ascom.* 7: 1–12; **Guého, E.; Leclerc, M.C.; Hoog, G.S. de; Dupont, B.** (1997). Molecular taxonomy and epidemiology of *Blastomyces* and *Histoplasma* species. *Mycoses* 40: 69–81; **Hoog, G.S. de; Bowman, B.; Graser, Y.; Haase, G.; El Fari, M.; Gerrits van den Ende, A.H.G.; Melzer-Krick, B.; Untereiner, W.A.** (1998). Molecular phylogeny and taxonomy of medically important fungi. *Medical Mycol.* 36: 52–56; **Peterson, S.; Sigler, L.** (1998). Molecular genetic variation in *Emmonsia crescens* and *Emmonsia parva*, etiologic agents of adiaspiromycosis, and their phylogenetic relationship to *Blastomyces dermatitidis* (*Ajello-*

myces dermatitidis) and other systemic fungal pathogens. *J. Clin. Microbiol.* 36: 2918–2925; **Sigler, L.; Flis, A.L.; Carmichael, J.W.** (1999). The genus *Uncinocarpus* (*Onygenaceae*) and its synonym *Brunneospora*: new concepts, combinations and connections to anamorphs in *Chrysosporium*, and further evidence of relationship with *Coccidioides immitis. Can. J. Bot.* 76: 1624–1636; **Sugiyama, M.; Ohara, A.; Mikawa, T.** (1999). Molecular phylogeny of onygenalean fungi based on small subunit ribosomal DNA (SSU rDNA) sequences. *Mycoscience* 40: 251–258; **Sugiyama, M.; Summerbell, R.C.; Mikawa, T.** (2002). Molecular phylogeny of onygenalean fungi based on small subunit (SSU) and large subunit (LSU) ribosomal DNA sequences. *Stud. Mycol.* 47: 5–23; **Untereiner, W.A.; Scott, J.A.; Naveau, F.A.; Currah, R.S.; Bachewich, J.** (2002). Phylogeny of *Ajellomyces*, *Polytolypa* and *Spiromastix* (*Onygenaceae*) inferred from rDNA sequence and nonmolecular data. *Stud. Mycol.* 47: 25–35; **Untereiner, W.A.; Scott, J.A.; Naveau, F.A.; Sigler, L.; Bachewich, J.; Angus, A.** (2004). The *Ajellomycetaceae*, a new family of vertebrate-associated Onygenales. *Mycologia* 96: 812–821.

Ophioparmaceae R.W. Rogers & Hafellner 1988
Lecanorales: Ascomycota

Thallus crustose. Ascomata apothecial, sessile, with a well-developed margin, disc reddish or brown. Interascal tissue of thick, sparingly branched, paraphyses with swollen apices. Asci with a well-developed J+ apical cap, without an ocular chamber, with an outer J+ gelatinized layer. Ascospores hyaline, elongate, transversely septate, without a sheath, coiled within the ascus. Anamorph pycnidial.

Significant Genera: *Ophioparma*.

Ophioparma ventosa, Argyll, Scotland; thallus and ascomata

Distribution: Boreal areas.

Economic Significance: None is known.

Ecology: Lichenized with green algae.

Notes: The family appears to be related to the *Umbilicariaceae*, as part of a basal group of the *Lecanoro-*

mycetes, but more research is needed. Links with the *Haematommataceae* have been assumed in the past, doubtless due to the red coloration of the ascomata in many species.

Ophioparma lapponica, Greenland; thallus and ascomata

References: **Kalb, K.; Staiger, B.** (1995). Rindenbewohnende Arten der Flechtengattung *Ophioparma* in Amerika. *Biblthca Lichenol.* 58: 191–198; **Printzen, C.; Rambold, G.** (1996). Two corticolous species of *Ophioparma* (*Lecanorales*) from East Asia. *Herzogia* 12: 23–29; **Rogers, R.W.; Hafellner, J.** (1988). *Haematomma* and *Ophioparma*: two superficially similar genera of lichenized fungi. *Lichenologist* 20: 167–174; **Skult, H.** (1997). Notes on the chemical and morphological variation of the lichen *Ophioparma ventosa* in East Fennoscandia. *Ann. bot. fenn.* 34: 291–297; **Wedin, M.; Wiklund, E.; Crewe, A.; Döring, H.; Ekman, S.; Nyberg, A.; Schmitt, I.; Lumbsch, H.T.** (2005). Phylogenetic relationships of *Lecanoromycetes* (*Ascomycota*) as revealed by sequences of mtSSU and nLSU rDNA sequence data. *Mycol. Res.* 109: 159–172.

Ophiostomataceae Nannf. 1932
Ophiostomatales: Ascomycota

Stromata absent. Ascomata perithecial, rarely cleistothecial, often aggregated, dark, fairly thin-walled, usually long-necked, lacking periphyses, with divergent ostiolar setae. Interascal tissue absent. Asci ? usually formed in chains from a fertile layer lining the ascomatal cavity, ± saccate, very thin-walled, evanescent, 8-spored. Ascospores small, hyaline, varied in shape, aseptate, often with eccentric wall thickening or sheaths. Anamorphs hyphomycetous, varied, often prominent, sometimes synnematous. Synnemata where present variable in form, the stalks often elongate, ranging from almost hyaline to ± black, freely branching in the fertile apical region to form lateral and terminal conidiogenous cells. Conidiogenous cells flask-shaped, sometimes elongate, with individual species apparently exhibiting percurrent, annellidic and sympodial proliferation. Conidia formed in slimy masses, small, hyaline, usually aseptate.

Significant Genera: *Ceratocystiopsis, Grosmannia, Ophiostoma*. Anamorphs: *Hyalorhinocladiella, Leptographium, Pesotum, Sporothrix*.

Ophiostoma novo-ulmi; ascomata formed from mating strains

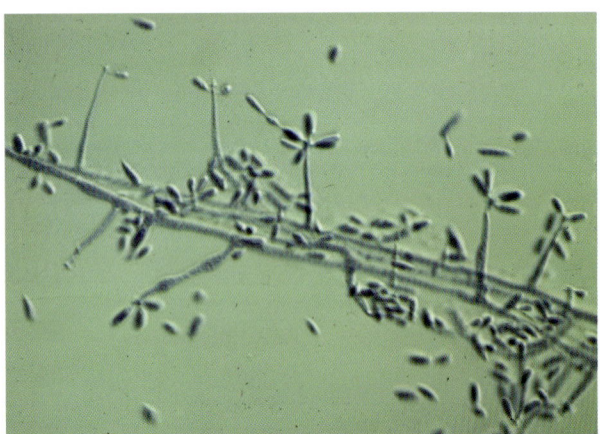

Sporothrix sp.; conidiophores and conidia

Distribution: Widespread, especially in temperate regions.

Economic Significance: Many species are important pathogens of conifers and broadleaved trees, causing sapstain and vascular wilt.

Ecology: At least most species are transmitted via insect vectors, with mycelial colonies developing in bark-boring beetle galleries and spreading through the vascular system; they are highly specific to their beetle vectors. In some cases spores and mycelium are transmitted in specialized organs, termed mycangia.

Notes: *Ophiostoma* species are the causal agents of Dutch elm disease, a devastating malady of *Ulmus* species that has significantly modified the entire landscape in the UK and other affected areas. Research on biological species concepts has identified at least three separate segregates, responsible for different historical epidemics. A few species of *Ophiostoma* and related *Sporothrix* anamorphs are systemic pathogens of humans.

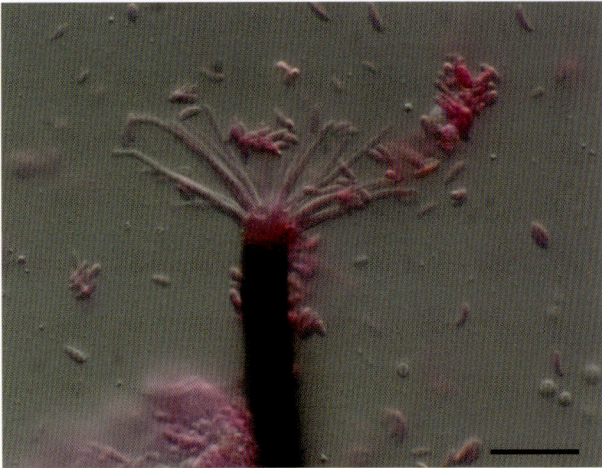

Ophiostoma novo-ulmi; ostiole with coronal setae

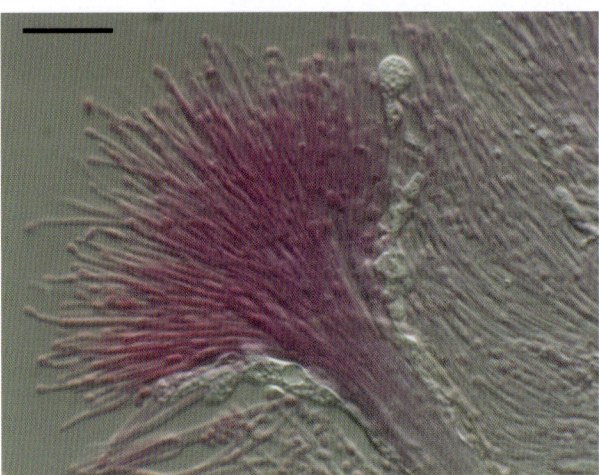

Ophiostoma novo-ulmi; synnematous anamorph

References: **Bates, M.R.; Buck, K.W.; Brasier, C.M.** (1993). Molecular relationships between *Ophiostoma ulmi* and the NAN and EAN races of *O. novo-ulmi* determined by restriction fragment length polymorphisms of nuclear DNA. *Mycol. Res.* 97: 449–455; **Benade, E.; Wingfield, M.J.; Van Wyk, P.S.** (1997). Conidium development in *Sporothrix* anamorphs of *Ophiostoma*. *Mycol. Res.* 101: 1108–1112; **Blackwell, M.; Jones, K.** (1997). Taxonomic diversity and interactions of insect-associated ascomycetes. *Biodiv. Cons.* 6: 689–699; **Brasier, C.M.** (1991). *Ophiostoma novo-ulmi* sp. nov., causative agent of current Dutch elm disease pandemics. *Mycopathologia* 115: 151–161; **Gorton, C.; Kim, S.H.; Henricot, B.; Webber, J.; Breuil, C.** (2004). Phylogenetic analysis of the bluestain fungus *Ophiostoma minus* based on partial ITS rDNA and β-tubulin gene sequences. *Mycol. Res.* 108: 759–765; **Grylls, B.T.; Seifert, K.A.** (1993). A synoptic key to species of *Ophiostoma, Ceratocystis* and *Ceratocystiopsis*. *Ceratocystis* and *Ophiostoma* Taxonomy, Ecology and Pathogenicity (St. Paul): 261–268; **Harrington, T.C.** (1988). *Leptographium* species, their distributions, hosts and insect vectors. *Leptographium Root Diseases on Conifers* (St. Paul): 1–39; **Hausner, G.; Reid, J.; Klassen, G.R.** (1993). On the phylogeny of *Ophiostoma, Ceratocystis* s.s., and *Microascus*, and relationships within *Ophiostoma* based on partial ribosomal DNA sequences. *Can. J. Bot.* 71: 1249–1265; **Samuels, G.J.** (1993). The case for distinguishing *Ceratocystis* and *Ophiostoma*. *Ceratocystis* and *Ophiostoma* Taxonomy, Ecology and Pathogenicity (St. Paul): 15–20; **Seifert, K.A.; Okada, G.** (1993). *Graphium* anamorphs of *Ophiostoma* species and similar anamorphs of other ascomycetes. *Ceratocystis* and *Ophiostoma* Taxonomy, Ecology and Pathogenic-

ity (St. Paul): 27–41; **Wyk, P.W.J. van; Wingfield, M.J.; Wyk, P.S. van** (1993). Ultrastructure of centrum and ascospore development in selected *Ceratocystis* and *Ophiostoma* species. *Ceratocystis* and *Ophiostoma* Taxonomy, Ecology and Pathogenicity (St. Paul): 133–138; **Zipfel, R.D.; Beer, Z.W. de; Jacobs, K.; Wingfield, B.D.; Wingfield, M.J.** (2006). Multi-gene phylogenies define *Ceratocystiopsis* and *Grosmannia* distinct from *Ophiostoma*. *Stud. Mycol.* 55: 75–97.

Orbiliaceae Nannf. 1932
Orbiliales: Ascomycota

Stromata absent. Ascomata apothecial, small, usually convex, brightly coloured or translucent; excipulum usually of thin-walled isodiametric cells, usually without marginal hairs; hymenium waxy, rarely setose. Interascal tissue of simple paraphyses, usually with knob-like apices. Asci very small, the apex truncate, with J– apical rings, often forked at the base. Ascospores small, hyaline, ovoid to ellipsoidal, often curved or even helical, mostly aseptate, sometimes ornamented. Anamorphs hyphomycetous, very varied in form, hyaline or brightly coloured. Fertile hyphae poorly differentiated, conidiogenous loci single or in clusters, terminal or as small side branches, then proliferating sympodially. Conidia extremely varied in form, cylindrical, ovoid, often elongate, sometimes curved or helical, usually septate, without mucous sheaths or appendages.

Significant Genera: *Orbilia*. Anamorphs: *Anguillospora*, *Arthrobotrys*, *Dactylella*, *Helicoön*, *Trinacrium*.

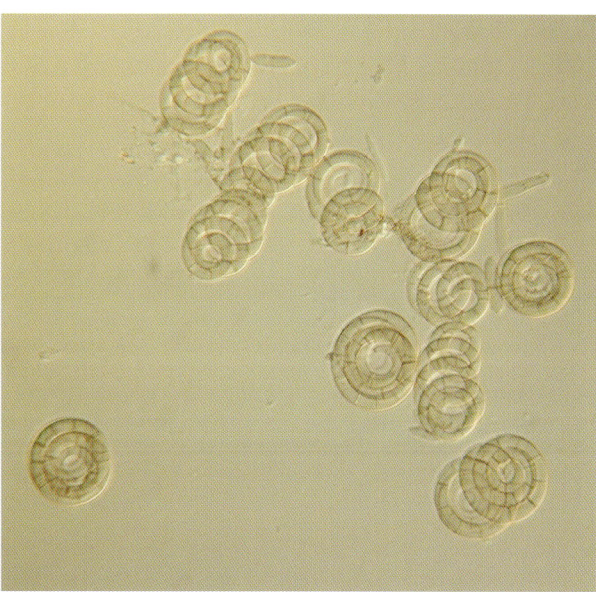
Helicoön pluriseptatum, Surrey, UK; conidia

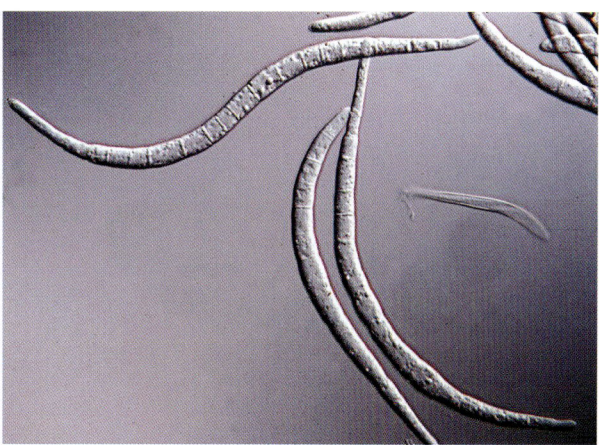
Anguillospora sp.; conidia

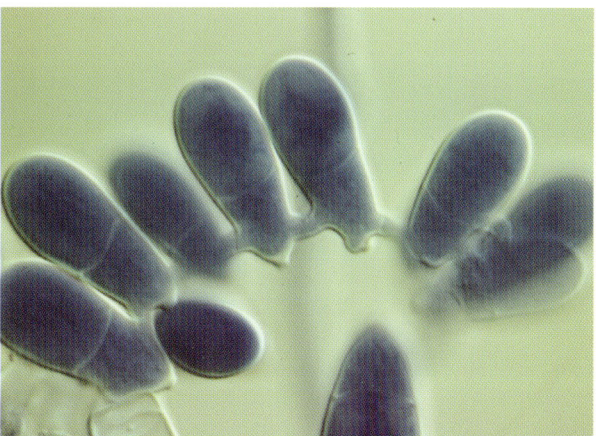
Arthrobotrys superba; conidia formed in chains

Orbilia luteorubella; ascomata on rotten wood

Distribution: Widespread, especially in temperate regions.

Economic Significance: Little is known. There have been studies on the use of some species as biocontrol agents, but products have not reached the market.

Ecology: Saprobic, found especially on wood in both wet and dry habitats. Many species are nematophagous, forming adhesive knobs, nets or ring-like traps in the presence of a food source; perhaps a strategy for life in low-nitrogen environments. There have been reports of associations with algae for a few species, but an established lichen association appears not to exist.

Notes: The *Orbiliaceae* was for long recognized as an outlier amongst the inoperculate discomycetes and molecular studies have confirmed its isolated position within the phylogeny of the *Ascomycota*.

References: **Eriksson, O.E.; Baral, H.-O.; Currah, R.S.; Hansen, K.; Kurtzman, C.P.; Rambold, G.; Laessøe, T. (eds)** (2003). Outline of *Ascomycota* – 2003. *Myconet* 9: 1–89; **Hagedorn, G.; Scholler, M.** (1999). A reevaluation of predatory orbiliaceous fungi. I. Phylogenetic analysis using rDNA sequence data. *Sydowia* 51: 27–48; **Harrington, F.A.; Pfister, D.H.; Potter, D.; Donoghue, M.J.** (1999). Phylogenetic studies within the *Pezizales*. I. 18S rRNA sequence data and classification. *Mycologia* 91: 41–50; **Liou, G.Y.; Tzean, S.S.** (1997). Phylogeny of the genus *Arthobotrys* and allied nematode-trapping fungi based on rDNA sequences. *Mycologia* 89: 876–884; **Pfister, D.H.** (1994). *Orbilia fimicola*, a nematophagous discomycete and its *Arthrobotrys* anamorph. *Mycologia* 86: 451–453; **Pfister, D.H.** (1997). Castor, Pollux and life histories of fungi. *Mycologia* 89: 1–23; **Spooner, B.M.** (1987). *Helotiales* of Australasia: *Geoglossaceae*, *Orbiliaceae*, *Sclerotiniaceae*, *Hyaloscyphaceae*. *Biblthca Mycol.* 116: 711 pp.; **Sugiyama, J.** (1998). Relatedness, phylogeny, and evolution of the fungi. *Mycoscience* 39: 487–511; **Tehler, A.G.; Farris, J.S.; Lipscomb, D.L.; Källersjö, M.** (2000). Phylogenetic analyses of the fungi based on large rDNA data sets. *Mycologia* 92: 459–474; **Webster, J.; Henrici, A.; Spooner, B.** (1998). *Orbilia fimicoloides* sp. nov., the teleomorph of *Dactylella* cf. *oxyspora*. *Mycol. Res.* 102: 99–102.

Pachnocybaceae Oberw. & R. Bauer 1989
Pachnocybales: Basidiomycota

Basidiomata stilboid, stipitate and capitate, smooth, not viscid or gelatinous. Fertile head ± globose, hyaline or pale yellowish, the entire surface composed of closely packed basidia. Stipe dark reddish-brown, composed of closely adherent thick-walled hyphae. Hyphae with simple septal pores, clamp connections absent, fertile hyphae repeatedly branched. Basidia clavate to cylindrical, aseptate, with basal clamp connections and an apical cluster of 4–6 minute sterigmata. Basidiospores ellipsoidal, hyaline or yellowish, thick-walled, smooth.

Significant Genera: *Pachnocybe*.

Distribution: Known from Europe and Canada.

Economic Significance: None is known.

Ecology: Presumably saprobic on decaying wood, detailed studies are not available.

Notes: At one time assumed to be a hyphomycetous anamorph, the type species *Pachnocybe ferruginea* is now known to be a meiotic basidiomycete with possible affinities to the *Platygloeaceae*. More phylogenetic studies are required.

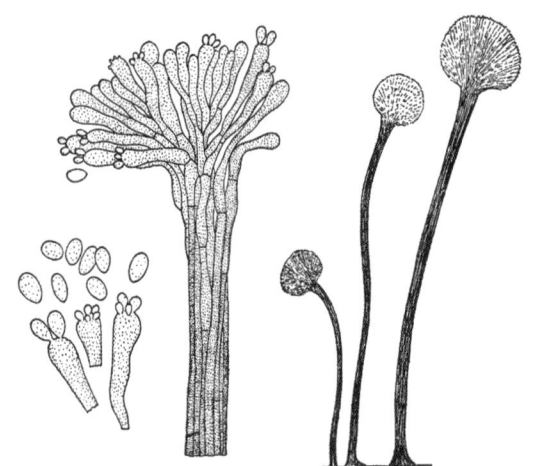

Pachnocybe ferruginea; stipitate basidioma and basidia

References: **Begerow, D.; Bauer, R.; Oberwinkler, F.** (1998). Phylogenetic studies on nuclear large subunit ribosomal DNA sequences of smut fungi and related taxa. *Can. J. Bot.* 75: 2045–2056; **Berres, M.E.; Szabo, L.J.; McLaughlin, D.J.** (1996). Phylogenetic relationships in auriculariaceous basidiomycetes based on 25S ribosomal DNA sequences. *Mycologia* 87: 821–840; **Kleven, N.L.; McLaughlin, D.J.** (1989). A light and electron microscopic study of the developmental cycle in the basidiomycete *Pachnocybe ferruginea*. *Can. J. Bot.* 67: 1336–1348; **Kropp, B.R.; Corden, M.E.** (1986). Morphology and taxonomy of *Pachnocybe ferruginea*. *Mycologia* 78: 334–342; **McLaughlin, D.J.; Doublés, J.C.; Lu, H.** (1990). Mitosis and the phylogeny of the basidiomycete *Pachnocybe ferruginea*. *Mycol. Soc. Amer. Newsl.* 41: 28; **Oberwinkler, F.; Bandoni, R.J.** (1982). A taxonomic survey of the gasteroid, auricularioid Heterobasidiomycetes. *Can. J. Bot.* 60: 1726–1750; **Oberwinkler, F.; Bauer, R.** (1989). The systematics of gasteroid, auricularioid Heterobasidiomycetes. *Sydowia* 41: 224–256.

Pachyascaceae Poelt ex P.M. Kirk, P.F. Cannon & J.C. David 2001
Uncertain position within Lecanoromycetes, Ascomycota

Thallus granular, much reduced. Ascomata apothecial, pulvinate, brownish, the peridium hardly developed; hymenium blueing in iodine. Interascal tissue of irregular anastomosing paraphyses, not swollen at the apices. Asci cylindrical, persistent, with distinct wall layers, not fissitunicate, with an enormously thickened apical cap which blues in iodine and a well-developed ocular chamber at least when immature, 12-spored. Ascospores brown, 1-septate, blueing in iodine, without a sheath. Anamorph unknown.

Significant Genera: *Pachyascus*.

Distribution: Arctic Europe.

Economic Significance: None is known.

Ecology: Lichenized with green algae.

Notes: No molecular data are available, the single genus included appearing distinctive morphologically and without obvious relatives.

References: **Poelt, J.** (1986). Über auf Moosen parasitierende Flechten. *Sydowia* 38: 241–254.

Pacisporaceae C. Walker, Błaszk., A. Schüssler & Schwarzott 2004
Diversisporales: Glomeromycota

Mycorrhizal species forming arbuscules and vesicles in roots (intraradical), anamorphic; the penetrating hyphae producing finely branched haustorial branches (arbuscules) or coils (pelotons) and vesicles. Spores hypogeous or partly epigeous, solitary, with a flexible internal wall, germination accompanied by the formation of a germination shield.

Significant Genera: *Pacispora*.

Distribution: Widespread, although undoubtedly more so than current records suggest.

Economic Significance: Enhancement of phosphorus (P) supply to the 'host' is a characteristic of AM (arbuscular mycorrhizal) infections. As such they have a significant influence, either directly or indirectly, on land plants including ecosystem variability and productivity.

Ecology: As mycorrhizal partners of land plants forming arbuscular mycorrhizas; the hosts of AM fungi are primarily found in P-limited tropical and temperate grasslands.

Notes: The genus *Gerdemannia* and family *Gerdemanniaceae* were introduced almost contemporaneously with *Pacispora*; the last name has priority and a new family name *Pacisporaceae* was therefore required as *Gerdemannia* was rendered illegitimate.

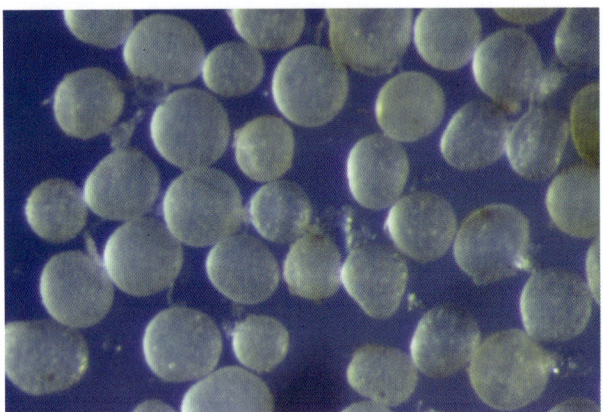
Pacispora scintillans; spores

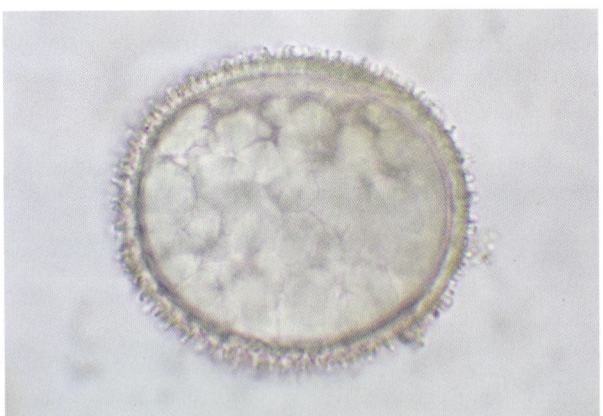
Pacispora chimonobambusae; spore

References: **Azcon-Aguilar, C.; Barea, J.M.** (1996). Arbuscular mycorrhizas and biological control of soil-borne plant pathogens – an overview of the mechanisms involved. *Mycorrhiza* 6: 457–464; **Oehl, F.; Sieverding, E.** (2004). *Pacispora*, a new vesicular arbuscular mycorrhizal fungal genus in the *Glomeromycetes*. *J. Appl. Bot., Angew. Bot.* 78: 72–82; **Redecker, D.** (2005). *Glomeromycota* Arbuscular mycorrhizal fungi and their relative(s). Version 01 July 2005. http://tolweb.org/Glomeromycota/28715/2005.07.01 in The Tree of Life Web Project, http://tolweb.org; **Schüßler, A.; Schwarzott, D.; Walker, C.** (2001). A new fungal phylum, the *Glomeromycota*: phylogeny and evolution. *Mycol. Res.* 105: 1413–1421; **van der Heijden, M.A.G.; Klironomos, J.N.; Ursic, M.; Moutoglis, P.; Streitwolf-Engel, R.; Boller, T.; Wiemken, A.; Sanders, I.R.** (1998). Mycorrhizal fungal diversity determines plant biodiversity, ecosystem variability and productivity. *Nature* Lond. 396: 69–72; **Walker, C.; Błaszkowski, J.; Schwarzott, D.; Schüssler, A.** (2004). *Gerdemannia* gen. nov., a genus separated from *Glomus*, and *Gerdemanniaceae* fam. nov., a new family in the *Glomeromycota*. *Mycol. Res.* 108: 707–718; **Walker, C.; Schüssler, A.** (2004). Nomenclatural clarifications and new taxa in the *Glomeromycota*. *Mycol. Res.* 108: 981–982.

Pannariaceae Tuck. 1872
Lecanorales: Ascomycota

Thallus crustose or foliose, usually dark grey. Ascomata apothecial, sessile, usually brown or black, usually with a well-developed margin composed of isodiametric cells. Interascal tissue of rigid, thick-walled, sparingly branched, paraphyses. Asci rather variable, usually with a well-developed apical cap and a J+ outer gelatinized layer, sometimes with a deeply staining tube-like apical ring. Ascospores usually small, aseptate, hyaline, sometimes ornamented. Anamorphs pycnidial, with bacilliform conidia.

Significant Genera: *Pannaria, Psoroma*.

Distribution: Widespread, especially in south temperate zones.

Psoroma hypnorum; thallus and ascomata

Protopannaria pezizoides; ascomata

Economic Significance: None is known.

Ecology: Lichenized with cyanobacteria, on tree bark, rocks etc., often amongst mosses.

Notes: The *Pannariaceae* comprises a well-defined group within the *Peltigerineae* (*Lecanorales*), clustering with the *Coccocarpiaceae* and *Collemataceae*. The taxonomy and position of *Fuscopannaria* needs further research.

Pannaria rubiginosa, Inverness-shire, Scotland; thalli and ascomata

References: **Ekman, S.; Joergensen, P.M.** (2002). Towards a molecular phylogeny for the lichen family *Pannariaceae* (*Lecanorales*, *Ascomycota*). *Can. J. Bot.* 80: 625–634; **Galloway, D.J.; James, P.W.** (1985). The lichen genus *Psoromidium* Stirton. *Lichenologist* 17: 173–188; **Galloway, D.J.; Jørgensen, P.M.** (1987). Studies in the lichen family *Pannariaceae* II. The genus *Leioderma* Nyl. *Lichenologist* 19: 345–400; **Henssen, A.** (1997). *Santessoniella*, a new cyanophilic genus of lichenized ascomycetes. *In* Tibell, L. & Hedberg, I. (eds), Lichen Studies Dedicated to Rolf Santesson. *Symb. bot. upsal.* 32: 75–93; **Jørgensen, M.** (1994). The lichen genus *Leioderma* Nyl. in Japan. *J. Jap. Bot.* 69: 383–386; **Jørgensen, P.M.** (1999). Studies in the lichen family *Pannariaceae* VIII. Seven new parmelioid lichens from New Zealand. *N.Z. Jl Bot.* 37: 257–268; **Jørgensen, P.M.** (2000). On the sorediate counterparts of the lichen *Fuscopannaria leucosticta*. *Bryologist* 103: 104–107; **Jørgensen, P.M.; Galloway, D.J.** (1989). Studies in the lichen family *Pannariaceae* III. The genus *Fuscoderma*, with additional notes and a revised key to *Leioderma*. *Lichenologist* 21: 295–301; **Jørgensen, P.M.; Wedin, M.** (1999). On *Psoroma* species from the southern hemisphere with cephalodia producing vegetative dispersal units. *Lichenologist* 31: 341–347; **Lumbsch, H.T.; Kothe, H.W.** (1992). Thallus surfaces in *Coccocarpiaceae* and *Pannariaceae* (lichenized *Ascomycetes*) viewed with scanning electron microscopy. *Mycotaxon* 43: 277–282; **Makhija, U.; Adawadkar, B.** (1999). The lichen genus *Parmeliella* (*Pannariaceae*) from the Andaman and Nicobar Islands of India. *Mycotaxon* 71: 323–334; **Øvstedal, D.O.; Smith, R.I.L.** (1993). A new *Pannaria* species from the Antarctic. *Cryptog. Bryol.-Lichénol.* 14: 337–340; **Wedin, M.; Wiklund, E.** (2004). The phylogenetic relationships of *Lecanorales* suborder *Peltigerineae* revisited. *Symb. bot. upsal.* 34: 469–475; **Wiklund, E.; Wedin, M.** (2003). The phylogenetic relationships of the cyanobacterial lichens in the *Lecanorales* suborder *Peltigerineae*. *Cladistics* 19: 419–431.

Papulosaceae Winka & O. Erikss. 2000
Uncertain position within Sordariomycetes, Ascomycota

Stromata absent. Ascomata perithecial, black-walled, deeply immersed in plant tissues, with a long periphysate neck. Interascal tissue composed of simple thin-walled lateral paraphyses. Asci formed in a basal fascicle, cylindrical, fairly thick-walled but not fissitunicate, persistent, with a simple large J+ apical ring, 8-spored. Ascospores ellipsoidal, hyaline, aseptate, thick-walled, verruculose, at first covered in a gelatinous sheath. Anamorph unknown.

Significant Genera: *Papulosa*.

Distribution: Known from the eastern USA.

Economic Significance: None is known.

Ecology: In *Juncus* stems, halophytic and probably saprobic.

Notes: The family appears isolated within the *Sordariomycetes*; preliminary molecular data indicate the *Ophiostomataceae* or the *Magnaporthaceae* as being the closest relative, but the families differ in many morphological and biological features.

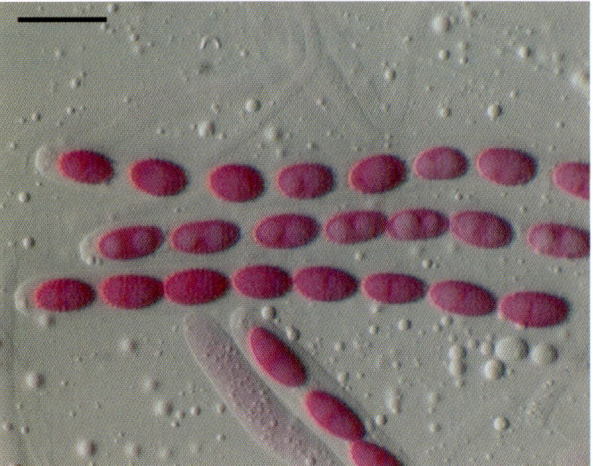

Papulosa amerospora, North Carolina, USA; asci and ascospores

References: **Winka, K.; Eriksson, O.E.** (2000). *Papulosa amerospora* accommodated in a new family (*Papulosaceae*, *Sordariomycetes*, *Ascomycota*) inferred from morphological and molecular data. *Mycoscience* 41: 97–103.

Paraglomeraceae Morton & Redecker 2001
Paraglomerales: Glomeromycota

Mycorrhizal species forming arbuscules and vesicles in roots (intraradical), anamorphic; the penetrating hyphae producing finely branched haustorial branches (arbuscules) or coils (pelotons) and vesicles. Spores hypogeous or partly epigeous, monomorphic and glomeroid (developing blastically from the tip of a sporogenous hypha); spore wall multilayed, with a thick and flexible inner layer (thin bi-layered flexible inner walls and germination orb absent).

Significant Genera: *Paraglomus*.

Distribution: Widespread, although undoubtedly more so than current records suggest.

Economic Significance: Enhancement of phosphorus (P) supply to the 'host' is a characteristic of AM (arbuscular mycorrhizal) infections. As such they have a significant influence, either directly or indirectly, on land plants including ecosystem variability and productivity.

Ecology: As mycorrhizal partners of land plants forming arbuscular mycorrhizas; the hosts of AM fungi are primarily found in P-limited tropical and temperate grasslands.

Notes: Considered (with *Archaeosporaceae*) to form the 'ancestral' lineage within the *Glomeromycota*.

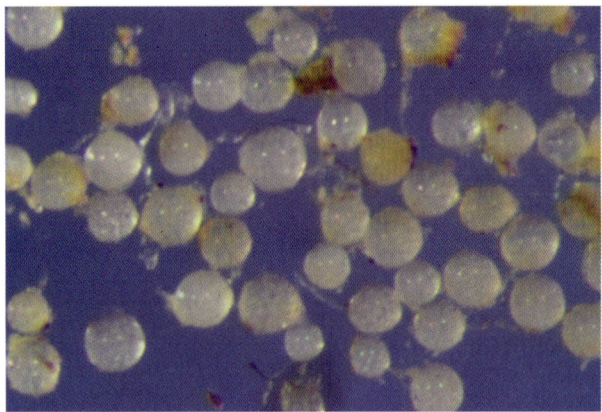

Paraglomus occultum; spores

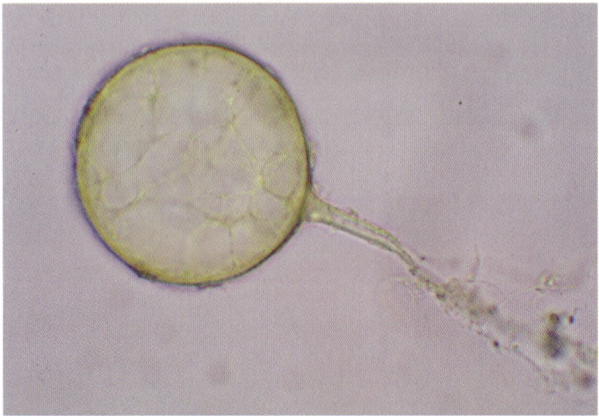

Paraglomus occultum; spore

References: **Azcon-Aguilar, C.; Barea, J.M.** (1996). Arbuscular mycorrhizas and biological control of soil-borne plant pathogens – an overview of the mechanisms involved. *Mycorrhiza* 6: 457–464; **Morton, J.B.; Redecker, D.** (2001). Two new families of *Glomales*, *Archaeosporaceae* and *Paraglomaceae*, with two new genera *Archaeospora* and *Paraglomus*, based on concordant molecular and morphological characters. *Mycologia* 93: 181–195; **Redecker, D.** (2005). *Glomeromycota* Arbuscular mycorrhizal fungi and their relative(s). Version 01 July 2005. http://tolweb.org/Glomeromycota/28715/2005.07.01 in The Tree of Life Web Project, http://tolweb.org; **Schüßler, A.; Schwarzott, D.; Walker, C.** (2001). A new fungal phylum, the *Glomeromycota*: phylogeny and evolution. *Mycol. Res.* 105: 1413–1421; **van der Heijden, M.A.G.; Klironomos, J.N.; Ursic, M.; Moutoglis, P.; Streitwolf-Engel, R.; Boller, T.; Wiemken, A.; Sanders, I.R.** (1998). Mycorrhizal fungal diversity determines plant biodiversity, ecosystem variability and productivity. *Nature* Lond. 396: 69–72.

Parmeliaceae Zenker 1827
Lecanorales: Ascomycota

Thallus foliose or fruticose, sometimes filamentous, corticate on both surfaces, usually with rhizoids, often brightly pigmented, often producing vegetative propagules of varied types. Ascomata apothecial, sessile, occasionally ± immersed or stalked, sometimes marginal, with a well-developed margin. Interascal tissue of sparingly branched paraphyses, sometimes pigmented, the apices sometimes swollen. Asci elongate or clavate with a well-developed J+ apical cap and a more weakly J+ plug. Ascospores varied, usually small, hyaline, aseptate, with or without a sheath. Pseudocyphellae sometimes present. Anamorphs where present pycnidial, ± immersed in the thallus, conidia varied in form but usually bacilliform or filiform.

Significant Genera: *Bryoria, Cetraria, Hypotrachyna, Menegazzia, Parmelia, Parmotrema, Usnea, Xanthoparmelia*.

Distribution: Cosmopolitan, found in a very wide range of climatic zones.

Usnea cirrosa, Arizona, USA; thallus and ascomata

Economic Significance: Little is known, although the wide range of metabolites formed might suggest that there is potential. Traditionally, species are used to dye wool and some are still used widely in the perfume industry.

Parmelina tiliacea, Salamanca, Spain; thallus and ascomata

Arctocetraria nigricans, Norway; thallus

Evernia prunastri, Vaucluse, France; thallus

Ecology: Lichenized with green algae, mostly growing on tree bark and rocks.

Notes: The *Parmeliaceae* is a very large group of lichenized fungi, with extensive variation in thallus form and chemistry that has led to instability in both generic and family concepts. There has been much recent research into molecular phylogenetics in the family. The data are still too sparse to identify a clear infrafamilial structure, although promising initial results have been obtained through analysis of mitochondrial DNA and there are some indications that the *Cetraria* group forms a well-supported clade.

References: **Adler, M.T.** (1990). An artificial key to the genera of the *Parmeliaceae* (*Lichenes, Ascomycotina*). *Mycotaxon* 38: 331–347; **Awasthi, D.D.** (1989). A key to the macrolichens of India and Nepal. *J. Hattori bot. Lab.* 65: 207–302; **Blanco, O.; Crespo, A.; Divakar, P.K.; Elix, J.A.; Lumbsch, H.T.** (2005). Molecular phylogeny of parmotremoid lichens (*Ascomycota, Parmeliaceae*). *Mycologia* 97: 150–159; **Blanco, O.; Crespo, A.; Ree, R.H.; Lumbsch, H.T.** (2006). Major clades of parmelioid lichens (*Parmeliaceae, Ascomycota*) and the evolution of their morphological and chemical diversity. *Mol. Phylogen. Evol.* 39: 52–69; **Crespo, A.; Blanco, O.; Hawksworth, D.L.** (2001). The potential of mitochondrial DNA for establishing phylogeny and stabilising generic concepts in the parmelioid lichens. *Taxon* 50: 807–819; **Crespo, A.; Bridge, P.D.; Hawksworth, D.L.; Grube, M.; Cubero, O.F.** (1999). Comparison of rRNA genotype frequencies of *Parmelia sulcata* from long established and recolonizing sites following sulphur dioxide amelioration. *Pl. Syst. Evol.* 217: 177–183; **Crespo, A.; Cubero, O.F.** (1998). A molecular approach to the circumscription and evaluation of some genera segregated from *Parmelia s. lat. Lichenologist* 30: 369–380; **Crespo, A.; Gavilán, R.; Elix, J.A.; Gutiérrez, G.** (1999). A comparison of morphological, chemical and molecular characters in some parmelioid genera. *Lichenologist* 31: 451–460; **Divakar, P.K.; Crespo, A.; Blanco, O.; Lumbsch, H.T.** (2006). Phylogenetic significance of morphological characters in the tropical *Hypotrachyna* clade of parmelioid lichens (*Parmeliaceae, Ascomycota*). *Mol. Phylogen. Evol.* 40: 448–458; **Elix, J.A.** (1996). A revision of the lichen genus *Relicina*. *Biblthca Lichenol.* 62: 150 pp.; **Grube, M.; Gutmann, B.; Rios, A. de los; Mattsson, J.-E.; Wedin, M.** (1999). An exceptional group-I intron-like insertion in the SSU rDNA of lichen mycobionts. *Curr. Genet.* 35: 536–541; **Ingold, C.T.** (1998). The basidium in *Ustilaginaceae*. *Mycologist* 12: 161–164; **Kärnefelt, I.; Emanuelsson, K.; Thell, A.** (1998). Anatomy and systematics of usneoid genera in the *Parmeliaceae*. *Nova Hedwigia* 67: 71–92; **Lumbsch, H.T.** (1998). The use of metabolic data in lichenology at the species and subspecific levels. *Lichenologist* 30: 357–367; **Mattsson, J.-E.; Wedin, M.** (1998). Phylogeny of the *Parmeliaceae* – DNA data versus morphological data. *Lichenologist* 30: 463–472; **Mattsson, J.-E.; Wedin, M.** (1999). A re-assessment of the family *Alectoriaceae*. *Lichenologist* 31: 431–440; **Thell, A.** (1999). Group I intron versus ITS sequences in phylogeny of cetrarioid lichens. *Lichenologist* 31: 441–449; **Thell, A.; Feuerer, T.; Kärnefelt, I.; Myllys, L.; Stenroos, S.** (2004). Monophyletic groups within the *Parmeliaceae* identified by ITS rDNA, β-tubulin and GADPH sequences. *Mycol. Prog.* 3: 297–314; **Thell, A.; Mattsson, J.-E.; Kärnefelt, I.** (1995). Lecanoralean ascus types in the lichenized families *Alectoriaceae* and *Parmeliaceae*. *Cryptog. bot.* 5: 120–127.

Parmulariaceae E. Müll. & Arx ex M.E. Barr 1979
Uncertain position within Dothideomycetes, Ascomycota

Superficial mycelium absent. Ascomata superficial or subcuticular, perithecial, developing from a more deeply immersed stromatic layer, sometimes as locules in a compound structure, strongly flattened, discoid or pulvinate, often elongate, opening by irregular disintegration or longitudinal splits; peridium composed of one or few layers of irregular or radiating cells. Interascal tissue of cellular pseudoparaphyses, often pigmented at the apices where they have broken free from the upper ascoma wall. Asci ovoid to clavate, with fissitunicate or rostrate dehiscence, immersed in J– gel. Ascospores hyaline to brown, septate, often with a sheath. Anamorphs poorly known, usually forming in locules within the stroma and probably spermatial.

Significant Genera: *Hysterostomella*, *Parmularia*, *Rhagadolobium*.

Rhagadolobium dicksoniifolium, New Zealand; ascostromata on living fern leaf

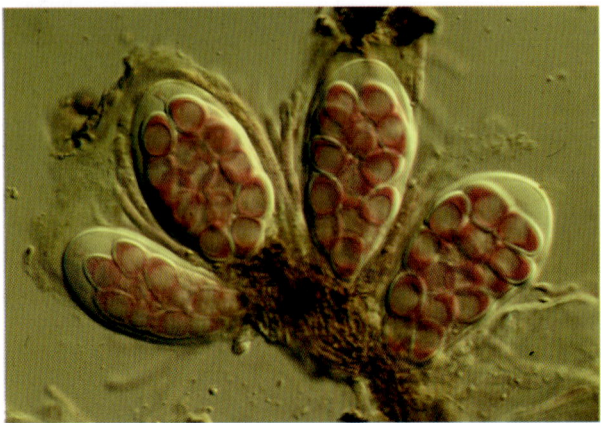
Cycloschizon fimbriatum, Kenya; asci and ascospores

Distribution: Most species have tropical distributions; a few occur in humid temperate zones.

Economic Significance: None is known.

Ecology: Biotrophic on living leaves and stems.

Notes: No molecular data are available. There are some morphological similarities with the *Asterinaceae*, which share the biotrophic mode of nutrition.

Englerodothis oleae, Mauritius; ascostroma on living leaf

References: **Butin, H.; Marmolejo, J.G.** (1988). *Coccodothis sphaeroidea* (Ascomycetes), parasito de *Juniperus flaccida* en el estado de Nuevo Leon, México. *Revta Mex. Micol.* 4: 9–12; **Inácio, C.A.** (2003). *A Monograph of the* Parmulariaceae [Thesis (PhD), University of London]: 417 pp.; **Inácio, C.A.; Cannon, P.F.** (2003). *Viegasella* and *Minterella*, two new genera of Parmulariaceae (Ascomycota), with notes on the species referred to *Schneepia*. *Mycol. Res.* 107: 82–92; **Sivanesan, A.; Sinha, A.R.P.** (1989). *Aldonata*, a new ascomycete genus in the Parmulariaceae. *Mycol. Res.* 92: 246–249; **Vasil'eva, L.N.** (1988). Sistematicheskoe polozhenie *Camarops polysperma* (Mont.) J.B.Miller i rod *Biscogniauxia* O.Kuntze na dal'nem vostoke [The taxonomic position of *Camarops polysperma* (Mont.) J.B. Miller and *Biscogniauxia* O. Kuntze in the Far East]. *Mikol. Fitopatol.* 22: 388–396; **Zhuang, W.Y.** (1988). The genus *Parencoelia* (Leotiaceae, Encoelioideae). *Mycotaxon* 32: 85–95.

Parodiellaceae Theiss. & Syd. ex M.E. Barr 1987
Pleosporales: Ascomycota

Ascomata superficial, perithecial, ± globose to pyriform with a thickened base, black, thick-walled and verrucose, strongly aggregated, with a well-developed apical lysigenous pore; peridium soft, composed of small pseudoparenchymatous cells. Interascal tissue of cellular pseudoparaphyses. Asci cylindrical, fissitunicate, with a well-developed ocular chamber, not blueing in iodine. Ascospores hyaline to brown, septate, striate, without a sheath. Anamorph coelomycetous. Conidiomata thick-walled, black. Conidiogenous cells ± cylindrical, proliferating percurrently. Conidia straight to falcate, hyaline, aseptate.

Significant Genera: *Parodiella*.

Distribution: Primarily tropical.

Economic Significance: Colonies frequently cover most of the leaf surface so must have a significant effect on growth. However, major crop plants are not affected.

Ecology: Biotrophic on leaves, mostly of *Leguminosae*.

Notes: Anamorph–teleomorph links are not fully investigated; here *Ascochytopsis* is assumed to be the anamorph of *Parodiella*. No molecular data are available.

Parodiella hedysari, Kenya; ascomata on leaf

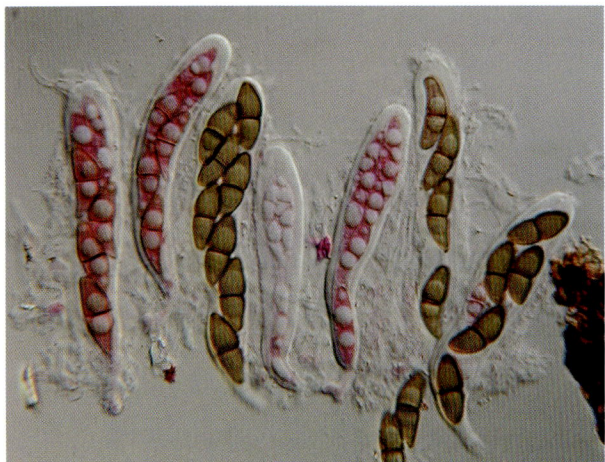

Parodiella hedysari, Kenya; asci and ascospores

References: **Cannon, P.F.** (2001). *Asterina nyanzae; Capnodium salicinum; Chaetothyrium guaraniticum; Cucurbitaria laburni; Dothidea sambuci; Hysterographium fraxini; Micropeltis ugandae; Parodiella hedysari; Tubeufia cerea; Vizella gomphispora.* IMI Descr. Fungi Bact. 141: [22] pp.

Parodiopsidaceae Toro 1952
Uncertain position within Dothideomycetes, Ascomycota

Mycelium superficial, dark, usually setose or hyphopodiate. Ascomata superficial, perithecial, sometimes stalked, globose, small, thin-walled, opening by breakdown of apical cells to form an irregular ostiole; peridium composed of 1–2 layers of dark pseudoparenchymatous cells, sometimes setose. Interascal tissue disintegrating at an early stage of development. Asci saccate, fissitunicate, sometimes J+ blue. Ascospores brown, septate, the septa sometimes thickened, sometimes ornamented, sometimes with a sheath or gelatinous appendages. Anamorphs hyphomycetous. Conidiophores dark, short, simple, usually poroid. Conidia solitary, elongate and often attenuated, thick-walled, septate.

Significant Genera: *Balladyna, Perisporiopsis*. Anamorphs: *Septoidium, Tretospora*.

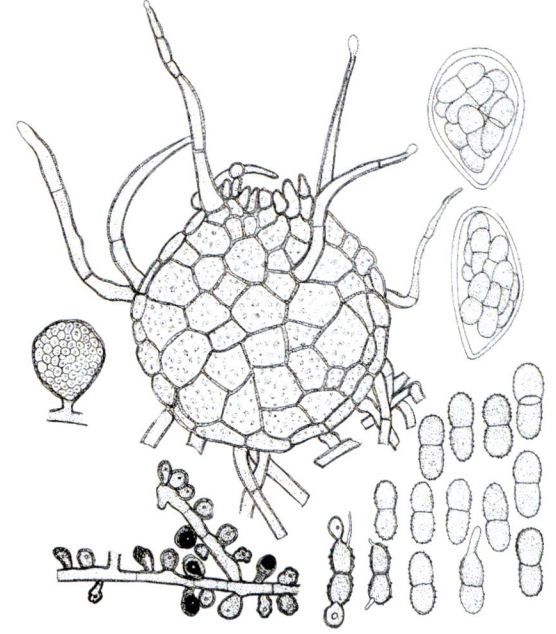

Balladynopsis ebbelsii, Tanzania; setose ascomata, asci, ascospores and appressoria

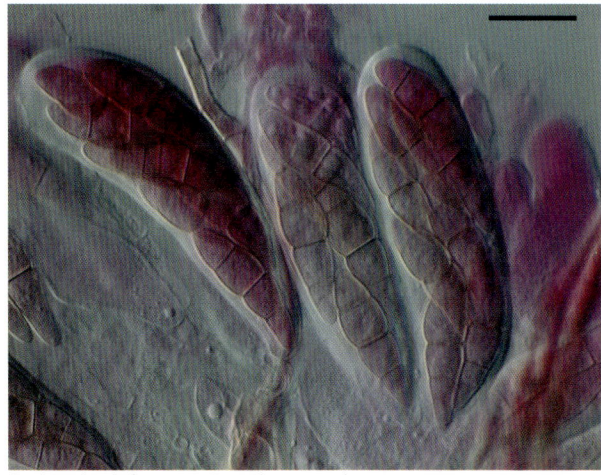

Perisporiopsis lantanae, Brazil; asci and ascospores

Distribution: Most species are tropical.

Economic Significance: Little is known; most species appear to have little deleterious effect on their hosts. *Pilgeriella anacardii* causes damage to cashew trees.

Ecology: Biotrophic on leaves or hyperparasitic on sooty moulds.

Notes: Possibly related to the *Capnodiaceae*, but little information on possible phylogenetic position is available.

References: **Arx, J.A. von; Müller, E.** (1984). Notes on some ascomycetes. *Sydowia* 37: 6–10; **Farr, M.L.** (1986). Notes on mostly neotropical fungi. *Sydowia* 38: 65–74; **Hughes, S.J.** (1993). *Meliolina* and its excluded species. *Mycol. Pap.* 166: 255 pp.; **Rodríguez Hernández, M.** (2000). *Leptomelia* (*Ascomycota, Parodiopsidaceae*): un nuevo registro para Cuba. *Revta Jardín bot. Nac., Univ. Habana* 21: 127–131; **Sivanesan, A.** (1981). *Balladynopsis, Balladynocallia* and *Alina. Mycol. Pap.* 146: 38 pp.

Ecology: Mostly saprobic on bark or wood.

Notes: An isolated family with distinctive morphological features. Many species have long-lived fruit bodies and are therefore frequently encountered during mycological survey.

References: **Bellemère, A.; Malherbe, M.-C.; Hafellner, J.** (1986). Les asques bituniqués du *Lecanidion atratum* (Hedw.) Rabenh. [=*Patellaria atrata* (Hedw.) Fr.] (*Lecanidiaceae*): étude ultrastructurale de la paroi au cours du développement et a la déhiscence. *Cryptog. Mycol.* 7: 113–147; **Kutorga, E.; Hawksworth, D.L.** (1997). A reassessment of the genera referred to the family *Patellariaceae* (*Ascomycota*). *Syst. Ascom.* 15: 1–110; **Liew, E.C.Y.; Aptroot, A.; Hyde, K.D.** (2000). Phylogenetic significance of the pseudoparaphyses in loculoascomycete taxonomy. *Mol. Phylogen. Evol.* 16: 392–402; **Magnes, M.; Scheuer, C.; Söderholm, U.** (1998). A new variety of *Baggea pachyascus* (*Patellariaceae, Ascomycota*) from Finland. *Mycotaxon* 68: 321–325; **Silva-Hanlin, D.M.; Hanlin, R.T.** (1999). Small subunit ribosomal RNA gene phylogeny of several loculoascomycetes and its taxonomic implications. *Mycol. Res.* 103: 153–160.

Patellariaceae Corda 1838
Patellariales: Ascomycota

Ascomata sessile, apothecial, round or elongated, usually closed when young but discoid when mature, with inrolled margins when dry; excipulum composed of carbonized pseudoparenchymatous tissue. Interascal tissue of trabeculate pseudoparaphyses, sometimes later developing true paraphyses, forming an epithecium; hymenial gel sometimes blueing in iodine. Asci thick-walled, cylindrical or clavate, fissitunicate, not blueing in iodine. Ascospores hyaline or brown, septate. Anamorphs coelomycetous.

Significant Genera: *Patellaria, Rhytidhysteron, Tryblidaria*.

Rhytidhysteron rufulum, India; ascomata on dead stem

Distribution: Widespread.

Economic Significance: None is known.

Paxillaceae Lotsy 1907
Boletales: Basidiomycota

Basidiomata pileate, stipitate, often tomentose, with a depressed centre and an involute margin, the stipe varied in position with a veil sometimes present. Clamp connections present. Hymenium poroid or lamellar, gills or pore walls decurrent, narrow with gills often repeatedly forked and sometimes anastomosing, pores where present irregular or angular, with cystidia usually present. Basidia small, clavate to capitate, usually 4-spored. Basidiospores ellipsoidal to fusiform or subreniform, yellow or brownish, smooth to punctate or spinose, thin-walled.

Significant Genera: *Gyrodon, Paxillus*.

Distribution: Widespread, especially in north temperate regions.

Economic Significance: Several *Paxillus* species are toxic, though some people appear to be able to eat them without ill effects.

Ecology: Species are ectomycorrizal, some with a broad range of woody plants and others strongly host-specific.

Notes: *Gyrodon*, with its poroid rather than lamellar hymenium, has been separated into the separate family *Gyrodontaceae*, but that family as originally circumscribed is clearly polyphyletic and its circumscription needs re-evaluation. Several gasteroid genera including *Melanogaster* may well also belong here. Other genera traditionally linked with *Paxillus*, such as the active wood-decaying *Tapinella*, have been shown to be quite separate using molecular characters.

Gyrodon lividus; basidiomata

Paxillus involutus; basidiomata

References: **Becerra, A.; Nouhra, E.; Daniele, G.; Domínguez, L.; McKay, D.** (2005). Ectomycorrhizas of *Cortinarius helodes* and *Gyrodon monticola* with *Alnus acuminata* from Argentina. *Mycorrhiza* 15: 7–15; **Binder, M.; Bresinsky, A.** (2002). Derivation of a polymorphic lineage of gasteromycetes from boletoid ancestors. *Mycologia* 94: 85–98; **Bresinsky, A.; Jarosch, M.; Fischer, M.; Schönberger, I.; Wittmann-Bresinsky, B.** (1999). Phylogenetic relationships within *Paxillus s.l.* (*Basidiomycetes, Boletales*): separation of a southern hemisphere genus. *Pl. Biol.* 1: 327–333; **Bruns, T.D.; Shefferson, R.P.** (2004). Evolutionary studies of ectomycorrhizal fungi: recent advances and future directions. *Can. J. Bot.* 82: 1122–1132; **Claridge, A.W.; Trappe, J.M.; Castellano, M.A.** (2001). Australasian truffle-like fungi. X. *Gymnopaxillus* (*Basidiomycota, Austropaxillaceae*). *Aust. Syst. Bot.* 14: 273–281; **Hahn, C.; Agerer, R.** (1999). [Studies on the *Paxillus involutus* complex]. *Nova Hedwigia* 69: 241–310; **Hönig, K.; Riefler, M.; Kottke, I.** (2000). Survey of *Paxillus involutus* (Batsch) Fr. inoculum and fruitbodies in a nursery by IGS-RLPs and IGS sequences. *Mycorrhiza* 9: 315–322; **Jarosch, M.; Bresinsky, A.** (1999). Speciation and phylogenetic distances within *Paxillus s. str.* (*Basidiomycetes, Boletales*). *Pl. Biol.* 1: 701–705; **Kretzer, A.M.; Bruns, T.D.** (1999). Use of *atp6* in fungal phylogenetics: an example from the *Boletales*. *Mol. Phylogen. Evol.* 13: 483–492.

Peltigeraceae Dumort. 1822
Peltigerales: Ascomycota

Thallus usually large and spreading, foliose, lobed, usually green or dark greyish, matt or shiny, sometimes with cephalodia, corticate only on the upper surface, the lower surface often tomentose and with conspicuous rhizoids. Ascomata apothecial, formed on the upper surface, laminar or marginal, with a thin covering layer which breaks apart at an early stage to expose the hymenium, flat or saddle-shaped, brown or black. Interascal tissue of simple conglutinate paraphyses, sometimes swollen at the tip. Asci fissitunicate with a thickened apex, a J+ apical ring and a more weakly J+ gelatinized outer layer. Ascospores hyaline to brown, elongate, transversely septate, sometimes ornamented. Anamorphs pycnidial, poorly known.

Significant Genera: *Peltigera, Solorina*.

Peltigera horizontalis, Wiltshire, UK; thallus and ascomata

Peltigera membranacea, Devon, UK; thallus and ascomata

Distribution: Widespread, especially in temperate regions but present from the tropics to Antarctica.

Economic Significance: Little is known. Many species are conspicuous and have been used as indicators of ecological continuity.

Ecology: Lichenized with cyanobacteria or green algae (then often with cyanobacterial cephalodia), on bare soil, in grass or on rocks or tree trunks.

Notes: This is one of four families of the *Peltigerales*, which constitute some of the most conspicuous and well-known lichen groups. The *Solorinaceae* was at one time separated from the *Peltigeraceae* using morphological criteria, but that is not strongly supported by molecular evidence and *Solorina* itself has been found to be polyphyletic.

Solorina saccata, Norway; thallus and ascomata

References: **Czeczuga, B.; Upreti, D.K.** (1992). Carotenoids in lichens of the *Lobaria* and *Peltigera* genera from India. *Bull. bot. Surv. India* 32: 80–85; **Eriksson, O.E.; Strand, Å.** (1995). Relationships of the genera *Nephroma*, *Peltigera* and *Solorina* (*Peltigerales*, *Ascomycota*) inferred from 18S rDNA sequences. *Syst. Ascom.* 14: 33–39; **Galloway, D.J.** (1991). Chemical evolution in the order *Peltigerales*: triterpenoids. *Symbiosis* 11: 327–344; **Galloway, D.J.** (1998). The lichen genus *Solorina* Ach. (*Peltigeraceae*, Lichenized *Ascomycotina*) in New Zealand. *Cryptog. Bryol.-Lichénol.* 19: 137–146; **Goffinet, B.; Bayer, R.J.** (1997). Characterization of mycobionts of photomorph pairs in the *Peltigerineae* (Lichenized *Ascomycetes*) based on internal transcribed spacer sequences of the nuclear ribosomal DNA. *Fungal Genetics Biol.* 21: 228–237; **Goward, T.; Goffinet, B.; Vitikainen, O.** (1995). Synopsis of the genus *Peltigera* (lichenized *Ascomycetes*) in British Columbia, with a key to the North American species. *Can. J. Bot.* 73: 91–111; **Hawksworth, D.L.** (1988). The variety of fungal-algal symbioses, their evolutionary significance, and the nature of lichens. *J. Linn. Soc. Bot.* 96: 3–20; **Holtan-Hartwig, J.** (1993). The lichen genus *Peltigera*, exclusive of the *P. canina* group, in Norway. *Sommerfeltia* 15: 1–77; **Miadlikowska, J.; Lutzoni, F.** (2000). Phylogenetic revision of the genus *Peltigera* (lichen-forming *Ascomycota*) based on morphological, chemical, and large subunit nuclear ribosomal DNA data. *Int. J. Pl. Sci.* 161: 925–958; **Miadlikowska, J.; Lutzoni, F.** (2004). Phylogenetic classification of peltigeralean fungi (*Peltigerales*, *Ascomycota*) based on ribosomal RNA small and large subunits. *Am. J. Bot.* 91: 449–464; **Miadlikowska, J.; Lutzoni, F.; Goward, T.; Zoller, S.; Posada, D.** (2003). New approach to an old problem: incorporating signal from gap-rich regions of ITS and rDNA large subunit into phylogenetic analyses to resolve the *Peltigera canina* species complex. *Mycologia* 95: 1181–1203; **Smith, R.I.L.; Øvstedal, D.O.** (1994). *Solorina spongiosa* in Antarctica: an extremely disjunct bipolar lichen. *Lichenologist* 26: 209–213; **Vitikainen, O.** (1994). Taxonomic revision of *Peltigera* (lichenized *Ascomycotina*) in Europe. *Acta bot. fenn.* 152: 1–96; **Wiklund, E.; Wedin, M.** (2003). The phylogenetic relationships of the cyanobacterial lichens in the *Lecanorales* suborder *Peltigerineae*. *Cladistics* 19: 419–431.

Peltulaceae Büdel 1986
Lichinales: Ascomycota

Thallus dark, clavate, peltate or ligulate, sometimes hollow, brown to black, not heavily gelatinized. Ascomata immersed in the thallus, initially ± perithecial in form but becoming ± apothecial, often formed from pycnidia, the peridium often not well-defined. Interascal tissue of simple or anastomosing paraphyses. Asci with a thickened apex when young, at least sometimes functionally bitunicate, often thin-walled when mature, polysporous, without well-defined apical structures, usually with a gelatinized layer. Anamorph pycnidial, immersed in the thallus.

Significant Genera: *Peltula*.

Peltula omphaliza, Norway; thalli

Peltula radicata, Saudi Arabia; ascomata on sandstone

Distribution: Widespread, especially in arid regions.

Economic Significance: None is known.

Ecology: Lichenized with cyanobacteria.

Notes: More research is needed but it appears that the *Peltulaceae* are an ancestral group within the *Lichinales*, with primarily rostrate rather than prototunicate asci as in most species of the *Lichinaceae*.

References: **Büdel, B.** (1987). Taxonomy and biology of the lichen genus *Peltula* Nyl. In Peveling, E. (ed.), Progress and Problems in Lichenology in the Eighties. *Biblthca Lichenol.* 25: 209–217; **Büdel, B.** (1990). Anatomical adaptions to the semiarid/arid environment in the lichen genus *Peltula*. *Biblthca Lichenol.* 38: 47–61; **Büdel, B.** (1995). The lichen genus *Neoheppia*. *Mycotaxon* 54: 137–145; **Filson, R.B.** (1988). The lichen genera *Heppia* and *Peltula* in Australia. *Muelleria* 6: 495–517; **Reeb, V.; Lutzoni, F.; Roux, C.** (2004). Contribution of *RPB2* to multilocus phylogenetic studies of the euascomycetes (*Pezizomycotina*, *Fungi*) with special emphasis on the lichen-forming *Acarosporaceae* and evolution of polyspory. *Mol. Phylogen. Evol.* 32: 1036–1060; **Upreti, D.K.; Büdel, B.** (1990). The lichen genera *Heppia* and *Peltula* in India. *J. Hattori bot. Lab.* 68: 279–284.

Peniophora quercina; basidiomata

Peniophora rufa; basidiomata

Peniophoraceae Lotsy 1907
Russulales: Basidiomycota

Basidiomata pulvinate to resupinate, effuse or reflexed, annual or perennial, waxy to leathery. Hyphal system monomitic, clamp connections usually present, with thin-walled refractile or encrusted cystidia. Hymenial surface almost always smooth. Basidia clavate, normally with 4 sterigmata and a basal clamp connection. Basidiospores large, ellipsoidal to allantoid, ± hyaline or pinkish, smooth, not staining in iodine.

Significant Genera: *Peniophora*.

Distribution: Cosmopolitan.

Economic Significance: Some *Peniophora* species may attack living trees, but economically significant damage is rare.

Ecology: Most species are saprobic, causing rots of standing and fallen wood. Many species are adapted to xeric conditions, being able to grow on rapidly drying dead attached branches.

Notes: Molecular studies suggest a wider definition for the family, including genera such as *Scytinostroma* and *Vararia*, but more research is needed.

References: **Binder, M.; Hibbett, D.S.; Larsson, K.-H.; Larsson, E.; Langer, E.; Langer, G.** (2005). The phylogenetic distribution of resupinate forms across the major clades of mushroom-forming fungi (*Homobasidiomycetes*). *Syst. Biodiv.* 3: 113–157; **Ginns, J.** (1998). Genera of the North American *Corticiaceae* sensu lato. *Mycologia* 90: 1–35; **Hallenberg, N.; Larsson, E.; Mahlapuu, M.** (1996). Phylogenetic studies in *Peniophora*. *Mycol. Res.* 100: 179–187; **Hjortstam, K.; Ryvarden, L.** (2004). Some new tropical genera and species of corticioid fungi (*Basidiomycotina, Aphyllophorales*). *Syn. Fung.* (Oslo) 18: 20–32; **Küffer, N.; Senn-Irlet, B.** (2005). Diversity and ecology of wood-inhabiting aphyllophoroid basidiomycetes on fallen woody debris in various forest types in Switzerland. *Mycol. Prog.* 4: 77–86; **Larsson, E.; Larsson, K.-H.** (2003). Phylogenetic relationships of russuloid basidiomycetes with emphasis on aphyllophoralean taxa. *Mycologia* 95: 1037–1065; **Larsson, K.-H.; Larsson, E.; Koljalg, U.** (2004). High phylogenetic diversity among corticioid homobasidiomycetes. *Mycol. Res.* 108: 983–1002; **Wu, S.H.; Chen, Z.C.** (1993). The genus *Duportella* Pat. (*Corticiaceae* s.l., *Basidiomycotina*) in Taiwan. *Bull. natn. Mus. Nat. Sci.* Taiwan 4: 101–111; **Yurchenko, E.O.** (2000). Key to the genus *Peniophora* (*Corticiaceae* s.l., *Basidiomycetes*) of Belorussia. *Mikol. Fitopatol.* 34: 37–41.

Pertusariaceae Körb. ex Körb. 1855
Pertusariales: Ascomycota

Thallus crustose, well-developed, often warted or fissured, usually greyish, the upper surface not corticate; cephalodia absent. Ascomata sessile or slightly sunken, apothecial or almost closed but with the hymenium usually exposed from an early stage, often with a very well-developed thalline margin. Interascal tissue of narrow, densely branched and anastomosed, pseudoparaphyses, the hymenium staining blue in iodine. Asci fissitunicate, short, widely cylindrical with a thick multilayered usually J+ wall, the apex often more strongly thickened, releasing spores through a ± vertical split, 2- to 8-spored. Ascospores large, aseptate, smooth, the wall often very thick and distinctly multilayered. Anamorphs pycnidial, producing filiform or bacillar conidia.

Significant Genera: *Pertusaria*.

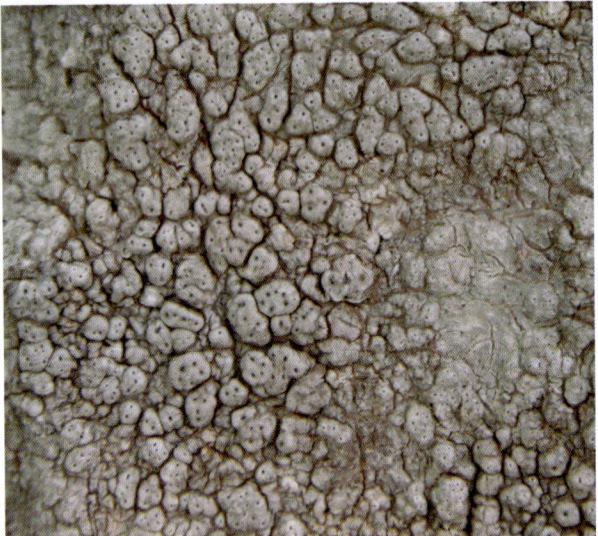
Pertusaria pertusa, Salamanca, Spain; thallus and ascomata

Pertusaria hemisphaerica, Salamanca, Spain; thallus

Distribution: Very widespread.

Economic Significance: None is known.

Ecology: Lichenized with green algae, on acid rocks and tree trunks and similar substrata.

Pertusaria tetrathalamia; thallus and ascomata

Notes: Species of the *Pertusariaceae* produce a particularly wide range of complex lichen metabolites and their presence correlates well with molecular data. The family appears natural and well circumscribed, now *Ochrolechia* has been separated into its own family.

References: **Archer, A.W.** (1997). The Lichen Genus *Pertusaria* in Australia. *Biblthca Lichenol.* 69: 249 pp.; **Lumbsch, H.T.; Dickhäuser, A.; Feige, G.B.** (1995). Systematic studies in the *Pertusariales* III. Taxonomic position of *Thamnochrolechia* (lichenized Ascomycetes). *Biblthca Lichenol.* 57: 355–361; **Lumbsch, H.T.; Nash, T.H.; Messuti, M.I.** (1999). A revision of *Pertusaria* species with hyaline ascospores in southwestern North America (*Pertusariales, Ascomycotina*). *Bryologist* 102: 215–239; **Lumbsch, H.T.; Schmitt, I.** (2001). Molecular data suggest that the lichen genus *Pertusaria* is not monophyletic. *Lichenologist* 33: 161–170; **Lumbsch, H.T.; Schmitt, I.; Palice, Z.; Wiklund, E.; Ekman, S.; Wedin, M.** (2004). Supraordinal phylogenetic relationships of *Lecanoromycetes* based on a Bayesian analysis of combined nuclear and mitochondrial sequences. *Mol. Phylogen. Evol.* 31: 822–832; **Schmitt, I.; Lumbsch, H.T.** (2004). Molecular phylogeny of the *Pertusariaceae* supports secondary chemistry as an important systematic character set in lichen-forming ascomycetes. *Mol. Phylogen. Evol.* 33: 43–55; **Schmitt, I.; Martín, M.P.; Türk, R.; Lumbsch, H.T.** (2003). Phylogenetic position of the genera *Varicellaria, Melanaria* and *Thamnochrolechia* (*Pertusariales*). *Biblthca Lichenol.* 86: 147–154; **Schmitt, I.; Yamamoto, Y.; Lumbsch, H.T.** (2006). Phylogeny of *Pertusariales* (*Ascomycotina*): resurrection of *Ochrolechiaceae* and new circumscription of *Megasporaceae*. *J. Hattori bot. Lab.* 100: 753–764; **Wedin, M.; Wiklund, E.; Crewe, A.; Döring, H.; Ekman, S.; Nyberg, A.; Schmitt, I.; Lumbsch, H.T.** (2005). Phylogenetic relationships of *Lecanoromycetes* (*Ascomycota*) as revealed by sequences of mtSSU and nLSU rDNA sequence data. *Mycol. Res.* 109: 159–172.

Pezizaceae Dumort. 1829
Pezizales: Ascomycota

Ascomata epigeous or hypogeous, large, apothecial or cleistothecial, formed from a single ascogonium or ascogonial coil, some shade of brown, violaceous, yellow etc. (without carotenoid pigments), usually sessile, generally flat or cupulate, without setose hairs; if cleistothecial, ascomata solid or chambered, hypogeous or emergent. Interascal tissue of simple or moniliform paraphyses, often swollen and pigmented at the apex, mostly absent in cleistothecial taxa. Asci cylindrical (sometimes globose in cleistothecial taxa), persistent, operculate, the wall blueing in iodine at least at the tip, ascal pore simple. Ascospores ellipsoidal to globose, hyaline or pale brown, smooth or ornamented, often guttulate, uninucleate. Anamorphs hyphomycetous, hyaline to brightly coloured, only known for a few genera; fertile hyphae wide and thin-walled, with a long stipe, dichotomously branched at the apex, with extensive sympodial proliferation, conidia hyaline, aseptate, thin-walled, usually formed on well-developed denticles.

Significant Genera: *Peziza, Plicaria*. Anamorphs: *Chromelosporium, Oedocephalum*.

Distribution: Widespread in temperate regions, a few species with tropical distributions.

Economic Significance: Little is known. Spores of some species survive heat sterilization and can cause problems in mushroom cultivation.

Ecology: Saprobic or ectomycorrhizal, on or in soil, leafmould etc., some species associated with dung or fire sites.

Notes: The *Pezizaceae* appears to be ± monophyletic, but traditional genera are poorly supported by molecular data, especially *Peziza* itself. The hypogeous habit seems to have evolved on at least three occasions. There are some indications that the family *Pezizaceae* is not closely related to other families currently accepted in the *Pezizales*, but more molecular data are needed.

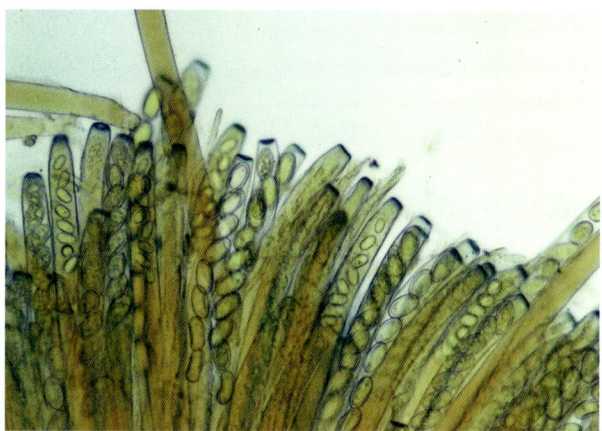

Peziza vesiculosa, Surrey, UK; asci in iodine showing opercula

Peziza cerea, Scotland; ascomata

Plicaria leiocarpa, Scotland; ascomata

Peziza sp., southern Chile; ascomata

References: **Bougher, N.L.; Lebel, T.** (2001). Sequestrate (trufflelike) fungi of Australia and New Zealand. *Aust. Syst. Bot.* 14: 439–484; **Doveri, F.; Cacialli, G.; Caroti, V.** (2000). [A guide to the identification of fimicolous *Pezizales* in Italy. Contribution to the study of fimicolous fungi – XXXII]. *Docums Mycol.* 30: 3–98; **Hansen, K.; Læssøe, T.; Pfister, D.H.** (2001). Phylogenetics of the *Pezizaceae*, with an emphasis on *Peziza*. *Mycologia* 93: 958–990; **Hansen, K.; Læssøe, T.; Pfister, D.H.** (2002). Phylogenetic diversity in the core group of *Peziza* inferred from ITS sequences and morphology. *Mycol. Res.* 106: 879–902; **Hansen, K.; LoBuglio, K.F.; Pfister, D.H.** (2005). Evolutionary relationships of the cup-fungus genus *Peziza* and *Pezizaceae* inferred from multiple nuclear genes: RPB2, β-tubulin and LSU rDNA. *Mol. Phylogen. Evol.* 36: 1–23; **Harrington, F.A.; Pfister, D.H.; Potter, D.; Donoghue, M.J.** (1999). Phylogenetic studies within the *Pezizales*. I. 18S rRNA sequence data and classification. *Mycologia* 91: 41–50;

Kimbrough, J.W.; Wu, C.-g.; Gibson, J.L. (1991). Ultrastructural evidence for a phylogenetic linkage of the truffle genus *Hydnobolites* to the *Pezizaceae* (Pezizales, Ascomycetes). *Bot. Gaz.* 152: 408–420; **Landvik, S.; Egger, K.N.; Schumacher, T.** (1997). Towards a subordinal classification of the *Pezizales* (Ascomycota): phylogenetic analyses of SSU rDNA sequences. *Nordic Jl Bot.* 17: 403–418; **Maia, L.C.; Yano, A.M.; Kimbrough, J.W.** (1996). Species of *Ascomycota* forming ectomycorrhizae. *Mycotaxon* 57: 371–390; **Moravec, J.** (1987). A taxonomic revision of the genus *Marcelleina*. *Mycotaxon* 30: 473–499; **Norman, J.E.; Egger, K.N.** (1997). Phylogeny of the genus *Plicaria* and its relationship to *Peziza* inferred from ribosomal DNA sequence analysis. *Mycologia* 88: 986–995; **Norman, J.E.; Egger, K.N.** (1999). Molecular phylogenetic analysis of *Peziza* and related genera. *Mycologia* 91: 820–829; **Tehler, A.; Little, D.P.; Farris, J.S.** (2003). The full-length phylogenetic tree from 1551 ribosomal sequences of chitinous fungi, *Fungi. Mycol. Res.* 107: 901–916.

Phacidiaceae Fr. 1849
Helotiales: Ascomycota

Stroma black, immersed (usually subcuticular), uniloculate, usually circular. Ascomata apothecial, without a separate wall, developing from central stromatal cells, opening by radial or longitudinal splits. Interascal tissue of simple, basally branched and anastomosing, apical paraphyses. Asci ± cylindrical, thin-walled, with a J+ apical ring. Ascospores hyaline, aseptate, without a sheath. Anamorphs coelomycetous, stromatic, locules immersed within host tissue. Conidiogenous cells cylindrical, proliferating percurrently. Conidia hyaline, cylindrical, usually with a gelatinous appendage.

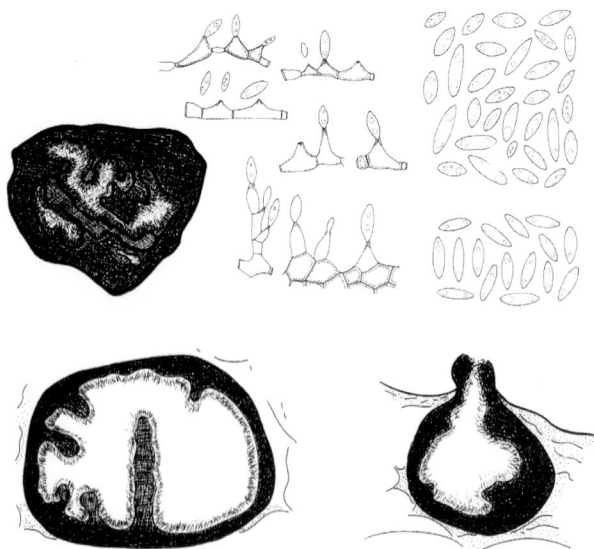

Phacidium coniferarum; conidiomata, conidiogenous cells and conidia

Significant Genera: *Phacidium*. Anamorph: *Ceuthospora*.

Distribution: Most species are found in temperate zones.

Economic Significance: A few species are minor pathogens of economically important forest trees.

Ecology: Parasitic or saprobic on leaves, especially of conifers.

Notes: Only limited molecular data are available, but the family may well be polyphyletic. Its delimitation is still uncertain.

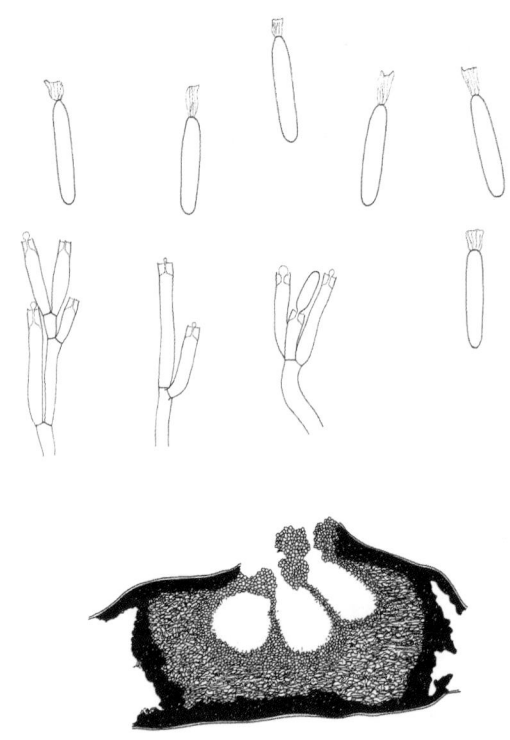

Ceuthospora lauri; conidioma, conidiogenous cells and conidia

References: **Carris, L.M.** (1990). Cranberry black rot fungi: *Allantophomopsis cytisporea* and *Allantophomopsis lycopodina*. *Can. J. Bot.* 68: 2283–2291; **DiCosmo, F.; Nag Raj, T.R.; Kendrick, W.B.** (1984). A revision of the *Phacidiaceae* and related anamorphs. *Mycotaxon* 21: 1–234; **Gernandt, D.S.; Platt, J.L.; Stone, J.K.; Spatafora, J.W.; Holst-Jensen, A.; Hamelin, R.C.; Kohn, L.M.** (2001). Phylogenetics of *Helotiales* and *Rhytismatales* based on partial small subunit nuclear ribosomal DNA sequences. *Mycologia* 93: 915–933; **Roll-Hansen, F.** (1987). *Phacidium infestans* and *Ph. abietis*. Hosts, especially *Abies* species in Norwegian nurseries. *Eur. J. For. Path.* 17: 311–315; **Roll-Hansen, F.** (1989). *Phacidium infestans*. A literature review. *Eur. J. For. Path.* 19: 237–250.

Phaeochoraceae K.D. Hyde, P.F. Cannon & M.E. Barr 1997
Phyllachorales: Ascomycota

Stromata immersed, sometimes conspicuously blackened, 1- to several-loculate, the ostiole not periphysate. Ascomata basically perithecial but often fusing to form a single cavity with columnar peridial elements. Interascal tissue of wide, thin-walled, evanescent paraphyses. Asci saccate or fusiform, long-stalked, evanescent. Ascospores aseptate, brown, sometimes with appendages, sometimes striate. Anamorphs coelomycetous, probably spermatial in function.

Significant Genera: *Phaeochora, Serenomyces*.

Distribution: Widespread, where the host family occurs.

Economic Significance: A few species cause cosmetic damage to ornamental palm species.

Ecology: Biotrophic on leaves of palms.

Notes: No molecular data are available and the placement of this family in the *Phyllachorales* is provisional.

Phaeochora steinheilii, Sicily; ascomata on living palm leaf

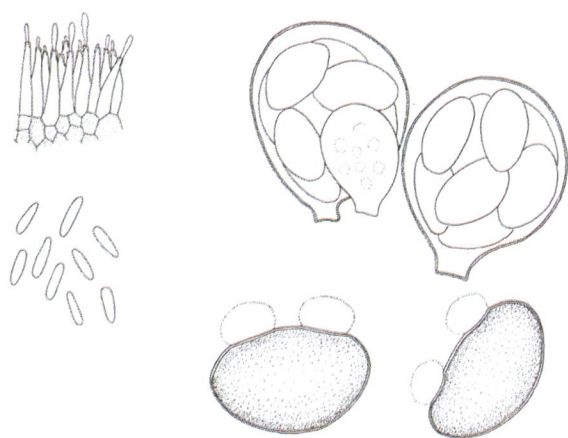

Phaeochora steinheilii; asci, ascospores and anamorph

References: **Barr, M.E.; Ohr, H.D.; Murphy, M.K.** (1989). The genus *Serenomyces* on Palms. *Mycologia* 81: 47–51; **Carrai, C.; D'Agliano, G.** (1992). *Phaeochora steinheilii* agente delle macchie di catrame su *Chamaerops*. *Inftore fitopatol.* 42: 17–20; **Hyde, K.D.; Cannon, P.F.** (1999). Fungi Causing Tar Spots on Palms. *Mycol. Pap.* 175: 114 pp.; **Hyde, K.D.; Cannon, P.F.; Barr, M.E.** (1997). *Phaeochoraceae*, a new ascomycete family from palms. *Syst. Ascom.* 15: 117–120.

Phaeosphaeriaceae M.E. Barr 1979
Pleosporales: Ascomycota

Ascomata perithecial, immersed to erumpent, gregarious or occasionally as locules in small stromata, ± globose, with a well-developed lysigenous, often periphysate, ostiole; peridium soft, composed of relatively small, thin-walled, pseudoparenchymatous cells. Interascal tissue sparse, of narrow cellular pseudoparaphyses. Asci cylindrical, fissitunicate, with a well-developed ocular chamber, not blueing in iodine. Ascospores brown, transversely septate, constricted only at the primary septum, occasionally muriform, sometimes with a sheath. Anamorphs coelomycetous, pycnidial in structure. Conidiogenous cells short, doliiform, annellidic when proliferating, formed from the inner wall cells. Conidia elongate, hyaline or pigmented, usually multiseptate.

Significant Genera: *Nodulosphaeria, Phaeosphaeria*. Anamorphs: *Phaeoseptoria, Stagonospora, Coniothyrium*.

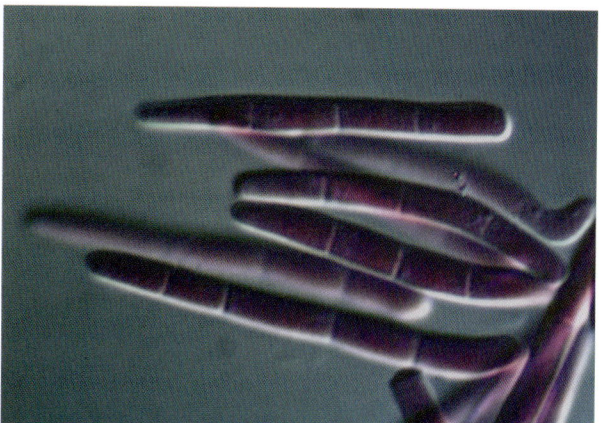

Phaeosphaeria avenaria, UK; conidia

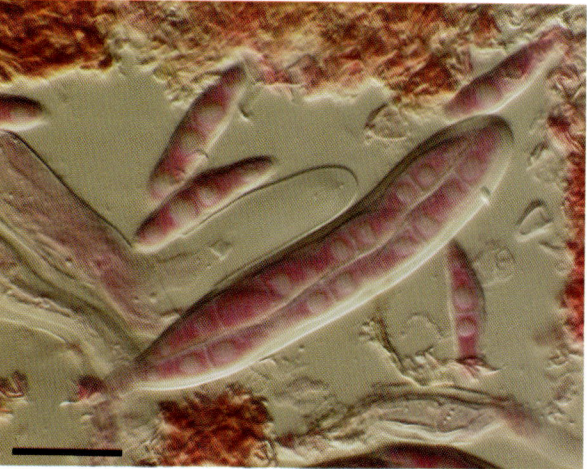

Phaeosphaeria macrosporidium, UK; asci and ascospores

Distribution: Cosmopolitan.

Economic Significance: A number of species are important seed-borne pathogens of cereals.

Ecology: Necrotrophic or saprobic on a wide range of plants, especially monocotyledons; a few species parasitic on other fungi.

Notes: Some outlying genera probably belong elsewhere and not all studies separate Phaeosphaeria and its close allies well from the Leptosphaeriaceae.

References: **Barr, M.E.** (1992). Additions to and notes on the Phaeosphaeriaceae (Pleosporales, Loculoascomycetes). Mycotaxon 43: 371–400; **Boise, J.R.** (1989). On Hadrospora, a new genus in the Phaeosphaeriaceae, and Byssothecium alpestris in the Dacampiaceae. Mem. N. Y. bot. Gdn 49: 308–310; **Câmara, M.P.S.; Palm, M.E.; Berkum, P. van; O'Neill, N.R.** (2002). Molecular phylogeny of Leptosphaeria and Phaeosphaeria. Mycologia 94: 630–640; **Câmara, M.P.S.; Palm, M.E.; Berkum, P. van; Stewart, E.L.** (2001). Systematics of Paraphaeosphaeria: a molecular and morphological approach. Mycol. Res. 105: 41–56; **Câmara, M.P.S.; Ramaley, A.W.; Castlebury, L.A.; Palm, M.E.** (2003). Neophaeosphaeria and Phaeosphaeriopsis, segregates of Paraphaeosphaeria. Mycol. Res. 107: 516–522; **Caten, C.E.** (1999). Molecular genetics of Stagonospora and Septoria. Septoria on Cereals, A Study of Pathosystems (Wallingford): 26–43; **Cunfer, B.M.; Ueng, P.P.** (1999). Taxonomy and identification of Septoria and Stagonospora species on small-grain cereals. A. Rev. Phytopath. 37: 267–284; **Khashnobish, A.; Shearer, C.A.** (1996). Phylogenetic relationships in some Leptosphaeria and Phaeosphaeria species. Mycol. Res. 100: 1355–1363; **Kodsueb, R.; Dhanasekraran, V.; Aptroot, A.; Lumyong, S.; McKenzie, E.H.C.; Hyde, K.D.; Jeewon, R.** (2006). The family Pleosporaceae: intergeneric relationships and phylogenetic perspectives based on sequence analyses of partial 28S rDNA. Mycologia 98: 571–583; **Leuchtmann, A.** (1984). Über Phaeosphaeria Miyake und andere bitunicate Ascomyceten mit mehrfach querseptierten Ascosporen. Sydowia 37: 75–194; **Liew, E.C.Y.; Aptroot, A.; Hyde, K.D.** (2000). Phylogenetic significance of the pseudoparaphyses in loculoascomycete taxonomy. Mol. Phylogen. Evol. 16: 392–402; **Ramaley, A.W.** (1997). New Paraphaeosphaeria species and their anamorphs. Mycotaxon 61: 347–358; **Shoemaker, R.A.** (1985). Canadian and some extralimital Nodulosphaeria and Entodesmium species. Can. J. Bot. 62: 2730–2753; **Shoemaker, R.A.; Babcock, C.E.** (1985). Canadian and some extralimital Paraphaeosphaeria species. Can. J. Bot. 63: 1284–1291; **Shoemaker, R.A.; Babcock, C.E.** (1989). Phaeosphaeria. Can. J. Bot. 67: 1500–1599.

Phaeotrichaceae Cain 1956
Pleosporales: Ascomycota

Ascomata perithecial or cleistothecial, black, thin-walled, where ostiolate with a well-developed, non-periphysate, pore, setose; peridium thin, composed of small pseudoparenchymatous cells. Interascal tissue absent or of ± evanescent cellular pseudoparaphyses. Asci saccate and evanescent or cylindrical and fissitunicate. Ascospores dark brown, septate, sometimes fragmenting, usually with terminal germ pores, sometimes with a sheath. Anamorphs unknown.

Significant Genera: Phaeotrichum, Trichodelitschia.

Distribution: Widespread.

Economic Significance: None is known.

Ecology: Saprobic, on herbivore dung.

Notes: An isolated family of Pleosporales, perhaps representing a basal lineage within the Dothideomycetes. The Sporormiaceae seem not to be related, despite similarities in ecology and morphology. More molecular data are needed.

Phaeotrichum cylindrosporum, Arizona, USA; ascomatal hairs

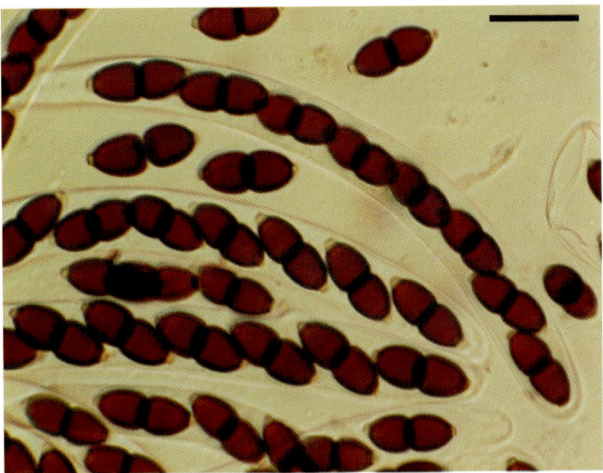
Trichodelitschia bisporula, Sweden; asci and ascospores

References: **Cain, R.F.** (1956). Studies of coprophilous ascomycetes II. Phaeotrichum, a new cleistocarpous genus in a new family, and its relationships. Can. J. Bot. 34: 675–687; **Ebersohn, C.; Eicker, A.** (1992). Trichodelitschia microspora, a new coprophilous species from South Africa. S. Afr. J. Bot. 58: 145–146; **Khan, R.S.; Krug, J.C.** (1994). A synopsis of the coprophilous ascomycetes of East Africa. Proceedings of the Thirteenth Plenary Meeting of AETFAT Zomba, Malawi, 2-11 April 1991, Vol. 1 (Malawi): 755–772; **Kruys, Å.; Eriksson, O.E.; Wedin, M.** (2006). Phylogenetic relationships of coprophilous Pleosporales (Dothideomycetes, Ascomycota), and the classification of some bitunicate taxa of unknown position. Mycol. Res. 110: 527–536; **Lumbsch, H.T.; Lindemuth, R.** (2001). Major lineages of Dothideomycetes (Ascomycota) inferred from SSU and LSU rDNA sequences. Mycol. Res. 105: 901–908.

Phaffomycetaceae Y. Yamada, H. Kawas., Nagats., Mikata & T. Seki 1999
Saccharomycetales: Ascomycota

Mycelium absent. Vegetative cells globose to cylindrical, proliferating by multilateral budding. Asci formed through transformation of diploid cells or by conjugation of independent cells, splitting apart to release the spores, 1- to 4-spored. Ascospores elongate or saturniform. Coenzyme Q-7 system present.

Significant Genera: *Phaffomyces*, *Starmera*.

Distribution: Widespread.

Economic Significance: None is known.

Ecology: Varied, but associated especially with necrotic lesions in cacti.

Notes: Very poorly known; no recent information is available.

References: **Yamada, Y.; Higashi, T.; Ando, S.; Mikata, K.** (1997). The phylogeny of strains of the genus *Pichia* Hansen (*Saccharomycetaceae*) based on the partial sequences of 18S ribosomal RNA: the proposals of *Phaffomyces* and *Starmera*, the new genera. *Bull. Fac. Agric. Shizuoka Univ.* 47: 23–35; **Yamada, Y.; Kawasaki, H.; Nagatsuka, Y.; Mikata, K.; Seki, T.** (1999). The phylogeny of the cactophilic yeasts based on the 18S ribosomal RNA gene sequences: the proposals of *Phaffomyces antillensis* and *Starmera caribaea*, new combinations. *Biosc., Biotechn., Biochem.* 63: 827–832.

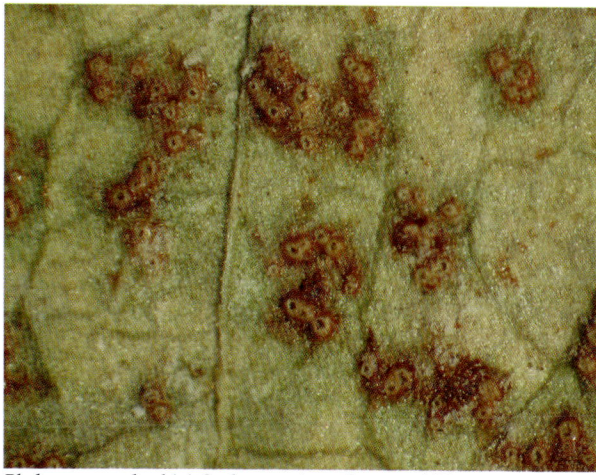
Phakopsora pachyrhizi, Sudan; aecia on leaves of *Vigna*

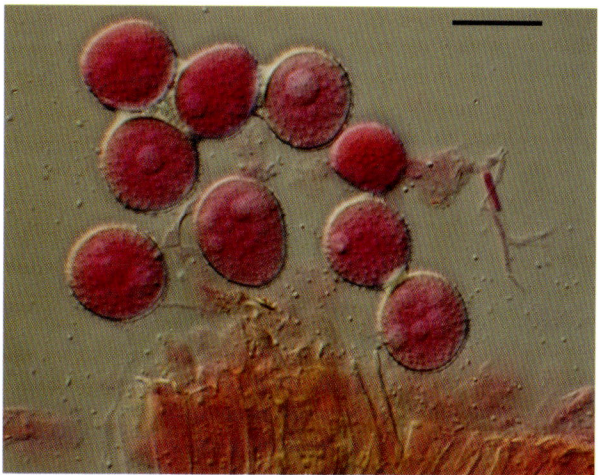
Cerotelium fici, Galápagos Islands; aeciospores

Phakopsoraceae Cummins & Hirats. f. 1983
Uredinales: Basidiomycota

Spermogonia conical to discoid, with discrete walls, subepidermal or subcuticular. Aecia mostly *Aecidium*-like or *Caeoma*-like, usually cupulate, with or without a distinct peridium. Aeciospores usually verrucose, formed in chains and separated by intercalary cells. Uredinia *Malupa*-like. Urediniospores, pedicellate, surrounded by varied sterile hyphal tissue. Telia erumpent or deeply embedded, with two to several layers of teliospores. Teliospores sessile, unicellular, ± adhering to each other and forming crusts, stalked crusts or thread-like columns, usually with a germ pore in each cell, germinating to form external basidia.

Significant Genera: *Cerotelium*, *Phakopsora*.

Distribution: Almost all species are tropical.

Phakopsora vitis; basidiospores

Economic Significance: *Phakopsora* species cause economically important diseases of soybean, grape and cotton, and *Cerotelium* species affect fig and cotton, among other plants.

Ecology: Biotrophic and heteroecious, affecting a wide range of broadleaved plants.

Notes: A poorly known family with little molecular phylogenetic data available; preliminary studies suggest that it is polyphyletic with *Kweilingia* requiring exclusion and perhaps related to the *Pileolariaceae*.

References: **Aime, M.C.** (2006). Towards resolving family-level relationships in rust fungi (*Uredinales*). *Mycoscience* 47: 112–122; **Cummins, G.B.; Hiratsuka, Y.** (2003). *Illustrated Genera of Rust Fungi* (St Paul): 225 pp.; **Frederick, R.D.; Snyder, C.L.; Peterson, G.L.; Bond, M.R.** (2002). Polymerase chain reaction assays for the detection and discrimination of the soybean rust pathogens *Phakopsora pachyrhizi* and *P. meibomiae*. *Phytopathology* 92: 217–227; **Gjærum, H.B.; Namaganda, M.; Lye, K.A.** (2000). Ugandan rust fungi 2. The genus *Phakopsora*. *Lidia* 5: 13–17; **Huseyin, E.S.; Selcuk, F.** (2004). Observations on the genera *Cerotelium*, *Melampsoridium* and *Pileolaria* (*Uredinales*) in Turkey. *Pakist. J. Bot.* 36: 203–207; **Ono, Y.** (2000). Taxonomy of the *Phakopsora ampelopsidis* species complex on vitaceous hosts in Asia including a new species, *P. euvitis*. *Mycologia* 92: 154–173; **Tessmann, D.J.; Dianese, J.C.; Genta, W.; Mio, L.L. May-de** (2004). Grape rust caused by *Phakopsora euvitis*, a new disease for Brazil. *Fitopatol. Brasil* 29: 338; **Wingfield, B.D.; Ericson, L.; Szaro, T.; Burdon, J.J.** (2004). Phylogenetic patterns in the *Uredinales*. *Australas. Pl. Path.* 33: 327–335.

Phallaceae Corda 1842
Phallales: Basidiomycota

Phallus impudicus, Hampshire, UK; basidioma

Basidiomata gasteroid, epigeous, either remaining truffle-like or with a fertile structure developing from a strongly gelatinous spherical or broadly ellipsoidal 'egg'. Fertile part often on a wide fleshy or spongy stalk, cylindrical or stellate or reticulate, sometimes brightly coloured, sometimes with a lattice-like veil-like indusium, branched or unbranched, the gleba internal or external. Hyphal system monomitic, generative hyphae gelatinized, with clamp connections. Gleba strongly gelatinized, deliquescent. Basidia small, evanescent, with 4–8 sterigmata. Basidiospores ellipsoidal to cylindrical, hyaline or pale brown, smooth, ± thin-walled, often truncate at the base.

Clathrus ruber; basidiomata

Aseroë rubra; basidioma

Significant Genera: *Aseroë, Clathrus, Mutinus, Phallus*.

Distribution: Widespread, especially tropical.

Economic Significance: Species of *Dictyophora* (sometimes considered a segregate of *Phallus*) are widely eaten in Oriental regions and are also valued as medicinal mushrooms.

Ecology: Saprobic, on soil or decaying wood, with spores dispersed by insects that are attracted to the fruit bodies by their distinctive foetid smell.

Notes: The Stinkhorns and their allies, characterized by a remarkable range of fruit body types. Several species have been reported as introductions into Europe.

References: **Baseia, I.G.; Gibertoni, T.B.; Maia, L.C.** (2003). *Phallus pygmaeus*, a new minute species from a Brazilian tropical rain forest. *Mycotaxon* 85: 77–79; **Beaton, G.; Malajczuk, N.** (1986). New species of *Gelopellis* and *Protubera* from Western Australia. *Trans. Br. mycol. Soc.* 87: 478–482; **Boa, E.** (1988). Edible stinkhorns?. *Mycologist* 2: 107–108; **Bougher, N.L.; Lebel, T.** (2001). Sequestrate (truffle-like) fungi of Australia and New Zealand. *Aust. Syst. Bot.* 14: 439–484; **Calonge, F.D.** (1985). El orden *Phallales* (*Gasteromycetes*) en España. I. *Phallaceae y Clathraceae*. *Boln Soc. Micol. Castell.* 10: 59–72; **Fan, L.; Liu, B.; Liu, Y.H.** (1994). The *Gasteromycetes* of China. *Beih. Nova Hedwigia* 108: 72 pp.; **Humpert, A.J.; Muench, E.L.; Giachini, A.J.; Castellano, M.A.; Spatafora, J.W.** (2001). Molecular phylogenetics of *Ramaria* and related genera: evidence from nuclear large subunit and mitochondrial small subunit rDNA sequences. *Mycologia* 93: 465–477; **Kreisel, H.** (1996). A preliminary survey of the genus *Phallus sensu lato*. *Czech Mycol.* 48: 273–281; **Marren, P.** (1995). Stinkhorns, earthstars and fungal flowers: the strange world of gasteromycete fungi. *British Wildlife* 6: 366–376; **Pine, E.M.; Hibbett, D.S.; Donoghue, M.J.** (1999). Phylogenetic relationships of cantharelloid and clavarioid *Homobasidiomycetes* based on mitochondrial and nuclear rDNA sequences. *Mycologia* 91: 944–963; **Stijve, T.** (1997). Close encounters with *Clathrus ruber*, the latticed stinkhorn. *Australas. Mycol. Newsl.* 16: 11–15.

Phaneromycetaceae Gamundí & Spinedi 1985
Uncertain position within Ascomycetes, Ascomycota

Phaneromyces macrosporus, Tierra del Fuego; vertical section through ascoma

Stroma absent. Ascomata apothecial, erumpent, flat to cupulate, brown or orange, with an outer wall of gelatinized hyphae with crystalline intrusions; hymenium not blueing in iodine. Interascal tissue of simple, filiform, paraphyses. Asci cylindric-clavate, thick-walled, ? sometimes fissitunicate. Ascospores multiseptate, hyaline.

Significant Genera: *Phaneromyces*.

Distribution: Known from temperate South America.

Economic Significance: None is known.

Ecology: Saprobic on dead wood.

Notes: Possibly allied to the *Ostropales*, but the asci are rather distinct. No molecular data are available.

References: **Gamundí, I.J.; Spinedi, H.A.** (1986). *Phaneromyces* Speg. & Hariot, a discomycetous genus of critical taxonomic position. *Sydowia* 38: 106–113; **Kutorga, E.; Hawksworth, D.L.** (1997). A reassessment of the genera referred to the family *Patellariaceae* (*Ascomycota*). *Syst. Ascom.* 15: 1–110.

Phelloriniaceae Ulbr. 1951
Agaricales: Basidiomycota

Basidiomata gasteroid, stipitate, initially hypogeous, sometimes covered by a veil, the remains of which persist at the base. Fertile part ± globose, glabrous or covered with large scales, stipe robust and ± woody; peridium thin, composed of gelatinized hyphae, releasing spores by disintegration and sometimes initially by apical fissures. Gleba powdery, initially white, with orange or brownish chambers separated by hyaline hyphal strands. Basidia persisting in bundles in the mature gleba, cylindrical to pyriform, eventually evanescent. Basidiospores globose, brown, thick-walled, warted or reticulate.

Significant Genera: *Dictyocephalos*, *Phellorinia*.

Distribution: Widespread but not frequently encountered.

Economic Significance: None is known.

Ecology: In soil in arid or semi-arid regions, presumably saprobic.

Notes: A poorly known family, probably related to the *Tulostomataceae*. *Chlamydopus* may also belong here but morphological features are quite distinct and further molecular studies are needed.

References: **Calonge, F.D.** (1998). *Gasteromycetes*, I. *Lycoperdales, Nidulariales, Phallales, Sclerodermatales, Tulostomatales*. *Fl. Mycol. Iberica* 3: 271 pp.; **Coetzee, J.C.; Eicker, A.; Van Wyck, A.E.** (1997). Taxonomic notes on the *Geastraceae, Tulostomataceae, Nidulariaceae* and *Sphaerobolaceae* (*Gasteromycetes*) *sensu* Bottomly, in southern Africa. *Bothalia* 27: 117–123; **Dios, M.M.; Moreno, G.; Altés, A.** (2002). *Dictyocephalos attenuatus* (*Agaricales, Phelloriniaceae*) new record from Argentina. *Mycotaxon* 84: 265–270; **Martín, M.P.; Hidalgo, E.; Altés, A.; Moreno, G.** (2000). Phylogenetic relationships in *Phelloriniaceae* (*Basidiomycotina*) based on ITS rDNA sequence analysis. *Cryptog. Mycol.* 21: 3–12; **Moreno, G.; Altés, A.; Kreisel, H.** (1997). Confirmation of the presence of *Dictyocephalos attenuatus* (*Gasteromycetes*) in Europe. *Mycotaxon* 64: 393–399; **Wright, J.E.; Moreno, G.; Altés, A.** (1993). *Dictyocephalos attenuatus* (*Gasteromycetes, Basidiomycotina*) new for Europe. *Cryptog. Mycol.* 14: 77–83.

Phillipsiellaceae Höhn. 1909
Uncertain position within Dothideomycetes, Ascomycota

Stromata absent. Ascomata small, scutate to pulvinate, possibly apothecial, greenish-black, sometimes surrounded by weakly developed mycelium; peridium scarcely developed, of a few rows of pseudoparenchymatous cells. Interascal tissue of wide pseudoparaphyses which are enlarged at the apex to form an epithecium-like layer. Asci saccate, sessile, thick-walled at least at the apex, with a wide ocular chamber, not fissitunicate. Ascospores hyaline to brown, simple, transversely septate or muriform. Anamorphs unknown.

Significant Genera: *Phillipsiella*.

Distribution: Known from the Neotropics.

Economic Significance: None is known.

Ecology: Saprobic, epiphytic on leaves, sometimes associated with leaf hairs.

Notes: A poorly known family of uncertain relationships. Synonymy with the *Saccardiaceae* has been proposed and possibly also with the *Schizothyriaceae*, though the latter family is itself highly heterogenous.

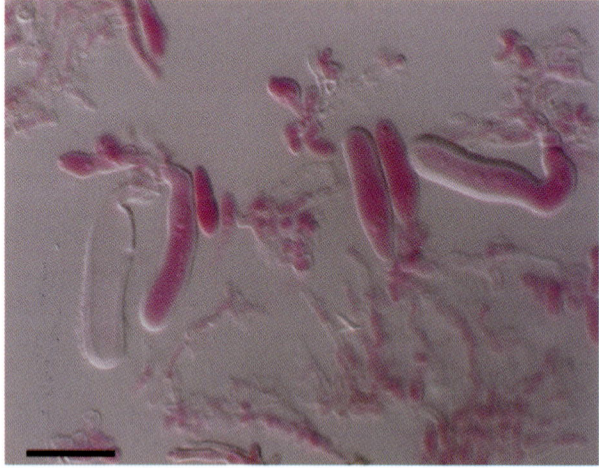

Phillipsiella sp., India; asci and ascospore

References: **Kohlmeyer, J.; Volkmann-Kohlmeyer, B.; Eriksson, O.E.** (1998). Fungi on *Juncus roemerianus*. 11. More new ascomycetes. *Can. J. Bot.* 76: 467–477; **Rossman, A.Y.; Mouchacca, J.; Samuels, G.J.** (1994). *Phillipsiella crescentiae* comb. nov. and a redescription of *P. atra*, type of the genus *Phillipsiella*. *Sydowia* 46: 66–74.

Phleogenaceae Gäum. 1926
Atractiellales: Basidiomycota

Basidiomata gasteroid, stipitate and capitate, the fertile head ± globose, ± smooth, not viscid or gelatinous, white or pale brown, the stipe cylindrical, compact, sometimes branched. Hyphae with simple septal pores, often thickened, clamp connections present, fertile hyphae bearing basidia on short lateral branches and interspersed with branched sterile hyphae. Basidia ± cylindrical, often curved or contorted, with three transverse septa, basidiospores developing laterally without clearly differentiated sterigmata. Basidiospores ovoid to reniform, brown, thick-walled, smooth.

Significant Genera: *Phleogena*.

Distribution: Widespread in both north and south temperate zones.

Economic Significance: None is known.

Ecology: Presumably saprobic, found on dead wood and bark.

Notes: A poorly known family with few (maybe only a single) species, separated from most of the 'stilboid' basidiomycetes by its clamp connections and transversely septate basidia. Its phylogenetic position requires further research.

Phleogena faginea; basidiomata

References: **Cook, P.** (1994). Profiles of fungi, 62. *Phleogena faginea*. *Mycologist* 8: 107; **Oberwinkler, F.; Bandoni, R.J.** (1982). A taxonomic survey of the gasteroid, auricularioid *Heterobasidiomycetes*. *Can. J. Bot.* 60: 1726–1750; **Oberwinkler, F.; Bauer, R.** (1989). The systematics of gasteroid, auricularioid *Heterobasidiomycetes*. *Sydowia* 41: 224–256.

Phlyctidaceae Poelt ex J.C. David & D. Hawksw. 1991
Lecanorales: Ascomycota

Thallus crustose, sometimes areolate. Ascomata apothecial, immersed, sometimes emergent, cupulate, thin-walled. Interascal tissue of branched and anastomosed paraphyses. Asci clavate, several-layered, with a J+ apical cap and an outer J+ gelatinized layer, 1- to 8-spored. Ascospores hyaline, transversely septate or muriform. Anamorphs pycnidial.

Phlyctis argena, Salamanca, Spain; sorediate thallus

Phlyctis agelaea, Devon, UK; ascomata and thallus

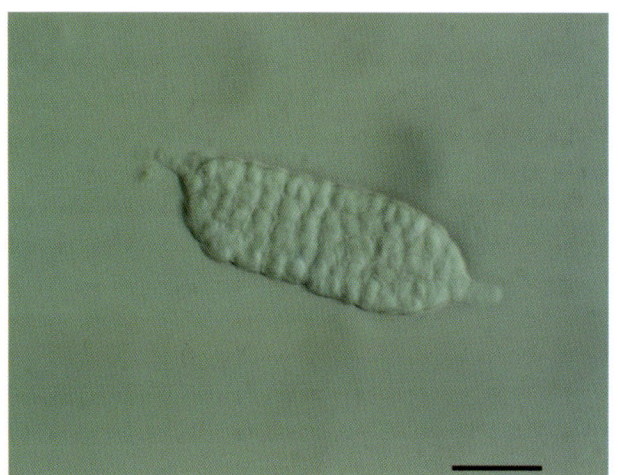
Phlyctis agelaea, Devon, UK; ascospore

Significant Genera: *Phlyctis*.

Distribution: Primarily tropical.

Economic Significance: None is known.

Ecology: Lichenized with green algae, usually on bark.

Notes: Apparently an isolated family, perhaps belonging to the *Ostropomycetidae* rather than the *Lecanoromycetidae*. More research is needed.

References: **Brusse, F.A.** (1987). *Psathyrophlyctis*, a new lichen genus from southern Africa. *Bothalia* 17: 182–184; **Galloway, D.J.; Guzmán Grimaldi, G.** (1988). A new species of *Phlyctis* from Chile. *Lichenologist* 20: 393–397; **Wedin, M.; Wiklund, E.; Crewe, A.; Döring, H.; Ekman, S.; Nyberg, A.; Schmitt, I.; Lumbsch, H.T.** (2005). Phylogenetic relationships of *Lecanoromycetes* (*Ascomycota*) as revealed by sequences of mtSSU and nLSU rDNA sequence data. *Mycol. Res.* 109: 159–172.

Phragmidiaceae Corda 1837
Uredinales: Basidiomycota

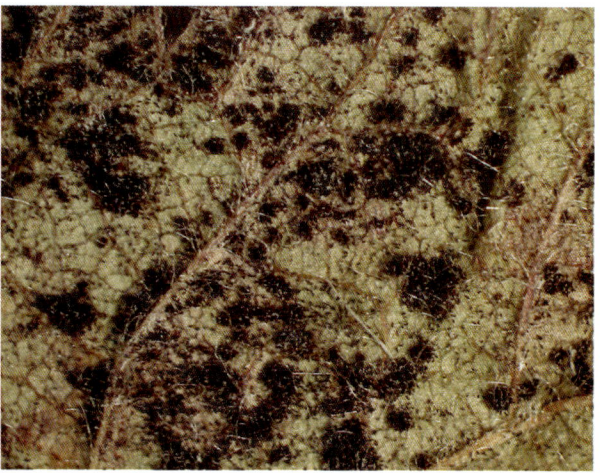
Phragmidium bulbosum, Sweden; telia

Spermogonia discoid, most often indeterminate and without discrete walls, subepidermal or subcuticular. Aecia usually *Caeoma*-like or *Petersonia*-like, lacking clear wall structures. Aeciospores formed in chains, with or without intercalary cells, verrucose or echinulate. Uredinia subepidermal or subcuticular. Urediniospores usually echinulate, pedicellate and often interspersed with paraphysis-like hyphae. Telia erumpent. Teliospores usually reddish or dark brown, pedicellate, 1-celled or transversely septate and with one or more germ pores in each cell, smooth, echinulate or verrucose, germinating to produce external basidia.

Significant Genera: *Hamaspora*, *Phragmidium*.

Distribution: Very widespread, especially in temperate climates.

Economic Significance: A number of *Phragmidium* species affect ornamental roses and *Phragmidium rubi-idaei* and *Kuehneola uredinis* can cause defoliation and stem blight of raspberry plants. Some species of *Phragmidium* have been investigated for their biocontrol potential of invasive species of *Rubus*.

Ecology: Biotrophic and autoecious, with almost all species associated with *Rosaceae*.

Notes: The family appears to be well-defined, though *Triphragmium* (the type of which is also on *Rosaceae*) may belong here; previous classifications have placed it in the *Raveneliaceae* or *Sphaerophragmiaceae*.

Mycotaxon 39: 249–256; **Ono, Y.** (2002). Life cycle and nuclear behavior in three rust fungi (*Uredinales*). *Mycoscience* 43: 37–45; **Ritz, C.M.; Maier, W.F.A.; Oberwinkler, F.; Wissemann, V.** (2005). Different evolutionary histories of two *Phragmidium* species infecting the same dog rose hosts. *Mycol. Res.* 109: 603–609; **Wahyuno, D.; Kakishima, M.; Ono, Y.** (2001). Morphological analyses of urediniospores and teliospores in seven *Phragmidium* species parasitic on ornamental roses. *Mycoscience* 42: 519–533; **Wahyuno, D.; Kakishima, M.; Ono, Y.** (2002). Aeciospore-surface structures of *Phragmidium* species parasitic on roses. *Mycoscience* 43: 159–167; **Wingfield, B.D.; Ericson, L.; Szaro, T.; Burdon, J.J.** (2004). Phylogenetic patterns in the *Uredinales*. *Australas. Pl. Path.* 33: 327–335.

Phragmoxenidiaceae Oberw. & R. Bauer 1990
Tremellales: Basidiomycota

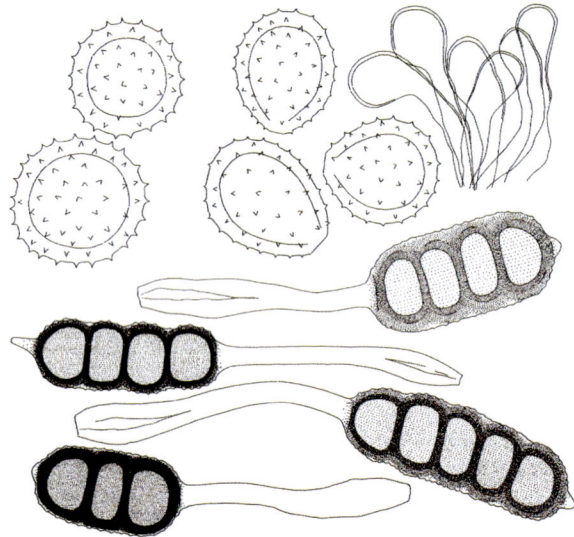

Phragmidium violaceum; aeciospores, urediniospores and paraphyses, and teliospores

Phragmidium bulbosum, China; teliospores

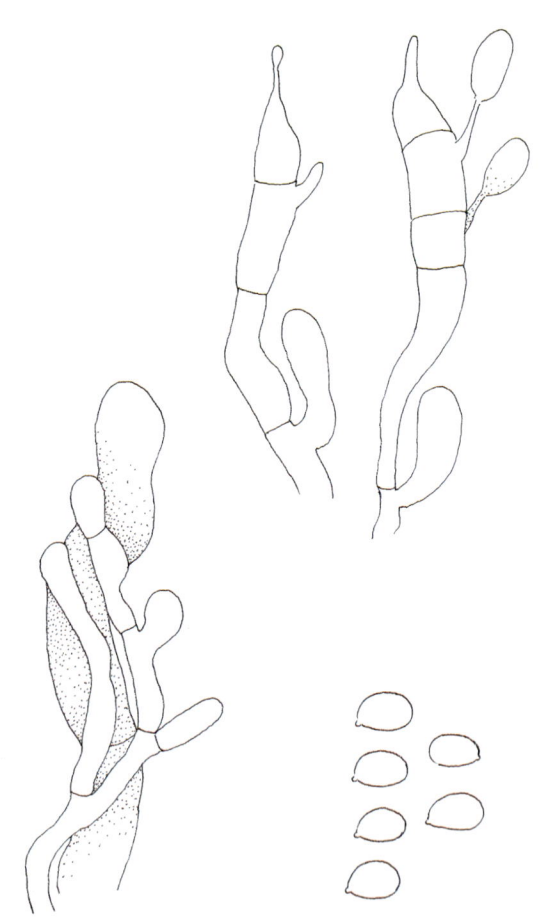

Phragmoxenidium mycophilum, Sweden; basidia and basidiospores

References: **Aime, M.C.** (2006). Towards resolving family-level relationships in rust fungi (*Uredinales*). *Mycoscience* 47: 112–122; **Cummins, G.B.; Hiratsuka, Y.** (2003). *Illustrated Genera of Rust Fungi* (St Paul): 225 pp.; **Gomez, D.R.; Evans, K.J.; Harvey, P.R.; Baker, J.; Barton, J.; Jourdan, M.; Morin, L.; Pennycook, S.R.; Scott, E.S.** (2006). Genetic diversity in the blackberry rust pathogen, *Phragmidium violaceum*, in Europe and Australasia as revealed by analysis of SAMPL. *Mycol. Res.* 110: 423–430; **Maier, W.; Begerow, D.; Weiss, M.; Oberwinkler, F.** (2003). Phylogeny of the rust fungi: an approach using nuclear large subunit ribosomal DNA sequences. *Can. J. Bot.* 81: 12–23; **McCain, J.W.; Hennen, J.F.** (1990). Taxonomic notes on *Frommeëlla* (*Uredinales*) – 1.

Basidiomata absent. Hyphae lacking clamp connections, septa with dolipores but lacking parenthosomes, cystidia absent. Basidia cylindrical, sometimes curved, with three lateral septa developing in the upper part, the sterigmata lateral at the distal end of the lower cells and terminal on the upper cell. Basidiospores subglobose to ellipsoidal, hyaline, thin-walled, smooth, not staining in iodine, germinating to form secondary spores.

Significant Genera: *Phragmoxenidium*.

Distribution: The only known species, *Phragmoxenidium mycophilum*, has been recorded several times from central and northern Europe.

Economic Significance: None is known.

Ecology: Parasitic in the hymenium of *Uthatobasidium fusisporum* (*Ceratobasidiaceae*).

Notes: The family is characterized by the unique combination of laterally septate basidia and hyphae with dolipores that lack parenthesomes. No molecular studies have been carried out.

References: **Oberwinkler, F.; Bauer, R.; Schneller J.** (1990). *Phragmoxenidium mycophilum* sp. nov., an unusual mycoparasitic heterobasidiomycete. *Syst. Appl. Microbiol.* 13: 186–191; **Roberts, P.** (1997). New and unusual Scandinavian heterobasidiomycetes. *Windahlia* 22: 15–22.

Phycomycetaceae Arx 1982
Mucorales: Zygomycota

Vegetative thallus a mycelium, complex fruit bodies absent. Sporophores (sporangiophores) arising from substratum mycelium or aerial hyphae, robust, always unbranched. Sporangia usually very large, multispored and columellate, specialized sporangiola absent; wall echinulate, diffluent, the sporangium forming a liquid droplet at maturity. Sporangiospores globose to ellipsoid, smooth, hyaline to yellowish. Zygospores globose to subglobose; wall brown to dark brown, opaque, ornamented with more or less coarse projections; suspensors tongs-like, smooth, more or less swollen, equal or unequal, with branched antler-like appendages; heterothallic or unknown.

Spinellus fusiger, Berkshire, UK; sporangiophores on old *Mycena* basidioma

Significant Genera: *Phycomyces*, *Spinellus*.

Distribution: Widespread but more common in temperate regions.

Economic Significance: None is known.

Ecology: Soil fungi but most readily isolated as a component of the early coprophilous succession, sometimes colonizing fruits of varying type lying on the ground or facultative parasite of fungi, for example, fruit bodies of *Tricholomataceae*.

Notes: No recent information is available.

References: **Benjamin, C.R.; Hesseltine, C.W.** (1959). Studies on the genus *Phycomyces*. *Mycologia* 51: 751–771; **Yamazaki, Y.; Ootaki, T.** (1996). Vegetative regeneration on sexual organs in *Phycomyces blakesleeanus*. *Mycoscience* 37: 269–275.

Phyllachoraceae Theiss. & Syd. 1915
Phyllachorales: Ascomycota

Phyllachora aff. *crotonis*, Galápagos Islands; colonies on living leaves of *Croton scouleri*

Stromata often well-developed, immersed in or erumpent from plant tissue, often clypeate, usually black but rarely brightly coloured, containing one to many ascomata. Ascomata perithecial, usually thin-walled, the ostioles periphysate; peridium usually composed of thin-walled compressed hyaline or brown tissue. Interascal tissue of simple rather wide thin-walled paraphyses, sometimes deliquescent. Asci ± cylindrical, thin-walled, persistent, usually with an inconspicuous I– apical ring. Ascospores usually hyaline and aseptate, rarely ornamented, often with a mucilaginous sheath. Anamorphs coelomycetous, with conidia usually formed in locules within the stroma. Conidiomata very variable in form, often strongly flattened or compressed by adjacent ascomata, occasionally ± pycnidial, the conidiomatal walls often undifferentiated. Conidiogenous cells ± cylindrical, often in clusters on short conidiophores, proliferating percurrently. Conidia small, aseptate, hyaline, filiform (and often curved) or bacillar.

Phyllachora ateleiae, Cuba; colonies on living leaves of *Ateleia gummifera*

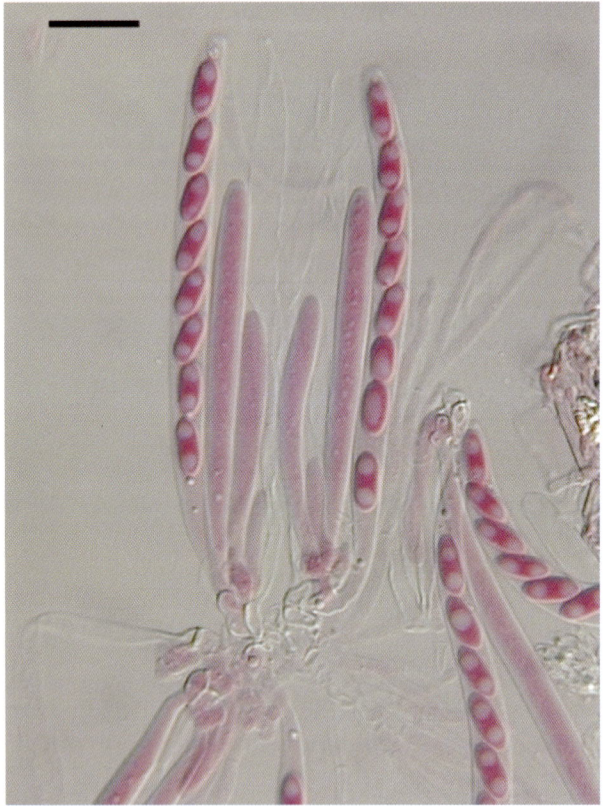

Fremitomyces punctatus, Kenya; asci and ascospores

Significant Genera: *Phyllachora*, *Polystigma*. Anamorphs: *Linochora*, *Rhodosticta*.

Distribution: Widespread, but diversity is especially concentrated in the tropics.

Economic Significance: Few species cause economically significant diseases of crops, although some of the cereal-inhabiting taxa contribute to crop loss via complexes of species including mycoparasites or secondary colonizers. *Polystigma* species cause premature leaf fall of woody species of *Rosaceae*.

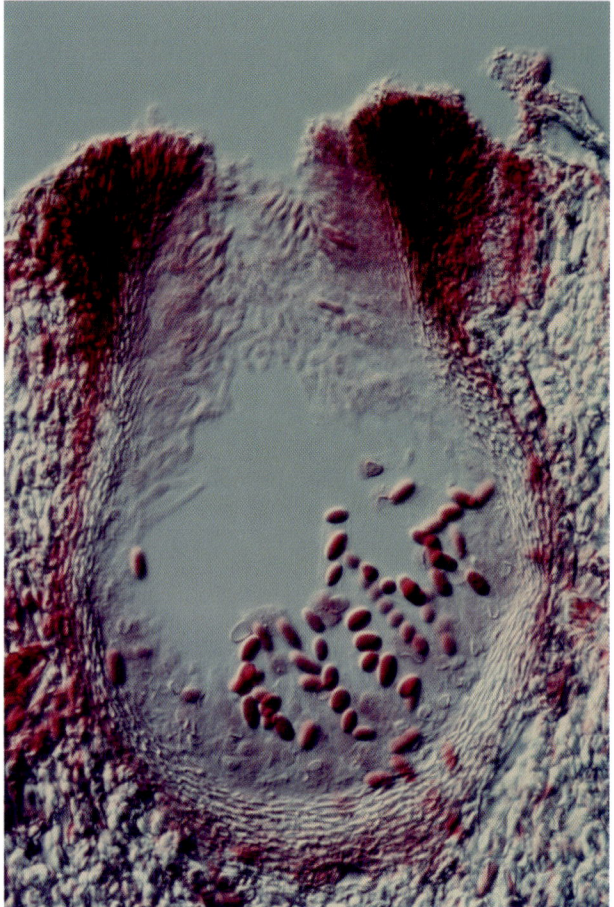

Stigmatula astragali; ascoma in vertical section

Ecology: Biotrophic in leaves and (rarely) stems, rarely causing necrosis. When this occurs it is usually the result of mycoparasitism. The conidial stages are almost always spermatial in function.

Notes: Almost no molecular data are available and the relationships of the family and the infrafamiliar classification are both tentative. At one time the *Glomerellaceae* were considered to belong with *Phyllachora* and its relatives, but there are significant ontogenetic and biological differences between the two groups. There are some indications that fungi of the *Phyllachoraceae* have coevolved with their angiosperm hosts.

References: **Cannon, P.F.** (1991). A revision of *Phyllachora* and some similar genera on the host family Leguminosae. *Mycol. Pap.* 163: 302 pp.; **Cannon, P.F.** (1994). Observations on coevolution of the *Phyllachoraceae* (Fungi: Ascomycota) with the Leguminosae. *Advances in Legume Systematics* Part 5. The Nitrogen Factor (Kew): 179–188; **Cannon, P.F.** (1996). Systematics and diversity of the *Phyllachoraceae* associated with Rosaceae, with a monograph of *Polystigma*. *Mycol. Res.* 100: 1409–1427; **Cannon, P.F.; Evans, H.C.** (1999). Biotrophic species of *Phyllachoraceae* associated with the angiosperm family Erythroxylaceae. *Mycol. Res.* 103: 577–590; **Hanlin, R.T.; Goh, T.-k.; Skarshaug, A.J.** (1992). A key to and descriptions of species assigned to *Ophiodothella*, based on the literature. *Mycotaxon* 44: 103–126; **Hyde, K.D.; Cannon, P.F.**

(1999). Fungi Causing Tar Spots on Palms. *Mycol. Pap.* 175: 114 pp.; **Malloch, D.; Mallik, A.** (1998). Taxonomy of *Orphnodactylis* [sic] *kalmiae* gen. et sp. nov. influenced by the hyperparasite *Didymosphaeria kalmiae*. *Can. J. Bot.* 76: 1265–1275; **Pearce, C.A.; Hyde, K.D.** (2006). Phyllachoraceae of Australia. *Fungal Diversity Res. Ser.* (Hong Kong) 17: 308 pp.; **Pearce, C.A.; Reddell, P.; Hyde, K.D.** (1999). A revision of *Phyllachora* (*Ascomycotina*) on hosts in the angiosperm family *Asclepiadaceae*, including *P. gloriana* sp. nov. on *Tylophora benthamii* from Australia. *Fungal Diversity* 3: 123–138; **Silva-Hanlin, D.M.W.; Hanlin, R.T.** (1998). The order *Phyllachorales*: taxonomic review. *Mycoscience* 39: 97–104.

Phyllobatheliaceae Bitter & F. Schill. 1927
Uncertain position within Dothideomycetes, Ascomycota

Thallus crustose, poorly developed. Ascomata perithecial, erumpent, sometimes with an involucrellum; the peridium thin-walled, ± hyaline but covered with dark melanized powdery material. Interascal tissue of simple paraphyses, not gelatinized. Asci saccate, ± thick-walled, fissitunicate, with a distinct ocular chamber, not blueing in iodine, 1- to 8-spored. Ascospores hyaline, muriform, without a sheath. Anamorphs pycnidial.

Significant Genera: *Phyllobathelium*.

Distribution: Tropical.

Economic Significance: None is known.

Ecology: Lichenized with green algae, foliicolous.

Notes: Affinities are uncertain and no molecular data are available.

References: **Aptroot, A.; Diederich, P.; Sérusiaux, E.; Sipman, H.J.M.** (1997). Lichens and Lichenicolous Fungi from New Guinea. *Biblthca Lichenol.* 64: 220 pp.; **Lücking, R.** (1997). Estado actual de las investigaciones sobre líquenes foliícolas en la región Neotrópica, con un análisis biogeográfico preliminar. *Trop. Bryol.* 13: 87–114; **Lücking, R.** (1998). Foliicolous lichens and their lichenicolous fungi collected during the Smithsonian International Cryptogamic Expedition to Guyana 1996. *Trop. Bryol.* 15: 45–76.

Physalacriaceae Corner 1970
Agaricales: Basidiomycota

Basidiomata stipitate, capitate, the fertile part globose or clavate, glabrous or pruinose, hollow, with a short central stipe. Hyphal system monomitic, clamp connections present, the hyphae not or slightly gelatinized, not staining in iodine. Hymenium covering the outer surface or in patches oriented towards the ground, the basidia sometimes accompanied by resiniferous cystidia. Basidia narrowly clavate with 2–4 sterigmata, often maturing sequentially. Basidiospores varied in size, ellipsoidal, fusiform, cylindrical or lacrimiform, hyaline, smooth, thin-walled, not staining in iodine.

Significant Genera: *Physalacria*.

Distribution: Widespread, though most species occur in the tropics, notably South-East Asia and Australasia.

Economic Significance: None is known. Of the other fungi potentially included in this family (see notes below), the widespread species *Flammulina velutipes* is a highly prized edible mushroom, cultivated forms of which are referred to as enoki or enokitake in Japan. Several *Armillaria* species are aggressive root pathogens of a wide range of woody plants, but others form mycorrhizas with orchids.

Ecology: Presumably saprobic, found on decaying leaves and wood. One genus is a parasite of marine algae.

Physalacria inflata; basidiomata

Notes: The *Physalacriaceae* has traditionally been restricted to a single reduced agaric genus, but molecular studies have indicated that *Physalacria* may be related to the well-known genera *Flammulina*, *Xerula* and possibly also *Armillaria*. Pending confirmation of these relationships, the family is here treated in its traditional sense. The genus *Paleoclavaria*, a fossil genus from amber, has been compared with *Physalacria*.

References: **Antonín, V.; Mossebo, D.C.** (2002). Two interesting central African collections of *Physalacria* (*Basidiomycetes*, *Agaricales*): *P. camerunensis* sp. nov. and the first African record of *P. tropica*. *Mycotaxon* 83: 419–424; **Berthier, J.** (1985). Les *Physalacriaceae* du globe. *Biblthca Mycol.* 98: 128 pp.; **Binder, M.; Hibbett, D.S.; Wang, Z.; Farnham, W.F.** (2006). Evolutionary relationships of *Mycaureola dilseae* (*Agaricales*), a basidiomycete pathogen of a subtidal rhodophyte. *Am. J. Bot.* 93: 547–556; **Dentinger, B.T.M.; McLaughlin, D.J.** (2006). Reconstructing the *Clavariaceae* using nuclear large subunit rDNA sequences and a new genus segregated from *Clavaria*. *Mycologia* 98: 746–762; **Horak, E.; Desjardin, D.E.** (1994). Reduced marasmioid and mycenoid agarics from Australasia. *Aust. Syst. Bot.* 7: 153–170; **Moncalvo, J.-M.; Vilgalys, R.; Redhead, S.A.; Johnson, J.E.; James, T.Y.; Aime, M.C.; Hofstetter, V.; Verduin, S.J.W.; Larsson, E.; Baroni, T.J.; Thorn, R.G.; Jacobsson, S.; Clemencon, H.; Miller, O.K.** (2002). One hundred and seventeen clades of euagarics. *Mol. Phylogen. Evol.* 23: 357–400; **Poinar, G.O.; Brown, A.E.** (2003). A non-gilled hymenomycete in Cretaceous amber. *Mycol. Res.* 107: 763–768; **Tanaka, I.; Doi, Y.; Hongo, T.** (2004). Two unusual species of *Physalacria* (*Basidiomycetes*, *Agaricales*) collected in New Zealand and Papua New Guinea during mycological expeditions by the National Science Museum, Tokyo. *Mycoscience* 45: 143–146; **Wilson, A.W.; Desjardin, D.E.** (2005). Phylogenetic relationships in the gymnopoid and marasmioid fungi (*Basidiomycetes*, euagarics clade). *Mycologia* 97: 667–679.

Physciaceae Zahlbr. 1898
Lecanorales: Ascomycota

Thallus crustose, foliose or fruticose. Ascomata apothecial, sessile, occasionally immersed, usually with a distinct margin. Interascal tissue composed of ± branched and anastomosing paraphyses, often with swollen and/or pigmented apices. Asci with a J+ apical cap, with a conical or funnel-shaped more weakly staining apical cushion, usually with a J+ gelatinized outer layer. Ascospores usually septate, usually dark brown with thickened walls and septa. Anamorphs pycnidial.

Heterodermia echinata; thallus and ascomata

Significant Genera: *Buellia*, *Heterodermia*, *Physcia*, *Rinodina*.

Distribution: Widespread.

Economic Significance: None is known.

Ecology: Lichenized with green algae, often on nitrogen-rich substrata.

Notes: A large and polymorphic family. The *Caliciaceae* almost certainly represents mazaedial analogues of the *Physciaceae*, and further research is needed to confirm its delimitation and major internal divisions. The *Buelliaceae* and *Pyxinaceae* were both subsumed into the *Physciaceae*, but should probably be resurrected at least at subfamily rank.

Physcia stellaris, Norway; thallus and ascomata

Buellia epigaea, Norway; thallus and ascomata

References: **Aptroot, A.** (1988). Lichens of Madagascar: the *Pyxinaceae* (syn. *Physciaceae*). *Cryptog. Bryol.-Lichénol.* 9: 141–147; **Aptroot, A.; Berendsen, W.** (1989). The nature of pruina on *Pyxinaceae* (*Lichenes*). *Proc. K. Ned. Akad. Wet. Ser. C, Biol. Med. Sci.* 92: 409–414; **Chen, J.B.; Wang, D.P.** (1999). The lichen family *Physciaceae* (*Ascomycota*) in China I. The genus *Anaptychia*. *Mycotaxon* 73: 335–342; **Crespo, A.; Blanco, O.; Llimona, X.; Ferencova, Z.; Hawksworth, D.L.** (2004). *Coscinocladium*, an overlooked endemic and monotypic Mediterranean lichen genus of *Physciaceae*, reinstated by molecular phylogenetic analysis. *Taxon* 53: 405–414; **Cubero, O.F.; Crespo, A.; Esslinger, T.L.; Lumbsch, H.T.** (2004). Molecular phylogeny of the genus *Physconia* (*Ascomycota*, *Lecanorales*) inferred from a Bayesian analysis of nuclear ITS rDNA sequences. *Mycol. Res.* 108: 498–505; **Dahlkild, A.; Kallersjo, M.; Lohtander, K.; Tehler, A.** (2001). Photobiont diversity in the *Physciaceae* (*Lecanorales*). *Bryologist* 104: 527–536; **Esslinger, T.L.** (2000). A key for the lichen genus *Physconia* in California, with descriptions for three new species occurring within the state. *Bull. Calif. Lichen Soc.* 7: 1–6; **Esslinger, T.L.; Bratt, C.** (1998). The *Heterodermia erinacea* group in North America, and a remarkable new disjunct distribution. *Lichenographia Thomsoniana, North American Lichenology in Honor of John W. Thomson* (Ithaca): 25–36; **Giralt, M.; Barbero, M.; Elix, J.A.** (2000). Notes on some corticolous and lignicolous *Buellia* species from the Iberian peninsula. *Lichenologist* 32: 105–128; **Giralt, M.; Llimona, X.** (1997). The saxicolous species of the genera *Rinodina* and *Rinodinella* lacking spot test reactions in the Iberian Peninsula. *Mycotaxon* 62: 175–224; **Grube, M.; Arup, U.** (2001). Molecular and morphological evolution in the *Physciaceae* (*Lecanorales*, lichenized *Ascomycotina*), with special emphasis on the genus *Rinodina*. *Lichenologist* 33: 63–72; **Helms, G.; Friedl, T.** (2003). Phylogenetic relationships of the *Physciaceae* inferred from rDNA sequence data and selected phenotypic characters. *Mycologia* 95: 1078–1099; **Lohtander, K; Källersjö, M.; Moberg, R.; Tehler, A.** (2000). The family *Physciaceae* in Fennoscandia: phylogeny inferred from ITS sequences. *Mycologia* 92: 728–735; **Lumbsch, H.T.; Schmitt, I.; Palice, Z.; Wiklund, E.; Ekman, S.; Wedin, M.** (2004). Supraordinal phylogenetic relationships of *Lecanoromycetes* based on a Bayesian analysis of combined nuclear and mitochondrial sequences. *Mol. Phylogen. Evol.* 31: 822–832; **Mayrhofer, H.; Kantvilas, G.; Ropin, K.** (2000). The corticolous species of the lichen genus *Rinodina* (*Physciaceae*) in temperate Australia. *Muelleria* 12: 169–194; **Moberg, R.; Nash, T.H.** (1999). The genus *Heterodermia* in the Sonoran Desert area. *Bryologist* 102: 1–14; **Nimis, P.L.; Tretiach, M.** (1997). A revision of *Tornabea*, a genus of fruticose lichens new to North America. *Bryologist* 100: 217–225; **Nordin, A.** (1997). Ascospore characters in *Physciaceae*: an ultrastructural study. *In* Tibell, L. & Hedberg, I. (eds), Lichen Studies Dedicated to Rolf Santesson. *Symb. bot. upsal.* 32: 195–208; **Obermayer, W.; Blaha, J.; Mayrhofer, H.** (2004). *Buellia centralis* and chemotypes of *Dimelaena oreina* in Tibet and other central Asian regions. *Symb. bot. upsal.* 34: 327–342; **Stenroos, S.K.; DePriest, P.T.** (1998). SSU rDNA phylogeny of cladoniiform lichens. *Am. J. Bot.* 85: 1548–1559; **Tibell, L.** (2003). *Tholurna dissimilis* and generic delimitations in *Caliciaceae* inferred from nuclear ITS and LSU rDNA phylogenies (*Lecanorales*, lichenized ascomycetes). *Mycol. Res.* 107: 1403–1418; **Wedin, M.; Baloch, E.; Grube, M.** (2004). Parsimony analyses of mtSSU and nITS rDNA sequences reveal the natural relationships of the lichen families *Physciaceae* and *Caliciaceae*. *Taxon* 51: 655–660; **Wedin, M.; Döring, H.; Nordin, A.; Tibell, L.** (2000). Small subunit rDNA phylogeny shows the lichen families *Caliciaceae* and *Physciaceae* (*Lecanorales*, *Ascomycotina*) to form a monophyletic group. *Can. J. Bot.* 78: 246–254.

Physodermataceae Sparrow 1952
Blastocladiales: Blastocladiomycota

Thallus of two types, a polycentric endobiotic system bearing septate, turbinate cells which form thick-walled resting spores, or monocentric with epibiotic zoosporangia producing zoospores or gamete-like structures; zoosporangia internally proliferous; resting spores disproportionately large, dark brown, spherical or ellipsoid, or assuming the shape of the host cell, oftenbearing pores through which protrude antler-like outgrowths, germinating by circumscissile or irregular fracture to form an elongate endosporangium with a broad apical papilla, this deliquescing to form a pore through which the single, large zoospore emerges.

Significant Genera: *Physoderma*.

Distribution: Widespread.

Economic Significance: *Physoderma maydis* is the causal agent of maize (*Zea mays*) brown spot. If the infections are severe and centred on the stem nodes lodging may result in a 25% or more loss of yield. Most other species are usually not of economic significance; *P. alfalfae* causes a similar disease (crown wart) on alfalfa (*Medicago*).

Ecology: Parasites of phanerogams, often with a host-specific or limited host range.

Notes: There are over 50 species in the family and are best known from both north temperate regions and from tropical and temperate Asia, probably reflecting a higher degree of recording and research in these areas. In maize, symptoms first appear as small yellowish spots on the leaf blade (sometimes in lateral bands), leaf sheath, stalk and, rarely, on the husks and tassel of the outer ear. The spots eventually turn a reddish to chocolate brown and merge to form irregular blotches. A brown powdery pustule is formed when the overlying tissue disintegrates. In alfalfa the galls form on the young crown buds with the mature resting spores released after decay of infected tissue.

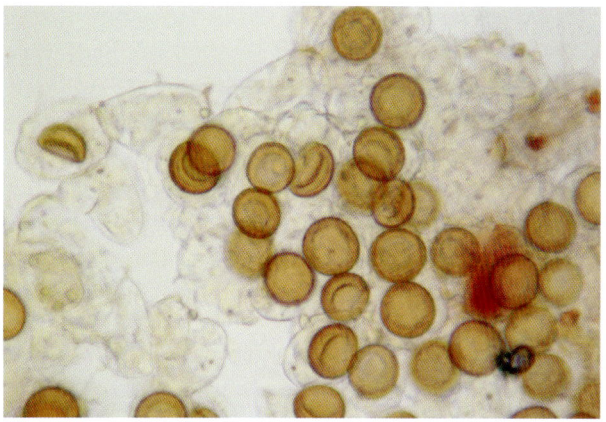

Physoderma asphodeli; resting spores

References: **Barr, D.J.S.** (2001). *Chytridiomycota*. *The Mycota* (Berlin) 7: 93–112.

Piedraiaceae Viégas ex M.E. Barr 1979
Uncertain position within Dothideomycetes, Ascomycota

Stromata small, black, multiloculate, erumpent; locules opening by irregular lysigenous breakdown of apical cells. Interascal tissue absent. Asci subglobose, one or few per locule, apparently bitunicate but with walls that break down at an early stage. Ascospores hyaline, aseptate, elongate-fusiform, with a gelatinous sheath which is strongly attenuated at each apex to form a filiform appendage. Anamorph unknown.

Significant Genera: *Piedraia*.

Distribution: Tropical, especially in South America.

Economic Significance: Causes the mycosis commonly referred to as black piedra.

Ecology: Keratinophilic, growing on hair.

Notes: Perhaps related to the *Myriangiales* but currently available phylogenies are inconclusive and more research is needed.

References: **Hoog, G.S. de; Guého, E.** (1998). Agents of white piedra, black piedra and tinea nigra. *Topley & Wilson's Microbiology and Microbial Infections* Edn 9. Vol. 4. Medical Mycology (London): 191–197; **Kane, J.; Summerbell, R.C.** (1999). *Trichophyton*, *Microsporum*, *Epidermophyton*, and agents of superficial mycoses. *Manual of Clinical Microbiology* (Washington): 1275–1294; **Lumbsch, H.T.; Lindemuth, R.** (2001). Major lineages of *Dothideomycetes* (*Ascomycota*) inferred from SSU and LSU rDNA sequences. *Mycol. Res.* 105: 901–908.

Pileolariaceae Cummins & Y. Hirats. 1983
Uredinales: Basidiomycota

Spermogonia conical or discoid, with discrete walls, subepidermal or subcuticular. Aecia and uredinia *Uredo*-like, erumpent, subepidermal, without distinct walls. Aeciospores and urediniospores morphologically similar, formed singly on short pedicels with varied ornamentation (verrucose, reticulate, ridged or helically marked) and usually zonate germ pores. Telia erumpent. Teliospores pedicellate, unicellular, pale or pigmented, thick-walled and often ornamented, rarely in clusters on common pedicels or accompanied by sterile cells, germ pore single, germinating to form an external basidium.

Significant Genera: *Pileolaria*, *Skierka*, *Uromycladium*.

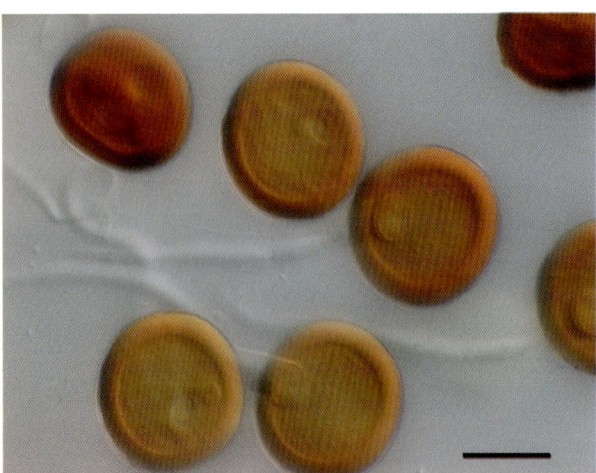

Pileolaria terebinthi, India; pedicellate teliospores

Distribution: Widespread, primarily in warm temperate or tropical zones.

Economic Significance: Some species attack economically important plants; *Pileolaria terebinthi* occurs on pistachio and *Atelocauda* and *Uromycladium* species on *Acacia*, but none are major pathogens.

Ecology: Biotrophic and autoecious, diverse especially on mimosoid legumes.

Notes: Molecular data are sparse, but there are some indications that the type genus *Pileolaria* is not closely related to the species associated with *Fabaceae*. *Atelocauda* has recently been subdivided and the segregate genus *Racospermomyces* appears to be evolutionarily distant from both *Pileolaria* and *Uromycladium*.

Uromycladium tepperianum, Victoria, Australia; urediniospores

References: **Aime, M.C.** (2006). Towards resolving family-level relationships in rust fungi (*Uredinales*). *Mycoscience* 47: 112–122; **Chen, W.Q.; Gardner, D.E.; Webb, D.T.** (1996). Biology and life cycle of *Atelocauda koae*, an unusual demicyclic rust. *Mycoscience* 37: 91–98; **Cummins, G.B.; Hiratsuka, Y.** (2003). *Illustrated Genera of Rust Fungi* (St Paul): 225 pp.; **Dianese, J.C.; Santos, L.T.P.; Medeiros, R.B.; Sanchez, M.** (1993). New *Skierka* species on *Cupania rugosa*, a native *Sapindaceae* from the cerrado. *Fitopatol. Brasil* 18: 342; **Gardner, D.E.; Hodges, C.S.** (1985). Spore surface morphology of Hawaiian *Acacia* rust fungi. *Mycologia* 77: 575–586; **Hiratsuka, Y.** (1988). Ontogeny and morphology of teliospores (probasidia) in *Uredinales* and their significance in taxonomy and phylogeny. *Mycotaxon* 31: 517–531; **Walker, J.** (2001). A revision of the genus *Atelocauda* (*Uredinales*) and description of *Racospermyces* gen. nov. for some rusts of *Acacia*. *Australas. Mycol.* 20: 3–28; **Wingfield, B.D.; Ericson, L.; Szaro, T.; Burdon, J.J.** (2004). Phylogenetic patterns in the *Uredinales*. *Australas. Pl. Path.* 33: 327–335.

Pilobolaceae Corda 1842
Mucorales: Zygomycota

Vegetative thallus a mycelium, complex fruit bodies absent. Sporophores (sporangiophores) arising either directly from submerged mycelium or usually singly from immersed, terminal or intercalary trophocysts; simple, stolons or rhizoids not produced; either phototropic or characterized by a period of rapid elongation immediately before sporangium maturity with or without an accompanying, thin-walled, subsporangial vesicle. Sporangia large, multispored and columellate, hemispherical or globose with a proximal flattened portion, specialized liberation mechanism present involving either forcible or passive (mediated by a thin zone in the wall) liberation of the entire sporangium; wall persistent, echinulate. Sporangiospores globose to ellipsoid or doliiform, smooth, hyaline to yellowish or pale brown. Zygospores globose to subglobose; wall brown to dark brown, opaque, ornamented with more or less coarse projections; suspensors tongs-like or opposed, smooth, more or less swollen, equal or unequal; heterothallic or presumed heterothallic.

Significant Genera: *Pilaira, Pilobolus, Utharomyces*.

Distribution: Widespread but some restricted to the tropics and subtropics.

Economic Significance: None is known.

Ecology: Coprophilous, frequently obligately so.

Notes: Originally considered to be monophyletic but recently analysis of molecular data has not confirmed this.

Pilobolus sp.; sporangia

References: **Cacialli, G.; Caroti, V.; Doveri, F.** (1999). Sul genere *Pilobolus* Tode 1784. Contributo allo studio dei funghi fimicoli – XXIX. *Funghi e Ambiente* 78–79: 13–25; **Hu, F.M.; Zheng, R.Y.; Chen, G.Q.** (1989). A redelimitation of the species of *Pilobolus*. *Mycosystema* 2: 111–133; **Kirk, P.M.; Benny, G.L.** (1980). The genus *Utharomyces* Boedijn (*Pilobolaceae*: Zygomycetes). *Trans. Br. mycol. Soc.* 75: 123–131; **Voigt, K.; Wöstemeyer, J.W.** (2001). Phylogeny and origin of 82 zygomycetes from all 54 genera of the *Mucorales* and *Mortierellales* based on combined analysis of actin and translation elongation factor EF-1α genes. *Gene* 270: 113–120.

Pilocarpaceae Zahlbr. 1905
Lecanorales: Ascomycota

Byssoloma leucoblepharum; ascomata on leaf

Byssoloma subundulatum; thalli on leaf

Thallus crustose. Ascomata apothecial, brightly coloured, with a poorly-developed margin composed of loosely intertwined hyphae, sometimes appearing woolly. Interascal tissue of branched and anastomosing paraphyses, usually immersed in gel, the apices not swollen. Asci with a well-developed J+ apical cap and a strongly J+ tube-like apical ring, usually without a well-developed ocular chamber, with an outer J+ gelatinized layer. Ascospores hyaline, often elongated, with one or more septa, usually without a sheath. Anamorphs pycnidial.

Significant Genera: *Byssoloma, Fellhanera*.

Distribution: Widespread, especially in the tropics.

Economic Significance: None is known.

Ecology: Lichenized with green algae. Many are foliicolous.

Notes: The family should perhaps be united with the *Micareaceae*; molecular data support this view.

Fellhanera bouteillei; thallus and ascomata

References: **Andersen, H.L.; Ekman, S.** (2004). Phylogeny of the *Micareaceae* inferred from nrSSU DNA sequences. *Lichenologist* 36: 27–35; **Andersen, H.L.; Ekman, S.** (2005). Disintegration of the *Micareaceae* (lichenized *Ascomycota*): a molecular phylogeny based on mitochondrial rDNA sequences. *Mycol. Res.* 109: 21–30; **Awasthi, D.D.; Mathur, R.** (1987). Species of the lichen genera *Bacidia, Badimia, Fellhanera* and *Mycobilimbia* from India. *Proc. Indian Acad. Sci.* Pl. Sci. 97: 481–503; **Cáceres, M.** (1999). A new foliicolous *Byssoloma* (lichenized ascomycetes: *Pilocarpaceae*) from the Atlantic rainforest in Pernambuco, Brazil. *Mycotaxon* 71: 383–386; **Kalb, K.; Vězda, A.** (1990). Die Flechtengattung *Byssoloma* in der Neotropis (eine taxonomisch-phytogeographische Studie). *Nova Hedwigia* 51: 435–451; **Lücking, R.** (1991). Neue Arten foliikoler Flechten aus Costa Rica, Zentralamerika. *Nova Hedwigia* 52: 267–304; **Lücking, R.; Lumbsch, H.T.; Elix, J.A.** (1994). Chemistry, anatomy and morphology of foliicolous species of *Fellhanera* and *Badimia* (lichenized *Ascomycotina*: *Lecanorales*). *Bot. Acta* 107: 393–401; **Thor, G.; Lücking, R.; Matsumoto, T.** (2000). The Foliicolous Lichens of Japan. *Symb. bot. upsal.* 32: 72 pp.

Piptocephalidaceae J. Schröt. 1886
Zoopagales: Zygomycota

Vegetative thallus a mycelium, complex fruit bodies absent. Sporophores (sporangiophores) erect, branched or unbranched, arising either directly from submerged mycelium or from the aerial hyphae. Sporangia (merosporangia) cylindrical, simple or branched basally, typically borne on a deciduous cell (head cell) which may be sterile or fertile, head cell rarely absent. Sporangiospores borne in a linear arrangement, usually cylindrical and smooth although appearing ornamented due to irregularities in the merosporangial wall. Zygospores warty, borne on tongs-like suspensors which are often spirally wound and with globose outgrowths.

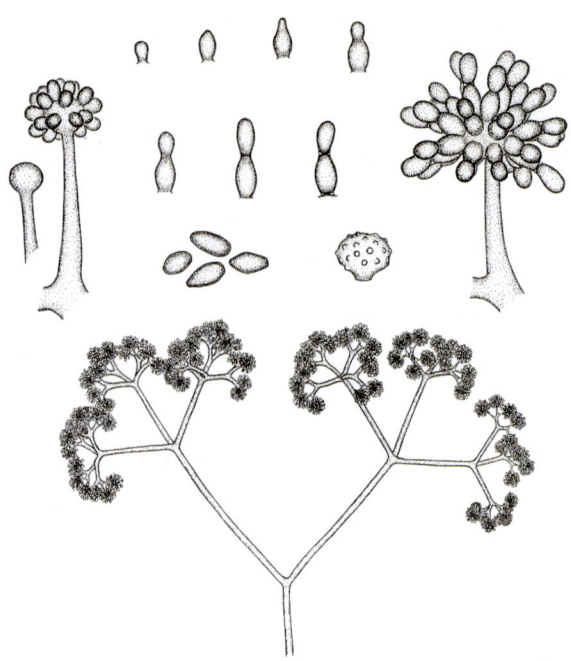

Piptocephalis lepidula, sporophore and sporangia. Reproduced with permission from *Aliso*, 1967.

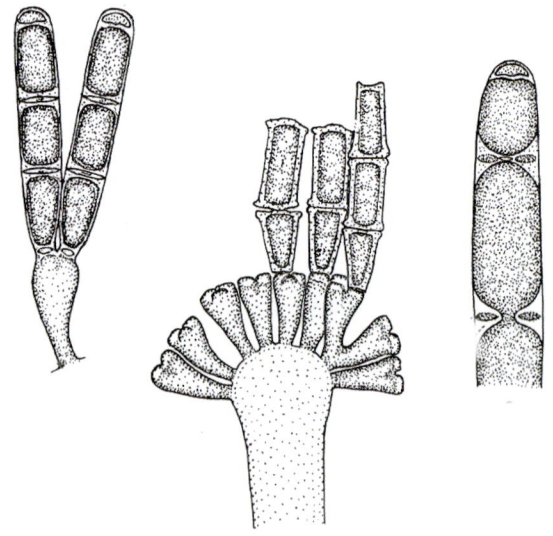

Syncephalis pycnosperma, sporangia. Reproduced with permission from *Aliso*, 1967.

Significant Genera: *Piptocephalis, Syncephalis*.

Distribution: Widespread.

Economic Significance: None is known.

Ecology: Obligate haustorial mycoparasites of *Mucorales* (rarely *Penicillium* and *Wynnea*).

Notes: Very poorly known; little recent information is available.

References: **Benjamin, R.K.** (1959). The Merosporangiferous *Mucorales*. *Aliso* 4: 321–433; **Benjamin, R.K.** (1963). Addenda to 'The Merosporangiferous *Mucorales*' II. *Aliso* 5: 273–288; **Tanabe, Y.; Saikawa, M.; Watanabe, M.M.; Sugiyama, J.** (2004). Molecular phylogeny of *Zygomycota* based on EF-1α and RPB1 sequences: limitations and utility of alternative markers to rDNA. *Mol. Phylogen. Evol.* 30: 438–449.

Pisolithaceae Ulbr. 1928
Boletales: Basidiomycota

Basidiomata gasteroid, epigeal, sessile or with a short, poorly defined, stipe composed of clustered mycelial cords, ± globose, white, yellowish or brown, smooth, the peridium thin and brittle, cracking irregularly to expose the spores. Gleba compound, with ± spherical or angular locules (peridioles) separated by plate-like layers of mycelial tissue, initially ± hyaline, becoming reddish-brown or almost black, powdery. Clamp connections present. Basidiospores globose, brown, thick-walled, with conspicuous spinose ornamentation.

Significant Genera: *Pisolithus*.

Distribution: Cosmopolitan. Molecular studies have confirmed that there have been multiple introductions of *Pisolithus* species from one part of the world to another, in association with their mycorrhizal partners.

Economic Significance: None is known.

Pisolithus tinctorius; basidioma

Ecology: On the ground, ectomycorrhizal with a range of woody plants.

Notes: There was considerable interest in *Pisolithus* as an apparently universal mycorrhizal species and its potential in forest growth and regeneration. Recent studies suggest that a number of species exist with restricted ecological requirements, reducing the potential for improvement on a global scale. The *Pisolithaceae* is combined with the *Sclerodermataceae* by some authors, but molecular studies confirm the two taxa are separate.

References: **Cairney, J.W.G.** (2002). *Pisolithus* – death of the pan-global super fungus. *New Phytol.* 153: 199–201; **Díez, J.; Anta, B.; Manjón, J.L.; Honrubia, M.** (2001). Genetic variability of *Pisolithus* isolates associated with native hosts and exotic eucalyptus in the western Mediterranean region. *New Phytol.* 149: 577–587; **Gill, M.; Watling, R.** (1986). The relationships of *Pisolithus* (*Sclerodermataceae*) to other fleshy fungi with particular reference to the occurrence and taxonomic significance of hydroxylated pulvinic acids. *Pl. Syst. Evol.* 154: 225–236; **Hitchcock, C.J.; Chambers, S.M.; Anderson, I.C.; Cairney, J.W.G.** (2003). Development of markers for simple sequence repeat-rich regions that discriminate between *Pisolithus albus* and *P. microcarpus*. *Mycol. Res.* 107: 699–706; **Junghans, D.T.; Gomes, E.A.; Guimarães, W.V.; Barros, E.G.; Araújo, E.F.** (1998). Genetic diversity of the ectomycorrhizal fungus *Pisolithus tinctorius* based on RAPD-PCR analysis. *Mycorrhiza* 7: 243–248; **Kanchanaprayudh, J.; Zhou, Z.H.; Yomyart, S.; Sihanonth, P.; Hogetsu, T.** (2003). Molecular phylogeny of ectomycorrhizal *Pisolithus* fungi associated with pine, dipterocarp, and eucalyptus trees in Thailand. *Mycoscience* 44: 287–294; **Kope, H.H.; Fortin, J.A.** (1990). Germination and comparative morphology of basidiospores of *Pisolithus arhizus*. *Mycologia* 82: 350–357; **Martin, F.; Díez, J.; Dell, B.; Delaruelle, C.** (2002). Phylogeography of the ectomycorrhizal *Pisolithus* species as inferred from nuclear ribosomal DNA ITS sequences. *New Phytol.* 153: 345–357; **Moyersoen, B.; Beever, R.E.; Martin, F.** (2003). Genetic diversity of *Pisolithus* in New Zealand indicates multiple long-distance dispersal from Australia. *New Phytol.* 160: 569–579; **Orlovich, D.A.; Ashford, A.E.** (1994). Structure and development of the dolipore septum in *Pisolithus tinctorius*. *Protoplasma* 178: 66–80.

Placynthiaceae Å.E. Dahl 1950
Peltigerales: Ascomycota

Thallus crustose to squamulose or minutely fruticose, with the mycobiont and photobiont evenly intermixed. Ascomata apothecial, on the upper surface, sessile, dark, without a well-developed thalline margin. Asci with a thickened apex, a J+ apical ring, an ocular chamber and a J+ outer gelatinized coat. Ascospores simple or transversely septate. Anamorphs pycnidial.

Placynthium nigrum, Norway; thallus and ascomata

Significant Genera: *Placynthium*.

Distribution: Primarily in north temperate regions.

Economic Significance: None is known.

Ecology: Lichenized with cyanobacteria, mainly on soil or rocks.

Notes: The family appears to be related to the *Collemataceae*; these two groups have been placed alongside the *Pannariaceae* in the *Peltigerales* suborder *Collematineae*.

Placynthium nigrum, Lancashire, UK; close-up of thallus

References: **Henssen, A.** (1997). *Santessoniella*, a new cyanophilic genus of lichenized ascomycetes. *In* Tibell, L. & Hedberg, I. (eds), Lichen Studies Dedicated to Rolf Santesson. *Symb. bot. upsal.* 32: 75–93; **Henssen, A.; Tønsberg, T.** (2000). *Spilonemella*, a new genus of cyanophilic lichens with species from North America and Japan (*Coccocarpiaceae*). *Bryologist* 103: 108–116; **Miadlikowska, J.; Lutzoni, F.** (2004). Phylogenetic classification of peltigeralean fungi (*Peltigerales*, *Ascomycota*) based on ribosomal RNA small and large subunits. *Am. J. Bot.* 91: 449–464; **Wedin, M.; Wiklund, E.** (2004). The phylogenetic relationships of *Lecanorales* suborder *Peltigerineae* revisited. *Symb. bot. upsal.* 34: 469–475; **Wiklund, E.; Wedin, M.** (2003). The phylogenetic relationships of the cyanobacterial lichens in the *Lecanorales* suborder *Peltigerineae*. *Cladistics* 19: 419–431.

Planistromellaceae M.E. Barr 1996
Dothideales: Ascomycota

Ascomata perithecial, multiloculate or uniloculate, immersed or erumpent, small, the ostioles periphysate, peridium of brown pseudoparenchymatous tissue. Interascal tissue of cellular pseudoparaphyses, frequently degenerating. Asci saccate, fissitunicate, with a distinct ocular chamber, not blueing in iodine. Ascospores hyaline or pale brown, aseptate or transversely septate, thin-walled, smooth, sometimes with a gelatinous coat. Anamorphs coelomycetous, often prominent.

Significant Genera: *Planistroma, Planistromella*. Anamorph: *Kellermania, Piptarthron*.

Distribution: Widespread but primarily known and studied in North America.

Economic Significance: None is known.

Ecology: Mostly necrotrophs and saprobes of plants, especially in xeric habitats.

Notes: The evidence appears weak for inclusion of *Microcyclus* in this family as is currently the case; that genus is in any case polyphyletic.

Planistroma yuccigenum, Arizona, USA; colonies in dead leaves of *Dasylirion wheeleri*

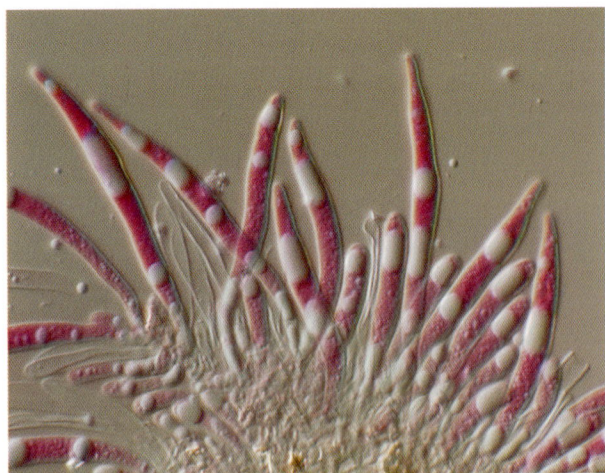

Planistroma yuccigenum, Arizona, USA; conidiophores and conidia

References: **Barr, M.E.** (1996). *Planistromellaceae*, a new family in the *Dothideales*. *Mycotaxon* 60: 433–442; **Ramaley, A.W.** (1995). New species of *Kellermania*, *Piptarthron*, *Planistroma*, and *Planistromella* from members of the *Agavaceae*. *Mycotaxon* 55: 255–268; **Ramaley, A.W.** (1998). New teleomorphs of the anamorphic genus *Kellermania*. *Mycotaxon* 66: 509–514.

Platygloeaceae Racib. 1909
Platygloeales, Basidiomycota

Basidiomata pezizoid, effused, pustular or intrahymenial, waxy, gelatinous or pellicular, non-fertile tissue often poorly developed. Hyphae hyaline with simple septal pores, thin-walled, usually with clamp connections and sometimes haustoria, cystidia absent. Basidia cylindrical, aseptate or becoming transversely septate, derived from a ± globose thin-walled cell with a basal clamp connection, the basidial cells producing small hyaline thin-walled basidiospores from (usually) single sterigmata. Anamorphs very diverse.

Significant Genera: *Helicobasidium, Helicogloea, Platygloea, Tuberculina.*

Distribution: Cosmopolitan.

Helicobasidium brebisonii; basidioma

Tuberculina persicina; parasitizing the rust *Tranzschelia prunispinosae* on leaves of *Anemone*

Economic Significance: *Helicobasidium mompa* and *H. brebissonii* and relatives cause violet root rot of a wide range of plants. *Tuberculina* species (see below) have been studied for their potential in biocontrol of rusts, though no products are currently available.

Ecology: Many species are mycoparasites; others are reported as saprobes but some of these are likely to be unrecognized parasites. *Helicobasidium* species are pathogens of plant roots but seem to have *Tuberculina* anamorphic stages that are parasitic on rusts. More research is needed into this apparently unique life strategy. *Tuberculina* species are traditionally studied in isolation from their meiotic stages and appear to be strongly host-specific.

Notes: Most recent phylogenies place this family in a basal position within the rusts, but it is polymorphic and certainly polyphyletic.

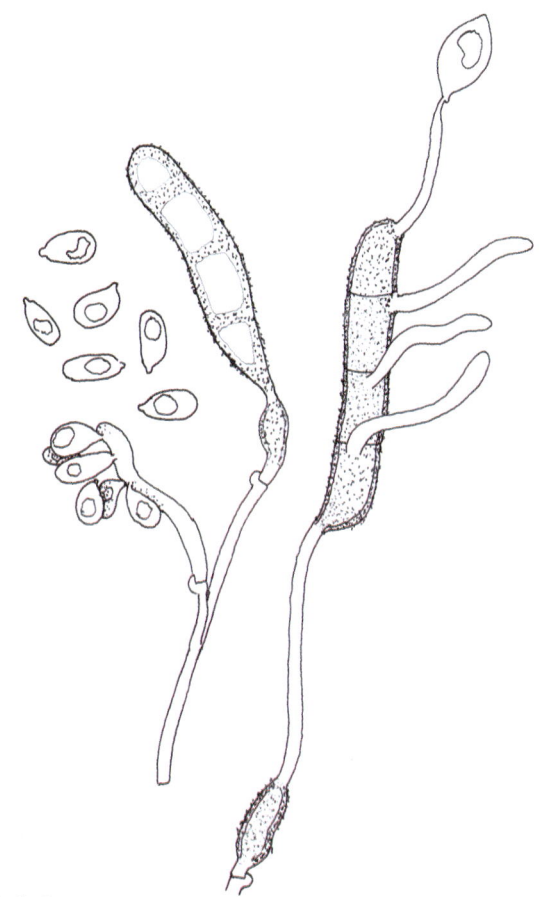

Spiculogloea occulta, Mallorca, Spain; basidia and basidiospores

References: **Bandoni, R.J.** (1998). On some species of *Mycogloea*. *Mycoscience* 39: 31–36; **Bauer, R.; Lutz, M.; Oberwinkler, F.** (2004). *Tuberculina*-rusts: a unique basidiomycetous interfungal cellular interaction with horizontal nuclear transfer. *Mycologia* 96: 960–967; **Berres, M.E.; Szabo, L.J.; McLaughlin, D.J.** (1996). Phylogenetic relationships in auriculariaceous basidiomycetes based on 25S ribosomal DNA sequences. *Mycologia* 87: 821–840; **Chen, C.J.; Oberwinkler, F.** (2000). *Helicogloea* species collected in Taiwan. *Mycotaxon* 76: 279–285; **Dueñas, M.** (2001). Iberian intrahymenial species of *Platygloeales, Tremellales* and *Tulasnellales*. *Nova Hedwigia* 72: 441–459; **Lutz, M.; Bauer, R.; Begerow, D.; Oberwinkler, F.** (2004). *Tuberculina–Thanatophytum/Rhizoctonia crocorum–Helicobasidium*: a unique mycoparasitic-phytoparasitic life strategy. *Mycol. Res.* 108: 227–238; **Lutz, M.; Bauer, R.; Begerow, D.; Oberwinkler, F.** (2004). *Tuberculina–Helicobasidium*: host specificity of the *Tuberculina*-stage reveals unexpected diversity within the group. *Mycologia* 96: 1316–1329; **Lutz, M.; Bauer, R.; Begerow, D.; Oberwinkler, F.; Triebel, D.** (2004). *Tuberculina*: rust relatives attack rusts. *Mycologia* 96: 614–626; **Nakamura, H.; Ikeda, K.-i.; Arakawa, M.; Akahira, T.; Matsumoto, N.** (2004). A comparative study of the violet root rot fungi, *Helicobasidium* *brebissonii* and *H. mompa*, from Japan. *Mycol. Res.* 108: 641–648; **Oberwinkler, F.; Bauer, R.; Tschen, J.** (1999). The mycoparasitism of *Platygloea bispora*. *Kew Bull.* 54: 763–769; **Wingfield, B.D.; Ericson, L.; Szaro, T.; Burdon, J.J.** (2004). Phylogenetic patterns in the Uredinales. *Australas. Pl. Path.* 33: 327–335.

Pleomassariaceae M.E. Barr 1979
Pleosporales: Ascomycota

Prosthemium stellare (teleomorph *Splanchnonema holoschistum*); conidia

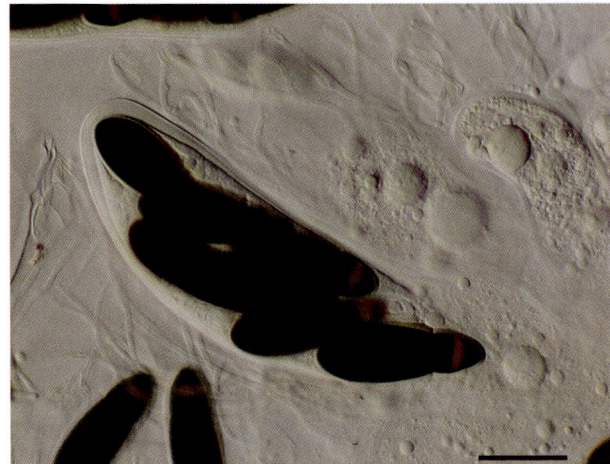

Splanchnonema scoriadeum, Russian Far East; ascus and ascospores

Splanchnonema scoriadeum, Russian Far East; ascomata

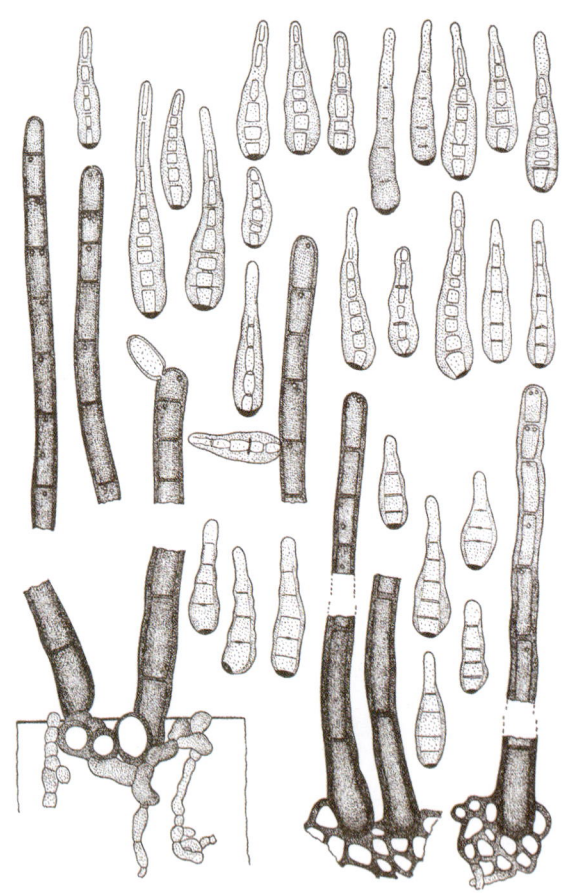

Helminthosporium solani; conidiophores and conidia

Ascomata immersed to erumpent, perithecial, often large, spherical or conical, black, with a well-developed lysigenous ostiole; peridium composed of large-celled pseudoparenchymatous tissue, narrower at apex and base. Interascal tissue of cellular pseudoparaphyses, immersed in gel. Asci cylindrical to clavate, fissitunicate, with a prominent ocular chamber, not blueing in iodine. Ascospores large, brown, transversely septate or muriform, sometimes distoseptate, with a large gelatinous sheath. Anamorphs coelomycetous with complex conidia or hyphomycetous with dark, thick-walled, conidia.

Significant Genera: *Asteromassaria*, *Splanchnonema*. Anamorph: *Dendryphiopsis*, *Helminthosporium*.

Distribution: Widespread in both temperate and tropical zones.

Economic Significance: A few species are weak pathogens of crop plants.

Ecology: Saprobic or necrotrophic on wood, bark and herbaceous material.

Notes: The family as currently circumscribed is almost certainly polyphyletic; few molecular data are available. Many species at one time assigned to *Helminthosporium* are now treated as anamorphs of *Pleosporaceae*.

References: **Liew, E.C.Y.; Aptroot, A.; Hyde, K.D.** (2000). Phylogenetic significance of the pseudoparaphyses in loculoascomycete taxonomy. *Mol. Phylogen. Evol.* 16: 392–402; **Lumbsch, H.T.; Lindemuth, R.** (2001). Major lineages of *Dothideomycetes* (*Ascomycota*) inferred from SSU and LSU rDNA sequences. *Mycol. Res.* 105: 901–908; **Shoemaker, R.A.; McLaughlin, J.; Greifenhagen, S.; Hambleton, S.** (2003). *Asteromassaria olivaceohirta*. Fungi Canadenses no. 345. *Can. J. Pl. Path.* 25: 384–386.

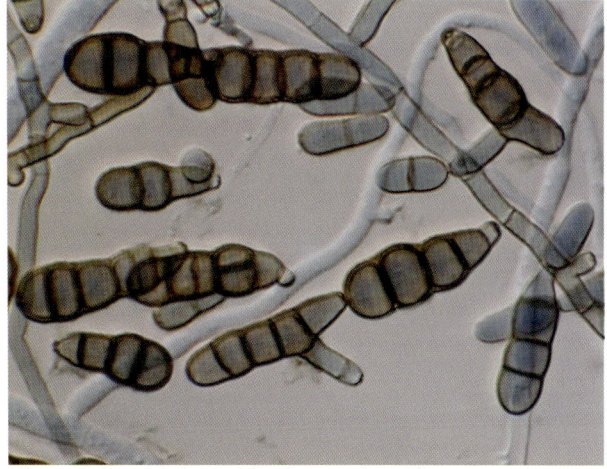

Alternaria brassicicola; conidiophore and conidia

Pleosporaceae Nitschke 1869
Pleosporales: Ascomycota

Ascomata perithecial, ± globose, thick-walled, immersed or erumpent, black, opening by a well-developed lysigenous ostiole, sometimes hairy or setose; peridium usually thick, with several layers of large thick-walled pseudoparenchymatous cells. Interascal tissue of cellular pseudoparaphyses. Asci ± cylindrical, fissitunicate, the inner wall often thickened in the apical region. Ascospores brown, septate, sometimes muriform, often with a gelatinous sheath. Anamorphs hyphomycetous.

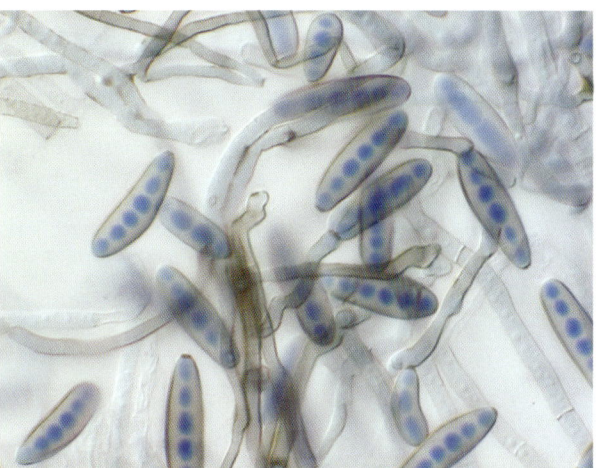

Cochliobolus hawaiiensis (*Bipolaris* anamorph); conidiophore and conidia

Significant Genera: *Cochliobolus, Lewia, Pleospora, Pyrenophora, Setosphaeria*. Anamorphs: *Alternaria, Bipolaris, Drechslera, Exserohilum, Stemphylium*.

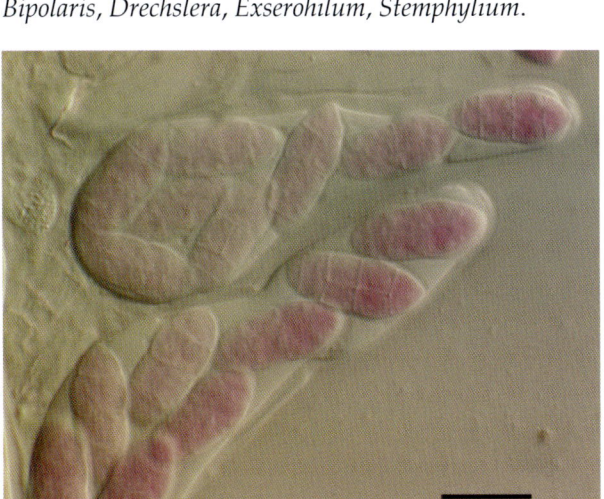

Leptosphaerulina trifolii, Bangladesh; asci and ascospores

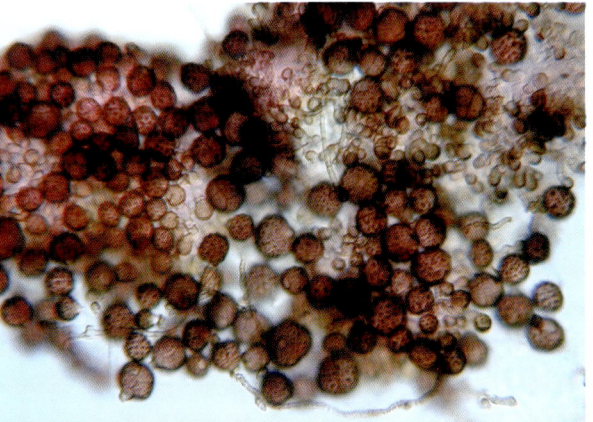

Epicoccum nigrum; colony with conidia

Distribution: Cosmopolitan.

Economic Significance: Many are important plant pathogens, known primarily as their anamorphs, and some species are mycotoxigenic. A few are implicated as medical pathogens.

Ecology: Necrotrophic pathogens and saprobes, especially associated with grasses. Many species are seed-borne.

Notes: The classification of several of the important pathogen groups within the *Pleosporaceae* has recently been reviewed and generic limits take into account both teleomorph and anamorph characters. These have been strongly supported by molecular evidence.

References: **Berbee, M.L.; Payne, B.P.; Zhang, G.J.; Roberts, R.G.; Turgeon, B.G.** (2003). Shared ITS DNA substitutions in isolates of opposite mating type reveal a recombining history for three presumed asexual species in the filamentous ascomycete genus *Alternaria*. *Mycol. Res.* 107: 169–182; **Berbee, M.L.; Pirseyedi, M.; Hubbard, S.** (1999). *Cochliobolus* phylogenetics and the origin of known, highly virulent pathogens, inferred from ITS and glyceraldehyde-3-phosphate dehydrogenase gene sequences. *Mycologia* 91: 964–977; **Borchardt, D.S.; Welz, H.G.; Geiger, H.H.** (1998). Genetic structure of *Setosphaeria turcica* populations in tropical and temperate climates. *Phytopathology* 88: 322–329; **Câmara, M.P.S.; O'Neill, N.R.; Berkum, P. van** (2002). Phylogeny of *Stemphylium* spp. based on ITS and glyceraldehyde-3-phosphate dehydrogenase gene sequences. *Mycologia* 94: 660–672; **Crous, P.W.; Janse, B.J.H.; Tunbridge, J.; Holz, G.** (1995). DNA homology between *Pyrenophora japonica* and *P. teres*. *Mycol. Res.* 99: 1098–1102; **Farr, D.F.; O'Neill, N.R.; Berkum, P.B. van** (2000). Morphological and molecular studies on *Dendryphion penicillatum* and *Pleospora papaveracea*, pathogens of *Papaver somniferum*. *Mycologia* 92: 145–153; **Kodsueb, R.; Dhanasekraran, V.; Aptroot, A.; Lumyong, S.; McKenzie, E.H.C.; Hyde, K.D.; Jeewon, R.** (2006). The family *Pleosporaceae*: intergeneric relationships and phylogenetic perspectives based on sequence analyses of partial 28S rDNA. *Mycologia* 98: 571–583; **Olivier, C.; Berbee, M.L.; Shoemaker, R.A.; Loria, R.** (2000). Molecular phylogenetic support from ribosomal DNA sequences for origin of *Helminthosporium* from *Leptosphaeria*-like loculoascomycete ancestors. *Mycologia* 92: 736–746; **Peever, T.L.; Su, G.; Carpenter-Boggs, L.; Timmer, L.W.** (2004). Molecular systematics of citrus-associated *Alternaria* species. *Mycologia* 96: 119–134; **Pryor, B.M.; Bigelow, D.M.** (2003). Molecular characterization of *Embellisia* and *Nimbya* species and their relationship to *Alternaria*, *Ulocladium* and *Stemphylium*. *Mycologia* 95: 1141–1154; **Pryor, B.M.; Gilbertson, R.L.** (2000). Molecular phylogenetic relationships amongst *Alternaria* species and related fungi based upon analysis of nuclear ITS and mt SSU rDNA sequences. *Mycol. Res.* 104: 1312–1321; **Serdani, M.; Kang, J.C.; Andersen, B.; Crous, P.W.** (2002). Characterisation of *Alternaria* species-groups associated with core rot of apples in South Africa. *Mycol. Res.* 106: 561–569; **Shimizu, K.; Tanaka, C.; Peng, Y.L.; Tsuda, M.** (1998). Phylogeny of *Bipolaris* inferred from nucleotide sequences of Brn1, a reductase gene involved in melanin biosynthesis. *J. gen. appl. Microbiol.* Tokyo 44: 251–258; **Shoemaker, R.A.** (1999). *Marielliottia*, a new genus of cereal and grass parasites segregated from *Drechslera*. *Can. J. Bot.* 76: 1558–1569; **Shoemaker, R.A.; Babcock, C.E.** (1987). *Wettsteinina*. *Can. J. Bot.* 65: 373–405; **Sivanesan, A.** (1987). Graminicolous species of *Bipolaris*, *Curvularia*, *Drechslera*, *Exserohilum* and their teleomorphs. *Mycol. Pap.* 158: 261 pp.; **Turgeon, B.G.; Berbee, M.L.** (1998). Evolution of pathogenic and reproductive strategies in *Cochliobolus* and related genera. *Molecular Genetics of Host-Specific Toxins in Plant Disease* (Dordrecht): 153–163; **Yun, S.H.; Berbee, M.L.; Yoder, O.C.; Turgeon, B.G.** (1999). Evolution of the fungal self-fertile reproductive life style from self-sterile ancestors. *Proc. natn Acad. Sci. U.S.A.* 96: 5592–5597.

Pleurostomataceae Réblová, L. Mostert, W. Gams & Crous 2004
Calosphaeriales: Ascomycota

Stromata absent. Ascomata perithecial, superficial or partially immersed, ± globose, papillate, black and smooth-walled; peridium leathery, two-layered. Interascal tissue lacking, periphyses present. Asci formed from proliferating ascogenous hyphae, thin-walled and unitunicate, saccate, the stalk remaining attached to the ascogenous hypha after dehiscence and sometimes eccentric, the apical part thickened but without any distinct structures, polysporous. Ascospores allantoid, hyaline, aseptate, thin-walled and smooth-walled, without a mucous sheath or appendages. Anamorph hyphomycetous, *Phialophora*-like.

Significant Genera: *Pleurostoma*. Anamorph: *Pleurostomophora*.

Distribution: Scattered throughout both tropical and temperate zones.

Economic Significance: None is known.

Ecology: Saprobic on rotting wood.

Notes: The family forms a sister group to the *Calosphaeriaceae*. Its multispored asci with stalks that remain attached after dehiscence are unparalleled.

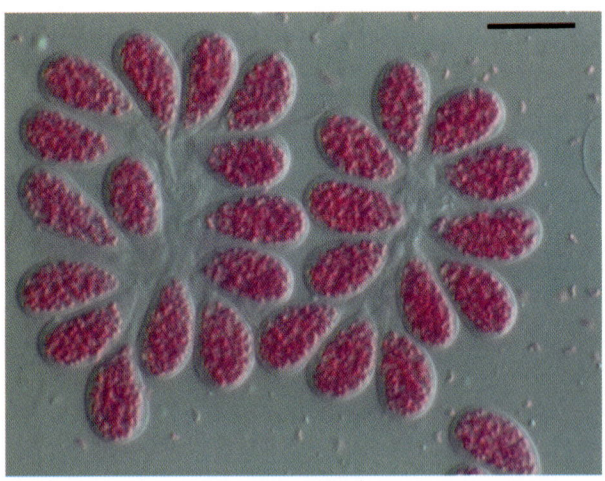

Pleurostoma ootheca, Argentina; asci and ascospores

References: **Barr, M.E.** (1985). Notes on the *Calosphaeriales*. *Mycologia* 77: 549–565; **Mostert, L.; Crous, P.W.; Groenewald, J.Z.; Gams, W.; Summerbell, R.C.** (2003). *Togninia* (*Calosphaeriales*) is confirmed as teleomorph of *Phaeoacremonium* by means of morphology, sexual compatibility and DNA phylogeny. *Mycologia* 95: 646–659; **Réblová, M.; Mostert, L.; Gams, W.; Crous, P.W.** (2004). New genera in the *Calosphaeriales*: *Togniniella* and its anamorph *Phaeocrella*, and *Calosphaeriophora* as anamorph of *Calosphaeria*. *Stud. Mycol.* 50: 533–550; **Romero, A.I.; Minter, D.W.** (1988). Fluorescence microscopy: an aid to the elucidation of ascomycete structures. *Trans. Br. mycol. Soc.* 90: 457–470; **Romero, A.I.; Samuels, G.J.** (1991). Studies on xylophilous fungi from Argentina. VI. *Ascomycotina* on *Eucalyptus viminalis* (Myrtaceae). *Beih. Sydowia* 43: 228–248; **Vijaykrishna, D.; Mostert, L.; Jeewon, R.; Gams, W.; Hyde, K.D.; Crous, P.W.** (2004). *Pleurostomophora*, an anamorph of *Pleurostoma* (*Calosphaeriales*), a new anamorph genus morphologically similar to *Phialophora*. *Stud. Mycol.* 50: 387–395.

Pleurotaceae Kühner 1980
Agaricales: Basidiomycota

Basidiomata pileate, mostly stipitate, fleshy or tough-textured, the stipes eccentric or lateral, the upper surface usually greyish or with blue or lilac tints, smooth. Hyphal system monomitic or dimitic, hyphae sometimes gelatinized, with clamp connections, not staining in iodine. Hymenium lamellar, veil sometimes present. Basidia clavate, usually with 4 sterigmata, often accompanied by thick-walled encrusted cystidia. Basidiospores hyaline, cylindrical, smooth or faintly punctate, ± thin-walled, varied in size.

Significant Genera: *Hohenbuehelia*, *Pleurotus*. Anamorph: *Nematoctonus*.

Distribution: Widespread in temperate and tropical zones.

Economic Significance: *Pleurotus ostreatus* and related species are widely cultivated, on straw, sawdust or other agroindustrial waste products. There has been research into the use of *Pleurotus* species in bioremediation of PCP-contaminated soils.

Hohenbuehelia atrocoerulea, UK; basidiomata

Pleurotus dryinus, Surrey, UK; basidiomata

Ecology: Saprobic on woody substrata, more rarely associated with roots of herbaceous plants. Many species are known to derive nutrition from nematodes, the hyphae with adhesive knobs or toxigenic secretory appendages.

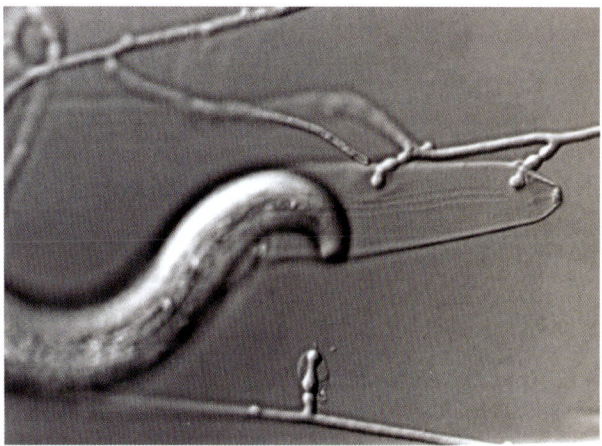
Nematoctonus sp.; hyphae parasitizing a nematode

Notes: A well-defined family, united by the nematophagous habit as well as morphological and molecular features. The *Tricholomataceae* may be a sister group, but that family needs further redefinition.

References: **Albertó, E.; Fazio, A.; Wright, J.E.** (1998). Reevaluation of *Hohenbuehelia nigra* and species with close affinities. *Mycologia* 90: 142–150; **Capelari, M.; Fungaro, M.H.P.** (2003). Determination of biological species and analysis of genetic variability by RAPD of isolates of *Pleurotus* subgenus *Coremiopleurotus*. *Mycol. Res.* 107: 1050–1054; **Corner, E.J.H.** (1994). On the agaric genera *Hohenbuehelia* and *Oudemansiella*. Part II: *Oudemansiella* Speg. *Gdns' Bull.* Singapore 46: 49–75; **Corner, E.J.H.** (1994). On the agaric genera *Hohenbuehelia* and *Oudemansiella*. Part I: *Hohenbuehelia*. *Gdns' Bull.* Singapore 46: 1–47; **De Gioia, T.; Sisto, D.; Rana, G.L.; Figliuolo, G.** (2005). Genetic structure of the *Pleurotus eryngii* species-complex. *Mycol. Res.* 109: 71–80; **Fazio, A.; Albertó, E.** (2001). Two new species of *Hohenbuehelia* from Argentina. *Mycotaxon* 77: 117–125; **Gonzalez, P.; Labarère, J.** (2000). Phylogenetic relationships of *Pleurotus* species according to the sequence and secondary structure of the mitochondrial small-subunit rRNA V4, V6 and V9 domains. *Microbiology* (Reading) 146: 209–221; **Hibbett, D.S.; Binder, M.** (2002). Evolution of complex fruiting-body morphologies in homobasidiomycetes. *Proc. R. Soc. Lond. B. Biol. Sci.* 269: 1963–1969; **Hibbett, D.S.; Thorn, R.G.** (1994). Nematode-trapping in *Pleurotus tuberregium*. *Mycologia* 86: 696–699; **Lechner, B.E.; Wright, J.E.; Albertó, E.** (2004). The genus *Pleurotus* in Argentina. *Mycologia* 96: 845–858; **Liou, G.Y.; Tzean, S.S.** (1997). Phylogeny of the genus *Arthobotrys* and allied nematode-trapping fungi based on rDNA sequences. *Mycologia* 89: 876–884; **Moncalvo, J.-M.; Vilgalys, R.; Redhead, S.A.; Johnson, J.E.; James, T.Y.; Aime, M.C.; Hofstetter, V.; Verduin, S.J.W.; Larsson, E.; Baroni, T.J.; Thorn, R.G.; Jacobsson, S.; Clemencon, H.; Miller, O.K.** (2002). One hundred and seventeen clades of euagarics. *Mol. Phylogen. Evol.* 23: 357–400; **Thorn, R.G.; Moncalvo, J.-M.; Reddy, C.A.; Vilgalys, R.** (2000). Phylogenetic analyses and the distribution of nematophagy support a monophyletic *Pleurotaceae* within the polyphyletic pleurotoid-lentinoid fungi. *Mycologia* 92: 241–252; **Vilgalys, R.** (1993). Evolution and taxonomy of the genus *Pleurotus* in Europe and North America. *Fungi and Lichens in the Baltic Region, Abstracts, Twelfth International Conference on Mycology and Lichenology* (Vilnius): 133; **Zervakis, G.I.** (1998). Mating competence and biological species within the subgenus *Coremiopleurotus*. *Mycologia* 90: 1063–1074;

Zervakis, G.I.; Moncalvo, J.-M.; Vilgalys, R. (2004). Molecular phylogeny, biogeography and speciation of the mushroom species *Pleurotus cystidiosus* and allied taxa. *Microbiology* (Reading) 150: 715–726.

Pluteaceae Kotl. & Pouzar 1972
Agaricales: Basidiomycota

Basidiomata typically stipitate and pileate, rarely secotioid and stalked or sessile. Stipe fleshy, sometimes developing from a bulbous basal volva, annular ring often present. Cap fleshy, often with a viscid and sometimes brightly coloured surface layer, fragments of the volva sometimes present. Hyphal system monomitic, hyphae with or without clamp connections, sometimes inflated, not staining in iodine. Hymenium lamellar, the gills ± free, often broad, usually lacking cystidia. Basidia often large, narrowly clavate, with 2 or 4 sterigmata. Basidiospores globose to cylindrical, usually smooth and thin-walled, hyaline or pinkish, sometimes staining blue in iodine.

Amanita fulva, Surrey, UK; basidioma

Significant Genera: *Amanita*, *Pluteus*, *Volvariella*.

Amanita spissa, Hampshire, UK; basidioma

Torrendia pulchella; basidioma

Amanita phalloides; basidiomata

Distribution: Widespread in both temperate and tropical zones.

Economic Significance: *Volvariella volvacea*, the paddy straw mushroom, is one of the most widely cultivated species, growing successfully on all kinds of lignocellulosic materials. Some species of *Amanita* are edible but others, including *A. phalloides* and *A. muscaria*, are highly poisonous and/or hallucinogenic.

Ecology: Forming ectomycorrhizas with roots of broadleaved trees, or saprobic on rotten wood, plant remains or humus.

Notes: The *Amanitaceae* and *Pluteaceae* have been separated by some authors and molecular evidence provides some support for this arrangement. However, the combined assemblage is clearly monophyletic and there is no unequivocal case for separation of the two taxa at family rather than subfamily level. The secotioid genus *Torrendia* has been confirmed as a member of the *Amanita caesarea* clade using molecular data.

References: **Banerjee, P.a; Sundberg, W.J.** (1995). The genus *Pluteus* section *Pluteus* (*Pluteaceae, Agaricales*) in the midwestern United States. *Mycotaxon* 53: 189–246; **Bhatt, R.P.; Miller, O.K.** (2004). *Amanita* subgenus *Lepidella* and related taxa in the southeastern United States. *Mem. N. Y. bot. Gdn* 89: 33–59; **Bougher, N.L.; Lebel, T.** (2002). Australasian sequestrate (truffle-like) fungi. XII. *Amarrendia* gen. nov.: an astipitate, sequestrate relative of *Torrendia* and *Amanita* (*Amanitaceae*) from Australia. *Aust. Syst. Bot.* 15: 513–525; **Drehmel, D.; Moncalvo, J.-M.; Vilgalys, R.** (1999). Molecular phylogeny of *Amanita* based on large-subunit ribosomal DNA sequences: implications for taxonomy and character evolution. *Mycologia* 91: 610–618; **Grgurinovic, C.A.** (2001). *Agaricales* in Australasia. *Aust. Syst. Bot.* 14: 395–406; **Miller, O.K.; Horak, E.** (1992). Observations on the genus *Torrendia* and a new species from Australia. *Mycologia* 84: 64–71; **Moncalvo, J.-M.; Vilgalys, R.; Redhead, S.A.; Johnson, J.E.; James, T.Y.; Aime, M.C.; Hofstetter, V.; Verduin, S.J.W.; Larsson, E.; Baroni, T.J.; Thorn, R.G.; Jacobsson, S.; Clemencon, H.; Miller, O.K.** (2002). One hundred and seventeen clades of euagarics. *Mol. Phylogen. Evol.* 23: 357–400; **Oda, T.; Tanaka, C.; Tsuda, M.** (2004). Molecular phylogeny and biogeography of the widely distributed *Amanita* species, *A. muscaria* and *A. pantherina*. *Mycol. Res.* 108: 885–896; **Seok, S.J.; Kim, Y.S.; Weon, H.Y.; Lee, K.H.; Park, K.M.; Min, K.H.; Yoo, K.H.** (2002). Taxonomic study on *Volvariella* in Korea. *Mycobiology* 30: 183–192; **Simmons, C.; Henkel, T.; Bas, C.** (2002). The genus *Amanita* in the Pakaraima Mountains of Guyana. *Persoonia* 17: 563–582; **Weiss, M.; Yang, Z.L.; Oberwinkler, F.** (1998). Molecular phylogenetic studies in the genus *Amanita*. *Can. J. Bot.* 76: 1170–1179; **Zhang, L.F.; Yang, J.B.; Yang, Z.L.** (2004). Molecular phylogeny of eastern Asian species of *Amanita* (*Agaricales, Basidiomycota*): taxonomic and biogeographic implications. *Fungal Diversity* 17: 219–238.

Pneumocystidaceae O.E. Eriksson. 1994
Pneumocystidales: Ascomycota

Mycelium absent. Vegetative cells thin-walled, irregularly shaped, uninucleate, with tubular extensions that enter invaginations in lung cells, dividing by fission; sometimes becoming thick-walled, ovoid and cyst-like (? transformed into asci), with 4–8 endogenously produced daughter cells (? ascospores), at first globose but becoming falcate.

Significant Genera: *Pneumocystis*.

Distribution: Cosmopolitan.

Economic Significance: Pneumonia caused by *Pneumocystis* infection is one of the principal causes of death in immunocompromized patients. The fungus is frequently present but non-symptomatic in otherwise healthy people.

Ecology: Extracellular parasites in lungs of mammals, especially humans and simians.

Notes: The family contains the single genus *Pneumocystis*, at one time considered to be a protozoan but now one of the basal members of the *Ascomycota*. At one stage only a single species, *Pneumocystis carinii*, was recognized, but recent research suggests that a wider evolutionary radiation may have taken place with species specific to many mammal groups.

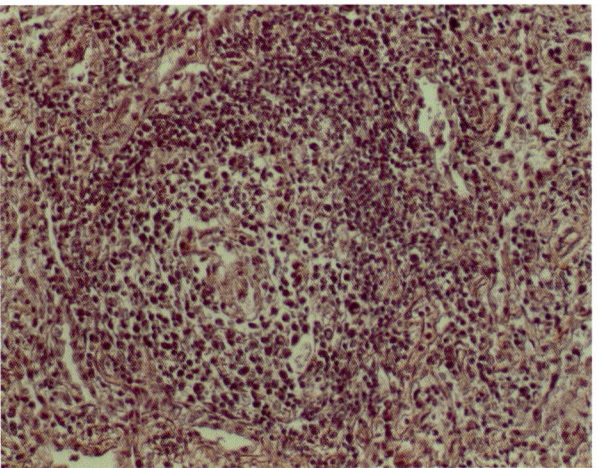

Pneumocystis carinii; cysts in lung tissue

References: **Chin, K.; Luttrell, T.D.; Roe, J.D.; Shadzi, S.; Wyder, M.A.; Kaneshiro, E.S.** (1999). Putative *Pneumocystis* dormant forms outside the mammalian host, and long-term culture derived from them: initial characterizations. *J. Eukary. Microbiol.* 46: 95S–99S; **Cushion, M.T.; Keely, S.P.; Stringer, J.R.** (2004). Molecular and phenotypic description of *Pneumocystis wakefieldiae* sp. nov., a new species in rats. *Mycologia* 96: 429–438; **Dei-Cas, E.; Mazars, E.; Aliouat, E.M.; Nevez, G.; Cailliez, J.C.; Camus, D.** (1998). The host specificity of *Pneumocystis carinii*. *J. Mycol. Médic.* 8: 1–6; **Demanche, C.; Berthelemy, M.; Petit, T.; Polack, B.; Wakefield, A.E.; Dei-Cas, E.; Guillot, J.** (2001). Phylogeny of *Pneumocystis carinii* from 18 primate species confirms host specificity and suggests coevolution. *J. Clin. Microbiol.* 39: 2126–2133; **Edman, J.C.; Kovacs, J.A.; Masur, H.; Santi, D.V.; Elwood, H.J.; Sogin, M.L.** (1988). Ribosomal RNA sequence shows *Pneumocystis carinii* to be a member of the Fungi. *Nature Lond.* 334: 519; **Eriksson, O.E.** (1994). *Pneumocystis carinii*, a parasite in lungs of mammals, referred to a new family and order (*Pneumocystidaceae, Pneumocystidales, Ascomycota*). *Syst. Ascom.* 13: 165–180; **Frenkel,**

J.K. (1999). *Pneumocystis* pneumonia, an immunodeficiency-dependent disease (IDD): a critical historical overview. *J. Eukary. Microbiol.* 46: 89S–92S; **Guillot, J.; Demanche, C.; Norris, K.; Wildschutte, H.; Wanert, F.; Berthelemy, M.; Tataine, S.; Dei-Cas, E.; Chermette, R.** (2004). Phylogenetic relationships among *Pneumocystis* from Asian macaques inferred from mitochondrial rRNA sequences. *Mol. Phylogen. Evol.* 31: 988–996; **Keely, S.P.; Fischer, J.M.; Cushion, M.T.; Stringer, J.R.** (2004). Phylogenetic identification of *Pneumocystis murina* sp. nov., a new species in laboratory mice. *Microbiology* (Reading) 150: 1153–1165; **Lu, J.J.; Bartlett, M.S.; Shaw, M.M.; Queener, S.F.; Smith, J.W.; Ortiz-Rivera, M.; Leibowitz, M.J.; Lee, C.H.** (1994). Typing of *Pneumocystis carinii* strains that infect humans based on nucleotide sequence variations of internal transcribed spacers of rRNA genes. *J. Clin. Microbiol.* 32: 2904–2912; **Mazars, E.; Dei-Cas, E.** (1998). Epidemiological and taxonomic impact of *Pneumocystis* diversity. *FEMS Immunol. Med. Microbiol.* 22: 75–80; **Norris, K.A.; Wildschutte, H.; Franko, J.; Board, K.F.** (2003). Genetic variation at the mitochondrial large-subunit rRNA locus of *Pneumocystis* isolates from simian immunodeficiency virus-infected Rhesus Macaques. *Clin. Diagn. Lab. Immunol.* 10: 1037–1042; **Palmer, R.J.; Settnes, O.P.; Lodal, J.; Wakefield, A.E.** (2000). Population structure of rat-derived *Pneumocystis carinii* in Danish wild rats. *Appl. Environm. Microbiol.* 66: 4954–4961; **Robberts, F.J.L.; Liebowitz, L.D.; Chalkley, L.J.** (2004). Genotyping and coalescent phylogenetic analysis of *Pneumocystis jiroveci* [sic] from South Africa. *J. Clin. Microbiol.* 42: 1505–1510; **Stringer, J.R.** (1998). The genome of *Pneumocystis carinii*. *FEMS Immunol. Med. Microbiol.* 22: 15–26; **Wakefield, A.E.; Stringer, J.R.; Tamburrini, E.; Dei-Cas, E.** (1998). Genetics, metabolism and host specificity of *Pneumocystis carinii*. *Medical Mycol.* 36: 183–193.

Podoscyphaceae D.A. Reid 1965
Polyporales: Basidiomycota

Cotylidia sp.; basidiomata

Basidiomata spathulate to funnel-shaped, often varied in form, the upper surface glabrous or hairy, tough and leathery, with a short wide stalk, developing singly or in clusters. Hyphal system monomitic, dimitic or trimitic, generative hyphae with thin walls and usually prominent clamp connections, skeletal hyphae hyaline, with thick refractive walls, thin-walled gloeocystidia often present. Hymenium smooth or weakly ridged, basidia sometimes accompanied by skeletal hyphae projecting through the hymenium. Basidia cylindrical to clavate, with 1–4 well-developed sterigmata. Basidiospores ± globose, ovoid or cylindrical, hyaline, thin-walled, smooth, not staining in iodine.

Significant Genera: *Cotylidia*, *Podoscypha*.

Podoscypha multizonata, Essex, UK; basidioma

Distribution: Widespread.

Economic Significance: None is known. *Podoscypha multizonata* has been proposed for Europe-wide protection under the Bern Convention.

Ecology: Growing on roots, stumps, buried wood and similar substrata.

Notes: Molecular data suggest that this family is highly polyphyletic, but further studies are needed to establish its true extent and relationships. Among other genera, *Cotylidia* may well not belong to this family, and the placement of *Caripia* is questionable.

References: **Chamuris, G.P.** (1988). The non-stipitate stereoid fungi in the northeastern United States and adjacent Canada. *Mycol. Mem.* 14: 247 pp.; **Chang, T.T.; Chiu, W.H.; Hua, J.** (1996). The Cultural Atlas of Wood-Inhabiting *Aphyllophorales* in Taiwan, vol. 1. *Mycol. Monogr.* 10: 126 pp.; **Douanla-Meli, C.; Langer, E.** (2004). A taxonomic study of the family *Podoscyphaceae* (*Basidiomycetes*), new species and new records in Cameroon. *Mycotaxon* 90: 323–335; **Ginns, J.** (1998). Genera of the North American *Corticiaceae* sensu lato. *Mycologia* 90: 1–35; **Legon, N.; Pegler, D.N.** (1996). Profiles of Fungi. 80. *Podoscypha multizonata* (Berk. & Br.) Pat.; 81. *Amylostereum chailletii* (Pers.: Fr.) Boidin; 82. *Pseudomerulius aureus* (Fr.) Jülich. *Mycologist* 10: 180; **Reid, D.A.** (1965). A monograph of the stipitate stereoid fungi. *Beih. Nova Hedwigia* 18: 484 pp.

Polyporaceae Fr. ex Corda 1839
Polyporales: Basidiomycota

Basidiomata polyporoid with a central, eccentric or lateral stalk, or resupinate; annual or perennial; fleshy, leathery or strongly woody, the upper surface smooth, tomentose or scaly, sometimes zonate. Hyphal system monomitic, dimitic or trimitic with skeleto-ligative hyphae, clamp connections frequent, cystidia absent. Hymenium tubular or lamellate, without setae. Basidia small, clavate, usually with 4 sterigmata. Basidiospores cylindrical to allantoid, thin-walled, hyaline, usually smooth, not staining in iodine.

Significant Genera: *Fomes, Lentinus, Polyporus, Poria, Trametes.*

Distribution: Cosmopolitan.

Economic Significance: Various taxa (e.g. species of *Lentinus, Lenzites, Polyporus* and *Poria*) cause economic damage to timber and rots of living hardwoods. A few (e.g. *Laetiporus sulphureus*) are edible.

Fomes fomentarius, Inverness-shire, Scotland; basidioma

Polyporus badius, Surrey, UK; basidioma

Ecology: Lignicolous, causing brown or white rot, mostly of broadleaved trees.

Trametes versicolor, Surrey, UK; basidiomata

Notes: A large and diverse family that will probably require subdivision following more research. The *Ganodermataceae* clusters within the core polyporoid clade in recent phylogenetic studies.

References: **Binder, M.; Hibbett, D.S.; Larsson, K.-H.; Larsson, E.; Langer, E.; Langer, G.** (2005). The phylogenetic distribution of resupinate forms across the major clades of mushroom-forming fungi (*Homobasidiomycetes*). *Syst. Biodiv.* 3: 113–157; **Carranza, J.; Ruiz-Boyer, A.** (2001). Cultural studies on some genera of *Basidiomycetes* (*Basidiomycota*) from Costa Rica. *Harvard Pap. Bot.* 6: 57–84; **Dai, Y.C.; Niemelä, T.; Kinnunen, J.** (2002). The polypore genera *Abundisporus* and *Perenniporia* (*Basidiomycota*) in China, with notes on *Haploporus*. *Ann. bot. fenn.* 39: 169–182; **Decock, C.; Ryvarden, L.** (2003). *Perenniporiella* gen. nov. segregated from *Perenniporia*, including a key to neotropical *Perenniporia* species with pileate basidiomes. *Mycol. Res.* 107: 93–103; **Hibbett, D.S.; Donoghue, M.J.** (2001). Analysis of character correlations among wood decay mechanisms, mating systems and substrate ranges in *Homobasidiomycetes*. *Syst. Biol.* 50: 215–242; **Hong, S.G.; Jeong, W.J.; Jung, H.S.** (2002). Amplification of mitochondrial small subunit ribosomal DNA of polypores and its potential for phylogenetic analysis. *Mycologia* 94: 823–833; **Kim, S.Y.; Jung, H.S.** (2001). Phylogenetic relationships of the *Polyporaceae* based on gene sequences of nuclear small subunit ribosomal RNAs. *Mycobiology* 29: 73–79; **Ko, K.S.; Jung, H.S.** (1999). Molecular phylogeny of *Trametes* and related genera. *Antonie van Leeuwenhoek* 75: 191–199; **Ko, K.S.; Jung, H.S.** (2002). Three nonorthologous ITS1 types are present in a polypore fungus *Trichaptum abietinum*. *Mol. Phylogen. Evol.* 23: 112–122; **Ko, K.S.; Jung, H.S.** (2002). Phylogenetic evaluation of *Polyporus* s. str. based on molecular sequences. *Mycotaxon* 82: 315–322; **Krüger, D.; Gargas, A.** (2004). The basidiomycete genus *Polyporus* – an emendation based on phylogeny and putative secondary structure of ribosomal RNA molecules. *Feddes Repert.* 115: 530–546; **Moncalvo, J.-M.; Vilgalys, R.; Redhead, S.A.; Johnson, J.E.; James, T.Y.; Aime, M.C.; Hofstetter, V.; Verduin, S.J.W.; Larsson, E.; Baroni, T.J.; Thorn, R.G.; Jacobsson, S.; Clemencon, H.; Miller, O.K.** (2002). One hundred and seventeen clades of euagarics. *Mol. Phylogen. Evol.* 23: 357–400; **Parmasto, E.; Hallenberg, N.** (2000). The genus *Abundisporus* (*Hymenomycetes, Basidiomycotina*). *Karstenia* 40: 129–138; **Silveira, R.M. Borges da; Saidman, B.O.; Wright, J.E.** (2003). *Polyporus* s. str. in southern South America: isoenzyme analysis. *Mycol. Res.* 107: 597–608; **Tomšovský, M.; Homolka, L.** (2004). Mating tests among geographically separated collections of the *Trametes versicolor* (Fr.) Pilát (*Basidiomycetes, Polyporales*) group. *Nova Hedwigia* 79: 425–431; **Valderrama, B.; Oliver, P.; Medrano-Soto, A.; Vazquez-Duhalt, R.** (2003). Evolutionary and structural diversity of fungal laccases. *Antonie van Leeuwenhoek* 84: 289–299.

Polystomellaceae Theiss. & P. Syd. 1915
Uncertain position within Dothideomycetes, Ascomycota

Stromata superficial or subcuticular, strongly flattened, black, multiloculate; composed of several layers of pseudoparenchymatous cells, without distinct ascomatal walls. Interascal tissue of cellular pseudoparaphyses, sometimes partially deliquescing at maturity. Asci cylindrical, ? fissitunicate, thick-walled especially at the apex but without clearly defined apical structures, usually 8-spored. Ascospores hyaline to brown, septate, without a sheath. Anamorphs poorly known.

Polystomella costaricensis, Ecuador; stromata on leaf

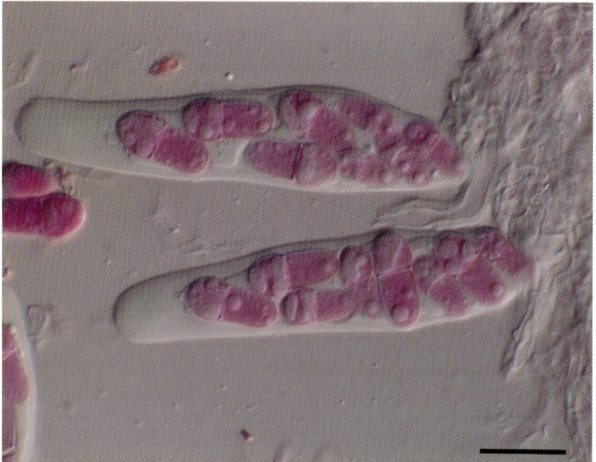
Polystomella costaricensis, Ecuador; asci and ascospores

Significant Genera: *Dothidella*, *Munkiella*.

Distribution: Confined to tropical zones, particularly of South America.

Economic Significance: None is known.

Ecology: Biotrophic on leaves.

Notes: Most species are poorly known and the affinities of the family are unclear. No molecular data are available.

References: **Swart, H.J.** (1987). Australian leaf-inhabiting fungi. XXV. *Dothidella inaequalis* and *Montagnella eucalypti*. *Trans. Br. mycol. Soc.* **89**: 483–488.

Porinaceae Rchb. 1828
Trichotheliales: Ascomycota

Thallus crustose, sometimes poorly developed. Ascomata perithecial, sometimes pale-walled and thin-walled, sometimes surrounded by a distinct involucrellum and/or covered by a thalline layer, the ostioles ± periphysate, sometimes surrounded by agglutinated radiating hairs. Interascal tissue of simple paraphyses. Asci cylindrical, thin-walled, not fissitunicate, sometimes with an inconspicuous refractive ring, not blueing in iodine. Ascospores hyaline, thin-walled or thick-walled, transversely septate or muriform, sometimes with a gelatinous sheath. Anamorphs coelomycetous.

Trichothelium argenteum; ascomata surrounded by agglutinated hairs

Porina simulans; thallus and ascomata

Significant Genera: *Porina, Trichothelium*.

Distribution: Widespread, especially in the wet tropics.

Economic Significance: None is known.

Porina epiphylla, French Guiana; asci, ascospores and paraphyses

Ecology: Lichenized with green algae, on living leaves, bark and siliceous rocks.

Notes: Little is known of the relationships of this family, as with many other groups primarily containing foliicolous lichens, and the division between the two principal genera is problematic. Currently available molecular data suggest placement within the *Ostropomycetidae*. The name *Trichotheliaceae* has also been used, incorrectly, for this family.

References: **Anon.** (2001). *Flora of Australia* (Melbourne) 58: 242 pp.; **Aptroot, A.; Diederich, P.; Sérusiaux, E.; Sipman, H.J.M.** (1997). Lichens and Lichenicolous Fungi from New Guinea. *Biblthca Lichenol.* 64: 220 pp.; **Baloch, E.; Grube, M.** (2006). Evolution and phylogenetic relationships within *Porinaceae* (*Ostropomycetidae*), focusing on foliicolous species. *Mycol. Res.* 110: 125–136; **Grube, M.; Baloch, E.; Lumbsch, H.T.** (2004). The phylogeny of *Porinaceae* (*Ostropomycetidae*) suggests a neotenic origin of perithecia in Lecanoromycetes. *Mycol. Res.* 108: 1111–1118; **Hafellner, J.; Kalb, K.** (1995). Studies in *Trichotheliales* ordo novus. *Biblthca Lichenol.* 57: 161–186; **Lücking, R.** (1996). Taxonomic studies in foliicolous species of the genus *Porina* I. The *Porina rufula* aggregate. *Bot. Acta* 109: 248–260; **Lücking, R.** (1998). Foliicolous lichens and their lichenicolous fungi collected during the Smithsonian International Cryptogamic Expedition to Guyana 1996. *Trop. Bryol.* 15: 45–76; **Lücking, R.** (2004). A revised key to foliicolous *Porinaceae* (*Ascomycota*: *Trichotheliales*). *Biblthca Lichenol.* 88: 409–426; **Lücking, R.; Cáceres, M.E.S.** (1999). New species or interesting records of foliicolous lichens. IV. *Porina pseudoapplanata* (lichenized ascomycetes: *Trichotheliaceae*), a remarkable new species with *Phyllophiale*-type isidia. *Lichenologist* 31: 349–358; **Lücking, R.; Cáceres, M.E.S.** (2004). Corticolous species of *Trichothelium* (*Ascomycota: Porinaceae*). *Mycol. Res.* 108: 571–575; **Lücking, R.; Ferraro, L.I.** (1997). New species or interesting records of foliicolous lichens. I. *Trichothelium argenteum* (Lichenized *Ascomycetes*: *Trichotheliaceae*). *Lichenologist* 29: 217–220; **Lücking, R.; Vĕzda, A.** (1998). Taxonomic studies in foliicolous species of the genus *Porina* (lichenized *Ascomycotina*: *Trichotheliaceae*) – II. The *Porina epiphylla* group. *Willdenowia* 28: 181–225; **McCarthy, P.M.** (1994). Corticolous species of *Porina* (lichenized *Ascomycotina*), *Trichotheliaceae* in Australia. II. *Nova Hedwigia* 59: 509–516; **McCarthy, P.M.; Elix, J.A.** (1996). *Myeloconis*, a new genus of pyrenocarpous lichens from the tropics. *Lichenologist* 28: 401–414; **McCarthy, P.M.; Malcolm, W.M.** (1997). The genera of *Trichotheliaceae*. *Lichenologist* 29: 1–8.

Porpidiaceae Hertel & Hafellner 1984
Lecanorales: Ascomycota

Thallus crustose. Ascomata apothecial, sessile or immersed, pale to black, sometimes irregularly shaped, with variably defined marginal tissue. Interascal tissue of branched and anastomosing paraphyses, often swollen at the apices and with a pigmented epithecium. Asci with a well-developed J+ apical cap, a conspicuous, strongly blueing, tube-like, apical ring and an outer J+ gelatinized layer. Ascospores hyaline, aseptate, often with a sheath. Anamorphs coelomycetous, pycnidial.

Significant Genera: *Bellemerea, Mycobilimbia, Porpidia*.

Bellemerea alpina, Australia; thallus and ascomata

Porpidia macrocarpa, Australia; thallus and ascomata

Distribution: Widespread, especially in south temperate regions.

Economic Significance: None is known.

Ecology: Lichenized with green algae, on rock and wood.

Notes: There is evidence to support the amalgamation of the *Porpidiaceae* with the *Lecideaceae*; species of *Lecidea* cluster within *Porpidia s.l.* in a recent phylogenetic study. However, further reappraisal of the morphological distinctions between the two families is required.

Porpidia cinereoatra, Harris, Scotland; thallus and ascomata

References: **Brodo, I.M.** (1995). *Koerberiella (Porpidiaceae, Ascomycotina)*, a new genus of lichens for North America. *Bryologist* 98: 609–611; **Brodo, I.M.; Hertel, H.** (1987). The lichen genus *Amygdalaria (Porpidiaceae)* in North America. *Herzogia* 7: 493–521; **Gowan, S.P.** (1989). A character analysis of the secondary products of the *Porpidiaceae* (lichenized *Ascomycotina*). *Syst. Bot.* 14: 77–90; **Gowan, S.P.** (1989). The lichen genus *Porpidia (Porpidiaceae)* in North America. *Bryologist* 92: 25–59; **Hafellner, J.** (1989). Die europäischen *Mycobilimbia*-Arten – eine erste Übersicht (lichenisierte *Ascomycetes, Lecanorales*). *Herzogia* 8: 53–59; **Hertel, H.** (1990). New records of lecideoid lichens from the Southern Hemisphere. *Mitt. bot. StSamml. Münch.* 28: 211–238; **Pietschmann, M.** (1990). Morphometrics of tubiform apical apparatus in *Lecideaceae, Micareaceae, Porpidiaceae* and allied families (lichenized *Ascomycetes, Lecanorales*): limitations and perspectives of statistical inference. *Nova Hedwigia* 51: 521–549; **Rambold, G.** (1989). A monograph of the saxicolous lecideoid lichens of Australia (excl. Tasmania). *Biblthca Lichenol.* 34: 345 pp.; **Rambold, G.; Hagedorn, G.** (1998). The distribution of selected diagnostic characters in the *Lecanorales*. *Lichenologist* 30: 473–487; **Reeb, V.; Lutzoni, F.; Roux, C.** (2004). Contribution of *RPB2* to multilocus phylogenetic studies of the euascomycetes (*Pezizomycotina, Fungi*) with special emphasis on the lichen-forming *Acarosporaceae* and evolution of polyspory. *Mol. Phylogen. Evol.* 32: 1036–1060.

Protogastraceae Zeller 1934
Boletales: Basidiomycota

Basidiomata gasteroid, minute, without rhizoids or other clear attachment structure, thin-walled, covered in loosely interwoven hyphae, forming a hollow sphere lined with hymenium. Basidia clavate to fusiform, 1- to 4-spored, inconspicuous and presumably evanescent. Basidiospores ovoid to ellipsoidal, smooth, thin-walled, yellowish, with a distinct exospore layer.

Significant Genera: *Protogaster*.

Distribution: Only known from the USA.

Economic Significance: None is known.

Ecology: Found on damp ground in burrows, associated with roots. Its nutritional status is unknown.

Notes: Very little is known about the single species *Protogaster rhizophilus* and its position in the *Boletales* requires confirmation.

References: **Reijnders, A.F.M.** (2000). A morphogenetic analysis of the basic characters of the gasteromycetes and their relation to other basidiomycetes. *Mycol. Res.* 104: 900–910; **Zeller, S.M.** (1934). *Protogaster*, representing a new order of the *Gasteromycetes*. *Ann. Mo. bot. Gdn* 21: 231–249; **Zeller, S.M.; Walker, L.B.** (1935). *Gasterella*, a new uniloculate gasteromycete. *Mycologia* 27: 573–579.

Protomycetaceae Gray 1821
Taphrinales: Ascomycota

Protomyces pachydermus, Essex, UK; a gall-forming species causing reddish discoloured stems and leaves of *Taraxacum*

Mycelium apparently diploid, growing intercellularly within plant tissue, sometimes forming thick-walled, smooth or ornamented, resting spores either throughout gall tissue or in a continuous subepidermal layer, the resting spores either converted directly into a spore sac, or rupturing to form the spore sac from an emergent internal membrane. Ascogenous cells ('spore sacs') formed either directly or via rupture and the subsequent emergence of an internal membrane, with a multinucleate protoplast which becomes oriented in a peripheral layer, the nuclei undergoing meiosis to form 4 small hyaline aseptate ascospores ('endospores'), the endospores discharged in a single mass. Anamorphs yeast-like.

Significant Genera: *Burenia*, *Protomyces*.

Distribution: Widespread in temperate zones.

Economic Significance: None is known.

Ecology: Biotrophic on leaves and stems, usually gall-forming.

Notes: A small and poorly studied group at the base of the *Ascomycota* phylogeny, close to the *Taphrinaceae*. *Saitoëlla* has been placed in the family, but molecular data suggest that it may be more closely related to the *Saccharomycetes*.

References: **Ahearn, D.G.; Sugiyama, J.; Simmons, R.B.** (1998). *Saitoella* S. Goto, Sugiyama, Hamamoto & Komagata. *The Yeasts, A Taxonomic Study* (Amsterdam): 600–601; **Döbbeler, P.** (1995). *Burenia myrrhidendri* spec. nov. (*Protomycetales*), ein bemerkenswerter biotropher ascomycet in den früchten einer baumförmigen umbellifere aus Costa Rica. *Nova Hedwigia* 60: 171–177; **Kurtzman, C.P.** (1998). *Protomyces* Unger. *The Yeasts, A Taxonomic Study* (Amsterdam): 353–357; **Kurtzman, C.P.; Blanz, P.A.** (1998). Ribosomal RNA/DNA sequence comparisons for assessing phylogenetic relationships. *The Yeasts, A Taxonomic Study* (Amsterdam): 69–74; **Nishida, H.; Sugiyama, J.** (1993). Phylogenetic relationships among *Taphrina*, *Saitoella*, and other higher fungi. *Mol. Biol. Evol.* 10: 431–436; **Sjamsuridzal, W.; Tajiri, Y.; Nishida, H.; Thuan, T.B.; Kawasaki, A.; Yokota, A.; Sugiyama, J.** (1997). Evolutionary relationships of members of the genera *Taphrina*, *Protomyces*, *Schizosaccharomyces*, and related taxa within the archiascomycetes: integrated analysis of genotypic and phenotypic characters. *Mycoscience* 38: 267–280; **Sugiyama, J.; Nagahama, T.; Nishida, H.** (1996). Fungal diversity and phylogeny with emphasis on 18S ribosomal DNA sequence divergence. *Microbial Diversity in Time and Space* Proceedings of the International Symposium, October 24-26, 1994, Tokyo, Japan (New York): 41–51.

Protoscyphaceae Kutorga & D. Hawksw. 1997
Uncertain position within Dothideomycetes, Ascomycota

Thallus absent, stroma absent or poorly developed. Ascomata not clearly delimited, ± perithecial, formed as pulvinate or discoid locules which later merge to form a compact hymenium, not blueing in iodine. Interascal tissue of filiform gelatinized pseudoparaphyses, forming an epithecium above the asci. Asci clavate, thick-walled, fissitunicate, with a distinct ocular chamber. Ascospores hyaline to pale brown, muriform. Anamorphs not known.

Significant Genera: *Protoscypha*.

Distribution: The only species is known from tropical Central and South America.

Economic Significance: None is known.

Ecology: Parasitic on *Coccodiella* species, biotrophic leaf-inhabiting fungi.

Notes: Placement of this family is problematic. The lack of well-defined ascomatal tissue and mycoparasitic nature might indicate affinities with the *Archontiales*, but ascus and interascal tissue structure differs. No molecular data are available.

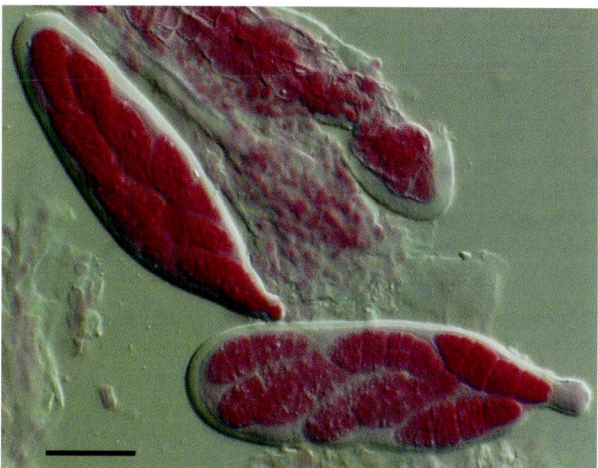

Protoscypha subtropica, Costa Rica; asci and ascospores

References: **Kutorga, E.; Hawksworth, D.L.** (1997). A reassessment of the genera referred to the family *Patellariaceae* (*Ascomycota*). *Syst. Ascom.* 15: 1–110; **Magnes, M.** (1997). Weltmonographie der *Triblidiaceae*. *Biblthca Mycol.* 165: 177 pp.

Protothelenellaceae Vězda, H. Mayrhofer & Poelt 1985
Uncertain position within Lecanoromycetes, Ascomycota

Thallus crustose and poorly developed or absent. Stromata absent. Ascomata intermediate between apothecial and perithecial, immersed, sometimes becoming erumpent, dark green to black, opening by a broad pore; peridium composed of interwoven hyphae in a gelatinous matrix. Interascal tissue of sparingly branched and anastomosing trabeculate pseudoparaphyses, becoming free at the apices, the apices not swollen; hymenium gelatinized, blueing in iodine; periphyses absent. Asci cylindrical, persistent, at first thick-walled but usually without separable layers, the apex somewhat thickened and blueing diffusely in iodine, with a J+ apical ring, splitting at the apex to release the spores. Ascospores hyaline, usually muriform, without a sheath. Anamorphs coelomycetous, pycnidial. Conidia bacilliform.

Significant Genera: *Protothelenella*.

Distribution: Widespread in north temperate regions.

Economic Significance: None is known.

Ecology: Saprobic on bark or lichenized with green algae, rarely lichenicolous or bryophilous.

Notes: A family of uncertain affinities, apparently allied to the *Thrombiaceae*, which has very similar ascus structure but differs in periphysis, ascospore and hymenial gel characteristics.

References: **David, J.C.** (1987). Studies on the genus *Epigloea*. *Syst. Ascom.* 6: 217–221; **Mayrhofer, H.** (1987). Ergänzende Studien zur Taxonomie der Gattung *Protothelenella*. *Herzogia* 7: 313–342; **Schmitt, I.; Mueller, G.; Lumbsch, H.T.** (2005). Ascoma morphology is homoplaseous and phylogenetically misleading in some pyrenocarpous lichens. *Mycologia* 97: 362–374; **Sherwood-Pike, M.A.; Boise, J.R.** (1986). Studies in lignicolous ascomycetes: *Xylopezia* and *Mycowinteria*. *Brittonia* 38: 35–44.

Psathyrellaceae Vilgalys, Moncalvo & Redhead 2001
Agaricales: Basidiomycota

Basidiomata pileate, often conical or campanulate, membranous to fleshy, glabrous, granular or scurfy, often becoming lacerate and/or deliquescent, the stipe central, veil absent or evanescent, sometimes annular. Clamp connections usually present. Gills free or attached, parallel-sided or wedge-shaped, usually dark-coloured, often deliquescent and often with large protruding cystidia. Basidia often short and broad, frequently dimorphic, trimorphic or tetramorphic. Basidiospores variously shaped, generally dark brown with thick and sometimes layered walls, usually with a well-developed germ pore, rarely ornamented.

Significant Genera: *Coprinellus, Coprinopsis, Parasola, Psathyrella*.

Distribution: Cosmopolitan.

Economic Significance: Several *Coprinopsis* species are edible and widely eaten, but are toxic when ingested with alcohol.

Ecology: Saprobic, in grass, on dung, rotten wood and similar substrata.

Notes: Many of these fungi have traditionally been classified in the *Coprinaceae*, but the type of *Coprinus* has been demonstrated to be distantly related to most of the other species in that genus and is now placed in the *Agaricaceae*. The *Bolbitiaceae* is the most closely related family to this group.

Coprinopsis atramentaria, Surrey, UK; basidioma

Coprinellus disseminatus, Devon, UK; basidiomata

Parasola kuehneri, Surrey, UK; basidioma

References: **Bogart, F. van de** (1979). The genus *Coprinus* in western North America, part III: section *Atramentarii*. *Mycotaxon* 10: 155–174; **Enderle, M.; Bender, H.** (1990). Studien zur Gattung *Coprinus* in der Bundesrepublik Deutschland. V. *Z. Mykol.* 56: 19–46; **Kits van Waveren, E.** (1985). The Dutch, French and British species of *Psathyrella*. *Persoonia* Suppl. 2: 300 pp.; **Moncalvo, J.-M.; Lutzoni, F.M.; Rehner, S.A.; Johnson, J.; Vilgalys, R.** (2000). Phylogenetic relationships of agaric fungi based on nuclear large subunit ribosomal DNA sequences. *Syst. Biol.* 49: 278–305; **Redhead, S.A.; Vilgalys, R.; Moncalvo, J.-M.; Johnson, J.; Hopple, J.S.** (2001). *Coprinus* Pers. and the disposition of *Coprinus* species *sensu lato*. *Taxon* 50: 203–241; **Smith, A.H.** (1972). The North American species of *Psathyrella*. *Mem. N. Y. bot. Gdn* 24: 633 pp.

Pseudeurotiaceae Malloch & Cain 1970
Uncertain position within Ascomycetes, Ascomycota

Stromata absent. Ascomata cleistothecial, formed from coiled initials, usually brown to black, the peridium pseudoparenchymatous, thin-walled, not cephalothecoid in structure. Interascal tissue absent. Asci irregularly disposed, ± globose, thin-walled, without apical structures, evanescent. Ascospores small, hyaline or brown, usually aseptate, smooth-walled, without germ pores. Anamorphs hyphomycetous, mostly *Sporothrix*-like but conidiogenous cells have only poorly defined denticles.

Significant Genera: *Pseudeurotium*. Anamorph: *Teberdinia*.

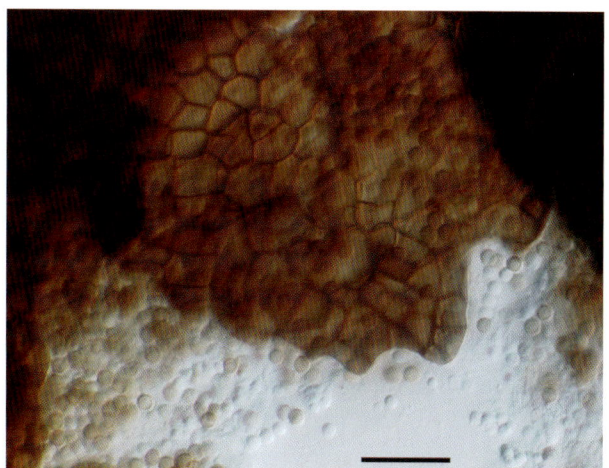

Pseudeurotium zonatum, Argentina; ascomatal wall and ascospores

Distribution: Widespread, primarily known from temperate regions.

Economic Significance: None is known.

Ecology: Saprobic on woody tissue and rotting vegetation, often isolated from soil.

Notes: A small and rather isolated family. *Connersia* and *Pleuroascus* are phylogenetically separate from *Pseudeurotium* and *Leuconeurospora* should be excluded.

References: **Arx, J.A. von; Figueras, M.J.; Guarro, J.** (1988). Sordariaceous ascomycetes without ascospore ejaculation. *Beih. Nova Hedwigia* 94: 104 pp.; **Sogonov, M.V.; Schroers, H.-J.; Gams, W.; Dijksterhuis, J.; Summerbell, R.C.** (2005). The hyphomycete *Teberdinia hygrophila* gen. nov., sp. nov. and related anamorphs of *Pseudeurotium* species. *Mycologia* 97: 695–709; **Suh, S.O.; Blackwell, M.** (1999). Molecular phylogeny of the cleistothecial fungi placed in *Cephalothecaceae* and *Pseudeurotiaceae*. *Mycologia* 91: 836–848.

Pseudoperisporiaceae Toro 1926
Uncertain position within Dothideomycetes, Ascomycota

Mycelium superficial or within surface layers of substratum. Ascomata superficial, perithecial, small, ± globose, with a clearly defined ostiole, sometimes setose; peridium thin-walled, composed of pseudoparenchymatous cells. Interascal tissue of narrowly cellular pseudoparaphyses, often deliquescing at maturity. Asci saccate, fissitunicate. Ascospores hyaline to brown, variously shaped, transversely septate. Anamorphs coelomycetous.

Significant Genera: *Epipolaeum, Phaeodimeriella, Wentiomyces*.

Distribution: Widespread, primarily in tropical zones.

Economic Significance: None is known.

Rasutoria abietis; colonies on *Abies* needles

Ecology: Saprobic or biotrophic on plant tissue, or on other fungi. Many parasitize sooty moulds. One species of *Raciborskiomyces* has been identified as a major component of soil mycobiota using molecular methods.

Notes: Species concepts are uncertain in many groups and the family is likely to be highly polyphyletic. Molecular data indicate a placement close to the *Capnodiaceae*, but more studies are needed.

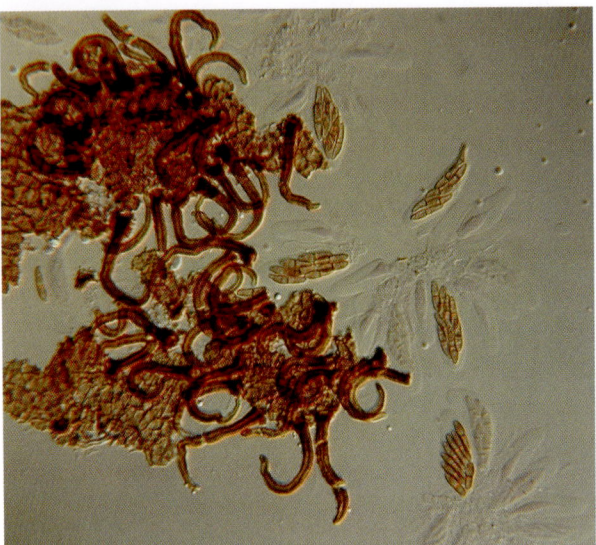

Phaeodimeriella sp., Kenya; ascoma, asci and ascospores

References: **Barr, M.E.** (1997). Notes on some 'dimeriaceous' fungi. *Mycotaxon* 64: 149–171; **Döbbeler, P.** (1987). *Ascomycetes* growing on *Polytrichum sexangulare*. *Arctic and Alpine Mycology* II (New York): 87–107; **Döbbeler, P.** (1997). Biodiversity of bryophilous ascomycetes. *Biodiv. Cons.* 6: 721–738; **Farr, M.L.** (1986). Notes on mostly neotropical fungi. *Sydowia* 38: 65–74; **Lumbsch, H.T.; Schmitt, I.; Lindemuth, R.; Miller, A.; Mangold, A.; Fernandez, F.; Huhndorf, S.** (2005). Performance of four ribosomal DNA regions to infer higher-level phylogenetic relationships of inoperculate euascomycetes (*Leotiomyceta*). *Mol. Phylogen. Evol.* 34: 512–524; **Sivanesan, A.** (1987). Studies on the genera *Dimeriellopsis, Hyalomeliolina* and *Nematostoma*. *Syst. Ascom.* 6: 201–212; **Valinsky, L.; Della Vedova, G.; Jiang, T.; Borneman, J.** (2002). Oligonucleotide fingerprinting of rRNA genes for analysis of fungal community composition. *Appl. Environm. Microbiol.* 68: 5999–6004.

Psoraceae Zahlbr. 1898
Lecanorales: Ascomycota

Thallus crustose or squamulose. Ascomata apothecial, sessile or immersed, flat or domed, sometimes on the thallus margin, without well-defined marginal tissue, pale or dark. Interascal tissue of usually wide paraphyses, occasionally anastomosing, with swollen apices and a crystalline epithecium, often agglutinated. Asci with a J+ apical cap, a usually well-developed strongly staining tubular ring and a sometimes very well-developed outer J+ gelatinized layer. Ascospores hyaline, aseptate, without a sheath. Anamorphs coelomycetous, pycnidial. Conidia usually bacilliform.

Significant Genera: *Psora*.

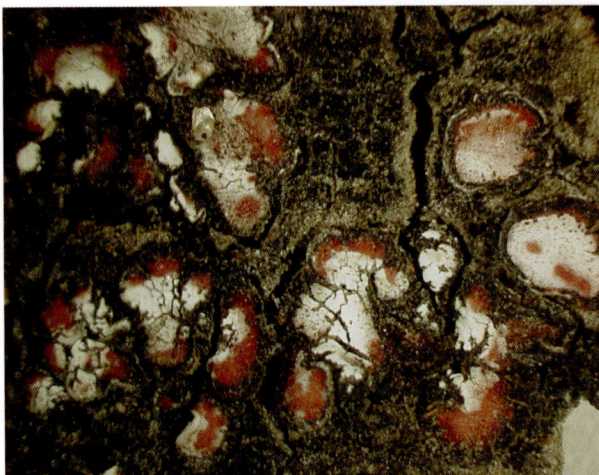

Psora decipiens, China; thallus and ascomata

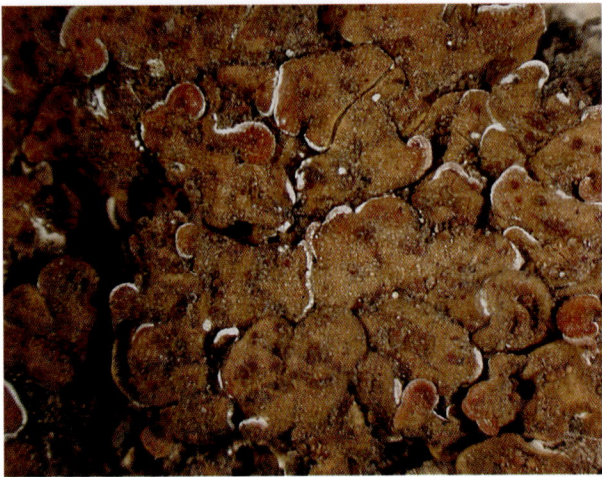

Psora rubiformis, New Mexico, USA; thallus

Distribution: Widespread.

Economic Significance: None is known.

Ecology: Lichenized with green algae, generally on calcareous soils or rock.

Notes: A small family, probably most closely related to the *Bacidiaceae*.

References: **Brusse, F.** (1985). *Glyphopeltis* (*Lecidiaceae*), a new lichen genus from southern Africa. *Lichenologist* 17: 267–268; **Buschbom, J.; Mueller, G.** (2004). Resolving evolutionary relationships in the lichen-forming genus *Porpidia* and related allies (*Porpidiaceae, Ascomycota*). *Mol. Phylogen. Evol.* 32: 66–82; **Hertel, H.; Rambold, G.** (1988). Cephalodiate Arten der Gattung *Lecidea* sensu lato (*Ascomycetes* lichenisati). *Pl. Syst. Evol.* 158: 289–312; **Lumbsch, H.T.; Kothe, H.W.** (1993). Die Entwicklung der Thalli von *Eremastrella crystallifera* und ihre taxonomische Bedeutung (*Psoraceae, Ascomycotina*). *Nova Hedwigia* 57: 19–32; **Pant, G.** (1989). A note on lichen genus *Protoblastenia* from India. *J. Bombay nat. Hist. Soc.* 85: 658–659; **Peršoh, D.; Beck, A.; Rambold, G.** (2004). The distribution of ascus types and photobiont selection in *Lecanoromycetes* (*Ascomycota*) against the background of a revised SSU nrDNA phylogeny. *Mycol. Prog.* 3: 103–121; **Timdal, E.** (1987). A revision of *Psora* (*Lecideaceae*) in North America. *Bryologist* 89: 253–275; **Timdal, E.** (1994). The ascus of *Glyphopeltis ligustica*. *Cryptog. Bryol.-Lichénol.* 15: 171–172.

Pterulaceae Corner 1970
Agaricales: Basidiomycota

Basidiomata clavarioid, filiform or clavate, usually stipitate, often dichotomously branched, negatively or positively geotropic, tough-textured or cartilaginous, the stipe sometimes setose or sclerotic. Hyphal system dimitic, skeletal hyphae hyaline, usually unbranched and indeterminate, sometimes poorly developed, clamp connections present or absent. Hymenium amphigenous, waxy, the basidia often interspersed with thick-walled or encrusted cystidia, often confined to the apical regions. Basidia large or small, clavate, with 2–4 sterigmata. Basidiospores often large, ellipsoidal to subglobose, hyaline, smooth and thin-walled or thick-walled.

Significant Genera: *Deflexula*, *Pterula*.

Distribution: Primarily found in the tropics, although there are some taxa from Europe and New Zealand.

Economic Significance: None is known.

Ecology: Primarily saprobic on rotten wood. One unnamed taxon is cultivated in the neotropics by ants of the *Apterostigma pilosum* group.

Notes: A poorly studied group whose position within the *Agaricales* needs further analysis.

References: **Cripps, C.L.; Caesar, T.J.** (1998). A conidia-forming basidiomycete in the *Pterulaceae*. *Mycotaxon* 69: 153–158; **Munkacsi, A.B.; Pan, J.J.; Villesen, P.; Mueller, U.G.; Blackwell, M.; McLaughlin, D.J.** (2004). Convergent coevolution in the domestication of coral mushrooms by fungus-growing ants. *Proc. R. Soc. Lond. B. Biol. Sci.* 271: 1777–1782; **Perez-Moreno, J.; Villarreal, L.** (1989). First report of the clavarioid genera *Lachnocladium* and *Pterula* (*Holobasidiomycetes*) from the Mexican tropic. *Micol. Neotrop. Aplic.* 2: 123–130; **Petersen, R.H.** (1988). The clavarioid fungi of New Zealand. *DSIR Bulletin* 236: 170 pp.; **Pine, E.M.; Hibbett, D.S.; Donoghue, M.J.** (1999). Phylogenetic relationships of cantharelloid and clavarioid *Homobasidiomycetes* based on mitochondrial and nuclear rDNA sequences. *Mycologia* 91: 944–963; **Roberts, P.** (1999). Clavarioid fungi from Korup National Park, Cameroon. *Kew Bull.* 54: 517–539; **Roberts, P.J.; Spooner, B.M.** (2000). Cantharelloid, clavarioid and thelephoroid fungi from Brunei Darussalam. *Kew Bull.* 55: 843–851.

Pucciniaceae Chevall. 1826
Uredinales: Basidiomycota

Spermogonia subepidermal, ampulliform, ostiolate with discrete walls. Aecia typically *Aecidium*-like, more rarely *Caeoma*-like or *Uredo*-like, with or without a distinct wall. Aeciospores usually verrucose, usually formed in chains. Uredinia *Uredo*-like or *Uredostilbe*-like, lacking a discrete wall or with laterally fused palisade-like walls, sometimes with paraphysis-like hyphae. Urediniospores verrucose or echinulate, pedicellate, germ pores variously arranged. Telia usually without discrete walls, rarely with a palisade-like peridium or separated into locules by sterile hyphal cells, sometimes with paraphyses. Teliospores formed singly, usually pedicellate and 1- or 2-celled, the septa transverse or oblique, each cell typically with a single germ pore, usually germinating to form an external basidium.

Gymnosporangium clavariiforme, Var, France; telia on branches of *Juniperus*

Significant Genera: *Gymnosporangium*, *Puccinia*, *Uromyces*.

Pterula multifida; basidiomata

Distribution: Cosmopolitan, well represented in all climatic zones.

Economic Significance: Many species are economically important parasites of crop plants. Cereal pathogens are particularly significant, including *Puccinia graminis*, *P. striiformis* and *P. triticina* on wheat, *P. melanocephala* on sugarcane and *P. purpurea* and *P. sorghi* on maize and sorghum. Other important crops affected include asparagus, beans, chysanthemum, citrus, cotton, groundnut, guava, onion, potato and sunflower.

Gymnosporangium cornutum, Sutherland, Scotland; aecia on leaf of *Sorbus aucuparia*

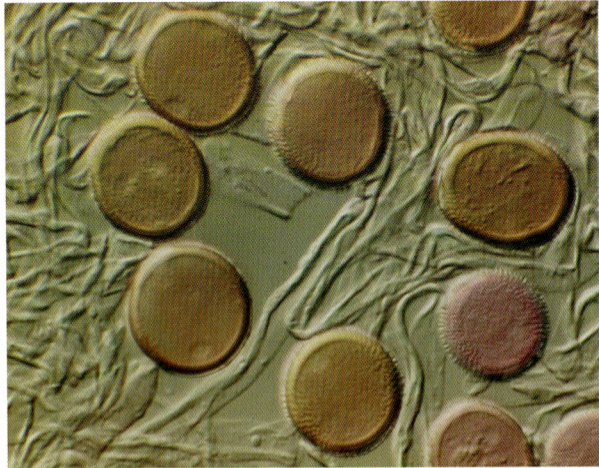

Puccinia mogiphanis, Galápagos Islands; urediniospores

Ecology: Biotrophic, heteroecious or autoecious, associated with leaf and stem tissue of a very wide range of plants.

Notes: An extremely diverse family; *Puccinia* alone is estimated to include 4000 described species. Host identity is unsurprisingly relied upon heavily for identification, but there is ample evidence for specificity in many species. There is strong molecular evidence that *Gymnosporangium* forms a separate clade from other genera of *Pucciniaceae*. Both *Puccinia* and *Uromyces* seem to be polyphyletic, suggesting that teliospore septation is not a significant character in evolutionary terms.

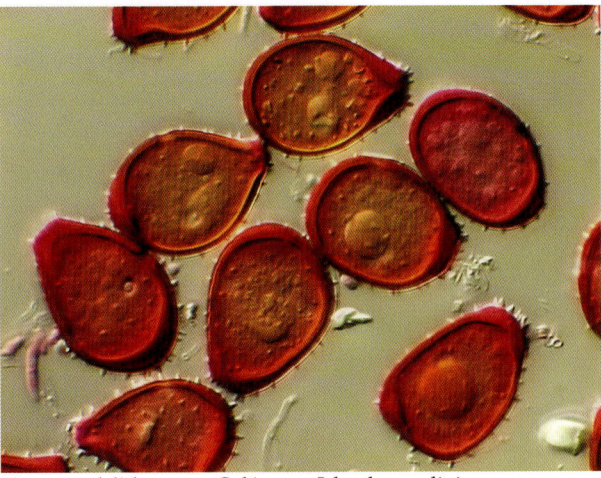

Uromyces dolichosporus, Galápagos Islands; urediniospores

References: **Aime, M.C.** (2006). Towards resolving family-level relationships in rust fungi (*Uredinales*). *Mycoscience* 47: 112–122; **Anikster, Y.; Eilam, T.; Bushnell, W.R.; Kosman, E.** (2005). Spore dimensions of *Puccinia* species of cereal hosts as determined by image analysis. *Mycologia* 97: 474–484; **Anikster, Y.; Szabo, L.J.; Eilam, T.; Manisterski, J.; Koike, S.T.; Bushnell, W.R.** (2004). Morphology, life cycle biology, and DNA sequence analysis of rust fungi on garlic and chives from California. *Phytopathology* 94: 569–577; **Araya, C.M.; Alleyne, A.T.; Steadman, J.R.; Eskridge, K.M.; Coyne, D.P.** (2004). Phenotypic and genotypic characterization of *Uromyces appendiculatus* from *Phaseolus vulgaris* in the Americas. *Pl. Dis.* 88: 830–836; **Chung, W.H.; Tsukiboshi, T.; Ono, Y.; Kakishima, M.** (2004). Morphological and phylogenetic analyses of *Uromyces appendiculatus* and *U. vignae* on legumes in Japan. *Mycoscience* 45: 233–244; **Chung, W.H.; Tsukiboshi, T.; Ono, Y.; Kakishima, M.** (2004). Phylogenetic analyses of *Uromyces viciae-fabae* and its varieties on *Vicia*, *Lathyrus*, and *Pisum* in Japan. *Mycoscience* 45: 1–8; **Cummins, G.B.; Hiratsuka, Y.** (2003). *Illustrated Genera of Rust Fungi* (St Paul): 225 pp.; **Kosman, E.; Pardes, E.; Anikster, Y.; Manisterski, J.; Yehuda, P.B.; Szabo, L.J.; Sharon, A.** (2004). Genetic variation and virulence on *Lr26* in *Puccinia triticina*. *Phytopathology* 94: 632–640; **Lee, S.K.; Kakishima, M.** (1999). Aeciospore surface structures of *Gymnosporangium* and *Roestelia* (*Uredinales*). *Mycoscience* 40: 109–120; **Maier, W.; Begerow, D.; Weiss, M.; Oberwinkler, F.** (2003). Phylogeny of the rust fungi: an approach using nuclear large subunit ribosomal DNA sequences. *Can. J. Bot.* 81: 12–23; **Maier, W.; Wingfield, B.D.; Mennicken, M.; Wingfield, M.J.** (2007). Polyphyly and two emerging lineages in the rust genera *Puccinia* and *Uromyces*. *Mycol. Res.* 111: 176-185; **Ono, Y.** (2002). The diversity of nuclear cycle in microcyclic rust fungi (*Uredinales*) and its ecological and evolutionary implications. *Mycoscience* 43: 421–439; **Van Der Merwe, M.; Ericson, L.; Walker, J.; Thrall, P.H.; Burdon, J.J.** (2007). Evolutionary relationships among species of *Puccinia* and *Uromyces* (*Pucciniaceae*, *Uredinales*) inferred from partial protein coding gene phylogenies. *Mycol. Res.* 111: 163-175; **Virtudazo, E.V.; Nakamura, H.; Kakishima, M.** (2001). Phylogenetic analysis of sugarcane rusts based on sequences of ITS, 5.8 S rDNA and D1/D2 regions of LSU rDNA. *J. Gen. Pl. Path.* 67: 28–26; **Weber, R.W.S.; Webster, J.; Engel, G.** (2003). Phylogenetic analysis of *Puccinia distincta* and *P. lagenophorae*, two closely related rust fungi causing epidemics on *Asteraceae* in Europe. *Mycol. Res.* 107: 15–24; **Wingfield, B.D.; Ericson, L.; Szaro, T.; Burdon, J.J.** (2004). Phylogenetic patterns in the *Uredinales*. *Australas.*

Pl. Path. 33: 327–335; **Wood, A.R.; Crous, P.W.** (2005). Morphological and molecular characterization of *Endophyllum* species on perennial asteraceous plants in South Africa. *Mycol. Res.* 109: 387–400.

Pucciniastraceae Gäum. ex Leppik 1972
Uredinales: Basidiomycota

Spermogonia subcuticular or subepidermal, ± globose, conical or discoid, without discrete walls. Aecia *Peridermium*-like or rarely *Milesia*-like, tubular or blister-like, with well-developed wall tissue. Aeciospores ornamented, often verrucose, formed in chains and interspersed with intercalary cells. Uredinia *Milesia*-like with usually well-developed walls, blister-like or domed, ostiolate. Urediniospores formed singly, hyaline or brown, ornamented (usually echinulate) with varied germ pore orientation and usually pedicellate. Telia intradermal or subepidermal, not erumpent, wall structures absent. Teliospores hyaline or brown, laterally adhering, formed in single layers, unicellular or with transverse septa, the cells sometimes with a single germ pore, germinating to penetrate the epidermis and form external basidia.

Significant Genera: *Milesina, Pucciniastrum*.

Distribution: Widespread in temperate climates.

Economic Significance: Some species occur on economically important conifers, but do not cause major diseases. *Pucciniastrum* species cause some damage to raspberry and ornamental *Fuchsia* plants.

Ecology: Biotrophic and heteroecious, the aecial stages all on conifer leaves, primarily on *Abies* species. Alternate hosts belong to a wide range of families.

Pucciniastrum goeppertianum, Colorado, USA; aecia

Notes: Molecular evidence suggests that this family is strongly polyphyletic, with at least three separate evolutionary clades. *Thekopsora* and some species of *Pucciniastrum* cluster with the *Coleosporiaceae* and *Cronartiaceae*; *Hyalopsora*, *Melampsoridium* and *Milesina* may also belong in this major clade.

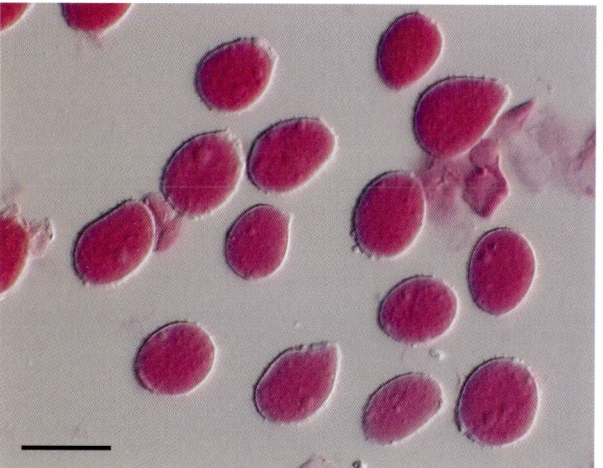

Pucciniastrum epilobii, Western Australia; urediniospores

Pucciniastrum areolatum; symptoms

References: **Aime, M.C.** (2006). Towards resolving family-level relationships in rust fungi (*Uredinales*). *Mycoscience* 47: 112–122; **Berndt, R.; Bauer, R.; Oberwinkler, F.** (1994). Ultrastructure of the host-parasite interface in the fern rusts *Milesia, Uredinopsis* and *Hyalopsora* (Pucciniastraceae, Uredinales). *Can. J. Bot.* 72: 1084–1094; **Berndt, R.; Oberwinkler, F.** (1997). Haustorial ultrastructure and morphology of *Melampsorella* and *Thekopsora areolata*. *Mycologia* 89: 698–705; **Cummins, G.B.; Hiratsuka, Y.** (2003). *Illustrated Genera of Rust Fungi* (St Paul): 225 pp.; **Kurkela, T.; Hanso, M.; Hantula, J.** (1999). Differentiating characteristics between *Melampsoridium* rusts infecting birch and alder leaves. *Mycologia* 91: 987–992; **Liang, Y.-m.; Tian, C.-m.; Kakishima, M.** (2006). Phylogenetic relationships on 14 morphologically similar species of *Pucciniastrum* in Japan based on rDNA sequence data. *Mycoscience* 47: 137–144; **Maier, W.; Begerow, D.; Weiss, M.; Oberwinkler, F.** (2003). Phylogeny of the rust fungi: an approach using nuclear large subunit ribosomal DNA sequences. *Can. J. Bot.* 81: 12–23; **Motokura, Y.; Kimishima, E.; Kimura, S.** (1999). Rust disease of raspberry caused by *Pucciniastrum americanum* intercepted in plant quarantine. *Res. Bull. Pl. Protect. Serv.* Japan 35: 103–108; **Yamaoka, Y.; Katsuya, K.** (1987). Axenic cultures of *Pucciniastrum agrimoniae, P. boehmeriae* and *P. coryli. Trans. Mycol. Soc. Japan* 28: 155–161.

Pucciniosiraceae Cummins & Y. Hirats. 1983
Uredinales: Basidiomycota

Spermogonia usually absent, where present subepidermal, ampulliform with discrete walls. Aecia and uredinia not produced. Telia with or without distinct wall structures, often *Aecidium*-like in gross appearance. Teliospores sessile, produced in basipetal succession, generally with intercalary cells, unicellular or bicellular, generally adhering to each other and often extruded as columns or threads, basidia external or internal.

Significant Genera: *Dietelia, Pucciniosira*.

Pucciniosira triumfettae, Trinidad; telia

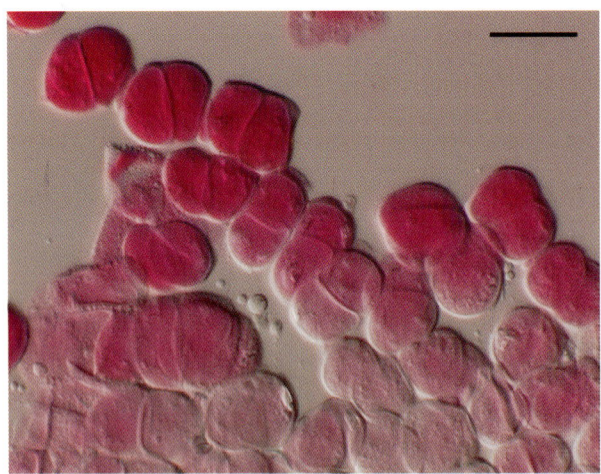
Pucciniosira triumfettae, Trinidad; teliospores

Distribution: Most species are tropical in distribution.

Economic Significance: Little is known; important crop plants are not affected.

Ecology: Biotrophic and autoecious, associated with a wide range of plant families.

Notes: An artificial assemblage of microcyclic rust fungi that are probably related to *Pucciniaceae*. No large-scale molecular phylogenetic studies are available to date, but preliminary studies support this conclusion.

References: **Aime, M.C.** (2006). Towards resolving family-level relationships in rust fungi (*Uredinales*). *Mycoscience* 47: 112–122; **Cummins, G.B.; Hiratsuka, Y.** (2003). *Illustrated Genera of Rust Fungi* (St Paul): 225 pp.; **Gjærum, H.B.; Namaganda, M.; Lye, K.A.** (2000). Ugandan rust fungi 3. The genera *Dietelia* and *Pucciniosira*. *Lidia* 5: 18–20; **Hernández, J.R.** (2000). *Baeodromus ranunculi*, a new rust on *Ranunculus* from Argentina, and a synopsis of *Baeodromus*. *Mycotaxon* 76: 329–336; **Wingfield, B.D.; Ericson, L.; Szaro, T.; Burdon, J.J.** (2004). Phylogenetic patterns in the *Uredinales*. *Australas. Pl. Path.* 33: 327–335.

Pyrenothricaceae Zahlbr. 1926
Uncertain position within Chaetothyriomycetes, Ascomycota

Thallus dark brown, indeterminate, gelatinous, composed of cyanobacterial filaments covered in anastomosing fungal hyphae. Ascomata perithecial, minute, pyriform, the ostiole small, ? lysigenous; peridium thin, dark, composed of pseudoparenchymatous cells. Interascal tissue lacking but with periphysoids extending into the perithecial chamber; hymenium J+ blue. Asci clavate to saccate, fissitunicate, with rostrate dehiscence. Ascospores dark grey, muriform. Anamorphs unknown.

Significant Genera: *Cyanoporina, Pyrenothrix*.

Distribution: Widespread, primarily tropical and subtropical.

Economic Significance: None is known.

Ecology: Associated with cyanobacteria, on leaves and bark.

Notes: Only three species are known from this family and its phylogenetic position has not been confirmed using molecular methods.

References: **Herrera-Campos, M. de los A.; Huhndorf, S.; Lücking, R.** (2005). The foliicolous lichen flora of Mexico IV: a new, foliicolous species of *Pyrenothrix* (*Chaetothyriales*: *Pyrenothricaceae*). *Mycologia* 97: 356–361; **Rogers, R.W.** (1992). Keys to the Australian lichen genera. *Flora of Australia* Vol. 54. Lichens – Introduction, *Lecanorales* 1 (Canberra): 65–94.

Pyrenulaceae Rabenh. 1870
Pyrenulales: Ascomycota

Thallus inconspicuous and usually immersed or absent. Ascomata perithecial, sometimes aggregated, globose to flattened, papillate, the ostioles sometimes lateral, sometimes surrounded by clypeus-like structures, the wall hyphal in construction. Interascal tissue initially of narrow anastomosing trabeculate pseudoparaphyses in an often faintly J+ gel, with true paraphyses subsequently formed from the basal layers. Asci cylindrical, fissitunicate, with a small or large ocular chamber, sometimes staining differentially, not blueing in iodine, without an apical ring, rarely thin-walled and evanescent, in which case the ascospores accumulating in a mazaedial mass. Ascospores hyaline or brown, transversely septate or muriform, the septa often thickened (distoseptate), rarely ornamented with grooves. Anamorphs coelomycetous. Conidiomata pycnidial, black. Conidia filiform, often curved.

Significant Genera: *Anthracothecium*, *Pyrenula*.

Distribution: Widespread, though most species are tropical.

Economic Significance: None is known.

Ecology: Usually lichenized with green algae (especially *Trentepohliaceae*), on bark.

Notes: A poorly known though speciose family, with molecular data indicating a position within the *Chaetothyriomycetes*. The *Requienellaceae* is now accepted as a distinct group which includes lichenized as well as non-lichenized species.

Pyrenula occidentalis, Argyll, Scotland; thallus and ascomata

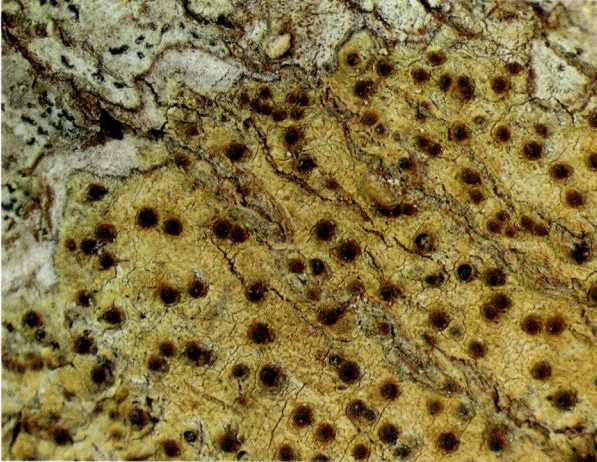

Pyrenula ochraceoflava; thallus and ascomata

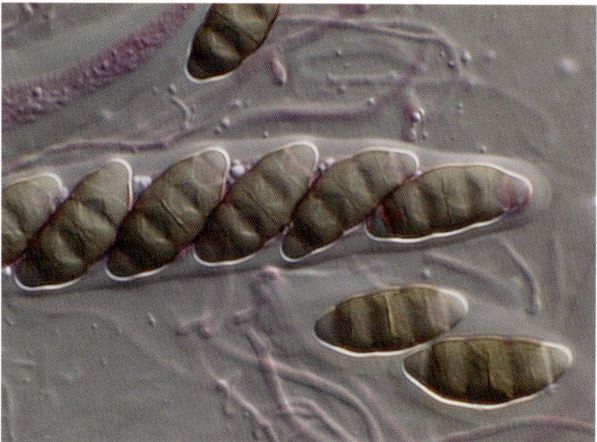

Pyrenula macrospora, UK; ascus and ascospores

References: **Aptroot, A.** (1991). Tropical pyrenocarpous lichens. A phylogenetic approach. *Tropical Lichens: Their Systematics, Conservation, and Ecology* (Oxford) 43: 253–273; **Aptroot, A.** (1991). A monograph of the *Pyrenulaceae* (excluding *Anthracothecium* and *Pyrenula*) and the *Requienellaceae*, with notes on the *Pleomassariaceae*, the *Trypetheliaceae* and *Mycomicrothelia* (lichenized and non-lichenized *Ascomycetes*. *Biblthca Lichenol.* 44: 178 pp.; **Bhattacharya, D.; Lutzoni, F.; Reeb, V.; Simon, D.; Fernandez, F.** (2000). Widespread occurrence of spliceosomal introns in the rDNA genes of ascomycetes. *Mol. Biol. Evol.* 17: 1971–1984; **Del Prado, R.; Schmitt, I.; Kautz, S.; Palice, Z.; Lücking, R.; Lumbsch, H.T.** (2006). Molecular data place *Trypetheliaceae* in *Dothideomycetes*. *Mycol. Res.* 110: 511–520; **Harada, H.** (1999). *Sulcopyrenula*, a new pyrenocarpous lichen genus (*Pyrenulaceae*, lichenized *Ascomycota*). *Lichenologist* 31: 567–573; **Harris, R.C.** (1989). A sketch of the family *Pyrenulaceae* (*Melanommatales*) in eastern North America. *Mem. N. Y. bot. Gdn* 49: 74–107; **Hawksworth, D.L.** (1986). The genus *Dipyrenis* Clem. *Nova Hedwigia* 43: 1–5; **Lumbsch, H.T.; Mangold, A.; Lücking, R.; Garcia, M.A.; Martín, M.P.** (2004). Phylogenetic position of the genera *Nadvornikia* and *Pyrgillus* (*Ascomycota*) based on molecular data. *Symb. bot. upsal.* 34: 9–17; **Tibell, L.** (1996). Caliciales. *Fl. Neotrop. Monogr.* 69: 78 pp.; **Upreti, D.K.** (1998). A key to the lichen genus *Pyrenula* from India, with nomenclatural notes. *Nova Hedwigia* 66: 557–576.

Pyronemataceae Corda 1842
Pezizales: Ascomycota

Ascomata apothecial or cleistothecial, sometimes on a well-developed hyphal mat, formed from clustered antheridia and ascogonia, small to large, discoid, cupulate or pulvinate, sessile or shortly stipitate, often brightly coloured due to carotenoid pigments, sometimes with hyaline or dark spinose hairs surrounding the hymenium; sometimes coalescing, if cleistothecial usually large, hypogeous or emergent, thick-walled, hollow, solid or chambered. Interascal tissue of simple or branched paraphyses, sometimes swollen and pigmented at the apices. Asci ± cylindrical, persistent, operculate, not blueing in iodine, with complex ascal pores, (sometimes ± globose in cleistothecial taxa). Ascospores usually ellipsoidal, hyaline or brown, smooth or ornamented, without a sheath, uninucleate, non-septate, guttulate or not. Anamorphs hyphomycetous.

Significant Genera: *Aleuria, Anthracobia, Geopora, Pyronema, Scutellinia, Trichophaea*.

Distribution: Cosmopolitan.

Economic Significance: *Pyronema* species are frequently encountered growing on damp plaster etc. in buildings.

Anthracobia melaloma, Scotland; ascomata

Geopyxis carbonaria, Scotland; ascomata

Ecology: Saprobic on soil, dung or rotten wood or mycorrhizal, some species hypogeous, others associated with fire sites.

Notes: Probably a polyphyletic family, but rather little molecular sequencing has been reported. Many of the mycorrhizal species appear to constitute significant components of forest soil mycobiota. The *Ascodesmidaceae* and *Glaziellaceae* may be synonyms.

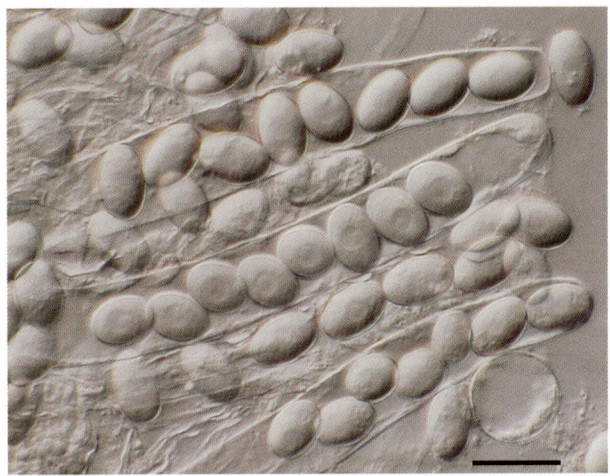
Pyronema domesticum, UK; asci and ascospores

Scutellinia scutellata, Argyll, Scotland; ascomata

References: **Benkert, D.** (1987). Beiträge zur Taxonomie der Gattung *Lamprospora* (*Pezizales*). *Z. Mykol*. 53: 195–271; **Hansen, K.; LoBuglio, K.F.; Pfister, D.H.** (2005). Evolutionary relationships of the cup-fungus genus *Peziza* and *Pezizaceae* inferred from multiple nuclear genes: RPB2, β-tubulin and LSU rDNA. *Mol. Phylogen. Evol*. 36: 1–23; **Hansen, K.; Perry, B.A.; Pfister, D.H.** (2006). Phylogenetic origins of two cleistothecial fungi, *Orbicula parietina* and *Lasiobolidium orbiculoides*, within the operculate discomycetes. *Mycologia* 97: 1023–1033; **Harrington, F.A.; Pfister, D.H.; Potter, D.; Donoghue, M.J.** (1999). Phylogenetic studies within the *Pezizales*. I. 18S rRNA sequence data and classification. *Mycologia* 91: 41–50; **Kimbrough, J.W.** (1989). Arguments towards restricting the limits of the *Pyronemataceae* (*Ascomycetes*, *Pezizales*). *Mem. N. Y. bot. Gdn* 49: 326–335; **Kimbrough, J.W.; Curry, K.J.** (1986). Septal structures in apothecial tissues of the tribe *Aleurieae* in the *Pyronemataceae* (*Pezizales*, *Ascomycetes*). *Mycologia* 78: 407–417; **Landvik, S.; Egger, K.N.; Schumacher, T.** (1997). Towards a subordinal classification of the *Pezizales* (*Ascomycota*): phylogenetic analyses of SSU rDNA sequences. *Nordic Jl Bot*. 17: 403–418; **Liu, C.Y.; Zhuang, W.Y.** (2006). Relationships among some members of the genus *Otidea* (*Pezizales*, *Pyronemataceae*). *Fungal Diversity* 23: 181–192; **Moore, R.T.** (1987). The genera of *Rhizoctonia*-like fungi: *Ascorhizoctonia*, *Ceratorhiza* gen. nov., *Epulorhiza* gen.

nov., *Moniliopsis*, and *Rhizoctonia*. *Mycotaxon* 29: 91–99; **Moravec, J.** (1989). A taxonomic revision of the genus *Cheilymenia* – 1. Species close to *Cheilymenia rubra*. *Mycotaxon* 36: 169–186; **Schumacher, T.** (1990). The genus *Scutellinia* (*Pyronemataceae*). *Op. bot.* 101: 107 pp.; **Zhuang, W.Y.; Korf, R.P.** (1986). A monograph of the genus *Aleurina* Massee (= *Jafneadelphus* Rifai). *Mycotaxon* 26: 361–400.

Pyxidiophoraceae G.R.W. Arnold 1971
Pyxidiophorales: Ascomycota

Stroma (thallus) absent. Ascomata perithecial, rarely cleistothecial, ± hyaline, often long-necked. Interascal tissue absent. Asci clavate, without an apical ring, very thin-walled, evanescent. Ascospores broadly fusiform to clavate, 1- to 3-septate, often caudate, with ± pigmented patches that probably have an attachment function. Anamorph if present *Chalara*-like or *Thaxteriola*.

Significant Genera: *Pyxidiophora*.

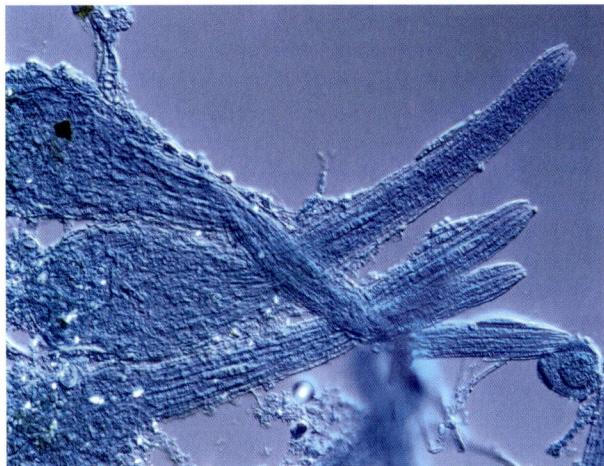

Pyxidiophora sp.; ascomata

Distribution: Known primarily from north temperate zones.

Economic Significance: None is known.

Ecology: Mainly coprophilous, usually associated with mites and other arthropods colonizing the substratum.

Notes: An enigmatic family that may occupy an isolated position within the *Laboulbeniomycetes*. The links between morphology and phylogeny in this subclass are difficult to resolve and more molecular data are required.

References: **Blackwell, M.; Bridges, J.R.; Moser, J.C.; Perry, T.J.** (1986). Hyperphoretic dispersal of a *Pyxidiophora* anamorph. *Science* N.Y. 232: 993–995; **Blackwell, M.; Malloch, D.W.** (1989). *Pyxidiophora*: life histories and arthropod associations of two species. *Can. J. Bot.* 67: 2552–2562; **Blackwell, M.; Malloch, D.W.** (1989). *Pyxidiophora* (*Pyxidiophoraceae*): a link between the *Laboulbeniales* and hyphal ascomycetes. *Mem. N. Y. bot. Gdn* 49: 23–32; **Blackwell, M.; Spatafora, J.W.; Malloch, D.; Taylor, J.W.** (1993). Consideration of higher taxonomic relationships involving *Pyxidiophora*. *Ceratocystis* and *Ophiostoma* Taxonomy, Ecology and Pathogenicity (St. Paul): 105–108; **Henk, D.A.; Weir, A.; Blackwell, M.** (2003). *Laboulbeniopsis termitarius*, an ectoparasite of termites newly recognized as a member of the *Laboulbeniomycetes*. *Mycologia* 95: 561–564; **Webster, J.; Hawksworth, D.L.** (1986). *Pyxidiophora spinulo-rostrata*, a new species with denticulate conidiophores from submerged twigs in south-west England. *Trans. Br. mycol. Soc.* 87: 77–79.

Quambalariaceae Z.W. Beer, Begerow & R. Bauer 2006
Microstromatales: Basidiomycota

Colonies stromatic, subcuticular or formed in substomatal cavities, wall structures poorly defined. Hyphae intercellular, with dolipore septa with swollen pore lips, enclosed on both sides by membrane caps. Telia absent. Conidiophores erumpent, single or clustered, hyaline, rarely branched, the fertile regions proliferating sympodially, weakly geniculate, the scars not thickened, usually with small denticles. Conidia generally clavate to cylindrical, hyaline, aseptate, smooth-walled, sometimes proliferating to form secondary conidia.

Significant Genera: *Quambalaria*.

Distribution: Species are known from Australia, South Africa and the Netherlands.

Economic Significance: *Quambalaria* infections have caused changes in preferred species of the host plant for commercial plantations in Australia.

Ecology: Known as minor pathogens of *Eucalyptus* and close relatives, and also as an isolate from human skin.

Notes: A distinct lineage of the *Microstromatales*, with unequivocal teleomorphic structures unknown. Species have some morphological similarities with the *Sporothrix* anamorphs of the *Ophiostomataceae*, and it is curious that that family also contains human as well as plant pathogens.

References: **Beer, Z.W. de; Begerow, D.; Bauer, R.; Pegg, G.S.; Crous, P.W.; Wingfield, M.J.** (2006). Phylogeny of the *Quambalariaceae* fam. nov., including important *Eucalyptus* pathogens in South Africa and Australia. *Stud. Mycol.* 55: 289–298; **Begerow, D.; Bauer, R.; Oberwinkler, F.** (2001). *Muribasidiospora*: Microstromatales or Exobasidiales?. *Mycol. Res.* 105: 798–810; **Simpson, J.A.** (2000). *Quambalaria*, a new genus of eucalypt pathogens. *Australas. Mycol.* 19: 57–62.

Radiomycetaceae Hesselt. & J.J. Ellis 1974
Mucorales: Zygomycota

Vegetative thallus a mycelium, complex fruit bodies absent. Sporophores (sporangiophores) arising from submerged mycelium, single or in pairs, simple or branched, rhizoids (at base) and stolons present, often well-developed, darkening with age; terminating in either spherical or clavate fertile vesicles bearing pedicellate sporangiola or a single sporangium. Sporangia unispored or multispored, pyriform to more or less spherical or lageniform and with a distinct spherical venter and long neck; columella absent or prominent and dome-shaped; wall echinulate (spines capitate) or appearing smooth, persistent or deliquescent; dehiscence simple by wall dissolution or fracture or through the neck following the dissolution of an apical mucilaginous plug. Sporangiospores broadly ellipsoid (sometimes with truncate ends), smooth, hyaline. Zygospores globose to subglobose; zygosporangial wall smooth and thin; suspensors opposed, bearing appendages; homothallic or unknown.

Significant Genera: *Radiomyces, Saksenaea*.

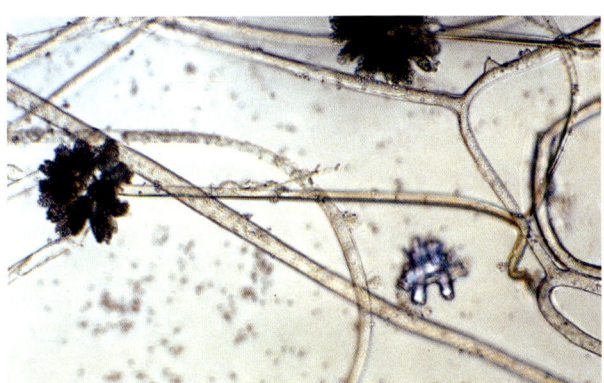

Radiomyces mexicanus; sporophore

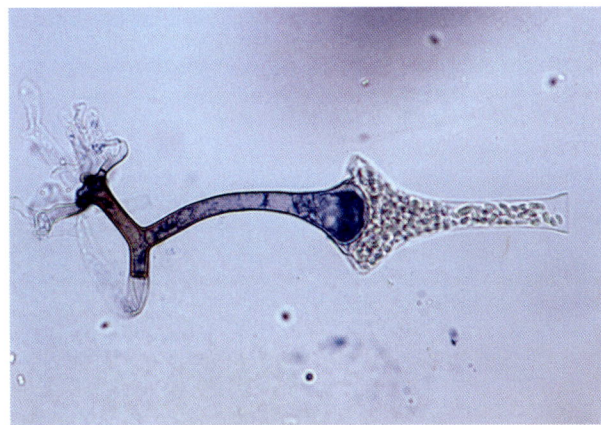

Saksenaea vasiformis; sporophore and sporangium

Distribution: Widespread, but especially in warm regions.

Economic Significance: None is known.

Ecology: Essentially soil fungi. Species of *Radiomyces* are relatively rare and most readily isolated as a component of the early coprophilous succession. *Saksenaea vasiformis* and *Apophysomyces elegans* are reported as emerging pathogens and infrequent causal agents of zygomycosis (such as severe cutaneous mucormycosis and osteomyelitis), usually due to traumatic implantation of the fungus.

Notes: This family is expanded to include two other genera of opportunistic pathogens of animals; the monotypic *Apophysomyces* (from *Mucoraceae*) and *Saksenaea* (from *Saksenaeaceae*).

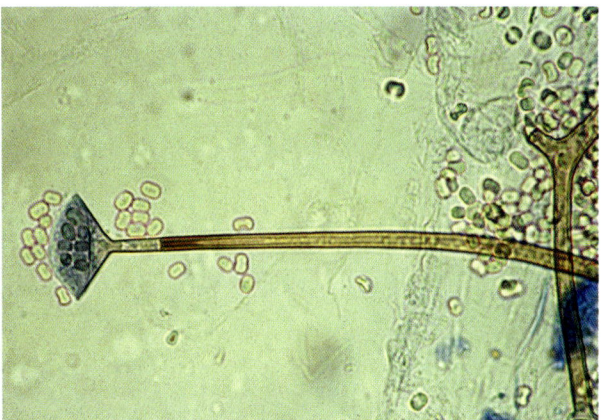

Apophysomyces elegans; sporophore

References: **Benny, G.L.; Benjamin, R.K.** (1992). The *Radiomycetaceae* (*Mucorales*; *Zygomycetes*). III. A new species of *Radiomyces*, and cladistic analysis and taxonomy of the family; with a discussion of evolutionary ordinal relationships in *Zygomycotina*. *Mycologia* 83: 713–735; **Chakrabarti, A.; Ghosh, A.; Prasad, G.S.; David, J.K.; Gupta, S.; Das, A.; Sakhuja, V.; Panda, N.K.; Singh, S.K.; Das, S.; Chakrabarti, T.** (2003). *Apophysomyces elegans*: an emerging zygomycete in India. *J. Clin. Microbiol.* 41: 783–788; **Kitz, D.J.; Embree, R.W.; Cazin, J.** (1980). *Radiomyces* a genus in the *Mucorales* pathogenic for mice. *Sabouraudia* 18: 115–121; **Meis, J.F.; Kullberg, B.J.; Pruszczynski, M.; Veth, R.P.** (1994). Severe osteomyelitis due to the zygomycete *Apophysomyces elegans*. *J. Clin. Microbiol.* 32: 3078–3081; **Ribes, J.A.; Vanover-Sams, C.L.; Baker, D.J.** (2000). *Zygomycetes* in human disease. *Clin. Microbiol. Rev.* 13: 236–301; **Tanphaichitr, V.S.; Chaiprasert, A.; Suvatte, V.; Thasnakorn, P.** (1990). Subcutaneous mucormycosis caused by *Saksenaea vasiformis* in a thalassaemic child: first case report in Thailand. *Mycoses* 33: 303–309.

Ramalinaceae C. Agardh 1821
Lecanorales: Ascomycota

Thallus fruticose, much-branched, erect or pendulous, the branches usually distinctly flattened (strap-shaped), smooth or ridged, greyish to yellow, not gelatinous, sometimes with pseudocyphellae and frequently with soralia. Ascomata apothecial, usually shortly stipitate, often on short lateral branches, usually pale, often pruinose. Interascal tissue of sparingly branched paraphyses. Asci elongate-clavate, with a J+ apical cap, and usually a well-developed non-staining apical cushion, without an outer J+ gelatinized layer. Ascospores hyaline, 1-septate, often curved, without a mucilaginous sheath. Anamorphs coelomycetous, pycnidial. Conidiomata immersed in the thallus, the ostiole sometimes blackened. Conidiogenous cells proliferating percurrently. Conidia hyaline, bacilliform.

Significant Genera: *Ramalina*.

Distribution: Widespread.

Economic Significance: None is known.

Ecology: Lichenized with green algae, on rocks, branches etc., often forming extensive carpets of thalli; often prominent in maritime communities.

Notes: Some molecular data suggest that this family should be merged within the *Bacidiaceae* (which has similar ascus morphology) and perhaps also the *Catillariaceae*, but perhaps those families are unacceptably heterogenous and more genes should be analysed. The genus *Ramalinora* was proposed as a crustose counterpart of *Ramalina*, but preliminary molecular data suggest that they are not closely related.

Ramalina fastigiata, Pembrokeshire, Wales; thallus with ascomata

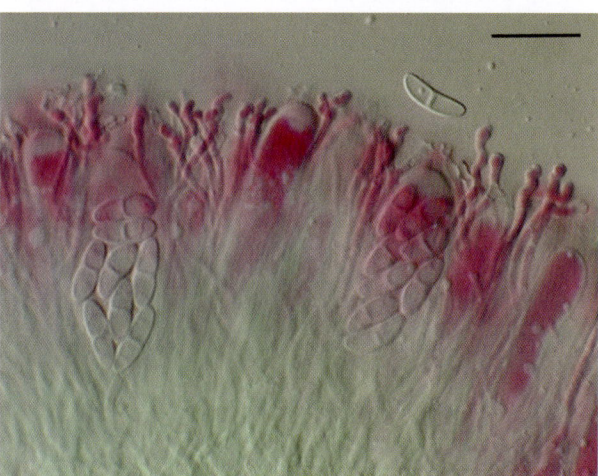
Ramalina siliquosa, Harris, Scotland; asci, ascospores and paraphyses

Ramalina spp., Carloway Broch, Lewis, Scotland; thalli

References: **Blanchon, D.J.; Braggins, J.E.; Stewart, A.** (1996). The lichen genus *Ramalina* in New Zealand. *J. Hattori bot. Lab.* 79: 43–98; **Cordeiro, L.M.C.; Iacomini, M.; Stocker-Wörgötter, E.** (2004). Culture studies and secondary compounds of six *Ramalina* species. *Mycol. Res.* 108: 489–497; **Culberson, C.F.; Culberson, W.L.; Johnson, A.** (1990). The *Ramalina americana* complex (*Ascomycotina, Ramalinaceae*): chemical and geographic correlations. *Bryologist* 93: 167–186; **Culberson, W.L.; Culberson, C.F.; Johnson, A.** (1993). Speciation in lichens of the *Ramalina siliquosa* complex (*Ascomycotina, Ramalinaceae*): gene flow and reproductive isolation. *Am. J. Bot.* 80: 1472–1481; **Groner, U.; LaGreca, S.** (1997). The 'Mediterranean' *Ramalina panizzei* north of the Alps: morphological, chemical and rDNA sequence data. *Lichenologist* 29: 441–454; **Kashiwadani, H.; Kalb, K.** (1993). The genus *Ramalina* in Brazil. *Lichenologist* 25: 1–31; **LaGreca, S.** (1999). A phylogenetic evaluation of the *Ramalina americana* chemotype complex (lichenized *Ascomycota, Ramalinaceae*) based on rDNA ITS sequence data. *Bryologist* 102: 602–618; **LaGreca, S.; Lumbsch, H.T.** (2001). No evidence from rDNA ITS sequence data for a placement of *Ramalinora* in the *Ramalinaceae*. *Lichenologist* 33: 172–176; **Lumbsch, H.T.** (1998). The use of metabolic data in lichenology at the species and subspecific levels. *Lichenologist* 30: 357–367; **Spjut, R.W.** (1996). *Niebla* and *Vermilacinia* (*Ramalinaceae*) from California and Baja California. *Sida Bot. Misc.* 14: 208 pp.; **Stevens, G.N.** (1987). The lichen genus *Ramalina* in Australia. *Bull. Br. Mus. nat. Hist.* Bot. 16: 107–223; **Stevens, G.N.** (1991). The tropical Pacific species of *Usnea* and *Ramalina* and their relationship to species in other parts of the world. *Tropical Lichens: Their Systematics, Conservation, and Ecology* (Oxford) 43: 47–67; **Stocker-Wörgotter.; Elix, J.A.; Grube, M.** (2004). Secondary chemistry of lichen-forming fungi: chemosyndromic variation and DNA-analyses of cultures and chemotypes in the *Ramalina farinacea* complex. *Bryologist* 107: 152–162.

Raveneliaceae Leppik 1972
Uredinales: Basidiomycota

Spermogonia discoid or conical, ostiolate, with distinct walls, subepidermal or subcuticular. Aecia *Aecidium*-like, *Caeoma*-like or *Uredo*-like, with or without a distinct wall and sometimes accompanied by paraphysis-like hyphae. Aeciospores typically verrucose or echinulate, solitary on pedicels or formed in chains with intercalary cells. Uredinia *Malupa*-like, *Calidion*-like or *Uredo*-like. Urediniospores usually surrounded by paraphysis-like hyphae, usually echinulate, pedicellate, germ pores usually well-developed, in various configurations. Telia erumpent, often with paraphysis-like hyphae. Teliospores pedicellate, either multiseptate or with several separate spores clustered on a shared simple or multicelled pedicel, when septate either vertically multiseptate or muriform, sometimes accompanied by hygroscopic cysts, thick-walled, smooth, verrucose, echinulate or with complex appendages, germ pores 1 or 2 in each cell, germinating to produce external basidia.

Significant Genera: *Ravenelia*, *Sphaerophragmium*.

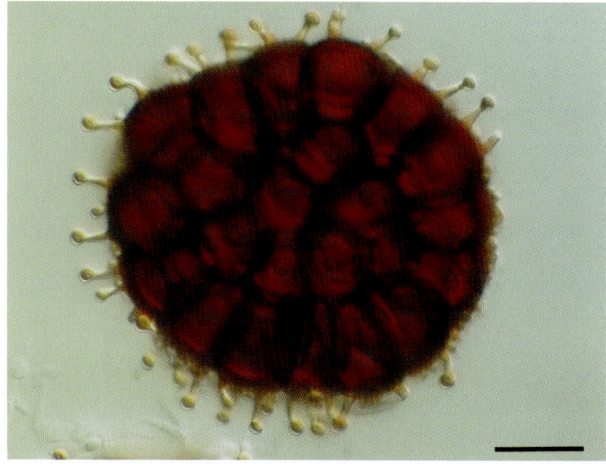

Ravenelia ornata, India; teliospore cluster

Distribution: Most species are tropical in distribution.

Economic Significance: A few species cause diseases of economically important trees, especially *Acacia* and *Emblica* species, but major losses are not reported. There has been interest in using *Diabole cubensis* to control invasive *Mimosa* species.

Ecology: Biotrophic, autoecious and associated particularly with the *Leguminosae* and *Rosaceae*.

Notes: A poorly known family that may well require reforming, but very few sequences are available. *Sphaerophragmium*, *Triphragmiopsis* and *Triphragmium* have in the past been separated into their own family.

Ravenelia mimosae-himalayanae; India; telia on leaves of *Mimosa himalayana*

Ravenelia mimosae-himalayanae; India; teliospore clusters and paraphyses

References: **Aime, M.C.** (2006). Towards resolving family-level relationships in rust fungi (*Uredinales*). *Mycoscience* 47: 112–122; **Berndt, B.** (1997). Morphology of haustoria of *Ravenelia* and *Kernkampella* spp. *Mycol. Res.* 101: 23–34; **Cummins, G.B.; Hiratsuka, Y.** (2003). *Illustrated Genera of Rust Fungi* (St Paul): 225 pp.; **Gjærum, H.B.; Namaganda, M.; Lye, K.A.** (2000). Ugandan rust fungi 11. The genus *Ravenelia*. *Lidia* 5: 91–96; **Hennen, J.F.; Figueiredo, M.B.; Carvalho, A.A. de** (2000). *Esalque holwayi* gen. et comb. nov., a rust of Brazilian ironwood (*Caesalpinia* species). *Mycologia* 92: 312–316; **Hennen, J.F.; Sotão, H.M.P.; Winkler Hennen, M.M.** (1998). The genus *Diorchidium* in the neotropics. *Mycologia* 90: 1079–1086; **Hernández, J.R.; Hennen, J.F.** (2002). The genus *Ravenelia* in Argentina. *Mycol. Res.* 106: 954–974; **Ono, Y.** (2002). The diversity of nuclear cycle in microcyclic rust fungi (*Uredinales*) and its ecological and evolutionary implications. *Mycoscience* 43: 421–439; **Reenen, M. van; Merwe, C.F. van der** (1992). The genus *Ravenelia* on *Acacia* in South Africa. *Phytophylactica* 24: 127; **Savile, D.B.O.** (1989). Raveneliaceae revisited. *Can. J. Bot.* 67: 2983–2994.

Requienellaceae Boise 1986
Pyrenulales: Ascomycota

Thallus absent or poorly developed and immersed. Ascomata immersed to superficial, perithecial, subglobose, papillate, black, thick-walled, often covered by a clypeate layer. Interascal tissue of sparsely septate, rarely branched, pseudoparaphyses. Asci cylindric-clavate, fissitunicate, with a broad cylindrical apical differentially staining region, not blueing in iodine. Ascospores dark brown, distoseptate. Anamorphs unknown.

Significant Genera: *Parapyrenis*, *Requienella*.

Distribution: Widespread.

Economic Significance: None is known.

Ecology: On wood and bark. Most species appear not to be lichenized.

Notes: Little is known of this family and no phylogenetic information is available. The distoseptate spores suggest placement within the *Pyrenulales*, but the interascal tissue is not typical of that order.

References: **Alias, S.A.; Hyde, K.D.; Jones, E.B.G.** (1996). *Pyrenographa xylographoides* from Malaysian and Australian mangroves. *Mycol. Res.* 100: 580–582; **Boise, J.** (1986). *Requienellaceae*, a new family of *Loculoascomycetes*. *Mycologia* 78: 37–41; **Harris, R.C.** (1989). A sketch of the family *Pyrenulaceae* (*Melanommatales*) in eastern North America. *Mem. N. Y. bot. Gdn* 49: 74–107; **Poonyth, A.D.; Hyde, K.D.; Aptroot, A.; Peerally, A.** (2000). *Mauritiana rhizophorae* gen. et sp. nov. (*Ascomycetes, Requienellaceae*), with a list of terrestrial saprobic mangrove fungi. *Fungal Diversity* 4: 101–116.

Rhamphosporaceae R. Bauer & Oberw. 1997
Doassansiales: Basidiomycota

Sori in living leaf and stem tissue, scattered or gregarious, becoming brown. Hyphae intracellular, haustoria present. Teliospores formed singly, usually lemon-shaped, hyaline or pale yellow, smooth or finely verruculose, formed on the branches of fertile hyphae, germinating directly to form basidia. Basidia cylindrical, septate with an apical cluster of 4–6 4-celled fertile branches, which each give rise to 2–3 basidiospores. Basidiospores filiform, anastomosing and proliferating to form secondary spores.

Significant Genera: *Rhamphospora*.

Distribution: Widespread in both north temperate and neotropic zones.

Economic Significance: None is known.

Ecology: Biotrophic, causing leaf and stem spots on *Nymphaeaceae* in freshwater habitats.

Notes: Apparently related to the *Doassansiaceae* but differing by the presence of haustoria in host cells.

References: **Bauer, R.; Oberwinkler, F.; Vánky, K.** (1997). Ultrastructural markers and systematics in smut fungi and allied taxa. *Can. J. Bot.* 75: 1273–1314; **Begerow, D.; Bauer, R.; Oberwinkler, F.** (1998). Phylogenetic studies on nuclear large subunit ribosomal DNA sequences of smut fungi and related taxa. *Can. J. Bot.* 75: 2045–2056; **Begerow, D.; Bauer, R.; Oberwinkler, F.** (2002). The *Exobasidiales*: an evolutionary hypothesis. *Mycol. Prog.* 1: 187–199; **Begerow, D.; Lutz, M.; Oberwinkler, F.** (2002). Implications of molecular characters for the phylogeny of the genus *Entyloma*. *Mycol. Res.* 106: 1392–1399; **Vánky, K.** (1987). Illustrated genera of smut fungi. *Cryptog. Stud.* 1: 159 pp.

Rhizinaceae Bonord. 1851
Pezizales: Ascomycota

Ascomata apothecial, epigeous, irregularly cupulate to discoid, the hymenial surface brown and often somewhat contorted, the stipe hardly developed. Interascal tissues of simple branched paraphyses, often swollen at the tip, with a brown mucous epithecium and dark thick-walled spine-like elements. Asci cylindrical, the apex rounded, not blueing in iodine. Ascospores hyaline, aseptate, verrucose, ? tetranucleate, with apical gelatinous appendages. Anamorphs unknown.

Significant Genera: *Rhizina*.

Distribution: Widespread in temperate zones.

Economic Significance: None is known.

Ecology: On soil and forest litter, also causing root rots of conifers.

Rhizina undulata; ascomata

Notes: *Rhizina* is the only genus assigned to this family. It has often been linked to the *Helvellaceae*, but is clearly distinct based on rDNA data.

References: **Abbott, S.P.; Currah, R.S.** (1997). The *Helvellaceae*: systematic revision and occurrence in northern and northwestern North America. *Mycotaxon* 62: 1–125; **Lygis, V.; Vasiliauskas, R.; Stenlid, J.** (2005). Clonality in the postfire root rot ascomycete *Rhizina undulata*. *Mycologia* 97: 788–792; **O'Donnell, K.; Cigelnik, E.; Weber, N.S.; Trappe, J.M.** (1997). Phylogenetic relationships among ascomycetous truffles and the true and false morels inferred from 18S and 28S ribosomal DNA sequence analysis. *Mycologia* 89: 48–65; **Weber, E.; Bresinsky, A.** (1992). Polyploidy in discomycetes. *Persoonia* Suppl. 14: 553–563.

Rhizocarpaceae M. Choisy ex Hafellner 1984
Lecanorales: Ascomycota

Thallus crustose or squamulose, often areolate or cracked, rarely absent, isidia rarely present. Ascomata apothecial, sometimes angular or elongated, sessile or immersed, flat or domed, black, marginal tissue variably developed. Interascal tissue of branched and anastomosed agglutinated paraphyses, not or slightly swollen at the apices, usually with a pigmented epithecium. Asci elongate-clavate, with a weakly J+ apical cap, with a discoid more strongly staining apical region and an outer J+ gelatinized layer, usually 8-spored. Ascospores hyaline or brown, 1-septate to muriform, usually with a sheath. Anamorphs coelomycetous, pycnidial. Conidia bacillar, elongate.

Significant Genera: *Rhizocarpon*.

Distribution: Widespread, but mainly in temperate and polar regions.

Rhizocarpon geographicum, Salamanca, Spain; thallus and ascomata

Economic Significance: None is known.

Ecology: Lichenized with green algae, usually on rock substrata. A few species are lichenicolous.

Notes: Molecular analysis confirms placement of the *Rhizocarpaceae* within the *Lecanorales*, but there is controversy as to its closest relatives. Most studies place it as a basal lineage, possibly with the *Fuscideaceae* as nearest neighbour. *Catolechia* may well not belong in the family. Many species have conspicuous thalli, and have been the subject of lichenometry studies.

Rhizocarpon sp., Salamanca, Spain; thallus and ascomata

References: **Buschbom, J.; Mueller, G.** (2004). Resolving evolutionary relationships in the lichen-forming genus *Porpidia* and related allies (*Porpidiaceae*, *Ascomycota*). *Mol. Phylogen. Evol.* 32: 66–82; **Feuerer, T.** (1991). Revision der europäischen Arten der Flechtengattung *Rhizocarpon* mit nichtgelbem Lager und vielzelligen Sporen. *Biblthca Lichenol.* 39: 218 pp.; **Grube, M.; Baloch, E.; Lumbsch, H.T.** (2004). The phylogeny of *Porinaceae* (*Ostropomycetidae*) suggests a neotenic origin of perithecia in *Lecanoromycetes*. *Mycol. Res.* 108: 1111–1118; **Ihlen, P.G.** (2004). Taxonomy of the non-yellow species of *Rhizocarpon* (*Rhizocarpaceae*, lichenized *Ascomycota*) in the Nordic countries, with hyaline and muriform ascospores. *Mycol. Res.* 108: 533–570; **Ihlen, P.G.; Ekman, S.** (2002). Outline of phylogeny and character evolution in *Rhizocarpon* (*Rhizocarpaceae*, lichenized *Ascomycota*) based on nuclear ITS and mitochondrial SSU ribosomal DNA sequences. *Biol. J. Linn. Soc.* 77: 535–546; **Poelt, J.** (1990). Parasitische Arten der Flechtengattung *Rhizocarpon*: eine weitere Übersicht. *Mitt. bot. StSamml. Münch.* 29: 515–538; **Rambold, G.; Meier, C.; Thamerus, M.** (1998). A comparative study on structure and functionality of asci in species of *Rhizocarpon* (*Lecanorales*, *Ascomycetes*). *Cryptog. Bryol.-Lichénol.* 19: 247–255; **Stenroos, S.K.; DePriest, P.T.** (1998). SSU rDNA phylogeny of cladoniiform lichens. *Am. J. Bot.* 85: 1548–1559; **Timdal, E.; Holtan-Hartwig, J.** (1988). A preliminary key to *Rhizocarpon* in Scandinavia. *Graphis Scripta* 2: 41–54; **Wedin, M.; Wiklund, E.; Crewe, A.; Döring, H.; Ekman, S.; Nyberg, A.; Schmitt, I.; Lumbsch, H.T.** (2005). Phylogenetic relationships of *Lecanoromycetes* (*Ascomycota*) as revealed by sequences of mtSSU and nLSU rDNA sequence data. *Mycol. Res.* 109: 159–172.

Rhizopogonaceae Gäum. & C.W. Dodge 1928
Boletales: Basidiomycota

Basidiomata gasteroid, often large, hypogeous, lacking a columella, more or less globose; peridium single-layered, composed of interwoven hyphal tissue or discrete hyphal strands, white, reddish-brown or black, many species bruising or staining green, blue, brown or black in KOH. Gleba loculate, sometimes staining grey, purple or almost black in iodine, with small irregular fertile chambers separated by gelatinized hyphal tissue; clamp connections absent. Basidia cylindrical to flask-shaped, soon evanescent, usually with 6–8 sterigmata. Basidiospores released passively, narrowly ellipsoidal to cylindrical, sometimes truncate, hyaline or pale brown, usually thin-walled, smooth, sometimes staining blue or purple in iodine.

Significant Genera: *Rhizopogon*.

Distribution: Widespread, mainly north temperate; perhaps introduced (with pines) in the southern hemisphere.

Economic Significance: None is known.

Ecology: Ectomycorrhizal with *Pinaceae*. A few species have been reported as associated with other plant hosts, but these may be incorrectly classified.

Notes: *Rhizopogon* in its current circumscription appears to be well-defined, monophyletic, and a sister group of the suilloid clade of the *Boletales*.

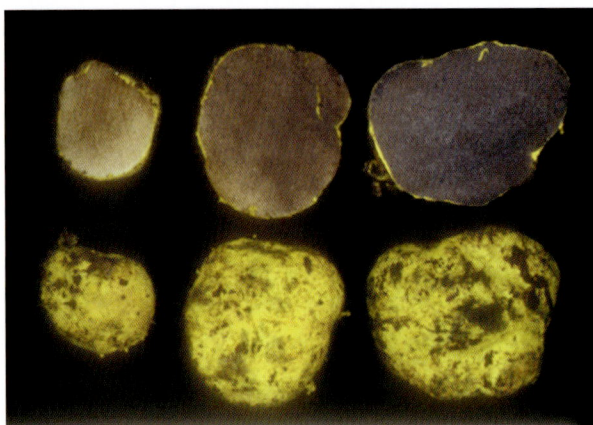

Rhizopogon truncatus; basidiomata

References: **Grubisha, L.C.; Trappe, J.M.; Molina, R.; Spatafora, J.W.** (2001). Biology of the ectomycorrhizal genus *Rhizopogon*. V. Phylogenetic relationships in the *Boletales* inferred from LSU rDNA sequences. *Mycologia* 93: 82–89; **Grubisha, L.C.; Trappe, J.M.; Molina, R.; Spatafora, J.W.** (2002). Biology of the ectomycorrhizal genus *Rhizopogon*. VI. Re-examination of infrageneric relationships inferred from phylogenetic analyses of ITS sequences. *Mycologia* 94: 607–619; **Jarosch, M.** (2001). *Biblthca Mycol.* 191: 158 pp.; **Johannesson, H.; Martín, M.P.** (1999). Cladistic analysis of European species of *Rhizopogon* (*Basidiomycotina*) based on morphological and molecular characters. *Mycotaxon* 71: 267–283; **Kretzer, A.M.; Dunham, S.; Molina, R.; Spatafora, J.W.** (2004). Microsatellite markers reveal the below ground distribution of genets in two species of *Rhizopogon* forming tuberculate ectomycorrhizas on Douglas fir. *New Phytol.* 161: 313–320; **Kretzer, A.M.; Luoma, D.L.; Molina, R.; Spatafora, J.W.** (2003). Taxonomy of the *Rhizopogon vinicolor* species complex based on analysis of ITS sequences and microsatellite loci. *Mycologia* 95: 480–487; **Martín, M.P.** (1996). The Genus *Rhizopogon* in Europe. *Edic. Espec. Soc. Catalana Micol.* 5: 173 pp.; **Martín, M.P.** (2001). The genus *Rhizopogon* in Finland. I. *R. abietis* and *R. ochraceorubens*. *Karstenia* 41: 23–24; **Martín, M.P.; Högberg, N.; Nylund, J.-E.** (1998). Molecular analysis confirms morphological reclassification of *Rhizopogon*. *Mycol. Res.* 102: 855–858; **Martín, M.P.; Raidl, S.** (2002). The taxonomic position of *Rhizopogon melanogastroides* (*Boletales*). *Mycotaxon* 84: 221–228; **Molina, R.; Trappe, J.M.; Grubisha, L.C.; Spatafora, J.W.** (1999). *Rhizopogon*. *Ectomycorrhizal Fungi* (Berlin): 129–161.

Rhynchogastremataceae Oberw. & B. Metzler 1989
Tremellales: Basidiomycota

Basidiomata absent. Hyphae separate or aggregated into synnemata, not gelatinized, with dolipore septa and clamp connections. Basidia formed in clusters with basal clamp connections, swollen at the base and with an elongated neck-like apical portion, partially cruciately septate at the apex, sterigmata not differentiated. Basidiospores discharged passively, yellow or brownish, thick-walled and verrucose, fused together in a tetrad by brown exudates. Anamorph yeast-like.

Significant Genera: *Rhynchogastrema*.

Distribution: Only recorded from Germany.

Economic Significance: None is known.

Ecology: Isolated from soil, apparently a facultative parasite of other fungi.

Notes: Only a single species is known; its affinities have not been studied using molecular methods. Septal pores and haustorial features suggest placement within the *Tremellales*.

References: **Metzler, B.; Oberwinkler, F.; Petzold, H.** (1989). *Rhynchogastrema* gen. nov. and *Rhynchogastremaceae* fam. nov. (*Tremellales*). *Syst. Appl. Microbiol.* 12: 280–287; **Sampaio, J.P.; Weiss, M.; Gadanho, M.; Bauer, R.** (2002). New taxa in the *Tremellales*: *Bulleribasidium oberjochense* gen. et sp. nov., *Papiliotrema bandonii* gen. et sp. nov. and *Fibulobasidium murrhardtense* sp. nov. *Mycologia* 94: 873–887.

Rhytismataceae Chevall. 1826
Rhytismatales: Ascomycota

Ascomata apothecial, immersed in plant tissue and single or aggregated with an often well-developed clypeate stroma, often elongated, opening by radial or longitudinal splits; upper wall usually black, thick-walled, composed of fungal and degraded plant cells; lower wall variously developed, sometimes black. Interascal tissue of simple paraphyses, sometimes anastomosing near the base, often swollen at the apex, often immersed in a gelatinous matrix, occasionally absent. Asci cylindrical, usually thin-walled, not fissitunicate, usually not thickened at the apex, occasionally with a minute refractive pore, very rarely blueing in iodine, often splitting irregularly. Ascospores almost always hyaline, usually aseptate, often elongate, often with mucous sheaths. Anamorphs coelomycetous, usually spermatial; conidiomata acervular or stromatic, with a palisade of sympodially proliferating conidiogenous cells and filiform or bacillar conidia.

Significant Genera: *Coccomyces*, *Lophodermium*, *Rhytisma*. Anamorphs: *Leptostroma*, *Melasmia*.

Distribution: Widespread, but especially prominent in temperate zones.

Rhytisma acerinum, Surrey, UK; ascomata

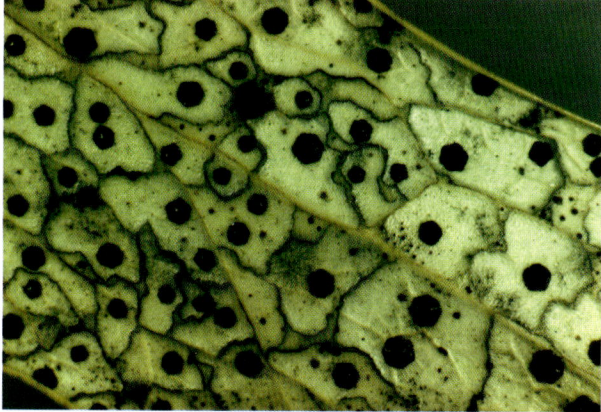
Coccomyces dentatus; colonies with ascomata

Economic Significance: A number of species are important plant pathogens, especially of conifers. The Christmas tree and other conifer production industries have been badly affected in North America by *Lophodermium seditiosum*; other species affect pines in the tropics.

Ecology: Biotrophic, necrotrophic or saprobic on leaves and bark, often initially as endophytes, causing tar spots, premature leaf fall and occasionally witches' broom symptoms.

Notes: The *Rhytismatales* have been treated as an independent order for many years, but molecular evidence suggest that the *Rhytismataceae* at least nests within the *Helotiales*. *Rhytisma acerinum* has been studied for use in air pollution monitoring, although the chemical factors influencing its distribution are unclear.

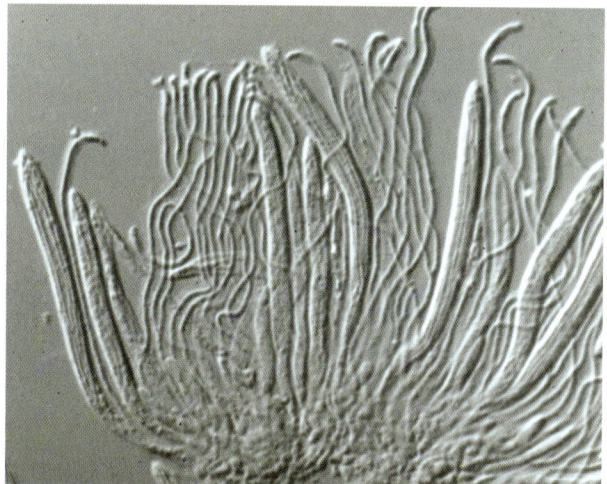
Lophodermium baculiferum; asci, ascospores and paraphyses

Rhytisma salicinum, Alberta, Canada; stromata on living leaves

References: **Cannon, P.F.; Minter, D.W.** (1986). The *Rhytismataceae* of the Indian Subcontinent. *Mycol. Pap.* 155: 123 pp.; **Deckert, R.J.; Melville, L.H., Peterson, R.P.** (2002). Structural features of a *Lophodermium* endophyte during the cryptic life-cycle phase in the foliage of *Pinus strobus*. *Mycol. Res.* 105: 991–997; **Ganley, R.J.; Brunsfeld, S.J.; Newcombe, G.** (2004). A community of unknown, endophytic fungi in western white pine. *Proc. natn Acad. Sci. U.S.A.* 101: 10107–10112; **Gernandt, D.S.; Platt, J.L.; Stone, J.K.; Spatafora, J.W.; Holst-Jensen, A.; Hamelin, R.C.; Kohn, L.M.** (2001). Phylogenetics of *Helotiales* and *Rhytismatales* based on partial small subunit nuclear ribosomal DNA sequences. *Mycologia* 93: 915–933; **Johnston, P.R.** (2000). *Rhytismatales* of Australia: the genus *Coccomyces*. *Aust. Syst. Bot.* 13: 199–243; **Johnston, P.R.** (2001). *Rhytismatales* of Australasia. *Aust. Syst. Bot.* 14: 377–384; **Johnston, P.R.** (2001). Monograph of the Monocotyledon-Inhabiting Species of *Lophodermium*. *Mycol. Pap.* 176: 239 pp.; **Minter, D.W.** (1981). *Lophodermium* on pines. *Mycol. Pap.* 147: 54 pp.; **Ortiz-García, S.; Gernandt, D.S.; Stone, J.K.; Johnston, P.R.;**

Chapela, I.H.; Salas-Lizana, R.; Alvarez-Buylla, E.R. (2003). Phylogenetics of *Lophodermium* from pine. *Mycologia* 95: 846–859; Sokolski, S.; Piché, Y.; Bérubé, J.A. (2004). *Lophodermium macci* sp. nov., a new species on senesced foliage of five-needle pines. *Mycologia* 96: 1261–1267; Stone, J.K.; Bacon, C.W.; White, J.F. (2000). An overview of endophytic microbes: endophytism defined. *Microbial Endophytes* (New York): 3–29.

Roccellaceae Chevall. 1826
Arthoniales: Ascomycota

Thallus crustose or fruticose; when crustose simple and often thin, when fruticose usually pruinose, sometimes with a thick and multilayered cortex. Ascomata with well-developed walls, apothecial or lirellate (round or elongated), rarely ± perithecial, often in groups of locules in a common receptacle. Asci rather variable but usually cylindric-clavate, thick-walled and fissitunicate, with a thickened apex and a distinct ocular chamber. Ascospores varied in shape, transversely and sometimes longitudinally septate, sometimes with an epispore. Anamorphs poorly known.

Significant Genera: *Enterographa*, *Opegrapha*, *Roccella*.

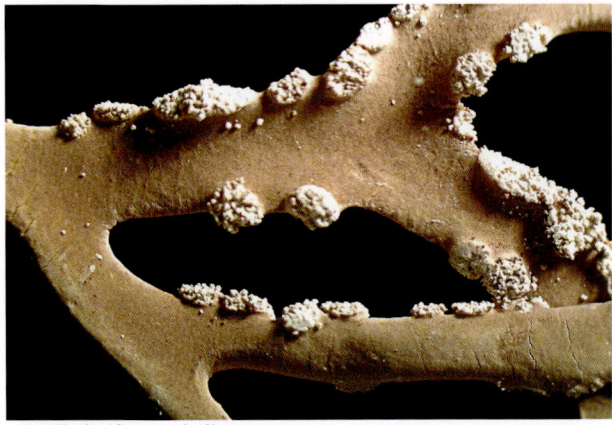
Roccella fuciformis; thallus

Opegrapha calcarea, Wiltshire, UK; ascomata developing from endolithic thallus

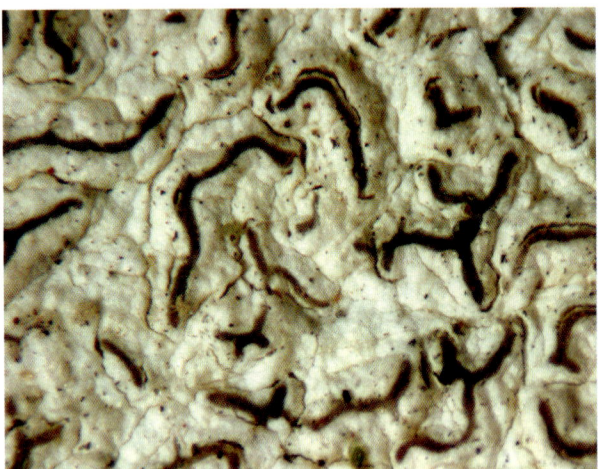
Enterographa anguinella; ascomata

Distribution: Cosmopolitan, with many species in coastal habitats.

Economic Significance: None is known.

Ecology: Mostly lichenized with green algae, on rocks or bark. Some species are lichenicolous.

Notes: The family is extremely diverse morphologically and more molecular studies are needed to establish its delimitation. The *Opegraphaceae* was placed into synonymy with the *Roccellaceae* some years ago, but studies of more genera are needed to confirm this placement.

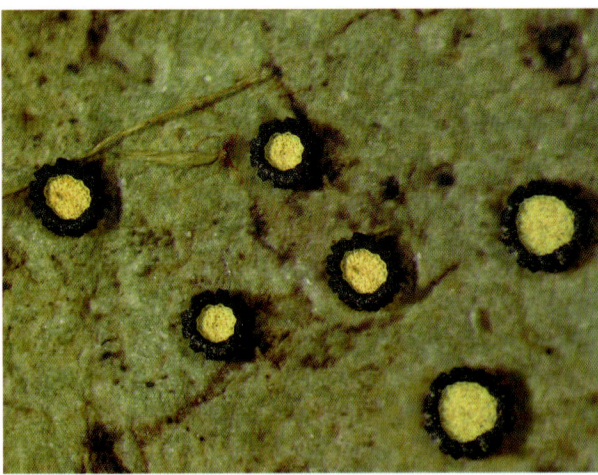
Cresponea melanocheloides; thallus and ascomata

References: Egea, J.M.; Torrente, P.; Manrique, E. (1993). The *Lecanactis grumulosa* group (*Opegraphaceae*) in the Mediterranean region. *Pl. Syst. Evol.* 187: 103–114; Ertz, D.; Christnach, C.; Wedin, M.; Diederich, P. (2005). A world monograph of the genus *Plectocarpon* (*Roccellaceae*, *Arthoniales*). *Biblthca Lichenol.* 91: 155 pp.; Feige, G.B.; Lumbsch, H.T.; Mies, B. (1993). Morphological and chemical changes in *Roccella* thalli infected by *Lecanactis grumulosa* (lichenized *Ascomycetes*, *Opegraphales*). *Cryptog. bot.* 3: 101–107; Follmann, G.; Peine, J. (1999). Camanchaca, a new genus of *Roccellaceae* (*Arthoniales*, *Loculoascomycetidae*) from the Atacama Desert, North Chile, South America. *J. Hattori bot. Lab.* 87: 259–276; Grube, M. (1998). Classification and phylogeny in the *Arthoniales* (Lichenized *Ascomycetes*). *Bryologist* 101: 377–391;

Henssen, A.; Thor, G. (1998). Studies in taxonomy and developmental morphology in *Chiodecton, Dichosporidium, Erythrodecton* and the new genus *Pulvinodecton* (*Arthoniales*, lichenized *Ascomycetes*). *Nordic Jl Bot.* 18: 95–120; **Lohtander, K.; Källersjö, M.; Tehler, A.** (1998). Dispersal strategies in *Roccellina capensis* (*Arthoniales*). *Lichenologist* 30: 341–350; **Lohtander, K.; Myllys, L.; Sundin, R.; Källersjö, M.; Tehler, A.** (1998). The species pair concept in the lichen *Dendrographa leucophaea* (*Arthoniales*): analyses based on ITS sequences. *Bryologist* 101: 404–411; **Myllys, L.; Källersjö, M.; Tehler, A.** (1998). A comparison of SSU rDNA data and morphological data in *Arthoniales* (*Euascomycetes*) phylogeny. *Bryologist* 101: 70–85; **Myllys, L.; Lohtander, K.; Källersjo, M.; Tehler, A.** (1999). Applicability of ITS data in *Roccellaceae* (*Arthoniales, Euascomycetes*) phylogeny. *Lichenologist* 31: 461–476; **Sparrius, L.B.** (2004). A monograph of *Enterographa* and *Sclerophyton*. *Biblthca Lichenol.* 89: 141 pp.; **Tehler, A.** (1990). A new approach to the phylogeny of *Euascomycetes* with a cladistic outline of *Arthoniales* focussing on *Roccellaceae*. *Can. J. Bot.* 68: 2458–2492; **Tehler, A.; Dahlkild, Å.; Eldenäs, P.; Feige, G.B.** (2004). The phylogeny and taxonomy of Macaronesian, European and Mediterranean *Roccella* (*Roccellaceae, Arthoniales*). *Symb. bot. upsal.* 34: 405–428; **Tehler, A.; Egea, J.M.** (1997). The phylogeny of *Lecanactis* (*Opegraphaceae*). *Lichenologist* 29: 397–414; **Tehler, A.; Lohtander, K.; Myllys, L.; Sundin, R.** (1997). On the identity of the genera *Hubbsia* and *Reinkella* (*Roccelaceae*). *In* Tibell, L. & Hedberg, I. (eds), Lichen Studies Dedicated to Rolf Santesson. *Symb. bot. upsal.* 32: 255–265; **Thor, G.** (1991). The lichen genus *Chiodecton* and five allied genera. *Op. bot.* 103: 92 pp.; **Torrente, P.; Egea, J.M.** (1989). La familia *Opegraphaceae* en el area Mediterránea de la Península Ibérica y Norte de Africa. *Biblthca Lichenol.* 32: 282 pp.

Roesleria subterranea, Kent, UK; ascoma

References: Redhead, S.A. (1985). *Roeslerina* gen. nov. (*Caliciales, Caliciaceae*), an ally of *Roesleria* and *Coniocybe*. *Can. J. Bot.* 62: 2514–2519; **Véghelyi, K.** (1989). The isolation and characteristics of *Roesleria hypogaea* Thüm. et Pass. *Acta phytopath. entom. Hung.* 24: 293–299; **Yao, Y.J.; Spooner, B.M.** (1999). *Roesleriaceae*, a new family of *Ascomycota*, and a new species of *Roeslerina*. *Kew Bull.* 54: 683–693.

Roesleriaceae Y.J. Yao & Spooner 1999
Uncertain position within Ascomycetes, Ascomycota

Stromata absent. Ascomata apothecial, stipitate, capitate, with excipulum poorly defined or lacking. Interascal tissue composed of narrow hypha-like paraphyses, sometimes interwoven and protruding beyond the hymenium. Asci cylindrical, thin-walled, ? sometimes in short chains, without apical structures, 8-spored, not blueing in iodine, evanescent. Ascospores lenticular to discoid, hyaline or greyish, smooth or finely ornamented, sometimes septate, accumulating in a mazaedium. Anamorphs not known.

Significant Genera: *Roesleria, Roeslerina*.

Distribution: Known from north temperate zones.

Economic Significance: One species is occasionally reported as a parasite of vines, fruit trees etc., but probably causes minimal economic damage.

Ecology: Hypogeous on plant roots, parasitic but sometimes becoming saprobic.

Notes: At one time included in the *Helotiales* and also placed as an ally of the calicioid lichens, but its evolutionary position remains enigmatic.

Russulaceae Lotsy 1907
Russulales: Basidiomycota

Basidiomata pileate and stipitate, the stipe central or rarely lateral and usually lacking a veil, or gasteroid (hypogeous or epigeous); usually fleshy and often brightly coloured. Flesh complex in organization, with islands of globose cells separated by filamentous (sometimes lactiferous) hyphae, not staining in iodine, usually with large hymenial cystidia arising from inner tissues. Hymenium usually lamellar, the gills free or decurrent. Basidia clavate, usually 4-spored. Basidiospores usually ± globose, hyaline to dark yellowish or rarely brown, with an outer ornamented layer that stains blue in iodine.

Russula sardonia, Hampshire, UK; basidioma

Significant Genera: *Arcangeliella, Lactarius, Macowanites, Russula, Zelleromyces*.

Distribution: Widespread; from Arctic and Antarctic regions through to the tropics, both Old World and New World.

Economic Significance: The edible species are an important source of income in some tropical developing countries, especially in Africa.

Ecology: Ectomycorhizal or saprobic, on the ground or associated with wood.

Lactarius chrysorrheus; basidiomata showing latex exudate

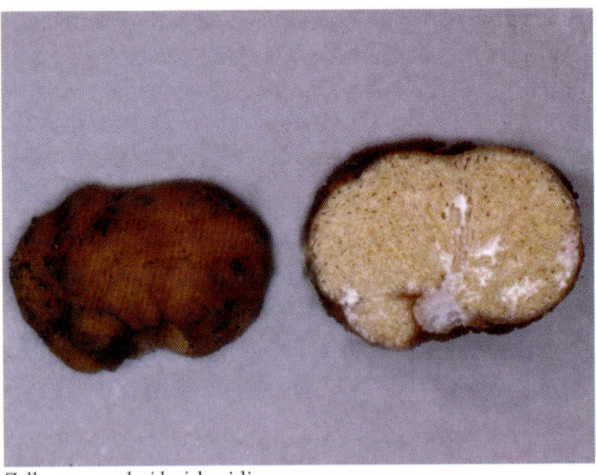

Zelleromyces claridgei; basidioma

Notes: The genera with gasteroid or hypogeous gasteroid fruit bodies have traditionally been separated from the agaricoid genera. However, recent molecular analysis has shown that the derived gasteroid form is relatively recent and appears to justify an integration of the gasteroid taxa in the agaricoid genera *Russula* and *Lactarius*. Further, as more tropical taxa are discovered the generic boundaries in the family are eroded. Thus, *Cystangium*, *Gymnomyces* and *Macowanites* are gasteroid species of *Russula* whilst *Arcangeliella* and *Zelleromyces* are gasteroid species of *Lactarius*.

References: **Binder, M.; Bresinsky, A.** (2002). Derivation of a polymorphic lineage of gasteromycetes from boletoid ancestors. *Mycologia* 94: 85–98; **Binder, M.; Hibbett, D.S.** (2002). Higher-level phylogenetic relationships of homobasidiomycetes (mushroom-forming fungi) inferred from four rDNA regions. *Mol. Phylogen. Evol.* 22: 76–90; **Buyck, B.** (1995). Towards a global and integrated approach on the taxonomy of *Russulales*. *Russulales News* 3: 3–17; **Buyck, B.; Horak, E.** (1999). New taxa of pleurotoid *Russulaceae*. *Mycologia* 91: 532–537; **Desjardin, D.E.** (2003). A unique ballistosporic hypogeous sequestrate *Lactarius* from California. *Mycologia* 95: 148–155; **Eberhardt, U.** (2002). Molecular kinship analyses of the agaricoid *Russulaceae*: correspondence with mycorrhizal anatomy and sporocarp features in the genus *Russula*. *Mycol. Prog.* 1: 201–223; **Larsson, E.; Larsson, K.-H.** (2003). Phylogenetic relationships of russuloid basidiomycetes with emphasis on aphyllophoralean taxa. *Mycologia* 95: 1037–1065; **Lebel, T.** (2003). Australasian sequestrate (truffle-like) fungi. XIV. *Gymnomyces* (*Russulales, Basidiomycota*). *Aust. Syst. Bot.* 16: 401–426; **Miller, S.L.; Buyck, B.** (2002). Molecular phylogeny of the genus *Russula* in Europe with a comparison of modern infrageneric classifications. *Mycol. Res.* 106: 259–276; **Miller, S.L.; Henkel, T.W.** (2004). Biology and molecular ecology of subiculate *Lactarius* species from Guyana. *Mem. N. Y. bot. Gdn* 89: 297–313; **Miller, S.L.; McClean, T.M.; Walker, J.F.; Buyck, B.** (2001). A molecular phylogeny of the *Russulales* including agaricoid, gasteroid and pleurotoid taxa. *Mycologia* 93: 344–354; **Moncalvo, J.-M.; Vilgalys, R.; Redhead, S.A.; Johnson, J.E.; James, T.Y.; Aime, M.C.; Hofstetter, V.; Verduin, S.J.W.; Larsson, E.; Baroni, T.J.; Thorn, R.G.; Jacobsson, S.; Clemencon, H.; Miller, O.K.** (2002). One hundred and seventeen clades of euagarics. *Mol. Phylogen. Evol.* 23: 357–400; **Shimono, Y.; Kato, M.; Takamatsu, S.** (2004). Molecular phylogeny of *Russulaceae* (*Basidiomycetes*; *Russulales*) inferred from the nucleotide sequences of nuclear large subunit rDNA. *Mycoscience* 45: 303–316.

Rutstroemiaceae Holst-Jensen, L.M. Kohn & T. Schumach. 1997
Helotiales: Ascomycota

Stromata present, composed of indeterminate substratal tissue permeated by undifferentiated thin-walled hyphae, sometimes with a black rind. Ascomata apothecial, stalked, brown to greenish, usually cupulate, without hairs. Interascal tissue of simple paraphyses. Asci usually with a J+ apical ring. Ascospores large or small, ellipsoidal, usually aseptate, hyaline or pale brown, often ± longitudinally symmetrical. Anamorphs spermatial or disseminative.

Significant Genera: *Lambertella, Lanzia, Rutstroemia*.

Distribution: Cosmopolitan, especially in temperate regions.

Economic Significance: None is known.

Ecology: Essentially saprobic on various plant parts, especially seeds and fruits, though initial infection may take place of living tissues.

Notes: The *Rutstroemiaceae* are phylogenetically distinct from the *Sclerotiniaceae* with which they have traditionally been placed. They are morphologically/ecologically distinguished through the development of sclerotia that are essentially modified plant organs rather than fungal tissues. *Poculum* is now recognized as a distinct genus and *Lambertella* has been shown to be polyphyletic with some species excluded from the family.

References: **Baral, H.O.** (1987). Der Apikalapparat der *Helotiales*. Eine lichtmikroskopische Studie über Arten mit Amyloidring. *Z. Mykol.* 53: 119–136; **Holst-Jensen, A.; Kohn, L.M.; Schumacher, T.** (1997). Nuclear rDNA phylogeny of the *Sclerotiniaceae*. *Mycologia* 89: 885–899; **Kohn, L.M.; Grenville, D.J.** (1989). Anatomy and histochemistry of stromatal anamorphs in the *Sclerotiniaceae*. *Can. J. Bot.* 67: 371–393; **Korf, R.P.; Zhuang, W.-y.** (1985). A synoptic key to the species of *Lambertella* (*Sclerotiniaceae*), with comments on a version prepared for TAXADAT, Anderegg's computer program. *Mycotaxon* 24: 361–386; **Schumacher, T.; Holst-Jensen, A.** (1997). A synopsis of the genus *Scleromitrula* (= *Verpatinia*) (*Ascomycota*: *Helotiales*: *Sclerotiniaceae*). *Mycoscience* 38: 55–69; **Spooner, B.M.** (1987). *Helotiales* of Australasia: *Geoglossaceae, Orbiliaceae, Sclerotiniaceae, Hyaloscyphaceae*. *Biblthca Mycol.* 116: 711 pp.; **Wang, Z.; Binder, M.; Schoch, C.L.; Johnston, P.R.; Spatafora, J.W.; Hibbett, D.S.** (2006). Evolution of helotialean fungi (*Leotiomycetes, Pezizomycotina*): a nuclear rDNA phylogeny. *Mol. Phylogen. Evol.* 41: 295–312; **Zhuang, W.Y.** (1996). The genera *Lambertella* and *Lanzia* (*Sclerotiniaceae*) in China. *Mycosystema* 8-9: 15–38; **Zhuang, W.Y.; Wang, Z.** (1997). Notes on sclerotiniaceous fungi on *Castanea* and *Castanopsis* in Asia. *Mycotaxon* 64: 449–454.

Rutstroemia firma, Surrey, UK; ascomata

Rutstroemia bulgarioides; ascomata

Saccardiaceae Höhn. 1909
Uncertain position within Dothideomycetes, Ascomycota

Mycelium superficial, sometimes setose, often inconspicuous. Stromata absent or small and basal. Ascomata superficial, perithecial, discoid, thin-walled, the upper part of the wall gelatinous, deliquescing to release the spores. Interascal tissue of usually narrow pseudoparaphyses, becoming free at the apex, the apices swollen and sometimes pigmented. Asci ± saccate, sessile, thick-walled, fissitunicate, without an apical ring. Ascospores hyaline or pale brown, muriform, without a sheath.

Significant Genera: *Johansonia*, *Saccardia*.

Distribution: Most species are tropical.

Economic Significance: None is known.

Ecology: Little is known, but species are superficial on living leaves and some appear to have biotrophic nutrition.

Notes: Very little is known of this family; a possible link with the *Myriangiales* has been postulated. No sequences are available.

References: Barr, M.E. (1987). New taxa and combinations in the *Loculoascomycetes*. *Mycotaxon* 29: 501–505; **Hsieh, W.H.; Chen, C.Y.; Sivanesan, A.** (1997). Some new ascomycetes from Taiwan. *Mycol. Res.* 101: 897–907; **Inácio, C.A.; Dianese, J.C.** (1998). Some foliicolous fungi on *Tabebuia* species. *Mycol. Res.* 102: 695–708.

Saccharomycetaceae G. Winter 1881
Saccharomycetales: Ascomycota

Mycelium ± absent. Vegetative cells reproducing by multilateral budding, ± ellipsoidal, without a mucous coating. Asci morphologically similar to vegetative cells, not in well-defined chains, ± globose, thin-walled, 1- to 4-spored, evanescent or semi-persistent, without an obvious discharge mechanism. Ascospores usually ± spherical, often ornamented with equatorial ridges etc. Fermentation positive, coenzyme system usually Q-6.

Significant Genera: *Pichia*, *Saccharomyces*.

Distribution: Cosmopolitan.

Economic Significance: *Saccharomyces cerevisiae* must be the single most economically important fungus in the world, with its role in bread making and alcohol production. Sequencing of its genome has led to a significant increase in interest in the functional genomics of fungi in general and yeasts are used extensively as model organisms and vehicles for transformants.

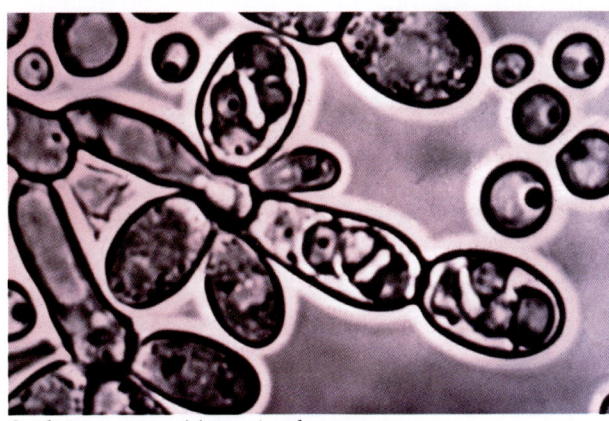

Saccharomyces cerevisiae; asci and ascospores

Ecology: Present in a very wide range of habitats, especially where carbohydrate sources are available, such as foodstuffs, nectar and other plant exudates.

Notes: The central family of the *Saccharomycetes*, recognizable especially by its fermentative capacity and budding nature.

References: Ando, S.; Mikata, K.; Tahara, Y.; Yamada, Y. (1996). Phylogenetic relationships of species of the genus *Saccharomyces* Meyen ex Reess deduced from partial base sequences of 18S and 26S ribosomal RNAs. *Biosc., Biotechn., Biochem.* 60: 1070–1075; **Ando, S.; Mikata, K.; Tahara, Y.; Yamada, Y.** (1996). Phylogenetic relationships of species of the genus *Kluyveromyces* van der Walt (*Saccharomycetaceae*) deduced from the partial sequences of 18S and 26S ribosomal RNAs. *Biosc., Biotechn., Biochem.* 60: 1063–1069; **Belloch, C.; Querol, A.; García, M.D.; Barrio, E.** (2000). Phylogeny of the genus *Kluyveromyces* inferred from the mitochondrial cytochrome-c oxidase II gene. *Int. J. Syst. Evol. Microbiol.* 50: 405–416; **Cliften, P.; Sudarsanam, P.; Desikan, A.; Fulton, L.; Fulton, B.; Majors, J.; Waterston, R.; Cohen, B.A.; Johnston, M.** (2003). Finding functional features in *Saccharomyces* genomes by phylogenetic footprinting. *Science* N.Y. 300: 71–76; **Dujon [*et al.*]** (2004). Genome evolution in yeasts. *Nature* Lond. 430: 35–44; **Edwards-Ingram, L.C.; Gent, M.E.; Hoyle, D.C.; Hayes, A.; Stateva, L.I.; Oliver, S.G.** (2004). Comparative Genomic Hybridization Provides New Insights Into the Molecular Taxonomy of the *Saccharomyces* Sensu Stricto Complex. *Genome Res.* 14: 1043–1051; **Fischer, G.; James, S.A.; Roberts, I.N.; Oliver, S.G.; Louis, E.J.** (2000). Chromosomal evolution in *Saccharomyces*. *Nature* Lond. 405: 451–454; **Hudson, J.R.; Dawson, E.P.; Rushing, K.L.; Jackson, C.H.; Lockshon, D.; Conover, D.; Lanciault, C.; Harris, J.R.; Simmons, S.J.; Rothstein, R.; Fields, S.** (1997). The complete set of predicted genes from *Saccharomyces cerevisiae* in a readily usable form. *Genome Res.* 7: 1169–1173; **James, S.A.; Cai, J.P.; Roberts, I.N.; Collins, M.D.** (1997). A phylogenetic analysis of the genus *Saccharomyces* based on 18S rRNA gene sequences: description of *Saccharomyces kunashirensis* sp. nov. and *Saccharomyces martiniae* sp. nov. *Int. J. Syst. Bacteriol.* 47: 453–460; **James, S.A.; Roberts, I.N.; Collins, M.D.** (1998). Phylogenetic heterogeneity of the genus *Williopsis* as revealed by 18S rRNA gene sequences. *Int. J. Syst. Bacteriol.* 48: 591–596; **Kurtzman, C.P.; Fell, J.W.** (1998). Definition, classification and nomenclature of the yeasts. *The Yeasts, A Taxonomic Study* (Amsterdam): 3–5; **Kurtzman, C.P.; Robnett, C.J.** (2003). Phylogenetic relationships among yeasts of the '*Saccharomyces* complex' determined from multigene sequence analyses. *FEMS Yeast Res.* 3: 417–432; **Mortimer, R.K.** (2000). Evolution and variation of the yeast (*Saccharomyces*) genome. *Genome Res.* 10: 403–409.

Saccharomycodaceae Kudrjanzev 1960
Saccharomycetales: Ascomycota

Mycelium poorly developed. Vegetative cells usually lemon-shaped, proliferating by bipolar budding. Asci either formed directly from an apparently undifferentiated vegetative cell, or by conjugation between a mother cell and its bud. Ascospores varied in form, sometimes brown, sometimes ornamented or with a median or eccentric flange. Fermentation usually positive.

Significant Genera: *Hanseniaspora, Nadsonia.*

Distribution: Cosmopolitan, in both tropical and temperate regions.

Economic Significance: Some species are economically important spoilage yeasts, others contribute to natural fermentation processes in food production, e.g. in coffee and cocoa.

Ecology: Isolated from a wide range of sources including spoiled food, decaying fruit and similar substrata.

Notes: Comparatively poorly known.

References: Boekhout, T.; Kurtzman, C.P.; O'Donnell, K.; Smith, M.T. (1994). Phylogeny of the yeast genera *Hanseniaspora* (anamorph *Kloeckera*), *Dekkera* (anamorph *Brettanomyces*), and *Eeniella* as inferred from partial 26S ribosomal DNA nucleotide sequences. *Int. J. Syst. Bacteriol.* 44: 781–786; **Cadez, N.; Poot, G.A.; Raspor, P.; Smith, M.T.** (2003). *Henseniaspora meyeri* sp. nov., *Hanseniaspora clermontiae* sp. nov., *Hanseniaspora lachancei* sp. nov. and *Hanseniaspora opuntiae* sp. nov., novel apiculate yeast species. *Int. J. Syst. Evol. Microbiol.* 53: 1671–1680; **Cadez, N.; Raspor, P.; de Cock, A,W.; Boekhout, T.; Smith, M.T.** (2002). Molecular identification and genetic diversity within species of the genera *Hanseniaspora* and *Kloeckera*. *FEMS Yeast Res.* 1: 279–289; **Esteve-Zarzoso, B.; Belloch, C.; Uruburu, F.; Querol, A.** (1999). Identification of yeasts by RFLP analysis of the 5.8S rRNA gene and the two ribosomal internal transcribed spacers. *Int. J. Syst. Bacteriol.* 49: 329–337; **Golubev, V.I.; Smith, M.T.; Poot, G.A.; Kock, J.L.F.** (1989). Species delineation in the genus *Nadsonia* Sydow. *Antonie van Leeuwenhoek* 55: 369–382; **Kurtzman, C.P.; Robnett, C.J.** (2003). Phylogenetic relationships among yeasts of the 'Saccharomyces complex' determined from multigene sequence analyses. *FEMS Yeast Res.* 3: 417–432; **Mikata, K.; Nakase, T.** (1997). Surface structure of ascospores of genus *Nadsonia* Sydow. *Microbiol. Culture Coll.* 13: 97–102; **Miller, M.W.; Phaff, H.J.** (1998). *Nadsonia* Sydow. *The Yeasts, A Taxonomic Study* (Amsterdam): 268–270; **Miller, M.W.; Phaff, H.J.** (1998). *Saccharomycodes* E.C. Hansen. *The Yeasts, A Taxonomic Study* (Amsterdam): 372–373; **Phaff, H.J.; Miller, M.W.** (1998). *Wickerhamia* Soneda. *The Yeasts, A Taxonomic Study* (Amsterdam): 409–410; **Simmons, R.B.; Ahearn, D.G.** (1985). Ascospore ornamentation in *Saccharomycodes ludwigii*. *Mycologia* 77: 660–662; **Smith, M.T.** (1998). *Hanseniaspora* Zikes. *The Yeasts, A Taxonomic Study* (Amsterdam): 214–220; **Tredoux, H.G.; Kock, J.L.F.; Lategan, P.M.** (1987). The use of cellular long-chain fatty acid composition in the identification of some yeasts association with the wine industry. *Syst. Appl. Microbiol.* 9: 299–306; **Yamada, Y.; Maeda, K.; Banno, I.** (1992). The phylogenetic relationships of the Q_6-equipped species in the teleomorphic apiculate yeast genera *Hanseniaspora, Nadsonia,* and *Saccharomycodes* based on the partial sequences of 18S and 26S ribosom al ribonucleic acids. *J. gen. appl. Microbiol. Tokyo* 38: 585–596.

Saccharomycopsidaceae Arx & Van der Walt 1987
Saccharomycetales: Ascomycota

Mycelium poorly to well-developed. Vegetative cells ± ellipsoidal, proliferating by multilateral budding, mycelial fragmentation also occurring. Asci ellipsoidal to ± globose, formed directly from vegetative cells, usually evanescent, usually 4-spored. Ascospores hyaline or pale brown, with ± equatorial flanges. Coenzyme system usually Q-8.

Significant Genera: *Saccharomycopsis.*

Distribution: Widespread.

Economic Significance: Sometimes isolated from foodstuffs or medical environments.

Ecology: Isolated from a very wide range of sources.

Notes: Comparatively poorly known.

References: Esteve-Zarzoso, B.; Belloch, C.; Uruburu, F.; Querol, A. (1999). Identification of yeasts by RFLP analysis of the 5.8S rRNA gene and the two ribosomal internal transcribed spacers. *Int. J. Syst. Bacteriol.* 49: 329–337; **Kurtzman, C.P.** (1999). Two new members of the *Saccharomycopsis* clade: *Saccharomycopsis microspora*, comb. nov. and *Candida lassenensis*, sp. nov. *Mycotaxon* 71: 241–250; **Kurtzman, C.P.; Robnett, C.J.** (1995). Molecular relationships among hyphal ascomycetous yeasts and yeastlike taxa. *Can. J. Bot.* 73: S824–S830; **Kurtzman, C.P.; Smith, M.T.** (1998). *Saccharomycopsis* Schiönning. *The Yeasts, A Taxonomic Study* (Amsterdam): 374–386; **Ota, A.; Morishita, H.** (1993). Effect of NaCl on the growth and morphology of *Saccharomycopsis fibuligera*. *Microbios* 73: 149–155; **Yamada, Y.; Matsuda, M.; Mikata, K.** (1996). The phylogeny of species of the genus *Saccharomycopsis* Schiönning [sic] (*Saccharomycetaceae*) based on the partial sequences of 18S and 26S ribosomal RNAs. *Biosc., Biotechn., Biochem.* 60: 1303–1307.

Sarcoscyphaceae Le Gal ex Eckblad 1968
Pezizales: Ascomycota

Ascomata apothecial, ± sessile or stalked, leathery or somewhat gelatinous, usually brightly coloured due to carotenoid pigments; excipulum composed of hyphal cells often embedded in a gelatinous matrix, sometimes with hyaline or rarely melanized hairs surrounding the hymenium. Interascal tissue composed of paraphyses, often anastomosing near the base and pigmented at the usually swollen apices. Asci cylindrical, persistent, with an often slightly subapical operculum, not blueing in iodine. Ascospores hyaline, smooth or ornamented, multinucleate. Anamorphs hyphomycetous.

Significant Genera: *Cookeina, Phillipsia, Sarcoscypha*.

Distribution: Cosmopolitan, prominent in tropical as well as temperate regions.

Economic Significance: None is known.

Ecology: Saprobic on soil and rotten wood.

Notes: Many species are large and brightly coloured and, therefore, frequently observed. *Cookeina* species are probably the most often encountered of all tropical discomycetes. The family seems to be well-defined in phylogenetic terms.

Cookeina tricholoma, Kenya; ascomata

Microstoma floccosum, Thailand; ascomata

References: **Butterfill, G.B.; Spooner, B.M.** (1995). *Sarcoscypha* (*Pezizales*) in Britain. *Mycologist* 9: 20–26; **Cabello, M.N.** (1988). Estudio sistemático del suborden *Sarcoscyphineae* (*Pezizales, Ascomycotina*) empleando técnicas numéricas. *Boln Soc. argent. Bot.* 25: 395–413; **Durrieu, G.; He, Q.B.; Chaumeton, J.-P.; Chauveau, C.** (1997). *Cookeina insititia*, a new *Discomycetes* from China. *Acta bot.*

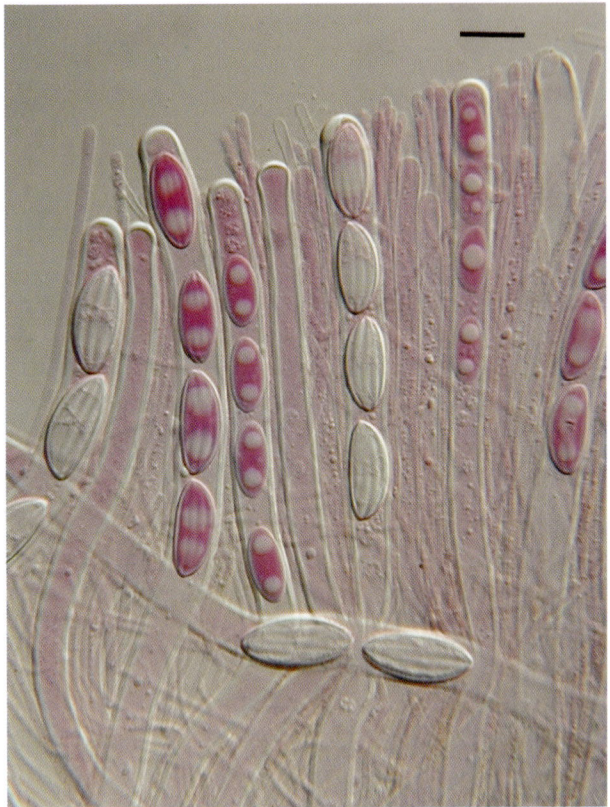
Phillipsia lutea, Guyana; asci, ascospores and paraphyses

Yunn. 19: 128–130; **Hansen, K.; Pfister, D.H.; Hibbett, D.S.** (1999). Phylogenetic relationships among species of *Phillipsia* inferred from molecular and morphological data. *Mycologia* 91: 299–314; **Harrington, F.A.** (1998). Relationships among *Sarcoscypha* species: evidence from molecular and morphological characters. *Mycologia* 90: 235–243; **Harrington, F.A.; Potter, D.** (1997). Phylogenetic relationships within *Sarcoscypha* based upon nucleotide sequences of the internal transcribed spacer of nuclear ribosomal DNA. *Mycologia* 89: 258–267; **Iturriaga, T.; Pfister, D.H.** (2006). A monograph of the genus *Cookeina* (*Ascomycota, Pezizales, Sarcoscyphaceae*). *Mycotaxon* 95: 137–180; **Landvik, S.; Egger, K.N.; Schumacher, T.** (1997). Towards a subordinal classification of the *Pezizales* (*Ascomycota*): phylogenetic analyses of SSU rDNA sequences. *Nordic Jl Bot.* 17: 403–418; **Li, L.T.; Kimbrough, J.W.** (1995). Septal structures in the *Sarcoscyphaceae* and *Sarcosomataceae* (*Pezizales*). *Int. J. Pl. Sci.* 156: 841–848; **Li, L.T.; Kimbrough, J.W.** (1996). Spore ontogeny in species of *Phillipsia* and *Wynnea* (*Pezizales*). *Can. J. Bot.* 74: 10–18; **Paden, J.W.** (1986). On the anamorph of *Phillipsia crispata*. *Mycotaxon* 25: 165–174; **Pfister, D.H.** (1989). *Kompsoscypha*: a new genus related to *Nanoscypha* (*Sarcoscyphaceae*). *Mem. N. Y. bot. Gdn* 49: 339–343; **Pfister, D.H.; Weinstein, R.N.; Iturriaga,T.** (2002). A phylogenetic study of the genus *Cookeina*. *Mycologia* 94: 673–682; **Wang, Z.; Zhuang, W.Y.** (1996). Taxonomic studies of the genus *Sarcoscypha* in China. *Mycosystema* 8-9: 39–52.

Sarcosomataceae Kobayasi 1937
Pezizales: Ascomycota

Ascomata large, apothecial, ± sessile, leathery or somewhat gelatinous, the disc pale or dark; peridium composed of intertwined hyphae often in a gelatinous matrix, dark brown, sometimes with dark brown setose hairs surrounding or interspersed with the hymenium. Interascal tissue of paraphyses, often anastomosing near the base. Asci cylindrical, persistent, with an often slightly subapical operculum. Ascospores hyaline, rarely ornamented, multinucleate, sometimes with a sheath. Anamorphs hyphomycetous.

Significant Genera: *Plectania*, *Sarcosoma*, *Urnula*.

Plectania chilensis; ascoma

Urnula craterium; ascomata

Distribution: Most species are temperate.

Economic Significance: Little is known, although there has been pharmaceutical interest in hexaketide metabolites produced by some species.

Ecology: Saprobic on rotten, sometimes buried, wood.

Notes: The family as currently circumscribed appears well-defined and evolutionarily isolated within the *Pezizales*.

References: **Bellemère, A.; Malherbe, M.C.; Chacun, H.; Meléndez-Howell, L.M.** (1990). L'étude ultrastructurale des asques et des ascospores de l'*Urnula helvelloides* Donadini, Berthet et Astier et les concepts d'asque subopérculé et de *Sarcosomataceae*. *Cryptog. Mycol.* 11: 203–238; **Benkert, D.** (1991). Bemerkenswerte Ascomyceten der DDR. 12. *Sarcoscyphaceae* und *Sarcosomataceae* (*Pezizales*). *Gleditschia* 19: 173–201; **Cao, J.Z.; Fan, L.; Liu, B.** (1992). Notes on the genus *Galiella* in China. *Mycologia* 84: 261–263; **Donadini, J.-C.** (1987). Etude des *Sarcoscyphaceae* ss. Le Gal (1) *Sarcosomataceae* et *Sarcoscyphaceae* ss. Korf. Le genre *Pseudoplectania* emend. nov. *P. ericae* sp. nov. (*Pezizales*). *Mycol. helv.* 2: 217–246; **Harrington, F.A.; Pfister, D.H.; Potter, D.; Donoghue, M.J.** (1999). Phylogenetic studies within the *Pezizales*. I. 18S rRNA sequence data and classification. *Mycologia* 91: 41–50; **Kopcke, B.; Weber, R.W.; Anke, H.** (2002). Galiellalactone and its biogenetic precursors as chemotaxonomic markers of the *Sarcosomataceae* (*Ascomycota*). *Phytochem.* 60: 709–714; **Landvik, S.; Egger, K.N.; Schumacher, T.** (1997). Towards a subordinal classification of the *Pezizales* (*Ascomycota*): phylogenetic analyses of SSU rDNA sequences. *Nordic Jl Bot.* 17: 403–418; **Li, L.T.; Kimbrough, J.W.** (1995). Spore wall ontogeny in *Pseudoplectania nigrella* and *Plectania nannfeldtii* (*Ascomycotina, Pezizales*). *Can. J. Bot.* 73: 1761–1767; **Li, L.T.; Kimbrough, J.W.** (1995). Septal structures in the *Sarcoscyphaceae* and *Sarcosomataceae* (*Pezizales*). *Int. J. Pl. Sci.* 156: 841–848; **Li, L.T.; Kimbrough, J.W.** (1996). Spore ontogeny of *Galiella rufa* (*Pezizales*). *Can. J. Bot.* 74: 1651–1656; **Zhuang, W.Y.; Wang, Z.** (1998). Sarcosomataceous discomycetes in China. *Mycotaxon* 67: 355–364.

Schizophyllaceae Quél. 1888
Agaricales: Basidiomycota

Basidioma usually pileate, typical tongue-shaped with a short eccentric stipe or sessile with a lateral attachment, rarely effuse, simple or densely imbricate, in some genera remaining cupulate or tubular, leathery or gelatinized, the cap grey, white or yellowish, smooth or hairy, the hairs rarely staining brown in iodine. Hyphal system monomitic, hyphae not inflated, usually with clamp connections. Hymenium initially smooth and waxy, becoming weakly tuberculate, radially ridged or splitting radially to form a series of lamellar plates, sometimes with cystidia. Basidia clavate, usually with 4 sterigmata. Basidiospores cylindrical, ovoid, ellipsoidal or allantoid, smooth and thin-walled, hyaline or yellowish-brown, not staining in iodine.

Significant Genera: *Schizophyllum*.

Distribution: Very widespread.

Economic Significance: *Schizophyllum commune* has been used extensively as a model organism for physiological and genetic studies It is also used for food, either directly or following fermentation to produce a cheese-like substance.

Ecology: Most species are saprobes on rotten wood, but *S. commune* is occasionally implicated as a pathogen or decay organism of woody plants or as an opportunistic invader of mammalian tissues.

Notes: Some molecular data indicate that the *Schizophyllaceae* and *Fistulinaceae* are related, but there are substantial morphological differences between the two families. *Henningsomyces* and *Rectipilus* were placed in the *Schizophyllaceae*, but the former at least is polyphyletic and neither appears closely related to *Schizophyllum*. The family should probably be restricted to that genus and *Auriculariopsis*.

Schizophyllum commune, NE China; basidiomata

References: **Bodensteiner, P.; Binder, M.; Moncalvo, J.-M.; Agerer, R.; Hibbett, D.S.** (2004). Phylogenetic relationships of cyphelloid homobasidiomycetes. *Mol. Phylogen. Evol.* 33: 501–515; **Buzina, W.; Lang-Loidolt, D.; Braun, H.; Freudenschuss, K.; Stammberger, H.** (2001). Development of molecular methods for identification of *Schizophyllum commune* from clinical samples. *J. Clin. Microbiol.* 39: 2391–2396; **Ginns, J.** (1998). Genera of the North American *Corticiaceae sensu lato*. *Mycologia* 90: 1–35; **James, T.Y.; Porter, D.; Hamrick, J.L.; Vilgalys, R.** (1999). Evidence for limited intercontinental gene flow in the cosmopolitan mushroom, *Schizophyllum commune*. *Evolution, Lancaster, Pa.* 53: 1665–1677; **James, T.Y.; Vilgalys, R.** (2001). Abundance and diversity of *Schizophyllum commune* spore clouds in the Caribbean detected by selective sampling. *Mol. Ecol.* 10: 471–479; **Kano, R.; Oomae, S.; Nakano, Y.; Minami, T.; Sukikara, M.; Nakayama, T.; Hasegawa, A.** (2002). First report on *Schizophyllum commune* from a dog. *J. Clin. Microbiol.* 40: 3535–3537; **Larsson, K.-H.; Larsson, E.; Koljalg, U.** (2004). High phylogenetic diversity among corticioid homobasidiomycetes. *Mycol. Res.* 108: 983–1002; **Moncalvo, J.-M.; Vilgalys, R.; Redhead, S.A.; Johnson, J.E.; James, T.Y.; Aime, M.C.; Hofstetter, V.; Verduin, S.J.W.; Larsson, E.; Baroni, T.J.; Thorn, R.G.; Jacobsson, S.; Clemencon, H.; Miller, O.K.** (2002). One hundred and seventeen clades of euagarics. *Mol. Phylogen. Evol.* 23: 357–400; **Sigler, L.; Estrada, S.; Montealegre, N.A.; Jaramillo, E.; Arango, M.; De Bedout, C.; Restrepo, A.** (1997). Maxillary sinusitis caused by *Schizophyllum commune* and experience with treatment. *J. Med. Vet. Mycol.* 35: 365–370; **Stalpers, J.A.** (1988). *Auriculariopsis* and the *Schizophyllales*. *Persoonia* 13: 495–504; **Wei, Y.L.; Dai, Y.C.** (2004). *Rectipilus*, a new genus to Chinese fungi flora. *Mycosystema* 23: 437–438.

Schizoporaceae Jülich 1982
Hymenochaetales: Basidiomycota

Basidiomata corticioid, resupinate, adnate, effuse, annual or perennial, white, yellowish or brown. Hymenium poroid with round or angular, often irregular pores, or with sparse to dense conical, or flat and irregular, spines. Hyphal system monomitic or dimitic, skeletal hyphae when present often poorly developed, generative hyphae with thin or thick walls, always with clamp connections. Cystidia usually present, varied in form, usually capitate with a globose encrusted or resinous tip. Basidia cylindrical to ± urniform, with 4 well-developed sterigmata. Basidiospores ellipsoidal, hyaline, thin-walled, smooth, not staining in iodine.

Significant Genera: *Hyphodontia*, *Schizopora*.

Distribution: Cosmopolitan.

Economic Significance: No significant decay problems are reported.

Ecology: Saprobic, causing white rots of standing and fallen wood of coniferous and broadleaved trees, rarely on herbaceous stems.

Notes: *Hyphodontia* is highly paraphyletic and needs further revision before the boundaries of this family can be confidently defined. *Oxyporus*, the other large genus currently placed in this family, may also need remodelling and repositioning. Provisionally the family is here restricted to *Schizopora* and a group of *Hyphodontia* species known to be close relatives. A longer-term solution could be merger with the *Hymenochaetaceae*.

Basidioradulum radula; basidioma

References: **Binder, M.; Hibbett, D.S.; Larsson, K.-H.; Larsson, E.; Langer, E.; Langer, G.** (2005). The phylogenetic distribution of resupinate forms across the major clades of mushroom-forming fungi (*Homobasidiomycetes*). *Syst. Biodiv.* 3: 113–157; **Ginns, J.** (1998). Genera of the North American *Corticiaceae sensu lato*. *Mycologia* 90: 1–35; **Hjortstam, K.; Ryvarden, L.** (2002). Studies in tropical corticioid fungi (*Basidiomycotina, Aphyllophorales*). *Aluta-*

ceodontia, *Botryodontia*, *Hyphodontia s.s.* and *Kneiffiella*. *Syn. Fung.* (Oslo) 15: 7–17; **Langer, E.** (1994). Die Gattung *Hyphodontia* John Eriksson. *Biblthca Mycol.* 154: 298 pp.; **Langer, E.** (1998). Evolution of *Hyphodontia* (*Corticiaceae*, *Basidiomycetes*) and related *Aphyllophorales* inferred from ribosomal DNA sequences. *Folia cryptog. Estonica* 33: 57–62; **Larsson, K.-H.; Larsson, E.; Koljalg, U.** (2004). High phylogenetic diversity among corticioid homobasidiomycetes. *Mycol. Res.* 108: 983–1002; **Martín, M.P.; Montón, C.** (2004). [Molecular identification of *Oxyporus latemarginatus* (*Basidiomycotina*) isolated from palms]. *Bol. Sanid. Veg., Plagas* 30: 93–96; **Paulus, B.; Hallenberg, N.; Buchanan, P.K.; Chambers, G.K.** (2000). A phylogenetic study of the genus *Schizopora* (*Basidiomycota*) based on ITS DNA sequences. *Mycol. Res.* 104: 1155–1163.

Schizosaccharomycetaceae Beij. ex Klöcker 1905
Schizosaccharomycetales: Ascomycota

Mycelium absent or poorly developed. Vegetative cells cylindrical with rounded ends, proliferating thallically by fission into 2 ± equal daughter cells. Asci formed by somatic conjugation of vegetative cells, often irregularly shaped, 4- to 8-spored. Ascospores globose to shortly cylindrical, blueing in iodine, smooth, without sheaths. Fermentation positive.

Significant Genera: *Schizosaccharomyces*.

Distribution: Cosmopolitan.

Economic Significance: Known as spoilage yeasts and used in fermentation processes, but most significant as a model organism for research.

Ecology: Isolated from a wide range of commodities, especially foods with high sugar content and various beverages.

Notes: Sequencing of the genome of *Schizosaccharomyces pombe*, completed in 2002, has underlined the depth of the division between this family and other ascomycotan yeasts. The genus, along with the segregate *Hasegawaea*, is now given its own class.

References: **Egel, R.** (1994). Regulation of meiosis and sporulation in *Schizosaccharomyces pombe*. *The Mycota* (Berlin) 1: 251–265; **Eriksson, O.E.; Svedskog, A.; Landvik, S.** (1993). Molecular evidence for the evolutionary hiatus between *Saccharomyces cerevisiae* and *Schizosaccharomyces pombe*. *Syst. Ascom.* 11: 119–162; **Esteve-Zarzoso, B.; Belloch, C.; Uruburu, F.; Querol, A.** (1999). Identification of yeasts by RFLP analysis of the 5.8S rRNA gene and the two ribosomal internal transcribed spacers. *Int. J. Syst. Bacteriol.* 49: 329–337; **Jeffery, J.; Kock, J.L.F.; Botha, A.; Coetzee, D.J.; Botes, P.J.** (1997). The value of lipid composition in the taxonomy of the *Schizosaccharomycetales*. *Antonie van Leeuwenhoek* 72: 327–335; **Kurtzman, C.P.; Blanz, P.A.** (1998). Ribosomal RNA/DNA sequence comparisons for assessing phylogenetic relationships. *The Yeasts, A Taxonomic Study* (Amsterdam): 69–74; **Mikata, K.; Yamada, Y.** (1999). The ubiquinone system in *Hasegawaea japonica* (Yukawa et Maki) Yamada et Banno: a new method for identifying ubiquinone homologs from yeast cells. *Res. Commun., Inst. Ferm.* Osaka 19: 41–46; **Munz, P.; Kohli, J.** (1995). *Schizosaccharomyces pombe*: genetic nomenclature and chromosome maps. *The Yeasts* Vol. 6. Yeast Genetics. Edn 2 (London): 583–588; **Sipiczki, M.** (1995). Phylogenesis of fission yeasts. Contradictions surrounding the origin of a century old genus. *Antonie van Leeuwenhoek* 68: 119–149; **Sugiyama, J.** (1998). Relatedness, phylogeny, and evolution of the fungi. *Mycoscience* 39: 487–511; **Vaughan-Martini, A.; Martini, A.** (1998). *Schizosaccharomyces* Lindner. *The Yeasts A Taxonomic Study* (Amsterdam): 391–394; **Wood, V. et al.** (2002). The genome sequence of *Schizosaccharomyces pombe*. *Nature* Lond. 415: 871–880.

Schizothyriaceae Höhn. ex Sacc., D. Sacc. & Traverso 1928
Microthyriales: Ascomycota

Ascomata thyrothecial, strongly flattened or crustose, rounded or elongate, opening by irregular splits; upper wall brown, composed of a single layer of ± epidermoid cells. Interascal tissue absent, or composed only of remnants of stromatal cells. Asci ± globose to saccate, fissitunicate, without an ocular chamber, not blueing in iodine. Ascospores hyaline to pale brown, transversely septate, without a sheath. Anamorphs unknown.

Significant Genera: *Schizothyrium*.

Distribution: Widespread, especially in the tropics.

Economic Significance: Very little is known; one species is implicated in sooty blotch disease of apples, but the economic impact is difficult to gauge.

Ecology: Mostly saprobic, epiphytic on leaves or stems.

Notes: A poorly understood family of uncertain affinities, and the principal genus is certainly polyphyletic. The assumed relationship with the *Microthyriales* may be due to parallel evolution rather than common ancestry.

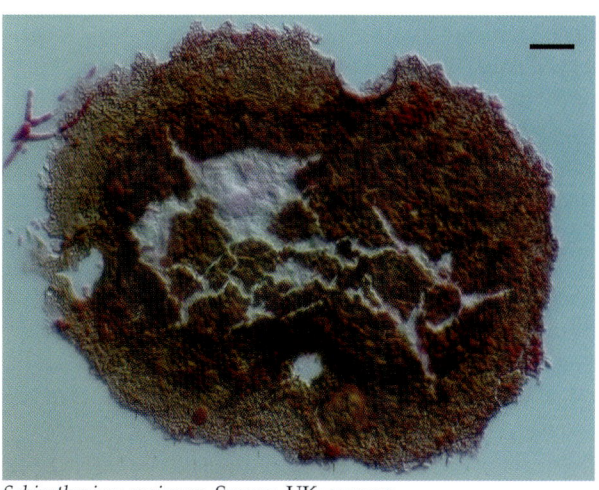

Schizothyrium speireum, Surrey, UK; ascoma

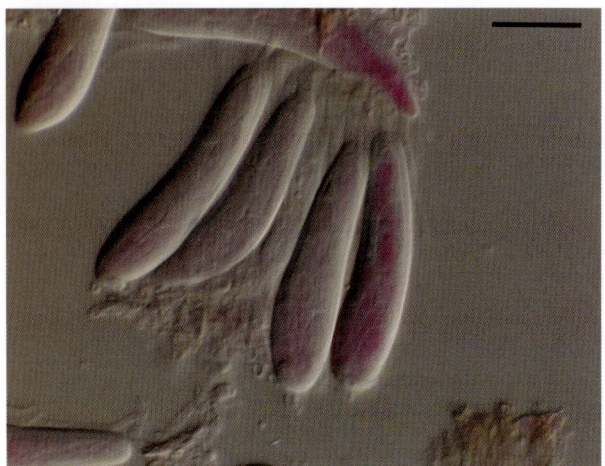

Schizothyrium speireum, India; asci, ascospores and interascal tissue

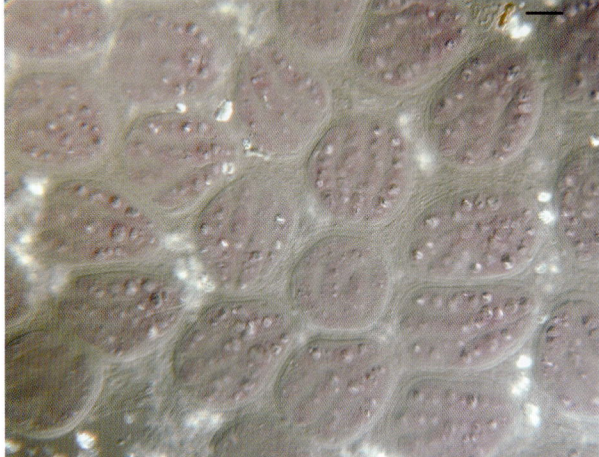

Schizothyrium scutelliforme, Sudan; asci and ascospores

References: **Farr, M.L.** (1987). Amazonian foliicolous fungi. IV. Some new and critical taxa in *Ascomycotina* and associated anamorphs. *Mycologia* 79: 97–116; **Williamson, S.M.; Sutton, T.B.** (2000). Sooty blotch and flyspeck of apple: etiology, biology and control. *Pl. Dis.* 84: 714–724.

Sclerodermataceae Corda 1842
Boletales: Basidiomycota

Basidiomata gasteroid, epigeous or semi-hypogeous, tuber-like, ± globose, sessile or rarely with a short stipe; peridium yellowish or brown, smooth, verrucose or squamose, not viscid or gelatinous, tough or cartilaginous, dehiscing by irregular fracture. Gleba becoming purplish or brown, initially interspersed with thin mycelial layers but finally becoming powdery. Capillitium absent. Basidia formed in small clusters without a true hymenium, ± clavate with 6–8 sterigmata, deliquescing at an early stage. Basidiospores brown, thick-walled, spinose or reticulate, not staining in iodine.

Significant Genera: *Scleroderma*.

Distribution: Widespread in both temperate and tropical regions.

Economic Significance: None is known.

Ecology: On the ground or associated with rotten wood, mostly ectomycorrhizal with woody plants.

Notes: Some authors combine this family with the *Astraeaceae*, *Calostomataceae* and *Pisolithaceae*.

Scleroderma citrinum, Surrey, UK; basidiomata

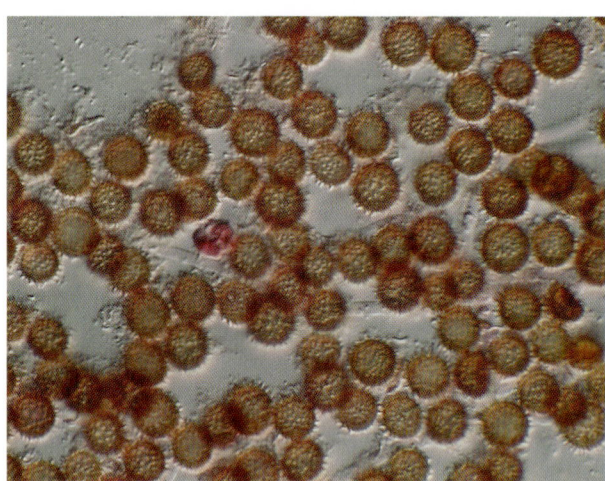

Scleroderma citrinum, Surrey, UK; basidiospores

Scleroderma verrucosum, Surrey, UK; dehiscent basidioma

References: **Binder, M.; Bresinsky, A.** (2002). Derivation of a polymorphic lineage of gasteromycetes from boletoid ancestors. *Mycologia* 94: 85–98; **Guzmán, G.; Ovrebo, C.L** (2000). New observations on sclerodermataceous fungi. *Mycologia* 92: 174–179; **Kasuya, T.; Guzmán, G.; Ramirez-Guillén, F.J.; Kato, T.** (2002). *Scleroderma laeve* (*Gasteromycetes, Sclerodermatales*), new to Japan. *Mycoscience* 43: 475–476; **Pegler, D.N.; Læssøe, T.; Spooner, B.M.** (1995). *British Puffballs, Earthstars and Stinkhorns* An Account of the British Gasteroid Fungi (Kew): 255 pp.; **Richter, D.L.** (1992). Six species of *Scleroderma* (*Gasteromycetes, Sclerodermatales*) described from pure cultures. *Mycotaxon* 45: 461–471; **Sims, K.P.; Sen, R.; Watling, R.; Jeffries, P.** (1999). Species and population structures of *Pisolithus* and *Scleroderma* identified by combined phenotypic and genomic marker analysis. *Mycol. Res.* 103: 449–458; **Sims, K.P.; Watling, R.; Jeffries, P.** (1995). A revised key to the genus *Scleroderma*. *Mycotaxon* 56: 403–420; **Watling, R.** (2006). The sclerodermatoid fungi. *Mycoscience* 47: 18–24; **Watling, R.; Sims, K.** (2004). Taxonomic and floristic notes on some larger Malaysian fungi. IV (*Scleroderma*). *Mem. N. Y. bot. Gdn* 89: 93–96.

Sclerotiniaceae Whetzel 1945
Helotiales: Ascomycota

Stromata present, as well-differentiated sclerotia or mummified host tissue. Ascomata apothecial, often long-stalked, usually brown, cupulate, without hairs, the stalk often darker, the wall sometimes gelatinized. Interascal tissue of simple paraphyses, without a mucous matrix. Asci cylindrical to cylindric-clavate, fairly thin-walled, usually with a J+ apical ring, releasing spores actively, usually 8-spored. Ascospores large or small, ellipsoidal, usually aseptate, hyaline or pale brown, often ± longitudinally symmetrical, rarely dimorphic, without a mucous sheath or appendages. Anamorphs often prominent and extremely varied in form, spermatial or disseminative. Conidial morphs hyphomycetous, forming hyaline aseptate conidia from sympodially proliferating branched conidiophores or in chains from sporodochium-like structures. Spermatia small, bacillar, produced in short chains from percurrently proliferating conidiogenous cells, sometimes in locules within sclerotia.

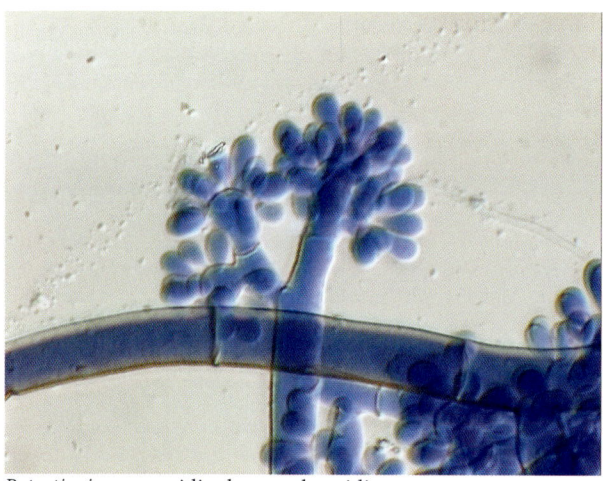
Botrytis cinerea; conidiophore and conidia

Significant Genera: *Botryotinia, Monilinia, Sclerotinia*. Anamorphs: *Botrytis, Monilia, Myrioconium*.

Distribution: Widespread, especially in temperate zones.

Dumontinia tuberosa, Scotland; ascomata

Botrytis tulipae; sclerotia in agar culture

Ciboria betulae, Scotland; ascomata

Economic Significance: Many important plant pathogens are included in the family.

Ecology: Primarily necrotrophs, infecting roots, leaves, fruits, seeds etc. and causing moulds, damping off etc. Sclerotia often overwinter and can survive for long periods in soil.

Notes: The *Sclerotiniaceae* is more tightly defined following separation of the *Rustroemiaceae*, but still contains several probably monophyletic units that appear to correlate with anamorph type. Further subdivision is probably needed, but more molecular data are needed.

References: **Anderson, J.B.; Kohn, L.B.** (1998). Genotyping, gene genealogies and genomics bring fungal population genetics above ground. *Trends in Ecology & Evolution* (Cambridge) 13: 444–449; **Anderson, J.B.; Kohn, L.M.** (1995). Clonality in soilborne, plant-pathogenic fungi. *A. Rev. Phytopath.* 33: 369–391; **Batra, L.R.** (1991). World species of *Monilinia* (*Fungi*): their ecology, biosystematics and control. *Mycol. Mem.* 16: 246 pp.; **Côté, M.-J.; Prud'homme, M.; Meldrum, A.J.; Tardif, M.-C.** (2004). Variations in sequence and occurrence of SSU rDNA group I introns in *Monilinia fructicola* isolates. *Mycologia* 96: 240–248; **Cubeta, M.A.; Cody, B.R.; Kohli, Y.; Kohn, L.M.** (1997). Clonality in *Sclerotinia sclerotiorum* on infected cabbage in eastern North Carolina. *Phytopathology* 87: 1000–1004; **Förster, H.; Adaskaveg, J.E.** (2000). Early brown rot infections in sweet cherry fruit are detected by *Monilinia*-specific DNA primers. *Phytopathology* 90: 171–178; **Fulton, C.E.; Leeuwen, G.C.M. van; Brown, A.E.** (1999). Genetic variation among and within *Monilinia* species causing brown rot of stone and pome fruits. *Eur. J. Pl. Path.* 105: 495–500; **Giraud, T.; Fortini, D.; Levis, C.; Lamarque, C.; Leroux, P.; LoBuglio, K.; Brygoo, Y.** (1999). Two sibling species of the *Botrytis cinerea* complex, *transposa* and *vacuma*, are found in sympatry on numerous host plants. *Phytopathology* 89: 967–973; **Holst-Jensen, A.; Kohn, L.M.; Jakobsen, K.S.; Schumacher, T.** (1997). Molecular phylogeny and evolution of *Monilinia* (*Sclerotiniaceae*) based on coding and noncoding rDNA sequences. *Am. J. Bot.* 84: 686–701; **Holst-Jensen, A.; Kohn, L.M.; Schumacher, T.** (1997). Nuclear rDNA phylogeny of the *Sclerotiniaceae*. *Mycologia* 89: 885–899; **Holst-Jensen, A.; Vaage, M.; Schumacher, T.** (1999). An approximation to the phylogeny of *Sclerotinia* and related genera. *Nordic Jl Bot.* 18: 705–719; **Holst-Jensen, A.; Vaage, M.; Schumacher, T.; Johansen, S.** (1999). Structural characteristics and possible horizontal transfer of Group I introns between closely related plant pathogenic fungi. *Mol. Biol. Evol.* 16: 114–126; **Kohn, L.M.** (1979). A monographic revision of the genus *Sclerotinia*. *Mycotaxon* 9: 365–444; **Kohn, L.M.; Grenville, D.J.** (1989). Ultrastructure of stromatal anamorphs in the *Sclerotiniaceae*. *Can. J. Bot.* 67: 394–406; **Schumacher, T.; Holst-Jensen, A.** (1997). A synopsis of the genus *Sclerotinia* (= *Verpatinia*) (*Ascomycota*: *Helotiales*: *Sclerotiniaceae*). *Mycoscience* 38: 55–69; **Schumacher, T.; Kohn, L.M.** (1985). A monographic revision of the genus *Myriosclerotinia*. *Can. J. Bot.* 63: 1610–1640; **Taylor, J.W.; Jacobson, D.J.; Fisher, M.C.** (1999). The evolution of asexual fungi: reproduction, speciation and classification. *A. Rev. Phytopath.* 37: 197–246; **Wang, Z.; Binder, M.; Schoch, C.L.; Johnston, P.R.; Spatafora, J.W.; Hibbett, D.S.** (2006). Evolution of helotialean fungi (*Leotiomycetes*, *Pezizomycotina*): a nuclear rDNA phylogeny. *Mol. Phylogen. Evol.* 41: 295–312; **Willetts, H.J.** (1997). Morphology, development and evolution of stromata/sclerotia and macroconidia of the *Sclerotiniaceae*. *Mycol. Res.* 101: 939–952.

Sebacinaceae K. Wells & Oberw. 1982
Sebacinales: Basidiomycota

Basidiomata pustulate, coralloid, tongue-like or funnel-shaped, gelatinous, cartilaginous or leathery. Hyphal system monomitic, hyphae brown-walled, without clamp connections, septa with dolipores and continuous parenthosomes, cystidia absent. Basidia ± globose to pyriform, without basal clamp connections, becoming longitudinally or cruciately septate to form 2 or 4 separate segments, the apical part elongate. Basidiospores ovoid, cylindrical or allantoid, hyaline, thin-walled, often germinating by repetition.

Significant Genera: *Sebacina*.

Distribution: Widespread.

Economic Significance: None is known.

Ecology: Terrestrial, forming mycorrhizas with a wide range of plants including orchids, *Ericaceae* and liverworts as well as ectomycorrhizas with conifers and angiosperm trees.

Notes: One of the basal families of the *Basidiomycota*, clearly distinct from the *Exidiaceae* (where the family has previously been placed) when sequences are examined. The diversity of mycorrhizal partners is remarkable and has led to suggestions that a mycorrhizal life strategy may be an ancestral condition for the basidiomycetes.

Sebacina incrustans; basidioma

References: **McCormick, M.K.; Whigham, D.F.; O'Neill, J.** (2004). Mycorrhizal diversity in photosynthetic terrestrial orchids. *New Phytol.* 163: 425–438; **Roberts, P.** (2003). *Sebacina concrescens* and *S. sparassoidea*: two conspicuous but neglected North American *Sebacina* species. *Sydowia* 55: 348–354; **Selosse, M.-A.; Bauer, R.; Moyersoen, B.** (2002). Basal hymenomycetes belonging to the *Sebacinaceae* are ectomycorrhizal on temperate deciduous trees. *New Phytol.* 155: 183–195; **Stalpers, J.A.; Andersen, T.F.** (1996). A synopsis of the taxonomy of teleomorphs connected with *Rhizoctonia* s.l. *Rhizoctonia Species, Taxonomy, Molecular Biology, Ecology, Pathology and Disease Control* (Dordrecht): 49–63; **Walker, J.F.; Parrent, J.L.** (2004). Molecular phylogenetic evidence for the mycorrhizal status of *Tremellodendron* (*Sebacinaceae*). *Mem. N. Y. bot. Gdn* 89: 291–296; **Weiss, M.; Oberwinkler, F.** (2001). Phylogenetic relationships in *Auriculariales* and related groups – hypotheses derived from nuclear ribosomal DNA sequences. *Mycol. Res.* 105: 403–415; **Weiss, M.; Selosse, M.-A.; Rexer, K.-H.; Urban, A.; Oberwinkler, F.** (2004). *Sebacinales*: a hitherto overlooked cosm of heterobasidiomycetes with a broad mycorrhizal potential. *Mycol. Res.* 108: 1003–1010.

Septobasidiaceae Racib. 1909
Septobasidiales: Basidiomycota

Basidiomata resupinate, effuse, not gelatinized, crustose or spongy in texture, often with a multilayered structure including chambers surrounding individual host insects. Haustoria present inside host cells, hyphae generally thin-walled, without clamp connections. Hymenial layer usually covering the outer surface, with basidia developing from thick-walled ± spherical fertile cells. Basidia elongate, usually cylindrical, with 1 to 3 transverse septa developing and usually with each cell forming a well-developed lateral sterigma. Basidiospores ellipsoidal to allantoid, hyaline, sometimes thick-walled, sometimes becoming septate, smooth-walled. Conidia formed in some species.

Significant Genera: *Septobasidium*.

Distribution: Widespread in the tropics and warm temperate regions.

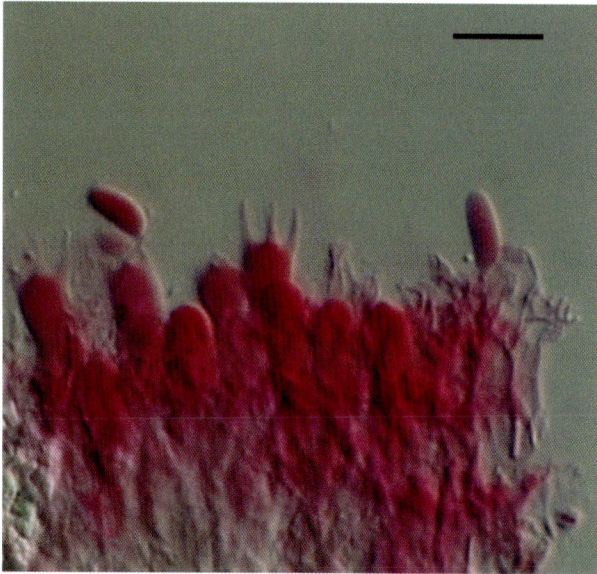
Septobasidium bogoriense, Malaysia; basidia and basidiospores

Septobasidium bogoriense, Malaysia; basidioma overgrowing scale insects

References: **Bandoni, R.J.** (1995). Dimorphic heterobasidiomycetes: taxonomy and parasitism. *Stud. Mycol.* 38: 13–27; **Begerow, D.; Bauer, R.; Oberwinkler, F.** (1998). Phylogenetic studies on nuclear large subunit ribosomal DNA sequences of smut fungi and related taxa. *Can. J. Bot.* 75: 2045–2056; **Gómez, L.D.; Henk, D.A.** (2004). Validation of the species of *Septobasidium* (*Basidiomycetes*) described by John N. Couch. *Lankesteriana* 4: 75–96; **Gómez, L.D.; Kisimova-Horovitz, L.** (2001). A new species of *Septobasidium* from Costa Rica. *Mycotaxon* 80: 255–259; **Henk, D.A.** (2005). New species of *Septobasidium* from southern Costa Rica and the southeastern United States. *Mycologia* 97: 908–913; **Wingfield, B.D.; Ericson, L.; Szaro, T.; Burdon, J.J.** (2004). Phylogenetic patterns in the *Uredinales*. *Australas. Pl. Path.* 33: 327–335.

Economic Significance: The combined fungal and insect colonies may cause significant damage to leaves and stems of living trees and shrubs. *Septobasidium* species have been examined for their potential in biological control, but successful products have not been developed to date.

Ecology: Parasitic on scale insect colonies; the association is antagonistic for many individual insects, but those not parasitized appear to be sheltered from attack by hyperparasitic insects. Species appear to be highly host-specific.

Notes: An isolated family, assumed to be a basal member of the *Urediniomycetes*. Detailed phylogenetic research has not so far been published.

Seuratiaceae Vuill. ex M.E. Barr 1987
Uncertain position within Ascomycetes, Ascomycota

Stromata superficial or absent. Ascomata crustose, dark, irregular in outline, the upper part deliquescing to expose the asci; peridium composed of pale chains of cells in a gelatinous matrix. Interascal tissue poorly developed, consisting of irregular interthecial strands. Asci scattered or in small groups, saccate, thick-walled, ? fissitunicate. Ascospores hyaline to brown, transversely septate, without a sheath. Anamorph usually hyphomycetous with elaborate lobed or torulose conidia.

Significant Genera: *Seuratia*. Anamorph: *Atichia*.

Distribution: Most species are known from the tropics.

Economic Significance: None is known.

Ecology: Saprobic and epiphytic on leaves.

Notes: Apparently an isolated family without close relatives. No molecular data are available.

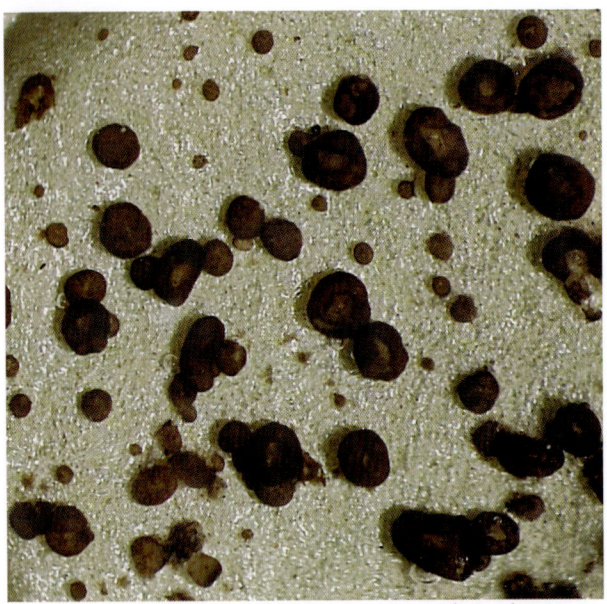

Seuratia millardetii, Tierra del Fuego; stromata on living leaves

Significant Genera: *Sigmoideomyces*.

Distribution: Widespread although monumentally poorly recorded.

Economic Significance: None is known.

Ecology: Parasitic on fungi.

Notes: Very poorly known; little recent information is available.

References: **Benny, G.L.; Benjamin, R.K.; Kirk, P.M.** (1992). A reevaluation of Cunninghamellaceae (Mucorales). Sigmoideomycetaceae fam. nov. and Reticulocephalis gen. nov.; cladistic analysis and description of two new species. Mycologia 84: 615–641; **Tanabe, Y.; O'Donnell, K.; Saikawa, M.; Sugiyama, J.** (2000). Molecular phylogeny of parasitic Zygomycota (Dimargaritales, Zoopagales) based on nuclear small subunit ribosomal DNA sequences. Mol. Phylogen. Evol. 16: 253–262.

Seuratia millardetii, Tierra del Fuego; asci and conidia

References: **Barr, M.E.** (1987). New taxa and combinations in the Loculoascomycetes. Mycotaxon 29: 501–505; **Rodríguez Hernández, M.; Camino Vilaró, M.C.** (1985). El género Seuratia (Ascomycotina) en Cuba. Revta Jardín bot. Nac., Univ. Habana 6: 61–63; **Sivanesan, A.; Hsieh, W.H.** (1995). A re-appraisal of the systematic status of the ascomycete genus Yoshinagaia. Mycol. Res. 99: 1295–1298; **Sutton, B.C.** (1986). Notes on Deuteromycetes. Sydowia 38: 324–338.

Sigmoideomycetaceae Benny, R.K. Benj. & P.M. Kirk 1992
Zoopagales: Zygomycota

Vegetative thallus a mycelium, complex fruit bodies absent. Fertile hyphae dichotomously branched, septate, coiled, either sessile or with a stalk-like region and thus raising the fertile head above the substratum, terminating in fertile vesicles or remaining sterile; spores pedicellate, borne on vesicles, ± hyaline, small, smothe-walled or more typically ornamented. Zygospores unknown.

Sirobasidiaceae Lindau 1897
Tremellales: Basidiomycota

Sirobasidium magnum, Singapore; habit drawing, basidia and basidiospores

Basidiomata small to medium-sized, usually erumpent, gelatinous, pustulate, pale or brightly coloured. Hyphal system monomitic, hyphae with clamp connections. Basidia formed in clusters or short chains, ± globose to ovoid, becoming obliquely or cruciately septate, sterigmata not differentiated. Basidiospores released actively, fusiform, aseptate, sometimes germinating to produce secondary spores.

Significant Genera: *Fibulobasidium, Sirobasidium*.

Distribution: Widespread, primarily tropical.

Economic Significance: None is known.

Ecology: On wood and bark, possibly parasitic on other fungi.

Notes: Molecular data indicate a relationship between *Sirobasidium* and *Fibulobasidium*; one clade of the polyphyletic genus *Tremella* may also belong here.

References: **Bandoni, R.J.** (1999). On an undescribed species of *Fibulobasidium*. *Can. J. Bot.* 76: 1540–1543; **Bandoni, R.J.; Boekhout, T.** (1998). Tremelloid genera with yeast phases: *Fibulobasidium* Bandoni, *Holtermannia* Saccardo & Traverso, *Sirobasidium* de Lagerheim & Patouillard, *Tremella* Persoon, *Trimorphomyces* Bandoni & Oberwinkler. *The Yeasts, A Taxonomic Study* (Amsterdam): 705–717; **Benkert, D.** (1991). *Xenolachne longicornis*, eine seltene Tremellacee in Deutschland (in memoriam Joe DUTY). *Mykol. MittBl.* 34: 79–82; **Fell, J.W.; Boekhout, T.; Fonseca, Á.; Scorzetti, G.; Statzell-Tallman, A.** (2000). Biodiversity and systematics of basidiomycetous yeasts as determined by large-subunit rDNA D1/D2 domain sequence analysis. *Int. J. Syst. Evol. Microbiol.* 50: 1351–1371; **Ingold, C.T.** (1995). Types of reproductive cell in *Exidia recisa* and *Sirobasidium intermediae*. *Mycol. Res.* 99: 1187–1190; **Sampaio, J.P.; Weiss, M.; Gadanho, M.; Bauer, R.** (2002). New taxa in the *Tremellales*: *Bulleribasidium oberjochense* gen. et sp. nov., *Papiliotrema bandonii* gen. et sp. nov. and *Fibulobasidium murrhardtense* sp. nov. *Mycologia* 94: 873–887.

Sistotremataceae Jülich 1982
Trechisporales: Basidiomycota

Basidiomata resupinate to pileate, rarely stipitate, sometimes very poorly developed, fragile, gelatinous or waxy. Hyphal system usually monomitic, hyphae inflated at the septa, clamp connections present or absent. Hymenium smooth, spiny, ridged or weakly poroid, cystidia where present thin-walled and conical or capitate, sometimes with refractive contents. Basidia urniform, pyriform or clavate to cylindrical with constricted mid portions, (2- to) 6- to 12-sterigmate, sometimes proliferating. Basidiospores globose, ellipsoidal or allantoid, rarely tetrahedral, hyaline, usually thin-walled, usually smooth, not staining in iodine.

Significant Genera: *Sistotrema, Trechispora*.

Distribution: Cosmopolitan.

Economic Significance: *Trechispora alnicola* has been reported as causing turfgrass disease, but the affinities of this species need further examination.

Ecology: Most species are saprobes on fallen wood, bark and similar substrata.

Notes: The family is largely united by its resupinate basidiomata and basidia with multiple sterigmata. Molecular data suggest a position within the cantharelloid clade and a somewhat different circumscription for the family, but further sequences are required before the classification can be stabilized.

Sistotrema confluens; basidioma

References: **Binder, M.; Hibbett, D.S.; Larsson, K.-H.; Larsson, E.; Langer, E.; Langer, G.** (2005). The phylogenetic distribution of resupinate forms across the major clades of mushroom-forming fungi (*Homobasidiomycetes*). *Syst. Biodiv.* 3: 113–157; **Ginns, J.** (1998). Genera of the North American *Corticiaceae* sensu lato. *Mycologia* 90: 1–35; **Greslebin, A.G.** (2001). *Sistotremateae* (*Corticiaceae, Aphyllophorales*) of the Patagonian Andes forests of Argentina. *Mycol. Res.* 105: 1392–1396; **Hibbett, D.S.; Binder, M.** (2002). Evolution of complex fruiting-body morphologies in homobasidiomycetes. *Proc. R. Soc. Lond.* B. Biol. Sci. 269: 1963–1969; **Küffer, N.; Senn-Irlet, B.** (2005). Diversity and ecology of wood-inhabiting aphyllophoroid basidiomycetes on fallen woody debris in various forest types in Switzerland. *Mycol. Prog.* 4: 77–86; **Larsson, K.-H.** (1994). Poroid species in *Trechispora* and the use of calcium oxalate crystals for species identification. *Mycol. Res.* 98: 1153–1172; **Larsson, K.-H.** (1996). New species and combinations in *Trechispora* (*Corticiaceae, Basidiomycotina*). *Nordic Jl Bot.* 16: 83–98; **Larsson, K.-H.** (1996). Taxonomy of *Trechispora farinacea* and proposed synonyms II. Species with a smooth hymenophore. *Nordic Jl Bot.* 16: 73–82; **Larsson, K.-H.; Larsson, E.; Koljalg, U.** (2004). High phylogenetic diversity among corticioid homobasidiomycetes. *Mycol. Res.* 108: 983–1002; **Ryvarden, L.** (1991). Genera of Polypores. Nomenclature and Taxonomy. *Syn. Fung.* (Oslo) 5: 363 pp.; **Wilkinson, H.T.** (1987). Association of *Trechispora alnicola* with yellow ring disease of *Poa pratensis*. *Can. J. Bot.* 65: 150–153.

Sordariaceae G. Winter 1885
Sordariales: Ascomycota

Stromata absent. Ascomata perithecial, dark, often leathery and thick-walled, usually ostiolate, the ostiole if present periphysate. Interascal tissue often present (of wide thin-walled paraphyses) but inconspicuous and usually evanescent. Asci large cylindrical, fairly thin-walled and not fissitunicate, rarely evanescent, usually with a small J– apical ring. Ascospores brown, aseptate or very rarely septate, sometimes ornamented with reticulations, pits or furrows, often with a gelatinous sheath but lacking elongate appendages. Anamorphs hyphomycetous, prominent in heterothallic species; either with conidia formed from rapidly growing fragmenting hyphae or bacillar spermatia produced from small percurrently proliferating ± cylindrical or flask-shaped conidiogenous cells.

Significant Genera: *Gelasinospora*, *Neurospora*, *Sordaria*. Anamorph: *Chrysonilia*.

Distribution: Very widespread.

Economic Significance: *Neurospora* species are resistant to high temperatures, spread very rapidly through the copious production of air-dispersed conidia, and may be found as contaminants of food processing and laboratories. They are also model organisms for genetic research.

Ecology: Saprobic, especially on dung and rotting vegetation but also found associated with foodstuffs.

Notes: Recent work suggests that the *Sordariaceae* nests within the *Lasiosphaeriaceae* and that *Gelasinospora* is polyphyletic with *Neurospora* (also polyphyletic) nesting within that clade.

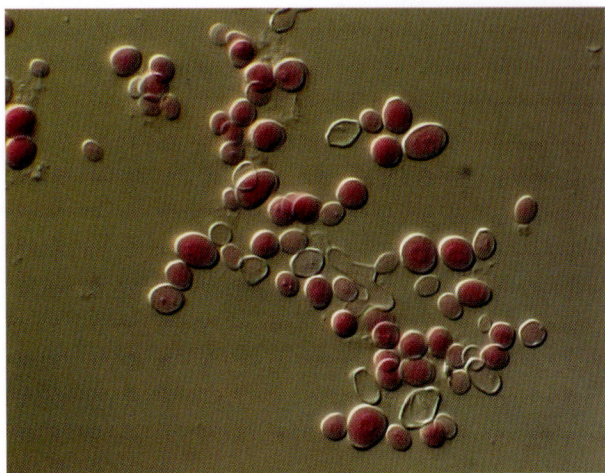

Neurospora sp.; conidia

Gelasinospora cerealis; ascus and ascospores

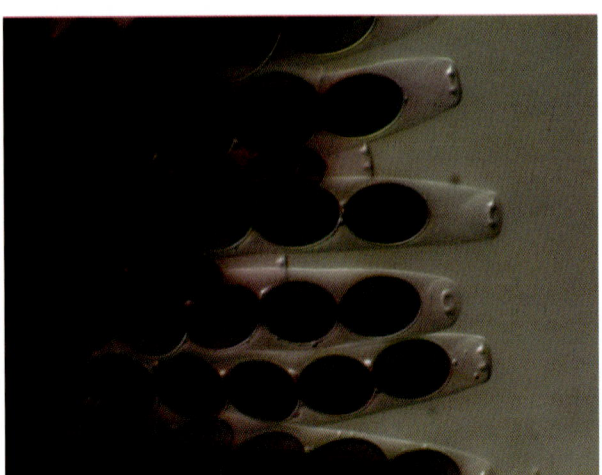

Sordaria fimicola; asci and ascospores

References: **Beatty, N.P.; Smith, M.L.; Glass, N.L.** (1994). Molecular characterization of mating-type loci in selected homothallic species of *Neurospora*, *Gelasinospora* and *Anixiella*. *Mycol. Res.* 98: 1309–1316; **Cai, L.; Jeewon, R.; Hyde, K.D.** (2006). Phylogenetic investigations of *Sordariaceae* based on multiple gene sequences and morphology. *Mycol. Res.* 110: 137–150; **Dettman, J.R.; Harbinski, F.M.; Taylor, J.W.** (2001). Ascospore morphology is a poor predictor of the phylogenetic relationships of *Neurospora* and *Gelasinospora*. *Fungal Genetics Biol.* 34: 49–61; **Dettman, J.R.; Jacobson, D.J.; Taylor, J.W.** (2006). Multilocus sequence data reveal extensive phylogenetic species diversity within the *Neurospora discreta* complex. *Mycologia* 98: 436–446; **Dettman, J.R.; Taylor, J.W.** (2004). Mutation and evolution of microsatellite loci in *Neurospora*. *Genetics* Bethesda 168: 1231–1248; **García, D.; Stchigel, A.M.; Cano, J.; Guarro, J; Hawksworth, D.L.** (2004). A synopsis and re-circumscription of *Neurospora* (syn. *Gelasinospora*) based on ultrastructural and 28S rDNA sequence data. *Mycol. Res.* 108: 1119–1142; **Glass, N.L.; Metzenberg, R.L.; Raju, N.B.** (1990). Homothallic *Sordariaceae* from nature: the absence of strains containing only the *a* mating type sequence. *Exp. Mycol.* 14: 274–289; **Huhndorf, S.M.; Miller, A.N.; Fernández, F.A.** (2004). Molecular systematics of the *Sordariales*: the order and the family *Lasiosphaeriaceae* revisited. *Mycologia* 96: 368–387; **Jacobson, D.J.; Powell, A.J.; Dettman, J.R.; Saenz, G.S.; Barton, M.M.; Hiltz, M.D.; Dvorachek, W.H.; Glass, N.L.; Taylor, J.W.; Natvig, D.O.** (2004). *Neurospora* in temperate forests of western North America. *Mycologia* 96: 66–74; **Krug, J.C.; Khan, R.S.; Jeng, R.S.** (1994). A new species of *Gelasinospora* with multiple germ pores. *Mycologia* 86: 250–253; **Perkins, D.D.** (1992). *Neurospora*: the organism behind the molecular revolution. *Genetics* Bethesda 130: 687–701; **Perkins, D.D.; Raju, N.B.** (1986). *Neurospora discreta*, a new heterothallic species defined by its crossing behaviour. *Exp.*

Mycol. 10: 323–338; **Perkins, D.D.; Turner, B.C.** (1988). *Neurospora* from natural populations: toward the population biology of a haploid eukaryote. *Exp. Mycol.* 12: 91–131; **Pöggeler, S.** (1999). Phylogenetic relationships between mating-type sequences from homothallic and heterothallic ascomycetes. *Curr. Genet.* 36: 222–231; **Read, N.D.; Beckett, A.** (1985). The anatomy of the mature perithecium in *Sordaria humana* and its significance for fungal multicellular development. *Can. J. Bot.* 63: 281–296; **Skupski, M.P.; Jackson, D.A.; Natvig, D.O.** (1997). Phylogenetic analysis of heterothallic *Neurospora* species. *Fungal Genetics Biol.* 21: 153–162.

Sorochytriaceae Dewel 1985
Blastocladiales: Blastocladiomycota

Colonial phase endobiotic; thallus polysporangiate through sequential septation, resulting segments forming rhizoids bearing sporangia giving rise to zoospores; a new colonial thallus developing intramatrically from encysted zoospores on host; if development extramatrical (or on agar) a branching, polycentric, rhizomycelial thallus results.

Significant Genera: *Sorochytrium*.

Distribution: Known only from the USA.

Economic Significance: None is known.

Ecology: Parasite of tardigrades.

Notes: Very poorly known; no recent information is available.

References: **Dewel, R.A.; Dewel, W.C.** (1990). The fine structure of the zoospore of *Sorochytrium milnesiophthora*. *Can. J. Bot.* 68: 1968–1977; **Dewel, R.A.; Joines, J.D.; Bond, J.J.** (1985). A new chytridiomycete parasitizing the tardigrade *Milnesium tardigradum*. *Can. J. Bot.* 63: 1525–1534.

Sparassidaceae Herter 1910
Polyporales: Basidiomycota

Basidiomata annual, coralloid, large, with flattened and often undulating branches, fleshy to tough-textured, usually yellowish or pale brown. Hyphal system monomitic, hyphae often with inflated sections, rarely gelatinized, clamp connections usually present, cystidia almost always absent. Hymenium amphigenous or restricted to the lower suface of branches, ± smooth, concolorous with the non-fertile parts. Basidia clavate or elongate, with 4 sterigmata. Basidiospores subglobose or ellipsoidal to ovoid, hyaline, smooth, thin-walled, not staining in iodine.

Significant Genera: *Sparassis*.

Distribution: Widespread, primarily in temperate biomes.

Economic Significance: *Sparassis crispa* is edible and wild-collected material is traded in some societies.

Ecology: Wood decay fungi causing brown rots, resulting in root or butt rot of living or dead trees.

Notes: Part of the polyporoid clade in its wide sense and apparently close to *Phaeolus* and *Grifola*. The former genus has simple polyporoid fruit bodies and shares brown rot capacity with *Sparassis*, the latter has complex fruit bodies but causes white rots.

Sparassis crispa, Surrey, UK; basidioma

References: **Burdsall, H.H.; Miller, O.K.** (1988). Neotypification of *Sparassis crispa*. *Mycotaxon* 31: 591–593; **Burdsall, H.H.; Miller, O.K.** (1988). Type studies and nomenclatural considerations in the genus *Sparassis*. *Mycotaxon* 31: 199–206; **Dai, Y.-C.; Wang, Z.; Binder, M.; Hibbett, D.S.** (2006). Phylogeny and a new species of *Sparassis* (*Polyporales*, *Basidiomycota*): evidence from mitochondrial atp6, nuclear rDNA and rpb2 genes. *Mycologia* 98: 584–592; **Desjardin, D.E.; Wang, Z.; Binder, M.; Hibbett, D.S.** (2004). *Sparassis cystidiosa* sp. nov. from Thailand is described using morphological and molecular data. *Mycologia* 96: 1010–1014; **Wang, Z.; Binder, M.; Dai, Y.C.; Hibbett, D.S.** (2004). Phylogenetic relationships of *Sparassis* inferred from nuclear and mitochondrial ribosomal DNA and RNA polymerase sequences. *Mycologia* 96: 1015–1029.

Spathulosporaceae Kohlm. 1973
Spathulosporales: Ascomycota

Thallus ± absent or with poorly developed intramatrical hyphae. Ascomata perithecial, superficial on the thallus, usually with dark sterile setose hairs; peridium black, relatively thin, 2-layered; hymenium extending over the lower part only of the inner wall of the ascoma. Interascal tissue present or not but ostiole periphysate. Asci ± clavate, thin-walled, evanescent. Ascospores hyaline, aseptate, with mucous appendages. Anamorph spermatial, with simple percurrently proliferating conidiogenous cells (antheridia) and conidia (spermatia) that sometimes have appendages.

Significant Genera: *Spathulospora*.

Spathulospora adelpha; ascoma with sterile hairs

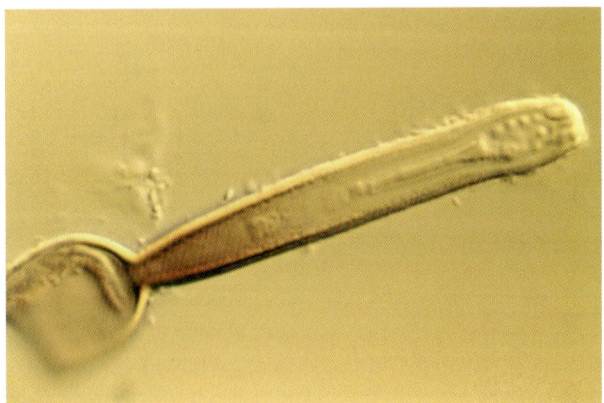
Spathulospora adelpha: tip of antheridium, spermatium with coiled apical filament

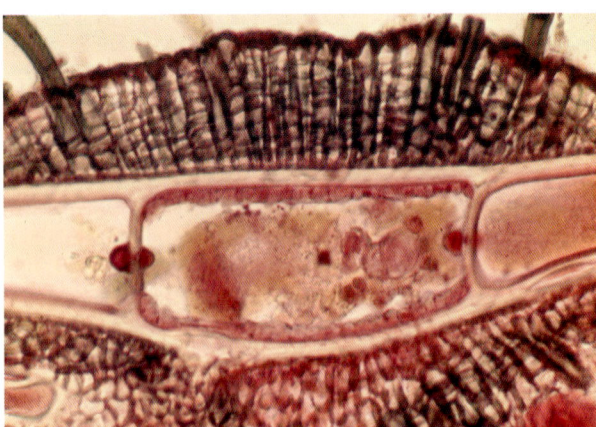
Spathulospora lanata; thallus around algal cells (*Ballia*) in longitudinal section

Distribution: Widespread in temperate zones, especially the South Pacific and South Atlantic.

Economic Significance: None is known.

Ecology: All species are obligate parasites on red algae in marine environments.

Notes: The family is still poorly known. It initially appeared to be phylogenetically isolated, but molecular evidence suggests that the family belongs with (or within) the *Lulworthiales*. The *Hispidocarpomycetaceae* may be closely related, but there are some differences in ascomatal development between the two families.

References: **Inderbitzin, P.; Lim, S.R.; Volkmann-Kohlmeyer, B.; Kohlmeyer, J.; Berbee, M.L.** (2004). The phylogenetic position of *Spathulospora* based on DNA sequences from dried herbarium material. *Mycol. Res.* 108: 737–748; **Kohlmeyer, J.** (1973). *Spathulosporales*, a new order and possible missing link between *Laboulbeniales* and pyrenomycetes. *Mycologia* 65: 614–647; **Kohlmeyer, J.; Kohlmeyer, E.** (1975). Biology and geographical distribution of *Spathulospora* species. *Mycologia* 67: 629–637; **Kohlmeyer, J.; Volkmann-Kohlmeyer, B.** (1991). Illustrated key to the filamentous higher marine fungi. *Bot. Mar.* 34: 1–61; **Nakagiri, A.; Ito, T.** (1997). *Retrostium amphiroae* gen. et sp. nov. inhabiting a marine red alga, *Amphiroa zonata*. *Mycologia* 89: 484–493.

Sphaerophoraceae Fr. 1831
Lecanorales: Ascomycota

Thallus foliose or fruticose, well-developed, sometimes with podetia or pseudopodetia. Ascomata apothecial, sessile, marginal or terminal, often ± enclosed by a thalline cup and a pigmented 'boundary layer' between generative and vegetative tissue. Interascal tissue absent. Asci cylindrical, thin-walled, evanescent or persistent (and then with tube-like apical structures). Ascospores brown, simple, ornamented with carbonized material, liberated in a black mazaedial mass. Anamorph pycnidial, producing ellipsoidal to filiform conidia from branched conidiophores with elongate conidiogenous cells.

Significant Genera: *Bunodophoron*, *Sphaerophorus*.

Sphaerophorus globosus, Harris, Scotland; fruticose thallus

Distribution: Widespread, diverse in south temperate regions.

Economic Significance: None is known.

Ecology: Lichenized with green algae, sometimes additionally with cyanobacteria in cephalodia, mostly on soil or rock.

Notes: The *Sphaerophoraceae* is well-defined in phylogenetic terms and is well separated from other familes of mazaediate lichens. Evanescent asci have evidently evolved on several occasions within the *Lecanorales*.

Bunodophoron melanocarpum; thallus

References: **Chen, J.B.** (1996). [The genus *Sphaerophorus* from China]. *Acta Mycol. Sin.* 15: 105–108; **Döring, H.; Henssen, A.; Wedin, M.** (1999). Ascoma development in *Neophyllis melacarpa* (*Lecanorales*, *Ascomycota*), with notes on the systematic position of the genus. *Aust. J. Bot.* 47: 783–794; **Döring, H.; Wedin, M.** (2000). Homology assessment of the boundary tissue in fruiting bodies of the lichen family *Sphaerophoraceae* (*Lecanorales*, *Ascomycota*). *Pl. Biol.* 2: 361–367; **Gargas, A.; DePriest, P.T.; Grube, M.; Tehler, A.** (1995). Multiple origins of lichen symbiosis in fungi suggested by SSU rDNA phylogeny. *Science* N.Y. 268: 1492–1495; **Henssen, A.; Döring, H.; Kantvilas, G.** (1992). *Austropeltum glareosum* gen. et sp. nov., a new lichen from mountain plateaux in Tasmania and New Zealand. *Bot. Acta* 105: 457–467; **Högnabba, F.; Wedin, M.** (2003). Molecular phylogeny of the *Sphaerophorus globosus* species complex. *Cladistics* 19: 224–232; **Sarrión, F.; Aragón, G.; Burgaz, A.R.** (1999). Studies on mazaediate lichens and calicioid fungi of the Iberian Peninsula. *Mycotaxon* 71: 169–198; **Stenroos, S.K.; DePriest, P.T.** (1998). SSU rDNA phylogeny of cladoniiform lichens. *Am. J. Bot.* 85: 1548–1559; **Tibell, L.** (1997). Anamorphs in mazaediate lichenized fungi and the *Mycocaliciaceae* ('*Caliciales s. lat.*'). *In* Tibell, L. & Hedberg, I. (eds), Lichen Studies Dedicated to Rolf Santesson. *Symb. bot. upsal.* 32: 291–322; **Wedin, M.** (1995). The lichen family *Sphaerophoraceae* (*Caliciales*, *Ascomycotina*) in temperate areas of the southern hemisphere. *Symb. bot. upsal.* 31: 102 pp.; **Wedin, M.; Döring, H.** (1999). The phylogenetic relationship of the *Sphaerophoraceae*, *Austropeltum* and *Neophyllis* (lichenized *Ascomycota*) inferred by SSU rDNA sequences. *Mycol. Res.* 103: 1131–1137; **Wedin, M.; Döring, H.; Ekman, S.** (2000). Molecular phylogeny of the lichen families *Cladoniaceae*, *Sphaerophoraceae*, and *Stereocaulaceae* (*Lecanorales*, *Ascomycotina*). *Lichenologist* 32: 171–187; **Wedin, M.; Tehler, A.; Gargas, A.** (1998). Phylogenetic relationships of *Sphaerophoraceae* (*Ascomycetes*) inferred from SSU rDNA sequences. *Pl. Syst. Evol.* 209: 75–83.

Sphinctrinaceae M. Choisy 1950
Mycocaliciales: Ascomycota

Thallus at least usually absent. Ascomata apothecial, often stalked, black, convex, with a well-developed margin. Interascal tissue absent. Asci cylindrical, thin-walled, rather tardily evanescent. Ascospores 0- to 1-septate, dark brown, ellipsoidal, with an ornamentation formed early in development beneath the plasmalemma. Anamorphs unknown.

Significant Genera: *Sphinctrina*.

Distribution: Most species have north temperate distributions.

Economic Significance: None is known.

Ecology: Lichenicolous or saprobic, rarely if ever lichenized.

Notes: The *Sphinctrinaceae* appears to be closely related to the *Mycocaliciaceae*; there are morphological and chemical similarities between the two families.

Sphinctrina turbinata; ascomata

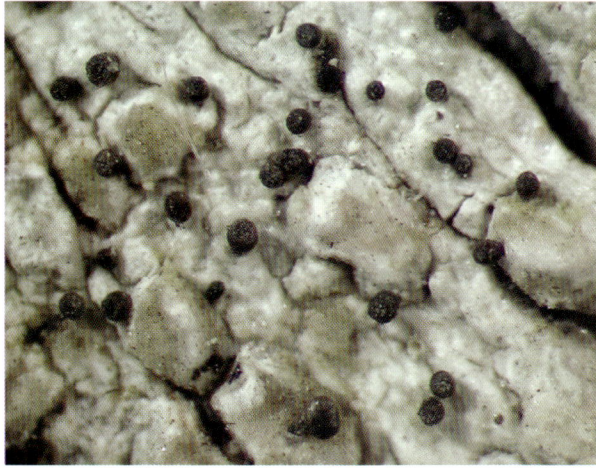

Sphinctrina tubiformis; ascomata

References: **Otto, P.; Krebs, G.** (1993). *Sphinctrina leucopoda* – ein seltener Flechtenparasit. *Boletus* 17: 97–100; **Sarrión, F.; Aragón,**

G.; Burgaz, A.R. (1999). Studies on mazaediate lichens and calicioid fungi of the Iberian Peninsula. *Mycotaxon* 71: 169–198; **Tibell, L.** (1987). Australasian *Caliciales*. *Symb. bot. upsal.* 27: 279 pp.; **Tibell, L.; Wedin, M.** (2000). *Mycocaliciales*, a new order for nonlichenized calicioid fungi. *Mycologia* 92: 577–581; **Wedin, M.; Tibell, L.** (1997). Phylogeny and evolution of *Caliciaceae, Mycocaliciaceae,* and *Sphinctrinaceae* (*Ascomycota*), with notes on the evolution of the prototunicate ascus. *Can. J. Bot.* 75: 1236–1242.

Spizellomycetaceae D.J.S. Barr 1980
Spizellomycetales: Chytridiomycota

Thallus eucarpic, monocentric; sporangium, or prosporangium, and resting spore endogenous within the zoospore cyst.

Significant Genera: *Spizellomyces*.

Distribution: Widespread although poorly recorded.

Economic Significance: None is known.

Ecology: Common in the soil and also as parasites of a range of soil organisms and plants.

Notes: Very poorly known; no recent information is available.

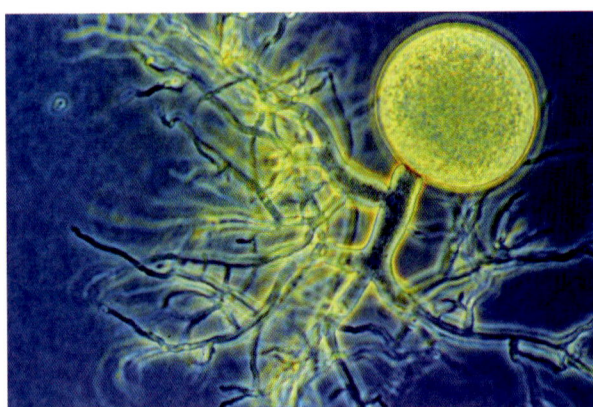

Spizellomyces sp.; sporangium

References: **Barr, D.J.S.** (2001). *Chytridiomycota. The Mycota* (Berlin) 7: 93–112; **Chen, S.F.; Hsu, M.L.; Chien, C.Y.** (2000). Some chytrids of Taiwan (III). *Bot. Bull. Acad. sin.* Taipei 41: 73–80; **Longcore, J.E.; Barr, D.J.S.; Désaulniers, N.** (1995). *Powellomyces*, a new genus in the *Spizellomycetales*. *Can. J. Bot.* 73: 1385–1390.

Sporidiobolaceae R.T. Moore 1980
Sporidiobolales: Basidiomycota

Colonies pinkish, yeast-like. Hyphae lacking haustoria, with colacosomes and septal pores. Teliospores ± globose, thin-walled or thick-walled, germinating to form elongate hypha-like transversely septate basidia with basidiospores formed at the distal end of each cell. Basidiospores small, hyaline, thin-walled, discharged passively. Anamorphs proliferating by budding and sometimes by ballistoconidia.

Significant Genera: *Rhodosporidium, Sporidiobolus*. Anamorphs: *Rhodotorula, Sporobolomyces*.

Distribution: Cosmopolitan. Some species are psychrophiles, occurring at high latitudes and altitudes.

Economic Significance: Some of the red yeasts in this family are widely encountered spoilage organisms and opportunistic human pathogens.

Ecology: Saprobic, found in an wide variety of habitats and especially in aquatic systems.

Notes: Many of the anamorphic species currently assigned to *Rhodotorula* and *Sporobolomyces* appear not to belong to this family.

References: **Bai, F.Y.; Zhao, J.H.; Takashima, M.; Jia, J.H.; Boekhout, T.; Nakase, T.** (2002). Reclassification of the *Sporobolomyces roseus* and *Sporidiobolus pararoseus* complexes, with the description of *Sporobolomyces phaffii* sp. nov. *Int. J. Syst. Evol. Microbiol.* 52: 2309–2314; **Biswas, S.K.; Yokoyama, K.; Nishimura, K.; Miyaji, M.** (2001). Molecular phylogenetics of the genus *Rhodotorula* and related basidiomycetous yeasts inferred from the mitochondrial cytochrome *b* gene. *Int. J. Syst. Evol. Microbiol.* 51: 1191–1199; **Boekhout, T.; Nakase, T.** (1998). *Sporobolomyces* Kluyver & van Niel. *The Yeasts, A Taxonomic Study* (Amsterdam): 828–843; **Fell, J.W.; Boekhout, T.; Fonseca, Á.; Scorzetti, G.; Statzell-Tallman, A.** (2000). Biodiversity and systematics of basidiomycetous yeasts as determined by large-subunit rDNA D1/D2 domain sequence analysis. *Int. J. Syst. Evol. Microbiol.* 50: 1351–1371; **Fell, J.W.; Scorzetti, G.; Statzell-Tallman, A.; Pinel, N.; Yarrow, D.** (2002). Recognition of the basidiomycetous yeast *Sporobolomyces ruberrimus* sp. nov. as a distinct species based on molecular and morphological analyses. *FEMS Yeast Res.* 1: 265–270; **Fell, J.W.; Statzell-Tallman, A.** (1998). *Rhodosporidium* Banno. *The Yeasts, A Taxonomic Study* (Amsterdam): 678–692; **Hamamoto, M.; Nakase, T.** (2000). Phylogenetic analysis of the ballistoconidium-forming yeast genus *Sporobolomyces* based on 18S rDNA sequences. *Int. J. Syst. Evol. Microbiol.* 50: 1373–1380; **Libkind, D.; Brizzio, S.; Ruffini, A.; Gadanho, M.; Broock, M. van; Sampaio, J.P.** (2003). Molecular characterization of carotenogenic yeasts from aquatic environments in Patagonia, Argentina. *Antonie van Leeuwenhoek* 84: 313–322; **Nagahama, T.; Hamamoto, M.; Nakase, T.; Takami, H.; Horikoshi, K.** (2001). Distribution and identification of red yeasts in deep-sea environments around the northwest Pacific Ocean. *Antonie van Leeuwenhoek* 80: 101–110; **Sampaio, J.P.; Gadanho, M.; Bauer, R.; Weiss, M.** (2003). Taxonomic studies in the *Microbotryomycetidae*: *Leucosporidium golubevii* sp. nov., *Leucosporidiella* gen. nov. and the new orders *Leucosporidiales* and *Sporidiobolales*. *Mycol. Prog.* 2: 53–68; **Scorzetti, G.; Fell, J.W.; Fonseca, A.; Statzell-Tallman, A.** (2002). Systematics of basidiomycetous yeasts: a comparison of large subunit D1/D2 and internal transcribed spacer rDNA regions. *FEMS Yeast Res.* 2: 495–517; **Statzell-Tallman, A.; Fell, J.W.** (1998). *Sporidiobolus* Nyland. *The Yeasts, A Taxonomic Study* (Amsterdam): 693–699; **Wang, Q.M.; Bai, F.Y.** (2004). Four new yeast species of the genus *Sporobolomyces* from plant leaves. *FEMS Yeast Res.* 4: 579–586.

Sporormiaceae Munk 1957
Pleosporales: Ascomycota

Stromata absent. Ascomata perithecial, black, thick-walled, sometimes hairy, with a well-developed lysigenous ostiole which is not periphysate. Interascal tissue copious, of cellular pseudoparaphyses. Asci cylindrical, fissitunicate, lacking complex apical structures. Ascospores usually dark brown, strongly constricted and fragmenting at the septa, usually with germ slits in each cell, occasionally ornamented, sometimes with a sheath. Anamorphs rare, coelomycetous, pycnidial, *Phoma*-like. Conidiogenous cells doliiform, proliferating percurrently. Conidia bacillar.

Significant Genera: *Preussia*, *Pycnidiophora*, *Sporormiella*.

Distribution: Cosmopolitan.

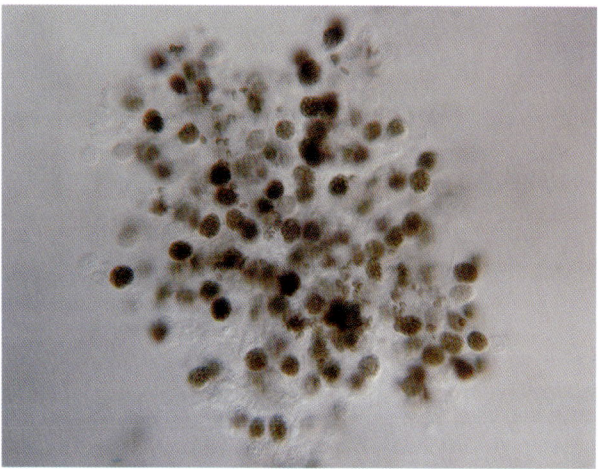

Pycnidiophora dispersa; asci and ascospores

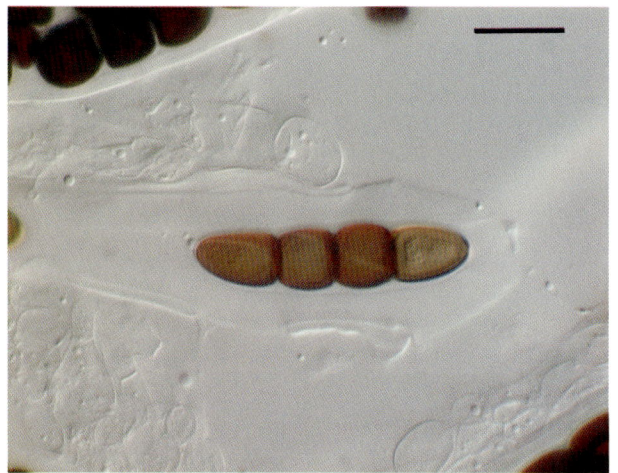

Sporormiella megalospora, Suffolk, UK; immature ascospore showing germ slits

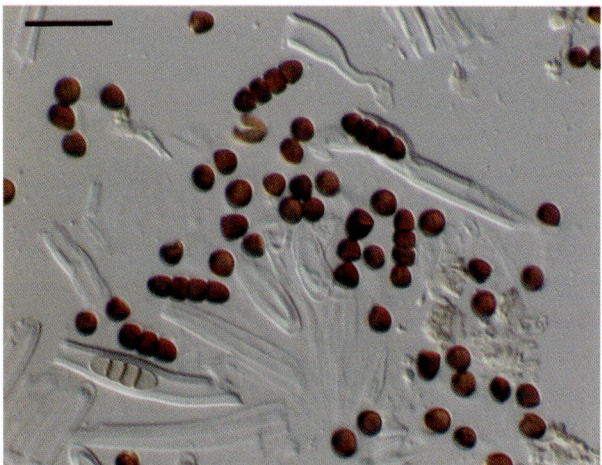

Sporormiella pulchella, Ontario, Canada; mature, mostly fragmented ascospores

Economic Significance: Little is known.

Ecology: Saprobic, associated with dung and rotting vegetation.

Notes: Species are frequently encountered as saprobes. Recent molecular data confirm this family as a constituent of the *Pleosporales* but separated from the *Delitschiaceae*, which occupies similar ecological niches and has many shared morphological characteristics.

References: **Arenal, F.; Platas, G.; Peláez, F.** (2004). Variability in spore length in some species of the genus *Preussia* (*Sporormiella*). *Mycotaxon* 89: 137–151; **Dissing, H.** (1992). Notes on the coprophilous pyrenomycete *Sporormia fimetaria*. *Persoonia* Suppl. 14: 389–394; **Guarro, J.; Abdullah, S.K.; Gené, J.; Al-Saadoon, A.H.** (1997). A new species of *Preussia* from submerged plant debris. *Mycol. Res.* 101: 305–308; **Hyde, K.D.; Steinke, T.S.** (1996). Two new species of *Delitschia* from submerged wood. *Mycoscience* 37: 99–102; **Kruys, Å.; Eriksson, O.E.; Wedin, M.** (2006). Phylogenetic relationships of coprophilous *Pleosporales* (*Dothideomycetes*, *Ascomycota*), and the classification of some bitunicate taxa of unknown position. *Mycol. Res.* 110: 527–536; **Liew, E.C.Y.; Aptroot, A.; Hyde, K.D.** (2000). Phylogenetic significance of the pseudoparaphyses in loculoascomycete taxonomy. *Mol. Phylogen. Evol.* 16: 392–402; **Silva-Hanlin, D.M.; Hanlin, R.T.** (1999). Small subunit ribosomal RNA gene phylogeny of several loculoascomycetes and its taxonomic implications. *Mycol. Res.* 103: 153–160.

Steccherinaceae Parmasto 1968
Polyporales: Basidiomycota

Basidiomata resupinate to reflexed, rarely pileate and stipitate, often effuse, usually adnate, membranous and tough-textured, the hymenial surface either with angular or labyrinthiform pores, or toothed, the teeth varied in form and development. Hyphal system dimitic, generative hyphae thin-walled, with or without clamp connections, skeletal hyphae thick-walled and sometimes encrusted. Basidia ± clavate, with 4 sterigmata and usually a basal clamp connection. Basidiospores ellipsoidal to ± cylindrical, thin-walled, smooth, not staining in iodine.

Significant Genera: *Antrodiella*, *Irpex*, *Steccherinum*.

Distribution: Widespread, especially in north temperate regions.

Economic Significance: None is known.

Ecology: Growing on dead wood causing white rot, sometimes replacing early colonizers.

Notes: A rather poorly defined family that will probably need reforming when further molecular data are available.

References: **Dai, Y.C.** (2004). Notes on the genus *Antrodiella* (*Basidiomycota, Aphyllophorales*) in China. *Mycotaxon* 89: 389–398; **Ginns, J.** (1998). Genera of the North American *Corticiaceae sensu lato*. *Mycologia* 90: 1–35; **Johannesson, H.; Renvall, P.; Stenlid, J.** (2000). Taxonomy of *Antrodiella* inferred from morphological and molecular data. *Mycol. Res.* 104: 92–99; **Kim, S.Y.; Jung, H.S.** (2002). Cladistic analysis of the *Polyporaceae* using morphological characters. *Mycotaxon* 82: 295–314; **Küffer, N.; Senn-Irlet, B.** (2005). Diversity and ecology of wood-inhabiting aphyllophoroid basidiomycetes on fallen woody debris in various forest types in Switzerland. *Mycol. Prog.* 4: 77–86; **Mossebo, D.C.; Ryvarden, L.** (2003). The genus *Mycorrhaphium* in Africa. *Mycotaxon* 88: 229–232.

Stephanospora caroticolor; basidiomata

References: **Beaton, G.; Pegler, D.N.; Young, T.W.K.** (1985). Gasteroid *Basidiomycota* of Victoria State, Australia: 5–7. *Kew Bull.* 40: 573–598; **Lebel, T.; Castellano, M.A.** (2002). Type studies of sequestrate *Russulales* II. Australian and New Zealand species related to *Russula*. *Mycologia* 94: 327–354; **Martín, M.P.; Raidl, S.; Tellería, M.T.** (2004). Molecular analyses confirm the relationship between *Stephanospora caroticolor* and *Lindtneria trachyspora*. *Mycotaxon* 90: 133–140; **Oberwinkler, F.; Horak, E.** (1979). *Stephanosporaceae* – eine neue Familie der *Basidiomycetes* mit aphyllophoralean und gastroiden Fruchtkörpern. *Pl. Syst. Evol.* 131: 157–164; **Pegler, D.N.; Young, T.W.K.** (1979). The gasteroid *Russulales*. *Trans. Br. mycol. Soc.* 72: 353–388.

Stephanosporaceae Oberw. & E. Horak 1979
Russulales: Basidiomycota

Basidiomata gasteroid or resupinate, often brightly coloured. Gasteroid forms ± globose, gleba loculate or labyrinthoid lacking a columella, with locules separated by tissue composed of inflated hyphae, clamp connections and cystidia absent, peridium glabrous or hairy, evanescent: resupinate forms poroid or with a smooth hymenium, otherwise similar anatomically. Basidia clavate, with 2 or 4 sterigmata, evanescent. Basidiospores ellipsoidal to pyriform, yellow or brown, spinose and with a prominent basal collar-like structure, not staining in iodine.

Significant Genera: *Lindtneria, Stephanospora*.

Distribution: Known from Eurasia and New Zealand.

Economic Significance: None is known.

Ecology: On the ground, associated with rotten wood or plant debris.

Notes: *Lindtneria* appears quite unlike *Stephanospora*, but molecular, chemical and anatomical data confirm their relationship. The systematic position of the *Stephanosporaceae* is uncertain; it clusters outside of the *Russulales* where it has traditionally been placed.

Stereaceae Pilát 1930
Russulales: Basidiomycota

Basidiomata appressed, effused-reflexed or discoid, rarely stalked, the outer surface often zoned, usually leathery or waxy. Hyphal system dimitic (rarely trimitic), typically differentiated into a hairy or hirsute outer skin, a closely woven cortex, and loose hyphae curving up to the hymenium, hyphae with simple septa, thick-walled cystidia common. Hymenium smooth to tuberculate or rugose. Basidia small and narrow, 4-spored, sometimes accompanied by ornamented paraphysis-like hyphae. Basidiospores ellipsoidal to cylindrical, hyaline, smooth, sometimes staining in iodine.

Stereum hirsutum, Devon, UK; basidiomata

Stereum gausapatum, Surrey, UK; basidiomata

Significant Genera: *Acanthophysium*, *Aleurodiscus*, *Stereum*.

Economic Significance: Species of *Aleurodiscus* sometimes cause economically significant disease in *Abies* and *Quercus*.

Ecology: Lignicolous or terrestrial (in leaf litter); typically saprobic.

Notes: Molecular evidence suggests that the family is generally well-circumscribed, but some remodelling will be required. Its position within the *Russulales* is confirmed.

Xylobolus frustulatus, Surrey, UK; basidioma

Distribution: Widespread.

References: **Binder, M.; Hibbett, D.S.** (2002). Higher-level phylogenetic relationships of homobasidiomycetes (mushroom-forming fungi) inferred from four rDNA regions. *Mol. Phylogen. Evol.* 22: 76–90; **Binder, M.; Hibbett, D.S.; Larsson, K.-H.; Larsson, E.; Langer, E.; Langer, G.** (2005). The phylogenetic distribution of resupinate forms across the major clades of mushroom-forming fungi (*Homobasidiomycetes*). *Syst. Biodiv.* 3: 113–157; **Ginns, J.** (1998). Genera of the North American *Corticiaceae sensu lato*. *Mycologia* 90: 1–35; **Küffer, N.; Senn-Irlet, B.** (2005). Diversity and ecology of wood-inhabiting aphyllophoroid basidiomycetes on fallen woody debris in various forest types in Switzerland. *Mycol. Prog.* 4: 77–86; **Larsson, E.; Hallenberg, N.** (2001). Species delimitation in the *Gloeocystidiellum porosum-clavuligerum* complex inferred from compatibility studies and nuclear rDNA sequence data. *Mycologia* 93: 907–914; **Larsson, E.; Larsson, K.-H.** (2003). Phylogenetic relationships of russuloid basidiomycetes with emphasis on aphyllophoralean taxa. *Mycologia* 95: 1037–1065; **Larsson, K.-H.; Larsson, E.; Koljalg, U.** (2004). High phylogenetic diversity among corticioid homobasidiomycetes. *Mycol. Res.* 108: 983–1002; **Miller, S.L.; McClean, T.M.; Walker, J.F.; Buyck, B.** (2001). A molecular phylogeny of the *Russulales* including agaricoid, gasteroid and pleurotoid taxa. *Mycologia* 93: 344–354; **Slippers, B.; Coutinho, T.A.; Wingfield, B.D.; Wingfield, M.J.** (2003). A review of the genus *Amylostereum* and its association with woodwasps. *S. Afr. J. Sci.* 99: 70–74; **Slippers, B.; Wingfield, B.D.; Coutinho, T.A.; Wingfield, M.J.** (2002). DNA sequence and RFLP data reflect geographical spread and relationships of *Amylostereum areolatum* and its insect vectors. *Mol. Ecol.* 11: 1845–1854; **Tabata, M.; Harrington, T.C.; Chen, W.; Abe, Y.** (2000). Molecular phylogeny of species in the genera *Amylostereum* and *Echinodontium*. *Mycoscience* 41: 585–593; **Wu, S.H.; Hibbett, D.S.; Binder, M.** (2001). Phylogenetic analyses of *Aleurodiscus s.l.* and allied genera. *Mycologia* 93: 720–731.

Stereocaulaceae Chevall. 1826
Lecanorales: Ascomycota

Stereocaulon aff. *tomentosum*, Bhutan; thallus and ascomata

Thallus composed of a crustose to squamulose primary thallus, often degenerating at an early stage, often with a solid erect shrubby secondary thallus. Ascomata apothecial, flat or domed, usually brown, without well-developed marginal tissues. Interascal tissue of simple, sparingly branched, paraphyses with hardly swollen apices. Asci with a strongly J+ apical cap with a central dark tubular structure, without an outer J+ gelatinized layer. Ascospores hyaline, often elongated, transversely septate, sometimes muriform. Cephalodia often present. Anamorph coelomycetous, pycnidial.

Significant Genera: *Stereocaulon*.

Distribution: Widespread, especially in temperate boreal and austral regions.

Economic Significance: None is known.

Ecology: Lichenized with green algae, the cephalodia with cyanobacteria; found on soil, rocks and wood. A number of species are tolerant of heavy metals.

Notes: *Muhria* has been shown to cluster within the genus *Stereocaulon*, while the recently described genus *Hertelidia* should be excluded. Many species of the sterile genera *Lepraria* and *Leproloma* also appear to be related to the *Stereocaulaceae*.

Stereocaulon aff. *vesuvianum*, Ecuador; vegetative thalli

References: **Ekman, S.; Tonsberg, T.** (2002). Most species of *Lepraria* and *Leproloma* form a monophyletic group closely related to *Stereocaulon*. *Mycol. Res.* 106: 1262–1276; **Fryday, A.M.; Coppins, B.J.** (1996). A new crustose *Stereocaulon* from the mountains of Scotland and Wales. *Lichenologist* 28: 513–519; **Högnabba, F.** (2006). Molecular phylogeny of the genus *Stereocaulon* (*Stereocaulaceae*, lichenized ascomycetes). *Mycol. Res.* 110: 1080–1092; **Jahns, H.M.; Kloöckner, P.a; Jørgensen, P.M.; Ott, S.** (1995). Development of thallus and ascocarps in *Stereocaulon tornensis*. *Biblthca Lichenol.* 58: 181–190; **Jørgensen, P.M.; Jahns, H.M.** (1987). *Muhria*, a remarkable new lichen genus from Scandinavia. *Notes R. bot. Gdn Edinb.* 44: 581–599; **Kivistö, L.** (1998). Taxonomy of *Stereocaulon paschale* and allied species in Finland. *Sauteria* 9: 25–36; **Printzen, C.; Kantvilas, G.** (2004). *Hertelidea*, genus novum Stereocaulacearum (*Ascomycetes* lichenisati). *Biblthca Lichenol.* 88: 539–553; **Sipman, H.J.M.** (1998). Notes on the lichen genus *Stereocaulon* in New Guinea. *Cryptog. Bryol.-Lichénol.* 19: 229–245; **Smith, R.I.L.; Øvstedal, D.O.** (1991). The lichen genus *Stereocaulon* in Antarctica and South Georgia. *Polar Biol.* 11: 91–102; **Stenroos, S.; DePriest, P.T.** (1998). Small insertions at a shared position in the SSU rDNA of *Lecanorales* (lichen-forming *Ascomycetes*). *Curr. Genet.* 33: 124–130; **Wedin, M.; Döring, H.** (1999). The phylogenetic relationship of the *Sphaerophoraceae*, *Austropeltum* and *Neophyllis* (lichenized *Ascomycota*) inferred by SSU rDNA sequences. *Mycol. Res.* 103: 1131–1137; **Wedin, M.; Döring, H.; Ekman, S.** (2000). Molecular phylogeny of the lichen families *Cladoniaceae*, *Sphaerophoraceae*, and *Stereocaulaceae* (*Lecanorales*, *Ascomycotina*). *Lichenologist* 32: 171–187.

Stictidaceae Fr. 1849
Ostropales: Ascomycota

Stroma poorly developed, restricted to intramatrical hyphae. Ascomata apothecial, often deeply immersed and almost perithecioid, usually with a well-developed margin formed largely of crystalline inclusions. Interascal tissue usually of simple paraphyses, sometimes branched and pigmented, sometimes swollen at the apices. Asci usually narrowly cylindrical with a strongly thickened apex. Ascospores often filiform, fragmenting. Anamorphs coelomycetous, pycnidial or ± acervular, opening with an irregular pore, inconspicuous. Conidiophores short. Conidiogenous cells in clusters, annellidic and proliferating percurrently. Conidia elongate (then sometimes fragmenting) or bacillar.

Significant Genera: *Schizoxylon*, *Stictis*.

Distribution: Very widespread, including boreal and austral zones.

Economic Significance: None is known.

Stictis stellata, UK; ascomata

Ecology: Saprobic on wood and stems, a few species are lichenized or lichenicolous.

Notes: The *Stictidaceae* appears to be a reasonably well-defined family and the *Ostropales* also appears to be a natural taxon, but more molecular data are needed. Preliminary results indicate that the genus *Stictis* is polyphyletic.

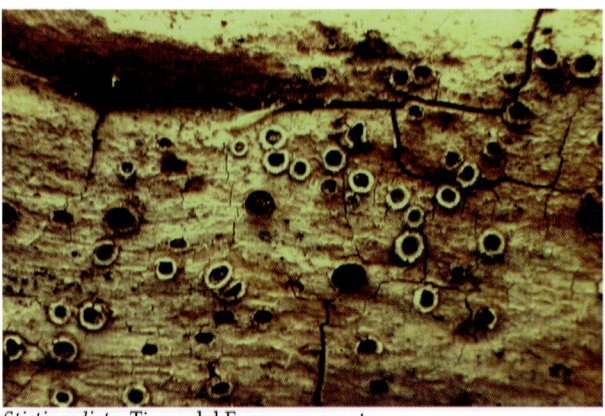
Stictis radiata, Tierra del Fuego; ascomata

References: **Johnston, P.R.** (1985). Anamorphs of the ostropalean genera *Schizoxylon* and *Acarosporina*. *Mycotaxon* 24: 349–360; **Lücking, R.; Stuart, B.L.; Lumbsch, H.T.** (2004). Phylogenetic relationships of *Gomphillaceae* and *Asterothyriaceae*: evidence from a combined Bayesian analysis of nuclear and mitochondrial sequences. *Mycologia* 96: 283–294; **Lumbsch, H.T.; Schmitt, I.; Palice, Z.; Wiklund, E.; Ekman, S.; Wedin, M.** (2004). Supraordinal phylogenetic relationships of *Lecanoromycetes* based on a Bayesian analysis of combined nuclear and mitochondrial sequences. *Mol. Phylogen. Evol.* 31: 822–832; **Wedin, M.; Döring, H.; Gilenstam, G.** (2006). *Stictis* s. lat. (*Ostropales, Ascomycota*) in northern Scandinavia, with a key and notes on morphological variation in relation to lifestyle. *Mycol. Res.* 110: 773–789.

Strigulaceae Zahlbr. 1898
Uncertain position within Chaetothyriomycetes, Ascomycota

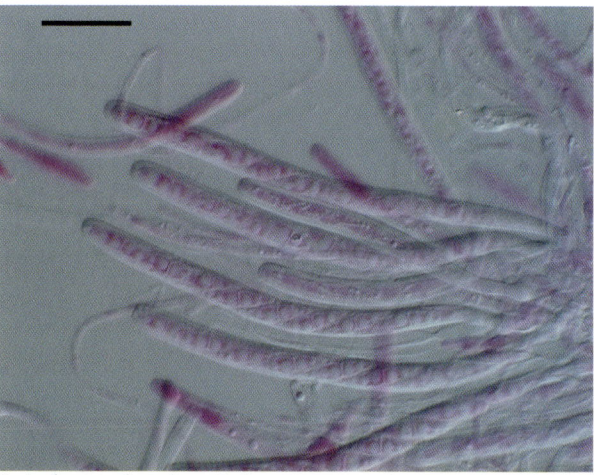

Strigula nemathora, Java, Indonesia; asci and ascospores

Thallus crustose, subcuticular, ± circular or emarginate, sometimes lobed. Ascomata perithecial, usually strongly flattened, immersed or erumpent, uniloculate, sometimes clypeate, ± hyaline to black. Interascal tissue of simple or sparingly branched paraphyses; hymenium blueing in iodine. Asci ± clavate, persistent, ± thin-walled, fissitunicate, with a distinct ocular chamber, not blueing in iodine. Ascospores hyaline, transversely septate, without a sheath. Anamorph coelomycetous, pycnidial.

Strigula macrocarpa; thallus and ascomata

Strigula nitidula, China; thallus and ascomata

Significant Genera: *Strigula*.

Distribution: Primarily tropical.

Economic Significance: None is known.

Ecology: Lichenized with green algae, especially species of *Cephaleuros*; usually on leaves.

Notes: The affinities of this family are uncertain, in common with most other groups of foliicolous lichens.

References: **Huhndorf, S.M.; Harris, R.C.** (1996). *Oletheriostrigula*, a new genus for *Massarina papulosa* (Fungi, Ascomycetes). *Brittonia* 48: 551–555; **Lücking, R.** (1991). Neue Arten foliikoler Flechten aus Costa Rica, Zentralamerika. *Nova Hedwigia* 52: 267–304; **Lücking, R.; Lücking, A.** (1995). Foliicolous lichens and bryophytes from Cocos Island, Costa Rica. A taxonomical and ecogeographical study, I. Lichens. *Herzogia* 11: 143–174; **Lücking, R.; Sérusiaux, E.; Maia, L.C.; Pereira, E.C.G.** (1998). A revision of the names of foliicolous lichenized fungi published by Batista and co-workers between 1960 and 1975. *Lichenologist* 30: 121–191; **McCarthy, P.M.; Streimann, H.; Elix, J.A.** (1996). New foliicolous species of *Strigula* from Lord Howe Island, Australia. *Lichenologist* 28: 239–244; **Roux, C.; Sérusiaux, E.; Bricaud, O.; Coppins, B.J.** (2004). Le genre *Strigula* (Lichens) en Europe et en Macaronésie. *Biblthca Lichenol.* 90: 96 pp.; **Schmitt, I.; Mueller, G.; Lumbsch, H.T.** (2005). Ascoma morphology is homoplaseous and phylogenetically misleading in some pyrenocarpous lichens. *Mycologia* 97: 362–374; **Sérusiaux, E.** (1998). Further observations on the lichen genus *Strigula* in New Zealand. *Bryologist* 101: 147–152.

Strophariaceae Singer & A.H. Sm. 1946
Agaricales: Basidiomycota

Basidiomata agaricoid or gasteroid; cap when present with an outer layer composed of closely adpressed hyphae that may be aggregated and encrusted with pigment to form scales or hairs, brightly coloured, usually yellow, brown or purplish. Hyphal system monomitic, always with clamp connections. Hymenium lamellate, the gills adnate to the stipe and sometimes sinuous. Basidia interspersed with clavate or mucronate cystidia that stain yellow in ammonia on the gill edges or over the entire lamellar surface. Basidiospores almost always smooth, yellow to brown or fuscous, thick-walled, with a broad germ pore at the hilar end.

Significant Genera: *Hypholoma*, *Pholiota*, *Psilocybe*.

Hypholoma fasciculare, Hampshire, UK; basidiomata on rotten log

Distribution: Cosmopolitan.

Economic Significance: *Pholiota nameko* is a popular edible mushroom in the Far East. *Psilocybe semilanceata* is notorious as a hallucinogenic mushroom, with substantial non-traditional industries supporting its consumption.

Pholiota squarrosa, Berkshire, UK; basidiomata on living tree

Ecology: Very variable; *Pholiota* species are lignicolous and can act as weak parasites while most other taxa are either saprobes in grassy habitats or are found in dung. Species are not ectomycorrhizal.

Notes: A polymorphic family that requires redefinition in the light of molecular evidence. *Psilocybe* has been shown to be polyphyletic and while the psilocybin-producing species may belong in this family, the non-hallucinogenic taxa occupy a separate clade.

References: **Boekhout, T.; Stalpers, J.; Verduin, S.J.W.; Rademaker, J.; Noordeloos, M.E.** (2002). Experimental taxonomic studies in *Psilocybe* sect. *Psilocybe*. *Mycol. Res.* 106: 1251–1261; **Bon, M.; Roux, P.** (2003). [Analytical key to the family Strophariaceae Singer & A.H. Smith]. *Docums Mycol.* 33: 3–54; **Bresinsky, A.; Binder, M.** (1998). [*Leratiomyces*, nom. nov. for a so far not validly described genus of the Strophariaceae (Agaricales) from New Caledonia]. *Z. Mykol.* 64: 79–82; **Clémençon, H.; Roffler, U.** (2003). The pseudosclerotia of the agaric *Stropharia luteonitens*. *Mycol. Prog.* 2: 235–238; **Guzmán, G.; Kasuya, T.** (2004). The known species of *Psilocybe* (Basidiomycotina, Agaricales, Strophariaceae) in Nepal. *Mycoscience* 45: 295–297; **Holec, J.** (1996). [A key to determination of the genus *Pholiota* and a survey of species known from the Czech Republic]. *Mykol. Listy* 57: 1–12; **Jacobsson, S.** (1991). *Pholiota* in northern Europe. *Windahlia* 19: 1–86; **Johnston, P.R.; Buchanan, P.K.** (1995). The genus *Psilocybe* (Agaricales) in New Zealand. *N.Z. Jl Bot.* 33: 379–388; **Kytövuori, I.** (1999). The *Stropharia semiglobata* group in NW Europe. *Karstenia* 39: 11–32; **Matsumoto, T.; Obatake, Y.; Fukumasa-Nakai, Y.; Nagasawa, E.** (2003). Phylogenetic position of *Pholiota nameko* in the genus *Pholiota* inferred from restriction analysis of ribosomal DNA. *Mycoscience* 44: 197–202; **Moncalvo, J.-M.; Vilgalys, R.; Redhead, S.A.; Johnson, J.E.; James, T.Y.; Aime, M.C.; Hofstetter, V.; Verduin, S.J.W.; Larsson, E.; Baroni, T.J.; Thorn, R.G.; Jacobsson, S.; Clemencon, H.; Miller, O.K.** (2002). One hundred and seventeen clades of euagarics. *Mol. Phylogen. Evol.* 23: 357–400; **Walther, G.; Garnica, S.; Weiss, M.** (2005). The systematic relevance of conidiogenesis modes in the gilled Agaricales. *Mycol. Res.* 109: 525–544.

Suillaceae Besl & Bresinsky 1997
Boletales: Basidiomycota

Basidiomata boletoid or rarely gasteroid, usually robust and fleshy; stipe when present stout, with a smooth or scabrous surface, with or without a ring; cap glabrous or squamulose, usually glutinous or viscid, the surface hyphae strongly gelatinized. Internal hyphae usually gelatinized, clamp connections almost always absent, cystidia usually present, often with resinous contents. Hymenium in boletoid forms sinuato-adnexed, always tubulate, usually yellow or olivaceous. Basidia narrowly clavate, 2- or 4-spored. Basidiospores generally small, ovoid to fusoid-cylindric or elongate, smooth, yellowish or brown.

Significant Genera: *Suillus*.

Distribution: Widespread, mainly in the north temperate area but also known from tropical Africa; perhaps introduced (with pines) in the southern hemisphere.

Economic Significance: Some species are edible and eaten by animals, including man; others are toxigenic.

Suillus variegatus, Surrey, UK; underside of basidioma showing pores

Suillus luteus, Andorra; basidiomata

Truncocolumella citrina; basidiomata

Ecology: Mycorrhizal, typically associated with conifers.

Notes: The autonomy of famillies in the suilloid clade (*Gomphidiaceae*, *Rhizopogonaceae*, *Suillaceae*) is weak and as such the three familes are maintained as distinct only for conventional reasons pending revision.

References: **Baura, G.; Szaro, T.M.; Bruns, T.D.** (1992). *Gastrosuillus laricinus* is a recent derivative of *Suillus grevillei*: molecular evidence. *Mycologia* 84: 592–597; **Besl, H.; Bresinsky, A.** (1997). Chemosystematics of *Suillaceae* and *Gomphidiaceae* (suborder *Suillineae*). *Pl. Syst. Evol.* 206: 223–242; **Binder, M.; Hibbett, D.S.** (2002). Higher-level phylogenetic relationships of homobasidiomycetes (mushroom-forming fungi) inferred from four rDNA regions. *Mol. Phylogen. Evol.* 22: 76–90; **Bruns, T.D.; Szaro, T.M.; Gardes, M.; Cullings, K.W.; Pan, J.J.; Taylor, D.L.; Horton, T.R.; Kretzer, A.; Garbelotto, M.; Li, Y.** (1998). A sequence database for the identification of ectomycorrhizal basidiomycetes by phylogenetic analysis. *Mol. Ecol.* 7: 257–272; **Dahlberg, A.; Finlay, R.D.** (1999). *Suillus. Ectomycorrhizal Fungi* (Berlin): 33–64; **Ding, M.R.; Wen, H.A.** (2003). Studies on *Suillus* from southwest China. *Nova Hedwigia* 76: 459–464; **Manian, S.; Sreenivasaprasad, S.; Bending, G.D.; Mills, P.R.** (2001). Genetic diversity and interrelationships among common European *Suillus* based on ribosomal DNA sequences. *FEMS Microbiol. Lett.* 204: 117–121; **Moncalvo, J.-M.; Vilgalys, R.; Redhead, S.A.; Johnson, J.E.; James, T.Y.; Aime, M.C.; Hofstetter, V.; Verduin, S.J.W.; Larsson, E.; Baroni, T.J.; Thorn, R.G.; Jacobsson, S.; Clemencon, H.; Miller, O.K.** (2002). One hundred and seventeen clades of euagarics. *Mol. Phylogen. Evol.* 23: 357–400; **Muller, L.A.H.; Lambaerts, M.; Vangronsveld, J.; Colpaert, J.V.** (2006). Isolation and characterization of microsatellite loci from the ectomycorrhizal basidiomycete *Suillus luteus*. *Mol. Ecol.* 6: 165–166; **Noordeloos, M.E.** (2000). [An introduction to the boletes – 1: characteristics and classification]. *Coolia* 43: 1–10; **Noordeloos, M.E.** (2000). [An introduction to the boletes – 2. From *Suillus* to *Paxillus*]. *Coolia* 43: 75–98; **Samson, J.; Fortin, J.A.** (1988). Structural characterization of *Fuscoboletinus* and *Suillus* ectomycorrhizae synthesized on *Larix laricina*. *Mycologia* 80: 382–392.

Syncephalastraceae Naumov ex R.K. Benj. 1959
Mucorales: Zygomycota

Vegetative thallus a mycelium, complex fruit bodies absent. Sporophores (sporangiophores) arising from aerial mycelium, simple or more commonly sympodially or racemosely branched, stolons and rhizoids present; terminating in spherical fertile vesicles bearing, over their entire surface, cylindrical sporangia (merosporangia). Sporangia few-spored, cylindrical, with a smooth and thin wall. Sporangiospores globose to subglobose, smooth, more or less hyaline. Chlamydospores often produced in vegetative hyphae. Zygospores globose to subglobose; zygosporangial wall warty; suspensors opposed, without appendages; heterothallic.

Significant Genera: *Syncephalastrum*.

Syncephalastrum racemosum; terminal vesicle, merosporangia

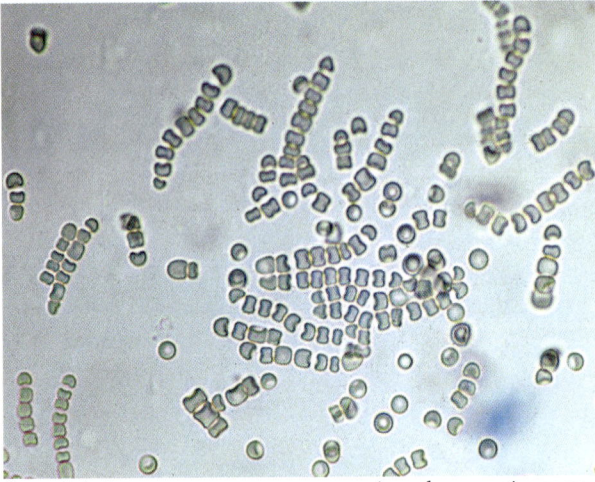

Syncephalastrum racemosum; merosporangia and sporangiospores

Distribution: Widespread, but more common in tropical and subtropical areas.

Economic Significance: None is known.

Ecology: Essentially soil fungi but most readily isolated as a component of the early coprophilous succession; rather common.

Notes: Considered a link between *Mucoraceae* and *Piptocephalidaceae* but analysis of molecular data does not support this view.

References: **Benjamin, R.K.** (1959). The Merosporangiferous Mucorales. *Aliso* 4: 321–433; **Zheng, R.Y.; Chen, G.Q.; Hu, F.M.** (1988). Monosporous varieties of *Syncephalastrum. Mycosystema* 1: 35–52.

Synchytriaceae J. Schröt. 1892
Chytridiales: Chytridiomycota

Thallus endobiotic, holocarpic, exogenous to the zoospore cyst, at maturity the whole developing into a sorus, prosorus or resting spore; resting spores formed by encystment of original thallus or following fusion of gametes from the resulting zygote, exospore thick-walled and either smooth, roughened or ridged, hyaline to reddish-brown or light to dark brown, functioning as a sporangium or prosorus on germination; zoospores and gametes ellipsoid to ovoid, usually with 1(–2) refringent globules, with a single posterior flagellum; inducing the development of galls in stems and roots (tubers) of susceptible hosts.

Significant Genera: *Synchytrium*.

Distribution: Widespread.

Economic Significance: The plant parasitic species of *Synchytrium* are important pathogens with *S. endobioticum*, the causal agent of potato wart disease, the most significant.

Ecology: Plant parasites or parasites of algae.

Notes: Over 100 species of *Synchytrium* have been described. The potato wart disease affects the tuber initials and tubers of potato (*Solanum tuberosum*), the only cultivated host, wild relatives in Mexico and, by artificial inoculation, tomato (*Lycopersicon esculentum*). The tuber intials become distorted and spongy but only the 'eyes' of mature tubers are infected and these develop the characteristic cauliflower-like outgrowths, initially white, turning green when exposed to light and eventually brown. The winter sporangia are mostly spherical and thick-walled.

References: **Barr, D.J.S.** (2001). Chytridiomycota. *The Mycota* (Berlin) 7: 93–112; **Hampson, M.C.; Yang, A.F.; Bal, A.K.** (1995). Ultrastructure of *Synchytrium endobioticum* resting spores and enhancement of germination using snails. *Mycologia* 86: 733–740; **Karling, J.S.** (1964). *Synchytrium* (New York): 470 pp.; **Percival, J.** (1910). Potato wart disease: the life history and cytology of *Synchytrium endobioticum. Zentbl. Bakt. ParasitKde* Abt. II 25: 440–447; **Raghavendra Rao, N.N.; Pavgi, M.S.** (1993). Life history of *Synchytrium* species parasitic on Cucurbitaceae. *Indian Phytopath.* 46: 36–43.

Taphrinaceae Gäum. 1928
Taphrinales: Ascomycota

Mycelium subcuticular or subepidermal, composed of dikaryotic ascogenous cells, vegetative tissue ± lacking. Ascomata absent. Interascal tissue absent. Asci formed either directly from ascogenous cells or with a separating stalk cell, forming a palisade on the surface of the host tissue, ± cylindrical, the end often truncate, ± persistent, usually 8-spored, the ascospores discharged simultaneously. Ascospores hyaline, aseptate, globose or ellipsoidal. Anamorph yeast-like, monokaryotic, formed from budding ascospores (including whilst within the ascus).

Significant Genera: *Taphrina*. Anamorph: *Lalaria*.

Distribution: Widespread, primarily in temperate regions.

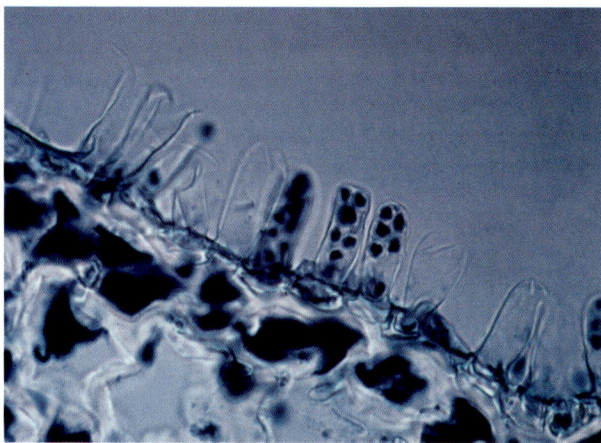
Taphrina deformans; asci and ascospores

Taphrina johansonii, Perthshire, Scotland; galls on *Populus tremula*

Economic Significance: A few species are significant plant pathogens, including the well-known peach leaf curl fungus *Taphrina deformans*.

Ecology: Biotrophic on plants, usually causing hyperplasia (galls, witches' brooms) or lesions.

Notes: There are strong indications of coevolution between *Taphrina* species and their host plants, with a number of evolutionary radiations strongly correlated with host genera. The *Protomycetaceae* is confirmed as a related family and both appear to be monophyletic, with the anamorphic genus *Saitoella* as a possible further representative of the *Taphrinales*.

References: **Bacigálová, K.** (1997). Species of *Taphrina* on *Betula* in Slovakia. *Czech Mycol.* 50: 107–118; **Inácio, J.; Rodrigues, M.G.; Sobral, P.; Fonseca, Á.** (2004). Characterisation and classification of phylloplane yeasts from Portugal related to the genus *Taphrina* and description of five novel *Lalaria* species. *FEMS Yeast Res.* 4: 541–555; **Kramer, C.L.** (1987). The *Taphrinales*. *Stud. Mycol.* 30: 151–166; **Moore, R.T.** (1998). *Lalaria* R.T. Moore. *The Yeasts, A Taxonomic Study* (Amsterdam): 582–591; **Nishida, H.; Sugiyama, J.** (1993). Phylogenetic relationships among *Taphrina*, *Saitoella*, and other higher fungi. *Mol. Biol. Evol.* 10: 431–436; **Rodrigues, M.G.; Fonseca, Á.** (2003). Molecular systematics of the dimorphic ascomycete genus *Taphrina*. *Int. J. Syst. Evol. Microbiol.* 53: 607–616; **Sjamsuridzal, W.; Tajiri, Y.; Nishida, H.; Thuan, T.B.; Kawasaki, A.; Yokota, A.; Sugiyama, J.** (1997). Evolutionary relationships of members of the genera *Taphrina*, *Protomyces*, *Schizosaccharomyces*, and related taxa within the archiascomycetes: integrated analysis of genotypic and phenotypic characters. *Mycoscience* 38: 267–280; **Sugiyama, J.** (1998). Relatedness, phylogeny, and evolution of the fungi. *Mycoscience* 39: 487–511.

Teloschistaceae Zahlbr. 1898
Teloschistales: Ascomycota

Thallus varied, ranging from crustose to ± foliose or fruticose, rhizines poorly developed or absent, mostly bright yellow, orange or red due to presence of anthraquinone pigments, other pigments rarely present; isidia and soredia sometimes present. Ascomata apothecial, disc-shaped or cup-shaped, rarely ciliate, also brightly coloured and usually with a well-developed thalline margin. Interascal tissue of simple paraphyses, often gelatinized and with swollen apices and a well-developed epithecial layer. Asci ± cylindrical, with a well-developed outer J+ layer, internal structures rudimentary or absent. Ascospores 0- to 3-septate, hyaline, the septa strongly thickened, polarilocular. Anamorph coelomycetous, pycnidial, with inconspicuous, sometimes multilocular conidiomata forming small hyaline ± bacillar conidia.

Significant Genera: *Caloplaca*, *Teloschistes*, *Xanthoria*.

Distribution: Cosmopolitan, particularly prevalent in temperate regions.

Economic Significance: None is known.

Ecology: Lichenized with green algae on well-lit rocks and bark, present particularly in nutrient-enriched habitats such as those affected by bird droppings.

Caloplaca aurantia, Aude, France; thallus and ascomata

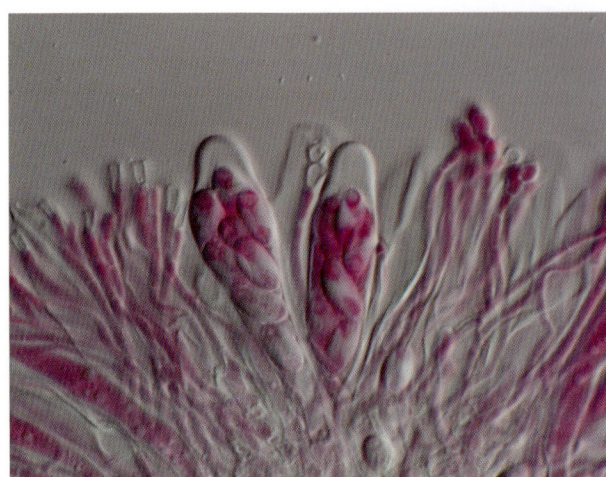

Xanthoria parietina, Barra, Scotland; asci, ascospores and paraphyses

Teloschistes chrysophthalmus, South Africa; thallus and ascomata

Notes: Species of the *Teloschistaceae* are some of the most prominent and well-known lichens, with their brightly coloured thalli; they are familiar components especially of maritime habitats. Surprisingly few data are available on the molecular phylogeny of the family, but its position within the wider *Lecanorales* clade is confirmed although evidence has been presented that the major genera, based primarily on thallus morphology, are polyphyletic.

Xanthoria elegans, Alberta, Canada; thalli and ascomata

References: **Almborn, O.** (1989). Revision of the lichen genus *Teloschistes* in central and southern Africa. *Nordic Jl Bot.* 8: 521–537; **Arup, U.** (1995). Littoral species of *Caloplaca* in North America: a summary and a key. *Bryologist* 98: 129–140; **Awasthi, D.D.** (1986). Macrolichen taxa of *Teloschistaceae* from India. *Proc. Indian Acad. Sci.* Pl. Sci. 96: 227–231; **Bellemère, A.; Hafellner, J.; Letrouit-Galinou, M.-A.** (1986). Ultrastructure et mode de déhiscence des asques chez les lichens des genres *Teloschistes* et *Apatoplaca* (*Teloschistaceae*). *Cryptog. Bryol.-Lichénol.* 7: 189–211; **Castello, M.** (1995). The lichen genus *Xanthoria* in Antarctica. *Cryptog. Bryol.-Lichénol.* 16: 79–87; **Franc, N.; Kärnefelt, E.I.** (1998). Phylogeny of *Xanthoria calcicola* and *X. parietina*, based on rDNA ITS sequences. *Graphis Scripta* 9: 49–54; **Gaya, E.; Lutzoni, F.; Zoller, S.; Navarro-Rosinés, P.** (2003). Phylogenetic study of *Fulgensia* and allied *Caloplaca* and *Xanthoria* species (*Teloschistaceae*, lichen-forming Ascomycota). *Am. J. Bot.* 90: 1095–1103; **Honegger, R.; Zippler, U.; Gansner, H.; Scherrer, S.** (2004). Mating systems in the genus *Xanthoria* (lichen-forming ascomycetes). *Mycol. Res.* 108: 480–488; **Kärnefelt, I.** (1988). Morphology and biogeography of saxicolous *Caloplaca* in southern Africa. *Monogr. Syst. Bot., Miss. Bot. Gdn* 25: 439–452; **Kärnefelt, I.** (1989). Morphology and phylogeny in the *Teloschistales*. *Cryptog. bot.* 1: 147–203; **Kärnefelt, I.** (1990). Isidiate taxa in the *Teloschistales* and their ecological and evolutionary significance. *Lichenologist* 22: 307–320; **Kondratyuk, S.; Kärnefelt, I.** (1997). Notes on *Xanthoria* Th. Fr. II. *Xanthoria poeltii*, a new lichen species from Europe. *Lichenologist* 29: 425–430; **Lindblom, L.** (1997). The genus *Xanthoria* (Fr.) Th. Fr. in North America. *J. Hattori bot. Lab.* 83: 75–172; **Lumbsch, H.T.** (1998). Taxonomic use of metabolic data in lichen-forming fungi. *Chemical Fungal Taxonomy* (New York): 345–387; **Lumbsch, H.T.; Schmitt, I.; Palice, Z.; Wiklund, E.; Ekman, S.; Wedin, M.** (2004). Supraordinal phylogenetic relationships of *Lecanoromycetes* based on a Bayesian analysis of combined nuclear and mitochondrial sequences. *Mol. Phylogen. Evol.* 31: 822–832; **Poelt, J.; Hinteregger, E.** (1993). Beiträge zur kenntnis der flechtenflora des Himalaya. VII die gattungen *Caloplaca, Fulgensia* und *Ioplaca* (mit englischem bestimmungsschlüssel). *Biblthca Lichenol.* 50: 247 pp.; **Søchting, U.; Lutzoni, F.** (2003). Molecular phylogenetic study at the generic boundary between the lichen-forming fungi *Caloplaca* and *Xanthoria* (Ascomycota, *Teloschistaceae*). *Mycol. Res.* 107: 1266–1276; **Westberg, M.; Kärnefelt, I.** (1998). The genus *Fulgensia* A. Massal. & De Not., a diverse group in the *Teloschistaceae*. *Lichenologist* 30: 515–532; **Wetmore, C.M.; Kärnefelt, E.I.** (1998). The lobate and subfruticose species of *Caloplaca* in North and central America. *Bryologist* 10: 230–255.

Terfeziaceae E. Fisch. 1897
Pezizales: Ascomycota

Ascomata large, cleistothecial, ± globose to turbinate, thick-walled, solid, the asci formed in marbled veins interspersed with sterile issue. Interascal tissue absent. Asci cylindrical to globose, indehiscent, sometimes blueing in iodine. Ascospores hyaline to pale brown, globose, sometimes ornamented, uninucleate. Anamorphs unknown.

Significant Genera: *Terfezia*.

Distribution: Widespread, especially in warm temperate regions.

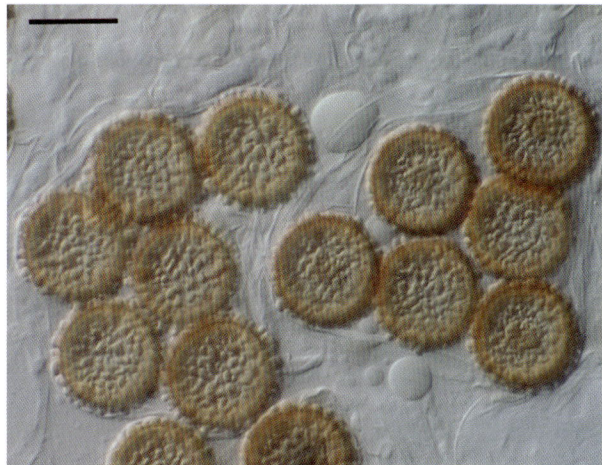

Terfezia boudieri, Iraq; asci and ascospores

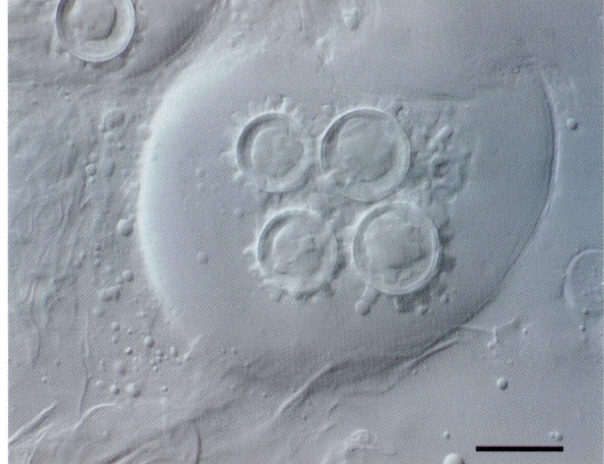

Terfezia arenaria, Caceres, Spain; immature ascus and ascospores

Economic Significance: Some species are used as food in arid regions of Africa and the Near East, though they are not prized highly in commercial terms.

Ecology: Hypogeous, sometimes emergent, mycorrhizal especially with woody plants of the *Cistaceae*. Most species occur in dry habitats.

Notes: One of several hypogeous groups within the *Pezizales*. There is some evidence that the *Pezizaceae* and *Terfeziaceae* are sister groups, but more data are needed.

References: **Abdullah, S.K.; Al-Issa, A.H.; Ewaz, J.O.; Al-Bader, S.M.** (1989). Taxonomy of edible hypogeous *Ascomycotina* of Iraq. *Int. J. Mycol. Lichenol.* 4: 9–21; **Alvarez, I.F.; Parladé, J.; Trappe, J.M.** (1993). *Loculotuber gennadii* gen. et comb. nov. and *Tuber multimaculatum* sp. nov. *Mycologia* 84: 926–929; **Diez, J.; Manjon, J.L.; Martin, F.** (2002). Molecular phylogeny of the mycorrhizal desert truffles (*Terfezia* and *Tirmania*), host specificity and edaphic tolerances. *Mycologia* 94: 247–259; **Ferdman, Y.; Aviram, S.; Roth-Bejerano, N.; Trappe, J.M.; Kagan-Zur, V.** (2005). Phylogenetic studies of *Terfezia pfeilii* and *Choiromyces echinulatus* (*Pezizales*) support new genera for southern African truffles: *Kalaharituber* and *Eremiomyces*. *Mycol. Res.* 109: 237–245; **Gutierrez, A.; Honrubia, M.; Morte, A.; Diaz, G.** (1996). Edible fungi adapted to arid and semi-arid areas. Molecular characterization and *in vitro* mycorrhization of micropropagated plantlets. *Cah. Opt. Méditerr.* 20: 139–144; **Khabar, L.; Najim, L.; Janex-Favre, M.C.; Parguey-Leduc, A.** (1994). L'ascocarpe de *Terfezia leonis* Tul. (Discomycètes, Tubérales). *Cryptog. Mycol.* 15: 187–206; **Kovacs, G.M.; Rudnoy, S.; Vagvolgyi, C.; Lasztity, D.; Racz, I.; Bratek, Z.** (2001). Intraspecific invariability of the internal transcribed spacer region of rDNA of the truffle *Terfezia terfezioides* in Europe. *Folia Microbiol.* Praha 46: 423–426; **Martin, F.; Costa, G.; Delaruelle, C.; Diez, J.** (1998). Genomic fingerprinting of ectomycorrhizal fungi by microsatellite-primed PCR. *Mycorrhiza Manual* Springer Lab Manual (Berlin): 463–474; **O'Donnell, K.; Cigelnik, E.; Weber, N.S.; Trappe, J.M.** (1997). Phylogenetic relationships among ascomycetous truffles and the true and false morels inferred from 18S and 28S ribosomal DNA sequence analysis. *Mycologia* 89: 48–65; **Percudani, R.; Trevisi, A.; Zambonelli, A.; Ottonello, S.** (1999). Molecular phylogeny of truffles (*Pezizales*: *Terfeziaceae*, *Tuberaceae*) derived from nuclear rDNA sequence analysis. *Mol. Phylogen. Evol.* 13: 169–180; **Taylor, F.W.; Thamage, D.M.; Baker, N.; Roth-Bejerano, N.; Kagan-Zur, V,** (1995). Notes on the Kalahari desert truffle, *Terfezia pfeilii*. *Mycol. Res.* 99: 874–878; **Trappe, J.M.** (1989). *Cazia flexiascus* gen. et sp. nov., a hypogeous fungus in the *Helvellaceae*. *Mem. N. Y. bot. Gdn* 49: 336–338.

Testudinaceae Arx 1971
Uncertain position within Dothideomycetes, Ascomycota

Stromata absent. Ascomata cleistothecial, black, thick-walled, fragmenting into predefined polygonal plates. Interascal tissue of branched evanescent pseudoparaphyses. Asci ± clavate, thick-walled when young, without apical structures, evanescent. Ascospores relatively small, brown, usually septate, mostly ornamented. Anamorphs unknown.

Significant Genera: *Testudina*, *Neotestudina*.

Distribution: Widespread, especially in xeric habitats.

Economic Significance: One species, *Neotestudina rosatii*, occurs as an opportunistic invader in medical environments.

Ecology: Mostly saprobes, isolated from soil, decaying plant tissues and similar substrata.

Notes: Limited molecular data are available, but it appears that the separation of this family from the *Zopfiaceae*, another family with cephalothecoid ascomata, is justified. It is probably polyphyletic.

Lepidosphaeria nicotiae, Western Sahara; ascospores

References: **Kruys, Å.; Eriksson, O.E.; Wedin, M.** (2006). Phylogenetic relationships of coprophilous *Pleosporales* (*Dothideomycetes*, *Ascomycota*), and the classification of some bitunicate taxa of unknown position. *Mycol. Res.* 110: 527–536; **LoBuglio, K.F.; Berbee, M.L.; Taylor, J.W.** (1996). Phylogenetic origins of the asexual mycorrhizal symbiont *Cenococcum geophilum* Fr. and other mycorrhizal fungi among the *Ascomycetes*. *Mol. Phylogen. Evol.* 6: 287–294; **Padhye, A.A.; McGinnis, M.R.** (1999). Fungi causing eumycotic mycetoma. *Manual of Clinical Microbiology* (Washington): 1318–1326; **Sivanesan, A.** (1991). IMI Descriptions of Fungi and Bacteria no. 1038. *Neotestudina rosatii*. *Mycopathologia* 114: 59–60.

Tetragoniomycetaceae Oberw. & Bandoni 1981
Tremellales: Basidiomycota

Basidiomata not developed. Hyphae with clamp connections, attached to host cells with haustoria. Fertile hyphae producing basidia at the apices of branches, with proliferation and slight elongation to produce new basidia. Basidia functioning as a dispersal/survival propagule, initially ellipsoidal to globose, becoming rhomboid and 4-celled, the cells arranged in a tetrad, with a thick ornamented outer wall, germinating directly to form dikaryotic hyphae. Basidiospores not produced.

Significant Genera: *Tetragoniomyces*.

Distribution: Known from Canada and northern Europe.

Economic Significance: None is known.

Ecology: Parasitic on other fungi, belonging to the *Rhizoctonia* complex.

Notes: The only species, *Tetragoniomyces uliginosa*, is a strange fungus that does not produce basidiospores. Its affinities are not well understood and placement in the *Tremellales* has been justified by septal ultrastructure and haustorial characters.

References: **Bandoni, R.J.** (1987). Taxonomic overview of the *Tremellales*. *Stud. Mycol.* 30: 87–110; **Clémençon, H.** (1990). *Tetragoniomyces uliginosus* (*Tremellales*) auf *Waitea nuda* (*Tulasnellales*). *Mycol. helv.* 4: 53–73; **Oberwinkler, F.** (1987). Heterobasidiomycetes with ontogenetic yeast-stages – systematic and phylogenetic aspects. *Stud. Mycol.* 30: 61–74; **Oberwinkler, F.; Bandoni, R.J.** (1981). *Tetragoniomyces* gen. nov. and *Tetragoniomycetaceae* fam. nov. (*Tremellales*). *Can. J. Bot.* 59: 1034–1040; **Put, K. van de; Van Ryckegem, G.; Antonissen, I.** (2000). Eerste vondsten van *Tetragoniomyces uliginosus* (P. Karst.) Oberw. & Bandoni in Vlaanderen. *Sterbeeckia* 19: 23–26.

Thelebolaceae Eckblad 1968
Thelebolales: Ascomycota

Stromata absent. Ascomata minute, ± globose or pulvinate, at least initially cleistothecial, the excipulum hyaline, poorly developed, ± glabrous. Interascal tissue poorly developed, composed of simple paraphyses. Asci ± ellipsoidal, often multispored, ± persistent, opening with a rather irregular vertical split. Ascospores usually small, hyaline, smooth or with ornamentation formed as an elaboration of an initially homogenous secondary wall layer. Anamorphs rarely encountered, where present hyphomycetous, *Sporothrix*-like.

Significant Genera: *Ascozonus*, *Thelebolus*.

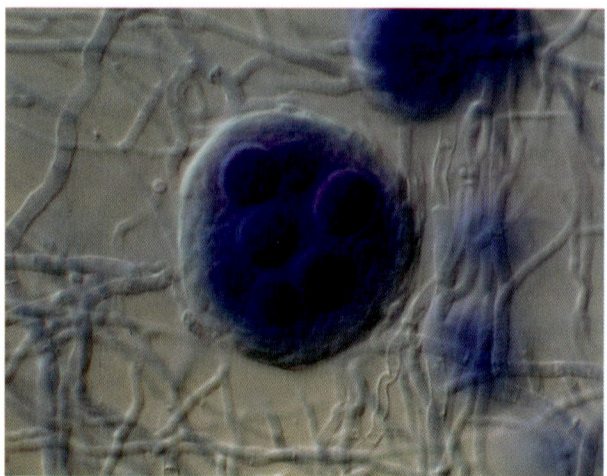

Thelebolus sp.; ascoma and asci

Distribution: Widespread, including boreal and austral regions.

Economic Significance: None is known.

Ecology: Saprobic, usually coprophilous.

Notes: At one time assumed to be highly reduced members of the *Pezizales*, molecular evidence suggests a closer link with the inoperculate discomycetes.

References: **Brummelin, J. van** (1998). Reconsideration of relationships within the *Thelebolaceae* based on ascus ultrastructure. *Persoonia* 16: 425–469; **Brummelin, J. van; Kristiansen, R.** (1998). Two rare coprophilous ascomycetes from Norway. *Persoonia* 17: 119–125; **Czymmek, K.J.; Klomparens, K.L.** (1992). The ultrastructure of ascosporogenesis in freeze-substituted *Thelebolus crustaceus*: enveloping membrane system and ascospore initial development. *Can. J. Bot.* 70: 1669–1683; **Landvik, S.; Kristiansen, R.; Schumacher, T.** (1998). Phylogenetic and structural studies in the *Thelebolaceae* (*Ascomycota*). *Mycoscience* 39: 49–56; **Momol, E.A.; Kimbrough, J.W.; Eriksson, O.E.** (1996). Phylogenetic relationships of *Thelebolus* indicated by 18S rDNA sequence analyses. *Syst. Ascom.* 14: 91–100; **Prokhorov, V.P.** (1998). The genera *Coprotus*, *Ascozonus*, *Thelebolus* and *Trichobolus* (*Pezizales*): the keys to identification of species. *Mikol. Fitopatol.* 32: 40–43; **Stchigel, A.M.; Cano, J.; MacCormack, W.; Guarro, J.** (2001). *Antarctomyces psychrophilus* gen. et sp. nov., a new ascomycete from Antarctica. *Mycol. Res.* 105: 377–382; **Wang, Y.Z.** (1999). The coprophilous discomycetes of Taiwan. *Bull. natn. Mus. Nat. Sci.* Taiwan 12: 49–74.

Thelenellaceae O.E. Erikss. ex H. Mayrhofer 1987
Uncertain position within Lecanoromycetes, Ascomycota

Thallus crustose, sometimes areolate. Ascomata perithecial, ± immersed, thick-walled, pale to dark brown, the ostiole sometimes periphysate. Interascal tissue of narrow, branched and anastomosed, pseudoparaphyses. Asci thick-walled, with separable wall layers, the apex sometimes thickened and with a small ocular chamber, not blueing in iodine. Ascospores hyaline to brown, thin-walled, muriform, the septa forming in a median position in each cell of the developing spore, without a sheath.

Significant Genera: *Julella*, *Thelenella*.

Distribution: Very widespread.

Economic Significance: None is known.

Ecology: Saprobic or lichenized with green algae, on rock and bark.

Notes: More molecular data are needed, but the family appears to be a sister group of the *Ostropales* and separation of the *Protothelenellaceae* is justified. The family may well be polyphyletic.

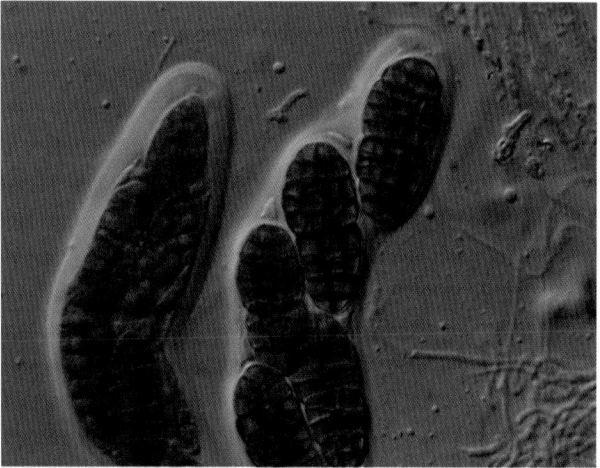

Julella avicenniae, Galápagos Islands; asci and ascospores

References: **Aptroot, A.** (1991). Tropical pyrenocarpous lichens. A phylogenetic approach. *Tropical Lichens: Their Systematics, Conservation, and Ecology* (Oxford) 43: 253–273; **Barr, M.E.** (1986). On *Julella*, *Delacourea*, and *Decaisnella*, three dictyosporous genera described by J.H.Fabre. *Sydowia* 38: 11–19; **Harada, H.** (1999). *Thelenella luridella* (lichenized *Ascomycota*, *Thelenellaceae*), newly found in Japan. *J. Nat. Hist. Mus. Inst.* Chiba 5: 91–95; **Kalb, K.** (1995). *Thelenella follmannii* sp. nov. (*Thelenellaceae*), eine neue corticole Flechtenart aus Jamaica (Westindien). *Flechten Follmann, Contributions to Lichenology in Honour of Gerhard Follmann* (Cologne): 249–253; **Mayrhofer, H.** (1987). Monographie der Flechtengattung *Thelenella*. *Biblthca Lichenol.* 26: 106 pp.; **Mayrhofer, H.; Poelt, J.** (1985). Die Flechtengattung *Microglaena* sensu Zahlbruckner in Europa. *Herzogia* 7: 13–79; **Schmitt, I.; Mueller, G.; Lumbsch, H.T.** (2005). Ascoma morphology is homoplaseous and phylogenetically misleading in some pyrenocarpous lichens. *Mycologia* 97: 362–374.

Thelephoraceae Chevall. 1826
Thelephorales: Basidiomycota

Basidiomata resupinate to flabelliform, typically thin-fleshed with a smooth or papillate hymenium, or not strictly resupinate with a smooth hymenium, rarely stalked with a toothed hymenium. Hyphal system monomitic or dimitic, clamp connections present or absent. Flesh where well-developed fibrous or leathery, usually dark, becoming greenish on treatment with KOH due to presence of thelephoric acid. Basidiospores globose to ellipsoidal or angular, often uneven in outline, usually ornamented, brownish to colourless (blue in one species), not staining in iodine, sometimes staining violet to dark blue in KOH.

Significant Genera: *Thelephora*, *Tomentella*.

Distribution: Widespread.

Economic Significance: None is known.

Ecology: On the ground, on soil, a few on wood. At least some species are ectomycorrhizal.

Notes: A fairly well-defined family, apparently paraphyletic as currently circumscribed with the *Bankeraceae* nesting within. That family differs primarily in morphological terms by its stalked basidiomata.

Thelephora anthocephala; basidiomata

Thelephora palmata; basidioma

References: **Ginns, J.** (1998). Genera of the North American *Corticiaceae sensu lato*. *Mycologia* 90: 1–35; **Hibbett, D.S.; Donoghue, M.J.** (2001). Analysis of character correlations among wood decay mechanisms, mating systems and substrate ranges in Homobasidiomycetes. *Syst. Biol.* 50: 215–242; **Jakucs, E.; Kovács, G.M.; Agerer, R.; Romsics, C.; Erös-Honti, Z.** (2005). Morphological-anatomical characterization and molecular identification of *Tomentella stuposa* ectomycorrhizae and related anatomotypes. *Mycorrhiza* 15: 247–258; **Kõljalg, U.** (1996). *Tomentella* (*Basidiomycota*) and related genera in temperate Eurasia. *Syn. Fung.* (Oslo) 9: 213 pp.; **Küffer, N.; Senn-Irlet, B.** (2005). Diversity and ecology of wood-inhabiting aphyllophoroid basidiomycetes on fallen woody debris in various forest types in Switzerland. *Mycol. Prog.* 4: 77–86; **Larsson, K.-H.; Larsson, E.; Koljalg, U.** (2004). High phylogenetic diversity among corticioid homobasidiomycetes. *Mycol. Res.* 108: 983–1002; **Ryvarden, L.** (2002). A note on the genus *Hydnodon* Banker. *Syn. Fung.* (Oslo) 15: 31–33; **Stalpers, J.A.** (1993). The aphyllophoraceous fungi I. Keys to the species of the *Thelephorales*. *Stud. Mycol.* 35: 168 pp.

Thelotremataceae Stizenb. 1862
Ostropales: Ascomycota

Thallus ± crustose. Ascomata apothecial, often deeply immersed and appearing urceolate or perithecial, usually with a well-developed thalline margin, carbonized or not. Interascal tissue of simple paraphyses. Asci usually with a well-developed apical cap and a distinct pore, rarely thin-walled and evanescent. Ascospores hyaline or brown, transversely septate, sometimes muriform, the septa often strongly thickened, rarely extruded in a mazaedial mass. Anamorph coelomycetous, pycnidial.

Significant Genera: *Diploschistes*, *Ocellularia*, *Thelotrema*.

Diploschistes muscorum, Norway; thallus and ascomata

Thelotrema lepadinum, Hampshire, UK; thallus and ascomata

Distribution: Primarily tropical, some species typical of arid regions.

Economic Significance: None is known.

Ecology: Lichenized with green algae, especially *Trentepohlia* species, on bark, rocks and soil.

Chapsa sp.; thallus and ascomata

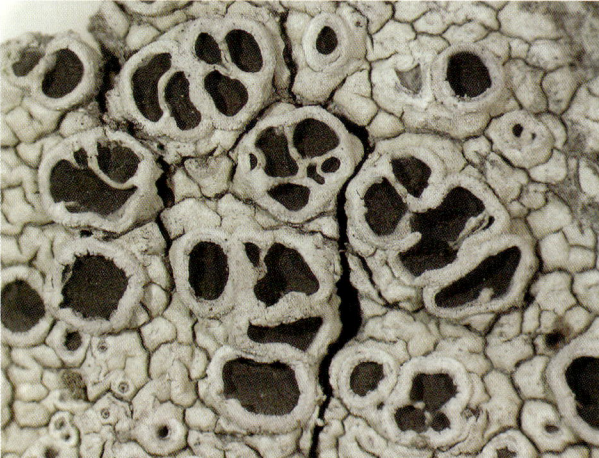

Diploschistes cinereocaesius; ascomata

Notes: Delineation of the principal genera needs reexamination according to molecular evidence, as does the relationship between this family and the *Graphidaceae*.

References: **Frisch, A.; Kalb, K.; Grube, M.** (2006). Contributions towards a new systematics of the lichen family *Thelotremataceae*. *Biblthca Lichenol.* 92: 556 pp.; **Kalb, K.; Staiger, B.; Elix, J.A.** (2004). A monograph of the lichen genus *Diorygma* – a first attempt. *Symb. bot. upsal.* 34: 133–181; **Kantvilas, G.; Vězda, A.** (2000). Studies on the lichen family *Thelotremataceae* in Tasmania. The genus *Chroodiscus* and its relatives. *Lichenologist* 32: 325–357; **Kauff, F.; Lutzoni, F.** (2002). Phylogeny of the *Gyalectales* and *Ostropales* (*Ascomycota, Fungi*): among and within order relationships based on nuclear ribosomal RNA small and large subunits. *Mol. Phylogen. Evol.* 25: 138–156; **Kauff, F.; Lutzoni, F.** (2002). Phylogeny of the *Gyalectales* and *Ostropales* (*Ascomycota, Fungi*): among and within order relationships based on nuclear ribosomal RNA small and large subunits. *Mol. Phylogen. Evol.* 25: 138–156; **Lumbsch, H.T.** (1993). Studien über die Flechtengattung *Diploschistes* I. *Nova Hedwigia* 56: 227–236; **Lumbsch, H.T.; Mangold, A.; Lücking, R.; Garcia, M.A.; Martín, M.P.** (2004). Phylogenetic position of the genera *Nadvornikia* and *Pyrgillus* (*Ascomycota*) based on molecular data. *Symb. bot. upsal.* 34: 9–17; **Lumbsch, H.T.; Schmitt, I.; Palice, Z.; Wiklund, E.; Ekman, S.; Wedin, M.** (2004). Supraordinal phylogenetic relationships of *Lecanoromycetes* based on a Bayesian analysis of combined nuclear and mitochondrial sequences. *Mol. Phylogen. Evol.* 31: 822–832; **Lumbsch, H.T.; Tehler, A.** (1998). A cladistic analysis of the genus *Diploschistes* (*Ascomycotina, Thelotremataceae*). *Bryologist* 101: 398–403; **Martín, M.P.; LaGreca, S.; Schmitt, I.; Lumbsch, H.T.**
(2003). Molecular phylogeny of *Diploschistes* inferred from ITS sequence data. *Lichenologist* 35: 27–32; **Matsumoto, T.** (2000). Taxonomic studies of the *Thelotremataceae* (*Graphidales*, lichenized *Ascomycota*) in Japan (1) Genus *Thelotrema*. *J. Hattori bot. Lab.* 88: 1–50; **Matsumoto, T.; Deguchi, H.** (1999). Pycnidial structures and genus concept in the *Thelotremataceae*. *Bryologist* 102: 86–91; **Staiger, B.; Kalb, K.; Grube, M.** (2006). Phylogeny and phenotypic variation in the lichen family *Graphidaceae* (*Ostropomycetidae, Ascomycota*). *Mycol. Res.* 110: 765–772.

Thrombiaceae Poelt & Vězda ex J.C. David & D. Hawksw. 1991

Uncertain position within Ascomycetes, Ascomycota

Thallus crustose, granular or film-like, ± gelatinized, often evanescent. Ascomata perithecial, immersed, the peridium dark, blueing in iodine towards the apex, thickened towards the ? periphysate ostiole, not clypeate. Interascal tissue of sparsely branched paraphyses. Asci cylindrical, persistent, thin-walled, with a well-developed J+ apical cap and a narrow cylindrical apical ring. Ascospores hyaline, aseptate, thin-walled, without a sheath. Anamorph not known.

Significant Genera: *Thrombium*.

Distribution: Known from north temperate regions.

Economic Significance: None is known.

Ecology: Lichenized with green algae (*Leptosira* species).

Notes: Related to and possibly synonymous with the *Protothelenellaceae*.

Thrombium epigaeum; thallus and ascomata

References: **Schmitt, I.; Mueller, G.; Lumbsch, H.T.** (2005). Ascoma morphology is homoplaseous and phylogenetically misleading in some pyrenocarpous lichens. *Mycologia* 97: 362–374.

Thyridiaceae J.Z. Yue & O.E. Erikss. 1987
Uncertain position within Sordariomycetes, Ascomycota

Stromata immersed to erumpent, usually yellow to brownish, soft-textured. Ascomata perithecial, ± globose to pyriform, often long-necked, thin-walled, the ostioles sometimes convergent, periphysate. Interascal tissue of thin-walled paraphyses and apical paraphyses. Asci cylindrical, persistent, without separable wall layers, thickened at the apex, with a J– apical ring. Ascospores at least partially brown, muriform, without a sheath. Anamorphs coelomycetous.

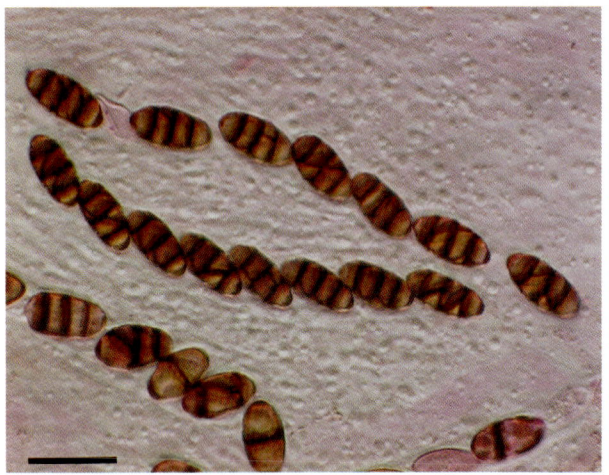

Thyridium vestitum, Yorkshire, UK; ascospores

Significant Genera: *Thyridium*.

Distribution: Widespread.

Economic Significance: None is known.

Ecology: Saprobic on wood and bark.

Notes: The family appears to be most closely related to the *Hypocreales*, with which it shares some wall morphology features, but more research is needed.

References: **Eriksson, O.E.; Yue, J.Z.** (1989). An amended description and disposition of the genus *Thyridium*. *Syst. Ascom.* 8: 9–16; **Leuchtmann, A.; Müller, E.** (1986). Über *Thyridium vestitum* und sein Anamorph. *Bot. Helv.* 96: 283–287; **Miller, A.N.; Huhndorf, S.M.** (2005). Multi-gene phylogenies indicate ascomal wall morphology is a better predictor of phylogenetic relationships than ascospore morphology in the *Sordariales* (*Ascomycota*, Fungi). *Mol. Phylogen. Evol.* 35: 60–75; **Samuels, G.J.; Rogerson, C.T.** (1989). *Endocreas lasiacidis* and *Sinosphaeria lasiacidis*, new tropical ascomycetes. *Stud. Mycol.* 31: 145–149; **Taylor, J.E.; Hyde, K.D.; Jones, E.B.G.** (1997). Fungi from palms. XXXV. *Thyridium chrysomallum* associated with *Archontophoenix alexandrae* (Palmae) cultivated in Hong Kong. *Sydowia* 49: 94–100.

Tilletiaceae J. Schröt. 1887
Tilletiales: Basidiomycota

Sori mostly forming in ovaries, forming so-called bunt balls composed of dark brown spores interspersed with sterile cells. Hyphae intercellular, with dolipore septa, lacking haustoria. Teliospores formed singly, yellowish to dark brown, almost always ornamented with spines, tubercules, ridges or reticulate patterns, usually surrounded by a hyaline gelatinous sheath or a mucous appendage, germinating to form an elongate cylindrical septate basidium bearing an apical whorl of 6–16 basidiospores. Basidiospores filiform, in some species copulating in pairs to give rise to infection hyphae or secondary spores. Anamorphs hyphomycetous, producing ballistoconidia.

Significant Genera: *Tilletia*.

Distribution: Cosmopolitan.

Tilletia laevis, Iraq; bunt balls in infected seeds

Tilletia laevis, Iraq; teliospores

Economic Significance: *Tilletia indica* causes karnal bunt disease of wheat, a highly important quarantine pathogen. Confusion between this and a similar species on ryegrass nearly led to complete suspension of the $5 billion US wheat export market in the mid 1990s. Other species also cause economically significant diseases of cereal crops.

Ecology: Biotrophic in ovaries, more rarely leaf and stem tissue, of grasses. *Erratomyces* occurs on legumes and appears to be a basal representative of the family.

Notes: Forming an isolated group within the smut fungi, these are the only members of the *Exobasidiomycetidae* with dolipore septa.

Tilletia controversa, Germany; teliospores

Tilletia narasimhanii, India; teliospores

References: **Anon.** (2004). Diagnostic protocols for regulated pests. PM 7/29(1). *Tilletia indica*. *Bulletin OEPP, EPPO Bulletin* 34: 219–227; **Begerow, D.; Bauer, R.; Oberwinkler, F.** (1998). Phylogenetic studies on nuclear large subunit ribosomal DNA sequences of smut fungi and related taxa. *Can. J. Bot.* 75: 2045–2056; **Castlebury, L.A.; Carris, L.M.; Vánky, K.** (2005). Phylogenetic analysis of *Tilletia* and allied genera in order *Tilletiales* (*Ustilaginomycetes*; *Exobasidiomycetidae*) based on large subunit nuclear DNA sequences. *Mycologia* 97: 888–900; **Chesmore, D.; Bernard, T.; Inman, A.J.; Bowyer, R.J.** (2003). Image analysis for the identification of the quarantine pest *Tilletia indica*. *Bulletin OEPP, EPPO Bulletin* 33: 495–499; **Cunnington, J.H.; Shivas, R.G.** (2004). The phylogenetic position of *Tilletia nigrifaciens*. *Australas. Mycol.* 22: 53–56; **Josefsen, L.; Christiansen, S.K.** (2002). PCR as a tool for the early detection and diagnosis of common bunt in wheat, caused by *Tilletia tritici*. *Mycol. Res.* 106: 1287–1292; **Levy, L.; Castlebury, L.A.; Carris, L.M.; Meyer, R.J.; Pimentel, G.** (2001). Internal transcribed spacer sequence-based phylogeny and polymerase chain reaction-restriction fragment length polmorphism differentiation of *Tilletia walkeri* and *T. indica*. *Phytopathology* 91: 935–940; **McDonald, J.G.; Wong, E.; White, G.P.** (2000). Differentiation of *Tilletia* species by rep-PCR genomic fingerprinting. *Pl. Dis.* 84: 1121–1125; **Palm, M.E.** (1999). Mycology and world trade: a view from the front line. *Mycologia* 91: 1–12; **Pimentel, G.; Carris, L.M.; Peever, T.L.** (2000). Characterization of interspecific hybrids between *Tilletia controversa* and *T. bromi*. *Mycologia* 92: 411–420; **Pimentel, G.; Peever, T.L.; Carris, L.M.** (2000). Genetic variation among natural populations of *Tilletia controversa* and *T. bromi*. *Phytopathology* 90: 376–383; **Tan, M.K.; Murray, G.M.** (2006). A molecular protocol using quenched FRET probes for the quarantine surveillance of *Tilletia indica*, the causal agent of karnal bunt of wheat. *Mycol. Res.* 110: 203–210; **Vánky, K.** (2004). Biodiversity and conservation of smut fungi (*Ustilaginomycetes* p.p., *Microbotryales*) reflected in Vánky, *Ustilaginales* exsiccata no. 1-1250. *Mycol. Balcanica* 1: 175–187; **Vánky, K.; McKenzie, E.H.C.** (2002). Smut Fungi of New Zealand. *Fungal Diversity Res. Ser.* (Hong Kong) 8: 259 pp.

Tilletiariaceae R.T. Moore 1980
Georgefischeriales: Basidiomycota

Sori formed in leaves or ovaries, bursting out of the plant tissues to expose masses of spores. Hyphae with simple poreless septa, haustoria absent. Teliospores subglobose or ovoid, thick-walled, sometimes with multiple wall layers, dark brown, smooth or spinose, sometimes aggregated into globose or polyhedral spore balls, germinating to produce a cylindrical, septate, sometimes branched basidium. Basidiospores discharged actively, formed laterally on well-developed sterigmata, clavate or allantoid, aseptate, sometimes germinating to form secondary spores.

Significant Genera: *Tilletiaria*, *Tolyposporella*.

Distribution: Widespread.

Economic Significance: None is known.

Ecology: *Tolyposporella* and *Phragmotaeinium* species are biotrophs, occurring in leaf or flower tissues of grasses. *Tilletiaria anomala* is only known in culture and its ecology needs investigation.

Notes: A small, poorly known smut group distinguished from the other families of the *Georgefischeriales* by its septate basidia.

References: **Bauer, R.; Begerow, D.; Nagler, A.; Oberwinkler, F.** (2001). The *Georgefischeriales*: a phylogenetic hypothesis. *Mycol. Res.* 105: 416–424; **Bauer, R.; Oberwinkler, F.; Vánky, K.** (1997). Ultrastructural markers and systematics in smut fungi and allied taxa. *Can. J. Bot.* 75: 1273–1314; **Begerow, D.; Bauer, R.; Oberwinkler, F.** (1998). Phylogenetic studies on nuclear large subunit ribosomal DNA sequences of smut fungi and related taxa. *Can. J. Bot.* 75: 2045–2056; **Boekhout, T.** (1998). *Tilletiaria* Bandoni & Johri. *The Yeasts, A Taxonomic Study* (Amsterdam): 703–704; **Ingold, C.T.** (1992). Discharge of ballistospores in *Tilletiaria anomala*. *Mycol. Res.* 96: 987–989; **Takashima, M.; Nakase, T.** (1996). A phylogenetic study of the genus *Tilletiopsis*, *Tilletiaria anomala* and related taxa based on the small subunit ribosomal DNA sequences. *J. gen. appl. Microbiol.* Tokyo 42: 421–429; **Vánky, K.** (1987). Illustrated genera of smut fungi. *Cryptog. Stud.* 1: 159 pp.

Togniniaceae Réblová, L. Mostert, W. Gams & Crous 2004
Calosphaeriales: Ascomycota

Stromata absent. Ascomata superficial to immersed, perithecial, ± globose, dark brown or black, long-necked, glabrous; peridium two-layered, leathery. Interascal tissue of abundant broad cellular paraphyses, sometimes branched and tapering. Ascogenous hyphae proliferating in a spicate arrangement. Asci elongate, clavate, the apex ± truncate and thickened but without distinct apical structures, sessile, breaking off near the base with the remnants remaining attached to the ascogenous hypha, 8-spored, not staining in iodine. Ascospores allantoid to ellipsoidal, hyaline, aseptate, without gelatinous sheath or appendages. Anamorphs hyphomycetous, pigmented. Conidiogenous cells proliferating percurrently, with flared collarettes. Conidia dimorphic.

Significant Genera: *Togninia*. Anamorph: *Phaeoacremonium*.

Distribution: Widespread in both temperate zones and also the tropics.

Economic Significance: Species (or species complexes) cause Petri and esca diseases of grapevines and have recently been recognized as one of the most serious of grape diseases.

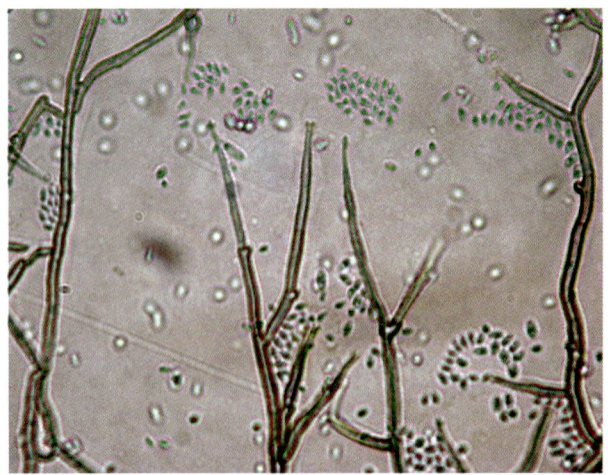

Phaeoacremonium parasiticum; conidiophores and conidia

Ecology: Species infect living parts of woody plants causing dieback and wilt, often with black oozing sap. Others are opportunistic pathogens of humans, typically causing deep-seated cysts.

Notes: Molecular studies indicate that this family is clearly separable from the remaining members of the *Calosphaeriales* as traditionally circumscribed, despite the distinctive spicate ascogenous hyphae. Relationships between that order and the *Diaporthales* need more study.

References: **Barr, M.E.** (1985). Notes on the *Calosphaeriales*. *Mycologia* 77: 549–565; **Fischer, M.; Edwards, J.; Cunnington, J.H.; Pascoe, I.G.** (2005). Basidiomycetous pathogens on grapevine: a new species from Australia – *Fomitiporia australiensis*. *Mycotaxon* 92: 85–96; **Mostert, L.; Crous, P.W.; Groenewald, J.Z.; Gams, W.; Summerbell, R.C.** (2003). *Togninia* (*Calosphaeriales*) is confirmed as teleomorph of *Phaeoacremonium* by means of morphology, sexual compatibility and DNA phylogeny. *Mycologia* 95: 646–659; **Mostert, L.; Groenewold, J.Z.; Summerbell, R.C.; Gams, W.; Crous, P.W.** (2006). Taxonomy and pathology of *Togninia* (*Diaporthales*) and its *Pheaoacremonium* anamorphs. *Stud. Mycol.* 54: 115 pp.; **Pascoe, I.G.; Edwards, J.; Cunnington, J.H.; Cottral, E.H.** (2004). Detection of the *Togninia* teleomorph of *Phaeoacremonium aleophilum* in Australia. *Phytopath. Mediterr.* 43: 51–58; **Réblová, M.; Mostert, L.; Gams, W.; Crous, P.W.** (2004). New genera in the *Calosphaeriales*: *Togniniella* and its anamorph *Phaeocrella*, and *Calosphaeriophora* as anamorph of *Calosphaeria*. *Stud. Mycol.* 50: 533–550; **Vijaykrishna, D.; Mostert, L.; Jeewon, R.; Gams, W.; Hyde, K.D.; Crous, P.W.** (2004). *Pleurostomophora*, an anamorph of *Pleurostoma* (*Calosphaeriales*), a new anamorph genus morphologically similar to *Phialophora*. *Stud. Mycol.* 50: 387–395.

Tremellaceae Fr. 1821
Tremellales: Basidiomycota

Basidiomata pustulate, cerebriform or irregularly lobed, sometimes lacking; when present gelatinous and often brightly coloured. Hyphal system monomitic, mostly with clamp connections and often haustoria, septa with dolipores. Hymenium composed of a compact layer of basidia, cystidia absent. Basidia formed at the apices of fertile hyphae with a basal clamp connection, often proliferating; variously shaped, with longitudinal, oblique or transverse septa, each cell typically producing one basidiospore, usually with a differentiated sterigma. Basidiospores usually discharged actively, globose or ellipsoidal, smooth, ± hyaline, thin-walled, not staining in iodine. Anamorphs yeast-like, proliferating by budding.

Significant Genera: *Tremella*. Anamorph: *Bullera*.

Distribution: Cosmopolitan.

Economic Significance: Some of the larger species are cultivated and eaten, especially in Oriental countries.

Ecology: Usually growing on woody substrata, often parasitic on other fungi.

Notes: The genus *Tremella* is clearly polyphyletic with species potentially belonging to at least four separate families. Further monographic study is required.

Tremella mesenterica, Devon, UK; basidiomata

Tremella globispora; basidiomata on old pyrenomycete fruit bodies

Tremella foliacea, Inverness-shire, Scotland; basidiomata

References: **Bandoni, R.; Ginns, J.** (1999). Notes on *Tremella mesenterica* and allied species. *Can. J. Bot.* 76: 1544–1557; **Bandoni, R.J.** (1987). Taxonomic overview of the *Tremellales*. *Stud. Mycol.* 30: 87–110; **Bandoni, R.J.; Boekhout, T.** (1998). Tremelloid genera with yeast phases: *Fibulobasidium* Bandoni, *Holtermannia* Saccardo & Traverso, *Sirobasidium* de Lagerheim & Patouillard, *Tremella* Persoon, *Trimorphomyces* Bandoni & Oberwinkler. *The Yeasts, A Taxonomic Study* (Amsterdam): 705–717; **Banno, I.; Yamada, Y.** (1998). *Fellomyces* Y. Yamada & Banno. *The Yeasts, A Taxonomic Study* (Amsterdam): 768–772; **Boekhout, T.; Nakase, T.** (1998). *Bullera* Derx. *The Yeasts, A Taxonomic Study* (Amsterdam): 731–741; **Chen, C.J.** (1998). Morphological and Molecular Studies in the Genus *Tremella*. *Biblthca Mycol.* 174: 225 pp.; **Chen, C.J.; Oberwinkler, F.; Chen, Z.C.** (1999). *Tremella occultifuroidea* sp. nov., a new mycoparasite of *Dacrymyces*. *Mycoscience* 40: 137–143; **Chen, C.J.; Oberwinkler, F.; Chen, Z.C.** (2001). Restudy of some type specimens of *Tremella* (I). *Mycotaxon* 77: 215–224; **Sampaio, J.P.; Weiss, M.; Gadanho, M.; Bauer, R.** (2002). New taxa in the *Tremellales*: *Bulleribasidium oberjochense* gen. et sp. nov., *Papiliotrema bandonii* gen. et sp. nov. and *Fibulobasidium murrhardtense* sp. nov. *Mycologia* 94: 873–887; **Scorzetti, G.; Fell, J.W.; Fonseca, A.; Statzell-Tallman, A.** (2002). Systematics of basidiomycetous yeasts: a comparison of large subunit D1/D2 and internal transcribed spacer rDNA regions. *FEMS Yeast Res.* 2: 495–517; **Weiss, M.; Oberwinkler, F.** (2001). Phylogenetic relationships in *Auriculariales* and related groups – hypotheses derived from nuclear ribosomal DNA sequences. *Mycol. Res.* 105: 403–415; **Yan, P.S.; Jiang, J.H.; Wang, D.C.; Luo, X.C.; Zhou, Q.** (2002). Molecular taxonomic relationships of *Auricularia* species inferred from RAPD markers. *Mycosystema* 21: 47–52; **Zugmaier, W.; Bauer, R.; Oberwinkler, F.** (1994). Mycoparasitism of some *Tremella* species. *Mycologia* 86: 49–56.

Tremellodendropsidaceae Jülich 1982
Tremellales: Basidiomycota

Basidiomata solitary or clustered, arbuscular with several ranks of branching, the branches sometimes flattened below, fleshy and cream, brownish or pinkish. Hyphal system monomitic, hyphae with clamp connections, cystidia absent. Basidia uruniform and long-stalked or elongate, completely or incompletely cruciately septate, with 4 well-developed sterigmata. Basidiospores subglobose to ellipsoidal, hyaline, smooth, not staining in iodine, sometimes germinating directly to produce secondary spores.

Significant Genera: *Tremellodendropsis*.

Distribution: Known from South-east Asia and Australasia.

Economic Significance: None is known.

Ecology: Terrestrial, presumably saprobic.

Notes: Preliminary molecular studies indicate that *Tremellodendropsis* (currently the only genus recognized in this family) belongs in the *Auriculariales*, and may be closely allied to *Protomerulius*.

Tremellodendropsis semivestita; basidiomata

References: **Petersen, R.H.** (1987). Notes on clavarioid fungi. XXI. New Zealand taxa of *Tremellodendropsis*. *Mycotaxon* 29: 45–65; **Weiss, M.; Oberwinkler, F.** (2001). Phylogenetic relationships in *Auriculariales* and related groups – hypotheses derived from nuclear ribosomal DNA sequences. *Mycol. Res.* 105: 403–415; **Wendland, I.** (1994). *Tremellodendropsis tuberosa* in Mecklenburg. *Boletus* 18: 102–104.

Triblidiaceae Rehm 1888
Triblidiales: Ascomycota

Stromata absent, the ascomata often strongly aggregated. Ascomata erumpent, ultimately ± apothecial, opening by irregular cracks in the upper wall; peridium black, composed of strongly melanized isodiametric cells. Interascal tissue of narrow trabeculate pseudoparaphyses, becoming free at the apex. Asci thin-walled, ± cylindrical, persistent, without separable wall layers, without distinct apical structures, sometimes becoming mature before the covering layer ruptures. Ascospores hyaline, multiseptate. Anamorph not known.

Significant Genera: *Triblidium*.

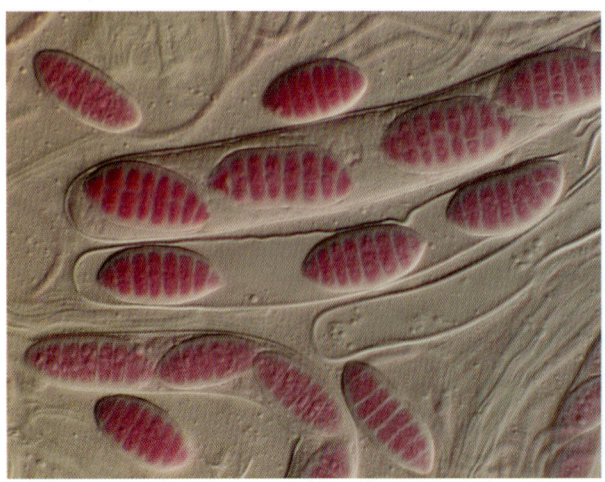
Triblidium craraense, Inverness-shire, Scotland; asci and ascospores

Distribution: Widespread.

Economic Significance: None is known.

Ecology: Apparently saprobic on bark.

Notes: The phylogenetic position of this family is uncertain; no sequences are available. A relationship with the *Rhytismatales* is possible, but there are differences especially in the interascal tissue.

References: **Eriksson, O.E.** (1992). *Huangshania verrucosa* gen. et sp. nov. (*Triblidiaceae*, *Triblidiales* ordo nov.). *Syst. Ascom.* 11: 1–10; **Magnes, M.** (1997). Weltmonographie der *Triblidiaceae*. *Biblthca Mycol.* 165: 177 pp.

Trichocomaceae E. Fisch. 1897
Eurotiales: Ascomycota

Mycelium well-developed, often brightly coloured (many species are greenish-grey), with specialized Hülle cells in some species. Stromata absent. Ascomata small, cleistothecial, rather varied in development; peridium varied, pseudoparenchymatous or hyphal, sometimes thick and sclerotioid, usually brightly coloured; ascogonia often coiled. Interascal tissue absent. Asci small, evanescent, ± globose, often formed in chains. Ascospores ± hyaline, usually bivalvate and often ornamented. Anamorphs hyphomycetous, often prominent, rarely synnematous. Conidiophores often with several orders of branching, sometimes weakly verrucose, with terminal clusters of conidiogenous cells. Conidiogenous cells flask-shaped, proliferating percurrently. Conidia in chains, frequently copious in quantity, small, aseptate, often thick-walled and usually ornamented.

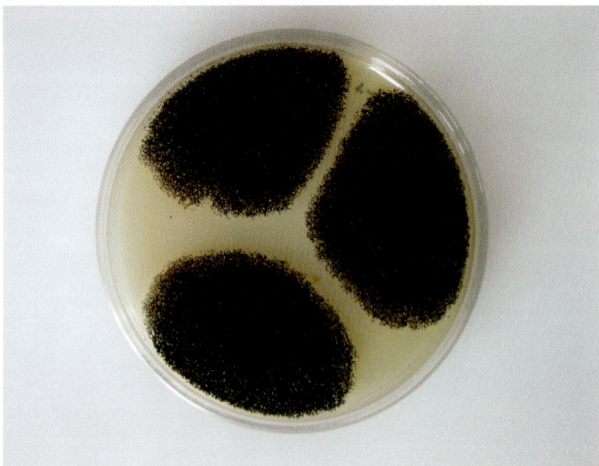
Aspergillus niger, UK; colonies on malt extract agar

Significant Genera: *Emericella*, *Eupenicillium*, *Eurotium*, *Talaromyces*. Anamorphs: *Aspergillus*, *Paecilomyces*, *Penicillium*

Distribution: Cosmopolitan.

Economic Significance: Famous for production of penicillins and related antibiotics, used in production of blue and Camembert-type cheeses, but almost equally prominent as biodeteriogens of food commodities. Production of toxic metabolites such as the aflatoxins, ochratoxins and patulins is a major concern in processed food industries. Some species are able to survive high or low temperatures and are sometimes present even in canned foods. Others are significant in human health, principally through allergenic effects but a few are direct pathogens.

Penicillium vulpinum; coremia

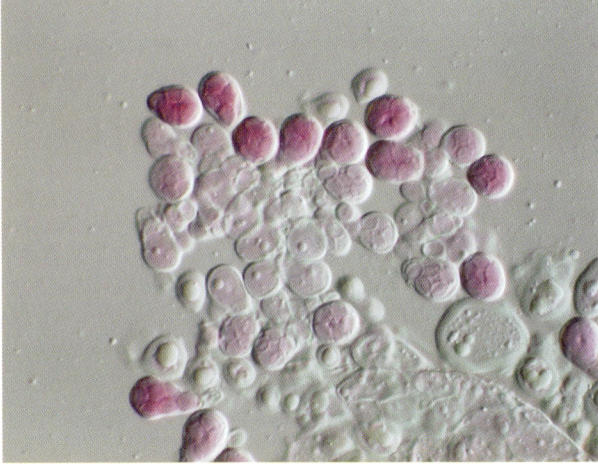
Eurotium chevalieri, England; asci and ascospores

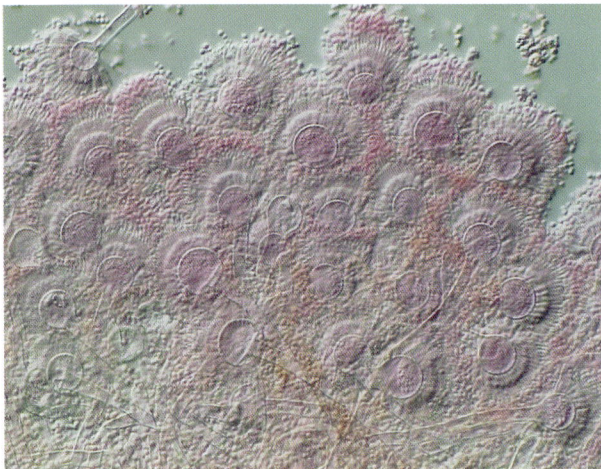
Aspergillus terreus, Scotland; conidiophores and conidia

Ecology: Saprobes with often aggressive colonization strategies, adaptable to extreme environments; ubiquitous in soil communities and extremely common associates of decaying plant material and foodstuffs.

Notes: The *Trichocomaceae* contains some of the most familiar fungi. The genome of *Aspergillus nidulans*, an important model organism for genetic research, has recently been sequenced.

References: **Domenech, J.; Prieto, A.; Barasoaín, I.; Gómez-Miranda, B.; Bernabé, M.; Leal, J.A.** (1999). Galactomannans from the cell walls of species of *Paecilomyces* sect. *Paecilomyces* and their teleomorphs as immunotaxonomic markers. *Microbiology* (Reading) 145: 2789–2796; **Frisvad, J.C.; Filtenborg, O.; Lund, F.; Samson, R.A.** (2000). The homogeneous species and series in subgenus *Penicillium* are related to mammal nutrition and excretion. *Integration of Modern Taxonomic Methods for Penicillium and Aspergillus Classification* (Amsterdam): 265–283; **Frisvad, J.C.; Samson, R.A.** (2000). *Neopetromyces* gen. nov. and an overview of teleomorphs of *Aspergillus* subgenus *Circumdati*. *Stud. Mycol.* 45: 201–207; **Frisvad, J.C.; Thrane, U.; Filtenborg, O.** (1998). Role and use of secondary metabolites in fungal taxonomy. *Chemical Fungal Taxonomy* (New York): 289–319; **Geiser, D.M.; Frisvad, J.C.; Taylor, J.W.** (1998). Evolutionary relationships in *Aspergillus* section *Fumigati* inferred from partial β-tubulin and hydrophobin DNA sequences. *Mycologia* 90: 831–845; **Hong, S.-B.; Go, S.-J.; Shin, H.-D.; Frisvad, J.C.; Samson, R.A.** (2006). Polyphasic taxonomy of *Aspergillus fumigatus* and related species. *Mycologia* 97: 1316–1329; **Horie, Y.; Fukihara, T.; Nishimura, K.; Taguchi, H.; Wang, D.L.; Li, R.Y.** (1997). New and interesting species of *Emericella* from Chinese soil. *Mycoscience* 37: 323–329; **Klich, M.A.** (2002). *Identification of Common Aspergillus Species* (Utrecht): 116 pp.; **Klich, M.A.; Cary, J.W.; Beltz, S.B.; Bennett, C.A.** (2003). Phylogenetic and morphological analysis of *Aspergillus ochraceoroseus*. *Mycologia* 95: 1252–1260; **Kozakiewicz, Z.** (1989). *Aspergillus* species on stored products. *Mycol. Pap.* 161: 188 pp.; **Kuraishi, H.; Itoh, M.; Katayama, Y.; Ito, T.; Hasegawa, A.; Sugiyama, J.** (2000). Ubiquinone systems in fungi. V. Distribution and taxonomic implications of ubiquinones in *Eurotiales*, *Onygenales* and the related plectomycete genera, except for *Aspergillus*, *Paecilomyces*, *Penicillium*, and their related teleomorphs. *Antonie van Leeuwenhoek* 77: 179–186; **Landvik, S.; Shailer, N.F.J.; Eriksson, O.E.** (1997). SSU rDNA sequence support for a close relationship between the *Elaphomycetales* and the *Eurotiales* and *Onygenales*. *Mycoscience* 37: 237–241; **Malloch, D.** (1986). The *Trichocomaceae*: relationships with other ascomycetes. *Advances in Penicillium and Aspergillus Systematics* (New York) 102: 365–382; **Ogawa, H.; Sugiyama, J.** (2000). Evolutionary relationships of the cleistothecial genera with *Penicillium*, *Geosmithia*, *Merimbla* and *Sarophorum* anamorphs as inferred from 18S rDNA sequence divergence. *Integration of Modern Taxonomic Methods for Penicillium and Aspergillus Classification* (Amsterdam): 149–161; **Ogawa, H.; Yoshimura, A.; Sugiyama, J.** (1997). Polyphyletic origins of species of the anamorphic genus *Geosmithia* and the relationships of the cleistothecial genera: evidence from 18S, 5.8S and 28S rDNA sequence analysis. *Mycologia* 89: 756–771; **Peterson, S.W.** (2000). Phylogenetic analysis of *Penicillium* species based on ITS and LSU-rDNA nucleotide sequences. *Integration of Modern Taxonomic Methods for Penicillium and Aspergillus Classification* (Amsterdam): 163–178; **Peterson, S.W.** (2000). Phylogenetic relationships in *Aspergillus* based on rDNA sequence analysis. *Integration of Modern Taxonomic Methods for Penicillium and Aspergillus Classification* (Amsterdam): 323–355; **Pitt, J.I.** (1995). Phylogeny in the genus *Penicillium*: a morphologist's perspective. *Can. J. Bot.* 73: S768–S777; **Pitt, J.I.; Samson, R.A.; Frisvad, J.C.** (2000). List of accepted species and their synonyms in the family *Trichocomaceae*.

Integration of Modern Taxonomic Methods for Penicillium and Aspergillus Classification (Amsterdam): 9–49; **Samson, R.A.; Frisvad, J.C.** (2004). *Penicillium* subgenus *Penicillium*: new taxonomic schemes, mycotoxins and other extrolites. *Stud. Mycol.* 49: 175–200; **Scott, J.; Untereiner, W.A.; Wong, B.; Straus, N.A.; Malloch, D.** (2004). Genotypic variation in *Penicillium chrysogenum* from indoor environments. *Mycologia* 96: 1095–1105; **Skouboe, P.; Frisvad, J.C.; Taylor, J.W.; Lauritsen, D.; Boysen, M.; Rossen, L.** (1999). Phylogenetic analysis of nucleotide sequences from the ITS region of terverticillate *Penicillium* species. *Mycol. Res.* 103: 873–881; **Tamura, M.; Kawahara, K.; Sugiyama, J.** (2000). Molecular phylogeny of *Aspergillus* and associated teleomorphs in the *Trichocomaceae* (*Eurotiales*). *Integration of Modern Taxonomic Methods for Penicillium and Aspergillus Classification* (Amsterdam): 357–372.

Tricholomataceae R. Heim ex Pouzar 1983

Agaricales: Basidiomycota

Basidiomata pileate and stipitate, usually large and sturdy, the cap viscid or not, glabrous to fibrillose or scaly, the stipe central, fleshy or fibrillose, veil absent (rarely with a web-like partial veil), often with a fleshy annular ring. Hyphal system monomitic, hyphae often with pigment-encrusted walls, not staining in iodine, clamp connections present or absent, cystidia usually absent. Hymenium lamellate, the gills pale, often sinuate. Basidia small, clavate, with 2 or 4 sterigmata. Basidiospores generally small, variously shaped (globose, ellipsoidal, fusiform or angular), hyaline or creamish, thin-walled, smooth, germ pore absent, very rarely staining in iodine.

Significant Genera: *Clitocybe, Tricholoma*.

Clitocybe fragrans, Surrey, UK; basidiomata

Distribution: Cosmopolitan, more prominent in temperate areas.

Economic Significance: *Tricholoma matsutake* is one of the most highly prized edible mushrooms in the world. It and closely related species are harvested from the wild throughout temperate Eurasia and North America and there is increasing concern over the need for conservation. A number of other species are also edible, but poisonous taxa also occur.

Lepista nuda, Berkshire, UK; basidioma

Ecology: *Tricholoma* species are ectomycorrhizal, occurring in coniferous and broadleaved forests.

Tricholomopsis rutilans, Surrey, UK; basidioma

Calyptella campanula, Sussex, UK; basidiomata

Notes: A very large and polymorphic family as defined in the *Dictionary of the Fungi* edn 9 that does not stand as a single unit when molecular evidence is taken into account. Several segregate families have been removed (e.g. *Hygrophoraceae, Marasmiaceae, Mycenaceae*) but the remaining cluster of genera is undoubtedly polyphyletic. Recognizable clades include those based on *Collybia – Clitocybe* and the *Lyophylleae* group (including *Termitomyces*). The description given here is primarily based on the type genus *Tricholoma*.

References: **Chapela, I.H.; Garbelotto, M.** (2004). Phylogeography and evolution in matsutake and close allies inferred by analyses of ITS sequences and AFLPs. *Mycologia* 96: 730–741; **Hwang, S.K.; Kim, J.G.** (2000). Secondary structural and phylogenetic implications of nuclear large subunit ribosomal RNA in the ectomycorrhizal fungus *Tricholoma matsutake*. *Curr. Microbiol.* 40: 250–256; **Kalamees, K.** (2001). Taxonomy and ecology of the *Tricholoma equestre* group in the Nordic and Baltic countries. *Folia cryptog. Estonica* 38: 13–23; **Lian, C.L.; Hogetsu, T.; Matsushita, N.; Guerin-Laguette, A.; Suzuki, K.; Yamada, A.** (2003). Development of microsatellite markers from an ectomycorrhizal fungus, *Tricholoma matsutake*, by an ISSR-suppression-PCR method. *Mycorrhiza* 13: 27–31; **Mankel, A.; Kost, G.; Kothe, E.** (1998). Re-evaluation of the phylogenetic relationship among species of the genus *Tricholoma*. *Microbiol. Res.* 153: 377–388; **Matsushita, N.; Kikuchi, K.; Sasaki, Y.; Guerin-Laguette, A.; Lapeyrie, F.; Vaario, L.M.; Intini, M.; Suzuki, K.** (2005). Genetic relationship of *Tricholoma matsutake* and *T. nauseosum* from the Northern Hemisphere based on analyses of ribosomal DNA spacer regions. *Mycoscience* 46: 90–96; **Moncalvo, J.-M.; Vilgalys, R.; Redhead, S.A.; Johnson, J.E.; James, T.Y.; Aime, M.C.; Hofstetter, V.; Verduin, S.J.W.; Larsson, E.; Baroni, T.J.; Thorn, R.G.; Jacobsson, S.; Clemencon, H.; Miller, O.K.** (2002). One hundred and seventeen clades of euagarics. *Mol. Phylogen. Evol.* 23: 357–400; **Murata, H.; Babasaki, K.** (2005). Intra- and inter-specific variations in the copy number of two types of retrotransposons from the ectomycorrhizal basidiomycete *Tricholoma matsutake*. *Mycorrhiza* 15: 381–386; **Noordeloos, M.E.** (1999). [An introduction to the genus *Tricholoma*]. *Coolia* 42: 163–182; **Suh, S.J.; Kim, I.H.; Nam, J.H.; Ghim, S.Y.; Bae, K.S.; Kim, J.G.** (2001). Cloning and phylogenetic analysis of chitin synthase genes from *Tricholoma matsutake*. *Mycobiology* 29: 179–182.

Trichosphaeriaceae G. Winter 1885
Trichosphaeriales: Ascomycota

Stromata absent, or reduced to a hyphal subiculum. Ascomata perithecial, superficial, often aggregated, ± globose, black, often thick-walled, usually setose, the ostiole papillate, periphysate; peridium composed of thick-walled angular cells, sometimes larger and more strongly pigmented towards the apex. Interascal tissue of narrow, persistent, thin-walled, true paraphyses formed from a basal meristem. Asci cylindrical, persistent, thin-walled, not fissitunicate, usually with a small J– apical ring. Ascospores variously shaped, hyaline to versicoloured, usually septate and rarely muriform, sometimes fragmenting at the septa, without germ pores, sometimes with a sheath. Anamorphs varied, hyphomycetous, with simple pigmented conidiophores and often with complex conidia.

Significant Genera: *Cresporhaphis, Trichosphaeria*.

Distribution: Cosmopolitan, best known from north temperate zones.

Economic Significance: None is known.

Ecology: Saprobic especially on wood and bark, occasionally on other fungi.

Notes: Preliminary information suggests that the *Trichosphaeriaceae* occupies a very isolated position within the *Sordariomycetes* and is not closely related to the *Annulatascaceae* as has been postulated.

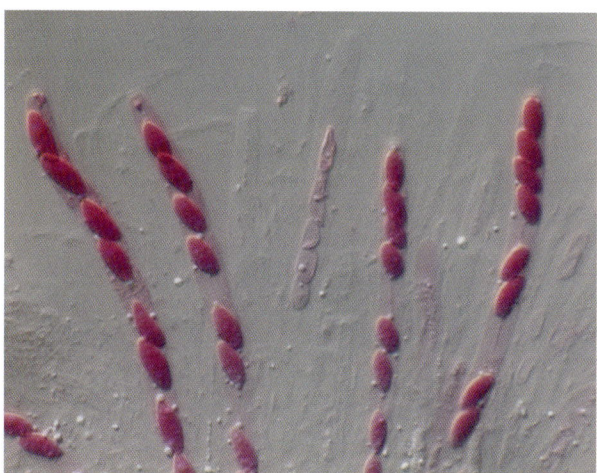

Trichosphaeria pilosa, Argyll, Scotland; asci and ascospores

References: **Campbell, J.; Shearer, C.A.** (2004). *Annulusmagnus* and *Ascitendus*, two new genera in the *Annulatascaceae*. *Mycologia* 96: 822–833; **Gams, W.** (2000). *Phialophora* and some similar morphologically little-differentiated anamorphs of divergent ascomycetes. *Stud. Mycol.* 45: 187–199; **Réblová, M.** (1999). Studies in *Chaetosphaeria* sensu lato II. *Coniobrevicolla* gen. & sp. nov. *Mycotaxon* 70: 421–429; **Réblová, M.** (1999). Studies in *Chaetosphaeria* sensu lato I. The genera *Chaetosphaerella* and *Tengiomyces* gen. nov. of the *Helminthosphaeriaceae*. *Mycotaxon* 70: 387–420; **Réblová, M.; Barr, M.E.; Samuels, G.J.** (1999). *Chaetosphaeriaceae*, a new family for *Chaetosphaeria* and its relatives. *Sydowia* 51: 49–70; **Réblová, M.; Seifert, K.A.** (2004). *Cryptadelphia* (*Trichosphaeriales*), a new genus for holomorphs with *Brachysporium* anamorphs and clarification of the taxonomic status of *Wallrothiella*. *Mycologia* 96: 343–367; **Seifert, K.A.** (1987). *Stromatographium* and *Acrostroma* gen. nov.: two tropical hyphomycete genera with distinctive synnema anatomies. *Can. J. Bot.* 65: 2196–2201.

Trypetheliaceae Zenker 1827
Pyrenulales: Ascomycota

Thallus crustose, often poorly developed or immersed. Ascomata globose, perithecial, sometimes aggregated within pseudostromatic tissues, papillate, the ostiole sometimes lateral. Interascal tissue of narrow branched and anastomosed trabeculate pseudoparaphyses. Asci cylindrical, fissitunicate, with a wide ocular chamber, without a refractive ring, not blueing in iodine. Ascospores usually hyaline, transversely septate or muriform, the septa often thickened, with a mucous sheath. Anamorphs coelomycetous.

Trypethelium aeneum; thallus and ascomata

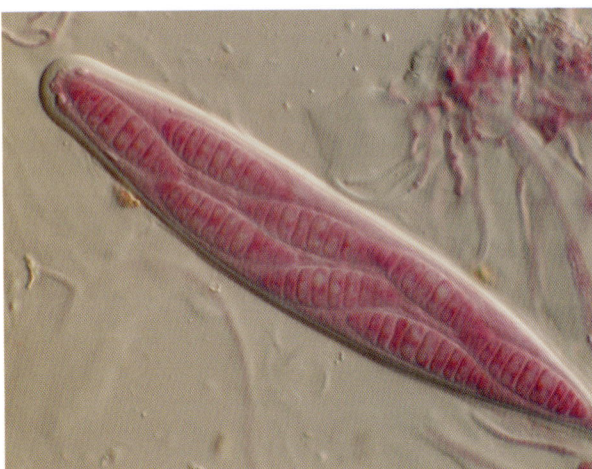
Trypethelium eluteriae, Queensland, Australia; ascus and ascospores

Significant Genera: *Astrothelium, Laurera, Trypethelium*.

Laurera purpurina; ascomata

Distribution: Cosmopolitan, but most species are found in the tropics.

Economic Significance: None is known.

Ecology: Lichenized with green algae, mostly on wood and bark.

Notes: Until recently this family has been assumed as related to the *Pyrenulaceae*, primarily based on shared ascospore characteristics, but molecular studies indicate that it belongs to the *Dothideomyces* and the *Pyrenulaceae* to the *Chaetothyriomycetes*.

Astrothelium galbineum; thallus and ascomata

References: **Del Prado, R.; Schmitt, I.; Kautz, S.; Palice, Z.; Lücking, R.; Lumbsch, H.T.** (2006). Molecular data place *Trypetheliaceae* in Dothideomycetes. *Mycol. Res.* 110: 511–520; **Harris, R.C.** (1986). The family Trypetheliaceae (Loculoascomycetes: lichenized Melanommatales) in Amazonian Brazil. *Acta Amazon.* Suppl. 14: 55–80; **Harris, R.C.** (1993). A revision of *Polymeridium* (Muell. Arg.) R.C. Harris (Trypetheliaceae). *Bolm Mus. paraense 'Emílio Goeldi' sér. bot.* 7: 619–644; **Harris, R.C.** (1998). A preliminary revision of *Pseudopyrenula* Müll. Arg. (lichenized Ascomycetes, Trypetheliaceae) with a redisposition of the names previously assigned to the genus. *Lichenographia Thomsoniana, North American Lichenology in Honor of John W. Thomson* (Ithaca): 133–148; **Makhija, U.; Patwardhan, P.G.** (1988). The lichen genus *Laurera* (family Trypetheliaceae) in India. *Mycotaxon* 31: 565–590; **Makhija, U.; Patwardhan, P.G.** (1993). A contribution to our knowledge of the lichen genus *Trypethelium* (family Trypetheliaceae). *J. Hattori bot. Lab.* 73: 183–219; **McCarthy, P.M.** (1995). *Laurera papillosa* (Trypetheliaceae), a remarkable new lichen from Papua New Guinea. *Lichenologist* 27: 310–314.

Tuberaceae Dumort. 1822
Pezizales: Ascomycota

Ascomata large, subterranean, cleistothecial, ± globose, thick-walled, solid, often ornamented, the asci formed in irregular marbled veins separated by sterile strands. Interascal tissue absent, at least at maturity. Asci formed apparently randomly within the fertile zones, ± globose, thick-walled but not appearing to be multilayered, indehiscent, usually less than 8-spored and frequently variable in this feature within individual ascomata. Ascospores brown, aseptate, usually strongly ornamented. Anamorphs unknown.

Significant Genera: *Tuber*.

Distribution: Widespread in temperate and tropical montane regions.

Economic Significance: *Tuber melanosporum* and *T. magnatum* are two of the most highly prized culinary fungi, with whole industries supporting their exploitation and considerable interest in their cultivation. Several other *Tuber* species are also used for food.

Ecology: Mycorrhizal with broadleaved trees, growing in the surface layer of soil and closely associated with roots. Highly specific pheromone-like chemicals are produced that play a role in dispersal by attracting mammals that uproot fruit bodies and distribute spores.

Notes: The *Tuberaceae* form a monophyletic clade within the *Pezizales*, related to but clearly distinct from the *Helvellaceae*. The subterranean life form has evolved at least four times within the order.

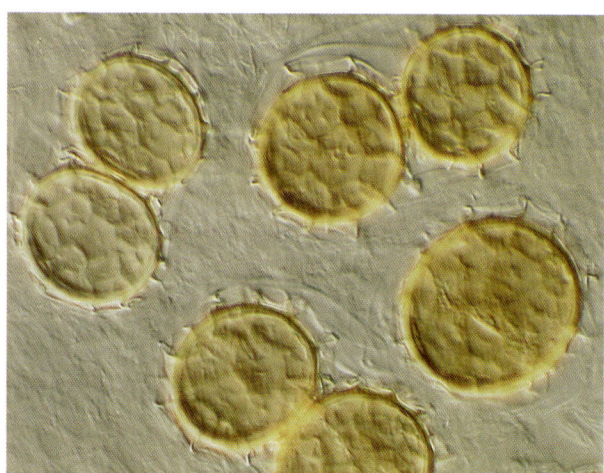

Tuber magnatum; asci and ascospores

References: **Amicucci, A.; Guidi, C.; Zambonelli, A.; Potenza, L.; Stocchi, V.** (2000). Multiplex PCR for the identification of white *Tuber* species. *FEMS Microbiol. Lett.* 189: 265–269; **Amicucci, A.; Zambonelli, A.; Giomaro, G.; Potenza, L.; Stocchi, V.** (1998). Identification of ectomycorrhizal fungi of the genus *Tuber* by species-specific ITS primers. *Mol. Ecol.* 7: 273–277; **Beaton, G.; Malajczuk, N.** (1986). A new species and a variety of *Labyrinthomyces* from Western Australia. *Trans. Br. mycol. Soc.* 86: 503–507; **Bertini, L.; Agostini, D.; Potenza, L.; Rossi, I.; Zeppa, S.; Zambonelli, A.; Stocchi, V.** (1998). Molecular markers for the identification of the ectomycorrhizal fungus *Tuber borchii*. *New Phytol.* 139: 565–570; **Comandini, O.; Pacioni, G.** (1997). Mycorrhizae of Asian black truffles, *Tuber himalayense* and *T. indicum*. *Mycotaxon* 63: 77–86; **Halász, K.; Bratek, Z.; Szegő, D.; Rudnóy, S.; Récz, I.; Lésztity, D.; Trappe, J.M.** (2005). Tests of species concepts of the small, white, European group of *Tuber* spp. based on morphology and rDNA ITS sequences with special reference to *Tuber rapaeodorum*. *Mycol. Prog.* 4: 281–290; **Harrington, F.A.; Pfister, D.H.; Potter, D.; Donoghue, M.J.** (1999). Phylogenetic studies within the *Pezizales*. I. 18S rRNA sequence data and classification. *Mycologia* 91: 41–50; **Iotti, M.; Amicucci, A.; Stocchi, V.; Zambonelli, A.** (2002). Morphological and molecular characterization of mycelia of some *Tuber* species in pure culture. *New Phytol.* 155: 499–505; **Landvik, S.; Egger, K.N.; Schumacher, T.** (1997). Towards a subordinal classification of the *Pezizales* (*Ascomycota*): phylogenetic analyses of SSU rDNA sequences. *Nordic Jl Bot.* 17: 403–418; **Mabru, D.; Douet, J.P.; Mouton, A.; Dupré, C.; Ricard, J.M.; Médina, B.; Castroviejo, M.; Chevalier, G.** (2004). PCR-RFLP using a SNP on the mitochondrial Lsu-rDNA as an easy method to differentiate *Tuber melanosporum* (Perigord truffle) and other truffle species in cans. *Int. J. Food Microbiol.* 94: 33–42; **Mello, A.; Cantisani, A.; Vizzini, A.; Bonfante, P.** (2002). Genetic variability of *Tuber uncinatum* and its relatedness to other black truffles. *Envir. Microbiol.* 4: 584–594; **Mello, A.; Garnero, L.; Bonfante, P.** (1999). Specific PCR primers as a reliable tool for the detection of white truffles in mycorrhizal roots. *New Phytol.* 141: 511–516; **Mello, A.; Vizzini, A.; Longato, S.; Rollo, F.; Bonfante, P.; Trappe, J.M.** (2000). *Tuber borchii* versus *Tuber maculatum*: neotype studies and DNA analyses. *Mycologia* 92: 326–331; **O'Donnell, K.; Cigelnik, E.; Weber, N.S.; Trappe, J.M.** (1997). Phylogenetic relationships among ascomycetous truffles and the true and false morels inferred from 18S and 28S ribosomal DNA sequence analysis. *Mycologia* 89: 48–65; **Paolocci, F.; Rubini, A.; Riccioni, C.; Topini, F.; Arcioni, S.** (2004). *Tuber aestivum* and *Tuber uncinatum*: two morphotypes or two species?. *FEMS Microbiol. Lett.* 235: 109–115; **Percudani, R.; Trevisi, A.; Zambonelli, A.; Ottonello, S.** (1999). Molecular phylogeny of truffles (*Pezizales*: *Terfeziaceae*, *Tuberaceae*) derived from nuclear rDNA sequence analysis. *Mol. Phylogen. Evol.* 13: 169–180; **Roux, C.; Séjalon-Delmas, N.; Martins, M.; Parguey-Leduc, A.; Dargent, R.; Bécard, G.** (1999). Phylogenetic relationships between European and Chinese truffles based on parsimony and distance analysis of ITS sequences. *FEMS Microbiol. Lett.* 180: 147–155; **Trappe, J.M.; Castellano, M.A.; Malajczuk, N.** (1992). Australasian truffle-like fungi. II. *Labyrinthomyces*, *Dingleya* and *Reddellomyces* gen. nov. (*Ascomycotina*). *Aust. Syst. Bot.* 5: 597–611; **Urban, A.; Neuner-Plattner, I.; Krisai-Greilhuber, I.; Haselwandter, K.** (2004). Molecular studies on terricolous microfungi reveal novel anamorphs of two *Tuber* species. *Mycol. Res.* 108: 749–758; **Zambonelli, A.; Rivetti, C.; Percudani, R.; Ottonello, S.** (2000). TuberKey: a DELTA-based tool for the description and interactive identification of truffles. *Mycotaxon* 74: 57–76.

Tubeufiaceae M.E. Barr 1979
Pleosporales: Ascomycota

Ascomata perithecial, sometimes aggregated into stromata, ± globose, thick-walled, many setose, opening by a well-developed lysigenous pore; peridium pale, fleshy, composed of small pseudoparenchymatous cells. Interascal tissue of cellular pseudoparaphyses. Asci cylindrical, fissitunicate, usually with a well-developed ocular chamber, not blueing in iodine. Ascospores hyaline or pale brown, transversely septate, rarely muriform, without a sheath. Anamorphs hyphomycetous, often helicosporous.

Significant Genera: *Podonectria*, *Tubeufia*. Anamorphs: *Helicoma*, *Helicomyces*.

Distribution: Widespread.

Economic Significance: A few species may have potential as biocontrol agents of insects, but there has been little or no research in this area.

Ecology: Biotrophic or saprobic, especially on scale insects or other fungi. The anamorphic forms are frequently aero-aquatic.

Tubeufia cerea, Arran, Scotland; ascomata on old pyrenomycete fungus

Podonectria cicadellidicola; ascomata on cicada larva

Notes: The *Tubeufiaceae* are unusual in the *Dothideomycetes* in that they have brightly coloured fleshy fruit bodies. The few phylogenetic analyses available are partially contradictory and more sequences are required to place the family with confidence.

Helicoma sp.; conidia

Tubeufia cerea; asci and ascospores

References: **Crane, J.L.; Shearer, C.A.; Barr, M.E.** (1998). A revision of *Boerlagiomyces* with notes and a key to the saprobic genera of *Tubeufiaceae*. *Can. J. Bot.* 76: 602–612; **Goos, R.D.** (1986). A review of the anamorph genus *Helicoma*. *Mycologia* 78: 744–761; **Goos, R.D.** (1987). Fungi with a twist: the helicosporous hyphomycetes. *Mycologia* 79: 1–22; **Goos, R.D.** (1989). On the anamorph genera *Helicosporium* and *Drepanospora*. *Mycologia* 81: 356–374; **Inderbitzin, P.; Landvik, S.; Abdel-Wahab, M.A.; Berbee, M.L.** (2001). *Aliquandostipitaceae*, a new family for two new tropical ascomycetes with unusually wide hyphae and dimorphic ascomata. *Am. J. Bot.* 88: 52–61; **Kodsueb, R.; Jeewon, R.; Vijaykrishna, D.; McKenzie, E.H.C.; Lumyong, P.; Lumyong, S.; Hyde, K.D.** (2006). Systematic revision of *Tubeufiaceae* based on morphological and molecular data. *Fungal Diversity* 21: 105–130; **Kodsueb, R.; Lumyong, S.; Lumyong, P.; McKenzie, E.H.C.; Ho, W.H.; Hyde, K.D.** (2004). *Acanthostigma* and *Tubeufia* species, including *T. claspisphaeria* sp. nov., from submerged wood in Hong Kong. *Mycologia* 96: 667–674; **Lumbsch, H.T.; Lindemuth, R.** (2001). Major lineages of *Dothideomycetes* (*Ascomycota*) inferred from SSU and LSU rDNA sequences. *Mycol. Res.* 105: 901–908; **Rodrigues, K.F.; Samuels, G.J.** (1994). *Letendraeopsis palmarum*, a new genus and species of loculoascomycetes. *Mycologia* 86: 254–258; **Rossman, A.Y.** (1987). The *Tubeufiaceae* and similar Loculoascomycetes. *Mycol. Pap.* 157: 71 pp.; **Scheuer, C.** (1991). *Taphrophila* (*Dothideales*: *Tubeufiaceae*) and two species of *Tubeufia* with dark setae. *Mycol. Res.* 95: 811–816; **Tsui, C.K.M.; Sivichai, S.; Berbee, M.L.** (2006). Molecular systematics of *Helicoma*, *Helicomyces* and *Helicosporium* and their teleomorphs inferred from rDNA sequences. *Mycologia* 98: 94–104; **Untereiner, W.A.; Straus, N.A.; Malloch, D.** (1995). A molecular-morphotaxonomic approach to the systematics of the *Herpotrichiellaceae* and allied black yeasts. *Mycol. Res.* 99: 897–913.

Tubulicrinaceae Jülich 1982
Polyporales: Basidiomycota

Basidiomata small or large, resupinate, usually thin and adnate, often inconspicuous. Hyphal system monomitic, hyphae thin-walled, with clamp connections. Hymenium smooth, hispid or pilose due to projecting cystidia. Cystidia thick-walled and conical, capitate or subulate, usually multi-rooted, sometimes encrusted and sometimes staining in iodine. Basidia ± cylindrical, attached laterally at the base to the fertile hypha with a basal clamp connections, with 4 sterigmata. Basidiospores globose, ellipsoidal, sigmoid or vermicular, smooth or ornamented, not staining in iodine.

Significant Genera: *Tubulicrinis*.

Distribution: Cosmopolitan, primarily from temperate regions.

Economic Significance: None is known.

Tubulicrinis subulatus; basidioma

Ecology: On dead wood and herbaceous plant material, some species commonly associated with ferns.

Notes: The family appears to be heterogenous, with representatives nesting within the hymenochaetoid and trechisporoid clades. Pending further molecular studies, its traditional circumscription is retained.

References: **Ginns, J.** (1998). Genera of the North American *Corticiaceae sensu lato*. *Mycologia* 90: 1–35; **Greslebin, A.G.; Rajchenberg, M.** (2003). Diversity of *Corticiaceae* sens. lat. in Patagonia, southern Argentina. *N.Z. Jl Bot.* 41: 437–446; **Hallenberg, N.** (1986). Cultural studies in *Tubulicrinis* and *Xenasmatella* (*Corticiaceae, Basidiomycetes*). *Mycotaxon* 27: 361–375; **Hjortstam, K.** (2001). *Tubulicrinaceae* – a survey of the genera and species. *Windahlia* 24: 1–14; **Küffer, N.; Senn-Irlet, B.** (2005). Diversity and ecology of wood-inhabiting aphyllophoroid basidiomycetes on fallen woody debris in various forest types in Switzerland. *Mycol. Prog.* 4: 77–86; **Larsson, K.-H.; Larsson, E.; Koljalg, U.** (2004). High phylogenetic diversity among corticioid homobasidiomycetes. *Mycol. Res.* 108: 983–1002; **Nakasone, K.K.** (1990). Cultural studies and identification of wood-inhabiting *Corticiaceae* and selected *Hymenomycetes* from North America. *Mycol. Mem.* 15: 412 pp.

Tulasnellaceae Juel 1897
Cantharellales: Basidiomycota

Basidiomata resupinate and effuse, typically waxy, gelatinized or ephemeral, usually pale. Hymenial surface smooth or warted. Hyphal system monomitic, hyphae with or without clamp connections, septa with dolipores and imperforate parenthesome, cystidia with refractive contents sometimes present. Basidia ± globose to clavate or pyriform, sterigmata initially subglobose with a tubular apex developing. Basidiospores ± globose, cylindrical, allantoid or sigmoid, hyaline, thin-walled, smooth, not staining in iodine, usually germinating to form secondary spores. Anamorphs *Rhizoctonia*-like.

Significant Genera: *Tulasnella*. Anamorph: *Epulorhiza*.

Tulasnella violea; basidioma

Distribution: Widespread.

Economic Significance: None is known.

Ecology: Typically found on decaying wood and other rotting vegetation, presumably saprobic in nutrition. Many anamorphic taxa have mycorrhizal associations with orchids. The linkages between teleomorph and anamorph need further study; some *Epulorhiza* species have associations with the *Sebacinaceae* rather than the *Tulasnellaceae*.

Notes: *Tulasnella* species appear to form a well-defined group, possibly close to the *Ceratobasidiaceae* and within the cantharelloid clade of euagarics rather than the *Tremellomycetes*. Further phylogenetic research is needed; there is some evidence to suggest that the family should be separated from *Cantharellales* at ordinal level.

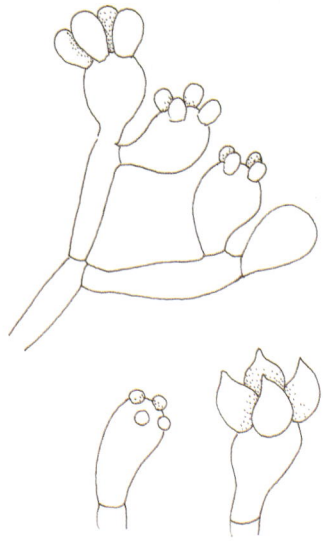

Tulasnella tomaculum, Devon, UK; basidia and basidiospores

References: **Ginns, J.** (1998). Genera of the North American *Corticiaceae sensu lato*. *Mycologia* 90: 1–35; **Gleason, F.H.; McGee, P.A.** (2002). Septal pore cap ultrastructure of fungi identified as *Epulorhiza* sp. (*sensu Sebacina*) isolated from Australian orchids. *Australas. Mycol.* 21: 12–15; **Greslebin, A.G.; Rajchenberg, M.** (2001). The genus *Tulasnella* with a new species in the Patagonian Andes forests of Argentina. *Mycol. Res.* 105: 1149–1151; **Hibbett et al.** (2007). A higher-level phylogenetic classification of the *Fungi*. *Mycol. Res.* 111(5): 509–548; **Kottke, I.; Beiter, A.; Weiss, M.; Haug, I.; Oberwinkler, F.; Nebel, M.** (2003). *Heterobasidiomycetes* form symbiotic associations with hepatics: *Jungermanniales* have sebacinoid mycobionts while *Aneura pinguis* (*Metzgeriales*) is associated with a *Tulasnella* species. *Mycol. Res.* 107: 957–968; **Kristiansen, K.A.; Freudenstein, J.V.; Rasmussen, F.N.; Rasmussen, H.N.** (2004). Molecular identification of mycorrhizal fungi in *Neuwiedia veratrifolia* (*Orchidaceae*). *Mol. Phylogen. Evol.* 33: 251–258; **Larsson, K.-H.; Larsson, E.; Koljalg, U.** (2004). High phylogenetic diversity among corticioid homobasidiomycetes. *Mycol. Res.* 108: 983–1002; **Ma, M.; Tan, T.K.; Wong, S.M.** (2003). Identification and molecular phylogeny of *Epulorhiza* isolates from tropical orchids. *Mycol. Res.* 107: 1041–1049; **McCormick, M.K.; Whigham, D.F.; O'Neill, J.** (2004). Mycorrhizal diversity in photosynthetic terrestrial orchids. *New Phytol.* 163: 425–438; **Moncalvo et al.** (2007). The cantharelloid clade: dealing with incongruent gene trees and phylogenetic reconstruction methods. *Mycologia* 98(6): 937–948; **Roberts, P.** (1994). Long-spored *Tulasnella* species from Devon, with additional notes on allantoid-spored species. *Mycol. Res.* 98: 1235–1244; **Stalpers, J.A.; Andersen, T.F.** (1996). A synopsis of the taxonomy of teleomorphs connected with *Rhizoctonia s.l. Rhizoctonia Species, Taxonomy, Molecular Biology, Ecology, Pathology and Disease Control* (Dordrecht): 49–63.

Tulostomataceae E. Fisch. 1900
Agaricales: Basidiomycota

Basidiomata gasteroid, epigeal at maturity, initially ± globose, volvate or not and then sometimes strongly gelatinized, with a long-stalked ± globose fertile head; peridium two-layered, dehiscent either by an apical pore (peristome) or pores, or irregular or circumscissile fragmentation. Capillitium usually well-developed, composed of hyaline to brown thick-walled hyphae, sometimes with helical or annulate thickening, sometimes inflated at the septa. Gleba not clearly loculate, powdery at maturity. Basidia narrowly clavate, usually with 4 lateral or terminal sterigmata. Basidiospores ± globose, brown or yellowish, thick-walled, usually verrucose or reticulate.

Significant Genera: *Battarrea*, *Tulostoma*.

Battarrea phalloides; basidiomata

Tulostoma brumale; basidiomata

Distribution: Widespread; known from most climatic zones but with greatest diversity in arid regions.

Economic Significance: None is known.

Ecology: Probably most species are saprobic, though a few have been reported as forming ectomycorrhizas.

Notes: *Battarrea* has been separated by many authors into its own family. Molecular data are sparse and while the group has been confirmed as clustering within the *Agaricales* its closest relatives are unclear.

References: **Barr, M.E.; Mathiassen, G.** (1998). Proposed redisposition of *Schizostoma vicinum* and a newly recognized taxon. *Mycotaxon* 69: 159–165; **Esqueda, M.; Herrera, T.; Pérez-Silva, E.; Aparicio, A.; Moreno, G.** (2002). Distribution of *Battarrea phalloides* in Mexico. *Mycotaxon* 82: 207–214; **Esqueda, M.; Moreno, G.; Pérez-Silva, E.; Sánchez, A.; Altés, A.** (2004). The genus *Tulostoma* in Sonora, Mexico. *Mycotaxon* 90: 409–422; **Hibbett, D.S.; Pine, E.M.; Langer, E.; Langer, G.; Donoghue, M.J.** (1996). Evolution of gilled mushrooms and puffballs inferred from ribosomal DNA sequences. *Proc. natn Acad. Sci. U.S.A.* 94: 12002–12006; **Jacobson, K.M.; Jacobson, P.J.; Miller, O.K.** (1999). The autecology of *Battarrea stevensii* in ephemeral rivers of Southwestern Africa. *Mycol. Res.* 103: 9–17; **Martín, M.P.; Hidalgo, E.; Altés, A.; Moreno, G.** (2000). Phylogenetic relationships in *Pheloriniaceae* (*Basidiomycotina*) based on ITS rDNA sequence analysis. *Cryptog. Mycol.* 21: 3–12; **Martín, M.P.; Johannesson, H.** (2000). *Battarrea phalloides* and *B. stevensii*, insight into a long-standing taxonomic puzzle. *Mycotaxon* 76: 67–75; **Masuya, H.; Asai, I.** (2004). Phylogenetic position of *Battarrea japonica* (Kawam.) Otani. *Bull. natn. Sci. Mus. Tokyo*, B 30: 9–13; **Moncalvo, J.-M.; Vilgalys, R.; Redhead, S.A.; Johnson, J.E.; James, T.Y.; Aime, M.C.; Hofstetter, V.; Verduin, S.J.W.; Larsson, E.; Baroni, T.J.; Thorn, R.G.; Jacobsson, S.; Clemencon, H.; Miller, O.K.** (2002). One hundred and seventeen clades of euagarics. *Mol. Phylogen. Evol.* 23: 357–400; **Wright, J.E.** (1987). The genus *Tulostoma* (*Gasteromycetes*) – a world monograph. *Biblthca Mycol.* 113: 338 pp.

Typhulaceae Jülich 1982
Agaricales: Basidiomycota

Basidiomata small, clavarioid, often arising from a sclerotium, filiform or cylindrical to clavate, with a differentiated head and stipe, often proliferating, smooth, pale and fleshy or cartilaginous. Hyphal system monomitic, clamp connections present, cystidia absent. Hymenium smooth, covering the upper part of the basidioma. Basidia clavate, short or elongate, with 2–8 well-developed sterigmata. Basidiospores usually ± reniform, smooth, hyaline and thin-walled.

Significant Genera: *Typhula*.

Distribution: Widespread, especially in north temperate zones.

Economic Significance: Some species of *Typhula* may cause extensive damage to turfgrass, especially following snow cover.

Ecology: Some species are saprobes on wood and rotting vegetation, others are pathogenic on roots and stems of grasses.

Notes: Preliminary molecular data suggest a link with the *Clavariaceae*, but further research is needed. *Typhula* itself is in need of revision.

Typhula erythropus; basidiomata

Typhula setipes; basidiomata

References: **Hsiang, T.; Matsumoto, N.; Millett, S.M.** (1999). Biology and management of *Typhula* snow moulds of turfgrass. *Pl. Dis.* 83: 788–798; **Hsiang, T.; Wu, C.R.** (2000). Genetic relationships of pathogenic *Typhula* species assessed by RAPD, ITS-RFLP and ITS sequencing. *Mycol. Res.* 104: 16–22; **Larsson, K.-H.; Larsson, E.; Koljalg, U.** (2004). High phylogenetic diversity among corticioid homobasidiomycetes. *Mycol. Res.* 108: 983–1002; **Matsumoto, N.; Tronsmo, A.M.; Shimanuki, T.** (1996). Genetic and biological characteristics of *Typhula ishikariensis* isolates from Norway. *Eur. J. Pl. Path.* 102: 431–439; **Metzler, B.** (1988). The conidia of *Typhula incarnata*. 2. Their function as spermatia. *Can. J. Bot.* 66: 1321–1324; **Pine, E.M.; Hibbett, D.S.; Donoghue, M.J.** (1999). Phylogenetic relationships of cantharelloid and clavarioid *Homobasidiomycetes* based on mitochondrial and nuclear rDNA sequences. *Mycologia* 91: 944–963; **Vergara, G.V.; Bughrara, S.S.; Jung, G.** (2004). Genetic variability of grey snow mould (*Typhula incarnata*). *Mycol. Res.* 108: 1283–1290; **Villarreal, L.; Pérez-Moreno, J.** (1991). The clavarioid fungi from Mexico, I.; Addition of the genera *Macrotyphula* and *Typhula*. *Micol. Neotrop. Aplic.* 4: 119–126; **Willetts, H.J.; Bullock, S.; Begg, E.; Matsumoto, N.** (1990). The structure and histochemistry of sclerotia of *Typhula incarnata*. *Can. J. Bot.* 68: 2083–2091.

Uleiellaceae Vánky 2001
Ustilaginales: Basidiomycota

Sori superficial in inflorescences, forming dark brown powdery masses of spores. Hyphae intracellular, covered by an electron-opaque matrix, septa lacking pores. Teliospores formed in clusters (apparently not true spore balls), with a thick reticulate outer layer, smooth, yellowish brown, germinating to form short septate branched basidia. Basidiospores formed successively at the apices of basidia, ovoid, hyaline, thin-walled.

Significant Genera: *Ulea*.

Distribution: Known from Brazil and Chile.

Economic Significance: None is known.

Ecology: Biotrophic in inflorescences of *Araucaria* species, causing significant disruption to organ development.

Notes: Affinities of this family are unclear and interpretation of morphological structures is difficult. The type genus *Uleiella* is a superfluous name for *Ulea*, published two years previously by the same author. The latter genus name has been considered a 'nomen subnudum', but conservation of *Uleiella* would be required to override its use.

References: **Butin, H.; Peredo, H.L.** (1986). Hongos parásitos en coníferas de América del Sur con especial referencia a Chile. *Biblthca Mycol.* 101: 100 pp.; **Vánky, K.** (1998). A survey of the spore-ball-forming smut fungi. *Mycol. Res.* 102: 513–526; **Vánky, K.** (2001). The emended *Ustilaginaceae* of the modern classificatory system for smut fungi. *Fungal Diversity* 6: 131–147.

Umbelopsidaceae W. Gams & W. Mey. 2003
Mucorales: Zygomycota

Vegetative thallus a mycelium, complex fruit bodies absent. Sporophores (sporangiophores) arising from submerged mycelium, densely cymose or verticillately branched, terminating in a sporangium, stolons or rhizoids not produced. Sporangia single-spored or multispored, globose or elongate, typically with columella small and rudimentary. Sporangiospores globose to ellipsoid, rounded or angular, rarely narrowly clavate and appearing appendaged, smooth, lightly pigmented. Zygospores unknown.

Significant Genera: *Umbelopsis*.

Distribution: Widespread.

Economic Significance: None is known.

Ecology: Essentially soil fungi.

Notes: This well-defined monogeneric family appears to occupy a basal position within the *Mucorales*; the species were originally placed in *Mortierella* (*Mortierellaceae*: *Mortierellales*).

References: **Kwanśa, H.; Ward, E.; Bateman, G.L.** (2006). Phylogenetic relationships among *Zygomycetes* from soil based on ITS1/2 rDNA sequences. *Mycol. Res.* 110: 501–510; **Mahoney, D.; Gams, W.; Meyer, W.; Starink-Willemse, M.** (2004). *Umbelopsis dimorpha* sp. nov., a link between *U. vinacea* and *U. versiformis*. *Mycol. Res.* 108: 107–111; **Meyer, W.; Gams, W.** (2003). Delimitation of *Umbelopsis* (*Mucorales*, *Umbelopsidaceae* fam. nov.) based on ITS sequence and RFLP data. *Mycol. Res.* 107: 339–350; **O'Donnell, K.; Lutzoni, F.M.; Ward, T.J.; Benny, G.L.** (2001). Evolutionary relationships among mucoralean fungi (*Zygomycota*): evidence for family polyphyly on a large scale. *Mycologia* 93: 286–296; **Sugiyama, M.; Tokumasu, S.; Gams, W.** (2003). *Umbelopsis gibberispora* sp. nov. from Japanese leaf litter and a clarification of *Micromucor ramannianus* var. *angulisporus*. *Mycoscience* 44: 217–226; **Tanabe, Y.; Saikawa, M.; Watanabe, M.M.; Sugiyama, J.** (2004). Molecular phylogeny of *Zygomycota* based on EF-1α and RPB1 sequences: limitations and utility of alternative markers to rDNA. *Mol. Phylogen. Evol.* 30: 438–449.

Umbilicariaceae Chevall. 1826
Umbilicariales: Ascomycota

Thallus large, foliose, usually lobed, greyish to black, leathery when hydrated, often warted, pustulate or areolate, usually with a central attachment, rhizoids often present. Ascomata apothecial, sessile or slightly stipitate, often convoluted, black, flat or domed, without well-developed marginal tissue. Interascal tissue of sparingly branched paraphyses, sometimes swollen at the apices. Asci cylindric-clavate, thick-walled, with a relatively small J+ apical cap and an outer layer of J+ gelatinized material, 1- to 8-spored. Ascospores hyaline to brown, variously septate. Anamorphs coelomycetous, pycnidial, sometimes multiloculate. Conidiogenous cells proliferating percurrently. Conidia catenate, small, cylindrical; thalloconidia often produced.

Significant Genera: *Lasallia*, *Umbilicaria*.

Distribution: Known especially from temperate, boreal and austral regions, but also in montane parts of warmer climatic zones.

Economic Significance: Little is known, although the thalli of some species are used as food in arctic regions and they have also been used as dye sources.

Lasallia rubiginosa, South Africa; thallus and ascomata

Lasallia pennsylvanica, Alberta, Canada; thallus and ascomata

Lasallia pustulata, Salamanca, Spain; thallus

Ecology: Lichenized with green algae, on rocks in maritime and upland regions, sometimes associated with nutrient-enriched habitats.

Notes: The *Umbilicariaceae* occupies an isolated position within the *Lecanoromycetes*, with no consistent relatives identifiable from phylogenetic studies. Species are often conspicuous members of lichen communities in boreal and austral areas.

Umbilicaria aff. *proboscidea*, Bhutan; thallus with gyrose ascomata

References: **Ascaso, C.; Sancho, L.G.; Valladares, F.** (1992). Fine structure of the thalloconidia of the lichen genus *Umbilicaria*. *Cryptog. Bryol.-Lichénol.* 13: 335–340; **Codogno, M.** (1995). The problem of geographic disjunctions in the *Umbilicariaceae* (lichens). *Nova Hedwigia* 60: 479–486; **Fahselt, D.; Alstrup, V.; Tavares, S.** (1995). Enzyme polymorphism in *Umbilicaria cylindrica* in northwestern Greenland. *Bryologist* 98: 118–122; **Fahselt, D.; Hageman, C.** (1994). Rhizine and upper thallus isozymes in umbilicate lichens. *Symbiosis* 16: 95–103; **Hageman, C.; Fahselt, D.** (1992). Relationships within the lichen family *Umbilicariaceae* based on enzyme electromorph data. *Lichenologist* 24: 91–100; **Hestmark, G.** (1990). Thalloconidia in the genus *Umbilicaria*. *Nordic Jl Bot.* 9: 547–574; **Hestmark, G.** (1992). Conidiogenesis in five species of *Umbilicaria*. *Mycol. Res.* 96: 1033–1043; **Hestmark, G.** (1997). Species diversity and reproductive strategies in the family *Umbilicariaceae* on high equatorial mountains – with remarks on global patterns. *Biblthca Lichenol.* 68: 195–202; **Hibbett et al.** (2007). A higher-level phylogenetic classification of the *Fungi*. *Mycol. Res.* 111(5): 509–548; **Ivanova, N.V.; DePriest, P.T.; Bobrova, V.K.; Troitsky, A.V.** (1999). Phylogenetic analysis of the lichen family *Umbilicariaceae* based on nuclear ITS1 and ITS2 rDNA sequences. *Lichenologist* 31: 477–489; **Narui, T.; Sawada, K.; Culberson, C.F.; Culberson, W.L.; Shibata, S.** (1999). Pustulan-type polysaccharides as a constant character of the *Umbilicariaceae* (lichenized *Ascomycotina*). *Bryologist* 102: 80–85; **Ott, S.; Brinkmann, M.; Wirtz, N.; Lumbsch, H.T.** (2004). Mitochondrial and nuclear ribosomal DNA data do not support the separation of the Antarctic lichens *Umbilicaria kappenii* and *Umbilicaria antarctica* as distinct species. *Lichenologist* 36: 227–234; **Reeb, V.; Lutzoni, F.; Roux, C.** (2004). Contribution of *RPB2* to multilocus phylogenetic studies of the euascomycetes (*Pezizomycotina*, *Fungi*) with special emphasis on the lichen-forming *Acarosporaceae* and evolution of polyspory. *Mol. Phylogen. Evol.* 32: 1036–1060; **Romeike, J.; Friedl, T.; Helms, G.; Ott, S.** (2002). Genetic diversity of algal and fungal partners in four species of *Umbilicaria* (lichenized *Ascomycetes*) along a transect of the Antarctic peninsula. *Mol. Biol. Evol.* 19: 1209–1217; **Sancho, L.G.** (1999). Clave para la determinación de la especies de la familia *Umbilicariaceae*. *Clementeana, Bol. Soc. Españ. Liquenol.* (SEL) 4: 42–47; **Wedin, M.; Wiklund, E.; Crewe, A.; Döring, H.; Ekman, S.; Nyberg, A.; Schmitt, I.; Lumbsch, H.T.** (2005). Phylogenetic relationships of *Lecanoromycetes* (*Ascomycota*) as revealed by sequences of mtSSU and nLSU rDNA sequence data. *Mycol. Res.* 109: 159–172.

Uncolaceae Buriticá 2000
Uredinales: Basidiomycota

Spermogonia ampulliform, without discrete cell walls, subepidermal. Aecia *Calidion*-like, peridium poorly developed, with the fertile cells surrounded by thick-walled incurving paraphysis-like hyphae, developing within stomata. Aeciospores echinulate, formed singly on short pedicels. Uredinia unknown. Telia developing from stomata, hyaline or yellowish. Teliospores unicellular, produced in short chains, becoming cuboid, hyaline, thick-walled, especially at the apex, with a single apical germ pore, germination unknown.

Significant Genera: *Uncol*.

Distribution: Only known from Colombia.

Economic Significance: None is known.

Ecology: Biotrophic on fern fronds (*Thelypteris* species).

Notes: Affinities are uncertain and no molecular data are available. Development within stomata is unusual, but may not have major evolutionary significance.

References: **Cummins, G.B.; Hiratsuka, Y.** (2003). *Illustrated Genera of Rust Fungi* (St Paul): 225 pp.

Urocystidaceae Begerow, R. Bauer & Oberw. 1998
Urocystidales: Basidiomycota

Sori usually in leaves and stems, often forming swellings or galls, containing a usually powdery mass of individual spores or more typically aggregated spore balls. Hyphae with simple septal pores with two membrane caps and two non-membranous inner plates closing the pore. Spore balls persistent, containing one or more fertile spores along with smaller and paler sterile cells. Teliospores subglobose to ovoid or elongate, often angular due to compression, reddish-brown or olivaceous, rarely septate, usually ± smooth, germinating to form a normally aseptate cylindrical basidium with an apical cluster of elongate or filiform basidiospores that fuse in pairs to form infection hyphae.

Significant Genera: *Urocystis*.

Distribution: Cosmopolitan.

Economic Significance: *Urocystis* species cause diseases of many economically important plants including wheat, rye, onion and ornamental flowers.

Ecology: Biotrophic in leaves and stems, more rarely inflorescences, roots etc. from a wide range of plant families.

Notes: The *Doassansiopsidaceae* is a sister group with weakly pigmented teliospores and associations with aquatic plants.

Tuburcinia leimbachii; sorus in stem of *Adonis*

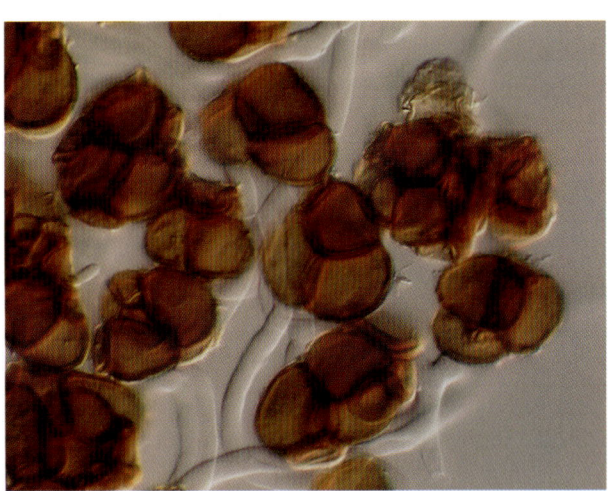
Urocystis irregularis, Norway; teliospores and paraphyses

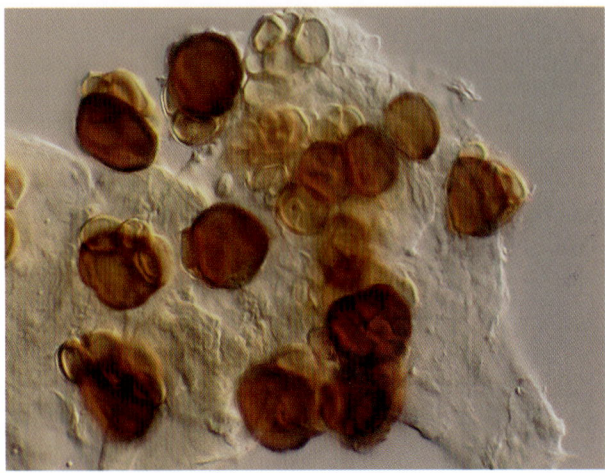
Urocystis anemones, Greece; teliospores

References: Bauer, R.; Oberwinkler, F.; Vánky, K. (1997). Ultrastructural markers and systematics in smut fungi and allied taxa. *Can. J. Bot.* 75: 1273–1314; **Begerow, D.; Bauer, R.; Oberwinkler, F.** (1998). Phylogenetic studies on nuclear large subunit ribosomal DNA sequences of smut fungi and related taxa. *Can. J. Bot.* 75: 2045–2056; **Denchev, C.M.** (2003). *Melanustilospora*, a new genus in the *Urocystales* (smut fungi). *Mycotaxon* 87: 475–477; **Ershad, D.** (2000). *Vankya*, a new genus of smut fungi. *Rostaniha* 1: 151–161; **Piepenbring, M.; Bauer, R.; Oberwinkler, F.** (1998). Teliospores of smut fungi. Teliospore connections, appendages, and germ pores studied by electron microscopy; phylogenetic discussion of characteristics of teliospores. *Protoplasma* 204: 202–218; **Vánky, K.** (1990). The genus *Mundkurella* (*Ustilaginales*). *Mycol. Res.* 94: 269–273; **Vánky, K.** (2001). The new classification of the smut fungi, exemplified by Australasian taxa. *Aust. Syst. Bot.* 14: 385–394; **Vánky, K.** (2004). Biodiversity and conservation of smut fungi (*Ustilaginomycetes p.p., Microbotryales*) reflected in Vánky, *Ustilaginales* exsiccata no. 1-1250. *Mycol. Balcanica* 1: 175–187.

Uropyxidaceae Cummins & Y. Hirats. 1983

Uredinales: Basidiomycota

Spermogonia conical or discoid, with discrete cellular walls, subepidermal or subcuticular. Aecia usually *Uredo*-like, with or without a clearly defined peridium, sometimes accompanied by paraphysis-like hyphae. Aeciospores echinulate or verrucose, borne single on pedicels or in chains. Uredinia *Malupa*-like, *Calidion*-like or *Uredo*-like, with or without paraphyses, rarely suprastomatal. Urediniospores borne singly on pedicels, usually echinulate, with scattered germ pores. Telia with or without paraphyses, subepidermal or suprastomatal. Teliospores usually with hygroscopic pedicels, very varied in morphology, 2-celled or multi-celled, the septa transverse, oblique or vertical, smooth, ornamented or appendaged, germinating from various parts of the teliospore to form external basidia.

Significant Genera: *Prospodium, Tranzschelia, Uropyxis*.

Distribution: Cosmopolitan, more diverse in warm regions.

Economic Significance: *Tranzschelia* species cause damage to rosaceous trees including plum, peach, apricot and almond. *Prospodium tuberculatum* is used in biocontrol of the invasive plant *Lantana camara* and *P. bicolor* is a pathogen in *Tabebuia* plantings.

Ecology: Biotrophic and autoecious or heteroecious, associated with many plant families.

Notes: A rather diverse family in morphological terms, requiring redefinition with the aid of molecular data. One preliminary study indicates an affinity with the *Chaconiaceae* and in another it clusters with the *Phakopsoraceae*.

Tranzschelia anemones; colonies on *Anemone nemorosa*

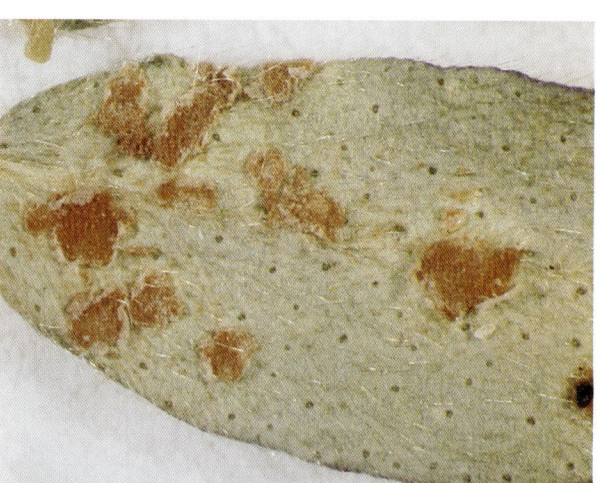
Uropyxis daleae, Mexico; uredinia

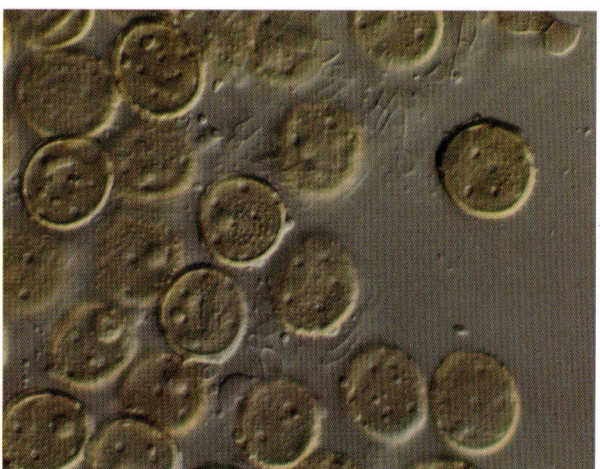
Uropyxis daleae, Mexico; urediniospores

References: Aime, M.C. (2006). Towards resolving family-level relationships in rust fungi (*Uredinales*). *Mycoscience* 47: 112–122; **Barreto, R.W.; Evans, H.C.; Ellison, C.A.** (1995). The mycobiota of the weed *Lantana camara* in Brazil, with particular reference to biological control. *Mycol. Res.* 99: 769–782; **Cummins, G.B.; Hiratsuka, Y.** (2003). *Illustrated Genera of Rust Fungi* (St Paul): 225 pp.; **Ferreira, F.A.; Hennen, J.F.** (1986). The life cycle, pathology, and taxonomy of the rust, *Prospodium bicolor* sp. nov., on yellow ipê, *Tabebuia serratifolia*, in Brazil. *Mycologia* 78: 795–803; **Hennen, J.F.; Sotão, H.M.P.** (1996). New species of *Uredinales* on *Bignoniaceae* from Brazil. *Sida* 17: 173–184; **López-Franco, R.M.; Hennen, J.F.** (1990). The genus *Tranzschelia* (*Uredinales*) in the Americas. *Syst. Bot.* 15: 560–591; **Maier, W.; Begerow, D.; Weiss, M.; Oberwinkler, F.** (2003). Phylogeny of the rust fungi: an approach using nuclear large subunit ribosomal DNA sequences. *Can. J. Bot.* 81: 12–23; **Ono, Y.** (1994). *Tranzschelia asiatica* sp. nov. and its taxonomic relationship to *Tranzschelia arthurii*. *Can. J. Bot.* 72: 1178–1186; **Wingfield, B.D.; Ericson, L.; Szaro, T.; Burdon, J.J.** (2004). Phylogenetic patterns in the *Uredinales*. *Australas. Pl. Path.* 33: 327–335.

Ustilaginaceae Tul. & C. Tul. 1847
Ustilaginales: Basidiomycota

Sori in inflorescences, leaves and stems, often bursting open at maturity to expose a dark brown powdery mass of spores, sometimes with a central columella and rarely also elater-like sterile hyphae. Hyphae intracellular, coated by an electron-opaque matrix, septa without pores. Teliospores formed singly, rarely in pairs, sometimes accompanied by sterile cells, ± globose, brown or olivaceous, usually ornamented (typically verrucose, echinulate or reticulate), germinating to produce hypha-like transversely septate basidia. Basidiospores formed laterally and terminally, elongate, hyaline, thin-walled.

Significant Genera: *Sporisorium*, *Ustilago*.

Ustilago maydis; sori in inflorescence of *Zea mays*

Distribution: Cosmopolitan.

Economic Significance: A number of species are important cereal pathogens, including *Sporisorium scitamineum* on sugarcane, *S. sorghi* on sorghum, *Ustilago avenae* on oats, *U. hordei* on barley, *U. krameri* on millet and *U. maydis* on maize. The latter species forms large galls in host tissue that are prized for food in the Neotropics; *Yenia esculenta* (on *Zizania*) is similarly eaten in the Orient.

Ecology: Biotrophic in living tissues of *Gramineae* and *Cyperaceae*.

Notes: Both of the principal genera in this family have been shown to be polyphyletic and further work on generic limits is needed. At least most of the members of this family reported on hosts other than grasses and sedges are likely to belong elsewhere.

Ustilago avenae, Germany; infected and uninfected inflorescences of *Arrhenatherum elatius*

Ustilago crus-galli, India; teliospores

References: **Austin, R.; Provart, N.J.; Sacadura, N.T.; Nugent, K.G.; Babu, M.; Saville, B.J.** (2004). A comparative genomic analysis of ESTs from *Ustilago maydis*. *Functional & Integrative Genomics* (Heidelberg) 4: 207–218; **Bauer, R.; Oberwinkler, F.; Vánky, K.** (1997). Ultrastructural markers and systematics in smut fungi and allied taxa. *Can. J. Bot.* 75: 1273–1314; **Begerow, D.; Bauer, R.; Oberwinkler, F.** (1998). Phylogenetic studies on nuclear large subunit ribosomal DNA sequences of smut fungi and related taxa. *Can. J. Bot.* 75: 2045–2056; **Jackson, A.P.** (2004). A reconciliation analysis of host switching in plant-fungal symbiosis. *Evolution, Lancaster, Pa.* 58: 1909–1923; **Menzies, J.G.; Bakkeren, G.; Matheson, F.; Procunier, J.D.; Woods, S.** (2003). Use of inter-simple sequence repeats and amplified fragment length polymorphisms to analyze genetic relationships among small grain-infecting species of *Ustilago*. *Phytopathology* 93: 167–175; **Piepenbring, M.; Stoll, M.; Oberwinkler, F.** (2002). The generic position of *Ustilago maydis*, *Ustilago scitaminea*, and *Ustilago esculenta* (*Ustilaginales*). *Mycol. Prog.* 1: 71–80; **Stoll, M.; Begerow, D.; Oberwinkler, F.** (2005). Molecular phylogeny of *Ustilago*, *Sporisorium*, and related taxa based on combined analyses of rDNA sequences. *Mycol. Res.* 109: 342–356; **Stoll, M.; Piepenbring, M.; Begerow, D.; Oberwinkler, F.** (2003). Molecular phylogeny of *Ustilago* and *Sporisorium* species (*Basidiomycota*, *Ustilaginales*) based on internal transcribed spacer (ITS) sequences. *Can. J. Bot.* 81: 976–984; **Vánky, K.** (2001). The new classification of the smut fungi, exemplified by Australasian taxa. *Aust. Syst. Bot.* 14: 385–394; **Vánky, K.** (2004). Biodiversity and conservation of smut fungi (*Ustilaginomycetes p.p.*, *Microbotryales*) reflected in Vánky, *Ustilaginales* exsiccata no. 1-1250. *Mycol. Balcanica* 1: 175–187.

Ustilentylomataceae R. Bauer & Oberw. 1997
Microbotryales: Basidiomycota

Sori forming in leaves or inflorescences, resulting in galls or scattered yellowish or brown spots, wall and columella absent. Mycelial septal pores simple, lacking caps; hyphae intercellular, haustoria lacking. Teliospores embedded in host tissue, single or aggregated into loose or tightly adhering balls, hyaline or yellowish to reddish brown, globose or angular due to compression, smooth, germinating to form a long 4-celled basidium (promycelium). Basidiospores produced laterally, sessile or on very short sterigmata, globose to ovoid, hyaline, thin-walled, smooth.

Significant Genera: *Ustilentyloma*.

Distribution: Known primarily from north temperate regions.

Economic Significance: None is known.

Ecology: Biotrophic and gall-forming on monocotyledonous plants.

Notes: The boundary between this family and the *Microbotryaceae* is not entirely clear; *Bauerago* may belong there rather than in the *Ustilentylomataceae*.

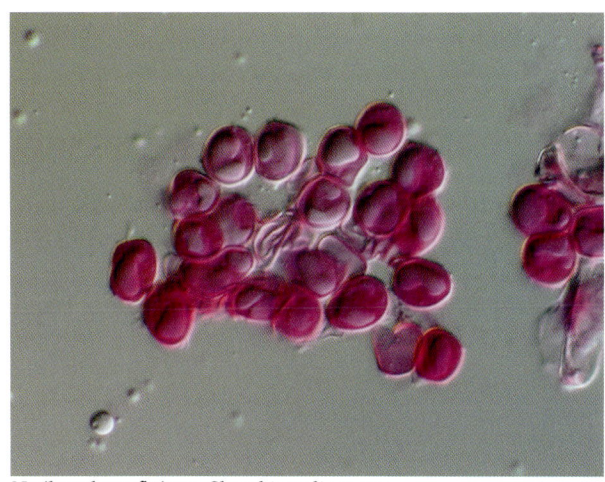

Ustilentyloma fluitans, Slovakia; teliospores

References: **Bauer, R.; Oberwinkler, F.; Vánky, K.** (1997). Ultrastructural markers and systematics in smut fungi and allied taxa. *Can. J. Bot.* 75: 1273–1314; **Begerow, D.; Bauer, R.; Oberwinkler, F.** (1998). Phylogenetic studies on nuclear large subunit ribosomal DNA sequences of smut fungi and related taxa. *Can. J. Bot.* 75: 2045–2056; **Denchev, C.M.** (1995). A comparative taxonomical investigation of *Ustilentyloma pleuropogonis* and *U. fluitans* (*Ustilaginales*). *Mycotaxon* 55: 243–254; **Kemler, M.; Goker, M.; Oberwinkler, F.; Begerow, D.** (2006). Implications of molecular characters for the phylogeny of the *Microbotryaceae* (*Basidiomycota*: *Urediniomycetes*). *BMC Evol. Biol.* 6: 35; **Piepenbring, M.** (2001). Smut fungi (*Ustilaginomycetes* and *Microbotryales*, *Basidiomycota*) in Panama. *Revta Biol. trop.* 49: 411–428; **Piepenbring, M.; Vánky, K.; Oberwinkler, F.** (1996). *Aurantiosporium*, a new genus for *Ustilago subnitens* (*Ustilaginales*). *Pl. Syst. Evol.* 199: 53–64; **Vánky, K.** (2001). The new classification of the smut fungi, exemplified by Australasian taxa. *Aust. Syst. Bot.* 14: 385–394; **Vánky, K.** (2004). Biodiversity and conservation of smut fungi (*Ustilaginomycetes p.p.*, *Microbotryales*) reflected in Vánky, *Ustilaginales* exsiccata no. 1-1250. *Mycol. Balcanica* 1: 175–187.

Valsaceae Tul. & C. Tul. 1861
Diaporthales: Ascomycota

Stromata varied in development, composed of a mixture of fungal and plant tissues, initially producing immersed stromatic conidiomata with ascomata following on the periphery. Ascomata perithecial, black, erect or oblique, often long-necked, the necks central or oblique, erumpent singly or converging through a stromatic disc. Asci numerous, clavate, with a single wall layer, the apex thickened and truncate with a large and conspicuous refractive J– ring, evanescent at the base, becoming free within the ascomatal cavity, 8- or poly-spored. Ascospores simple, ± hyaline, allantoid, without gelatinous sheath or appendages. Anamorphs coelomycetous. Conidiomata immersed in stromatic tissue, often convoluted. Conidiogenous cells lining the entire inner cavity, ± cylindrical, proliferating percurrently. Conidia usually small, hyaline, aseptate, allantoid.

Significant Genera: *Leucostoma*, *Valsa*. Anamorph: *Cytospora*.

Distribution: Widespread, especially in temperate regions.

Valsa ambiens, Norway; outer view of stromata

Economic Significance: Many species are parasitic on economically important plants, including important canker pathogens of fruit trees.

Ecology: Saprobes, endophytes and weak to virulent parasites of woody plants.

Notes: Following recent research, the *Valsaceae* is restricted to a small group of genera related to *Valsa* itself and the *Diaporthaceae* is now separated as an independent family. Traditional characters used to distinguish families of the *Diaporthales* are found to be unreliable based on evidence from molecular methods and there are a number of well-supported clades without formal names.

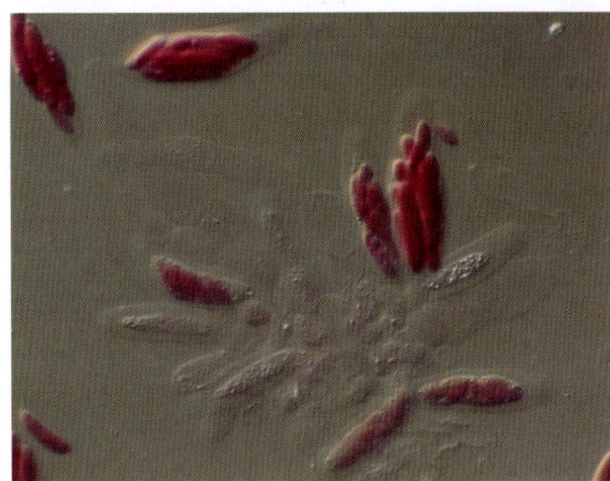

Valsa ambiens, Norway; asci and paraphyses

References: **Adams, G.C.; Surve-Iyer, R.S.; Iezzoni, A.F.** (2002). Ribosomal DNA sequence divergence and group I introns within the *Leucostoma* species *L. cinctum*, *L. persoonii*, and *L. parapersoonii* sp. nov., ascomycetes that cause cytospora canker of fruit trees. *Mycologia* 94: 947–967; **Adams, G.C.; Wingfield, M.J.; Common, R.; Roux, J.** (2005). Phylogenetic relationships and morphology of *Cytospora* species and related teleomorphs (*Ascomycota*, *Diaporthales*, *Valsaceae*) from *Eucalyptus*. *Stud. Mycol.* 52: 146 pp.; **Castlebury, L.A.; Rossman, A.Y.; Jaklitsch, W.J.; Vasilyeva, L.N.** (2002). A preliminary overview of the *Diaporthales* based on large subunit nuclear ribosomal DNA sequences. *Mycologia* 94: 1017–1031; **Gille, A.** (1990). Bestimmung der an Kirschgehölzen vorkommenden *Cytospora*-Arten. *Arch. phytopath. Pflanz.* 26: 237–245; **Spielman, L.J.** (1985). A monograph of *Valsa* on hardwoods in North America. *Can. J. Bot.* 63: 1355–1378; **Wang, D.C.; Iezzoni, A.; Adams, G.** (1998). Genetic heterogeneity of *Leucostoma* species in Michigan peach orchards. *Phytopathology* 88: 376–381.

Venturiaceae E. Müll. & Arx ex M.E. Barr 1979
Pleosporales: Ascomycota

Stromata variably developed, immersed or ± superficial, sometimes composed of a well-developed subiculum. Ascomata perithecial, aggregated on or in the stroma, becoming superficial, usually small, ± globose, often setose or hairy, opening by a well-defined lysigenous pore; peridium composed of ± small pseudoparenchymatous cells. Interascal tissue of narrow cellular pseudoparaphyses, sometimes evanescent. Asci ± cylindrical, fissitunicate, with a well-defined ocular chamber, not blueing in iodine, usually 8-spored. Ascospores variously pigmented, usually asymmetrical, 1-septate, sometimes with a gelatinous sheath. Anamorphs usually prominent, hyphomycetous. Conidiophores pigmented, formed singly or in clusters from vegetative cells, irregular in form, often branched. Conidiogenous cells, either terminal and sometimes intercalary, with multiple ± pigmented fertile loci, proliferating sympodially, or clavate to irregularly annellate and formed from a small basal stroma. Conidia irregular, aseptate or septate, brown, borne in either weakly branched chains or non-catenate.

Significant Genera: *Apiosporina, Gibbera, Venturia*. Anamorphs: *Fusicladium, Spilocaea*.

Distribution: Widespread, especially in temperate regions.

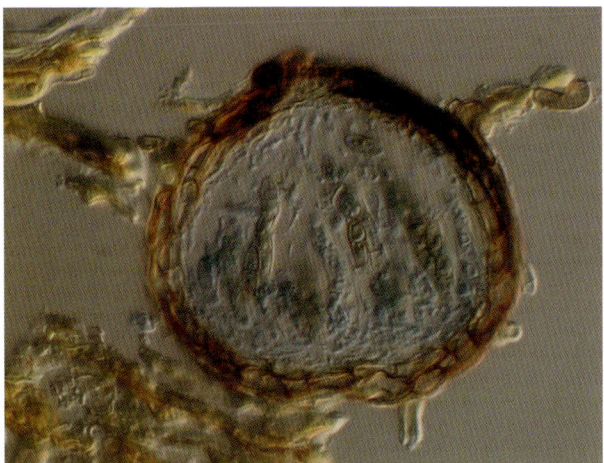

Venturia ribis, Finland; vertical section through ascoma

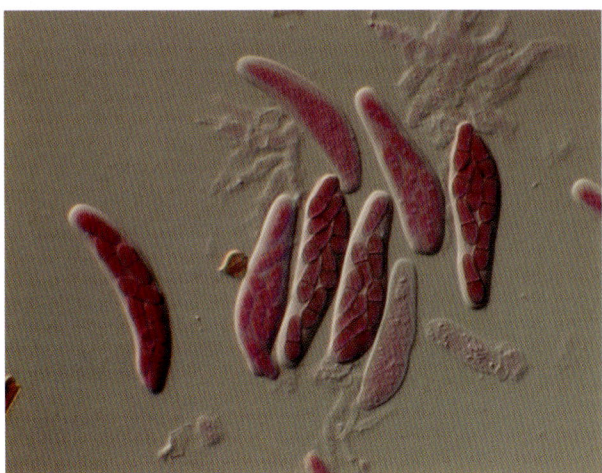

Venturia ribis, Finland; asci and ascospores

Economic Significance: Some are significant plant pathogens, causing a range of symptoms including scab, canker, premature leaf fall and witches' broom.

Ecology: Mostly endophytes and weak parasites of aerial plant parts, including bark and wood, leaf and fruit tissue; rarely biotrophic. Little is known except for the important pathogens.

Notes: The family is poorly known and more molecular research is needed. Some species have *Cladosporium*-like anamorphs, but these have been demonstrated to be only distantly related to the main *Cladosporium* clade.

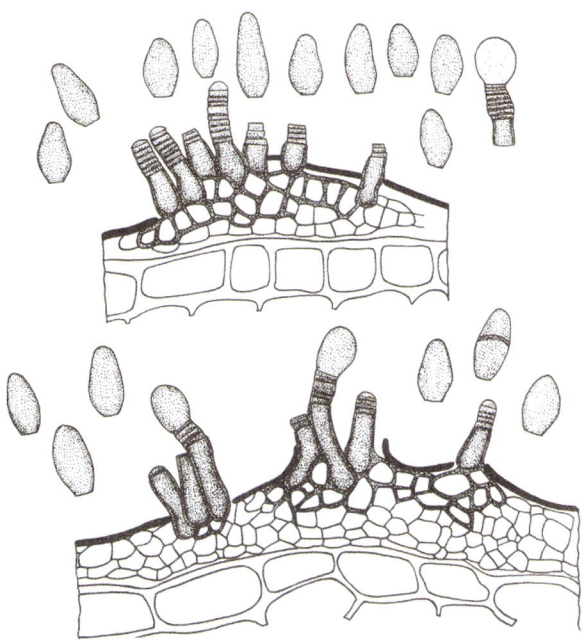

Spilocaea pyracanthae; conidiophores and conidia

References: **Barr, M.E.** (1989). The *Venturiaceae* in North America: revisions and additions. *Sydowia* 41: 25–40; **Braun, U.; Crous, P.W.; Dugan, F.; Hoog, G.S. de** (2003). Phylogeny and taxonomy of *Cladosporium*-like hyphomycetes, including *Davidiella* gen. nov., the teleomorph of *Cladosporium s. str.*. *Mycol. Prog.* 2: 3–18; **Carris, L.M.; Poole, A.P.** (1993). A new species of *Protoventuria* on leaves of *Vaccinium macrocarpon*. *Mycologia* 85: 93–99; **Hughes, S.J.; Arnold, G.** (1989). *Antennatula cubensis* sp. nov. and its *Hormisciomyces* synanamorph. *Mem. N. Y. bot. Gdn* 49: 198–201; **Hughes, S.J.; Seifert, K.A.** (1998). The hyphomycete genus *Heterosporiopsis* Petrak. *Sydowia* 50: 192–199; **Ishii, H.; Yanase, H.** (2000). *Venturia nashicola*, the scab fungus of Japanese and Chinese pears: a species distinct from *V. pirina*. *Mycol. Res.* 104: 755–759; **Newcombe, G.** (2003). Native *Venturia inopina* sp. nov., specific to *Populus trichocarpa* and its hybrids. *Mycol. Res.* 107: 108–116; **Pascoe, I.G.; Sutton, B.C.** (1990). *Protoventuria parahebicola* sp. nov. (*Venturiaceae*), the teleomorph of *Fusicladium veronicae* on *Parahebe perfoliata*. *Aust. Syst. Bot.* 3: 281–285; **Samuels, G.J.; Barr, M.E.; Rogerson, C.T.** (1988). *Xenomeris saccifolii* and *Gibbera sphyrospermi*, new tropical species of the *Venturiaceae* (Fungi, Pleosporales). *Brittonia* 40: 392–397; **Schnabel, G.; Schnabel, E.L.; Jones, A.L.** (1999). Characterization of ribosomal DNA from *Venturia inaequalis* and its phylogenetic relationship to rDNA from other tree-fruit *Venturia* species. *Phytopathology* 89: 100–108; **Sivanesan, A.** (1984). *Bitunicate Ascomycetes and their Anamorphs* (Vaduz): 700 pp.; **Tenzer, I.; Gessler, C.** (1999). Genetic diversity of *Venturia inaequalis* across Europe. *Eur. J. Pl. Path.* 105: 545–552; **Untereiner, W.A.; Naveau, F.A.** (1999). Molecular systematics of the *Herpotrichiellaceae* with an assessment of the phylogenetic positions of *Exophiala dermatitidis* and *Phialophora americana*. *Mycologia* 91: 67–83.

Verrucariaceae Zenker 1827
Verrucariales: Ascomycota

Thallus usually well-developed, variable in form including crustose, foliose and squamulose types, occasionally filmy or immersed within the substratum, greenish, brownish, grey or black. Ascomata perithecial, often clypeate with a well-developed involucrellum, sometimes strongly aggregated, sometimes thin-walled or with the apical part thicker, the ostiolar canal lines with periphyses. Interascal tissue of narrow gelatinized paraphysis-like hyphae, sometimes evanescent, without apical paraphyses. Asci cylindrical to clavate, with two distinct wall layers but dehiscence sometimes obscure and sometimes evanescent, the apical region sometimes blueing in iodine, usually 8-spored. Ascospores colourless to dark brown, variously septate, often with a gelatinous sheath. Anamorphs coelomycetous, pycnidial, immersed in the thallus. Conidiogenous cells short, proliferating percurrently. Conidia bacilliform.

Significant Genera: *Endocarpon*, *Polyblastia*, *Staurothele*, *Verrucaria*.

Verrucaria maura, Barra, Scotland; thallus and ascomata

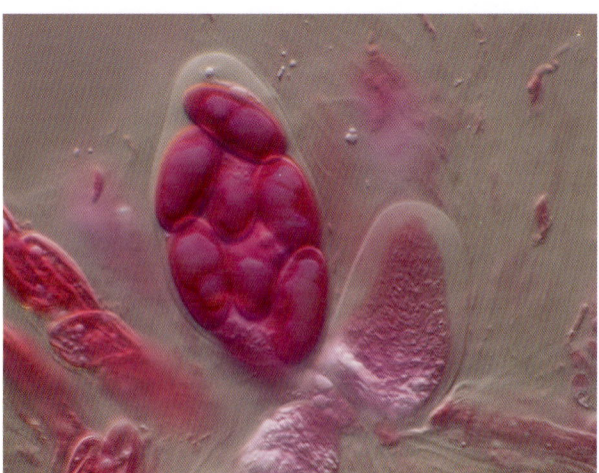
Verrucaria aethiobola, Ireland; asci and ascospores

Distribution: Cosmopolitan, prominent especially in temperate zones.

Economic Significance: None is known.

Ecology: Lichenized with green algae, on rocks, soil etc., more rarely on wood and bark.

Notes: The family is poorly researched and generic concepts need revision. *Verrucaria* species with black thalli form a prominent part of the temperate maritime lichen community.

Dermatocarpon miniatum, Arizona, USA; thalli

Normandina pulchella; thalli

References: **Aptroot, A.** (1991). A conspectus of *Normandina* (*Verrucariaceae*, lichenized *Ascomycetes*). Willdenowia Beih. 21: 263–267; **Aptroot, A.** (1991). Tropical pyrenocarpous lichens. A phylogenetic approach. *Tropical Lichens: Their Systematics, Conservation, and Ecology* (Oxford) 43: 253–273; **Breuss, O.** (1996). Revision der Flechtengattung *Placidiopsis* (*Verrucariaceae*). Öst. Z. Pilzk. 5: 65–94; **Breuss, O.** (1998). On the taxonomy of 'Catapyrenium' plumbeum (lichenized *Ascomycetes*, *Verrucariaceae*). Annln naturh. Mus. Wien Ser. B, Bot. Zool. 100: 671–676; **Grube, M.** (1999). Epifluorescence studies of the ascus in *Verrucariales* (lichenized *Ascomycotina*). Nova Hedwigia 68: 241–249; **Harada, H.** (1996). Taxonomic notes on the lichen family *Verrucariaceae* in Japan (X). *Thelidium radiatum* Harada sp. nov. Hikobia 12: 133–136; **Heiðmarsson, S.** (2003). Molecular study of *Dermatocarpon miniatum* (*Verrucariales*) and allied taxa. Mycol. Res. 107: 459–468; **Thomson, J.W.** (1987). The lichen genera *Catapyrenium* and *Placidiopsis* in North America. Bryologist 90: 27–39; **Thomson, J.W.** (1991). The lichen genus *Staurothele* in North America. Bryologist 94: 351–367.

Vezdaeaceae Poelt & Vězda ex J.C. David & D. Hawksw. 1991
Lecanorales: Ascomycota

Thallus crustose, initially subcuticular. Ascomata apothecial, rounded or irregular, sessile or short-stalked, marginal tissue poorly developed, formed from paraphysis-like filaments. Interascal tissue of branched and anastomosing hyphae, sometimes entwining asci. Asci with a J+ apical cap and a well-developed more darkly staining tubular ring. Ascospores hyaline, sometimes septate, sometimes ornamented. Anamorph hyphomycetous. Conidia formed from thalline hyphae.

Significant Genera: *Vezdaea*.

Distribution: Most have European distributions, though *Vezdaea* species have also been reported from Australia.

Economic Significance: None is known.

Ecology: Lichenized with green algae (*Leptosira* species).

Notes: An apparently isolated family for which no molecular data are available.

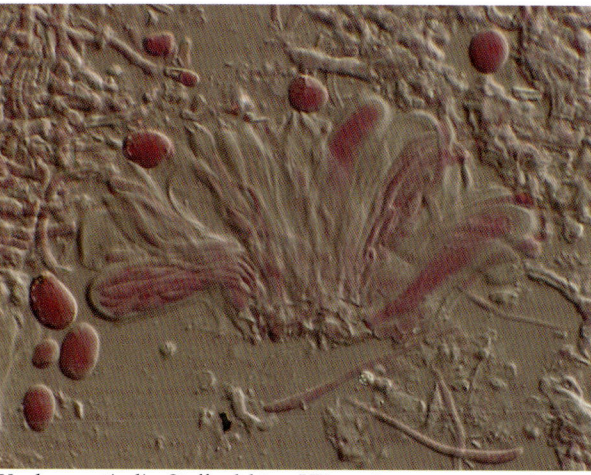
Vezdaea aestivalis, Staffordshire, UK; asci, ascospores and paraphyses

References: **Coppins, B.J.** (1987). The genus *Vezdaea* in the British Isles. *Lichenologist* 19: 167–176; **Giralt, M.; Poelt, J.; Suanjak, M.** (1993). Die Flechtengattung *Vezdaea* mit *V. cobria* spec. nov. *Herzogia* 9: 715–724; **Hawksworth, D.L.** (1988). The variety of fungal-algal symbioses, their evolutionary significance, and the nature of lichens. *J. Linn. Soc. Bot.* 96: 3–20; **Scheidegger, C.** (1995). Reproductive strategies in *Vezdaea* (*Lecanorales*, lichenized *Ascomycetes*): a low-temperature scanning electron microscopy study of a ruderal species. *Cryptog. bot.* 5: 163–171.

Vezdaea aestivalis; thallus and ascomata

Vezdaea dawsoniae; ascomata

Vialaeaceae P.F. Cannon 1995
Uncertain position within Sordariomycetidae, Ascomycota

Stromata absent or poorly developed. Ascomata perithecial, black, ± immersed, long-necked, the necks central, surrounded by a clypeus; peridium with occasional pores in the cell walls. Interascal tissue of short, wide, very thin-walled, paraphyses, evanescent at an early stage. Asci cylindrical, persistent, thin-walled, with a complex, partially J+, apical structure. Ascospores elongated, hyaline, strongly isthmoid, septate, without a sheath.

Significant Genera: *Vialaea*.

Distribution: Known from both temperate and tropical regions.

Economic Significance: One species is known to attack mango trees, but no significant economic damage is caused.

Ecology: Both known species are weak parasites of twigs.

Notes: No sequences are available for this family and its placement is, therefore, very uncertain. The isthmoid ascospores are diagnostic.

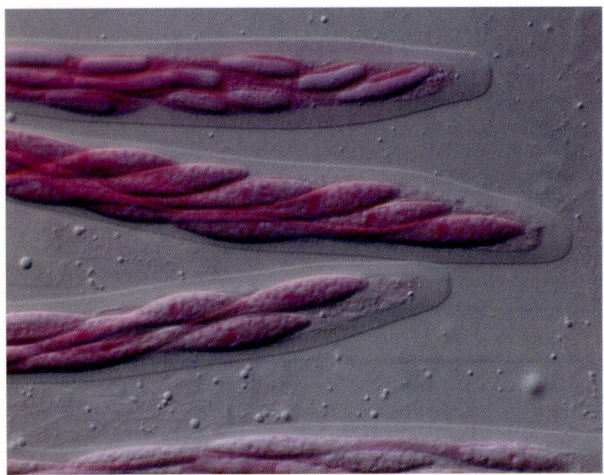

Vialaea insculpta, UK; asci and ascospores

References: **Cannon, P.F.** (1995). Studies on fungi with isthmoid ascospores: the genus *Vialaea*, with the description of the new family Vialaeaceae. *Mycol. Res.* 99: 367–373; **Redlin, S.C.** (1989). Observations of *Vialaea insculpta* (*Amphisphaeriaceae*). *Sydowia* 41: 296–307.

Vibrisseaceae Locq. 1972
Helotiales: Ascomycota

Stromata absent. Ascomata apothecial, flat or pulvinate, the disc pale or brightly coloured, sessile or long-stalked; excipulum composed of thin-walled angular cells, narrow, usually dark, without hairs. Interascal tissue of simple paraphyses, sometimes swollen at the apices. Asci elongated, with a thick J– apical cap penetrated by a narrow pore. Ascospores filiform, hyaline, multiseptate, fragmenting. Anamorphs hyphomycetous, aero-aquatic.

Significant Genera: *Vibrissea*.

Vibrissea guernisacii, UK; ascomata

Distribution: Cosmopolitan, best known from north temperate regions.

Economic Significance: None is known.

Ecology: Saprobic, usually on wood and bark immersed in fresh water. Some species may be root endophytes.

Notes: Preliminary studies indicate that the *Vibrisseaceae* might cluster within the much larger family Helotiaceae, but the latter assemblage is almost certainly polyphyletic itself and needs much more detailed molecular analysis. *Vibrissea* and its allies appears to be linked phylogenetically with *Mollisia* and *Tapesia* (traditionally placed in the *Dermateaceae*) but more research is needed to clarify the relationships.

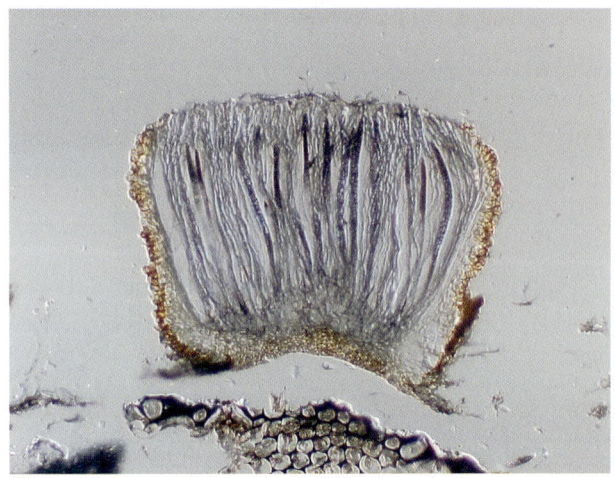

Vibrissea sporogyra, UK; vertical section through ascoma

References: **Baral, H.O.** (1987). Der Apikalapparat der *Helotiales*. Eine lichtmikroskopische Studie über Arten mit Amyloidring. *Z. Mykol.* 53: 119–136; **Hamad, S.R.; Webster, J.** (1988). *Anavirga dendromorpha*, anamorph of *Apostemidium torrenticola*. *Sydowia* 40: 60–64; **Iturriaga, T.** (1997). *Vibrissea pfisteri*, a new species with an unusual ecology. *Mycotaxon* 61: 215–221; **Kohn, L.M.** (1989). *Chlorovibrissea* (*Helotiales, Leotiaceae*), a new genus of austral discomycetes. *Mem. N. Y. bot. Gdn* 49: 112–118; **Wang, Z.; Binder, M.; Hibbett, D.S.** (2006). Life history and systematics of the aquatic discomycete *Mitrula* (*Helotiales, Ascomycota*) based on cultural, morphological and molecular studies. *Am. J. Bot.* 92: 1565–1574; **Wang, Z.; Binder, M.; Schoch, C.L.; Johnston, P.R.; Spatafora, J.W.; Hibbett, D.S.** (2006). Evolution of helotialean fungi (*Leotiomycetes, Pezizomycotina*): a nuclear rDNA phylogeny. *Mol. Phylogen. Evol.* 41: 295–312.

Vizellaceae H.J. Swart 1971
Uncertain position within Dothideomycetidae, Ascomycota

Mycelium superficial, the hyphae hyaline, cylindrical, with thickened brown septa. Ascomata strongly flattened, perithecial, intra- or subcuticular, the upper wall composed of a single layer of pseudoparenchymatous cells, without a clearly defined ostiole. Interascal tissue absent or poorly developed, sometimes with a central columella. Asci ovoid to saccate, probably evanescent. Ascospores brown, usually aseptate or with a small appendage-like cell, with a conspicuous hyaline band, without a sheath.

Significant Genera: *Vizella*.

Distribution: Tropical and subtropical regions on all continents.

Economic Significance: None is known.

Ecology: Apparently biotrophic on leaves, but nutritional status has not been intensively studied.

Notes: Affinities of this family are uncertain and no molecular data are available. The conspicuous hyaline bands (areas lacking pigmentation) around the ascospores probably function to aid germination but a number of unrelated families of the *Ascomycota* show these structures.

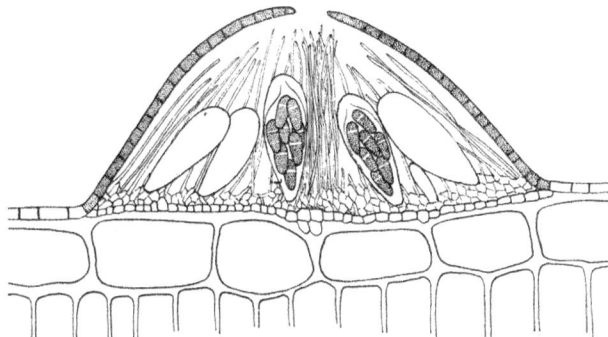

Entopeltis tetrorchidii; vertical section through ascoma

References: **Cannon, P.F.** (2001). *Asterina nyanzae; Capnodium salicinum; Chaetothyrium guaraniticum; Cucurbitaria laburni; Dothidea sambuci; Hysterographium fraxini; Micropeltis ugandae; Parodiella hedysari; Tubeufia cerea; Vizella gomphispora*. IMI Descr. Fungi Bact. 141: [22] pp.; **Farr, M.L.** (1987). Amazonian foliicolous fungi. IV. Some new and critical taxa in *Ascomycotina* and associated anamorphs. Mycologia 79: 97–116; **Gadgil, P.D.** (1995). Mycological records 4: *Vizella tunicata* sp. nov. N.Z. Jl For. Sci. 25: 107–110; **Sivanesan, A.; Sutton, B.C.** (1985). Microfungi on *Xanthorrhoea*. Trans. Br. mycol. Soc. 85: 239–255; **Taylor, J.E.; Crous, P.W.** (1998). *Coleroa senniana; Helicosingula leucadendri; Leptosphaeria protearum; Mycosphaerella jonkershoekensis, M. bellula; Teratosphaeria maculiformis, T. fibrillosa; Trimmatostroma macowanii; Vizella interrupta*. IMI Descr. Fungi Bact. 135: [26] pp..

Volvocisporiaceae Begerow, R. Bauer & Oberw. 2001
Microstromatales: Basidiomycota

Sori formed in substomatal cavities, wall structures not defined. Hyphae intercellular, with simple pores with ± rounded pore lips enclosed on both sides by membrance caps. Telia absent. Basidia derived from fertile hyphae or sac-like structures that are sometimes surrounded by sterile hyphal tissue, erumpent through host stomata, clavate or elongate, with a cluster of 6–8 sterigmata at the apex. Basidiospores discharged passively, hyaline, globose and muriform. Anamorph unknown.

Significant Genera: *Volvocisporium*.

Distribution: Only reported from India.

Economic Significance: None is known.

Ecology: Parasitic on leaves of *Triumfetta* (*Tiliaceae*).

Notes: The only species has unusual muriform basidia with an outer layer of cells surrounding an inner core. Recognition of the family implies acceptance that the *Microstromataceae* is paraphyletic according to some analyses.

References: **Beer, Z.W. de; Begerow, D.; Bauer, R.; Pegg, G.S.; Crous, P.W.; Wingfield, M.J.** (2006). Phylogeny of the *Quambalariaceae* fam. nov., including important *Eucalyptus* pathogens in South Africa and Australia. Stud. Mycol. 55: 289–298; **Begerow, D.; Bauer, R.; Oberwinkler, F.** (2001). *Muribasidiospora*: Microstromatales or Exobasidiales?. Mycol. Res. 105: 798–810.

Wallemiaceae R.T. Moore 1996
Wallemiales: Basidiomycota

Thallus, stromata and basidiomata absent. Hyphae with dolipore septa and vesiculate parenthosomes. Colonies small and brown, composed of clusters of subhyaline to pigmented fertile hyphae, becoming septate near the apex. These cells subdivide and fragment into four cylindrical daughter cells, which become rounded, thick-walled and verrucose or minutely spinose, adhering in short chains that disperse as a powdery mass. Chlamydospore-like structures known in one species.

Significant Genera: *Wallemia*.

Distribution: Widely distributed but sometimes overlooked as colonies tend to become overgrown on substrata with low water tension.

Economic Significance: *Wallemia* species are spoilage organisms occurring on dried fish and other desiccated products including dried fruits, jams, marzipan, rice and spices. Toxigenic strains are known, and subcutaneous infections and allergenic effects have been reported.

Ecology: Saprobic, capable of growth over wide ranges of water tension and resistant to acid conditions.

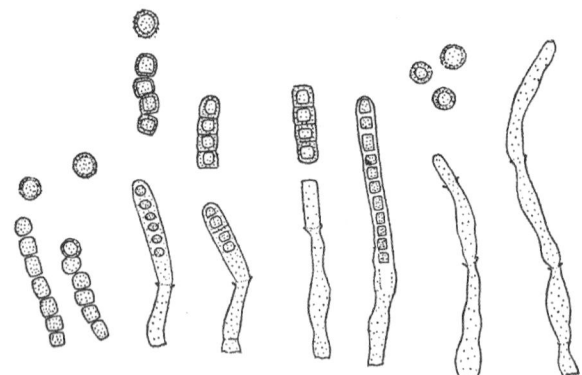

Wallemia sebi; fertile hyphae and chains of spores

Notes: There has been speculation that *Wallemia* spores are teleomorphic in origin and not conidia, but to date no definitive studies have been made. The family is extremely isolated in evolutionary terms and has no close relatives; the three species currently recognized may belong to at least two separate genera.

References: **Hashmi, M.H.; Morgan-Jones, G.** (1973). Conidium ontogeny in hyphomycetes. The meristem arthrospores of *Wallemia sebi*. *Can. J. Bot.* 51: 1669–1671; **Moore, R.T.** (1986). A note on *Wallemia sebi*. *Antonie van Leeuwenhoek* 52: 183–187; **Moore, R.T.** (1996). The dolipore/parenthosome septum in modern taxonomy. *Rhizoctonia Species, Taxonomy, Molecular Biology, Ecology, Pathology and Disease Control* (Dordrecht): 13–35; **Zalar, P.; Hoog, G.S. de; Schroers, H.-J.; Frank, J.M.; Gunde-Cimerman, N.** (2005). Taxonomy and phylogeny of the xerophilic genus *Wallemia* (*Wallemiomycetes* and *Wallemiales*, cl. Et ord. nov.). *Antonie van Leeuwenhoek* 87: 311–328.

Websdaneaceae Vánky 2001
Ustilaginales: Basidiomycota

Sori in stems and leaves causing hypertrophy, erumpent to expose a black powdery mass of loosely adhering spore balls. Hyphae intracellular; peridium, columella and sterile cells lacking. Spore balls subglobose to irregularly elongate. Teliospores varied in size and shape, often polyhedral and/or with irregular acute processes, olivaceous, the wall uneven in thickness, faintly verrucose, germinating to form basidia. Basidia filiform, (1-) 2- to 4-celled, with basidiospores formed successively on well-developed sterigmata. Basidiospores fusiform or ovoid, thin-walled, hyaline, producing secondary spores by budding.

Significant Genera: *Websdanea*.

Distribution: Only known from Western Australia.

Economic Significance: None is known.

Ecology: Biotrophic in green tissues of *Lyginia* (*Restionaceae*), causing sterility of affected plants.

Notes: Preliminary molecular data suggests that the family is isolated within the *Ustilaginales*, but its affinities need further research. Only one species has been described.

References: **Piepenbring, M.; Begerow, D.; Oberwinkler, F.** (1999). Molecular sequence data assess the value of morphological characteristics for a phylogenetic classification of species of *Cintractia*. *Mycologia* 91: 485–498; **Vánky, K.** (1997). *Websdanea*, a new genus of smut fungi. *Mycotaxon* 65: 183–190; **Vánky, K.** (2001). The new classification of the smut fungi, exemplified by Australasian taxa. *Aust. Syst. Bot.* 14: 385–394; **Vánky, K.** (2001). The emended *Ustilaginaceae* of the modern classificatory system for smut fungi. *Fungal Diversity* 6: 131–147; **Vánky, K.** (2004). Biodiversity and conservation of smut fungi (*Ustilaginomycetes* p.p., *Microbotryales*) reflected in Vánky, *Ustilaginales* exsiccata no. 1-1250. *Mycol. Balcanica* 1: 175–187.

Xanthopyreniaceae Zahlbr. 1926
Pyrenulales: Ascomycota

Thallus crustose or absent (frequently lichenicolous). Ascomata solitary, perithecial, globose, wall pale to usually dark, most strongly pigmented in the cell walls, ostiole not periphysate. Hymenial gel not blueing in iodine. Interascal tissue of anastomosing trabeculate pseudoparaphyses. Asci persistent, with two functional wall layers, fissitunicate, usually subcylindrical, not blueing in iodine, often with a small ocular chamber. Ascospores hyaline or becoming brownish, simple to usually transversely 1-septate, sometimes becoming roughened with age, often with a thin mucous sheath. Anamorphs coelomycetous, pycnidial. Conidia ellipsoidal.

Significant Genera: *Collemopsidium*, *Pyrenocollema*.

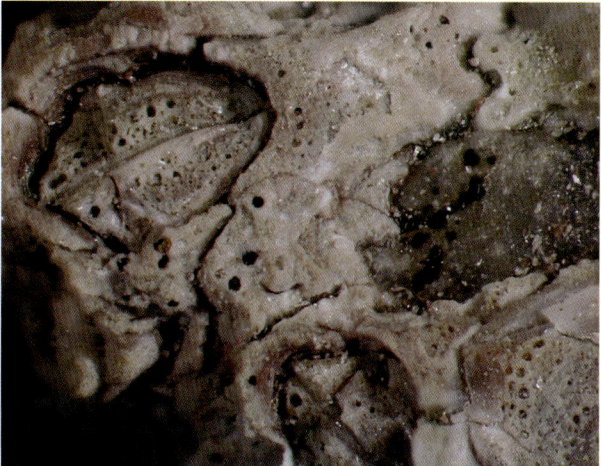
Pyrenocollema halodytes, Devon, UK; ascomata on barnacle

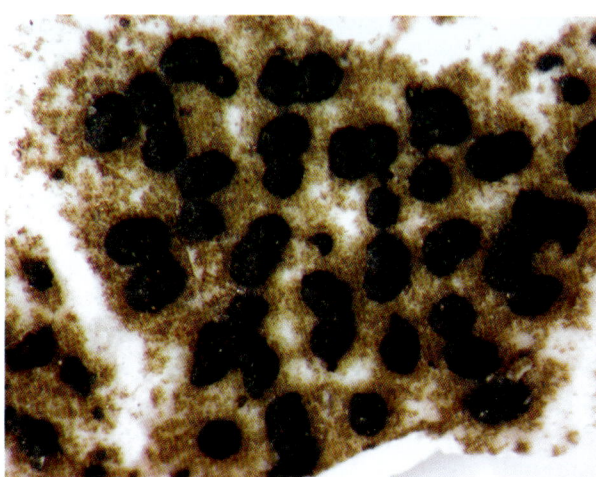
Pyrenocollema halodytes; ascomata on barnacle

Distribution: Widespread especially in aquatic environments, both freshwater and marine.

Economic Significance: None is known.

Ecology: Lichenized with cyanobacteria or green algae, or non-lichenized but then often lichenicolous; marine species primarily occurring on calcareous rock or shells of marine organisms.

Notes: Circumscription of the family is uncertain and application of several generic names is controversial.

References: **Calatayud, V.; Triebel, D.; Pérez-Ortega, S.** (2007). *Zwackhiomyces cervinae*, a new lichenicolous fungus (*Xanthopyreniaceae* on *Acarospora*, with a key to the known species of the genus. *Lichenologist* 39: 129-134; **Fałtynowicz, W.; Sągin, B.** (1995). *Pyrenocollema halodytes*, a new lichen species in Poland. *Acta Mycologica* Warszawa 30: 147–150; **Harada, H.** (1999). *Pyrenocollema japonicum*, a new freshwater species of pyrenocarpous lichen from Japan. *Bryologist* 102: 50–52; **Hawksworth, D.L.** (1988). The variety of fungal-algal symbioses, their evolutionary significance, and the nature of lichens. *J. Linn. Soc. Bot.* 96: 3–20; **McCarthy, P.M.; Kantvilas, G.** (1999). *Pyrenocollema montanum*, a new species from Tasmania. *Lichenologist* 31: 227–230; **Mohr, F.; Ekman, S.; Heegaard, E.** (2004). Evolution and taxonomy of the marine *Collemopsidium* species (lichenized *Ascomycota*) in north-west Europe. *Mycol. Res.* 108: 515–532; **Santesson, R.** (1992). *Pyrenocollema elegans*, a new marine lichen. *Lichenologist* 24: 7–11; **Schultz, M.; Porembski, S.; Büdel, B.** (2000). Diversity of rock-inhabiting cyanobacterial lichens: studies on granite inselbergs along the Orinoco and in Guyana. *Pl. Biol.* 2: 482–494.

Xenasmataceae Oberw. 1966
Polyporales: Basidiomycota

Basidiomata resupinate, waxy or gelatinous, effuse and often inconspicuous. Hyphal system monomitic, hyphae often gelatinized, cystidia present or absent. Hymenium smooth. Basidia widely cylindrical to ± urniform, attached laterally at the base to the fertile hypha, clamp connections lacking. Basidiospores hyaline, often staining in iodine.

Significant Genera: *Phlebiella*, *Xenasma*, *Xenasmatella*.

Distribution: Widespread, primarily known from temperate zones.

Economic Significance: None is known.

Ecology: Saprobic on fallen wood and similar substrata.

Notes: Few sequences are available, but the family appears to contain at least two quite separate lineages and its position is unclear.

References: **Ginns, J.** (1998). Genera of the North American *Corticiaceae sensu lato*. *Mycologia* 90: 1–35; **Greslebin, A.G.; Rajchenberg, M.** (2003). Diversity of *Corticiaceae sens. lat.* in Patagonia, southern Argentina. *N.Z. Jl Bot.* 41: 437–446; **Hallenberg, N.** (1986). Cultural studies in *Tubulicrinis* and *Xenasmatella* (*Corticiaceae*, *Basidiomycetes*). *Mycotaxon* 27: 361–375; **Hjortstam, K.; Larsson, K.-H.** (1987). Additions to *Phlebiella* (*Corticiaceae*,

Basidiomycetes), with notes on *Xenasma* and *Sistotrema*. *Mycotaxon* 29: 315–319; **Larsson, K.-H.; Larsson, E.; Koljalg, U.** (2004). High phylogenetic diversity among corticioid homobasidiomycetes. *Mycol. Res.* 108: 983–1002; **Lin, S.H.; Chen, Z.C.** (1990). The *Corticiaceae* and the resupinate *Hydnaceae* of Taiwan. *Taiwania* 35: 69–111; **Maekawa, N.** (1993). Taxonomic study of Japanese *Corticiaceae* (*Aphyllophorales*) I. *Rep. Tottori Mycol. Inst.* 31: 1–149; **Nakasone, K.K.** (1990). Cultural studies and identification of wood-inhabiting *Corticiaceae* and selected *Hymenomycetes* from North America. *Mycol. Mem.* 15: 412 pp.

Xylariaceae Tul. & C. Tul. 1861
Xylariales: Ascomycota

Stromata usually well-developed, extremely varied in morphology, generally composed of fungal tissue only, globose to applanate, sometimes stipitate and branched, occasionally reduced to a clypeus or a narrow shell surrounding individual ascomata, sometimes perennial, usually black, the internal tissue white or concolorous with the surface. Ascomata perithecial, black, globose, the ostiole periphysate, immersed within stromatic tissue, the ostiolar region often papillate or umbilicate. Interascal tissue well-developed, of narrow ± thick-walled paraphyses. Asci cylindrical, persistent, ± thick-walled but without separable layers, almost always with a large ± complex J+ apical structure, usually 8-spored. Ascospores usually dark brown, aseptate (sometimes with a hyaline daughter cell), with a germ slit, sometimes with a mucous sheath. Anamorphs hyphomycetous, often formed on the surface of stromatic tissues. Conidiophores occasionally synnematous or simple but often with complex branching. Conidiogenous cells irregular, proliferating sympodially. Conidia small, usually brown, aseptate, seceding ± rhexolytically.

Significant Genera: *Camillea*, *Hypoxylon*, *Rosellinia*, *Xylaria*. Anamorphs: *Dematophora*, *Nodulisporium*.

Camillea mucronata, Galápagos Islands; detail of ostiolar region of stromata

Hypoxylon rutilum, Surrey, UK; stromata

Thamnomyces camerunensis, Kenya; stromata

Distribution: Cosmopolitan.

Economic Significance: A few species are economically important pathogens of roots and stem bases, especially in the tropics.

Ecology: Saprobes or weak pathogens, many endophytic, mainly in wood and bark, occasionally in soil and dung.

Xylaria longipes, Surrey, UK; stromata

Notes: Many stromata are long-persistent, forming a prominent component of the mycota in local surveys. Generic concepts have been heavily researched with a number of segregates recognized within the *Hypoxylon* complex but further molecular phylogenetic studies are needed. *Daldinia* appears to cluster within the main *Hypoxylon* clade.

References: **Callan, B.E.; Rogers, J.D.** (1993). A synoptic key to *Xylaria* species from continental United States and Canada based on cultural and anamorphic features. *Mycotaxon* 46: 141–154; **Chapela, I.H.; Petrini, O.; Bielser, G.** (1993). The physiology of ascospore eclosion in *Hypoxylon fragiforme*: mechanisms in the early recognition and establishment of an endophytic symbiosis. *Mycol. Res.* 97: 157–162; **Collado, J.; Platas, G.; Peláez, F.** (2001). Identification of an endophytic *Nodulisporium* sp. from *Quercus ilex* in central Spain as the anamorph of *Biscogniauxia mediterranea* by rDNA sequence analysis and effect of different ecological factors on distribution of the fungus. *Mycologia* 93: 875–886; **González, F.S.M.; Rogers, J.D.** (1989). A preliminary account of *Xylaria* of Mexico. *Mycotaxon* 34: 283–373; **Granmo, A.; Hammelev, D.; Knudsen, H.; Laessøe, T.; Sasa, M.; Whalley, A.J.S.** (1989). The genera *Biscogniauxia* and *Hypoxylon* (*Sphaeriales*) in the Nordic countries. *Op. bot.* 100: 59–84; **Hsieh, H.-M.; Ju, Y.-M.; Rogers, J.D.** (2005). Molecular phylogeny of *Hypoxylona* and closely related genera. *Mycologia* 97: 844–865; **Johannesson, H.; Laessøe, T.; Stenlid, J.** (2000). Molecular and morphological investigation of *Daldinia* in northern Europe. *Mycol. Res.* 104: 275–280; **Ju, Y.M.; Rogers, J.D.** (1996). A revision of the genus *Hypoxylon*. *Mycol. Mem.* 20: 365 pp.; **Ju, Y.M.; Rogers, J.D.; San Martín, F.** (1997). A revision of the genus *Daldinia*. *Mycotaxon* 61: 243–293; **Ju, Y.M.; Rogers, J.D.; San Martín, F.; Granmo, A.** (1998). The genus *Biscogniauxia*. *Mycotaxon* 66: 1–98; **Laessøe, T.** (1999). The *Xylaria comosa* complex. *Kew Bull.* 54: 605–619; **Laessøe, T.; Rogers, J.D.; Whalley, A.J.S.** (1989). *Camillea, Jongiella* and light-spored species of *Hypoxylon*. *Mycol. Res.* 93: 121–155; **Laessøe, T.; Spooner, B.M.** (1994). *Rosellinia* & *Astrocystis* (*Xylariaceae*): new species and generic concepts. *Kew Bull.* 49: 1–70; **Lee, J.S.; Ko, K.S.; Jung, H.S.** (2000). Phylogenetic analysis of *Xylaria* based on nuclear ribosomal ITS1-5.8S-ITS2 sequences. *FEMS Microbiol. Lett.* 187: 89–93; **Petrini, L.; Petrini, O.** (1986). Xylariaceous fungi as endophytes. *Sydowia* 38: 216–234; **Platas, G.; Ruibal, C.; Collado, J.; Platas, G.; Ruibal, C.; Collado, J.** (2004). Size and sequence heterogeneity in the ITS1 of *Xylaria hypoxylon* isolates. *Mycol. Res.* 108: 71–75; **Rogers, J.D.; Ju, Y.M.** (1998). The genus *Kretzschmaria*. *Mycotaxon* 68: 345–393; **Rogers, J.D.; Ju, Y.M.; San Martín, F.** (1998). *Podosordaria*: a redefinition based on cultural studies of the type species, *P. mexicana*, and two new species. *Mycotaxon* 67: 61–72; **Sánchez-Ballesteros, J.; González, V.; Salazar, O.; Acero, J.; Portal, M.A.; Julián, M.; Rubio, V.; Bills, G.F.; Polishook, J.D.; Platas, G.; Mochales, S.; Peláez, F.** (2000). Phylogenetic study of *Hypoxylon* and related genera based on ribosomal ITS sequences. *Mycologia* 92: 964–977; **Stadler, M.; Ju, Y.M.; Rogers, J.D.** (2004). Chemotaxonomy of *Entonaema, Rhopalostroma* and other *Xylariaceae*. *Mycol. Res.* 108: 239–256; **Triebel, D.; Peršoh, D.; Wollweber, H.; Stadler, M.** (2005). Phylogenetic relationships among *Daldinia, Entonaema* and *Hypoxylon* as inferred from ITS nrDNA analyses of *Xylariales*. *Nova Hedwigia* 80: 25–43; **Whalley, A.J.S** (1996). The xylariaceous way of life. *Mycol. Res.* 100: 897–922; **Whalley, A.J.S.; Edwards, R.L.** (1995). Secondary metabolites and systematic arrangement within the *Xylariaceae*. *Can. J. Bot.* 73: S802–S810.

Zoopagaceae Drechsler 1938
Zoopagales: Zygomycota

Mycelium non-septate, produced outside the host, with adhesive material on hyphae to trap prey and forming haustoria at the point of contact. Spores (conidia) simple or appendaged, either solitary, formed successively or formed in relatively short chains on short hyphae (unknown in one genus where only terminal, intercalary or lateral chlamydospores are formed). Zygospores minute, polyhedral to warty, borne on somewhat parallel, randomly arranged or spirally twisted suspensors.

Significant Genera: *Zoopage*.

Distribution: Widespread although under-recorded.

Economic Significance: None is known.

Ecology: Predacious parasites of nematodes, amoebae and other small terrestrial animals (one genus parasitic on algae).

Notes: Almost all our knowledge of these fungi is derived from work published during the 1930s.

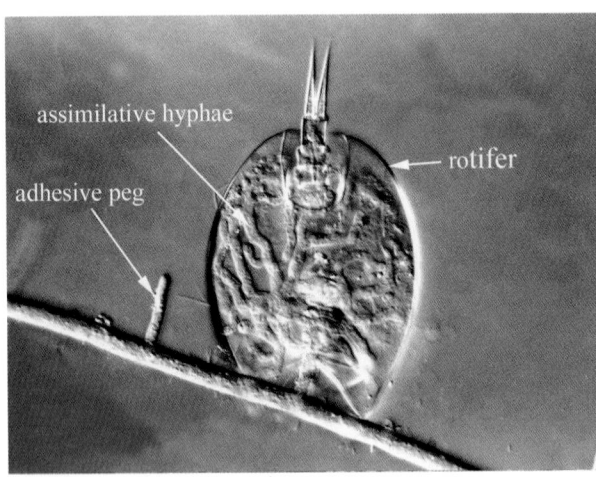

Zoophagus sp.; fungus in rotifer

References: **Barron, G.L.** (2004). Fungal parasites and predators of rotifers, nematodes, and other invertebrates. *Biodiversity of Fungi, Inventory and Monitoring Methods* (Amsterdam): 435–450; **Duddington, C.L.** (1973). Zoopagales. *In* Ainsworth, G.C.; Sparrow, F.K.; Sussman, A.S. (Eds). *The Fungi* (London) IVB: 231–234; **Dyal, R.** (1973). Key to *Phycomycetes* predaceous or parasitic in nematodes or amoebae I. Zoopagales. *Sydowia* 27: 293–301; **Tanabe, Y.; O'Donnell, K.; Saikawa, M.; Sugiyama, J.** (2000). Molecular phylogeny of parasitic Zygomycota (Dimargaritales, Zoopagales) based on nuclear small subunit ribosomal DNA sequences. *Mol. Phylogen. Evol.* 16: 253–262.

Zopfiaceae G. Arnaud ex D. Hawksw. 1992
Uncertain position within Dothideomycetes, Ascomycota

Ascomata cleistothecial, usually superficial, ± globose, black, usually non-ostiolate; peridium usually thick-walled. Interascal tissue of trabecular evanescent pseudoparaphyses. Asci globose to saccate, evanescent. Ascospores ellipsoidal, 1-septate, dark brown, often ornamented, without a sheath. Anamorphs unknown.

Significant Genera: *Caryospora*, *Zopfia*.

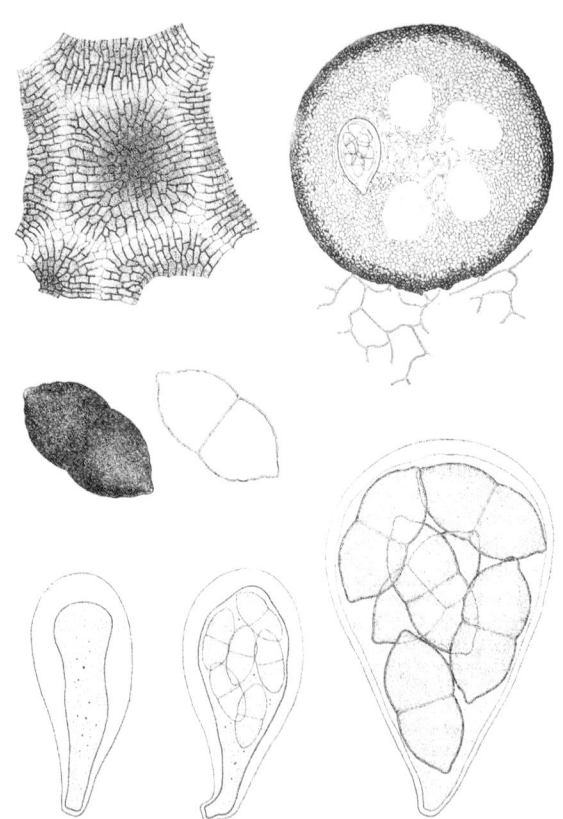

Zopfia rhizophila; ascoma, asci and ascospores

Distribution: Widespread.

Economic Significance: None is known.

Ecology: Apparently saprobic, found especially on rhizomes and roots. A few species are found in marine environments.

Notes: One of a number of families with cleistothecial (cephalothecial) ascomata, the walls of which fragment into polygonal plates. Molecular data are sparse but those available confirm separation from the *Testudinaceae*, which has superficially similar ascomata. The family appears to be polyphyletic.

References: **Hyde, K.D.** (1989). *Caryospora mangrovei* sp. nov. and notes on marine fungi from Thailand. *Trans. Mycol. Soc. Japan* 30: 333–341; **Kruys, Å.; Eriksson, O.E.; Wedin, M.** (2006). Phylogenetic relationships of coprophilous *Pleosporales* (*Dothideomycetes*, *Ascomycota*), and the classification of some bitunicate taxa of unknown position. *Mycol. Res.* 110: 527–536; **LoBuglio, K.F.; Berbee, M.L.; Taylor, J.W.** (1996). Phylogenetic origins of the asexual mycorrhizal symbiont *Cenococcum geophilum* Fr. and other mycorrhizal fungi among the *Ascomycetes*. *Mol. Phylogen. Evol.* 6: 287–294; **Ranghoo, V.M.; Hyde, K.D.** (1999). *Ascomauritiana lignicola* gen. et sp. nov., an ascomycete from submerged wood in Mauritius. *Mycol. Res.* 103: 938–942.

GLOSSARY

a- (**an-**) (prefix), not having; not; as in acaudate, anaerobe, aniso-.

ab- (prefix), position away from.

abaxial (of a basidiospore), the side away from the long axis of the basidium (Corner, 1948); cf. adaxial.

aberrant, an organism that deviates in one or more ways from the norm.

abhymenial, opposite the spore-producing surface.

abjection, the separating of a spore from a sporophore or sterigma by an act of the fungus.

abjunction, the cutting off of a spore from a hypha by a septum.

abraded (of lichen thalli), having the surface worn; eroded.

abrupt, as if cut off transversely; truncate.

abscission, separating by disappearance of a joining layer or wall, as of conidia from a conidiogenous cell.

absorb, to obtain food by taking up water and dissolved substances across a membrane; cf. ingest.

abstriction, abjunction and then abscission, esp. by constriction.

acantha, a sharp pointed process; a spine.

acanthocyte, spiny cell produced on a short branch from the vegetative mycelium of *Stropharia* spp. (Farr, *Mycotaxon* **11**: 241, 1980).

acanthohyphidium, see hyphidium.

acanthophysis, see hyphidium.

acaudate, not having a tail.

accumbent, resting against anything.

acellular, not divided into cells, e.g. a myxomycete plasmodium.

acephalous, not having a head.

acerose, needle-like and stiff; like a pine needle.

acervate, massed up; heaped; growth in heaps or groups.

acervulus (pl. **-i**; adj. **-lar**), a ± saucer-shaped conidioma (embedded in host tissue) in which the hymenium of conidiogenous cells develops on the floor of the cavity from a pseudoparenchymatous stroma beneath an integument of host tissue which ruptures at maturity; acervular conidioma.

acetabuliform, saucer-like in form.

achroic (**achromatic, achrous**), having no colour or pigment.

acicular, slender and pointed; needle-shaped.

acidiphilous (**acidophilous, acidophilic**), growing on or in conditions of low hydrogen ion concentration; e.g. *Scytalidium acidophilum* with an optimum pH for growth of 3, with good growth even at pH 1 (Miller *et al.*, *Internat. Biodet.* **20**: 27, 1984); also used of lichens on peaty soils or bark of a pH below 5.

acro- (combining form), at the end; apical; terminal.

acroauxic (of conidiophores), growth in length restricted to the apical region.

acrochroic, coloured specially in the hyphal tips at the growing point.

acrogenous, development at the apex.

acronema, extension of flagellum tip containing the two central microtubules but none of the nine peripheral elements.

acropetal, (1) describes chains of conidia in which the youngest is at the apex, basifugal; cf. basipetal; (2) a pattern of apical growth.

acropleurogenous, formed at the end and on the sides.

acrospore, an apical spore.

acrosporogenous (of conidial maturation), cells delimited and maturing in sequence from base to apex as the tip of the conidium expands (Luttrell, 1963).

acroton, a spinule in lichens bearing side branches.

actinogyrose (**actinogyr**) (of apothecia), disc gyrose and having no proper margin.

aculeate, having narrow spines.

aculeolate, having spine-like processes.

acuminate, gradually narrowing to a point.

acute, (1) pointed; (2) less than a right angle.

adaxial (of a basidiospore), the side next to the long axis of the basidium, usually that with the apiculus (Corner, 1948); cf. abaxial.

adelphogamy, pseudomictic copulation of mother and daughter cells, as in some yeasts (Gäumann & Dodge, 1928: 13).

adenose, having glands; gland-like.

adhesive disc, see holdfast.

adiaspore, a large spherical chlamydospore produced in the lungs of animals by the enlargement of an inhaled conidium of *Emmonsia* spp.; causing adiaspiromycosis. *Chrysosporium pruinosum* produces similar spores in culture (Carmichael, *CJB* **40**: 1167, 1962).

adnate (of lamellae or tubes), joined to the stipe; if lamellae, proximal end not notched (cf. sinuate); sometimes restricted to lamellae widely joined to the stipe (cf. adnexed); (of pellicle, scales, etc.), tightly fixed to the surface.

adnexed (of lamellae), narrowly joined to the stipe (cf. adnate); an ambiguous term.

addressed, see appressed.

adspersed, of wide distribution; scattered.

aduncate, bent; hooked; crooked.

adventitious septum, see septum.

adventive branching (of fruticose lichens), branching not of the normal pattern; e.g. regenerate branches produced after damage to the original branches in *Cladonia*.

aecidioid teliospore, see teliospore.

aecidioid urediniospores, see urediniospore.

aecidiospore, see aeciospore.

aecidium, see aecium.

aeciospore, (aecidiospores, plasmogamospores), produced in aecia (sing. -ium; aecidiosori), are unicellular, non-repeating vegetative spores, usually resulting from dikaryotization (and thus usually associated with pycnia), which germinate to give dikaryotic mycelium. Aeciospores (aecial aeciospore (I¹), Laundon) are typically catenulate, thin-walled and verrucose but sometimes they resemble typical urediniospores when they are designated **uredinioid aeciospores** by Cummins (= aecial urediniospores (II¹), Laundon; primary uredospores, Winter).

aeciotelium, see teliospore.

aecium, see aeciospore.

aequi-hymeniiferous (of hymenial development in agarics), having basidia which mature and shed their spores evenly over the surface of each lamella; the non-*Coprinus* type (Buller, *Researches* **2**: 19, 1922); cf. inaequi-hymeniiferous.

aerobe, an organism needing free oxygen for growth; cf. anaerobe.

aerogenic, describes an organism that produces detectable gas during the breakdown of carbohydrate.

aerole (of lichens), a scale-like area on the thallus delimited by cracks or depressions.

aetiology, the science of the causes of disease; etiology (Amer.).

agamic (**agamous**), asexual.

agaricicolous, living on agarics.

agglutinate, fixed together as if with glue.

aggregate, (1) (in taxonomy; '**agg.**' or '**aggr.**'), rarely used in mycology, has been used for groups of closely related morphospecies only distinguishable with difficulty; (2) (in descriptions), near together, crowded.

akinete, (1) a non-motile reproductive structure; (2) a resting cell.

alate, winged.

alepidote, having no scales or scurf; smooth.

aleuriospore (obsol.), formerly used for a thick-walled and pigmented but sometimes thin-walled and hyaline conidium developed from the blown-out end of a conidiogenous cell or hyphal branch from which it secedes with difficulty, as in *Aleurisma, Mycogone, Microsporum*; 'chlamydospore' sensu Hughes (1953); gangliospore. Since introduced by Vuillemin (1911), aleuriospore has been used in various senses, see Mason (1933, 1937) and Barron (1968), and finally rejected as a confused term (Kendrick, *Taxonomy of Fungi imperfecti*, 1971).

algal-layer (of lichen thalli), the photobiont-containing layer (usually between the upper cortex and the medulla) of the thallus.

algicolous, living on algae; **- fungi** see van Donk & Brumsz (*in* Reisser (Ed.), *Algae and symbiosis*: 567, 1992; review), algae.

aliform, wing-like in form.

alkaphilic, used of organisms growing well at high pH values; e.g. *Fusarium* sp. at pH 10 (Hiura & Tanimura, *in* Horrikoshi & Grant (Eds), *Superbugs: microorganisms in extreme environments*: 287, 1991).

allantoid (esp. of spores), slightly curved with rounded ends; sausage-like in form.

allergy An acquired, specific, altered capacity to react. It is acquired by exposure to allergenic particles; the sensitivity acquired from a single exposure is specific to one or to closely related species, although multiple exposures may result in multiple sensitivities; and subsequent re-exposure results in an altered capacity to react or allergic reaction. The form of that reaction depends on the nature of the allergenic particle, for instance, its size and chemical characteristics, the immunological reactivity of the subject and the circumstances of exposure. The two forms of allergy of most concern in this context are an immediate reaction, characterized by rhinitis and hay fever-like symptoms and a late reaction, characterized by alveolitis or pneumonitis. Fungal spores have been implicated as causative agents of both types of allergic reaction. Rhinitis and asthma are caused by normal everyday exposure to airborne allergens in subjects who are constitutionally predisposed (atopic) and who produce specific IgE antibodies against the allergen. Symptoms occur within a few minutes of exposure and may be provoked by 10^4 spores/m^{-3} air, or fewer, typically of fungi with spores larger than 10 µm. The spores may be components of the normal air spora, including *Cladosporium*, *Alternaria* and *Didymella*, or they may be associated with work environments, for instance cereal rusts and smuts and *Verticillium lecanii* when harvesting, *Agaricus bisporus* and *Boletus edulis* spores when preparing mushroom soup, and *Aspergillus flavus* and *A. awamori* from surface fermentations. Asthma may also be associated with exposure to fungal enzymes during their production. Allergic alveolitis occurs in non-atopic subjects after intense exposures to spores, typically 10^6-10^{10} spores/m^{-3}. At least 10^8 spores/m^{-3} may be required for sensitization but species differ in their antigenicity. Symptoms occur about 4 h after exposure and persist for 24-36 h if there is no further exposure. They include influenza-like symptoms, feverishness, chills, a dry cough, breathlessness and weight loss. With repeated exposure, breathlessness becomes increasingly severe and eventually permanent lung damage may occur with fibrosis and the increased load on the heart may lead to death. Specific IgG antibodies develop and may be an aid to diagnosis although implication of a fungus in the disease may require further tests. The disease is typically occupational and associated with poorly stored agricultural products. The classic form is farmer's lung, usually caused by thermophilic actinomycetes but sometimes by fungi, including *Aspergillus flavus*, *A. versicolor* and *Eurotium rubrum* (syn. *Aspergillus umbrosus*). Other forms of allergic alveolitis include cheese-washer's lung (*Penicillium casei*), malt-worker's lung (*Aspergillus clavatus*, *A. fumigatus*), suberosis (*Penicillium frequentans*), maple-bark stripper's lung (*Cryptostroma corticale*), sawmill worker's lung (*Rhizopus rhizopodiformis*, *Penicillium* spp., *Aspergillus fumigatus*, *Trichoderma viride*), sequoiosis (*Aureobasidium pullulans*, *Graphium* spp.), mushroom picker's lung (*Pleurotus ostreatus*, *Pholiota nameko*, *Aspergillus fumigatus*, *Cephalotrichum stemonitis*, and allergic alveolitis from citric acid fermentations (*Aspergillus fumigatus*, *A. niger*, *Penicillium* spp.). Mouldy lichens have also been reported to cause allergic alveolitis.

Allergic skin reactions may be caused by spores of the *Arthrinium arundinis* state of *Apiospora montagnei* in workers cutting the canes of *Arundo donax* in France, by contact with lichens in wood-cutters and people using lichens in decorations (Richardson, *in* Galun (Ed.), *CRC Handbook of lichenology* **3**: 98, 1988; review), and secondary to dermatophyte infections (see mycid).

For further information, see Pepys (*Hypersensitivity diseases of the lungs due to fungi and organic dusts*, 1969), Wilken-Jensen & Gravesen (*Atlas of moulds in Europe causing respiratory allergy*, 1984), Lacey (*in* Hawksworth (Ed.), *Frontiers in mycology*: 157, 1991), Lacey & Crook (*Ann. occup. Hyg.* **32**: 515, 1988), Lacey & Dutkiewicz (*J. Aerosol Sci.*, 1994).

alliaceous, having a taste or smell of onions or garlic; cepaceous.

allochronic, occurring at different time periods, e.g. contemporary and fossil specimens.

allochrous (**allochroous**), changing from one colour to another.

allochthonous, transported to the place where found; not indigenous; cf. autochthonous.

allocyst, a chlamydospore-like structure in *Flammula gummosa* (Kühner, 1946).

allopatric, occurring in different geographical regions; cf. sympatric.

alpha-spore (**A-spore**, **α-spore**), a fertile, fusoid to oblong, biguttulate spore of an anamorph of the *Valsaceae* (*Phomopsis*); cf. beta-spore.

alternate host, one or other of the two unlike hosts of an heteroecious rust.

alutaceous, the colour of buff leather.

alveola, (1) a small surface cavity or hollow; (2) a pore of a polypore (obsol.).

alveolate, marked with ± 6-sided (honey-comb-like) hollows; faveolate.

AM, arbuscular mycorrhiza; see mycorrhiza.

amend, the act and result of making an alteration, not necessarily to correct a fault or error; cf. emend.

amerospore, a 1-celled (i.e. non-septate) spore with a length/width ratio ‹ 15:1 (cf. scolecospore); if elongated, axis single and not curved through more than 180° (cf. helicospore); any protuberances ‹ $1/4$ spore body length (cf. staurospore).

ametoecious, see autoecious.

amixis, see heterothallism.

amoeboid, not having a cell wall and changing in form, like an amoeba.

amphi- (prefix), the two (sorts, sides).

amphigenous, making growth all round or on two sides.

amphimixis, copulation of two cells and nuclei which are not near relations, e.g. egg and sperm; cf. apomixis, automixis and pseudomixis.

amphispore, a second, special type of urediniospore.

amphithallism, see homothallism.

amphithecium, the thalline margin of an apothecium (L).

amphitrichous (amphitrichiate), having one flagellum at each pole.

amplectant, covering; embracing.

ampliate, made greater; enlarged.

ampoule hypha, see hypha.

ampulla, (1) the swollen tip of a conidiogenous cell which produces synchronous blastic conidia (as in *Gonatobotrytum*); (2) a conidiophore which develops a number of short branches or discrete conidiogenous cells (as in *Aspergillus*).

ampulliform, flask-like in form.

amyloid (of asci, spores, etc.), stained blue by iodine (see Iodine, Stains); cf. dextrinoid. See Dodd & McCracken (*Mycol.* **64**: 1341, 1972; nature of fungal starch), amylomycan.

amylomycan, a name proposed for the I+ blue or red compounds associated with asci (Common, *Mycotaxon* **41**: 67, 1991).

an-, see a-.

anaerobe, an organism able to grow without free oxygen. An **obligate -** grows only without free oxygen; a **facultative -** grows with or without free oxygen. See Zehnder (Ed.) (*Biology of anaerobic micro-organisms*, 1988); cf. aerobe.

analogous, showing a resemblance in form, structure, or function which is not considered to be evidence of evolutionary relatedness; cf. homologous.

anamorph, (1) (of shapes), a deformed figure appearing in proportion when correctly viewed; (2) (of fungi), asexual ('imperfect') form or morph (e.g. that characterized only by presence or absence of conidia).

anaphylaxis, manifestation of a change (immediate hypersensitivity) in a living animal from the uniting of an antibody with its antigen which may result in the death of the animal; cf. allergy.

anaphysis, a thread-like conidiophore persisting in apothecia of *Ephebe*.

anastomosing, joining irregularly to give a vein-like network.

anastomosis (pl. **anastomoses**), the fusion between branches of the same or different hyphae (or other structures) to make a network.

androphore, a branch forming antheridia, as in *Pyronema*.

anemophilous (of spores), taken about by air currents.

aneuploid, having a chromosome number which is not a multiple of the haploid set.

angio- (of a sporocarp), closed at least till the spores are mature; cf. endoangiocarpous, gymnoangiocarpous, hemiangiocarpous and cleistocarp.

angiocarpous (of a basidiome), hymenial surface at first exposed but later covered by an incurving pileus margin and/or excrescences from the stipe (Singer, 1975: 26); also used in a parallel way for Ascomycota.

-angium (**-ange**, suffix), a structure having no opening; a cavity.

anguilluliform, worm-like or eel-like in form.

angular septum, see septum.

angustate, narrowed.

anheliophilous, preferring diffuse light; cf. heliophilous.

aniso- (prefix), unequal.

anisogamy, the copulation of gametes of unlike form or physiology, i.e. of **-gametes**; heterogamy; cf. isogamy.

anisokont, having flagella of unequal length; heterokont.

anisospory, having spores of more than one kind.

anisotomic dichotomic branching, branching where one dichotomy becomes stouter and forms a main stem so that the other branch of the dichotomy appears to be lateral, as in *Alectoria ochroleuca* ; cf. isotomic dichotomic branching.

annellate (of asci), ones with a thickened apical pore (e.g. Leotiales); see ascus; **annellations**; see annellidic.

annellidic (of conidiogenesis), holoblastic conidiogenesis in which the conidiogenous cell (**annellide**, annellophore) by repeated enteroblastic percurrent proliferation produces a basipetal sequence of conidia (**annelloconidia**, annellospores) leaving the distal end marked by transverse bands (**annellations**).

annellophore, see annellidic.

annular, ring-like; ring-like arrangement.

annulus, (1) (of basidiomata), a ring-like partial veil, or part of it, round the stipe after expansion of the pileus; hymenial veil; apical veil; ring; an **-** near the top of the stipe is **superior** (an **armilla**, fide Gäumann & Dodge, 1928: 453), one lower down, **inferior**; (2) (in *Papulospora*), the ring of cells around a bulbil; (3) (of asci), the apical ring; anneau apicale; (4) (in *Alternaria*), thickening in apices of conidiogenous cells, fide Campbell (*Arch. Mikrobiol.* **69**: 60, 1970).

anoderm, having no skin.

anterior, (1) at or in the direction of the front; (2) (of lamellae), the end at the edge of the pileus.

antheridium (pl. **-a**, **antherid**), the male gametangium, either formed from a haplophase thallus, or in which meiosis occurs after delimitation.

antherozoid, a motile male cell; a sperm.

anthracobiontic, obligately inhabiting burnt areas; **anthracophilous**, sporulation favoured by burnt areas; **anthracophobic**, sporulation suppressed or checked on burnt areas; **anthracoxenous**, incidence and growth not affected by burnt areas (Moser, 1949).

anthropophilic (of dermatophytes, etc.), preferentially pathogenic for man; cf. zoophilic.

anti- (in combination), against.

antibiosis, antagonism between two organisms resulting in one overcoming the other.

anticlinal, perpendicular to the surface; cf. periclinal.

antrorse, directed upwards or forwards.

apical, at the end (or **apex**); **- granule**, a deeply staining granule at the hyphal apex, esp. in *Basidiomycetes*; the 'Spitzenkorper' of Brunswik (1924); **- veil**, see annulus; **- wall building**, see wall building.

apiculate, having an apiculus.

apiculus (of a spore), a short projection at one end; a projection by which it was fixed to the sterigma (Josserand); apicule; hilar appendage.

apileate, having no pileus; resupinate.

apiosporous (of 2-celled spores), where one cell is markedly smaller then the other.

aplanetism, the condition of having non-motile spores in place of zoospores.

aplanogamete, a non-motile gamete.

aplanospore, (1) a naked, amoeboid or non-amoeboid mobile cell; (2) a sporangiospore.

aplerotic, of an oospore which occupies ‹ 60% of the oogonial volume (Shahzad *et al.*, *Bot. J. Linn. Soc.* **108**: 143, 1992).

apobasidium, see basidium.

apocyte, multinucleate cell in which the multinucleate condition is accidental, transitory or secondary. See coenocyte.

apodial, having no stalk; sessile.

apogamy, the apomictic development of diploid cells.

apomixis (adj. **apomictic**), the development of sexual cells into spores, etc., without being fertilized; cf. amphimixis, automixis, and pseudomixis.

apophysis, a swelling or a swollen filament, e.g. at the end of a sporangiophore below the sporangium in *Mucorales* (cf. columella) or on the stem of some species of *Geastrum*; (in basidiomycetes), the swelling at the tip of a sterigma from which the basidiospore develops and which becomes the hilar appendage (q.v.).

apoplastic, movement of substances via the cell walls, not entering the living cell; cf. symplastic.

apothecium (pl. **-ia**), a cup-like or saucer-like ascoma in which the hymenium is exposed at maturity, sessile or stipitate, the stipes sometimes lichenized (podetium; q.v.). See the following for terminology of anatomical structures of apothecia: Degelius (*Sym. bot. upsal.* **13** (2), 1954; tabulation of terms), Korf (*Sci. Rep. Yokohama nat. Univ.* II **7**: 7, 1958; *in* Ainsworth *et al.* (Eds), *The Fungi* **4A**: 249, 1973), Letrouit-Galinou (*Bryologist* **71**: 297, 1969), Maas Geesteranus (*Blumea* **6**: 41, 1947), Sheard (*Lichenologist* **3**: 328, 1967).

appendage, a process (outgrowth) of any sort. For coelomycete conidial appendage terminology see Nag Raj (*Coelomycetous anamorphs*, 1993).

appendiculate, (1) (of an agaric basidioma), having the edge of the expanded pileus fringed with tooth-like remains of the veil, as in *Psathyrella candolleana*; (2) (of a spore), having one or more setulae.

applanate, flattened.

apposed, placed in apposition; side by side or next to each other.

appressed (**adpressed**), closely flattened down.

appressorium (pl. **-ia**), a swelling on a germ-tube or hypha, esp. for attachment in an early stage of infection, as in certain *Uredinales* and in *Colletotrichum*; the '.... expression of the genotype during the final phase of germination', whether or not morphologically differentiated from vegetative hyphae, as long as the structure adheres to and penetrates the host (Emmett & Parbery, *Ann. Rev. Phytopath.* **13**: 146, 1975); the term hyphopodium (q.v.) is probably best treated as a synonym.

arachnoid, covered with, or formed of, delicate hairs or fibres; araneose.

araneose (**araneous**), see arachnoid.

arboricolous, growing on trees.

arbuscle (**arbuscule**), see mycorrhiza.

archaeascus, see ascus.

archicarp (of ascomycetes), the cell, hypha, or coil which later becomes the ascoma or part of it.

archontosome, an electron-dense body occurring near nuclei at all stages from crozier formation to the development of young ascospores in *Xylaria polymorpha*. See Beckett & Crawford (*J. gen. Microbiol.* **63**: 269, 1970).

arcuate, arc-like.

ardella, a small spot-like apothecium, as in the lichen *Arthonia*.

ardosiaceous (ardesiaceous), slate-coloured.

areolate, having division by cracks into small areas.

arescent, becoming crustose on drying.

arid, dry.

armilla, see annulus.

armillate, edged; fringed; frilled.

arrect, stiffly upright.

arthric (of conidiogenesis), thallic conidiogenesis characterized by the conversion of a pre-existing, determinate hyphal element into a conidium (**arthroconidium**, thallic-arthroconidium, arthrospore), as in *Geotrichum*. See arthrocatenate.

arthro- (prefix), jointed.

arthrocatenate (of thalloconidia), formed in chains by the simultaneous or random fragmentation of a hypha.

arthroconidium (pl. **-ia**), see arthric.

arthrospore, (1) see arthric; (2) a specialized uninucleate cell functioning as a spore and derived from the disarticulation of cells of a formerly vegetative branch (*Asellariales*).

arthrosterigma (of lichens), a septate conidiophore (spermatiophore) (obsol.).

articulated, jointed.

ascending (**ascendent**) (of an annulus), having the free edge above attached, cf. descending; (of conidiophores), curving up, cf. erect; (of lamellae), on a cone-like or an unexpanded pileus.

ascigerous, having asci.

ascigerous centrum, the special tissue which produces the asci and hamathecium.

asco- (prefix), pertaining to an ascus.

ascocarp, see ascoma.

ascoconidiophore, the phialide bearing an ascoconidium in *Ascoconidium* (Seaver, *Mycol.* **34**: 412, 1942).

ascoconidium, a conidium formed directly from an ascospore, esp. when still within the ascus (e.g. *Claussenomyces*).

ascogenous (ascogenic), ascus-producing or -supporting.

ascogonium, the cell or group of cells in *Ascomycota* fertilized by a sexual act.

ascoma (pl. **ascomata**), an ascus-containing structure, ascocarp.

ascomycete, one of the *Ascomycota*.

ascoparaphysis see paraphysis.

ascophore, (1) an ascus-producing hypha, esp. the stalk-like hyphae supporting asci in *Cephaloascus*; (2) apothecium (obsol.).

ascophyte, hypothetical autotrophic ancestor of the *Ascomycota* (Cain, 1972), cf. basidiophyte.

ascoplasm, epiplasm (q.v.).

ascospore, a spore produced in an ascus by 'free cell formation'; the ascospore wall is multilayered, it consists of an outer **perispore**, an intermediary layer, the proper wall (**epispore**) and sometimes an internal **endospore**; major differences in which layers are thickened, folded or pigmented can give rise to considerable variation even in a single family (e.g. *Lasiosphaeriaceae*); see Bellemère (*in* Hawksworth (Ed.), *Ascomycete systematics*: 111, 1994), basidiospore, spore wall.

ascostome, a pore in the apex of an ascus (obsol.).

ascostroma, a stroma in or on which asci are produced, usually restricted to groups with ascolocular ontogeny.

ascus (pl. **asci**), term introduced by Nees (*Syst. Pilze*: 164, 1817) for the typically sac-like cell (first figured in *Pertusaria* by Micheli in 1729) characteristic of *Ascomycota*, in which (after karyogamy and meiosis) ascospores (generally 8) are produced by 'free cell formation'. Asci vary considerably in structure, and work in the last two decades has shown previous separation into only 2–3 categories (e.g. **bitunicate, prototunicate, unitunicate**) to be an over simplification. Sherwood (1981) illustrated 9 main types distinguishable by light microscopy (reproduced on p. 36 of edn 7 of this *Dictionary*): prototunivate, bitunicate, astropalean, annellate, hypodermataceous, pseudoperculate, operculate, lecanoralean, and verrucariod). Eriksson (1981) distinguished 7 types of dehiscence in bitunicate asci with an ectotunica and distinct endotunica (see p. 37 of edn 7). These classifications mask a much wider range of variation; Bellemère (1994) recognized 3 predehiscence types and 11 dehiscence categories. The details of the asci are stressed in ascomycete systematics, esp. in lichen-forming orders where reactions with iodine are emphasized (q.v.) (Hafellner, 1984).

Bitunicate asci with two functional wall layers; those splitting at discharge (fissitunicate; 'jack-in-the-box') had been correlated with an ascolocular ontogeny by Luttrell (1951). Reynolds (1989) critically examined this paradigm and found the term to be applied to different ascus types and that an exclusive link to ascostromatic fungi could not be upheld; he also introduced the term extenditunicate for asci which extend without any splitting of the wall layers (Reynolds, *Cryptog. Mycol.* **10**: 305, 1989).

Much variation depends on the modifications in the various wall layers, especially the thickness of the walls and the *c* and *d* layers, and the details of apical differentiation (Bellemère, 1994). Caution is needed in comparing ascus staining reactions (see iodine) and structures in the absence of ultrastructural data.

Also encountered are **- crown** (annular thickenings in *Phyllachora*), and **- plug** (thickening in the apex through which the spores are forcibly discharged).

ascus plug, thickening in the apex through which the spores are forcibly discharged.

aseptate, having no cross walls.

aseptic, free from damaging microorganisms.

asexual, without sex organs or sex spores; vegetative.

asperate, rough with projections or points.

aspergilliform (of a sporulating structure), resembling that of an *Aspergillus* conidiophore.

aspergilloma, a 'fungus ball' composed principally of hyphae of *Aspergillus*, found in a pre-existing cavity (esp. in an upper lobe of the lung) or a bronchus, which usually has a relatively benign or asymptomatic effect; cf. aspergillosis.

aspergillosis, any disease in humans or animals caused by *Aspergillus* (esp. *A. fumigatus*); esp. common in birds (see Chute *et al.* (1971), Ainsworth & Austwick (1973) under Medical and veterinary mycology). See Austwick, (*in* Raper & Fennell, Eds, *The genus Aspergillus*, 1965), Bossche *et al.* (Eds) (*Aspergillus and aspergillosis*, 1988).

asperulate, delicately asperate.

aspicilioid (of lecanorine apothecia), more or less immersed in the thallus, at least when young.

A-spore, see alpha-spore.

asporogenic (**asporogenous**), not forming spores.

assimilative, (1) taking in; (2) (of hyphae) having to do with the growth phase before reproduction; non-reproductive; vegetative.

astatocoenocytic (of nuclear behaviour in basidiomycetes), haplont mycelium cells coenocytic, diplont binucleate but coenocytic and without clamps when aeration insufficient, basidioma binucleate; in contrast to **holocoenocytic** (haplont and diplont coenocytic, only developing basidium binucleate), **heterocytic** (haplont regularly coenocytic), and the **normal** condition when the haplont is uninucleate, the diplont binucleate (Boidin, *in* Petersen (Ed.), *Evolution in the higher basidiomycetes*: 129, 1971).

asterophysis, see seta.

asteroseta, (1) see cystidium; (2) see seta.

astomate (astomous), lacking an ostiole.

asymmetric (of spores), having one side flattened or concave.

atomate, having a powdered surface.

attenuate, (1) narrowed; (2) (of a pathogen), having lowered pathogenicity or virulence.

atypical, not normal.

auleate (of gasteromycete basidiomata), a closed basidioma in which pleated plates of trama project into the glebal cavity from top and sides. See Dring (1973); after Kreisel (1969).

autecology, ecological studies on a single species and its relationship to the biological and physiochemical aspects of its environment.

aut-eu-form, an autoecious rust having all the spore stages.

authentic (of specimens, cultures, etc.), identified by the author of the name of the taxon to which they are referred.

auto- (prefix), self-inducing, -producing, etc.

autobasidium, see basidium.

autochthonous, (1) indigenous; cf. allochthonous; (2) (of soil organisms), continuously active, as opposed to **zymogenous** organisms which become active when a suitable substrate becomes available (Winogradsky, 1924); cf. exochthonous (Park, 1957).

autodeliquescent (of lamellae and pileus of *Coprinus*), becoming liquid by **-digestion**.

autoecious, completing the life cycle on one host (esp. of rusts; cf. heteroecious) or group of closely related collateral hosts; ametoecious (de Bary).

autogamy, the fusion of nuclei in pairs within the female organ, without cell fusion having taken place.

autolysis, self-digestion of a cell or tissue by endogenous enzymes.

automictic sexual reproduction, karyogamy between daughter nuclei of different meioses in the same gametangium (Dick, 1972).

automixis, self-fertilization by the fusion of two closely related sexual cells or nuclei; cf. amphi-, apo-, pseudomixis.

autotroph (adj. **autotrophic**) (of a living organism), one not using organic compounds as primary sources of energy, i.e. using energy from light or inorganic reactions as do green plants, lichen-forming fungi, and the photosynthetic iron and sulphur bacteria. See Fry & Peel (Eds) (*Autotrophic micro-organisms*, 1954), Lees (*Biochemistry of autotrophic bacteria*, 1955); cf. heterotrophic.

autotropism, an avoidance (−) response between neighbouring hyphae which in part is responsible for the spacing of hyphae at the colony margin (Trinci, *in* Jennings & Rayner (Eds), *The ecology and physiology of the fungal mycelium*: 23, 1984).

auxotroph, a biochemical mutant which will only grow on the minimal medium after the addition of one or more specific substances.

aversion, the inhibition of growth at the adjacent edges of colonies of microorganisms, esp. in a culture of one species; cf. antagonism; barrage.

axenic (of cultures), consisting of one organism; uncontaminated; a pure culture; cf. gnotobiotic.

axeny, inhospitality; 'passive' as opposed to 'active' resistance of a plant to a pathogen (Gäumann, 1946).

axial canal (- mass), see ascus.

axoneme, the main core of a flagellum composed of 2 central microtubules surrounded by 9 double microtubules.

azotodesmic nitrogen-fixing (Pike & Carroll, *in* Alexopoulos & Mims, *Introductory mycology*, edn 3, 1980).

azygospore, a parthenogenetic zygospore; characteristic of some *Mucorales*. See Benjamin (*Aliso* **5**: 235, 1963; list).

baccate, soft throughout like a berry.

bacillar (**bacilliform**), rod-like in form.

bacteriostatic, of a substance, or a concentration of a bactericide, which will not let growth of bacteria take place but which is not bactericidal.

bactobiont, see photobiont.

baculate, (1) (**baculiform**) (of spores), rod-shaped; (2) (of surface ornamentation), rod-shaped.

balanoid, acorn-shaped.

ballistoconidium (pl. **-ia**), a forcibly abjected conidium.

ballistospore, a forcibly abjected basidiospore. See Nyland (*Mycol.* **41**: 688, 1950); **- discharge**, see Buller (*Researches* **2–6**, 1922–34), Olive (*Science, N.Y.* **146**: 524, 1964), Ingold (*Friesia* **9**: 66, 1969; review).

barbate, having one or more groups of hairs; bearded.

barrage, the space between two mycelia which have an aversion for one another (Vandendries & Brodie, 1933); cf. zone lines.

basal body, (1) (of *Blastocladiaceae*), the part of the thallus fixed to the substratum by rhizoids at the lower end (Indoh, 1940); (2), see blepharoplast.

basal frill (of a spore), the apical part of a conidiogenous cell, or basal part of a cell which is carried away with the detached conidium following rhexolytic secession (Sutton, *CJB* **45**: 1251, 1967); cf. marginal frill.

basauxic (of conidiophores), elongating by a basal growing point (Hughes, *CJB* **31**: 650, 1953). See meristem arthrospore.

basidiocarp, see basidioma.

basidiograph, the straight-line graph obtained by plotting the ratio of the length (*l*) to the width (*w*) against the length of the basidia of a species of agaric (Corner, 1947: 214); cf. sporograph.

basidiole, a basidium-like hymenial element that lacks sterigmata because it is either young or permanently sterile; best restricted to immature basidia fide Singer (1962).

basidioma (pl. **-ata**), a basidium-producing organ; basidiome (Donk, *Taxon* **18**: 666, 1969); basidiocarp; carpophore; fruit-body; hymenophore; sporophore.

basidiophytes, Cain's (1972) term for hypothetical autotrophic ancestors of basidiomycetes; cf. ascophyte.

basidiospore, a propagative cell (typically a ballistospore but in gasteromycetes a statismospore) containing one or two haploid nuclei produced, after meiosis, on a basidium. The colour, form and ornamentation of the basidiospore are fundamental to basidiomycete classification and an essential part of any specific description. Greater use of electron microscopy has revealed an increasing complexity of the wall layers or teguments.

basidium (pl. **-ia**), (1) the cell or organ, diagnostic for basidiomycetes, from which, after karyogamy and meiosis, basidiospores (generally 4) are produced externally each on an extension (sterigma, q.v.) of its wall; (2) a conidiophore or phialide (obsol.).

The confused terminology applied to basidia (sense 1) and their parts has been traced by Clémençon (*Z. Mykol.* **54**: 3, 1988) and is analyzed by Talbot (*TBMS* **61**: 497, 1973) whose recommended usage (basically that of Donk) and synonymy is adopted in the series of definitions which follow: **pro-**, the morphological part or developmental stage of the basidium in which karyogamy occurs; primary basidial cell; probasidial cyst; **hypo-** (Martin) p.p.; teliospore of *Uredinales*. **meta-**, (1) the morphological part or developmental stage in which meiosis occurs; hypo- (Martin) p.p.; **epi-** (Martin) p.p.; promycelium of *Uredinales*. (When the whole meta- includes pro- remnants the distal and functional part may be distinguished as a **pario-** (Talbot, 1973).) (2) See proto- below. **holo-**, a basidium (e.g. of *Agaricus*) in which the meta- is not divided by primary septa (see septum) but may become adventitiously septate (see septum) (Talbot, *Taxon* **17**: 625, 1968). A holo- may be a **sticho-**, cylindrical, with nuclear spindles longitudinal and at different levels, or a **chiasto-**, clavate, with nuclear spindles across the basidium and at the same level. **phragmo-**, a basidium in which the meta- is divided by primary septa, usually cruciate (e.g. *Tremella*) or transverse (e.g. *Auricularia*) (Talbot, 1968).

Among other terms applied to basidia are: **apo-**, one with non-apiculate spores borne symmetrically on the sterigmata and not forcibly discharged (Rogers, *Mycol.* **39**: 558, 1947); **auto-**, one with spores borne asymmetrically and forcibly discharged; **endo-**, one developing within the basidioma, as in gasteromycetes; **epi-**, Martin's term for protosterigma, see sterigma; **hetero-**, a basidium of the *Heterobasidiomycetes*, usually a phragmo-; **homo-**, a basidium of the *Homobasidiomycetes*, usually a holo-; **hypo-**, = pro- (Donk), meta- (Martin); (of *Septobasidium*) = basidium (Martin); **pleuro-**, one relatively broad at the base and with bifurcated spreading 'roots', as in *Pleurobasidium* (Donk); **proto-**, a primitive basidium; the opposite of meta- in the sense of changed or degenerate basidium; **repeto-**, see Chadefaud (*Rev. mycol.* **39**: 173, 1975); **sclero-**, the thick-walled, encysted, gemma-like probasidium of the *Uredinales* (teliospore) and the *Auriculariales* (Janchen, 1923). See also Wells & Wells (*Basidium and basidiocarp evolution, cytology, function and development* 1982).

basifugal, development from the base up, acropetal.

basionym (basinym, basonym) (in nomenclature), the name-bringing or epithet-bringing synonym on which a new transfer or new combination is based. Donk (*Bull. Jard. bot. Buitenz.* sér. 3 **18**: 274, 1949) uses **isonym** for a name derived from a basionym. Cf. synisonym, synonym.

basipetal, describes a chain of conidia in which new spores are formed at the base, the oldest at the apex, cf. acropetal.

basocatenate (of conidia), formed in chains with the youngest conidium at the basal or proximal end of the chain.

beaded (of a lamella), having a line of small drops of liquid on the edge.

beak (of ascoma or conidioma), an elongated neck through which the spores are discharged. See rostrum.

behind (of lamellae), the end nearest the stipe.

beta-spore (**B-spore**, **β-spore**), a fertile, usually hamate, spore of an anamorph of the *Valsaceae* (*Phomopsis*); cf. alpha-spore.

bi- (prefix), twice; having two; two-.

biallelic (of an incompatibility system), having 2 alleles per locus; cf. multiallelic.

biatorine (of apothecia), of the lecideine type, pale or more or less coloured and soft in consistency.

bibulous (of the surface of a pileus), able to take up water.

bicampanulate, like two bells arranged mouth to mouth.

biconic (of spores), like two cones attached base to base.

bifarious, in two lines or series; distichous.

bifid, having a crack or division near the middle; forked.

bifurcate, to divide or divided into two parts or branches.

bilabiate, (1) two-lipped; (2) (of asci), ones in which the ectotunica splits in a lip-like manner to expose the endotunica (e.g. *Pertusaria*).

bilaminate, two-layered.

bilateral, see trama.

binate, in two parts.

binding hyphae, see hyphal analysis.

binucleate-phase, the dikaryo-phase.

bio- (prefix), pertaining to life.

biocide, a substance which kills living organisms; cf. biostat.

biogenous, living on another living organism; parasitic.

biologic form (Marshall Ward) or **race** (Klebahn), see physiologic race.

biont, a living organism; commonly used as a suffix to a word indicating the nature or position of the biont.

biophagous, see biogenous.

biophilous, see biogenous.

biostat, a substance which causes living organisms to stop growing; cf. biocide.

biotroph (adj. **-trophic**), an obligate parasite (cf. necrotroph, saprotroph), growing on another organism, in intimate association with its cytoplasm.

biotype, (1) (Scheibe) = physiologic race; (2) one individual; a group of individuals having a like genetic make up (Christensen & Rodenhiser, *Bot. Rev.* **6**: 389, 1940; Waterhouse & Watson, *Proc. Linn. Soc. NSW* **66**: 269, 1941).

bipartite, having division into two.

bipolar, (1) (of spore), at the two ends (poles); (2) (of an incompatibility system), having 1 locus; unifactorial; cf. tetrapolar; (3) occurring in both Arctic and Antarctic regions.

bird's nest fungi, the *Nidulariaceae*.

biseriate (**biserial**), in two series.

bitunicate, (1) having two walls; (2) (of asci), with two functional layers, that may or may not rupture or extend at discharge; see ascus.

biuncinate, two-hooked.

bivalvate, (1) (of spores), lens-shaped and having a hyaline rim, as in *Arthrinium*; (2) (of asci), see ascus.

biverticillate (of a penicillus), having branching at two levels, i.e. having metulae bearing phialides.

blasteniospore, a polarilocular (q.v.) ascospore.

blastic (of conidiogenesis), one of the two basic sorts of conidiogenesis (cf. thallic), characterized by a marked enlargement of a recognizable conidial initial *before* the initial is delimited by a septum. The conidium is differentiated from *part* of a cell (Kendrick, 1971: 255); **entero-**, when the inner wall (see tretic) or neither wall (see phialidic) of the blastic conidiogenous cell contributes to the formation of the **conidium** (blastic conidium) (cf. holoblastic); **holo-**, when both outer and inner walls of the blastic conidiogenous cell contribute to the formation of the conidium (cf. enteroblastic and see annellidic); **mono-**, when a conidiogenous cell has only one conidiogenous locus; **poly-**, when a conidiogenous cell has several conidiogenous loci.

blastidium, a lichen propagule produced by the budding of thalli in a yeast-like manner (Poelt, *Flora, Jena* **169**: 23, 1980).

blastocatenate (of blastoconidia), formed in chains with the youngest at the apical or distal end of the chain.

blastoconidium, a blastic (q.v.) conidium.

blastospore, a spore formed by marked enlargement of a recognizable conidium initial before the initial is delimited by a septum. The conidium differentiates from part of the cell. See Kendrick (Ed.) (*Taxonomy of fungi imperfecti*, 1971).

blematogen (**blematogen layer**), the undifferentiated tissue which becomes the universal veil in agarics (Atkinson, *Am. J. Bot.* **1**: 3, 1914).

blepharoplast (of zoospores), the basal body or granule (**kinetosome**) from which arise the longitudinal fibres constituting the axoneme of a flagellum; joined to the nucleus by a **rhizoplast**.

boletinoid (of hymenophores), having a structure intermediate between pores and gills.

bombysine, like silk.

booted, see peronate.

botryo-aleuriospore, one of an apical cluster of aleuriospores developed basipetally from the conidiogenous cells.

botryoblastospore, clusters of conidia borne on the swollen apex (ampulla) of a conidiogenous cell, arising synchronously or asynchronously, either singly or in chains.

botryose, racemose; grouped like grapes.

botuliform, cylindrical with rounded ends; sausage-like in form; see allantoid.

bourrelet, see ascus.

brachy- (prefix), short.

brachycyclic, with aeciospores (q.v.) missing from the life-cycle, but replaced by aecidioid urediniospores (q.v.).

brachyform, with aeciospores (q.v.) missing from the life-cycle.

brachymeiosis (obsol.), a third division, once claimed to occur in the ascus.

breathing pore, raised aperture in the upper cortex of *Parmelia exasperata* from the medulla to the exterior.

brevicollate, short-necked.

bridging hypha, a branch hypha joining two other hyphae (Buller, *Researches* **4**: 152, 1931); **- species** or **- host**, a plant by which a specialized parasite went, in Marshall Ward's opinion (*Ann. Bot.* **15**: 560, 1902; *Ann. Myc.* **1**: 132, 1903; but see Bean *et al.*, *Ann. Bot.* N.S. **18**: 129, 1954), from a susceptible to a resistant host.

bromatia, the rounded swellings at the ends of hyphae of ant fungi (see Insects and fungi) which are used by the ants as food.

broom cells (of agarics), cells bearing apical appendages to give a broom-like appearance on pileus or edge of lamella, as in *Marasmius rotula*; cellules en brosse (Singer, 1962: 62).

B-spore, see beta-spore; cf. alpha spore.

buckle, see clamp-connection.

budding, a process of multiplication in 1-celled fungi or in spores, in which there is a development of a new cell from a small outgrowth; cf. fission.

bulbil, a discrete, compact, multicellular, thalloidic propagule initiated in one of several ways but always homogeneous throughout development, with all cells acropetally produced and expanding more or less synchronously to many times (e.g. 4–10×) the diameter of the colourless, thin-walled hyphae from which they arise; pseudoparenchymatous at least at maturity, and lacking internal differentiation (found in certain basidiomycetes such as *Burgoa* and *Minimedusa*, distinguished from sclerotia; see Weresub & LeClair, *CJB* **49**: 2203, 1971); lichenized in *Multiclavula vernalis* (Poelt & Obermayer, *Herzogia* **8**: 289, 1990).

bulbillate (of a stipe), having a small or not clearly marked bulb at the base.

bulbillosis, the condition, in *Agaricales*, in which basidiome sporulation is suppressed and the basidial function taken on by bulbils, as in *Rhacophyllus* (Singer, 1962: 27).

bulbous, (1) bulb-like; (2) (of a stipe), having a swelling at the base.

bullate, (1) having bubble-like or blister-like swellings; (2) (of a pileus), having a rounded projection at the centre.

bursiculate, bag-like.

bursiform, bag-like.

byssisede, see byssoid.

byssoid, cotton-like; made up of delicate threads; floccose. **- lichens**, see Egea *et al.* (*Lichenologist* **27**: 351, 1995; in *Arthoniales*), Hafellner & Vězda (*Nova Hedw.* **55**: 183, 1992; key 17 gen. thalli).

cadavericole, an organism living on corpses.

caducous (of spores, etc.), falling off readily, deciduous.

caeoma (pl. **caeomata**), an aecium as in *Caeoma*, i.e. without peridial cells and with or without paraphyses.

caeomatoid (of aecia), resembling caeomata; sometimes, incorrectly, 'caeomoid'.

Caesar's mushroom, basidioma of the edible *Amanita caesarea*.

caespitose (**cespitose**), in groups or tufts like grass; cf. gregarious.

caespitulus (pl. **caespituli**), a tuft of spores.

calcarate, having a projection or spur.

calcareous, containing lime.

calceiform, shoe-like in form.

calceolate, see calceiform.

calcicolous, an organism (**calcicole**) growing on substrates rich in calcium; esp. of spp. on limestone or chalky rocks or soils.

callose, hard or thick and sometimes rough.

callosities (of fungi), wall thickenings associated with the penetration of fungicolous parasites (Swart, *TBMS* **64**: 511, 1975). See papillae.

calvescent, becoming bare or bald.

calvous, naked, bare.

calyciform, cup-like.

calycular, cup-like.

calyptra, a cap or hood.

campanulate, bell-like in form.

campestroid, agarics having a pileus with a diam. : stipe ratio of 1 or ›1. See Freeman (*Mycotaxon* **8**: 1, 1979); cf. placomycetoid.

campylidium (pl. **-ia**), helmet-shaped conidiomata occurring in various, mainly foliicolous, tropical lichenized genera (e.g. *Badimia*, *Loflammia*, *Sporopodium*); the name *Pyrenotrichum* (*Ascomycetes*, inc. sed.) has been applied to many of these conidiomata. See Sérusiaux (*Lichenologist* **18**: 1, 1986).

canal, sometimes applied to the pore connecting the two cells of a polarilocular spore.

canaliculate, having longitudinal grooves.

cancellate, reticulate; like a network, as the basidioma of *Clathrus*.

candidiasis, a cosmop. disease of humans (including thrush, mouget, etc.) and animals caused by species of *Candida* (syn. *Monilia* auct.), esp. *C. albicans*; moniliasis; candidosis. See Winner & Hurley (*Candida albicans*, 1964; (Eds), *Symposium on Candida infections*, 1966), Odds (*Candida and candidosis*, edn. 2, 1988), Turnbay *et al.* (Eds) (*Candida and candidamycosis*, 1991).

candidosis, see candidiasis.

canescent, becoming hoary or grey.

caninoid venation, see veins.

cap, see pileus.

capillaceous, see capilliform.

capilliconidium, a secondary conidium produced on a long capillary tube in *Entomophthorales*.

capilliform, hair-like; thread-like; capillaceous.

capillitium (of myxomycetes and gasteromycetes), a mass of sterile, thread-like elements, tubes or fibres among the spores.

capitate, having a well-formed head.

capitate-fastigiate (of macrolichens), having a thallus cortex of erect, parallel hyphae terminated by swollen and pigmented apical cells.

capitellum, a little head.

capitulum, a stalked globose apical apothecium, as in the *Caliciales*; cf. mazaedium.

capsule, a hyaline gelatinous sheath surrounding the cell of certain yeasts and bacteria.

carbonaceous, dark-coloured and readily broken; charcoal-like or cinder-like.

carbonicolous, living on burnt ground; pyrophilous (q.v.).

carinate, keeled; boat-like.

cariose, decayed.

carioso-cancellate, becoming latticed by decay.

carminophilic (of basidia), becoming densely granular (= siderophilous (or carminophilous) granulation) after treatment with aceto-carmine stain.

carnose (**carnous**), fleshy.

carnulose, somewhat fleshy.

carpogenous, living on fruit.

carpogonium (generally of algae, sometimes of fungi, e.g. *Erysiphaceae*), the female sex organ.

carpophore, (1) stalk of the sporocarp; (2) sometimes (esp. in France) = basidioma.

carpophoroid, a sterile carpophore-like body, in agarics, of unknown function (Singer, 1962: 22).

carrier, an organism harbouring a parasite without itself showing disease (Anon., *TBMS* **33**: 154, 1950).

cartilaginous, firm and tough but readily bent.

cartilaginous layer, sometimes applied to the sterome in *Cladonia* and the chondroid axis (q.v.) in *Usnea*.

caryo-, see karyo-.

cassideous, helmet-shaped.

catahymenium, see hymenium.

cataphyses, pseudoparaphyses (Groenhart, *Persoonia* **4**: 11, 1965); see hamathecium.

cata-species, a species where spermatia (q.v.) are absent.

catathecium (pl. **-ia**), a flattened ascoma, having the wall more or less radial in structure, and with a basal plate, e.g. *Trichothyrina*; cf. thyriothecium.

catenate (**catenulate**), in chains or end-to-end series. See arthrocatenate, basocatenate, blastocatenate.

catenophysis, a persistent chain of utricular, thin-walled cells formed by the vertical separation of the pseudoparenchyma in the centrum of certain ascomycetes, e.g. some *Halosphaeriaceae* (see Kohlmeyer & Kohlmeyer, *Mycol.* **63**: 857, 1971).

catenulate, see catenate.

catenuliform, chain-like.

catothecium, see catathecium.

cauda, tail; tail-like appendage.

caudate, having a tail.

caulescent, having a stem; becoming stemmed.

caulicolous, living on herbaceous stems; **caulicole**, a fungus which does this.

caulocystidium, see cystidium.

cauloplane, the stem surface.

cavernose, having hollows or cavities.

cavernula (pl. **-ae**), cavity; esp. the cavities in the lower cortex of *Cavernularia*.

cecidium, a gall (q.v.); caused by an animal (**zoo-**), esp. an insect; caused by a fungus (**myco-**).

cell, a unit of cytoplasm containing one or more nuclei, limited by a membrane (the **cell membrane**), and, in fungi, usually enclosed by a wall. Adjacent fungal cells do not necessarily contain individual protoplasts (q.v.) because cytoplasm and nuclei may pass from one cell to another.

cellular, composed of or comprising cells; cf. hyphal.

cellulin, a chitan-glucan complex which occurs as granules in the cells and plugs (**- plugs**) at hyphal constrictions in *Leptomitales* (see Lee *et al.*, *Mycol.* **68**: 87, 1976).

cellulolytic fungi, fungi able to utilize cellulose-containing materials (incl. plant cellulose, paper, cloth, etc.), e.g. *Chaetomiaceae* (Chahal & Wang, *Mycol.* **70**: 160, 1978). Cellophane or filter paper is often used in the culture of these fungi.

ceno-, see coeno-.

central body (of ascomycetes), the cell structure (central apparatus) from which astral rays emanate and initiate a cleavage of the cytoplasm (Harper, 1905). See Lindegren *et al.* (*Can. J. Genet. Cytol.* **7**: 37, 1965).

centric (**central**) (of a stipe), at the centre of the pileus; (of oogonium of *Saprolegniaceae*), having one or two layers of fat droplets surrounding the central cytoplasm in contrast to **sub-**, having the cytoplasm surrounded by one layer of droplets on one side, and by two or three layers on the other and **ex-**, having one large drop or a lunate row of droplets on one side (Coker, 1923) cf. excentric.

centrifugal, from the centre outwards.

centriole, short-cylindrical or barrel-shaped cell organelle 300–500 nm long × 150 nm diam. (kinetosome lacking an axoneme).

centripetal, towards the centre.

centrum, the structures within an ascoma, i.e. the asci and interascal tissue (the hamathecium); for types of centrum organization see Reynolds (Ed.) (*Ascomycete systematics*, 1981), and under *Ascomycota*.

cep, basidioma of the edible *Boletus edulis*; **fungi suilli** of Pliny and other classical writers.

cepaceous, see alliaceous.

cephalodium (pl. **-ia**), a delimited region within (**internal -**), or a warty, squamulose, or fruticose structure on the surface of, a lichen thallus containing a photobiont different from that characteristic of the rest of the thallus. Generally cephalodia contain cyanobacteria (e.g. *Nostoc*) whilst the rest of the thallus contains a green alga (e.g. *Trebouxia*). *Nostoc* cephalodia fix atmospheric nitrogen (see Millbank & Kershaw, *New Phytol.* **68**: 721, 1969). Known in about 400 lichenized spp. in diverse orders; see paracephalodium, photomorph, phycotype.

cephalothecoid, fragmenting along a series of predefined suture lines, as in ascomata of *Cephalotheca*.

ceraceous (**cereous**), wax-like.

ceranoid, having horn-like branches.

cerebriform, brain-like; convoluted.

cernuous, hanging down; drooping; nodding.

cespitose, see caespitose.

chalaroplectenchyma, see plectenchyma.

chartaceous, paper-like.

chasmothecium (pl. **-ia**), a closed fruitbody having no predefined opening but with the asci arranged in a single basal fascicle, e.g. an ascoma of *Erysiphe* (Braun *et al.* in Belanger *et al.* (Eds). *The Powdery Mildews: A comprehensive Tretise*, 2000); cf. cleistothecium.

cheilocystidium, see cystidium.

cheiroid, see chiroid.

chemical race, a group of chemically differentiated individuals or populations and not of any particular taxonomic rank (i.e. a chemical race may be a species, var., or chemotype); **- strain**, used as an infraspecific taxonomic rank in lichenized taxa distinguished only by chemical characters (Lamb, *Nature* **168**: 38, 1951).

chemosyndrome, a biogenetically meaningful set of major and minor natural metabolic products produced by a species (Culberson & Culberson, *Syst. Bot.* **1**: 325, 1977).

chemotaxis, a taxis (q.v.) in response to a chemical stimulus. Gooday (*in* Carlile (Ed.), *Primitive sensory and communication systems*, 155, 1975; chemotaxis in fungi), Bonner (*Mycol.* **69**: 443, 1977; in cellular slime moulds).

chemotropism, a reaction to a chemical, e.g. oxygen (Robinson, *New Phytologist* **72**: 1349, 1973), a hormone (Banbury, *J. Exper. Bot.* **6**: 235, 1975) or (in the case of water moulds but not other fungi) nutrients (Musgrove *et al.*, *J. Gen. Microbiol.* **101**: 65, 1977).

chemotype, a group of chemically differentiated individuals of a species of unknown or of no taxonomic significance.

chiastobasidium, see basidium.

chimeroid (of lichen thalli), two species with different types of algae previously regarded as distinct joined together as a composite ('chimeroid') thallus showing these 'photomorphs' (q.v.; phycosymbiodemes, photosymbiodeme, phycotypes) are the result of interactions between a single mycobiont and distinct photobionts, see phycotype.

chionophilous, see nitrophilous.

chiroid (**cheiroid**), shaped like a hand with the fingers together and not divergent, e.g. the conidia of *Cheiromyces*; cf. digitate, palmate.

chitinoclastic, chitin decomposing.

chitosan, a partially deacetylated form of chitin characteristic of zygomycetes; **chitases**, enzymes (EC 3.2.1.99) from fungi and bacteria able to hydrolyse chitosan (Monaghan *et al.*, *Nature* (*New Biol.*) **245**: 78, 1973).

chitosome, a small spheroidal structure (40–70 nm diam.), containing chitin synthetase zymogen, found in many fungi (Bracker *et al.*, *Proc. Nat. Acad. Sci. USA* **73**: 4570, 1976).

chlamydocyst, a two-walled resting zoosporangium of *Blastocladiaceae* within a hypha.

chlamydospore, an asexual 1-celled spore (primarily for perennation, not dissemination) originating endogenously and singly within part of a pre-existing cell, by the contraction of the protoplast and possessing an inner secondary and often thickened hyaline or brown wall, usually impregnated with hydrophobic material. Originally proposed by de Bary in 1859 for *Asterophora* anamorphs. See Griffiths (*Nova Hedw.* **25**: 503, 1974; origin, structure, function), Hughes (*in* Arai (Ed.), *Filamentous microorganisms*: 1, 1985; definition, occurrence).

chlorophycophilous (of fungi), lichenized with a green photobiont (Pike & Carroll, *in* Alexopoulos & Mims, *Introductory mycology*, edn 3, 1980).

chondroid axis, the cartilaginous axis occupying the central portion of the medulla in *Usnea*.

chromo- (combining form), colour.

chromogen, a stain-producing organism (Erlich, 1941).

chromogenesis, colour production.

chromogenic (chromogenous), colour-producing.

chromophilous, deeply staining.

chryseous, golden yellow.

chrysocystidium, see cystidium.

chrysogonidia, photobiont cells of *Trentepohlia* (obsol.).

cicatricose, with longitudinal ridges.

cicatrized (of conidiogenous cells and conidia), having thickened scars.

ciliate, edged with hairs.

ciliatulate, thinly ciliate.

cilium (pl. **-ia**), (1) an appendage of animal cells, e.g. protozoa; sometimes used for the flagellum (q.v.) of a zoospore; (2) a hair-like out-growth, e.g. from the edge of an apothecium or lichen thallus.

-cillin (suffix), for penicillins; derivatives of carboxy-6-amino-penicillanic acid (*WHO Chron.* **17**: 400, 1963).

cincinnate (cincinnal), rolled round; curled.

cingulate, edged all round.

circadian, pertaining to a day, e.g. a 24-hr rhythm; cf. diel, diurnal.

circinate, twisted round; coiled.

circum- (prefix), all round; round about.

circumcinct, having a band around the middle.

circumscissile, opening or cracking along a circle.

cirrate (cirrose), rolled round (curled) or becoming so.

cirrus (cirrhus), a curl-like tuft; a tendril-like mass or 'spore horn' of forced-out spores.

cisternal ring, a ring-like arrangement of the endoplasmic reticulum which appears to bud and give rise to vesicles.

clamp-connection (-connexion (obsol.), clamp, clamp-cell) (of basidiomycetes), a hyphal outgrowth which, at cell division, makes a connection between the resulting two cells by fusion with the lower; buckle; nodose septum; by-pass hypha.

clathrate (clathroid), like a network; latticed.

clava, a club-like fruiting structure, e.g. of *Cordyceps*.

clavate, (1) club-like; narrowing in the direction of the base; (2) (of stipes of agarics), narrowing to the apex; cf. obclavate. See also clava.

clavulate, somewhat club-like.

clavus, the sclerotium of ergot (obsol.).

cleistocarp, see cleistothecium.

cleistothecium (pl. -ia), a closed fruitbody having no predefined opening; in ascomycota with asci not regularly arranged, e.g. an ascoma of *Thielavia*; cf. chasmothecium.

closterospore (in dermatology), a multinucleate phragmospore, as in *Trichophyton* (obsol.).

cluster-cup, see aecium.

clypeate, having a clypeus.

clypeus, a shield-like stromatic growth, with or without host tissue, over one or more ascomata or conidiomata.

co- (prefix), together.

coacervate, massed (heaped), together.

coadnate, united, cohering, connate.

coalescent, joined together.

coarctate, crushed together, crowded, constricted.

cochleariform, spoon-like in form.

cochleate, shell-like in form; twisted like a shell.

coeno- (prefix), living together, e.g. multinucleate.

coenocentrum (of oomycetes), a small deeply staining body at the centre of the multinucleate oosphere to which the egg-nucleus goes.

coenocyte (adj. coenocytic), a multinucleate mass of protoplasm; (adj., of fungi), non-cellular, in the sense of non-septate; Vuillemin (1912) used coenocyte for a cell usually multinucleate and apocyte for one temporarily or secondarily multinucleate; coenocytium originally used for a structure resulting from nuclear division not followed by cytoplasmic cleavage, in contrast to a syncytium, a multinucleate structure resulting from the fusion of several protoplasts.

coenogametes, multinucleate gametangia (q.v.) which, upon fusion, give a coenozygote.

coenozygote, see coenogametes.

colacosome a cell organelle.

collabent, falling in; collapsing.

collarette, a cup-shaped structure at the apex of a conidiogenous cell.

collariate, having a collar; see collarette.

collarium, the ring of tissue to which the proximal ends of remote lamellae are attached as in *Marasmius* spp., etc.

collateral host, see autoecious.

colliculose (colliculous), with rounded swellings; blistered.

collulum, the neck of a conidiogenous cell (Zaleski, 1927).

colony (of bacteria and yeasts), a mass of individuals, generally of one species, living together; (of mycelial fungi), a group of hyphae (frequently with spores) which, if from one spore or cell, may be one individual.

colour. Corner (*Clavaria*, 1950) proposed the following series of terms for the more precise description of pigmentation of basidiomata and hyphae: **achroic**, without true pigmentation; **euchroic**, having true pigmentation as opposed to **epichroic**, discoloration due to injury; **hysterochroic**, slowly discoloured from base to apex in old age; euchroism may be **acrochroic**, coloured specially in the hyphal tips at the growing point, **metachroic**, changing colour through the appearance of a new pigment in maturer tissue, **ectochroic**, pigment on the outside of the hypha, **mesochroic**, pigment in the hyphal wall, or **endochroic**, pigment inside the cell. The last may be subdivided according to whether the pigment is diffused in the cytoplasm (**cytochroic**), in the cell vacuoles (**cystochroic**) or in oil drops (**lipochroic**); cf. Pigments.

columella, a sterile central axis within a mature fruit-body which may be uni- or multicellular, unbranched or branched, of fungal or host origin; (of gasteromycetes, after Cunningham) **axile -**, when an axis in the gleba; **dendroid -**, having lateral branches, as in *Gymnoglossum*; **percurrent -**, joining the peridium at the apex of gleba; **pseudo-**, embryonic tissue in the mature peridium of *Geastrum*; **simple -**, not branched, as in *Secotium*.

comate, having hairs; shaggy.

commensalism, a form of symbiosis (q.v.).

commissure, a closing join; a seam.

commixt, mixed with; intermingled.

community, any phytosociological taxon.

comose, having hairs in groups or tufts.

compaginate, joined tightly together.

companion cell, contiguous donor gametangium of *Olpidiopsidales*.

compatible (of mating types, strains, etc.), able to be cross-mated; cross-fertile; **hemi-**, a homokaryon compatible with 1 of the 2 components of the dikaryon.

complanate, flat; smooth.

complex, sometimes used to designate a group of closely related species.

complicate, bent upon itself.

compound, made up of a number of parts.

compressed (of a stipe), flattened transversely.

concatenate, in chains; catenulate.

concave (esp. of a pileus), hollowed out; basin-like.

concentric bodies, ultrastructures found in many lichenized fungi and also in some other fungi such as

Rhopographus, Sphaerotheca, Cercospora, Pseudopeziza, Sphaceloma; 'elliptical bodies'. See Griffiths & Greenwood (*Arch. Microbiol.* **87**: 285, 1972), Pons *et al.* (*TBMS* **83**: 181, 1984), Beilharz (*TBMS* **84**: 79, 1985), Meyer (*Mycol.* **79**: 44, 1987; in granules in *Allomyces*).

conceptacle, any hollow structure producing spores or spermatia.

conchate (**conchiform**), like a bivalve shell.

concolorous, of one colour.

concrescent, becoming joined.

concrete, joined by growth.

conditioning, the process by which fungi enzymically soften up substrata such as dead leaves before detritivorous animals can eat them.

conducting hypha, see hypha.

conferted, near together; crowded.

confervoid, composed of loose filaments or cells.

confluent, (1) (of sori, etc.), coming together; running into one another; (2) (of the flesh of a stipe), continuous with the trama of the pileus.

congeneric, one of two or more taxa considered to belong to one genus; cf. synonym.

congested, very near together.

conglobate, (1) massed into a ball; (2) (of the bases of stipes), together making a fleshy mass.

conglutinate, glued together; esp. of paraphyses in *Lecanorales*.

conidange, a small pycnidium (in a lichen thallus) having no stout wall (des Abbayes, 1951) (obsol.).

conidiangium, pycnidium (obsol.).

conidiogenesis, the process of conidium formation. Concepts of conidiogenesis have increasingly been used in anamorphic fungal systematics since Hughes (*CJB* **31**: 577, 1953) classified some hyphomycetes according to the different methods by which conidia develop from conidiophores and the ways in which conidiophores (and conidiogenous cells) grow before, during and after conidia are produced. The historical development of this approach has been covered by Kendrick (Ed.) (*Taxonomy of fungi imperfecti*, 1971), for hyphomycetes, by Sutton (*in* Ainsworth *et al.* (Eds), *The Fungi* **4A**: 513, 1973) for coelomycetes, and by Vobis (*Bibl. Lich.* **14**, 1980) for lichenized pycnidia. Cole & Samson (*Patterns of development in conidial fungi*, 1979) emphasized the contribution of ultrastructural data to developmental concepts, whilst Minter *et al.* (*TBMS* **79**: 75, 1982; **80**: 39, 1983; **81**: 109, 1983) reassessed optical and electron microscopy observations and demonstrated a continuum of developmental processes. See wall building.

conidiogenous, producing conidia; **- cell**, any cell from or within which a conidium is directly produced; **- locus**, the place on a conidiogenous cell at which a conidium arises; Kendrick (1971: 258).

conidiole, a small conidium, esp. one on another; a secondary conidium, as in *Empusa*.

conidioma (pl. **-ata**), a specialized multi-hyphal, conidia-bearing structure (Kendrick & Nag Raj *in* Kendrick (Ed.), *The whole fungus* **1**: 51, 1979). See acervulus, pycnidium, sporodochium, synnema (all obsol. nouns, but used adjectivally, e.g. acervular conidioma); cf. conidiophore.

conidiophore, a simple or branched hypha (a **fertile hypha**) bearing or consisting of conidiogenous cells from which conidia are produced; sometimes used when describing reduced structures for the conidiogenous cell.

conidium (pl. **-ia**), a specialized, non-motile (cf. zoospore), asexual spore, usually caducous, not developed by cytoplasmic cleavage (cf. sporangiospore) or free-cell formation (cf. ascospore); in certain *Oomycota* produced through the incomplete development of zoosporangia which fall off and germinate to produce a germination tube. See Sutton (*TBMS* **86**: 1, 1986; derivation etc.); **- initial**, a cell, or part of a cell from which a conidium develops; **- ontogeny**, conidiogenesis (q.v.).

conjugate, joined; in twos; **- nuclei**, two nuclei in one cell which undergo division (**- division**) at the same time, as in basidiomycetes.

conjugation, copulation (q.v.), esp. isogamic copulation; **- tube**, a tube between two copulating cells.

connate, joined by growth.

connective, see disjunctor.

connective hyphae, hyphae of the connective tissue of the context (Fayod, 1889).

connivent, (1) touching but not organically joined; (2) (of a pileus margin), touching the stipe.

consortium, a form of symbiosis (q.v.) in which two or more organisms live together in an interdependent way (obsol.).

conspecific, of two or more taxa considered to be one species; cf. synonym.

constipate, crowded together.

contaminated, (1) bearing, or intermixed with, a pathogen, as spores on seeds, fungi in soil (c.f. infection and infested); (2) (of cultures), not pure.

context (of hymenomycetes), the hyphal mass between the superior surface and the subhymenium or the trama of basidiocarps.

contiguous, touching; joining.

contingent, touching.

continuous, (1) (of spores, hyphae, etc.), having no septa; (2) (of a stipe), one with the tissue of the pileus or peridium; (3) (of cultures), see culture.

convergence, describes two organisms with many characters in common but which are descended from widely separate origins. Difficult to distinguish from parallelism due to intermediate conditions.

convex (of a pileus), equally rounded; broadly obtuse; **convexo-expanded**, having the edge bent over; **convexo-plane**, convex when young, flat after expansion.

copulants, Kniep's name for copulating structures of like form.

copulation, the fusion of sexual elements; conjugation; **gametangial -**, the fusion of two sexual organs; **heterogamic -**, the fusion of gametes morphologically unlike; **isogamic -**, the fusion of gametes morphologically like; conjugation in the narrower sense; **planogamic -**, the fusion of motile gametes to give a motile zygote (a planozygote); cf. merogamy.

coralloid, much branched; like coral in form; esp. basidiomata of *Clavaria*; cf. forate.

corbiculae, protective structures forming a stroma around the telia of certain rusts (Kuhnholtz-Lordat, *Bull. mens. Acad. Sci. Lett. Montpellier* **71**: 91, 1942; see *RAM* **26**: 468, 1947); paraphysis; pseudoparaphysis.

coremium, see synnema.
coriacellate, somewhat coriaceous.
coriaceous, like leather in texture.
corium, see spore wall.
corneous, (1) horn-like in texture; (2) (of a substance), like horn.
corniform, shaped like a horn.
cornute, (1) horned; horn-like in form; (2) (of aecia), long and tube-like, as in *Gymnosporangium*.
coronate, crowned.
corrugate, wrinkled.
cortex, a more or less thick outer covering, cf. derm.; **epi-** (q.v.); **corticate**, having a cortex.
corticate, see cortex.
corticolous, living on bark; **corticole**, an organism which does this.
cortina (of agarics), a partial veil (or part of one), frequently web-like, covering the mature gills.
corymbose, arranged in clusters.
coscinocystidium, a cystidium projecting as a pseudocystidium.
coscinoid, a pitted conducting element in *Linderomyces*.
costate, veined or ribbed.
costiferous, see hypha.
cotyliform, plate-like or wheel-like with an upturned edge.
crateriform, cup-like or crater-like in form.
crenate, having the edge toothed with rounded teeth.
crenulate, delicately crenate.
crescentic, see lunate.
cribose (**cribriform**), having a network like a sieve.
crispate, curled and twisted.
crista, tubular, pouch-like or shelf-like inwardly directed fold of the inner membrane of a mitochondrion; site of ATP production during aerobic metabolism.
cristate, crested.
crowded (of gills), very close together; conferted.
crozier, the hook of an ascogenous hypha before ascus-development; ascus crook.
cruciate, (1) in the form of a cross; (2) (of basidial septa), vertical and at right angles.
crust, a general term for a hard surface layer, esp. of a sporocarp; crustose.
crustose (**crustaceous**), crust-like; used for lichens having a thallus stretching over and firmly fixed to the substratum by the whole of their lower surface; such thalli generally lack rhizinae and a lower cortex.
crypta, a sleeve-like formation around a tree root (esp. evergreens) in tropics and subtropics developed by certain agarics (Singer, 1962: 20).
cryptic, inconspicuous or hidden.
cryptoendolithic (of organisms, esp. lichens), surviving at low temperatures through modification of the thallus so that it can exist inside rock between rock crystals (Friedmann, *Science, N.Y.* **215**: 1045, 1982). Cf. endolithic.
ctenoid, comb-like.
cucullate, hood-like or cowl-like in form.
culmicolous, living on stems, esp. those of grasses; caulicolous; **culmicole**, an organism which does this.
culture, a growth of one organism or of a group of organisms for the purpose of experiment (esp. of microorganisms on laboratory media) or sometimes for trade (e.g. a **mushroom -**); **continuous -**, one in which the culture medium is simultaneously added and withdrawn (harvested) so that the volume remains constant; see Calcott (*Continuous culture of cells*, 2 vols, 1981); **enrichment -**, a culture which favours the growth of the desired organism in a mixed culture or population; **pure -**, a - of one sort of organism; **- medium**, see medium; **type -**, see type.
cumulate, massed together; heaped up.
cuneate, see cuneiform.
cuneiform (**cuneate**), thinner at one end than the other; wedge or axe-blade shaped.
cupulate, cup-like in form, as e.g. in conidiomata.
cuspidate (e.g. of a pileus or cystidium), having a well-marked sharp outgrowth or point at the top.
cutis (**cuticle**) (of basidiomata), the outer layer consisting of compressed hyphae parallel to the surface; the upper and lower layers of the **cutis** are sometimes distinguished as **epi-** and **sub-**. See Shaffer (*Brittonia* **22**: 230, 1970; cuticular terminology in *Russula*), Singer (*Agaricales in Modern Taxonomy* edn 4: 69, 1986); also pellis.
cyanescent, becoming blue.
cyanobiont, see photobiont.
cyanophilous (of spores, etc.), readily absorbing a blue stain such as cotton blue or gentian violet.
cyanophycophilous (of fungi), ones lichenized with a cyanobacterium (Pike & Carroll, *in* Alexopoulos & Mims, *Introductory mycology*, edn 3, 1980); see photobiont.
cyanotrophic (of fungi, esp. lichen-forming spp.), obtaining nutrients (esp. nitrates fixed from the atmosphere) by forming regular connections to free-living or ± lichenized cyanobacteria; see Poelt & Mayrhofer (*Pl. Syst. Evol.* **158**: 265, 1988); see also cephalodium.
cyathiform, like a cup, a little wider at the top than at the bottom, and sometimes stalked.
cyclosis, cytoplasmic streaming; characteristic of eukaryotes.
cymbiform, boat-shaped; navicular.
cymose, a more or less flat-topped cluster.
cyphella (pl. **-ae**), a break in the lower (rarely upper) cortex of a lichen thallus which is roundish or ovate and in section appears as a cup-like structure lined with a layer of loosely connected, frequently globular, cells formed from the medulla, characteristic of *Sticta*; also used for pores open to the halliomedulla in *Oropogon*.
cyst, (1) an encysted cell (? the product of meiosis after karyogamy), usually aggregated into a cystosorus which germinates to produce a zoospore (*Plasmodiophorales*); (2) an encysted zoospore (planospore) which becomes a gametangium (*Blastocladiales*); **macro-** (of *Myxomycota*), an encysted aggregate of myxoamoebae; the resting form of a young plasmodium; the alternative to the sorocarp in some cellular slime moulds (dictyostelids) (Nickerson & Raper, *Am. J. Bot.* **60**: 190, 1973); propagule, especially the walled structure of the encysted zoospore (*Peronosporales*); resting spores of chrysomonads etc.; **micro-** (of *Myxomycota*), an encysted myxamoeba or swarm spore; **spore-**, a cell, hollow organ or sac-like structure enclosing a mass of protoplasm containing spores as in the *Ascosphaeraceae* (Skou, 1982).

cystesium, a cell which differentiates to adhere to a cystidium arising from the opposite hymenium (Horner & Moore, *TBMS* **88**: 488, 1987).

cystidiole (of hymenomycetes), a simple hymenial cell of about the same diameter as the basidia but remaining sterile and protruding beyond the hymenial surface.

cystidium (pl. **-ia**), a sterile body, frequently of distinctive shape, occurring at any surface of a basidioma, particularly the hymenium from which it frequently projects. Cystidia have been classified and named according to their: (1) *origin*: **hymenial -** (**tramal -**), originating from hymenial (tramal) hyphae; **pseudo-**, derived from a conducting element, filamentous to fusoid, oily contents, embedded not projecting; **coscino-**, see coscinoid; **skeleto-**, the apical part of a skeletal hypha (frequently ± inflated) projecting into or through the hymenium; false seta; **macro-**, arising deep in the trama in Lactario-Russulae; **hypho-**, hypha-like, derived from generative hyphae. (2) *position* (first by Buller, **2**): on the pileus surface (**pileo-**, **dermato-**, Fayod); at the edge (**cheilo-**), side (**pleuro-**), or within (**endo-**) a lamella; on the stipe (**caulo-**). (3) *form*: **lepto-**, smooth, thin-walled; **lampro-**, thick-walled, with or without encrustation (**setiform lampro-**, awl-shaped, wall pigmented; **asteroseta**, a radially branched lampro-; **microsclerid**, a versiform, endolampro-; **lyo-**, cylindrical to conical, very thick-walled, abruptly thin-walled at apex, not encrusted, colourless, as in *Tubulicrinis* (Donk, 1956); **monilioid gloeo-**, (torulose gloeo- (Bourdot & Galzin, 1928); moniliform paraphysis (Burt, 1918); pseudophysis; **schizo-** (Nikolayeve, 1956, 1961), monilioid, frequently with a beaded apex (as in *Hericiaceae* and *Corticiaceae*). (4) *contents*: **gloeo-**, thin-walled, usually irregular, contents hyaline or yellowish and highly refractile; **chryso-**, like lepto- but with highly staining contents; **hypo-**, (Larsen & Burdsall, *Mem. N.Y. bot. Gdn* **28**: 123, 1976); **oleo-**, having an oily resinous exudate; **pseudo-**, see (1) above. See also hyphidium, seta. Reviews of cystidia include Romagnesi (*Rev. Mycol., Paris* **9** (suppl.): 4, 1944), Talbot (*Bothalia* **6**: 249, 1954), Lentz (*Bot. Rev.* **20**: 135, 1954), Smith (*in* Ainsworth & Sussman (Eds), *The fungi* **2**: 151, 1966), Price (*Nova Hedw.* **24**: 515, 1975; types in polypores).

cystochroic, with pigment in the cell vacuoles.

cystosorus (of *Chytridiales*), a group of united cysts or resting spores.

cystospore, (1) an encysted zoospore formed at the exit to the zoosporangium and germinating to produce a new zoospore (planospore) as in *Achlya* (*Oomycetes*), *Achlyogeton* and *Achlyella* (*Chytridiomycetes*); (2) (of *Amoebidiales*) spores released from an encysted amoeboid cell.

cytochroic, with pigment diffused in the cytoplasm.

cytolysis, breaking up or solution of the cell wall.

cytoskeleton, intracellular network of protein filaments that is insoluble in non-ionic detergents.

dacryoid, having one end rounded and the other more or less pointed; pear-like or tear-like in form.

dactyloid, finger-like.

datum (pl. **data**), facts, figures, information and observations. Often used adjectivally, e.g. data bank, data matrix, database.

decay, the destruction of plant or animal material by fungi and other microorganisms.

deciduous (of spores, etc.), falling away at maturity; shed, either with (e.g. teliospores) or without (e.g. urediniospores) a fragment of the pedicel or sporophore; cf. persistent.

declinate, bent or curved down or forwards.

declivate (**declivous**), sloping.

decolourate, colourless.

decorticate, having no cortex.

decumbent, resting on the substratum with the ends turned up.

decurrent (of lamellae), running down the stipe.

decurved (of the pileus edge), bent down.

decussate (of lichen thalli), having the surface divided and crossed by dark lines.

dediploidization (in ascomycetes and basidiomycetes), the making of haploid cells (or hyphae) by a dikaryotic diploid mycelium or cell.

dehiscence papilla, morphologically and ultrastructurally distinct protuberance on an undischarged sporangium that becomes converted into an exit tube.

dehiscent (**dehiscing**) (of asci or fruit-bodies), opening when mature, by pores or by becoming broken into parts.

deliquescent, becoming liquid, e.g. after maturing.

deltoid, triangular in shape.

dematiaceous (of mycelium, spores, etc.), pigmented, more or less darkly; cf. moniliaceous.

-deme (suffix), a neutral term, always used with a prefix, and denoting any group of individuals within a taxon (q.v.; usually a species); first proposed by Gilmour & Gregor (*Nature* **144**: 333, 1939); occasionally used in mycology, e.g. agamodeme (predominantly apomictic), photosymbiodeme (of lichen thalli with different photobionts).

demicyclic, see microcyclic.

dendritic, irregularly branched; tree-like; dendroid.

dendrohyphidium, see hyphidium.

dendroid, tree-like in form; dendritic.

dendrophysis, see hyphidium.

denigrate, blackened.

dense body vesicle (**DBV**), cytoplasmic vesicle, associated with phosphorylated glucan metabolism, found in a variety of TEM morphological states ranging from a single (or several) electron-opaque core in an amorphous matrix to a highly structured, myelin-like arrangement of alternating electron-opaque and electron-translucent layers.

dentate, toothed; cf. denticulate.

denticle, a small tooth-like projection esp. one on which a spore is borne.

denticulate, having small teeth; cf. dentate.

denuded, uncovered or glabrous by loss of scales, etc.

depauperate, poorly developed.

dependent, hanging down.

deplanate, flat.

depressed (1) (of a pileus), having the middle lower than the edge; (2) (of lamellae), sinuate (q.v.).

derived, of a character that has changed from the form in which it appeared in an ancestor.

derm (**dermium**) (of basidiomata), an outer layer in which the hyphae are perpendicular to the surface (Lowag, 1941); cf. cortex.

dermatocyst (**dermatocystidium**), see cystidium (1).

dermatophyte, a fungus parasitizing keratinized tissue (hair, skin, nails) of humans and animals and causing **dermatophytosis** (pl. **-es**) (ringworm, tinea). These fungi, which are typically anamorphic fungi (hyphomycetes) with teleomorphs in the *Arthrodermataceae* (*Onygenales*), have frequently been treated as a special group the 'Dermatophytes' or Ringworm Fungi. Ringworm is cosmop. and dermatophytes, or non- or weakly pathogenic dermatophyte-like fungi, occur widely in soil and other keratin-containing substrata such as bird's nests. There are 3 main gen. (*Epidermophyton*, *Microsporum*, *Trichophyton* distinguished by characteristic macroconidia), + c. 30 syn., c. 40 spp. (1,000 names).

dermatophytid, a pustular allergic eruption (**id-reaction**) of the skin at a distance from a primary infection by a dermatophyte.

dermis (of lichens), the limiting layer of a thallus (obsol.).

descending (**descendant**) (of an annulus), having the free edge below the attached (cf. ascending).

determinate, (1) clearly marked; definite; (2) (of conidiophores), growth ceasing with the production of terminal conidia.

detersile (of villosity), removable so that the surface becomes bare.

detoxification, the conversion of a toxin (e.g. an inhibitory phytoalexin) to non-toxic (non-inhibitory) products.

deuteroconidium (of dermatophytes), a spore-like cell, the outcome of the division of a hemispore (protoconidium).

deuterogamy, the condition in which other processes replace fusion of gametes, as in some fungi, the macroalgae, and phanerogams; secondary pairing.

dextrinoid (of spores, etc.), stained yellowish-brown or reddish-brown by Melzer's iodine (see iodine); pseudoamyloid (Singer); cf. amyloid.

diageotropism, tendency to horizontal growth in relation to the earth surface.

diaphanous, transparent or nearly so.

diaspore, (1) any unit of dissemination, e.g. a spore, fragment of mycelium, sclerotium (Sernander, 1927); (2) (of lichens), particularly applied to vegetative propagules. See hormocyst, isidium, soredium.

dicaryon, see dikaryon.

dichohyphidium, see hyphidium.

dichophysis, see hyphidium.

dichotomous, branching, frequently successively, into two more or less equal arms.

dictydine granules, see plasmodic granules.

dictyochlamydospore, a non-deciduous multicelled chlamydospore composed of an outer wall separable from the walls of the component cells which are rather easily separated from each other, as in some *Phoma* spp. formerly ascribed to *Peyronellaea* (Luedemann, *Mycol.* **51**: 778, 1961).

dictyoporospore, a deciduous, multicelled porospore the component cells of which are firmly united and not enclosed by an outer wall, as in *Alternaria* (Luedemann, *Mycol.* **51**: 778, 1961).

dictyoseptate, having transverse and longitudinal cross walls (septa), like layers of cement between bricks; muriform.

dictyosomes, ± spherical vesicles associated with the edges of the membrane-bound sacs (cisternae) which constitute the golgi apparatus in *Oomycota* and other fungi as shown by electron microscopy.

dictyosporangium, a septate sporangium, as in *Dictyuchus*.

dictyospore, differs from an amerospore (q.v.) by being divided by intersecting septa in more than one plane; muriform spore.

didymospore, differs from an amerospore (q.v.) in having one transverse septum.

diel, a 24-hr periodicity; cf. circadian, diurnal.

differential hosts, the special species or cultivars of host plants the reactions of which are used for determining physiologic races.

diffluent, breaking up in water.

diffract (of a pileus surface), cracked into small areas; areolate.

diffuse, widely or loosely spreading and having no distinct margin; **- wall building**, see wall building.

digitate, with deep radiating divisions, finger-like; cf. chiroid, digitate.

dikaryon (adj. **dikaryotic**), a cell having two genetically distinct haploid nuclei.

dikaryoparaphysis, see hyphidium.

dikaryotization, the conversion of a homokaryon into a dikaryon typically by the fusion of 2 compatible homokaryons; **illegitimate -**, the sporadic occurrence of a dikaryon in non-compatible di-mon matings.

dilacerate, torn asunder.

dimerous (of basidia), having a constriction between the probasidium and the metabasidium, as in *Brachybasidium*.

dimidiate, (1) shield-like; appearing to lack one half, or having one half very much smaller than the other; (2) (of a pileus), without a stalk and semi-circular; (3) (of lamellae), stretching only halfway to the stipe; (4) (of an ascomatal wall), having the outer wall covering only the top part.

dimitic, see hyphal analysis.

dimixis, see heterothallism.

dimorphic, having two forms; esp. of *Histoplasma*, *Sporothrix*, and other pathogens of humans and animals which have yeast and mycelial habits. See Romano (*in* Ainsworth & Sussman (Eds), *The Fungi* **2**: 181, 1966; review), Szaniszlo (Ed.) (*Fungal dimorphism ....*, 1985), San-Blas (*Handb. Appl. Mycol.* **2**. *Humans, animals and insects*: 459, 1991; molec. aspects).

dioecism (adj. **dioecious**), the condition in which the male and female sex structures are on different thalli, e.g. in certain *Laboulbeniales*; also reported in *Lecidea verruca* where the 'male' thalli are mostly smaller (Poelt, *Pl. Syst. Evol.* **135**: 81, 1980); cf. monoecism, heterothallism.

diorchidioid (of teliospores), 2-celled and with septum longitudinal.

diphycophilous, fungi lichenized with both a green and a blue-green photobiont (Pike & Carroll, *in* Alexopoulos & Mims, *Introductory mycology*, edn 3, 1980).

diplo- (prefix), two; twice; double.

diploconidium, a binucleate conidium.
diplodioecious, see heterophytic.
diploheteroecious, see heterophytic.
diploid, (1) (of a nucleus), having the 2*n* number of chromosomes; (2) (of a cell), having the 2*n* number of chromosomes in one (synkaryotic, 2*n*) or two (dikaryotic, *n* + *n*) nuclei; (3) (of a mycelium), made up of dikaryotic diploid cells.
diploidization, the process by which a haploid cell (or mycelium) becomes a diploid (dikaryotic) cell (mycelium) having conjugate nuclei (Buller, 1941); cf. heterokaryotise.
diplokaryon, see synkaryon.
diplomonoecious, see homophytic.
diplont, the thallus of the diploid stage; the sporophyte.
diplophase, the part of a life-history in which the cells are diploid.
diplostichous, in two lines or groups.
diplostromatic, see stroma.
diplosynoecious, see homophytic.
direct (of fruit-body development), cell enlargement occurring at the same time as cell division; in **indirect** development cell enlargement mainly occurs after the period of cell division (Corner, 1950).
disc, (1) (**disk**) (of discomycetes), the round, plate-like or curved spore-producing part of the ascoma; (2) (of a pileus), the central part of the top surface.
disciform, round and flat.
discocarp, an ascoma in which the hymenium is uncovered when the asci and spores are mature; an apothecium.
discoid, flat and circular; resembling a disc.
discolourous, of a different colour, as of the two surfaces of a foliose lichen thallus.
discothecium, an ascostroma resembling an apothecium but bearing cylindrical bitunicate asci and differing from a hysterothecium by the weathering away of the covering layer (Korf, *Mycol.* **54**: 25, 1962).
discrete, (1) separate; not joining; (2) (of a conidiogenous cell), not subtended by a conidiophore; cf. integrated.
discrete body, a non-functional cleistothecial initial of a dermatophyte in culture; pseudocleistothecium.
disjunctor, a cell or projection, sometimes having a short existence, developing through the pores of septal lamellae of adjoining conidia in a chain (e.g. in *Monilinia*); a connective. See Batra (*Mycol.* **80**: 660, 1988).
disk, see disc.
dispersal spore, a spore disseminated by wind, water, or other agent; diaspore.
dispore, one of the spores of a 2-spored basidium as opposed to a **tetraspore**, one of the spores of a 4-spored basidium (Corner, 1947); cf. monospore.
dissepiment, a partition, e.g. that between the pores of a polypore.
dissociation, Leonian's name for mutation or saltation.
distal, situated away from either the centre of a body or the point of origin; terminal; cf. proximal.
distichous, in two lines.
distoseptate (of septation), having the individual cells each surrounded by a sac-like wall distinct from the outer wall, as in *Drechslera* (Luttrell, *Mycol.* **55**: 672, 1963); cf. euseptate.
diurnal, in daylight hours; cf. circadian, diel.

divaricate, divergent at right angles.
diverticulum, a pocket-like side branch, as on mycelium of *Pythium*.
dolabrate (**dolabriform**), hatchet-like in form.
dolabriform, having the shape of the head of an axe or cleaver.
dolicho- (in Greek combinations), long.
dolichospore, a long spore.
doliiform, barrel-like in form.
dolipore septum, a septum of a dikaryotic basidiomycete hypha which flares out in the middle portion forming a barrel-shaped structure with open ends as shown by electron microscopy; see Markham (*MR* **98**: 1089, 1994; review), Moore (*in* Hawksworth (Ed.), *Identification and characterization of pest organisms*: 249, 1994), Moore & McAlear (*Am. J. Bot.* **49**: 86, 1962); septal pore swelling; cf. parenthesome.
dorsal, back or upper surface; the surface facing away from the axis, cf. ventral; sometimes used for the upper surface of foliose lichens.
druse, a stellate cluster of large crystals in a lichen thallus.
dry spore, a spore that becomes separated without slime from the cell producing it (Mason, 1937); cf. slime spore.
dual phenomenon (in anamorphic fungi), the condition in which a fungus is made up of two culturally different elements or individuals (Hansen, *Mycol.* **30**: 442, 1938; Hansen & Snyder, *Am. J. Bot.* **30**: 419, 1943).
dual propagule (in a lichen), one comprising elements of both the fungal and the photosynthetic partner (e.g. isidium, soredium).
duplex (of the context), in two layers, that adjacent to the lamellae or tubes being harder than the one over it.
duvet (of dermatophytes), a soft, thick layer of hyphae like brushed-up cloth.
dysgonic (of dermatophytes), growing more slowly in culture, frequently with less aerial mycelium, than a normal, **eugonic**, strain. See Johnstone & La Touche (*TBMS* **39**: 442, 1956).
dystrophic, inadequately nourished; cf. eutrophic, oligotrophic.
e- (prefix) (**ex-**), from; out of; without; not having.
eccentric, see excentric.
echinate (dim. **echinulate**) (of spores, etc.), having sharply pointed spines; spinose.
echinulate, see echinate.
eclosion, an explosive series of movements which results in the release of a germinating inner spore from a rigid exosporium, as in *Hypoxylon fragiforme* (Chapela *et al.*, *CJB* **68**: 2571, 1990).
ecorticate, having no cortex.
ecotype, part of a population of a species showing morphological, chemical, or physiological characteristics which appear to be genetically determined and correlated with particular ecological conditions, but which are not considered of major taxonomic significance.
ectal, outer; outermost.
ectal excipulum (of ascomata), the outer layers including the subhymenium in a non-lichenized apothecium, sometimes multi-layered; see excipulum.
ecto- (prefix), outside.
ectoascus, the outer wall of a fissitunicate (q.v.) ascus, as in *Lecanidion*.

ectochroic, with pigment on the outside of the hypha.
ectomycorrhiza, see mycorrhiza.
ectoparasite, a parasite living on the outside of its host.
ectoplacodial, see stroma.
ectospore, (1) an exogenous spore; (2) a basidiospore (obsol.).
ectostroma, see stroma.
ectothecal (of ascomycetes), having the hymenium exposed.
ectothrix, living on the surface of hair.
ectotroph (adj. **ectotrophic**), see mycorrhiza.
ectotropic, curving out.
ectotunica, the outer wall of a bitunicate ascus.
edaphic, pertaining to the soil.
effete, (1) overmature, exhausted; (2) (of fruiting bodies), empty.
effigurate (of lichen thalli), obscurely lobed.
efflorescent, bursting out of.
effuse, stretched out flat, esp. as a film-like growth.
effused-reflexed (of *Basidiomycetes*), stretched out over the substratum but turned up at the edge to make a pileus.
egg, (1) the female gamete; (2) (of phalloids, *Amanita*, etc.), the young basidioma before the volva is broken.
eguttulate without oil-like drops (**guttules**) inside.
elater, (1) a free capillitium-thread, e.g. in myxomycetes and *Farysia*; (2) a body with spiral or annular markings in the gleba of *Battarrea*.
ellipsoidal (of spores, etc.), elliptical in optical section.
emarginate, (1) (of lamellae), see sinuate; (2) (of apothecia), lacking a thalline exciple (excipulum thallinum) (lichens) or a raised proper exciple (excipulum proprium).
emend, to correct an error; cf. amend.
endemic, native to one country or geographical region.
endo- (prefix), inside.
endoascospores, spore-like cells produced within ascospores (see Morgan-Jones, *CJB* **51**: 493, 1972).
endoascus, the often extensible inner wall layers of a bitunicate (q.v.) ascus.
endobasidial (of a conidiophore in a lichenized pycnidium), having a secondary sporing branch (obsol.); cf. exobasidial.
endobasidium, see basidium.
endobiotic, making growth inside living organisms.
endocarpinoid (of lichenized perithecia), sunk into the tissues of the thallus, as in *Endocarpon* (obsol.).
endocarpous (of gasteromycetes, etc.), having the mature hymenium covered over; angiocarpous.
endochroic, with pigment inside the cell.
endocommensal, an organism living as a commensal (see symbiosis) inside another (e.g. *Trichomycetes* in the gut of *Insecta*).
endoconidium, a conidium formed inside a hypha, e.g. as in *Thielaviopsis basicola*.
endocyanosis, the inclusion of cyanobacteria inside the cells of another organism; in fungi, *Geosiphon*.
endocyclic, see microcyclic.
endocystidium, see cystidium.
endoectothrix, making growth in and on a hair.
endogenous, (1) living inside; (2) immersed in the substratum; (3) undergoing development within.

endogonidium, a gonidium having its development inside a receptable or gonidangium (obsol.).
endohypha, see hypha.
endokapylic, a thallus of a lichenicolous fungus in which no morphologically distinct lichenized structure is formed (Poelt & Vĕzda, 1984; Rambold & Triebel, *Bibl. Lich.* **48**, 1992).
endolithic, in stone (Kobluk & Kahle, *Bull. Can. Pet. Geol.* **25**: 208, 1977; fungi, bibliogr.); cf. epilithic. See also cryptoendolithic.
endomycobiont, a fungal biont in a symbiosis completely immersed in the tissues of the host (e.g. the fungal partner in a mycophycobiosis); an inhabitant (see symbiosis); also used of certain mycorrhizas.
endomycorrhiza, see mycorrhiza.
endo-operculation (of sporangial dehiscence in chytrids), operculum forced off and carried away by the emerging sporogenous contents (cf. exo-operculation).
endoparasite, a parasite living inside its host.
endoperidermal, within the periderm (Lambright & Tucker, *Bryologist* **83**: 170, 1980); endophloeodal + hypophloeodal.
endoperidium, the inner layer of the peridium.
endophloeodic (**endophloeodal**, **endophloeic**) (of the thallus of a crustaceous lichen), almost entirely immersed in bark.
endophyllous, living within (i.e., below the cuticle) leaves.
endopropagule, a propagule produced inside the body (medical mycology).
endosaprophytism, the destruction of an alga by the fungus in a lichen (Elenkin).
endosclerotium, a sclerotium of endogenous origin.
endospore, (1) the inner wall of a spore (see ascospore, spore wall); (2) an endogenous spore, e.g. a sporangiospore.
endosymbiont, an organism which lives in mutualistic symbiosis within the cells of another organism; the inhabitant.
endothrix, living inside a hair.
endotrophic, see mycorrhiza.
endotunica, endoascus (q.v.).
endozoic, living inside an animal.
enphytotic, a plant disease of which the damage is constant from year to year; cf. epiphytotic.
ensate (**ensiform**), narrow and pointed; sword-like in form.
enteroblastic, see blastic.
entheogen, a plant (or fungus) substance used by humans in prehistory associated with religous feelings; see ethnomycology, hallucinogenic fungi, soma.
entire (of edges of lamellae, etc.), not torn; having no teeth.
ento- (prefix), inside.
entomo- (prefix), of *Insecta*.
entomogenous, living in or on insects, esp. as pathogens.
entomophilous (of fungi), having spores distributed by insects.
entoparasitic, parasitic inside the host.
entoplacodial, see stroma.
entostroma, see stroma.
epapillate, having no papillae.
epi- (prefix), upon.

epibasidium, see basidium.
epibiotic, living on the surface of another organism.
epibryophilous, growing over bryophytes.
epichroic, discoloration due to injury.
epicortex, a thin polysaccharide-like layer over the surface of the cellular upper cortex in thalli of some *Parmeliaceae* visible by SEM (Hawksworth, *in* Hale, *Smithson. Contr. bot.* **10**: 5, 1973) and which may have regular pores functioning in gas exchange (Hale, *Lichenologist* **13**: 1, 1981). See Hyvärinen (*Lichenologist* **24**: 267, 1992; environmental induction), Lumbsch & Kothe (*Mycotaxon* **43**: 277, 1992; *Coccocarpiaceae, Pannariaceae*).
epicutis, see cutis.
epiflora, surface flora; sometimes applied (incorrectly) to the microbiota on seed surfaces; the epibiota.
epigeal (**epigean**, **epigeic**, **epigeous**), on the earth.
epigeic (of lichens), not attached to any substrate but blowing about on the surface of the ground.
epigenous, growing on the surface.
epigynous, having the antheridium above the oogonium on one hypha.
epihymenium, a thin layer of interwoven hyphae on the surface of the hymenium (Corner, 1950); cf. epithecium.
epikapylic, a thallus of a lichenicolous fungus in which a morphologically distinct lichenized structure is formed (Poelt & Vězda, 1984).
epilithic, living on the surface of stones; cf. endolithic.
epinecral layer, see necral layer.
epiphloeodal, living upon bark.
epiphragm, the membrane over the young fruit-body in the *Nidulariaceae*.
epiphyllous, on the upper surface of a leaf; foliicolous.
epiphyte, a plant living on another, but not as a parasite.
epiphytic (adj.), frequently = corticolous.
epiphytotic, an epidemic among plants; cf. enphytotic.
epiplasm (of an ascus), the cytoplasm not used up in the 'free cell formation' of ascospores.
epispore, see ascospore, spore wall.
episporium, see spore wall.
epistroma, see stroma.
epithalline, of a falsely thalline apothecial edge in lichenized fungi.
epithecium, tissue at the surface of an apothecium formed by the branching of the ends of the paraphyses above the asci; cf. epihymenium, pseudoepithecium.
epithelium, see cutis.
epitunica, see exosporium; spore wall.
epixylic (**epixylous**), living on wood; lignicolous.
epizoic, living on animals.
epizootic, an epidemic among animals.
epruinose, having no pruina.
equal (of a stipe), having the same diameter throughout.
erect, upright; straight, not curved, up.
erinaceous, prickly like a hedgehog.
erose (of a lamella, etc.), having delicate tooth-like projections from the edge.
erratic (of lichen thalli), not fixed to the substratum and often blowing around, e.g. *Chondropsis semiviridis*, *Sphaerothallia esculenta* ('manna'); epigaeic; vagrant; wandering lichens.
erumpent, bursting through the surface of the substratum; cf. perrumpent.

eseptate, see aseptate.
esorediate (**esorediose**), having no soredia.
ethnomycology, mycology as a branch of ethnology. See Wasson & Wasson (*Mushrooms, Russia and history*, 2 vols. 1957), and the writings by Wasson and Heim on hallucinogenic fungi; also Lowy (*Mycol.* **64**: 816, 1972, Maya codices; **66**: 188, 1974, *Amanita muscaria* and the thunderbolt legend; *Revista Interam. Rev.* **2**: 405, 1972, **5**: 110, 1975, **10**: 94, 1980, mushrooms and religion), Redlinger (Ed.) (*The sacred mushroom seeker*, 1990), soma.
etiology, see aetiology.
eu- (prefix), true; sometimes used, but wrongly, for the subgenus or section including the type species of the generic name of which it is an infrageneric taxon.
eucarpic, developing reproductive structures on limited portions of the thallus; residual nucleate protoplasm remaining and capable of further mitotic growth and regeneration.
euchroic, having true pigmentation; c.f. epichroic.
eucortex (of lichens), a cortex composed of well-differentiated tissue.
euform, of species with all spore forms present in the lifecycle.
eugonic, see dysgonic.
eugonidium, a bright green lichen photobiont (e.g. *Trebouxia*) (obsol.).
euhymenium, see hymenium.
eukaryote (adj. **eukaryotic**), one of the *Eukaryota*; cf. prokaryote.
eumorphic, well formed.
eumycetoma, see mycetoma.
euseptate (of conidial septation), having cells separated by multilayered walls of similar structure to lateral walls, as in *Pyricularia* (Luttrell, *Mycol.* **55**: 672, 1963); cf. distoseptate.
eustroma, see stroma.
euthecium, an ascoma (cleistothecium, perithecium, apothecium) of an euascomycete; cf. pseudothecium.
euthyplectenchyma, see plectenchyma.
eutrophic, rich in nutrients; cf. dystrophic, oligotrophic.
eutrophication, nutrient enrichment, usually used when directly or indirectly caused by human influences.
eutypoid, having groups of perithecia in a stroma with the ostioles vertical and breaking through the surface individually; cf. valsoid.
evanescent, having a short existence; fugacious.
everted, turned inside out.
exasperate, roughened with hard projecting points.
excavate, hollow out.
excentric (**eccentric**), (1) one sided; (2) (of a stipe), at one side or not in the centre of the pileus; cf. centric.
exciple, see excipulum.
excipulum (of ascomata), tissue or tissues containing the hymenium in an apothecium, or forming the walls of a perithecium; cf. **ectal -, medullary -. - proprium**, non-lichenized excipular tissue forming the margins of an apothecium of a lichenized fungus; **- thallinum**, lichenized excipular tissue of a lecanorine apothecium, external to an excipulum proprium (which may be much reduced), usually with a structure like that of the vegetative lichen thallus. See references under apothecium, tissue types.

exhabitant, see symbiosis.

exigynous, having the antheridial stalk arising directly from the oogonial cell above the basal septum.

exit tube, extension of the sporangium, produced prior to or during sporangial discharge, which enables sporangial contents to be released outside the host or substrate.

exo- (prefix), outside.

exobasidial, (1) having the basidia uncovered; (2) separated by a wall from the basidium; (3) (of a conidiophore in a lichenized pycnidium sporophore; obsol.), having no secondary branch (Steiner); cf. endobasidial.

exochthonous (of soil organisms), invaders ill-adapted to live in soil (Park, 1957); cf autochthonous.

exogenization, an hypothetical process whereby endogenously formed spores become exogenously formed: a mechanism proposed to support the evolution of basidiomycetes from the ancestral ascomycetes (Clemençon, *Persoonia* **9**: 363, 1977).

exogenous, undergoing development outside.

exolete (of perithecia, pycnidia, etc.), long over-mature; empty.

exo-operculation (of sporangial dehiscence in chytrids), the operculum is hinged to the rim of the pore; 'true operculation'; cf. endo-operculation.

exoperidium, the outer layer of the peridium.

exopropagule, a propagule formed outside the body (medical mycology).

exospore, see spore wall.

exosporium, see spore wall.

exotic, (1) (adj.), of another country; not indigenous; (2) (n), an - organism.

expallant (of a pileus), becoming pale on drying.

explanate, spread out.

explosive (of asci), see ascus.

exserted, sticking out; protruding (e.g. a mature ascus of *Ascobolus*).

extendituniate, see ascus.

extramatrical, (1) living on or near the surface of the matrix or substratum; (2) VAM structures (mycelium, spores) developing outside roots of a phytobiont.

extrusome, membrane-bound structure derived from vesicle of the Golgi system and anchored to the cell membrane by proteinaceous particles; contents extruded in respose to stimuli.

fabiform, see reniform.

facultative, (1) sometimes; not necessarily; not obligate (q.v.); (2) (of a parasite), having the power of living as a saprobe; able to be cultured on laboratory media; (3) **- synonym**, see synonym.

falcate (**falciform**), curved like the blade of a scythe or sickle.

falciphore, see falx.

false membrane (of a smut), a tissue of sterile fungal cells limiting the sorus, as in *Sphacelotheca*.

falx, a 'fertile hypha' or conidiophore of *Zygosporium*, having the form of a bill-hook. Falces may be sessile or on special hyphae or **falciphores** (Mason, 1941).

farctate (of a stipe), having the centre softer than the outer layer; stuffed.

farinaceous (**farinose**), like meal in form or smell.

fasciate (**fasciated**), massed or joined side by side.

fascicle, (1) (esp. of hyphae), a little group or bundle; (2) (of books or exsiccatae), one part, or collection of separate leaves, of a work issued in parts.

fasciculate, having growth in fascicles; **- basidium**, see basidium.

fastigiate, having parallel, massed, upright branches.

fastigiate cortex (of lichens), made up of parallel hyphae at right angles to the axis of the thallus; cf. fibrous cortex.

fatiscent, cracked or falling apart.

faveolate (**favose**), honeycombed; alveolate.

favic chandeliers, dichotomously branched, swollen, hyphal tips, growing submerged from the edge of the colony of *Trichophyton schoenleinii*.

favoid, like a honeycomb.

fenestrate, (1) having windows or openings; (2) (of spores), muriform.

fertile hypha, see conidiophore.

fertilization, the fusion of sex nuclei; **- tube/hypha** hypha developing from an antheridial gametic cell; passing through the antheridium wall and bridging the gap between non-contiguous gametangia to penetrate the oogonium.

fibril, (1) a very small fibre; (2) (in *Usnea*), short simple branches perpendicular to the main branches.

fibrillar surface coat, fibrous component attached to the flagellar membrane and covering the entire surface of the flagellum.

fibrillose, covered with silk-like fibres.

fibrosin a refractive substance found in some fungi.

fibrous cortex (of lichens), made up of loosely woven distinct hyphae parallel with the long axis of the thallus; cf. fastigiate cortex.

filamentous, (1) thread-like; filamentose; (2) (of lichens), the photobiont forms a filament of cells which is surrounded by hyphae or cells of the mycobiont (e.g. *Cystocoleus*, *Racodium*, *Coenogonium*).

filiform, thread-like.

fimbriate, edged; delicately toothed; fringed; cf. fimbrillate.

fimbrillate, having a very small fringe; cf. fimbriate.

fimicolous, living on animal droppings; coprophilous.

fission, (1) becoming two by division of the complete organism; cf. budding; (2) (of conidial liberation), secession by the separation of a double septum; cf. fracture, lysis.

fissitunicate, see ascus.

fistular (**fistulose**), hollow, like a pipe.

flabellate (**flabelliform**), like a fan; in the form of a half-circle.

flaccid, not stiff; limp; flabby.

flagellum (pl. **flagella**), cylindrical extension of a eukaryotic cell, bounded by a plasma membrane and containing an axoneme; two types can be distinguished by electron microscopy, the **whiplash** with a smooth continuous surface (as in *Chytridiomycota*) and the **tinsel**, characteristic of *Hyphochytriomycota*, with the surface covered with hair-like processes (**mastigonemes** or **flimmers**); **-ar apparatus**, complex consisting of one or more basal bodies which may bear flagella, may have microtubular and fibrous roots associated with their bases; **-ar fibrous roots**, roots composed of a bundle of filaments, frequently appearing cross-striated; **-ar hairs**, filamentous appendages usually arranged in one or more rows but not covering the entire surface of a flagellum; **-ar scales**, organic structures of discrete size and shape, often covering the whole surface of the flagellum external to the plasma membrane, usually assembled in the dictyosome. See Barr (*Mycol.* **84**: 1, 1992; terminology flagellar apparatus). See also axoneme, blepharoplast; and cf. cilium.

flesh, the trama, esp. of the pileus of an agaric or bolete.

fleshy (of sporocarp), soft, not cartilaginous-like or wood-like.

flexuous (**flexuose**), wavy.

flexuous hyphae (of *Uredinales*), an unbranched or branched haploid hyphal projection from a pycnium, which may be diploidized by a pycniospore of opposite 'sex' (Craigie, see *Nature* **141**: 33, 1938); cf. receptive body.

flocci, cotton-like groups or tufts.

floccose, cottony; byssoid.

flocculent (of a liquid-culture), having small masses of cells throughout or as a deposit.

flocculose, delicately cottony.

flora, (1) the plants of a particular geographical area or habitat; (2) a description, catalogue or list of all or some groups of plants in a particular area. Formerly applied to fungi and lichens (i.e. **fungus -**, **lichen -**), but as fungi are not plants the term mycobiota (q.v.) is preferred.

fluorescent, giving out light when placed in ultraviolet (or other) radiation.

flush (of fungal growth), the sudden development of a large quantity of mycelium or a periodic surge of basidiomata emergence, esp. in mushroom cultures.

foliicolous, living on leaves; **- lichens**, see Santesson (*Symb. bot. upsal.* **12** (1), 1952; monogr.), Farkas & Sipman (*Trop. bryol.* **7**: 93, 1993; checklist 482 spp., bibliogr.), 81; bibliogr. 83 papers 1952-85), Lücking (*Beih. Nova Hedw.* **104**: 1, 1992; keys 228 spp. Costa Rica), Ferraro (*Bonplandia* **5**: 191, 1983; S. Am.), Santesson & Tibell (*Austrobaileya* **2**: 529, 1988; 66 spp. Australia), Sérusiaux (*Bot. J. Linn. Soc.* **100**: 87, 1989; review).

foliole, a small leaf-like excrescence on the surface of a foliose lichen.

foliose, (1) leaf-like; (2) (of lichens), having a layered (stratose) thallus, usually with a lower cortex, and attached to the substratum either by rhizines or at the base, but not by the whole lower surface (e.g. *Parmelia*, *Peltigera*).

foot cell, (1) a basal cell supporting the conidiophore in *Aspergillus*; (2) the basal cell of the conidium in *Fusarium*. See Sutton (*TBMS* **86**: 1, 1986; occurrence in anamorphic fungi).

forate (of 'gasteromycete' basidiomata), invagination of the primordial tissue resulting in a series of pits; the type of development generally known as 'coralloid' of which it is the opposite (Dring, 1973).

-form (suffix), shape.

fornicate, arched; (of *Geastrum*), having the fibrous and fleshy layers of the fruit body becoming arched over the cup-like mycelial layer.

foveate, having small holes or cavities; pitted.

foveolate, delicately pitted; dimpled; dim. of foveate.

fracture (of conidial liberation), secession involving the rupture of the wall of an adjacent vegetative or degenerate cell at a point removed from the septum; cf. fission, lysis, rhexolytic.

fragmentation spores, conidia produced by hyphae breaking up into separate cells.

free (of lamellae or tubes), not joined to the stipe; cf. remote, seceding.

free cell formation, the process by which the 8 nuclei, each with some adjacent cytoplasm, are cut off by walls in the immature ascus to become ascospores.

friable, readily powdered.

fructicolous, living on fruit.

fructification, see fruit-body.

fruit-body (fructification), a general term for spore-bearing organs in both macrofungi and microfungi. The more precise terms ascoma (ascocarp), basidioma (basidiocarp), conidioma, sporocarp, etc. are preferred usage.

fruticolous, living on shrubs.

fruticose, (1) shrub-like; (2) (of lichens), having an upright or hanging thallus of radiate structure (e.g. *Cladonia*, *Ramalina*, *Usnea*).

fruticulose (of lichens), having a minutely shrubby habit (e.g. *Ephebe*, *Polychidium*).

fugacious, see evanescent.

fulcrum (of lichens), a conidiophore within a pycnidium (obsol.).

fungal, (1) (n.), a fungus (obsol.); (2) (adj.), see fungous.

fungicidal, able to kill fungus spores or mycelium.

fungiform, mushroom-shaped.

fungistasis, (1) see mycostasis; (2) (**fungistatic**) (of a substance, or of a concentration of a fungicide), inhibiting fungus growth but not fungicidal; cf. genestasis.

fungivorous, using fungi as food; **fungivore**, an organism which does this.

fungoid, fungus-like.

fungology, mycology (obsol.).

fungoma, fungus ball formation.

fungophobia, a horror or dread of fungi (Hay, *British Fungi*: 6 , 1887) (obsol.).

fungous, of, or having to do with, fungi; fungal.

funicular, cord-like.

funicular cord (**funiculus**), the cord of hyphae by which the peridioles in *Nidulariaceae* (e.g. *Cyathus*) are at first fixed to the inner wall of the peridium; see splash cup.

funiculose (of hyphae), aggregated into rope-like strands, plectonematogenous (q.v.).

funoid, composed of rope-like strands or fibres; funicular (q.v.).

furcate, forked.

furfuraceous, covered with bran-like particles; scurfy.

fuscous, dusky; too brown for a grey (Corner, 1958).

fuseau, a fusoid macroconidium of a dermatophyte (e.g. *Microsporum*); a spindle (obsol.).

fusiform, spindle-like; narrowing toward the ends.

fusoid, somewhat fusiform.

fuzzy coat, the outer gelatinous coat of an ascus, esp. of one staining blue in iodine; see ascus.

galeate, hooded; hat-shaped or helmet-shaped.

galvanotropism, a reaction to an electrical field (McGillivray & Gow, *J. Gen. Microbiol.* **132**: 2515, 1986).

gametangium (pl. **-ia**) (gametange), cell containing gametes or gametic nuclei; the gametangium may initially be diploid and is the site of meiosis, or haploid. See zygangium.

gamete, naked uninucleate haploid cell with the sole function of fusing with another gamete to produce a zygote; sometimes used for the sex-nuclei of coenogametes.

gametogenesis, the development of gametes.

gametophyte, a haploid or sexual plant; haplont or haplophase; cf. sporophyte.

gametothallus, a thallus producing gametes; cf. sporothallus.

gamma particle, a DNA-containing cytoplasmic organelle in the zoospore of *Blastocladiella emersonii* (Myers & Cantino, *The gamma particle*, 1974; Barstow & Lovett, *Mycol.* **67**: 518, 1975).

gangliform, having knots; knotted.

gangliospore, Subramanian's (*Curr. Sci.* **31**: 410, 1962) term for aleuriospore in the sense of 'holoblastic conidium'.

gasteroconidium, see gasterospore.

gasterospore (**gasteroconidium**), a thick-walled, globose, chlamydospore of *Ganoderma*; probably apomictic (see Bose, *Mycol.* **25**: 432, 1933; *Sydowia, Beih.* **1**: 176, 1957).

gel tissue, a mixture of gel and hyphae found in members of the *Leotiales* and *Tremellales*; the gel may arise either by direct secretion or by disintegration of hyphae: see Moore (*Mycol.* **57**: 114, 1965; *Am. J. Bot.* **52**: 389, 1965, ontogenesis; *Stain Technol.* **40**: 23, 1965, staining); cf. gliatope.

gelatin, product obtained by boiling collagen, soluble in water above *c.* 40°C. Gels of *c.* 4–12% used to test ability of some microorganisms to liquefy or hydrolyse gelatin.

gelatinous, jelly-like; used for the hyphae of tissues which become partly dissolved and glutinous with moisture.

gemma (pl. **-ae**), (1) an asexual propagule borne singly or in chains at the ends of hyphae, referred to in older literature as a chlamydospore (*Saprolegniaceae*); (2) another term for oidia in *Basidiomycetes* (Gäumann, *The Fungi*: 449, 1928), rejected for this usage by Kendrick & Watling (*in* Kendrick (Ed.), *The whole fungus* **2**: 477, 1979).

generative hyphae, see hyphal analysis.

genestasis, inhibition of sporulation; **genistat**, a substance preventing or reducing sporulation in fungi without materially affecting vegetative growth; 'antisporulator' (Horsfall, 1947); cf. fungistatic.

geniculate, bent like a knee.

genistat, see genestasis.

genocentric, see reproductocentric.

geofungi, soil fungi (Cooke, 1963).

geophilous, earth loving, e.g. of fungi having underground fruit bodies; cf. terricolous.

geotropism, a reaction to gravity, e.g. sporangiophores of *Mucorales* and stipes, gills and tubes of fruit bodies of *Basidiomycota* (Moore, *New Phytol.* **117**: 3, 1991).

germ pore, a differentiated, frequently apical area, or hollow, in a spore wall (esp. in rusts) through which a **germ tube** (a germination hyphae) may come out; see Melendez-Howell (*Ann. Sci. nat. Bot. sér.* 12 **8**: 487, 1967; germ pore of basidiospores).

germ slit, a thin area of spore wall usually orientated along the long axis of the spore. In *Bryothele mira* the germ slit is transverse (Dobbeler, *Nova Hedw.* **66**: 337, 1998).

germ tube, a germination hyphae which is formed by a germinating spore.

germicide, a substance causing destruction of microorganisms.

germination by repetition, producing secondary spores in place of germ tubes, as in *Heterobasidiomycetes* and *Sporobolomyces*.

gibber, gibbous (q.v.).

gibbous (of a pileus), having a swelling or wide umbo, or having a convex top and a flat underside; gibber, gibbose.

gill (of an agaric), commonly used in English for lamella (q.v.) which is to be preferred as a more international term; **- fungi**, members of the *Agaricales*.

gilvous, pale yellow.

glabrous, smooth; not hairy.

glaireous, slimy.

glaucous, having a bluish-grey waxy bloom.

gleba, the sporing tissue in an angiocarpous sporocarp, esp. of gasteromycetes and hypogeous *Pezizales*; **glebal mass**, the projectile of *Sphaerobolus*.

glebula, a rounded process from a lichen thallus.

gleocystidium, see cystidium.

gleoplerous hyphae (oil hyphae), hyphae with very long cells (or unicellular), with numerous oil drops in the plasma. See Jülich (*in* Gams (Ed.), *Kleine Kryptogamenflora* **II**(7), 1984).

gliatope, a site of heavy gel production (Moore, *Am. J. Bot.* **52**: 391, 1965). See gel tissue.

globoid (**globose, globular, globulose**), spherical or almost so.

glochidiate, covered with barbed bristles.

gloeocystidium, see cystidium.

glomerulus (**glomerule**), a clump or cluster; frequently used for clusters of photobiont cells in lichens.

glossoid cell, elongate (tongue-shaped) cell containing an elaborate extrusome (*Haptoglossa*).

glucans, one of the main constituents of fungal walls. **R-glucans** are alkali insoluble; **S-glucans** are soluble.

gluten, (1) a substance on the surface of some agarics, etc., which is sticky when wet; (2) spore mass in *Phallus*.

glutinous, sticky; made up of, or covered with, gluten.

glypholecine, having particularly labyrinth-like lirella as in *Glypholecia*.

gnotobiotic (of cultures), ones in which all the living components are known; cf. axenic.

golgi body. Dictyosome with large numbers of cisternae which may be visible using light microscopy after staining.

gongylidius (pl. **gongylidia**), a bulbous structure developed by fungi cultivated by termites.

gonidial layer, photobiont layer in a lichen thallus (obsol.).

gonidimium, a hymenial alga (obsol.).

gonidium, photobiont (obsol.).

gonimium, a cyanobacterial cell in a lichen thallus (obsol.).

goniocyst (**goniocystula**), a group of algal cells derived from a single cell surrounded by a hyphal envelope forming a roundish structure which is not a soralium (e.g. the vegetative thallus '*Botrydina vulgaris*'). See Sérusiaux (*Lichenologist* **17**: 1, 1985).

goniocystangium, cup-like structure bearing goniocysts (q.v.) on foliicolous species of *Catillaria* (Vězda, 1980) and *Opegrapha* (Santesson, *Svensk Naturv.* 1968: 176, 1968; Sérusiaux, 1985).

goniosporous, having angled spores.

gonosphere, a zoospore of the *Chytridiales* (obsol.).

gonotocont, the organ in which meiosis takes place.

gossamers, fine, floating mycelial nets produced by fungi on media lacking added carbon (Wainwright, *Mycologist* **1**: 182, 1987).

granular (**granulate**, **granulose**) (of a surface), covered with very small particles.

granuloma, a nodule of firm tissue formed as a reaction to chronic irritation.

graphium (pl. **-ia**), the synnema of *Graphium*.

gregarious, in companies or groups but not joined together.

guttate, (1) having tear-like drops; (2) (of a pileus), marked as if by drops of liquid.

guttulate (of spores), having one or more oil-like drops (**guttules**) inside.

gymnocarpous (of a sporocarp), open, with the hymenium appearing and developing to maturity exposed and not enclosed; cf. angiocarpous.

gymnoplast, see protoplast.

gymnothecium, an ascoma in which the peridium is a loose hyphal network, typical of *Gymnoascaceae*.

gynoecium (pl. **-a**), the female gametangium, formed from a haplophase thallus.

gynophore (of *Pyronemataceae*), the multinucleate female structure undergoing development.

gyrate (**gyrose**), curved to the back and to the front in turn; folded and wavy; convoluted like a brain; (of an apothecioid ascoma), concentrically folded, e.g. *Umbilicaria*.

habitat, natural place of occurrence of an organism.

haerangium, the sporulating organ of certain ascomycetes (e.g. *Fugascus*, *Ceratostomella*), classed by Falck as *Herangiomycetes*, in which the 8 ascospores developed from the **octophore** are contained by a membrane and surrounded by a circle of hairs (the **tentacle**) around the ostiole of the perithecium (Falck, 1947).

hair (in *Agaricales*), one of the hair-shaped epicuticular elements forming a pilose covering or down under a lens and not homologous with a cystidium, pseudoparaphysis, or seta, e.g. in *Lachnella*, *Crinipellis* (as restricted by Singer, 1962: 61).

haline, found near the sea shore (obsol.).

hallucinogenic fungi (**teonanácatl**), basidiomata (of *Psilocybe* (*P. mexicana*, etc.; see Singer and Singer & Smith, *Mycol.* **50**: 239, 1958), *Stropharia*, *Paneolus*, *Lycoperdon cruciatum*, *L. mixtecorum*, etc.) eaten by Mexican Indians during magical ceremonies to induce cerebral effects. A crystalline active principle has been isolated and named **psilocybin** (**psilocin**).

Lit.: Heim, Wasson *et al.* (*Les champignons hallucinogènes du Mexique*, 1958; *Nouvelles investigations sur les champignons hallucinogènes*, 1967 [reprinted from *Archiv. Mus. Nat. d'Hist. Nat.* sér. 7 **6**, **9**]). See ethnomycology, soma.

halmophagous (of ectotrophic mycorrhiza), having a mantle and a Hartig net (Burgeff, 1943).

halonate, (1) (of a leaf-spot), having concentric rings; one of the 'frog-eye' type; (2) (of a spore), having a transparent coat around it.

halophilic, tolerating salt; living in salt water; *Dendryphiella salina* is esp. well studied and there appears to be no salt accumulation in its vacuoles (see Jennings, in Rodriguez-Valera (Ed.), *General and applied aspects of halophilic microorganisms*: 107, 1991; review, bibliogr.).

hamate (**hamose**, **hamous**), hooked; unciate; cf. hamulate.

hamathecium (Eriksson, *Opera Bot.* **60**: 15, 1981), a neutral term for all kinds of hyphae or other tissues between asci, or projecting into the locule or ostiole of ascomata; usually of carpocentral origin; interascal tissues. Eriksson recognized seven categories:
(A) **Interascal pseudoparenchyma**, carpocentral tissues unchanged or compressed between developing asci; e.g. *Wettsteinina*.
(B) **Paraphyses**, hyphae originating from the base of the cavity, usually unbranched and not anastomosed; e.g. *Pyrenula*, *Xylaria*.
(C) **Paraphysoids** (trabecular pseudoparaphyses; tinophyses), interascal or pre-ascal tissue stretching and coming to resemble pseudoparaphyses; often only remotely septate, anastomosing and very narrow (see Barr, *Mycol.* **71**: 935, 1979); e.g. *Patellaria*, *Melanomma*.
(D) **Pseudoparaphyses** (cellular pseudoparaphyses; cataphyses), hyphae originating above the level of the asci and growing downwards between the developing asci, finally becoming attached to the base of the cavity and often also then free in the upper part; often regularly septate, branched and anastomosing and broader; e.g. *Pleospora*.
(E) **Periphysoids**, short hyphae originating above the level of the developing asci but not reaching the base of the cavity; e.g. *Nectria*, *Metacapnodium*.
(F) **Periphyses**, hyphae confined to the ostiolar canal; unbranched and not anastomosing; can occur in conjunction with (B), (D) or (E); e.g. *Gibberella*, *Pyrenula*.
(G) **Hamathecial tissue absent** (not figured), e.g. *Dothidea*.

hamulate (**hamulose**), having little hooks; cf. hamate.

haplo- (prefix), one only; single.

haploconidium (of *Tremellales*), a uninucleate conidium.

haplodioicious, see heterothallism.

haplogonidia (**haplogonimia**), gonidia (gonimia) in ones, not in groups (obsol.).

haploheteroecious, see heterothallism.

haploid, (1) (of a nucleus), having the n number of chromosomes; (2) (of a cell), having 1 haploid nucleus; (3) (of a mycelium), made up of haploid cells.

haplomonoecious, see homothallism.

haplont, the thallus of the haplophase; the gametophyte.

haplophase, the part of the life history in which the cells are haploid.

haplostromatic, see stroma.

haplosynoecious, see homothallism.

hapteron, (1) an aerial organ of attachment of some fruticose lichens (e.g. *Alectoria sarmentosa* subsp. *vexillifera*) formed by a secondary branch which becomes attached to the substratum; (2) attachment organ at base of a funicular cord in *Nidulariaceae*.

haptonema, filamentous appendage (usually coiled) consisting of the plasma membrane, a sheath of endoplasmic reticulum, and a core of several microtubules anchored near the kinetosome.

hastate, like a spear-head or arrow-head in form.

haustorial cap, an electron-dense, cap-like mass at the end of a lobe of the haustorial apparatus of *Exobasidium camelliae* (Mims, *Mycol.* **74**: 188, 1982).

haustorium (pl. **-ia**), a special hyphal branch, esp. one within a living cell of the host, for absorption of food (see Karling, *Am. J. Bot.* **19**: 41, 1932). Honegger (*New Phytol.* **103**: 785, 1986) distinguishes three main types of fungus–plant cell interactions: (1) **wall-to-wall apposition** with no penetration; (2) **intracellular haustoria** where the fungus penetrates into the plant cell, with or without the formation of special sheath, neckband, or collar; (3) **intraparietal haustoria** where penetration is restricted to the wall layers (common in some groups of lichens).

helicospore, a non-septate or septate spore, with a single (usually elongated) axis curved through at least 180° but may describe one or more complete rotations, in two or three dimensions (cf. amerospore, scolecospore); any protuberances, other than setulae, $<1/4$ spore body length (cf. staurospore). See Goos (*Mycol.* **79**: 4, 1987; terminology).

heliophilous, preferring direct sunlight; cf. anheliophilous.

heliozooid, amoeba-like, but having well-marked ray-like pseudopodia.

helminthoid, worm-like in form; vermiform.

helotism, the physiologic relation of alga to fungus in a lichen (obsol.).

hemi- (prefix), half; in part; cf. semi-.

hemiamyloidity The red colouration when iodine solutions are applied to the hymenium of most *Ascomycota* (Baral, *Mycotaxon* **29**: 399, 1987).

hemiangiocarpous (of a sporocarp), opening before quite mature.

hemiascospore, ascospore of a hemiascus.

hemiascus, the atypical multispored ascus of the *Hemiasci*.

hemicompatible, see compatible.

hemifissitunicate, see ascus.

hemiform, of species with no spermatia (q.v.) or aeciospores (q.v.) in the life-cycle.

hemiparasite, a facultative parasite.

hemispore (esp. of dermatophytes), (1) a cell at the end of a filament, which later becomes by division deuteroconidia; protoconidium (after Vuillemin); (2) one of the two cells produced by a primary trans-septum in an ascospore (Eriksson, *Ark. Bot.* II **6**: 339, 1967), see septum.

hepaticolous, growing on liverworts (*Hepaticae*); cf. muscicolous.

herbarium (pl. **-ia**), (1) a collection of dried plants; (2) the place in which such a collection is stored. Often also used for dried reference collections of fungi, especially when curated along with plant specimens.

herbicolous, living on herbs.

hetero- (prefix), other; not normal; different.

heterobasidium, see basidium.

heterocytic, see astatocoenocytic.

heteroecious (n., **heteroecism**), undergoing different parasitic stages on two unlike hosts, as in the *Uredinales*.

heterogametes, gametes of different form.

heterogamy, the copulation of heterogametes; cf. isogamy.

heterokaryosis (adj. **heterokaryotic**), (1) (of cells), the condition of having two or more genetically different nuclei, sometimes as a result of anastomosis, cf. dikaryotic; (2) (of mycelia), being made up of heterokaryotic cells; see Parmeter *et al.* (*Ann. Rev. Phytopath.* **1**: 51, 1963; in plant pathogenic fungi), Caten & Jinks (*TBMS* **49**: 81, 1966; occurrence in nature).

heterokaryotic, (1) having two or more slightly ($<5\%$) genetically different nuclei in common cytoplasm (fungi); (2) showing nuclear dimorphism (protists).

heterokaryotic, (1) having two or more slightly ($<5\%$) genetically different nuclei in common cytoplasm (fungi); (2) showing nuclear dimorphism (protists).

heterokaryotise (of rusts and pyrenomycetes), fusion of haploid structures of opposite sex which does not give a conjugate arrangement of the nuclei, cf. diploidization.

heteromerous, (1) (of a lichen thallus), having the mycobiont and photobiont in well-marked layers, usually between the medulla and the upper cortex; (2) (of trama in *Russulaceae*), having sphaerocyst nests among filamentous hyphae; cf. homoiomerous.

heteromixis, see heterothallism.

heteromorphic (**heteromorphous**), (1) having variation from normal structure; (2) having organs of different length; (3) (of agaric lamella edge), sterile due to the pressure of cystidia; cf. homomorphous.

heterophytic, the equivalent in the sporophyte generation of dioecious in the gametophyte generation (Blakeslee, *Bot. Gaz.* **42**: 161, 1906); cf. homophytic.

heteroplastic, see heterokaryotic.

heterospory, (1) having asexually produced spores of more than one kind (de Bary, 1887: 496) (obsol.); (2) having spores which differ in the mating type (+ or –) in heterothallic fungi (e.g. in *Mucorales*) (Blakeslee, *Bot. Gaz.* **42**: 161, 1906); (3) polymorphism of basidiospores in *Agaricales* associated with extreme conditions (Heim, *Rev. Mycol.* **8**: 32, 1943).

heterothallism (adj. **heterothallic**), condition of sexual reproduction in which 'conjugation is possible only through the interaction of different thalli' (Blakeslee, 1904). Heterothallism and homothallism (q.v.) were first applied by Blakeslee to the methods of zygospore formation in the *Mucorales*, and he considered the terms to correspond to dioecism and monoecism for higher plants. Heterothallism has, however, been used as the equivalent of haplodioecism or dioecism (as in *Dictyuchus monosporus*, where male and female organs are produced on different individuals), and self-sterility or self-incompatibility (as in *Ascobolus magnificus*, where male and female organs are developed on one individual). Whitehouse (*Biol. Rev.* **24**: 411, 1949) has distinguished the first type (haplodioecism) as '**morphological -**', and the second (haploid incompatibility) as '**physiological -**'. Physiological heterothallism may be determined either by two allelomorphs at one locus or by multiple allelomorphs at one or two loci (Whitehouse, *New Phytol.* **48**: 212, 1949). Multiple-allelomorph physiological heterothallism is characteristic of the hymenomycetes and gasteromycetes, among which Whitehouse has estimated 35% are heterothallic and bipolar (with one locus), 55% are heterothallic and tetrapolar (with two loci). Pontecorvo (*Adv. Genetics* **5**: 194, 1953) designated as **relative -** the formation of crossed asci in excess of 50% by the combination of certain homothallic strains of *Aspergillus nidulans*. Drayton & Groves (*Mycol.* **44**: 132, 1952) would restrict heterothallism to morphological heterothallism, Korf (*Nature* **170**: 534, 1952) would use heterothallism in a wide physiological sense and would distinguish a morphologically heterothallic organism as being both haplo-dioecious (dioecious) and heterothallic.

Burnett (*New Phytol.* **55**: 50, 1956) proposed the following terminology for mating systems of fungi: **heteromixis** (adj. heteromictic) for fusion of genetically different nuclei which includes dimixis when 2 types of nuclei (= heterothallism sensu Blakeslee, morphological, and 2-allelomorph physiological heterothallism), **diaphoromixis** when more than 2 types of nuclei (= multiple allelomorph physiological heterothallism), and **homoheteromixis** = secondary homothallism, amphithallism; **homomixis** (adj. homomictic) = homothallism; **amixis** (adj. amictic), apomixis in haploid organisms.

heterotroph (adj. **heterotrophic**) (of living organisms), using organic compounds as primary sources of energy, cf. autotrophic.

heteroxenous, having more than one host.

heterozygous, having heterokaryosis resulting from the fusion of gametes.

hiascent, becoming wide open.

hilar appendage (of basidiospores), the small wart-like or cone-like projection which connects the spore with the sterigma; sterigmatal appendage (Smith); apicule (Josserand); cf. apophysis.

hilum, a mark or scar, esp. that on a spore at the point of attachment to a conidiogenous cell or sterigma.

himantioid (of mycelium), in spreading fan-like cords, as in *Himantia*.

hirsute, having long hairs.

hirtose (**hirtous**), having hairs; hirsute.

hispid, having hairs or bristles.

hispidulous, somewhat, or delicately; cf. hispid.

histogenous, (1) produced from tissue; (2) (of spores), produced from hyphae or cells, without conidiogenous cells.

histolysis, the disappearance or solution of a wall or tissue.

hoary (esp. of a pileus or stipe), covered thickly with silk-like hairs; canescent.

holdfast, a process from the thallus for attachment, e.g. appressorium, hyphopodium, stigmatopodium, and stomatopodium; cf. rhizoid, hapteron, haustorium.

holo- (prefix), all; whole; entire.

holobasidium, see basidium.

holoblastic, see blastic.

holocarpic, having all the thallus used for the fruit-body.

holocarpous (of lichen thalli), ones formed by colonies of a free-living photobiont being invaded by a mycobiont and developing directly into fruiting bodies (Henssen, *Lichenologist* **18**: 51, 1986).

holocoenocytic, see astatocoenocytic.

hologamy, the condition in which all the thallus becomes a gametangium, i.e. there is fusion between two mature individuals as in *Polyphagus*.

holomorph, the whole fungus in all its morphs and phases; cf. anamorph, teleomorph.

holomorphum, two or more accepted species, each comprising teleomorph and anamorph, but scarcely distinguishable in one of the morphs (Tribe, *Bull. BMS* **17**: 94, 1983).

holophyte, a physiologically self-supporting green plant.

holosaprophyte, a true saprophyte (Johow).

holosporous (of conidial maturation), the conidium approaches its final size and shape before delimiting cells and maturing as a whole (Luttrell, 1963).

homeostasis, the maintenance of constant chemical and physical conditions within a living organism. See Whittenbury *et al.* (Eds) (*Homeostatic mechanisms in micro-organisms*, 1987).

homo- (prefix), one and the same.

homobasidium, see basidium.

homobium, a self-supporting association of a fungus and an alga, as in lichens.

homohetromixis, see heterothallism.

homohylic vesicle, sporangial vesicle, the wall of which is continuous with, and of the same material as the wall layer, or one of the wall layers of the sporangium. See also plasmamembranic vesicle, precipitative vesicle.

homoiomerous, (1) (of a lichen thallus), having the mycobiont and phycobiont evenly intermixed throughout the thallus, as in *Collema*; (2) (of trama in agarics), composed of hyphal tissue only; cf. heteromerous.

homokaryotic, having genetically identical nuclei, as in a line of isolates without variation (Brierley, *Ann. appl. Biol.* **18**: 429, 1931).

homokaryotic diplospory, incorporation of one or more homogenetic diploid nuclei in an oospore formed in response to abortive meioses in the oogonium (Dick, 1972).

homologous, showing a resemblance in form or structure, but not necessarily function, which is considered to be evidence of evolutionary relatedness; cf. analogous. See Poelt (*in* Hawksworth (Ed.), *Frontiers in mycology*: 85, 1991; homologous characters in lichens).

homomixis, haplo-monoecism or monoecism. Whitehouse (*Biol. Rev.* **24**: 428, 1949) recognized two types: **primary homothallism** such as occurs in a homokaryotic individual and **secondary homothallism** (= pseudo heterothallism or facultative heterothallism (Dodge), **amphithallism**, Lange, 1952) as in a thallus derived from one heterokaryotic spore containing nuclei of compatible mating types; cf. heterothallism.

homomorphous (of an agaric lamella edge), hymenium on edge not differentiated from that on the faces; cf. heteromorphous.

homophytic, the equivalent in the sporophyte generation of monoecious in the gametophyte generation (Blakeslee, *Bot. Gaz.* **42**: 161, 1906); cf heterophytic.

homoplasmic, see homokaryotic.

homospory, (1) having asexually produced spores of only one kind (= isospory); (2) having spores which are not differentiated according to mating type (c.f. heterospory).

homothallism (adj. **homothallic**), the condition in which sexual reproduction can occur without the interaction of two differing thalli.

homozygous, with identical alleles at the same locus on homologous chromosomes.

hormocyst, a propagule or diaspore produced in a special **hormocystangium** comprising a few cyanobacterial cells and fungal hyphae; produced by a few gelatinous lichens, e.g. *Lempholemma cladodes*, *L. vesiculiferum*. See Degelius (*Svensk bot. Tidsk.* **39**: 419, 1945), Henssen (*Lichenologist* **4**: 99, 1969).

hormocystangium, see hormocyst.

host, (1) a living organism harbouring a parasite; frequently in the sense of 'suscept' (q.v.); sometimes, as in 'host index', in a general sense covering certain substrata; (2) a commercial computer operator keeping databases other than their own on its machines and making them available to others for on-line searching for a fee.

hülle cells, terminal or intercalary thick-walled cells which occur in large numbers in association with the ascomata of, e.g. *Aspergillus nidulans*.

humicolous, living in or on decaying organic matter, soil.

hyaline, transparent or nearly so; translucent; frequently used in the sense of colourless.

hyalo- (prefix) (of spores), hyaline or brightly coloured, esp. for groups of anamorphic fungi (**hyalodidymae**, etc.).

hybrid, the result of a cross between organisms belong to different taxa which yield viable progeny. It is unclear how common the process is in nature with fungi; as with plants, hybridization can occur at levels from genus down; e.g. *Saccharomycopsis fibuligera* × *Yarrowia lipolytica* (Nga et al., *J. gen. Microbiol.* **138**: 223, 1992), *Cladonia grayi* × *C. mero-chlorophaea* (Culberson et al., *Am. J. Bot.* **75**: 1135, 1988); a **sexual -**. A **mechanical -** (in lichens) is formed by the growth together of propagules from different genera, species or genotypes to form a single thallus not involving sexual crossing (Hawksworth, *in* Street (Ed.), *Essays in plant taxonomy*: 211, 1978; see Lichens). The existing rules of nomenclature for hybrids and grafts are available to name both kinds of progeny (Hawksworth, *Internat. Lich. Newsl.* **21**: 59, 1988).

hydrofungi, aquatic fungi (Cooke, 1963).

hydrogenosome, membrane-bound organelle found in some anaerobic fungi and protozoa involved in the production of hydrogen and carbon dioxide.

hydrotropism, a reaction to water.

hygrophanous, having a water-soaked appearance when wet.

hygrophilous, preferring a moist habitat.

hygroscopic, (1) becoming soft in wet air, hard in dry; (2) (of a sporocarp), opening and discharging spores in dry air.

hymenial algae (or gonidia), algal cells in the hymenium of a lichenized ascomycete, e.g. *Endocarpon*, *Staurothele*, *Thelendia*; **- cystidium**, see cystidium; **- veil**, see annulus.

hymeniderm (of basidiomata), an outer layer composed of an unstratified layer of single cells or hyphal tips; see derm.

hymenium, the spore-bearing layer of a fruit-body; (of basidiomycetes) **euhymenium** (Donk, *Persoonia* **3**: 210, 1964), a hymenium in which the basidia and their sterile homologues are the first elements to be formed, as a palisade; in a **static** (non-thickening) **euhymenium** the exhausted basidia are replaced at the same level by intercalation; in a **thickening euhymenium** the tramal hyphae grow between the exhausted basidia to form a new hymenium above the old, as in the *Cantharellaceae*; **catahymenium** (Lemke, *CJB* **42**: 218, 1964), a hymenium in which hyphidia are the first-formed elements and the basidia embedded at various levels elongate to reach the surface and do not form a palisade; cf. thecium, apothecium.

hymenophore, a spore-bearing structure, esp. a basidioma, or that part of it bearing the hymenium; cf. sporophore.

hymenopodium (**hymenopode**), tissue under the hymenium; subhymenium or hypothecium.

hyper- (prefix), above.

hyperepiphyllous growing on epiphyllous or foliicolous lichens or bryophytes, esp. in tropical rain forests.

hyperparasite, a parasite growing on another parasite (obsol.); fungicolous fungi is a preferable general term where parasitism has not been established.

hyperplasia, over-development of some sort (e.g. swellings, galls, witches' brooms) as a reaction to a disease-producing agent; cf. hypoplasia.

hypersaprophyte, a saprophyte only found on substrates invaded by other saprophytes, e.g. *Herpotrichiellaceae*, *Lasiosphaeria*, *Nectria sanguinea*, etc. (Munk, *Sydowia, Beih.* **1**, 1957).

hypersensitivity, increased sensitivity, as in the condition in which there is death of the host tissue at the point of attack by a pathogen, so that the infection does not spread; esp. of reaction to rusts for which the word was first used by Stakman; intolerance.

hypertonic (of culture media), having an osmotic pressure higher than that of the organism cultured; cf. hypotonic.

hypertrophy (of organs, etc.), the state of having growth greater than normal.

hypertrophyte, a parasitic fungus causing hyperplasia in a plant (Wakker).

hypha (pl. **hyphae**), one of the filaments of a mycelium; Vuillemin restricted the term to septate filaments; cf. siphon; **ampoule -**, a swollen hypha as in certain basidiomycetes; **arboriform -**, much branched skeletal - of *Ganoderma*; **ascogenous -**, a dikaryotic hypha from which an ascus develops; **binding -**, see Hyphal analysis; **costiferous -**, transverse ribs (costae) on the inner surface of the hyphal wall of the gill trama in *Paxillus involutus* and *P. filamentosus*, non-amyloid, stained by Congo red and calcofluor but not toluidine blue; **endo-** (intrahyphal **-**), vegetative or fertile element initiated by the differentiation within a - from the innermost wall layer (Cole & Samson, 1979); **inflated -**, one in which cells behind the growing apex enlarge and cause the apparent rapid rate of growth characteristic of most agaric and gasteromycete basidiocarps; in an **uninflated -** no change of cell size occurs as in most polypores; **mediate -** and **mycelial -**, see Corner (*TBMS* **17**: 54, 1932); **oleiferous -**, do not carry latex (cf. lactifer) but frequently resinous substances (Singer, 1960: 34); **oiliferous -**, a submerged hypha of an endolithic lichen having torulose, guttulate cells; see des Abbayes (*Traité de Lichenologie*, 1951); **racquet -**, one of racket cells (q.v.); **skeletal -**, see Hyphal analysis; **stuffing -**, see Corner (*TBMS* **17**: 54, 1932). See flexuous hypha, hyphal analysis, hyphal peg, textura.

hyphal, composed of or comprising hyphae; cf. cellular.

hyphal analysis, a procedure by which the development and structure of basidiomata can be investigated, providing important taxonomic criteria. Three main types of hyphal systems of increasing complexity were recognized by Corner (*TBMS* **17**: 51, 1932) (Figs 16, 17): **monomitic**, having hyphae of one kind (generative hyphae which are branched, septate, with or without clamp-connexions, thin-walled to thick-walled, and of unlimited length; they give rise both to other hyphal types and to the hymenium) (Teixeira, *Mycol.* **52**: 30, 1961; gen. hyphae of polypores); **dimitic**, having hyphae of two kinds (generative and skeletal hyphae which are thick-walled, aseptate, and of limited length, with thin-walled apices, generally unbranched but when terminal they can develop arboriform branching or taper) or generative and binding (see below); **trimitic**, having hyphae of three kinds (generative, skeletal and binding (or ligative) hyphae which are aseptate, thick-walled, much branched, either *Bovista*-type with tapering branches or coralloid; they bind the skeletal and generative hyphae together). In *Polyporaceae* and *Lentinaceae*, intercalary skeletal hyphae can give rise to ligative branching, the entire element being termed a skeleto-binding cell (Corner, 1981) or skeleto-ligative hypha (Pegler, 1983). Corner also recognizes **sarco-dimitic** (in which the skeletal hyphae are replaced by thick-walled, long, inflating fusiform elements) and **sarcotrimitic** (in which the generative hyphae give both thick-walled inflated elements similar to binding hyphae but septate) types.
Most soft and fleshy basidiomata are monomitic, with hyphae which are generally inflated (most agaricoid and clavarioid fungi). Hard and tough basidiomata may be monomitic with the generative hyphae developing thickened walls, dimitic, with skeletal hyphae (e.g. *Phellinus* spp.) or (esp. when perennial) trimitic (e.g. *Fomes, Ganoderma, Microporus xanthopus*). Every species has a well-defined and constant construction, which is maintained regardless of changes in the external morphology of the basidioma due to environmental conditions, hence the importance of hyphal analysis in taxonomy.

hyphal fusions, see anastomosis. Vegetative hyphae of a mycelium may fuse, forming an interconnected network. Fusion between hyphae of different mycelia is controlled by genetic systems which determine sexual (see sex) or vegetative compatibility (q.v.).

hyphal net ('**Hyphenfilz**'), organ of attachment in some squamulose (placodioid) lichens (e.g. *Psora decipiens*) where a delicately branched reticulate net penetrates the substrate (see Poelt & Baumgärtner, *Öst. bot. Z.* **111**: 1, 1964; rhizinose strand).

hyphal peg (of basidiomata), a bunch of somewhat interwoven hyphae extending from the trama (where it originates) to the hymenium from which it may project (Singer); (of hyphae), projection from a hypha for fusion (Buller), peg-hypha.

hyphal rhizoid, a hypha acting as a rhizoid.

hyphidium (pl. **-ia**), (paraphysis, pseudoparaphysis, paraphysoid, dikaryoparaphysis, and pseudophysis sensu Singer (1962) are syn. or near syn.), a little, or strongly, modified terminal hypha in the hymenium of hymenomycetes. Donk (*Persoonia* **3**: 229, 1964) distinguished; **haplo-** (simple **-**), unmodified, unbranched or little branched; **dendro-** (dendrophysis), irregularly strongly branched; **dicho-** (dichophysis), repeatedly dichotomously branched; **acantho-** (acanthophysis; bottle-brush paraphysis (Burt, 1918)), having pin-like outgrowths near the apex; in *Corticiaceae* may be botryose, clavate, coralloid, or cylindrical; cf. cystidium.

hyphocystidium, see cystidium.

hyphoid, (1) like hyphae; cobwebby; (2) (of aecia of *Dasyspora*), having aeciospores on hyphal projections from stomata (Arthur).

hyphophore, erect stalked peltate asexual sporophores in the *Asterothyriaceae* (e.g. *Echinoplaca, Gyalideopsis, Tricharia*). See Sérusiaux & De Sloover (*Veröff. Geobot. Inst. ETH, Rübel* **91**: 260, 1986; types), Vězda (*Čas. slez. Muz. Silesiae A* **22**: 67, 1973; *Folia geobot. phytotax., Praha* **14**: 43, 1979).

hyphopodium (pl. **-ia**), a short branch of one or two cells on epiphytic mycelium of *Meliolales*, etc.; in a **capitate -** the end is rounded, = appressorium (fide Mibey & Hawksworth, *SA* **14**: 25, 1995); **mucronate -** = conidiogenous cell (fide Hughes, *CJB* **59**: 1514, 1981). A **stigmatopodium (stigmopodium)** is a hyphopodium in which the end cell or **stigmatocyst** has a haustorium (Arnaud). A stigmatocyst in a hypha is a **node cell**; cf. Doidge (*Bothalia* **4**: 273, 1942). See also Walker (*Mycotaxon* **11**: 1, 1980).

hypnocyst, an *Alternaria*-like group of cells (Chippindale, *TBMS* **14**: 203, 1929; Griffiths, *Nova Hedw.* **25**: 511, 1974).

hypnospora, a resting spore.

hypo- (prefix), under.

hypobasidium, see basidium.

hypochnoid, having effused, resupinate, dry, rather loosely intertwined hyphae, as in *Tomentella* (formerly *Hypochnus*).

hypocreacous, fleshy and brightly coloured, like *Hypocrea*.

hypocystidium, see cystidium.

hypodermataceous (of asci), ones which are essentially unitunicate, lack any apical thickening, and which discharge the spores through a narrow pore; see ascus.

hypogean (**hypogeal, hypogeic, hypogeous**), in the earth.

hypogenous, produced lower down.

hypogyny (adj. **hypogynous**), the condition of having the antheridium under the oogonium and on the same hypha.

hypolithic, see endolithic.

hyponecral, see necral layer.

hyponym, a name only; one having no description or reference to a specimen.

hypoparasite, a hidden parasite; a pathogen dispersed along with another pathogen, such as a mycovirus in an *Ophiostoma ulmi* population.

hypophloeodal, under the periderm or bark; endophloeodal; subcutical; within the bark.

hypophyllous, on the under surface of a leaf.

hypoplasia, the state of having growth less than normal, cf. hyperplasia.

hypostroma, see stroma.

hypothallus, (1) (of lichens), the first hyphae of the thallus to grow, usually used of a crustaceous lichen which has no photobiont cells or cortex; = prothallus (prothallus), fide Maas Geesteranus (*Blumea* **6**: 47, 1947) who restricts hypothallus to the spongy tissue on the underside of the thallus in *Anzia*, *Pannaria* and *Pannoparmelia*, but see spongiostratum; (2) (of myxomycetes), the thin layer on the surface of the substratum not used up in sporangial development; Ross (*Mycol.* **65**: 477, 1973) distinguished epihypothallic and subhypothallic development.

hypothecium, medullary excipulum; the hyphal layer under the subhymenium in an apothecium; sometimes used indiscriminately for all tissues below the hymenium (including the subhymenium).

hypotonic (of culture media), having an osmotic pressure lower than that of the organism cultured, cf. hypertonic.

hysteriaceous (**hysterioid**, **hysteriiform**), long and cleft, like the **hysterothecium** (ascoma) of the *Hysteriaceae*; lirellate.

hysterochroic, slowly discoloured from base to apex in old age.

hysterophyte, a saprophyte (obsol.).

hysterothecium, an elongated ascoma like that of the *Hysteriaceae* with a slit-like line of dehiscence.

id-reaction, see dermatophytid.

imbricate (of pilei, scales, squamules, etc.), partly covering one another like the tiles on a roof.

immaculate, not spotted.

immarginate, having no well-defined edge.

immersed, embedded in the substratum.

immobolisation, the controlled, intentional, attachment of fungal cells in fermentation technology (Webb, *Mycologist* **3**: 163, 1989); cf. biomass support particles.

imperforate, having no opening.

inaequi-hymeniiferous (of hymenial development in agarics), having basidia which mature and shed their spores in zones; the coprinus type (Buller, *Researches* **2**: 19, 1922); cf. aequi-hymeniiferous.

incised, as if cut into; esp. of a pileus margin or lobes of a foliose lichen thallus.

incompatible, (1) (of sex), unable to be cross-mated due to mating type or fertility barriers; (2) (of vegetative mycelia), unable to form a stable heterokaryon due to genetic differences at one or more vegetative compatibility (vc, het) loci. See vegetative compatibility.

incrassate, made thick.

incrusted (of hyphae), having matter excreted on the walls (Corner, 1950).

incubation period, the time between inoculation and the development of visible symptoms.

indefinite, not sharply limited.

indehiscent (of sporocarps, sporangia, etc.), not opening, or with no special method of opening.

indeterminate, (1) having the edge not well-defined, esp. of fruit-bodies and leaf-spots; (2) (of conidiophores), continuing growth indefinitely.

indigenous, natural to a country or region; native.

indirect (of fruit-body development), see direct.

individualism in fungi, mechanisms may exist in nature to 'define' individuals involving co-operative (hyphal fusions, heterokaryosis) and individualistic methods. See Todd & Rayner (*Sci. Progr.* **66**: 331, 1980), incompatible.

indumentum, a covering, such as hairs, etc.

indurated, made hard.

indusium, (1) cover; (2) (of phalloids), a net-like structure hanging from the top of the stipe under the pileus.

inermous, having no spines or prickles.

infarctate, solid; turgid.

infect (of a pathogen), to enter and establish a pathogenic relationship with an organism; to enter and persist in a carrier (*TBMS* **33**: 155, 1950); to make an attack on an organism; (of an agent), to make infection of an organism take place; **-ed** (of an organism), attacked by a pathogen, cf. contaminated; **-ion**, the act of infecting; **-ious** (of diseases), resulting from infection; sometimes used in the sense of able to be handed on by touch (contagious) or by inoculum; **-ive** (of a pathogen), able to make an attack on a living organism; (of a vector, medium, etc.), having the power of effecting the transmission of a pathogen.

inferior (of an annulus), low down on the stipe.

infested, attacked by animals, esp. insects; sometimes used of fungi in soil or other substrata in the sense of 'contaminated'.

infissitunicate (Dughi, *C. r. hebd. Séanc. Acad. Sci., Paris* **243**: 750, 1956), see ascus.

inflated hypha, see hypha.

inflexed (of pileus margin), turned down.

infundibuliform, funnel-like in form.

ingest, to obtain food by engulfing it; see phagotrophic; cf. absorb.

inhabitant, see symbiosis.

innate, bedded in; immersed.

inoculate, to put a microorganism, or a substance containing one, into an organism or a substratum.

inoculation, the act of inoculating [of (an organism or substratum)] *with* (the inoculum); *of* (the inoculum) *into* (an organism or substratum); *by* (an agent or method).

inoculum, the substance, generally a pathogen, used for inoculating.

inoculum potential (of a fungus or other microorganism), the energy of growth available for colonization of a substratum at the surface of the substratum to be colonized (Garrett, 1956).

inoperculate (of an ascus or sporangium), opening by an irregular apical split to discharge the spores, see ascus; cf. operculate.

inordinate, in no order.

inquinant, stained; blackened; dirty (obsol.).

insititious, of inserted nature, introduced from without.

inspissate, made thick.

integrated (of conidiogenous cells), incorporated in the main axis or branches of the conidiophore; cf. discrete.

inter- (prefix), between; among.

interascal tissue, see hamathecium.

interascicular parenchyma, the paraphysis-like hyphae or paraphysoidal interthecial fibres (Stevens) (obsol.).

interbiotic, living as a parasite on or near one or more living organisms, as certain rhizoidal chytrids.

intercalary, (1) (of growth), between the apex and the base; (2) (of cells, spores, etc.), between two cells.

intercellular, between cells.

interspace (of a pileus), the space between the lamellae.

interthecial, between asci.

intervenose (of a pileus), veined in the interspaces.

intra- (prefix), within; inside.

intracellular, within the cell.

intrahyphal, see hypha.

intramatrical, living in the matrix or substratum.

intramatrical spores, an alternative name for vesicles produced in host roots by most endomycorrhizal fungi.

intraparietal, within a wall or walls (e.g. of crystals amongst the tissues of an exciple).

intricate cortex (of lichen thalli), made up of hyphae twisted together; cf. textura intricata.

introrse, in the direction of the central axis; inwards.

intumescence, a swelling.

invaginated, covered by a sheath.

involucrellum, tissue forming the upper part of a perithecioid ascoma surrounding the true exciple, not involving host or substrate materials (cf. clypeus) and generally dimidiate (e.g. *Verrucaria*).

involute, rolled in.

iodine (I, J), used as Lugol's solution (I 0.5g, KI 1.5 g, water 100 ml), in potassium iodide (**IKI**; I 1%, KI 3%), and formerly often Melzer's reagent giving blue, red, lavender or violet colours seen best after pre-treatment with potassium hydroxide in spores, asci, hymenial tissues, etc. Reactions can vary according to the kind of iodine solution used, its concentration, the age of the material, and the nature of any pretreatment; all need to be reported when referring to such tests. Extensively used in the systematics of lichenized and non-lichenized fungi since Nylander (*Flora, Jena* **48**: 465, 1865). See Baral (*Mycotaxon* **29**: 399, 1987; caution in use), Common (*Mycotaxon* **41**: 67, 1991; review I+ materials), Kohn & Korf (*Mycotaxon* **3**: 165, 1975; pretreatment), Nannfeldt (*TBMS* **67**: 283, 1976; ascus plugs), amylomycan, amyloid, dextrinoid.

ionomidotic reaction, release of a dark pigment into aqueous potassium hydroxide (KOH) mounts; an important taxonomic criterion in some dark-coloured *Leotiales* apothecia (e.g. *Claussenomyces*; Oullette & Korf, *Mycotaxon* **10**: 255, 1979).

irpicoid, having teeth, or becoming toothed, as in *Irpex*.

isidiate, having isidia.

isidiiferous, of a lichen thallus bearing isidia.

isidium (pl. **-ia**), a photobiont-containing protuberance of the cortex in lichens which may be warty, cylindrical, clavate, scale-like, coralloid, simple, or branched; occurring directly on the thallus (e.g. *Peltigera praetextata*, *Pseudevernia furfuracea*); may become sorediate (e.g. *Lobaria pulmonaria*). See Bailey (*in* Brown *et al. Lichenology: progress and problems*: 214, 1976; review), Du Rietz (*Svensk bot. Tidskr.* **18**: 141, 1924; classification), Kershaw & Millbank (*Lichenologist* **4**: 214, 1970; rôle), Puymaly (*Botaniste* **48**: 237, 1965).

iso- (prefix), equal.

isodiametric with equal diameters: having diameters or axes of equal length.

isogamete, one of two sex cells like in form.

isogamy, the conjugation of isogametes; cf. anisogamy.

isohaplont, a haplont of cells having genotypically like nuclei (Kniep, 1928), cf. miktohaplont.

isomorphic, like in form but unlike in structure.

isonym, a homotypic synonym; see basionym.

isoplanogamete, one of two motile sex cells like in form.

isospory, see homospory.

isotomic dichotomous branching, branching in which both dichotomies are about the same thickness and length so that the dichotomic pattern is visible even in older parts of the thallus, as in *Cladonia evansii*, cf. anisotomic dichotomic branching.

isthmospore, a spore comprising two or more cells interconnected by a much narrower region, as in the ascospores of *Vialaea* (Cannon, *MR* **99**: 367, 1995); a conidium composed of four more thick-walled cells separated by thin-walled cells as in *Isthmospora* (Hughes, *Mycol. Pap.* **50**, 1953).

isthmus, (1) the narrower or thinner-walled portion of an isthmospore (q.v.); (2) the thickened medial perforated septum of a polarilocular ascospore.

isthmus disarticulation, protoplasmic retraction during spore formation, with the secretion of an endosporic wall membrane.

ixo- (prefix), sticky.

ixocutis (of a pileus), a slimy cuticle.

ixotrichoderm (**ixotrichodermium**) (of a pileus), a trichodermium composed of gelatinized hyphae (Snell, 1939; for a discussion, see Shaffer, *Mycol.* **58**: 486, 1966).

jack-in-the-box discharge, see ascus.

juvenescence, the process of maturing at a stage of development normally immature.

karyochorisis, somatic nuclear division resulting from a constriction of the nuclear membrane (Moore, *Z. Zellforsch.* **63**: 921, 1964).

karyogamy, the fusion of two sex nuclei after cell fusion, i.e. after plasmogamy.

karyotype, the size and number of chromosomes in an organism. Generally determined by deduction from mating studies, microscopically or from specialized electrophoresis methods. See McCluskey & Mills (*Mol. Pl.-Microbe Interactions* **3**: 366, 1990), pulsed field gel-electrophoresis.

katothecium, see catathecium.

keratinophylic, (1) capable of decomposing keratin; (2) of many fungi causing superficial mycoses in humans; see dermatophyte.

kinetid, flagellar apparatus including kinetosomes and their associated tubules and fibres.

kinetosome, intracellular, non-membrane bound organelles; microtubular cylinders (0.2 µm diam.) organized in the 9 (A, B C- microtubules) + 0 pattern and with axoneme extensions in the 9 (A, B- microtubules) + 2 pattern.

labiate, having lips; lip-like.
labium, a lip.
labriform, lip-shaped; frequently used for terminal soralia of lichens having this shape.
labyrinthiform, having the form of a labyrinth, intricate.
laccate, polished; varnished; shining.
lacerate, as if roughly cut or torn.
lacinia, a delicate branch of a foliose lichen thallus having an anatomical structure typical of foliose lichens.
laciniate (of an edge, etc.), as if cut into delicate bands.
lacrimiform (**lacrimoid lacrymiform**, **lacrymoid**), like a tear drop.
lacteous, like milk.
lactescent, becoming like milk.
lactifer, a latex-carrying hypha (Singer, 1960: 33).
lactiferous, having a milk-like juice.
lactiferous hypha, see hypha.
lacuna, a hole or hollow.
lacunose, having lacunae.
laevigate, smooth.
lageniform, swollen at the base, narrowed at the top; like a Florence flask.
lamella (pl. **lamellae**) (of an agaric), one of the characteristic hymenium-covered vertical plates on the underside of the pileus; gill.
lamellate, (1) having lamellae; (2) made up of thin plates.
lamellula (pl. **lamellulae**), a small lamella which runs from the edge of the pileus towards the stipe, as in *Russula*.
lamina (pl. **laminae**), (1) blade; (2) the main part of the thallus in foliose lichens; (3) epithecium + hymenium + subhymenium in an apothecium (Hertel, *Beih. Nova Hedw.* **24**, 1967); (4) (of leaves) the flat surface; -l, on the lamina.
lamprocystidium, see cystidium.
lanate, like wool; covered with short hair-like processes.
lanceolate, see acerose.
languid, feeble; hanging down.
lanose, see lanate.
lanuginose, see lanate; see also nematogenous.
lateral, at the side.
latticed, cross-barred; like a network.
lecanoralean, (1) (of asci), of asci which are essentially bitunicate in structure, generally thick-walled with an especially strongly thickened apex, and in which discharge is by a rostrate eversion of the endoascus; see ascus; (2) (of apothecioid ascomata), see lecanorine.
lecanorine (of an apothecium), having an excipulum (q.v.) thallinum (e.g. *Lecanora*, *Parmelia*); lecanoroid.
lecideine (of an apothecium), one having no excipulum thallinum (q.v.) and the margin usually consisting only of the excipulum proprium (e.g. *Lecidea*, *Bacidia*).
lecythiform, like a stoppered bottle; ninepin-shaped.
leiodisc (of an apothecium), having a smooth glazed disc.
leiosporous, having smooth spores.
lentic, habitat still water (lakes); cf. lotic.
lenticular, like a double convex lens in form.
lentiginose (**lentiginous**), having very small spots as though freckled.
lepidote, covered with small scales.
leprose (of lichens), having the surface or the whole thallus entirely dissolved into soredia (e.g. *Lepraria*).
leptocystidium, see cystidium.

leptodermatous (of hyphae), having the outer wall thinner than the lumen; cf. mesodermatous.
lepto-form, see leptospore.
leptogonidium, a photobiont consisting of small-sized cells (obsol.); cf. gonidium.
leptospore (of *Uredinales*), a teliospore (q.v.) adapted for immediate germination without a dormant period.
leptotichous (of tissue), thin-walled.
lesion, a wound; a well-marked but limited diseased-area.
leucosporous, having spores white in the mass.
levigate, see laevigate.
ligative hyphae, Pouzar's term for binding hyphae; see hyphal analysis.
ligneous (**lignose**), wood-like.
lignicolous, living on or in wood.
lignituber, see papillae.
ligulate (**liguliform**), flat and narrow; strap-like in form; lorate.
limbate, (1) edged with another colour; (2) (of a volva), adnate to base of stipe and having a narrow, free, membranous margin (Bas, 1969).
limoniform, lemon-like in form.
linear, long and narrow.
lineolate, marked with lines.
linguiform, see lingulate.
lingulate, tongue-like in form.
lipochroic, with pigment in oil drops.
lipsanenchyma, primordial tissue of a basidioma, other than the universal veil, covering the hymenium (Reijnders, 1963; Singer, 1962: 29).
lirella, a long, narrow apothecium as in *Graphis* and *Hysterium*.
litho- (prefix), pertaining to rocks.
lithophyte, a plant living on rocks; for fungi living on rocks see saxicolous.
lithophytic, the habit of a lithophyte (q.v.).
lithotroph, utilizing rocks as nourishment.
littoral, growing on sea or lake shores.
lituate, forked, and having the points turned out a little.
lobate, lobed.
lobulate, having small lobes.
locule (**loculus**), a cavity, esp. one in a stroma.
lomasome, a vesicle derived from an intracytoplasmic membrane (Marchant *et al.*, *New Phytol.* **66**: 623, 1967); cf. plasmalemmasome.
longicollous, having long beaks or necks.
longiseptum, see septum.
longuinose, see lanate.
lophotrichous (**lophotrichate**), having several flagella at one or both ends.
lorate, like a narrow band; strap-like in form; ligulate.
lotic, habitat running water (streams); cf. lentic.
lumen, the central cavity of a cell or other structure.
lunate, like a new moon; crescentic.
luteous, yellow.
lymabiont, an organism only found in sewage.
lymaphile, an organism commonly found in sewage (Cooke, *Sydowia*, *Beih.* **1**, 1957).
lymaphobe, an organism never found in sewage (Cooke, *Sydowia*, *Beih.* **1**, 1957).
lymaxene, an organism rarely found in sewage (Cooke, *Sydowia*, *Beih.* **1**, 1957).
lyocystidium, see cystidium.

lysigenous, formed by the breaking down of cells; cf. schizogenous.

lysis, (1) dissolution of a cell, e.g. by a lysin; (2) (of conidial liberation), secession by the dissolution of the wall of the adjacent cell; cf. fission, fracture.

macaedium, see mazaedium.

macro- (prefix), long, but commonly used in the sense of mega- (q.v.).

macrocephalic, see septum.

macroconidium, (1) the larger, and generally more diagnostic conidium of a fungus which has microconidia (and sometimes also mesoconidia) in addition; (2) (infrequent), a long large conidium.

macrocyclic, of species with all spore forms present in the life-cycle.

macrocyst, see cyst.

macronematous (of conidiophores), morphologically different from vegetative hyphae (cf. micronematous).

macrophylline (of foliose lichens), having large lobes (obsol.).

macroscopic, visible without a lens.

macrospore, a large spore when there are spores of two sizes.

maculate, spotted; blotched.

maculicole (adj. **-icolous**), an organism living on spots, e.g. leaf spots.

malaceoid venation, see veins.

malacoid, like mucilage.

mammiform, breast-like in form.

mantle, a compact layer of hyphae enclosing short feeder roots of ectomycorrhizal plants, connected to the Hartig net on the inside, and to the extramatrical hyphae (q.v.) on the outside; acts as a nutrient sink.

marcescent (of basidiomata), withering, drying up *in situ*; cf. putrescent.

marginal frill (of a conidium), the periclinal wall left attached to a spore after secession; cf. basal frill.

marginal veil (of agarics), an incurving proliferation of the margin of the pileus which protects the developing hymenium; cf. partial veil.

marginate, (1) having a well-marked edge; (2) (of basal bulb of agaric stipe), having a gutter-like rim as in *Leucocortinarius bulbiger*.

margo proprius, see excipulum (proprium), proper margin.

margo thallinus, see excipulum (thallinum), thalline margin.

mastigoneme, see flagellum.

mastigopod (of myxomycetes), a swarm cell (obsol.).

matrix, (1) the substratum in or on which an organism is living; (2) mucilaginous material in which conidia and some ascospores are produced, influences dissemination, survival, germination etc. See Louis & Cooke (*TBMS* **84**: 661, 1985).

mazaedium, a spore mass formed by an ascoma, as in *Caliciales* and *Onygenaceae*, in which the spores, generally with sterile elements, become free from the asci as a dry, loose powdery mass on the fruiting surface.

medallion clamp, a clamp connection with a space between the main hypha and the hook.

medulla, (1) (of lichen thalli), the loose layer of hyphae below the cortex and algal layer; (2) (of sporocarps of macromycetes), the part composed mainly or entirely of longitudinal hyphae.

medullary excipulum (of ascomata), tissue below the generative layer in an apothecium; hypothecium. See excipulum.

mega- (prefix), of great size; large; cf. macro-.

megaspore, see macrospore.

meiocyte, a cell in which meiosis takes place; cf. gonotokont.

meiophase, the part of a life cycle in which a diploid nucleus undergoes reduction.

meiosporangium, a thick-walled diploid sporangium of certain *Blastocladiales* producing uninucleate, haploid zoospores (**meiospores**, q.v.) (Emerson, 1950); see Dick (*Mycologist* **1**: 166, 1987); cf. mitosporangium.

meiospore, (1) a spore from a meiosporangium (q.v.); (2) (of ascomycetes and basidiomycetes), a basidiospore or ascospore which is the product of meiosis (see Kendrick & Watling, *in* Kendrick (Ed.), *The whole fungus* **2**: 473, 1979).

meiotangium, for the sporangium or gametangium in which meiosis occurs. See Corner (*TBMS* **15**: 336, 1931).

melanized, containing dark brown pigments.

melanosporous, black-spored.

melophase, the part of a life cycle in which a diploid nucleus undergoes reduction.

membranous (**membranaceous**), like a thin skin or parchment.

memnospore, a spore remaining at its place of origin (Gregory, *in* Madelin (Ed.), *The fungus spore*, 1966); cf. xenospore.

merenchyma, see plectenchyma.

merismatoid (of a pileus), made up of smaller pilei.

merispore, see sporidesm.

meristem arthrospore, one of the chain of conidia maturing in basipetal succession and originating by apical wall building at the tip of the condiogenous cell; **- blastospore**, a conidium arising either apically or laterally from a conidiogenous cell which elongates through ring wall building at the base (a basauxic conidiophore; Hughes, 1953).

meristematic (of conidiophores), see wall building.

meristogenous (of pycnidia, etc.), formed by growth and division of one hypha; **symphogenous**, formed from a number of hyphae. See Sutton (*in* Ainsworth *et al.* (Eds), *The Fungi* **4A**: 1973).

merogamy, copulation between special sex cells or gametes.

merosporangium (pl. **-ia**), (of *Zygomycetes*), a cylindrical outgrowth from the swollen end of a sporangiophore in which a chain-like series of sporangiospores is generally produced. See Benjamin (*Mycol.* **58**: 1, 1966; review, *in* Kendrick (Ed.), *The whole fungus* **2**: 573, 1979).

mesochroic, with pigment in the hyphal wall.

mesodermatous (of hyphae), having the outer wall and lumen of about the same thickness; cf. leptodermatous.

mesophile, see thermophily.

mesospore, (1) a 1-celled teliospore among 2-celled ones; (2) an amphispore (obsol.); (3) the middle layer of a three-layered spore wall.

meta- (prefix), changed in form or position; between; with; after.

metabasidium, see basidium.

metabiosis, the association of two organisms acting or living one after the other; cf. synergism.

metacellulose, a cellulose in certain fungi.

metachroic, changing colour through the appearance of a new pigment in maturer tissue.

metachromic, giving a red reaction to cresyl blue. See Singer (*The Agaricales (mushrooms) in modern taxonomy*: 77, 1951).

metaphysis (obsol.), used by Petrak; see paraphysis.

metaplasm, see epiplasm.

metathallus, assimilative (photobiont-containing) part of a lichen thallus, esp. where there is also a prothallus (q.v.).

metoecious (obsol.), used by de Bary; see heteroecious.

metula, a conidiophore branch having phialides, e.g. of *Penicillium* and *Aspergillus* (obsol.).

metuloid, an encrusted cystidium thick-walled at maturity, as in *Peniophora*.

micaceous (of a pileus surface), covered with bright particles.

micro- (prefix), small; one-thousandth (Système International d'Unités); see micrometre.

microaerophilic, making best growth under lowered oxygen pressure.

microbe, a microorganism (q.v.).

microbial (adj.), pertaining to microbes (q.v.).

microbiota, all the microorganisms present in the area or habitat specificied, including algae, bacteria and protozoa as well as fungi; see mycobiota.

microbodies, see peroxisome.

microcephalic, see septum.

microconidium, (1) the smaller conidium of a fungus which also has macroconidia; (2) a spermatium (q.v.).

microculture, a culture of an organism under the microscope, as in a hanging drop.

microcyclic (of conidiation), germination of spores by the direct function of the conidia without the intervention of mycelial growth (Hanlin, *Mycoscience* **35**: 113, 1994); (of rusts; endocyclic, demicyclic), species with either no aeciospores nor urediniospore, or similar species where the teliospore is replace with a aecidioid teliospore (q.v.).

microcyst, see cyst.

microcytospore, (1) an encysted zoospore (planospore); (2) an encysted gametangium (*Blastocladiales*).

microendospore, minute cytoplasmic particles behaving like spores in *Ophiostoma ulmi* (Ouellette & Gagnon, *CJB* **38**: 235, 1960).

microflora, sometimes used, inappropriately, for all the microorganisms present in a specified site or habitat; see microbiota, mycobiota.

microform, of species with no aeciospore (q.v.) or urediniospore (q.v.) in the life-cycle.

microgonidia, very small green bodies in lichen hyphae (Minks) (obsol.).

microlichens, lichens in which the whole of their morphological characteristics can be seen only with a magnifier equal or larger than ×10. See Messuti (*Br. Lich. Soc. Bull.* **73**: 49, 1993).

micrometre, one-thousandth of a millimetre (0.001 mm; 1 μm) (Système International d'Unités); formerly often as 'μ'.

micron, see micrometre.

micronematous (**micronemeous**), (1) having hyphae of small diameter; (2) (of conidiophores) similar morphologically to vegetative hyphae (cf. Micronemeae).

microorganism, an organism which belongs to a phylum many members of which either cannot be seen with the unaided eye or require microscopic examination and/or growth in pure culture for their identification; a microbe; includes all unicellular prokaryotes and eukaryotes, and also some multicellular eukaryotes, i.e. microscopic algae, bacteria, fungi (including yeasts), protozoa and viruses; sometimes used (incorrectly) only for prokaryotes (bacteria and viruses). See Cowan (*A dictionary of microbial taxonomy*: 162, 1978; inappropriateness of term), Zavarzin (*in* Allsopp *et al.* (Eds), *Microbial diversity and ecosystem function*: 17, 1995; concept).

microphylline (of lichen thalli), composed of minute lobes or scales.

microsclerid, see cystidium.

microsclerotium, a very small sclerotium, as in *Verticillium dahliae*; pseudosclerotium.

microsporangium, secondary sporangium formed from a zoospore cyst, either without any mitoses, or with very few mitoses.

microspore, (1) a small spore, where there are spores of two sizes; (2) a spore from a microsporangium (q.v.).

miktohaplont, a haplont made up of cells having genotypically different nuclei (Kniep, 1928); cf. isohaplont.

mischoblastiomorph (of ascospores), ones similar to the polarilocular type but either without a septum or only with an incomplete septum.

mitochondrion, membrane-bound intracellular organelle containing enzymes and electron transport chains for oxidative respiration of organic acids and the concomitant production of ATP. Possesses DNA, messenger RNA and small ribosomes and thus capable of protein synthesis.

mitosporangium, a thin-walled diploid sporangium of certain *Blastocladiales* producing uninucleate diploid zoospores (**mitospores**) (Emerson, 1950); cf. meiosporangium.

mitospore, (1) a spore from a mitosporangium (q.v.); (2) (of ascomycetes and basidiomycetes), any non-basidiosporous or non-ascosporous propagule (see Kendrick & Watling, *in* Kendrick (Ed.), *The whole fungus* **2**: 473, 1979).

mitrate (**mitriform**), mitre-like in form.

mitriform, see mitrate.

mollicute, see mycoplasma.

monandrous (of oospores), formed when only one functioning antheridium is present, cf. polyandrous.

monaxial, having one individual stem or axis.

moniliaceous (of mycelium, spores, etc.), hyaline or brightly coloured; mucedinaceous; cf. dematiaceous.

moniliform (**monilioid**), having swellings at regular intervals like a string of beads.

mono- (prefix), one.

monoblastic (of a conidiogenous cell), producing a blastic conidium at one locus.

monocarpic (of *Exobasidium* infections), circumscribed and annual (Nannfeldt, 1981: 10), cf. polycarpic, surculicolous.

monocentric (of a chytrid thallus), having one centre of growth and development; see polycentric and cf. reproductocentric.

monocephalic (adj. **monocephalous**), 1-headed.

monoclinous, having the antheridium on the oogonial stalk; cf. androgynous.

monoecism (adj. **monoecious**), having the male and female sex organs on one thallus; cf. dioecism, heterothallism, homothallism.

monogeocentric, see reproductocentric.

monokaryon, see haplont.

monokaryotic, having genetically identical haploid nuclei; cf. dikaryon.

monomitic, see hyphal analysis.

monomorphic, having one structure or form; not pleomorphic.

monomycelial (of an isolate), from one spore or hyphal tip.

mononematous (of conidiophores), solitary or in tufts or loose fascicles; cf. synnematous.

monophagous, see monophagy.

monophagy (adj. **monophagous**) (of *Chytridiales*), the condition of having the thallus in one host cell; the opposite of polyphagy, in which the thallus branches invade more than one host cell.

monophialidic (of a conidiogenous cell), having one locus through which conidia are produced; cf. polyphialidic.

monophyllous (of foliose lichens), having a single leaf-like thallus.

monoplanetism (of zoospores in oomycetes), the condition of having one motile phase, with no resting period.

monopodial, a type of branching in which a persistant main axis gives off branches, one at a time and frequently in alternate or spiral series.

monoreproductocentric, see reproductocentric.

monospermous, 1-spored.

monospore, used by Corner (*New Phytol.* **46**: 195, 1947) for the one spore maturing on a 2-spored basidium normally bearing 2 dispores (q.v.).

monosporic (adj. **monosporous**), 1-spored.

monostichous, in one line or series.

monotretic, see tretic.

monoverticillate (of a penicillus), composed of phialides only.

moriform, like a mulberry (*Morus*) fruit in form.

morph, form.

morphotype, a group of morphologically differentiated individuals of a species of unknown or of no taxonomic significance.

mucedinous, white or pale in colour and mould-like; mucedinoid.

mucilaginous, sticky when wet; slimy.

mucronate, pointed; ending in a short, sharp point.

multi- (prefix), a great number; many; much.

multiallelic (of an incompatibility system), having more than 2 alleles per locus; cf. biallelic.

multifid, having division into a number of parts or lobes.

multiguttulate, containing many oil-like drops.

multiperforate, of a septum with many pores connecting compartments. See Cole & Samson (*Patterns of development in conidial fungi*, 1979).

multiseptate, having a number of septa.

multisporous, having a number of spores.

multivesicular bodies, small vesicles limited by a membrane which in *Sclerotinia fructigena* originate from the endoplasmic reticulum and are possibly related to extracellular enzyme secretion (Calonge *et al.*, *J. gen. Microbiol.* **55**: 177, 1969).

Munk pores. Small (c. 1 μm) pores, each surrounded by a ring of thickening, between cells of the ascoma wall in the *Nitschkeaceae*.

muricate, rough with short, hard outgrowths.

muriculate, delicately muricate.

muriform (of spores), see dictyospore.

muriform cell, a thick-walled, dark, muriform cell (frequently referred to as a sclerotic cell or body), found in tissues affected by chromoblastomycosis (Matsumoto *et al.*, *Mycol.* **76**: 244, 1984).

muscicolous, growing on *Musci*; cf. hepaticolous.

musiform, banana-shaped (basidiospores in *Exobasidium*, fide Nannfeldt, *Symb. bot. upsal.* **23**: 27, 1981).

muticate (adj. **muticous**), having no point; not sharp at the ends.

mutualism, persistent and intimate association between organisms of different size in which the larger organism (the host) utilizes novel or enhanced properties possessed by the smaller partner(s) (symbionts), e.g. lichens, mycorrhizas. See Douglas & Smith (in Smith, *Pap. Proc. R. Soc. Tasmania* **123**: 1, 1989), symbiosis.

myc- (**mycet-**, **myceto-**, **myco-**) (prefix), pertaining to fungi.

mycangium (pl. **-ia**), a sac or cup-shaped fungal repository of ectodermal origin located in or on an ambrosia beetle (Batra, *Trans. Kansas Acad. Sci.* **66**: 226, 1963).

mycelial cord, a discrete filamentous aggregation of hyphae which, in contrast to a rhizomorph (q.v.), has no apical meristem; syrrotia. Thompson & Rayner (*TBMS* **78**: 193, 1982) prefer not to use 'mycelial strand' for such structures.

mycelial muff, a subterranean hyphal system surrounding a living root (Buscott & Roux, *TBMS* **89**: 249, 1987).

mycelium (pl. **-ia**), a mass of hyphae; the thallus of a fungus; 'spawn'; **mycelioid**, like mycelium. See Jennings & Rayner (Eds) (*The ecology and physiology of the fungal mycelium*, 1984), Gregory (*TBMS* **82**: 1, 1984; review).

myceloconidium, see stylospore.

mycetal, a fungus or a lichen (obsol.).

-mycetes (suffix), indicating the rank of a fungal class.

mycetobionts, fungus-dependent, used of obligate fungus-feeding arthropods.

mycetocyte, see mycetosome.

mycetoma (maduramycosis, madura foot), a disease, esp. tropical, of the foot or other part of man resulting in tumefactions and characterized by mycotic granules ('grains') in the infected tissues. Although a clinical entity, many different fungi (**eumycetoma**) and actinomycetes (**actinomycetoma**) are involved. Mycetomas can be roughly classified according to whether the grains are white or yellow (*Nocardia madurae*, *Pseudallescheria boydii*, *Aspergillus* spp., etc.), red (*Streptomyces pelletieri*, *S. somaliensis*) or black (*Madurella myce-*

tomatis, etc.). See Mahgoub & Murray (*Mycetoma*, 1973).

mycetophagy, see mycophagy.

mycetophiles, fungus-loving, facultative fungus-feeding arthropods.

mycetophilous, see mycophilic.

mycetosome, a sac-like structure in the gut of Anobiid beetles lined with cells (**mycetocytes**) containing yeast cells.

mycid, a secondary effect (manifested as eczema, urticaria, etc.) which is an allergic reaction to spores or toxin of a dermatophyte; dermatophytid. Cf. mycosis (2). A mycid may be a trichophytid (caused by *Trichophyton*); microsporid (*Microsporum*); epidermophytid (*Epidermophyton*).

-mycin (suffix), the recommended ending for names coined for antibiotics derived from actinomycetes.

myco- (prefix), pertaining to fungi.

mycobiont, the fungal component of a lichen (Scott, *Nature* **179**: 486, 1957); cf. phycobiont, photobiont.

mycobiota, (1) the total fungal inventory of the area under consideration (e.g. all the species present); (2) the fungal mass present (e.g. in a soil sample).

mycocecidium, see cecidium.

mycoclena (orig. 'micoclena'), term coined by Peyronel (1922) for the 'fungus mantle' of an ectotrophic mycorrhiza having a loose structure; cf. mycoderm.

mycoderm, coined by Ziegenspeck (1929) for a compact, tissue-like, ectotrophic mycorrhiza; cf. mycolena.

mycoflora, see mycobiota (a more appropriate term as fungi are not plants).

mycogenous, coming from, or living on, fungi.

mycohaemia (**mycohemia**), a condition in which fungi are present in the blood stream.

mycolatry, the worship of fungi (introduced by Wasson, 1980).

mycoliths, sand grains bound together with mycelium, forming structures 5–6 cm long, esp. by *Melanospora tulasnei*.

mycologist, one engaged in the pursuit of mycology.

mycology, the scientific study of fungi.

mycolysis, the lysis of a fungus.

mycomysticism, mystical state induced by eating hallucinogenic fungi.

mycoparasitism, the parasitism of one fungus by another (the **mycoparasite**); preferable to hyperparasitism which has been used for the same phenomenon.

mycopathology, the study of disease caused by fungi.

mycophage, (1) a mycophagist; (2) a phage-like antibacterial substance produced by certain actinomycetes.

mycophagist, an eater of fungi.

mycophagy (adj. **mycophagous**), (1) the use of fungi as food; mycetophagy; (2) the lysis of a fungus by a phage.

mycophilic, (1) fond of fungi (or mushrooms); mycetophilous; (2) growing on fungi.

mycophobia, fear of mushrooms.

mycophthorous (of a fungus), parasitic on another fungus; see mycoparasitism.

mycophycobiosis, an obligate symbiosis between a systemic (inhabitant) marine fungus and a marine alga in which the alga is the exhabitant and dominates (Kohlmeyer & Kohlmeyer, *Bot. Mar.* **15**: 109, 1972), e.g. *Mycosphaerella ascophylli* on *Ascophyllum nodosum*; cf. Lichens.

mycoplasm, a symbiotic phase of rust fungus and host protoplasm (Eriksson; now taken as an error); see also mycoplasma.

mycoplasma, (1) an intimate relationship between plant-invading fungi or other microorganisms and their host cells (Frank, *Ber. Deutsch. bot. Ges.* **7**: 332, 1889) (obsol.); (2) bacterium-like organisms without a cell wall living inside cells of a host, mollicutes (Krass & Gardner, *Internat. J. Syst. Bact.* **23**: 62, 1973).

mycorrhiza (pl. **mycorrhizas, mycorrhizae**) (fungus root). A symbiotic, non-pathogenic or feebly or weakly pathogenic association of a fungus and the roots of a plant. Found in the majority, perhaps 85% of plant species. The resulting dual organism (cf. lichen) was first described in detail by Frank (1895) who observed it on the roots of the major tree species of temperate forests. Frank (1887) then recorded two types of mycorrhiza, which he called: (a) **ectotrophic** (the characteristic mycorrhiza of temperate and boreal forest trees with different basidiomycetes esp. spp. of *Amanita*, *Boletus*, *Cortinarius*, *Russula*, *Suillus*, and some ascomycetes, e.g. *Tuber*), in which the fungus forms a sheath on the surface of the root from which hyphae extend outward into the soil and inwards between the outer cortical cells with which they interface to form a 'Hartig net'; and: (b) **endotrophic** (e.g. of orchid-basidiomycete and ericoid-ascomycete associations) in which the fungal hyphae enter the cortical cells of the root, enveloped by the plasmalemma of host.

The mycorrhiza now known to be most widely distributed both through the plant kingdom (see Trappe, in Safir *et al.*, 1987), and geographically, is a type of endo-infection formed by *Glomeromycota* and called vesicular-arbuscular (**VA**, or **VAM**). In this, the penetrating hyphae produce finely branched haustorial branches (arbuscules) or coils (pelotons) and vesicles (Schlicht, 1889; Gallaud, 1905; Mosse, 1956). The taxonomic status of VA fungi is reviewed by Morton & Benny (*Mycotaxon* **37**: 471, 1990).

Current classifications of mycorrhizal types avoid use of the suffix 'trophic' and recognize the following categories: **Ectomycorrhiza**, **Vesicular-arbuscular** (occasionally simply 'arbuscular'; AM), **Ericoid**, **Orchid**, **Arbutoid** and **Monotropoid** (Lewis, *Biol. Rev.* **48**: 261, 1973; Read, *CJB* **61**: 985, 1983).

Early development of some plants, especially those with very small seeds (e.g. orchids) are completely dependent upon fungal infection. This dependence may be retained where, as in *Monotropaceae*, and some members of *Orchidaceae* and *Gentianaceae*, plants lack chlorophyll throughout their lives, and hence continue to require all carbon and most mineral supplies from their fungal symbiont. In green plants the main function of infection differs according to mycorrhizal type. Enhancement of phosphorus (P) supply to the 'host' is characteristic of VA infections, that of nitrogen (N) of ericoid infections, while both N and P supplies can be supplemented by ecto, ectendo and probably orchid infections. The global distribution of mycorrhizal types largely reflects these functional attributes, selection favouring hosts of VA fungi in primarily P limited tropical and temperate grasslands, ectomycorrhizal plants in predominantly N and P limited temperate and boreal forests, and ericoid hosts in N-limited tundra (Read, *Experientia* **47**: 376, 1991). In each of these circumstances host plants show differing levels of responsiveness to infection depending upon the extent of nutrient limitation and the structure of their root systems. Those with evolutionarily advanced 'fibrous' root systems, e.g. grasses, are less responsive than those with primitive coarsely branched so called 'magnolioid' roots (Baylis, see Sanders *et al.*, 1975).

mycosclerid, see cystidium.

mycosin, a nitrogenous substance like animal chitin in the cell wall of fungi.

mycosis (pl. **mycoses**), (1) a fungus disease of humans, animals, or, rarely, plants (e.g. tracheomycosis). Mycoses are frequently named after the part attacked (**broncho-**, respiratory tract; **dermato-**, skin; **onycho-**, nails; **oto-**, ear; **pneumo-**, lungs), or the pathogen **blasto-** (q.v.) (*Blastomyces*); **coccidioido-**, coccidioidal granuloma (*Coccidioides immitis*); (2) the first limited infection of a dermatophyte; cf. mycid. Nomenclature, see Odds (*J. med. vet. Mycol.* **30**: 21, 1992).

mycostasis (adj. **mycostatic**), inhibition of fungal growth; fungistasis (fungistatic); sporostasis; **mycostatic** in soil, see Dobbs *et al.* (*Nature* **172**: 197, 1953), Parkinson & Waid (Eds) (*The ecology of fungi*, 130, 1960).

mycostratum, see perispore; spore wall.

mycosymbiont, see mycobiont.

mycosymbiosis, symbiosis of two or more fungi (Vainio, 1921; cf. *Gonidiomyces*).

mycothallus (pl. **-lli**), a mutualistic symbiosis of a fungus with a hepatic (liverwort) or fern gametophyte (Boullard, *Syllogeus* **19**: 1, 1979; *in* Pirozynski & Hawksworth (Eds), *Coevolution of fungi with plants and animals*: 107, 1988).

mycotic (esp. of disease), caused by fungi.

mycotoxin, see phytotoxic mycotoxins.

mycotroph, a fungus which obtains its nutrients from another fungus; cf. mycoparasitism.

mycotrophic (of plants), having mycorrhiza.

myriosporous, having many spores; cf. oligosporous.

myrmecophilous (of fungi), being a covering or food for ants.

mytiliform, like a mussel shell in form.

myxarioid (of basidia), having a stalk-like portion separated by a wall from the globose metabasidial portion, as in *Myxarium* (Donk, *Persoonia* **4**: 232, 1966).

myxomyceticolous, growing on myxomycetes.

myxosporium, see perisporium; spore wall.

nacreous, like mother-of-pearl.

napiform, turnip-like in form.

nassace (**nasse**), the finger-like protrusion of the inner part of a bitunicate ascus into the inner tunica; internal apical beak; see ascus.

nasse, see nassace.

navel, see umbilicus.

navicular (**naviculate**), boat-like in form; cymbiform.

necral layer, a layer of horny dead fungal hyphae with indistinct lumina in or near the cortex of lichens; **epinecral layer** if above the algal layer, **hyponecral layer** if below.

necrophagous, saprobic.

necrophyte, an organism living on dead material (Münch); cf. perthophyte, saprophyte.

necrosis, death of plant cells, esp. when resulting in the tissue becoming dark in colour; commonly a symptom of fungus infection.

necrotroph, a parasite that derives its energy from dead cells of the host (Thrower, *Phytopath. Z.* **56**: 258, 1966); cf. biotroph.

nematogenous, of conidiogenous cells arising at all levels from single hyphae; lanose. See Gams (*Cephalosporium-artige Schimmelpilze*, 1971).

nematophagous, nematode-feeding. Nematophagous fungi are also reported as fossils (Jansson & Poinar, *TBMS* **87**: 471, 1986).

nemoral, living in woods or groves.

neo- (prefix), new.

neosexual, see protosexual.

neotony, a process in which normal development of cells, except those involved in reproduction, is arrested; results in a sexually mature organism with juvenile features; see Kreisel (*in* Hawksworth (Ed.), *Frontiers in mycology*: 69, 1991; in fungal phylogeny); cf. pedogenesis.

nervicolous, living on veins of leaves or stems.

neurotoxin, a toxin which affects the nervous system.

nidose (**nidorose**), having an unpleasant smell.

nidulant, lying free in a cavity.

nimbospore, a spore having a gelatinous, apparently many-layered wall, e.g. *Histoplasma capsulatum* (Nielsen & Evans, *J. Bact.* **68**: 261, 1954).

nitid (adj. **nitidous**), smooth and clear; lustrous.

nitrophilous, having a preference for habitats rich in nitrogen; chionophilous; **nitrophobous**, having a preference for habitats poor in nitrogen.

node cell, see hyphopodium.

nodose-septum, see clamp-connection.

nodular bodies (of dermatophytes), rounded bodies made up of massed hyphae.

nodulose (of spores), having broad-based, blunt, wart-like excrescences.

nodum (pl. **noda**) (in phytosociology), particular well-defined plant communities.

nonfissitunicate (of asci), ones in which discharge does not involve a separation of the wall layers; see ascus.

non-target organisms, organisms found with or near those being treated with chemical or biological control agents.

notate (of surfaces), marked by straight or curved lines.

nubilated, cloudy and semi-opaque as viewed by transmitted light.

nuclear cap (of *Blastocladiaceae*), a body at one side of the nucleus of a zoospore or gamete.

nurse cells (in *Scleroderma*), hyphae supplying food material to spores which have come away from the basidia.

nutant, nodding.

nutriocyte (of *Ascosphaera*), the inflated part of the ascogonium which eventually develops into a spore cyst (Spiltoir & Olive, *Mycol.* **47**: 240, 1955).

ob- (prefix), inversely or oppositely.

obclavate, inversely clavate (widest at the base); cf. clavate.

oblate, with the equatorial diameter greater than the polar diameter; **- spheroidal**, shaped like the earth.

obligate, (1) necessary; essential; (2) (of a parasite), living as a parasite in nature, sometimes of one that has not been cultured on laboratory media, cf. facultative; see synonym.

oblique septum, see septum.

oblong (of spores), twice as long as wide and having somewhat truncate ends; **- ellipsoid** (of spores), rounded-oblong; having long sides parallel and ends almost hemispherical.

obovate, inversely ovate.

obovoid, inversely ovoid.

obpyriform, the reverse of pyriform.

obsolete, (1) (of organs or parts), rudimentary or absent; (2) (of terms), no longer in use.

obsubulate, very narrow; pointed at the base and a little wider at the tip.

obtuse, (1) rounded or blunt; (2) greater than a right angle.

occluded, closed; often used of the lumina of hyphae or pseudoparenchymatous cells.

ocellate, having rounded marks, like eyes.

ocellus, an eyespot functioning as a lens and concentrating light rays on a sensitive spot.

ochrosporous, having yellow or yellow-brown spores.

octo- (in combinations), 8.

octophore, see haerangium.

octopolar (of incompatability systems), having 3 loci, as in *Psathyrella coprobia* (Jurand & Kemp, *Genetical Res., Cambr.* **22**: 125, 1973); cf. tetrapolar.

octospore, one spore of an 8-spored ascus.

octosporous, producing spores in 8s.

ocular chamber, see ascus.

odontoid, tooth-like; dentate.

oedocephaloid, having a swelling at the end or tip, as conidiophores of *Oedocephalum* and *Cunninghamella*.

-oid (suffix), like; having the form of. Most of the many mycological terms ending in this suffix (e.g. achlyoid, as in *Achlya*; daedaleoid, as *Daedalea*) have not been compiled in this *Dictionary*.

oidiophore, a structure producing oidia.

oidiospore, see oidium.

oidium (pl. **-ia**), (1) spermatia formed on hyphal branches, esp. in heterothallic hymenomycetes; (2) flat-ended conidia formed by the breaking up (usually centripetally) of a hypha into cells, as in *Geotrichum candidum*; arthrospore; (3) a mildew.

oidization, dikaryotization by the fusion of an oidium with a haploid hypha.

oleocystidium, see cystidium.

oleoso-locular (of spores), having cells like drops of oil.

oligosporous, having few spores; cf. myriosporous.

oligotrophic, poor in nutrients; cf. dystrophic, eutrophic.

omnivorous (of parasites), attacking a number of different hosts.

omphalodisc, (1) an orbicular conical-shaped disk; (2) (of *Umbilicaria* [*Omphalodiscus*]), an apothecium with a central knob of sterile hyphae.

oogamy, heterogamy in which the gametes are a non-motile egg and a small, motile sperm.

operculate (of an ascus or sporangium), opening by an apical lid to discharge the spores, as in the ascus of the *Pezizales*; see ascus; cf. inoperculate.

operculum, a cover or lid.

opisthokont, having one or more flagella at the posterior end.

opportunistic (of fungi), normally saprobic and frequently common but on occasion able to cause disease in or grow on a host (esp. of humans or other animals), rendered susceptible by some predisposing factor(s).

opsis-form, of species with no urediniospore (q.v.) in the life-cycle.

orbilla (pl. **orbillae**), an apothecium (obsol.), (Sprengel, *Intr. study cryptog.*, 1807).

orculiform, see polarilocular.

ornamented (of organs, esp. spores), having the surface marked or sculptured with lines, wrinkles, warts, striations, ridges, reticulations, fibrils, scales etc.; not smooth. See basidiospore.

ornithocoprophilous, preferring habitats rich in bird droppings.

orthidium (obsol.), conidiomata of *Pyrenotrichum*, sometimes incorrectly listed as if a generic name. See Hawksworth (*Bull. Br. Mus. nat. Hist., Bot.* **9**: 59, 1981).

orthotrophy, Corda's (1842) term for the condition in which a basidiospore primordium develops at the apex of the apophysis in contrast to **heterotrophy** when the primordium develops laterally.

oscule (obsol.), a pore of a rust spore (Tulasne).

-osis (suffix), condition of; state caused by.

osmophily (adj. **-ilic**, **-ilous**), making growth under conditions of high osmotic pressure, as in *Xeromyces* and some yeasts on conc. sugar solutions. See Moustafa (*Can. J. Microbiol.* **21**: 1573, 1975; osmophilous spp. of Kuwait).

osmotolerant, capable of growing in high osmotic pressure, e.g. some yeasts and filamentous fungi on concentrated sugar solutions.

osmotrophic, exhibiting absorptive nutrition as do fungi.

ostiole (sing. **ostiolum**), the schizogenous, paraphysis-lined cavity, ending in a pore, in the papilla or neck of a perithecium (Miller, *Mycol.* **20**: 196, 1928); any pore by which spores are freed from an ascigerous or pycnidial fruit-body. See v. Arx (*K. ned. Akad. Wet.*, C, **76**: 289, 1973; taxonomic importance in *Ascomycota*).

ostropalean (of asci), ones essentially unitunicate in structure and with a thickened apex penetrated by a narrow pore; see ascus.

oval, widely elliptical.

ovariicolous, living in ovaries.

ovate (of a surface [or sometimes a solid]), **ovoid** (of a solid), like a hen's egg with the narrower end at the top.

oxydated (of crustose lichens), having thalli tinged rust-red by iron oxides; oxydized.

pachydermatous, (1) thick-skinned; (2) (of hyphae), having the outer wall thicker than the lumen.

pachypleurous, thick-walled.

palisade plectenchyma, plectenchyma in the cortex of a lichen thallus composed of hyphae arranged perpendicular to the surface.

palisade-cells (of lichens), the end cells of the hyphae of a fastigiate cortex.

palisoderm (of basidiomata), an outer layer composed of several strata of cells or hyphal tips; see derm.

pallid, light-coloured; pale.

pallisadoplectenchyma, see plectenchyma.
palmate, having lobes radiating from a common centre but not extending to the point of insertion; cf. chiroid, digitate.
paludal, living in wet places (marshes).
pannose (**panniform**), having the appearance of felt or woollen cloth.
papilionaceous, variegated; mottled; marked with different colours as the lamellae of certain *Panaeolus* spp.
papilla, a small rounded process.
papillae (of host tissue), localized wall thickenings on the inner surface of plant cell walls at sites penetrated by fungi; callosity, lignituber, callus, 'sheath', are whole or part synonyms (Aist, *Ann. Rev. Phytopath.* **14**: 145, 1976).
papillate, having a papilla (q.v.).
papulose, covered with pimples or pustules.
papulospore, asexual spore in e.g. *Papulaspora sepedonioides* (Weresub & LeClair, *CJB* **49**: 2203, 1971).
paracapillitium (of *Lycoperdales*), a capillitium composed of thin-walled, hyaline, septate hyphae in contrast to a true capillitium of thick-walled, brown, aseptate hyphae (Kreisel, 1962).
paracephalodium (pl. **-ia**), a hyphal mat covering cyanobacteria arising from a squamulose lichen thallus which has a green alga as the photobiont (Poelt & Mayrhofer, *Pl. Syst. Evol.* **158**: 265, 1988).
paramorph, a neutral term proposed by Huxley (*The new systematics*: 37, 1940) for any form differing from the mean of the group; advocated for use in fossil fungi lacking characters essential for their proper placement, but believed to have affinities with particular non-fossil groups by Reynolds (*Mycotaxon* **23**: 141, 1985).
paraphysis (pl. **paraphyses**), a sterile, upward-growing, basally attached hyphal element in a hymenium, esp. in ascomycetes where they are generally filiform, unbranched or branched, and the free ends frequently make an epithecium over the asci, see hamathecium; in basidiomycetes, see hyphidium; **apical paraphyses**, the downward-growing hyphae with free tips in the centrum of hypocrealean fungi (Luttrell, *TBMS* **48**: 135, 1965), periphysoids; **asco-**, a multicellular diploid storage hypha originating from the base of the ascus in *Erysiphaceae* (Speer, *Sydowia* **27**: 1, 1976); **paraphysate**, having paraphyses.
paraphysoid, (1) (of ascomycetes), see hamathecium; (2) (of basidiomycetes), a sterile accessory hymenial structure; see basidiole, cystidiole, hyphidium; **- network**, branched and anastomosing paraphysoids surrounding asci in some ascolocular ascomycetes.
paraplectenchyma, plectenchyma (q.v.) composed of cells which have isodiametric lumina and unthickened walls (see Yoshimura & Shimada, *Bull. Kochi Gakuen Jun. Coll.* **11**: 13, 1980).
parasexual cycle, a mechanism discovered by Pontecorvo & Roper in 1952 in filamentous fungi by which recombination of hereditary properties is based not on sexual reproduction (meiosis) but on the mitotic cycle. The essential features of the process are: (1) the production of diploid nuclei in a heterokaryotic haploid mycelium; (2) the multiplication of the diploid nuclei along with haploid nuclei in a heterokaryotic mycelium; (3) the sorting out of a diploid homokaryon; (4) segregation and recombination by crossing-over at mitosis; (5) haploidization of the diploid nuclei.

The results are similar to those achieved by meiosis but instead of a regular sequence of events in time as in the meiotic cycle, in the parasexual cycle the various processes may all be occurring at one time in one mycelium, at rates which have been estimated. The parasexual cycle may (as in *Aspergillus nidulans*) or may not (*A. niger*) be accompanied by a sexual cycle. Although the details of the mechanism of the parasexual cycle are unknown, Pontecorvo and his school obtained data from the parasexual cycle for mapping the 8 chromosomes of *Aspergillus nidulans*, results which showed good agreement with those for the same fungus obtained by the analysis of the sexual cycle. See Pontecorvo *et al.* (*Advances in Genetics* **5**: 141, 1953), Pontecorvo & Käfer (*Advances in Genetics* **9**: 71, 1958). For other aspects see Pontecorvo (*Ann. Rev. Microbiol.* **10**: 393, 1956), Pontecorvo *et al.* (*J. gen. Microbiol.* **8**: 198, 1953; *Aspergillus niger*, **11**: 94, 1954; *Penicillium chrysogenum*), Buxton (*J. gen. Microbiol.* **15**: 133, 1956; *Fusarium oxysporum*). See protosexual.
parasite, an organism living on or in, and obtaining its nutrients from, its host, another living organism; a biotroph or necrotroph; cf. parasymbiont, pathogen, symbiosis.
parasoredium, soredium-like structure, originally used for a structured plectenchyma in the upper thallus layer of the *Umbilicaria hirsuta* aggr. (see Codogno *et al.*, *Pl. Syst. Evol.* **165**: 55, 1989).
parasymbiont (obsol.), a fungus or lichen living on a lichen thallus but not causing any obvious damage (Zopf, *Ber. dtsch. bot. Ges.* **15**: 90, 1897); a commensalistic or gall-forming lichenicolous fungus (Hawksworth, *Bot. J. Linn. Soc.* **96**: 3, 1988).
parathecium, (1) (of apothecia), the outside hyphal layer, esp. if darker in colour (obsol.); (2) ectal excipulum (q.v.).
parenthesome, a curved double membrane (which may be perforate, imperforate, or vesiculate; see Moore, *Bot. Marina* **23**: 362, 1980) on each side of a dolipore septum (Moore & McAlear, *Am. J. Bot.* **49**: 86, 1962); 'dolipore'; septal pore cap.
parietal, fixed to the wall, e.g. of asci in a perithecium.
pariobasidium, see basidium.
parmuliform, shield-shaped with the margins slightly upturned.
part spore, one of the 1-celled spores resulting from the breaking up of a 2 or more celled ascospore.
parthenogamy, the state of an oospore formed with a diploid nucleus resulting from restitution at telophase I or telophase II of meiosis (note that the parthenogametic state cannot be presumed from the absence of an antheridium).
parthenogenesis, the apomitic development of haploid cells (Gäumann).
parthenomixis, see parthenogamy.
parthenospore, an oospore (aboospore) or zygospore (azygospore) produced by parthenogenesis.
partial veil (or **inner veil**) (of agarics), a layer of tissue, developed from the stipe, which joins the stipe to the pileus edge during hymenium development, and

which later may become an annulus or cortina; = velum (Persoon).

patelliform, like a round plate having a well-marked edge.

patent, stretching out; spreading.

pathogen, a parasite able to cause disease in a particular host or range of hosts (*TBMS* **33**: 155, 1950); **-ic**, disease-causing or able to be so; **-icity**, the condition of being pathogenic.

pectinate, like the teeth of a comb.

pectinate hyphae, comb-like hyphae of e.g. *Microsporum audouinii*.

pedicel, a small stalk.

pedicellate, having a pedicel.

pedogamy, pseudomixis between mature and immature cells, e.g. copulation between a yeast mother cell and its bud.

pedogenesis, (1) reproduction in young or immature organisms; (2) soil formation.

peg, see hyphal peg.

pellet, a three-dimensional colony in a liquid culture (particularly a shaken culture).

pellicle, (1) outermost living layer lying below any non-living secreted material, containing plasma membrane plus underlying epiplasm or other membranes and may show ridges, folds or distinct crests; (2) a growth on the surface of a liquid culture; (3) (of agaric basidiomata), a detachable, skin-like cuticle of the pileus.

pellicular veil, a very thin partial veil of a sporophore not having a stipe (Singer, 1962); cf. cortina.

pelliculose, like a thin crust, as is the hymenial layer in *Thelephoraceae*.

pellis, the cellular cortical layers, not belonging to the veils, of a basidioma (Bas, *Persoonia* **5**: 327, 1969); cuticle. Bas distinguished different layers of the pellis as **supra-**, **medio-**, and **sub-**, and different topographies of the pellis as **pilei-**, etc.

pellucid-striate (of a pileus), having a somewhat transparent top so that the gills are seen through it as rays.

peloton, see mycorrhiza.

peltate, in the form of a round plate with a stalk from the centre of the underside.

pendant, (1) (of a fruticose lichen), one hanging downwards; (2) (in an ascus) see ascus.

penicillate, like a little brush.

penicillus, the brush-like conidiogenous apparatus of *Penicillium* and related genera composed of a stipe bearing a tuft of conidiogenous cells and other elements formerly on the rami and metulae.

percurrent, (1) extending throughout the entire length, as of the columella of a gasteromycete basidioma; (2) growing through in the direction of the long axis, as of a conidial germ tube emerging through the hilum or of a proliferation growing through the tip of the conidiogenous cell (Luttrell, 1963).

perennial, living for a number of years.

perforation lysis, the process by which degradation of resistant fungus propagules in soil is initiated (Old & Wong, *Soil Biol. Biochem.* **8**: 285, 1976; review).

pergameneous (**pergamenous**, **pergamentaceous**), like paper.

periclinal, curved in the direction of, or parallel to, the surface or the circumference; cf. anticlinal; **- thickening**, zone of increased material surrounding the protoplasmic channel at the apex of a 'phialide' (see Sutton, *The Coelomycetes*, 1980).

peridermium, an aecium as in the form genus *Peridermium*.

peridiole (**peridiolum**) (esp. of *Nidulariaceae*), a division of the gleba having a separate wall, frequently acting as a unit for distribution.

peridium, the wall or limiting membrane of a sporangium or other fruit-body, an excipule; **peridial cells** (esp. of aecia), the cells of the peridium.

perifulcrium, the wall of a pycnidium in a lichen thallus (obsol.).

periphysis (pl. **periphyses**), an upward-pointing hypha inside, or near, the ostiole of a perithecium, pycnidium, or pycnium; see hamathecium.

periphysoid, see hamathecium.

periphyton, the 'assemblage of organisms growing upon free surfaces of submerged objects in water and covering them with a slimy coat.' Young (1945) fide Cooke (*Bot. Rev.* **22**: 616, 1956).

perispore (**perisporium**), sheath outside the true spore wall. See Harmaja (*Karstenia* **14**: 123, 1974; cyanophilic in *Pezizales*); ascospore, spore wall.

perisporial sac, a perispore forming a loose envelope around the spore as in *Coprinus* sp.

perisporium see perispore.

peristome, an edging round an opening, esp. of basidiomata of certain gasteromycetes.

perithecium (pl. **-ia**), a subglobose or flask-like ostiolate ascoma; sometimes limited to ascohymenial types formed from the development of an ascogonium (not of stromatic origin), but now widely used as a general term regardless of the ontogenetic type. See Cherepanova (*Vestn. Leningrad Univ. Biol.* **3**: 39, 1986; types and evol. pathways).

peritrichous (**peritrichiate**), having hairs or flagella all over the surface.

peronate, sheathed; having a boot or covering, esp. of the lower part of a stipe covered by a volva or veil.

peroxisome, one of the subcellular organelles which have indispensable functions in the metabolism of *n*-alkalenes, fatty acids, methanol, and several nitrogen-containing compounds in eukaryotic microorganisms. See Tanaka & Ueda (*MR* **97**: 1025, 1993).

perrumpent, breaking through; cf. erumpent.

persistent, (1) (of interascal tissues) still evident at maturity; (2) (of spores), non-deciduous; (3) (of teliospore pedicels), remaining firmly attached to the spore after liberation.

perthophyte (**perthotroph**), a necrophyte on dead tissues of living hosts (Münch); cf. saprophyte.

pervious (of lichenized scyphi), open or perforate basally.

petrophilous, see saxicolous.

phaeo- (prefix), dark-coloured or swarthy, esp. of spores.

phagosome, a membrane surrounding an endosymbiont to form a distinctive structure, as *Nostoc*-containing vesicles of *Geosiphon*.

phagotrophic, feeding by ingestion, engulfing food.

phalacrogenous, of conidiogenous cells arising at the same level from single hyphae to form a turf-like layer; velvety. See Gams (*Cephalosporium-artige Schimmelpilze*, 1971).

phellophagy, ability to attack cork cells (Speer, *Mycotaxon* **21**: 235, 1984).

pheromone, a substance secreted to the outside by an individual and received by a second individual of the same species, in which it induces a specific reaction, e.g. a definite behaviour or developmental process (Karlson & Luscher, *Nature* **183**: 55, 1959).

phialide (after Vuillemin), a cell which develops one or more (the **polyphialide** of Hughes, *Mycol. Pap.* **45**, 1951) open-ended conidiogenous loci from which a basipetal succession of conidia, **phialospores**, develops without an increase in length of the phialide itself (Hughes, *loc. cit.*); cf. annellophore; sterigma. In some fungi, e.g. *Acremonium*, the phialide may be the conidiophore; more frequently the phialide is either an end cell of a conidiophore or attached to a conidiophore (or **phialophore**). See Roquebert (*Rev. Myc.* **40**: 417, 1976; review), terminus phialospore, and Minter *et al.* (*TBMS* **81**: 109, 1983; history).

phialidic (of conidiogenesis, obsol.), the sort of conidiogenesis in which each conidium (**phialoconidium**, phialidic conidium, phialospore) originates by the laying down of new wall material not from existing walls or layers of the wall of the conidiogenous cell (**phialide**). A *basipetal* succession of conidia is formed from a *fixed* conidiogenous locus (cf. tretic). **mono-, poly-,** (of phialides), producing conidia through a single opening or a sympodial irregular or synchronous succession of openings, respectively, in the conidiogenous cell wall.

phialophore, see phialide.

phialospore, see phialide.

phoenicoid fungi, fungi growing amongst the ashes of former fires (Carpenter & Trappe, *Mycotaxon* **23**: 203, 1985).

photo- (prefix), pertaining to light.

photobiont, a photosynthetic symbiont in a lichen which may be a eukaryotic alga (phycobiont), see Gärtner (*in* Reisser (Ed.), *Algae and symbiosis*: 325, 1992; review systematics), or a cyanobacterium (bactobiont, cyanobiont) (Ahmadjian, *Internat. Lich. Newsl.* **15** (2): 19, 1982), Büdel (*in* Reisser (Ed.), *Algae and symbiosis*: 301, 1992; review systematics).

photomorph, an organism whose form is determined by the nature of its photosynthesis (Laundon, *Taxon* **44**: 387, 1995); see also phycotype, phototype, photosymbiodeme, phycosymbiodeme.

photophilous, having a preference for well-illuminated habitats; cf. heliophilous, anheliophilous.

photophobous, having a preference for shaded habitats.

photosporogenic, requiring light for sporogenesis.

photosymbiodeme, a replacement term for phycosymbiodeme to allow for one biont being a cyanobacterium (Stocker-Wörgötter & Türk, *Crypt. Bot.* **4**: 300, 1994); a photomorph.

photosymbiodeme, a replacement term for phycosymbiodeme to allow for one biont being a cyanobacterium (Stocker-Wörgötter & Türk, *Crypt. Bot.* **4**: 300, 1994); a photomorph. See Lichens.

phototaxis, movement (e.g. of zoospores) influenced by light.

phototropism, a reaction of conidiophores, sporangiophores, asci, stipes etc. to light (Bergman *et al.*, *Bact. Rev.* **33**: 99, 1969).

phragmobasidium, see basidium.

phragmospore, differs from an amerospore (q.v.) and didymospore (q.v.) in having 2 to many transverse septa.

phyco- (prefix), pertaining to algae.

phycobiont, the algal partner in a lichen (Scott, *Nature* **179**: 486, 1957), photobiont (q.v.); cf. mycobiont.

phycolichenes (obsol.), lichens in which the vegetative thallus morphology is determined by the photobiont and which are of uncertain systematic position as the sporocarps are unknown (e.g. *Cystocoleus*, *Racodium*).

phycophilous, growing with or on algae.

phycosymbiodeme, joined lichen thalli with a single mycobiont but different photobionts (Renner & Galloway, *Mycotaxon* **16**: 197, 1982); photomorph, phycotype.

phycosymbiont, phycobiont.

phycotrophic (of fungi), obtaining nutrients from algae (Dobbs, *Lichenologist* **4**: 323, 1970).

phycotype, see type.

phyllidium, lichen propagule formed by abstriction of a leaf-like or scale-like portion of the thallus. See Poelt (*Flora, Jena* **169**: 23, 1980).

phyllocladia, the granular, verrucose, coralloid, squamuliform, digitate, peltate, or foliaceous parts of the thallus of *Stereocaulon* which contain the photobiont.

phylloplane, the leaf surface; Last & Deighton (*TBMS* **48**: 83, 1965), the non-parasitic biotas of the leaf surface.

phyllosphere, the zone immediately surrounding a leaf; frequently used in the sense of phylloplane (q.v.). See Preece & Dickinson (Eds) (*Ecology of leaf surface microorganisms*, 1970), Dickinson & Preece (Eds) (*Microbiology of aerial plant surfaces*, 1976), Blakeman (Ed.) (*Microbial ecology of the phylloplane*, 1981), Fokkema & van der Heuvel (Eds) (*Microbiology of the phyllosphere*, 1986); cf. rhizoplane; spermoplane.

physiologic race, one of a group of forms alike in morphology but unlike in certain cultural, physiological, biochemical, pathological, or other characters.

phytoalexin, a metabolite produced by a plant in response to infection by a fungus or other pathogen (or by an abiotic factor) inhibitory to the invading pathogen. Reviews: Cruickshank, *Ann. Rev. Phytopath.* **1**: 351, 1963; Kuć, *Ann. Rev. Phytopath.* **10**: 207, 1972; Ingram, *Bot. Rev.* **38**: 343, 1972.

Phytoalexins include: capsidiol, glyceollin, ipomearone, kievitone, phaseollin, pisatin, rishitin, wyerone.

phytotoxic mycotoxins, (toxins injurious to plants; see phytotoxin (2) under toxin). May be host specific (e.g. helminthosporoside, phytoalternarins, victorin) or non-host specific (e.g. alternaric acid, baccatin, culmomarasmin, diaporthin, fusaric acid, fusicoccin, helminthosporal, ophiobolin (cochliobolin), skyrin, tentoxin). Reviews: Pringle & Scheffer (*Ann. Rev. Phytopath.* **2**: 133, 1964), Wright (*Ann. Rev. Microbiol.* **22**: 269, 1968), Wood *et al.* (Eds) (*Phytotoxins in plant disease*, 1972), Strobel (*Ann. Rev. Pl. Physiol.* **25**: 541, 1974), Durbin (Ed.) (*Toxins in plant diseases*, 1981).

pileate, having a pileus.

pileipellis, see pellis.

pileocystidium, see cystidium.

pileolus, a small pileus.

pileus, the hymenium-supporting part of the basidioma of non-resupinate *Basidiomycetes*; cap.

pilose, covered with hairs.

pionnotes (of *Fusarium*), a spore mass having a fat-like or grease-like appearance; **pseudo-**, see sporodochium.

piricularin, a phytotoxin from *Pyricularia oryzae* (Togashi *et al.*, *Ann. phytopath. Soc. Japan* **25**: 142, 1960).

piriform, see pyriform.

placodioid (of a crustose lichen thallus), disc-shaped with plicate lobes at the circumference.

placodiomorph, a 2-celled spore with a thickened septum which may or may not have a pore; cf. polarilocular.

placodium, see stroma, thyriothecium.

placoid, see placodioid.

placomycetoid, pileus with diam : stipe ratio ‹1; cf. campestroid.

plage, (1) a smooth, paler-coloured, or colourless spot on a surface; (2) (of basidiospores), esp. a smooth spot above the hilar appendage.

plane, flat.

planoconvex, a flat zygote.

planocyte (**planont**), a motile cell.

planogamete, a motile gamete; zoogamete.

planont, see planocyte.

planospore, a zoospore.

planozygote, (1) a motile zygote; (2) a flat zygote (of a pileus or spore), convex but somewhat flat.

plaque, a clear area in a bacterial colony caused by localized viral lysis; also applied to similar areas in fungal cultures (e.g. Riegman & Wessels, *TBMS* **75**: 325, 1980).

plasma membrane, see plasmalemma.

plasmalemma, outer membrane composed of phospholipids and proteins which surrounds a cell and regulates exchange of materials between the cell and its environment; the limiting (boundary) membrane of the cytoplasm; cell, cytoplasmic, or plasma membrane.

plasmalemmasome, an intracytoplasmic vesicle (formed by invagination of the plasmalemma) filled with tubular diverticula; cf. lomasome.

plasmamembranic vesicle, sporangial vesicle, the membrane of which is the same membrane bounding the protoplasm within the sporangium, c.f. homohylic vesicle, precipitative vesicle.

plasmodesma (pl. **-ta**), an isthmus-like strand of protoplasm connecting adjacent cells.

plasmodic granules, very small, dark-coloured particles on the surface of the peridium, and frequently of the spores, in the *Cribrariaceae*; dictydine granules.

plasmogamospore, aeciospore (Laundon, *TBMS* **58**: 345, 1972).

plasmogamy, fusion of two cells or plasmodial cytoplasms without karyogamy (nuclear fusion) or a precursor to karyogamy.

platygonidia, photobionts occurring in stellately or orbicular spreading colonies (e.g. *Cephaleuros*) (obsol.).

platyphyllous, broadly lobed.

platysmoid, (1) (of lichen thalli), loosely attached foliose thalli with ascending lobes, as in *Platismatia* (obsol.); (2) (of tissues), a scleroplectenchyma in which the hyphae are brown (Yoshimura & Shimada, *Bull. Kochi Gakuen Jun. Coll.* **11**: 13, 1980).

plectenchyma, a thick tissue formed by hyphae becoming twisted and fixed together; synchyma (Vuillemin); it is **prosenchyma** (**proso-**) when the hyphal elements are seen to be hyphae; **pseudoparenchyma** (**para-**), when they are not; Vuillemin (1912) distinguished as **merenchyma** tissue derived by cell division in several planes, and Degelius (1954) used **euthyplectenchyma** for hyphal tissue with no cellular structure. Yoshimura & Shimada (*Bull. Kochi Gakuen Jun. Coll.* **11**: 13, 1980) key nine categories of plectenchyma, **chalaroplectenchyma** (hyphal walls not united, lumina wide), **pallisadoplectenchyma** (hyphae parallel, not coherent, with intercellular spaces), **platysmoid** (a scleroplectenchyma with brown hyphal walls), **scleroplectenchyma** (cell-walls thickened, lumina narrow), **scleroprosoplectenchyma** (hyphae parallel, cohering, walls thick, lumina narrow), and **serioplectenchyma** (hyphae parallel, cohering, walls not thick, lumina wide).

plectonematogenous, of conidiogenous cells arising from rope-like strands of interwoven hyphae (not from single hyphae), the strands intertwined and not synnematous; funiculose. See Gams (*Cephalosporium-artige Schimmelpilze*, 1971).

pleiosporous, having many spores.

pleioxeny, the condition of plurivorous parasitism.

pleoanamorphy, having more than one anamorph. See Hennebert (*in* Sugiyama (Ed.), *Pleomorphic fungi: the diversity and its taxonomic implications*, 1987).

pleomorphic, (1) of fungi having more than one independent form or spore-stage in the life cycle, especially of holomorphs comprising a teleomorph and one or more anamorphs; polymorphic; see Savile (*Mycol.* **61**: 1161, 1970), Sugiyama (Ed.) (*Pleomorphic fungi: their diversity and its taxonomic implications*, 1987); (2) (of dermatophytes), changes due to 'degeneration' in culture.

pleomorphism, the condition of being pleomorphic.

pleont, any one of the two or more states of a pleomorphic fungus (Delphino, 1887).

plerotic, of an oospore which completely fills the oogonial cavity; but also of an oospore which occupies ›65% of the oogonium volume (Dick *et al.*, 1992).

pleuracrogenous, formed at the end and on the sides.

pleuro- (in combination), side, at the side.

pleurobasidium, see basidium.

pleurocystidium, see cystidium.

pleurogenous, formed on the side.

pleurosporous, having spores on the sides, e.g. a basidium of the *Uredinales*.

plicate, folded into pleats; **plica**, a pleat.

plug (in an ascus), see ascus.

plurilocular, (1) (of ascospores), many-celled; (2) (of stromata), having several locules.

plurivorous, attacking a number of hosts or substrates; not specialized.

pneumomycosis, see mycosis.

pocket rot, localized rotting of trunks of trees or roots by wood-destroying fungi.

podetium (pl. **-ia**), lichenized stem-like portion (stipe, or discopodium) bearing the hymenial discs and sometimes conidiomata in a fruticose apothecium (Ahti,

Lichenologist **14**: 105, 1982), esp. as in *Cladonia*; cf. pseudopodetium.
polar (of bacteria, spores, etc.), at the ends or poles.
polar-diblastic, see polarilocular.
polaribilocular, polarilocular (q.v.) with two cells.
polarilocular (of ascospores), bicellular and the two cells separated by a central perforated septum; orculiform, polaribilocular, polar-diblastic. See Sheard (*Lichenologist* **3**: 328, 1967).
poleophilous, town loving; sometimes used of lichens which thrive in urban areas (e.g. *Lecanora conizaeoides* in Eur.).
poly- (prefix), a great number; many.
polyascous, having many asci; esp. having the asci in one hymenium, not separated by sterile bands.
polyblastic (of conidiogenous cell), producing blastic conidia at several conidiogenous loci, either synchronously or irregularly.
polycarpic (of *Exobasidium* infections), systemic (or circumscribed) and perennial (Nannfeldt, 1981: 15); cf. monocarpic.
polycentric, having a number of centres of growth and development and more than one reproductive organ, as in the *Cladochytriaceae* (Karling, *Mycol.* **36**: 528, 1934); see monocentric; cf. reproductocentric.
polycephalous, many-headed.
polychotomous, having an apex dividing simultaneously into more than two branches (Corner, 1950).
polydactyloid venation, see veins.
polyenergid, coenocytic.
polymorphic, having different forms; pleomorphic.
polymorphic species, species with a series of intergrading morphological features; e.g. resulting from inbreeding or automictic sexual reproduction.
polyphagous, see polyphagy.
polyphagy (adj. **polyphagous**), see monophagy.
polyphialidic (of a conidiogenous cell), having more than one conidiogenous locus at which conidia are produced; cf. monophialidic.
polyphyllous (of foliose lichen thalli), having many connected leaf-like lobes.
polyplanetism, sequence of two or more motile flagellate phases with interspersed mobile aplanosporic phases in the zoosporic part of the life history; the aplanosporic phase may be naked or as a walled cyst; motile phases may be monomorphic or dimorphic.
polysidia, specialized dual vegetative propagules in certain *Pyxine* spp. formed in depressions at the tips of coral-like structures (**polysidiangia**) which recall isidia but are not themselves propagules (Kalb, *Bibl. Lich.* **24**, 1987).
polysporous, many-spored.
polytomous, dividing into many branches, usually at one node or point.
polytretic, see tretic.
polyxeny, see pleioxeny.
pore, (1) a small opening, as in tretic (q.v.) conidiogenesis; (2) in **- fungi** (*Polyporaceae* and *Boletaceae*), the mouth of a tube.
poricidal (of asci), see ascus.
poriform, pore-like.
poroconidium, see tretic.
porospore, see tretic.
posterior, (1) at or in the direction of the back; (2) (of a lamella), the end at or near the stipe.
praemorse (of the stipe base), as if broken off; truncate.
precipitative vesicle, sporangial vesicle, the membrane of which is formed after extrusion of naked sporangial protoplasm, whether partially cleaved into planonts or not, possibly as a precipitation reaction between periprotoplasmic colloids and the environment; amorphous or fibrillar, c.f. homohylic vesicle, plasmamembranic vesicle.
primary, first; first-formed; **- homothallism**, see homothallism; **- mycelium** (of basidiomycetes), the haploid mycelium from a basidiospore; **- septum**, see septum; **- squamules**, the first formed squamules of *Cladonia* from which the podetia arise; **- universal veil**, see protoblem; **- uredo** (uredium, uredinium, q.v.).
primordial, first in order of appearance; pertaining to the earliest stages of development; **- covering** or **cuticle** = blematogen; **- hypha**, intensely coloured hyphae of the epicutis in *Russula* (Melzer, 1934); **- shaft**, the monaxial basidioma initial, as in *Clavariaceae* (Corner, 1950); **- tissue**, undifferentiated tissue of a basidioma initial; cf. lipsanenchyma; **- veil**, protoblem; **primordium**, the earliest stage of development of an organ.
primospore, a spore very like a cell of the thallus (MacMillan).
probasidium, see basidium.
pro-diploidization hypha, a hypha which may be diploidized, cf. flexuous hypha.
progametangium (of *Mucorales*), a hyphal branch forming a gametangium and suspensor cell.
progamones, a group of sex hormones of zygomycetes (Reschke & Plempel, *Z. Pflanzenphysiol.* **67**: 343, 1972).
prohybrid, a mycelium having additional nuclei from hyphal fusions (after Dodge, *Mycol.* **28**: 407, 1936).
prokaryote (adj. **-otic**), an organism lacking membrane-limited nuclei and not exhibiting mitosis, e.g. bacteria; cf. eukaryote.
prolate (of a spore, sporocarp, etc.), elongated in the direction of the poles; **- spheroidal**, only slightly so. See subglobose.
proliferation, successive development of new parts, esp. of new sporangia within the old wall in *Oomycota*, or of new wall material in conidiogenous cells.
promycelium, Tulasne's term for the germ tube of the teliospore (*Uredinales*) or ustilospore (*Ustilaginales*) from which **promycelial spores** (Plowright) (sporidia) are produced. The teliospore has been interpreted as a probasidium, the **-** as a metabasidium (after septation a phragmobasidium) and the ballistisporous promycelial spores basidiospores, as have the non-ballistisporic smut sporidia (see Donk, *K. ned. Akad. Wet. C* **76**: 109, 1973).
proper exciple (**margin**), see excipulum (proprium).
prophialide, see metula; primary sterigma.
prosenchyma, see plectenchyma.
prosoplectenchyma, see plectenchyma.
prosorus (of *Chytridiales*), a cell giving a group of sporangia (the sorus).
proteoglycan, an antitumour metabolite from *Coriolus pubescens* active against sarcoma-180.

proteophilous fungi, fungi associated with ammonia-rich (e.g. urea affected) soils (Sagara, *Trans. mycol. Soc. Japan* **14**: 41, 1973).

proterospore, a spore formed at the start of the sporulation period in *Ganoderma* and able to germinate easily without passing through the gut of a fly larva (Nuss, *Pl. Syst. Evol.* **141**: 53, 1982).

prothallus, see hypothallus.

prothecium, a primitive or rudimentary perithecium, as in the *Gymnoascaceae*.

proto- (prefix), primitive; primordial.

protoaecium, a haploid structure which, after diploidization, becomes a fruiting structure.

protobasidium, see basidium.

protoblem, a loose flocculent mycelial layer covering the universal veil, as in *Amanita*; primordial veil.

protoconidium, see hemispore.

protogonidium, the first of a series of gonidia (obsol.).

protohymenial, having a primitive hymenium (Maire).

protoperithecium, a young but walled perithecium before ascus formation (Ellis, *Mycol.* **51**: 416, 1960).

protoplast, traditionally the totality of the living cell constituents, whether walled or not, but now frequently used for the cell protoplasm after experimental removal of the cell wall, a usage to which Brenner *et al.* (*Nature* **181**: 1713, 1958) proposed to restrict the term. Fungal protoplasts are proving to have value in the study of cell organelles, biochemistry, genetics; they also have potential applications in biotechnology, esp. since protoplasts from different strains or even species can be fused (first achieved in 1975) into an aggregate protoplast and produce a heterokaryon with changed properties. See Villanueva (*in* Ainsworth & Sussman (Eds), *The Fungi* **2**: 3, 1966), Perberdy (*Sci. Progr.* **60**: 73, 1972; review methods), Villanueva *et al.* (Eds) (*Yeast, mould and protoplasts*, 1973), Perberdy *et al.* (Eds) (*Microbial and plant protoplasts*, 1976), Peberdy & Ferenczy (Eds) (*Fungal protoplasts. Applications in biochemistry and genetics*, 1985), Peberdy (*Microbiol. Sci.* **4**: 108, 1987; review). **sphaeroplast** a - enclosed by a modified or fragmentary cell wall [**gymnoplast** was used by Küster (1935) for plant cells without a cell wall and Frey-Wyssling (*Nature* **216**: 516, 1967) prefered **semi-gymnoplast** to sphaeroplast.].

protosexual (of yeasts or other organisms), having diploid or dikaryotic cells which produce haploid or unisexual cells in the absence of fructifications or sexual spores; in contrast to parasexual (q.v.), which is redefined to cover organisms having both protosexual and sexual cycles, and **neo-sexual**, for organisms having a sexual but not a protosexual cycle (Wickerham, *Mycol.* **56**: 254, 1964, cf. **57**: 134, 1965, **58**: 943, 1967).

protospore, (1) the multinucleated mass of cytoplasm cut out by primary cleavage planes, to be followed by further cleavage to form the uninucleate spores of *Phycomyces* and other *Mucorales* (Harper), and the sporangiospores of *Coccidioides*; (2) (of *Synchytriaceae*), a 1-nucleate portion of protoplasm which becomes the sporangium.

protosterigma, see basidium.

protothecium, an incompletely differentiated ascoma containing neither asci nor ascospores (Shoemaker, *CJB* **34**: 641, 1955).

prototunicate (of asci), basically unitunicate, but with the wall lysing at or before maturity and lacking differentiated apical structures (e.g. *Saccharomycetales*); such asci may develop in an hymenium or be distributed randomly in the interior of the ascoma. See ascus.

protouredinium, see protoaecium.

protuberate (of conidia), having short projections.

proximal, situated close to either the centre of a body or the point of origin; cf. distal.

pruinose, having a frost-like or flour-like surface covering of **pruina**. Often caused by calcium oxalate hydrates on lichen thalli; see Wadsten & Moberg (*Lichenologist* **17**: 239, 1985).

pseudo- (prefix), false; spurious.

pseudoamyloid, see dextrinoid.

pseudoangiocarpous (of a basidioma), hymenial surface at first exposed but later covered by an incurving pileus margin and/or excrescences from the stipe (Singer, 1975: 26).

pseudocleistothecium, see discrete body.

pseudocortex (of lichen thalli), a false cortex, used for the outer layer of the pseudopodetia in *Pycnothelia papillaria*.

pseudocyphella (pl. **-e**), an opening in the cortex of lichens where the medulla is exposed to the open air but lacking specialized cells surrounding the cavity; they provide valuable taxonomic characters in e.g. *Alectoria*, *Bryoria*, *Pseudocyphellaria*.

pseudocystidium, (1) (in agarics), see cystidium; (2) (of *Entomophthora*), the organ penetrating the insect cuticle allowing conidiophores to emerge.

pseudodiblastic (of ascospores), having oil-drops at the poles so that they superficially resemble polarilocular spores (q.v.).

pseudoepithecium, an amorphous or granular layer overlying paraphyses in an apothecium and in which their tips are immersed, but not forming a separate tissue.

pseudofissitunicate (of asci), see ascus.

pseudoheterothallism, see heterothallism.

pseudoidia, separated hyphal cells able to be germinated (Bensaude).

pseudoisidium (pl. **-ia**), (1) an outgrowth from the surface of a lichen thallus resembling an isidium (e.g. *Gyalideopsis*; see Vězda, *Folia geobot. phytotax.* **14**: 48, 1979); (2) isidium without photosynthetic cells in *Pseudocyphellaria* (see Galloway, *Bull. Br. Mus. nat. Hist., Bot.* **17**: 1, 1988).

pseudomixis (**-gamy**), the type of fertilization in which the copulating elements are not special sexual cells; cf. amphimixis

pseudomorph, a stroma made up of plant parts kept together by plectenchyma.

pseudomycelium (of *Candida*, etc.), loosely united, catenulate groups of cells (see Zobl, *RAM* **23**: 177, 1944).

pseudomycorrhiza, see mycorrhiza.

pseudooperculate (of asci), ones which are essentially unitunicate in structure and with a thickened apical cap which splits completely away at discharge (e.g. *Odontotrema*); see ascus.

pseudoparaphysis (pseudoparaphyses; of ascomycetes), see hamathecium; (of basidiomycetes), see hyphidium.

pseudoparenchyma, see plectenchyma.

pseudoperidium, a false peridium; covering membrane of the aecium in the *Uredinales*.

pseudoperithecium (of *Laboulbeniales*), a perithecium-like structure in which the asci and spores become free.

pseudophialide, a cell bearing a sporangiolum in the *Kickxellaceae*.

pseudophyse, see cystidium.

pseudophysis, see hyphidium.

pseudopionnotes, see sporodochium.

pseudopodetium (pl. **-ia**), a lichenized podetium-like structure of vegetative origin, ascogonia arising on this not on the pre-formed granular or squamulose thallus initials (e.g. *Cladia*, *Stereocaulon*).

pseudopycnidium (obsol.), a pycnidium-like structure of hyphal tissue, as in certain anamorphic fungi.

pseudorhiza, a rooting base, as in *Collybia radicata* (Buller, 4).

pseudosclerotium, a compacted mass of intermixed substratum (soil, stones, etc.) held together by mycelium, as in *Polyporus tuberaster* (see stone-fungus; also zone lines).

pseudoseptum, (1) (obsol.) a protoplasmic or vacuolar membrane looking like a septum (= distoseptum, distoseptate) as in *Corynespora*; (2) (of *Blastocladiales*), a septum having pores.

pseudosetae (false setae), the upturned free-ends of context hyphae in the hymenium of *Duportella*.

pseudospore, (1) (of *Acrasiales*), an encysted myxamoeba; (2) (of *Ustilaginales*), a basidiospore (obsol.); (3) a chlamydospore, as in *Rhizoctonia rubi*.

pseudostem (of gasteromycete basidiomata), spongy tissue in which hyphae are not orientated parallel to the stipe axis (Dring, 1973).

pseudostroma, a stroma formed of thalline tissue and remnants of host tissue (see Eriksson, *Opera bot.* **60**: 14, 1981), an aggregation of perithecial ascomata into a pustule some partly of bark cells altered by the fungus (see Johnson, *Ann. Mo bot. Gdn* **27**: 31, 1940), a coelomycetous conidioma of fungal and host tissue (see Sutton, *The Coelomycetes*, 1980)); cf. substroma.

pseudothecium, (1) an ascostromatic ascoma having asci in numerous unwalled locules, as in loculoascomycetes; cf. eutheocium; (2) a protoperithecium.

psychrophile, see thermophily.

psychrotolerant, growing at temperatures below 10°C (opt. below 20°C).

pterate, having wings; alate.

ptyophagous (of endotrophic mycorrhiza), the young hyphae rupturing and extruding plasmal masses (**ptyosomes**) which are digested by the host cells (Burgeff, 1924); **tolypophagous**, the penetrating hyphae killed and digested by the host (Burgeff, 1924); **thamnisophagous**, forming haustorial arbuscles which are finally digested by the host (Burgeff, 1938); cf. halmophagous.

pubescent, having soft hairs.

puff-ball, basidioma of the *Lycoperdales*.

puffing, a phenomenon in which thousands of asci in an apothecial ascoma discharge their ascospores simultaneously, producing a visible cloud.

pullulation, budding, as in yeasts.

pulsed field gel-electrophoresis (PFGE), a term used to describe a number of different techniques for separating large pieces of DNA, e.g. chromosomes. Different proprietary systems available, e.g. CHEF, FIGE, PFGE, but all involve applying electric field to an electrophoresis gel as a series of pulses or variations rather than a single continuous constant field. See Boekhout et al. (*Ant. v. Leeuwenhoek* **63**: 157, 1993).

pulveraceo-delitescent, covered with a layer of powdery granules.

pulverulent, powdered; as if powdered over.

pulvinate, cushion-like in form.

punctate, marked with very small spots; **puncta**, small spots.

punctiform (of rust sori, bacterial colonies, etc.), very small, but seen without a lens.

pustule, a blister-like, frequently erumpent, spot or sporemass.

putrescent (of basidiomata), decaying, rotting; cf. marcescent.

pycnidiospore, a conidium in or from a pycnidium (obsol.).

pycnidium (pl. **-ia**), a frequently ± flask-shaped conidioma of fungal tissue with a circular or longitudinal ostiole, the inner surface of which is lined entirely or partially by conidiogenous cells; pycnidial conidioma.

pycniospore (pl. **-a**), (of *Uredinales*), a spore from a pycnium; spermatium; sometimes used in error for pycnidiospore.

pycnium (in *Uredinales*), the pycnidium-like haploid fruit-body, or spermogonium. See Hiratsuka & Cummins (*Mycol.* **55**: 487, 1963), Savile (*Mycol.* **63**: 1089, 1971).

pycnoascocarp, an ascoma arising from a pycnidial conidioma.

pycnoconidium, see pycnidiospore (obsol.).

pycnogonidium, see pycnidiospore, pycniospore, or stylospore (obsol.).

pycnosclerotium, a more or less hard-walled structure resembling a pycnidial conidioma but having no spores.

pycnosis (of *Microthyriaceae*), the process by which a part of the stroma is arched up and becomes thick while an ascigerous hymenium is formed under it.

pycnospore, formerly occasionally used for pycniospore or pycnidiospore (obsol.).

pycnostroma, see stroma.

pycnothecium (of *Microthyriaceae*), an ascoma formed by pycnosis.

pycnothyrium, a superficial flattened shield-shaped conidioma with radiate upper and sometimes lower walls; pycnothyrial conidioma. See Hughes (*Mycol. Pap.* **50**: 7, 1953). Characteristic of *Microthyriaceae*.

pyrenium, a pyrenomycete ascoma (obsol.).

pyrenocarp, a perithecium (s.l., q.v.); used colloquially as a term for any fungus with a perithecium-like ascoma.

pyrenocarpous, see pyrenocarp.

pyriform, pear-like in form; cf. obpyriform.

pyrophilous, growing on burnt ground, steam-sterilized soil etc.; carbonicolous; **- fungi**, fireplace fungi, phoenicoid fungi (see Ramsbottom, *Mushrooms and toadstools*: 231, 1953; list of some larger fungi, Webster et al., *TBMS* **47**: 445, 1964; discomycetes, e.g. *Pyronema*).

pyroxylophilous, living on burnt wood.

pyxidate, provided with a lid, pertaining to, of having the character of a box; box-like.

quadrangular, see rhomboidal.

quellkörper, a mucilaginous mass of thick-walled cells within the ascoma of *Nitschkeaceae*; believed to induce rupture of the ascoma.

racemose, arranged along an elongated hypha, typically unilaterally.

rachis, a geniculate or zig-zag holoblastic extension of a conidiogenous cell (as in *Tritirachium*) resulting from sympodial conidiogenous cell development; **rachiform** (of conidiogenous cells), having a rachis; cf. raduliform.

racket (racquette) cell (of dermatophytes), a hyphal cell having a swelling at one end; cf. hypha.

racquette cell, see racket cell.

radial (of lichen thalli), radially symmetrical in transverse section (e.g. *Alectoria, Bryoria, Coelocaulon, Usnea*).

radiate, spreading from a centre.

radicating (of stipes), like a root; rooting.

radula spore (**radulaspore, radulospore**), one of the slimy spores borne over the surface of ascospores as in *Nectria coryli* while still in the ascus; **dry - -** = sympodulospore (Mason, 1933, 1937).

raduliform (of conidiogenous cells), the elongating conidiogenous axis resulting from holoblastic sympodial development, clavate or somewhat inflated rather than zig-zag; cf. rachiform.

ramarioid, with a form similar to that of the basidioma of *Ramaria*.

ramicolous, living on branches.

ramoconidium, an apical branch of a conidiophore which secedes and functions as a conidium, as in *Cladosporium* and *Subramaniomyces*.

ramus, (1) (pl. **-i**), a branch (Lat.); (2) (of a penicillus), a cell bearing a verticil of 'metulae' and phialides.

rangiferoid, branched like a reindeer's horn.

raphe (of *Chaetomella*), the longitudinal dehiscence mechanism. See Di Cosmo & Cole (*CJB* **58**: 1129, 1980; ultrastr. development).

raphides, needle-shaped crystals; as hyphae inside some lichen thalli.

receptacle, an axis having one or more organs, as the stem in *Phallales*; any hymenium-supporting structure.

receptive body, a small branched or unbranched process from the stroma (as in *Stromatinia gladioli*) able to be 'spermatized' by microconidia (see Drayton, *Mycol.* **26**: 46, 1934); **receptive hypha** = flexuous hypha, trichogyne, and possibly other like structures.

recognition (of mutualistic symbionts), the process by which two compatible potential symbionts initiate the relationship.

recurved (of a pileus), convexo-expanded (q.v.).

reflexed (of an edge), turned up or back.

remote (of lamellae), proximal end free and at some distance from the stipe; cf. free, seceding.

reniform, kidney-like in form; fabiform.

repand (of a pileus), having a waved edge which is turned back.

repeating spore, a spore which gives rise to the same type of mycelium as that on which it developed.

repent prostrate, creeping.

repetition (spore germination by), producing a new spore like the first.

reproductocentric (of *Chytridiales*), having development of one (**mono-, monogenocentric**) or more reproductive structures at the centre of gravity of the thallus; genocentric. See Karling (*Am. J. Bot.* **19**: 54, 1932); cf. monocentric and polycentric.

resistance, the power of an organism to overcome, completely or in some degree, the effect of a pathogen or other damaging factor. **acquired -**, a non-inherited resistance response in a normally susceptible host following a predisposing treatment. See immunity, axeny.

resistant sporangium, see meiosporangium.

resting sporangium, see meiosporangium.

resting spore, (1) a spore germinating after a resting period (frequently after overwintering), as does an oospore or a teliospore; a 'winter spore'; (2) an encysted zygote formed after the fusion of gametes, meiosis occuring on germination to produce planospores which encyst to produce gametes (*Blastocladiales*).

resupinate (of basidiomata), flat on the substrate with the hymenium on the outer side.

reticulate, like a net; netted.

retraction septum, see septum (adventitious).

retroneme, a bipartite tubular hair.

retrorse, backward.

revolute, having the edge rolled back or up.

rhagadiose, having deep cracks.

rhexolytic, secession of conidia involving the circumscissile splitting of the periclinal wall of the cell below the basal conidial septum (Cole & Samson, *Patterns of development in conidial fungi*, 1979) rather than the septum itself; fracture; cf. schizolytic.

rhizina (pl. **-ae**), a root-like hair or thread; the attachment organs of many foliose lichens (e.g. *Parmelia*); rhizine; they may be divided into several types, for details see Gyelnik (*Bot. Közl.* **24**: 122, 1927), Hale & Kurokawa (*Contr. U.S. natn Herb.* **36**: 122, 1964), Hannemann (*Bibl. Lich.* **1**, 1973); cf. hyphal net.

rhizinose-strand ('Rhizinenstränge'), a rhizine-like organ of attachment in squamulose (placodioid) lichens (e.g. *Toninia*), which is tough and much branched. At least 3 types occur. See Poelt & Baumgartner (*Öst. bot. Z.* **111**: 1, 1964), Hannemann (*Bibl. Lich.* **1**, 1973); cf. hyphal net.

rhizoid, a root-like structure consisting of anucleate filaments; branched, extension of a chytrid thallus acting as a feeding organ (Karling, *Am. J. Bot.* **19**: 44, 1932), cf. haustorium, rhizina, and holdfast; **-al**, having, or made up of, rhizoids.

rhizomorph, a root-like aggregation of hyphae having a well-defined apical meristem (cf. mycelial cord) and frequently differentiated into a rind of small dark-coloured cells surrounding a central core of elongated colourless cells. See Snider (*Mycol.* **51**: 693, 1961; *Armillaria mellea*), Jacques-Felix (*BSMF* **83**: 1, 1967; *Marasmius*, **84**: 161, 1969; agarics).

rhizomycelium, a rhizoidal system which resembles mycelium, e.g. thallus of the *Cladochytriaceae* (Karling, *Am. J. Bot.* **19**: 53, 1932).

rhizoplane, the surface of a root. See Lynch (Ed.) (*The rhizosphere*, 1990).

rhizoplast, see blepharoplast.

rhizosphere, the region immediately surrounding a root and influenced by its presences; the microbiota is frequently richer than that of the soil away from a root. See Katznelson *et al.* (*Bot. Rev.* **14**: 543, 1948).

rhodosporous, having light-red spores.

rhomboidal, resembling an equilateral not right-angled parallelogram (a rhomboid); quadrangular.

rhynchosporous, having beaked spores.

rimose (1) cracked; (2) (of a pileus surface), cracked; originally, cracked in all directions (the recommended usage); frequently, cracked by radial fissures as in *Inocybe*; cf. rimulose.

rimulose, having small cracks; cf. rimose.

rind, sometimes used for the firm outer layer of a rhizomorph or other organ; cortex (q.v.).

ring, (1) see annulus; (2) (of liquid cultures, esp. of bacteria), growth at the surface, sticking to the glass; - **wall building**, see wall building.

ringworm, see dermatophyte.

rivulose, marked with lines like little rivers.

rodlet, structural unit of conidial and some hyphal walls composed of particles *c.* 50 Å diam. arranged in linear series (Hess *et al.*, *Mycol.* **60**: 290, 1968).

roestelioid (of an aecium), long and tube-like, as in *Gymnosporangium*.

roridous, covered with drops of liquid like dew.

rostrate, (1) beaked; (2) (of asci), see ascus; bent tip of macroconidia of *Microsporum canis* and other anamorphic fungi.

rostrum, any beak-like process.

rostrupioid (of *Uredinales*), having teliospores as in *Rostrupia*.

rosulate, in a rosette.

ruderal, (1) living in waste places; (2) (of fungi) having a high growth rate, rapidly germinating spores, and a short life expectancy due to exhaustion of the available nutrients; cf. zymogenous (see autochthonous).

rugose, wrinkled; cf. rugulose.

rugulose, delicately wrinkled; cf. rugose.

rumposome, an organelle in zoospores of certain *Chytridiomycota* located close to the cell wall; tooth-like in section and honey-comb like in surface view; see *Rumpomycetes*.

rupestral (**rupestrine**), living on walls or rocks; cf. saxicolous.

saccate, like a sac or bag.

saltation (of fungi), mutation; dissociation.

saprobe (**saprogen**, **saprotroph**), an organism using dead organic material as food, and commonly causing its decay (saprobe is the preferred term for fungi); **saprobic**, **saprogen-ic** (**-ous**), **saprophilous**, **saprotrophic** (adj.). See Hudson (*Fungal saprophytism*, edn 2, 1980), saprophyte.

saprogen, a saprobe (q.v.).

saprophyte, a plant feeding by external digestion of dead organic matter; commonly misapplied to fungi where saprobe (q.v.) is the preferred term.

saprotroph, (1) a saprobe (q.v.); (2) a necrophyte on dead material which is not part of a living host (Münch), cf. perthophyte.

sarciniform, bundle-like, as the dictyospore of *Stemphylium botryosum*.

sarcodimitic, see hyphal analysis.

sarcotrimitic, see hyphal analysis.

satratoxins, toxins of *Stachybotrys atra* (syn. *S. alternans*); the cause of stachybotryotoxicosis in farm animals and humans.

saturnine (**saturniform**) (of ascospores), having a flat edge round the middle (as in some *Hansenula* spp.).

saxicolous, growing on rocks. A few fungi are able to live on saxicolous substrata (Kobluk & Kahle, *Bull. Can. Pet. Geol.* **25**: 208, 1977; bibliogr.); *Lichenothelia* is primarily saxicolous. Rocks are one of the main substrata for lichens and the lit. on the latter is vast. Lichens can also change the mineral composition of rocks through the action of oxalic acid, etching being visible by SEM (see Syers & Iskandar, in Ahmadjian & Hale (Eds), *The Lichens*: 225, 1974, Jones *et al.*, *Lichenologist* **12**: 277, 1980).

scabrid, rough with delicate and irregular projections.

scabrous, rough.

scar, (1) (of yeasts), **bud -**, on parent cell; **birth -**, on daughter cell; (2) (of anamorphic fungi), at conidiogenous locus and conidial base/apex, left after secession of conidium.

scariose, thin; paper-like.

schizidium, a propagule formed by upper layers of a lichen thallus splitting off as scale-like segments from the main lobes (e.g. the lobule-like structures in *Fulgensia bracteata* subsp. *deformis*). See Poelt (*Flora, Jena* **159**: 23, 1980).

schizobiont, bacteria once considered to be additional symbionts of lichens.

schizogenous, formed by cracking or splitting, cf. lysigenous.

schizogony, the process of division of a schizont.

schizolytic, secession of conidia involving a splitting of the delimiting septum so that one half of the crosswall becomes the base of the seceding conidium and the other half remains at the apex of the conidiogenous cell (Cole & Samson, *Patterns of development in conidial fungi*, 1979); cf. rhexolytic.

schizont, a vegetative thallus, having no wall, which undergoes simple or multiple division.

scissile (of the flesh of a pileus), separating into horizontal layers.

sclerified, becoming thick walled, as in sclerenchyma.

sclerobasidium, see basidium.

sclerocarps, sclerotium-like modified ascomata permanently lacking a sexual capacity and acting as sclerotia, as in *Varicosporina ramulosa* (Kohlmeyer & Charles, *CJB* **59**: 1787, 1981).

scleroplectenchyma, plectenchyma (q.v.) composed of very thick-walled conglutinate cells. See Yoshimura & Shimada (*Bull. Kochi Gakuen Jun. Coll.* **11**: 13, 1990); stereome.

scleroprosoplectenchyma, see plectenchyma.

sclerothionine, a plant growth-promoting metabolic product of *Sclerotinia libertiana* (Matsu & Satomura, *Agr. biol. Chem.* **32**: 611, 1968).

sclerotic cell, see muriform cell.

sclerotium, (1) a firm, frequently rounded, mass of hyphae with or without the addition of host tissue or soil, normally having no spores in or on it (cf. bulbil, stroma); see Willetts (*Biol. Rev.* **46**: 387, 1971; survival,

47: 515, 1972; morphogenesis), Willetts & Bullock (*MR* **96**: 801, 1992; developm. biology). A sclerotium may give rise to a fruit body, a stroma (as in ergot), or mycelium. (2) (of *Myxomycetes*), the firm, resting condition of a plasmodium.

scobiculate, in fine grains, like sawdust.

scolecospore, a spore resembling an amerospore (q.v.) but with or without septa and a length/width ratio › 15:1; cf. amerospore.

scorpioid, a branching system in which the laterals are curved so that they all appear to arise from one side of the main stem, as in *Cladonia arbuscula*.

scrobiculate, (1) roughened; resembling sawdust; (2) (of lichens), coarsely pitted; foveolate.

scrupose, rough with very small hard points.

scutate, like a round plate or shield.

scutellum, see thyriothecium.

scutulum (pl. **-a**), cup-like crust or mat of hyphae produced in the follicles of the scalp or the body in infections by *Trichophyton schoenleinii*.

scyphoid, cup-like.

scyphus, a cup-like apex of a lichenized podetium, as in *Cladonia fimbriata*.

seceding, (1) (of lamellae), at first joined to the stipe (adnate), then free (q.v.); separating from the stipe (cf. seceding); (2) (of conidia), at first attached to the conidiogenous cell, then separating by schizolysis or rhexolysis; secession.

secondary mycelium (of basidiomycetes), the dikaryotic mycelium resulting from plasmogamy in the primary mycelium (q.v.); mycelium developed from the base of a basidioma (Corner).

secondary spores (of basidiomycetes), spores other than basidiospores.

secotioid (of basidiomycetes), the margin of the pileus does not break free from the stipe (or if it does the pileus never fully expands), lamellae convoluted and anastomosed, basidiospores not ballistosporic; phylogeny of a **- syndrome**, see Thiers (*Mycol.* **76**: 1, 1984).

secund, having parts directed to one side only; cf. scorpioid.

segment (of a dictyospore), a part of the spore cut off by an A-trans-septum (Eriksson, *Opera bot.* **60**, 1981); see septum.

seiospore, a dry dispersal spore.

semi- (prefix), half; cf. hemi-.

semifissitunicate, see ascus.

semigymnoplast, see protoplast.

semimacronematous, only slightly different from vegetative hyphae; frequently ascending, rarely erect. See macronematous.

senescence (of fungi), the degeneration which makes indefinite propagation of certain fungi in culture impossible. See Holliday (*Nature* **221**: 1224, 1969; *Neurospora*, *Podospora*).

sensitive, reacting with severe symptoms to the attack of a given pathogen (*TBMS* **33**: 155, 1950); **sensitivity**, the tendency of an organism attacked by a disease to give more or less strong symptoms; sensibility (Wilbrink).

separating, see seceding; **- cell**, a cell between a conidium and a conidiogenous cell, involved in rhexolytic secession (q.v.).

septal pore apparatus, see dolipore septum (**- - swelling**), parenthesome (**- - cap**); **- plug**, an occlusion of a septal pore.

septate, having septa.

septum (pl. **septa**), a cell wall or partition; Talbot (*Taxon* **17**: 622, 1968) distinguishes **primary -**, a septum formed in direct association with nuclear division (by constriction, mitosis, or meiosis) separating the daughter cells and having a pore (see Markham, *MR* **98**: 1089, 1994; review) which may be modified as a dolipore (q.v.; in basidiomycetes) or be associated with Woronin bodies (in ascomycetes; see Kimbrough, *in* Hawksworth, 1994: 127, ultrastr.), and **adventitious -** (retraction or retaining septum, 'cloison de retrait'), a septum formed in the absence of, or independently of, nuclear division, esp. in association with the movement of cytoplasm from one part of the fungus to another; primary septa are characteristic of higher fungi, adventitious septa of lower fungi where nuclear division is by constriction; **angular -** (Eriksson, *Ark. Bot.* II **6**: 339, 1967), oblique septum; **longiseptum**, a longitudinal septum within a spore (Reynolds, *Mycol.* **63**: 1173, 1971); **oblique -**, a septum within a segment of a spore arising at an oblique angle to that delimiting the segment (Eriksson, *Opera bot.* **60**, 1981); **trans-**, a transverse septum within a spore (Reynolds, 1971), which may be an **A trans-septum** (forming a segment) or a **B trans-septum** (laid down in a segment after division by a longiseptum; never in macrocephalic spores) (Eriksson, 1981). Ascospores in which the septation proceeds from the primary septum towards the poles, so that immature spores have longer end cells are termed **macrocephalic**; those in which it proceeds by each trans-septum dividing a segment into two of ± equal lengths are **microcephalic** (Eriksson, 1967, 1981), Curry & Kimbrough (*Mycol.* **75**: 781, 1983; *Pezizaceae*). See also euseptate, distoseptate, multiperforate (septum).

sequestrate, fungal fruit-bodies which have evolved from having exposed hymenia and forcibly discharged spores to a closed or even hypogeous habit in which spores are retained in the fruit-body until it decays or is eaten by an animal vector. Many sequestrate taxa can be clearly recognized as being derived from specific spore-discharging ancestors, e.g. *Rhizopogon* from *Suillus*.

sericeous, like silk.

serioplectenchyma, see plectenchyma.

serous (of latex), like serum; watery; opalescent.

serrate, edged with teeth, like a saw.

serrulate, delicately toothed.

sessile, having no stem.

seta (pl. **-ae**) (Lat., a bristle), (1) a stiff hair, generally thick-walled and dark in colour; in hyphomycetes, see Dev Rao (*in* Subramanian (Ed.), *Taxonomy of fungi*: 397, 1984); (2) (in hymenomycetes), a sterile hyphal end, thick-walled, darkening in KOH sol., found frequently projecting from the hymenium in xanthochroic basidiomata. Lentz (1954) distinguished 'seta', 'embedded seta', and 'stellate seta' (asteroseta). Smith (1966) treated the last as a cystidium; see cystidium (3).

setaceous (Lat., *setaceus*), bristle-like; cf. setose.

setiform, see setaceous.

setose (Lat. setosus; bristly), covered with bristles; cf. setaceous, setulose.

setula (Lat., a little bristle), (1) a delicate hair-like appendage arising from a conidium, as in *Dinemasporium*; (2) (in hymenomycetes), a thick-walled, pigmented, terminal element of a tramal cystidium.

setule (in hymenomycetes), a thin-walled, rarely pigmented, usually lageniform cystidium on the pileus or stipe.

setulose (Lat., setulosus), covered with fine bristles or hairs; cf. setaceous.

shape, configuration or form, total effect produced by the outline of the structure.

sicyospore, a thick-walled storage cell (obsol.).

sigmoid, curved like the letter 'S'.

sikyotic cell (the 'Schröpfkopf-Zelle' of Burgeff, 1924), the terminal cell of a *Parasitella simplex* hypha by which the parasite anchors itself to its host (*Absidia glauca*). After several days this organ differentiates into a **sikospore** (sikyotic spore). See Kellner *et al.* (*Mycologist* **5**: 120, 1991).

simblospore, Langeron's (1945) term for zoospore (q.v.).

simple, unbranched; having no divisions.

sinuate, (1) (of lamellae), notched at the proximal end at junction with stipe (cf. adnate); emarginate; (2) (of an edge), undulating.

siphon (obsol.), an aseptate hypha (Vuillemin, 1912).

skeletal hypha, see hypha.

skeletal hyphae, see hyphal analysis.

skeletocystidium, see cystidium (1).

skiophilous, shade loving; cf. photophobous.

slime, a wet, generally sticky, substance; mucus; **- flux**, a thick liquid from the stems or branches of trees made up of, or having a connection with, fungi and bacteria (Ogilvie, *TBMS* **9**: 167, 1924; Stautz, *Phytopath. Z.* **3**: 163, 1931); **- moulds**, the *Acrasiomycota*, *Dictyosteliomycota*, and *Myxomycota*; **- spore**, a spore that becomes separated with slime from the cell producing it (Mason, 1937), cf. dry spore.

slug (in slime moulds), the aggregated pseudoplasmodium of *Dictyosteliomycota*.

soleiform, shaped like the sole of a shoe, i.e. elongate-ellipsoid, with one cell much larger and broader than the other.

solopathogenic (of a smut such as *Ustilago zeae*), a pathogenic monosporidial line (Christensen).

soma, (1) body, excluding reproductive parts or phase; (2) in Aryan religion, *Amanita muscaria* (Wasson, *Soma: the divine mushroom of immortality*, 1968, *The wondrous mushroom*, 1980); **-tic**, pertaining to the soma. See also ethnomycology, hallucinogenic fungi.

somatogamy, fusion of somatic (vegetative) cells or hyphae involving plasmogamy but not karyogamy.

soralium (pl. **-ia**), decorticate portions of a lichen thallus where **soredia** are located. Usually formed from medullary tissues thrusting upwards through the cortical layers and so sometimes with the chemistry of the medulla rather than of the cortex. Soralia may be **diffuse** (the upper surface of the lichen becoming a continuous soredial mass), or **delimited** (i.e. confined to well-defined areas), and can be classified according to where they originate. They can arise on tubercles (**tuberculate -**) or as fissures (**fissural -**) in some genera (Hawksworth, *Lichenologist* **5**: 181, 1972). Soralia can arise at the tips of isidia, and in some taxa the soralia can contain a mixture of soredia and isidia-like structures.

soredium (pl. **-ia**), a non-corticate combination of phycobiont cells and fungal hyphae having the appearance of a powdery granule, and capable of reproducing a lichen vegetatively. For their liberation and dispersal see Bailey (*J. Linn. Soc., Bot.* **59**: 479, 1966, *Revue bryol. lichén.* **36**: 314, 1969, *in* Brown *et al.* (Eds), *Lichenology: progress & problems*: 215, 1976); cf. soralium.

sorus, a fruiting structure in certain fungi, esp. the spore mass in *Uredinales* and *Ustilaginales*; a group of fruit-bodies, as in *Synchytriaceae*.

spalted (of wood), with dark zone lines (q.v.) due to the interactions of different fungal colonies of the same species; as caused, e.g. by *Diatrype disciformis*; generally brittle, short-grained, and easily breaking through decay.

sparassoid, composed of interlaced flabelliform branches forming ball-like structures recalling *Sparassis* basidiomata. Known also in *Pezizales* (Korf, *Rept Tottori mycol. Inst.* **10**: 389, 1973).

spathulate, like a spoon in form.

sperm, a male sex cell, typically motile; **-atiophore**, a spermatia-producing or -supporting structure; **-atium** (pl. **-a**), a 'sex' (+ or -) cell, e.g. a pycniospore; a microconidium in discomycetes and pyrenomycetes; a non-motile gamete, as in *Laboulbeniales*; **-atization**, the placing of spermatia on structures (receptive hyphae, etc.) for diploidization; **-odochidium**, a fruit-body having spermodochia in a lysigenous cavity in the suscept tissue (Whetzel, *Mycol.* **35**: 337, 1943); **-odochium**, a spermogonium having no wall (Whetzel, *Mycol.* **29**: 135, 1937); **-ogonium** (**-agone**, **-agonium**), a walled structure in which spermatia are produced, as in ascomycetes; a pycnium of a rust; a lichen pycnidium (obsol.).

spermatiospore, see spermatium.

spermatium, (pl. **-a**), see sperm.

spermoplane, the surface of a seed; **spermosphere**, the microhabitat around a seed in soil (Verona, *Ann. Inst. Pasteur* **105**: 75, 1963); cf. phylloplane; rhizoplane.

sphacelium, the structure forming conidia in *Claviceps* from which the sclerotium develops (Fredrickson *et al.*, *MR* **95**: 1101, 1991).

sphaerocysts, globose cells in tissues of fungi, e.g. *Russula* and *Lactarius*.

sphaeroplast, see protoplast.

spheridium, see capitulum.

spheroplast, see protoplast.

spherule, (1) a sporangium-like structure in *Coccidioides* (Baker & Mrak, *Am. J. trop. Med.* **21**: 589, 1941); (2) a multinucleate cell of a resting myxomycete plasmodium.

spicate, bearing a spike.

spicule (**spiculum**) (Tulasne; obol.), see sterigma.

spiculospore, a spore formed at the tip of a pointed structure often elongate and so resembling a spike, as in *Hirsutella* and *Akanthomyces* (Subramanian, *Curr. Sci.* **31**: 410, 1962).

spiculum (pl. **spicula**), see sterigma.

spilodium, introduced by Lettau (*Beih. Feddes Rep.* **69**: 62, 1932) for the minute round blackish structures on the thallus of *Dirina massiliensis* caused by *Milospium graphideorum*.

spindle, see fuseau.

spindle pole body, microtubule organizing centre, functionally homologous to the animal cell centrosome.

spindle-organ, see turbinate cell.

spine, a narrow sharply pointed process; **spinule**, a small spine (in lichens); **spiny**, having spines; **spinose** (dim. **spinulose**), delicately spiny.

spiral hypha, a hypha ending in a spiral or helical coil, as in *Trichophyton* (Davidson & Gregory, *TBMS* **21**: 98, 1937).

spitzenkörper, a labile vesicle-rich and actin-rich structure present in growing tips of most, but not all, fungi, whose behaviour correlates with tip growth rate and direction (Grove & Bracker, *J. Bact.* **104**: 989, 1970; López-Franco & Bracker, *Protoplasma* **195**: 90, 1996; Bourett & Howard, *Protoplasma* **163**: 199, 1991).

splash (splashing) cup, an open cup-like structure (as in *Cladonia*, *Cyathus*, and the liverwort *Marchantia*), from which the reproductive bodies are discharged by falling drops of water. See Brodie (*CJB* **29**: 593, 1951); bird's nest fungi.

spongiostratum, used for the hypothallus of *Anzia* and *Pannoparmelia* (Hannemann, *Bibl. Lich.* **1**, 1973). See Henssen & Dobelmann (*Bibl. Lich.* **25**: 103, 1987).

spora, the spore content of a particular place or ecological niche.

sporabola, the curve made by a basidiospore after discharge from its sterigma (Buller, *Researches* **1**; Ingold, 1971: 111).

sporangial vesicle, vesicle produced at the mouth of the sporangium during planont maturation and discharge. See homohylic vesicle, plasmamembranic vesicle, precipitative vesicle.

sporangiocyst (of *Chytridiales*), a resting sporangium (A. Fischer); cf. cystosorus.

sporangiolum (pl. **-a**, **sporangiole**), (1) (of *Mucorales*), a small sporangium with or without a columella, generally having a small number of spores; (2) a degenerating arbuscule (Janse, 1897) (obsol.).

sporangiophore, thallus element (usually morphologically differentiated) subtending one or more sporangia.

sporangiosorus, group of spherical sporangia fused together and formed from a single plasmodium; also one or more lobed sporangia formed from a single plasmodium.

sporangiospore, walled spore produced in a sporangium; **primary infestation -** (of *Eccrinales*), 1–4-nucleate, thick-walled spore which serves to transmit an infestation from one host individual to another after passage through the gut; **secondary infestation -** (of *Eccrinales*), multinucleate, thin-walled spore which germinates in the same gut as where they were produced.

sporangium (pl. **-ia**, **sporange**), an organ enclosing endogenously generated spore(s), the walls of the spore(s) not being derived from the supporting or containing structure.

spore, a general term for a reproductive structure in fungi, bacteria, and cryptogamic plants. In fungi, a differentiated morphological form which may be: (a) specialized for dissemination; (b) produced in response to, and resistant to, adverse conditions; and/or (c) produced during or as a result of a sexual or asexual reproductive process. Commonly 1-celled, but in fungi frequently a multicelled structure (e.g. phragmospore, spore ball, etc.) which is in effect a group of 1-celled spores because every cell may produce one or more germ tubes. Thick-walled or thin-walled, pigmented or not, motile or non-motile.

More attention has been given to the spore than to any other fungal structure. Spore morphology (e.g. flagellation of zoospores) and development (e.g. of ascospores and basidiospores; sexual and asexual spores) provide basic taxonomic criteria and biologically spores may be differentiated into groups disseminated by wind, water, insects and other animals, etc., and those which allow a fungus to survive conditions unfavourable for growth (e.g. resting spores), although one type of spore may serve several functions (Sutton, *TBMS* **86**: 1, 1986). Vuillemin (1912) defined spores morphologically (e.g. motile and non-motile) and biologically and borrowed or coined terms for his different spore types. Most of these terms have never been in current use but the naming of spores continues (see Spore terminology).

spore ball, a unit of dispersal comprised of a more or less firmly aggregated group of spores (e.g. *Sorosporium*, *Tolyposporium*) or spores and sterile cells (e.g. *Urocystis*).

spore horn, see cirrus.

spore specific gravity, Buller (**1**: 153, 1909) determined the Specific Gravity (SG) of various agaric spores as 1•02–1•43.

spore wall. Conventional or electron microscopy shows the spore wall to be layered. The terminology of these layers by different authors is somewhat confused (see the comparison by Payak, 1964: 33). Five layers (the spore wall proper; **eusporium**) which have been distinguished are, from within outwards: (1) **endosporium** (endospore, corium), which is usually thin and is the last to develop during sporogenesis; (2) **episporium**, the thick, fundamental layer which determines the shape of the spore; (3) **exosporium** (exospore, epitunica, trachytectum, tunica), a layer derived from (2) but chemically distinct and frequently responsible for the ornamentation; (4) **perisporium** (= mucostratum, myxosporium), a layer, frequently fugacious, enveloping the whole spore and limited by (5), the hardly visible **ectosporium** (sporothecium). On this disappearance of (4) and (5) (3) is the outer spore layer. (1)-(3) are thus the spore wall proper; (4) and (5) of extrasporal origin.

sporidesm (sporodesm), a compound spore or spore-ball, the components of which are merispores. See teleutosporodesm.

sporidiolae, spore-like bodies produced inside **sporidiomata**, perithecium-like structures, in *Kathistes* (Malloch & Blackwell, *CJB* **68**: 1712, 1990).

sporidiole, a small spore.

sporidioma (pl. **-omata**), a sporidium (sporidia) containing structure.

sporidium (pl. **-ia**), (1) a basidiospore of the *Uredinales* and *Ustilaginales* or, in the latter, any spore other than an ustilospore; (2) ascospore (obsol.).

sporocarp, a general term for spore-bearing organs; fruit-body (q.v.). Used esp. of *Acrasiomycota*, *Myxomycota* and *Endogonaceae*.

sporocladium (pl. **-ia**), a special sporogenous branch in the *Kickxellaceae*.

sporocyst, a cyst producing asexual spores.

sporodochium (pl. **-ia**), conidioma, typical of the *Tuberculariaceae* in which the spore mass is supported by a superficial cushion-like (pulvinate) mass of short conidiophores and pseudoparenchyma; **pionnote -** (**pseudopionnotes**, Sherbakoff, 1915) (of *Fusarium*), minute sporodochia near the surface of the substrate having no stroma, the spores forming a continuous slimy layer; cf. acervulus, pionnotes.

sporogenesis, spore development.

sporogenous, producing, having or supporting spores; cf. conidiogenous.

sporograph, the straight-line graph obtained by plotting the ratio (E) of the length (D) to width (d) against the length of the basidiospores of a species of agaric (Corner, *New Phytol.* **46**: 196, 1947); cf. Q.

sporoma (pl. **-omata**), a multicellular structure specially developed to produce spores.

sporont, a thallus on which spores will be produced.

sporophagous, feeding on spores, as in certain thrips species on fungus spores (Ananthakrishnan *et al.*, *Proc. Ind. Acad. Sci. Anim. Ser.* **92**: 95, 1983), where the dimensions of the mouth-parts are related to the sizes of the spores eaten (Ananthakrishnan & Dhileepan, *Proc. Ind. Acad. Sci. Anim. Ser.* **93**: 243, 1984); species lists (Ananthakrishnan *et al.*, 1984).

sporophore, (1) a spore-producing or -supporting structure, esp. a conidiophore; (2) (of macrofungi), ascoma, basidioma; cf. hymenophore; a basidium (sensu Berkeley).

sporoplasm, the spore-producing protoplasm within the epiplasm in a sporangium or ascus (Guilliermond).

sporostasis (adj. **sporostatic**), inhibition of spore germination; cf. mycostasis; sporostatic products of fungi, see Robinson *et al.* (*TBMS* **51**: 113, 1968).

sporothallus, a thallus producing spores; cf. gametothallus.

sporothecium, (1) the tip of a basidium bearing basidiospores when the basidiospores are sessile (Clémençon, *Z. Pilzk.* **36**: 113, 1970); (2) see spore wall.

sporotrichosis, a lymphatic disease in humans and animals (*Sporothrix schenckii*; see de Beurmann & Gougerot, *Les Sporotrichoses*, 1912), Norden (*Acta Path. Microbiol. Scand., Suppl.* **84**, 1951).

squamose, having scales.

squamule, a small scale.

squamulose, (1) having small scales; (2) growth form of a lichen thallus.

squarrose, rough with scales.

stalagmoid (of spores, **stalagmospores**), like a long tear or drop.

statismospore, a spore not forcibly discharged; cf. ballistospore.

statolon, an antiviral substance (which induces interferon formation) from *Penicillium stoloniferum*; the active principle of which is considered to be RNA of viral origin (Banks *et al.*, *Nature* **218**: 542, 1968).

staurospore (**stauroconidium**), a non-septate or septate spore with more than one axis; axes not curved through more than 180° (cf. helicospore); protuberances present and $›1/4$ spore body length (cf. amerospore).

stellate, like a star in form; **- seta**, a compound seta having several radiating arms; asterophysis.

stephanocyst, a structure, typically bicellular (basal cell cup-like, terminal cell globose), found in certain basidiomycetes. See Burdsall (*Mycol.* **61**: 915, 1969).

stereome (of lichens), a scleroplectenchyma which forms the main supporting tissue of the thallus, as in *Alectoria*, *Bryoria*, and *Cladonia*.

sterigma (pl. **-ata**), (1) (of a basidium, q.v.), an extension of the metabasidium composed of a basal filamentous or inflated part (the **proto-**; epibasidium) and an apical spore-bearing projection (the **spiculum**); (2) (of *Aspergillus*, etc.) [a usage not recommended], phialide (**secondary -**); prophialide (**primary -**); metula; (3) (of lichens), a spermatiophore (Nylander).

sterile, (1) not producing spores or a sporocarp; (2) free from living microorganisms; sterilized.

stichobasidium, see basidium.

stigmatocyst, see hyphopodium.

stigmatopodium (**stigmopodium**), see hyphopodium.

stilbaceous (obsol.), having synnemata; synnematous (q.v.).

stilboid, a sterile, basidioma-like structure (as in *Mycena citricolor* and other agarics) which functions as a propagule (Singer, 1962: 25); gemma (Buller); cf. carpophoroid.

stipe, a stalk.

stipitate, stalked.

stock (of basidiomycetes), a dikaryotic mycelium (fide Raper, 1966); cf. strain.

stolon, a 'runner', as in *Rhizopus*.

stomatopodium (**stomopodium**), a hyphal branch (an appressorium; cf. hyphopodium) or 'plug' above or in a stoma.

stone-fungus, the hard pseudosclerotium of *Polyporus tuberaster*; Pietraia fungaia. On being watered, an edible basidioma is produced. The Canadian tuckahoe is the same species (Vanterpool & Macrae, *CJB* **29**: 147, 1951).

strand plectenchyma, plectenchyma in strands forming supporting tissues in a lichen thallus.

stratose thallus, a lichen thallus having the tissue in horizontal layers.

striate, marked with delicate lines, grooves or ridges.

strigose, rough with sharp-pointed hairs; hispid.

strobiliform, like a fir-cone in form.

stroma (pl. **stromata**), a mass or matrix of vegetative hyphae, with or without tissue of the host or substrate, sometimes sclerotium-like in form, in or on which spores or fruit bodies bearing spores are produced. Many ascomycetes (esp. *Xylariales*) and anamorphic fungi have stromata; a few *Uredinales* and other fungi. **ecto-** (**epi-**, Fuisting), a **-**, normally conidial, formed in the periderm and frequently breaking through the bark; **endo-** (**ento-**) (**hypho-**, Fuisting), a perithecial **-** formed under the **ecto-**; **eu-**, one of fungal tissue only;

pseudo- see under pseudostroma. Ruhland's name for the ostiolar disc is **placodium**, forms (as in *Diatrype*) from the **endo-** being **ento-placodial**, those (at least in part) from the **ecto-** being **ecto-placodial**. A species having ectostroma and endostroma is **diplostromatic**; one with only one **haplostromatic**. See Miller (*Mycol.* **20**: 188, 1928); cf. sclerotium.

stromatolites (lichen), laminar calcretes formed abiotically in rock and sometimes wrongly interpreted as fossils (Klappa, *Sediment. Petrol.* **49**: 387, 1979).

stuffed (of a stipe), having the inside of a different structure from that of the outer layer.

stupose, of tissue formed from hyphae which are not gelatinized.

stylospore, (1) a spore on a pedicel or hypha, esp. a urediniospore (obsol.); (2) an elongated pycnidiospore (obsol.); (3) the sporangiolum (the 'Stielgemmen' of Linnemann, 1941) of *Mortierella*.

suaveolent, having a sweet smell.

sub- (prefix), under; below; frequently in the sense of approximating to the condition qualified, slightly, somewhat.

subcentric, see centric.

subcutis, see cutis.

subglobose, not quite spherical.

subhymenium, generative tissue below the hymenium; sometimes used as equivalent to medullary exciple or hypothecium.

subiculum (subicule), a net-, wool-, or crust-like growth of mycelium under fruit-bodies.

suboperculate, see ascus.

substrate (substratum), although these two terms are frequently treated as synonyms by mycologists both have useful special senses: (1) **substrate**: (in enzymology) is applied to the substance on which an enzyme acts and in microbiology to the substances (e.g. culture medium constituents) utilized by a microorganism for growth in distinction to the material; (2) **substratum** (pl. **substrata**): (in ecology), the material on which an organism is growing or to which it is attached.

substroma, pseudostroma (q.v.) in which the vegetative hyphae of the host predominate (Johnston, *Ann. Mo. bot. Gdn* **27**: 31, 1940).

subulate, slender and tapering to a point; awl-shaped.

subuniversal veil, see protoblem.

sulcate, grooved.

sulcus, a furrow or groove.

summer spore, a spore germinating without resting, frequently living only a short time; cf. resting spore.

superficial, on the surface of the substratum.

superior (of an annulus), near the top of the stipe.

supine (of fructifications), closely applied to the substratum.

suprahilar plage (of basidiospores, esp. of *Lactarius* and *Russula*), the area above the hilar appendage on which the eusporial ornamentation is lacking or reduced (Kühner, 1926).

surculicolous (of *Exobasidium* infections), monocarpic (q.v.) and systemic in annual shoots (Nannfeldt, 1981).

suscept, a living organism which is **susceptible** to (able to be attacked by; non-immune to) a given disease, pathogen, or toxin.

suspensor, a hypha supporting a gamete, gametangium, or esp. a zygospore.

swarm-cell (of *Myxomycetes* and some *Chytridiales*), a motile cell acting, before or after division, as an isogamete.

swarm-spore (swarmer), see zoospore.

sycosis, a fungus disease of the hair follicles; esp. of the face; ringworm of the beard.

sym-, see **syn-**.

symbiosis, associations between unlike organisms, generally ones persisting for long periods (relative to the generation time of the interacting organisms); apparently first used by the lichenologist A.B. Frank in 1877 (often credited to de Bary, 1879) who later coined the word mycorrhiza (q.v.). At times equated with mutualism (q.v.), but correctly also covering parasitic (harmful) and commensalistic (unharmful) associations (Ahmadjian & Paracer, 1987).

There has been debate as to which partner might be regarded as host in mutualistic symbioses (e.g. de Bary regarded the algae in lichens as the host, and Douglas, 1994, the fungus as host to the algae); in order to circumvent this controversy, Law & Lewis (1983) used the neutral terms **exhabitant** (the organism forming the outer tissues) and **inhabitant** (the enclosed organism).

In lichens, Poelt (*Abstr. IMC2*, 1977) recognized: **two-membered -**, one alga + one fungus, **three-membered -**, one alga + two fungi (lichenicolous fungi; q.v.), or two algae + one fungus (cephalodium; q.v.); and **four-membered -**, two algae + two fungi (lichenicolous lichens); Hawksworth (*Bot. J. Linn. Soc.* **96**: 3, 1988) used '-biont' (q.v.) instead and discussed further fungal/algal interactions (see also Rambold & Triebel, *Bibl. Lich.* 48, 1992).

sympatric, occurring in the same geographical region; cf. allopatric.

symphogenous, see meristogenous.

symplastic, entering living cells; cf. apoplastic.

sympodial (of conidiogenous cells), characterized by continued growth, after the main axis has produced a terminal spore, by the development of a succession of apices each of which originates below and to one side of the previous apex.

sympodioconidium (sympodulospore), a spore produced on a sympodula.

sympodula, a sympodial conidiogenous cell.

syn- (sym-) (in compounds), growing together; adhesion; aggregation.

synanamorph, any one of two or more anamorphs which have the same teleomorph.

synaptonemal complex, proteinaceous, longitudinally aligned structure which usually unites homologous chromosomes during the prophase of meiosis.

synascus, the gametangium of *Ascosphaera* (Varitchak, 1933).

synchronospore, a spore produced simultaneously with other neighbouring spores.

syncytium, see coenocyte.

synergism, two organisms or environmental factors acting simultaneously to effect a change greater than either could alone; e.g. the increase of fungicidal value in certain mixtures of fungicides, fungicides and non-toxic materials, or of air pollutants; cf. metabiosis.

syngamy, fertilization; the fusion of male and female cells to form a zygote.

synisonym, one of two or more names having the same basionym (Donk, *Persoonia* **1**: 175, 1960).

synkaryon, a diploid zygote nucleus.

synkaryotic (of a nucleus), having 2*n* chromosomes.

synnema, (pl. **synnemata**), a conidioma composed of a more or less compacted group of erect and sometimes fused conidiophores bearing conidia at the apex only or on both apex and sides. Seifert (*Stud. Mycol.* **27**: 1, 1985) distinguished 3 types: **determinate -**, with a terminal, non-elongated conidiogenous zone, growth ceasing after sporulation has begun, e.g. *Stilbella*; **indeterminate -**, with an elongated fertile zone, sometimes covering the whole conidioma, growth continuing after sporulation, e.g. *Doratomyces*; **compound -**, branched in which determinate or indeterminate branches are formed on a branched or unbranched axis, e.g. *Tilachlidiopsis*. Anatomical stipe types found in the three groups include **parallel -**, of primarily parallel hyphae; **intricate -**, of primarily or entirely textura intricata; **basistromatic -**, with well-defined basal stromata; **amphistromatic -**, well-defined basal stroma, stipe of parallel hyphae and an apical dome of textura angularis to globulosa with conidiogenous cells; **cupulate -**, conidiogenous zone concave.

synnema coremium, sometimes used for synnemata with looser fascicles as in 'coremioid' spp. of *Penicillium* and *Aspergillus* (obsol.).

synnematous (**synnematogenous**), having synnemata. See Seifert (*Stud. Mycol.* **27**: 1, 1985, *Stilbella* and allies, *Mem. N.Y. bot. Gdn* **59**: 109, 1990, keys).

synonym, another name for a species or group, esp. a later or illegitimate name not currently employed for the taxon. If two or more names are based on the same type they are **homotypic** (**nomenclatural**, 'obligate', ≡) synonyms, if on different types they may be **heterotypic** (**taxonomic**, 'facultative', =) synonyms.

synzoospore, multinucleate zoospore with many sets of flagella.

syrrotium (pl. **-ia**), term coined by Falk (1912) for the mycelial cord of *Merulius*. See Thompson (*in* Jennings & Rayner (Eds), *The ecology and physiology of the fungal mycelium*: 185, 1984).

systemic, (1) (of a parasite), spreading throughout the host; (2) (of a fungicide), absorbed, esp. by the roots, and translocated to other parts of the plant.

tartareous, having a thick rough crumbling surface.

taxis (frequently a suffix), a movement of a plasmodium or zoospore as a reaction to a one-sided stimulus; + (positive) when the movement is in the direction of the stimulus, – (negative) when away from the stimulus. The following tactic movements of zoospores have been described; (1) **chemotaxis**, in response to root exudates (*Pythium* spp.) and amino acids (*Allomyces* spp.); (2) **gravitaxis** (*Phytophthora palmivora*); (3) **electrotaxis** (*Phytophthora palmivora*; Morris *et al.*, *Plant Cell & Environment* **15**: 645, 1992).

taxon (pl. **taxa**, **taxons**), a taxonomic group of any rank (Code, Art. 1). See Lam (*Taxon* **6**: 213, 1957; history and usage).

teleblem (**teleoblema**), see universal veil.

teleomorph, the sexual ('perfect') form or morph (e.g. that characterized by ascomata or basidiomata).

teleutosorus, see telium.

teleutospore, see teliospore.

teleutosporodesm, Donk's (*Proc. Kon. Nederl. Akad. Wetensch.* C **75**: 385, 1972) term for teliospore (obsol.).

teliospore (teleutospores, teleutosporodesma, winter spores, black rust spores), produced in **telia** (sing. -ium; teleutosori; ontogeny and morphology, see Hiratsuka, *Mycotaxon* **31**: 517, 1988), are basidia-producing spores. Telia and teliospores, which characterize the teleomorph of rust fungi, show wide morphological variation but typically teliospores are resting spores, 2- or more celled, sessile or pedicellate but not deciduous, and the thick wall is variously ornamented. Rarely they resemble typical aeciospores when they are designated **aecidioid teliospores** by Cummins (= telial aeciospores (I^{III}), Laundon). Teliospores that germinate immediately, especially in species of genera that usually show dormancy, may be termed leptospores.

telium (pl. **-ia**), a sorus producing teliospores.

tempeh (**tempé**), an oriental food composed of *Rhizopus oligosporus* fermented soybeans (Hesseltine, *Mycol.* **57**: 154, 1965).

tenacle, see haerangium.

terebrate, having scattered perforations.

terebrator (of lichens), a trichogyne (Lindau).

terete, cylindrical; frequently circular in section but narrowing to one end.

terminus (**phialo**) **spore**, a phialospore of a 1-spored phialide, i.e. one terminating the growth of the phialide (Mason, 1933).

terrestrial, growing on land as opposed to in water; cf. terricolous.

terricolous, growing on the ground; cf. terrestrial.

terverticillate (of a penicillus), having branching at three levels, i.e. having rami bearing metulae and phialides.

tessellate, marked with a mosaic design; chequered.

tetra (prefix), four; **-cytes**, the spores resulting from meiosis; **-d**, a group of four; **-polar** (of incompatibility systems), having 2 loci; bifactorial; cf. bipolar; **-spore**, see dispore; **-tomic**, 4-times furcate at one node.

tetrachotomous, with four branches arising from the same point.

texospore, ascospore coated with a layer of cells of paraphysal origin, as in *Texosporium* (Tibell & Hofsten, *Mycol.* **60**: 557, 1968).

textura, see tissue types.

thalamium, asci + hamathecium (obsol.).

thallic (of conidiogenesis), one of the two basic sorts (cf. blastic) in which any enlargement of the recognizable conidial initial occurs *after* the initial has been delimited by one or more septa. The conidium is differentiated from a *whole* cell. **entero-**, thallic conidiogenesis in which the outer wall of the sporogenous cell is not involved in the formation of the spore wall (as for sporangiospores).

thalline exciple (margin), see excipulum (thallinum).

thalloconidium, a propagule produced and seceded directly from the lower cortex and(or) rhizines of certain *Umbilicaria* spp.; similar structures arise from the prothallus in *Protoparmelia*, *Rhizoplaca* and *Sporastatia* (Poelt & Obermayer, *Herzogia* **8**: 273, 1990); thalloconidia are dark brown, smooth to rugged, with 2–3 wall layers, and consist of one to 2500 cells. See Hestmark (*Nord. Jl Bot.* **9**: 547, 1990; ultrastr., occurrence); see also thallyles.

thallodic, of, pertaining to, or belonging to a thallus (Weresub & LeClair, *CJB* **49**: 2203, 1971).

thallospore, (1) an asexual spore having neither conidiophore nor conidiogenous cell, or one which is not separated from the hypha or conidiogenous cell producing it; i.e. an arthrospore, blastospore, or chlamydospore (and aleuriospore) (after Vuillemin; see Mason, 1933); (2) a thalloconidium (q.v.).

thallus, the vegetative body of a thallophyte; **heteromerous-**, a layered thallus; **homoiomerous-**, an unlayered thallus.

thallyles, minute thallus-like propagules produced on the underside of certain *Umbilicaria* thalli (Krog & Swinscow, *Nordic Jl Bot.* **6**: 75, 1986); see also thalloconidia.

thamniscophagous, see ptyophagous.

theca, see ascus (obsol.).

thecaspore, see ascospore (obsol.).

thecium, the part of an apothecium containing the asci between the epithecium and hypothecium; sometimes used for the whole sporocarp or as equivalent to hymenium.

thermodury, withstanding high temperature, esp. when in a dormant state, e.g. as spores; cf. thermophily.

thermophily, making active growth at high temperature; cf. thermodury. Fungi may be classified as **thermophiles** (adj. **-ilic**), growth at 20–50+°C (opt. 40–50+°C). See Cooney & Emerson (*Thermophilic fungi*, 1964; descriptions), Emerson (*in* Ainsworth & Sussman (Eds), *The Fungi* **3**: 105, 1968), Crisan (*Mycol.* **65**: 1170, 1973; concepts), Bilaĭ & Zakharchenko (*Opredelitel' Termofill'nykh Gribov*, 1987; keys, illustr. 38 spp.); **thermotolerant fungi**, e.g. *Aspergillus fumigatus*, *Absidia ramosa*, max. *c.* 50°C, min. well below 20°C; **mesophiles** (adj. **-ilic**), growth 10-40°C (opt. 20-35°C); **psychrophiles** (adj. **-ilic**), growth below 10°C (opt. below 20°C).

thigmotropism, a reaction of germ tubes and hyphae to plant and other sufaces (Dickson, *Phytopath. Z.* **66**: 38, 1969, Kwon & Hoch, *Exp. Mycol.* **15**: 116, 1991).

tholus, see ascus.

thryptogen (**thryptophyte**), an organism increasing the sensitivity of a suscept to outside factors, e.g. to cold (Langer, 1936).

thyriothecium, an inverted flattened ascoma, having the wall ('scutellum', 'placodium') more or less radial in structure, and lacking a basal plate, e.g. *Microthyrium*; cf. catathecium.

thyrsus (pl. **thyrsi**), (1) a type of inflorescence (Bot.); (2) the densely branched apices of some lichens, e.g. *Cladonia stellaris*.

thysanoblastic (of conidiogenesis), when 'the whole of the upper surface of the conidiogenous cell takes part in the process of conidium formation and secession, and both schizolysis and rhexolysis occur alternately in successively seceding conidia' (Roux & van Warmelo, *MR* **92**: 225, 1989).

tichus, peripheral layer of cells of perithecial walls forming a dark protective layer as in *Pleospora herbarum* (Groenhart, *Persoonia* **4**: 11, 1965).

tinophyses (Groenhart, *Persoonia* **4**: 11, 1965), paraphysoids; see hamathecium.

tissue types, Korf distinguished the types of hyphal tissues in discomycetes as different **textura**s and this is now applied to all ascomycetes and coelomycetes. Tissue (textura) types (from Korf, *Sci. Rep. Yokohama nat. Univ.* II **7**: 13, 1958; which is derived from Starbäck, 1895). See also Dargan (*Nova Hedw.* **44**: 489, 1987; *Xylariaceae*), plectenchyma.

tolypophagous, see ptyophagous.

tomentose, having a covering of soft, matted hairs (a **tomentum**; downy.

tonophily (adj. **-ilic**, **-ilous**), the ability to grow under conditions of high osmotic pressure.

torsive, spirally twisted.

torulose (**torulous**), cylindrical but having swellings at intervals; moniliform.

totipotent, bisexual.

toxin, a non-enzymic metabolite of one organism which is injurious to another; **myco-**, a toxin produced by a fungus, esp. one affecting humans or animals; **patho-** (Wheeler & Luke, *Ann. Rev. Phytopath.* **17**: 223, 1963), see vivotoxin; **phyto-**, (1) a toxin produced by a plant (cf. phytoalexins); (2) (frequently, but better avoided), a toxin injurious to plants (see phytotoxic mycotoxins); **vivo-**, a toxin 'produced in the infected host by the pathogen and/or its host, which functions in the production of disease, but is not itself the initial inciting agent' (Dimond & Waggoner, *Phytopath.* **43**: 229, 1953); pathotoxin.

toxiphilous, favouring a polluted habitat (e.g. *Lecanora conizaeoides* in area of high sulphur dioxide pollution), cf. poleophilous; **toxiphobous**, not tolerating such a habitat, e.g. *Usnea* spp.); **toxitolerant**, tolerant of toxins.

trabecula (pl. **-ae**; adj. **-ate**), (1) a lamella primordium; (2) (of *Gymnoglossum* and other gasteromycetes), plates of undifferentiated primordial tissue in the developing gleba forming a branch of a dendroid columella; (3) (of pseudoparaphyses), paraphysoids, tinophyses, see hamathecium.

trachytectum, see exosporium; spore wall.

trama, the layer(s) of hyphae in the central part of a lamella of an agaric, a spine of *Hydnaceae*, or the dissepiment between pores in a polypore; cf. context.

trans-septum, see septum.

tree hair, (1) lichens, esp. fruticose spp. (*Bryoria*, *Usnea*), etc. growing on tree trunks (obsol.); (2) the lichen *Pseudevernia furfuracea* (mousse d'arbre); a source of perfume; cf. oak-moss.

tremelloid, (1) like jelly or wet gelatin; gelatinous; (2) *Tremella*-like.

tretic (of conidiogenesis), the sort of conidiogenesis in which each conidium (**tretoconidium**, tretic conidium, poroconidium, porospore) is delimited by an extension of the inner wall of the conidiogenous cell. Tretoconidia are solitary or in acropetal chains (cf. phialidic). **mono-**, **poly-**, (of conidiogenous cells), producing tretoconidia by the extrusion of the inner wall through one or several channels, respectively.

tri- (in combination), three, triple.
trichidium, see sterigma.
trichocyst, a subpellicular organelle of many ciliates and dinoflagellates; sometimes an offensive weapon able to disable prey, sometimes an anchoring device.
trichoderm (of basidiomata), an outer layer composed of hair-like elements projecting from the surface (Furtado, *Mycol.* **57**: 599, 1965). See derm.
trichogyne, the receptive hypha of the female organ, esp. in certain ascomycetes.
trichospore, a caducous, dehiscent, monosporous sporangium with basal appendages characteristic of the *Harpellales* (Moss & Lichtwardt, *CJB* **54**: 2346, 1976).
trichotomous, with three branches arising from the same point.
trimerous, in threes.
trimitic, see hyphal analysis.
triquetrous, three-edged, three-cornered.
tristichous, in three rows.
troglobiotic, living in caves.
troop, a group of sporocarps (esp. basidiomata), generally from one mycelium.
trophocyst (of *Pilobolus*), a hyphal swelling from which a sporangiophore is produced.
trophogonium (**trophogone**) (of ascomycetes), an antheridium of which the only use is supplying food (Dangeard).
truncate, ending abruptly, as though with the end cut off horizontally.
tubercule (**tubercle**), a small wart-like process; **tuberculate**, having tubercles, syn. of punctate.
tumid, swollen; inflated.
tunic, see exospore.
tunica, a coat, esp. a thin white membrane round the peridiole in most species of the *Nidulariaceae*. See also spore wall (2) and basidiospore.
turbid, not clear; cloudy.
turbinate, like a top in form. **- organ** or **cell** (of *Cladochytriaceae*), a swelling on the vegetative thallus (see Karling, *Am. J. Bot.* **18**: 528, 1931); spindle-organ.
turgid, tightly swollen.
type (in nomenclature), the element on which the descriptive matter fulfilling the conditions of valid publication of a scientific name is based, or is considered to have been based, and which fixes the application of the name; e.g. a family name on a genus, a generic name on a species (a **- species**; see also nomen species), a specific name generally a **- specimen**, which may be a slide, sometimes on a **- culture** (incorrectly for fungi if still metabolically active), a Figure, or a description. Numerous terms with the suffix '-type' have been used in nomenclature, both formally and informally (*see* Hawksworth, *A draft glossary of terms used in bionomenclature*, [IUBS Monogr. 9], 1994), and only a selection of those most used by mycologists are included here: **epi-** a specimen or illustration used to serve as an interpretive type where the existing type material is inadequate for the precise application of the name; **ex-type**, out of the type, used especially for living cultures where the holotype is a dried culture or one preserved in a metabolically inactive state; **holo-**, the single element on which the describing author based a name; **iso-**, a duplicate or part of the type collection (other than the holotype) [(in immunology), part of the imunoglobin molecule from the mouse used in the characterization of sera]; [**histo-**, a reaction between different types of cells]; **lecto-**, an element selected in a later work from the original material where no holotype was designated; **mono-**, the only species included in a genus when first described; **neo-**, specimen or other material designated as nomenclatural type when all the original material is missing; **para-**, any specimen other than the holotype on which the first account of a species or other group is based; **patho-**, see pathovar; **phyco-**, each of the morphologically distinct structures derived by symbiosis between a single mycobiont and different photobionts (Swinscow, *Lichenologist* **9**: 89, 1977; see Lichens); **syn-** one of several elements cited by an author when originally proposing a name but where no holotype was selected ; **topo-**, a later collection from the original locality; **typo-**, the specimen used to prepare an illustration where the latter is the type.
ulcerose, ulcer-like.
umbilicate, having a small hollow; esp. of a pileus having a hollow on the top above the stipe.
umbilicus, (1) the central hold-fast occurring in some foliose lichens (e.g. *Umbilicaria*), navel, umbo; (2) the pore in the perispore of an ascospore (Eriksson, *Opera bot.* **60**, 1981).
umbo, a central swelling like the boss at the centre of a shield; esp. one on top of a pileus above the stipe; see also umbilicus; **-nate**, having an umbo.
uncinate (**uncate**), (1) hooked; (2) of gill insertion, near sinuate (q.v.).
under cortex, lower cortex in foliose lichens.
undulate, wavy.
ungulate, shaped like a horse's hoof.
unialgal (of cultures of lichen photobionts), ones in which a single algal species is present but which may also contain bacteria, fungi, or other organisms.
unipolar, at one end only (esp. of a bacterial cell).
uniseriate, in one row.
universal veil (of agarics and gasteromycetes), a layer of tissue covering the basidioma while development takes place; teleblem; blematogen; cf. volva; **primary - -** = protoblem.
unorientated, not arranged in any particular direction.
unstratified (of lichen thalli), not layered; homoiomerous (q.v.).
urceolate, pitcher-like in form.
urediniospore (**uredinospore, urediospore**), summer spores, red rust spores), repeating vegetative spores (which give urediniospores again or teliospores), usually on dikaryotic mycelium, in **uredinia** (uredosori, uredia; morphological types of uredinia, see Sathe, *Kavaka* **5**: 59, 1977; Hiratsuka & Sato, *in* Scott & Chakravorty (Eds), 1982). Typical urediniospores are unicellular, pedicellate, deciduous, with the pigmented echinulate wall showing two or more germ pores. Rarely they resemble typical aeciospores when they are designated **aecidioid urediniospores** by Cummins (= uredinial aeciospores (I^{II}), Laundon). **Amphispores** (II^{II};X) or resting urediniospores are produced by some rusts. These spores generally have thicker and darker walls than normal urediniospores.

uredinium (pl. **-ia, uredium, uredosorus**), see urediniospore.

uredoconidium (of *Cumminsiella*), see Kuhnholtz-Lordat (*RAM* **26**: 469, 1947).

urniform, having the form of an urn.

ustilospore, Donk's (*K. ned. Akad. Wet.* C **76**: 111, 1973) name for a smut spore; **ustospore** (Moore, *Ant. v. Leeuwenhoek* **38**: 579, 1972).

ustospore, see ustilospore.

utricle, the bladder-like covering of certain fungi, e.g. *Dendrogaster*.

utriform, bag-like.

vagant (of lichens), see vagrant.

vagrant (of lichens), unattached; erratic; vagant. See Rosentreter (*Bryologist* **96**: 333, 1993; N. Am.).

valsoid, having groups of perithecia with their beaks pointing inward (convergent), or even parallel to the surface, as in *Valsa*; cf. eutypoid.

VAM fungi, vesicular-arbuscular mycorrhizal fungi; VA fungi; see mycorrhiza.

variolarioid, having powdery or granular tubercules.

vegetative, see assimilative.

veil, see annulus (**apical -, hymenial -**), cortina, **marginal -, partial - (inner -), pellicular -,** protoblem (**primordial -**), **universal -**.

veins (of lichens), strands of tissue on the lower surface of foliose lichens, esp. *Peltigera* where they may replace a lower cortex. Gyelnik (*Bot. Közlemén.* **24**: 122, 1927) distinguished 2 types: **caninoid -** where the strands are separated to the tips of the lobes; **polydactyloid -** where the strands are confluent towards the tips of the lobes. Maas Geesteranus also recognized **malaceoid -**, where the undersurface has a few whitish interstices faintly indicating venation.

velar, pertaining to a veil.

velum, see veil.

velutinate (velutinous), thickly covered with delicate hairs; like velvet; see phalacrogenous.

venose, having veins.

ventral, front, or lower surface; the surface facing the axis, cf. dorsal; frequently used for the lower surface of foliose lichens.

ventricose, swelling out in the middle or at one side; inflated.

vermiform (vermicular), worm-like.

verruca (pl. **verrucae**), a wart-like swelling.

verrucarioid (of asci), see ascus.

verrucose, having small rounded processes or 'warts'.

verruculose, delicately verrucose.

versiform, of different forms; changing form with age.

vertex, (1) the top of an organ; (2) pileus (obsol.).

verticillate, having parts in rings (**verticils**); whorled.

vesicle, (1) a bladder-like sac; (2) (of *Aspergillus*), the swollen apex of the conidiophore; (3) (of *Pythium*), the evanescent extra-sporangial structure in which zoospores are differentiated; **homohylic -**, sporangial vesicle, the wall of which is continuous with, and of the same material as the wall layer, or one of the wall layers of the sporangium; **vesiculose**, made from or full of vesicles.

vesicular bodies, thin-walled vesicles in the subhymenium of certain hymenomycetes (mostly *Thelephoraceae*) (Overholts); vesicular-arbuscular type of mycorrhiza, see mycorrhiza.

viable, living; able to make growth.

villi (sing. **villus**), long soft hairs.

villose (villous), covered with villi, which are not matted; cf. tomentose.

vinescent, turning wine-red.

virescent, turning green.

virgate, banded; streaked.

virose, (1) poisonous; (2) having a strong and unpleasant smell.

virulence, the degree or measure of pathogenicity; **virulent**, strongly pathogenic.

viscid, slimy, sticky, glutinous, lubricous, mucilaginous, viscous; cf. gelatinous, ixo-.

viteline, yellow like egg yolk.

vittate, having longitudinal lines, bands, or ridges.

volutin, a reserve material of fungi, esp. yeasts, seen as electron-dense granules; metachromatic polymetaphosphate material. See Nagel (*Bot. Rev.* **14**: 174, 1948).

volva (of agarics and gasteromycetes), the cup-like lower part of the universal veil round the base of the mature stipe or receptacle; sometimes = universal veil, which is the preferred usage, fide Bas (*Persoonia* **5**: 304, 1969), who should be consulted for details of volva types and terminology.

wall building, descriptive of hyphal growth in which cell wall material is produced by certain ultrastructural secretory bodies in the cytoplasm. Three types of wall building may be distinguished: **apical -** in which the bodies are concentrated at the hyphal tip, producing new wall by distal growth, forming a cylindrical hypha in which the youngest wall material is at the tip; **ring -**, in which the bodies are concentrated adjacent to the cell wall at some point below the tip, in the shape of an imaginary ring, producing new wall by proximal growth, forming a cylindrical hypha in which the youngest wall material is always at the base; **diffuse -**, in which the bodies occur throughout the cytoplasm at a low concentration, producing lateral growth (i.e. swelling of the cylindrical hypha) by alteration of pre-existing wall.

The terms assume special significance in conidial development, where they have been used to clarify the concepts of thallic and blastic. Apical wall building occurs in *Geniculosporium*, *Cladosporium* and *Scopulariopsis*, and 'phialides' where conidia are produced in gummy masses (e.g. *Trichoderma*) or false chains (e.g. *Mariannaea*). Ring wall building occurs in 'phialides' with conidia in true chains (e.g. *Penicillium*, *Chalara*), in so-called meristem arthrospores (e.g. *Wallemia*) and in conidiogenous cells of basauxic fungi (e.g. *Arthrinium*). Diffuse wall building occurs simultaneously with or shortly after apical or ring wall building in most of the preceding examples, but its occurrence is much delayed or even absent in thallic development (e.g. *Geotrichum*). Wall building is a preferable term to meristem which implies growth by cell division rather than within a single cell. See Minter *et al.* (*TBMS* **79**: 75, 1982, **80**: 39, 1983, **81**: 109, 1983).

wandering lichens, lichens with an epigeic habit (e.g. *Parmelia afrorevoluta*; see Paulson & Hastings, *Knowledge* **37**: 319, 1914).

winter spore, a resting spore for overwintering, e.g. a teliospore of *Puccinia graminis*. See teliospore.

xanthochroic (of a hymenomycete basidioma), having a reddish-brown or yellowish-brown context which darkens on treatment with KOH.

xenobiotic, (1) a chemical not normally synthesized or metabolized by living organisms, e.g. a manufactured drug; (2) chemical waste or other pollutant toxic to a living organism.

xenospore, a spore dispersed from its place of origin (Gregory, *in* Madelin, 1966); cf. memnospore.

xero- (prefix), dry; drought.

xerophilic, favouring habitats in which water is not available; either living in desert conditions or where water is not generally available because of the physiological status of cells (c.f. water activity).

xerophyte, a plant of dry habitats; sometimes incorrectly applied to fungi.

xerotolerant, able to grow under dry conditions (Pitt *in* Duckworth (Ed.), *Water relations of foods*, 1975).

xylogenous, living on wood.

xyloma (of *Dothideales*), a sclerotium-like body producing sporogenous structures inside (de Bary) (obsol.).

xylostromata, sheets of mycelium as in *Xylostroma*.

zeorine (of apothecia), like those of *Zeora*.

zeugite, the organ in which fertilization is completed and the dikaryophase ends; e.g. an ascus or a basidium.

zonate, having concentric lines often forming alternating pale and darker zones near the margins; used of crustose lichen thalli, polypore surfaces, etc.

zonation (of cultures), regular concentric variation of texture, pigmentation or sporulation frequently associated with fluctuations (esp. diurnal) in light, temperature, or other factors; 'Liesegang' phenomenon. See Bisby (*Mycol.* **17**: 89, 1925), Hein (*Am. J. Bot.* **17**: 143, 1930), Kafi & Tarr (*TBMS* **46**: 549, 1964). **ecological -**.

zone lines, narrow, dark brown, or black, lines (pseudosclerotia) or plates (pseudosclerotial plates) in decayed wood (esp. hardwoods) generally caused by fungi (Lopez *et al.*, *TBMS* **64**: 465, 1975); see also spalted (wood).

zoogametes, a motile gamete; planogametes.

zoogloea (of bacteria), a colony embedded in a slimy substance.

zoogonidium, (1) ? = zoospore (q.v.), (2) an aplanospore of a photobiont within the thallus of a lichen (obsol.).

zoophilic (of dermatophytes, etc.), preferentially pathogenic for animals; cf. anthrophilic.

zoosporangium (pl. **-ia**, **zoosporange**), a sporangium producing zoospores.

zoospore, a motile sporangiospore, i.e. one having flagella; swarm spore; swarmer; simblospore; planospore; planont; cf. swarm-cell; see Waterhouse (*TBMS* **45**: 1, 1962), Fuller (*Mycol.* **69**: 1, 1977), Lange & Olson (*Dansk bot. Arkiv.* **33**, 1979; uniflagellate zoospores).

zygangium, gametangium of a zygomycete.

zygoconidium (pl. **-ia**), asexual propagule formed from the fusion or conjoining of two conidia generated simultaneously from adjacent conidiogenous loci. Kendrick & Watling (*The Whole Fungus*: 543, 1979) applied the term isthmospore (q.v.) to this structure and although similar in appearance, it is distinct in the nature of its origin. Zygoconidia are known in a number of genera, mainly basidiomycete anamorphs (e.g. *Anastomyces*, *Christiansenia* and *Zygogloea*). The term was first introduced by Boidin (*Bull. Soc. Linn. Lyon* **39**: 132, 1970) but more recently taken up by others following Oberwinkler & Bandoni (*Norw. J. Bot.* **2**: 501, 1982).

zygophore (of *Mucorales*), a special hyphal branch producing copulation branches.

zygosporangium (pl. **-ia**), see sporangium.

zygospore, the resting spore resulting from the conjugation of isogametes or (in *Zygomycetes*), from the fusion of like gametangia.

zygote, the result of fusion of two gametes; a cell in which two nuclei of opposite sex have undergone fusion (Buller, 1941).

zymogenous, ferment producing. See autochthonous.

zymogram, (1) the pattern of bands obtained by electrophoretic enzyme analysis; (2) a tabulation of carbohydrate fermentations test results.

ACCEPTED FAMILIES AND FREQUENTLY CITED SYNONYMS

Absidiaceae Arx, *Sydowia* **35**: 21 (1982) = **Mucoraceae**.
Acanthophysiaceae Boidin, Mugnier & Canales, *Mycotaxon* **66**: 486 (1998) = **Stereaceae**.
Acarosporaceae Zahlbr., *Nat. Pflanzenfam.* (Leipzig) **1**(1*): 150 (1906).
Acaulosporaceae J.B. Morton & Benny, *Mycotaxon* **37**: 479 (1990).
Achaetomiaceae Mukerji, *Taxonomy of Fungi.* (*Proc. int. Symp. Madras, 1973*) Part 2 (Madras) **1**: 261 (1978) = **Chaetomiaceae**.
Acrocordiaceae Oksner ex M.E. Barr, *Mycotaxon* **29**: 505 (1987) = **Monoblastiaceae**.
Acrospermaceae Fuckel, *Jb. nassau. Ver. Naturk.* **23-24**: 92 (1870).
Actinopeltaceae Petr., *Annls mycol.* **22**: 55 (1924) = **Melanconidaceae**.
Adelococcaceae Triebel, *Sendtnera* **1**: 278 (1993).
Agaricaceae Chevall., *Fl. gén. env. Paris* (Paris): 121 (1826).
Agaricostilbaceae Oberw. & R. Bauer, *Sydowia* **41**: 240 (1989).
Agyriaceae Corda, *Icon. fung.* (Prague) **2**: 36 (1838).
Ajellomycetaceae Unter., J.A. Scott & Sigler, *Mycologia* **96**(4): 819 (2004).
Albatrellaceae (Pouzar) Nuss, *Hoppea* **39**: 174 (1980).
Alectoriaceae (Hue) Tomas., *Archo bot. Sist. Fito-geogr. Genet.* **25**: 235 (1949) = **Parmeliaceae**.
Aleuriaceae Arpin, *Bull. mens. Soc. linn. Lyon* **38**(suppl.): 128 (1969) = **Pyronemataceae**.
Aleurismataceae Vuill., *Bull. Séanc. Soc. Sci. Nancy* Sér. 3 **12**: 413 (1911) = **Hypocreaceae**.
Aleurodiscaceae Jülich, *Biblthca Mycol.* **85**: 354 (1982) = **Stereaceae**.
Aliquandostipitaceae Inderb., *Am. J. Bot.* **88**(1): 54 (2001).
Alphitomorphaceae Corda, *Anleit. Stud. Mykol.* Prag: 120 (1842) = **Erysiphaceae**.
Alternariaceae Earle, *Repert. mic. uomo*: 505 (1934) = **Pleosporaceae**.
Amanitaceae R. Heim ex Pouzar, *Česká Mykol.* **37**: 173 (1983) = **Pluteaceae**.
Amauroascaceae Arx, *Persoonia* **13**(3): 284 (1987) = **Onygenaceae**.
Amorphothecaceae Parbery, *Aust. J. Bot.* **17**: 345 (1969).
Amphisphaeriaceae G. Winter, *Rabenh. Krypt.-Fl.* (Leipzig) **1**(2): 259 (1885).
Amylariaceae Corner, *Beih. Nova Hedwigia* **33**: 5 (1970) = **Bondarzewiaceae**.
Amylocorticaceae Jülich, *Biblthca Mycol.* **85**: 354 (1982) = **Atheliaceae**.
Amylosporaceae Jülich, *Biblthca Mycol.* **85**: 354 (1982) = **Bondarzewiaceae**.
Amylostereaceae Boidin, Mugnier & Canales, *Mycotaxon* **66**: 487 (1998) = **Stereaceae**.
Anamylopsoraceae Lumbsch & Lunke, *Pl. Syst. Evol.* **198**(3-4): 285 (1995).
Anaptychiaceae Körb., *Parerga lichenol.* (Breslau): 19 (1859) = **Physciaceae**.
Ancylistaceae J. Schröt., *Nat. Pflanzenfam.* (Leipzig) **1**(1): 92 (1893).
Annulatascaceae S.W. Wong, K.D. Hyde & E.B.G. Jones, *Syst. Ascom.* **16**(1-2): 18 (1998).
Antennulariaceae Locq., *Mycol. gén. struct.* (Paris): 190 (1984) = **Venturiaceae**.
Antennulariellaceae Woron., *Annls mycol.* **23**: 178 (1925).
Anthracoideaceae Denchev, *Mycotaxon* **65**: 413 (1997).
Aphanopsidaceae Printzen & Rambold, *Lichenologist* **27**(2): 100 (1995).
Aphelariaceae Corner, *Beih. Nova Hedwigia* **33**: 5 (1970).
Apiosporaceae K.D. Hyde, J. Fröhl., Joanne E. Taylor & M.E. Barr, *Sydowia* **50**(1): 23 (1998).
Aporpiaceae Bondartsev & Bondartseva, *Bot. Zh. SSSR* **45**: 1694 (1960).
Appendicisporaceae C. Walker, Vestberg & A. Schüssler, *Mycol. Res.* **111**: [in press] (2007).
Arachniaceae Coker & Couch, *Gasteromycetes E. U.S. Canada* (Chapel Hill): 144 (1928) = **Lycoperdaceae**.
Archaeosporaceae Morton & Redecker, *Mycologia* **93**(1): 182 (2001).
Arctomiaceae Th. Fr., *Nova Acta R. Soc. Scient. upsal.* Ser. 3 **3**: 287 (1860).
Argynnaceae Shearer & J.L. Crane, *Trans. Br. mycol. Soc.* **75**: 193 (1980).
Armatellaceae Hosag., *Sydowia* **55**(2): 165 (2003) = **Meliolaceae**.
Arthoniaceae Rchb., *Deut. Bot. Herb.-Buch*: 13 (1841).
Arthopyreniaceae Walt. Watson, *New Phytol.* **28**: 107 (1929).
Arthriniaceae (Sacc.) Nann., *Repert. mic. uomo*: 484 (1934) = **Apiosporaceae**.
Arthrobotryaceae Corda, *Icon. fung.* (Prague) **5**: 14 (1842) = **Orbiliaceae**.
Arthrodermataceae Locq. ex Currah, *Mycotaxon* **24**: 36 (1985).
Arthrorhaphidaceae Poelt & Hafellner, *Phyton* Horn **17**: 220 (1976).
Ascobolaceae Boud. ex Sacc., *Bot. Zbl.* **18**: 219 (1884).
Ascocorticiaceae J. Schröt., *Krypt.-Fl. Schlesien* (Breslau) **3**(2): 15 (1893).
Ascodesmidaceae J. Schröt., *Krypt.-Fl. Schlesien* (Breslau) **3**(2): 31 (1893).

Ascodichaenaceae D. Hawksw. & Sherwood, *Mycotaxon* **16**: 262 (1982).
Ascoideaceae J. Schröt., *Nat. Pflanzenfam.* (Leipzig) **1**(1): 145 (1894).
Ascoporiaceae Kutorga & D. Hawksw., *Syst. Ascom.* **15**(1-2): 25 (1997).
Ascosphaeraceae Olive & Spiltoir, *Mycologia* **47**: 242 (1955).
Asellariaceae Manier ex Manier & Lichtw., *Annls Sci. Nat. Bot.*, sér. 12 **9**: 526 (1968).
Aspergillaceae Link, *Abh. dt. Akad. Wiss. Berlin*: 165 (1826) ['1824'] = **Trichocomaceae**.
Aspidotheliaceae Räsänen ex J.C. David & D. Hawksw., *Syst. Ascom.* **10**(1): 13 (1991).
Asterinaceae Hansf., *Mycol. Pap.* **15**: 188 (1946).
Asterodontaceae Parmasto, *Folia cryptog. Estonica* **37**: 55 (2001) ['2000'] = **Hymenochaetaceae**.
Asteromellaceae Melnik, *Mikol. Fitopatol.* **20**(2): 101 (1986) = **Didymosphaeriaceae**.
Asterostromataceae (Donk) Pouzar, *Česká Mykol.* **37**: 173 (1983) = **Lachnocladiaceae**.
Asterothyriaceae Walt. Watson ex R. Sant., *Symb. bot. upsal.* **12**(1): 316 (1952).
Astraeaceae Zeller ex Jülich, *Biblthca Mycol.* **85**: 355 (1982).
Astrocystidiaceae Hara, *Bot. Mag.* Tokyo **27**: 473 (1913) = **Xylariaceae**.
Astrotheliaceae Zahlbr., *Syllabus*. Edn 2 (Berlin): 46 (1898) = **Trypetheliaceae**.
Atelosaccharomycetaceae Guillierm., *Clef dichot. déterm. Levures*: 11 (1928) = **Filobasidiaceae**.
Atheliaceae Jülich, *Biblthca Mycol.* **85**: 355 (1982).
Atichiaceae Racib., *Parasit. Alg. Pilze Java's* (Jakarta) **3**: 41 (1900) = **Seuratiaceae**.
Atractiellaceae R.T. Moore, *Mycotaxon* **59**: 8 (1996) = **Hoehnelomycetaceae**.
Atractogloeaceae Oberw. & R. Bauer, *Sydowia* **41**: 245 (1989).
Aulographaceae Luttr. ex P.M. Kirk, P.F. Cannon & J.C. David, *Ainsworth & Bisby's Dictionary of the Fungi*. Edn 9 (Wallingford): ix (2001).
Aureobasidiaceae Cif., *Man. Mic. Med.* (Pavia) **1**: 178 (1958) = **Dothioraceae**.
Auriculariaceae Fr., *Epicr. syst. mycol.* (Upsaliae): 530 (1838).
Auriculariopsidaceae Jülich, *Biblthca Mycol.* **85**: 355 (1982) = **Schizophyllaceae**.
Auriscalpiaceae Maas Geest., *Proc. K. Ned. Akad. Wet.* Ser. C, Biol. Med. Sci. **66**: 426 (1963).
Bacidiaceae Walt. Watson, *New Phytol.* **28**: 27 (1929).
Bactrosporaceae Rabenh., *Krypt.-Fl. Sachsen* (Leipzig) **2**: 60 (1870) = **Roccellaceae**.
Baeomycetaceae Dumort., *Anal. fam. pl.* (Tournay): 71 (1829).
Bagliettoaceae Servít, *Českoslov. Lišejn. Čeledi Verrucariaceae*: 17 (1954) = **Verrucariaceae**.
Balsamiaceae E. Fisch., *Nat. Pflanzenfam.* (Leipzig) **1**(1): 288 (1897) = **Helvellaceae**.
Bankeraceae Donk, *Persoonia* **1**: 405 (1961).
Basidiobolaceae Engl. & E. Gilg, *Syllabus*. Edn 9 & 10 (Berlin): 45 (1924).
Batistiaceae Samuels & K.F. Rodrigues, *Mycologia* **81**(1): 54 (1989).
Battarreaceae Corda, *Anleit. Stud. Mykol.* Prag: 118 (1842) = **Tulostomataceae**.
Beenakiaceae Jülich, *Biblthca Mycol.* **85**: 356 (1982) = **Gomphaceae**.
Bertiaceae Smyk, *Ukr. bot. Zh.* **38**(6): 47 (1981) = **Nitschkiaceae**.
Biatoraceae A. Massal. ex Stizenb., *Ber. Tät. St Gall. naturw. Ges.*: 163 (1862) = **Bacidiaceae**.
Biatorellaceae M. Choisy ex Hafellner & Casares, *Nova Hedwigia* **55**(3-4): 316 (1992).
Bionectriaceae Samuels & Rossman, *Stud. Mycol.* **42**: 15 (1999).
Bjerkanderaceae Jülich, *Biblthca Mycol.* **85**: 356 (1982) = **Hapalopilaceae**.
Blastocladiaceae H.E. Petersen, *Bot. Tidsskr.* **29**: 357 (1909).
Blumeriaceae V.P. Gelyuta, *Biol. Zh. Armenii* **41**(5): 356 (1988) = **Erysiphaceae**.
Bolbitiaceae Singer, *Pap. Mich. Acad. Sci.* **32**: 147 (1948).
Boletaceae Chevall., *Fl. gén. env. Paris* (Paris): 248 (1826).
Boletellaceae Jülich, *Biblthca Mycol.* **85**: 357 (1982) = **Boletaceae**.
Boletinellaceae P.M. Kirk, P.F. Cannon & J.C. David, *Ainsworth & Bisby's Dictionary of the Fungi*. Edn 9 (Wallingford): ix (2001).
Boletoideae Schultz, *Nat. Syst. Pflanzenr.* (Berlin): 250 (1832) = **Boletaceae**.
Boletopsidaceae Bondartsev & Singer ex Jülich, *Biblthca Mycol.* **85**: 357 (1982) = **Bankeraceae**.
Boliniaceae Rick, *Broteria* ser. bot. **25**: 65 (1931).
Bondarzewiaceae Kotl. & Pouzar, *Česká Mykol.* **11**: 163 (1957).
Boreostereaceae Jülich, *Biblthca Mycol.* **85**: 357 (1982).
Botrydiaceae Lindl., *Veg. kingd.* (London): 41 (1846) = **Sclerotiniaceae**.
Botryobasidiaceae (Parmasto) Jülich, *Biblthca Mycol.* **85**: 357 (1982).
Botryoconiaceae Cif. & Vegni, *Riv. Patol. veg., Pavia* sér. 3 **3**(2): 96 (1963) = **Cryptobasidiaceae**.

Botryohypochnaceae (Parmasto) Jülich, *Biblthca Mycol.* **85**: 358 (1982) = **Botryobasidiaceae**.
Botryosphaeriaceae Theiss. & P. Syd., *Annls mycol.* **16**: 16 (1918).
Brachybasidiaceae Gäum., *Vergl. Morph. Pilze* (Jena): 489 (1926).
Brauniellaceae Singer, *Boln Soc. argent. Bot.* **10**: 66 (1962) = **Hymenogasteraceae**.
Brefeldiellaceae (Theiss.) E. Müll. & Arx, *Beitr. Kryptfl. Schweiz* **11**(2): 148 (1962).
Brigantiaeaceae Hafellner & Bellem., *Nova Hedwigia* **35**: 246 (1982) ['1981'].
Broomeiaceae Zeller, *Mycologia* **40**: 647 (1948).
Buelliaceae Zahlbr., *Nat. Pflanzenfam.* (Leipzig) **1**(1*): 230 (1907) = **Physciaceae**.
Bulgariaceae Fr., *Summa veg. Scand.* (Stockholm) **2**: 235 (1849).
Byssocorticaceae Jülich, *Biblthca Mycol.* **85**: 358 (1982) = **Atheliaceae**.
Byssolomataceae Zahlbr., *Nat. Pflanzenfam..* Edn 2 (Leipzig) **8**: 133 (1926) = **Pilocarpaceae**.
Cainiaceae J.C. Krug, *Sydowia* **30**: 123 (1978).
Caliciaceae Chevall., *Fl. gén. env. Paris* (Paris): 314 (1826).
Caloceraceae Rea, *Brit. basidiomyc.* (Cambridge): 740 (1922) = **Dacrymycetaceae**.
Caloplacaceae Zahlbr., *Nat. Pflanzenfam.* (Leipzig) **1**(1*): 226 (1907) = **Teloschistaceae**.
Caloporaceae Bondartseva, *Mikol. Fitopatol.* **17**(4): 269 (1983) = **Polyporaceae**.
Calosphaeriaceae Munk, *Dansk bot. Ark.* **17**(1): 278 (1957).
Calostomataceae E. Fisch., *Nat. Pflanzenfam.* (Leipzig) **1**(1**): 339 (1900).
Calycidiaceae Elenkin, *Izv. glav. bot. Sada SSSR* **28**(3-4): 267 (1929).
Camptobasidiaceae R.T. Moore, *Mycotaxon* **59**: 8 (1996) = **Platygloeaceae**.
Candelariaceae Hakul., *Ann. bot. Soc. Zool.-Bot. Fenn. Vanamo* **27**(3): 11 (1954).
Cantharellaceae J. Schröt., *Krypt.-Fl. Schlesien* (Breslau) **3**(1): 413 (1888).
Capnodiaceae (Sacc.) Höhn. ex Theiss., *Verh. zool.-bot. Ges. Wein* **66**: 363 (1916).
Carbomycetaceae Trappe, *Trans. Br. mycol. Soc.* **57**: 87 (1971).
Catabotrydaceae Petr. ex M.E. Barr, *Mycotaxon* **39**: 83 (1990).
Catathelasmataceae Wasser, *Agarikovye Griby SSSR* (Kiev): 29 (1985) = **Tricholomataceae**.
Catenariaceae (Sparrow) Couch, *Mycologia* **37**: 187 (1945).
Catillariaceae Hafellner, *Beih. Nova Hedwigia* **79**: 271 (1984).
Catinariaceae Hale ex Hafellner, *Beih. Nova Hedwigia* **79**: 272 (1984) = **Bacidiaceae**.
Caulochytriaceae Subram., *Curr. Sci.* **43**: 723 (1974).
Cejpomycetaceae Jülich, *Biblthca Mycol.* **85**: 359 (1982) = **Ceratobasidiaceae**.
Celidiaceae (A. Massal.) J. Schröt., *Krypt.-Fl. Schlesien* (Breslau) **3**(2): 122 (1893) = **Arthoniaceae**.
Cenangiaceae Rehm, *Rabenh. Krypt.-Fl.* (Leipzig) **1**(3): 213 (1888) = **Helotiaceae**.
Cenomycetaceae Chevall., *Fl. gén. env. Paris* (Paris): 585 (1826) = **Cladoniaceae**.
Cephaloascaceae L.R. Batra, *Tech. Bull. U.S. Dep. Agric.* **1469**: 3 (1973).
Cephalocladiaceae Corda, *Icon. fung.* (Prague) **2**: 11 (1838) = **Sclerotiniaceae**.
Cephalothecaceae Höhn., *Annls mycol.* **15**: 362 (1917).
Ceratobasidiaceae G.W. Martin, *Lloydia* **11**: 114 (1948).
Ceratocystidaceae Locq., *Syn. gen. fung.* (Paris): [1] (1972).
Ceratomycetaceae S. Colla, *Fl. ital. crypt.* (Florence) **1**(16): 134 (1934).
Ceratosporiaceae (Ferraris) Nann., *Repert. mic. uomo*: 507 (1934) = **Amphisphaeriaceae**.
Ceratostomataceae G. Winter, *Rabenh. Krypt.-Fl.* (Leipzig) **1**(2): 247 (1885).
Cercosporaceae Nann., *Repert. mic. uomo*: 507 (1934) = **Mycosphaerellaceae**.
Cercosporellaceae Nann., *Repert. mic. uomo*: 473 (1934) = **Mycosphaerellaceae**.
Cerinomycetaceae Jülich, *Biblthca Mycol.* **85**: 358 (1982).
Cetrariaceae Schaer., *Enum. critic. lich. europ.* (Bern): 12 (1850) = **Parmeliaceae**.
Chaconiaceae Cummins & Y. Hirats., *Illustr. Gen. Rust Fungi. rev. edit.* (St. Paul): 14 (1983).
Chadefaudiellaceae Faurel & Schotter ex Benny & Kimbr., *Mycotaxon* **12**: 46 (1980).
Chaetocladiaceae A. Fisch., *Rabenh. Krypt.-Fl.* (Leipzig) **1**(4): 283 (1892) = **Mucoraceae**.
Chaetodermataceae Jülich, *Biblthca Mycol.* **85**: 359 (1982) = **Stereaceae**.
Chaetomiaceae G. Winter, *Rabenh. Krypt.-Fl.* (Leipzig) **1**(2): 153 (1885).
Chaetoporellaceae Jülich, *Biblthca Mycol.* **85**: 359 (1982) = **Schizoporaceae**.
Chaetosphaerellaceae Huhndorf, A.N. Mill. & F.A. Fernández, *Mycol. Res.* **108**(12): 1387 (2004).
Chaetosphaeriaceae Réblová, M.E. Barr & Samuels, *Sydowia* **51**(1): 56 (1999).
Chaetothyriaceae Hansf. ex M.E. Barr, *Mycologia* **71**: 943 (1979).
Chamonixiaceae Jülich, *Biblthca Mycol.* **85**: 359 (1982) = **Boletaceae**.

Chiodectonaceae Zahlbr., *Nat. Pflanzenfam.* (Leipzig) **1**(1*): 102 (1905) = **Roccellaceae**.
Chionosphaeraceae Oberw. & Bandoni, *Can. J. Bot.* **60**(9): 1732 (1982).
Chlamydozymaceae Wick., *Mycologia* **56**: 257 (1964) = **Metschnikowiaceae**.
Chloridiaceae (Sacc.) Nann., *Repert. mic. uomo*: 496 (1934) = **Chaetosphaeriaceae**.
Choanephoraceae J. Schröt., *Nat. Pflanzenfam.* (Leipzig) **1**(1): 131 (1894).
Christianseniaceae F. Rath, *Atti Soc. ital. Sci. nat. Mus. Civico Storia nat. Milano* **132**(2): 17 (1991).
Chrysoconiaceae Jülich, *Biblthca Mycol.* **85**: 360 (1982) = **Coniophoraceae**.
Chrysoglutenaceae Jatta, *Fl. ital. crypt.* (Florence) **1**(3): 778 (1911) = **Nectriaceae**.
Chrysomyxaceae Gäum. ex Leppik, *Ann. bot. fenn.* **9**: 139 (1972) = **Coleosporiaceae**.
Chrysothricaceae Zahlbr., *Nat. Pflanzenfam.* (Leipzig) **1**(1*): 117 (1905).
Chytridiaceae Nowak., *Akad. umiejetnošci Krakowie. Wydzíat mat.-przyród.* Pamietník **4**: 174 (1878).
Ciglidiaceae Chevall., *Fl. gén. env. Paris* (Paris): 382 (1826) = **Pucciniaceae**.
Cintractiaceae Vánky, *Mycotaxon* **32**: 344 (2000).
Cintractiellaceae Vánky, *Fungal Diversity* **13**: 172 (2003) = **Ustilaginaceae**.
Cladiaceae Filson, *J. Hattori bot. Lab.* **49**: 12 (1981) = **Cladoniaceae**.
Cladochytriaceae J. Schröt., *Nat. Pflanzenfam.* (Leipzig) **1**(1): 80 (1892).
Cladoniaceae Zenker, *Pharmaceutische Waarenkunde* (Eisenach) **1**: 124 (1827).
Cladosporiaceae (Mathieu) Nann., *Repert. mic. uomo*: 404 (1934) = **Mycosphaerellaceae**.
Clasterosporiaceae (Sacc.) Nann., *Repert. mic. uomo*: 498 (1934) = **Magnaporthaceae**.
Clathraceae Chevall., *Fl. gén. env. Paris* (Paris): 120 (1826) = **Phallaceae**.
Claustulaceae G. Cunn., *Proc. Linn. Soc. N. S. W.* **56**: 198 (1931) = **Phallaceae**.
Clavariaceae Chevall., *Fl. gén. env. Paris* (Paris): 102 (1826).
Clavariadelphaceae Corner, *Beih. Nova Hedwigia* **33**: 6 (1970) = **Gomphaceae**.
Clavarichaetaceae Jülich, *Biblthca Mycol.* **85**: 360 (1982) = **Hymenochaetaceae**.
Clavicipitaceae (Lindau) O.E. Erikss., *Mycotaxon* **15**: 224 (1982).
Clavicoronaceae Corner, *Beih. Nova Hedwigia* **33**: 6 (1970) = **Auriscalpiaceae**.
Clavulinaceae (Donk) Donk, *Beih. Nova Hedwigia* **1**(4): 407 (1961).
Climacodontaceae Jülich, *Biblthca Mycol.* **85**: 360 (1982) = **Meruliaceae**.
Clintamraceae Vánky, *Fungal Diversity* **95**(3): 142 (2001).
Clypeosphaeriaceae G. Winter, *Rabenh. Krypt.-Fl.* (Leipzig) **1**(2): 554 (1886).
Coccocarpiaceae (Mont. ex Müll. Berol.) Henssen, *Syst. Ascom.* **5**: 314 (1986).
Coccodiniaceae Höhn. ex O.E. Erikss., *Op. bot. Soc. bot. Lund* **60**: 42 (1981).
Coccoideaceae Henn. ex Sacc. & D. Sacc., *Syll. fung.* (Abellini) **17**: 860 (1905).
Coccotremataceae Henssen ex J.C. David & D. Hawksw., *Syst. Ascom.* **10**(1): 14 (1991).
Cochlonemataceae Dudd., *The Fungi* (London) **4B**: 233 (1974).
Coelomomycetaceae Couch ex Couch, *J. Elisha Mitchell scient. Soc.* **78**: 135 (1962).
Coenogoniaceae (Fr.) Stizenb., *Ber. Tät. St Gall. naturw. Ges.*: 140 (1862) = **Gyalectaceae**.
Coleosporiaceae Dietel, *Nat. Pflanzenfam.* (Leipzig) **1**(1**): 548 (1900).
Collemataceae Zenker, *Pharmaceutische Waarenkunde* (Eisenach) **1**(3): 124 (1827).
Collemataceae Zenker, *Pharmaceutische Waarenkunde* (Eisenach) **1**: 124 (1827) = **Collemataceae**.
Coltriciaceae Jülich, *Biblthca Mycol.* **85**: 361 (1982) = **Hymenochaetaceae**.
Completoriaceae Humber, *Mycotaxon* **34**(2): 453 (1989).
Coniocarpaceae Rchb., *Deut. Bot. Herb.-Buch* **1**(2): xxviii (1841) = **Arthoniaceae**.
Coniochaetaceae Malloch & Cain, *Can. J. Bot.* **49**: 878 (1971).
Coniocybaceae Rchb., *Handb. nat. Pfl.-Syst.* (Dresden): 132 (1837).
Coniophoraceae Ulbr., *Krypt.-Fl. Anfäng.* (Berlin) **1**(3): 120 (1928).
Coniothyriaceae W.B. Cooke, *Revta Biol., Lisb.* **12**: 289 (1983) ['1980-1983'] = **Leptosphaeriaceae**.
Conopleaceae Chevall., *Fl. gén. env. Paris* (Paris): 38 (1826) = **Sarcosomataceae**.
Cookellaceae Höhn. ex Sacc. & Trotter, *Syll. fung.* (Abellini) **22**: 585 (1913).
Coprinaceae Overeem & Weese, *Icon. Fung. Malay., Heft VI: Coprinaceae* (St. Gravenhage): 3 (1924) = **Agaricaceae**.
Coprinaceae Gäum., *Vergl. Morph. Pilze* (Jena): 530 (1926) = **Agaricaceae**.
Coraceae Tomas. ex Tomas., *Archo bot. Sist. Fito-geogr. Genet.* **26**: 104 (1950) = **Atheliaceae**.
Cordieritidaceae (Sacc.) Sacc., *Syll. fung.* (Abellini) **8**: 810 (1889) = **Helotiaceae**.
Coremiaceae (Ferraris) Nann., *Repert. mic. uomo*: 511 (1934) = **Trichocomaceae**.
Coriolaceae (Imazeki) Singer, *Publções Inst. Micol. Recife* **304**: 6 (1961) = **Polyporaceae**.

Corneromycetaceae Jülich, *Persoonia* **10**: 336 (1979) = **Coniophoraceae**.
Corniculariaceae Schaer., *Enum. critic. lich. europ.* (Bern): 4 (1850) = **Parmeliaceae**.
Coronophoraceae Höhn., *Sber. Akad. Wiss. Wien Math.-naturw. Kl.*, Abt. 1 **116**: 624 (1907) = **Nitschkiaceae**.
Corticiaceae Herter, *Krypt.-Fl. Brandenburg* (Leipzig) **6**: 70 (1910).
Cortinariaceae R. Heim ex Pouzar, *Česká Mykol.* **37**: 174 (1983).
Cortinariaceae (Fayod) R. Heim, *Trab. Mus. Ciènc. nat. Barcelona* sèr. bot. **15**(3): 115 (1934) = **Cortinariaceae**.
Coryneaceae Corda, *Icon. fung.* (Prague) **3**: 36 (1839) = **Melanconidaceae**.
Coryneliaceae Sacc. ex Berl. & Voglino, *Syll. fung.*. Addit. (Abellini): 193 (1886).
Corynesporascaceae Sivan., *Mycol. Res.* **100**(7): 786 (1996).
Corynitaceae Kalchbr., *Értek. termész. Köréb. Magy. tudom. Akad.* **10**(17): 14 (1880) = **Phallaceae**.
Craterellaceae Herter, *Krypt.-Fl. Brandenburg* (Leipzig) **6**: 141 (1910) = **Cantharellaceae**.
Crepidotaceae Singer, *Lilloa* **22**: 584 (1951).
Cribbeaceae Singer, J.E. Wright & E. Horak, *Darwiniana* **12**: 611 (1963) = **Cortinariaceae**.
Cristiniaceae Jülich, *Biblthca Mycol.* **85**: 361 (1982) = **Atheliaceae**.
Crocyniaceae M. Choisy ex Hafellner, *Beih. Nova Hedwigia* **79**: 274 (1984).
Cronartiaceae Dietel, *Nat. Pflanzenfam.* (Leipzig) **1**(1**): 548 (1900).
Cryphonectriaceae Gryzenh. & M.J. Wingf., *Mycologia* **98**(2): 246 (2006).
Cryptobasidiaceae Malençon ex Donk, *Reinwardtia* **4**: 114 (1956).
Cryptococcaceae Kütz. ex Castell. & Chalm., *Man. trop. med.*. 3rd Edn (London): 1070 (1919) = **Filobasidiaceae**.
Cryptomycocolacaceae Oberw. & R. Bauer, *Mycologia* **82**(6): 672 (1990).
Cryptoporaceae Jülich, *Biblthca Mycol.* **85**: 361 (1982) = **Polyporaceae**.
Cryptotheciaceae A.L. Sm., *Trans. Br. mycol. Soc.* **11**: 190 (1926) = **Arthoniaceae**.
Cryptotheliaceae Walt. Watson, *New Phytol.* **28**: 113 (1929) = **Trypetheliaceae**.
Cucurbitariaceae G. Winter, *Rabenh. Krypt.-Fl.* (Leipzig) **1**(2): 308 (1885).
Cudoniaceae P.F. Cannon, *Ainsworth & Bisby's Dictionary of the Fungi*. Edn 9 (Wallingford): ix (2001).
Cuniculitremaceae J.P. Sampaio & R. Kirschner, *Antonie van Leeuwenhoek* **80**(2): 155 (2001).
Cunninghamellaceae Naumov ex R.K. Benj., *Aliso* **4**: 415 (1959).
Cycloschizaceae Locq., D. Pons & Sal.-Cheb., *Cah. Micropaleontol.* **1**: 117 (1981) = **Parmulariaceae**.
Cylindrobasidiaceae Jülich, *Biblthca Mycol.* **85**: 362 (1982) = **Meruliaceae**.
Cypheliaceae Zahlbr., *Nat. Pflanzenfam.* (Leipzig) **1**(1*): 83 (1903) = **Caliciaceae**.
Cyphellaceae Lotsy, *Vortr. bot. Stammesgesch.* **1**: 695 (1907).
Cystobasidiaceae Gäum., *Vergl. Morph. Pilze* (Jena): 411 (1926) = **Platygloeaceae**.
Cystofilobasidiaceae Well & Bandoni, *The Mycota* (Berlin) **7**(B): 113 (2001).
Cystostereaceae Jülich, *Biblthca Mycol.* **85**: 362 (1982).
Cytidiaceae Jülich, *Biblthca Mycol.* **85**: 363 (1982) = **Corticiaceae**.
Cytisporei Fr., *Syst. orb. veg.* (Lund) **1**: 118 (1825) = **Valsaceae**.
Cyttariaceae Speg., *Boln Acad. nac. Cienc. Córdoba* **11**: 253 (1887).
Dacampiaceae Körb., *Syst. lich. germ.* (Breslau): 322 (1855).
Dacrymycetaceae J. Schröt., *Krypt.-Fl. Schlesien* (Breslau) **3**(1): 399 (1888).
Dacryobolaceae Jülich, *Biblthca Mycol.* **85**: 363 (1982) = **Meruliaceae**.
Dactyliaceae (Sacc.) Nann., *Repert. mic. uomo*: 472 (1934) = **Orbiliaceae**.
Dactylosporaceae Bellem. & Hafellner, *Cryptog. Mycol.* **3**: 79 (1982).
Daedaleaceae Jülich, *Biblthca Mycol.* **85**: 363 (1982) = **Fomitopsidaceae**.
Delitschiaceae M.E. Barr, *Mycotaxon* **76**: 109 (2000).
Dendrosphaeraceae Cif. ex Benny & Kimbr., *Mycotaxon* **12**: 22 (1980) = **Trichocomaceae**.
Dendryphiaceae Corda, *Icon. fung.* (Prague) **4**: 32 (1840) = **Pleosporaceae**.
Dentinaceae Kotl. & Pouzar, *Česká Mykol.* **26**: 217 (1972) = **Hydnaceae**.
Dermateaceae Fr., *Summa veg. Scand.* (Stockholm) **2**: 345 (1849).
Dermatocarpaceae (Eschw.) Stizenb., *Ber. Tät. St Gall. naturw. Ges.*: 149 (1862) ['1861/2'] = **Verrucariaceae**.
Dermatosoraceae Vánky, *Fungal Diversity* **237**(2): 135 (2001).
Diachanthodaceae Jülich, *Biblthca Mycol.* **85**: 363 (1982) = **Meripilaceae**.
Diademaceae Shoemaker & C.E. Babc., *Can. J. Bot.* **70**(8): 1618 (1992).
Diaporthaceae Höhn. ex Wehm., *Am. J. Bot.* **13**: 638 (1926).
Diatrypaceae Nitschke, *Verh. naturh. Ver. preuss. Rheinl.* **26**: 73 (1869).
Dicantharellaceae Jülich, *Biblthca Mycol.* **85**: 364 (1982) = **Lachnocladiaceae**.

Dicellomycetaceae Parmasto, *Eesti NSV Tead. Akad. Toim.* Biol. seer **17**: 226 (1968) = **Brachybasidiaceae**.
Dichaenaceae Fr., *Summa veg. Scand.* (Stockholm) **2**: 380 (1849) = **Ascodichaenaceae**.
Dichostereaceae Jülich, *Biblthca Mycol.* **85**: 364 (1982) = **Lachnocladiaceae**.
Dicranophoraceae J.H. Mirza, *Mucor. Pakistan* (Faisalabad): 19 (1979) = **Mucoraceae**.
Dictyolaceae Gäum., *Vergl. Morph. Pilze* (Jena): 511 (1926) = **Tricholomataceae**.
Dictyonemataceae Tomas. ex Tomas., *Archo bot. Sist. Fito-geogr. Genet.* **26**: 104 (1950) = **Atheliaceae**.
Didymobotryaceae Nann., *Repert. mic. uomo*: 515 (1934) = **Clavicipitaceae**.
Didymocladiaceae Nann., *Repert. mic. uomo*: 471 (1934) = **Hypocreaceae**.
Didymosphaeriaceae Munk, *Dansk bot. Ark.* **15**(2): 128 (1953).
Digitatisporaceae Jülich, *Biblthca Mycol.* **85**: 365 (1982) = **Atheliaceae**.
Dimargaritaceae R.K. Benj., *Aliso* **4**: 364 (1959).
Dimorphocystidiaceae Jülich, *Biblthca Mycol.* **85**: 365 (1982) = **Pterulaceae**.
Dimorphomycetaceae S. Colla, *Fl. ital. crypt.* (Florence) **1**(16): 50 (1934) = **Laboulbeniaceae**.
Diplocladiaceae Nann., *Repert. mic. uomo*: 470 (1934) = **Hypocreaceae**.
Diplocystaceae Kreisel, *Feddes Repert.* **85**: 334 (1974).
Diplodermaceae Fr., *Summa veg. Scand.* (Stockholm) **2**: 440 (1849) = **Astraeaceae**.
Diploschistaceae Zahlbr., *Nat. Pflanzenfam.* (Leipzig) **1**(1*): 121 (1905) = **Thelotremataceae**.
Dipodascaceae Engl. & E. Gilg, *Syllabus*. Edn 9 & 10 (Berlin): 59 (1924).
Diporothecaceae Mibey & D. Hawksw., *Syst. Ascom.* **14**(1): 27 (1995).
Dirinaceae Zahlbr., *Nat. Pflanzenfam.* (Leipzig) **1**(1*): 105 (1905) = **Roccellaceae**.
Discinaceae Benedix, *Z. Pilzk.* **27**: 100 (1961).
Discomycetes Fr., *Summa veg. Scand.* (Stockholm) **2**: 343 (1849) = **Helvellaceae**.
Diversisporaceae C. Walker & A. Schüssler, *Mycol. Res.* **108**(9): 981 (2004).
Doassansiaceae (Azbukina & Karatygin) R.T. Moore ex P.M. Kirk, P.F. Cannon & J.C. David, *Ainsworth & Bisby's Dictionary of the Fungi.* Edn 9 (Wallingford): ix (2001).
Doassansiopsidaceae Begerow, R. Bauer & Oberw., *Can. J. Bot.* **75**(12): 2052 (1998) ['1997'].
Dothideaceae Chevall., *Fl. gén. env. Paris* (Paris): 446 (1826).
Dothioraceae Theiss. & P. Syd., *Annls mycol.* **15**: 444 (1918) ['1917'].
Duportellaceae Jülich, *Biblthca Mycol.* **85**: 365 (1982) = **Peniophoraceae**.
Eballistraceae R. Bauer, Begerow, A. Nagler & Oberw., *Mycol. Res.* **105**(4): 423 (2001).
Ecchynaceae Rea, *Brit. basidiomyc.* (Cambridge): 16 (1922) = **Phleogenaceae**.
Echinochaetaceae Jülich, *Biblthca Mycol.* **85**: 366 (1982) = **Polyporaceae**.
Echinodontiaceae Donk, *Persoonia* **1**: 405 (1961).
Ectolechiaceae (Vain.) Zahlbr., *Nat. Pflanzenfam.* (Leipzig) **1**(1*): 122 (1905).
Eigleraceae Hafellner, *Beih. Nova Hedwigia* **79**: 276 (1984) = **Hymeneliaceae**.
Elaphomycetaceae Tul. ex Paol., *Syll. fung.* (Abellini) **8**: 863 (1889).
Elasmomycetaceae Locq. ex Pegler & T.W.K. Young, *Trans. Br. mycol. Soc.* **72**: 367 (1979) = **Russulaceae**.
Elixiaceae Lumbsch, *J. Hattori bot. Lab.* **83**: 62 (1997).
Elsinoaceae Höhn. ex Sacc. & Trotter, *Syll. fung.* (Abellini) **22**: 584 (1913).
Empusaceae Clem. & Shear, *Gen. fung..* Edn 2 (Minneapolis): 37 (1931) = **Entomophthoraceae**.
Endocarpaceae Rchb., *Consp. Regni Veget.* (Leipzig): 21 (1828) = **Verrucariaceae**.
Endochytriaceae Sparrow ex D.J.S. Barr, *Can. J. Bot.* **58**: 2390 (1980).
Endogonaceae Paol., *Syll. fung.* (Abellini) **8**: 905 (1889).
Endomycetaceae J. Schröt., *Krypt.-Fl. Schlesien* (Breslau) **3**(2): 208 (1893).
Endophyllaceae Dietel, *Nat. Pflanzenfam.* (Leipzig) **1**(1**): 35 (1897) = **Pucciniaceae**.
Endopyreniaceae Zahlbr., *Syllabus*. Edn 2 (Berlin): 46 (1898) = **Verrucariaceae**.
Englerulaceae Henn., *Hedwigia* Beibl. **43**: 353-354 (1904).
Entolomataceae Kotl. & Pouzar, *Česká Mykol.* **26**: 218 (1972).
Entomophthoraceae Nowak., *Bot. Ztg.* **35**: 35 (1877).
Entomosporiaceae W.B. Cooke, *Revta Biol., Lisb.* **12**: 290 (1983) = **Dermateaceae**.
Entophlyctidaceae Whiffen, *Farlowia* **1**: 591 (1944) = **Endochytriaceae**.
Entorrhizaceae R. Bauer & Oberw., *Can. J. Bot.* **75**(8): 1311 (1997).
Entylomataceae R. Bauer & Oberw., *Can. J. Bot.* **75**(8): 1312 (1997).
Entylomellaceae Cif., *Acad. Republ. Pop. Rom. Amag. Tr. Săvul.*: 175 (1959) = **Tilletiaceae**.
Eoterfeziaceae G.F. Atk., *Bot. Gaz.* **34**: 40 (1902).
Ephebaceae Th. Fr., *Nova Acta R. Soc. Scient. upsal.* Ser. 3 **3**: 289 (1860) = **Lichinaceae**.

Epigloeaceae Zahlbr., *Nat. Pflanzenfam.* (Leipzig) **1**(1*): 53 (1903).
Epitheliaceae Jülich, *Biblthca Mycol.* **85**: 366 (1982).
Eremascaceae Engl. & E. Gilg, *Syllabus*. Edn 9 & 10 (Berlin): 59 (1924).
Eremomycetaceae Malloch & Cain, *Can. J. Bot.* **49**: 847 (1971).
Eremotheciaceae Kurtzman, *J. Industr. Microbiol.* **14**(6): 527 (1995).
Erinaceae Quél., *Enchir. fung.* (Paris): 188 (1886) = **Bankeraceae**.
Erysiphaceae Tul. & C. Tul., *Select. fung. carpol.* (Paris) **1**: 191 (1861).
Euantennariaceae S. Hughes & Corlett ex S. Hughes, *N.Z. Jl Bot.* **10**: 238 (1972).
Euceratomycetaceae I.I. Tav., *Mycotaxon* **11**: 488 (1980).
Eurotiaceae Clem. & Shear, *Gen. fung..* Edn 2 (Minneapolis): 50 (1931) = **Trichocomaceae**.
Everniaceae (Hue) Tomas., *Archo bot. Sist. Fito-geogr. Genet.* **1**(1): 238 (1897) = **Parmeliaceae**.
Excipulaceae Bonord., *Handb. Allgem. mykol.* (Stuttgart): 225 (1851) = **Dermateaceae**.
Exidiaceae R.T. Moore, *Mycologia* **70**: 1016 (1978).
Exobasidiaceae J. Schröt., *Krypt.-Fl. Schlesien* (Breslau) **3**(1): 413 (1888).
Fabosporaceae E.K. Novák & Zsolt, *Acta bot. hung.* **7**: 99 (1961) = **Saccharomycopsidaceae**.
Faerberiaceae Pouzar, *Česká Mykol.* **37**: 174 (1983) = **Polyporaceae**.
Farysiaceae Vánky, *Fungal Diversity* **95**(3): 143 (2001).
Favolaschiaceae (Murrill) Singer, *Beih. Nova Hedwigia* **29**: 380 (1969) = **Marasmiaceae**.
Fenestellaceae M.E. Barr, *Mycologia* **71**: 952 (1979).
Filobasidiaceae L.S. Olive, *J. Elisha Mitchell scient. Soc.* **84**: 261 (1968).
Fimetariaceae D.A. Griffiths & Seaver, *N. Amer. Fl.* (New York) **3**(1): 65 (1910) = **Sordariaceae**.
Fistulinaceae Lotsy, *Vortr. bot. Stammesgesch.* **1**: 695 (1907).
Fomitaceae Jülich, *Biblthca Mycol.* **85**: 367 (1982) = **Polyporaceae**.
Fomitopsidaceae Jülich, *Biblthca Mycol.* **85**: 367 (1982).
Fusariaceae (Ferraris) Nann., *Repert. mic. uomo*: 518 (1934) = **Nectriaceae**.
Fuscideaceae Hafellner, *Beih. Nova Hedwigia* **79**: 278 (1984).
Galactiniaceae Berthet ex Le Gal, *Bull. Soc. mycol. Fr.* **85**: 10 (1969) = **Pezizaceae**.
Ganodermataceae (Donk) Donk, *Bull. bot. Gdns Buitenz.* Ser. III **17**: 474 (1948).
Gasterellaceae Zeller, *Mycologia* **40**: 639 (1948).
Gasteromycetes Fr., *Summa veg. Scand.* (Stockholm) **2**: 429 (1849) = **Phallaceae**.
Gastroboletaceae Singer, *Boln Soc. argent. Bot.* **10**: 57 (1962) = **Boletaceae**.
Gastrosporiaceae Pilát, *Bull. Soc. mycol. Fr.* **50**: 46 (1934).
Gautieriaceae Zeller, *Mycologia* **40**: 666 (1948) = **Gomphaceae**.
Geastraceae Corda, *Anleit. Stud. Mykol.* Prag: 104 (1842).
Gelatinodiscaceae S.E. Carp., *Mycotaxon* **3**: 231 (1976) = **Helotiaceae**.
Gelopellaceae Zeller, *Mycologia* **31**: 20 (1939) = **Phallaceae**.
Geminaginaceae Vánky, *Fungal Diversity* **95**(3): 141 (2001).
Geneaceae Trappe, *Mycotaxon* **9**: 330 (1979) = **Pyronemataceae**.
Geoglossaceae Corda, *Icon. fung.* (Prague) **2**: 35 (1838).
Georgefischeriaceae R. Bauer, Begerow & Oberw., *Can. J. Bot.* **75**(8): 1312 (1997).
Georgefisheriaceae R. Bauer, Begerow & Oberw., *Can. J. Bot.* **75**(8): 1312 (1997) = **Georgefischeriaceae**.
Geosiphonaceae Engl. & E. Gilg, *Syllabus*. Edn 9 & 10 (Berlin): 24 (1924).
Gigaspermaceae Jülich, *Biblthca Mycol.* **85**: 367 (1982).
Gigasporaceae J.B. Morton & Benny, *Mycotaxon* **37**: 483 (1990).
Gilbertellaceae Benny, *Mycologia* **83**(2): 150 (1991) = **Choanephoraceae**.
Gjaerumiaceae R. Bauer, M. Lutz & Oberw., *Mycol. Res.* **109**(11): 1257 (2005).
Glaziellaceae J.L. Gibson, *Mycologia* **78**: 953 (1986).
Glenosporaceae Nann., *Repert. mic. uomo*: 423 (1934) = **Septobasidiaceae**.
Gloeocystidiellaceae (Parmasto) Jülich, *Biblthca Mycol.* **85**: 368 (1982) = **Stereaceae**.
Gloeoheppiaceae Henssen, *Lichenologist* **27**(4): 266 (1995).
Gloeomucronaceae Jülich, *Biblthca Mycol.* **85**: 368 (1982) = **Hydnaceae**.
Gloeophyllaceae Jülich, *Biblthca Mycol.* **85**: 368 (1982).
Gloeosporidiellaceae Melnik, *Mikol. Fitopatol.* **20**: 101 (1986) = **Dermateaceae**.
Gloiothelaceae Boidin, Mugnier & Canales, *Mycotaxon* **66**: 487 (1998) = **Stereaceae**.
Glomeraceae Piroz. & Dalpé, *Symbiosis* **7**(1): 19 (1989).
Glomerellaceae Locq., *Mycol. gén. struct.* (Paris): 175 (1984).

Glomerillaceae Norman, *Bot. Notiser*: 193 (1869) ['1868'] = **Verrucariaceae**.
Glomosporiaceae Cif., *Quad. Lab. crittogam., Pavia* **27**: 1 (1963).
Glyphidaceae (Th. Fr.) Stizenb., *Ber. Tät. St Gall. naturw. Ges.*: 151 (1862) = **Graphidaceae**.
Gnomoniaceae G. Winter, *Rabenh. Krypt.-Fl.* (Leipzig) **1**(2): 570 (1886).
Gomphaceae Donk, *Persoonia* **1**: 406 (1961).
Gomphidiaceae Maire ex Jülich, *Biblthca Mycol.* **85**: 369 (1982).
Gomphillaceae Walt. Watson ex Hafellner, *Beih. Nova Hedwigia* **79**: 280 (1984).
Gonapodyaceae H.E. Petersen ex P.M. Kirk, P.F. Cannon & J.C. David, *Ainsworth & Bisby's Dictionary of the Fungi*. Edn 9 (Wallingford): ix (2001).
Gonatobotrydiaceae (Sacc.) Nann., *Repert. mic. uomo*: 172 (1934) = **Ceratostomataceae**.
Grammotheleaceae Jülich, *Biblthca Mycol.* **85**: 369 (1982).
Graphidaceae Dumort., *Comment. bot.* (Tournay): 69 (1822).
Graphinellaceae Walt. Watson, *New Phytol.* **28**: 98 (1929) = **Odontotremataceae**.
Graphiolaceae Clem. & Shear, *Gen. fung.*. Edn 2 (Minneapolis): 156 (1931).
Graphostromataceae M.E. Barr, J.D. Rogers & Y.M. Ju, *Mycotaxon* **48**: 533 (1993).
Grifolaceae Jülich, *Biblthca Mycol.* **85**: 369 (1982) = **Meripilaceae**.
Gyalectaceae Stizenb., *Ber. Tät. St Gall. naturw. Ges.*: 158 (1862).
Gymnoascaceae Baran., *Bot. Ztg.* **30**: 158 (1872).
Gypsoplacaceae Timdal, *Biblthca Lichenol.* **38**: 423 (1990).
Gyrodontaceae (Singer) Heinem., *Bull. Jard. bot. État Brux.* **21**: 228 (1951) = **Paxillaceae**.
Gyrophoraceae Zenker, *Pharmaceutische Waarenkunde* (Eisenach): 124 (1827) = **Umbilicariaceae**.
Gyroporaceae Locq., *Mycol. gén. struct.* (Paris): 122 (1984).
Haddowiaceae Jülich, *Biblthca Mycol.* **85**: 370 (1982) = **Ganodermataceae**.
Haematommataceae Hafellner, *Beih. Nova Hedwigia* **79**: 281 (1984).
Halosphaeriaceae E. Müll. & Arx ex Kohlm., *Can. J. Bot.* **50**: 1951 (1972).
Hansenulaceae E.K. Novák & Zsolt, *Acta bot. hung.* **7**: 98 (1961) = **Saccharomycetaceae**.
Hapalopilaceae Jülich, *Biblthca Mycol.* **85**: 370 (1982).
Haplographiaceae (Sacc.) E. Castell. & Chalm., *Man. trop. med.*. 3rd Edn (London): 1070 (1919) = **Hyaloscyphaceae**.
Haplomycetes Fr., *Summa veg. Scand.* (Stockholm) **2**: 485 (1849) = **Mucoraceae**.
Haploporaceae Jülich, *Biblthca Mycol.* **85**: 370 (1982) = **Polyporaceae**.
Harpellaceae L. Léger & Duboscq ex P.M. Kirk & P.F. Cannon, in Cannon & Kirk, *Fungal Families of the World* **(Wallingford)**: 152 (2007).
Harpidiaceae Vězda ex Hafellner, *Beih. Nova Hedwigia* **79**: 283 (1984) = **Lichinaceae**.
Harpochytriaceae Wille, *Petermanns Mitt.* Ergänz. no. **131**: 371 (1900).
Helicocephalidaceae Boedijn, *Sydowia* **12**: 355 (1959).
Helicomycetaceae Nann., *Repert. mic. uomo*: 473 (1934) = **Tubeufiaceae**.
Helicosporiaceae Nann., *Repert. mic. uomo*: 507 (1934) = **Tubeufiaceae**.
Helminthosphaeriaceae Samuels, Cand. & Magni, *Mycologia* **89**(1): 144 (1997).
Helminthosporiaceae Corda, *Icon. fung.* (Prague) **1**: 12 (1837) = **Pleomassariaceae**.
Helocarpaceae Hafellner, *Beih. Nova Hedwigia* **79**: 285 (1984) = **Micareaceae**.
Helotiaceae Rehm, *Rabenh. Krypt.-Fl.* (Leipzig) **1**(3): 647 (1886).
Helvellaceae Fr., *Syst. mycol.* (Lundae) **2**: 1 (1823).
Hemiascosporiaceae L.R. Batra, *Mycologia* **65**: 795 (1973) = **Ascodesmidaceae**.
Hemigasteraceae Gäum. & C.W. Dodge, *Comp. Morph. Fungi* (London): 466 (1928).
Hemihysteriaceae (Speg.) Sacc. & Traverso, *Annls mycol.* **5**: 318 (1907) = **Asterinaceae**.
Hemiphacidiaceae Korf, *Mycologia* **54**: 26 (1962).
Heppiaceae Zahlbr., *Nat. Pflanzenfam.* (Leipzig) **1**(1*): 176 (1906).
Hericiaceae Donk, *Persoonia* **3**: 269 (1964).
Herpomycetaceae (Thaxt.) I.I. Tav., *Mycotaxon* **13**: 469 (1981).
Herpothallaceae Tomas. ex Tomas., *Archo bot. Sist. Fito-geogr. Genet.* **26**: 103 (1950) = **Arthoniaceae**.
Herpotrichiellaceae Munk, *Dansk bot. Ark.* **15**(2): 131 (1953).
Heteroacanthellaceae P. Roberts, *Mycologist* **12**(4): 147 (1998) = **Ceratobasidiaceae**.
Heterobasidiaceae Jülich, *Biblthca Mycol.* **85**: 371 (1982) = **Bondarzewiaceae**.
Heterodeaceae Filson, *Lichenologist* **10**: 15 (1978).
Heterogastridiaceae Oberw. & R. Bauer, *Mycologia* **82**(1): 57 (1990).

Heteroscyphaceae Jülich, *Persoonia* **10**: 336 (1979) = **Exidiaceae**.
Heterosphaeriaceae Rehm, *Rabenh. Krypt.-Fl.* (Leipzig) **1**(3): 198 (1888) = **Helotiaceae**.
Hispidicarpomycetaceae Nakagiri, *Mycologia* **85**(4): 649 (1993).
Histoplasmataceae Redaelli & Cif., *Boll. Sez. ital. Soc. int. Microbiol.* **6**(19): 377 (1934) = **Onygenaceae**.
Hoehnelomycetaceae Jülich, *Biblthca Mycol.* **85**: 371 (1982).
Hormodendraceae Nann., *Repert. mic. uomo*: 399 (1934) = **Mycosphaerellaceae**.
Humariaceae Velen., *Monogr. Discom. Bohem.* (Prague): 319 (1934) = **Pyronemataceae**.
Hyaloriaceae Lindau, *Nat. Pflanzenfam.* (Leipzig) **1**(1**): 95 (1897).
Hyaloscyphaceae Nannf., *Nova Acta R. Soc. Scient. upsal.* Ser. 4 **8**(2): 258 (1932).
Hybogasteraceae Jülich, *Biblthca Mycol.* **85**: 372 (1982).
Hydnaceae Chevall., *Fl. gén. env. Paris* (Paris): 270 (1826).
Hydnangiaceae Gäum. & C.W. Dodge, *Comp. Morph. Fungi* (London): 485 (1928).
Hydnodontaceae Jülich, *Biblthca Mycol.* **85**: 372 (1982) = **Thelephoraceae**.
Hydnotryaceae M. Lange, *Dansk bot. Ark.* **16**: 27 (1956) = **Discinaceae**.
Hygrophoraceae Lotsy, *Vortr. bot. Stammesgesch.*: 706 (1907).
Hygrophoropsidaceae Kühner, *Bull. mens. Soc. linn. Lyon* **49**(num. spec.): 900 (1980).
Hymenangiaceae Corda, *Icon. fung.* (Prague) **5**: 28 (1842) = **Hymenogasteraceae**.
Hymeneliaceae Körb., *Syst. lich. germ.* (Breslau): 327 (1855).
Hymenochaetaceae Imazeki & Toki, *Bull. Govt Forest Exp. Stn Meguro* **67**: 24 (1954).
Hymenogasteraceae Vittad., *Monogr. Tuberac.* (Milano): 11 (1831).
Hymenogrammaceae Jülich, *Biblthca Mycol.* **85**: 372 (1982) = **Grammotheleaceae**.
Hymenomycetes Fr., *Epicr. syst. mycol.* (Upsaliae): v (1836) = **Agaricaceae**.
Hyphodermataceae Jülich, *Biblthca Mycol.* **85**: 373 (1982).
Hypochnaceae J. Schröt., *Krypt.-Fl. Schlesien* (Breslau) **3**(1): 415 (1888) = **Arthoniaceae**.
Hypochnellaceae Jülich, *Biblthca Mycol.* **85**: 373 (1982) = **Atheliaceae**.
Hypocreaceae De Not., *G. bot. ital.* n.s. **2**: 48 (1844).
Hypodermataceae Rehm, *Rabenh. Krypt.-Fl.* (Leipzig) **1**(3): 28 (1887) = **Rhytismataceae**.
Hypogaeaceae E. Horak, *Sydowia* **17**: 299 (1964) = **Agaricaceae**.
Hypogymniaceae Poelt ex Elix, *Brunonia* **2**: 176 (1980) = **Parmeliaceae**.
Hypomycetaceae (Lindau) Earle, *Contr. U.S. natnl. Herb.* **6**: 168 (1901) = **Hypocreaceae**.
Hyponectriaceae Petr., *Annls mycol.* **21**: 305 (1923).
Hypoxylaceae DC., *Fl. Franç..* Edn 3 (Paris) **2**: 280 (1805) = **Xylariaceae**.
Hypsostromataceae Huhndorf, *Mycologia* **86**(2): 266 (1994).
Hysterangiaceae E. Fisch., *Nat. Pflanzenfam.* (Leipzig) **1**(1**): 304 (1899).
Hysteriaceae Chevall., *Fl. gén. env. Paris* (Paris): 432 (1826).
Icmadophilaceae Triebel, *Biblthca Lichenol.* **53**: 227 (1993).
Imbricariaceae Chevall., *Fl. gén. env. Paris* (Paris): 618 (1826) = **Physciaceae**.
Incrustoporiaceae Jülich, *Biblthca Mycol.* **85**: 373 (1982) = **Polyporaceae**.
Inocybaceae Jülich, *Biblthca Mycol.* **85**: 374 (1982).
Inonotaceae Fiasson & Niemelä, *Karstenia* **24**(1): 23 (1984) = **Hymenochaetaceae**.
Iodophanaceae Prokhorov, *Mikol. Fitopatol.* **27**(2): 25 (1993) = **Pezizaceae**.
Irpicaceae Spirin & Zmitr., *Mycena* **3**: 48 (2003) = **Steccherinaceae**.
Ischnodermataceae Jülich, *Biblthca Mycol.* **85**: 374 (1982) = **Hapalopilaceae**.
Isidiaceae Rchb., *Deut. Bot. Herb.-Buch*: 14 (1841) = **Pertusariaceae**.
Ixechinaceae (R. Heim) Guzmán, *Boln. Soc. mex. Micol.* **8**: 59 (1974) = **Boletaceae**.
Karstenellaceae Harmaja, *Karstenia* **14**: 110 (1974).
Kathistaceae Malloch & M. Blackw., *Can. J. Bot.* **68**(8): 1719 (1990).
Kickxellaceae Linder, *Farlowia* **1**: 56 (1943).
Koerberiellaceae Hafellner, *Beih. Nova Hedwigia* **79**: 286 (1984) = **Porpidiaceae**.
Koralionastetaceae Kohlm. & Volkm.-Kohlm., *Mycologia* **79**(5): 764 (1987).
Kriegeriellaceae M.E. Barr, *Mycotaxon* **29**: 502 (1987) = **Pleosporaceae**.
Krieglsteineraceae Pouzar, *Beitr. Kenntn. Pilze Mitteleur.* **3**: 404 (1987) = **Platygloeaceae**.
Laboulbeniaceae G. Winter, *Rabenh. Krypt.-Fl.* (Leipzig) **1**(2): 918 (1886).
Laboulbeniaceae heterothallicae S. Colla, *Fl. ital. crypt.* (Florence) **1**(6): 55 (1934) = **Laboulbeniaceae**.
Lachneaceae Velen., *Monogr. Discom. Bohem.* (Prague): 300 (1934) = **Pyronemataceae**.
Lachnocladiaceae D.A. Reid, *Beih. Nova Hedwigia* **18**: 45 (1965).

Lactariaceae Gäum., *Vergl. Morph. Pilze* (Jena): 529 (1926) = **Russulaceae**.
Laetiporaceae Jülich, *Biblthca Mycol.* **85**: 375 (1982) = **Polyporaceae**.
Lahmiaceae O.E. Erikss., *Mycotaxon* **27**: 357 (1986).
Laricifomitaceae Jülich, *Biblthca Mycol.* **85**: 375 (1982) = **Fomitopsidaceae**.
Lasiolomataceae Hafellner, *Beih. Nova Hedwigia* **79**: 287 (1984) = **Ectolechiaceae**.
Lasiosphaeriaceae Nannf., *Nova Acta R. Soc. Scient. upsal.* Ser. 4 **8**(2): 50 (1932).
Lautosporaceae Kohlm., Volkm.-Kohlm. & O.E. Erikss., *Bot. Mar.* **38**(2): 169 (1995).
Lecanactidaceae Stizenb., *Ber. Tät. St Gall. naturw. Ges.*: 155 (1862) = **Graphidaceae**.
Lecaniaceae Walt. Watson, *New Phytol.* **28**: 23 (1929) = **Bacidiaceae**.
Lecanidiaceae O.E. Erikss., *Op. bot. Soc. bot. Lund* **60**: 78 (1981) = **Patellariaceae**.
Lecanoraceae Körb., *Syst. lich. germ.* (Breslau): 104 (1855).
Lecideaceae Chevall., *Fl. gén. env. Paris* (Paris): 549 (1826).
Lecidomataceae Hafellner, *Beih. Nova Hedwigia* **79**: 299 (1984) = **Psoraceae**.
Legeriomycetaceae Pouzar, *Folia geobot. phytotax.* **7**: 319 (1972).
Lentariaceae Jülich, *Biblthca Mycol.* **85**: 375 (1982) = **Gomphaceae**.
Lenzitopsidaceae Jülich, *Biblthca Mycol.* **85**: 376 (1982) = **Thelephoraceae**.
Leotiaceae Corda, *Icon. fung.* (Prague) **5**: 37 (1842).
Leptogiaceae Körb., *Syst. lich. germ.* (Breslau): 416 (1855) = **Collemataceae**.
Leptopeltidaceae Höhn. ex Trotter, *Syll. fung.* (Abellini) **24**: 1255 (1928).
Leptosphaeriaceae M.E. Barr, *Mycotaxon* **29**: 503 (1987).
Leptostromataceae Sacc., *Syll. fung.* (Abellini) **3**: 625 (1884) = **Rhytismataceae**.
Letrouitiaceae Hafellner & Bellem., *Nova Hedwigia* **35**: 382 (1982) ['1981'].
Leucocoprinaceae (Singer) Jülich, *Biblthca Mycol.* **85**: 376 (1982) = **Agaricaceae**.
Leucogastraceae Moreau ex Fogel, *Can. J. Bot.* **57**: 1723 (1979).
Leucosporidiaceae Jülich, *Biblthca Mycol.* **85**: 377 (1982).
Leveillulaceae V.P. Gelyuta, *Biol. Zh. Armenii* **41**(5): 357 (1988) = **Erysiphaceae**.
Lichenotheliaceae Henssen, *Syst. Ascom.* **5**: 137 (1986).
Lichinaceae Nyl., *Mém. Soc. natn. Sci. nat. math. Cherbourg* **2**: 8 (1854).
Licrostromataceae Jülich, *Biblthca Mycol.* **85**: 377 (1982) = **Corticiaceae**.
Limnoperdaceae G.A. Escobar, *Mycologia* **68**: 878 (1976) = **Cyphellaceae**.
Lindtneriaceae Jülich, *Biblthca Mycol.* **85**: 377 (1982) = **Corticiaceae**.
Lipomycetaceae E.K. Novák & Zsolt, *Acta bot. hung.* **7**: 97 (1961).
Litschauerellaceae Jülich, *Biblthca Mycol.* **85**: 377 (1982) = **Tubulicrinaceae**.
Lobariaceae Chevall., *Fl. gén. env. Paris* (Paris): 609 (1826).
Lobiolataceae C. Agardh, *Aphor. bot.* (Lund): 92 (1821) = **Peltigeraceae**.
Lopadiaceae Hafellner, *Beih. Nova Hedwigia* **79**: 300 (1984) = **Ectolechiaceae**.
Lophiostomataceae Sacc., *Syll. fung.* (Abellini) **2**: 672 (1883).
Lophotrichaceae Seth, *Nova Hedwigia* **19**: 592 (1971) = **Microascaceae**.
Loramycetaceae Dennis ex Digby & Goos, *Mycologia* **79**: 829 (1988) ['1987'].
Loxosporaceae Kalb & Staiger, *Biblthca Lichenol.* **59**: 13 (1995).
Lulworthiaceae Kohlm., Spatafora & Volkm.-Kohlm., *Mycologia* **30-32**: 456 (2000).
Lycoperdaceae Chevall., *Fl. gén. env. Paris* (Paris): 348 (1826).
Lysuraceae Corda, *Icon. fung.* (Prague) **5**: 28 (1842) = **Phallaceae**.
Macrosporiaceae (Lindau) Nann., *Repert. mic. uomo*: 503 (1934) = **Pleosporaceae**.
Magnaporthaceae P.F. Cannon, *Syst. Ascom.* **13**(1): 26 (1994).
Marasmiaceae Roze ex Kühner, *Bull. mens. Soc. linn. Lyon* **49**: 76 (1980).
Massalongiaceae Wedin, P.M. Jørg. & Wiklund, *Lichenologist* **39**(1): 63 (2007).
Massariaceae Nitschke, *Verh. naturh. Ver. preuss. Rheinl.* **26**: 73 (1869).
Massarinaceae Munk, *Friesia* **5**: 305 (1956) = **Lophiostomataceae**.
Massariovalsaceae Hara, *Bot. Mag. Tokyo* **27**: 474 (1913) = **Melanconidaceae**.
Mastodiaceae Zahlbr., *Nat. Pflanzenfam.* (Leipzig) **1**(1*): 240 (1907).
Medeolariaceae Korf, *Mycotaxon* **15**: 231 (1982).
Megalariaceae Hafellner, *Beih. Nova Hedwigia* **79**: 302 (1984).
Megalosporaceae Vězda ex Hafellner & Bellem., *Nova Hedwigia* **35**: 216 (1982) ['1981'].
Megasporaceae Lumbsch, *J. Hattori bot. Lab.* **75**: 302 (1994).
Meiorganaceae R. Heim ex Jülich, *Biblthca Mycol.* **85**: 378 (1982) = **Coniophoraceae**.

Melampsoraceae Dietel, *Nat. Pflanzenfam.* (Leipzig) **1**(1**): 38 (1897).
Melanconiaceae Corda, *Icon. fung.* (Prague) **5**: 33 (1842) = **Melanconidaceae**.
Melanconidaceae G. Winter, *Rabenh. Krypt.-Fl.* (Leipzig) **1**(2): 764 (1886).
Melaniellaceae R. Bauer, Vánky, Begerow & Oberw., *Mycologia* **4**(no. 3): 482 (1999).
Melanogastraceae E. Fisch., *Nat. Pflanzenfam.*. Edn 2 (Leipzig) **7**(a): 9 (1933).
Melanommataceae G. Winter, *Rabenh. Krypt.-Fl.* (Leipzig) **1**(2): 220 (1885).
Melanopsichiaceae Vánky, *Fungal Diversity* **95**(3): 140 (2001).
Melanotaeniaceae Begerow, R. Bauer & Oberw., *Can. J. Bot.* **75**(12): 2053 (1998) ['1997'].
Melaspileaceae Walt. Watson, *New Phytol.* **28**: 94 (1929).
Meliolaceae G.W. Martin ex Hansf., *Mycol. Pap.* **15**: 23 (1946).
Meliolinaceae S. Hughes, *Mycol. Pap.* **166**: 176 (1993).
Melogrammataceae G. Winter, *Rabenh. Krypt.-Fl.* (Leipzig) **1**(2): 797 (1886) = **Melanconidaceae**.
Meripilaceae Jülich, *Biblthca Mycol.* **85**: 378 (1982).
Meristacraceae Humber, *Mycotaxon* **34**(2): 456 (1989).
Merolpidiaceae A. Fisch., *Rabenh. Krypt.-Fl.* (Leipzig) **1**(4): 45 (1892) = **Synchytriaceae**.
Meruliaceae P. Karst., *Revue mycol.* Toulouse **3**(9): 19 (1881).
Mesnieraceae Arx & E. Müll., *Stud. Mycol.* **9**: 94 (1975).
Mesophelliaceae (G. Cunn.) Jülich, *Biblthca Mycol.* **85**: 379 (1982).
Metacapnodiaceae S. Hughes & Corlett, *N.Z. Jl Bot.* **10**: 239 (1972).
Metschnikowiaceae T. Kamieński, *Trudy S. Petersb. Obschch. Est. Otd. Bot.* **30**: 345 (1899).
Micareaceae Vězda ex Hafellner, *Beih. Nova Hedwigia* **79**: 306 (1984).
Microascaceae Luttr. ex Malloch, *Mycologia* **62**: 734 (1970).
Microbotryaceae R.T. Moore, *Mycotaxon* **59**: 17 (1996).
Microcaliciaceae Tibell, *Beih. Nova Hedwigia* **79**: 701 (1984).
Microglaenaceae Servít, *Československ. Lišejn. Čeledi Verrucariaceae*: 17 (1954) = **Thelenellaceae**.
Micromycopsidaceae Subram., *Curr. Sci.* **43**: 723 (1974) = **Synchytriaceae**.
Micropeltidaceae Clem. & Shear, *Gen. fung.*. Edn 2 (Minneapolis): 100 (1931).
Microporaceae Jülich, *Biblthca Mycol.* **85**: 379 (1982) = **Polyporaceae**.
Microstromataceae Jülich, *Biblthca Mycol.* **85**: 379 (1982).
Microtheliopsidaceae O.E. Erikss., *Op. bot. Soc. bot. Lund* **60**: 97 (1981).
Microthyriaceae Sacc., *Syll. fung.* (Abellini) **2**: 658 (1883).
Mikronegeriaceae Cummins & Y. Hirats., *Illustr. Gen. Rust Fungi. rev. edit.* (St. Paul): 13 (1983).
Miltideaceae Hafellner, *Beih. Nova Hedwigia* **79**: 308 (1984).
Mitrulaceae Rchb., *Consp. Regni Veget.* (Leipzig): 13 (1828) = **Sclerotiniaceae**.
Mixiaceae C.L. Kramer, *Stud. Mycol.* **30**: 159 (1987).
Mollisiaceae Rehm, *Rabenh. Krypt.-Fl.* (Leipzig) **1**(3): 503 (1891) = **Dermateaceae**.
Monascaceae J. Schröt., *Nat. Pflanzenfam.* (Leipzig) **1**(1): 148 (1894).
Moniliaceae Dumort., *Comment. bot.* (Tournay): 96 (1822) = **Sclerotiniaceae**.
Monoblastiaceae Walt. Watson, *New Phytol.* **28**: 106 (1929).
Monoblepharidaceae A. Fisch., *Rabenh. Krypt.-Fl.* (Leipzig) **1**(A): 378 (1892).
Monolpidiaceae A. Fisch., *Rabenh. Krypt.-Fl.* (Leipzig) **1**(4): 20 (1892) = **Olpidiaceae**.
Monotosporaceae Vuill., *Bull. Soc. mycol. Fr.* **28**: 118 (1912) = **Hysteriaceae**.
Montagneaceae Singer, *Revue mycol.* Toulouse **40**: 61 (1976) = **Agaricaceae**.
Morchellaceae Rchb., *Pflanzenreich* (Leipzig): 2 (1834).
Moriolaceae Zahlbr., *Nat. Pflanzenfam.* (Leipzig) **1**(1*): 52 (1903).
Mortierellaceae A. Fisch., *Rabenh. Krypt.-Fl.* (Leipzig) **1**(4): 268 (1892).
Mucoraceae Dumort., *Comment. bot.* (Tournay): 69 (1822).
Munkiellaceae (Theiss. & P. Syd.) Luttr., *The Fungi* (London) **IV A**: 154 (1973) = **Polystomellaceae**.
Muribasidiosporaceae Kamat & Rajendren, *Mycologia* **61**: 1159 (1969) = **Exobasidiaceae**.
Mycenaceae Roze, *Bull. Soc. bot. Fr. Act. bot.* **23**: 109 (1876).
Mycenastraceae Zeller, *Mycologia* **40**: 648 (1948).
Mycobilimbiaceae Hafellner, *Beih. Nova Hedwigia* **79**: 308 (1984) = **Porpidiaceae**.
Mycoblastaceae Hafellner, *Beih. Nova Hedwigia* **79**: 310 (1984).
Mycoboniaceae Jülich, *Biblthca Mycol.* **85**: 380 (1982) = **Boreostereaceae**.
Mycocaliciaceae Alf. Schmidt, *Mitt. Staatsinst. Allg. Bot. Hamburg* **13**: 127 (1970).
Mycogalopsidaceae Gjurašin, *Acta bot. Inst. bot., Zagreb* **1**: 12 (1925) = **Pyronemataceae**.

Mycoporaceae Zahlbr., *Nat. Pflanzenfam.* (Leipzig) **1**(1*): 77 (1903).
Mycorrhaphiaceae Jülich, *Biblthca Mycol.* **85**: 380 (1982) = **Steccherinaceae**.
Mycosphaerellaceae Lindau, *Nat. Pflanzenfam.* (Leipzig) **1**(1): 421 (1897).
Mycosyringaceae R. Bauer & Oberw., *Can. J. Bot.* **75**(8): 1312 (1997).
Mycotyphaceae Benny & R.K. Benj., *Mycotaxon* **22**: 122 (1985).
Myeloconidiaceae P.M. McCarthy, *Flora of Australia* (Melbourne) **58**(A): 227 (2001).
Myelospermataceae K.D. Hyde & S.W. Wong, *Mycol. Res.* **44**(1): 349 (1999).
Myriangiaceae Nyl., *Mém. Soc. natn. Sci. nat. math. Cherbourg* **2**: 9 (1854).
Mytilinidiaceae Kirschst., *Verh. bot. Ver. Prov. Brandenb.* **66**: 28 (1924).
Myxariaceae Jülich, *Biblthca Mycol.* **85**: 380 (1982) = **Exidiaceae**.
Myxomyriangiaceae (Theiss.) Theiss., *Annls mycol.* **15**: 438 (1917) = **Elsinoaceae**.
Myxonemei Bonord., *Handb. Allgem. mykol.* (Stuttgart): 148 (1851) = **Pucciniaceae**.
Myxophacidiaceae Petr., *Annls mycol.* **20**: 302 (1922) = **Ascodichaenaceae**.
Myxothyriaceae Höhn., *Sber. Akad. Wiss. Wien Math.-naturw. Kl.*, Abt. 1 **118**: 1168 (1909) = **Asterinaceae**.
Myxotrichaceae Locq. ex Currah, *Mycotaxon* **24**: 103 (1985).
Myxotrichellaceae (Ferraris) Nann., *Repert. mic. uomo*: 496 (1934) = **Myxotrichaceae**.
Naetrocymbaceae Höhn. ex R.C. Harris, *More Florida Lichens. Incl. 10 Cent Tour Pyrenol.* (New York): 59 (1995).
Nectaromycetaceae Cif. & Redaelli, *Annls mycol.* **27**: 251 (1929) = **Metschnikowiaceae**.
Nectriaceae Tul. & C. Tul., *Select. fung. carpol.* (Paris) **3**: 3 (1865).
Nemariaceae Navas, *Broteria* ser. bot. **9**: 74 (1910) = **Roccellaceae**.
Nematosporaceae E.K. Novák & Zsolt, *Acta bot. hung.* **7**: 99 (1961) = **Eremotheciaceae**.
Neocallimastigaceae I.B. Heath, *Can. J. Bot.* **61**: 304 (1983).
Neolectaceae Redhead, *Can. J. Bot.* **55**: 305 (1977).
Neozygitaceae Ben Ze'ev, R.G. Kenneth & Uziel, *Mycotaxon* **28**(2): 321 (1987).
Nephromataceae Wetmore ex J.C. David & D. Hawksw., *Syst. Ascom.* **10**(1): 15 (1991).
Nesolechiaceae Arnold, *Flora Jena* **41**: 701 (1858) = **Parmeliaceae**.
Niaceae Jülich, *Biblthca Mycol.* **85**: 381 (1982).
Nidulariaceae Dumort., *Comment. bot.* (Tournay): 69 (1822).
Niessliaceae Kirschst., *Annls mycol.* **37**: 89 (1939).
Nigrofomitaceae Jülich, *Biblthca Mycol.* **85**: 381 (1982) = **Polyporaceae**.
Nitschkiaceae (Fitzp.) Nannf., *Nova Acta R. Soc. Scient. upsal.* Ser. 4 **8**(2): 56 (1932).
Obryzaceae Körb., *Syst. lich. germ.* (Breslau): 427 (1855).
Ochrolechiaceae R.C. Harris ex Lumbsch & I. Schmitt, *J. Hattori bot. Lab.* **100**: 760 (2006).
Octavianiaceae Locq. ex Pegler & T.W.K. Young, *Trans. Br. mycol. Soc.* **72**: 379 (1979).
Odontotremataceae D. Hawksw. & Sherwood, *Mycotaxon* **16**: 263 (1982).
Oedogoniomycetaceae D.J.S. Barr, *Handbook of Protoctista* (Boston): 463 (1990).
Oidiaceae Link, *Abh. dt. Akad. Wiss. Berlin*: 2 (1826) ['1824'] = **Erysiphaceae**.
Oliveoniaceae P. Roberts, *Folia cryptog. Estonica* **33**: 128 (1998).
Olpidiaceae J. Schröt., *Krypt.-Fl. Schlesien* (Breslau) **3**(1): 180 (1889).
Omphalariaceae Körb., *Syst. lich. germ.* (Breslau): 422 (1855) = **Lichinaceae**.
Omphalotaceae Bresinsky, *Pl. Syst. Evol.* **150**(1-2): 113 (1985).
Onygenaceae Berk., *Intr. crypt. bot.* (London): 272 (1857).
Opegraphaceae Stizenb., *Ber. Tät. St Gall. naturw. Ges.*: 153 (1862) = **Roccellaceae**.
Ophioparmaceae R.W. Rogers & Hafellner, *Lichenologist* **20**(2): 172 (1988).
Ophiostomataceae Nannf., *Nova Acta R. Soc. Scient. upsal.* Ser. 4 **8**(2): 30 (1932).
Oplotheciaceae Bat. & Cif., *Saccardoa* **2**: 242 (1963) = **Trichosphaeriaceae**.
Orbiliaceae Nannf., *Nova Acta R. Soc. Scient. upsal.* Ser. 4 **8**(2): 250 (1932).
Orphniosporaceae Bellem. & Hafellner, *Beih. Nova Hedwigia* **79**: 312 (1984) = **Fuscideaceae**.
Ostracodermataceae Malençon, *Bull. Soc. mycol. Fr.* **80**: 208 (1964) = **Pezizaceae**.
Ostropaceae Rehm, *Rabenh. Krypt.-Fl.* (Leipzig) **1**(3): 185 (1888) = **Stictidaceae**.
Otideaceae Eckblad, *Nytt Mag. Bot.* **15**: 82 (1968) = **Pyronemataceae**.
Pachnocybaceae Oberw. & R. Bauer, *Sydowia* **41**: 244 (1989).
Pachyascaceae Poelt ex P.M. Kirk, P.F. Cannon & J.C. David, *Ainsworth & Bisby's Dictionary of the Fungi.* Edn 9 (Wallingford): ix (2001).
Pachykytosporaceae Jülich, *Biblthca Mycol.* **85**: 382 (1982) = **Polyporaceae**.

Pacisporaceae C. Walker, Błaszk., A. Schüssler & Schwarzott, *Mycol. Res.* **108**(9): 981 (2004).
Pannariaceae Tuck., *Gen. lich.* (Amherst): xii (1872).
Papulosaceae Winka & O. Erikss., *Mycoscience* **13**(1): 102 (2000).
Papyrodiscaceae Jülich, *Biblthca Mycol.* **85**: 382 (1982) = **Corticiaceae**.
Paracoccidioidaceae Redaelli & Cif., *Memorie R. Accad. Ital.* **8**(12): 595 (1937) = **Ajellomycetaceae**.
Paraglomeraceae Morton & Redecker, *Mycologia* **93**(1): 188 (2001).
Paraphelariaceae Jülich, *Biblthca Mycol.* **85**: 383 (1982) = **Auriculariaceae**.
Paratheliaceae Zahlbr., *Nat. Pflanzenfam.* (Leipzig) **1**(1*): 71 (1903) = **Verrucariaceae**.
Parmeliaceae Zenker, *Pharmaceutische Waarenkunde* (Eisenach): 124 (1827).
Parmulariaceae E. Müll. & Arx ex M.E. Barr, *Mycologia* **71**: 944 (1979).
Parodiellaceae Theiss. & Syd. ex M.E. Barr, *Mycotaxon* **29**: 503 (1987).
Parodiopsidaceae Toro, *J. Agric. Univ. P. Rico* **36**: 66 (1952).
Patellariaceae Corda, *Icon. fung.* (Prague) **2**: 37 (1838).
Patouillardinaceae Jülich, *Biblthca Mycol.* **85**: 383 (1982) = **Exidiaceae**.
Paxillaceae Lotsy, *Vortr. bot. Stammesgesch.* **1**: 706 (1907).
Peltideaceae Körb., *Lichenogr. germ.* (Breslau): 4 (1846) = **Peltigeraceae**.
Peltigeraceae Dumort., *Comment. bot.* (Tournay): 68 (1822).
Peltulaceae Büdel, *Syst. Ascom.* **5**: 149 (1986).
Peniophoraceae Lotsy, *Vortr. bot. Stammesgesch.* **1**: 687 (1907).
Perenniporiaceae Jülich, *Biblthca Mycol.* **85**: 383 (1982) = **Polyporaceae**.
Pertusariaceae Körb. ex Körb., *Syst. lich. germ.* (Breslau): 377 (1855).
Peyritschiellaceae Thaxt., *Mem. Am. Acad. Arts Sci.* **13**: 236 (1908) = **Laboulbeniaceae**.
Pezizaceae Dumort., *Anal. fam. pl.* (Tournay): 72 (1829).
Pezizellaceae Velen., *Monogr. Discom. Bohem.* (Prague): 154 (1934) = **Helotiaceae**.
Phacidiaceae Fr., *Summa veg. Scand.* (Stockholm) **2**: 367 (1849).
Phaeochoraceae K.D. Hyde, P.F. Cannon & M.E. Barr, *Syst. Ascom.* **15**(1-2): 118 (1997).
Phaeolaceae Jülich, *Biblthca Mycol.* **85**: 384 (1982) = **Polyporaceae**.
Phaeosphaeriaceae M.E. Barr, *Mycologia* **71**: 948 (1979).
Phaeotrichaceae Cain, *Can. J. Bot.* **34**: 676 (1956).
Phaffomycetaceae Y. Yamada, H. Kawas., Nagats., Mikata & T. Seki, *Biosc., Biotechn., Biochem.* **67**(2): 831 (1999).
Phakopsoraceae (Arthur) Cummins & Hirats. f., *Illustr. Gen. Rust Fungi. rev. edit.* (St. Paul): 13 (1983).
Phallaceae Corda, *Icon. fung.* (Prague) **5**: 29 (1842).
Phaneromycetaceae Gamundí & Spinedi, *Sydowia* **38**: 112 (1985).
Phelloriniaceae Ulbr., *Ber. bayer. bot. Ges.* **64**: 264 (1951).
Phillipsiellaceae Höhn., *Sber. Akad. Wiss. Wien Math.-naturw. Kl., Abt. 1* **118**: 361 (1909).
Phlebiaceae Jülich, *Biblthca Mycol.* **85**: 385 (1982) = **Meruliaceae**.
Phleogenaceae Gäum., *Vergl. Morph. Pilze* (Jena): 417 (1926).
Phlyctidaceae Poelt ex J.C. David & D. Hawksw., *Syst. Ascom.* **10**(1): 15 (1991).
Phragmidiaceae Corda, *Icon. fung.* (Prague) **1**: 6 (1837).
Phragmoxenidiaceae Oberw. & R. Bauer, *Syst. Appl. Microbiol.* **13**(2): 190 (1990).
Phycomycetaceae Arx, *Sydowia* **35**: 22 (1982).
Phylaciaceae Speer, *Bull. Soc. mycol. Fr.* **96**: 138 (1980) = **Xylariaceae**.
Phylacteriaceae Imazeki, *Mycologia* **42**: 558 (1953) = **Thelephoraceae**.
Phyllachoraceae Theiss. & Syd., *Annls mycol.* **13**: 168 (1915).
Phylliscaceae Th. Fr., *Nova Acta R. Soc. Scient. upsal.* Ser. 3 **3**: 288 (1860) = **Lichinaceae**.
Phyllobatheliaceae Bitter & F. Schill., *Hedwigia Beibl.* **67**: 272 (1927).
Phyllopsoraceae Zahlbr., *Nat. Pflanzenfam.* (Leipzig) **1**(1*): 138 (1905) = **Bacidiaceae**.
Phyllostictaceae Fr., *Summa veg. Scand.* (Stockholm) **2**: 420 (1849) = **Botryosphaeriaceae**.
Phymatosphaeriaceae Speg., *An. Soc. cient. argent.* **26**: 57 (1888) = **Myriangiaceae**.
Physalacriaceae Corner, *Beih. Nova Hedwigia* **33**: 10 (1970).
Physciaceae Zahlbr., *Syllabus.* Edn 2 (Berlin): 46 (1898).
Physmaceae Walt. Watson, *New Phytol.* **28**: 36 (1929) = **Collemataceae**.
Physodermataceae Sparrow, *Mycologia* **44**: 768 (1952).
Physosporellaceae Höhn. ex M.E. Barr, *Mycologia* **68**: 615 (1976) = **Hyponectriaceae**.
Piedraiaceae Viégas ex Cif., Bat. & S. Camposa{2}, *Publções Inst. Micol. Recife* **45**(1-6): 7 (1956) = **Piedraiaceae**.

Piedraiaceae Viégas ex M.E. Barr, *Mycologia* **77**: 939 (1979).
Pileolariaceae (Arthur) Cummins & Y. Hirats., *Illustr. Gen. Rust Fungi. rev. edit.* (St. Paul): 14 (1983).
Pilobolaceae Corda, *Anleit. Stud. Mykol.* Prag: 17 (1842).
Pilocarpaceae Zahlbr., *Nat. Pflanzenfam.* (Leipzig) **1**(1*): 116 (1905).
Pilodermataceae Jülich, *Biblthca Mycol.* **85**: 385 (1982) = **Atheliaceae**.
Pilophoraceae Stizenb., *Ber. Tät. St Gall. naturw. Ges.*: 166 (1862) = **Cladoniaceae**.
Piptocephalidaceae J. Schröt., *Krypt.-Fl. Schlesien* (Breslau) **3**(1): 215 (1886).
Piptoporaceae Jülich, *Biblthca Mycol.* **85**: 385 (1982) = **Fomitopsidaceae**.
Pisocarpiaceae Corda, *Icon. fung.* (Prague) **2**: 24 (1838) = **Pisolithaceae**.
Pisolithaceae Ulbr., *Krypt.-Fl. Anfäng.* (Berlin) **1**(3): 413 (1928).
Pithoascaceae Benny & Kimbr., *Mycotaxon* **12**: 45 (1980) = **Microascaceae**.
Placolecidaceae Hafellner, *Beih. Nova Hedwigia* **79**: 317 (1984) = **Catillariaceae**.
Placynthiaceae Å.E. Dahl, *Meddr Grønland Biosc.* **150**(2): 49 (1950).
Planistromellaceae M.E. Barr, *Mycotaxon* **60**: 433 (1996).
Planoroccellaceae Elenkin, *Izv. glav. bot. Sada SSSR* **28**(3-4): 267 (1929) = **Roccellaceae**.
Platygloeaceae Racib., *Bull. int. Acad. Sci. Lett. Cracovie* Cl. sci. math. nat. Sér. B, sci. nat. **3**: 346 (1909).
Platystomaceae J. Schröt., *Krypt.-Fl. Schlesien* (Breslau) **3**(2): 323 (1894) = **Lophiostomataceae**.
Plectodiscellaceae Woron., *Mykol. Zentbl.* **4**: 232 (1914) = **Elsinoaceae**.
Pleomassariaceae M.E. Barr, *Mycologia* **71**: 949 (1979).
Pleosporaceae Nitschke, *Verh. naturh. Ver. preuss. Rheinl.* **26**: 74 (1869).
Pleurostomataceae Réblová, L. Mostert, W. Gams & Crous, *Stud. Mycol.* **50**: 540 (2004).
Pleurotaceae Kühner, *Bull. mens. Soc. linn. Lyon* **49**: 184 (1980).
Pleurotremataceae Walt. Watson, *New Phytol.* **28**: 113 (1929) = **Pyrenulaceae**.
Plicariaceae Velen., *Monogr. Discom. Bohem.* (Prague): 342 (1934) = **Pezizaceae**.
Plicaturaceae Jülich, *Biblthca Mycol.* **85**: 386 (1982) = **Atheliaceae**.
Ploettnerulaceae Kirschst., *Verh. bot. Ver. Prov. Brandenb.* **66**: 27 (1924) = **Dermateaceae**.
Pluteaceae Kotl. & Pouzar, *Česká Mykol.* **26**: 218 (1972).
Pneumocystidaceae O.E. Erikss., *Syst. Ascom.* **13**(2): 170 (1994).
Podaxaceae Corda, *Icon. fung.* (Prague) **5**: 24 (1842) = **Agaricaceae**.
Podoscyphaceae D.A. Reid, *Beih. Nova Hedwigia* **18**: 43 (1965).
Polyactidaceae Corda, *Icon. fung.* (Prague) **2**: 14 (1838) = **Sclerotiniaceae**.
Polyporaceae Fr. ex Corda, *Icon. fung.* (Prague) **3**: 49 (1839).
Polystictaceae Rea, *Brit. basidiomyc.* (Cambridge): 10 (1922) = **Hymenochaetaceae**.
Polystigmataceae Höhn. ex Nannf., *Nova Acta R. Soc. Scient. upsal.* Ser. 4 **8**(2): 51 (1932) = **Phyllachoraceae**.
Polystomellaceae Theiss. & P. Syd., *Annls mycol.* **13**: 158 (1915).
Poriaceae Locq., *Bull. Jard. bot. État Brux.* **27**: 561 (1957) = **Polyporaceae**.
Porinaceae Walt. Watson, *New Phytol.* **28**: 109 (1929) = **Pertusariaceae**.
Porinaceae Rchb., *Consp. Regni Veget.* (Leipzig): 20 (1828).
Porocyphaceae Körb., *Syst. lich. germ.* (Breslau): 425 (1855) = **Lichinaceae**.
Porodisculaceae Jülich, *Biblthca Mycol.* **85**: 386 (1982) = **Polyporaceae**.
Porotheleaceae Murrill, *Mycologia* **8**: 56 (1916) = **Schizophyllaceae**.
Porpidiaceae Hertel & Hafellner, *Beih. Nova Hedwigia* **79**: 318 (1984).
Protogastraceae Zeller, *Ann. Mo. bot. Gdn* **21**: 235 (1934).
Protomycetaceae Gray, *Nat. Arr. Brit. Pl.* (London) **1**: 532 (1821).
Protophallaceae Zeller, *Mycologia* **31**: 22 (1939) = **Phallaceae**.
Protoscyphaceae Kutorga & D. Hawksw., *Syst. Ascom.* **15**(1-2): 70 (1997).
Protothelenellaceae Vězda, H. Mayrhofer & Poelt, *Herzogia* **7**(1-2): 26 (1985).
Psathyrellaceae (Kühner) Vilgalys, Moncalvo & Redhead, *Taxon* **50**(1): 226 (2001).
Pseudeurotiaceae Malloch & Cain, *Can. J. Bot.* **48**: 1815 (1970).
Pseudohiatulaceae (Singer) Grgur., *Australas. Mycol.* **19**(2): 73 (2000) = **Tricholomataceae**.
Pseudoperisporiaceae Toro, *Scient. Surv. P. Rico* **8**(1): 40 (1926).
Pseudophacidiaceae Rehm, *Rabenh. Krypt.-Fl.* (Leipzig) **1**(3): 60 (1887) = **Ascodichaenaceae**.
Pseudophysciaceae (Hue) Tomas., *Archo bot. Sist. Fito-geogr. Genet.* **25**: 238 (1949) = **Physciaceae**.
Pseudorhizinaceae Harmaja, *Karstenia* **14**: 14 (1974) = **Discinaceae**.
Pseudosphaeriaceae Höhn., *Sber. Akad. Wiss. Wien Math.-naturw. Kl.*, Abt. 1 **116**: 129 (1907) = **Pleosporaceae**.

Pseudovalsaceae M.E. Barr, *Mycol. Mem.* **7**: 151 (1978) = **Melanconidaceae**.
Psiloniaceae Corda, *Icon. fung.* (Prague) **1**: 16 (1837) = **Nectriaceae**.
Psoraceae Zahlbr., *Syllabus*. Edn 2 (Berlin): 44 (1898).
Psorulaceae Hafellner, *Beih. Nova Hedwigia* **79**: 326 (1984) = **Psoraceae**.
Pterulaceae Corner, *Beih. Nova Hedwigia* **33**: 10 (1970).
Pterygellaceae Jülich, *Biblthca Mycol.* **85**: 386 (1982) = **Cantharellaceae**.
Pucciniaceae Chevall., *Fl. gén. env. Paris* (Paris): 413 (1826).
Pucciniastraceae Gäum. ex Leppik, *Ann. bot. fenn.* **9**: 139 (1972).
Pucciniosiraceae (Dietel) Cummins & Y. Hirats., *Illustr. Gen. Rust Fungi. rev. edit.* (St. Paul): 15 (1983).
Pulverariaceae Schltdl., *Fl. berol.* (Berlin) **2**: 99 (1824) = **Chrysothricaceae**.
Punctulariaceae Donk, *Persoonia* **3**: 287 (1964) = **Corticiaceae**.
Pyrenidiaceae Zahlbr., *Syllabus*. Edn 2 (Berlin): 46 (1898) = **Dacampiaceae**.
Pyrenogastraceae Jülich, *Biblthca Mycol.* **85**: 387 (1982) = **Geastraceae**.
Pyrenomycetes Fr., *Summa veg. Scand.* (Stockholm) **2**: 375 (1849) = **Xylariaceae**.
Pyrenophoraceae M.E. Barr, *Mycologia* **71**: 948 (1979) = **Pleosporaceae**.
Pyrenopsidaceae Th. Fr., *Nova Acta R. Soc. Scient. upsal.* Ser. 3 **3**: 285 (1860) = **Lichinaceae**.
Pyrenothamniaceae Zahlbr., *Syllabus*. Edn 2 (Berlin): 46 (1898) = **Verrucariaceae**.
Pyrenothricaceae Zahlbr., *Nat. Pflanzenfam.*. Edn 2 (Leipzig) **8**: 91 (1926).
Pyrenulaceae Rabenh., *Krypt.-Fl. Sachsen* (Leipzig) **2**: 42 (1870).
Pyronemataceae Corda, *Anleit. Stud. Mykol.* Prag: 149 (1842).
Pyxidiophoraceae G.R.W. Arnold, *Z. Pilzk.* **37**: 191 (1971).
Quambalariaceae Z.W. Beer, Begerow & R. Bauer, *Stud. Mycol.* **55**: 295 (2006).
Radiomycetaceae Hesselt. & J.J. Ellis, *Mycologia* **66**: 91 (1974).
Radulaceae Gäum., *Vergl. Morph. Pilze* (Jena): 511 (1926) = **Valsaceae**.
Ramaleaceae Elenkin, *Izv. glav. bot. Sada SSSR* **28**(3-4): 267 (1929) = **Cladoniaceae**.
Ramalinaceae C. Agardh, *Aphor. bot.* (Lund): 93 (1821).
Ramariaceae Corner, *Beih. Nova Hedwigia* **33**: 11 (1970) = **Gomphaceae**.
Ramulariaceae (Sacc.) Nann., *Repert. mic. uomo*: 472 (1934) = **Mycosphaerellaceae**.
Raveneliaceae Leppik, *Ann. bot. fenn.* **9**: 139 (1972).
Repetobasidiaceae Jülich, *Biblthca Mycol.* **85**: 388 (1982) = **Sistotremataceae**.
Requienellaceae Boise, *Mycologia* **78**: 37 (1986).
Rhamphosporaceae R. Bauer & Oberw., *Can. J. Bot.* **75**(8): 1312 (1997).
Rhizidiaceae J. Schröt., *Krypt.-Fl. Schlesien* (Breslau) **3**(1): 189 (1886) = **Chytridiaceae**.
Rhizinaceae Bonord., *Handb. Allgem. mykol.* (Stuttgart): 200 (1851).
Rhizocarpaceae M. Choisy ex Hafellner, *Beih. Nova Hedwigia* **79**: 327 (1984).
Rhizomorphaceae Chevall., *Fl. gén. env. Paris* (Paris): 506 (1826) = **Marasmiaceae**.
Rhizophlyctaceae H.E. Petersen, *Bot. Tidsskr.* **29**: 357 (1909) = **Spizellomycetaceae**.
Rhizopogonaceae Gäum. & C.W. Dodge, *Comp. Morph. Fungi* (London): 468 (1928).
Rhizothyriaceae Tehon ex Bat. & Cif., *Mycopath. Mycol. appl.* **11**: 71 (1959) = **Helotiaceae**.
Rhodophyllaceae Singer, *Lilloa* **22**: 601 (1951) = **Entolomataceae**.
Rhopalosporaceae Hafellner, *Beih. Nova Hedwigia* **79**: 334 (1984) = **Fuscideaceae**.
Rhynchogastremataceae Oberw. & B. Metzler, *Syst. Appl. Microbiol.* **12**(3): 283 (1989).
Rhytismataceae Chevall., *Fl. gén. env. Paris* (Paris): 439 (1826).
Richoniellaceae Jülich, *Biblthca Mycol.* **85**: 388 (1982) = **Entolomataceae**.
Rigidoporaceae Jülich, *Biblthca Mycol.* **85**: 388 (1982) = **Meripilaceae**.
Rimulariaceae Hafellner, *Beih. Nova Hedwigia* **79**: 331 (1984) = **Agyriaceae**.
Roccellaceae Chevall., *Fl. gén. env. Paris* (Paris): 604 (1826).
Roccellinastraceae Hafellner, *Beih. Nova Hedwigia* **79**: 332 (1984) = **Micareaceae**.
Roesleriaceae Yao & Spooner, *Kew Bull.* **106**(7): 684 (1999).
Russulaceae Lotsy, *Vortr. bot. Stammesgesch.*: 708 (1907).
Ruthiaceae Bat., J.A. Lima & M.A. Tatlasse{?}, *Publções Inst. Micol. Recife* **251**: 5 (1962) = **Strigulaceae**.
Rutstroemiaceae Holst-Jensen, L.M. Kohn & T. Schumach., *Mycologia* **89**(6): 895 (1997).
Saccardiaceae Höhn., *Sber. Akad. Wiss. Wien Math.-naturw. Kl., Abt. 1* **118**: 369 (1909).
Saccharomycetaceae G. Winter, *Rabenh. Krypt.-Fl.* (Leipzig) **1**(1): 58 (1881).
Saccharomycodaceae Kudrjanzev, *System. Hefen* (Berlin): 270 (1960).
Saccharomycopsidaceae Arx & Van der Walt, *Stud. Mycol.* **30**: 174 (1987).

Saccoblastiaceae Jülich, *Biblthca Mycol.* **85**: 389 (1982) = **Platygloeaceae**.
Saccotheciaceae Bonord., *Abh. Mykol.* (Halle): 82 (1864) = **Dothioraceae**.
Sarcinellaceae (Ferraris) Nann., *Repert. mic. uomo*: 506 (1934) = **Englerulaceae**.
Sarcodontaceae Bondartsev & Singer ex Singer, *Beih. Nova Hedwigia* **77**: 20 (1983) = **Bankeraceae**.
Sarcoscyphaceae Le Gal ex Eckblad, *Nytt Mag. Bot.* **15**: 103 (1968).
Sarcosomataceae Kobayasi, *J. Jap. Bot.* **13**: 516 (1937).
Sarcostromellaceae Boedijn, *Persoonia* **1**: 16 (1959) = **Boliniaceae**.
Sarrameanaceae Hafellner, *Beih. Nova Hedwigia* **79**: 336 (1984) = **Fuscideaceae**.
Schadoniaceae Hafellner, *Beih. Nova Hedwigia* **79**: 337 (1984) = **Bacidiaceae**.
Schaereriaceae M. Choisy ex Hafellner, *Beih. Nova Hedwigia* **79**: 399 (1984) = **Agyriaceae**.
Schizophyllaceae Quél., *Fl. mycol. France* (Paris): 365 (1888).
Schizoporaceae Jülich, *Biblthca Mycol.* **85**: 389 (1982).
Schizosaccharomycetaceae Beij. ex Klöcker, *Lafar's Handb. Techn. Mykol.* Edn 2 **4**: 189 (1905).
Schizosporaceae Dietel, *Nat. Pflanzenfam.* (Leipzig) **1**(1**): 37 (1897) = **Pucciniosiraceae**.
Schizothyriaceae Höhn. ex Trotter, Sacc., D. Sacc. & Traverso, *Syll. fung.* (Abellini) **24**: 1254 (1928).
Sclerodermataceae Corda, *Icon. fung.* (Prague): 23 (1842).
Sclerophoraceae Körb., *Lichenogr. germ.* (Breslau): 4 (1846) = **Coniocybaceae**.
Sclerotiaceae Dumort., *Comment. bot.* (Tournay): 69 (1822) = **Typhulaceae**.
Sclerotiniaceae Whetzel, *Mycologia* **37**: 652 (1945).
Scoliciosporaceae Hafellner, *Beih. Nova Hedwigia* **79**: 340 (1984) = **Lecanoraceae**.
Scortechiniaceae Huhndorf, A.N. Mill. & F.A. Fernández, *Mycol. Res.* **108**(12): 1387 (2004) = **Nitschkiaceae**.
Scytinopogonaceae Jülich, *Biblthca Mycol.* **85**: 389 (1982) = **Clavariaceae**.
Sebacinaceae K. Wells & Oberw., *Mycologia* **74**: 329 (1982).
Secotiaceae Tul., *Annls Sci. Nat. Bot.*, sér. 3 **4**: 176 (1845) = **Agaricaceae**.
Sepedoniaceae Fr., *Syst. mycol.* (Lundae) **3**(2): 436 (1832) = **Hypocreaceae**.
Septobasidiaceae Racib., *Bull. int. Acad. Sci. Lett. Cracovie* Cl. sci. math. nat. Sér. B, sci. nat. **3**: 359 (1909).
Septocylindriaceae (Sacc.) Nann., *Repert. mic. uomo*: 471 (1934) = **Mycosphaerellaceae**.
Septoriaceae W.B. Cooke, *Revta Biol., Lisb.* **12**(12): 289 (1983) = **Mycosphaerellaceae**.
Serpulaceae Jarosch & Bresinsky, *Biblthca Mycol.* **191**: 90 (2001) = **Coniophoraceae**.
Seuratiaceae Vuill. ex M.E. Barr, *Mycotaxon* **29**: 501 (1987).
Sigmoideomycetaceae Benny, R.K. Benj. & P.M. Kirk, *Mycologia* **84**(5): 620 (1992).
Siphulaceae Rchb., *Handb. nat. Pfl.-Syst.* (Dresden): 132 (1837) = **Icmadophilaceae**.
Sirobasidiaceae Lindau, *Nat. Pflanzenfam.* (Leipzig): 89 (1897).
Sistotremataceae Jülich, *Biblthca Mycol.* **85**: 390 (1982).
Solorinellaceae Vězda & Poelt, *Phyton* Horn **30**(1): 48 (1990) = **Asterothyriaceae**.
Sordariaceae G. Winter, *Rabenh. Krypt.-Fl.* (Leipzig) **1**(2): 162 (1885).
Sorochytriaceae Dewel, *Can. J. Bot.* **63**(9): 1532 (1985).
Sparassidaceae Herter, *Krypt.-Fl. Brandenburg* (Leipzig) **6**: 167 (1910).
Sparsitubaceae Jülich, *Biblthca Mycol.* **85**: 390 (1982) = **Polyporaceae**.
Spathulosporaceae Kohlm., *Mycologia* **65**: 615 (1973).
Sphaerellaceae Nitschke, *Verh. naturh. Ver. preuss. Rheinl.* **26**: 74 (1869) = **Mycosphaerellaceae**.
Sphaeriaceae Fr., *Syst. orb. veg.* (Lund) **1**: 103 (1825) = **Xylariaceae**.
Sphaerioidaceae Sacc., *Syll. fung.* (Abellini) **3**: 1 (1884) = **Botryosphaeriaceae**.
Sphaerobolaceae J. Schröt., *Krypt.-Fl. Schlesien* (Breslau) **3**(1): 688 (1889) = **Geastraceae**.
Sphaerophoraceae Fr., *Lich. eur. reform.* (Lund): 7 (1831).
Sphaerophoropsidaceae Elenkin, *Izv. glav. bot. Sada SSSR* **28**(3-4): 267 (1929) = **Cladoniaceae**.
Sphaerophragmiaceae Cummins & Y. Hirats., *Illustr. Gen. Rust Fungi. rev. edit.* (St. Paul): 15 (1983) = **Raveneliaceae**.
Sphaerosomataceae Bail, *Syst. Pilze* (Bonn) **2**: 65 (1858) = **Pyronemataceae**.
Sphinctrinaceae M. Choisy, *Bull. mens. Soc. linn. Lyon* **19**: 65 (1950).
Spilomataceae Cheval., *Fl. gén. env. Paris* (Paris): 580 (1826) = **Agyriaceae**.
Spizellomycetaceae D.J.S. Barr, *Can. J. Bot.* **58**(22): 2384 (1980).
Splanchnomycetaceae Corda, *Icon. fung.* (Prague) **5**: 26 (1842) = **Rhizopogonaceae**.
Sporidiobolaceae R.T. Moore, *Bot. Mar.* **23**: (1980).
Sporocadaceae Corda, *Icon. fung.* (Prague) **5**: 34 (1842) = **Amphisphaeriaceae**.
Sporochytriaceae A. Fisch., *Rabenh. Krypt.-Fl.* (Leipzig) **1**(4): 85 (1892) = **Chytridiaceae**.

Sporormiaceae Munk, *Dansk bot. Ark.* **17**(1): 450 (1957).
Sporoschismataceae (Sacc.) Nann., *Repert. mic. uomo*: 501 (1934) = **Chaetosphaeriaceae**.
Squamarinaceae Hafellner, *Beih. Nova Hedwigia* **79**: 342 (1984) = **Bacidiaceae**.
Staurotheleaceae Servít, *Českoslov. Lišejn. Čeledi Verrucariaceae*: 17BB (1954) = **Verrucariaceae**.
Steccherinaceae Parmasto, *Consp. System. Corticiac.* (Tartu): 169 (1968).
Stegasphaeriaceae Syd. & P. Syd., *Annls mycol.* **14**: 364 (1916) = **Mesnieraceae**.
Stephanosporaceae Oberw. & E. Horak, *Pl. Syst. Evol.* **131**: 162 (1979).
Stephanothecaceae Petr., *Annls mycol.* **29**: 345 (1931) = **Elsinoaceae**.
Stereaceae Pilát, *Hedwigia* Beibl. **70**: 34 (1930).
Stereocaulaceae Chevall., *Fl. gén. env. Paris* (Paris): 596 (1826).
Stichoclavariaceae Ulbr., *Krypt.-Fl. Anfäng.* (Berlin) **1**(3): 82 (1928) = **Clavariaceae**.
Stictaceae (Nyl.) Stizenb., *Ber. Tät. St Gall. naturw. Ges.*: 174 (1862) = **Lobariaceae**.
Stictidaceae Fr., *Summa veg. Scand.* (Stockholm) **2**: 345 (1849).
Strigulaceae Zahlbr., *Syllabus.* Edn 2 (Berlin): 46 (1898).
Strobilomycetaceae E.-J. Gilbert, *Bolets* (Paris): 105 (1931) = **Boletaceae**.
Strophariaceae Singer & A.H. Sm., *Mycologia* **38**: 503 (1946).
Subulicystidiaceae Jülich, *Biblthca Mycol.* **85**: 391 (1982) = **Hyphodermataceae**.
Suillaceae (Singer) Besl & Bresinsky, *Pl. Syst. Evol.* **206**(1-4): 239 (1997).
Sydowiellaceae Lar.N. Vassiljeva, *Pirenomits. Lokuloaskomits. Severa Dal'nego Vostoka* (Leningrad): 210 (1987) = **Melanconidaceae**.
Syncephalastraceae Naumov ex R.K. Benj., *Aliso* **4**: 327 (1959).
Synchytriaceae J. Schröt., *Nat. Pflanzenfam.* (Leipzig) **1**(1): 71 (1892).
Taiwanascaceae Sivan. & H.S. Chang, *Mycol. Res.* **101**(2): 178 (1997) = **Niessliaceae**.
Taphrinaceae Gäum., *Comp. Morph. Fungi* (London): 161 (1928).
Tapinellaceae C. Hahn, *Sendtnera* **6**: 122 (1999) = **Hygrophoropsidaceae**.
Teloschistaceae Zahlbr., *Syllabus.* Edn 2 (Berlin): 45 (1898).
Tephromelataceae Hafellner, *Beih. Nova Hedwigia* **79**: 344 (1984) = **Bacidiaceae**.
Terfeziaceae E. Fisch., *Nat. Pflanzenfam.* (Leipzig) **1**(1): 312 (1897).
Testudinaceae Arx, *Persoonia* **6**: 366 (1971).
Tetragoniomycetaceae Oberw. & Bandoni, *Can. J. Bot.* **59**: 1039 (1981).
Thamnidiaceae Fitzp., *Lower fung.* (New York): 242 (1930) = **Mucoraceae**.
Thelebolaceae (Brumm.) Eckblad, *Nytt Mag. Bot.* **15**: 22 (1968).
Thelenellaceae H. Mayrhofer, *Biblthca Lichenol.* **26**: 16 (1987).
Thelephoraceae Chevall., *Fl. gén. env. Paris* (Paris): 84 (1826).
Thelidiaceae Walt. Watson, *New Phytol.* **28**: 107 (1929) = **Verrucariaceae**.
Thelotremataceae (Nyl.) Stizenb., *Ber. Tät. St Gall. naturw. Ges.*: 167 (1862).
Thermoascaceae Apinis, *Trans. Br. mycol. Soc.* **50**: 581 (1967) = **Trichocomaceae**.
Tholurnaceae Elenkin, *Izv. glav. bot. Sada SSSR* **28**(3-4): 267 (1929) = **Caliciaceae**.
Thrombiaceae Poelt & Vězda ex J.C. David & D. Hawksw., *Syst. Ascom.* **10**(1): 16 (1991).
Thyridiaceae J.Z. Yue & O.E. Eriksś., *Syst. Ascom.* **6**(2): 233 (1987).
Tilletiaceae J. Schröt., *Krypt.-Fl. Schlesien* (Breslau) **3**(1): 276 (1887).
Tilletiariaceae R.T. Moore, *Bot. Mar.* **23**: 23 (1980).
Togniniaceae Réblová, L. Mostert, W. Gams & Crous, *Stud. Mycol.* **50**: 540 (2004).
Tomentellaceae Warm., *Haandb. syst. Bot..* Edn 2 (Kjøbenhavn): 110 (1890) = **Thelephoraceae**.
Torrendiaceae Jülich, *Biblthca Mycol.* **85**: 392 (1982) = **Pluteaceae**.
Trametaceae Boidin, Mugnier & Canales, *Mycotaxon* **66**: 487 (1998) = **Polyporaceae**.
Trapeliaceae M. Choisy ex Hertel, *Vortr. GesGeb. Bot.* n.f. **4**: 181 (1970) = **Agyriaceae**.
Tremellaceae Fr., *Syst. mycol.* (Lundae) **1**: lv (1821).
Tremellodendropsidaceae Jülich, *Biblthca Mycol.* **85**: 392 (1982).
Tremellogastraceae Zeller, *Mycologia* **40**: 662 (1948) = **Sclerodermataceae**.
Tremoleciaceae Hafellner, *Beih. Nova Hedwigia* **79**: 346 (1984) = **Hymeneliaceae**.
Triblidiaceae Rehm, *Rabenh. Krypt.-Fl.* (Leipzig) **1**(2): 191 (1888).
Trichocomaceae E. Fisch., *Nat. Pflanzenfam.* (Leipzig) **1**(1): 310 (1897).
Tricholomataceae R. Heim ex Pouzar, *Česká Mykol.* **37**: 175 (1983).
Tricholomataceae (Fayod) R. Heim, *Trab. Mus. Ciènc. nat. Barcelona* sèr. bot. **15**(3): 88 (1934) = **Tricholomataceae**.

Trichopeltidaceae Theiss., *Zentbl. Bakt. ParasitKde* Abt. II **39**: 629 (1913) = **Microthyriaceae**.
Trichopeltinaceae (Theiss. & P. Syd.) Bat., C.A.A. Costa & Cif., *Atti Ist. bot. Univ. Lab. crittog. Pavia* sér. 5 **15**: 37 (1958) = **Microthyriaceae**.
Trichophytonaceae Vuill. ex E.K. Novák & Galgoczy, *Acta bot. hung.* **15**: 124 (1969) = **Arthrodermataceae**.
Trichosphaeriaceae G. Winter, *Rabenh. Krypt.-Fl.* (Leipzig) **1**(2): 191 (1885).
Trichothallaceae Bat. & Cif., *Mycopath. Mycol. appl.* **11**: 95 (1959) = **Euantennariaceae**.
Trichothyriaceae Theiss., *Beih. bot. Zbl.* Abt. 1 **2**: 3 (1914) = **Microthyriaceae**.
Trinacriaceae Nann., *Repert. mic. uomo*: 473 (1934) = **Orbiliaceae**.
Triposporiaceae (Ferraris) Nann., *Repert. mic. uomo*: 507 (1934) = **Asterinaceae**.
Triposporiopsidaceae S. Hughes, *Mycologia* **68**: 712 (1976) = **Capnodiaceae**.
Tripterosporaceae Cain, *Can. J. Bot.* **34**: 699 (1956) = **Lasiosphaeriaceae**.
Truncocolumellaceae Agerer, *Sendtnera* **6**: 38 (1999) = **Suillaceae**.
Trypetheliaceae Zenker, *Pharmaceutische Waarenkunde* (Eisenach): 123 (1827).
Tuberaceae Dumort., *Comment. bot.* (Tournay): 69 (1822).
Tuberculariaceae Fr., *Syst. orb. veg.* (Lund) **1**: 169 (1825) = **Nectriaceae**.
Tubeufiaceae M.E. Barr, *Mycologia* **71**: 948 (1979).
Tubulicrinaceae Jülich, *Biblthca Mycol.* **85**: 392 (1982).
Tulasnellaceae Juel, *Bih. K. svenska VetenskAkad. Handl.* **23**, 3(12): 21 (1897).
Tulostomataceae E. Fisch., *Nat. Pflanzenfam.* (Leipzig) **1**(1**): 342 (1900).
Tylosporaceae Jülich, *Biblthca Mycol.* **85**: 393 (1982) = **Atheliaceae**.
Typhulaceae Jülich, *Biblthca Mycol.* **85**: 393 (1982).
Uleiellaceae Vánky, *Fungal Diversity* **241**(2): 139 (2001).
Umbelopsidaceae W. Gams & W. Mey., *Mycol. Res.* **107**(3): 348 (2003).
Umbilicariaceae Chevall., *Fl. gén. env. Paris* (Paris): 640 (1826).
Uncolaceae Buriticá, *Revta Acad. colomb. cienc. exact. fís. nat.* **24**(no. 90): 113 (2000).
Urocystaceae Begerow, R. Bauer & Oberw., *Can. J. Bot.* **75**(12): 2052 (1998) ['1997'].
Urophlyctaceae Hadar, *C. r. hebd. Séanc. Acad. Sci., Paris* sér. III **29**(15 février): 330 (1982) = **Physodermataceae**.
Urophlyctidaceae Hadar, *C. r. hebd. Séanc. Acad. Sci., Paris* sér. III **294**(15 février): 330 (1982) = **Physodermataceae**.
Uropyxidaceae (Arthur) Cummins & Y. Hirats., *Illustr. Gen. Rust Fungi. rev. edit.* (St. Paul): 14 (1983).
Usneaceae Zenker, *Pharmaceutische Waarenkunde* (Eisenach): 124 (1827) = **Parmeliaceae**.
Ustilaginaceae Tul. & C. Tul., *Annls Sci. Nat. Bot.*, sér. 3 **7**: 14 (1847).
Ustilentylomataceae R. Bauer & Oberw., *Can. J. Bot.* **75**(8): 1311 (1997).
Valsaceae Tul. & C. Tul., *Select. fung. carpol.* (Paris) **1**: 180 (1861).
Venturiaceae E. Müll. & Arx ex M.E. Barr, *Mycologia* **71**: 947 (1979).
Verrucariaceae Zenker, *Pharmaceutische Waarenkunde* (Eisenach): 123 (1827).
Vezdaeaceae Poelt & Vězda ex J.C. David & D. Hawksw., *Syst. Ascom.* **10**(1): 16 (1991).
Vialaeaceae P.F. Cannon, *Mycol. Res.* **99**(3): 368 (1995).
Vibrisseaceae Locq., *Syn. gen. fung.* (Paris): [1] (1972).
Vibrisseaceae Korf, *Mycosystema* **3**: 23 (1990) = **Vibrisseaceae**.
Vizellaceae H.J. Swart, *Trans. Br. mycol. Soc.* **57**: 456 (1971).
Volutellaceae (Ferraris) Nann., *Repert. mic. uomo*: 517 (1934) = **Nectriaceae**.
Volvocisporiaceae Begerow, R. Bauer & Oberw., *Mycol. Res.* **10**(7): 809 (2001).
Wallemiaceae R.T. Moore, *Rhizoctonia Species. Taxonomy, Molecular Biology, Ecology, Pathology and Disease Control* (Dordrecht): 20 (1996).
Websdaneaceae Vánky, *Fungal Diversity* **239**(1): 135 (2001).
Wrightoporiaceae Jülich, *Biblthca Mycol.* **85**: 393 (1982) = **Bondarzewiaceae**.
Xanthopsorellaceae Hafellner, *Beih. Nova Hedwigia* **79**: 348 (1984) = **Catillariaceae**.
Xanthopyreniaceae Zahlbr., *Nat. Pflanzenfam.*. Edn 2 (Leipzig) **8**: 91 (1926).
Xenasmataceae Oberw., *Sydowia* **19**: 25 (1966).
Xerocomaceae (Singer) Pegler & T.W.K. Young, *Trans. Br. mycol. Soc.* **76**: 112 (1981) = **Boletaceae**.
Xylariaceae Tul. & C. Tul., *Select. fung. carpol.* (Paris) **2**: 3 (1861).
Xylographaceae Tuck., *Syn. N. Amer. lich.* (Boston) **2**: 110 (1888) = **Agyriaceae**.
Xylophagaceae Murrill, *Torreya* **3**: 7 (1903) = **Coniophoraceae**.
Zaghouaniaceae Syd. & P. Syd., *Monogr. Uredin.* (Lipsiae) **3**: 586 (1915) = **Pucciniaceae**.

Zerovaemycetaceae Gorovij, *Dopov. Akad. Nauk URSR* Ser. B(8): 745 (1977) = **Psathyrellaceae**.
Zodiomycetaceae (Thaxt.) Nann., *Repert. mic. uomo*: 263 (1934) = **Laboulbeniaceae**.
Zoopagaceae Drechsler, *Mycologia* **30**: 154 (1938).
Zopfiaceae G. Arnaud ex D. Hawksw., *Syst. Ascom.* **11**(1): 77 (1992).

DATE DUE